Parts 260 to 265
Revised as of July 1, 2010

# Protection of Environment

Containing a codification of documents
of general applicability and future effect

As of July 1, 2010

*With Ancillaries*

Published by
Office of the Federal Register
National Archives and Records
Administration

A Special Edition of the Federal Register

Code of Federal Regulations

MW01534503

# U.S. GOVERNMENT OFFICIAL EDITION NOTICE

### Legal Status and Use of Seals and Logos

The seal of the National Archives and Records Administration (NARA) authenticates the Code of Federal Regulations (CFR) as the official codification of Federal regulations established under the Federal Register Act. Under the provisions of 44 U.S.C. 1507, the contents of the CFR, a special edition of the Federal Register, shall be judicially noticed. The CFR is prima facie evidence of the original documents published in the Federal Register (44 U.S.C. 1510).

It is prohibited to use NARA's official seal and the stylized Code of Federal Regulations logo on any republication of this material without the express, written permission of the Archivist of the United States or the Archivist's designee. Any person using NARA's official seals and logos in a manner inconsistent with the provisions of 36 CFR part 1200 is subject to the penalties specified in 18 U.S.C. 506, 701, and 1017.

### Use of ISBN Prefix

This is the Official U.S. Government edition of this publication and is herein identified to certify its authenticity. Use of the 0–16 ISBN prefix is for U.S. Government Printing Office Official Editions only. The Superintendent of Documents of the U.S. Government Printing Office requests that any reprinted edition clearly be labeled as a copy of the authentic work with a new ISBN.

**GPO** U.S. GOVERNMENT PRINTING OFFICE

U.S. Superintendent of Documents • Washington, DC 20402–0001

http://bookstore.gpo.gov

Phone: toll-free (866) 512-1800; DC area (202) 512-1800

# Table of Contents

*Cite this Code:*   **CFR**

*To cite the regulations in
this volume use title,
part and section num-
ber. Thus, 40 CFR 260.1
refers to title 40, part
260, section 1.*

# Explanation

The Code of Federal Regulations is a codification of the general and permanent rules published in the Federal Register by the Executive departments and agencies of the Federal Government. The Code is divided into 50 titles which represent broad areas subject to Federal regulation. Each title is divided into chapters which usually bear the name of the issuing agency. Each chapter is further subdivided into parts covering specific regulatory areas.

Each volume of the Code is revised at least once each calendar year and issued on a quarterly basis approximately as follows:

Title 1 through Title 16.................................................as of January 1
Title 17 through Title 27...............................................as of April 1
Title 28 through Title 41...............................................as of July 1
Title 42 through Title 50...............................................as of October 1

The appropriate revision date is printed on the cover of each volume.

## LEGAL STATUS

The contents of the Federal Register are required to be judicially noticed (44 U.S.C. 1507). The Code of Federal Regulations is prima facie evidence of the text of the original documents (44 U.S.C. 1510).

## HOW TO USE THE CODE OF FEDERAL REGULATIONS

The Code of Federal Regulations is kept up to date by the individual issues of the Federal Register. These two publications must be used together to determine the latest version of any given rule.

To determine whether a Code volume has been amended since its revision date (in this case, July 1, 2010), consult the "List of CFR Sections Affected (LSA)," which is issued monthly, and the "Cumulative List of Parts Affected," which appears in the Reader Aids section of the daily Federal Register. These two lists will identify the Federal Register page number of the latest amendment of any given rule.

## EFFECTIVE AND EXPIRATION DATES

Each volume of the Code contains amendments published in the Federal Register since the last revision of that volume of the Code. Source citations for the regulations are referred to by volume number and page number of the Federal Register and date of publication. Publication dates and effective dates are usually not the same and care must be exercised by the user in determining the actual effective date. In instances where the effective date is beyond the cut-off date for the Code a note has been inserted to reflect the future effective date. In those instances where a regulation published in the Federal Register states a date certain for expiration, an appropriate note will be inserted following the text.

## OMB CONTROL NUMBERS

The Paperwork Reduction Act of 1980 (Pub. L. 96–511) requires Federal agencies to display an OMB control number with their information collection request.

Many agencies have begun publishing numerous OMB control numbers as amendments to existing regulations in the CFR. These OMB numbers are placed as close as possible to the applicable recordkeeping or reporting requirements.

## OBSOLETE PROVISIONS

Provisions that become obsolete before the revision date stated on the cover of each volume are not carried. Code users may find the text of provisions in effect on a given date in the past by using the appropriate numerical list of sections affected. For the period before January 1, 2001, consult either the List of CFR Sections Affected, 1949–1963, 1964–1972, 1973–1985, or 1986–2000, published in eleven separate volumes. For the period beginning January 1, 2001, a "List of CFR Sections Affected" is published at the end of each CFR volume.

## "[RESERVED]" TERMINOLOGY

The term "[Reserved]" is used as a place holder within the Code of Federal Regulations. An agency may add regulatory information at a "[Reserved]" location at any time. Occasionally "[Reserved]" is used editorially to indicate that a portion of the CFR was left vacant and not accidentally dropped due to a printing or computer error.

## INCORPORATION BY REFERENCE

*What is incorporation by reference?* Incorporation by reference was established by statute and allows Federal agencies to meet the requirement to publish regulations in the Federal Register by referring to materials already published elsewhere. For an incorporation to be valid, the Director of the Federal Register must approve it. The legal effect of incorporation by reference is that the material is treated as if it were published in full in the Federal Register (5 U.S.C. 552(a)). This material, like any other properly issued regulation, has the force of law.

*What is a proper incorporation by reference?* The Director of the Federal Register will approve an incorporation by reference only when the requirements of 1 CFR part 51 are met. Some of the elements on which approval is based are:

(a) The incorporation will substantially reduce the volume of material published in the Federal Register.

(b) The matter incorporated is in fact available to the extent necessary to afford fairness and uniformity in the administrative process.

(c) The incorporating document is drafted and submitted for publication in accordance with 1 CFR part 51.

*What if the material incorporated by reference cannot be found?* If you have any problem locating or obtaining a copy of material listed as an approved incorporation by reference, please contact the agency that issued the regulation containing that incorporation. If, after contacting the agency, you find the material is not available, please notify the Director of the Federal Register, National Archives and Records Administration, 8601 Adelphi Road, College Park, MD 20740-6001, or call 202-741-6010.

## CFR INDEXES AND TABULAR GUIDES

A subject index to the Code of Federal Regulations is contained in a separate volume, revised annually as of January 1, entitled CFR INDEX AND FINDING AIDS. This volume contains the Parallel Table of Authorities and Rules. A list of CFR titles, chapters, subchapters, and parts and an alphabetical list of agencies publishing in the CFR are also included in this volume.

An index to the text of "Title 3—The President" is carried within that volume.

The Federal Register Index is issued monthly in cumulative form. This index is based on a consolidation of the "Contents" entries in the daily Federal Register.

A List of CFR Sections Affected (LSA) is published monthly, keyed to the revision dates of the 50 CFR titles.

## REPUBLICATION OF MATERIAL

There are no restrictions on the republication of material appearing in the Code of Federal Regulations.

## INQUIRIES

For a legal interpretation or explanation of any regulation in this volume, contact the issuing agency. The issuing agency's name appears at the top of odd-numbered pages.

For inquiries concerning CFR reference assistance, call 202–741–6000 or write to the Director, Office of the Federal Register, National Archives and Records Administration, 8601 Adelphi Road, College Park, MD 20740-6001 or e-mail fedreg.info@nara.gov.

## SALES

The Government Printing Office (GPO) processes all sales and distribution of the CFR. For payment by credit card, call toll-free, 866-512-1800, or DC area, 202-512-1800, M-F 8 a.m. to 4 p.m. e.s.t. or fax your order to 202-512-2250, 24 hours a day. For payment by check, write to: US Government Printing Office – New Orders, P.O. Box 979050, St. Louis, MO 63197-9000. For GPO Customer Service call 202-512-1803.

## ELECTRONIC SERVICES

The full text of the Code of Federal Regulations, the LSA (List of CFR Sections Affected), The United States Government Manual, the Federal Register, Public Laws, Public Papers, Daily Compilation of Presidential Documents and the Privacy Act Compilation are available in electronic format via *Federalregister.gov*. For more information, contact Electronic Information Dissemination Services, U.S. Government Printing Office. Phone 202-512-1530, or 888-293-6498 (toll-free). E-mail, *gpoaccess@gpo.gov*.

The Office of the Federal Register also offers a free service on the National Archives and Records Administration's (NARA) World Wide Web site for public law numbers, Federal Register finding aids, and related information. Connect to NARA's web site at *www.archives.gov/federal-register*. The NARA site also contains links to GPO Access.

RAYMOND A. MOSLEY,
*Director,*
*Office of the Federal Register.*
*July 1, 2010.*

# THIS TITLE

Title 40—PROTECTION OF ENVIRONMENT is composed of thirty-two volumes. The parts in these volumes are arranged in the following order: parts 1–49, parts 50–51, part 52 (52.01–52.1018), part 52 (52.1019–end of part 52), parts 53–59, part 60 (60.1–end of part 60, sections), part 60 (Appendices), parts 61–62, part 63 (63.1–63.599), part 63 (63.600–63.1199), part 63 (63.1200–63.1439), part 63 (63.1440–63.6175), part 63 (63.6580–63.8830), part 63 (63.8980–end of part 63) parts 64–71, parts 72–80, parts 81–84, part 85–§86.599–99, part 86 (86.600–1–end of part 86), parts 87–99, parts 100–135, parts 136–149, parts 150–189, parts 190–259, parts 260–265, parts 266–299, parts 300–399, parts 400–424, parts 425–699, parts 700–789, parts 790–999, and part 1000 to end. The contents of these volumes represent all current regulations codified under this title of the CFR as of July 1, 2010.

Chapter I—Environmental Protection Agency appears in all thirty-two volumes. Regulations issued by the Council on Environmental Quality, including an Index to Parts 1500 through 1508, appear in the volume containing part 1000 to end. The OMB control numbers for title 40 appear in §9.1 of this chapter.

For this volume, Michele Bugenhagen was Chief Editor. The Code of Federal Regulations publication program is under the direction of Michael L. White, assisted by Ann Worley.

# Title 40—Protection of Environment

(This book contains parts 260 to 265)

---

1

# CHAPTER I—ENVIRONMENTAL PROTECTION AGENCY (CONTINUED)

---

EDITORIAL NOTE: Nomenclature changes to chapter I appear at 65 FR 47324, 47325, Aug. 2, 2000, and 66 FR 34375, 34376, June 28, 2001.

## SUBCHAPTER I—SOLID WASTES (CONTINUED)

# SUBCHAPTER I—SOLID WASTES (CONTINUED)

## PART 260—HAZARDOUS WASTE MANAGEMENT SYSTEM: GENERAL

### Subpart A—General

Sec.
260.1 Purpose, scope, and applicability.
260.2 Availability of information; confidentiality of information.
260.3 Use of number and gender.

### Subpart B—Definitions

260.10 Definitions.
260.11 References.

### Subpart C—Rulemaking Petitions

260.20 General.
260.21 Petitions for equivalent testing or analytical methods.
260.22 Petitions to amend part 261 to exclude a waste produced at a particular facility.
260.23 Petitions to amend 40 CFR part 273 to include additional hazardous wastes.
260.30 Non-waste determinations and variances from classification as a solid waste.
260.31 Standards and criteria for variances from classification as a solid waste.
260.32 Variances to be classified as a boiler.
260.33 Procedures for variances from classification as a solid waste, for variances to be classified as a boiler, or for non-waste determinations.
260.34 Standards and criteria for non-waste determinations.
260.40 Additional regulation of certain hazardous waste recycling activities on a case-by-case basis.
260.41 Procedures for case-by-case regulation of hazardous waste recycling activities.
260.42 Notification requirement for hazardous secondary materials.
260.43 Legitimate recycling of hazardous secondary materials regulated under § 260.34, § 261.2(a)(2)(ii), and § 261.4(a)(23), (24), or (25).

AUTHORITY: 42 U.S.C. 6905, 6912(a), 6921–6927, 6930, 6934, 6935, 6937, 6938, 6939, and 6974.

SOURCE: 45 FR 33073, May 19, 1980, unless otherwise noted.

## Subpart A—General

### § 260.1 Purpose, scope, and applicability.

(a) This part provides definitions of terms, general standards, and overview information applicable to parts 260 through 265 and 268 of this chapter.

(b) In this part: (1) Section 260.2 sets forth the rules that EPA will use in making information it receives available to the public and sets forth the requirements that generators, transporters, or owners or operators of treatment, storage, or disposal facilities must follow to assert claims of business confidentiality with respect to information that is submitted to EPA under parts 260 through 265 and 268 of this chapter.

(2) Section 260.3 establishes rules of grammatical construction for parts 260 through 265 and 268 of this chapter.

(3) Section 260.10 defines terms which are used in parts 260 through 265 and 268 of this chapter.

(4) Section 260.20 establishes procedures for petitioning EPA to amend, modify, or revoke any provision of parts 260 through 265 and 268 of this chapter and establishes procedures governing EPA's action on such petitions.

(5) Section 260.21 establishes procedures for petitioning EPA to approve testing methods as equivalent to those prescribed in parts 261, 264, or 265 of this chapter.

(6) Section 260.22 establishes procedures for petitioning EPA to amend subpart D of part 261 to exclude a waste from a particular facility.

[45 FR 33073, May 19, 1980, as amended at 51 FR 40636, Nov. 7, 1986]

### § 260.2 Availability of information; confidentiality of information.

(a) Any information provided to EPA under parts 260 through 265 and 268 of this chapter will be made available to the public to the extent and in the manner authorized by the Freedom of Information Act, 5 U.S.C. section 552, section 3007(b) of RCRA and EPA regulations implementing the Freedom of Information Act and section 3007(b), part 2 of this chapter, as applicable.

(b) Any person who submits information to EPA in accordance with parts 260 through 266 and 268 of this chapter may assert a claim of business confidentiality covering part or all of that

5

information by following the procedures set forth in §2.203(b) of this chapter. Information covered by such a claim will be disclosed by EPA only to the extent, and by means of the procedures, set forth in part 2, subpart B, of this chapter except that information required by §262.53(a) and §262.83 that is submitted in a notification of intent to export a hazardous waste will be provided to the U.S. Department of State and the appropriate authorities in the transit and receiving or importing countries regardless of any claims of confidentiality. However, if no such claim accompanies the information when it is received by EPA, it may be made available to the public without further notice to the person submitting it.

[45 FR 33073, May 19, 1980, as amended at 51 FR 28682, Aug. 8, 1986; 51 FR 40636, Nov. 7, 1986; 61 FR 16309, Apr. 12, 1996]

## § 260.3  Use of number and gender.

As used in parts 260 through 265 and 268 of this chapter:

(a) Words in the masculine gender also include the feminine and neuter genders; and

(b) Words in the singular include the plural; and

(c) Words in the plural include the singular.

[45 FR 33073, May 19, 1980, as amended at 51 FR 40636, Nov. 7, 1986]

## Subpart B—Definitions

### § 260.10  Definitions.

When used in parts 260 through 273 of this chapter, the following terms have the meanings given below:

*Above ground tank* means a device meeting the definition of "tank" in §260.10 and that is situated in such a way that the entire surface area of the tank is completely above the plane of the adjacent surrounding surface and the entire surface area of the tank (including the tank bottom) is able to be visually inspected.

*Act* or *RCRA* means the Solid Waste Disposal Act, as amended by the Resource Conservation and Recovery Act of 1976, as amended, 42 U.S.C. section 6901 et seq.

*Active life* of a facility means the period from the initial receipt of hazardous waste at the facility until the Regional Administrator receives certification of final closure.

*Active portion* means that portion of a facility where treatment, storage, or disposal operations are being or have been conducted after the effective date of part 261 of this chapter and which is not a closed portion. (See also "closed portion" and "inactive portion".)

*Administrator* means the Administrator of the Environmental Protection Agency, or his designee.

*Ancillary equipment* means any device including, but not limited to, such devices as piping, fittings, flanges, valves, and pumps, that is used to distribute, meter, or control the flow of hazardous waste from its point of generation to a storage or treatment tank(s), between hazardous waste storage and treatment tanks to a point of disposal onsite, or to a point of shipment for disposal off-site.

*Aquifer* means a geologic formation, group of formations, or part of a formation capable of yielding a significant amount of ground water to wells or springs.

*Authorized representative* means the person responsible for the overall operation of a facility or an operational unit (i.e., part of a facility), e.g., the plant manager, superintendent or person of equivalent responsibility.

*Battery* means a device consisting of one or more electrically connected electrochemical cells which is designed to receive, store, and deliver electric energy. An electrochemical cell is a system consisting of an anode, cathode, and an electrolyte, plus such connections (electrical and mechanical) as may be needed to allow the cell to deliver or receive electrical energy. The term battery also includes an intact, unbroken battery from which the electrolyte has been removed.

*Boiler* means an enclosed device using controlled flame combustion and having the following characteristics:

(1)(i) The unit must have physical provisions for recovering and exporting thermal energy in the form of steam, heated fluids, or heated gases; and

(ii) The unit's combustion chamber and primary energy recovery sections(s) must be of integral design. To be of integral design, the combustion chamber and the primary energy recovery section(s) (such as waterwalls and superheaters) must be physically formed into one manufactured or assembled unit. A unit in which the combustion chamber and the primary energy recovery section(s) are joined only by ducts or connections carrying flue gas is not integrally designed; however, secondary energy recovery equipment (such as economizers or air preheaters) need not be physically formed into the same unit as the combustion chamber and the primary energy recovery section. The following units are not precluded from being boilers solely because they are not of integral design: process heaters (units that transfer energy directly to a process stream), and fluidized bed combustion units; and

(iii) While in operation, the unit must maintain a thermal energy recovery efficiency of at least 60 percent, calculated in terms of the recovered energy compared with the thermal value of the fuel; and

(iv) The unit must export and utilize at least 75 percent of the recovered energy, calculated on an annual basis. In this calculation, no credit shall be given for recovered heat used internally in the same unit. (Examples of internal use are the preheating of fuel or combustion air, and the driving of induced or forced draft fans or feedwater pumps); or

(2) The unit is one which the Regional Administrator has determined, on a case-by-case basis, to be a boiler, after considering the standards in §260.32.

*Carbon regeneration unit* means any enclosed thermal treatment device used to regenerate spent activated carbon.

*Cathode ray tube or CRT* means a vacuum tube, composed primarily of glass, which is the visual or video display component of an electronic device. A used, intact CRT means a CRT whose vacuum has not been released. A used, broken CRT means glass removed from its housing or casing whose vacuum has been released.

*Certification* means a statement of professional opinion based upon knowledge and belief.

*Closed portion* means that portion of a facility which an owner or operator has closed in accordance with the approved facility closure plan and all applicable closure requirements. (See also "active portion" and "inactive portion".)

*Component* means either the tank or ancillary equipment of a tank system.

*Confined aquifer* means an aquifer bounded above and below by impermeable beds or by beds of distinctly lower permeability than that of the aquifer itself; an aquifer containing confined ground water.

*Container* means any portable device in which a material is stored, transported, treated, disposed of, or otherwise handled.

*Containment building* means a hazardous waste management unit that is used to store or treat hazardous waste under the provisions of subpart DD of parts 264 or 265 of this chapter.

*Contingency plan* means a document setting out an organized, planned, and coordinated course of action to be followed in case of a fire, explosion, or release of hazardous waste or hazardous waste constituents which could threaten human health or the environment.

*Corrosion expert* means a person who, by reason of his knowledge of the physical sciences and the principles of engineering and mathematics, acquired by a professional education and related practical experience, is qualified to engage in the practice of corrosion control on buried or submerged metal piping systems and metal tanks. Such a person must be certified as being qualified by the National Association of Corrosion Engineers (NACE) or be a registered professional engineer who has certification or licensing that includes education and experience in corrosion control on buried or submerged metal piping systems and metal tanks.

*CRT collector* means a person who receives used, intact CRTs for recycling, repair, resale, or donation.

*CRT glass manufacturer* means an operation or part of an operation that uses a furnace to manufacture CRT glass.

*CRT processing* means conducting all of the following activities:

(1) Receiving broken or intact CRTs; and

(2) Intentionally breaking intact CRTs or further breaking or separating broken CRTs; and

(3) Sorting or otherwise managing glass removed from CRT monitors.

*Designated facility* means:

(1) A hazardous waste treatment, storage, or disposal facility which:

(i) Has received a permit (or interim status) in accordance with the requirements of parts 270 and 124 of this chapter;

(ii) Has received a permit (or interim status) from a State authorized in accordance with part 271 of this chapter; or

(iii) Is regulated under § 261.6(c)(2) or subpart F of part 266 of this chapter; and

(iv) That has been designated on the manifest by the generator pursuant to § 262.20.

(2) *Designated facility* also means a generator site designated on the manifest to receive its waste as a return shipment from a facility that has rejected the waste in accordance with § 264.72(f) or § 265.72(f) of this chapter.

(3) If a waste is destined to a facility in an authorized State which has not yet obtained authorization to regulate that particular waste as hazardous, then the designated facility must be a facility allowed by the receiving State to accept such waste.

*Destination facility* means a facility that treats, disposes of, or recycles a particular category of universal waste, except those management activities described in paragraphs (a) and (c) of §§ 273.13 and 273.33 of this chapter. A facility at which a particular category of universal waste is only accumulated, is not a destination facility for purposes of managing that category of universal waste.

*Dike* means an embankment or ridge of either natural or man-made materials used to prevent the movement of liquids, sludges, solids, or other materials.

*Dioxins and furans (D/F)* means tetra, penta, hexa, hepta, and octa-chlorinated dibenzo dioxins and furans.

*Discharge* or *hazardous waste discharge* means the accidental or intentional spilling, leaking, pumping, pouring, emitting, emptying, or dumping of hazardous waste into or on any land or water.

*Disposal* means the discharge, deposit, injection, dumping, spilling, leaking, or placing of any solid waste or hazardous waste into or on any land or water so that such solid waste or hazardous waste or any constituent thereof may enter the environment or be emitted into the air or discharged into any waters, including ground waters.

*Disposal facility* means a facility or part of a facility at which hazardous waste is intentionally placed into or on any land or water, and at which waste will remain after closure. The term disposal facility does not include a corrective action management unit into which remediation wastes are placed.

*Drip pad* is an engineered structure consisting of a curbed, free-draining base, constructed of non-earthen materials and designed to convey preservative kick-back or drippage from treated wood, precipitation, and surface water run-on to an associated collection system at wood preserving plants.

*Elementary neutralization unit* means a device which:

(1) Is used for neutralizing wastes that are hazardous only because they exhibit the corrosivity characteristic defined in § 261.22 of this chapter, or they are listed in subpart D of part 261 of the chapter only for this reason; and

(2) Meets the definition of tank, tank system, container, transport vehicle, or vessel in § 260.10 of this chapter.

*EPA hazardous waste number* means the number assigned by EPA to each hazardous waste listed in part 261, subpart D, of this chapter and to each characteristic identified in part 261, subpart C, of this chapter.

*EPA identification number* means the number assigned by EPA to each generator, transporter, and treatment, storage, or disposal facility.

*EPA region* means the states and territories found in any one of the following ten regions:

Region I—Maine, Vermont, New Hampshire, Massachusetts, Connecticut, and Rhode Island.

Region II—New York, New Jersey, Commonwealth of Puerto Rico, and the U.S. Virgin Islands.

Region III—Pennsylvania, Delaware, Maryland, West Virginia, Virginia, and the District of Columbia.

Region IV—Kentucky, Tennessee, North Carolina, Mississippi, Alabama, Georgia, South Carolina, and Florida.

Region V—Minnesota, Wisconsin, Illinois, Michigan, Indiana and Ohio.

Region VI—New Mexico, Oklahoma, Arkansas, Louisiana, and Texas.

Region VII—Nebraska, Kansas, Missouri, and Iowa.

Region VIII—Montana, Wyoming, North Dakota, South Dakota, Utah, and Colorado.

Region IX—California, Nevada, Arizona, Hawaii, Guam, American Samoa, Commonwealth of the Northern Mariana Islands.

Region X—Washington, Oregon, Idaho, and Alaska.

*Equivalent method* means any testing or analytical method approved by the Administrator under §§260.20 and 260.21.

*Existing hazardous waste management (HWM) facility* or *existing facility* means a facility which was in operation or for which construction commenced on or before November 19, 1980. A facility has commenced construction if:

(1) The owner or operator has obtained the Federal, State and local approvals or permits necessary to begin physical construction; and either

(2)(i) A continuous on-site, physical construction program has begun; or

(ii) The owner or operator has entered into contractual obligations— which cannot be cancelled or modified without substantial loss—for physical construction of the facility to be completed within a reasonable time.

*Existing portion* means that land surface area of an existing waste management unit, included in the original Part A permit application, on which wastes have been placed prior to the issuance of a permit.

*Existing tank system* or *existing component* means a tank system or component that is used for the storage or treatment of hazardous waste and that is in operation, or for which installation has commenced on or prior to July 14, 1986. Installation will be considered to have commenced if the owner or operator has obtained all Federal, State, and local approvals or permits necessary to begin physical construction of the site or installation of the tank system and if either (1) a continuous on-site physical construction or installation program has begun, or (2) the owner or operator has entered into contractual obligations—which cannot be canceled or modified without substantial loss—for physical construction of the site or installation of the tank system to be completed within a reasonable time.

*Explosives or munitions emergency* means a situation involving the suspected or detected presence of unexploded ordnance (UXO), damaged or deteriorated explosives or munitions, an improvised explosive device (IED), other potentially explosive material or device, or other potentially harmful military chemical munitions or device, that creates an actual or potential imminent threat to human health, including safety, or the environment, including property, as determined by an explosives or munitions emergency response specialist. Such situations may require immediate and expeditious action by an explosives or munitions emergency response specialist to control, mitigate, or eliminate the threat.

*Explosives or munitions emergency response* means all immediate response activities by an explosives and munitions emergency response specialist to control, mitigate, or eliminate the actual or potential threat encountered during an explosives or munitions emergency. An explosives or munitions emergency response may include in-place render-safe procedures, treatment or destruction of the explosives or munitions and/or transporting those items to another location to be rendered safe, treated, or destroyed. Any reasonable delay in the completion of an explosives or munitions emergency response caused by a necessary, unforeseen, or uncontrollable circumstance will not terminate the explosives or munitions emergency. Explosives and munitions emergency responses can occur on either public or private lands and are not limited to responses at RCRA facilities.

*Explosives or munitions emergency response specialist* means an individual trained in chemical or conventional munitions or explosives handling, transportation, render-safe procedures, or destruction techniques. Explosives or munitions emergency response specialists include Department of Defense

(DOD) emergency explosive ordnance disposal (EOD), technical escort unit (TEU), and DOD-certified civilian or contractor personnel; and other Federal, State, or local government, or civilian personnel similarly trained in explosives or munitions emergency responses.

*Facility* means:

(1) All contiguous land, and structures, other appurtenances, and improvements on the land, used for treating, storing, or disposing of hazardous waste, or for managing hazardous secondary materials prior to reclamation. A facility may consist of several treatment, storage, or disposal operational units (e.g., one or more landfills, surface impoundments, or combinations of them).

(2) For the purpose of implementing corrective action under 40 CFR 264.101 or 267.101, all contiguous property under the control of the owner or operator seeking a permit under Subtitle C of RCRA. This definition also applies to facilities implementing corrective action under RCRA Section 3008(h).

(3) Notwithstanding paragraph (2) of this definition, a remediation waste management site is not a facility that is subject to 40 CFR 264.101, but is subject to corrective action requirements if the site is located within such a facility.

*Federal agency* means any department, agency, or other instrumentality of the Federal Government, any independent agency or establishment of the Federal Government including any Government corporation, and the Government Printing Office.

*Federal, State and local approvals or permits necessary to begin physical construction* means permits and approvals required under Federal, State or local hazardous waste control statutes, regulations or ordinances.

*Final closure* means the closure of all hazardous waste management units at the facility in accordance with all applicable closure requirements so that hazardous waste management activities under parts 264 and 265 of this chapter are no longer conducted at the facility unless subject to the provisions in § 262.34.

*Food-chain crops* means tobacco, crops grown for human consumption, and crops grown for feed for animals whose products are consumed by humans.

*Free liquids* means liquids which readily separate from the solid portion of a waste under ambient temperature and pressure.

*Freeboard* means the vertical distance between the top of a tank or surface impoundment dike, and the surface of the waste contained therein.

*Gasification.* For the purpose of complying with 40 CFR 261.4(a)(12)(i), gasification is a process, conducted in an enclosed device or system, designed and operated to process petroleum feedstock, including oil-bearing hazardous secondary materials through a series of highly controlled steps utilizing thermal decomposition, limited oxidation, and gas cleaning to yield a synthesis gas composed primarily of hydrogen and carbon monoxide gas.

*Generator* means any person, by site, whose act or process produces hazardous waste identified or listed in part 261 of this chapter or whose act first causes a hazardous waste to become subject to regulation.

*Ground water* means water below the land surface in a zone of saturation.

*Hazardous secondary material* means a secondary material (e.g., spent material, by-product, or sludge) that, when discarded, would be identified as hazardous waste under part 261 of this chapter.

*Hazardous secondary material generated and reclaimed under the control of the generator* means:

(1) That such material is generated and reclaimed at the generating facility (for purposes of this definition, generating facility means all contiguous property owned, leased, or otherwise controlled by the hazardous secondary material generator); or

(2) That such material is generated and reclaimed at different facilities, if the reclaiming facility is controlled by the generator or if both the generating facility and the reclaiming facility are controlled by a person as defined in § 260.10, and if the generator provides one of the following certifications: "on behalf of [insert generator facility name], I certify that this facility will send the indicated hazardous secondary material to [insert reclaimer facility

name], which is controlled by [insert generator facility name] and that [insert the name of either facility] has acknowledged full responsibility for the safe management of the hazardous secondary material," or "on behalf of [insert generator facility name] I certify that this facility will send the indicated hazardous secondary material to [insert reclaimer facility name], that both facilities are under common control, and that [insert name of either facility] has acknowledged full responsibility for the safe management of the hazardous secondary material." For purposes of this paragraph, "control" means the power to direct the policies of the facility, whether by the ownership of stock, voting rights, or otherwise, except that contractors who operate facilities on behalf of a different person as defined in §260.10 shall not be deemed to "control" such facilities, or

(3) That such material is generated pursuant to a written contract between a tolling contractor and a toll manufacturer and is reclaimed by the tolling contractor, if the tolling contractor certifies the following: "On behalf of [insert tolling contractor name], I certify that [insert tolling contractor name], has a written contract with [insert toll manufacturer name] to manufacture [insert name of product or intermediate] which is made from specified unused materials, and that [insert tolling contractor name] will reclaim the hazardous secondary materials generated during this manufacture. On behalf of [insert tolling contractor name], I also certify that [insert tolling contractor name] retains ownership of, and responsibility for, the hazardous secondary materials that are generated during the course of the manufacture, including any releases of hazardous secondary materials that occur during the manufacturing process. For purposes of this paragraph, tolling contractor means a person who arranges for the production of a product or intermediate made from specified unused materials through a written contract with a toll manufacturer. Toll manufacturer means a person who produces a product or intermediate made from specified unused materials pursuant to a written contract with a tolling contractor.

*Hazardous secondary material generator* means any person whose act or process produces hazardous secondary materials at the generating facility. For purposes of this paragraph, "generating facility" means all contiguous property owned, leased, or otherwise controlled by the hazardous secondary material generator. For the purposes of §261.2(a)(2)(ii) and §261.4(a)(23), a facility that collects hazardous secondary materials from other persons is not the hazardous secondary material generator.

*Hazardous waste* means a hazardous waste as defined in §261.3 of this chapter.

*Hazardous waste constituent* means a constituent that caused the Administrator to list the hazardous waste in part 261, subpart D, of this chapter, or a constituent listed in table 1 of §261.24 of this chapter.

*Hazardous waste management unit* is a contiguous area of land on or in which hazardous waste is placed, or the largest area in which there is significant likelihood of mixing hazardous waste constituents in the same area. Examples of hazardous waste management units include a surface impoundment, a waste pile, a land treatment area, a landfill cell, an incinerator, a tank and its associated piping and underlying containment system and a container storage area. A container alone does not constitute a unit; the unit includes containers and the land or pad upon which they are placed.

*In operation* refers to a facility which is treating, storing, or disposing of hazardous waste.

*Inactive portion* means that portion of a facility which is not operated after the effective date of part 261 of this chapter. (See also "active portion" and "closed portion".)

*Incinerator* means any enclosed device that:

(1) Uses controlled flame combustion and neither meets the criteria for classification as a boiler, sludge dryer, or carbon regeneration unit, nor is listed as an industrial furnace; or

(2) Meets the definition of infrared incinerator or plasma arc incinerator.

*Incompatible waste* means a hazardous waste which is unsuitable for:

11

(1) Placement in a particular device or facility because it may cause corrosion or decay of containment materials (e.g., container inner liners or tank walls); or

(2) Commingling with another waste or material under uncontrolled conditions because the commingling might produce heat or pressure, fire or explosion, violent reaction, toxic dusts, mists, fumes, or gases, or flammable fumes or gases.

(See appendix V of parts 264 and 265 of this chapter for examples.)

*Individual generation site* means the contiguous site at or on which one or more hazardous wastes are generated. An individual generation site, such as a large manufacturing plant, may have one or more sources of hazardous waste but is considered a single or individual generation site if the site or property is contiguous.

*Industrial furnace* means any of the following enclosed devices that are integral components of manufacturing processes and that use thermal treatment to accomplish recovery of materials or energy:

(1) Cement kilns

(2) Lime kilns

(3) Aggregate kilns

(4) Phosphate kilns

(5) Coke ovens

(6) Blast furnaces

(7) Smelting, melting and refining furnaces (including pyrometallurgical devices such as cupolas, reverberator furnaces, sintering machine, roasters, and foundry furnaces)

(8) Titanium dioxide chloride process oxidation reactors

(9) Methane reforming furnaces

(10) Pulping liquor recovery furnaces

(11) Combustion devices used in the recovery of sulfur values from spent sulfuric acid

(12) Halogen acid furnaces (HAFs) for the production of acid from halogenated hazardous waste generated by chemical production facilities where the furnace is located on the site of a chemical production facility, the acid product has a halogen acid content of at least 3%, the acid product is used in a manufacturing process, and, except for hazardous waste burned as fuel, hazardous waste fed to the furnace has a minimum halogen content of 20% as-generated.

(13) Such other devices as the Administrator may, after notice and comment, add to this list on the basis of one or more of the following factors:

(i) The design and use of the device primarily to accomplish recovery of material products;

(ii) The use of the device to burn or reduce raw materials to make a material product;

(iii) The use of the device to burn or reduce secondary materials as effective substitutes for raw materials, in processes using raw materials as principal feedstocks;

(iv) The use of the device to burn or reduce secondary materials as ingredients in an industrial process to make a material product;

(v) The use of the device in common industrial practice to produce a material product; and

(vi) Other factors, as appropriate.

*Infrared incinerator* means any enclosed device that uses electric powered resistance heaters as a source of radiant heat followed by an afterburner using controlled flame combustion and which is not listed as an industrial furnace.

*Inground tank* means a device meeting the definition of "tank" in § 260.10 whereby a portion of the tank wall is situated to any degree within the ground, thereby preventing visual inspection of that external surface area of the tank that is in the ground.

*Injection well* means a well into which fluids are injected. (See also "underground injection".)

*Inner liner* means a continuous layer of material placed inside a tank or container which protects the construction materials of the tank or container from the contained waste or reagents used to treat the waste.

*Installation inspector* means a person who, by reason of his knowledge of the physical sciences and the principles of engineering, acquired by a professional education and related practical experience, is qualified to supervise the installation of tank systems.

*Intermediate facility* means any facility that stores hazardous secondary materials for more than 10 days, other than a hazardous secondary material generator or reclaimer of such material.

*International shipment* means the transportation of hazardous waste into or out of the jurisdiction of the United States.

*Lamp*, also referred to as "universal waste lamp", is defined as the bulb or tube portion of an electric lighting device. A lamp is specifically designed to produce radiant energy, most often in the ultraviolet, visible, and infra-red regions of the electromagnetic spectrum. Examples of common universal waste electric lamps include, but are not limited to, fluorescent, high intensity discharge, neon, mercury vapor, high pressure sodium, and metal halide lamps.

*Land-based unit* means an area where hazardous secondary materials are placed in or on the land before recycling. This definition does not include land-based production units.

*Landfill* means a disposal facility or part of a facility where hazardous waste is placed in or on land and which is not a pile, a land treatment facility, a surface impoundment, an underground injection well, a salt dome formation, a salt bed formation, an underground mine, a cave, or a corrective action management unit.

*Landfill cell* means a discrete volume of a hazardous waste landfill which uses a liner to provide isolation of wastes from adjacent cells or wastes. Examples of landfill cells are trenches and pits.

*Land treatment facility* means a facility or part of a facility at which hazardous waste is applied onto or incorporated into the soil surface; such facilities are disposal facilities if the waste will remain after closure.

*Leachate* means any liquid, including any suspended components in the liquid, that has percolated through or drained from hazardous waste.

*Leak-detection system* means a system capable of detecting the failure of either the primary or secondary containment structure or the presence of a release of hazardous waste or accumulated liquid in the secondary containment structure. Such a system must employ operational controls (e.g., daily visual inspections for releases into the secondary containment system of aboveground tanks) or consist of an interstitial monitoring device designed to detect continuously and automatically the failure of the primary or secondary containment structure or the presence of a release of hazardous waste into the secondary containment structure.

*Liner* means a continuous layer of natural or man-made materials, beneath or on the sides of a surface impoundment, landfill, or landfill cell, which restricts the downward or lateral escape of hazardous waste, hazardous waste constituents, or leachate.

*Management* or *hazardous waste management* means the systematic control of the collection, source separation, storage, transportation, processing, treatment, recovery, and disposal of hazardous waste.

*Manifest* means: The shipping document EPA Form 8700–22 (including, if necessary, EPA Form 8700–22A), originated and signed by the generator or offeror in accordance with the instructions in the appendix to 40 CFR part 262 and the applicable requirements of 40 CFR parts 262 through 265.

*Manifest tracking number* means: The alphanumeric identification number (*i.e.*, a unique three letter suffix preceded by nine numerical digits), which is pre-printed in Item 4 of the Manifest by a registered source.

*Mercury-containing equipment* means a device or part of a device (including thermostats, but excluding batteries and lamps) that contains elemental mercury integral to its function.

*Military munitions* means all ammunition products and components produced or used by or for the U.S. Department of Defense or the U.S. Armed Services for national defense and security, including military munitions under the control of the Department of Defense, the U.S. Coast Guard, the U.S. Department of Energy (DOE), and National Guard personnel. The term military munitions includes: confined gaseous, liquid, and solid propellants, explosives, pyrotechnics, chemical and riot control agents, smokes, and incendiaries used by DOD components, including bulk explosives and chemical warfare agents, chemical munitions, rockets, guided and ballistic missiles, bombs, warheads, mortar rounds, artillery ammunition, small arms ammunition, grenades, mines, torpedoes, depth charges, cluster munitions and dispensers, demolition charges, and devices and components thereof. Military munitions do not include wholly inert items, improvised explosive devices, and nuclear weapons, nuclear devices, and nuclear components thereof. However, the term does include non-nuclear components of nuclear devices, managed under DOE's nuclear weapons program after all required sanitization operations under the Atomic Energy Act of 1954, as amended, have been completed.

*Mining overburden returned to the mine site* means any material overlying an economic mineral deposit which is removed to gain access to that deposit and is then used for reclamation of a surface mine.

*Miscellaneous unit* means a hazardous waste management unit where hazardous waste is treated, stored, or disposed of and that is not a container, tank, surface impoundment, pile, land treatment unit, landfill, incinerator, boiler, industrial furnace, underground injection well with appropriate technical standards under part 146 of this chapter, containment building, corrective action management unit, unit eligible for a research, development, and demonstration permit under 40 CFR 270.65, or staging pile.

*Movement* means that hazardous waste transported to a facility in an individual vehicle.

*New hazardous waste management facility* or *new facility* means a facility which began operation, or for which construction commenced after November 19, 1980. (See also "Existing hazardous waste management facility".)

*New tank system* or *new tank component* means a tank system or component that will be used for the storage or treatment of hazardous waste and for which installation has commenced after July 14, 1986; except, however, for purposes of §264.193(g)(2) and §265.193(g)(2), a new tank system is one for which construction commences after July 14, 1986. (See also "existing tank system.")

*On ground tank* means a device meeting the definition of "tank" in §260.10 and that is situated in such a way that the bottom of the tank is on the same level as the adjacent surrounding surface so that the external tank bottom cannot be visually inspected.

*On-site* means the same or geographically contiguous property which may be divided by public or private right-of-way, provided the entrance and exit between the properties is at a cross-roads intersection, and access is by crossing as opposed to going along, the right-of-way. Non-contiguous properties owned by the same person but connected by a right-of-way which he controls and to which the public does not have access, is also considered on-site property.

*Open burning* means the combustion of any material without the following characteristics:

(1) Control of combustion air to maintain adequate temperature for efficient combustion,

(2) Containment of the combustion-reaction in an enclosed device to provide sufficient residence time and mixing for complete combustion, and

(3) Control of emission of the gaseous combustion products.

(See also "incineration" and "thermal treatment".)

*Operator* means the person responsible for the overall operation of a facility.

*Owner* means the person who owns a facility or part of a facility.

*Partial closure* means the closure of a hazardous waste management unit in accordance with the applicable closure requirements of parts 264 and 265 of this chapter at a facility that contains other active hazardous waste management units. For example, partial closure may include the closure of a tank (including its associated piping and underlying containment systems), landfill cell, surface impoundment, waste pile, or other hazardous waste management unit, while other units of the same facility continue to operate.

*Performance Track member facility* means a facility that has been accepted by EPA for membership in the National Environmental Performance Track Program and is still a member of the Program. The National Environmental Performance Track Program is a voluntary, facility based, program for top environmental performers. Facility members must demonstrate a good record of compliance, past success in achieving environmental goals, and commit to future specific quantified environmental goals, environmental management systems, local community outreach, and annual reporting of measurable results.

*Person* means an individual, trust, firm, joint stock company, Federal Agency, corporation (including a government corporation), partnership, association, State, municipality, commission, political subdivision of a State, or any interstate body.

*Personnel* or *facility personnel* means all persons who work at, or oversee the operations of, a hazardous waste facility, and whose actions or failure to act may result in noncompliance with the requirements of part 264 or 265 of this chapter.

*Pesticide* means any substance or mixture of substances intended for preventing, destroying, repelling, or mitigating any pest, or intended for use as a plant regulator, defoliant, or desiccant, other than any article that:

(1) Is a new animal drug under FFDCA section 201(w), or

(2) Is an animal drug that has been determined by regulation of the Secretary of Health and Human Services not to be a new animal drug, or

(3) Is an animal feed under FFDCA section 201(x) that bears or contains any substances described by paragraph (1) or (2) of this definition.

*Pile* means any non-containerized accumulation of solid, nonflowing hazardous waste that is used for treatment or storage and that is not a containment building.

*Plasma arc incinerator* means any enclosed device using a high intensity electrical discharge or arc as a source of heat followed by an afterburner using controlled flame combustion and which is not listed as an industrial furnace.

*Point source* means any discernible, confined, and discrete conveyance, including, but not limited to any pipe, ditch, channel, tunnel, conduit, well, discrete fissure, container, rolling stock, concentrated animal feeding operation, or vessel or other floating craft, from which pollutants are or may be discharged. This term does not include return flows from irrigated agriculture.

*Publicly owned treatment works* or *POTW* means any device or system used in the treatment (including recycling and reclamation) of municipal sewage or industrial wastes of a liquid nature which is owned by a "State" or "municipality" (as defined by section 502(4) of the CWA). This definition includes sewers, pipes, or other conveyances only if they convey wastewater to a POTW providing treatment.

*Qualified Ground-Water Scientist* means a scientist or engineer who has received a baccalaureate or post-graduate degree in the natural sciences or engineering, and has sufficient training and experience in ground-water hydrology and related fields as may be demonstrated by state registration, professional certifications, or completion of accredited university courses that enable that individual to make sound professional judgements regarding ground-water monitoring and contaminant fate and transport.

*Regional Administrator* means the Regional Administrator for the EPA Region in which the facility is located, or his designee.

*Remediation waste* means all solid and hazardous wastes, and all media (including ground water, surface water, soils, and sediments) and debris, that are managed for implementing cleanup.

*Remediation waste management site* means a facility where an owner or operator is or will be treating, storing or disposing of hazardous remediation wastes. A remediation waste management site is not a facility that is subject to corrective action under 40 CFR 264.101, but is subject to corrective action requirements if the site is located in such a facility.

*Replacement unit* means a landfill, surface impoundment, or waste pile unit (1) from which all or substantially all of the waste is removed, and (2) that is subsequently reused to treat, store, or dispose of hazardous waste. "Replacement unit" does not apply to a unit from which waste is removed during closure, if the subsequent reuse solely involves the disposal of waste from that unit and other closing units or corrective action areas at the facility, in accordance with an approved closure plan or EPA or State approved corrective action.

*Representative sample* means a sample of a universe or whole (e.g., waste pile, lagoon, ground water) which can be expected to exhibit the average properties of the universe or whole.

*Run-off* means any rainwater, leachate, or other liquid that drains over land from any part of a facility.

*Run-on* means any rainwater, leachate, or other liquid that drains over land onto any part of a facility.

*Saturated zone* or *zone of saturation* means that part of the earth's crust in which all voids are filled with water.

*Sludge* means any solid, semi-solid, or liquid waste generated from a municipal, commercial, or industrial wastewater treatment plant, water supply treatment plant, or air pollution control facility exclusive of the treated effluent from a wastewater treatment plant.

*Sludge dryer* means any enclosed thermal treatment device that is used to dehydrate sludge and that has a maximum total thermal input, excluding the heating value of the sludge itself, of 2,500 Btu/lb of sludge treated on a wet-weight basis.

*Small Quantity Generator* means a generator who generates less than 1000 kg of hazardous waste in a calendar month.

*Solid waste* means a solid waste as defined in § 261.2 of this chapter.

*Sorbent* means a material that is used to soak up free liquids by either adsorption or absorption, or both. *Sorb* means to either adsorb or absorb, or both.

*Staging pile* means an accumulation of solid, non-flowing remediation waste (as defined in this section) that is not a containment building and that is used only during remedial operations for temporary storage at a facility. Staging piles must be designated by the Director according to the requirements of 40 CFR 264.554.

*State* means any of the several States, the District of Columbia, the Commonwealth of Puerto Rico, the Virgin Islands, Guam, American Samoa, and the Commonwealth of the Northern Mariana Islands.

*Storage* means the holding of hazardous waste for a temporary period, at the end of which the hazardous waste is treated, disposed of, or stored elsewhere.

*Sump* means any pit or reservoir that meets the definition of tank and those troughs/trenches connected to it that serve to collect hazardous waste for transport to hazardous waste storage, treatment, or disposal facilities; except that as used in the landfill, surface impoundment, and waste pile rules, "sump" means any lined pit or reservoir that serves to collect liquids drained from a leachate collection and removal system or leak detection system for subsequent removal from the system.

*Surface impoundment* or *impoundment* means a facility or part of a facility which is a natural topographic depression, man-made excavation, or diked area formed primarily of earthen materials (although it may be lined with man-made materials), which is designed to hold an accumulation of liquid wastes or wastes containing free liquids, and which is not an injection well. Examples of surface impoundments are holding, storage, settling, and aeration pits, ponds, and lagoons.

*Tank* means a stationary device, designed to contain an accumulation of hazardous waste which is constructed primarily of non-earthen materials (e.g., wood, concrete, steel, plastic) which provide structural support.

*Tank system* means a hazardous waste storage or treatment tank and its associated ancillary equipment and containment system.

*TEQ* means toxicity equivalence, the international method of relating the toxicity of various dioxin/furan congeners to the toxicity of 2,3,7,8-tetrachlorodibenzo-p-dioxin.

*Thermal treatment* means the treatment of hazardous waste in a device which uses elevated temperatures as the primary means to change the chemical, physical, or biological character or composition of the hazardous waste. Examples of thermal treatment processes are incineration, molten salt, pyrolysis, calcination, wet air oxidation, and microwave discharge. (See also "incinerator" and "open burning".)

16

*Thermostat* means a temperature control device that contains metallic mercury in an ampule attached to a bimetal sensing element, and mercury-containing ampules that have been removed from these temperature control devices in compliance with the requirements of 40 CFR 273.13(c)(2) or 273.33(c)(2).

*Totally enclosed treatment facility* means a facility for the treatment of hazardous waste which is directly connected to an industrial production process and which is constructed and operated in a manner which prevents the release of any hazardous waste or any constituent thereof into the environment during treatment. An example is a pipe in which waste acid is neutralized.

*Transfer facility* means any transportation-related facility, including loading docks, parking areas, storage areas and other similar areas where shipments of hazardous waste or hazardous secondary materials are held during the normal course of transportation.

*Transport vehicle* means a motor vehicle or rail car used for the transportation of cargo by any mode. Each cargo-carrying body (trailer, railroad freight car, etc.) is a separate transport vehicle.

*Transportation* means the movement of hazardous waste by air, rail, highway, or water.

*Transporter* means a person engaged in the offsite transportation of hazardous waste by air, rail, highway, or water.

*Treatability Study* means a study in which a hazardous waste is subjected to a treatment process to determine: (1) Whether the waste is amenable to the treatment process, (2) what pretreatment (if any) is required, (3) the optimal process conditions needed to achieve the desired treatment, (4) the efficiency of a treatment process for a specific waste or wastes, or (5) the characteristics and volumes of residuals from a particular treatment process. Also included in this definition for the purpose of the §261.4 (e) and (f) exemptions are liner compatibility, corrosion, and other material compatibility studies and toxicological and health effects studies. A "treatability study" is not a means to commercially treat or dispose of hazardous waste.

*Treatment* means any method, technique, or process, including neutralization, designed to change the physical, chemical, or biological character or composition of any hazardous waste so as to neutralize such waste, or so as to recover energy or material resources from the waste, or so as to render such waste non-hazardous, or less hazardous; safer to transport, store, or dispose of; or amenable for recovery, amenable for storage, or reduced in volume.

*Treatment zone* means a soil area of the unsaturated zone of a land treatment unit within which hazardous constituents are degraded, transformed, or immobilized.

*Underground injection* means the subsurface emplacement of fluids through a bored, drilled or driven well; or through a dug well, where the depth of the dug well is greater than the largest surface dimension. (See also "injection well".)

*Underground tank* means a device meeting the definition of "tank" in §260.10 whose entire surface area is totally below the surface of and covered by the ground.

*Unfit-for use tank system* means a tank system that has been determined through an integrity assessment or other inspection to be no longer capable of storing or treating hazardous waste without posing a threat of release of hazardous waste to the environment.

*United States* means the 50 States, the District of Columbia, the Commonwealth of Puerto Rico, the U.S. Virgin Islands, Guam, American Samoa, and the Commonwealth of the Northern Mariana Islands.

*Universal waste* means any of the following hazardous wastes that are managed under the universal waste requirements of part 273 of this chapter:

(1) Batteries as described in §273.2 of this chapter;

(2) Pesticides as described in §273.3 of this chapter;

(3) Mercury-containing equipment as described in §273.4 of this chapter; and

(4) Lamps as described in §273.5 of this chapter.

*Universal Waste Handler:*

(1) Means:

17

(i) A generator (as defined in this section) of universal waste; or

(ii) The owner or operator of a facility, including all contiguous property, that receives universal waste from other universal waste handlers, accumulates universal waste, and sends universal waste to another universal waste handler, to a destination facility, or to a foreign destination.

(2) Does not mean:

(i) A person who treats (except under the provisions of 40 CFR 273.13 (a) or (c), or 273.33 (a) or (c)), disposes of, or recycles universal waste; or

(ii) A person engaged in the off-site transportation of universal waste by air, rail, highway, or water, including a universal waste transfer facility.

*Universal Waste Transporter* means a person engaged in the off-site transportation of universal waste by air, rail, highway, or water.

*Unsaturated zone* or *zone of aeration* means the zone between the land surface and the water table.

*Uppermost aquifer* means the geologic formation nearest the natural ground surface that is an aquifer, as well as lower aquifers that are hydraulically interconnected with this aquifer within the facility's property boundary.

*Used oil* means any oil that has been refined from crude oil, or any synthetic oil, that has been used and as a result of such use is contaminated by physical or chemical impurities.

*Vessel* includes every description of watercraft, used or capable of being used as a means of transportation on the water.

*Wastewater treatment unit* means a device which:

(1) Is part of a wastewater treatment facility that is subject to regulation under either section 402 or 307(b) of the Clean Water Act; and

(2) Receives and treats or stores an influent wastewater that is a hazardous waste as defined in § 261.3 of this chapter, or that generates and accumulates a wastewater treatment sludge that is a hazardous waste as defined in § 261.3 of this chapter, or treats or stores a wastewater treatment sludge which is a hazardous waste as defined in § 261.3 of this Chapter; and

(3) Meets the definition of tank or tank system in § 260.10 of this chapter.

*Water (bulk shipment)* means the bulk transportation of hazardous waste which is loaded or carried on board a vessel without containers or labels.

*Well* means any shaft or pit dug or bored into the earth, generally of a cylindrical form, and often walled with bricks or tubing to prevent the earth from caving in.

*Well injection:* (See "underground injection".)

*Zone of engineering control* means an area under the control of the owner/operator that, upon detection of a hazardous waste release, can be readily cleaned up prior to the release of hazardous waste or hazardous constituents to ground water or surface water.

[45 FR 33073, May 19, 1980]

EDITORIAL NOTE: For FEDERAL REGISTER citations affecting § 260.10, see the List of CFR Sections Affected, which appears in the Finding Aids section of the printed volume and on GPO Access.

## § 260.11   References.

(a) When used in parts 260 through 268 and 278 of this chapter, the following publications are incorporated by reference. These incorporations by reference were approved by the Director of the Federal Register pursuant to 5 U.S.C. 552(a) and 1 CFR part 51. These materials are incorporated as they exist on the date of approval and a notice of any change in these materials will be published in the FEDERAL REGISTER. Copies may be inspected at the Library, U.S. Environmental Protection Agency, 1200 Pennsylvania Ave., NW. (3403T), Washington, DC 20460, *libraryhq@epa.gov;* or at the National Archives and Records Administration (NARA). For information on the availability of this material at NARA, call 202–741–6030, or go to: *http:// www.archives.gov/federal_register/ code_of_federal_regulations/ ibr_locations.html.*

(b) The following materials are available for purchase from the American Society for Testing and Materials, 100 Barr Harbor Drive, P.O. Box C700, West Conshohocken, PA 19428–2959.

(1) ASTM D–93–79 or D–93–80, "Standard Test Methods for Flash Point by Pensky-Martens Closed Cup Tester," IBR approved for § 261.21.

(2) ASTM D–1946–82, "Standard Method for Analysis of Reformed Gas by Gas Chromatography," IBR approved for §§ 264.1033, 265.1033.

(3) ASTM D 2267–88, "Standard Test Method for Aromatics in Light Naphthas and Aviation Gasolines by Gas Chromatography," IBR approved for § 264.1063.

(4) ASTM D 2382–83, "Standard Test Method for Heat of Combustion of Hydrocarbon Fuels by Bomb Calorimeter (High-Precision Method)," IBR approved for §§ 264.1033, 265.1033.

(5) ASTM D 2879–92, "Standard Test Method for Vapor Pressure—Temperature Relationship and Initial Decomposition Temperature of Liquids by Isoteniscope," IBR approved for § 265.1084.

(6) ASTM D–3278–78, "Standard Test Methods for Flash Point for Liquids by Setaflash Closed Tester," IBR approved for § 261.21(a).

(7) ASTM E 168–88, "Standard Practices for General Techniques of Infrared Quantitative Analysis," IBR approved for § 264.1063.

(8) ASTM E 169–87, "Standard Practices for General Techniques of Ultraviolet-Visible Quantitative Analysis," IBR approved for § 264.1063.

(9) ASTM E 260–85, "Standard Practice for Packed Column Gas Chromatography," IBR approved for § 264.1063.

(10) ASTM E 926–88, "Standard Test Methods for Preparing Refuse-Derived Fuel (RDF) Samples for Analyses of Metals," Test Method C—Bomb, Acid Digestion Method.

(c) The following materials are available for purchase from the National Technical Information Service, 5285 Port Royal Road, Springfield, VA 22161; or for purchase from the Superintendent of Documents, U.S. Government Printing Office, Washington, DC 20402, (202) 512–1800.

(1) "APTI Course 415: Control of Gaseous Emissions," EPA Publication EPA–450/2–81–005, December 1981, IBR approved for §§ 264.1035 and 265.1035.

(2) Method 1664, Revision A, n-Hexane Extractable Material (HEM; Oil and Grease) and Silica Gel Treated n-Hexane Extractable Material (SGT-HEM; Non-polar Material) by Extraction and Gravimetry, PB99–121949, IBR approved for part 261, appendix IX.

(3) The following methods as published in the test methods compendium known as "Test Methods for Evaluating Solid Waste, Physical/Chemical Methods," EPA Publication SW–846, Third Edition. A suffix of "A" in the method number indicates revision one (the method has been revised once). A suffix of "B" in the method number indicates revision two (the method has been revised twice). A suffix of "C" in the method number indicates revision three (the method has been revised three times). A suffix of "D" in the method number indicates revision four (the method has been revised four times).

(i) Method 0010, dated September 1986 and in the Basic Manual, IBR approved for part 261, appendix IX.

(ii) Method 0020, dated September 1986 and in the Basic Manual, IBR approved for part 261, appendix IX.

(iii) Method 0030, dated September 1986 and in the Basic Manual, IBR approved for part 261, appendix IX.

(iv) Method 1320, dated September 1986 and in the Basic Manual, IBR approved for part 261, appendix IX.

(v) Method 1311, dated September 1992 and in Update I, IBR approved for part 261, appendix IX, and §§ 261.24, 268.7, 268.40.

(vi) Method 1330A, dated September 1992 and in Update I, IBR approved for part 261, appendix IX.

(vii) Method 1312 dated September 1994 and in Update III, IBR approved for part 261, appendix IX and § 278.3(b)(1).

(viii) Method 0011, dated December 1996 and in Update III, IBR approved for part 261, appendix IX, and part 266, appendix IX.

(ix) Method 0023A, dated December 1996 and in Update III, IBR approved for part 261, appendix IX, part 266, appendix IX, and § 266.104.

(x) Method 0031, dated December 1996 and in Update III, IBR approved for part 261, appendix IX.

(xi) Method 0040, dated December 1996 and in Update III, IBR approved for part 261, appendix IX.

(xii) Method 0050, dated December 1996 and in Update III, IBR approved for part 261, appendix IX, part 266, appendix IX, and § 266.107.

(xiii) Method 0051, dated December 1996 and in Update III, IBR approved for part 261, appendix IX, part 266, appendix IX, and § 266.107.

(xiv) Method 0060, dated December 1996 and in Update III, IBR approved for part 261, appendix IX, § 266.106, and part 266, appendix IX.

(xv) Method 0061, dated December 1996 and in Update III, IBR approved for part 261, appendix IX, § 266.106, and part 266, appendix IX.

(xvi) Method 9071B, dated April 1998 and in Update IIIA, IBR approved for part 261, appendix IX.

(xvii) Method 1010A, dated November 2004 and in Update IIIB, IBR approved for part 261, appendix IX.

(xviii) Method 1020B, dated November 2004 and in Update IIIB, IBR approved for part 261, appendix IX.

(xix) Method 1110A, dated November 2004 and in Update IIIB, IBR approved for § 261.22 and part 261, appendix IX.

(xx) Method 1310B, dated November 2004 and in Update IIIB, IBR approved for part 261, appendix IX.

(xxi) Method 9010C, dated November 2004 and in Update IIIB, IBR approved for part 261, appendix IX and §§ 268.40, 268.44, 268.48.

(xxii) Method 9012B, dated November 2004 and in Update IIIB, IBR approved for part 261, appendix IX and §§ 268.40, 268.44, 268.48.

(xxiii) Method 9040C, dated November 2004 and in Update IIIB, IBR approved for part 261, appendix IX and § 261.22.

(xxiv) Method 9045D, dated November 2004 and in Update IIIB, IBR approved for part 261, appendix IX.

(xxv) Method 9060A, dated November 2004 and in Update IIIB, IBR approved for part 261, appendix IX, and §§ 264.1034, 264.1063, 265.1034, 265.1063.

(xxvi) Method 9070A, dated November 2004 and in Update IIIB, IBR approved for part 261, appendix IX.

(xxvii) Method 9095B, dated November 2004 and in Update IIIB, IBR approved, part 261, appendix IX, and §§ 264.190, 264.314, 265.190, 265.314, 265.1081, 267.190(a), 268.32.

(d) The following materials are available for purchase from the National Fire Protection Association, 1 Batterymarch Park, P.O. Box 9101, Quincy, MA 02269–9101.

(1) "Flammable and Combustible Liquids Code" (1977 or 1981), IBR approved for §§ 264.198, 265.198, 267.202(b).

(2) [Reserved]

(e) The following materials are available for purchase from the American Petroleum Institute, 1220 L Street, Northwest, Washington, DC 20005.

(1) API Publication 2517, Third Edition, February 1989, "Evaporative Loss from External Floating-Roof Tanks," IBR approved for § 265.1084.

(2) [Reserved]

(f) The following materials are available for purchase from the Environmental Protection Agency, Research Triangle Park, NC.

(1) "Screening Procedures for Estimating the Air Quality Impact of Stationary Sources, Revised", October 1992, EPA Publication No. EPA–450/R–92–019, IBR approved for part 266, appendix IX.

(2) [Reserved]

(g) The following materials are available for purchase from the Organisation for Economic Co-operation and Development, Environment Direcorate, 2 rue Andre Pascal, 75775 Paris Cedex 16, France.

(1) OECD Green List of Wastes (revised May 1994), Amber List of Wastes and Red List of Wastes (both revised May 1993) as set forth in Appendix 3, Appendix 4 and Appendix 5, respectively, to the OECD Council Decision C(92)39/FINAL (Concerning the Control of Transfrontier Movements of Wastes Destined for Recovery Operations), IBR approved for 262.89 of this chapter.

(2) [Reserved]

[70 FR 34560, June 14, 2005, as amended at 70 FR 53453, Sept. 8, 2005; 70 FR 59575, Oct. 12, 2005; 72 FR 39352, July 18, 2007]

## Subpart C—Rulemaking Petitions

### § 260.20  General.

(a) Any person may petition the Administrator to modify or revoke any provision in parts 260 through 266, 268 and 273 of this chapter. This section sets forth general requirements which apply to all such petitions. Section 260.21 sets forth additional requirements for petitions to add a testing or analytical method to part 261, 264 or 265 of this chapter. Section 260.22 sets forth additional requirements for petitions to exclude a waste or waste-derived material at a particular facility from § 261.3 of this chapter or the lists of hazardous wastes in subpart D of part 261 of this chapter. Section 260.23 sets forth additional requirements for petitions to amend part 273 of this chapter to include additional hazardous wastes or categories of hazardous waste as universal waste.

(b) Each petition must be submitted to the Administrator by certified mail and must include:

(1) The petitioner's name and address;

(2) A statement of the petitioner's interest in the proposed action;

(3) A description of the proposed action, including (where appropriate) suggested regulatory language; and

(4) A statement of the need and justification for the proposed action, including any supporting tests, studies, or other information.

(c) The Administrator will make a tentative decision to grant or deny a petition and will publish notice of such tentative decision, either in the form of an advanced notice of proposed rulemaking, a proposed rule, or a tentative determination to deny the petition, in the FEDERAL REGISTER for written public comment.

(d) Upon the written request of any interested person, the Administrator may, at his discretion, hold an informal public hearing to consider oral comments on the tentative decision. A person requesting a hearing must state the issues to be raised and explain why written comments would not suffice to communicate the person's views. The Administrator may in any case decide on his own motion to hold an informal public hearing.

(e) After evaluating all public comments the Administrator will make a final decision by publishing in the FEDERAL REGISTER a regulatory amendment or a denial of the petition.

[45 FR 33073, May 19, 1980, as amended at 51 FR 40636, Nov. 7, 1986; 57 FR 38564, Aug. 25, 1992; 60 FR 25540, May 11, 1995]

## § 260.21 Petitions for equivalent testing or analytical methods.

(a) Any person seeking to add a testing or analytical method to part 261, 264, or 265 of this chapter may petition for a regulatory amendment under this section and § 260.20. To be successful, the person must demonstrate to the satisfaction of the Administrator that the proposed method is equal to or superior to the corresponding method prescribed in part 261, 264, or 265 of this chapter, in terms of its sensitivity, accuracy, and precision (i.e., reproducibility).

(b) Each petition must include, in addition to the information required by § 260.20(b):

(1) A full description of the proposed method, including all procedural steps and equipment used in the method;

(2) A description of the types of wastes or waste matrices for which the proposed method may be used;

(3) Comparative results obtained from using the proposed method with those obtained from using the relevant or corresponding methods prescribed in part 261, 264, or 265 of this chapter;

(4) An assessment of any factors which may interfere with, or limit the use of, the proposed method; and

(5) A description of the quality control procedures necessary to ensure the sensitivity, accuracy and precision of the proposed method.

(c) After receiving a petition for an equivalent method, the Administrator may request any additional information on the proposed method which he may reasonably require to evaluate the method.

(d) If the Administrator amends the regulations to permit use of a new testing method, the method will be incorporated by reference in § 260.11 and added to "Test Methods for Evaluating Solid Waste, Physical/Chemical Methods," EPA Publication SW–846, U.S. Environmental Protection Agency, Office of Resource Conservation and Recovery, Washington, DC 20460.

[45 FR 33073, May 19, 1980, as amended at 49 FR 47391, Dec. 4, 1984; 70 FR 34561, June 14, 2005; 74 FR 30230, June 25, 2009]

## § 260.22 Petitions to amend part 261 to exclude a waste produced at a particular facility.

(a) Any person seeking to exclude a waste at a particular generating facility from the lists in subpart D of part 261 may petition for a regulatory amendment under this section and §260.20. To be successful:

(1) The petitioner must demonstrate to the satisfaction of the Administrator that the waste produced by a particular generating facility does not meet any of the criteria under which the waste was listed as a hazardous or an acutely hazardous waste; and

(2) Based on a complete application, the Administrator must determine, where he has a reasonable basis to believe that factors (including additional constituents) other than those for which the waste was listed could cause the waste to be a hazardous waste, that such factors do not warrant retaining the waste as a hazardous waste. A waste which is so excluded, however, still may be a hazardous waste by operation of subpart C of part 261.

(b) The procedures in this Section and §260.20 may also be used to petition the Administrator for a regulatory amendment to exclude from §261.3(a)(2)(ii) or (c), a waste which is described in these Sections and is either a waste listed in subpart D, or is derived from a waste listed in subpart D. This exclusion may only be issued for a particular generating, storage, treatment, or disposal facility. The petitioner must make the same demonstration as required by paragraph (a) of this section. Where the waste is a mixture of solid waste and one or more listed hazardous wastes or is derived from one or more hazardous wastes, his demonstration must be made with respect to the waste mixture as a whole; analyses must be conducted for not only those constituents for which the listed waste contained in the mixture was listed as hazardous, but also for factors (including additional constituents) that could cause the waste mixture to be a hazardous waste. A waste which is so excluded may still be a hazardous waste by operation of subpart C of part 261.

(c) If the waste is listed with codes "I", "C", "R", or "E", in subpart D,

(1) The petitioner must show that the waste does not exhibit the relevant characteristic for which the waste was listed as defined in §261.21, §261.22, §261.23, or §261.24 using any applicable methods prescribed therein. The petitioner also must show that the waste

does not exhibit any of the other characteristics defined in §261.21, §261.22, §261.23, or §261.24 using any applicable methods prescribed therein;

(2) Based on a complete application, the Administrator must determine, where he has a reasonable basis to believe that factors (including additional constituents) other than those for which the waste was listed could cause the waste to be hazardous waste, that such factors do not warrant retaining the waste as a hazardous waste. A waste which is so excluded, however, still may be a hazardous waste by operation of subpart C of part 261.

(d) If the waste is listed with code "T" in subpart D,

(1) The petitioner must demonstrate that the waste:

(i) Does not contain the constituent or constituents (as defined in Appendix VII of part 261 of this chapter) that caused the Administrator to list the waste; or

(ii) Although containing one or more of the hazardous constituents (as defined in appendix VII of part 261) that caused the Administrator to list the waste, does not meet the criterion of §261.11(a)(3) when considering the factors used by the Administrator in §261.11(a)(3) (i) through (xi) under which the waste was listed as hazardous; and

(2) Based on a complete application, the Administrator must determine, where he has a reasonable basis to believe that factors (including additional constituents) other than those for which the waste was listed could cause the waste to be a hazardous waste, that such factors do not warrant retaining the waste as a hazardous waste; and

(3) The petitioner must demonstrate that the waste does not exhibit any of the characteristics defined in §261.21, §261.22, §261.23, and §261.24 using any applicable methods prescribed therein;

(4) A waste which is so excluded, however, still may be a hazardous waste by operation of subpart C of part 261.

(e) If the waste is listed with the code "H" in subpart D,

(1) The petitioner must demonstrate that the waste does not meet the criterion of §261.11(a)(2); and

(2) Based on a complete application, the Administrator must determine, where he has a reasonable basis to believe that additional factors (including additional constituents) other than those for which the waste was listed could cause the waste to be a hazardous waste, that such factors do not warrant retaining the waste as a hazardous waste; and

(3) The petitioner must demonstrate that the waste does not exhibit any of the characteristics defined in §261.21, §261.22, §261.23, and §261.24 using any applicable methods prescribed therein;

(4) A waste which is so excluded, however, still may be a hazardous waste by operation of subpart C of part 261.

(f) [Reserved for listing radioactive wastes.]

(g) [Reserved for listing infectious wastes.]

(h) Demonstration samples must consist of enough representative samples, but in no case less than four samples, taken over a period of time sufficient to represent the variability or the uniformity of the waste.

(i) Each petition must include, in addition to the information required by §260.20(b):

(1) The name and address of the laboratory facility performing the sampling or tests of the waste;

(2) The names and qualifications of the persons sampling and testing the waste;

(3) The dates of sampling and testing;

(4) The location of the generating facility;

(5) A description of the manufacturing processes or other operations and feed materials producing the waste and an assessment of whether such processes, operations, or feed materials can or might produce a waste that is not covered by the demonstration;

(6) A description of the waste and an estimate of the average and maximum monthly and annual quantities of waste covered by the demonstration;

(7) Pertinent data on and discussion of the factors delineated in the respective criterion for listing a hazardous waste, where the demonstration is based on the factors in §261.11(a)(3);

(8) A description of the methodologies and equipment used to obtain the representative samples;

(9) A description of the sample handling and preparation techniques, including techniques used for extraction, containerization and preservation of the samples;

(10) A description of the tests performed (including results);

(11) The names and model numbers of the instruments used in performing the tests; and

(12) The following statement signed by the generator of the waste or his authorized representative:

I certify under penalty of law that I have personally examined and am familiar with the information submitted in this demonstration and all attached documents, and that, based on my inquiry of those individuals immediately responsible for obtaining the information, I believe that the submitted information is true, accurate, and complete. I am aware that there are significant penalties for submitting false information, including the possibility of fine and imprisonment.

(j) After receiving a petition for an exclusion, the Administrator may request any additional information which he may reasonably require to evaluate the petition.

(k) An exclusion will only apply to the waste generated at the individual facility covered by the demonstration and will not apply to waste from any other facility.

(l) The Administrator may exclude only part of the waste for which the demonstration is submitted where he has reason to believe that variability of the waste justifies a partial exclusion.

[45 FR 33073, May 19, 1980, as amended at 50 FR 28742, July 15, 1985; 54 FR 27116, June 27, 1989; 58 FR 46049, Aug. 31, 1994; 70 FR 34561, June 14, 2005; 71 FR 40258, July 14, 2006]

**§260.23 Petitions to amend 40 CFR part 273 to include additional hazardous wastes.**

(a) Any person seeking to add a hazardous waste or a category of hazardous waste to the universal waste regulations of part 273 of this chapter may petition for a regulatory amendment under this section, 40 CFR 260.20, and subpart G of 40 CFR part 273.

(b) To be successful, the petitioner must demonstrate to the satisfaction of the Administrator that regulation under the universal waste regulations of 40 CFR part 273: Is appropriate for the waste or category of waste; will improve management practices for the waste or category of waste; and will improve implementation of the hazardous waste program. The petition must include the information required by 40 CFR 260.20(b). The petition should also address as many of the factors listed in 40 CFR 273.81 as are appropriate for the waste or category of waste addressed in the petition.

(c) The Administrator will grant or deny a petition using the factors listed in 40 CFR 273.81. The decision will be based on the weight of evidence showing that regulation under 40 CFR part 273 is appropriate for the waste or category of waste, will improve management practices for the waste or category of waste, and will improve implementation of the hazardous waste program.

(d) The Administrator may request additional information needed to evaluate the merits of the petition.

[60 FR 25540, May 11, 1995]

### § 260.30  Non-waste determinations and variances from classification as a solid waste.

In accordance with the standards and criteria in § 260.31 and § 260.34 and the procedures in § 260.33, the Administrator may determine on a case-by-case basis that the following recycled materials are not solid wastes:

(a) Materials that are accumulated speculatively without sufficient amounts being recycled (as defined in § 261.1(c)(8) of this chapter);

(b) Materials that are reclaimed and then reused within the original production process in which they were generated;

(c) Materials that have been reclaimed but must be reclaimed further before the materials are completely recovered.

(d) Hazardous secondary materials that are reclaimed in a continuous industrial process; and

(e) Hazardous secondary materials that are indistinguishable in all relevant aspects from a product or intermediate.

[50 FR 661, Jan. 4, 1985; 50 FR 14219, Apr. 11, 1985, as amended at 59 FR 48041, Sept. 19, 1994; 73 FR 64758, Oct. 30, 2008]

### § 260.31  Standards and criteria for variances from classification as a solid waste.

(a) The Administrator may grant requests for a variance from classifying as a solid waste those materials that are accumulated speculatively without sufficient amounts being recycled if the applicant demonstrates that sufficient amounts of the material will be recycled or transferred for recycling in the following year. If a variance is granted, it is valid only for the following year, but can be renewed, on an annual basis, by filing a new application. The Administrator's decision will be based on the following criteria:

(1) The manner in which the material is expected to be recycled, when the material is expected to be recycled, and whether this expected disposition is likely to occur (for example, because of past practice, market factors, the nature of the material, or contractual arrangements for recycling);

(2) The reason that the applicant has accumulated the material for one or more years without recycling 75 percent of the volume accumulated at the beginning of the year;

(3) The quantity of material already accumulated and the quantity expected to be generated and accumulated before the material is recycled;

(4) The extent to which the material is handled to minimize loss;

(5) Other relevant factors.

(b) The Administrator may grant requests for a variance from classifying as a solid waste those materials that are reclaimed and then reused as feedstock within the original production process in which the materials were generated if the reclamation operation is an essential part of the production process. This determination will be based on the following criteria:

(1) How economically viable the production process would be if it were to use virgin materials, rather than reclaimed materials;

(2) The extent to which the material is handled before reclamation to minimize loss;

(3) The time periods between generating the material and its reclamation, and between reclamation and return to the original primary production process;

(4) The location of the reclamation operation in relation to the production process;

(5) Whether the reclaimed material is used for the purpose for which it was originally produced when it is returned to the original process, and whether it is returned to the process in substantially its original form;

(6) Whether the person who generates the material also reclaims it;

(7) Other relevant factors.

(c) The Regional Administrator may grant requests for a variance from classifying as a solid waste those materials that have been reclaimed but must be reclaimed further before recovery is completed if, after initial reclamation, the resulting material is commodity-like (even though it is not yet a commercial product, and has to be reclaimed further). This determination will be based on the following factors:

(1) The degree of processing the material has undergone and the degree of further processing that is required;

(2) The value of the material after it has been reclaimed;

(3) The degree to which the reclaimed material is like an analogous raw material;

(4) The extent to which an end market for the reclaimed material is guaranteed;

(5) The extent to which the reclaimed material is handled to minimize loss;

(6) Other relevant factors.

[50 FR 662, Jan. 4, 1985, as amended at 59 FR 48041, Sept. 19, 1994; 71 FR 16902, Apr. 4, 2006]

§ 260.32 Variances to be classified as a boiler.

In accordance with the standards and criteria in §260.10 (definition of "boiler"), and the procedures in §260.33, the Administrator may determine on a case-by-case basis that certain enclosed devices using controlled flame combustion are boilers, even though they do not otherwise meet the defini-

tion of boiler contained in §260.10, after considering the following criteria:

(a) The extent to which the unit has provisions for recovering and exporting thermal energy in the form of steam, heated fluids, or heated gases; and

(b) The extent to which the combustion chamber and energy recovery equipment are of integral design; and

(c) The efficiency of energy recovery, calculated in terms of the recovered energy compared with the thermal value of the fuel; and

(d) The extent to which exported energy is utilized; and

(e) The extent to which the device is in common and customary use as a "boiler" functioning primarily to produce steam, heated fluids, or heated gases; and

(f) Other factors, as appropriate.

[50 FR 662, Jan. 4, 1985, as amended at 59 FR 48041, Sept. 19, 1994]

§ 260.33 Procedures for variances from classification as a solid waste, for variances to be classified as a boiler, or for non-waste determinations.

The Administrator will use the following procedures in evaluating applications for variances from classification as a solid waste, applications to classify particular enclosed controlled flame combustion devices as boilers, or applications for non-waste determinations.

(a) The applicant must apply to the Administrator for the variance or non-waste determination. The application must address the relevant criteria contained in §260.31, §260.32, or §260.34, as applicable.

(b) The Administrator will evaluate the application and issue a draft notice tentatively granting or denying the application. Notification of this tentative decision will be provided by newspaper advertisement or radio broadcast in the locality where the recycler is located. The Administrator will accept comment on the tentative decision for 30 days, and may also hold a public hearing upon request or at his discretion. The Administrator will issue a final decision after receipt of comments and after the hearing (if any).

(c) For non-waste determinations, in the event of a change in circumstances that affect how a hazardous secondary material meets the relevant criteria contained in § 260.34 upon which a non-waste determination has been based, the applicant must re-apply to the Administrator for a formal determination that the hazardous secondary material continues to meet the relevant criteria and therefore is not a solid waste.

[59 FR 48041, Sept. 19, 1994, as amended at 73 FR 64758, Oct. 30, 2008]

### § 260.34  Standards and criteria for non-waste determinations.

(a) An applicant may apply to the Administrator for a formal determination that a hazardous secondary material is not discarded and therefore not a solid waste. The determinations will be based on the criteria contained in paragraphs (b) or (c) of this section, as applicable. If an application is denied, the hazardous secondary material might still be eligible for a solid waste variance or exclusion (for example, one of the solid waste variances under § 260.31). Determinations may also be granted by the State if the State is either authorized for this provision or if the following conditions are met:

(1) The State determines the hazardous secondary material meets the criteria in paragraphs (b) or (c) of this section, as applicable;

(2) The State requests that EPA review its determination; and

(3) EPA approves the State determination.

(b) The Administrator may grant a non-waste determination for hazardous secondary material which is reclaimed in a continuous industrial process if the applicant demonstrates that the hazardous secondary material is a part of the production process and is not discarded. The determination will be based on whether the hazardous secondary material is legitimately recycled as specified in § 260.43 and on the following criteria:

(1) The extent that the management of the hazardous secondary material is part of the continuous primary production process and is not waste treatment;

(2) Whether the capacity of the production process would use the hazardous secondary material in a reasonable time frame and ensure that the hazardous secondary material will not

be abandoned (for example, based on past practices, market factors, the nature of the hazardous secondary material, or any contractual arrangements);

(3) Whether the hazardous constituents in the hazardous secondary material are reclaimed rather than released to the air, water or land at significantly higher levels from either a statistical or from a health and environmental risk perspective than would otherwise be released by the production process; and

(4) Other relevant factors that demonstrate the hazardous secondary material is not discarded.

(c) The Administrator may grant a non-waste determination for hazardous secondary material which is indistinguishable in all relevant aspects from a product or intermediate if the applicant demonstrates that the hazardous secondary material is comparable to a product or intermediate and is not discarded. The determination will be based on whether the hazardous secondary material is legitimately recycled as specified in § 260.43 and on the following criteria:

(1) Whether market participants treat the hazardous secondary material as a product or intermediate rather than a waste (for example, based on the current positive value of the hazardous secondary material, stability of demand, or any contractual arrangements);

(2) Whether the chemical and physical identity of the hazardous secondary material is comparable to commercial products or intermediates;

(3) Whether the capacity of the market would use the hazardous secondary material in a reasonable time frame and ensure that the hazardous secondary material will not be abandoned (for example, based on past practices, market factors, the nature of the hazardous secondary material, or any contractual arrangements);

(4) Whether the hazardous constituents in the hazardous secondary material are reclaimed rather than released to the air, water or land at significantly higher levels from either a statistical or from a health and environmental risk perspective than would otherwise be released by the production process; and

(5) Other relevant factors that demonstrate the hazardous secondary material is not discarded.

[73 FR 64758, Oct. 30, 2008]

### §260.40 Additional regulation of certain hazardous waste recycling activities on a case-by-case basis.

(a) The Regional Administrator may decide on a case-by-case basis that persons accumulating or storing the recyclable materials described in §261.6(a)(2)(iii) of this chapter should be regulated under §261.6 (b) and (c) of this chapter. The basis for this decision is that the materials are being accumulated or stored in a manner that does not protect human health and the environment because the materials or their toxic constituents have not been adequately contained, or because the materials being accumulated or stored together are incompatible. In making this decision, the Regional Administrator will consider the following factors:

(1) The types of materials accumulated or stored and the amounts accumulated or stored;

(2) The method of accumulation or storage;

(3) The length of time the materials have been accumulated or stored before being reclaimed;

(4) Whether any contaminants are being released into the environment, or are likely to be so released; and

(5) Other relevant factors.

(b) [Reserved]

The procedures for this decision are set forth in §260.41 of this chapter.

[50 FR 662, Jan. 4, 1985, as amended at 71 FR 40258, July 14, 2006]

### §260.41 Procedures for case-by-case regulation of hazardous waste recycling activities.

The Regional Administrator will use the following procedures when determining whether to regulate hazardous waste recycling activities described in §261.6(a)(2)(iii) under the provisions of §261.6 (b) and (c), rather than under the provisions of subpart F of part 266 of this chapter.

(a) If a generator is accumulating the waste, the Regional Administrator will issue a notice setting forth the factual basis for the decision and stating that the person must comply with the applicable requirements of subparts A, C, D, and E of part 262 of this chapter. The notice will become final within 30 days, unless the person served requests a public hearing to challenge the decision. Upon receiving such a request, the Regional Administrator will hold a public hearing. The Regional Administrator will provide notice of the hearing to the public and allow public participation at the hearing. The Regional Administrator will issue a final order after the hearing stating whether or not compliance with part 262 is required. The order becomes effective 30 days after service of the decision unless the Regional Administrator specifies a later date or unless review by the Administrator is requested. The order may be appealed to the Administrator by any person who participated in the public hearing. The Administrator may choose to grant or to deny the appeal. Final Agency action occurs when a final order is issued and Agency review procedures are exhausted.

(b) If the person is accumulating the recyclable material as a storage facility, the notice will state that the person must obtain a permit in accordance with all applicable provisions of parts 270 and 124 of this chapter. The owner or operator of the facility must apply for a permit within no less than 60 days and no more than six months of notice, as specified in the notice. If the owner or operator of the facility wishes to challenge the Regional Administrator's decision, he may do so in his permit application, in a public hearing held on the draft permit, or in comments filed on the draft permit or on the notice of intent to deny the permit. The fact sheet accompanying the permit will specify the reasons for the Agency's determination. The question of whether the Regional Administrator's decision was proper will remain open for consideration during the public comment period discussed under §124.11 of this chapter and in any subsequent hearing.

[50 FR 663, Jan. 4, 1985, as amended at 71 FR 40258, July 14, 2006]

### §260.42 Notification requirement for hazardous secondary materials.

(a) Hazardous secondary material generators, tolling contractors, toll manufacturers, reclaimers, and intermediate facilities managing hazardous secondary materials which are excluded from regulation under § 261.2(a)(2)(ii), § 261.4(a)(23), (24), or (25) must send a notification prior to operating under the exclusion(s) and by March 1 of each even numbered year thereafter to the Regional Administrator using EPA Form 8700–12 that includes the following information:

(1) The name, address, and EPA ID number (if applicable) of the facility;

(2) The name and telephone number of a contact person;

(3) The NAICS code of the facility;

(4) The exclusion under which the hazardous secondary materials will be managed (e.g., § 261.2(a)(2)(ii), § 261.4(a)(23), (24), and/or (25));

(5) For reclaimers and intermediate facilities managing hazardous secondary materials in accordance with § 261.4(a)(24) or (25), whether the reclaimer or intermediate facility has financial assurance (not applicable for persons managing hazardous secondary materials generated and reclaimed under the control of the generator);

(6) When the facility expects to begin managing the hazardous secondary materials in accordance with the exclusion;

(7) A list of hazardous secondary materials that will be managed according to the exclusion (reported as the EPA hazardous waste numbers that would apply if the hazardous secondary materials were managed as hazardous wastes);

(8) For each hazardous secondary material, whether the hazardous secondary material, or any portion thereof, will be managed in a land-based unit;

(9) The quantity of each hazardous secondary material to be managed annually; and

(10) The certification (included in EPA Form 8700–12) signed and dated by an authorized representative of the facility.

(b) If a hazardous secondary material generator, tolling contractor, toll manufacturer, reclaimer or intermediate facility has submitted a notification, but then subsequently stops managing hazardous secondary materials in accordance with the exclusion(s), the facility must notify the Regional Administrator within thirty (30) days using EPA Form 8700–12. For purposes of this section, a facility has stopped managing hazardous secondary materials if the facility no longer generates, manages and/or reclaims hazardous secondary materials under the exclusion(s) and does not expect to manage any amount of hazardous secondary materials for at least one year.

[73 FR 64759, Oct. 30, 2008]

§ 260.43  Legitimate recycling of hazardous secondary materials regulated under § 260.34, § 261.2(a)(2)(ii), and § 261.4(a)(23), (24), or (25).

(a) Persons regulated under § 260.34 or claiming to be excluded from hazardous waste regulation under § 261.2(a)(2)(ii), § 261.4(a)(23), (24), or (25) because they are engaged in reclamation must be able to demonstrate that the recycling is legitimate. Hazardous secondary material that is not legitimately recycled is discarded material and is a solid waste. In determining if their recycling is legitimate, persons must address the requirements of § 260.43(b) and must consider the requirements of § 260.43(c) below.

(b) Legitimate recycling must involve a hazardous secondary material that provides a useful contribution to the recycling process or to a product or intermediate of the recycling process, and the recycling process must produce a valuable product or intermediate.

(1) The hazardous secondary material provides a useful contribution if it

(i) Contributes valuable ingredients to a product or intermediate; or

(ii) Replaces a catalyst or carrier in the recycling process; or

(iii) Is the source of a valuable constituent recovered in the recycling process; or

(iv) Is recovered or regenerated by the recycling process; or

(v) Is used as an effective substitute for a commercial product.

(2) The product or intermediate is valuable if it is

(i) Sold to a third party; or

(ii) Used by the recycler or the generator as an effective substitute for a commercial product or as an ingredient or intermediate in an industrial process.

(c) The following factors must be considered in making a determination as to the overall legitimacy of a specific recycling activity.

(1) The generator and the recycler should manage the hazardous secondary material as a valuable commodity. Where there is an analogous raw material, the hazardous secondary material should be managed, at a minimum, in a manner consistent with the management of the raw material. Where there is no analogous raw material, the hazardous secondary material should be contained. Hazardous secondary materials that are released to the environment and are not recovered immediately are discarded.

(2) The product of the recycling process does not

(i) Contain significant concentrations of any hazardous constituents found in Appendix VIII of part 261 that are not found in analogous products; or

(ii) Contain concentrations of any hazardous constituents found in Appendix VIII of part 261 at levels that are significantly elevated from those found in analogous products; or

(iii) Exhibit a hazardous characteristic (as defined in part 261 subpart C) that analogous products do not exhibit.

(3) In making a determination that a hazardous secondary material is legitimately recycled, persons must evaluate all factors and consider legitimacy as a whole. If, after careful evaluation of these other considerations, one or both of the factors are not met, then this fact may be an indication that the material is not legitimately recycled.

However, the factors in this paragraph do not have to be met for the recycling to be considered legitimate. In evaluating the extent to which these factors are met and in determining whether a process that does not meet one or both of these factors is still legitimate, persons can consider the protectiveness of the storage methods, exposure from toxics in the product, the bioavailability of the toxics in the product, and other relevant considerations.

[73 FR 64759, Oct. 30, 2008]

# PART 261—IDENTIFICATION AND LISTING OF HAZARDOUS WASTE

## Subpart A—General

## Subpart B—Criteria for Identifying the Characteristics of Hazardous Waste and for Listing Hazardous Waste

## Subpart C—Characteristics of Hazardous Waste

## Subpart D—Lists of Hazardous Wastes

## Subpart E—Exclusions/Exemptions

## Subparts F-G [Reserved]

## Subpart H —Financial Requirements for Management of Excluded Hazardous Secondary Materials

AUTHORITY: 42 U.S.C. 6905, 6912(a), 6921, 6922, 6924(y) and 6938.

SOURCE: 45 FR 33119, May 19, 1980, unless otherwise noted.

## Subpart A—General

### § 261.1  Purpose and scope.

(a) This part identifies those solid wastes which are subject to regulation as hazardous wastes under parts 262 through 265, 268, and parts 270, 271, and 124 of this chapter and which are subject to the notification requirements of section 3010 of RCRA. In this part:

(1) Subpart A defines the terms "solid waste" and "hazardous waste", identifies those wastes which are excluded from regulation under parts 262 through 266, 268 and 270 and establishes special management requirements for hazardous waste produced by conditionally exempt small quantity generators and hazardous waste which is recycled.

(2) Subpart B sets forth the criteria used by EPA to identify characteristics

of hazardous waste and to list particular hazardous wastes.

(3) Subpart C identifies characteristics of hazardous waste.

(4) Subpart D lists particular hazardous wastes.

(b)(1) The definition of solid waste contained in this part applies only to wastes that also are hazardous for purposes of the regulations implementing subtitle C of RCRA. For example, it does not apply to materials (such as non-hazardous scrap, paper, textiles, or rubber) that are not otherwise hazardous wastes and that are recycled.

(2) This part identifies only some of the materials which are solid wastes and hazardous wastes under sections 3007, 3013, and 7003 of RCRA. A material which is not defined as a solid waste in this part, or is not a hazardous waste identified or listed in this part, is still a solid waste and a hazardous waste for purposes of these sections if:

(i) In the case of sections 3007 and 3013, EPA has reason to believe that the material may be a solid waste within the meaning of section 1004(27) of RCRA and a hazardous waste within the meaning of section 1004(5) of RCRA; or

(ii) In the case of section 7003, the statutory elements are established.

(c) For the purposes of §§ 261.2 and 261.6:

(1) A "spent material" is any material that has been used and as a result of contamination can no longer serve the purpose for which it was produced without processing;

(2) "Sludge" has the same meaning used in § 260.10 of this chapter;

(3) A "by-product" is a material that is not one of the primary products of a production process and is not solely or separately produced by the production process. Examples are process residues such as slags or distillation column bottoms. The term does not include a co-product that is produced for the general public's use and is ordinarily used in the form it is produced by the process.

(4) A material is "reclaimed" if it is processed to recover a usable product, or if it is regenerated. Examples are recovery of lead values from spent batteries and regeneration of spent solvents. In addition, for purposes of

§§ 261.2(a)(2)(ii), 261.4(a)(23), and 261.4(a)(24) smelting, melting and refining furnaces are considered to be solely engaged in metals reclamation if the metal recovery from the hazardous secondary materials meets the same requirements as those specified for metals recovery from hazardous waste found in § 266.100(d)(1)–(3) of this chapter, and if the residuals meet the requirements specified in § 266.112 of this chapter.

(5) A material is "used or reused" if it is either:

(i) Employed as an ingredient (including use as an intermediate) in an industrial process to make a product (for example, distillation bottoms from one process used as feedstock in another process). However, a material will not satisfy this condition if distinct components of the material are recovered as separate end products (as when metals are recovered from metal-containing secondary materials); or

(ii) Employed in a particular function or application as an effective substitute for a commercial product (for example, spent pickle liquor used as phosphorous precipitant and sludge conditioner in wastewater treatment).

(6) "Scrap metal" is bits and pieces of metal parts (e.g.,) bars, turnings, rods, sheets, wire) or metal pieces that may be combined together with bolts or soldering (e.g., radiators, scrap automobiles, railroad box cars), which when worn or superfluous can be recycled.

(7) A material is "recycled" if it is used, reused, or reclaimed.

(8) A material is "accumulated speculatively" if it is accumulated before being recycled. A material is not accumulated speculatively, however, if the person accumulating it can show that the material is potentially recyclable and has a feasible means of being recycled; and that—during the calendar year (commencing on January 1)—the amount of material that is recycled, or transferred to a different site for recycling, equals at least 75 percent by weight or volume of the amount of that material accumulated at the beginning of the period. In calculating the percentage of turnover, the 75 percent requirement is to be applied to each material of the same type (e.g., slags from a single smelting process) that is recy-

cled in the same way (*i.e.*, from which the same material is recovered or that is used in the same way). Materials accumulating in units that would be exempt from regulation under § 261.4(c) are not to be included in making the calculation. (Materials that are already defined as solid wastes also are not to be included in making the calculation.) Materials are no longer in this category once they are removed from accumulation for recycling, however.

(9) "Excluded scrap metal" is processed scrap metal, unprocessed home scrap metal, and unprocessed prompt scrap metal.

(10) "Processed scrap metal" is scrap metal which has been manually or physically altered to either separate it into distinct materials to enhance economic value or to improve the handling of materials. Processed scrap metal includes, but is not limited to scrap metal which has been baled, shredded, sheared, chopped, crushed, flattened, cut, melted, or separated by metal type (i.e., sorted), and, fines, drosses and related materials which have been agglomerated. (Note: shredded circuit boards being sent for recycling are not considered processed scrap metal. They are covered under the exclusion from the definition of solid waste for shredded circuit boards being recycled (§ 261.4(a)(14)).

(11) "Home scrap metal" is scrap metal as generated by steel mills, foundries, and refineries such as turnings, cuttings, punchings, and borings.

(12) "Prompt scrap metal" is scrap metal as generated by the metal working/fabrication industries and includes such scrap metal as turnings, cuttings, punchings, and borings. Prompt scrap is also known as industrial or new scrap metal.

[45 FR 33119, May 19, 1980, as amended at 48 FR 14293, Apr. 1, 1983; 50 FR 663, Jan. 4, 1985; 51 FR 10174, Mar. 24, 1986; 51 FR 40636, Nov. 7, 1986; 62 FR 26018, May 12, 1997; 73 FR 64760, Oct. 30, 2008; 75 FR 13001, Mar. 18, 2010]

### § 261.2 Definition of solid waste.

(a)(1) A *solid waste* is any discarded material that is not excluded under § 261.4(a) or that is not excluded by a variance granted under §§ 260.30 and

260.31 or that is not excluded by a non-waste determination under §§ 260.30 and 260.34.

(2)(i) A *discarded material* is any material which is:

(A) Abandoned, as explained in paragraph (b) of this section; or

(B) Recycled, as explained in paragraph (c) of this section; or

(C) Considered inherently waste-like, as explained in paragraph (d) of this section; or

(D) A military munition identified as a solid waste in § 266.202.

(ii) A hazardous secondary material is not discarded if it is generated and reclaimed under the control of the generator as defined in § 260.10, it is not speculatively accumulated as defined in § 261.1(c)(8), it is handled only in non-land-based units and is contained in such units, it is generated and reclaimed within the United States and its territories, it is not otherwise subject to material-specific management conditions under § 261.4(a) when reclaimed, it is not a spent lead acid battery (see § 266.80 and § 273.2), it does not meet the listing description for K171 or K172 in § 261.32, and the reclamation of the material is legitimate, as specified under § 260.43. (See also the notification requirements of § 260.42). (For hazardous secondary materials managed in land-based units, see § 261.4(a)(23)).

(b) Materials are solid waste if they are *abandoned* by being:

(1) Disposed of; or

(2) Burned or incinerated; or

(3) Accumulated, stored, or treated (but not recycled) before or in lieu of being abandoned by being disposed of, burned, or incinerated.

(c) Materials are solid wastes if they are *recycled*—or accumulated, stored, or treated before recycling—as specified in paragraphs (c)(1) through (4) of this section.

(1) *Used in a manner constituting disposal.* (i) Materials noted with a "*" in Column 1 of Table 1 are solid wastes when they are:

(A) Applied to or placed on the land in a manner that constitutes disposal; or

(B) Used to produce products that are applied to or placed on the land or are otherwise contained in products that are applied to or placed on the land (in which cases the product itself remains a solid waste).

(ii) However, commercial chemical products listed in § 261.33 are not solid wastes if they are applied to the land and that is their ordinary manner of use.

(2) *Burning for energy recovery.* (i) Materials noted with a "*" in column 2 of Table 1 are solid wastes when they are:

(A) Burned to recover energy;

(B) Used to produce a fuel or are otherwise contained in fuels (in which cases the fuel itself remains a solid waste).

(ii) However, commercial chemical products listed in § 261.33 are not solid wastes if they are themselves fuels.

(3) *Reclaimed.* Materials noted with a "—" in column 3 of Table 1 are not solid wastes when reclaimed. Materials noted with an "*" in column 3 of Table 1 are solid wastes when reclaimed unless they meet the requirements of §§ 261.2(a)(2)(ii), or 261.4(a)(17), or 261.4(a)(23), or 261.4(a)(24) or 261.4(a)(25).

(4) *Accumulated speculatively.* Materials noted with a "*" in column 4 of Table 1 are solid wastes when accumulated speculatively.

TABLE 1

| | Use constituting disposal (§ 261.2(c)(1)) | Energy recovery/ fuel (§ 261.2(c)(2)) | Reclamation (261.2(c)(3)), except as provided in §§ 261.2(a)(2)(ii), 261.4(a)(17), 261.4(a)(23), 261.4(a)(24), or 261.4(a)(25) | Speculative accumulation (§ 261.2(c)(4)) |
|---|---|---|---|---|
| | 1 | 2 | 3 | 4 |
| Spent Materials | (*) | (*) | (*) | (*) |
| Sludges (listed in 40 CFR Part 261.31 or 261.32) | (*) | (*) | (*) | (*) |
| Sludges exhibiting a characteristic of hazardous waste | (*) | (*) | — | (*) |

TABLE 1—Continued

| | Use constituting disposal (§261.2(c)(1)) | Energy recovery/fuel (§261.2(c)(2)) | Reclamation (261.2(c)(3)), except as provided in §§261.2(a)(2)(ii), 261.4(a)(17), 261.4(a)(23), 261.4(a)(24), or 261.4(a)(25) | Speculative accumulation (§261.2(c)(4)) |
|---|---|---|---|---|
| | 1 | 2 | 3 | 4 |
| By-products (listed in 40 CFR 261.31 or 261.32) | (*) | (*) | (*) | (*) |
| By-products exhibiting a characteristic of hazardous waste | (*) | (*) | — | (*) |
| Commercial chemical products listed in 40 CFR 261.33 | (*) | (*) | — | — |
| Scrap metal that is not excluded under §261.4(a)(13) | (*) | (*) | (*) | (*) |

Note: The terms "spent materials," "sludges," "by-products," and "scrap metal" and "processed scrap metal" are defined in §261.1.

(d) *Inherently waste-like materials.* The following materials are solid wastes when they are recycled in any manner:

(1) Hazardous Waste Nos. F020, F021 (unless used as an ingredient to make a product at the site of generation), F022, F023, F026, and F028.

(2) Secondary materials fed to a halogen acid furnace that exhibit a characteristic of a hazardous waste or are listed as a hazardous waste as defined in subparts C or D of this part, except for brominated material that meets the following criteria:

(i) The material must contain a bromine concentration of at least 45%; and

(ii) The material must contain less than a total of 1% of toxic organic compounds listed in appendix VIII; and

(iii) The material is processed continually on-site in the halogen acid furnace via direct conveyance (hard piping).

(3) The Administrator will use the following criteria to add wastes to that list:

(i)(A) The materials are ordinarily disposed of, burned, or incinerated; or

(B) The materials contain toxic constituents listed in appendix VIII of part 261 and these constituents are not ordinarily found in raw materials or products for which the materials substitute (or are found in raw materials or products in smaller concentrations) and are not used or reused during the recycling process; and

(ii) The material may pose a substantial hazard to human health and the environment when recycled.

(e) *Materials that are not solid waste when recycled.* (1) Materials are not solid wastes when they can be shown to be recycled by being:

(i) Used or reused as ingredients in an industrial process to make a product, provided the materials are not being reclaimed; or

(ii) Used or reused as effective substitutes for commercial products; or

(iii) Returned to the original process from which they are generated, without first being reclaimed or land disposed. The material must be returned as a substitute for feedstock materials. In cases where the original process to which the material is returned is a secondary process, the materials must be managed such that there is no placement on the land. In cases where the materials are generated and reclaimed within the primary mineral processing industry, the conditions of the exclusion found at §261.4(a)(17) apply rather than this paragraph.

(2) The following materials are solid wastes, even if the recycling involves use, reuse, or return to the original process (described in paragraphs (e)(1) (i) through (iii) of this section):

(i) Materials used in a manner constituting disposal, or used to produce products that are applied to the land; or

33

(ii) Materials burned for energy recovery, used to produce a fuel, or contained in fuels; or

(iii) Materials accumulated speculatively; or

(iv) Materials listed in paragraphs (d)(1) and (d)(2) of this section.

(f) *Documentation of claims that materials are not solid wastes or are conditionally exempt from regulation.* Respondents in actions to enforce regulations implementing subtitle C of RCRA who raise a claim that a certain material is not a solid waste, or is conditionally exempt from regulation, must demonstrate that there is a known market or disposition for the material, and that they meet the terms of the exclusion or exemption. In doing so, they must provide appropriate documentation (such as contracts showing that a second person uses the material as an ingredient in a production process) to demonstrate that the material is not a waste, or is exempt from regulation. In addition, owners or operators of facilities claiming that they actually are recycling materials must show that they have the necessary equipment to do so.

[50 FR 664, Jan. 4, 1985, as amended at 50 FR 33542, Aug. 20, 1985; 56 FR 7206, Feb. 21, 1991; 56 FR 32688, July 17, 1991; 56 FR 42512, Aug. 27, 1991; 57 FR 38564, Aug. 25, 1992; 59 FR 48042, Sept. 19, 1994; 62 FR 6651, Feb. 12, 1997; 62 FR 26019, May 12, 1997; 63 FR 28636, May 26, 1998; 64 FR 24513, May 11, 1999; 67 FR 11253, Mar. 13, 2002; 71 FR 40258, July 14, 2006; 73 FR 64760, Oct. 30, 2008; 75 FR 13001, Mar. 18, 2010]

## § 261.3   Definition of hazardous waste.

(a) A solid waste, as defined in § 261.2, is a hazardous waste if:

(1) It is not excluded from regulation as a hazardous waste under § 261.4(b); and

(2) It meets any of the following criteria:

(i) It exhibits any of the characteristics of hazardous waste identified in subpart C of this part. However, any mixture of a waste from the extraction, beneficiation, and processing of ores and minerals excluded under § 261.4(b)(7) and any other solid waste exhibiting a characteristic of hazardous waste under subpart C is a hazardous waste only if it exhibits a characteristic that would not have been exhibited by the excluded waste alone if such mixture had not occurred, or if it

continues to exhibit any of the characteristics exhibited by the non-excluded wastes prior to mixture. Further, for the purposes of applying the Toxicity Characteristic to such mixtures, the mixture is also a hazardous waste if it exceeds the maximum concentration for any contaminant listed in table 1 to § 261.24 that would not have been exceeded by the excluded waste alone if the mixture had not occurred or if it continues to exceed the maximum concentration for any contaminant exceeded by the nonexempt waste prior to mixture.

(ii) It is listed in subpart D of this part and has not been excluded from the lists in subpart D of this part under §§ 260.20 and 260.22 of this chapter.

(iii) [Reserved]

(iv) It is a mixture of solid waste and one or more hazardous wastes listed in subpart D of this part and has not been excluded from paragraph (a)(2) of this section under §§ 260.20 and 260.22, paragraph (g) of this section, or paragraph (h) of this section; however, the following mixtures of solid wastes and hazardous wastes listed in subpart D of this part are not hazardous wastes (except by application of paragraph (a)(2)(i) or (ii) of this section) if the generator can demonstrate that the mixture consists of wastewater the discharge of which is subject to regulation under either section 402 or section 307(b) of the Clean Water Act (including wastewater at facilities which have eliminated the discharge of wastewater) and;

(A) One or more of the following spent solvents listed in § 261.31—benzene, carbon tetrachloride, tetrachloroethylene, trichloroethylene or the scrubber waters derived-from the combustion of these spent solvents—*Provided,* That the maximum total weekly usage of these solvents (other than the amounts that can be demonstrated not to be discharged to wastewater) divided by the average weekly flow of wastewater into the headworks of the facility's wastewater treatment or pretreatment system does not exceed 1 part per million, OR the total measured concentration of these solvents entering the headworks of the facility's wastewater treatment system (at facilities subject to regulation under the Clean Air Act, as amended, at 40 CFR parts 60, 61, or 63, or at facilities subject to an enforceable limit in a federal operating permit that minimizes fugitive emissions), does not exceed 1 part per million on an average weekly basis. Any facility that uses benzene as a solvent and claims this exemption must use an aerated biological wastewater treatment system and must use only lined surface impoundments or tanks prior to secondary clarification in the wastewater treatment system. Facilities that choose to measure concentration levels must file a copy of their sampling and analysis plan with the Regional Administrator, or State Director, as the context requires, or an authorized representative ("Director" as defined in 40 CFR 270.2). A facility must file a copy of a revised sampling and analysis plan only if the initial plan is rendered inaccurate by changes in the facility's operations. The sampling and analysis plan must include the monitoring point location (headworks), the sampling frequency and methodology, and a list of constituents to be monitored. A facility is eligible for the direct monitoring option once they receive confirmation that the sampling and analysis plan has been received by the Director. The Director may reject the sampling and analysis plan if he/she finds that, the sampling and analysis plan fails to include the above information; or the plan parameters would not enable the facility to calculate the weekly average concentration of these chemicals accurately. If the Director rejects the sampling and analysis plan or if the Director finds that the facility is not following the sampling and analysis plan, the Director shall notify the facility to cease the use of the direct monitoring option until such time as the bases for rejection are corrected; or

(B) One or more of the following spent solvents listed in §261.31-methylene chloride, 1,1,1-trichloroethane, chlorobenzene, o-dichlorobenzene, cresols, cresylic acid, nitrobenzene, toluene, methyl ethyl ketone, carbon disulfide, isobutanol, pyridine, spent chlorofluorocarbon solvents, 2-ethoxyethanol, or the scrubber waters derived-from the combustion of these spent solvents—*Provided* That the maximum total weekly usage of these solvents (other than the amounts that can be demonstrated not to be discharged to wastewater) divided by the average weekly flow of wastewater into the headworks of the facility's wastewater treatment or pretreatment system does not exceed 25 parts per million, OR the total measured concentration of these solvents entering the headworks of the facility's wastewater treatment system (at facilities subject to regulation under the Clean Air Act as amended, at 40 CFR parts 60, 61, or 63, or at facilities subject to an enforceable limit in a federal operating permit that minimizes fugitive emissions), does not exceed 25 parts per million on an average weekly basis. Facilities that choose to measure concentration levels must file a copy of their sampling and analysis plan with the Regional Administrator, or State Director, as the context requires, or an authorized representative ("Director" as defined in 40 CFR 270.2). A facility must file a copy of a revised sampling and analysis plan only if the initial plan is rendered inaccurate by changes in the facility's operations. The sampling and analysis plan must include the monitoring point location (headworks), the sampling frequency and methodology, and a list of constituents to be monitored. A facility is eligible for the direct monitoring option once they receive confirmation that the sampling and analysis plan has been received by the Director. The Director may reject the sampling and analysis plan if he/she finds that, the sampling and analysis plan fails to include the above information; or the plan parameters would not enable the facility to calculate the weekly average concentration of these chemicals accurately. If the Director rejects the sampling and analysis plan or if the Director finds that the facility is not following the sampling and analysis plan, the Director shall notify the facility to cease the use of the direct monitoring option until such time as the bases for rejection are corrected; or

(C) One of the following wastes listed in §261.32, provided that the wastes are discharged to the refinery oil recovery sewer before primary oil/water/solids separation—heat exchanger bundle cleaning sludge from the petroleum refining industry (EPA Hazardous Waste No. K050), crude oil storage tank sediment from petroleum refining operations (EPA Hazardous Waste No. K169), clarified slurry oil tank sediment and/or in-line filter/separation solids from petroleum refining operations (EPA Hazardous Waste No. K170), spent hydrotreating catalyst (EPA Hazardous Waste No. K171), and spent hydrorefining catalyst (EPA Hazardous Waste No. K172); or

(D) A discarded hazardous waste, commercial chemical product, or chemical intermediate listed in §§ 261.31 through 261.33, arising from *de minimis* losses of these materials. For purposes of this paragraph (a)(2)(iv)(D), *de minimis* losses are inadvertent releases to a wastewater treatment system, including those from normal material handling operations (e.g., spills from the unloading or transfer of materials from bins or other containers, leaks from pipes, valves or other devices used to transfer materials); minor leaks of process equipment, storage tanks or containers; leaks from well maintained pump packings and seals; sample purgings; relief device discharges; discharges from safety showers and rinsing and cleaning of personal safety equipment; and rinsate from empty containers or from containers that are rendered empty by that rinsing. Any manufacturing facility that claims an exemption for *de minimis* quantities of wastes listed in §§ 261.31 through 261.32, or any nonmanufacturing facility that claims an exemption for *de minimis* quantities of wastes listed in subpart D of this part must either have eliminated the discharge of wastewaters or have included in its Clean Water Act permit application or submission to its pretreatment control authority the constituents for which each waste was listed (in 40 CFR 261 appendix VII) of this part; and the constituents in the table "Treatment Standards for Hazardous Wastes" in 40 CFR 268.40 for which each waste has a treatment standard (*i.e.*, Land Disposal Restriction constituents). A facility is eligible to claim the exemption once the permit writer or control authority has been notified of possible *de minimis* releases via the Clean Water Act permit

application or the pretreatment control authority submission. A copy of the Clean Water permit application or the submission to the pretreatment control authority must be placed in the facility's on-site files; or

(E) Wastewater resulting from laboratory operations containing toxic (T) wastes listed in subpart D of this part, Provided, That the annualized average flow of laboratory wastewater does not exceed one percent of total wastewater flow into the headworks of the facility's wastewater treatment or pretreatment system or provided the wastes, combined annualized average concentration does not exceed one part per million in the headworks of the facility's wastewater treatment or pretreatment facility. Toxic (T) wastes used in laboratories that are demonstrated not to be discharged to wastewater are not to be included in this calculation; or

(F) One or more of the following wastes listed in § 261.32—wastewaters from the production of carbamates and carbamoyl oximes (EPA Hazardous Waste No. K157)—*Provided* that the maximum weekly usage of formaldehyde, methyl chloride, methylene chloride, and triethylamine (including all amounts that cannot be demonstrated to be reacted in the process, destroyed through treatment, or is recovered, *i.e.*, what is discharged or volatilized) divided by the average weekly flow of process wastewater prior to any dilution into the headworks of the facility's wastewater treatment system does not exceed a total of 5 parts per million by weight OR the total measured concentration of these chemicals entering the headworks of the facility's wastewater treatment system (at facilities subject to regulation under the Clean Air Act as amended, at 40 CFR parts 60, 61, or 63, or at facilities subject to an enforceable limit in a federal operating permit that minimizes fugitive emissions), does not exceed 5 parts per million on an average weekly basis. Facilities that choose to measure concentration levels must file copy of their sampling and analysis plan with the Regional Administrator, or State Director, as the context requires, or an authorized representative ("Director" as defined in 40 CFR 270.2). A facility must file a copy of a revised sampling and analysis plan only if the initial plan is rendered inaccurate by changes in the facility's operations. The sampling and analysis plan must include the monitoring point location (headworks), the sampling frequency and methodology, and a list of constituents to be monitored. A facility is eligible for the direct monitoring option once they receive confirmation that the sampling and analysis plan has been received by the Director. The Director may reject the sampling and analysis plan if he/she finds that, the sampling and analysis plan fails to include the above information; or the plan parameters would not enable the facility to calculate the weekly average concentration of these chemicals accurately. If the Director rejects the sampling and analysis plan or if the Director finds that the facility is not following the sampling and analysis plan, the Director shall notify the facility to cease the use of the direct monitoring option until such time as the bases for rejection are corrected; or

(G) Wastewaters derived-from the treatment of one or more of the following wastes listed in § 261.32—organic waste (including heavy ends, still bottoms, light ends, spent solvents, filtrates, and decantates) from the production of carbamates and carbamoyl oximes (EPA Hazardous Waste No. K156).—*Provided*, that the maximum concentration of formaldehyde, methyl chloride, methylene chloride, and triethylamine prior to any dilutions into the headworks of the facility's wastewater treatment system does not exceed a total of 5 milligrams per liter OR the total measured concentration of these chemicals entering the headworks of the facility's wastewater treatment system (at facilities subject to regulation under the Clean Air Act as amended, at 40 CFR parts 60, 61, or 63, or at facilities subject to an enforceable limit in a federal operating permit that minimizes fugitive emissions), does not exceed 5 milligrams per liter on an average weekly basis. Facilities that choose to measure concentration levels must file copy of their sampling and analysis plan with the Regional Administrator, or State Director, as the context requires, or an

authorized representative ("Director" as defined in 40 CFR 270.2). A facility must file a copy of a revised sampling and analysis plan only if the initial plan is rendered inaccurate by changes in the facility's operations. The sampling and analysis plan must include the monitoring point location (headworks), the sampling frequency and methodology, and a list of constituents to be monitored. A facility is eligible for the direct monitoring option once they receive confirmation that the sampling and analysis plan has been received by the Director. The Director may reject the sampling and analysis plan if he/she finds that, the sampling and analysis plan fails to include the above information; or the plan parameters would not enable the facility to calculate the weekly average concentration of these chemicals accurately. If the Director rejects the sampling and analysis plan or if the Director finds that the facility is not following the sampling and analysis plan, the Director shall notify the facility to cease the use of the direct monitoring option until such time as the bases for rejection are corrected.

(v) *Rebuttable presumption for used oil.* Used oil containing more than 1000 ppm total halogens is presumed to be a hazardous waste because it has been mixed with halogenated hazardous waste listed in subpart D of part 261 of this chapter. Persons may rebut this presumption by demonstrating that the used oil does not contain hazardous waste (for example, to show that the used oil does not contain significant concentrations of halogenated hazardous constituents listed in appendix VIII of part 261 of this chapter).

(b) A solid waste which is not excluded from regulation under paragraph (a)(1) of this section becomes a hazardous waste when any of the following events occur:

(1) In the case of a waste listed in subpart D of this part, when the waste first meets the listing description set forth in subpart D of this part.

(2) In the case of a mixture of solid waste and one or more listed hazardous wastes, when a hazardous waste listed in subpart D is first added to the solid waste.

(3) In the case of any other waste (including a waste mixture), when the waste exhibits any of the characteristics identified in subpart C of this part.

(c) Unless and until it meets the criteria of paragraph (d) of this section:

(1) A hazardous waste will remain a hazardous waste.

(2)(i) Except as otherwise provided in paragraph (c)(2)(ii), (g) or (h) of this section, any solid waste generated from the treatment, storage, or disposal of a hazardous waste, including any sludge, spill residue, ash emission control dust, or leachate (but not including precipitation run-off) is a hazardous waste. (However, materials that are reclaimed from solid wastes and that are used beneficially are not solid wastes and hence are not hazardous wastes under this provision unless the reclaimed material is burned for energy recovery or used in a manner constituting disposal.)

(ii) The following solid wastes are not hazardous even though they are generated from the treatment, storage, or disposal of a hazardous waste, unless they exhibit one or more of the characteristics of hazardous waste:

(A) Waste pickle liquor sludge generated by lime stabilization of spent pickle liquor from the iron and steel industry (SIC Codes 331 and 332).

(B) Waste from burning any of the materials exempted from regulation by § 261.6(a)(3)(iii) and (iv).

(C)(1) Nonwastewater residues, such as slag, resulting from high temperature metals recovery (HTMR) processing of K061, K062 or F006 waste, in units identified as rotary kilns, flame reactors, electric furnaces, plasma arc furnaces, slag reactors, rotary hearth furnace/electric furnace combinations or industrial furnaces (as defined in paragraphs (6), (7), and (13) of the definition for "Industrial furnace" in 40 CFR 260.10), that are disposed in subtitle D units, provided that these residues meet the generic exclusion levels identified in the tables in this paragraph for all constituents, and exhibit no characteristics of hazardous waste. Testing requirements must be incorporated in a facility's waste analysis plan or a generator's self-implementing waste analysis plan; at a minimum, composite samples of residues must be collected and analyzed quarterly and/or when the process or operation generating the waste changes. Persons claiming this exclusion in an enforcement action will have the burden of proving by clear and convincing evidence that the material meets all of the exclusion requirements.

| Constituent | Maximum for any single composite sample—TCLP (mg/l) |
|---|---|
| Generic exclusion levels for K061 and K062 nonwastewater HTMR residues | |
| Antimony | 0.10 |
| Arsenic | 0.50 |
| Barium | 7.6 |
| Beryllium | 0.010 |
| Cadmium | 0.050 |
| Chromium (total) | 0.33 |
| Lead | 0.15 |
| Mercury | 0.009 |
| Nickel | 1.0 |
| Selenium | 0.16 |
| Silver | 0.30 |
| Thallium | 0.020 |
| Zinc | 70 |
| Generic exclusion levels for F006 nonwastewater HTMR residues | |
| Antimony | 0.10 |
| Arsenic | 0.50 |
| Barium | 7.6 |
| Beryllium | 0.010 |
| Cadmium | 0.050 |
| Chromium (total) | 0.33 |
| Cyanide (total) (mg/kg) | 1.8 |
| Lead | 0.15 |
| Mercury | 0.009 |
| Nickel | 1.0 |
| Selenium | 0.16 |
| Silver | 0.30 |
| Thallium | 0.020 |
| Zinc | 70 |

(2) A one-time notification and certification must be placed in the facility's files and sent to the EPA region or authorized state for K061, K062 or F006 HTMR residues that meet the generic exclusion levels for all constituents and do not exhibit any characteristics that are sent to subtitle D units. The notification and certification that is placed in the generators or treaters files must be updated if the process or operation generating the waste changes and/or if the subtitle D unit receiving the waste changes. However, the generator or treater need only notify the EPA region or an authorized state on an annual basis if such changes occur. Such notification and certification should be sent to the EPA region or authorized state by the end of

the calendar year, but no later than December 31. The notification must include the following information: The name and address of the subtitle D unit receiving the waste shipments; the EPA Hazardous Waste Number(s) and treatability group(s) at the initial point of generation; and, the treatment standards applicable to the waste at the initial point of generation. The certification must be signed by an authorized representative and must state as follows: "I certify under penalty of law that the generic exclusion levels for all constituents have been met without impermissible dilution and that no characteristic of hazardous waste is exhibited. I am aware that there are significant penalties for submitting a false certification, including the possibility of fine and imprisonment."

(D) Biological treatment sludge from the treatment of one of the following wastes listed in §261.32—organic waste (including heavy ends, still bottoms, light ends, spent solvents, filtrates, and decantates) from the production of carbamates and carbamoyl oximes (EPA Hazardous Waste No. K156), and wastewaters from the production of carbamates and carbamoyl oximes (EPA Hazardous Waste No. K157).

(E) Catalyst inert support media separated from one of the following wastes listed in §261.32—Spent hydrotreating catalyst (EPA Hazardous Waste No. K171), and Spent hydrorefining catalyst (EPA Hazardous Waste No. K172).

(d) Any solid waste described in paragraph (c) of this section is not a hazardous waste if it meets the following criteria:

(1) In the case of any solid waste, it does not exhibit any of the characteristics of hazardous waste identified in subpart C of this part. (However, wastes that exhibit a characteristic at the point of generation may still be subject to the requirements of part 268, even if they no longer exhibit a characteristic at the point of land disposal.)

(2) In the case of a waste which is a listed waste under subpart D of this part, contains a waste listed under subpart D of this part or is derived from a waste listed in subpart D of this part, it also has been excluded from paragraph (c) of this section under §§260.20 and 260.22 of this chapter.

(e) [Reserved]

(f) Notwithstanding paragraphs (a) through (d) of this section and provided the debris as defined in part 268 of this chapter does not exhibit a characteristic identified at subpart C of this part, the following materials are not subject to regulation under 40 CFR parts 260, 261 to 266, 268, or 270:

(1) Hazardous debris as defined in part 268 of this chapter that has been treated using one of the required extraction or destruction technologies specified in Table 1 of §268.45 of this chapter; persons claiming this exclusion in an enforcement action will have the burden of proving by clear and convincing evidence that the material meets all of the exclusion requirements; or

(2) Debris as defined in part 268 of this chapter that the Regional Administrator, considering the extent of contamination, has determined is no longer contaminated with hazardous waste.

(g)(1) A hazardous waste that is listed in subpart D of this part solely because it exhibits one or more characteristics of ignitability as defined under §261.21, corrosivity as defined under §261.22, or reactivity as defined under §261.23 is not a hazardous waste, if the waste no longer exhibits any characteristic of hazardous waste identified in subpart C of this part.

(2) The exclusion described in paragraph (g)(1) of this section also pertains to:

(i) Any mixture of a solid waste and a hazardous waste listed in subpart D of this part solely because it exhibits the characteristics of ignitability, corrosivity, or reactivity as regulated under paragraph (a)(2)(iv) of this section; and

(ii) Any solid waste generated from treating, storing, or disposing of a hazardous waste listed in subpart D of this part solely because it exhibits the characteristics of ignitability, corrosivity, or reactivity as regulated under paragraph (c)(2)(i) of this section.

(3) Wastes excluded under this section are subject to part 268 of this chapter (as applicable), even if they no longer exhibit a characteristic at the point of land disposal.

(4) Any mixture of a solid waste excluded from regulation under § 261.4(b)(7) and a hazardous waste listed in subpart D of this part solely because it exhibits one or more of the characteristics of ignitability, corrosivity, or reactivity as regulated under paragraph (a)(2)(iv) of this section is not a hazardous waste, if the mixture no longer exhibits any characteristic of hazardous waste identified in subpart C of this part for which the hazardous waste listed in subpart D of this part was listed.

(h)(1) Hazardous waste containing radioactive waste is no longer a hazardous waste when it meets the eligibility criteria and conditions of 40 CFR part 266, Subpart N ("eligible radioactive mixed waste").

(2) The exemption described in paragraph (h)(1) of this section also pertains to:

(i) Any mixture of a solid waste and an eligible radioactive mixed waste; and

(ii) Any solid waste generated from treating, storing, or disposing of an eligible radioactive mixed waste.

(3) Waste exempted under this section must meet the eligibility criteria and specified conditions in 40 CFR 266.225 and 40 CFR 266.230 (for storage and treatment) and in 40 CFR 266.310 and 40 CFR 266.315 (for transportation and disposal). Waste that fails to satisfy these eligibility criteria and conditions is regulated as hazardous waste.

[57 FR 7632, Mar. 3, 1992; 57 FR 23063, June 1, 1992, as amended at 57 FR 37263, Aug. 18, 1992; 57 FR 41611, Sept. 10, 1992; 57 FR 49279, Oct. 30, 1992; 59 FR 38545, July 28, 1994; 60 FR 7848, Feb. 9, 1995; 63 FR 28637, May 26, 1998; 63 FR 42184, Aug. 6, 1998; 66 FR 27297, May 16, 2001; 66 FR 50333, Oct. 3, 2001; 70 FR 34561, June 14, 2005; 70 FR 57784, Oct. 4, 2005; 71 FR 40258, July 14, 2006]

## § 261.4  Exclusions.

(a) *Materials which are not solid wastes.* The following materials are not solid wastes for the purpose of this part:

(1)(i) Domestic sewage; and

(ii) Any mixture of domestic sewage and other wastes that passes through a sewer system to a publicly-owned treatment works for treatment. "Domestic sewage" means untreated sanitary wastes that pass through a sewer system.

(2) Industrial wastewater discharges that are point source discharges subject to regulation under section 402 of the Clean Water Act, as amended.

[*Comment:* This exclusion applies only to the actual point source discharge. It does not exclude industrial wastewaters while they are being collected, stored or treated before discharge, nor does it exclude sludges that are generated by industrial wastewater treatment.]

(3) Irrigation return flows.

(4) Source, special nuclear or by-product material as defined by the Atomic Energy Act of 1954, as amended, 42 U.S.C. 2011 *et seq.*

(5) Materials subjected to in-situ mining techniques which are not removed from the ground as part of the extraction process.

(6) Pulping liquors (*i.e.*, black liquor) that are reclaimed in a pulping liquor recovery furnace and then reused in the pulping process, unless it is accumulated speculatively as defined in § 261.1(c) of this chapter.

(7) Spent sulfuric acid used to produce virgin sulfuric acid, unless it is accumulated speculatively as defined in § 261.1(c) of this chapter.

(8) Secondary materials that are reclaimed and returned to the original process or processes in which they were generated where they are reused in the production process provided:

(i) Only tank storage is involved, and the entire process through completion of reclamation is closed by being entirely connected with pipes or other comparable enclosed means of conveyance;

(ii) Reclamation does not involve controlled flame combustion (such as occurs in boilers, industrial furnaces, or incinerators);

(iii) The secondary materials are never accumulated in such tanks for over twelve months without being reclaimed; and

(iv) The reclaimed material is not used to produce a fuel, or used to produce products that are used in a manner constituting disposal.

(9)(i) Spent wood preserving solutions that have been reclaimed and are reused for their original intended purpose; and

(ii) Wastewaters from the wood preserving process that have been reclaimed and are reused to treat wood.

(iii) Prior to reuse, the wood preserving wastewaters and spent wood preserving solutions described in paragraphs (a)(9)(i) and (a)(9)(ii) of this section, so long as they meet all of the following conditions:

(A) The wood preserving wastewaters and spent wood preserving solutions are reused on-site at water borne plants in the production process for their original intended purpose;

(B) Prior to reuse, the wastewaters and spent wood preserving solutions are managed to prevent release to either land or groundwater or both;

(C) Any unit used to manage wastewaters and/or spent wood preserving solutions prior to reuse can be visually or otherwise determined to prevent such releases;

(D) Any drip pad used to manage the wastewaters and/or spent wood preserving solutions prior to reuse complies with the standards in part 265, subpart W of this chapter, regardless of whether the plant generates a total of less than 100 kg/month of hazardous waste; and

(E) Prior to operating pursuant to this exclusion, the plant owner or operator prepares a one-time notification stating that the plant intends to claim the exclusion, giving the date on which the plant intends to begin operating under the exclusion, and containing the following language: "I have read the applicable regulation establishing an exclusion for wood preserving wastewaters and spent wood preserving solutions and understand it requires me to comply at all times with the conditions set out in the regulation." The plant must maintain a copy of that document in its on-site records until closure of the facility. The exclusion applies so long as the plant meets all of the conditions. If the plant goes out of compliance with any condition, it may apply to the appropriate Regional Administrator or state Director for reinstatement. The Regional Administrator or state Director may reinstate the exclusion upon finding that the plant has returned to compliance with all conditions and that the violations are not likely to recur.

(10) EPA Hazardous Waste Nos. K060, K087, K141, K142, K143, K144, K145, K147, and K148, and any wastes from the coke by-products processes that are hazardous only because they exhibit the Toxicity Characteristic (TC) specified in section 261.24 of this part when, subsequent to generation, these materials are recycled to coke ovens, to the tar recovery process as a feedstock to produce coal tar, or mixed with coal tar prior to the tar's sale or refining. This exclusion is conditioned on there being no land disposal of the wastes from the point they are generated to the point they are recycled to coke ovens or tar recovery or refining processes, or mixed with coal tar.

(11) Nonwastewater splash condenser dross residue from the treatment of K061 in high temperature metals recovery units, provided it is shipped in drums (if shipped) and not land disposed before recovery.

(12)(i) Oil-bearing hazardous secondary materials (i.e., sludges, byproducts, or spent materials) that are generated at a petroleum refinery (SIC code 2911) and are inserted into the petroleum refining process (SIC code 2911—including, but not limited to, distillation, catalytic cracking, fractionation, gasification (as defined in 40 CFR 260.10) or thermal cracking units (i.e., cokers)) unless the material is placed on the land, or speculatively accumulated before being so recycled. Materials inserted into thermal cracking units are excluded under this paragraph, provided that the coke product also does not exhibit a characteristic of hazardous waste. Oil-bearing hazardous secondary materials may be inserted into the same petroleum refinery where they are generated, or sent directly to another petroleum refinery and still be excluded under this provision. Except as provided in paragraph (a)(12)(ii) of this section, oil-bearing hazardous secondary materials generated elsewhere in the petroleum industry (i.e., from sources other than petroleum refineries) are not excluded under this section. Residuals generated from processing or recycling materials excluded under this paragraph (a)(12)(i), where such materials as generated would have otherwise met a listing under subpart D of this part, are designated as F037 listed wastes when disposed of or intended for disposal.

(ii) Recovered oil that is recycled in the same manner and with the same conditions as described in paragraph (a)(12)(i) of this section. Recovered oil is oil that has been reclaimed from secondary materials (including wastewater) generated from normal petroleum industry practices, including refining, exploration and production, bulk storage, and transportation incident thereto (SIC codes 1311, 1321, 1381, 1382, 1389, 2911, 4612, 4613, 4922, 4923, 4789, 5171, and 5172.) Recovered oil does not include oil-bearing hazardous wastes listed in subpart D of this part; however, oil recovered from such wastes may be considered recovered oil. Recovered oil does not include used oil as defined in 40 CFR 279.1.

(13) Excluded scrap metal (processed scrap metal, unprocessed home scrap metal, and unprocessed prompt scrap metal) being recycled.

(14) Shredded circuit boards being recycled provided that they are:

(i) Stored in containers sufficient to prevent a release to the environment prior to recovery; and

(ii) Free of mercury switches, mercury relays and nickel-cadmium batteries and lithium batteries.

(15) Condensates derived from the overhead gases from kraft mill steam strippers that are used to comply with 40 CFR 63.446(e). The exemption applies only to combustion at the mill generating the condensates.

(16) Comparable fuels or comparable syngas fuels that meet the requirements of § 261.38.

(17) Spent materials (as defined in § 261.1) (other than hazardous wastes listed in subpart D of this part) generated within the primary mineral processing industry from which minerals, acids, cyanide, water, or other values are recovered by mineral processing or by beneficiation, provided that:

(i) The spent material is legitimately recycled to recover minerals, acids, cyanide, water or other values;

(ii) The spent material is not accumulated speculatively;

(iii) Except as provided in paragraph (a)(17)(iv) of this section, the spent material is stored in tanks, containers, or

buildings meeting the following minimum integrity standards: a building must be an engineered structure with a floor, walls, and a roof all of which are made of non-earthen materials providing structural support (except smelter buildings may have partially earthen floors provided the secondary material is stored on the non-earthen portion), and have a roof suitable for diverting rainwater away from the foundation; a tank must be free standing, not be a surface impoundment (as defined in 40 CFR 260.10), and be manufactured of a material suitable for containment of its contents; a container must be free standing and be manufactured of a material suitable for containment of its contents. If tanks or containers contain any particulate which may be subject to wind dispersal, the owner/operator must operate these units in a manner which controls fugitive dust. Tanks, containers, and buildings must be designed, constructed and operated to prevent significant releases to the environment of these materials.

(iv) The Regional Administrator or State Director may make a site-specific determination, after public review and comment, that only solid mineral processing spent material may be placed on pads rather than tanks containers, or buildings. Solid mineral processing spent materials do not contain any free liquid. The decision-maker must affirm that pads are designed, constructed and operated to prevent significant releases of the secondary material into the environment. Pads must provide the same degree of containment afforded by the non-RCRA tanks, containers and buildings eligible for exclusion.

(A) The decision-maker must also consider if storage on pads poses the potential for significant releases via groundwater, surface water, and air exposure pathways. Factors to be considered for assessing the groundwater, surface water, air exposure pathways are: The volume and physical and chemical properties of the secondary material, including its potential for migration off the pad; the potential for human or environmental exposure to hazardous constituents migrating from the pad via each exposure pathway, and

42

the possibility and extent of harm to human and environmental receptors via each exposure pathway.

(B) Pads must meet the following minimum standards: Be designed of non-earthen material that is compatible with the chemical nature of the mineral processing spent material, capable of withstanding physical stresses associated with placement and removal, have run on/runoff controls, be operated in a manner which controls fugitive dust, and have integrity assurance through inspections and maintenance programs.

(C) Before making a determination under this paragraph, the Regional Administrator or State Director must provide notice and the opportunity for comment to all persons potentially interested in the determination. This can be accomplished by placing notice of this action in major local newspapers, or broadcasting notice over local radio stations.

(v) The owner or operator provides notice to the Regional Administrator or State Director providing the following information: The types of materials to be recycled; the type and location of the storage units and recycling processes; and the annual quantities expected to be placed in land-based units. This notification must be updated when there is a change in the type of materials recycled or the location of the recycling process.

(vi) For purposes of paragraph (b)(7) of this section, mineral processing spent materials must be the result of mineral processing and may not include any listed hazardous wastes. Listed hazardous wastes and characteristic hazardous wastes generated by non-mineral processing industries are not eligible for the conditional exclusion from the definition of solid waste.

(18) Petrochemical recovered oil from an associated organic chemical manufacturing facility, where the oil is to be inserted into the petroleum refining process (SIC code 2911) along with normal petroleum refinery process streams, provided:

(i) The oil is hazardous only because it exhibits the characteristic of ignitability (as defined in §261.21) and/or

toxicity for benzene (§261.24, waste code D018); and

(ii) The oil generated by the organic chemical manufacturing facility is not placed on the land, or speculatively accumulated before being recycled into the petroleum refining process. An "associated organic chemical manufacturing facility" is a facility where the primary SIC code is 2869, but where operations may also include SIC codes 2821, 2822, and 2865; and is physically co-located with a petroleum refinery; and where the petroleum refinery to which the oil being recycled is returned also provides hydrocarbon feedstocks to the organic chemical manufacturing facility. "Petrochemical recovered oil" is oil that has been reclaimed from secondary materials (i.e., sludges, byproducts, or spent materials, including wastewater) from normal organic chemical manufacturing operations, as well as oil recovered from organic chemical manufacturing processes.

(19) Spent caustic solutions from petroleum refining liquid treating processes used as a feedstock to produce cresylic or naphthenic acid unless the material is placed on the land, or accumulated speculatively as defined in §261.1(c).

(20) Hazardous secondary materials used to make zinc fertilizers, provided that the following conditions specified are satisfied:

(i) Hazardous secondary materials used to make zinc micronutrient fertilizers must not be accumulated speculatively, as defined in §261.1 (c)(8).

(ii) Generators and intermediate handlers of zinc-bearing hazardous secondary materials that are to be incorporated into zinc fertilizers must:

(A) Submit a one-time notice to the Regional Administrator or State Director in whose jurisdiction the exclusion is being claimed, which contains the name, address and EPA ID number of the generator or intermediate handler facility, provides a brief description of the secondary material that will be subject to the exclusion, and identifies when the manufacturer intends to begin managing excluded, zinc-bearing hazardous secondary materials under the conditions specified in this paragraph (a)(20).

(B) Store the excluded secondary material in tanks, containers, or buildings that are constructed and maintained in a way that prevents releases of the secondary materials into the environment. At a minimum, any building used for this purpose must be an engineered structure made of non-earthen materials that provide structural support, and must have a floor, walls and a roof that prevent wind dispersal and contact with rainwater. Tanks used for this purpose must be structurally sound and, if outdoors, must have roofs or covers that prevent contact with wind and rain. Containers used for this purpose must be kept closed except when it is necessary to add or remove material, and must be in sound condition. Containers that are stored outdoors must be managed within storage areas that:

(*1*) Have containment structures or systems sufficiently impervious to contain leaks, spills and accumulated precipitation; and

(*2*) Provide for effective drainage and removal of leaks, spills and accumulated precipitation; and

(*3*) Prevent run-on into the containment system.

(C) With each off-site shipment of excluded hazardous secondary materials, provide written notice to the receiving facility that the material is subject to the conditions of this paragraph (a)(20).

(D) Maintain at the generator's or intermediate handlers's facility for no less than three years records of all shipments of excluded hazardous secondary materials. For each shipment these records must at a minimum contain the following information:

(*1*) Name of the transporter and date of the shipment;

(*2*) Name and address of the facility that received the excluded material, and documentation confirming receipt of the shipment; and

(*3*) Type and quantity of excluded secondary material in each shipment.

(iii) Manufacturers of zinc fertilizers or zinc fertilizer ingredients made from excluded hazardous secondary materials must:

(A) Store excluded hazardous secondary materials in accordance with the storage requirements for generators and intermediate handlers, as specified in paragraph (a)(20)(ii)(B) of this section.

(B) Submit a one-time notification to the Regional Administrator or State Director that, at a minimum, specifies the name, address and EPA ID number of the manufacturing facility, and identifies when the manufacturer intends to begin managing excluded, zinc-bearing hazardous secondary materials under the conditions specified in this paragraph (a)(20).

(C) Maintain for a minimum of three years records of all shipments of excluded hazardous secondary materials received by the manufacturer, which must at a minimum identify for each shipment the name and address of the generating facility, name of transporter and date the materials were received, the quantity received, and a brief description of the industrial process that generated the material.

(D) Submit to the Regional Administrator or State Director an annual report that identifies the total quantities of all excluded hazardous secondary materials that were used to manufacture zinc fertilizers or zinc fertilizer ingredients in the previous year, the name and address of each generating facility, and the industrial process(s) from which they were generated.

(iv) Nothing in this section preempts, overrides or otherwise negates the provision in § 262.11 of this chapter, which requires any person who generates a solid waste to determine if that waste is a hazardous waste.

(v) Interim status and permitted storage units that have been used to store only zinc-bearing hazardous wastes prior to the submission of the one-time notice described in paragraph (a)(20)(ii)(A) of this section, and that afterward will be used only to store hazardous secondary materials excluded under this paragraph, are not subject to the closure requirements of 40 CFR Parts 264 and 265.

(21) Zinc fertilizers made from hazardous wastes, or hazardous secondary materials that are excluded under paragraph (a)(20) of this section, provided that:

(i) The fertilizers meet the following contaminant limits:

(A) For metal contaminants:

| Constituent | Maximum Allowable Total Concentration in Fertilizer, per Unit (1%) of Zinc (ppm) |
|---|---|
| Arsenic | 0.3 |
| Cadmium | 1.4 |
| Chromium | 0.6 |
| Lead | 2.8 |
| Mercury | 0.3 |

(B) For dioxin contaminants the fertilizer must contain no more than eight (8) parts per trillion of dioxin, measured as toxic equivalent (TEQ).

(ii) The manufacturer performs sampling and analysis of the fertilizer product to determine compliance with the contaminant limits for metals no less than every six months, and for dioxins no less than every twelve months. Testing must also be performed whenever changes occur to manufacturing processes or ingredients that could significantly affect the amounts of contaminants in the fertilizer product. The manufacturer may use any reliable analytical method to demonstrate that no constituent of concern is present in the product at concentrations above the applicable limits. It is the responsibility of the manufacturer to ensure that the sampling and analysis are unbiased, precise, and representative of the product(s) introduced into commerce.

(iii) The manufacturer maintains for no less than three years records of all sampling and analyses performed for purposes of determining compliance with the requirements of paragraph (a)(21)(ii) of this section. Such records must at a minimum include:

(A) The dates and times product samples were taken, and the dates the samples were analyzed;

(B) The names and qualifications of the person(s) taking the samples;

(C) A description of the methods and equipment used to take the samples;

(D) The name and address of the laboratory facility at which analyses of the samples were performed;

(E) A description of the analytical methods used, including any cleanup and sample preparation methods; and

(F) All laboratory analytical results used to determine compliance with the contaminant limits specified in this paragraph (a)(21).

(22) Used cathode ray tubes (CRTs)

(i) Used, intact CRTs as defined in §260.10 of this chapter are not solid wastes within the United States unless they are disposed, or unless they are speculatively accumulated as defined in §261.1(c)(8) by CRT collectors or glass processors.

(ii) Used, intact CRTs as defined in §260.10 of this chapter are not solid wastes when exported for recycling provided that they meet the requirements of §261.40.

(iii) Used, broken CRTs as defined in §260.10 of this chapter are not solid wastes provided that they meet the requirements of §261.39.

(iv) Glass removed from CRTs is not a solid waste provided that it meets the requirements of §261.39(c).

(23) Hazardous secondary material generated and reclaimed within the United States or its territories and managed in land-based units as defined in §260.10 of this chapter is not a solid waste provided that:

(i) The material is contained;

(ii) The material is a hazardous secondary material generated and reclaimed under the control of the generator, as defined in §260.10;

(iii) The material is not speculatively accumulated, as defined in §261.1(c)(8);

(iv) The material is not otherwise subject to material-specific management conditions under paragraph (a) of this section when reclaimed, it is not a spent lead acid battery (see §266.80 and §273.2 of this chapter), and it does not meet the listing description for K171 or K172 in §261.32;

(v) The reclamation of the material is legitimate, as specified under §260.43 of this chapter; and

(vi) In addition, persons claiming the exclusion under this paragraph (a)(23) must provide notification as required by §260.42 of this chapter. (For hazardous secondary material managed in a non-land-based unit, see §261.2(a)(2)(ii)).

(24) Hazardous secondary material that is generated and then transferred to another person for the purpose of reclamation is not a solid waste, provided that:

(i) The material is not speculatively accumulated, as defined in § 261.1(c)(8);

(ii) The material is not handled by any person or facility other than the hazardous secondary material generator, the transporter, an intermediate facility or a reclaimer, and, while in transport, is not stored for more than 10 days at a transfer facility, as defined in § 260.10 of this chapter, and is packaged according to applicable Department of Transportation regulations at 49 CFR Parts 173, 178, and 179 while in transport;

(iii) The material is not otherwise subject to material-specific management conditions under paragraph (a) of this section when reclaimed, it is not a spent lead-acid battery (see § 266.80 and § 273.2 of this chapter), and it does not meet the listing description for K171 or K172 in § 261.32;

(iv) The reclamation of the material is legitimate, as specified under § 260.43 of this chapter;

(v) The hazardous secondary material generator satisfies all of the following conditions:

(A) The material must be contained.

(B) Prior to arranging for transport of hazardous secondary materials to a reclamation facility (or facilities) where the management of the hazardous secondary materials is not addressed under a RCRA Part B permit or interim status standards, the hazardous secondary material generator must make reasonable efforts to ensure that each reclaimer intends to properly and legitimately reclaim the hazardous secondary material and not discard it, and that each reclaimer will manage the hazardous secondary material in a manner that is protective of human health and the environment. If the hazardous secondary material will be passing through an intermediate facility where the management of the hazardous secondary materials is not addressed under a RCRA Part B permit or interim status standards, the hazardous secondary material generator must make contractual arrangements with the intermediate facility to ensure that the hazardous secondary material is sent to the reclamation facility identified by the hazardous secondary material generator, and the hazardous secondary material gener-

ator must perform reasonable efforts to ensure that the intermediate facility will manage the hazardous secondary material in a manner that is protective of human health and the environment. Reasonable efforts must be repeated at a minimum of every three years for the hazardous secondary material generator to claim the exclusion and to send the hazardous secondary materials to each reclaimer and any intermediate facility. In making these reasonable efforts, the generator may use any credible evidence available, including information gathered by the hazardous secondary material generator, provided by the reclaimer or intermediate facility, and/or provided by a third party. The hazardous secondary material generator must affirmatively answer all of the following questions for each reclamation facility and any intermediate facility:

(1) Does the available information indicate that the reclamation process is legitimate pursuant to § 260.43 of this chapter? In answering this question, the hazardous secondary material generator can rely on their existing knowledge of the physical and chemical properties of the hazardous secondary material, as well as information from other sources (e.g., the reclamation facility, audit reports, etc.) about the reclamation process. (By responding to this question, the hazardous secondary material generator has also satisfied its requirement in § 260.43(a) of this chapter to be able to demonstrate that the recycling is legitimate).

(2) Does the publicly available information indicate that the reclamation facility and any intermediate facility that is used by the hazardous secondary material generator notified the appropriate authorities of hazardous secondary materials reclamation activities pursuant to § 260.42 of this chapter and have they notified the appropriate authorities that the financial assurance condition is satisfied per paragraph (a)(24)(vi)(F) of this section? In answering these questions, the hazardous secondary material generator can rely on the available information documenting the reclamation facility's

and any intermediate facility's compliance with the notification requirements per §260.42 of this chapter, including the requirement in §260.42(a)(5) to notify EPA whether the reclaimer or intermediate facility has financial assurance.

(3) Does publicly available information indicate that the reclamation facility or any intermediate facility that is used by the hazardous secondary material generator has not had any formal enforcement actions taken against the facility in the previous three years for violations of the RCRA hazardous waste regulations and has not been classified as a significant non-complier with RCRA Subtitle C? In answering this question, the hazardous secondary material generator can rely on the publicly available information from EPA or the state. If the reclamation facility or any intermediate facility that is used by the hazardous secondary material generator has had a formal enforcement action taken against the facility in the previous three years for violations of the RCRA hazardous waste regulations and has been classified as a significant non-complier with RCRA Subtitle C, does the hazardous secondary material generator have credible evidence that the facilities will manage the hazardous secondary materials properly? In answering this question, the hazardous secondary material generator can obtain additional information from EPA, the state, or the facility itself that the facility has addressed the violations, taken remedial steps to address the violations and prevent future violations, or that the violations are not relevant to the proper management of the hazardous secondary materials.

(4) Does the available information indicate that the reclamation facility and any intermediate facility that is used by the hazardous secondary material generator have the equipment and trained personnel to safely recycle the hazardous secondary material? In answering this question, the generator may rely on a description by the reclamation facility or by an independent third party of the equipment and trained personnel to be used to recycle the generator's hazardous secondary material.

(5) If residuals are generated from the reclamation of the excluded hazardous secondary materials, does the reclamation facility have the permits required (if any) to manage the residuals? If not, does the reclamation facility have a contract with an appropriately permitted facility to dispose of the residuals? If not, does the hazardous secondary material generator have credible evidence that the residuals will be managed in a manner that is protective of human health and the environment? In answering these questions, the hazardous secondary material generator can rely on publicly available information from EPA or the state, or information provided by the facility itself.

(C) The hazardous secondary material generator must maintain for a minimum of three years documentation and certification that reasonable efforts were made for each reclamation facility and, if applicable, intermediate facility where the management of the hazardous secondary materials is not addressed under a RCRA Part B permit or interim status standards prior to transferring hazardous secondary material. Documentation and certification must be made available upon request by a regulatory authority within 72 hours, or within a longer period of time as specified by the regulatory authority. The certification statement must:

(1) Include the printed name and official title of an authorized representative of the hazardous secondary material generator company, the authorized representative's signature, and the date signed;

(2) Incorporate the following language: "I hereby certify in good faith and to the best of my knowledge that, prior to arranging for transport of excluded hazardous secondary materials to [insert name(s) of reclamation facility and any intermediate facility], reasonable efforts were made in accordance with §261.4(a)(24)(v)(B) to ensure that the hazardous secondary materials would be recycled legitimately, and otherwise managed in a manner that is protective of human health and the environment, and that such efforts were based on current and accurate information."

(D) The hazardous secondary material generator must maintain at the generating facility for no less than three (3) years records of all off-site shipments of hazardous secondary materials. For each shipment, these records must, at a minimum, contain the following information:

(*1*) Name of the transporter and date of the shipment;

(*2*) Name and address of each reclaimer and, if applicable, the name and address of each intermediate facility to which the hazardous secondary material was sent;

(*3*) The type and quantity of hazardous secondary material in the shipment.

(E) The hazardous secondary material generator must maintain at the generating facility for no less than three (3) years confirmations of receipt from each reclaimer and, if applicable, each intermediate facility for all off-site shipments of hazardous secondary materials. Confirmations of receipt must include the name and address of the reclaimer (or intermediate facility), the type and quantity of the hazardous secondary materials received and the date which the hazardous secondary materials were received. This requirement may be satisfied by routine business records (e.g., financial records, bills of lading, copies of DOT shipping papers, or electronic confirmations of receipt); and

(vi) Reclaimers of hazardous secondary material excluded from regulation under this exclusion and intermediate facilities as defined in § 260.10 of this chapter satisfy all of the following conditions:

(A) The reclaimer and intermediate facility must maintain at its facility for no less than three (3) years records of all shipments of hazardous secondary material that were received at the facility and, if applicable, for all shipments of hazardous secondary materials that were received and subsequently sent off-site from the facility for further reclamation. For each shipment, these records must at a minimum contain the following information:

(*1*) Name of the transporter and date of the shipment;

(*2*) Name and address of the hazardous secondary material generator and, if applicable, the name and address of the reclaimer or intermediate facility which the hazardous secondary materials were received from;

(*3*) The type and quantity of hazardous secondary material in the shipment; and

(*4*) For hazardous secondary materials that, after being received by the reclaimer or intermediate facility, were subsequently transferred off-site for further reclamation, the name and address of the (subsequent) reclaimer and, if applicable, the name and address of each intermediate facility to which the hazardous secondary material was sent.

(B) The intermediate facility must send the hazardous secondary material to the reclaimer(s) designated by the hazardous secondary materials generator.

(C) The reclaimer and intermediate facility must send to the hazardous secondary material generator confirmations of receipt for all off-site shipments of hazardous secondary materials. Confirmations of receipt must include the name and address of the reclaimer (or intermediate facility), the type and quantity of the hazardous secondary materials received and the date which the hazardous secondary materials were received. This requirement may be satisfied by routine business records (e.g., financial records, bills of lading, copies of DOT shipping papers, or electronic confirmations of receipt).

(D) The reclaimer and intermediate facility must manage the hazardous secondary material in a manner that is at least as protective as that employed for analogous raw material and must be contained. An "analogous raw material" is a raw material for which a hazardous secondary material is a substitute and serves the same function and has similar physical and chemical properties as the hazardous secondary material.

(E) Any residuals that are generated from reclamation processes will be managed in a manner that is protective of human health and the environment. If any residuals exhibit a hazardous characteristic according to subpart C

of 40 CFR part 261, or if they themselves are specifically listed in subpart
D of 40 CFR part 261, such residuals are
hazardous wastes and must be managed
in accordance with the applicable requirements of 40 CFR parts 260 through
272.

(F) The reclaimer and intermediate
facility has financial assurance as required under subpart H of 40 CFR part
261.

(vii) In addition, all persons claiming
the exclusion under this paragraph
(a)(24) of this section must provide notification as required under §260.42 of
this chapter.

(25) Hazardous secondary material
that is exported from the United States
and reclaimed at a reclamation facility
located in a foreign country is not a
solid waste, provided that the hazardous secondary material generator
complies with the applicable requirements of paragraph (a)(24)(i)–(v) of this
section (excepting paragraph
(a)(v)(B)(2) of this section for foreign
reclaimers and foreign intermediate facilities), and that the hazardous secondary material generator also complies with the following requirements:

(i) Notify EPA of an intended export
before the hazardous secondary material is scheduled to leave the United
States. A complete notification must
be submitted at least sixty (60) days before the initial shipment is intended to
be shipped off-site. This notification
may cover export activities extending
over a twelve (12) month or lesser period. The notification must be in writing, signed by the hazardous secondary
material generator, and include the
following information:

(A) Name, mailing address, telephone
number and EPA ID number (if applicable) of the hazardous secondary material generator;

(B) A description of the hazardous
secondary material and the EPA hazardous waste number that would apply
if the hazardous secondary material
was managed as hazardous waste and
the U.S. DOT proper shipping name,
hazard class and ID number (UN/NA)
for each hazardous secondary material
as identified in 49 CFR parts 171
through 177;

(C) The estimated frequency or rate
at which the hazardous secondary ma-

terial is to be exported and the period
of time over which the hazardous secondary material is to be exported;

(D) The estimated total quantity of
hazardous secondary material;

(E) All points of entry to and departure from each foreign country through
which the hazardous secondary material will pass;

(F) A description of the means by
which each shipment of the hazardous
secondary material will be transported
(e.g., mode of transportation vehicle
(air, highway, rail, water, etc.), type(s)
of container (drums, boxes, tanks,
etc.));

(G) A description of the manner in
which the hazardous secondary material will be reclaimed in the receiving
country;

(H) The name and address of the reclaimer, any intermediate facility and
any alternate reclaimer and intermediate facilities; and

(I) The name of any transit countries
through which the hazardous secondary material will be sent and a description of the approximate length of
time it will remain in such countries
and the nature of its handling while
there (for purposes of this section, the
terms "Acknowledgement of Consent",
"receiving country" and "transit country" are used as defined in 40 CFR
262.51 with the exception that the
terms in this section refer to hazardous
secondary materials, rather than hazardous waste):

(ii) Notifications submitted by mail
should be sent to the following mailing
address: Office of Enforcement and
Compliance Assurance, Office of Federal Activities, International Compliance Assurance Division, (Mail Code
2254A), Environmental Protection
Agency, 1200 Pennsylvania Ave., NW.,
Washington, DC 20460. Hand-delivered
notifications should be delivered to: Office of Enforcement and Compliance
Assurance, Office of Federal Activities,
International Compliance Assurance
Division, Environmental Protection
Agency, Ariel Rios Bldg., Room 6144,
12th St. and Pennsylvania Ave., NW.,
Washington, DC 20004. In both cases,
the following shall be prominently dis-
played on the front of the envelope:
"Attention: Notification of Intent to
Export."

(iii) Except for changes to the telephone number in paragraph (a)(25)(i)(A) of this section and decreases in the quantity of hazardous secondary material indicated pursuant to paragraph (a)(25)(i)(D) of this section, when the conditions specified on the original notification change (including any exceedance of the estimate of the quantity of hazardous secondary material specified in the original notification), the hazardous secondary material generator must provide EPA with a written renotification of the change. The shipment cannot take place until consent of the receiving country to the changes (except for changes to paragraph (a)(25)(i)(I) of this section and in the ports of entry to and departure from transit countries pursuant to paragraphs (a)(25)(i)(E) of this section) has been obtained and the hazardous secondary material generator receives from EPA an Acknowledgment of Consent reflecting the receiving country's consent to the changes.

(iv) Upon request by EPA, the hazardous secondary material generator shall furnish to EPA any additional information which a receiving country requests in order to respond to a notification.

(v) EPA will provide a complete notification to the receiving country and any transit countries. A notification is complete when EPA receives a notification which EPA determines satisfies the requirements of paragraph (a)(25)(i) of this section. Where a claim of confidentiality is asserted with respect to any notification information required by paragraph (a)(25)(i) of this section, EPA may find the notification not complete until any such claim is resolved in accordance with 40 CFR 260.2.

(vi) The export of hazardous secondary material under this paragraph (a)(25) is prohibited unless the receiving country consents to the intended export. When the receiving country consents in writing to the receipt of the hazardous secondary material, EPA will send an Acknowledgment of Consent to the hazardous secondary material generator. Where the receiving country objects to receipt of the hazardous secondary material or withdraws a prior consent, EPA will notify the hazardous secondary material gen-

erator in writing. EPA will also notify the hazardous secondary material generator of any responses from transit countries.

(vii) For exports to OECD Member countries, the receiving country may respond to the notification using tacit consent. If no objection has been lodged by any receiving country or transit countries to a notification provided pursuant to paragraph (a)(25)(i) of this section within thirty (30) days after the date of issuance of the acknowledgement of receipt of notification by the competent authority of the receiving country, the transboundary movement may commence. In such cases, EPA will send an Acknowledgment of Consent to inform the hazardous secondary material generator that the receiving country and any relevant transit countries have not objected to the shipment, and are thus presumed to have consented tacitly. Tacit consent expires one (1) calendar year after the close of the thirty (30) day period; renotification and renewal of all consents is required for exports after that date.

(viii) A copy of the Acknowledgment of Consent must accompany the shipment. The shipment must conform to the terms of the Acknowledgment of Consent.

(ix) If a shipment cannot be delivered for any reason to the reclaimer, intermediate facility or the alternate reclaimer or alternate intermediate facility, the hazardous secondary material generator must re-notify EPA of a change in the conditions of the original notification to allow shipment to a new reclaimer in accordance with paragraph (iii) of this section and obtain another Acknowledgment of Consent.

(x) Hazardous secondary material generators must keep a copy of each notification of intent to export and each Acknowledgment of Consent for a period of three years following receipt of the Acknowledgment of Consent.

(xi) Hazardous secondary material generators must file with the Administrator no later than March 1 of each year, a report summarizing the types, quantities, frequency and ultimate destination of all hazardous secondary materials exported during the previous alendar year. Annual reports sub-mitted by mail should be sent to the ollowing address: Office of Enforce-ment and Compliance Assurance, Office f Federal Activities, International Compliance Assurance Division (Mail Code 2254A), Environmental Protection Agency, 1200 Pennsylvania Ave., NW., Washington, DC 20460. Hand-delivered reports should be delivered to: Office of Enforcement and Compliance Assur-ance, Office of Federal Activities, International Compliance Assurance Division, Environmental Protection Agency, Ariel Rios Bldg., Room 6144, 12th St. and Pennsylvania Ave., NW., Washington, DC 20004. Such reports must include the following informa-tion:

(A) Name, mailing and site address, and EPA ID number (if applicable) of the hazardous secondary material gen-erator;

(B) The calendar year covered by the report;

(C) The name and site address of each reclaimer and intermediate facility;

(D) By reclaimer and intermediate facility, for each hazardous secondary material exported, a description of the hazardous secondary material and the EPA hazardous waste number that would apply if the hazardous secondary material was managed as hazardous waste, DOT hazard class, the name and U.S. EPA ID number (where applicable) for each transporter used, the total amount of hazardous secondary mate-rial shipped and the number of ship-ments pursuant to each notification;

(E) A certification signed by the haz-ardous secondary material generator which states: "I certify under penalty of law that I have personally examined and am familiar with the information submitted in this and all attached doc-uments, and that based on my inquiry of those individuals immediately re-sponsible for obtaining the informa-tion, I believe that the submitted infor-mation is true, accurate, and complete. I am aware that there are significant penalties for submitting false informa-tion including the possibility of fine and imprisonment."

(xii) All persons claiming an exclu-sion under this paragraph (a)(25) must provide notification as required by §260.42 of this chapter.

(b) *Solid wastes which are not haz-ardous wastes.* The following solid wastes are not hazardous wastes:

(1) Household waste, including house-hold waste that has been collected, transported, stored, treated, disposed, recovered (e.g., refuse-derived fuel) or reused. "Household waste" means any material (including garbage, trash and sanitary wastes in septic tanks) de-rived from households (including single and multiple residences, hotels and motels, bunkhouses, ranger stations, crew quarters, campgrounds, picnic grounds and day-use recreation areas). A resource recovery facility managing municipal solid waste shall not be deemed to be treating, storing, dis-posing of, or otherwise managing haz-ardous wastes for the purposes of regu-lation under this subtitle, if such facil-ity:

(i) Receives and burns only

(A) Household waste (from single and multiple dwellings, hotels, motels, and other residential sources) and

(B) Solid waste from commercial or industrial sources that does not con-tain hazardous waste; and

(ii) Such facility does not accept haz-ardous wastes and the owner or oper-ator of such facility has established contractual requirements or other ap-propriate notification or inspection procedures to assure that hazardous wastes are not received at or burned in such facility.

(2) Solid wastes generated by any of the following and which are returned to the soils as fertilizers:

(i) The growing and harvesting of ag-ricultural crops.

(ii) The raising of animals, including animal manures.

(3) Mining overburden returned to the mine site.

(4) Fly ash waste, bottom ash waste, slag waste, and flue gas emission con-trol waste, generated primarily from the combusion of coal or other fossil fuels, except as provided by §266.112 of this chapter for facilities that burn or process hazardous waste.

(5) Drilling fluids, produced waters, and other wastes associated with the exploration, development, or produc-tion of crude oil, natural gas or geo-thermal energy.

(6)(i) Wastes which fail the test for the Toxicity Characteristic because chromium is present or are listed in subpart D due to the presence of chromium, which do not fail the test for the Toxicity Characteristic for any other constituent or are not listed due to the presence of any other constituent, and which do not fail the test for any other characteristic, if it is shown by a waste generator or by waste generators that:

(A) The chromium in the waste is exclusively (or nearly exclusively) trivalent chromium; and

(B) The waste is generated from an industrial process which uses trivalent chromium exclusively (or nearly exclusively) and the process does not generate hexavalent chromium; and

(C) The waste is typically and frequently managed in non-oxidizing environments.

(ii) Specific wastes which meet the standard in paragraphs (b)(6)(i) (A), (B), and (C) (so long as they do not fail the test for the toxicity characteristic for any other constituent, and do not exhibit any other characteristic) are:

(A) Chrome (blue) trimmings generated by the following subcategories of the leather tanning and finishing industry; hair pulp/chrome tan/retan/wet finish; hair save/chrome tan/retan/wet finish; retan/wet finish; no beamhouse; through-the-blue; and shearling.

(B) Chrome (blue) shavings generated by the following subcategories of the leather tanning and finishing industry: Hair pulp/chrome tan/retan/wet finish; hair save/chrome tan/retan/wet finish; retan/wet finish; no beamhouse; through-the-blue; and shearling.

(C) Buffing dust generated by the following subcategories of the leather tanning and finishing industry; hair pulp/chrome tan/retan/wet finish; hair save/chrome tan/retan/wet finish; retan/wet finish; no beamhouse; through-the-blue.

(D) Sewer screenings generated by the following subcategories of the leather tanning and finishing industry: Hair pulp/chrome tan/retan/wet finish; hair save/chrome tan/retan/wet finish; retan/wet finish; no beamhouse; through-the-blue; and shearling.

(E) Wastewater treatment sludges generated by the following subcat-egories of the leather tanning and fin-ishing industry: Hair pulp/chrome tan retan/wet finish; hair save/chrome tan retan/wet finish; retan/wet finish; no beamhouse; through-the-blue; an shearling.

(F) Wastewater treatment sludge generated by the following subcat-egories of the leather tanning and fin-ishing industry: Hair pulp/chrome tan retan/wet finish; hair save/chrome tan retan/wet finish; and through-the-blue

(G) Waste scrap leather from the leather tanning industry, the shoe manufacturing industry, and other leather product manufacturing indus-tries.

(H) Wastewater treatment sludges from the production of $TiO_2$ pigment using chromium-bearing ores by the chloride process.

(7) Solid waste from the extraction beneficiation, and processing of ores and minerals (including coal, phos-phate rock, and overburden from the mining of uranium ore), except as pro-vided by § 266.112 of this chapter for fa-cilities that burn or process hazardous waste.

(i) For purposes of § 261.4(b)(7 beneficiation of ores and minerals is restricted to the following activities; crushing; grinding; washing; dissolu-tion; crystallization; filtration; sort-ing; sizing; drying; sintering; pelletizing; briquetting; calcining to remove water and/or carbon dioxide; roasting, autoclaving, and/or chlorination in preparation for leach-ing (except where the roasting (and/or autoclaving and/or chlorination)/leach-ing sequence produces a final or inter-mediate product that does not undergo further beneficiation or processing; gravity concentration; magnetic sepa-ration; electrostatic separation; flota-tion; ion exchange; solvent extraction; electrowinning; precipitation; amal-gamation; and heap, dump, vat, tank, and in situ leaching.

(ii) For the purposes of § 261.4(b)(7), solid waste from the processing of ores and minerals includes only the fol-lowing wastes as generated:

(A) Slag from primary copper proc-essing;

(B) Slag from primary lead proc-essing;

(C) Red and brown muds from bauxite refining;

(D) Phosphogypsum from phosphoric acid production;

(E) Slag from elemental phosphorus production;

(F) Gasifier ash from coal gasification;

(G) Process wastewater from coal gasification;

(H) Calcium sulfate wastewater treatment plant sludge from primary copper processing;

(I) Slag tailings from primary copper processing;

(J) Fluorogypsum from hydrofluoric acid production;

(K) Process wastewater from hydrofluoric acid production;

(L) Air pollution control dust/sludge from iron blast furnaces;

(M) Iron blast furnace slag;

(N) Treated residue from roasting/leaching of chrome ore;

(O) Process wastewater from primary magnesium processing by the anhydrous process;

(P) Process wastewater from phosphoric acid production;

(Q) Basic oxygen furnace and open hearth furnace air pollution control dust/sludge from carbon steel production;

(R) Basic oxygen furnace and open hearth furnace slag from carbon steel production;

(S) Chloride process waste solids from titanium tetrachloride production;

(T) Slag from primary zinc processing.

(iii) A residue derived from co-processing mineral processing secondary materials with normal beneficiation raw materials or with normal mineral processing raw materials remains excluded under paragraph (b) of this section if the owner or operator:

(A) Processes at least 50 percent by weight normal beneficiation raw materials or normal mineral processing raw materials; and,

(B) Legitimately reclaims the secondary mineral processing materials.

(8) Cement kiln dust waste, except as provided by §266.112 of this chapter for facilities that burn or process hazardous waste.

(9) Solid waste which consists of discarded arsenical-treated wood or wood products which fails the test for the Toxicity Characteristic for Hazardous Waste Codes D004 through D017 and which is not a hazardous waste for any other reason if the waste is generated by persons who utilize the arsenical-treated wood and wood products for these materials' intended end use.

(10) Petroleum-contaminated media and debris that fail the test for the Toxicity Characteristic of §261.24 (Hazardous Waste Codes D018 through D043 only) and are subject to the corrective action regulations under part 280 of this chapter.

(11) Injected groundwater that is hazardous only because it exhibits the Toxicity Characteristic (Hazardous Waste Codes D018 through D043 only) in §261.24 of this part that is reinjected through an underground injection well pursuant to free phase hydrocarbon recovery operations undertaken at petroleum refineries, petroleum marketing terminals, petroleum bulk plants, petroleum pipelines, and petroleum transportation spill sites until January 25, 1993. This extension applies to recovery operations in existence, or for which contracts have been issued, on or before March 25, 1991. For groundwater returned through infiltration galleries from such operations at petroleum refineries, marketing terminals, and bulk plants, until [insert date six months after publication]. New operations involving injection wells (beginning after March 25, 1991) will qualify for this compliance date extension (until January 25, 1993) only if:

(i) Operations are performed pursuant to a written state agreement that includes a provision to assess the groundwater and the need for further remediation once the free phase recovery is completed; and

(ii) A copy of the written agreement has been submitted to: Waste Identification Branch (5304), U.S. Environmental Protection Agency, 1200 Pennsylvania Ave., NW., Washington, DC 20460.

(12) Used chlorofluorocarbon refrigerants from totally enclosed heat transfer equipment, including mobile

air conditioning systems, mobile refrigeration, and commercial and industrial air conditioning and refrigeration systems that use chlorofluorocarbons as the heat transfer fluid in a refrigeration cycle, provided the refrigerant is reclaimed for further use.

(13) Non-terne plated used oil filters that are not mixed with wastes listed in subpart D of this part if these oil filters have been gravity hot-drained using one of the following methods:

(i) Puncturing the filter anti-drain back valve or the filter dome end and hot-draining;

(ii) Hot-draining and crushing;

(iii) Dismantling and hot-draining; or

(iv) Any other equivalent hot-draining method that will remove used oil.

(14) Used oil re-refining distillation bottoms that are used as feedstock to manufacture asphalt products.

(15) Leachate or gas condensate collected from landfills where certain solid wastes have been disposed, provided that:

(i) The solid wastes disposed would meet one or more of the listing descriptions for Hazardous Waste Codes K169, K170, K171, K172, K174, K175, K176, K177, K178 and K181 if these wastes had been generated after the effective date of the listing;

(ii) The solid wastes described in paragraph (b)(15)(i) of this section were disposed prior to the effective date of the listing;

(iii) The leachate or gas condensate do not exhibit any characteristic of hazardous waste nor are derived from any other listed hazardous waste;

(iv) Discharge of the leachate or gas condensate, including leachate or gas condensate transferred from the landfill to a POTW by truck, rail, or dedicated pipe, is subject to regulation under sections 307(b) or 402 of the Clean Water Act.

(v) As of February 13, 2001, leachate or gas condensate derived from K169–K172 is no longer exempt if it is stored or managed in a surface impoundment prior to discharge. As of November 21, 2003, leachate or gas condensate derived from K176, K177, and K178 is no longer exempt if it is stored or managed in a surface impoundment prior to discharge. After February 26, 2007, leachate or gas condensate derived

from K181 will no longer be exempt if it is stored or managed in a surface impoundment prior to discharge. There is one exception: if the surface impoundment is used to temporarily store leachate or gas condensate in response to an emergency situation (e.g., shut down of wastewater treatment system), provided the impoundment has a double liner, and provided the leachate or gas condensate is removed from the impoundment and continues to be managed in compliance with the conditions of this paragraph (b)(15)(v) after the emergency ends.

(16) [Reserved]

(17) Solid waste that would otherwise meet the definition of low-level mixed wastes (LLMW) pursuant to § 266.210 of this chapter that is generated at the Ortho-McNeil Pharmaceutical, Inc. (OMP Spring House) research and development facility in Spring House, Pennsylvania and treated on-site using a bench-scale high temperature catalytic oxidation unit is not a hazardous waste provided that:

(i) The total volume of LLMW generated and treated is no greater than 50 liters/year, (ii) OMP Spring House submits a written report to the EPA Region III office once every six months beginning six months after June 27, 2005, that must contain the following:

(A) Analysis demonstrating the destruction and removal efficiency of the treatment technology for all organic components of the wastestream,

(B) Analysis demonstrating the capture efficiencies of the treatment technology for all radioactive components of the wastestream and an estimate of the amount of radioactivity released during the reporting period,

(C) Analysis (including concentrations of constituents, including inorganic constituents, present and radioactivity) of the wastestream prior to and after treatment,

(D) Volume of the wastestream being treated per batch, as well as a total for the duration of the reporting period, and

(E) Final disposition of the radioactive residuals from the treatment of the wastestream.

(iii) OMP Spring House makes no significant changes to the design or operation of the high temperature catalytic oxidation unit or the wastestream.

(iv) This exclusion will remain in affect for 5 years from June 27, 2005.

(c) *Hazardous wastes which are exempted from certain regulations.* A hazardous waste which is generated in a product or raw material storage tank, a product or raw material transport vehicle or vessel, a product or raw material pipeline, or in a manufacturing process unit or an associated non-waste-treatment-manufacturing unit, is not subject to regulation under parts 262 through 265, 268, 270, 271 and 124 of this chapter or to the notification requirements of section 3010 of RCRA until it exits the unit in which it was generated, unless the unit is a surface impoundment, or unless the hazardous waste remains in the unit more than 90 days after the unit ceases to be operated for manufacturing, or for storage or transportation of product or raw materials.

(d) *Samples.* (1) Except as provided in paragraph (d)(2) of this section, a sample of solid waste or a sample of water, soil, or air, which is collected for the sole purpose of testing to determine its characteristics or composition, is not subject to any requirements of this part or parts 262 through 268 or part 270 or part 124 of this chapter or to the notification requirements of section 3010 of RCRA, when:

(i) The sample is being transported to a laboratory for the purpose of testing; or

(ii) The sample is being transported back to the sample collector after testing; or

(iii) The sample is being stored by the sample collector before transport to a laboratory for testing; or

(iv) The sample is being stored in a laboratory before testing; or

(v) The sample is being stored in a laboratory after testing but before it is returned to the sample collector; or

(vi) The sample is being stored temporarily in the laboratory after testing for a specific purpose (for example, until conclusion of a court case or enforcement action where further testing of the sample may be necessary).

(2) In order to qualify for the exemption in paragraphs (d)(1) (i) and (ii) of this section, a sample collector shipping samples to a laboratory and a laboratory returning samples to a sample collector must:

(i) Comply with U.S. Department of Transportation (DOT), U.S. Postal Service (USPS), or any other applicable shipping requirements; or

(ii) Comply with the following requirements if the sample collector determines that DOT, USPS, or other shipping requirements do not apply to the shipment of the sample:

(A) Assure that the following information accompanies the sample:

(*1*) The sample collector's name, mailing address, and telephone number;

(*2*) The laboratory's name, mailing address, and telephone number;

(*3*) The quantity of the sample;

(*4*) The date of shipment; and

(*5*) A description of the sample.

(B) Package the sample so that it does not leak, spill, or vaporize from its packaging.

(3) This exemption does not apply if the laboratory determines that the waste is hazardous but the laboratory is no longer meeting any of the conditions stated in paragraph (d)(1) of this section.

(e) *Treatability Study Samples.* (1) Except as provided in paragraph (e)(2) of this section, persons who generate or collect samples for the purpose of conducting treatability studies as defined in section 260.10, are not subject to any requirement of parts 261 through 263 of this chapter or to the notification requirements of Section 3010 of RCRA, nor are such samples included in the quantity determinations of §261.5 and §262.34(d) when:

(i) The sample is being collected and prepared for transportation by the generator or sample collector; or

(ii) The sample is being accumulated or stored by the generator or sample collector prior to transportation to a laboratory or testing facility; or

(iii) The sample is being transported to the laboratory or testing facility for the purpose of conducting a treatability study.

(2) The exemption in paragraph (e)(1) of this section is applicable to samples

of hazardous waste being collected and shipped for the purpose of conducting treatability studies provided that:

(i) The generator or sample collector uses (in "treatability studies") no more than 10,000 kg of media contaminated with non-acute hazardous waste, 1000 kg of non-acute hazardous waste other than contaminated media, 1 kg of acute hazardous waste, 2500 kg of media contaminated with acute hazardous waste for each process being evaluated for each generated waste stream; and

(ii) The mass of each sample shipment does not exceed 10,000 kg; the 10,000 kg quantity may be all media contaminated with non-acute hazardous waste, or may include 2500 kg of media contaminated with acute hazardous waste, 1000 kg of hazardous waste, and 1 kg of acute hazardous waste; and

(iii) The sample must be packaged so that it will not leak, spill, or vaporize from its packaging during shipment and the requirements of paragraph A or B of this subparagraph are met.

(A) The transportation of each sample shipment complies with U.S. Department of Transportation (DOT), U.S. Postal Service (USPS), or any other applicable shipping requirements; or

(B) If the DOT, USPS, or other shipping requirements do not apply to the shipment of the sample, the following information must accompany the sample:

(1) The name, mailing address, and telephone number of the originator of the sample;

(2) The name, address, and telephone number of the facility that will perform the treatability study;

(3) The quantity of the sample;

(4) The date of shipment; and

(5) A description of the sample, including its EPA Hazardous Waste Number.

(iv) The sample is shipped to a laboratory or testing facility which is exempt under § 261.4(f) or has an appropriate RCRA permit or interim status.

(v) The generator or sample collector maintains the following records for a period ending 3 years after completion of the treatability study:

(A) Copies of the shipping documents;

(B) A copy of the contract with the facility conducting the treatability study;

(C) Documentation showing:

(1) The amount of waste shipped under this exemption;

(2) The name, address, and EPA identification number of the laboratory or testing facility that received the waste;

(3) The date the shipment was made; and

(4) Whether or not unused samples and residues were returned to the generator.

(vi) The generator reports the information required under paragraph (e)(2)(v)(C) of this section in its biennial report.

(3) The Regional Administrator may grant requests on a case-by-case basis for up to an additional two years for treatability studies involving bioremediation. The Regional Administrator may grant requests on a case-by-case basis for quantity limits in excess of those specified in paragraphs (e)(2) (i) and (ii) and (f)(4) of this section, for up to an additional 5000 kg of media contaminated with non-acute hazardous waste, 500 kg of non-acute hazardous waste, 2500 kg of media contaminated with acute hazardous waste and 1 kg of acute hazardous waste:

(i) In response to requests for authorization to ship, store and conduct treatability studies on additional quantities in advance of commencing treatability studies. Factors to be considered in reviewing such requests include the nature of the technology, the type of process (e.g., batch versus continuous), size of the unit undergoing testing (particularly in relation to scale-up considerations), the time/quantity of material required to reach steady state operating conditions, or test design considerations such as mass balance calculations.

(ii) In response to requests for authorization to ship, store and conduct treatability studies on additional quantities after initiation or completion of initial treatability studies, when: There has been an equipment or mechanical failure during the conduct of a treatability study; there is a need to verify the results of a previously conducted treatability study; there is a need to study and analyze alternative techniques within a previously evaluated treatment process; or there is a need to do further evaluation of an ongoing treatability study to determine final specifications for treatment.

(iii) The additional quantities and timeframes allowed in paragraph (e)(3) (i) and (ii) of this section are subject to all the provisions in paragraphs (e) (1) and (e)(2) (iii) through (vi) of this section. The generator or sample collector must apply to the Regional Administrator in the Region where the sample is collected and provide in writing the following information:

(A) The reason why the generator or sample collector requires additional time or quantity of sample for treatability study evaluation and the additional time or quantity needed;

(B) Documentation accounting for all samples of hazardous waste from the waste stream which have been sent for or undergone treatability studies including the date each previous sample from the waste stream was shipped, the quantity of each previous shipment, the laboratory or testing facility to which it was shipped, what treatability study processes were conducted on each sample shipped, and the available results on each treatability study;

(C) A description of the technical modifications or change in specifications which will be evaluated and the expected results;

(D) If such further study is being required due to equipment or mechanical failure, the applicant must include information regarding the reason for the failure or breakdown and also include what procedures or equipment improvements have been made to protect against further breakdowns; and

(E) Such other information that the Regional Administrator considers necessary.

(f) *Samples Undergoing Treatability Studies at Laboratories and Testing Facilities.* Samples undergoing treatability studies and the laboratory or testing facility conducting such treatability studies (to the extent such facilities are not otherwise subject to RCRA requirements) are not subject to any requirement of this part, part 124, parts 262–266, 268, and 270, or to the notification requirements of Section 3010

of RCRA provided that the conditions of paragraphs (f) (1) through (11) of this section are met. A mobile treatment unit (MTU) may qualify as a testing facility subject to paragraphs (f) (1) through (11) of this section. Where a group of MTUs are located at the same site, the limitations specified in (f) (1) through (11) of this section apply to the entire group of MTUs collectively as if the group were one MTU.

(1) No less than 45 days before conducting treatability studies, the facility notifies the Regional Administrator, or State Director (if located in an authorized State), in writing that it intends to conduct treatability studies under this paragraph.

(2) The laboratory or testing facility conducting the treatability study has an EPA identification number.

(3) No more than a total of 10,000 kg of "as received" media contaminated with non-acute hazardous waste, 2500 kg of media contaminated with acute hazardous waste or 250 kg of other "as received" hazardous waste is subject to initiation of treatment in all treatability studies in any single day. "As received" waste refers to the waste as received in the shipment from the generator or sample collector.

(4) The quantity of "as received" hazardous waste stored at the facility for the purpose of evaluation in treatability studies does not exceed 10,000 kg, the total of which can include 10,000 kg of media contaminated with non-acute hazardous waste, 2500 kg of media contaminated with acute hazardous waste, 1000 kg of non-acute hazardous wastes other than contaminated media, and 1 kg of acute hazardous waste. This quantity limitation does not include treatment materials (including nonhazardous solid waste) added to "as received" hazardous waste.

(5) No more than 90 days have elapsed since the treatability study for the sample was completed, or no more than one year (two years for treatability studies involving bioremediation) have elapsed since the generator or sample collector shipped the sample to the laboratory or testing facility, whichever date first occurs. Up to 500 kg of treated material from a particular waste stream from treatability studies may

be archived for future evaluation up to five years from the date of initial receipt. Quantities of materials archived are counted against the total storage limit for the facility.

(6) The treatability study does not involve the placement of hazardous waste on the land or open burning of hazardous waste.

(7) The facility maintains records for 3 years following completion of each study that show compliance with the treatment rate limits and the storage time and quantity limits. The following specific information must be included for each treatability study conducted:

(i) The name, address, and EPA identification number of the generator or sample collector of each waste sample;

(ii) The date the shipment was received;

(iii) The quantity of waste accepted;

(iv) The quantity of "as received" waste in storage each day;

(v) The date the treatment study was initiated and the amount of "as received" waste introduced to treatment each day;

(vi) The date the treatability study was concluded;

(vii) The date any unused sample or residues generated from the treatability study were returned to the generator or sample collector or, if sent to a designated facility, the name of the facility and the EPA identification number.

(8) The facility keeps, on-site, a copy of the treatability study contract and all shipping papers associated with the transport of treatability study samples to and from the facility for a period ending 3 years from the completion date of each treatability study.

(9) The facility prepares and submits a report to the Regional Administrator, or state Director (if located in an authorized state), by March 15 of each year, that includes the following information for the previous calendar year:

(i) The name, address, and EPA identification number of the facility conducting the treatability studies;

(ii) The types (by process) of treatability studies conducted;

(iii) The names and addresses of persons for whom studies have been conducted (including their EPA identification numbers);

(iv) The total quantity of waste in storage each day;

(v) The quantity and types of waste subjected to treatability studies;

(vi) When each treatability study was conducted;

(vii) The final disposition of residues and unused sample from each treatability study.

(10) The facility determines whether any unused sample or residues generated by the treatability study are hazardous waste under § 261.3 and, if so, are subject to parts 261 through 268, and part 270 of this chapter, unless the residues and unused samples are returned to the sample originator under the § 261.4(e) exemption.

(11) The facility notifies the Regional Administrator, or State Director (if located in an authorized State), by letter when the facility is no longer planning to conduct any treatability studies at the site.

(g) *Dredged material that is not a hazardous waste.* Dredged material that is subject to the requirements of a permit that has been issued under 404 of the Federal Water Pollution Control Act (33 U.S.C.1344) or section 103 of the Marine Protection, Research, and Sanctuaries Act of 1972 (33 U.S.C. 1413) is not a hazardous waste. For this paragraph (g), the following definitions apply:

(1) The term *dredged material* has the same meaning as defined in 40 CFR 232.2;

(2) The term *permit* means:

(i) A permit issued by the U.S. Army Corps of Engineers (Corps) or an approved State under section 404 of the Federal Water Pollution Control Act (33 U.S.C. 1344);

(ii) A permit issued by the Corps under section 103 of the Marine Protection, Research, and Sanctuaries Act of 1972 (33 U.S.C. 1413); or

(iii) In the case of Corps civil works projects, the administrative equivalent of the permits referred to in paragraphs (g)(2)(i) and (ii) of this section, as provided for in Corps regulations (for example, see 33 CFR 336.1, 336.2, and 337.6).

[45 FR 33119, May 19, 1980]

EDITORIAL NOTE: For FEDERAL REGISTER citations affecting §261.4, see the List of CFR Sections Affected, which appears in the Finding Aids section of the printed volume and on GPO Access.

## §261.5 Special requirements for hazardous waste generated by conditionally exempt small quantity generators.

(a) A generator is a conditionally exempt small quantity generator in a calendar month if he generates no more than 100 kilograms of hazardous waste in that month.

(b) Except for those wastes identified in paragraphs (e), (f), (g), and (j) of this section, a conditionally exempt small quantity generator's hazardous wastes are not subject to regulation under parts 262 through 268, and parts 270 and 124 of this chapter, and the notification requirements of section 3010 of RCRA, provided the generator complies with the requirements of paragraphs (f), (g), and (j) of this section.

(c) When making the quantity determinations of this part and 40 CFR part 262, the generator must include all hazardous waste that it generates, except hazardous waste that:

(1) Is exempt from regulation under 40 CFR 261.4(c) through (f), 261.6(a)(3), 261.7(a)(1), or 261.8; or

(2) Is managed immediately upon generation only in on-site elementary neutralization units, wastewater treatment units, or totally enclosed treatment facilities as defined in 40 CFR 260.10; or

(3) Is recycled, without prior storage or accumulation, only in an on-site process subject to regulation under 40 CFR 261.6(c)(2); or

(4) Is used oil managed under the requirements of 40 CFR 261.6(a)(4) and 40 CFR part 279; or

(5) Is spent lead-acid batteries managed under the requirements of 40 CFR part 266, subpart G; or

(6) Is universal waste managed under 40 CFR 261.9 and 40 CFR part 273;

(7) Is a hazardous waste that is an unused commercial chemical product (listed in 40 CFR part 261, subpart D or exhibiting one or more characteristics in 40 CFR part 261, subpart C) that is generated solely as a result of a laboratory clean-out conducted at an eligible academic entity pursuant to §262.213.

For purposes of this provision, the term eligible academic entity shall have the meaning as defined in §262.200 of Part 262.

(d) In determining the quantity of hazardous waste generated, a generator need not include:

(1) Hazardous waste when it is removed from on-site storage; or

(2) Hazardous waste produced by on-site treatment (including reclamation) of his hazardous waste, so long as the hazardous waste that is treated was counted once; or

(3) Spent materials that are generated, reclaimed, and subsequently reused on-site, so long as such spent materials have been counted once.

(e) If a generator generates acute hazardous waste in a calendar month in quantities greater than set forth below, all quantities of that acute hazardous waste are subject to full regulation under parts 262 through 268, and parts 270 and 124 of this chapter, and the notification requirements of section 3010 of RCRA:

(1) A total of one kilogram of acute hazardous wastes listed in §§261.31 or 261.33(e).

(2) A total of 100 kilograms of any residue or contaminated soil, waste, or other debris resulting from the clean-up of a spill, into or on any land or water, of any acute hazardous wastes listed in §§261.31, or 261.33(e).

NOTE TO PARAGRAPH (E): "Full regulation" means those regulations applicable to generators of 1,000 kg or greater of hazardous waste in a calendar month.

(f) In order for acute hazardous wastes generated by a generator of acute hazardous wastes in quantities equal to or less than those set forth in paragraphs (e)(1) or (e)(2) of this section to be excluded from full regulation under this section, the generator must comply with the following requirements:

(1) Section 262.11 of this chapter;

(2) The generator may accumulate acute hazardous waste on-site. If he accumulates at any time acute hazardous wastes in quantities greater than those set forth in paragraph (e)(1) or (e)(2) of this section, all of those accumulated wastes are subject to regulation under parts 262 through 268, and parts 270 and

124 of this chapter, and the applicable notification requirements of section 3010 of RCRA. The time period of §262.34(a) of this chapter, for accumulation of wastes on-site, begins when the accumulated wastes exceed the applicable exclusion limit;

(3) A conditionally exempt small quantity generator may either treat or dispose of his acute hazardous waste in an on-site facility or ensure delivery to an off-site treatment, storage, or disposal facility, either of which, if located in the U.S., is:

(i) Permitted under part 270 of this chapter;

(ii) In interim status under parts 270 and 265 of this chapter;

(iii) Authorized to manage hazardous waste by a State with a hazardous waste management program approved under part 271 of this chapter;

(iv) Permitted, licensed, or registered by a State to manage municipal solid waste and, if managed in a municipal solid waste landfill is subject to Part 258 of this chapter;

(v) Permitted, licensed, or registered by a State to manage non-municipal non-hazardous waste and, if managed in a non-municipal non-hazardous waste disposal unit after January 1, 1998, is subject to the requirements in §§257.5 through 257.30 of this chapter; or

(vi) A facility which:

(A) Beneficially uses or reuses, or legitimately recycles or reclaims its waste; or

(B) Treats its waste prior to beneficial use or reuse, or legitimate recycling or reclamation; or

(vii) For universal waste managed under part 273 of this chapter, a universal waste handler or destination facility subject to the requirements of part 273 of this chapter.

(g) In order for hazardous waste generated by a conditionally exempt small quantity generator in quantities of 100 kilograms or less of hazardous waste during a calendar month to be excluded from full regulation under this section, the generator must comply with the following requirements:

(1) Section 262.11 of this chapter;

(2) The conditionally exempt small quantity generator may accumulate hazardous waste on-site. If he accumulates at any time 1,000 kilograms or

greater of his hazardous wastes, all of those accumulated wastes are subject to regulation under the special provisions of part 262 applicable to generators of greater than 100 kg and less than 1000 kg of hazardous waste in a calendar month as well as the requirements of parts 263 through 268, and parts 270 and 124 of this chapter, and the applicable notification requirements of section 3010 of RCRA. The time period of §262.34(d) for accumulation of wastes on-site begins for a conditionally exempt small quantity generator when the accumulated wastes equal or exceed 1000 kilograms;

(3) A conditionally exempt small quantity generator may either treat or dispose of his hazardous waste in an on-site facility or ensure delivery to an off-site treatment, storage or disposal facility, either of which, if located in the U.S., is:

(i) Permitted under part 270 of this chapter;

(ii) In interim status under parts 270 and 265 of this chapter;

(iii) Authorized to manage hazardous waste by a State with a hazardous waste management program approved under part 271 of this chapter;

(iv) Permitted, licensed, or registered by a State to manage municipal solid waste and, if managed in a municipal solid waste landfill is subject to Part 258 of this chapter;

(v) Permitted, licensed, or registered by a State to manage non-municipal non-hazardous waste and, if managed in a non-municipal non-hazardous waste disposal unit after January 1, 1998, is subject to the requirements in §§257.5 through 257.30 of this chapter; or

(vi) A facility which:

(A) Beneficially uses or reuses, or legitimately recycles or reclaims its waste; or

(B) Treats its waste prior to beneficial use or reuse, or legitimate recycling or reclamation; or

(vii) For universal waste managed under part 273 of this chapter, a universal waste handler or destination facility subject to the requirements of part 273 of this chapter.

(h) Hazardous waste subject to the reduced requirements of this section may be mixed with non-hazardous waste and remain subject to these reduced requirements even though the resultant mixture exceeds the quantity limitations identified in this section, unless the mixture meets any of the characteristics of hazardous waste identified in subpart C.

(i) If any person mixes a solid waste with a hazardous waste that exceeds a quantity exclusion level of this section, the mixture is subject to full regulation.

(j) If a conditionally exempt small quantity generator's wastes are mixed with used oil, the mixture is subject to part 279 of this chapter. Any material produced from such a mixture by processing, blending, or other treatment is also so regulated.

[51 FR 10174, Mar. 24, 1986, as amended at 51 FR 28682, Aug. 8, 1986; 51 FR 40637, Nov. 7, 1986; 53 FR 27163, July 19, 1988; 58 FR 26424, May 3, 1993; 60 FR 25541, May 11, 1995; 61 FR 34278, July 1, 1996; 63 FR 24968, May 6, 1998; 63 FR 37782, July 14, 1998; 68 FR 44665, July 30, 2003; 73 FR 72954, Dec. 1, 2008; 75 FR 13001, Mar. 18, 2010]

### §261.6 Requirements for recyclable materials.

(a)(1) Hazardous wastes that are recycled are subject to the requirements for generators, transporters, and storage facilities of paragraphs (b) and (c) of this section, except for the materials listed in paragraphs (a)(2) and (a)(3) of this section. Hazardous wastes that are recycled will be known as "recyclable materials."

(2) The following recyclable materials are not subject to the requirements of this section but are regulated under subparts C through N of part 266 of this chapter and all applicable provisions in parts 268, 270, and 124 of this chapter.

(i) Recyclable materials used in a manner constituting disposal (40 CFR part 266, subpart C);

(ii) Hazardous wastes burned (as defined in section 266.100(a)) in boilers and industrial furnaces that are not regulated under subpart O of part 264 or 265 of this chapter (40 CFR part 266, subpart H);

(iii) Recyclable materials from which precious metals are reclaimed (40 CFR part 266, subpart F);

(iv) Spent lead-acid batteries that are being reclaimed (40 CFR part 266, subpart G).

(3) The following recyclable materials are not subject to regulation under parts 262 through parts 268, 270 or 124 of this chapter, and are not subject to the notification requirements of section 3010 of RCRA:

(i) Industrial ethyl alcohol that is reclaimed except that, unless provided otherwise in an international agreement as specified in §262.58:

(A) A person initiating a shipment for reclamation in a foreign country, and any intermediary arranging for the shipment, must comply with the requirements applicable to a primary exporter in §§262.53, 262.56 (a)(1)–(4), (6), and (b), and 262.57, export such materials only upon consent of the receiving country and in conformance with the EPA Acknowledgment of Consent as defined in subpart E of part 262, and provide a copy of the EPA Acknowledgment of Consent to the shipment to the transporter transporting the shipment for export;

(B) Transporters transporting a shipment for export may not accept a shipment if he knows the shipment does not conform to the EPA Acknowledgment of Consent, must ensure that a copy of the EPA Acknowledgment of Consent accompanies the shipment and must ensure that it is delivered to the facility designated by the person initiating the shipment.

(ii) Scrap metal that is not excluded under §261.4(a)(13);

(iii) Fuels produced from the refining of oil-bearing hazardous waste along with normal process streams at a petroleum refining facility if such wastes result from normal petroleum refining, production, and transportation practices (this exemption does not apply to fuels produced from oil recovered from oil-bearing hazardous waste, where such recovered oil is already excluded under §261.4(a)(12);

(iv)(A) Hazardous waste fuel produced from oil-bearing hazardous wastes from petroleum refining, production, or transportation practices, or produced from oil reclaimed from such hazardous wastes, where such hazardous wastes are reintroduced into a process that does not use distillation or does not produce products from crude oil so long as the resulting fuel meets the used oil specification under § 279.11 of this chapter and so long as no other hazardous wastes are used to produce the hazardous waste fuel;

(B) Hazardous waste fuel produced from oil-bearing hazardous waste from petroleum refining production, and transportation practices, where such hazardous wastes are reintroduced into a refining process after a point at which contaminants are removed, so long as the fuel meets the used oil fuel specification under § 279.11 of this chapter; and

(C) Oil reclaimed from oil-bearing hazardous wastes from petroleum refining, production, and transportation practices, which reclaimed oil is burned as a fuel without reintroduction to a refining process, so long as the reclaimed oil meets the used oil fuel specification under § 279.11 of this chapter.

(4) Used oil that is recycled and is also a hazardous waste solely because it exhibits a hazardous characteristic is not subject to the requirements of parts 260 through 268 of this chapter, but is regulated under part 279 of this chapter. Used oil that is recycled includes any used oil which is reused, following its original use, for any purpose (including the purpose for which the oil was originally used). Such term includes, but is not limited to, oil which is re-refined, reclaimed, burned for energy recovery, or reprocessed.

(5) Hazardous waste that is exported to or imported from designated member countries of the Organization for Economic Cooperation and Development (OECD) (as defined in § 262.58(a)(1)) for purpose of recovery is subject to the requirements of 40 CFR part 262, subpart H, if it is subject to either the Federal manifesting requirements of 40 CFR Part 262, to the universal waste management standards of 40 CFR Part 273, or to State requirements analogous to 40 CFR Part 273.

(b) Generators and transporters of recyclable materials are subject to the applicable requirements of parts 262 and 263 of this chapter and the notification requirements under section 3010 of RCRA, except as provided in paragraph (a) of this section.

(c) (1) Owners and operators of facilities that store recyclable materials before they are recycled are regulated under all applicable provisions of subparts A though L, AA, BB, and CC of parts 264 and 265, and under parts 124, 266, 267, 268, and 270 of this chapter and the notification requirements under section 3010 of RCRA, except as provided in paragraph (a) of this section. (The recycling process itself is exempt from regulation except as provided in § 261.6(d).)

(2) Owners or operators of facilities that recycle recyclable materials without storing them before they are recycled are subject to the following requirements, except as provided in paragraph (a) of this section:

(i) Notification requirements under section 3010 of RCRA;

(ii) Sections 265.71 and 265.72 (dealing with the use of the manifest and manifest discrepancies) of this chapter.

(iii) Section 261.6(d) of this chapter.

(d) Owners or operators of facilities subject to RCRA permitting requirements with hazardous waste management units that recycle hazardous wastes are subject to the requirements of subparts AA and BB of part 264, 265 or 267 of this chapter.

[50 FR 49203, Nov. 29, 1985]

EDITORIAL NOTE: For FEDERAL REGISTER citations affecting § 261.6, see the List of CFR Sections Affected, which appears in the Finding Aids section of the printed volume and on GPO Access.

## § 261.7 Residues of hazardous waste in empty containers.

(a)(1) Any hazardous waste remaining in either: an empty container; or an inner liner removed from an empty container, as defined in paragraph (b) of this section, is not subject to regulation under parts 261 through 268, 270, or 124 this chapter or to the notification requirements of section 3010 of RCRA.

(2) Any hazardous waste in either a container that is not empty or an inner liner removed from a container that is not empty, as defined in paragraph (b) of this section, is subject to regulation under parts 261 through 268, 270 and 124 of this chapter and to the notification requirements of section 3010 of RCRA.

(b)(1) A container or an inner liner removed from a container that has held any hazardous waste, except a waste that is a compressed gas or that is identified as an acute hazardous waste listed in §§261.31 or 261.33(e) of this chapter is empty if:

(i) All wastes have been removed that can be removed using the practices commonly employed to remove materials from that type of container, e.g., pouring, pumping, and aspirating, *and*

(ii) No more than 2.5 centimeters (one inch) of residue remain on the bottom of the container or inner liner, *or*

(iii)(A) No more than 3 percent by weight of the total capacity of the container remains in the container or inner liner if the container is less than or equal to 119 gallons in size; or

(B) No more than 0.3 percent by weight of the total capacity of the container remains in the container or inner liner if the container is greater than 119 gallons in size.

(2) A container that has held a hazardous waste that is a compressed gas is empty when the pressure in the container approaches atmospheric.

(3) A container or an inner liner removed from a container that has held an acute hazardous waste listed in §§261.31 or 261.33(e) is empty if:

(i) The container or inner liner has been triple rinsed using a solvent capable of removing the commercial chemical product or manufacturing chemical intermediate;

(ii) The container or inner liner has been cleaned by another method that has been shown in the scientific literature, or by tests conducted by the generator, to achieve equivalent removal; or

(iii) In the case of a container, the inner liner that prevented contact of the commercial chemical product or manufacturing chemical intermediate with the container, has been removed.

[45 FR 78529, Nov. 25, 1980, as amended at 47 FR 36097, Aug. 18, 1982; 48 FR 14294, Apr. 1, 1983; 50 FR 1999, Jan. 14, 1985; 51 FR 40637, Nov. 7, 1986; 70 FR 10815, Mar. 4, 2005; 70 FR 53453, Sept. 8, 2005; 75 FR 13002, Mar. 18, 2010]

### §261.8 PCB wastes regulated under Toxic Substance Control Act.

The disposal of PCB-containing dielectric fluid and electric equipment containing such fluid authorized for use and regulated under part 761 of this chapter and that are hazardous only because they fail the test for the Toxicity Characteristic (Hazardous Waste Codes D018 through D043 only) are exempt from regulation under parts 261 through 265, and parts 268, 270, and 124 of this chapter, and the notification requirements of section 3010 of RCRA.

[55 FR 11862, Mar. 29, 1990]

### §261.9 Requirements for Universal Waste.

The wastes listed in this section are exempt from regulation under parts 262 through 270 of this chapter except as specified in part 273 of this chapter and, therefore are not fully regulated as hazardous waste. The wastes listed in this section are subject to regulation under 40 CFR part 273:

(a) Batteries as described in 40 CFR 273.2;

(b) Pesticides as described in §273.3 of this chapter;

(c) Mercury-containing equipment as described in §273.4 of this chapter; and

(d) Lamps as described in §273.5 of this chapter.

[60 FR 25541, May 11, 1995, as amended at 64 FR 36487, July 6, 1999; 70 FR 45520, Aug. 5, 2005]

## Subpart B—Criteria for Identifying the Characteristics of Hazardous Waste and for Listing Hazardous Waste

### §261.10 Criteria for identifying the characteristics of hazardous waste.

(a) The Administrator shall identify and define a characteristic of hazardous waste in subpart C only upon determining that:

(1) A solid waste that exhibits the characteristic may:

(i) Cause, or significantly contribute to, an increase in mortality or an increase in serious irreversible, or incapacitating reversible, illness; or

(ii) Pose a substantial present or potential hazard to human health or the environment when it is improperly treated, stored, transported, disposed of or otherwise managed; and

(2) The characteristic can be:

(i) Measured by an available standardized test method which is reasonably within the capability of generators of solid waste or private sector laboratories that are available to serve generators of solid waste; or

(ii) Reasonably detected by generators of solid waste through their knowledge of their waste.

(b) [Reserved]

## § 261.11  Criteria for listing hazardous waste.

(a) The Administrator shall list a solid waste as a hazardous waste only upon determining that the solid waste meets one of the following criteria:

(1) It exhibits any of the characteristics of hazardous waste identified in subpart C.

(2) It has been found to be fatal to humans in low doses or, in the absence of data on human toxicity, it has been shown in studies to have an oral LD 50 toxicity (rat) of less than 50 milligrams per kilogram, an inhalation LC 50 toxicity (rat) of less than 2 milligrams per liter, or a dermal LD 50 toxicity (rabbit) of less than 200 milligrams per kilogram or is otherwise capable of causing or significantly contributing to an increase in serious irreversible, or incapacitating reversible, illness. (Waste listed in accordance with these criteria will be designated Acute Hazardous Waste.)

(3) It contains any of the toxic constituents listed in appendix VIII and, after considering the following factors, the Administrator concludes that the waste is capable of posing a substantial present or potential hazard to human health or the environment when improperly treated, stored, transported or disposed of, or otherwise managed:

(i) The nature of the toxicity presented by the constituent.

(ii) The concentration of the constituent in the waste.

(iii) The potential of the constituent or any toxic degradation product of the constituent to migrate from the waste into the environment under the types of improper management considered in paragraph (a)(3)(vii) of this section.

(iv) The persistence of the constituent or any toxic degradation product of the constituent.

(v) The potential for the constituent or any toxic degradation product of the constituent to degrade into non-harmful constituents and the rate of degradation.

(vi) The degree to which the constituent or any degradation product of the constituent bioaccumulates in ecosystems.

(vii) The plausible types of improper management to which the waste could be subjected.

(viii) The quantities of the waste generated at individual generation sites or on a regional or national basis.

(ix) The nature and severity of the human health and environmental damage that has occurred as a result of the improper management of wastes containing the constituent.

(x) Action taken by other governmental agencies or regulatory programs based on the health or environmental hazard posed by the waste or waste constituent.

(xi) Such other factors as may be appropriate.

Substances will be listed on appendix VIII only if they have been shown in scientific studies to have toxic, carcinogenic, mutagenic or teratogenic effects on humans or other life forms.

(Wastes listed in accordance with these criteria will be designated Toxic wastes.)

(b) The Administrator may list classes or types of solid waste as hazardous waste if he has reason to believe that individual wastes, within the class or type of waste, typically or frequently are hazardous under the definition of hazardous waste found in section 1004(5) of the Act.

(c) The Administrator will use the criteria for listing specified in this section to establish the exclusion limits referred to in § 261.5(c).

[45 FR 33119, May 19, 1980, as amended at 55 FR 18726, May 4, 1990; 57 FR 14, Jan. 2, 1992]

## Subpart C—Characteristics of Hazardous Waste

### § 261.20  General.

(a) A solid waste, as defined in § 261.2, which is not excluded from regulation as a hazardous waste under § 261.4(b), is a hazardous waste if it exhibits any of the characteristics identified in this subpart.

[*Comment:* §262.11 of this chapter sets forth the generator's responsibility to determine whether his waste exhibits one or more of the characteristics identified in this subpart]

(b) A hazardous waste which is identified by a characteristic in this subpart is assigned every EPA Hazardous Waste Number that is applicable as set forth in this subpart. This number must be used in complying with the notification requirements of section 3010 of the Act and all applicable recordkeeping and reporting requirements under parts 262 through 265, 268, and 270 of this chapter.

(c) For purposes of this subpart, the Administrator will consider a sample obtained using any of the applicable sampling methods specified in appendix I to be a representative sample within the meaning of part 260 of this chapter.

[*Comment:* Since the appendix I sampling methods are not being formally adopted by the Administrator, a person who desires to employ an alternative sampling method is not required to demonstrate the equivalency of his method under the procedures set forth in §§260.20 and 260.21.]

[45 FR 33119, May 19, 1980, as amended at 51 FR 40636, Nov. 7, 1986; 55 FR 22684, June 1, 1990; 56 FR 3876, Jan. 31, 1991]

## §261.21 Characteristic of ignitability.

(a) A solid waste exhibits the characteristic of ignitability if a representative sample of the waste has any of the following properties:

(1) It is a liquid, other than an aqueous solution containing less than 24 percent alcohol by volume and has flash point less than 60 °C (140 °F), as determined by a Pensky-Martens Closed Cup Tester, using the test method specified in ASTM Standard D 93–79 or D 93–80 (incorporated by reference, see §260.11), or a Setaflash Closed Cup Tester, using the test method specified in ASTM Standard D 3278–78 (incorporated by reference, see §260.11).

(2) It is not a liquid and is capable, under standard temperature and pressure, of causing fire through friction, absorption of moisture or spontaneous chemical changes and, when ignited, burns so vigorously and persistently that it creates a hazard.

(3) It is an ignitable compressed gas.

(i) The term "compressed gas" shall designate any material or mixture having in the container an absolute pressure exceeding 40 p.s.i. at 70 °F or, regardless of the pressure at 70 °F, having an absolute pressure exceeding 104 p.s.i. at 130 °F; or any liquid flammable material having a vapor pressure exceeding 40 p.s.i. absolute at 100 °F as determined by ASTM Test D–323.

(ii) A compressed gas shall be characterized as ignitable if any one of the following occurs:

(A) Either a mixture of 13 percent or less (by volume) with air forms a flammable mixture or the flammable range with air is wider than 12 percent regardless of the lower limit. These limits shall be determined at atmospheric temperature and pressure. The method of sampling and test procedure shall be acceptable to the Bureau of Explosives and approved by the director, Pipeline and Hazardous Materials Technology, U.S. Department of Transportation (see Note 2).

(B) Using the Bureau of Explosives' Flame Projection Apparatus (see Note 1), the flame projects more than 18 inches beyond the ignition source with valve opened fully, or, the flame flashes back and burns at the valve with any degree of valve opening.

(C) Using the Bureau of Explosives' Open Drum Apparatus (see Note 1), there is any significant propagation of flame away from the ignition source.

(D) Using the Bureau of Explosives' Closed Drum Apparatus (see Note 1), there is any explosion of the vapor-air mixture in the drum.

(4) It is an oxidizer. An oxidizer for the purpose of this subchapter is a substance such as a chlorate, permanganate, inorganic peroxide, or a nitrate, that yields oxygen readily to stimulate the combustion of organic matter (see Note 4).

(i) An organic compound containing the bivalent -O-O- structure and which may be considered a derivative of hydrogen peroxide where one or more of the hydrogen atoms have been replaced by organic radicals must be classed as an organic peroxide unless:

(A) The material meets the definition of a Class A explosive or a Class B explosive, as defined in §261.23(a)(8), in which case it must be classed as an explosive,

(B) The material is forbidden to be offered for transportation according to 49 CFR 172.101 and 49 CFR 173.21,

(C) It is determined that the predominant hazard of the material containing an organic peroxide is other than that of an organic peroxide, or

(D) According to data on file with the Pipeline and Hazardous Materials Safety Administration in the U.S. Department of Transportation (see Note 3), it has been determined that the material does not present a hazard in transportation.

(b) A solid waste that exhibits the characteristic of ignitability has the EPA Hazardous Waste Number of D001.

NOTE 1: A description of the Bureau of Explosives' Flame Projection Apparatus, Open Drum Apparatus, Closed Drum Apparatus, and method of tests may be procured from the Bureau of Explosives.

NOTE 2: As part of a U.S. Department of Transportation (DOT) reorganization, the Office of Hazardous Materials Technology (OHMT), which was the office listed in the 1980 publication of 49 CFR 173.300 for the purposes of approving sampling and test procedures for a flammable gas, ceased operations on February 20, 2005. OHMT programs have moved to the Pipeline and Hazardous Materials Safety Administration (PHMSA) in the DOT.

NOTE 3: As part of a U.S. Department of Transportation (DOT) reorganization, the Research and Special Programs Administration (RSPA), which was the office listed in the 1980 publication of 49 CFR 173.151a for the purposes of determining that a material does not present a hazard in transport, ceased operations on February 20, 2005. RSPA programs have moved to the Pipeline and Hazardous Materials Safety Administration (PHMSA) in the DOT.

NOTE 4: The DOT regulatory definition of an oxidizer was contained in §173.151 of 49 CFR, and the definition of an organic peroxide was contained in paragraph 173.151a. An organic peroxide is a type of oxidizer.

[45 FR 33119, May 19, 1980, as amended at 46 FR 35247, July 7, 1981; 55 FR 22684, June 1, 1990; 70 FR 34561, June 14, 2005; 71 FR 40259, July 14, 2006]

### § 261.22  Characteristic of corrosivity.

(a) A solid waste exhibits the characteristic of corrosivity if a representative sample of the waste has either of the following properties:

(1) It is aqueous and has a pH less than or equal to 2 or greater than or equal to 12.5, as determined by a pH meter using Method 9040C in "Test Methods for Evaluating Solid Waste, Physical/Chemical Methods," EPA Publication SW-846, as incorporated by reference in § 260.11 of this chapter.

(2) It is a liquid and corrodes steel (SAE 1020) at a rate greater than 6.35 mm (0.250 inch) per year at a test temperature of 55 °C (130 °F) as determined by Method 1110A in "Test Methods for Evaluating Solid Waste, Physical/Chemical Methods," EPA Publication SW-846, and as incorporated by reference in § 260.11 of this chapter.

(b) A solid waste that exhibits the characteristic of corrosivity has the EPA Hazardous Waste Number of D002.

[45 FR 33119, May 19, 1980, as amended at 46 FR 35247, July 7, 1981; 55 FR 22684, June 1, 1990; 58 FR 46049, Aug. 31, 1993; 70 FR 34561, June 14, 2005]

### § 261.23  Characteristic of reactivity.

(a) A solid waste exhibits the characteristic of reactivity if a representative sample of the waste has any of the following properties:

(1) It is normally unstable and readily undergoes violent change without detonating.

(2) It reacts violently with water.

(3) It forms potentially explosive mixtures with water.

(4) When mixed with water, it generates toxic gases, vapors or fumes in a quantity sufficient to present a danger to human health or the environment.

(5) It is a cyanide or sulfide bearing waste which, when exposed to pH conditions between 2 and 12.5, can generate toxic gases, vapors or fumes in a quantity sufficient to present a danger to human health or the environment.

(6) It is capable of detonation or explosive reaction if it is subjected to a strong initiating source or if heated under confinement.

(7) It is readily capable of detonation or explosive decomposition or reaction at standard temperature and pressure.

(8) It is a forbidden explosive as defined in 49 CFR 173.54, or is a Division 1.1, 1.2 or 1.3 explosive as defined in 49 CFR 173.50 and 173.53.

(b) A solid waste that exhibits the characteristic of reactivity has the EPA Hazardous Waste Number of D003.

[45 FR 33119, May 19, 1980, as amended at 55 FR 22684, June 1, 1990; 75 FR 13002, Mar. 18, 2010]

## § 261.24 Toxicity characteristic.

(a) A solid waste (except manufactured gas plant waste) exhibits the characteristic of toxicity if, using the Toxicity Characteristic Leaching Procedure, test Method 1311 in "Test Methods for Evaluating Solid Waste, Physical/Chemical Methods," EPA Publication SW-846, as incorporated by reference in § 260.11 of this chapter, the extract from a representative sample of the waste contains any of the contaminants listed in table 1 at the concentration equal to or greater than the respective value given in that table. Where the waste contains less than 0.5 percent filterable solids, the waste itself, after filtering using the methodology outlined in Method 1311, is considered to be the extract for the purpose of this section.

(b) A solid waste that exhibits the characteristic of toxicity has the EPA Hazardous Waste Number specified in Table 1 which corresponds to the toxic contaminant causing it to be hazardous.

TABLE 1—MAXIMUM CONCENTRATION OF CONTAMINANTS FOR THE TOXICITY CHARACTERISTIC

| EPA HW No. [1] | Contaminant | CAS No. [2] | Regulatory Level (mg/L) |
|---|---|---|---|
| D004 | Arsenic | 7440–38–2 | 5.0 |
| D005 | Barium | 7440–39–3 | 100.0 |
| D018 | Benzene | 71–43–2 | 0.5 |
| D006 | Cadmium | 7440–43–9 | 1.0 |
| D019 | Carbon tetrachloride | 56–23–5 | 0.5 |
| D020 | Chlordane | 57–74–9 | 0.03 |
| D021 | Chlorobenzene | 108–90–7 | 100.0 |
| D022 | Chloroform | 67–66–3 | 6.0 |
| D007 | Chromium | 7440–47–3 | 5.0 |
| D023 | o-Cresol | 95–48–7 | [4] 200.0 |
| D024 | m-Cresol | 108–39–4 | [4] 200.0 |
| D025 | p-Cresol | 106–44–5 | [4] 200.0 |
| D026 | Cresol | | [4] 200.0 |
| D016 | 2,4-D | 94–75–7 | 10.0 |
| D027 | 1,4-Dichlorobenzene | 106–46–7 | 7.5 |
| D028 | 1,2-Dichloroethane | 107–06–2 | 0.5 |
| D029 | 1,1-Dichloroethylene | 75–35–4 | 0.7 |
| D030 | 2,4-Dinitrotoluene | 121–14–2 | [3] 0.13 |
| D012 | Endrin | 72–20–8 | 0.02 |
| D031 | Heptachlor (and its epoxide) | 76–44–8 | 0.008 |
| D032 | Hexachlorobenzene | 118–74–1 | [3] 0.13 |
| D033 | Hexachlorobutadiene | 87–68–3 | 0.5 |
| D034 | Hexachloroethane | 67–72–1 | 3.0 |
| D008 | Lead | 7439–92–1 | 5.0 |

TABLE 1—MAXIMUM CONCENTRATION OF CONTAMINANTS FOR THE TOXICITY CHARACTERISTIC

| EPA HW No. [1] | Contaminant | CAS No. [2] | Regulatory Level (mg/L) |
|---|---|---|---|
| D013 | Lindane | 58–89–9 | 0.4 |
| D009 | Mercury | 7439–97–6 | 0.2 |
| D014 | Methoxychlor | 72–43–5 | 10.0 |
| D035 | Methyl ethyl ketone | 78–93–3 | 200.0 |
| D036 | Nitrobenzene | 98–95–3 | 2.0 |
| D037 | Pentrachlorophenol | 87–86–5 | 100.0 |
| D038 | Pyridine | 110–86–1 | [3] 5.0 |
| D010 | Selenium | 7782–49–2 | 1.0 |
| D011 | Silver | 7440–22–4 | 5.0 |
| D039 | Tetrachloroethylene | 127–18–4 | 0.7 |
| D015 | Toxaphene | 8001–35–2 | 0.5 |
| D040 | Trichloroethylene | 79–01–6 | 0.5 |
| D041 | 2,4,5-Trichlorophenol | 95–95–4 | 400.0 |
| D042 | 2,4,6-Trichlorophenol | 88–06–2 | 2.0 |
| D017 | 2,4,5-TP (Silvex) | 93–72–1 | 1.0 |
| D043 | Vinyl chloride | 75–01–4 | 0.2 |

[1] Hazardous waste number.
[2] Chemical abstracts service number.
[3] Quantitation limit is greater than the calculated regulatory level. The quantitation limit therefore becomes the regulatory level.
[4] If o-, m-, and p-Cresol concentrations cannot be differentiated, the total cresol (D026) concentration is used. The regulatory level of total cresol is 200 mg/l.

[55 FR 11862, Mar. 29, 1990, as amended at 55 FR 22684, June 1, 1990; 55 FR 26987, June 29, 1990; 58 FR 46049, Aug. 31, 1993; 67 FR 11254, Mar. 13, 2002; 71 FR 40259, July 14, 2006]

# Subpart D—Lists of Hazardous Wastes

## § 261.30 General.

(a) A solid waste is a hazardous waste if it is listed in this subpart, unless it has been excluded from this list under §§ 260.20 and 260.22.

(b) The Administrator will indicate his basis for listing the classes or types of wastes listed in this subpart by employing one or more of the following Hazard Codes:

Ignitable Waste ........................ (I)
Corrosive Waste ....................... (C)
Reactive Waste ........................ (R)
Toxicity Characteristic Waste ... (E)
Acute Hazardous Waste ............ (H)
Toxic Waste ............................. (T)

Appendix VII identifies the constituent which caused the Administrator to list the waste as a Toxicity Characteristic Waste (E) or Toxic Waste (T) in §§ 261.31 and 261.32.

(c) Each hazardous waste listed in this subpart is assigned an EPA Hazardous Waste Number which precedes the name of the waste. This number

67

must be used in complying with the notification requirements of Section 3010 of the Act and certain recordkeeping and reporting requirements under parts 262 through 265, 267, 268, and 270 of this chapter.

(d) The following hazardous wastes listed in § 261.31 are subject to the exclusion limits for acutely hazardous wastes established in § 261.5: EPA Hazardous Wastes Nos. F020, F021, F022, F023, F026 and F027.

[45 FR 33119, May 19, 1980, as amended at 48 FR 14294, Apr. 1, 1983; 50 FR 2000, Jan. 14, 1985; 51 FR 40636, Nov. 7, 1986; 55 FR 11863, Mar. 29, 1990; 75 FR 13002, Mar. 18, 2010]

## § 261.31 Hazardous wastes from non-specific sources.

(a) The following solid wastes are listed hazardous wastes from non-specific sources unless they are excluded under §§ 260.20 and 260.22 and listed in appendix IX.

| Industry and EPA hazardous waste No. | Hazardous waste | Hazard code |
|---|---|---|
| Generic: | | |
| F001 | The following spent halogenated solvents used in degreasing: Tetrachloroethylene, trichloroethylene, methylene chloride, 1,1,1-trichloroethane, carbon tetrachloride, and chlorinated fluorocarbons; all spent solvent mixtures/blends used in degreasing containing, before use, a total of ten percent or more (by volume) of one or more of the above halogenated solvents or those solvents listed in F002, F004, and F005; and still bottoms from the recovery of these spent solvents and spent solvent mixtures. | (T) |
| F002 | The following spent halogenated solvents: Tetrachloroethylene, methylene chloride, trichloroethylene, 1,1,1-trichloroethane, chlorobenzene, 1,1,2-trichloro-1,2,2-trifluoroethane, ortho-dichlorobenzene, trichlorofluoromethane, and 1,1,2-trichloroethane; all spent solvent mixtures/blends containing, before use, a total of ten percent or more (by volume) of one or more of the above halogenated solvents or those listed in F001, F004, or F005; and still bottoms from the recovery of these spent solvents and spent solvent mixtures. | (T) |
| F003 | The following spent non-halogenated solvents: Xylene, acetone, ethyl acetate, ethyl benzene, ethyl ether, methyl isobutyl ketone, n-butyl alcohol, cyclohexanone, and methanol; all spent solvent mixtures/blends containing, before use, only the above spent non-halogenated solvents; and all spent solvent mixtures/blends containing, before use, one or more of the above non-halogenated solvents, and, a total of ten percent or more (by volume) of one or more of those solvents listed in F001, F002, F004, and F005; and still bottoms from the recovery of these spent solvents and spent solvent mixtures. | (I)* |
| F004 | The following spent non-halogenated solvents: Cresols and cresylic acid, and nitrobenzene; all spent solvent mixtures/blends containing, before use, a total of ten percent or more (by volume) of one or more of the above non-halogenated solvents or those solvents listed in F001, F002, and F005; and still bottoms from the recovery of these spent solvents and spent solvent mixtures. | (T) |
| F005 | The following spent non-halogenated solvents: Toluene, methyl ethyl ketone, carbon disulfide, isobutanol, pyridine, benzene, 2-ethoxyethanol, and 2-nitropropane; all spent solvent mixtures/blends containing, before use, a total of ten percent or more (by volume) of one or more of the above non-halogenated solvents or those solvents listed in F001, F002, or F004; and still bottoms from the recovery of these spent solvents and spent solvent mixtures. | (I,T) |
| F006 | Wastewater treatment sludges from electroplating operations except from the following processes: (1) Sulfuric acid anodizing of aluminum; (2) tin plating on carbon steel; (3) zinc plating (segregated basis) on carbon steel; (4) aluminum or zinc-aluminum plating on carbon steel; (5) cleaning/stripping associated with tin, zinc and aluminum plating on carbon steel; and (6) chemical etching and milling of aluminum. | (T) |
| F007 | Spent cyanide plating bath solutions from electroplating operations | (R, T) |
| F008 | Plating bath residues from the bottom of plating baths from electroplating operations where cyanides are used in the process. | (R, T) |
| F009 | Spent stripping and cleaning bath solutions from electroplating operations where cyanides are used in the process. | (R, T) |
| F010 | Quenching bath residues from oil baths from metal heat treating operations where cyanides are used in the process. | (R, T) |
| F011 | Spent cyanide solutions from salt bath pot cleaning from metal heat treating operations. | (R, T) |
| F012 | Quenching waste water treatment sludges from metal heat treating operations where cyanides are used in the process. | (T) |

| Industry and EPA hazardous waste No. | Hazardous waste | Hazard code |
|---|---|---|
| F019 ..................................... | Wastewater treatment sludges from the chemical conversion coating of aluminum except from zirconium phosphating in aluminum can washing when such phosphating is an exclusive conversion coating process. Wastewater treatment sludges from the manufacturing of motor vehicles using a zinc phosphating process will not be subject to this listing at the point of generation if the wastes are not placed outside on the land prior to shipment to a landfill for disposal and are either: disposed in a Subtitle D municipal or industrial landfill unit that is equipped with a single clay liner and is permitted, licensed or otherwise authorized by the state; or disposed in a landfill unit subject to, or otherwise meeting, the landfill requirements in §258.40, §264.301 or §265.301. For the purposes of this listing, motor vehicle manufacturing is defined in paragraph (b)(4)(i) of this section and (b)(4)(ii) of this section describes the recordkeeping requirements for motor vehicle manufacturing facilities. | (T) |
| F020 ..................................... | Wastes (except wastewater and spent carbon from hydrogen chloride purification) from the production or manufacturing use (as a reactant, chemical intermediate, or component in a formulating process) of tri- or tetrachlorophenol, or of intermediates used to produce their pesticide derivatives. (This listing does not include wastes from the production of Hexachlorophene from highly purified 2,4,5-trichlorophenol.). | (H) |
| F021 ..................................... | Wastes (except wastewater and spent carbon from hydrogen chloride purification) from the production or manufacturing use (as a reactant, chemical intermediate, or component in a formulating process) of pentachlorophenol, or of intermediates used to produce its derivatives. | (H) |
| F022 ..................................... | Wastes (except wastewater and spent carbon from hydrogen chloride purification) from the manufacturing use (as a reactant, chemical intermediate, or component in a formulating process) of tetra-, penta-, or hexachlorobenzenes under alkaline conditions. | (H) |
| F023 ..................................... | Wastes (except wastewater and spent carbon from hydrogen chloride purification) from the production of materials on equipment previously used for the production or manufacturing use (as a reactant, chemical intermediate, or component in a formulating process) of tri- and tetrachlorophenols. (This listing does not include wastes from equipment used only for the production or use of Hexachlorophene from highly purified 2,4,5-trichlorophenol.). | (H) |
| F024 ..................................... | Process wastes, including but not limited to, distillation residues, heavy ends, tars, and reactor clean-out wastes, from the production of certain chlorinated aliphatic hydrocarbons by free radical catalyzed processes. These chlorinated aliphatic hydrocarbons are those having carbon chain lengths ranging from one to and including five, with varying amounts and positions of chlorine substitution. (This listing does not include wastewaters, wastewater treatment sludges, spent catalysts, and wastes listed in §261.31 or §261.32.). | (T) |
| F025 ..................................... | Condensed light ends, spent filters and filter aids, and spent desiccant wastes from the production of certain chlorinated aliphatic hydrocarbons, by free radical catalyzed processes. These chlorinated aliphatic hydrocarbons are those having carbon chain lengths ranging from one to and including five, with varying amounts and positions of chlorine substitution. | (T) |
| F026 ..................................... | Wastes (except wastewater and spent carbon from hydrogen chloride purification) from the production of materials on equipment previously used for the manufacturing use (as a reactant, chemical intermediate, or component in a formulating process) of tetra-, penta-, or hexachlorobenzene under alkaline conditions. | (H) |
| F027 ..................................... | Discarded unused formulations containing tri-, tetra-, or pentachlorophenol or discarded unused formulations containing compounds derived from these chlorophenols. (This listing does not include formulations containing Hexachlorophene sythesized from prepurified 2,4,5-trichlorophenol as the sole component.). | (H) |
| F028 ..................................... | Residues resulting from the incineration or thermal treatment of soil contaminated with EPA Hazardous Waste Nos. F020, F021, F022, F023, F026, and F027. | (T) |
| F032 ..................................... | Wastewaters (except those that have not come into contact with process contaminants), process residuals, preservative drippage, and spent formulations from wood preserving processes generated at plants that currently use or have previously used chlorophenolic formulations (except potentially cross-contaminated wastes that have had the F032 waste code deleted in accordance with §261.35 of this chapter or potentially cross-contaminated wastes that are otherwise currently regulated as hazardous wastes (i.e., F034 or F035), and where the generator does not resume or initiate use of chlorophenolic formulations). This listing does not include K001 bottom sediment sludge from the treatment of wastewater from wood preserving processes that use creosote and/or pentachlorophenol. | (T) |
| F034 ..................................... | Wastewaters (except those that have not come into contact with process contaminants), process residuals, preservative drippage, and spent formulations from wood preserving processes generated at plants that use creosote formulations. This listing does not include K001 bottom sediment sludge from the treatment of wastewater from wood preserving processes that use creosote and/or pentachlorophenol. | (T) |

| Industry and EPA hazardous waste No. | Hazardous waste | Hazard code |
|---|---|---|
| F035 ................................ | Wastewaters (except those that have not come into contact with process contaminants), process residuals, preservative drippage, and spent formulations from wood preserving processes generated at plants that use inorganic preservatives containing arsenic or chromium. This listing does not include K001 bottom sediment sludge from the treatment of wastewater from wood preserving processes that use creosote and/or pentachlorophenol. | (T) |
| F037 ................................ | Petroleum refinery primary oil/water/solids separation sludge—Any sludge generated from the gravitational separation of oil/water/solids during the storage or treatment of process wastewaters and oily cooling wastewaters from petroleum refineries. Such sludges include, but are not limited to, those generated in oil/water/solids separators; tanks and impoundments; ditches and other conveyances; sumps; and stormwater units receiving dry weather flow. Sludge generated in stormwater units that do not receive dry weather flow, sludges generated from non-contact once-through cooling waters segregated for treatment from other process or oily cooling waters, sludges generated in aggressive biological treatment units as defined in § 261.31(b)(2) (including sludges generated in one or more additional units after wastewaters have been treated in aggressive biological treatment units) and K051 wastes are not included in this listing. This listing does include residuals generated from processing or recycling oil-bearing hazardous secondary materials excluded under § 261.4(a)(12)(i), if those residuals are to be disposed of. | (T) |
| F038 ................................ | Petroleum refinery secondary (emulsified) oil/water/solids separation sludge—Any sludge and/or float generated from the physical and/or chemical separation of oil/water/solids in process wastewaters and oily cooling wastewaters from petroleum refineries. Such wastes include, but are not limited to, all sludges and floats generated in: induced air flotation (IAF) units, tanks and impoundments, and all sludges generated in DAF units. Sludges generated in stormwater units that do not receive dry weather flow, sludges generated from non-contact once-through cooling waters segregated for treatment from other process or oily cooling waters, sludges and floats generated in aggressive biological treatment units as defined in § 261.31(b)(2) (including sludges and floats generated in one or more additional units after wastewaters have been treated in aggressive biological treatment units) and F037, K048, and K051 wastes are not included in this listing. | (T) |
| F039 ................................ | Leachate (liquids that have percolated through land disposed wastes) resulting from the disposal of more than one restricted waste classified as hazardous under subpart D of this part. (Leachate resulting from the disposal of one or more of the following EPA Hazardous Wastes and no other Hazardous Wastes retains its EPA Hazardous Waste Number(s): F020, F021, F022, F026, F027, and/or F028.). | (T) |

*(I,T) should be used to specify mixtures that are ignitable and contain toxic constituents.

(b) Listing Specific Definitions: (1) For the purposes of the F037 and F038 listings, oil/water/solids is defined as oil and/or water and/or solids.(2) (i) For the purposes of the F037 and F038 listings, aggressive biological treatment units are defined as units which employ one of the following four treatment methods: activated sludge; trickling filter; rotating biological contactor for the continuous accelerated biological oxidation of wastewaters; or high-rate aeration. High-rate aeration is a system of surface impoundments or tanks, in which intense mechanical aeration is used to completely mix the wastes, enhance biological activity, and (A) the units employ a minimum of 6 hp per million gallons of treatment volume; and either (B) the hydraulic retention time of the unit is no longer than 5 days; or (C) the hydraulic retention time is no longer than 30 days and the unit does not generate a sludge

that is a hazardous waste by the Toxicity Characteristic.

(ii) Generators and treatment, storage and disposal facilities have the burden of proving that their sludges are exempt from listing as F037 and F038 wastes under this definition. Generators and treatment, storage and disposal facilities must maintain, in their operating or other onsite records, documents and data sufficient to prove that: (A) the unit is an aggressive biological treatment unit as defined in this subsection; and (B) the sludges sought to be exempted from the definitions of F037 and/or F038 were actually generated in the aggressive biological treatment unit.

(3) (i) For the purposes of the F037 listing, sludges are considered to be generated at the moment of deposition in the unit, where deposition is defined as at least a temporary cessation of lateral particle movement.

(ii) For the purposes of the F038 listing, (A) sludges are considered to be generated at the moment of deposition in the unit, where deposition is defined as at least a temporary cessation of lateral particle movement and (B) floats are considered to be generated at the moment they are formed in the top of the unit.

(4) For the purposes of the F019 listing, the following apply to wastewater treatment sludges from the manufacturing of motor vehicles using a zinc phosphating process.

(i) Motor vehicle manufacturing is defined to include the manufacture of automobiles and light trucks/utility vehicles (including light duty vans, pick-up trucks, minivans, and sport utility vehicles). Facilities must be engaged in manufacturing complete vehicles (body and chassis or unibody) or chassis only.

(ii) Generators must maintain in their on-site records documentation

and information sufficient to prove that the wastewater treatment sludges to be exempted from the F019 listing meet the conditions of the listing. These records must include: the volume of waste generated and disposed of off site; documentation showing when the waste volumes were generated and sent off site; the name and address of the receiving facility; and documentation confirming receipt of the waste by the receiving facility. Generators must maintain these documents on site for no less than three years. The retention period for the documentation is automatically extended during the course of any enforcement action or as requested by the Regional Administrator or the state regulatory authority.

[46 FR 4617, Jan. 16, 1981]

EDITORIAL NOTE: For FEDERAL REGISTER citations affecting §261.31, see the List of CFR Sections Affected, which appears in the Finding Aids section of the printed volume and on GPO Access.

## §261.32   Hazardous wastes from specific sources.

(a)The following solid wastes are listed hazardous wastes from specific sources unless they are excluded under §§260.20 and 260.22 and listed in appendix IX.

| Industry and EPA hazardous waste No. | Hazardous waste | Hazard code |
|---|---|---|
| Wood preservation: K001 ..... | Bottom sediment sludge from the treatment of wastewaters from wood preserving processes that use creosote and/or pentachlorophenol. | (T) |
| Inorganic pigments: | | |
| K002 | Wastewater treatment sludge from the production of chrome yellow and orange pigments. | (T) |
| K003 | Wastewater treatment sludge from the production of molybdate orange pigments ...... | (T) |
| K004 | Wastewater treatment sludge from the production of zinc yellow pigments ................ | (T) |
| K005 | Wastewater treatment sludge from the production of chrome green pigments ............. | (T) |
| K006 | Wastewater treatment sludge from the production of chrome oxide green pigments (anhydrous and hydrated). | (T) |
| K007 | Wastewater treatment sludge from the production of iron blue pigments ................. | (T) |
| K008 | Oven residue from the production of chrome oxide green pigments ...................... | (T) |
| Organic chemicals: | | |
| K009 | Distillation bottoms from the production of acetaldehyde from ethylene ................. | (T) |
| K010 | Distillation side cuts from the production of acetaldehyde from ethylene ................ | (T) |
| K011 | Bottom stream from the wastewater stripper in the production of acrylonitrile ............ | (R, T) |
| K013 | Bottom stream from the acetonitrile column in the production of acrylonitrile ........... | (R, T) |
| K014 | Bottoms from the acetonitrile purification column in the production of acrylonitrile ...... | (T) |
| K015 | Still bottoms from the distillation of benzyl chloride ................................ | (T) |
| K016 | Heavy ends or distillation residues from the production of carbon tetrachloride ......... | (T) |
| K017 | Heavy ends (still bottoms) from the purification column in the production of epichlorohydrin. | (T) |
| K018 | Heavy ends from the fractionation column in ethyl chloride production .................. | (T) |
| K019 | Heavy ends from the distillation of ethylene dichloride in ethylene dichloride production. | (T) |
| K020 | Heavy ends from the distillation of vinyl chloride in vinyl chloride monomer production | (T) |
| K021 | Aqueous spent antimony catalyst waste from fluoromethanes production ................. | (T) |
| K022 | Distillation bottom tars from the production of phenol/acetone from cumene ............. | (T) |
| K023 | Distillation light ends from the production of phthalic anhydride from naphthalene ....... | (T) |
| K024 | Distillation bottoms from the production of phthalic anhydride from naphthalene ........ | (T) |
| K025 | Distillation bottoms from the production of nitrobenzene by the nitration of benzene ... | (T) |
| K026 | Stripping still tails from the production of methy ethyl pyridines ...................... | (T) |
| K027 | Centrifuge and distillation residues from toluene diisocyanate production ............... | (R, T) |

| Industry and EPA hazardous waste No. | Hazardous waste | Hazard code |
|---|---|---|
| K028 | Spent catalyst from the hydrochlorinator reactor in the production of 1,1,1-trichloroethane. | (T) |
| K029 | Waste from the product steam stripper in the production of 1,1,1-trichloroethane | (T) |
| K030 | Column bottoms or heavy ends from the combined production of trichloroethylene and perchloroethylene. | (T) |
| K083 | Distillation bottoms from aniline production | (T) |
| K085 | Distillation or fractionation column bottoms from the production of chlorobenzenes | (T) |
| K093 | Distillation light ends from the production of phthalic anhydride from ortho-xylene | (T) |
| K094 | Distillation bottoms from the production of phthalic anhydride from ortho-xylene | (T) |
| K095 | Distillation bottoms from the production of 1,1,1-trichloroethane | (T) |
| K096 | Heavy ends from the heavy ends column from the production of 1,1,1-trichloroethane | (T) |
| K103 | Process residues from aniline extraction from the production of aniline | (T) |
| K104 | Combined wastewater streams generated from nitrobenzene/aniline production | (T) |
| K105 | Separated aqueous stream from the reactor product washing step in the production of chlorobenzenes. | (T) |
| K107 | Column bottoms from product separation from the production of 1,1-dimethylhydrazine (UDMH) from carboxylic acid hydrazines. | (C,T) |
| K108 | Condensed column overheads from product separation and condensed reactor vent gases from the production of 1,1-dimethylhydrazine (UDMH) from carboxylic acid hydrazides. | (I,T) |
| K109 | Spent filter cartridges from product purification from the production of 1,1-dimethylhydrazine (UDMH) from carboxylic acid hydrazides. | (T) |
| K110 | Condensed column overheads from intermediate separation from the production of 1,1-dimethylhydrazine (UDMH) from carboxylic acid hydrazides. | (T) |
| K111 | Product washwaters from the production of dinitrotoluene via nitration of toluene | (C,T) |
| K112 | Reaction by-product water from the drying column in the production of toluenediamine via hydrogenation of dinitrotoluene. | (T) |
| K113 | Condensed liquid light ends from the purification of toluenediamine in the production of toluenediamine via hydrogenation of dinitrotoluene. | (T) |
| K114 | Vicinals from the purification of toluenediamine in the production of toluenediamine via hydrogenation of dinitrotoluene. | (T) |
| K115 | Heavy ends from the purification of toluenediamine in the production of toluenediamine via hydrogenation of dinitrotoluene. | (T) |
| K116 | Organic condensate from the solvent recovery column in the production of toluene diisocyanate via phosgenation of toluenediamine. | (T) |
| K117 | Wastewater from the reactor vent gas scrubber in the production of ethylene dibromide via bromination of ethene. | (T) |
| K118 | Spent adsorbent solids from purification of ethylene dibromide in the production of ethylene dibromide via bromination of ethene. | (T) |
| K136 | Still bottoms from the purification of ethylene dibromide in the production of ethylene dibromide via bromination of ethene. | (T) |
| K149 | Distillation bottoms from the production of alpha- (or methyl-) chlorinated toluenes, ring-chlorinated toluenes, benzoyl chlorides, and compounds with mixtures of these functional groups, (This waste does not include still bottoms from the distillation of benzyl chloride.). | (T) |
| K150 | Organic residuals, excluding spent carbon adsorbent, from the spent chlorine gas and hydrochloric acid recovery processes associated with the production of alpha- (or methyl-) chlorinated toluenes, ring-chlorinated toluenes, benzoyl chlorides, and compounds with mixtures of these functional groups. | (T) |
| K151 | Wastewater treatment sludges, excluding neutralization and biological sludges, generated during the treatment of wastewaters from the production of alpha- (or methyl-) chlorinated toluenes, ring-chlorinated toluenes, benzoyl chlorides, and compounds with mixtures of these functional groups. | (T) |
| K156 | Organic waste (including heavy ends, still bottoms, light ends, spent solvents, filtrates, and decantates) from the production of carbamates and carbamoyl oximes. (This listing does not apply to wastes generated from the manufacture of 3-iodo-2-propynyl n-butylcarbamate.). | (T) |
| K157 | Wastewaters (including scrubber waters, condenser waters, washwaters, and separation waters) from the production of carbamates and carbamoyl oximes. (This listing does not apply to wastes generated from the manufacture of 3-iodo-2-propynyl n-butylcarbamate.). | (T) |
| K158 | Bag house dusts and filter/separation solids from the production of carbamates and carbamoyl oximes. (This listing does not apply to wastes generated from the manufacture of 3-iodo-2-propynyl n-butylcarbamate.). | (T) |
| K159 | Organics from the treatment of thiocarbamate wastes | (T) |
| K161 | Purification solids (including filtration, evaporation, and centrifugation solids), bag house dust and floor sweepings from the production of dithiocarbamate acids and their salts. (This listing does not include K125 or K126.). | (R,T) |

| Industry and EPA hazardous waste No. | Hazardous waste | Hazard code |
|---|---|---|
| K174 ................................. | Wastewater treatment sludges from the production of ethylene dichloride or vinyl chloride monomer (including sludges that result from commingled ethylene dichloride or vinyl chloride monomer wastewater and other wastewater), unless the sludges meet the following conditions: (i) they are disposed of in a subtitle C or non-hazardous landfill licensed or permitted by the state or federal government; (ii) they are not otherwise placed on the land prior to final disposal; and (iii) the generator maintains documentation demonstrating that the waste was either disposed of in an on-site landfill or consigned to a transporter or disposal facility that provided a written commitment to dispose of the waste in an off-site landfill. Respondents in any action brought to enforce the requirements of subtitle C must, upon a showing by the government that the respondent managed wastewater treatment sludges from the production of vinyl chloride monomer or ethylene dichloride, demonstrate that they meet the terms of the exclusion set forth above. In doing so, they must provide appropriate documentation (e.g., contracts between the generator and the landfill owner/operator, invoices documenting delivery of waste to landfill, etc.) that the terms of the exclusion were met. | (T) |
| K175 ................................. | Wastewater treatment sludges from the production of vinyl chloride monomer using mercuric chloride catalyst in an acetylene-based process. | (T) |
| K181 ................................. | Nonwastewaters from the production of dyes and/or pigments (including nonwastewaters commingled at the point of generation with nonwastewaters from other processes) that, at the point of generation, contain mass loadings of any of the constituents identified in paragraph (c) of this section that are equal to or greater than the corresponding paragraph (c) levels, as determined on a calendar year basis. These wastes will not be hazardous if the nonwastewaters are: (i) disposed in a Subtitle D landfill unit subject to the design criteria in §258.40, (ii) disposed in a Subtitle C landfill unit subject to either §264.301 or §265.301, (iii) disposed in other Subtitle D landfill units that meet the design criteria in §258.40, §264.301, or §265.301, or (iv) treated in a combustion unit that is permitted under Subtitle C, or an onsite combustion unit that is permitted under the Clean Air Act. For the purposes of this listing, dyes and/or pigments production is defined in paragraph (b)(1) of this section. Paragraph (d) of this section describes the process for demonstrating that a facility's nonwastewaters are not K181. This listing does not apply to wastes that are otherwise identified as hazardous under §§261.21–261.24 and 261.31–261.33 at the point of generation. Also, the listing does not apply to wastes generated before any annual mass loading limit is met. | (T) |
| Inorganic chemicals: | | |
| K071 ................................. | Brine purification muds from the mercury cell process in chlorine production, where separately prepurified brine is not used. | (T) |
| K073 ................................. | Chlorinated hydrocarbon waste from the purification step of the diaphragm cell process using graphite anodes in chlorine production. | (T) |
| K106 ................................. | Wastewater treatment sludge from the mercury cell process in chlorine production .... | (T) |
| K176 ................................. | Baghouse filters from the production of antimony oxide, including filters from the production of intermediates (e.g., antimony metal or crude antimony oxide). | (E) |
| K177 ................................. | Slag from the production of antimony oxide that is speculatively accumulated or disposed, including slag from the production of intermediates (e.g., antimony metal or crude antimony oxide). | (T) |
| K178 ................................. | Residues from manufacturing and manufacturing-site storage of ferric chloride from acids formed during the production of titanium dioxide using the chloride-ilmenite process. | (T) |
| Pesticides: | | |
| K031 ................................. | By-product salts generated in the production of MSMA and cacodylic acid ................. | (T) |
| K032 ................................. | Wastewater treatment sludge from the production of chlordane ................................. | (T) |
| K033 ................................. | Wastewater and scrub water from the chlorination of cyclopentadiene in the production of chlordane. | (T) |
| K034 ................................. | Filter solids from the filtration of hexachlorocyclopentadiene in the production of chlordane. | (T) |
| K035 ................................. | Wastewater treatment sludges generated in the production of creosote ..................... | (T) |
| K036 ................................. | Still bottoms from toluene reclamation distillation in the production of disulfoton ........ | (T) |
| K037 ................................. | Wastewater treatment sludges from the production of disulfoton ............................... | (T) |
| K038 ................................. | Wastewater from the washing and stripping of phorate production ........................... | (T) |
| K039 ................................. | Filter cake from the filtration of diethylphosphorodithioic acid in the production of phorate. | (T) |
| K040 ................................. | Wastewater treatment sludge from the production of phorate ................................... | (T) |
| K041 ................................. | Wastewater treatment sludge from the production of toxaphene ............................... | (T) |
| K042 ................................. | Heavy ends or distillation residues from the distillation of tetrachlorobenzene in the production of 2,4,5-T. | (T) |
| K043 ................................. | 2,6-Dichlorophenol waste from the production of 2,4-D ........................................... | (T) |
| K097 ................................. | Vacuum stripper discharge from the chlordane chlorinator in the production of chlordane. | (T) |
| K098 ................................. | Untreated process wastewater from the production of toxaphene .............................. | (T) |
| K099 ................................. | Untreated wastewater from the production of 2,4-D .............................................. | (T) |
| K123 ................................. | Process wastewater (including supernates, filtrates, and washwaters) from the production of ethylenebisdithiocarbamic acid and its salt. | (T) |

| Industry and EPA hazardous waste No. | Hazardous waste | Hazard code |
|---|---|---|
| K124 | Reactor vent scrubber water from the production of ethylenebisdithiocarbamic acid and its salts. | (C, T) |
| K125 | Filtration, evaporation, and centrifugation solids from the production of ethylenebisdithiocarbamic acid and its salts. | (T) |
| K126 | Baghouse dust and floor sweepings in milling and packaging operations from the production or formulation of ethylenebisdithiocarbamic acid and its salts. | (T) |
| K131 | Wastewater from the reactor and spent sulfuric acid from the acid dryer from the production of methyl bromide. | (C, T) |
| K132 | Spent absorbent and wastewater separator solids from the production of methyl bromide. | (T) |
| Explosives: | | |
| K044 | Wastewater treatment sludges from the manufacturing and processing of explosives | (R) |
| K045 | Spent carbon from the treatment of wastewater containing explosives | (R) |
| K046 | Wastewater treatment sludges from the manufacturing, formulation and loading of lead-based initiating compounds. | (T) |
| K047 | Pink/red water from TNT operations | (R) |
| Petroleum refining: | | |
| K048 | Dissolved air flotation (DAF) float from the petroleum refining industry | (T) |
| K049 | Slop oil emulsion solids from the petroleum refining industry | (T) |
| K050 | Heat exchanger bundle cleaning sludge from the petroleum refining industry | (T) |
| K051 | API separator sludge from the petroleum refining industry | (T) |
| K052 | Tank bottoms (leaded) from the petroleum refining industry | (T) |
| K169 | Crude oil storage tank sediment from petroleum refining operations | (T) |
| K170 | Clarified slurry oil tank sediment and/or in-line filter/separation solids from petroleum refining operations. | (T) |
| K171 | Spent Hydrotreating catalyst from petroleum refining operations, including guard beds used to desulfurize feeds to other catalytic reactors (this listing does not include inert support media). | (I,T) |
| K172 | Spent Hydrorefining catalyst from petroleum refining operations, including guard beds used to desulfurize feeds to other catalytic reactors (this listing does not include inert support media). | (I,T) |
| Iron and steel: | | |
| K061 | Emission control dust/sludge from the primary production of steel in electric furnaces | (T) |
| K062 | Spent pickle liquor generated by steel finishing operations of facilities within the iron and steel industry (SIC Codes 331 and 332). | (C,T) |
| Primary aluminum: | | |
| K088 | Spent potliners from primary aluminum reduction | (T) |
| Secondary lead: | | |
| K069 | Emission control dust/sludge from secondary lead smelting. (NOTE: This listing is stayed administratively for sludge generated from secondary acid scrubber systems. The stay will remain in effect until further administrative action is taken. If EPA takes further action effecting this stay, EPA will publish a notice of the action in the FEDERAL REGISTER). | (T) |
| K100 | Waste leaching solution from acid leaching of emission control dust/sludge from secondary lead smelting. | (T) |
| Veterinary pharmaceuticals: | | |
| K084 | Wastewater treatment sludges generated during the production of veterinary pharmaceuticals from arsenic or organo-arsenic compounds. | (T) |
| K101 | Distillation tar residues from the distillation of aniline-based compounds in the production of veterinary pharmaceuticals from arsenic or organo-arsenic compounds. | (T) |
| K102 | Residue from the use of activated carbon for decolorization in the production of veterinary pharmaceuticals from arsenic or organo-arsenic compounds. | (T) |
| Ink formulation: | | |
| K086 | Solvent washes and sludges, caustic washes and sludges, or water washes and sludges from cleaning tubs and equipment used in the formulation of ink from pigments, driers, soaps, and stabilizers containing chromium and lead. | (T) |
| Coking: | | |
| K060 | Ammonia still lime sludge from coking operations | (T) |
| K087 | Decanter tank tar sludge from coking operations | (T) |
| K141 | Process residues from the recovery of coal tar, including, but not limited to, collecting sump residues from the production of coke from coal or the recovery of coke by-products produced from coal. This listing does not include K087 (decanter tank tar sludges from coking operations). | (T) |
| K142 | Tar storage tank residues from the production of coke from coal or from the recovery of coke by-products produced from coal. | (T) |
| K143 | Process residues from the recovery of light oil, including, but not limited to, those generated in stills, decanters, and wash oil recovery units from the recovery of coke by-products produced from coal. | (T) |
| K144 | Wastewater sump residues from light oil refining, including, but not limited to, intercepting or contamination sump sludges from the recovery of coke by-products produced from coal. | (T) |
| K145 | Residues from naphthalene collection and recovery operations from the recovery of coke by-products produced from coal. | (T) |

| Industry and EPA hazardous waste No. | Hazardous waste | Hazard code |
|---|---|---|
| K147 ................................. | Tar storage tank residues from coal tar refining ............................................................ | (T) |
| K148 ................................. | Residues from coal tar distillation, including but not limited to, still bottoms ................. | (T) |

(b) *Listing Specific Definitions:* (1) For the purposes of the K181 listing, dyes and/or pigments production is defined to include manufacture of the following product classes: dyes, pigments, or FDA certified colors that are classified as azo, triarylmethane, perylene or anthraquinone classes. Azo products include azo, monoazo, diazo, triazo, polyazo, azoic, benzidine, and pyrazolone products. Triarylmethane products include both triarylmethane and triphenylmethane products. Wastes that are not generated at a dyes and/or pigments manufacturing site, such as wastes from the offsite use, formulation, and packaging of dyes and/or pigments, are not included in the K181 listing.

(c) *K181 Listing Levels.* Nonwastewaters containing constituents in amounts equal to or exceeding the following levels during any calendar year are subject to the K181 listing, unless the conditions in the K181 listing are met.

| Constituent | Chemical abstracts No. | Mass levels (kg/yr) |
|---|---|---|
| Aniline ................................. | 62–53–3 | 9,300 |
| o-Anisidine ........................... | 90–04–0 | 110 |
| 4-Chloroaniline ...................... | 106–47–8 | 4,800 |
| p-Cresidine ........................... | 120–71–8 | 660 |
| 2,4-Dimethylaniline ................ | 95–68–1 | 100 |
| 1,2-Phenylenediamine ........... | 95–54–5 | 710 |
| 1,3-Phenylenediamine ........... | 108–45–2 | 1,200 |

(d) *Procedures for demonstrating that dyes and/or pigment nonwastewaters are not K181.* The procedures described in paragraphs (d)(1)–(d)(3) and (d)(5) of this section establish when nonwastewaters from the production of dyes/pigments would not be hazardous (these procedures apply to wastes that are not disposed in landfill units or treated in combustion units as specified in paragraph (a) of this section). If the nonwastewaters are disposed in landfill units or treated in combustion units as described in paragraph (a) of this section, then the nonwastewaters are not hazardous. In order to demonstrate that it is meeting the landfill disposal or combustion conditions contained in the K181 listing description, the generator must maintain documentation as described in paragraph (d)(4) of this section.

(1) *Determination based on no K181 constituents.* Generators that have knowledge (e.g., knowledge of constituents in wastes based on prior sampling and analysis data and/or information about raw materials used, production processes used, and reaction and degradation products formed) that their wastes contain none of the K181 constituents (*see* paragraph (c) of this section) can use their knowledge to determine that their waste is not K181. The generator must document the basis for all such determinations on an annual basis and keep each annual documentation for three years.

(2) *Determination for generated quantities of 1,000 MT/yr or less for wastes that contain K181 constituents.* If the total annual quantity of dyes and/or pigment nonwastewaters generated is 1,000 metric tons or less, the generator can use knowledge of the wastes (e.g., knowledge of constituents in wastes based on prior analytical data and/or information about raw materials used, production processes used, and reaction and degradation products formed) to conclude that annual mass loadings for the K181 constituents are below the listing levels of paragraph (c) of this section. To make this determination, the generator must:

(i) Each year document the basis for determining that the annual quantity of nonwastewaters expected to be generated will be less than 1,000 metric tons.

(ii) Track the actual quantity of nonwastewaters generated from January 1 through December 31 of each year. If, at any time within the year, the actual waste quantity exceeds 1,000 metric tons, the generator must comply with the requirements of paragraph (d)(3) of this section for the remainder of the year.

75

(iii) Keep a running total of the K181 constituent mass loadings over the course of the calendar year.

(iv) Keep the following records on site for the three most recent calendar years in which the hazardous waste determinations are made:

(A) The quantity of dyes and/or pigment nonwastewaters generated.

(B) The relevant process information used.

(C) The calculations performed to determine annual total mass loadings for each K181 constituent in the nonwastewaters during the year.

(3) *Determination for generated quantities greater than 1,000 MT/yr for wastes that contain K181 constituents.* If the total annual quantity of dyes and/or pigment nonwastewaters generated is greater than 1,000 metric tons, the generator must perform all of the steps described in paragraphs ((d)(3)(i)–(d)(3)(xi) of this section) in order to make a determination that its waste is not K181.

(i) Determine which K181 constituents (see paragraph (c) of this section) are reasonably expected to be present in the wastes based on knowledge of the wastes (e.g., based on prior sampling and analysis data and/or information about raw materials used, production processes used, and reaction and degradation products formed).

(ii) If 1,2-phenylenediamine is present in the wastes, the generator can use either knowledge or sampling and analysis procedures to determine the level of this constituent in the wastes. For determinations based on use of knowledge, the generator must comply with the procedures for using knowledge described in paragraph (d)(2) of this section and keep the records described in paragraph (d)(2)(iv) of this section. For determinations based on sampling and analysis, the generator must comply with the sampling and analysis and recordkeeping requirements described below in this section.

(iii) Develop a waste sampling and analysis plan (or modify an existing plan) to collect and analyze representative waste samples for the K181 constituents reasonably expected to be present in the wastes. At a minimum, the plan must include:

(A) A discussion of the number of samples needed to characterize the wastes fully;

(B) The planned sample collection method to obtain representative waste samples;

(C) A discussion of how the sampling plan accounts for potential temporal and spatial variability of the wastes.

(D) A detailed description of the test methods to be used, including sample preparation, clean up (if necessary), and determinative methods.

(iv) Collect and analyze samples in accordance with the waste sampling and analysis plan.

(A) The sampling and analysis must be unbiased, precise, and representative of the wastes.

(B) The analytical measurements must be sufficiently sensitive, accurate and precise to support any claim that the constituent mass loadings are below the listing levels of paragraph (c) of this section.

(v) Record the analytical results.

(vi) Record the waste quantity represented by the sampling and analysis results.

(vii) Calculate constituent-specific mass loadings (product of concentrations and waste quantity).

(viii) Keep a running total of the K181 constituent mass loadings over the course of the calendar year.

(ix) Determine whether the mass of any of the K181 constituents listed in paragraph (c) of this section generated between January 1 and December 31 of any year is below the K181 listing levels.

(x) Keep the following records on site for the three most recent calendar years in which the hazardous waste determinations are made:

(A) The sampling and analysis plan.

(B) The sampling and analysis results (including QA/QC data)

(C) The quantity of dyes and/or pigment nonwastewaters generated.

(D) The calculations performed to determine annual mass loadings.

(xi) Nonhazardous waste determinations must be conducted annually to verify that the wastes remain nonhazardous.

(A) The annual testing requirements are suspended after three consecutive successful annual demonstrations that the wastes are nonhazardous. The generator can then use knowledge of the wastes to support subsequent annual determinations.

(B) The annual testing requirements are reinstated if the manufacturing or waste treatment processes generating the wastes are significantly altered, resulting in an increase of the potential for the wastes to exceed the listing levels.

(C) If the annual testing requirements are suspended, the generator must keep records of the process knowledge information used to support a nonhazardous determination. If testing is reinstated, a description of the process change must be retained.

(4) *Recordkeeping for the landfill disposal and combustion exemptions.* For the purposes of meeting the landfill disposal and combustion condition set out in the K181 listing description, the generator must maintain on site for three years documentation demonstrating that each shipment of waste was received by a landfill unit that is subject to or meets the landfill design standards set out in the listing description, or was treated in combustion units as specified in the listing description.

(5) *Waste holding and handling.* During the interim period, from the point of generation to completion of the hazardous waste determination, the generator is responsible for storing the wastes appropriately. If the wastes are determined to be hazardous and the generator has not complied with the subtitle C requirements during the interim period, the generator could be subject to an enforcement action for improper management.

[46 FR 4618, Jan. 16, 1981]

EDITORIAL NOTE: For FEDERAL REGISTER citations affecting §261.32, see the List of CFR Sections Affected, which appears in the Finding Aids section of the printed volume and on GPO Access.

## §261.33 Discarded commercial chemical products, off-specification species, container residues, and spill residues thereof.

The following materials or items are hazardous wastes if and when they are discarded or intended to be discarded as described in §261.2(a)(2)(i), when they are mixed with waste oil or used oil or other material and applied to the land for dust suppression or road treatment, when they are otherwise applied to the land in lieu of their original intended use or when they are contained in products that are applied to the land in lieu of their original intended use, or when, in lieu of their original intended use, they are produced for use as (or as a component of) a fuel, distributed for use as a fuel, or burned as a fuel.

(a) Any commercial chemical product, or manufacturing chemical intermediate having the generic name listed in paragraph (e) or (f) of this section.

(b) Any off-specification commercial chemical product or manufacturing chemical intermediate which, if it met specifications, would have the generic name listed in paragraph (e) or (f) of this section.

(c) Any residue remaining in a container or in an inner liner removed from a container that has held any commercial chemical product or manufacturing chemical intermediate having the generic name listed in paragraphs (e) or (f) of this section, unless the container is empty as defined in §261.7(b) of this chapter.

[*Comment:* Unless the residue is being beneficially used or reused, or legitimately recycled or reclaimed; or being accumulated, stored, transported or treated prior to such use, re-use, recycling or reclamation, EPA considers the residue to be intended for discard, and thus, a hazardous waste. An example of a legitimate re-use of the residue would be where the residue remains in the container and the container is used to hold the same commercial chemical product or manufacturing chemical intermediate it previously held. An example of the discard of the residue would be where the drum is sent to a drum reconditioner who reconditions the drum but discards the residue.]

(d) Any residue or contaminated soil, water or other debris resulting from the cleanup of a spill into or on any land or water of any commercial chemical product or manufacturing chemical intermediate having the generic name listed in paragraph (e) or (f) of this section, or any residue or contaminated soil, water or other debris resulting from the cleanup of a spill, into or on any land or water, of any off-specification chemical product and manufacturing chemical intermediate which, if it met specifications, would have the generic name listed in paragraph (e) or (f) of this section.

[Comment: The phrase "commercial chemical product or manufacturing chemical intermediate having the generic name listed in . . ." refers to a chemical substance which is manufactured or formulated for commercial or manufacturing use which consists of the commercially pure grade of the chemical, any technical grades of the chemical that are produced or marketed, and all formulations in which the chemical is the sole active ingredient. It does not refer to a material, such as a manufacturing process waste, that contains any of the substances listed in paragraph (e) or (f). Where a manufacturing process waste is deemed to be a hazardous waste because it contains a substance listed in paragraph (e) or (f), such waste will be listed in either § 261.31 or § 261.32 or will be identified as a hazardous waste by the characteristics set forth in subpart C of this part.]

(e) The commercial chemical products, manufacturing chemical intermediates or off-specification commercial chemical products or manufacturing chemical intermediates referred to in paragraphs (a) through (d) of this section, are identified as acute hazardous wastes (H) and are subject to the small quantity exclusion defined in § 261.5(e).

[Comment: For the convenience of the regulated community the primary hazardous properties of these materials have been indicated by the letters T (Toxicity), and R (Reactivity). Absence of a letter indicates that the compound only is listed for acute toxicity. Wastes are first listed in alphabetical order by substance and then listed again in numerical order by Hazardous Waste Number.]

These wastes and their corresponding EPA Hazardous Waste Numbers are:

| Hazardous waste No. | Chemical abstracts No. | Substance |
|---|---|---|
| P023 | 107–20–0 | Acetaldehyde, chloro- |
| P002 | 591–08–2 | Acetamide, N-(aminothioxomethyl)- |
| P057 | 640–19–7 | Acetamide, 2-fluoro- |
| P058 | 62–74–8 | Acetic acid, fluoro-, sodium salt |
| P002 | 591–08–2 | 1-Acetyl-2-thiourea |
| P003 | 107–02–8 | Acrolein |
| P070 | 116–06–3 | Aldicarb |
| P203 | 1646–88–4 | Aldicarb sulfone. |
| P004 | 309–00–2 | Aldrin |
| P005 | 107–18–6 | Allyl alcohol |
| P006 | 20859–73–8 | Aluminum phosphide (R,T) |
| P007 | 2763–96–4 | 5-(Aminomethyl)-3-isoxazolol |
| P008 | 504–24–5 | 4-Aminopyridine |
| P009 | 131–74–8 | Ammonium picrate (R) |
| P119 | 7803–55–6 | Ammonium vanadate |
| P099 | 506–61–6 | Argentate(1-), bis(cyano-C)-, potassium |
| P010 | 7778–39–4 | Arsenic acid $H_3 AsO_4$ |
| P012 | 1327–53–3 | Arsenic oxide $As_2 O_3$ |
| P011 | 1303–28–2 | Arsenic oxide $As_2 O_5$ |
| P011 | 1303–28–2 | Arsenic pentoxide |
| P012 | 1327–53–3 | Arsenic trioxide |
| P038 | 692–42–2 | Arsine, diethyl- |
| P036 | 696–28–6 | Arsonous dichloride, phenyl- |
| P054 | 151–56–4 | Aziridine |
| P067 | 75–55–8 | Aziridine, 2-methyl- |
| P013 | 542–62–1 | Barium cyanide |
| P024 | 106–47–8 | Benzenamine, 4-chloro- |
| P077 | 100–01–6 | Benzenamine, 4-nitro- |
| P028 | 100–44–7 | Benzene, (chloromethyl)- |
| P042 | 51–43–4 | 1,2-Benzenediol, 4-[1-hydroxy-2-(methylamino)ethyl]-, (R)- |
| P046 | 122–09–8 | Benzeneethanamine, alpha,alpha-dimethyl- |
| P014 | 108–98–5 | Benzenethiol |
| P127 | 1563–66–2 | 7-Benzofuranol, 2,3-dihydro-2,2-dimethyl-, methylcarbamate. |
| P188 | 57–64–7 | Benzoic acid, 2-hydroxy-, compd. with (3aS-cis)-1,2,3,3a,8,8a-hexahydro-1,3a,8-trimethylpyrrolo[2,3-b]indol-5-yl methylcarbamate ester (1:1). |
| P001 | [1]81–81–2 | 2H-1-Benzopyran-2-one, 4-hydroxy-3-(3-oxo-1-phenylbutyl)-, & salts, when present at concentrations greater than 0.3% |
| P028 | 100–44–7 | Benzyl chloride |
| P015 | 7440–41–7 | Beryllium powder |
| P017 | 598–31–2 | Bromoacetone |
| P018 | 357–57–3 | Brucine |
| P045 | 39196–18–4 | 2-Butanone, 3,3-dimethyl-1-(methylthio)-, O-[(methylamino)carbonyl] oxime |

| Haz-ardous waste No. | Chemical abstracts No. | Substance |
|---|---|---|
| P021 | 592–01–8 | Calcium cyanide |
| P021 | 592–01–8 | Calcium cyanide Ca(CN)$_2$ |
| P189 | 55285–14–8 | Carbamic acid, [(dibutylamino)- thio]methyl-, 2,3-dihydro-2,2-dimethyl- 7-benzofuranyl ester. |
| P191 | 644–64–4 | Carbamic acid, dimethyl-, 1-[(dimethyl-amino)carbonyl]- 5-methyl-1H- pyrazol-3-yl ester. |
| P192 | 119–38–0 | Carbamic acid, dimethyl-, 3-methyl-1- (1-methylethyl)-1H- pyrazol-5-yl ester. |
| P190 | 1129–41–5 | Carbamic acid, methyl-, 3-methylphenyl ester. |
| P127 | 1563–66–2 | Carbofuran. |
| P022 | 75–15–0 | Carbon disulfide |
| P095 | 75–44–5 | Carbonic dichloride |
| P189 | 55285–14–8 | Carbosulfan. |
| P023 | 107–20–0 | Chloroacetaldehyde |
| P024 | 106–47–8 | p-Chloroaniline |
| P026 | 5344–82–1 | 1-(o-Chlorophenyl)thiourea |
| P027 | 542–76–7 | 3-Chloropropionitrile |
| P029 | 544–92–3 | Copper cyanide |
| P029 | 544–92–3 | Copper cyanide Cu(CN) |
| P202 | 64–00–6 | m-Cumenyl methylcarbamate. |
| P030 | ..................... | Cyanides (soluble cyanide salts), not otherwise specified |
| P031 | 460–19–5 | Cyanogen |
| P033 | 506–77–4 | Cyanogen chloride |
| P033 | 506–77–4 | Cyanogen chloride (CN)Cl |
| P034 | 131–89–5 | 2-Cyclohexyl-4,6-dinitrophenol |
| P016 | 542–88–1 | Dichloromethyl ether |
| P036 | 696–28–6 | Dichlorophenylarsine |
| P037 | 60–57–1 | Dieldrin |
| P038 | 692–42–2 | Diethylarsine |
| P041 | 311–45–5 | Diethyl-p-nitrophenyl phosphate |
| P040 | 297–97–2 | O,O-Diethyl O-pyrazinyl phosphorothioate |
| P043 | 55–91–4 | Diisopropylfluorophosphate (DFP) |
| P004 | 309–00–2 | 1,4,5,8-Dimethanonaphthalene, 1,2,3,4,10,10-hexa- chloro-1,4,4a,5,8,8a,-hexahydro-, (1alpha,4alpha,4abeta,5alpha,8alpha,8abeta)- |
| P060 | 465–73–6 | 1,4,5,8-Dimethanonaphthalene, 1,2,3,4,10,10-hexa- chloro-1,4,4a,5,8,8a-hexahydro-, (1alpha,4alpha,4abeta,5beta,8beta,8abeta)- |
| P037 | 60–57–1 | 2,7:3,6-Dimethanonaphth[2,3-b]oxirene, 3,4,5,6,9,9-hexachloro-1a,2,2a,3,6,6a,7,7a-octahydro-, (1aalpha,2beta,2aalpha,3beta,6beta,6aalpha,7beta, 7aalpha)- |
| P051 | [1] 72–20–8 | 2,7:3,6-Dimethanonaphth [2,3-b]oxirene, 3,4,5,6,9,9-hexachloro-1a,2,2a,3,6,6a,7,7a-octahydro-, (1aalpha,2beta,2abeta,3alpha,6alpha,6abeta,7beta, 7aalpha)-, & metabolites |
| P044 | 60–51–5 | Dimethoate |
| P046 | 122–09–8 | alpha,alpha-Dimethylphenethylamine |
| P191 | 644–64–4 | Dimetilan. |
| P047 | [1] 534–52–1 | 4,6-Dinitro-o-cresol, & salts |
| P048 | 51–28–5 | 2,4-Dinitrophenol |
| P020 | 88–85–7 | Dinoseb |
| P085 | 152–16–9 | Diphosphoramide, octamethyl- |
| P111 | 107–49–3 | Diphosphoric acid, tetraethyl ester |
| P039 | 298–04–4 | Disulfoton |
| P049 | 541–53–7 | Dithiobiuret |
| P185 | 26419–73–8 | 1,3-Dithiolane-2-carboxaldehyde, 2,4-dimethyl-, O- [(methylamino)- carbonyl]oxime. |
| P050 | 115–29–7 | Endosulfan |
| P088 | 145–73–3 | Endothall |
| P051 | 72–20–8 | Endrin |
| P051 | 72–20–8 | Endrin, & metabolites |
| P042 | 51–43–4 | Epinephrine |
| P031 | 460–19–5 | Ethanedinitrile |
| P194 | 23135–22–0 | Ethanimidothioic acid, 2-(dimethylamino)-N-[[(methylamino) carbonyl]oxy]-2-oxo-, methyl ester. |
| P066 | 16752–77–5 | Ethanimidothioic acid, N-[[(methylamino)carbonyl]oxy]-, methyl ester |
| P101 | 107–12–0 | Ethyl cyanide |
| P054 | 151–56–4 | Ethyleneimine |
| P097 | 52–85–7 | Famphur |
| P056 | 7782–41–4 | Fluorine |
| P057 | 640–19–7 | Fluoroacetamide |
| P058 | 62–74–8 | Fluoroacetic acid, sodium salt |
| P198 | 23422–53–9 | Formetanate hydrochloride. |
| P197 | 17702–57–7 | Formparanate. |
| P065 | 628–86–4 | Fulminic acid, mercury(2+) salt (R,T) |
| P059 | 76–44–8 | Heptachlor |
| P062 | 757–58–4 | Hexaethyl tetraphosphate |
| P116 | 79–19–6 | Hydrazinecarbothioamide |
| P068 | 60–34–4 | Hydrazine, methyl- |

79

| Hazardous waste No. | Chemical abstracts No. | Substance |
|---|---|---|
| P063 | 74–90–8 | Hydrocyanic acid |
| P063 | 74–90–8 | Hydrogen cyanide |
| P096 | 7803–51–2 | Hydrogen phosphide |
| P060 | 465–73–6 | Isodrin |
| P192 | 119–38–0 | Isolan. |
| P202 | 64–00–6 | 3-Isopropylphenyl N-methylcarbamate. |
| P007 | 2763–96–4 | 3(2H)-Isoxazolone, 5-(aminomethyl)- |
| P196 | 15339–36–3 | Manganese, bis(dimethylcarbamodithioato-S,S′)-, |
| P196 | 15339–36–3 | Manganese dimethyldithiocarbamate. |
| P092 | 62–38–4 | Mercury, (acetato-O)phenyl- |
| P065 | 628–86–4 | Mercury fulminate (R,T) |
| P082 | 62–75–9 | Methanamine, N-methyl-N-nitroso- |
| P064 | 624–83–9 | Methane, isocyanato- |
| P016 | 542–88–1 | Methane, oxybis[chloro- |
| P112 | 509–14–8 | Methane, tetranitro- (R) |
| P118 | 75–70–7 | Methanethiol, trichloro- |
| P198 | 23422–53–9 | Methanimidamide, N,N-dimethyl-N′-[3-[[(methylamino)-carbonyl]oxy]phenyl]-, monohydrochloride. |
| P197 | 17702–57–7 | Methanimidamide, N,N-dimethyl-N′-[2-methyl-4-[[(methylamino)carbonyl]oxy]phenyl]- |
| P050 | 115–29–7 | 6,9-Methano-2,4,3-benzodioxathiepin, 6,7,8,9,10,10-hexachloro-1,5,5a,6,9,9a-hexahydro-, 3-oxide |
| P059 | 76–44–8 | 4,7-Methano-1H-indene, 1,4,5,6,7,8,8-heptachloro-3a,4,7,7a-tetrahydro- |
| P199 | 2032–65–7 | Methiocarb. |
| P066 | 16752–77–5 | Methomyl |
| P068 | 60–34–4 | Methyl hydrazine |
| P064 | 624–83–9 | Methyl isocyanate |
| P069 | 75–86–5 | 2-Methyllactonitrile |
| P071 | 298–00–0 | Methyl parathion |
| P190 | 1129–41–5 | Metolcarb. |
| P128 | 315–8–4 | Mexacarbate. |
| P072 | 86–88–4 | alpha-Naphthylthiourea |
| P073 | 13463–39–3 | Nickel carbonyl |
| P073 | 13463–39–3 | Nickel carbonyl Ni(CO)₄, (T-4)- |
| P074 | 557–19–7 | Nickel cyanide |
| P074 | 557–19–7 | Nickel cyanide Ni(CN)₂ |
| P075 | ¹54–11–5 | Nicotine, & salts |
| P076 | 10102–43–9 | Nitric oxide |
| P077 | 100–01–6 | p-Nitroaniline |
| P078 | 10102–44–0 | Nitrogen dioxide |
| P076 | 10102–43–9 | Nitrogen oxide NO |
| P078 | 10102–44–0 | Nitrogen oxide NO₂ |
| P081 | 55–63–0 | Nitroglycerine (R) |
| P082 | 62–75–9 | N-Nitrosodimethylamine |
| P084 | 4549–40–0 | N-Nitrosomethylvinylamine |
| P085 | 152–16–9 | Octamethylpyrophosphoramide |
| P087 | 20816–12–0 | Osmium oxide OsO₄, (T-4)- |
| P087 | 20816–12–0 | Osmium tetroxide |
| P088 | 145–73–3 | 7-Oxabicyclo[2.2.1]heptane-2,3-dicarboxylic acid |
| P194 | 23135–22–0 | Oxamyl. |
| P089 | 56–38–2 | Parathion |
| P034 | 131–89–5 | Phenol, 2-cyclohexyl-4,6-dinitro- |
| P048 | 51–28–5 | Phenol, 2,4-dinitro- |
| P047 | ¹534–52–1 | Phenol, 2-methyl-4,6-dinitro-, & salts |
| P020 | 88–85–7 | Phenol, 2-(1-methylpropyl)-4,6-dinitro- |
| P009 | 131–74–8 | Phenol, 2,4,6-trinitro-, ammonium salt (R) |
| P128 | 315–18–4 | Phenol, 4-(dimethylamino)-3,5-dimethyl-, methylcarbamate (ester). |
| P199 | 2032–65–7 | Phenol, (3,5-dimethyl-4-(methylthio)-, methylcarbamate |
| P202 | 64–00–6 | Phenol, 3-(1-methylethyl)-, methyl carbamate. |
| P201 | 2631–37–0 | Phenol, 3-methyl-5-(1-methylethyl)-, methyl carbamate. |
| P092 | 62–38–4 | Phenylmercury acetate |
| P093 | 103–85–5 | Phenylthiourea |
| P094 | 298–02–2 | Phorate |
| P095 | 75–44–5 | Phosgene |
| P096 | 7803–51–2 | Phosphine |
| P041 | 311–45–5 | Phosphoric acid, diethyl 4-nitrophenyl ester |
| P039 | 298–04–4 | Phosphorodithioic acid, O,O-diethyl S-[2-(ethylthio)ethyl] ester |
| P094 | 298–02–2 | Phosphorodithioic acid, O,O-diethyl S-[(ethylthio)methyl] ester |
| P044 | 60–51–5 | Phosphorodithioic acid, O,O-dimethyl S-[2-(methylamino)-2-oxoethyl] ester |
| P043 | 55–91–4 | Phosphorofluoridic acid, bis(1-methylethyl) ester |
| P089 | 56–38–2 | Phosphorothioic acid, O,O-diethyl O-(4-nitrophenyl) ester |

| Haz-ardous waste No. | Chemical abstracts No. | Substance |
|---|---|---|
| P040 | 297–97–2 | Phosphorothioic acid, O,O-diethyl O-pyrazinyl ester |
| P097 | 52–85–7 | Phosphorothioic acid, O-[4-[(dimethylamino)sulfonyl]phenyl] O,O-dimethyl ester |
| P071 | 298–00–0 | Phosphorothioic acid, O,O,-dimethyl O-(4-nitrophenyl) ester |
| P204 | 57–47–6 | Physostigmine. |
| P188 | 57–64–7 | Physostigmine salicylate. |
| P110 | 78–00–2 | Plumbane, tetraethyl- |
| P098 | 151–50–8 | Potassium cyanide |
| P098 | 151–50–8 | Potassium cyanide K(CN) |
| P099 | 506–61–6 | Potassium silver cyanide |
| P201 | 2631–37–0 | Promecarb |
| P070 | 116–06–3 | Propanal, 2-methyl-2-(methylthio)-, O-[(methylamino)carbonyl]oxime |
| P203 | 1646–88–4 | Propanal, 2-methyl-2-(methyl-sulfonyl)-, O-[(methylamino)carbonyl] oxime. |
| P101 | 107–12–0 | Propanenitrile |
| P027 | 542–76–7 | Propanenitrile, 3-chloro- |
| P069 | 75–86–5 | Propanenitrile, 2-hydroxy-2-methyl- |
| P081 | 55–63–0 | 1,2,3-Propanetriol, trinitrate (R) |
| P017 | 598–31–2 | 2-Propanone, 1-bromo- |
| P102 | 107–19–7 | Propargyl alcohol |
| P003 | 107–02–8 | 2-Propenal |
| P005 | 107–18–6 | 2-Propen-1-ol |
| P067 | 75–55–8 | 1,2-Propylenimine |
| P102 | 107–19–7 | 2-Propyn-1-ol |
| P008 | 504–24–5 | 4-Pyridinamine |
| P075 | [1]54–11–5 | Pyridine, 3-(1-methyl-2-pyrrolidinyl)-, (S)-, & salts |
| P204 | 57–47–6 | Pyrrolo[2,3-b]indol-5-ol, 1,2,3,3a,8,8a-hexahydro-1,3a,8-trimethyl-, methylcarbamate (ester), (3aS-cis)-. |
| P114 | 12039–52–0 | Selenious acid, dithallium(1+) salt |
| P103 | 630–10–4 | Selenourea |
| P104 | 506–64–9 | Silver cyanide |
| P104 | 506–64–9 | Silver cyanide Ag(CN) |
| P105 | 26628–22–8 | Sodium azide |
| P106 | 143–33–9 | Sodium cyanide |
| P106 | 143–33–9 | Sodium cyanide Na(CN) |
| P108 | [1]57–24–9 | Strychnidin-10-one, & salts |
| P018 | 357–57–3 | Strychnidin-10-one, 2,3-dimethoxy- |
| P108 | [1]57–24–9 | Strychnine, & salts |
| P115 | 7446–18–6 | Sulfuric acid, dithallium(1+) salt |
| P109 | 3689–24–5 | Tetraethyldithiopyrophosphate |
| P110 | 78–00–2 | Tetraethyl lead |
| P111 | 107–49–3 | Tetraethyl pyrophosphate |
| P112 | 509–14–8 | Tetranitromethane (R) |
| P062 | 757–58–4 | Tetraphosphoric acid, hexaethyl ester |
| P113 | 1314–32–5 | Thallic oxide |
| P113 | 1314–32–5 | Thallium oxide $Tl_2$ $O_3$ |
| P114 | 12039–52–0 | Thallium(I) selenite |
| P115 | 7446–18–6 | Thallium(I) sulfate |
| P109 | 3689–24–5 | Thiodiphosphoric acid, tetraethyl ester |
| P045 | 39196–18–4 | Thiofanox |
| P049 | 541–53–7 | Thioimidodicarbonic diamide [($H_2$ N)C(S)]$_2$ NH |
| P014 | 108–98–5 | Thiophenol |
| P116 | 79–19–6 | Thiosemicarbazide |
| P026 | 5344–82–1 | Thiourea, (?-chlorophenyl) |
| P072 | 86–88–4 | Thiourea, 1-naphthalenyl- |
| P093 | 103–85–5 | Thiourea, phenyl- |
| P185 | 26419–73–8 | Tirpate. |
| P123 | 8001–35–2 | Toxaphene |
| P118 | 75–70–7 | Trichloromethanethiol |
| P119 | 7803–55–6 | Vanadic acid, ammonium salt |
| P120 | 1314–62–1 | Vanadium oxide $V_2$ $O_5$ |
| P120 | 1314–62–1 | Vanadium pentoxide |
| P084 | 4549–40–0 | Vinylamine, N-methyl-N-nitroso- |
| P001 | [1]81–81–2 | Warfarin, & salts, when present at concentrations greater than 0.3% |
| P205 | 137–30–4 | Zinc, bis(dimethylcarbamodithioato-S,S')-, |
| P121 | 557–21–1 | Zinc cyanide |
| P121 | 557–21–1 | Zinc cyanide Zn(CN)$_2$ |
| P122 | 1314–84–7 | Zinc phosphide $Zn_3$ $P_2$, when present at concentrations greater than 10% (R,T) |
| P205 | 137–30–4 | Ziram. |
| P001 | [1]81–81–2 | 2H-1-Benzopyran-2-one, 4-hydroxy-3-(3-oxo-1-phenylbutyl)-, & salts, when present at concentrations greater than 0.3% |
| P001 | [1]81–81–2 | Warfarin, & salts, when present at concentrations greater than 0.3% |

| Haz-ardous waste No. | Chemical abstracts No. | Substance |
|---|---|---|
| P002 | 591–08–2 | Acetamide, -(aminothioxomethyl)- |
| P002 | 591–08–2 | 1-Acetyl-2-thiourea |
| P003 | 107–02–8 | Acrolein |
| P003 | 107–02–8 | 2-Propenal |
| P004 | 309–00–2 | Aldrin |
| P004 | 309–00–2 | 1,4,5,8-Dimethanonaphthalene, 1,2,3,4,10,10-hexa-chloro-1,4,4a,5,8,8a,-hexahydro-, (1alpha,4alpha,4abeta,5alpha,8alpha,8abeta)- |
| P005 | 107–18–6 | Allyl alcohol |
| P005 | 107–18–6 | 2-Propen-1-ol |
| P006 | 20859–73–8 | Aluminum phosphide (R,T) |
| P007 | 2763–96–4 | 5-(Aminomethyl)-3-isoxazolol |
| P007 | 2763–96–4 | 3(2H)-Isoxazolone, 5-(aminomethyl)- |
| P008 | 504–24–5 | 4-Aminopyridine |
| P008 | 504–24–5 | 4-Pyridinamine |
| P009 | 131–74–8 | Ammonium picrate (R) |
| P009 | 131–74–8 | Phenol, 2,4,6-trinitro-, ammonium salt (R) |
| P010 | 7778–39–4 | Arsenic acid $H_3$ $AsO_4$ |
| P011 | 1303–28–2 | Arsenic oxide $As_2$ $O_5$ |
| P011 | 1303–28–2 | Arsenic pentoxide |
| P012 | 1327–53–3 | Arsenic oxide $As_2$ $O_3$ |
| P012 | 1327–53–3 | Arsenic trioxide |
| P013 | 542–62–1 | Barium cyanide |
| P014 | 108–98–5 | Benzenethiol |
| P014 | 108–98–5 | Thiophenol |
| P015 | 7440–41–7 | Beryllium powder |
| P016 | 542–88–1 | Dichloromethyl ether |
| P016 | 542–88–1 | Methane, oxybis[chloro- |
| P017 | 598–31–2 | Bromoacetone |
| P017 | 598–31–2 | 2-Propanone, 1-bromo- |
| P018 | 357–57–3 | Brucine |
| P018 | 357–57–3 | Strychnidin-10-one, 2,3-dimethoxy- |
| P020 | 88–85–7 | Dinoseb |
| P020 | 88–85–7 | Phenol, 2-(1-methylpropyl)-4,6-dinitro- |
| P021 | 592–01–8 | Calcium cyanide |
| P021 | 592–01–8 | Calcium cyanide Ca(CN)₂ |
| P022 | 75–15–0 | Carbon disulfide |
| P023 | 107–20–0 | Acetaldehyde, chloro- |
| P023 | 107–20–0 | Chloroacetaldehyde |
| P024 | 106–47–8 | Benzenamine, 4-chloro- |
| P024 | 106–47–8 | p-Chloroaniline |
| P026 | 5344–82–1 | 1-(o-Chlorophenyl)thiourea |
| P026 | 5344–82–1 | Thiourea, (2-chlorophenyl)- |
| P027 | 542–76–7 | 3-Chloropropionitrile |
| P027 | 542–76–7 | Propanenitrile, 3-chloro- |
| P028 | 100–44–7 | Benzene, (chloromethyl)- |
| P028 | 100–44–7 | Benzyl chloride |
| P029 | 544–92–3 | Copper cyanide |
| P029 | 544–92–3 | Copper cyanide Cu(CN) |
| P030 | .................... | Cyanides (soluble cyanide salts), not otherwise specified |
| P031 | 460–19–5 | Cyanogen |
| P031 | 460–19–5 | Ethanedinitrile |
| P033 | 506–77–4 | Cyanogen chloride |
| P033 | 506–77–4 | Cyanogen chloride (CN)Cl |
| P034 | 131–89–5 | 2-Cyclohexyl-4,6-dinitrophenol |
| P034 | 131–89–5 | Phenol, 2-cyclohexyl-4,6-dinitro- |
| P036 | 696–28–6 | Arsonous dichloride, phenyl- |
| P036 | 696–28–6 | Dichlorophenylarsine |
| P037 | 60–57–1 | Dieldrin |
| P037 | 60–57–1 | 2,7:3,6-Dimethanonaphth[2,3-b]oxirene, 3,4,5,6,9,9-hexachloro-1a,2,2a,3,6,6a,7,7a-octahydro-, (1aalpha,2beta,2aalpha,3beta,6beta,6aalpha,7beta, 7aalpha)- |
| P038 | 692–42–2 | Arsine, diethyl- |
| P038 | 692–42–2 | Diethylarsine |
| P039 | 298–04–4 | Disulfoton |
| P039 | 298–04–4 | Phosphorodithioic acid, O,O-diethyl S-[2-(ethylthio)ethyl] ester |
| P040 | 297–97–2 | O,O-Diethyl O-pyrazinyl phosphorothioate |
| P040 | 297–97–2 | Phosphorothioic acid, O,O-diethyl O-pyrazinyl ester |
| P041 | 311–45–5 | Diethyl-p-nitrophenyl phosphate |
| P041 | 311–45–5 | Phosphoric acid, diethyl 4-nitrophenyl ester |
| P042 | 51–43–4 | 1,2-Benzenediol, 4-[1-hydroxy-2-(methylamino)ethyl]-, (R)- |
| P042 | 51–43–4 | Epinephrine |
| P043 | 55–91–4 | Diisopropylfluorophosphate (DFP) |
| P043 | 55–91–4 | Phosphorofluoridic acid, bis(1-methylethyl) ester |

| Haz-ardous waste No. | Chemical ab-stracts No. | Substance |
|---|---|---|
| P044 | 60–51–5 | Dimethoate |
| P044 | 60–51–5 | Phosphorodithioic acid, O,O-dimethyl S-[2-(methyl amino)-2-oxoethyl] ester |
| P045 | 39196–18–4 | 2-Butanone, 3,3-dimethyl-1-(methylthio)-, O-[(methylamino)carbonyl] oxime |
| P045 | 39196–18–4 | Thiofanox |
| P046 | 122–09–8 | Benzeneethanamine, alpha,alpha-dimethyl- |
| P046 | 122–09–8 | alpha,alpha-Dimethylphenethylamine |
| P047 | ¹534–52–1 | 4,6-Dinitro-o-cresol, & salts |
| P047 | ¹534–52–1 | Phenol, 2-methyl-4,6-dinitro-, & salts |
| P048 | 51–28–5 | 2,4-Dinitrophenol |
| P048 | 51–28–5 | Phenol, 2,4-dinitro- |
| P049 | 541–53–7 | Dithiobiuret |
| P049 | 541–53–7 | Thioimidodicarbonic diamide [(H₂ N)C(S)]₂ NH |
| P050 | 115–29–7 | Endosulfan |
| P050 | 115–29–7 | 6,9-Methano-2,4,3-benzodioxathiepin, 6,7,8,9,10,10-hexachloro-1,5,5a,6,9,9a-hexahydro-, 3-oxide |
| P051 | ¹72–20–8 | 2,7:3,6-Dimethanonaphth [2,3-b]oxirene, 3,4,5,6,9,9-hexachloro-1a,2,2a,3,6,6a,7,7a-octahydro-, (1aalpha,2beta,2abeta,3alpha,6alpha,6abeta,7beta, 7aalpha)-, & metabolites |
| P051 | 72–20–8 | Endrin |
| P051 | 72–20–8 | Endrin, & metabolites |
| P054 | 151–56–4 | Aziridine |
| P054 | 151–56–4 | Ethyleneimine |
| P056 | 7782–41–4 | Fluorine |
| P057 | 640–19–7 | Acetamide, 2-fluoro- |
| P057 | 640–19–7 | Fluoroacetamide |
| P058 | 62–74–8 | Acetic acid, fluoro-, sodium salt |
| P058 | 62–74–8 | Fluoroacetic acid, sodium salt |
| P059 | 76–44–8 | Heptachlor |
| P059 | 76–44–8 | 4,7-Methano-1H-indene, 1,4,5,6,7,8,8-heptachloro-3a,4,7,7a-tetrahydro- |
| P060 | 465–73–6 | 1,4,5,8-Dimethanonaphthalene, 1,2,3,4,10,10-hexa-chloro-1,4,4a,5,8,8a-hexahydro-, (1alpha,4alpha,4abeta,5beta,8beta,8abeta)- |
| P060 | 465–73–6 | Isodrin |
| P062 | 757–58–4 | Hexaethyl tetraphosphate |
| P062 | 757–58–4 | Tetraphosphoric acid, hexaethyl ester |
| P063 | 74–90–8 | Hydrocyanic acid |
| P063 | 74–90–8 | Hydrogen cyanide |
| P064 | 624–83–9 | Methane, isocyanato- |
| P064 | 624–83–9 | Methyl isocyanate |
| P065 | 628–86–4 | Fulminic acid, mercury(2+) salt (R,T) |
| P065 | 628–86–4 | Mercury fulminate (R,T) |
| P066 | 16752–77–5 | Ethanimidothioic acid, N-[[(methylamino)carbonyl]oxy]-, methyl ester |
| P066 | 16752–77–5 | Methomyl |
| P067 | 75–55–8 | Aziridine, 2-methyl- |
| P067 | 75–55–8 | 1,2-Propylenimine |
| P068 | 60–34–4 | Hydrazine, methyl- |
| P068 | 60–34–4 | Methyl hydrazine |
| P069 | 75–86–5 | 2-Methyllactonitrile |
| P069 | 75–86–5 | Propanenitrile, 2-hydroxy-2-methyl- |
| P070 | 116–06–3 | Aldicarb |
| P070 | 116–06–3 | Propanal, 2-methyl-2-(methylthio)-, O-[(methylamino)carbonyl]oxime |
| P071 | 298–00–0 | Methyl parathion |
| P071 | 298–00–0 | Phosphorothioic acid, O,O,-dimethyl O-(4-nitrophenyl) ester |
| P072 | 86–88–4 | alpha-Naphthylthiourea |
| P072 | 86–88–4 | Thiourea, 1-naphthalenyl- |
| P073 | 13463–39–3 | Nickel carbonyl |
| P073 | 13463–39–3 | Nickel carbonyl Ni(CO)₄, (T-4)- |
| P074 | 557 19–7 | Nickel cyanide |
| P074 | 557–19–7 | Nickel cyanide Ni(CN)₂ |
| P075 | ¹54–11–5 | Nicotine, & salts |
| P075 | ¹54–11–5 | Pyridine, 3-(1-methyl-2-pyrrolidinyl)-, (S)-, & salts |
| P076 | 10102–43–9 | Nitric oxide |
| P076 | 10102–43–9 | Nitrogen oxide NO |
| P077 | 100–01–6 | Benzenamine, 4-nitro- |
| P077 | 100–01–6 | p-Nitroaniline |
| P078 | 10102–44–0 | Nitrogen dioxide |
| P078 | 10102–44–0 | Nitrogen oxide NO₂ |
| P081 | 55–63–0 | Nitroglycerine (R) |
| P081 | 55–63–0 | 1,2,3-Propanetriol, trinitrate (R) |
| P082 | 62–75–9 | Methanamine, -methyl-N-nitroso- |
| P082 | 62–75–9 | N-Nitrosodimethylamine |
| P084 | 4549–40–0 | N-Nitrosomethylvinylamine |
| P084 | 4549–40–0 | Vinylamine, -methyl-N-nitroso- |
| P085 | 152–16–9 | Diphosphoramide, octamethyl- |
| P085 | 152–16–9 | Octamethylpyrophosphoramide |

83

| Haz-ardous waste No. | Chemical ab-stracts No. | Substance |
|---|---|---|
| P087 | 20816–12–0 | Osmium oxide OsO$_4$, (T-4)- |
| P087 | 20816–12–0 | Osmium tetroxide |
| P088 | 145–73–3 | Endothall |
| P088 | 145–73–3 | 7-Oxabicyclo[2.2.1]heptane-2,3-dicarboxylic acid |
| P089 | 56–38–2 | Parathion |
| P089 | 56–38–2 | Phosphorothioic acid, O,O-diethyl O-(4-nitrophenyl) ester |
| P092 | 62–38–4 | Mercury, (acetato-O)phenyl- |
| P092 | 62–38–4 | Phenylmercury acetate |
| P093 | 103–85–5 | Phenylthiourea |
| P093 | 103–85–5 | Thiourea, phenyl- |
| P094 | 298–02–2 | Phorate |
| P094 | 298–02–2 | Phosphorodithioic acid, O,O-diethyl S-[(ethylthio)methyl] ester |
| P095 | 75–44–5 | Carbonic dichloride |
| P095 | 75–44–5 | Phosgene |
| P096 | 7803–51–2 | Hydrogen phosphide |
| P096 | 7803–51–2 | Phosphine |
| P097 | 52–85–7 | Famphur |
| P097 | 52–85–7 | Phosphorothioic acid, O-[4-[(dimethylamino)sulfonyl]phenyl] O,O-dimethyl ester |
| P098 | 151–50–8 | Potassium cyanide |
| P098 | 151–50–8 | Potassium cyanide K(CN) |
| P099 | 506–61–6 | Argentate(1-), bis(cyano-C)-, potassium |
| P099 | 506–61–6 | Potassium silver cyanide |
| P101 | 107–12–0 | Ethyl cyanide |
| P101 | 107–12–0 | Propanenitrile |
| P102 | 107–19–7 | Propargyl alcohol |
| P102 | 107–19–7 | 2-Propyn-1-ol |
| P103 | 630–10–4 | Selenourea |
| P104 | 506–64–9 | Silver cyanide |
| P104 | 506–64–9 | Silver cyanide Ag(CN) |
| P105 | 26628–22–8 | Sodium azide |
| P106 | 143–33–9 | Sodium cyanide |
| P106 | 143–33–9 | Sodium cyanide Na(CN) |
| P108 | [1] 157–24–9 | Strychnidin-10-one, & salts |
| P108 | [1] 157–24–9 | Strychnine, & salts |
| P109 | 3689–24–5 | Tetraethyldithiopyrophosphate |
| P109 | 3689–24–5 | Thiodiphosphoric acid, tetraethyl ester |
| P110 | 78–00–2 | Plumbane, tetraethyl- |
| P110 | 78–00–2 | Tetraethyl lead |
| P111 | 107–49–3 | Diphosphoric acid, tetraethyl ester |
| P111 | 107–49–3 | Tetraethyl pyrophosphate |
| P112 | 509–14–8 | Methane, tetranitro-(R) |
| P112 | 509–14–8 | Tetranitromethane (R) |
| P113 | 1314–32–5 | Thallic oxide |
| P113 | 1314–32–5 | Thallium oxide Tl$_2$ O$_3$ |
| P114 | 12039–52–0 | Selenious acid, dithallium(1+) salt |
| P114 | 12039–52–0 | Tetraethyldithiopyrophosphate |
| P115 | 7446–18–6 | Thiodiphosphoric acid, tetraethyl ester |
| P115 | 7446–18–6 | Plumbane, tetraethyl- |
| P116 | 79–19–6 | Tetraethyl lead |
| P116 | 79–19–6 | Thiosemicarbazide |
| P118 | 75–70–7 | Methanethiol, trichloro- |
| P118 | 75–70–7 | Trichloromethanethiol |
| P119 | 7803–55–6 | Ammonium vanadate |
| P119 | 7803–55–6 | Vanadic acid, ammonium salt |
| P120 | 1314–62–1 | Vanadium oxide V$_2$O$_5$ |
| P120 | 1314–62–1 | Vanadium pentoxide |
| P121 | 557–21–1 | Zinc cyanide |
| P121 | 557–21–1 | Zinc cyanide Zn(CN)$_2$ |
| P122 | 1314–84–7 | Zinc phosphide Zn$_3$ P$_2$, when present at concentrations greater than 10% (R,T) |
| P123 | 8001–35–2 | Toxaphene |
| P127 | 1563–66–2 | 7-Benzofuranol, 2,3-dihydro-2,2-dimethyl-, methylcarbamate. |
| P127 | 1563–66–2 | Carbofuran |
| P128 | 315–8–4 | Mexacarbate |
| P128 | 315–18–4 | Phenol, 4-(dimethylamino)-3,5-dimethyl-, methylcarbamate (ester) |
| P185 | 26419–73–8 | 1,3-Dithiolane-2-carboxaldehyde, 2,4-dimethyl-, O-[(methylamino)-carbonyl]oxime. |
| P185 | 26419–73–8 | Tirpate |
| P188 | 57–64–7 | Benzoic acid, 2-hydroxy-, compd. with (3aS-cis)-1,2,3,3a,8,8a-hexahydro-1,3a,8-trimethylpyrrolo[2,3-b]indol-5-yl methylcarbamate ester (1:1) |
| P188 | 57–64–7 | Physostigmine salicylate |
| P189 | 55285–14–8 | Carbamic acid, [(dibutylamino)-thio]methyl-, 2,3-dihydro-2,2-dimethyl-7-benzofuranyl ester |
| P189 | 55285–14–8 | Carbosulfan |
| P190 | 1129–41–5 | Carbamic acid, methyl-, 3-methylphenyl ester |

| Haz-rdous waste No. | Chemical ab-stracts No. | Substance |
|---|---|---|
| 190 | 1129–41–5 | Metolcarb |
| 191 | 644–64–4 | Carbamic acid, dimethyl-, 1-[(dimethyl-amino)carbonyl]-5-methyl-1H-pyrazol-3-yl ester |
| 191 | 644–64–4 | Dimetilan |
| 192 | 119–38–0 | Carbamic acid, dimethyl-, 3-methyl-1-(1-methylethyl)-1H-pyrazol-5-yl ester |
| 192 | 119–38–0 | Isolan |
| 194 | 23135–22–0 | Ethanimidthioic acid, 2-(dimethylamino)-N-[[(methylamino) carbonyl]oxy]-2-oxo-, methyl ester |
| 194 | 23135–22–0 | Oxamyl |
| 196 | 15339–36–3 | Manganese, bis(dimethylcarbamodithioato-S,S')-, |
| 196 | 15339–36–3 | Manganese dimethyldithiocarbamate |
| 197 | 17702–57–7 | Formparanate |
| 197 | 17702–57–7 | Methanimidamide, N,N-dimethyl-N'-[2-methyl-4-[[(methylamino)carbonyl]oxy]phenyl]- |
| 198 | 23422–53–9 | Formetanate hydrochloride |
| 198 | 23422–53–9 | Methanimidamide, N,N-dimethyl-N'-[3-[[(methylamino)-carbonyl]oxy]phenyl]-monohydrochloride |
| 199 | 2032–65–7 | Methiocarb |
| 199 | 2032–65–7 | Phenol, (3,5-dimethyl-4-(methylthio)-, methylcarbamate |
| 201 | 2631–37–0 | Phenol, 3-methyl-5-(1-methylethyl)-, methyl carbamate |
| 201 | 2631–37–0 | Promecarb |
| 202 | 64–00–6 | m-Cumenyl methylcarbamate |
| 202 | 64–00–6 | 3-Isopropylphenyl N-methylcarbamate |
| 202 | 64–00–6 | Phenol, 3-(1-methylethyl)-, methyl carbamate |
| 203 | 1646–88–4 | Aldicarb sulfone |
| 203 | 1646–88–4 | Propanal, 2-methyl-2-(methyl-sulfonyl)-, O-[(methylamino)carbonyl] oxime |
| 204 | 57–47–6 | Physostigmine |
| 204 | 57–47–6 | Pyrrolo[2,3-b]indol-5-ol, 1,2,3,3a,8,8a-hexahydro-1,3a,8-trimethyl-, methylcarbamate (ester), (3aS-cis)- |
| 205 | 137–30–4 | Zinc, bis(dimethylcarbamodithioato-S,S')-, |
| 205 | 137–30–4 | Ziram |

[1] CAS Number given for parent compound only.

(f) The commercial chemical products, manufacturing chemical intermediates, or off-specification commercial chemical products referred to in paragraphs (a) through (d) of this section, are identified as toxic wastes (T), unless otherwise designated and are subject to the small quantity generator exclusion defined in §261.5 (a) and (g).

*Comment:* For the convenience of the regulated community, the primary hazardous properties of these materials have been indicated by the letters T (Toxicity), R (Reactivity), I (Ignitability) and C (Corrosivity). Absence of a letter indicates that the compound is only listed for toxicity. Wastes are first listed in alphabetical order by substance and then listed again in numerical order by Hazardous Waste Number.]

These wastes and their corresponding EPA Hazardous Waste Numbers are:

| Haz-ardous waste No. | Chemical ab-stracts No. | Substance |
|---|---|---|
| U394 | 30558–43–1 | A2213. |
| U001 | 75–07–0 | Acetaldehyde (I) |
| U034 | 75–87–6 | Acetaldehyde, trichloro- |
| U187 | 62–44–2 | Acetamide, N-(4-ethoxyphenyl)- |
| U005 | 53–96–3 | Acetamide, N-9H-fluoren-2-yl- |
| U240 | [1] 94–75–7 | Acetic acid, (2,4-dichlorophenoxy)-, salts & esters |
| U112 | 141–78–6 | Acetic acid ethyl ester (I) |
| U144 | 301–04–2 | Acetic acid, lead(2+) salt |
| U214 | 563–68–8 | Acetic acid, thallium(1+) salt |
| see F027 | 93–76–5 | Acetic acid, (2,4,5-trichlorophenoxy)- |
| U002 | 67–64–1 | Acetone (I) |
| U003 | 75–05–8 | Acetonitrile (I,T) |
| U004 | 98–86–2 | Acetophenone |
| U005 | 53–96–3 | 2-Acetylaminofluorene |
| U006 | 75–36–5 | Acetyl chloride (C,R,T) |
| U007 | 79–06–1 | Acrylamide |
| U008 | 79–10–7 | Acrylic acid (I) |
| U009 | 107–13–1 | Acrylonitrile |
| U011 | 61–82–5 | Amitrole |
| U012 | 62–53–3 | Aniline (I,T) |

85

| Haz-ardous waste No. | Chemical ab-stracts No. | Substance |
|---|---|---|
| U136 | 75–60–5 | Arsinic acid, dimethyl- |
| U014 | 492–80–8 | Auramine |
| U015 | 115–02–6 | Azaserine |
| U010 | 50–07–7 | Azirino[2′,3′:3,4]pyrrolo[1,2-a]indole-4,7-dione, 6-amino-8-[[(aminocarbonyl)oxy]methyl]-1,1a,2,8,8a,8b-hexahydro-8a-methoxy-5-methyl-, [1aS-(1aalpha, 8beta,8aalpha,8balpha)]- |
| U280 | 101–27–9 | Barban. |
| U278 | 22781–23–3 | Bendiocarb. |
| U364 | 22961–82–6 | Bendiocarb phenol. |
| U271 | 17804–35–2 | Benomyl. |
| U157 | 56–49–5 | Benz[j]aceanthrylene, 1,2-dihydro-3-methyl- |
| U016 | 225–51–4 | Benz[c]acridine |
| U017 | 98–87–3 | Benzal chloride |
| U192 | 23950–58–5 | Benzamide, 3,5-dichloro-N-(1,1-dimethyl-2-propynyl)- |
| U018 | 56–55–3 | Benz[a]anthracene |
| U094 | 57–97–6 | Benz[a]anthracene, 7,12-dimethyl- |
| U012 | 62–53–3 | Benzenamine (I,T) |
| U014 | 492–80–8 | Benzenamine, 4,4′-carbonimidoylbis[N,N-dimethyl- |
| U049 | 3165–93–3 | Benzenamine, 4-chloro-2-methyl-, hydrochloride |
| U093 | 60–11–7 | Benzenamine, N,N-dimethyl-4-(phenylazo)- |
| U328 | 95–53–4 | Benzenamine, 2-methyl- |
| U353 | 106–49–0 | Benzenamine, 4-methyl- |
| U158 | 101–14–4 | Benzenamine, 4,4′-methylenebis[2-chloro- |
| U222 | 636–21–5 | Benzenamine, 2-methyl-, hydrochloride |
| U181 | 99–55–8 | Benzenamine, 2-methyl-5-nitro- |
| U019 | 71–43–2 | Benzene (I,T) |
| U038 | 510–15–6 | Benzeneacetic acid, 4-chloro-alpha-(4-chlorophenyl)-alpha-hydroxy-, ethyl ester |
| U030 | 101–55–3 | Benzene, 1-bromo-4-phenoxy- |
| U035 | 305–03–3 | Benzenebutanoic acid, 4-[bis(2-chloroethyl)amino]- |
| U037 | 108–90–7 | Benzene, chloro- |
| U221 | 25376–45–8 | Benzenediamine, ar-methyl- |
| U028 | 117–81–7 | 1,2-Benzenedicarboxylic acid, bis(2-ethylhexyl) ester |
| U069 | 84–74–2 | 1,2-Benzenedicarboxylic acid, dibutyl ester |
| U088 | 84–66–2 | 1,2-Benzenedicarboxylic acid, diethyl ester |
| U102 | 131–11–3 | 1,2-Benzenedicarboxylic acid, dimethyl ester |
| U107 | 117–84–0 | 1,2-Benzenedicarboxylic acid, dioctyl ester |
| U070 | 95–50–1 | Benzene, 1,2-dichloro- |
| U071 | 541–73–1 | Benzene, 1,3-dichloro- |
| U072 | 106–46–7 | Benzene, 1,4-dichloro- |
| U060 | 72–54–8 | Benzene, 1,1′-(2,2-dichloroethylidene)bis[4-chloro- |
| U017 | 98–87–3 | Benzene, (dichloromethyl)- |
| U223 | 26471–62–5 | Benzene, 1,3-diisocyanatomethyl- (R,T) |
| U239 | 1330–20–7 | Benzene, dimethyl- (I) |
| U201 | 108–46–3 | 1,3-Benzenediol |
| U127 | 118–74–1 | Benzene, hexachloro- |
| U056 | 110–82–7 | Benzene, hexahydro- (I) |
| U220 | 108–88–3 | Benzene, methyl- |
| U105 | 121–14–2 | Benzene, 1-methyl-2,4-dinitro- |
| U106 | 606–20–2 | Benzene, 2-methyl-1,3-dinitro- |
| U055 | 98–82–8 | Benzene, (1-methylethyl)- (I) |
| U169 | 98–95–3 | Benzene, nitro- |
| U183 | 608–93–5 | Benzene, pentachloro- |
| U185 | 82–68–8 | Benzene, pentachloronitro- |
| U020 | 98–09–9 | Benzenesulfonic acid chloride (C,R) |
| U020 | 98–09–9 | Benzenesulfonyl chloride (C,R) |
| U207 | 95–94–3 | Benzene, 1,2,4,5-tetrachloro- |
| U061 | 50–29–3 | Benzene, 1,1′-(2,2,2-trichloroethylidene)bis[4-chloro- |
| U247 | 72–43–5 | Benzene, 1,1′-(2,2,2-trichloroethylidene)bis[4- methoxy- |
| U023 | 98–07–7 | Benzene, (trichloromethyl)- |
| U234 | 99–35–4 | Benzene, 1,3,5-trinitro- |
| U021 | 92–87–5 | Benzidine |
| U202 | [1]81–07–2 | 1,2-Benzisothiazol-3(2H)-one, 1,1-dioxide, & salts |
| U278 | 22781–23–3 | 1,3-Benzodioxol-4-ol, 2,2-dimethyl-, methyl carbamate. |
| U364 | 22961–82–6 | 1,3-Benzodioxol-4-ol, 2,2-dimethyl-, |
| U203 | 94–59–7 | 1,3-Benzodioxole, 5-(2-propenyl)- |
| U141 | 120–58–1 | 1,3-Benzodioxole, 5-(1-propenyl)- |
| U367 | 1563–38–8 | 7-Benzofuranol, 2,3-dihydro-2,2-dimethyl- |
| U090 | 94–58–6 | 1,3-Benzodioxole, 5-propyl- |
| U064 | 189–55–9 | Benzo[rst]pentaphene |
| U248 | [1]81–81–2 | 2H-1-Benzopyran-2-one, 4-hydroxy-3-(3-oxo-1-phenyl-butyl)-, & salts, when present at concentrations of 0.3% or less |
| U022 | 50–32–8 | Benzo[a]pyrene |
| U197 | 106–51–4 | p-Benzoquinone |

| Hazardous waste No. | Chemical abstracts No. | Substance |
|---|---|---|
| U023 | 98–07–7 | Benzotrichloride (C,R,T) |
| U085 | 1464–53–5 | 2,2′-Bioxirane |
| U021 | 92–87–5 | [1,1′-Biphenyl]-4,4′-diamine |
| U073 | 91–94–1 | [1,1′-Biphenyl]-4,4′-diamine, 3,3′-dichloro- |
| U091 | 119–90–4 | [1,1′-Biphenyl]-4,4′-diamine, 3,3′-dimethoxy- |
| U095 | 119–93–7 | [1,1′-Biphenyl]-4,4′-diamine, 3,3′-dimethyl- |
| U225 | 75–25–2 | Bromoform |
| U030 | 101–55–3 | 4-Bromophenyl phenyl ether |
| U128 | 87–68–3 | 1,3-Butadiene, 1,1,2,3,4,4-hexachloro- |
| U172 | 924–16–3 | 1-Butanamine, N-butyl-N-nitroso- |
| U031 | 71–36–3 | 1-Butanol (I) |
| U159 | 78–93–3 | 2-Butanone (I,T) |
| U160 | 1338–23–4 | 2-Butanone, peroxide (R,T) |
| U053 | 4170–30–3 | 2-Butenal |
| U074 | 764–41–0 | 2-Butene, 1,4-dichloro- (I,T) |
| U143 | 303–34–4 | 2-Butenoic acid, 2-methyl-, 7-[[2,3-dihydroxy-2-(1-methoxyethyl)-3-methyl-1-oxobutoxy]methyl]-2,3,5,7a-tetrahydro-1H-pyrrolizin-1-yl ester, [1S-[1alpha(Z),7(2S*,3R*),7aalpha]]- |
| U031 | 71–36–3 | n-Butyl alcohol (I) |
| U136 | 75–60–5 | Cacodylic acid |
| U032 | 13765–19–0 | Calcium chromate |
| U372 | 10605–21–7 | Carbamic acid, 1H-benzimidazol-2-yl, methyl ester. |
| U271 | 17804–35–2 | Carbamic acid, [1-[(butylamino)carbonyl]-1H-benzimidazol-2-yl]-, methyl ester. |
| U280 | 101–27–9 | Carbamic acid, (3-chlorophenyl)-, 4-chloro-2-butynyl ester. |
| U238 | 51–79–6 | Carbamic acid, ethyl ester |
| U178 | 615–53–2 | Carbamic acid, methylnitroso-, ethyl ester |
| U373 | 122–42–9 | Carbamic acid, phenyl-, 1-methylethyl ester. |
| U409 | 23564–05–8 | Carbamic acid, [1,2-phenylenebis (iminocarbonothioyl)]bis-, dimethyl ester. |
| U097 | 79–44–7 | Carbamic chloride, dimethyl- |
| U389 | 2303–17–5 | Carbamothioic acid, bis(1-methylethyl)-, S-(2,3,3-trichloro-2-propenyl) ester. |
| U387 | 52888–80–9 | Carbamothioic acid, dipropyl-, S-(phenylmethyl) ester. |
| U114 | [1] 111–54–6 | Carbamodithioic acid, 1,2-ethanediylbis-, salts & esters |
| U062 | 2303–16–4 | Carbamothioic acid, bis(1-methylethyl)-, S-(2,3-dichloro-2-propenyl) ester |
| U279 | 63–25–2 | Carbaryl. |
| U372 | 10605–21–7 | Carbendazim. |
| U367 | 1563–38–8 | Carbofuran phenol. |
| U215 | 6533–73–9 | Carbonic acid, dithallium(1+) salt |
| U033 | 353–50–4 | Carbonic difluoride |
| U156 | 79–22–1 | Carbonochloridic acid, methyl ester (I,T) |
| U033 | 353–50–4 | Carbon oxyfluoride (R,T) |
| U211 | 56–23–5 | Carbon tetrachloride |
| U034 | 75–87–6 | Chloral |
| U035 | 305–03–3 | Chlorambucil |
| U036 | 57–74–9 | Chlordane, alpha & gamma isomers |
| U026 | 494–03–1 | Chlornaphazin |
| U037 | 108–90–7 | Chlorobenzene |
| U038 | 510–15–6 | Chlorobenzilate |
| U039 | 59–50–7 | p-Chloro-m-cresol |
| U042 | 110–75–8 | 2-Chloroethyl vinyl ether |
| U044 | 67–66–3 | Chloroform |
| U046 | 107–30–2 | Chloromethyl methyl ether |
| U047 | 91–58–7 | beta-Chloronaphthalene |
| U048 | 95–57–8 | o-Chlorophenol |
| U049 | 3165–93–3 | 4-Chloro-o-toluidine, hydrochloride |
| U032 | 13765–19–0 | Chromic acid $H_2$ $CrO_4$, calcium salt |
| U050 | 218–01–9 | Chrysene |
| U051 | .................... | Creosote |
| U052 | 1319–77–3 | Cresol (Cresylic acid) |
| U053 | 4170–30–3 | Crotonaldehyde |
| U055 | 98–82–8 | Cumene (I) |
| U246 | 506–68–3 | Cyanogen bromide (CN)Br |
| U197 | 106–51–4 | 2,5-Cyclohexadiene-1,4-dione |
| U056 | 110–82–7 | Cyclohexane (I) |
| U129 | 58–89–9 | Cyclohexane, 1,2,3,4,5,6-hexachloro-, (1alpha,2alpha,3beta,4alpha,5alpha,6beta)- |
| U057 | 108–94–1 | Cyclohexanone (I) |
| U130 | 77–47–4 | 1,3-Cyclopentadiene, 1,2,3,4,5,5-hexachloro- |
| U058 | 50–18–0 | Cyclophosphamide |
| U240 | [1] 94–75–7 | 2,4-D, salts & esters |
| U059 | 20830–81–3 | Daunomycin |

| Hazardous waste No. | Chemical abstracts No. | Substance |
|---|---|---|
| U060 | 72–54–8 | DDD |
| U061 | 50–29–3 | DDT |
| U062 | 2303–16–4 | Diallate |
| U063 | 53–70–3 | Dibenz[a,h]anthracene |
| U064 | 189–55–9 | Dibenzo[a,i]pyrene |
| U066 | 96–12–8 | 1,2-Dibromo-3-chloropropane |
| U069 | 84–74–2 | Dibutyl phthalate |
| U070 | 95–50–1 | o-Dichlorobenzene |
| U071 | 541–73–1 | m-Dichlorobenzene |
| U072 | 106–46–7 | p-Dichlorobenzene |
| U073 | 91–94–1 | 3,3′-Dichlorobenzidine |
| U074 | 764–41–0 | 1,4-Dichloro-2-butene (I,T) |
| U075 | 75–71–8 | Dichlorodifluoromethane |
| U078 | 75–35–4 | 1,1-Dichloroethylene |
| U079 | 156–60–5 | 1,2-Dichloroethylene |
| U025 | 111–44–4 | Dichloroethyl ether |
| U027 | 108–60–1 | Dichloroisopropyl ether |
| U024 | 111–91–1 | Dichloromethoxy ethane |
| U081 | 120–83–2 | 2,4-Dichlorophenol |
| U082 | 87–65–0 | 2,6-Dichlorophenol |
| U084 | 542–75–6 | 1,3-Dichloropropene |
| U085 | 1464–53–5 | 1,2:3,4-Diepoxybutane (I,T) |
| U108 | 123–91–1 | 1,4-Diethyleneoxide |
| U028 | 117–81–7 | Diethylhexyl phthalate |
| U395 | 5952–26–1 | Diethylene glycol, dicarbamate. |
| U086 | 1615–80–1 | N,N′-Diethylhydrazine |
| U087 | 3288–58–2 | O,O-Diethyl S-methyl dithiophosphate |
| U088 | 84–66–2 | Diethyl phthalate |
| U089 | 56–53–1 | Diethylstilbesterol |
| U090 | 94–58–6 | Dihydrosafrole |
| U091 | 119–90–4 | 3,3′-Dimethoxybenzidine |
| U092 | 124–40–3 | Dimethylamine (I) |
| U093 | 60–11–7 | p-Dimethylaminoazobenzene |
| U094 | 57–97–6 | 7,12-Dimethylbenz[a]anthracene |
| U095 | 119–93–7 | 3,3′-Dimethylbenzidine |
| U096 | 80–15–9 | alpha,alpha-Dimethylbenzylhydroperoxide (R) |
| U097 | 79–44–7 | Dimethylcarbamoyl chloride |
| U098 | 57–14–7 | 1,1-Dimethylhydrazine |
| U099 | 540–73–8 | 1,2-Dimethylhydrazine |
| U101 | 105–67–9 | 2,4-Dimethylphenol |
| U102 | 131–11–3 | Dimethyl phthalate |
| U103 | 77–78–1 | Dimethyl sulfate |
| U105 | 121–14–2 | 2,4-Dinitrotoluene |
| U106 | 606–20–2 | 2,6-Dinitrotoluene |
| U107 | 117–84–0 | Di-n-octyl phthalate |
| U108 | 123–91–1 | 1,4-Dioxane |
| U109 | 122–66–7 | 1,2-Diphenylhydrazine |
| U110 | 142–84–7 | Dipropylamine (I) |
| U111 | 621–64–7 | Di-n-propylnitrosamine |
| U041 | 106–89–8 | Epichlorohydrin |
| U001 | 75–07–0 | Ethanal (I) |
| U404 | 121–44–8 | Ethanamine, N,N-diethyl- |
| U174 | 55–18–5 | Ethanamine, N-ethyl-N-nitroso- |
| U155 | 91–80–5 | 1,2-Ethanediamine, N,N-dimethyl-N′-2-pyridinyl-N′-(2-thienylmethyl)- |
| U067 | 106–93–4 | Ethane, 1,2-dibromo- |
| U076 | 75–34–3 | Ethane, 1,1-dichloro- |
| U077 | 107–06–2 | Ethane, 1,2-dichloro- |
| U131 | 67–72–1 | Ethane, hexachloro- |
| U024 | 111–91–1 | Ethane, 1,1′-[methylenebis(oxy)]bis[2-chloro- |
| U117 | 60–29–7 | Ethane, 1,1′-oxybis-(I) |
| U025 | 111–44–4 | Ethane, 1,1′-oxybis[2-chloro- |
| U184 | 76–01–7 | Ethane, pentachloro- |
| U208 | 630–20–6 | Ethane, 1,1,1,2-tetrachloro- |
| U209 | 79–34–5 | Ethane, 1,1,2,2-tetrachloro- |
| U218 | 62–55–5 | Ethanethioamide |
| U226 | 71–55–6 | Ethane, 1,1,1-trichloro- |
| U227 | 79–00–5 | Ethane, 1,1,2-trichloro- |
| U410 | 59669–26–0 | Ethanimidothioic acid, N,N′- [thiobis[(methylimino)carbonyloxy]]bis-, dimethyl ester |
| U394 | 30558–43–1 | Ethanimidothioic acid, 2-(dimethylamino)-N-hydroxy-2-oxo-, methyl ester. |
| U359 | 110–80–5 | Ethanol, 2-ethoxy- |
| U173 | 1116–54–7 | Ethanol, 2,2′-(nitrosoimino)bis- |
| U395 | 5952–26–1 | Ethanol, 2,2′-oxybis-, dicarbamate. |

| Haz-ardous waste No. | Chemical abstracts No. | Substance |
|---|---|---|
| U004 | 98–86–2 | Ethanone, 1-phenyl- |
| U043 | 75–01–4 | Ethene, chloro- |
| U042 | 110–75–8 | Ethene, (2-chloroethoxy)- |
| U078 | 75–35–4 | Ethene, 1,1-dichloro- |
| U079 | 156–60–5 | Ethene, 1,2-dichloro-, (E)- |
| U210 | 127–18–4 | Ethene, tetrachloro- |
| U228 | 79–01–6 | Ethene, trichloro- |
| U112 | 141–78–6 | Ethyl acetate (I) |
| U113 | 140–88–5 | Ethyl acrylate (I) |
| U238 | 51–79–6 | Ethyl carbamate (urethane) |
| U117 | 60–29–7 | Ethyl ether (I) |
| U114 | ¹111–54–6 | Ethylenebisdithiocarbamic acid, salts & esters |
| U067 | 106–93–4 | Ethylene dibromide |
| U077 | 107–06–2 | Ethylene dichloride |
| U359 | 110–80–5 | Ethylene glycol monoethyl ether |
| U115 | 75–21–8 | Ethylene oxide (I,T) |
| U116 | 96–45–7 | Ethylenethiourea |
| U076 | 75–34–3 | Ethylidene dichloride |
| U118 | 97–63–2 | Ethyl methacrylate |
| U119 | 62–50–0 | Ethyl methanesulfonate |
| U120 | 206–44–0 | Fluoranthene |
| U122 | 50–00–0 | Formaldehyde |
| U123 | 64–18–6 | Formic acid (C,T) |
| U124 | 110–00–9 | Furan (I) |
| U125 | 98–01–1 | 2-Furancarboxaldehyde (I) |
| U147 | 108–31–6 | 2,5-Furandione |
| U213 | 109–99–9 | Furan, tetrahydro-(I) |
| U125 | 98–01–1 | Furfural (I) |
| U124 | 110–00–9 | Furfuran (I) |
| U206 | 18883–66–4 | Glucopyranose, 2-deoxy-2-(3-methyl-3-nitrosoureido)-, D- |
| U206 | 18883–66–4 | D-Glucose, 2-deoxy-2-[[(methylnitrosoamino)-carbonyl]amino]- |
| U126 | 765–34–4 | Glycidylaldehyde |
| U163 | 70–25–7 | Guanidine, N-methyl-N′-nitro-N-nitroso- |
| U127 | 118–74–1 | Hexachlorobenzene |
| U128 | 87–68–3 | Hexachlorobutadiene |
| U130 | 77–47–4 | Hexachlorocyclopentadiene |
| U131 | 67–72–1 | Hexachloroethane |
| U132 | 70–30–4 | Hexachlorophene |
| U243 | 1888–71–7 | Hexachloropropene |
| U133 | 302–01–2 | Hydrazine (R,T) |
| U086 | 1615–80–1 | Hydrazine, 1,2-diethyl- |
| U098 | 57–14–7 | Hydrazine, 1,1-dimethyl- |
| U099 | 540–73–8 | Hydrazine, 1,2-dimethyl- |
| U109 | 122–66–7 | Hydrazine, 1,2-diphenyl- |
| U134 | 7664–39–3 | Hydrofluoric acid (C,T) |
| U134 | 7664–39–3 | Hydrogen fluoride (C,T) |
| U135 | 7783–06–4 | Hydrogen sulfide |
| U135 | 7783–06–4 | Hydrogen sulfide H₂ S |
| U096 | 80–15–9 | Hydroperoxide, 1-methyl-1-phenylethyl- (R) |
| U116 | 96–45–7 | 2-Imidazolidinethione |
| U137 | 193–39–5 | Indeno[1,2,3-cd]pyrene |
| U190 | 85–44–9 | 1,3-Isobenzofurandione |
| U140 | 78–83–1 | Isobutyl alcohol (I,T) |
| U141 | 120–58–1 | Isosafrole |
| U142 | 143–50–0 | Kepone |
| U143 | 303–34–4 | Lasiocarpine |
| U144 | 301–04–2 | Lead acetate |
| U146 | 1335–32–6 | Lead, bis(acetato-O)tetrahydroxytri- |
| U145 | 7446–27–7 | Lead phosphate |
| U146 | 1335–32–6 | Lead subacetate |
| U129 | 58–89–9 | Lindane |
| U163 | 70–25–7 | MNNG |
| U147 | 108–31–6 | Maleic anhydride |
| U148 | 123–33–1 | Maleic hydrazide |
| U149 | 109–77–3 | Malononitrile |
| U150 | 148–82–3 | Melphalan |
| U151 | 7439–97–6 | Mercury |
| U152 | 126–98–7 | Methacrylonitrile (I, T) |
| U092 | 124–40–3 | Methanamine, N-methyl- (I) |
| U029 | 74–83–9 | Methane, bromo- |
| U045 | 74–87–3 | Methane, chloro- (I, T) |

| Haz-ardous waste No. | Chemical abstracts No. | Substance |
|---|---|---|
| U046 | 107–30–2 | Methane, chloromethoxy- |
| U068 | 74–95–3 | Methane, dibromo- |
| U080 | 75–09–2 | Methane, dichloro- |
| U075 | 75–71–8 | Methane, dichlorodifluoro- |
| U138 | 74–88–4 | Methane, iodo- |
| U119 | 62–50–0 | Methanesulfonic acid, ethyl ester |
| U211 | 56–23–5 | Methane, tetrachloro- |
| U153 | 74–93–1 | Methanethiol (I, T) |
| U225 | 75–25–2 | Methane, tribromo- |
| U044 | 67–66–3 | Methane, trichloro- |
| U121 | 75–69–4 | Methane, trichlorofluoro- |
| U036 | 57–74–9 | 4,7-Methano-1H-indene, 1,2,4,5,6,7,8,8-octachloro-2,3,3a,4,7,7a-hexahydro- |
| U154 | 67–56–1 | Methanol (I) |
| U155 | 91–80–5 | Methapyrilene |
| U142 | 143–50–0 | 1,3,4-Metheno-2H-cyclobuta[cd]pentalen-2-one, 1,1a,3,3a,4,5,5,5a,5b,6-decachlorooctahydro- |
| U247 | 72–43–5 | Methoxychlor |
| U154 | 67–56–1 | Methyl alcohol (I) |
| U029 | 74–83–9 | Methyl bromide |
| U186 | 504–60–9 | 1-Methylbutadiene (I) |
| U045 | 74–87–3 | Methyl chloride (I,T) |
| U156 | 79–22–1 | Methyl chlorocarbonate (I,T) |
| U226 | 71–55–6 | Methyl chloroform |
| U157 | 56–49–5 | 3-Methylcholanthrene |
| U158 | 101–14–4 | 4,4′-Methylenebis(2-chloroaniline) |
| U068 | 74–95–3 | Methylene bromide |
| U080 | 75–09–2 | Methylene chloride |
| U159 | 78–93–3 | Methyl ethyl ketone (MEK) (I,T) |
| U160 | 1338–23–4 | Methyl ethyl ketone peroxide (R,T) |
| U138 | 74–88–4 | Methyl iodide |
| U161 | 108–10–1 | Methyl isobutyl ketone (I) |
| U162 | 80–62–6 | Methyl methacrylate (I,T) |
| U161 | 108–10–1 | 4-Methyl-2-pentanone (I) |
| U164 | 56–04–2 | Methylthiouracil |
| U010 | 50–07–7 | Mitomycin C |
| U059 | 20830–81–3 | 5,12-Naphthacenedione, 8-acetyl-10-[(3-amino-2,3,6-trideoxy)-alpha-L-lyxo-hexopyranosyl)oxy]-7,8,9,10-tetrahydro-6,8,11-trihydroxy-1-methoxy-, (8S-cis)- |
| U167 | 134–32–7 | 1-Naphthalenamine |
| U168 | 91–59–8 | 2-Naphthalenamine |
| U026 | 494–03–1 | Naphthalenamine, N,N′-bis(2-chloroethyl)- |
| U165 | 91–20–3 | Naphthalene |
| U047 | 91–58–7 | Naphthalene, 2-chloro- |
| U166 | 130–15–4 | 1,4-Naphthalenedione |
| U236 | 72–57–1 | 2,7-Naphthalenedisulfonic acid, 3,3′-[(3,3′-dimethyl[1,1′-biphenyl]-4,4′-diyl)bis(azo)bis[5-amino-4-hydroxy]-, tetrasodium salt |
| U279 | 63–25–2 | 1-Naphthalenol, methylcarbamate. |
| U166 | 130–15–4 | 1,4-Naphthoquinone |
| U167 | 134–32–7 | alpha-Naphthylamine |
| U168 | 91–59–8 | beta-Naphthylamine |
| U217 | 10102–45–1 | Nitric acid, thallium(1+) salt |
| U169 | 98–95–3 | Nitrobenzene (I,T) |
| U170 | 100–02–7 | p-Nitrophenol |
| U171 | 79–46–9 | 2-Nitropropane (I,T) |
| U172 | 924–16–3 | N-Nitrosodi-n-butylamine |
| U173 | 1116–54–7 | N-Nitrosodiethanolamine |
| U174 | 55–18–5 | N-Nitrosodiethylamine |
| U176 | 759–73–9 | N-Nitroso-N-ethylurea |
| U177 | 684–93–5 | N-Nitroso-N-methylurea |
| U178 | 615–53–2 | N-Nitroso-N-methylurethane |
| U179 | 100–75–4 | N-Nitrosopiperidine |
| U180 | 930–55–2 | N-Nitrosopyrrolidine |
| U181 | 99–55–8 | 5-Nitro-o-toluidine |
| U193 | 1120–71–4 | 1,2-Oxathiolane, 2,2-dioxide |
| U058 | 50–18–0 | 2H-1,3,2-Oxazaphosphorin-2-amine, N,N-bis(2-chloroethyl)tetrahydro-, 2-oxide |
| U115 | 75–21–8 | Oxirane (I,T) |
| U126 | 765–34–4 | Oxiranecarboxyaldehyde |
| U041 | 106–89–8 | Oxirane, (chloromethyl)- |
| U182 | 123–63–7 | Paraldehyde |
| U183 | 608–93–5 | Pentachlorobenzene |
| U184 | 76–01–7 | Pentachloroethane |
| U185 | 82–68–8 | Pentachloronitrobenzene (PCNB) |

| Haz-ardous waste No. | Chemical abstracts No. | Substance |
|---|---|---|
| See F027 | 87–86–5 | Pentachlorophenol |
| U161 | 108–10–1 | Pentanol, 4-methyl- |
| U186 | 504–60–9 | 1,3-Pentadiene (I) |
| U187 | 62–44–2 | Phenacetin |
| U188 | 108–95–2 | Phenol |
| U048 | 95–57–8 | Phenol, 2-chloro- |
| U039 | 59–50–7 | Phenol, 4-chloro-3-methyl- |
| U081 | 120–83–2 | Phenol, 2,4-dichloro- |
| U082 | 87–65–0 | Phenol, 2,6-dichloro- |
| U089 | 56–53–1 | Phenol, 4,4′-(1,2-diethyl-1,2-ethenediyl)bis-, (E)- |
| U101 | 105–67–9 | Phenol, 2,4-dimethyl- |
| U052 | 1319–77–3 | Phenol, methyl- |
| U132 | 70–30–4 | Phenol, 2,2′-methylenebis[3,4,6-trichloro- |
| U411 | 114–26–1 | Phenol, 2-(1-methylethoxy)-, methylcarbamate. |
| U170 | 100–02–7 | Phenol, 4-nitro- |
| See F027 | 87–86–5 | Phenol, pentachloro- |
| See F027 | 58–90–2 | Phenol, 2,3,4,6-tetrachloro- |
| See F027 | 95–95–4 | Phenol, 2,4,5-trichloro- |
| See F027 | 88–06–2 | Phenol, 2,4,6-trichloro- |
| U150 | 148–82–3 | L-Phenylalanine, 4-[bis(2-chloroethyl)amino]- |
| U145 | 7446–27–7 | Phosphoric acid, lead(2+) salt (2:3) |
| U087 | 3288–58–2 | Phosphorodithioic acid, O,O-diethyl S-methyl ester |
| U189 | 1314–80–3 | Phosphorus sulfide (R) |
| U190 | 85–44–9 | Phthalic anhydride |
| U191 | 109–06–8 | 2-Picoline |
| U179 | 100–75–4 | Piperidine, 1-nitroso- |
| U192 | 23950–58–5 | Pronamide |
| U194 | 107–10–8 | 1-Propanamine (I,T) |
| U111 | 621–64–7 | 1-Propanamine, N-nitroso-N-propyl- |
| U110 | 142–84–7 | 1-Propanamine, N-propyl- (I) |
| U066 | 96–12–8 | Propane, 1,2-dibromo-3-chloro- |
| U083 | 78–87–5 | Propane, 1,2-dichloro- |
| U149 | 109–77–3 | Propanedinitrile |
| U171 | 79–46–9 | Propane, 2-nitro- (I,T) |
| U027 | 108–60–1 | Propane, 2,2′-oxybis[2-chloro- |
| U193 | 1120–71–4 | 1,3-Propane sultone |
| See F027 | 93–72–1 | Propanoic acid, 2-(2,4,5-trichlorophenoxy)- |
| U235 | 126–72–7 | 1-Propanol, 2,3-dibromo-, phosphate (3:1) |
| U140 | 78–83–1 | 1-Propanol, 2-methyl- (I,T) |
| U002 | 67–64–1 | 2-Propanone (I) |
| U007 | 79–06–1 | 2-Propenamide |
| U084 | 542–75–6 | 1-Propene, 1,3-dichloro- |
| U243 | 1888–71–7 | 1-Propene, 1,1,2,3,3,3-hexachloro- |
| U009 | 107–13–1 | 2-Propenenitrile |
| U152 | 126–98–7 | 2-Propenenitrile, 2-methyl- (I,T) |
| U008 | 79–10–7 | 2-Propenoic acid (I) |
| U113 | 140–88–5 | 2-Propenoic acid, ethyl ester (I) |
| U118 | 97–63–2 | 2-Propenoic acid, 2-methyl , ethyl ester |
| U162 | 80–62–6 | 2-Propenoic acid, 2-methyl-, methyl ester (I,T) |
| U373 | 122–42–9 | Propham. |
| U411 | 114–26–1 | Propoxur. |
| U387 | 52888–80–9 | Prosulfocarb. |
| U194 | 107–10–8 | n-Propylamine (I,T) |
| U083 | 78–87–5 | Propylene dichloride |
| U148 | 123–33–1 | 3,6-Pyridazinedione, 1,2-dihydro- |
| U196 | 110–86–1 | Pyridine |
| U191 | 109–06–8 | Pyridine, 2-methyl- |
| U237 | 66–75–1 | 2,4-(1H,3H)-Pyrimidinedione, 5-[bis(2-chloroethyl)amino]- |
| U164 | 56–04–2 | 4(1H)-Pyrimidinone, 2,3-dihydro-6-methyl-2-thioxo- |
| U180 | 930–55–2 | Pyrrolidine, 1-nitroso- |
| U200 | 50–55–5 | Reserpine |
| U201 | 108–46–3 | Resorcinol |
| U202 | ¹81–07–2 | Saccharin, & salts |
| U203 | 94–59–7 | Safrole |
| U204 | 7783–00–8 | Selenious acid |

| Haz-ardous waste No. | Chemical ab-stracts No. | Substance |
|---|---|---|
| U204 | 7783–00–8 | Selenium dioxide |
| U205 | 7488–56–4 | Selenium sulfide |
| U205 | 7488–56–4 | Selenium sulfide SeS$_2$ (R,T) |
| U015 | 115–02–6 | L-Serine, diazoacetate (ester) |
| See F027 | 93–72–1 | Silvex (2,4,5-TP) |
| U206 | 18883–66–4 | Streptozotocin |
| U103 | 77–78–1 | Sulfuric acid, dimethyl ester |
| U189 | 1314–80–3 | Sulfur phosphide (R) |
| See F027 | 93–76–5 | 2,4,5-T |
| U207 | 95–94–3 | 1,2,4,5-Tetrachlorobenzene |
| U208 | 630–20–6 | 1,1,1,2-Tetrachloroethane |
| U209 | 79–34–5 | 1,1,2,2-Tetrachloroethane |
| U210 | 127–18–4 | Tetrachloroethylene |
| See F027 | 58–90–2 | 2,3,4,6-Tetrachlorophenol |
| U213 | 109–99–9 | Tetrahydrofuran (I) |
| U214 | 563–68–8 | Thallium(I) acetate |
| U215 | 6533–73–9 | Thallium(I) carbonate |
| U216 | 7791–12–0 | Thallium(I) chloride |
| U216 | 7791–12–0 | thallium chloride TlCl |
| U217 | 10102–45–1 | Thallium(I) nitrate |
| U218 | 62–55–5 | Thioacetamide |
| U410 | 59669–26–0 | Thiodicarb. |
| U153 | 74–93–1 | Thiomethanol (I,T) |
| U244 | 137–26–8 | Thioperoxydicarbonic diamide [(H$_2$ N)C(S)]$_2$ S$_2$, tetramethyl- |
| U409 | 23564–05–8 | Thiophanate-methyl. |
| U219 | 62–56–6 | Thiourea |
| U244 | 137–26–8 | Thiram |
| U220 | 108–88–3 | Toluene |
| U221 | 25376–45–8 | Toluenediamine |
| U223 | 26471–62–5 | Toluene diisocyanate (R,T) |
| U328 | 95–53–4 | o-Toluidine |
| U353 | 106–49–0 | p-Toluidine |
| U222 | 636–21–5 | o-Toluidine hydrochloride |
| U389 | 2303–17–5 | Triallate. |
| U011 | 61–82–5 | 1H-1,2,4-Triazol-3-amine |
| U226 | 71–55–6 | 1,1,1-Trichloroethane |
| U227 | 79–00–5 | 1,1,2-Trichloroethane |
| U228 | 79–01–6 | Trichloroethylene |
| U121 | 75–69–4 | Trichloromonofluoromethane |
| See F027 | 95–95–4 | 2,4,5-Trichlorophenol |
| See F027 | 88–06–2 | 2,4,6-Trichlorophenol |
| U404 | 121–44–8 | Triethylamine. |
| U234 | 99–35–4 | 1,3,5-Trinitrobenzene (R,T) |
| U182 | 123–63–7 | 1,3,5-Trioxane, 2,4,6-trimethyl- |
| U235 | 126–72–7 | Tris(2,3-dibromopropyl) phosphate |
| U236 | 72–57–1 | Trypan blue |
| U237 | 66–75–1 | Uracil mustard |
| U176 | 759–73–9 | Urea, N-ethyl-N-nitroso- |
| U177 | 684–93–5 | Urea, N-methyl-N-nitroso- |
| U043 | 75–01–4 | Vinyl chloride |
| U248 | [1] 81–81–2 | Warfarin, & salts, when present at concentrations of 0.3% or less |
| U239 | 1330–20–7 | Xylene (I) |
| U200 | 50–55–5 | Yohimban-16-carboxylic acid, 11,17-dimethoxy-18-[(3,4,5-trimethoxybenzoyl)oxy]-, methyl ester, (3beta,16beta,17alpha,18beta,20alpha)- |
| U249 | 1314–84–7 | Zinc phosphide Zn$_3$ P$_2$, when present at concentrations of 10% or less |
| U001 | 75–07–0 | Acetaldehyde (I) |
| U001 | 75–07–0 | Ethanal (I) |
| U002 | 67–64–1 | Acetone (I) |
| U002 | 67–64–1 | 2-Propanone (I) |
| U003 | 75–05–8 | Acetonitrile (I,T) |
| U004 | 98–86–2 | Acetophenone |
| U004 | 98–86–2 | Ethanone, 1-phenyl- |
| U005 | 53–96–3 | Acetamide, -9H-fluoren-2-yl- |
| U005 | 53–96–3 | 2-Acetylaminofluorene |
| U006 | 75–36–5 | Acetyl chloride (C,R,T) |
| U007 | 79–06–1 | Acrylamide |
| U007 | 79–06–1 | 2-Propenamide |

| Haz-ardous waste No. | Chemical abstracts No. | Substance |
|---|---|---|
| U008 | 79–10–7 | Acrylic acid (I) |
| U008 | 79–10–7 | 2-Propenoic acid (I) |
| U009 | 107–13–1 | Acrylonitrile |
| U009 | 107–13–1 | 2-Propenenitrile |
| U010 | 50–07–7 | Azirino[2',3':3,4]pyrrolo[1,2-a]indole-4,7-dione, 6-amino-8-[[(aminocarbonyl)oxy]methyl]-1,1a,2,8,8a,8b-hexahydro-8a-methoxy-5-methyl-, [1aS-(1aalpha, 8beta,8aalpha,8balpha)]- |
| U010 | 50–07–7 | Mitomycin C |
| U011 | 61–82–5 | Amitrole |
| U011 | 61–82–5 | 1H-1,2,4-Triazol-3-amine |
| U012 | 62–53–3 | Aniline (I,T) |
| U012 | 62–53–3 | Benzenamine (I,T) |
| U014 | 492–80–8 | Auramine |
| U014 | 492–80–8 | Benzenamine, 4,4'-carbonimidoylbis[N,N-dimethyl- |
| U015 | 115–02–6 | Azaserine |
| U015 | 115–02–6 | L-Serine, diazoacetate (ester) |
| U016 | 225–51–4 | Benz[c]acridine |
| U017 | 98–87–3 | Benzal chloride |
| U017 | 98–87–3 | Benzene, (dichloromethyl)- |
| U018 | 56–55–3 | Benz[a]anthracene |
| U019 | 71–43–2 | Benzene (I,T) |
| U020 | 98–09–9 | Benzenesulfonic acid chloride (C,R) |
| U020 | 98–09–9 | Benzenesulfonyl chloride (C,R) |
| U021 | 92–87–5 | Benzidine |
| U021 | 92–87–5 | [1,1'-Biphenyl]-4,4'-diamine |
| U022 | 50–32–8 | Benzo[a]pyrene |
| U023 | 98–07–7 | Benzene, (trichloromethyl)- |
| U023 | 98–07–7 | Benzotrichloride (C,R,T) |
| U024 | 111–91–1 | Dichloromethoxy ethane |
| U024 | 111–91–1 | Ethane, 1,1'-[methylenebis(oxy)]bis[2-chloro- |
| U025 | 111–44–4 | Dichloroethyl ether |
| U025 | 111–44–4 | Ethane, 1,1'-oxybis[2-chloro- |
| U026 | 494–03–1 | Chlornaphazin |
| U026 | 494–03–1 | Naphthalenamine, N,N'-bis(2-chloroethyl)- |
| U027 | 108–60–1 | Dichloroisopropyl ether |
| U027 | 108–60–1 | Propane, 2,2'-oxybis[2-chloro- |
| U028 | 117–81–7 | 1,2-Benzenedicarboxylic acid, bis(2-ethylhexyl) ester |
| U028 | 117–81–7 | Diethylhexyl phthalate |
| U029 | 74–83–9 | Methane, bromo- |
| U029 | 74–83–9 | Methyl bromide |
| U030 | 101–55–3 | Benzene, 1-bromo-4-phenoxy- |
| U030 | 101–55–3 | 4-Bromophenyl phenyl ether |
| U031 | 71–36–3 | 1-Butanol (I) |
| U031 | 71–36–3 | n-Butyl alcohol (I) |
| U032 | 13765–19–0 | Calcium chromate |
| U032 | 13765–19–0 | Chromic acid H$_2$ CrO$_4$, calcium salt |
| U033 | 353–50–4 | Carbonic difluoride |
| U033 | 353–50–4 | Carbon oxyfluoride (R,T) |
| U034 | 75–87–6 | Acetaldehyde, trichloro- |
| U034 | 75–87–6 | Chloral |
| U035 | 305–03–3 | Benzenebutanoic acid, 4-[bis(2-chloroethyl)amino]- |
| U035 | 305–03–3 | Chlorambucil |
| U036 | 57–74–9 | Chlordane, alpha & gamma isomers |
| U036 | 57–74–9 | 4,7-Methano-1H-indene, 1,2,4,5,6,7,8,8-octachloro-2,3,3a,4,7,7a-hexahydro- |
| U037 | 108–90–7 | Benzene, chloro- |
| U037 | 108–90–7 | Chlorobenzene |
| U038 | 510–15–6 | Benzeneacetic acid, 4-chloro-alpha-(4-chlorophenyl)-alpha-hydroxy-, ethyl ester |
| U038 | 510–15–6 | Chlorobenzilate |
| U039 | 59–50–7 | p-Chloro-m-cresol |
| U039 | 59–50–7 | Phenol, 4-chloro-3-methyl- |
| U041 | 106–89–8 | Epichlorohydrin |
| U041 | 106–89–8 | Oxirane, (chloromethyl)- |
| U042 | 110–75–8 | 2-Chloroethyl vinyl ether |
| U042 | 110–75–8 | Ethene, (2-chloroethoxy)- |
| U043 | 75–01–4 | Ethene, chloro- |
| U043 | 75–01–4 | Vinyl chloride |
| U044 | 67–66–3 | Chloroform |
| U044 | 67–66–3 | Methane, trichloro- |
| U045 | 74–87–3 | Methane, chloro- (I,T) |
| U045 | 74–87–3 | Methyl chloride (I,T) |
| U046 | 107–30–2 | Chloromethyl methyl ether |
| U046 | 107–30–2 | Methane, chloromethoxy- |
| U047 | 91–58–7 | beta-Chloronaphthalene |

| Haz-ardous waste No. | Chemical ab-stracts No. | Substance |
|---|---|---|
| U047 | 91–58–7 | Naphthalene, 2-chloro- |
| U048 | 95–57–8 | o-Chlorophenol |
| U048 | 95–57–8 | Phenol, 2-chloro- |
| U049 | 3165–93–3 | Benzenamine, 4-chloro-2-methyl-, hydrochloride |
| U049 | 3165–93–3 | 4-Chloro-o-toluidine, hydrochloride |
| U050 | 218–01–9 | Chrysene |
| U051 | ..................... | Creosote |
| U052 | 1319–77–3 | Cresol (Cresylic acid) |
| U052 | 1319–77–3 | Phenol, methyl- |
| U053 | 4170–30–3 | 2-Butenal |
| U053 | 4170–30–3 | Crotonaldehyde |
| U055 | 98–82–8 | Benzene, (1-methylethyl)-(I) |
| U055 | 98–82–8 | Cumene (I) |
| U056 | 110–82–7 | Benzene, hexahydro-(I) |
| U056 | 110–82–7 | Cyclohexane (I) |
| U057 | 108–94–1 | Cyclohexanone (I) |
| U058 | 50–18–0 | Cyclophosphamide |
| U058 | 50–18–0 | 2H-1,3,2-Oxazaphosphorin-2-amine, N,N-bis(2-chloroethyl)tetrahydro-, 2-oxide |
| U059 | 20830–81–3 | Daunomycin |
| U059 | 20830–81–3 | 5,12-Naphthacenedione, 8-acetyl-10-[(3-amino-2,3,6-trideoxy)-alpha-L-lyxo-hexopyranosyl)oxy]-7,8,9,10-tetrahydro-6,8,11-trihydroxy-1-methoxy-, (8S-cis)- |
| U060 | 72–54–8 | Benzene, 1,1′-(2,2-dichloroethylidene)bis[4-chloro- |
| U060 | 72–54–8 | DDD |
| U061 | 50–29–3 | Benzene, 1,1′-(2,2,2-trichloroethylidene)bis[4-chloro- |
| U061 | 50–29–3 | DDT |
| U062 | 2303–16–4 | Carbamothioic acid, bis(1-methylethyl)-, S-(2,3-di chloro-2-propenyl) ester |
| U062 | 2303–16–4 | Diallate |
| U063 | 53–70–3 | Dibenz[a,h]anthracene |
| U064 | 189–55–9 | Benzo[rst]pentaphene |
| U064 | 189–55–9 | Dibenzo[a,i]pyrene |
| U066 | 96–12–8 | 1,2-Dibromo-3-chloropropane |
| U066 | 96–12–8 | Propane, 1,2-dibromo-3-chloro- |
| U067 | 106–93–4 | Ethane, 1,2-dibromo- |
| U067 | 106–93–4 | Ethylene dibromide |
| U068 | 74–95–3 | Methane, dibromo- |
| U068 | 74–95–3 | Methylene bromide |
| U069 | 84–74–2 | 1,2-Benzenedicarboxylic acid, dibutyl ester |
| U069 | 84–74–2 | Dibutyl phthalate |
| U070 | 95–50–1 | Benzene, 1,2-dichloro- |
| U070 | 95–50–1 | o-Dichlorobenzene |
| U071 | 541–73–1 | Benzene, 1,3-dichloro- |
| U071 | 541–73–1 | m-Dichlorobenzene |
| U072 | 106–46–7 | Benzene, 1,4-dichloro- |
| U072 | 106–46–7 | p-Dichlorobenzene |
| U073 | 91–94–1 | [1,1′-Biphenyl]-4,4′-diamine, 3,3′-dichloro- |
| U073 | 91–94–1 | 3,3′-Dichlorobenzidine |
| U074 | 764–41–0 | 2-Butene, 1,4-dichloro-(I,T) |
| U074 | 764–41–0 | 1,4-Dichloro-2-butene (I,T) |
| U075 | 75–71–8 | Dichlorodifluoromethane |
| U075 | 75–71–8 | Methane, dichlorodifluoro- |
| U076 | 75–34–3 | Ethane, 1,1-dichloro- |
| U076 | 75–34–3 | Ethylidene dichloride |
| U077 | 107–06–2 | Ethane, 1,2-dichloro- |
| U077 | 107–06–2 | Ethylene dichloride |
| U078 | 75–35–4 | 1,1-Dichloroethylene |
| U078 | 75–35–4 | Ethene, 1,1-dichloro- |
| U079 | 156–60–5 | 1,2-Dichloroethylene |
| U079 | 156–60–5 | Ethene, 1,2-dichloro-, (E)- |
| U080 | 75–09–2 | Methane, dichloro- |
| U080 | 75–09–2 | Methylene chloride |
| U081 | 120–83–2 | 2,4-Dichlorophenol |
| U081 | 120–83–2 | Phenol, 2,4-dichloro- |
| U082 | 87–65–0 | 2,6-Dichlorophenol |
| U082 | 87–65–0 | Phenol, 2,6-dichloro- |
| U083 | 78–87–5 | Propane, 1,2-dichloro- |
| U083 | 78–87–5 | Propylene dichloride |
| U084 | 542–75–6 | 1,3-Dichloropropene |
| U084 | 542–75–6 | 1-Propene, 1,3-dichloro- |
| U085 | 1464–53–5 | 2,2′-Bioxirane |
| U085 | 1464–53–5 | 1,2:3,4-Diepoxybutane (I,T) |
| U086 | 1615–80–1 | N,N′-Diethylhydrazine |
| U086 | 1615–80–1 | Hydrazine, 1,2-diethyl- |

94

| Haz-ardous waste No. | Chemical abstracts No. | Substance |
|---|---|---|
| U087 | 3288–58–2 | O,O-Diethyl S-methyl dithiophosphate |
| U087 | 3288–58–2 | Phosphorodithioic acid, O,O-diethyl S-methyl ester |
| U088 | 84–66–2 | 1,2-Benzenedicarboxylic acid, diethyl ester |
| U088 | 84–66–2 | Diethyl phthalate |
| U089 | 56–53–1 | Diethylstilbesterol |
| U089 | 56–53–1 | Phenol, 4,4′-(1,2-diethyl-1,2-ethenediyl)bis-, (E)- |
| U090 | 94–58–6 | 1,3-Benzodioxole, 5-propyl- |
| U090 | 94–58–6 | Dihydrosafrole |
| U091 | 119–90–4 | [1,1′-Biphenyl]-4,4′-diamine, 3,3′-dimethoxy- |
| U091 | 119–90–4 | 3,3′-Dimethoxybenzidine |
| U092 | 124–40–3 | Dimethylamine (I) |
| U092 | 124–40–3 | Methanamine, -methyl-(I) |
| U093 | 60–11–7 | Benzenamine, N,N-dimethyl-4-(phenylazo)- |
| U093 | 60–11–7 | p-Dimethylaminoazobenzene |
| U094 | 57–97–6 | Benz[a]anthracene, 7,12-dimethyl- |
| U094 | 57–97–6 | 7,12-Dimethylbenz[a]anthracene |
| U095 | 119–93–7 | [1,1′-Biphenyl]-4,4′-diamine, 3,3′-dimethyl- |
| U095 | 119–93–7 | 3,3′-Dimethylbenzidine |
| U096 | 80–15–9 | alpha,alpha-Dimethylbenzylhydroperoxide (R) |
| U096 | 80–15–9 | Hydroperoxide, 1-methyl-1-phenylethyl-(R) |
| U097 | 79–44–7 | Carbamic chloride, dimethyl- |
| U097 | 79–44–7 | Dimethylcarbamoyl chloride |
| U098 | 57–14–7 | 1,1-Dimethylhydrazine |
| U098 | 57–14–7 | Hydrazine, 1,1-dimethyl- |
| U099 | 540–73–8 | 1,2-Dimethylhydrazine |
| U099 | 540–73–8 | Hydrazine, 1,2-dimethyl- |
| U101 | 105–67–9 | 2,4-Dimethylphenol |
| U101 | 105–67–9 | Phenol, 2,4-dimethyl- |
| U102 | 131–11–3 | 1,2-Benzenedicarboxylic acid, dimethyl ester |
| U102 | 131–11–3 | Dimethyl phthalate |
| U103 | 77–78–1 | Dimethyl sulfate |
| U103 | 77–78–1 | Sulfuric acid, dimethyl ester |
| U105 | 121–14–2 | Benzene, 1-methyl-2,4-dinitro- |
| U105 | 121–14–2 | 2,4-Dinitrotoluene |
| U106 | 606–20–2 | Benzene, 2-methyl-1,3-dinitro- |
| U106 | 606–20–2 | 2,6-Dinitrotoluene |
| U107 | 117–84–0 | 1,2-Benzenedicarboxylic acid, dioctyl ester |
| U107 | 117–84–0 | Di-n-octyl phthalate |
| U108 | 123–91–1 | 1,4-Diethyleneoxide |
| U108 | 123–91–1 | 1,4-Dioxane |
| U109 | 122–66–7 | 1,2-Diphenylhydrazine |
| U109 | 122–66–7 | Hydrazine, 1,2-diphenyl- |
| U110 | 142–84–7 | Dipropylamine (I) |
| U110 | 142–84–7 | 1-Propanamine, N-propyl-(I) |
| U111 | 621–64–7 | Di-n-propylnitrosamine |
| U111 | 621–64–7 | 1-Propanamine, N-nitroso-N-propyl- |
| U112 | 141–78–6 | Acetic acid ethyl ester (I) |
| U112 | 141–78–6 | Ethyl acetate (I) |
| U113 | 140–88–5 | 2-Propenoic acid, ethyl ester (I) |
| U113 | 140–88–5 | Ethyl acrylate (I) |
| U114 | [1]111–54–6 | Carbamodithioic acid, 1,2-ethanediylbis-, salts & esters |
| U114 | [1]111–54–6 | Ethylenebisdithiocarbamic acid, salts & esters |
| U115 | 75–21–8 | Ethylene oxide (I,T) |
| U115 | 75–21–8 | Oxirane (I,T) |
| U116 | 96–45–7 | Ethylenethiourea |
| U116 | 96–45–7 | 2-Imidazolidinethione |
| U117 | 60–29–7 | Ethane, 1,1′-oxybis-(I) |
| U117 | 60–29–7 | Ethyl ether (I) |
| U118 | 97–63–2 | Ethyl methacrylate |
| U118 | 97–63–2 | 2-Propenoic acid, 2-methyl-, ethyl ester |
| U119 | 62–50–0 | Ethyl methanesulfonate |
| U119 | 62–50–0 | Methanesulfonic acid, ethyl ester |
| U120 | 206–44–0 | Fluoranthene |
| U121 | 75–69–4 | Methane, trichlorofluoro- |
| U121 | 75–69–4 | Trichloromonofluoromethane |
| U122 | 50–00–0 | Formaldehyde |
| U123 | 64–18–6 | Formic acid (C,T) |
| U124 | 110–00–9 | Furan (I) |
| U124 | 110–00–9 | Furfuran (I) |
| U125 | 98–01–1 | 2-Furancarboxaldehyde (I) |
| U125 | 98–01–1 | Furfural (I) |
| U126 | 765–34–4 | Glycidylaldehyde |

95

| Hazardous waste No. | Chemical abstracts No. | Substance |
|---|---|---|
| U126 | 765–34–4 | Oxiranecarboxyaldehyde |
| U127 | 118–74–1 | Benzene, hexachloro- |
| U127 | 118–74–1 | Hexachlorobenzene |
| U128 | 87–68–3 | 1,3-Butadiene, 1,1,2,3,4,4-hexachloro- |
| U128 | 87–68–3 | Hexachlorobutadiene |
| U129 | 58–89–9 | Cyclohexane, 1,2,3,4,5,6-hexachloro-, (1alpha,2alpha,3beta,4alpha,5alpha,6beta)- |
| U129 | 58–89–9 | Lindane |
| U130 | 77–47–4 | 1,3-Cyclopentadiene, 1,2,3,4,5,5-hexachloro- |
| U130 | 77–47–4 | Hexachlorocyclopentadiene |
| U131 | 67–72–1 | Ethane, hexachloro- |
| U131 | 67–72–1 | Hexachloroethane |
| U132 | 70–30–4 | Hexachlorophene |
| U132 | 70–30–4 | Phenol, 2,2′-methylenebis[3,4,6-trichloro- |
| U133 | 302–01–2 | Hydrazine (R,T) |
| U134 | 7664–39–3 | Hydrofluoric acid (C,T) |
| U134 | 7664–39–3 | Hydrogen fluoride (C,T) |
| U135 | 7783–06–4 | Hydrogen sulfide |
| U135 | 7783–06–4 | Hydrogen sulfide $H_2S$ |
| U136 | 75–60–5 | Arsinic acid, dimethyl- |
| U136 | 75–60–5 | Cacodylic acid |
| U137 | 193–39–5 | Indeno[1,2,3-cd]pyrene |
| U138 | 74–88–4 | Methane, iodo- |
| U138 | 74–88–4 | Methyl iodide |
| U140 | 78–83–1 | Isobutyl alcohol (I,T) |
| U140 | 78–83–1 | 1-Propanol, 2-methyl- (I,T) |
| U141 | 120–58–1 | 1,3-Benzodioxole, 5-(1-propenyl)- |
| U141 | 120–58–1 | Isosafrole |
| U142 | 143–50–0 | Kepone |
| U142 | 143–50–0 | 1,3,4-Metheno-2H-cyclobuta[cd]pentalen-2-one, 1,1a,3,3a,4,5,5,5a,5b,6-decachlorooctahydro- |
| U143 | 303–34–4 | 2-Butenoic acid, 2-methyl-, 7-[[2,3-dihydroxy-2-(1-methoxyethyl)-3-methyl-1-oxobutoxy]methyl]-2,3,5,7a-tetrahydro-1H-pyrrolizin-1-yl ester, [1S-[1alpha(Z),7(2S*,3R*),7aalpha]]- |
| U143 | 303–34–4 | Lasiocarpine |
| U144 | 301–04–2 | Acetic acid, lead(2+) salt |
| U144 | 301–04–2 | Lead acetate |
| U145 | 7446–27–7 | Lead phosphate |
| U145 | 7446–27–7 | Phosphoric acid, lead(2+) salt (2:3) |
| U146 | 1335–32–6 | Lead, bis(acetato-O)tetrahydroxytri- |
| U146 | 1335–32–6 | Lead subacetate |
| U147 | 108–31–6 | 2,5-Furandione |
| U147 | 108–31–6 | Maleic anhydride |
| U148 | 123–33–1 | Maleic hydrazide |
| U148 | 123–33–1 | 3,6-Pyridazinedione, 1,2-dihydro- |
| U149 | 109–77–3 | Malononitrile |
| U149 | 109–77–3 | Propanedinitrile |
| U150 | 148–82–3 | Melphalan |
| U150 | 148–82–3 | L-Phenylalanine, 4-[bis(2-chloroethyl)amino]- |
| U151 | 7439–97–6 | Mercury |
| U152 | 126–98–7 | Methacrylonitrile (I,T) |
| U152 | 126–98–7 | 2-Propenenitrile, 2-methyl- (I,T) |
| U153 | 74–93–1 | Methanethiol (I,T) |
| U153 | 74–93–1 | Thiomethanol (I,T) |
| U154 | 67–56–1 | Methanol (I) |
| U154 | 67–56–1 | Methyl alcohol (I) |
| U155 | 91–80–5 | 1,2-Ethanediamine, N,N-dimethyl-N′-2-pyridinyl-N′-(2-thienylmethyl)- |
| U155 | 91–80–5 | Methapyrilene |
| U156 | 79–22–1 | Carbonochloridic acid, methyl ester (I,T) |
| U156 | 79–22–1 | Methyl chlorocarbonate (I,T) |
| U157 | 56–49–5 | Benz[j]aceanthrylene, 1,2-dihydro-3-methyl- |
| U157 | 56–49–5 | 3-Methylcholanthrene |
| U158 | 101–14–4 | Benzenamine, 4,4′-methylenebis[2-chloro- |
| U158 | 101–14–4 | 4,4′-Methylenebis(2-chloroaniline) |
| U159 | 78–93–3 | 2-Butanone (I,T) |
| U159 | 78–93–3 | Methyl ethyl ketone (MEK) (I,T) |
| U160 | 1338–23–4 | 2-Butanone, peroxide (R,T) |
| U160 | 1338–23–4 | Methyl ethyl ketone peroxide (R,T) |
| U161 | 108–10–1 | Methyl isobutyl ketone (I) |
| U161 | 108–10–1 | 4-Methyl-2-pentanone (I) |
| U161 | .108–10–1 | Pentanol, 4-methyl- |
| U162 | 80–62–6 | Methyl methacrylate (I,T) |
| U162 | 80–62–6 | 2-Propenoic acid, 2-methyl-, methyl ester (I,T) |
| U163 | 70–25–7 | Guanidine, -methyl-N′-nitro-N-nitroso- |
| U163 | 70–25–7 | MNNG |

| Haz-ardous waste No. | Chemical ab-stracts No. | Substance |
|---|---|---|
| U164 | 56–04–2 | Methylthiouracil |
| U164 | 56–04–2 | 4(1H)-Pyrimidinone, 2,3-dihydro-6-methyl-2-thioxo- |
| U165 | 91–20–3 | Naphthalene |
| U166 | 130–15–4 | 1,4-Naphthalenedione |
| U166 | 130–15–4 | 1,4-Naphthoquinone |
| U167 | 134–32–7 | 1-Naphthalenamine |
| U167 | 134–32–7 | alpha-Naphthylamine |
| U168 | 91–59–8 | 2-Naphthalenamine |
| U168 | 91–59–8 | beta-Naphthylamine |
| U169 | 98–95–3 | Benzene, nitro- |
| U169 | 98–95–3 | Nitrobenzene (I,T) |
| U170 | 100–02–7 | p-Nitrophenol |
| U170 | 100–02–7 | Phenol, 4-nitro- |
| U171 | 79–46–9 | 2-Nitropropane (I,T) |
| U171 | 79–46–9 | Propane, 2-nitro- (I,T) |
| U172 | 924–16–3 | 1-Butanamine, N-butyl-N-nitroso- |
| U172 | 924–16–3 | N-Nitrosodi-n-butylamine |
| U173 | 1116–54–7 | Ethanol, 2,2′-(nitrosoimino)bis- |
| U173 | 1116–54–7 | N-Nitrosodiethanolamine |
| U174 | 55–18–5 | Ethanamine, -ethyl-N-nitroso- |
| U174 | 55–18–5 | N-Nitrosodiethylamine |
| U176 | 759–73–9 | N-Nitroso-N-ethylurea |
| U176 | 759–73–9 | Urea, N-ethyl-N-nitroso- |
| U177 | 684–93–5 | N-Nitroso-N-methylurea |
| U177 | 684–93–5 | Urea, N-methyl-N-nitroso- |
| U178 | 615–53–2 | Carbamic acid, methylnitroso-, ethyl ester |
| U178 | 615–53–2 | N-Nitroso-N-methylurethane |
| U179 | 100–75–4 | N-Nitrosopiperidine |
| U179 | 100–75–4 | Piperidine, 1-nitroso- |
| U180 | 930–55–2 | N-Nitrosopyrrolidine |
| U180 | 930–55–2 | Pyrrolidine, 1-nitroso- |
| U181 | 99–55–8 | Benzenamine, 2-methyl-5-nitro- |
| U181 | 99–55–8 | 5-Nitro-o-toluidine |
| U182 | 123–63–7 | 1,3,5-Trioxane, 2,4,6-trimethyl- |
| U182 | 123–63–7 | Paraldehyde |
| U183 | 608–93–5 | Benzene, pentachloro- |
| U183 | 608–93–5 | Pentachlorobenzene |
| U184 | 76–01–7 | Ethane, pentachloro- |
| U184 | 76–01–7 | Pentachloroethane |
| U185 | 82–68–8 | Benzene, pentachloronitro- |
| U185 | 82–68–8 | Pentachloronitrobenzene (PCNB) |
| U186 | 504–60–9 | 1-Methylbutadiene (I) |
| U186 | 504–60–9 | 1,3-Pentadiene (I) |
| U187 | 62–44–2 | Acetamide, -(4-ethoxyphenyl)- |
| U187 | 62–44–2 | Phenacetin |
| U188 | 108–95–2 | Phenol |
| U189 | 1314–80–3 | Phosphorus sulfide (R) |
| U189 | 1314–80–3 | Sulfur phosphide (R) |
| U190 | 85–44–9 | 1,3-Isobenzofurandione |
| U190 | 85–44–9 | Phthalic anhydride |
| U191 | 109–06–8 | 2-Picoline |
| U191 | 109–06–8 | Pyridine, 2-methyl- |
| U192 | 23950–58–5 | Benzamide, 3,5-dichloro-N-(1,1-dimethyl-2-propynyl)- |
| U192 | 23950–58–5 | Pronamide |
| U193 | 1120–71–4 | 1,2-Oxathiolane, 2,2-dioxide |
| U193 | 1120–71–4 | 1,3-Propane sultone |
| U194 | 107–10–8 | 1-Propanamine (I,T) |
| U194 | 107–10–8 | n-Propylamine (I,T) |
| U196 | 110–86–1 | Pyridine |
| U197 | 106–51–4 | p-Benzoquinone |
| U197 | 106–51–4 | 2,5-Cyclohexadiene-1,4-dione |
| U200 | 50–55–5 | Reserpine |
| U200 | 50–55–5 | Yohimban-16-carboxylic acid, 11,17-dimethoxy-18-[(3,4,5-trimethoxybenzoyl)oxy]-, methyl ester,(3beta,16beta,17alpha,18beta,20alpha)- |
| U201 | 108–46–3 | 1,3-Benzenediol |
| U201 | 108–46–3 | Resorcinol |
| U202 | 181–07–2 | 1,2-Benzisothiazol-3(2H)-one, 1,1-dioxide, & salts |
| U202 | 181–07–2 | Saccharin, & salts |
| U203 | 94–59–7 | 1,3-Benzodioxole, 5-(2-propenyl)- |
| U203 | 94–59–7 | Safrole |
| U204 | 7783–00–8 | Selenious acid |
| U204 | 7783–00–8 | Selenium dioxide |

| Haz-ardous waste No. | Chemical abstracts No. | Substance |
|---|---|---|
| U205 | 7488–56–4 | Selenium sulfide |
| U205 | 7488–56–4 | Selenium sulfide SeS$_2$ (R,T) |
| U206 | 18883–66–4 | Glucopyranose, 2-deoxy-2-(3-methyl-3-nitrosoureido)-, D- |
| U206 | 18883–66–4 | D-Glucose, 2-deoxy-2-[[(methylnitrosoamino)-carbonyl]amino]- |
| U206 | 18883–66–4 | Streptozotocin |
| U207 | 95–94–3 | Benzene, 1,2,4,5-tetrachloro- |
| U207 | 95–94–3 | 1,2,4,5-Tetrachlorobenzene |
| U208 | 630–20–6 | Ethane, 1,1,1,2-tetrachloro- |
| U208 | 630–20–6 | 1,1,1,2-Tetrachloroethane |
| U209 | 79–34–5 | Ethane, 1,1,2,2-tetrachloro- |
| U209 | 79–34–5 | 1,1,2,2-Tetrachloroethane |
| U210 | 127–18–4 | Ethene, tetrachloro- |
| U210 | 127–18–4 | Tetrachloroethylene |
| U211 | 56–23–5 | Carbon tetrachloride |
| U211 | 56–23–5 | Methane, tetrachloro- |
| U213 | 109–99–9 | Furan, tetrahydro-(I) |
| U213 | 109–99–9 | Tetrahydrofuran (I) |
| U214 | 563–68–8 | Acetic acid, thallium(1+) salt |
| U214 | 563–68–8 | Thallium(I) acetate |
| U215 | 6533–73–9 | Carbonic acid, dithallium(1+) salt |
| U215 | 6533–73–9 | Thallium(I) carbonate |
| U216 | 7791–12–0 | Thallium(I) chloride |
| U216 | 7791–12–0 | Thallium chloride TlCl |
| U217 | 10102–45–1 | Nitric acid, thallium(1+) salt |
| U217 | 10102–45–1 | Thallium(I) nitrate |
| U218 | 62–55–5 | Ethanethioamide |
| U218 | 62–55–5 | Thioacetamide |
| U219 | 62–56–6 | Thiourea |
| U220 | 108–88–3 | Benzene, methyl- |
| U220 | 108–88–3 | Toluene |
| U221 | 25376–45–8 | Benzenediamine, ar-methyl- |
| U221 | 25376–45–8 | Toluenediamine |
| U222 | 636–21–5 | Benzenamine, 2-methyl-, hydrochloride |
| U222 | 636–21–5 | o-Toluidine hydrochloride |
| U223 | 26471–62–5 | Benzene, 1,3-diisocyanatomethyl- (R,T) |
| U223 | 26471–62–5 | Toluene diisocyanate (R,T) |
| U225 | 75–25–2 | Bromoform |
| U225 | 75–25–2 | Methane, tribromo- |
| U226 | 71–55–6 | Ethane, 1,1,1-trichloro- |
| U226 | 71–55–6 | Methyl chloroform |
| U226 | 71–55–6 | 1,1,1-Trichloroethane |
| U227 | 79–00–5 | Ethane, 1,1,2-trichloro- |
| U227 | 79–00–5 | 1,1,2-Trichloroethane |
| U228 | 79–01–6 | Ethene, trichloro- |
| U228 | 79–01–6 | Trichloroethylene |
| U234 | 99–35–4 | Benzene, 1,3,5-trinitro- |
| U234 | 99–35–4 | 1,3,5-Trinitrobenzene (R,T) |
| U235 | 126–72–7 | 1-Propanol, 2,3-dibromo-, phosphate (3:1) |
| U235 | 126–72–7 | Tris(2,3-dibromopropyl) phosphate |
| U236 | 72–57–1 | 2,7-Naphthalenedisulfonic acid, 3,3′-[(3,3′-dimethyl[1,1′-biphenyl]-4,4′-diyl)bis(azo)bis[5-amino-4-hy-droxy]-, tetrasodium salt |
| U236 | 72–57–1 | Trypan blue |
| U237 | 66–75–1 | 2,4-(1H,3H)-Pyrimidinedione, 5-[bis(2-chloroethyl)amino]- |
| U237 | 66–75–1 | Uracil mustard |
| U238 | 51–79–6 | Carbamic acid, ethyl ester |
| U238 | 51–79–6 | Ethyl carbamate (urethane) |
| U239 | 1330–20–7 | Benzene, dimethyl- (I,T) |
| U239 | 1330–20–7 | Xylene (I) |
| U240 | [1] 94–75–7 | Acetic acid, (2,4-dichlorophenoxy)-, salts & esters |
| U240 | [1] 94–75–7 | 2,4-D, salts & esters |
| U243 | 1888–71–7 | Hexachloropropene |
| U243 | 1888–71–7 | 1-Propene, 1,1,2,3,3,3-hexachloro- |
| U244 | 137–26–8 | Thioperoxydicarbonic diamide [(H$_2$N)C(S)]$_2$ S$_2$, tetramethyl- |
| U244 | 137–26–8 | Thiram |
| U246 | 506–68–3 | Cyanogen bromide (CN)Br |
| U247 | 72–43–5 | Benzene, 1,1′-(2,2,2-trichloroethylidene)bis[4- methoxy- |
| U247 | 72–43–5 | Methoxychlor |
| U248 | [1] 81–81–2 | 2H-1-Benzopyran-2-one, 4-hydroxy-3-(3-oxo-1-phenyl-butyl)-, & salts, when present at concentrations of 0.3% or less |
| U248 | [1] 81–81–2 | Warfarin, & salts, when present at concentrations of 0.3% or less |
| U249 | 1314–84–7 | Zinc phosphide Zn$_3$ P$_2$, when present at concentrations of 10% or less |
| U271 | 17804–35–2 | Benomyl |

| Haz- ardous waste No. | Chemical ab- stracts No. | Substance |
|---|---|---|
| U271 | 17804–35–2 | Carbamic acid, [1-[(butylamino)carbonyl]-1H-benzimidazol-2-yl]-, methyl ester |
| U278 | 22781–23–3 | Bendiocarb |
| U278 | 22781–23–3 | 1,3-Benzodioxol-4-ol, 2,2-dimethyl-, methyl carbamate |
| U279 | 63–25–2 | Carbaryl |
| U279 | 63–25–2 | 1-Naphthalenol, methylcarbamate |
| U280 | 101–27–9 | Barban |
| U280 | 101–27–9 | Carbamic acid, (3-chlorophenyl)-, 4-chloro-2-butynyl ester |
| U328 | 95–53–4 | Benzenamine, 2-methyl- |
| U328 | 95–53–4 | o-Toluidine |
| U353 | 106–49–0 | Benzenamine, 4-methyl- |
| U353 | 106–49–0 | p-Toluidine |
| U359 | 110–80–5 | Ethanol, 2-ethoxy- |
| U359 | 110–80–5 | Ethylene glycol monoethyl ether |
| U364 | 22961–82–6 | Bendiocarb phenol |
| U364 | 22961–82–6 | 1,3-Benzodioxol-4-ol, 2,2-dimethyl-, |
| U367 | 1563–38–8 | 7-Benzofuranol, 2,3-dihydro-2,2-dimethyl- |
| U367 | 1563–38–8 | Carbofuran phenol |
| U372 | 10605–21–7 | Carbamic acid, 1H-benzimidazol-2-yl, methyl ester |
| U372 | 10605–21–7 | Carbendazim |
| U373 | 122–42–9 | Carbamic acid, phenyl-, 1-methylethyl ester |
| U373 | 122–42–9 | Propham |
| U387 | 52888–80–9 | Carbamothioic acid, dipropyl-, S-(phenylmethyl) ester |
| U387 | 52888–80–9 | Prosulfocarb |
| U389 | 2303–17–5 | Carbamothioic acid, bis(1-methylethyl)-, S-(2,3,3-trichloro-2-propenyl) ester |
| U389 | 2303–17–5 | Triallate |
| U394 | 30558–43–1 | A2213 |
| U394 | 30558–43–1 | Ethanimidothioic acid, 2-(dimethylamino)-N-hydroxy-2-oxo-, methyl ester |
| U395 | 5952–26–1 | Diethylene glycol, dicarbamate |
| U395 | 5952–26–1 | Ethanol, 2,2'-oxybis-, dicarbamate |
| U404 | 121–44–8 | Ethanamine, N,N-diethyl- |
| U404 | 121–44–8 | Triethylamine |
| U409 | 23564–05–8 | Carbamic acid, [1,2-phenylenebis (iminocarbonothioyl)]bis-, dimethyl ester |
| U409 | 23564–05–8 | Thiophanate-methyl |
| U410 | 59669–26–0 | Ethanimidothioic acid, N,N'-[thiobis[(methylimino)carbonyloxy]]bis-, dimethyl ester |
| U410 | 59669–26–0 | Thiodicarb |
| U411 | 114–26–1 | Phenol, 2-(1-methylethoxy)-, methylcarbamate |
| U411 | 114–26–1 | Propoxur |
| See F027 | 93–76–5 | Acetic acid, (2,4,5-trichlorophenoxy)- |
| See F027 | 87–86–5 | Pentachlorophenol |
| See F027 | 87–86–5 | Phenol, pentachloro- |
| See F027 | 58–90–2 | Phenol, 2,3,4,6-tetrachloro- |
| See F027 | 95–95–4 | Phenol, 2,4,5-trichloro- |
| See F027 | 88–06–2 | Phenol, 2,4,6-trichloro- |
| See F027 | 93–72–1 | Propanoic acid, 2-(2,4,5-trichlorophenoxy)- |
| See F027 | 93–72–1 | Silvex (2,4,5-TP) |
| See F027 | 93–76–5 | 2,4,5-T |
| See F027 | 58–90–2 | 2,3,4,6-Tetrachlorophenol |
| See F027 | 95–95–4 | 2,4,5-Trichlorophenol |
| See F027 | 88–06–2 | 2,4,6-Trichlorophenol |

[1] CAS Number given for parent compound only.

[45 FR 78529, 78541, Nov. 25, 1980]

EDITORIAL NOTE: For FEDERAL REGISTER citations affecting §261.33, see the List of CFR Sections Affected, which appears in the Finding Aids section of the printed volume and on GPO Access.

99

## § 261.35  Deletion of certain hazardous waste codes following equipment cleaning and replacement.

(a) Wastes from wood preserving processes at plants that do not resume or initiate use of chlorophenolic preservatives will not meet the listing definition of F032 once the generator has met all of the requirements of paragraphs (b) and (c) of this section. These wastes may, however, continue to meet another hazardous waste listing description or may exhibit one or more of the hazardous waste characteristics.

(b) Generators must either clean or replace all process equipment that may have come into contact with chlorophenolic formulations or constituents thereof, including, but not limited to, treatment cylinders, sumps, tanks, piping systems, drip pads, fork lifts, and trams, in a manner that minimizes or eliminates the escape of hazardous waste or constituents, leachate, contaminated drippage, or hazardous waste decomposition products to the ground water, surface water, or atmosphere.

(1) Generators shall do one of the following:

(i) Prepare and follow an equipment cleaning plan and clean equipment in accordance with this section;

(ii) Prepare and follow an equipment replacement plan and replace equipment in accordance with this section; or

(iii) Document cleaning and replacement in accordance with this section, carried out after termination of use of chlorophenolic preservations.

(2) Cleaning Requirements.

(i) Prepare and sign a written equipment cleaning plan that describes:

(A) The equipment to be cleaned;

(B) How the equipment will be cleaned;

(C) The solvent to be used in cleaning;

(D) How solvent rinses will be tested; and

(E) How cleaning residues will be disposed.

(ii) Equipment must be cleaned as follows:

(A) Remove all visible residues from process equipment;

(B) Rinse process equipment with an appropriate solvent until dioxins and dibenzofurans are not detected in the final solvent rinse.

(iii) Analytical requirements.

(A) Rinses must be tested by using an appropriate method.

(B) "Not detected" means at or below the following lower method calibration limits (MCLs): The 2,3,7,8-TCDD-based MCL—0.01 parts per trillion (ppt), sample weight of 1000 g, IS spiking level of 1 ppt, final extraction volume of 10–50 µL. For other congeners—multiply the values by 1 for TCDF/PeCDD/PeCDF, by 2.5 for HxCDD/HxCDF/HpCDD/HpCDF, and by 5 for OCDD/OCDF.

(iv) The generator must manage all residues from the cleaning process as F032 waste.

(3) Replacement requirements.

(i) Prepare and sign a written equipment replacement plan that describes:

(A) The equipment to be replaced;

(B) How the equipment will be replaced; and

(C) How the equipment will be disposed.

(ii) The generator must manage the discarded equipment as F032 waste.

(4) Documentation requirements.

(i) Document that previous equipment cleaning and/or replacement was performed in accordance with this section and occurred after cessation of use of chlorophenolic preservatives.

(c) The generator must maintain the following records documenting the cleaning and replacement as part of the facility's operating record:

(1) The name and address of the facility;

(2) Formulations previously used and the date on which their use ceased in each process at the plant;

(3) Formulations currently used in each process at the plant;

(4) The equipment cleaning or replacement plan;

(5) The name and address of any persons who conducted the cleaning and replacement;

(6) The dates on which cleaning and replacement were accomplished;

(7) The dates of sampling and testing;

(8) A description of the sample handling and preparation techniques, including techniques used for extraction, containerization, preservation, and chain-of-custody of the samples;

(9) A description of the tests performed, the date the tests were performed, and the results of the tests;

(10) The name and model numbers of the instrument(s) used in performing the tests;

(11) QA/QC documentation; and

(12) The following statement signed by the generator or his authorized representative:

I certify under penalty of law that all process equipment required to be cleaned or replaced under 40 CFR 261.35 was cleaned or replaced as represented in the equipment cleaning and replacement plan and accompanying documentation. I am aware that there are significant penalties for providing false information, including the possibility of fine or imprisonment.

[55 FR 50482, Dec. 6, 1990, as amended at 56 FR 30195, July 1, 1991; 70 FR 34561, June 14, 2005]

## Subpart E—Exclusions/Exemptions

SOURCE: 71 FR 42948, July 28, 2006, unless otherwise noted.

### §261.38 Exclusion of comparable fuel and syngas fuel.

(a) *Specifications for excluded fuels.* Wastes that meet the specifications for comparable fuel or syngas fuel under paragraphs (a)(1) or (a)(2) of this section, respectively, and the other requirements of this section, are not solid wastes.

(1) *Comparable fuel specifications.*—(i) *Physical specifications.*—(A) *Heating value.* The heating value must exceed 5,000 Btu/lbs. (11,500 J/g).

(B) *Viscosity.* The viscosity must not exceed: 50 cS, as-fired.

(ii) *Constituent specifications.* For compounds listed in Table 1 to this section, the specification levels and, where non-detect is the specification, minimum required detection limits are: (*see* Table 1 of this section).

(2) *Synthesis gas fuel specifications.*— Synthesis gas fuel (*i.e.,* syngas fuel) that is generated from hazardous waste must:

(i) Have a minimum Btu value of 100 Btu/Scf;

(ii) Contain less than 1 ppmv of total halogen;

(iii) Contain less than 300 ppmv of total nitrogen other than diatomic nitrogen ($N_2$);

(iv) Contain less than 200 ppmv of hydrogen sulfide; and

(v) Contain less than 1 ppmv of each hazardous constituent in the target list of appendix VIII constituents of this part.

(3) *Blending to meet the specifications.* (i) Hazardous waste shall not be blended to meet the comparable fuel specification under paragraph (a)(1) of this section, except as provided by paragraph (a)(3)(ii) of this section:

(ii) *Blending to meet the viscosity specification.* A hazardous waste blended to meet the viscosity specification for comparable fuel shall:

(A) As generated and prior to any blending, manipulation, or processing, meet the constituent and heating value specifications of paragraphs (a)(1)(i)(A) and (a)(1)(ii) of this section;

(B) Be blended at a facility that is subject to the applicable requirements of parts 264, 265, or 267 or §262.34 of this chapter; and

(C) Not violate the dilution prohibition of paragraph (a)(6) of this section.

(4) *Treatment to meet the comparable fuel specifications.* (i) A hazardous waste may be treated to meet the specifications for comparable fuel set forth in paragraph (a)(1) of this section provided the treatment:

(A) Destroys or removes the constituents listed in the specification or raises the heating value by removing or destroying hazardous constituents or materials;

(B) Is performed at a facility that is subject to the applicable requirements of parts 264, 265, or 267, or §262.34 of this chapter; and

(C) Does not violate the dilution prohibition of paragraph (a)(6) of this section.

(ii) Residuals resulting from the treatment of a hazardous waste listed in subpart D of this part to generate a comparable fuel remain a hazardous waste.

(5) *Generation of a syngas fuel.* (i) A syngas fuel can be generated from the processing of hazardous wastes to meet the exclusion specifications of paragraph (a)(2) of this section provided the processing:

(A) Destroys or removes the constituents listed in the specification or raises the heating value by removing or destroying constituents or materials;

(B) Is performed at a facility that is subject to the applicable requirements of parts 264, 265, or 267, or § 262.34 of this chapter or is an exempt recycling unit pursuant to § 261.6(c); and

(C) Does not violate the dilution prohibition of paragraph (a)(6) of this section.

(ii) Residuals resulting from the treatment of a hazardous waste listed in subpart D of this part to generate a syngas fuel remain a hazardous waste.

(6) *Dilution prohibition.* No generator, transporter, handler, or owner or operator of a treatment, storage, or disposal facility shall in any way dilute a hazardous waste to meet the specifications of paragraphs (a)(1)(i)(A) or (a)(1)(ii) of this section for comparable fuel, or paragraph (a)(2) of this section for syngas.

(b) *Implementation.*—(1) *General.*—(i) Wastes that meet the specifications provided by paragraph (a) of this section for comparable fuel or syngas fuel are excluded from the definition of solid waste provided that the conditions under this section are met. For purposes of this section, such materials are called excluded fuel; the person claiming and qualifying for the exclusion is called the excluded fuel generator and the person burning the excluded fuel is called the excluded fuel burner.

(ii) The person who generates the excluded fuel must claim the exclusion by complying with the conditions of this section and keeping records necessary to document compliance with those conditions.

(2) *Notices.* (i) *Notices to State RCRA and CAA Directors in authorized States or regional RCRA and CAA Directors in unauthorized States.* (A) The generator must submit a one-time notice, except as provided by paragraph (b)(2)(i)(C) of this section, to the Regional or State RCRA and CAA Directors, in whose jurisdiction the exclusion is being claimed and where the excluded fuel will be burned, certifying compliance with the conditions of the exclusion and providing the following documentation:

(1) The name, address, and RCRA ID number of the person/facility claiming the exclusion;

(2) The applicable EPA Hazardous Waste Code(s) that would otherwise apply to the excluded fuel;

(3) The name and address of the units meeting the requirements of paragraphs (b)(3) and (c) of this section, that will burn the excluded fuel;

(4) An estimate of the average and maximum monthly and annual quantity of material for which an exclusion would be claimed, except as provided by paragraph (b)(2)(i)(C) of this section; and

(5) The following statement, which shall be signed and submitted by the person claiming the exclusion or his authorized representative:

Under penalty of criminal and civil prosecution for making or submitting false statements, representations, or omissions, I certify that the requirements of 40 CFR 261.38 have been met for all comparable fuels identified in this notification. Copies of the records and information required at 40 CFR 261.38(b)(8) are available at the generator's facility. Based on my inquiry of the individuals immediately responsible for obtaining the information, the information is, to the best of my knowledge and belief, true, accurate, and complete. I am aware that there are significant penalties for submitting false information, including the possibility of fine and imprisonment for knowing violations.

(B) If there is a substantive change in the information provided in the notice required under this paragraph, the generator must submit a revised notification.

(C) Excluded fuel generators must include an estimate of the average and maximum monthly and annual quantity of material for which an exclusion would be claimed only in notices submitted after December 19, 2008 for newly excluded fuel or for revised notices as required by paragraph (b)(2)(i)(B) of this section.

(ii) *Public notice.* Prior to burning an excluded fuel, the burner must publish in a major newspaper of general circulation local to the site where the fuel will be burned, a notice entitled "Notification of Burning a Fuel Excluded Under the Resource Conservation and Recovery Act" and containing the following information:

(A) Name, address, and RCRA ID number of the generating facility(ies);

(B) Name and address of the burner and identification of the unit(s) that will burn the excluded fuel;

(C) A brief, general description of the manufacturing, treatment, or other process generating the excluded fuel;

(D) An estimate of the average and maximum monthly and annual quantity of the excluded fuel to be burned; and

(E) Name and mailing address of the Regional or State Directors to whom the generator submitted a claim for the exclusion.

(3) *Burning.* The exclusion applies only if the fuel is burned in the following units that also shall be subject to Federal/State/local air emission requirements, including all applicable requirements implementing section 112 of the Clean Air Act:

(i) Industrial furnaces as defined in §260.10 of this chapter;

(ii) Boilers, as defined in §260.10 of this chapter, that are further defined as follows:

(A) Industrial boilers located on the site of a facility engaged in a manufacturing process where substances are transformed into new products, including the component parts of products, by mechanical or chemical processes; or

(B) Utility boilers used to produce electric power, steam, heated or cooled air, or other gases or fluids for sale;

(iii) Hazardous waste incinerators subject to regulation under subpart O of parts 264 or 265 of this chapter and applicable CAA MACT standards.

(iv) Gas turbines used to produce electric power, steam, heated or cooled air, or other gases or fluids for sale.

(4) *Fuel analysis plan for generators.* The generator of an excluded fuel shall develop and follow a written fuel analysis plan which describes the procedures for sampling and analysis of the material to be excluded. The plan shall be followed and retained at the site of the generator claiming the exclusion.

(i) At a minimum, the plan must specify:

(A) The parameters for which each excluded fuel will be analyzed and the rationale for the selection of those parameters;

(B) The test methods which will be used to test for these parameters;

(C) The sampling method which will be used to obtain a representative sample of the excluded fuel to be analyzed;

(D) The frequency with which the initial analysis of the excluded fuel will be reviewed or repeated to ensure that the analysis is accurate and up to date; and

(E) If process knowledge is used in the determination, any information prepared by the generator in making such determination.

(ii) For each analysis, the generator shall document the following:

(A) The dates and times that samples were obtained, and the dates the samples were analyzed;

(B) The names and qualifications of the person(s) who obtained the samples;

(C) A description of the temporal and spatial locations of the samples;

(D) The name and address of the laboratory facility at which analyses of the samples were performed;

(E) A description of the analytical methods used, including any clean-up and sample preparation methods;

(F) All quantitation limits achieved and all other quality control results for the analysis (including method blanks, duplicate analyses, matrix spikes, *etc.*), laboratory quality assurance data, and the description of any deviations from analytical methods written in the plan or from any other activity written in the plan which occurred;

(G) All laboratory results demonstrating whether the exclusion specifications have been met; and

(H) All laboratory documentation that support the analytical results, unless a contract between the claimant and the laboratory provides for the documentation to be maintained by the laboratory for the period specified in paragraph (b)(9) of this section and also provides for the availability of the documentation to the claimant upon request.

(iii) Syngas fuel generators shall submit for approval, prior to performing sampling, analysis, or any management of an excluded syngas fuel, a fuel analysis plan containing the elements of paragraph (b)(4)(i) of this section to the appropriate regulatory authority.

The approval of fuel analysis plans must be stated in writing and received by the facility prior to sampling and analysis to demonstrate the exclusion of a syngas. The approval of the fuel analysis plan may contain such provisions and conditions as the regulatory authority deems appropriate.

(5) *Excluded fuel sampling and analysis.* (i) *General.* For wastes for which an exclusion is claimed under the specifications provided by paragraphs (a)(1) or (a)(2) of this section, the generator of the waste must test for all the constituents in appendix VIII to this part, except those that the generator determines, based on testing or knowledge, should not be present in the fuel. The generator is required to document the basis of each determination that a constituent with an applicable specification should not be present. The generator may not determine that any of the following categories of constituents with a specification in Table 1 to this section should not be present:

(A) A constituent that triggered the toxicity characteristic for the constituents that were the basis for listing the hazardous secondary material as a hazardous waste, or constituents for which there is a treatment standard for the waste code in 40 CFR 268.40;

(B) A constituent detected in previous analysis of the waste;

(C) Constituents introduced into the process that generates the waste; or

(D) Constituents that are byproducts or side reactions to the process that generates the waste.

NOTE TO PARAGRAPH (B)(5): Any claim under this section must be valid and accurate for all hazardous constituents; a determination not to test for a hazardous constituent will not shield a generator from liability should that constituent later be found in the excluded fuel above the exclusion specifications.

(ii) *Use of process knowledge.* For each waste for which the comparable fuel or syngas exclusion is claimed where the generator of the excluded fuel is not the original generator of the hazardous waste, the generator of the excluded fuel may not use process knowledge pursuant to paragraph (b)(5)(i) of this section and must test to determine that all of the constituent specifications of paragraphs (a)(1) and (a)(2) of this section, as applicable, have been met.

(iii) The excluded fuel generator may use any reliable analytical method to demonstrate that no constituent of concern is present at concentrations above the specification levels. It is the responsibility of the generator to ensure that the sampling and analysis are unbiased, precise, and representative of the excluded fuel. For the fuel to be eligible for exclusion, a generator must demonstrate that:

(A) The 95% upper confidence limit of the mean concentration for each constituent of concern is not above the specification level; and

(B) The analyses could have detected the presence of the constituent at or below the specification level.

(iv) Nothing in this paragraph preempts, overrides or otherwise negates the provision in § 262.11 of this chapter, which requires any person who generates a solid waste to determine if that waste is a hazardous waste.

(v) In an enforcement action, the burden of proof to establish conformance with the exclusion specification shall be on the generator claiming the exclusion.

(vi) The generator must conduct sampling and analysis in accordance with the fuel analysis plan developed under paragraph (b)(4) of this section.

(vii) *Viscosity condition for comparable fuel.* (A) Excluded comparable fuel that has not been blended to meet the kinematic viscosity specification shall be analyzed as-generated.

(B) If hazardous waste is blended to meet the kinematic viscosity specification for comparable fuel, the generator shall:

(*1*) Analyze the hazardous waste as-generated to ensure that it meets the constituent and heating value specifications of paragraph (a)(1) of this section; and

(*2*) After blending, analyze the fuel again to ensure that the blended fuel meets all comparable fuel specifications.

(viii) Excluded fuel must be re-tested, at a minimum, annually and must be retested after a process change that could change its chemical or physical properties in a manner than may affect conformance with the specifications.

(6) (Reserved)

(7) *Speculative accumulation.* Excluded fuel must not be accumulated speculatively, as defined in §261.1(c)(8).

(8) *Operating record.* The generator must maintain an operating record on site containing the following information:

(i) All information required to be submitted to the implementing authority as part of the notification of the claim:

(A) The owner/operator name, address, and RCRA ID number of the person claiming the exclusion;

(B) For each excluded fuel, the EPA Hazardous Waste Codes that would be applicable if the material were discarded; and

(C) The certification signed by the person claiming the exclusion or his authorized representative.

(ii) A brief description of the process that generated the excluded fuel. If the comparable fuel generator is not the generator of the original hazardous waste, provide a brief description of the process that generated the hazardous waste;

(iii) The monthly and annual quantities of each fuel claimed to be excluded;

(iv) Documentation for any claim that a constituent is not present in the excluded fuel as required under paragraph (b)(5)(i) of this section;

(v) The results of all analyses and all detection limits achieved as required under paragraph (b)(4) of this section;

(vi) If the comparable fuel was generated through treatment or blending, documentation of compliance with the applicable provisions of paragraphs (a)(3) and (a)(4) of this section;

(vii) If the excluded fuel is to be shipped off-site, a certification from the burner as required under paragraph (b)(10) of this section;

(viii) The fuel analysis plan and documentation of all sampling and analysis results as required by paragraph (b)(4) of this section; and

(ix) If the generator ships excluded fuel off-site for burning, the generator must retain for each shipment the following information on-site:

(A) The name and address of the facility receiving the excluded fuel for burning;

(B) The quantity of excluded fuel shipped and delivered;

(C) The date of shipment or delivery;

(D) A cross-reference to the record of excluded fuel analysis or other information used to make the determination that the excluded fuel meets the specifications as required under paragraph (b)(4) of this section; and

(E) A one-time certification by the burner as required under paragraph (b)(10) of this section.

(9) *Records retention.* Records must be maintained for a period of three years.

(10) *Burner certification to the generator.* Prior to submitting a notification to the State and Regional Directors, a generator of excluded fuel who intends to ship the excluded fuel off-site for burning must obtain a one-time written, signed statement from the burner:

(i) Certifying that the excluded fuel will only be burned in an industrial furnace, industrial boiler, utility boiler, or hazardous waste incinerator, as required under paragraph (b)(3) of this section;

(ii) Identifying the name and address of the facility that will burn the excluded fuel; and

(iii) Certifying that the State in which the burner is located is authorized to exclude wastes as excluded fuel under the provisions of this section.

(11) *Ineligible waste codes.* Wastes that are listed as hazardous waste because of the presence of dioxins or furans, as set out in appendix VII of this part, are not eligible for these exclusions, and any fuel produced from or otherwise containing these wastes remains a hazardous waste subject to the full RCRA hazardous waste management requirements.

(12) *Regulatory status of boiler residues.* Burning excluded fuel that was otherwise a hazardous waste listed under §§261.31 through 261.33 does not subject boiler residues, including bottom ash and emission control residues, to regulation as derived-from hazardous wastes.

(13) *Residues in containers and tank systems upon cessation of operations.* (i) Liquid and accumulated solid residues that remain in a container or tank system for more than 90 days after the container or tank system ceases to be

operated for storage or transport of excluded fuel product are subject to regulation under parts 262 through 265, 267, 268, 270, 271, and 124 of this chapter.

(ii) Liquid and accumulated solid residues that are removed from a container or tank system after the container or tank system ceases to be operated for storage or transport of excluded fuel product are solid wastes subject to regulation as hazardous waste if the waste exhibits a characteristic of hazardous waste under §§ 261.21 through 261.24 or if the fuel were otherwise a hazardous waste listed under §§ 261.31 through 261.33 when the exclusion was claimed.

(iii) Liquid and accumulated solid residues that are removed from a container or tank system and which do not meet the specifications for exclusion under paragraphs (a)(1) or (a)(2) of this section are solid wastes subject to regulation as hazardous waste if:

(A) The waste exhibits a characteristic of hazardous waste under §§ 261.21 through 261.24; or

(B) The fuel were otherwise a hazardous waste listed under §§ 261.31 through 261.33. The hazardous waste code for the listed waste applies to these liquid and accumulated solid residues.

(14) *Waiver of RCRA Closure Requirements.* Interim status and permitted storage and combustion units, and generator storage units exempt from the permit requirements under § 262.34 of this chapter, are not subject to the closure requirements of 40 CFR Parts 264, 265, and 267 provided that the storage and combustion unit has been used to manage only hazardous waste that is subsequently excluded under the conditions of this section, and that afterward will be used only to manage fuel excluded under this section.

(15) *Spills and leaks.* (i) Excluded fuel that is spilled or leaked and that therefore no longer meets the conditions of the exclusion is discarded and must be managed as a hazardous waste if it exhibits a characteristic of hazardous waste under §§ 261.21 through 261.24 or if the fuel were otherwise a hazardous waste listed in §§ 261.31 through 261.33.

(ii) For excluded fuel that would have otherwise been a hazardous waste listed in §§ 261.31 through 261.33 and which is spilled or leaked, the hazardous waste code for the listed waste applies to the spilled or leaked material.

(16) Nothing in this section preempts, overrides, or otherwise negates the provisions in CERCLA Section 103, which establish reporting obligations for releases of hazardous substances, or the Department of Transportation requirements for hazardous materials in 49 CFR parts 171 through 180.

(c) *Failure to comply with the conditions of the exclusion.* An excluded fuel loses its exclusion if any person managing the fuel fails to comply with the conditions of the exclusion under this section, and the material must be managed as hazardous waste from the point of generation. In such situations, EPA or an authorized State agency may take enforcement action under RCRA section 3008(a).

Table 1 to § 261.38--Detection and Detection Limit Values for Comparable Fuel Specification

| Chemical name | CAS No. | Concentration Limit (mg/kg at 10,000 Btu/lb) | Minimum Required Detection Limit (mg/kg) |
|---|---|---|---|
| Total Nitrogen as N............................. | NA | 4900 | .......... |
| Total Halogens as Cl.............................. | NA | 540 | .......... |
| Total Organic Halogens as Cl...................... | NA | (a) | .......... |
| Polychlorinated biphenyls, total [Aroclors, total] | 1336-36-3 | ND | 1.4 |
| Cyanide, total.................................... | 57-12-5 | ND | 1 |
| Metals: | | | |
| Antimony, total............................... | 7440-36-0 | 12 | .......... |
| Arsenic, total................................. | 7440-38-2 | 0.23 | .......... |
| Barium, total.................................. | 7440-39-3 | 23 | .......... |
| Beryllium, total............................... | 7440-41-7 | 1.2 | .......... |
| Cadmium, total................................ | 7440-43-9 | 1.2 | .......... |
| Chromium, total............................... | 7440-47-3 | 2.3 | .......... |
| Cobalt.......................................... | 7440-48-4 | 4.6 | .......... |
| Lead, total..................................... | 7439-92-1 | 31 | .......... |
| Manganese..................................... | 7439-96-5 | 1.2 | .......... |
| Mercury, total................................. | 7439-97-6 | 0.25 | .......... |
| Nickel, total................................... | 7440-02-0 | 58 | .......... |
| Selenium, total................................ | 7782-49-2 | 0.23 | .......... |
| Silver, total.................................... | 7440-22-4 | 2.3 | .......... |
| Thallium, total................................. | 7440-28-0 | 23 | .......... |
| Hydrocarbons: | | | |
| Benzo[a]anthracene............................ | 56-55-3 | 2400 | .......... |
| Benzene........................................ | 71-43-2 | 4100 | .......... |
| Benzo[b]fluoranthene........................... | 205-99-2 | 2400 | .......... |
| Benzo[k]fluoranthene........................... | 207-08-9 | 2400 | .......... |
| Benzo[a]pyrene................................. | 50-32-8 | 2400 | .......... |
| Chrysene....................................... | 218-01-9 | 2400 | .......... |
| Dibenzo[a,h]anthracene........................ | 52-70-3 | 2400 | .......... |
| 7,12-Dimethylbenz[a]anthracene................ | 57-97-6 | 2400 | .......... |
| Fluoranthene................................... | 206-44-0 | 2400 | .......... |
| Indeno(1,2,3-cd)pyrene........................ | 193-39-5 | 2400 | .......... |
| 3-Methylcholanthrene.......................... | 56-49-5 | 2400 | .......... |
| Naphthalene................................... | 91-20-3 | 3200 | .......... |
| Toluene........................................ | 108-88-3 | 36000 | .......... |
| Oxygenates: | | | |
| Acetophenone................................. | 98-86-1 | 2400 | .......... |
| Acrolein........................................ | 107-02-8 | 39 | .......... |
| Allyl alcohol.................................... | 107-18-6 | 30 | .......... |
| Bis(2-ethylhexyl)phthalate [Di-2-ethylhexyl phthalate] | 117-81-7 | 2400 | .......... |
| Butyl benzyl phthalate......................... | 85-68-7 | 2400 | .......... |
| o-Cresol [2-Methyl phenol].................... | 95-48-7 | 2400 | .......... |
| m-Cresol [3-Methyl phenol]................... | 108-39-4 | 2400 | .......... |
| p-Cresol [4-Methyl phenol].................... | 106-44-5 | 2400 | .......... |
| Di-n-butyl phthalate........................... | 84-74-2 | 2400 | .......... |

| | | | |
|---|---|---|---|
| Diethyl phthalate............................. | 84-66-2 | 2400 | .......... |
| 2,4-Dimethylphenol............................. | 105-67-9 | 2400 | .......... |
| Dimethyl phthalate............................. | 131-11-3 | 2400 | .......... |
| Di-n-octyl phthalate........................... | 117-84-0 | 2400 | .......... |
| Endothall...................................... | 145-73-3 | 100 | .......... |
| Ethyl methacrylate............................. | 97-63-2 | 39 | .......... |
| 2-Ethoxyethanol [Ethylene glycol monoethyl ether] | 110-80-5 | 100 | .......... |
| Isobutyl alcohol............................... | 78-83-1 | 39 | .......... |
| Isosafrole..................................... | 120-58-1 | 2400 | .......... |
| Methyl ethyl ketone [2-Butanone]............... | 78-93-3 | 39 | .......... |
| Methyl methacrylate............................ | 80-62-6 | 39 | .......... |
| 1,4-Naphthoquinone............................. | 130-15-4 | 2400 | .......... |
| Phenol......................................... | 108-95-2 | 2400 | .......... |
| Propargyl alcohol [2-Propyn-1-ol].............. | 107-19-7 | 30 | .......... |
| Safrole........................................ | 94-59-7 | 2400 | .......... |
| Sulfonated Organics: | | | |
| Carbon disulfide............................... | 75-15-0 | ND | 39 |
| Disulfoton..................................... | 298-04-4 | ND | 2400 |
| Ethyl methanesulfonate......................... | 62-50-0 | ND | 2400 |
| Methyl methanesulfonate........................ | 66-27-3 | ND | 2400 |
| Phorate........................................ | 298-02-2 | ND | 2400 |
| 1,3-Propane sultone............................ | 1120-71-4 | ND | 100 |
| Tetraethyldithiopyrophosphate [Sulfotepp]...... | 3689-24-5 | ND | 2400 |
| Thiophenol [Benzenethiol]...................... | 108-98-5 | ND | 30 |
| O,O,O-Triethyl phosphorothioate................ | 126-68-1 | ND | 2400 |
| Nitrogenated Organics: | | | |
| Acetonitrile [Methyl cyanide].................. | 75-05-8 | ND | 39 |
| 2-Acetylaminofluorene [2-AAF].................. | 53-96-3 | ND | 2400 |
| Acrylonitrile.................................. | 107-13-1 | ND | 39 |
| 4-Aminobiphenyl................................ | 92-67-1 | ND | 2400 |
| 4-Aminopyridine................................ | 504-24-5 | ND | 100 |
| Aniline........................................ | 62-53-3 | ND | 2400 |
| Benzidine...................................... | 92-87-5 | ND | 2400 |
| Dibenz[a,j]acridine............................ | 224-42-0 | ND | 2400 |
| O,O-Diethyl O-pyrazinyl phosphorothioate [Thionazin] | 297-97-2 | ND | 2400 |
| Dimethoate..................................... | 60-51-5 | ND | 2400 |
| p-(Dimethylamino) azobenzene [4-Dimethylaminoazobenzene] | 60-11-7 | ND | 2400 |
| 3,3[prime]-Dimethylbenzidine................... | 119-93-7 | ND | 2400 |
| α,α-Dimethylphenethylamine......... | 122-09-8 | ND | 2400 |
| 3,3[prime]-Dimethoxybenzidine.................. | 119-90-4 | ND | 100 |
| 1,3-Dinitrobenzene [m-Dinitrobenzene].......... | 99-65-0 | ND | 2400 |
| 4,6-Dinitro-o-cresol........................... | 534-52-1 | ND | 2400 |
| 2,4-Dinitrophenol.............................. | 51-28-5 | ND | 2400 |
| 2,4-Dinitrotoluene............................. | 121-14-2 | ND | 2400 |
| 2,6-Dinitrotoluene............................. | 606-20-2 | ND | 2400 |
| Dinoseb [2-sec-Butyl-4,6-dinitrophenol]........ | 88-85-7 | ND | 2400 |
| Diphenylamine.................................. | 122-39-4 | ND | 2400 |
| Ethyl carbamate [Urethane].................... | 51-79-6 | ND | 100 |
| Ethylenethiourea (2-Imidazolidinethione)....... | 96-45-7 | ND | 110 |

| | | | |
|---|---|---|---|
| Famphur............................................ | 52-85-7 | ND | 2400 |
| Methacrylonitrile.............................. | 126-98-7 | ND | 39 |
| Methapyrilene................................... | 91-80-5 | ND | 2400 |
| Methomyl.......................................... | 16752-77-5 | ND | 57 |
| 2-Methyllactonitrile, [Acetone cyanohydrin].... | 75-86-5 | ND | 100 |
| Methyl parathion.............................. | 298-00-0 | ND | 2400 |
| MNNG (N-Metyl-N-nitroso-N[prime]-nitroguanidine) | 70-25-7 | ND | 110 |
| 1-Naphthylamine, [α-Naphthylamine]...... | 134-32-7 | ND | 2400 |
| 2-Naphthylamine, [β-Naphthylamine]........ | 91-59-8 | ND | 2400 |
| Nicotine...................................... | 54-11-5 | ND | 100 |
| 4-Nitroaniline, [p-Nitroaniline]............... | 100-01-6 | ND | 2400 |
| Nitrobenzene..................................... | 98-96-3 | ND | 2400 |
| p-Nitrophenol, [p-Nitrophenol]................ | 100-02-7 | ND | 2400 |
| 5-Nitro-o-toluidine............................ | 99-55-8 | ND | 2400 |
| N-Nitrosodi-n-butylamine...................... | 924-16-3 | ND | 2400 |
| N-Nitrosodiethylamine.......................... | 55-18-5 | ND | 2400 |
| N-Nitrosodiphenylamine, [Diphenylnitrosamine].. | 86-30-6 | ND | 2400 |
| N-Nitroso-N-methylethylamine................. | 10595-95-6 | ND | 2400 |
| N-Nitrosomorpholine.......................... | 59-89-2 | ND | 2400 |
| N-Nitrosopiperidine........................... | 100-75-4 | ND | 2400 |
| N-Nitrosopyrrolidine.......................... | 930-55-2 | ND | 2400 |
| 2-Nitropropane................................. | 79-46-9 | ND | 2400 |
| Parathion....................................... | 56-38-2 | ND | 2400 |
| Phenacetin...................................... | 62-44-2 | ND | 2400 |
| 1,4-Phenylene diamine, [p-Phenylenediamine].... | 106-50-3 | ND | 2400 |
| N-Phenylthiourea.............................. | 103-85-5 | ND | 57 |
| 2-Picoline [alpha-Picoline]................... | 109-06-8 | ND | 2400 |
| Propylthioracil, [6-Propyl-2-thiouracil]....... | 51-52-5 | ND | 100 |
| Pyridine........................................ | 110-86-1 | ND | 2400 |
| Strychnine...................................... | 57-24-9 | ND | 100 |
| Thioacetamide.................................. | 62-55-5 | ND | 57 |
| Thiofanox....................................... | 39196-18-4 | ND | 100 |
| Thiourea........................................ | 62-56-6 | ND | 57 |
| Toluene-2,4-diamine [2,4-Diaminotoluene]...... | 95-80-7 | ND | 57 |
| Toluene-2,6-diamine [2,6-Diaminotoluene]....... | 823-40-5 | ND | 57 |
| o-Toluidine..................................... | 95-53-4 | ND | 2400 |
| p-Toluidine..................................... | 106-49-0 | ND | 100 |
| 1,3,5-Trinitrobenzene, [sym-Trinitobenzene].... | 99-35-4 | ND | 2400 |
| Halogenated Organics: | | | |
| Allyl chloride.............................. | 107-05-1 | ND | 39 |
| Aramite........................................ | 140-57-8 | ND | 2400 |
| Benzal chloride [Dichloromethyl benzene]....... | 98-87-3 | ND | 100 |
| Benzyl chloride............................... | 100-44-77 | ND | 100 |
| bis(2-Chloroethyl)ether [Dichoroethyl ether]... | 111-44-4 | ND | 2400 |
| Bromoform [Tribromomothane]................... | 75-25-2 | ND | 39 |
| Bromomethane [Methyl bromide]................. | 74-83-9 | ND | 39 |
| 4-Bromophenyl phenyl ether [p-Bromo diphenyl ether] | 101-55-3 | ND | 2400 |
| Carbon tetrachloride......................... | 56-23-5 | ND | 39 |
| Chlordane....................................... | 57-74-9 | ND | 14 |

109

| | | | |
|---|---|---|---|
| p-Chloroaniline.............................. | 106-47-8 | ND | 2400 |
| Chlorobenzene................................. | 108-90-7 | ND | 39 |
| Chlorobenzilate............................... | 510-15-6 | ND | 2400 |
| p-Chloro-m-cresol........................... | 59-50-7 | ND | 2400 |
| 2-Chloroethyl vinyl ether...................... | 110-75-8 | ND | 39 |
| Chloroform.................................... | 67-66-3 | ND | 39 |
| Chloromethane [Methyl chloride]............... | 74-87-3 | ND | 39 |
| 2-Chloronaphthalene [beta-Chloronaphthalene]... | 91-58-7 | ND | 2400 |
| 2-Chlorophenol [o-Chlorophenol]................ | 95-57-8 | ND | 2400 |
| Chloroprene [2-Chloro-1,3-butadiene].......... | 1126-99-8 | ND | 39 |
| 2,4-D [2,4-Dichlorophenoxyacetic acid]........ | 94-75-7 | ND | 7 |
| Diallate...................................... | 2303-16-4 | ND | 3400 |
| 1,2-Dibromo-3-chloropropane.................. | 96-12-8 | ND | 39 |
| 1,2-Dichlorobenzene [o-Dichlorobenzene]....... | 95-50-1 | ND | 2400 |
| 1,3-Dichlorobenzene [m-Dichlorobenzene]....... | 541-73-1 | ND | 2400 |
| 1,4-Dichlorobenzene [p-Dichlorobenzene]....... | 106-46-7 | ND | 2400 |
| 3,3[prime]-Dichlorobenzidine.................. | 91-94-1 | ND | 2400 |
| Dichlorodifluoromethane [CFC-12]............... | 75-71-8 | ND | 39 |
| 1,2-Dichloroethane [Ethylene dichloride]....... | 107-06-2 | ND | 39 |
| 1,1-Dichloroethylene [Vinylidene chloride]..... | 75-35-4 | ND | 39 |
| Dichloromethoxy ethane [Bis(2-chloroethoxy)methane] | 111-91-1 | ND | 2400 |
| 2,4-Dichlorophenol........................... | 120-83-2 | ND | 2400 |
| 2,6-Dichlorophenol........................... | 87-65-0 | ND | 2400 |
| 1,2-Dichloropropane [Propylene dichloride]..... | 78-87-5 | ND | 39 |
| cis-1,3-Dichloropropylene...................... | 10061-01-5 | ND | 39 |
| trans-1,3-Dichloropropylene.................... | 10061-02-6 | ND | 39 |
| 1,3-Dichloro-2-propanol....................... | 96-23-1 | ND | 30 |
| Endosulfan I.................................. | 959-98-8 | ND | 1.4 |
| Endosulfan II................................. | 33213-65-9 | ND | 1.4 |
| Endrin........................................ | 72-20-8 | ND | 1.4 |
| Endrin aldehyde.............................. | 7421-93-4 | ND | 1.4 |
| Endrin Ketone................................ | 53494-70-5 | ND | 1.4 |
| Epichlorohydrin [1-Chloro-2,3-epoxy propane]... | 106-89-8 | ND | 30 |
| Ethylidene dichloride [1,1-Dichloroethane]..... | 75-34-3 | ND | 39 |
| 2-Fluoroacetamide............................ | 640-19-7 | ND | 100 |
| Heptachlor.................................... | 76-44-8 | ND | 1.4 |
| Heptachlor epoxide........................... | 1024-57-3 | ND | 2.8 |
| Hexachlorobenzene........................... | 118-74-1 | ND | 2400 |
| Hexachloro-1,3-butadiene [Hexachlorobutadiene]. | 87-68-3 | ND | 2400 |
| Hexachlorocyclopentadiene.................... | 77-47-4 | ND | 2400 |
| Hexachloroethane............................ | 67-72-1 | ND | 2400 |
| Hexachlorophene............................. | 70-30-4 | ND | 59000 |
| Hexachloropropene [Hexachloropropylene]........ | 1888-71-7 | ND | 2400 |
| Isodrin....................................... | 465-73-6 | ND | 2400 |
| Kepone [Chlordecone].......................... | 143-50-0 | ND | 4700 |
| Lindane [gamma-BHC] [gamma-Hexachlorocyclohexane]....... | 58-89-9 | ND | 1.4 |
| Methylene chloride [Dichloromethane].......... | 75-09-2 | ND | 39 |
| 4,4[prime]-Methylene-bis(2-chloroaniline)...... | 101-14-4 | ND | 100 |
| Methyl iodide [Iodomethane].................... | 74-88-4 | ND | 39 |

| | | | |
|---|---|---|---|
| Pentachlorobenzene............................ | 608-93-5 | ND | 2400 |
| Pentachloroethane............................. | 76-01-7 | ND | 39 |
| Pentachloronitrobenzene [PCNB] [Quintobenzene] [Quintozene]. | 82-68-8 | ND | 2400 |
| Pentachlorophenol.............................. | 87-88-5 | ND | 2400 |
| Pronamide....................................... | 23950-58-5 | ND | 2400 |
| Silvex [2,4,5-Trichlorophenoxypropionic acid].. | 93-72-1 | ND | 7 |
| 2,3,7,8-Tetrachlorodibenzo-p-dioxin [2,3,7,8-TCDD] | 1746-01-6 | ND | 30 |
| 1,2,4,5-Tetrachlorobenzene..................... | 95-94-3 | ND | 2400 |
| 1,1,2,2-Tetrachloroethane..................... | 79-35-4 | ND | 39 |
| Tetrachloroethylene [Perchloroethylene]........ | 127-18-4 | ND | 39 |
| 2,3,4,6-Tetrachlorophenol...................... | 58-90-2 | ND | 2400 |
| 1,2,4-Trichlorobenzene........................ | 120-82-1 | ND | 2400 |
| 1,1,1-Trichloroethane [Methyl chloroform]...... | 71-56-6 | ND | 39 |
| 1,1,2-Trichloroethane [Vinyl trichloride]...... | 79-00-5 | ND | 39 |
| Trichloroethylene.............................. | 79-01-6 | ND | 39 |
| Trichlorofluoromethane [Trichlormonofluoromethane]........... | 75-69-4 | ND | 39 |
| 2,4,5-Trichlorophenol.......................... | 95-95-4 | ND | 2400 |
| 2,4,6-Trichlorophenol.......................... | 88-06-2 | ND | 2400 |
| 1,2,3-Trichloropropane......................... | 96-18-4 | ND | 39 |
| Vinyl Chloride................................. | 75-01-4 | ND | 39 |

Notes:

NA--Not Applicable.

ND--Nondetect.

([a]) 25 or individual halogenated organics listed below.

[75 FR 33716, Jun. 15, 2010]

§ 261.39 Conditional Exclusion for Used, Broken Cathode Ray Tubes (CRTs) and Processed CRT Glass Undergoing Recycling.

Used, broken CRTs are not solid wastes if they meet the following conditions:

(a) *Prior to processing:* These materials are not solid wastes if they are destined for recycling and if they meet the following requirements:

(1) *Storage.* The broken CRTs must be either:

(i) Stored in a building with a roof, floor, and walls, or

(ii) Placed in a container (*i.e.,* a package or a vehicle) that is constructed, filled, and closed to minimize releases to the environment of CRT glass (including fine solid materials).

(2) *Labeling.* Each container in which the used, broken CRT is contained must be labeled or marked clearly with one of the following phrases: "Used cathode ray tube(s)-contains leaded glass " or "Leaded glass from televisions or computers." It must also be labeled: "Do not mix with other glass materials."

(3) *Transportation.* The used, broken CRTs must be transported in a container meeting the requirements of paragraphs (a)(1)(ii) and (2) of this section.

(4) *Speculative accumulation and use constituting disposal.* The used, broken CRTs are subject to the limitations on speculative accumulation as defined in paragraph (c)(8) of this section. If they are used in a manner constituting disposal, they must comply with the applicable requirements of part 266, subpart C instead of the requirements of this section.

(5) *Exports.* In addition to the applicable conditions specified in paragraphs (a)(1)–(4) of this section, exporters of used, broken CRTs must comply with the following requirements:

(i) Notify EPA of an intended export before the CRTs are scheduled to leave the United States. A complete notification should be submitted sixty (60) days before the initial shipment is intended to be shipped off-site. This notification may cover export activities extending over a twelve (12) month or lesser period. The notification must be in writing, signed by the exporter, and include the following information:

(A) Name, mailing address, telephone number and EPA ID number (if applicable) of the exporter of the CRTs.

(B) The estimated frequency or rate at which the CRTs are to be exported and the period of time over which they are to be exported.

(C) The estimated total quantity of CRTs specified in kilograms.

(D) All points of entry to and departure from each foreign country through which the CRTs will pass.

(E) A description of the means by which each shipment of the CRTs will be transported (*e.g.*, mode of transportation vehicle (air, highway, rail, water, etc.), type(s) of container (drums, boxes, tanks, etc.)).

(F) The name and address of the recycler and any alternate recycler.

(G) A description of the manner in which the CRTs will be recycled in the foreign country that will be receiving the CRTs.

(H) The name of any transit country through which the CRTs will be sent and a description of the approximate length of time the CRTs will remain in such country and the nature of their handling while there.

(ii) Notifications submitted by mail should be sent to the following mailing address: Office of Enforcement and Compliance Assurance, Office of Federal Activities, International Compliance Assurance Division, (Mail Code 2254A), Environmental Protection Agency, 1200 Pennsylvania Ave., NW., Washington, DC 20460. Hand-delivered notifications should be sent to: Office of Enforcement and Compliance Assurance, Office of Federal Activities, International Compliance Assurance Division, (Mail Code 2254A), Environmental Protection Agency, Ariel Rios Bldg., Room 6144, 1200 Pennsylvania Ave., NW., Washington, DC. In both cases, the following shall be prominently displayed on the front of the envelope: "Attention: Notification of Intent to Export CRTs."

(iii) Upon request by EPA, the exporter shall furnish to EPA any additional information which a receiving country requests in order to respond to a notification.

(iv) EPA will provide a complete notification to the receiving country and any transit countries. A notification is complete when EPA receives a notification which EPA determines satisfies the requirements of paragraph (a)(5)(i) of this section. Where a claim of confidentiality is asserted with respect to any notification information required by paragraph (a)(5)(i) of this section, EPA may find the notification not complete until any such claim is resolved in accordance with 40 CFR 260.2.

(v) The export of CRTs is prohibited unless the receiving country consents to the intended export. When the receiving country consents in writing to the receipt of the CRTs, EPA will forward an Acknowledgment of Consent to Export CRTs to the exporter. Where the receiving country objects to receipt of the CRTs or withdraws a prior consent, EPA will notify the exporter in writing. EPA will also notify the exporter of any responses from transit countries.

(vi) When the conditions specified on the original notification change, the exporter must provide EPA with a written renotification of the change, except for changes to the telephone number in paragraph (a)(5)(i)(A) of this section and decreases in the quantity indicated pursuant to paragraph (a)(5)(i)(C) of this section. The shipment cannot take place until consent of the receiving country to the changes has been obtained (except for changes to information about points of entry and departure and transit countries pursuant to paragraphs (a)(5)(i)(D) and (a)(5)(i)(H) of this section) and the exporter of CRTs receives from EPA a copy of the Acknowledgment of Consent to Export CRTs reflecting the receiving country's consent to the changes.

(vii) A copy of the Acknowledgment of Consent to Export CRTs must accompany the shipment of CRTs. The shipment must conform to the terms of the Acknowledgment.

(viii) If a shipment of CRTs cannot be delivered for any reason to the recycler or the alternate recycler, the exporter of CRTs must renotify EPA of a change in the conditions of the original notification to allow shipment to a new recycler in accordance with paragraph (a)(5)(vi) of this section and obtain another Acknowledgment of Consent to Export CRTs.

(ix) Exporters must keep copies of notifications and Acknowledgments of Consent to Export CRTs for a period of three years following receipt of the Acknowledgment.

(b) *Requirements for used CRT processing:* Used, broken CRTs undergoing CRT processing as defined in §260.10 of this chapter are not solid wastes if they meet the following requirements:

(1) *Storage.* Used, broken CRTs undergoing processing are subject to the requirement of paragraph (a)(4) of this section.

(2) *Processing.*

(i) All activities specified in paragraphs (2) and (3) of the definition of "CRT processing" in §260.10 of this chapter must be performed within a building with a roof, floor, and walls; and

(ii) No activities may be performed that use temperatures high enough to volatilize lead from CRTs.

(c) *Processed CRT glass sent to CRT glass making or lead smelting:* Glass from used CRTs that is destined for recycling at a CRT glass manufacturer or a lead smelter after processing is not a solid waste unless it is speculatively accumulated as defined in §261.1(c)(8).

(d) *Use constituting disposal:* Glass from used CRTs that is used in a manner constituting disposal must comply with the requirements of 40 CFR part 266, subpart C instead of the requirements of this section.

**§261.40  Conditional  Exclusion  for Used, Intact Cathode Ray Tubes (CRTs) Exported for Recycling.**

Used, intact CRTs exported for recycling are not solid wastes if they meet the notice and consent conditions of §261.39(a)(5), and if they are not speculatively accumulated as defined in §261.1(c)(8).

**§261.41  Notification  and  Record-keeping for Used, Intact Cathode Ray Tubes (CRTs) Exported for Reuse.**

(a) Persons who export used, intact CRTs for reuse must send a one-time notification to the Regional Administrator. The notification must include a statement that the notifier plans to export used, intact CRTs for reuse, the notifier's name, address, and EPA ID number (if applicable) and the name and phone number of a contact person.

(b) Persons who export used, intact CRTs for reuse must keep copies of normal business records, such as contracts, demonstrating that each shipment of exported CRTs will be reused. This documentation must be retained for a period of at least three years from the date the CRTs were exported.

## Subparts F–G [Reserved]

## Subpart H—Financial Requirements for Management of Excluded Hazardous Secondary Materials

SOURCE: 73 FR 64764, Oct. 30, 2008, unless otherwise noted.

**§261.140  Applicability.**

(a) The requirements of this subpart apply to owners or operators of reclamation and intermediate facilities managing hazardous secondary materials excluded under 40 CFR §261.4(a)(24), except as provided otherwise in this section.

(b) States and the Federal government are exempt from the financial assurance requirements of this subpart.

**§261.141  Definitions of terms as used in this subpart.**

The terms defined in §265.141(d), (f), (g), and (h) of this chapter have the same meaning in this subpart as they do in §265.141 of this chapter.

**§261.142  Cost estimate.**

(a) The owner or operator must have a detailed written estimate, in current dollars, of the cost of disposing of any hazardous secondary material as listed or characteristic hazardous waste, and the potential cost of closing the facility as a treatment, storage, and disposal facility.

(1) The estimate must equal the cost of conducting the activities described in paragraph (a) of this section at the point when the extent and manner of the facility's operation would make these activities the most expensive; and

(2) The cost estimate must be based on the costs to the owner or operator of

hiring a third party to conduct these activities. A third party is a party who is neither a parent nor a subsidiary of the owner or operator. (See definition of parent corporation in § 265.141(d) of this chapter.) The owner or operator may use costs for on-site disposal in accordance with applicable requirements if he can demonstrate that on-site disposal capacity will exist at all times over the life of the facility.

(3) The cost estimate may not incorporate any salvage value that may be realized with the sale of hazardous secondary materials, or hazardous or nonhazardous wastes if applicable under § 265.5113(d) of this chapter, facility structures or equipment, land, or other assets associated with the facility.

(4) The owner or operator may not incorporate a zero cost for hazardous secondary materials, or hazardous or nonhazardous wastes if applicable under § 265.5113(d) of this chapter that might have economic value.

(b) During the active life of the facility, the owner or operator must adjust the cost estimate for inflation within 60 days prior to the anniversary date of the establishment of the financial instrument(s) used to comply with § 261.143. For owners and operators using the financial test or corporate guarantee, the cost estimate must be updated for inflation within 30 days after the close of the firm's fiscal year and before submission of updated information to the Regional Administrator as specified in § 261.143(e)(3). The adjustment may be made by recalculating the cost estimate in current dollars, or by using an inflation factor derived from the most recent Implicit Price Deflator for Gross National Product published by the U.S. Department of Commerce in its Survey of Current Business, as specified in paragraphs (b)(1) and (2) of this section. The inflation factor is the result of dividing the latest published annual Deflator by the Deflator for the previous year.

(1) The first adjustment is made by multiplying the cost estimate by the inflation factor. The result is the adjusted cost estimate.

(2) Subsequent adjustments are made by multiplying the latest adjusted cost estimate by the latest inflation factor.

(c) During the active life of the facility, the owner or operator must revise the cost estimate no later than 30 days after a change in a facility's operating plan or design that would increase the costs of conducting the activities described in paragraph (a) or no later than 60 days after an unexpected event which increases the cost of conducting the activities described in paragraph (a) of this section. The revised cost estimate must be adjusted for inflation as specified in paragraph (b) of this section.

(d) The owner or operator must keep the following at the facility during the operating life of the facility: The latest cost estimate prepared in accordance with paragraphs (a) and (c) and, when this estimate has been adjusted in accordance with paragraph (b), the latest adjusted cost estimate.

### § 261.143  Financial assurance condition.

Per § 261.4(a)(24)(vi)(F) of this chapter, an owner or operator of a reclamation or intermediate facility must have financial assurance as a condition of the exclusion as required under § 261.4(a)(24) of this chapter. He must choose from the options as specified in paragraphs (a) through (e) of this section.

(a) *Trust fund.* (1) An owner or operator may satisfy the requirements of this section by establishing a trust fund which conforms to the requirements of this paragraph and submitting an originally signed duplicate of the trust agreement to the Regional Administrator. The trustee must be an entity which has the authority to act as a trustee and whose trust operations are regulated and examined by a Federal or State agency.

(2) The wording of the trust agreement must be identical to the wording specified in § 261.151(a)(1), and the trust agreement must be accompanied by a formal certification of acknowledgment (for example, see § 261.151(a)(2)). Schedule A of the trust agreement must be updated within 60 days after a change in the amount of the current cost estimate covered by the agreement.

(3) The trust fund must be funded for the full amount of the current cost estimate before it may be relied upon to satisfy the requirements of this section.

(4) Whenever the current cost estimate changes, the owner or operator must compare the new estimate with the trustee's most recent annual valuation of the trust fund. If the value of the fund is less than the amount of the new estimate, the owner or operator, within 60 days after the change in the cost estimate, must either deposit an amount into the fund so that its value after this deposit at least equals the amount of the current cost estimate, or obtain other financial assurance as specified in this section to cover the difference.

(5) If the value of the trust fund is greater than the total amount of the current cost estimate, the owner or operator may submit a written request to the Regional Administrator for release of the amount in excess of the current cost estimate.

(6) If an owner or operator substitutes other financial assurance as specified in this section for all or part of the trust fund, he may submit a written request to the Regional Administrator for release of the amount in excess of the current cost estimate covered by the trust fund.

(7) Within 60 days after receiving a request from the owner or operator for release of funds as specified in paragraph (a) (5) or (6) of this section, the Regional Administrator will instruct the trustee to release to the owner or operator such funds as the Regional Administrator specifies in writing. If the owner or operator begins final closure under subpart G of 40 CFR part 264 or 265, an owner or operator may request reimbursements for partial or final closure expenditures by submitting itemized bills to the Regional Administrator. The owner or operator may request reimbursements for partial closure only if sufficient funds are remaining in the trust fund to cover the maximum costs of closing the facility over its remaining operating life. No later than 60 days after receiving bills for partial or final closure activities, the Regional Administrator will instruct the trustee to make reim-

bursements in those amounts as the Regional Administrator specifies in writing, if the Regional Administrator determines that the partial or final closure expenditures are in accordance with the approved closure plan, or otherwise justified. If the Regional Administrator has reason to believe that the maximum cost of closure over the remaining life of the facility will be significantly greater than the value of the trust fund, he may withhold reimbursements of such amounts as he deems prudent until he determines, in accordance with §265.143(i) that the owner or operator is no longer required to maintain financial assurance for final closure of the facility. If the Regional Administrator does not instruct the trustee to make such reimbursements, he will provide to the owner or operator a detailed written statement of reasons.

(8) The Regional Administrator will agree to termination of the trust when:

(i) An owner or operator substitutes alternate financial assurance as specified in this section; or

(ii) The Regional Administrator releases the owner or operator from the requirements of this section in accordance with paragraph (i) of this section.

(b) *Surety bond guaranteeing payment into a trust fund.* (1) An owner or operator may satisfy the requirements of this section by obtaining a surety bond which conforms to the requirements of this paragraph and submitting the bond to the Regional Administrator. The surety company issuing the bond must, at a minimum, be among those listed as acceptable sureties on Federal bonds in Circular 570 of the U.S. Department of the Treasury.

(2) The wording of the surety bond must be identical to the wording specified in §261.151(b).

(3) The owner or operator who uses a surety bond to satisfy the requirements of this section must also establish a standby trust fund. Under the terms of the bond, all payments made thereunder will be deposited by the surety directly into the standby trust fund in accordance with instructions from the Regional Administrator. This standby trust fund must meet the requirements specified in paragraph (a) of this section, except that:

(i) An originally signed duplicate of the trust agreement must be submitted to the Regional Administrator with the surety bond; and

(ii) Until the standby trust fund is funded pursuant to the requirements of this section, the following are not required by these regulations:

(A) Payments into the trust fund as specified in paragraph (a) of this section;

(B) Updating of Schedule A of the trust agreement (see § 261.151(a)) to show current cost estimates;

(C) Annual valuations as required by the trust agreement; and

(D) Notices of nonpayment as required by the trust agreement.

(4) The bond must guarantee that the owner or operator will:

(i) Fund the standby trust fund in an amount equal to the penal sum of the bond before loss of the exclusion under § 261.4(a)(24) of this chapter or

(ii) Fund the standby trust fund in an amount equal to the penal sum within 15 days after an administrative order to begin closure issued by the Regional Administrator becomes final, or within 15 days after an order to begin closure is issued by a U.S. district court or other court of competent jurisdiction; or

(iii) Provide alternate financial assurance as specified in this section, and obtain the Regional Administrator's written approval of the assurance provided, within 90 days after receipt by both the owner or operator and the Regional Administrator of a notice of cancellation of the bond from the surety.

(5) Under the terms of the bond, the surety will become liable on the bond obligation when the owner or operator fails to perform as guaranteed by the bond.

(6) The penal sum of the bond must be in an amount at least equal to the current cost estimate, except as provided in paragraph (f) of this section.

(7) Whenever the current cost estimate increases to an amount greater than the penal sum, the owner or operator, within 60 days after the increase, must either cause the penal sum to be increased to an amount at least equal to the current cost estimate and submit evidence of such increase to the Regional Administrator, or obtain other financial assurance as specified in this section to cover the increase. Whenever the current cost estimate decreases, the penal sum may be reduced to the amount of the current cost estimate following written approval by the Regional Administrator.

(8) Under the terms of the bond, the surety may cancel the bond by sending notice of cancellation by certified mail to the owner or operator and to the Regional Administrator. Cancellation may not occur, however, during the 120 days beginning on the date of receipt of the notice of cancellation by both the owner or operator and the Regional Administrator, as evidenced by the return receipts.

(9) The owner or operator may cancel the bond if the Regional Administrator has given prior written consent based on his receipt of evidence of alternate financial assurance as specified in this section.

(c) *Letter of credit.* (1) An owner or operator may satisfy the requirements of this section by obtaining an irrevocable standby letter of credit which conforms to the requirements of this paragraph and submitting the letter to the Regional Administrator. The issuing institution must be an entity which has the authority to issue letters of credit and whose letter-of-credit operations are regulated and examined by a Federal or State agency.

(2) The wording of the letter of credit must be identical to the wording specified in § 261.151(c).

(3) An owner or operator who uses a letter of credit to satisfy the requirements of this section must also establish a standby trust fund. Under the terms of the letter of credit, all amounts paid pursuant to a draft by the Regional Administrator will be deposited by the issuing institution directly into the standby trust fund in accordance with instructions from the Regional Administrator. This standby trust fund must meet the requirements of the trust fund specified in paragraph (a) of this section, except that:

(i) An originally signed duplicate of the trust agreement must be submitted to the Regional Administrator with the letter of credit; and

(ii) Unless the standby trust fund is funded pursuant to the requirements of this section, the following are not required by these regulations:

(A) Payments into the trust fund as specified in paragraph (a) of this section;

(B) Updating of Schedule A of the trust agreement (see §261.151(a)) to show current cost estimates;

(C) Annual valuations as required by the trust agreement; and

(D) Notices of nonpayment as required by the trust agreement.

(4) The letter of credit must be accompanied by a letter from the owner or operator referring to the letter of credit by number, issuing institution, and date, and providing the following information: The EPA Identification Number (if any issued), name, and address of the facility, and the amount of funds assured for the facility by the letter of credit.

(5) The letter of credit must be irrevocable and issued for a period of at least 1 year. The letter of credit must provide that the expiration date will be automatically extended for a period of at least 1 year unless, at least 120 days before the current expiration date, the issuing institution notifies both the owner or operator and the Regional Administrator by certified mail of a decision not to extend the expiration date. Under the terms of the letter of credit, the 120 days will begin on the date when both the owner or operator and the Regional Administrator have received the notice, as evidenced by the return receipts.

(6) The letter of credit must be issued in an amount at least equal to the current cost estimate, except as provided in paragraph (f) of this section.

(7) Whenever the current cost estimate increases to an amount greater than the amount of the credit, the owner or operator, within 60 days after the increase, must either cause the amount of the credit to be increased so that it at least equals the current cost estimate and submit evidence of such increase to the Regional Administrator, or obtain other financial assurance as specified in this section to cover the increase. Whenever the current cost estimate decreases, the amount of the credit may be reduced to the amount of the current cost estimate following written approval by the Regional Administrator.

(8) Following a determination by the Regional Administrator that the hazardous secondary materials do not meet the conditions of the exclusion under §261.4(a)(24), the Regional Administrator may draw on the letter of credit.

(9) If the owner or operator does not establish alternate financial assurance as specified in this section and obtain written approval of such alternate assurance from the Regional Administrator within 90 days after receipt by both the owner or operator and the Regional Administrator of a notice from the issuing institution that it has decided not to extend the letter of credit beyond the current expiration date, the Regional Administrator will draw on the letter of credit. The Regional Administrator may delay the drawing if the issuing institution grants an extension of the term of the credit. During the last 30 days of any such extension the Regional Administrator will draw on the letter of credit if the owner or operator has failed to provide alternate financial assurance as specified in this section and obtain written approval of such assurance from the Regional Administrator.

(10) The Regional Administrator will return the letter of credit to the issuing institution for termination when:

(i) An owner or operator substitutes alternate financial assurance as specified in this section; or

(ii) The Regional Administrator releases the owner or operator from the requirements of this section in accordance with paragraph (i) of this section.

(d) *Insurance.* (1) An owner or operator may satisfy the requirements of this section by obtaining insurance which conforms to the requirements of this paragraph and submitting a certificate of such insurance to the Regional Administrator At a minimum, the insurer must be licensed to transact the business of insurance, or eligible to provide insurance as an excess or surplus lines insurer, in one or more States.

117

(2) The wording of the certificate of insurance must be identical to the wording specified in § 261.151(d).

(3) The insurance policy must be issued for a face amount at least equal to the current cost estimate, except as provided in paragraph (f) of this section. The term "face amount" means the total amount the insurer is obligated to pay under the policy. Actual payments by the insurer will not change the face amount, although the insurer's future liability will be lowered by the amount of the payments.

(4) The insurance policy must guarantee that funds will be available whenever needed to pay the cost of removal of all hazardous secondary materials from the unit, to pay the cost of decontamination of the unit, to pay the costs of the performance of activities required under subpart G of 40 CFR parts 264 or 265, as applicable, for the facilities covered by this policy. The policy must also guarantee that once funds are needed, the insurer will be responsible for paying out funds, up to an amount equal to the face amount of the policy, upon the direction of the Regional Administrator, to such party or parties as the Regional Administrator specifies.

(5) After beginning partial or final closure under 40 CFR parts 264 or 265, as applicable, an owner or operator or any other authorized person may request reimbursements for closure expenditures by submitting itemized bills to the Regional Administrator. The owner or operator may request reimbursements only if the remaining value of the policy is sufficient to cover the maximum costs of closing the facility over its remaining operating life. Within 60 days after receiving bills for closure activities, the Regional Administrator will instruct the insurer to make reimbursements in such amounts as the Regional Administrator specifies in writing if the Regional Administrator determines that the expenditures are in accordance with the approved plan or otherwise justified. If the Regional Administrator has reason to believe that the maximum cost over the remaining life of the facility will be significantly greater than the face amount of the policy, he may withhold reimbursement of such amounts as he deems prudent until he determines, in accordance with paragraph (h) of this section, that the owner or operator is no longer required to maintain financial assurance for the particular facility. If the Regional Administrator does not instruct the insurer to make such reimbursements, he will provide to the owner or operator a detailed written statement of reasons.

(6) The owner or operator must maintain the policy in full force and effect until the Regional Administrator consents to termination of the policy by the owner or operator as specified in paragraph (i)(10) of this section. Failure to pay the premium, without substitution of alternate financial assurance as specified in this section, will constitute a significant violation of these regulations warranting such remedy as the Regional Administrator deems necessary. Such violation will be deemed to begin upon receipt by the Regional Administrator of a notice of future cancellation, termination, or failure to renew due to nonpayment of the premium, rather than upon the date of expiration.

(7) Each policy must contain a provision allowing assignment of the policy to a successor owner or operator. Such assignment may be conditional upon consent of the insurer, provided such consent is not unreasonably refused.

(8) The policy must provide that the insurer may not cancel, terminate, or fail to renew the policy except for failure to pay the premium. The automatic renewal of the policy must, at a minimum, provide the insured with the option of renewal at the face amount of the expiring policy. If there is a failure to pay the premium, the insurer may elect to cancel, terminate, or fail to renew the policy by sending notice by certified mail to the owner or operator and the Regional Administrator. Cancellation, termination, or failure to renew may not occur, however, during the 120 days beginning with the date of receipt of the notice by both the Regional Administrator and the owner or operator, as evidenced by the return receipts. Cancellation, termination, or failure to renew may not occur and the policy will remain in full force and effect in the event that on or before the date of expiration:

(i) The Regional Administrator deems the facility abandoned; or

(ii) Conditional exclusion or interim status is lost, terminated, or revoked; or

(iii) Closure is ordered by the Regional Administrator or a U.S. district court or other court of competent jurisdiction; or

(iv) The owner or operator is named as debtor in a voluntary or involuntary proceeding under Title 11 (Bankruptcy), U.S. Code; or

(v) The premium due is paid.

(9) Whenever the current cost estimate increases to an amount greater than the face amount of the policy, the owner or operator, within 60 days after the increase, must either cause the face amount to be increased to an amount at least equal to the current cost estimate and submit evidence of such increase to the Regional Administrator, or obtain other financial assurance as specified in this section to cover the increase. Whenever the current cost estimate decreases, the face amount may be reduced to the amount of the current cost estimate following written approval by the Regional Administrator.

(10) The Regional Administrator will give written consent to the owner or operator that he may terminate the insurance policy when:

(i) An owner or operator substitutes alternate financial assurance as specified in this section; or

(ii) The Regional Administrator releases the owner or operator from the requirements of this section in accordance with paragraph (i) of this section.

(e) *Financial test and corporate guarantee.* (1) An owner or operator may satisfy the requirements of this section by demonstrating that he passes a financial test as specified in this paragraph. To pass this test the owner or operator must meet the criteria of either paragraph (e)(1) (i) or (ii) of this section:

(i) The owner or operator must have:

(A) Two of the following three ratios: A ratio of total liabilities to net worth less than 2.0; a ratio of the sum of net income plus depreciation, depletion, and amortization to total liabilities greater than 0.1; and a ratio of current

assets to current liabilities greater than 1.5; and

(B) Net working capital and tangible net worth each at least six times the sum of the current cost estimates and the current plugging and abandonment cost estimates; and

(C) Tangible net worth of at least $10 million; and

(D) Assets located in the United States amounting to at least 90 percent of total assets or at least six times the sum of the current cost estimates and the current plugging and abandonment cost estimates.

(ii) The owner or operator must have:

(A) A current rating for his most recent bond issuance of AAA, AA, A, or BBB as issued by Standard and Poor's or Aaa, Aa, A, or Baa as issued by Moody's; and

(B) Tangible net worth at least six times the sum of the current cost estimates and the current plugging and abandonment cost estimates; and

(C) Tangible net worth of at least $10 million; and

(D) Assets located in the United States amounting to at least 90 percent of total assets or at least six times the sum of the current cost estimates and the current plugging and abandonment cost estimates.

(2) The phrase "current cost estimates" as used in paragraph (e)(1) of this section refers to the cost estimates required to be shown in paragraphs 1–4 of the letter from the owner's or operator's chief financial officer (§261.151(e)). The phrase "current plugging and abandonment cost estimates" as used in paragraph (e)(1) of this section refers to the cost estimates required to be shown in paragraphs 1–4 of the letter from the owner's or operator's chief financial officer (§144.70(f) of this chapter).

(3) To demonstrate that he meets this test, the owner or operator must submit the following items to the Regional Administrator:

(i) A letter signed by the owner's or operator's chief financial officer and worded as specified in §261.151(e); and

(ii) A copy of the independent certified public accountant's report on examination of the owner's or operator's financial statements for the latest completed fiscal year; and

(iii) If the chief financial officer's letter providing evidence of financial assurance includes financial data showing that the owner or operator satisfies paragraph (e)(1)(i) of this section that are different from the data in the audited financial statements referred to in paragraph (e)(3)(ii)of this section or any other audited financial statement or data filed with the SEC, then a special report from the owner's or operator's independent certified public accountant to the owner or operator is required. The special report shall be based upon an agreed upon procedures engagement in accordance with professional auditing standards and shall describe the procedures performed in comparing the data in the chief financial officer's letter derived from the independently audited, year-end financial statements for the latest fiscal year with the amounts in such financial statements, the findings of the comparison, and the reasons for any differences.

(4) The owner or operator may obtain an extension of the time allowed for submission of the documents specified in paragraph (e)(3) of this section if the fiscal year of the owner or operator ends during the 90 days prior to the effective date of these regulations and if the year-end financial statements for that fiscal year will be audited by an independent certified public accountant. The extension will end no later than 90 days after the end of the owner's or operator's fiscal year. To obtain the extension, the owner's or operator's chief financial officer must send, by the effective date of these regulations, a letter to the Regional Administrator of each Region in which the owner's or operator's facilities to be covered by the financial test are located. This letter from the chief financial officer must:

(i) Request the extension;

(ii) Certify that he has grounds to believe that the owner or operator meets the criteria of the financial test;

(iii) Specify for each facility to be covered by the test the EPA Identification Number (if any issued), name, address, and current cost estimates to be covered by the test;

(iv) Specify the date ending the owner's or operator's last complete fiscal year before the effective date of these regulations in this subpart;

(v) Specify the date, no later than 90 days after the end of such fiscal year, when he will submit the documents specified in paragraph (e)(3) of this section; and

(vi) Certify that the year-end financial statements of the owner or operator for such fiscal year will be audited by an independent certified public accountant.

(5) After the initial submission of items specified in paragraph (e)(3) of this section, the owner or operator must send updated information to the Regional Administrator within 90 days after the close of each succeeding fiscal year. This information must consist of all three items specified in paragraph (e)(3) of this section.

(6) If the owner or operator no longer meets the requirements of paragraph (e)(1) of this section, he must send notice to the Regional Administrator of intent to establish alternate financial assurance as specified in this section. The notice must be sent by certified mail within 90 days after the end of the fiscal year for which the year-end financial data show that the owner or operator no longer meets the requirements. The owner or operator must provide the alternate financial assurance within 120 days after the end of such fiscal year.

(7) The Regional Administrator may, based on a reasonable belief that the owner or operator may no longer meet the requirements of paragraph (e)(1) of this section, require reports of financial condition at any time from the owner or operator in addition to those specified in paragraph (e)(3) of this section. If the Regional Administrator finds, on the basis of such reports or other information, that the owner or operator no longer meets the requirements of paragraph (e)(1) of this section, the owner or operator must provide alternate financial assurance as specified in this section within 30 days after notification of such a finding.

(8) The Regional Administrator may disallow use of this test on the basis of qualifications in the opinion expressed by the independent certified public accountant in his report on examination of the owner's or operator's financial

statements (see paragraph (e)(3)(ii) of this section). An adverse opinion or a disclaimer of opinion will be cause for disallowance. The Regional Administrator will evaluate other qualifications on an individual basis. The owner or operator must provide alternate financial assurance as specified in this section within 30 days after notification of the disallowance.

(9) The owner or operator is no longer required to submit the items specified in paragraph (e)(3) of this section when:

(i) An owner or operator substitutes alternate financial assurance as specified in this section; or

(ii) The Regional Administrator releases the owner or operator from the requirements of this section in accordance with paragraph (i) of this section.

(10) An owner or operator may meet the requirements of this section by obtaining a written guarantee. The guarantor must be the direct or higher-tier parent corporation of the owner or operator, a firm whose parent corporation is also the parent corporation of the owner or operator, or a firm with a "substantial business relationship" with the owner or operator. The guarantor must meet the requirements for owners or operators in paragraphs (e)(1) through (8) of this section and must comply with the terms of the guarantee. The wording of the guarantee must be identical to the wording specified in §261.151(g)(1). A certified copy of the guarantee must accompany the items sent to the Regional Administrator as specified in paragraph (e)(3) of this section. One of these items must be the letter from the guarantor's chief financial officer. If the guarantor's parent corporation is also the parent corporation of the owner or operator, the letter must describe the value received in consideration of the guarantee. If the guarantor is a firm with a "substantial business relationship" with the owner or operator, this letter must describe this "substantial business relationship" and the value received in consideration of the guarantee. The terms of the guarantee must provide that:

(i) Following a determination by the Regional Administrator that the hazardous secondary materials at the owner or operator's facility covered by this guarantee do not meet the conditions of the exclusion under §261.4(a)(24) of this chapter, the guarantor will dispose of any hazardous secondary material as hazardous waste and close the facility in accordance with closure requirements found in parts 264 or 265 of this chapter, as applicable, or establish a trust fund as specified in paragraph (a) of this section in the name of the owner or operator in the amount of the current cost estimate.

(ii) The corporate guarantee will remain in force unless the guarantor sends notice of cancellation by certified mail to the owner or operator and to the Regional Administrator. Cancellation may not occur, however, during the 120 days beginning on the date of receipt of the notice of cancellation by both the owner or operator and the Regional Administrator, as evidenced by the return receipts.

(iii) If the owner or operator fails to provide alternate financial assurance as specified in this section and obtain the written approval of such alternate assurance from the Regional Administrator within 90 days after receipt by both the owner or operator and the Regional Administrator of a notice of cancellation of the corporate guarantee from the guarantor, the guarantor will provide such alternate financial assurance in the name of the owner or operator.

(f) *Use of multiple financial mechanisms.* An owner or operator may satisfy the requirements of this section by establishing more than one financial mechanism per facility. These mechanisms are limited to trust funds, surety bonds, letters of credit, and insurance. The mechanisms must be as specified in paragraphs (a) through (d) of this section, respectively, of this section, except that it is the combination of mechanisms, rather than the single mechanism, which must provide financial assurance for an amount at least equal to the current cost estimate. If an owner or operator uses a trust fund in combination with a surety bond or a letter of credit, he may use the trust fund as the standby trust fund for the other mechanisms. A single standby trust fund may be established for two or more mechanisms. The Regional Administrator may use any or all of the mechanisms to provide for the facility.

(g) *Use of a financial mechanism for multiple facilities.* An owner or operator may use a financial assurance mechanism specified in this section to meet the requirements of this section for more than one facility. Evidence of financial assurance submitted to the Regional Administrator must include a list showing, for each facility, the EPA Identification Number (if any issued), name, address, and the amount of funds assured by the mechanism. If the facilities covered by the mechanism are in more than one Region, identical evidence of financial assurance must be submitted to and maintained with the Regional Administrators of all such Regions. The amount of funds available through the mechanism must be no less than the sum of funds that would be available if a separate mechanism had been established and maintained for each facility. In directing funds available through the mechanism for any of the facilities covered by the mechanism, the Regional Administrator may direct only the amount of funds designated for that facility, unless the owner or operator agrees to the use of additional funds available under the mechanism.

(h) *Removal and Decontamination Plan for Release* (1) An owner or operator of a reclamation facility or an intermediate facility who wishes to be released from his financial assurance obligations under § 261.4(a)(24)(vi)(F) of this chapter must submit a plan for removing all hazardous secondary material residues to the Regional Administrator at least 180 days prior to the date on which he expects to cease to operate under the exclusion.

(2) The plan must include, at least:

(A) For each hazardous secondary materials storage unit subject to financial assurance requirements under § 261.4(a)(24)(vi)(F), a description of how all excluded hazardous secondary materials will be recycled or sent for recycling, and how all residues, contaminated containment systems (liners, etc), contaminated soils, subsoils, structures, and equipment will be removed or decontaminated as necessary

to protect human health and the environment, and

(B) A detailed description of the steps necessary to remove or decontaminate all hazardous secondary material residues and contaminated containment system components, equipment, structures, and soils including, but not limited to, procedures for cleaning equipment and removing contaminated soils, methods for sampling and testing surrounding soils, and criteria for determining the extent of decontamination necessary to protect human health and the environment; and

(C) A detailed description of any other activities necessary to protect human health and the environment during this timeframe, including, but not limited to, leachate collection, run-on and run-off control, etc; and

(D) A schedule for conducting the activities described which, at a minimum, includes the total time required to remove all excluded hazardous secondary materials for recycling and decontaminate all units subject to financial assurance under § 261.4(a)(24)(vi)(F) and the time required for intervening activities which will allow tracking of the progress of decontamination.

(3) The Regional Administrator will provide the owner or operator and the public, through a newspaper notice, the opportunity to submit written comments on the plan and request modifications to the plan no later than 30 days from the date of the notice. He will also, in response to a request or at his discretion, hold a public hearing whenever such a hearing might clarify one or more issues concerning the plan. The Regional Administrator will give public notice of the hearing at least 30 days before it occurs. (Public notice of the hearing may be given at the same time as notice of the opportunity for the public to submit written comments, and the two notices may be combined.) The Regional Administrator will approve, modify, or disapprove the plan within 90 days of its receipt. If the Regional Administrator does not approve the plan, he shall provide the owner or operator with a detailed written statement of reasons for the refusal and the owner or operator must modify the plan or submit a new plan for approval within 30 days after

receiving such written statement. The Regional Administrator will approve or modify this plan in writing within 60 days. If the Regional Administrator modifies the plan, this modified plan becomes the approved plan. The Regional Administrator must assure that the approved plan is consistent with paragraph (h) of this section. A copy of the modified plan with a detailed statement of reasons for the modifications must be mailed to the owner or operator.

(4) Within 60 days of completion of the activities described for each hazardous secondary materials management unit, the owner or operator must submit to the Regional Administrator, by registered mail, a certification that all hazardous secondary materials have been removed from the unit and the unit has been decontaminated in accordance with the specifications in the approved plan. The certification must be signed by the owner or operator and by a qualified Professional Engineer. Documentation supporting the Professional Engineer's certification must be furnished to the Regional Administrator, upon request, until he releases the owner or operator from the financial assurance requirements for §261.4(a)(24)(vi)(F).

(i) *Release of the owner or operator from the requirements of this section.* Within 60 days after receiving certifications from the owner or operator and a qualified Professional Engineer that all hazardous secondary materials have been removed from the facility or a unit at the facility and the facility or a unit has been decontaminated in accordance with the approved plan per paragraph (h), the Regional Administrator will notify the owner or operator in writing that he is no longer required under §261.4(a)(24)(vi)(F) to maintain financial assurance for that facility or a unit at the facility, unless the Regional Administrator has reason to believe that all hazardous secondary materials have not been removed from the facility or unit at a facility or that the facility or unit has not been decontaminated in accordance with the approved plan. The Regional Administrator shall provide the owner or operator a detailed written statement of any such reason to believe that all haz-

ardous secondary materials have not been removed from the unit or that the unit has not been decontaminated in accordance with the approved plan.

## §§261.144–261.146 [Reserved]

## §261.147 Liability requirements.

(a) *Coverage for sudden accidental occurrences.* An owner or operator of a hazardous secondary material reclamation facility or an intermediate facility subject to financial assurance requirements under §261.4(a)(24)(vi)(F) of this chapter, or a group of such facilities, must demonstrate financial responsibility for bodily injury and property damage to third parties caused by sudden accidental occurrences arising from operations of the facility or group of facilities. The owner or operator must have and maintain liability coverage for sudden accidental occurrences in the amount of at least $1 million per occurrence with an annual aggregate of at least $2 million, exclusive of legal defense costs. This liability coverage may be demonstrated as specified in paragraphs (a) (1), (2), (3), (4), (5), or (6) of this section:

(1) An owner or operator may demonstrate the required liability coverage by having liability insurance as specified in this paragraph.

(i) Each insurance policy must be amended by attachment of the Hazardous Secondary Material Facility Liability Endorsement, or evidenced by a Certificate of Liability Insurance. The wording of the endorsement must be identical to the wording specified in §261.151(h). The wording of the certificate of insurance must be identical to the wording specified in §261.151(i). The owner or operator must submit a signed duplicate original of the endorsement or the certificate of insurance to the Regional Administrator, or Regional Administrators if the facilities are located in more than one Region. If requested by a Regional Administrator, the owner or operator must provide a signed duplicate original of the insurance policy.

(ii) Each insurance policy must be issued by an insurer which, at a minimum, is licensed to transact the business of insurance, or eligible to provide insurance as an excess or surplus lines insurer, in one or more States.

(2) An owner or operator may meet the requirements of this section by passing a financial test or using the guarantee for liability coverage as specified in paragraphs (f) and (g) of this section.

(3) An owner or operator may meet the requirements of this section by obtaining a letter of credit for liability coverage as specified in paragraph (h) of this section.

(4) An owner or operator may meet the requirements of this section by obtaining a surety bond for liability coverage as specified in paragraph (i) of this section.

(5) An owner or operator may meet the requirements of this section by obtaining a trust fund for liability coverage as specified in paragraph (j) of this section.

(6) An owner or operator may demonstrate the required liability coverage through the use of combinations of insurance, financial test, guarantee, letter of credit, surety bond, and trust fund, except that the owner or operator may not combine a financial test covering part of the liability coverage requirement with a guarantee unless the financial statement of the owner or operator is not consolidated with the financial statement of the guarantor. The amounts of coverage demonstrated must total at least the minimum amounts required by this section. If the owner or operator demonstrates the required coverage through the use of a combination of financial assurances under this paragraph, the owner or operator shall specify at least one such assurance as "primary" coverage and shall specify other assurance as "excess" coverage.

(7) An owner or operator shall notify the Regional Administrator in writing within 30 days whenever:

(i) A claim results in a reduction in the amount of financial assurance for liability coverage provided by a financial instrument authorized in paragraphs (a)(1) through (a)(6) of this section; or

(ii) A Certification of Valid Claim for bodily injury or property damages caused by a sudden or non-sudden accidental occurrence arising from the operation of a hazardous secondary material reclamation facility or intermediate facility is entered between the owner or operator and third-party claimant for liability coverage under paragraphs (a)(1) through (a)(6) of this section; or

(iii) A final court order establishing a judgment for bodily injury or property damage caused by a sudden or non-sudden accidental occurrence arising from the operation of a hazardous secondary material reclamation facility or intermediate facility is issued against the owner or operator or an instrument that is providing financial assurance for liability coverage under paragraphs (a)(1) through (a)(6) of this section.

(b) Coverage for nonsudden accidental occurrences. An owner or operator of a hazardous secondary material reclamation facility or intermediate facility with land-based units, as defined in § 260.10 of this chapter, which are used to manage hazardous secondary materials excluded under § 261.4(a)(24) of this chapter or a group of such facilities, must demonstrate financial responsibility for bodily injury and property damage to third parties caused by nonsudden accidental occurrences arising from operations of the facility or group of facilities. The owner or operator must have and maintain liability coverage for nonsudden accidental occurrences in the amount of at least $3 million per occurrence with an annual aggregate of at least $6 million, exclusive of legal defense costs. An owner or operator who must meet the requirements of this section may combine the required per-occurrence coverage levels for sudden and nonsudden accidental occurrences into a single per-occurrence level, and combine the required annual aggregate coverage levels for sudden and nonsudden accidental occurrences into a single annual aggregate level. Owners or operators who combine coverage levels for sudden and nonsudden accidental occurrences must maintain liability coverage in the amount of at least $4 million per occurrence and $8 million annual aggregate. This liability coverage may be demonstrated as specified in paragraph (b)(1), (2), (3), (4), (5), or (6) of this section:

(1) An owner or operator may demonstrate the required liability coverage by having liability insurance as specified in this paragraph.

(i) Each insurance policy must be amended by attachment of the Hazardous Secondary Material Facility Liability Endorsement or evidenced by a Certificate of Liability Insurance. The wording of the endorsement must be identical to the wording specified in §261.151(h). The wording of the certificate of insurance must be identical to the wording specified in §261.151(i). The owner or operator must submit a signed duplicate original of the endorsement or the certificate of insurance to the Regional Administrator, or Regional Administrators if the facilities are located in more than one Region. If requested by a Regional Administrator, the owner or operator must provide a signed duplicate original of the insurance policy.

(ii) Each insurance policy must be issued by an insurer which, at a minimum, is licensed to transact the business of insurance, or eligible to provide insurance as an excess or surplus lines insurer, in one or more States.

(2) An owner or operator may meet the requirements of this section by passing a financial test or using the guarantee for liability coverage as specified in paragraphs (f) and (g) of this section.

(3) An owner or operator may meet the requirements of this section by obtaining a letter of credit for liability coverage as specified in paragraph (h) of this section.

(4) An owner or operator may meet the requirements of this section by obtaining a surety bond for liability coverage as specified in paragraph (i) of this section.

(5) An owner or operator may meet the requirements of this section by obtaining a trust fund for liability coverage as specified in paragraph (j) of this section.

(6) An owner or operator may demonstrate the required liability coverage through the use of combinations of insurance, financial test, guarantee, letter of credit, surety bond, and trust fund, except that the owner or operator may not combine a financial test covering part of the liability coverage re-quirement with a guarantee unless the financial statement of the owner or operator is not consolidated with the financial statement of the guarantor. The amounts of coverage demonstrated must total at least the minimum amounts required by this section. If the owner or operator demonstrates the required coverage through the use of a combination of financial assurances under this paragraph, the owner or operator shall specify at least one such assurance as "primary" coverage and shall specify other assurance as "excess" coverage.

(7) An owner or operator shall notify the Regional Administrator in writing within 30 days whenever:

(i) A claim results in a reduction in the amount of financial assurance for liability coverage provided by a financial instrument authorized in paragraphs (b)(1) through (b)(6) of this section; or

(ii) A Certification of Valid Claim for bodily injury or property damages caused by a sudden or non-sudden accidental occurrence arising from the operation of a hazardous secondary material treatment and/or storage facility is entered between the owner or operator and third-party claimant for liability coverage under paragraphs (b)(1) through (b)(6) of this section; or

(iii) A final court order establishing a judgment for bodily injury or property damage caused by a sudden or non-sudden accidental occurrence arising from the operation of a hazardous secondary material treatment and/or storage facility is issued against the owner or operator or an instrument that is providing financial assurance for liability coverage under paragraphs (b)(1) through (b)(6) of this section.

(c) *Request for variance.* If an owner or operator can demonstrate to the satisfaction of the Regional Administrator that the levels of financial responsibility required by paragraph (a) or (b) of this section are not consistent with the degree and duration of risk associated with treatment and/or storage at the facility or group of facilities, the owner or operator may obtain a variance from the Regional Administrator. The request for a variance must be submitted in writing to the Regional Administrator. If granted, the variance will take the form of an adjusted level of required liability coverage, such level to be based on the Regional Administrator's assessment of the degree and duration of risk associated with the ownership or operation of the facility or group of facilities. The Regional Administrator may require an owner or operator who requests a variance to provide such technical and engineering information as is deemed necessary by the Regional Administrator to determine a level of financial responsibility other than that required by paragraph (a) or (b) of this section.

(d) *Adjustments by the Regional Administrator.* If the Regional Administrator determines that the levels of financial responsibility required by paragraph (a) or (b) of this section are not consistent with the degree and duration of risk associated with treatment and/or storage at the facility or group of facilities, the Regional Administrator may adjust the level of financial responsibility required under paragraph (a) or (b) of this section as may be necessary to protect human health and the environment. This adjusted level will be based on the Regional Administrator's assessment of the degree and duration of risk associated with the ownership or operation of the facility or group of facilities. In addition, if the Regional Administrator determines that there is a significant risk to human health and the environment from nonsudden accidental occurrences resulting from the operations of a facility that is not a surface impoundment, pile, or land treatment facility, he may require that an owner or operator of the facility comply with paragraph (b) of this section. An owner or operator must furnish to the Regional Administrator, within a reasonable time, any information which the Regional Administrator requests to determine whether cause exists for such adjustments of level or type of coverage.

(e) *Period of coverage.* Within 60 days after receiving certifications from the owner or operator and a qualified Professional Engineer that all hazardous secondary materials have been removed from the facility or a unit at the facility and the facility or a unit has been decontaminated in accordance with the approved plan per §261.143(h),

the Regional Administrator will notify the owner or operator in writing that he is no longer required under §261.4(a)(24)(vi)(F) to maintain liability coverage for that facility or a unit at the facility, unless the Regional Administrator has reason to believe that that all hazardous secondary materials have not been removed from the facility or unit at a facility or that the facility or unit has not been decontaminated in accordance with the approved plan.

(f) *Financial test for liability coverage.* (1) An owner or operator may satisfy the requirements of this section by demonstrating that he passes a financial test as specified in this paragraph. To pass this test the owner or operator must meet the criteria of paragraph (f)(1) (i) or (ii) of this section:

(i) The owner or operator must have:

(A) Net working capital and tangible net worth each at least six times the amount of liability coverage to be demonstrated by this test; and

(B) Tangible net worth of at least $10 million; and

(C) Assets in the United States amounting to either:

(*1*) At least 90 percent of his total assets; or

(*2*) at least six times the amount of liability coverage to be demonstrated by this test.

(ii) The owner or operator must have:

(A) A current rating for his most recent bond issuance of AAA, AA, A, or BBB as issued by Standard and Poor's, or Aaa, Aa, A, or Baa as issued by Moody's; and

(B) Tangible net worth of at least $10 million; and

(C) Tangible net worth at least six times the amount of liability coverage to be demonstrated by this test; and

(D) Assets in the United States amounting to either:

(*1*) At least 90 percent of his total assets; or

(*2*) at least six times the amount of liability coverage to be demonstrated by this test.

(2) The phrase "amount of liability coverage" as used in paragraph (f)(1) of this section refers to the annual aggregate amounts for which coverage is required under paragraphs (a) and (b) of this section and the annual aggregate

amounts for which coverage is required under paragraphs (a) and (b) of 40 CFR 264.147 and 265.147.

(3) To demonstrate that he meets this test, the owner or operator must submit the following three items to the Regional Administrator:

(i) A letter signed by the owner's or operator's chief financial officer and worded as specified in §261.151(f). If an owner or operator is using the financial test to demonstrate both assurance as specified by §261.143(e), *and* liability coverage, he must submit the letter specified in §261.151(f) to cover both forms of financial responsibility; a separate letter as specified in §261.151(e) is not required.

(ii) A copy of the independent certified public accountant's report on examination of the owner's or operator's financial statements for the latest completed fiscal year.

(iii) If the chief financial officer's letter providing evidence of financial assurance includes financial data showing that the owner or operator satisfies paragraph (f)(1)(i) of this section that are different from the data in the audited financial statements referred to in paragraph (f)(3)(ii) of this section or any other audited financial statement or data filed with the SEC, then a special report from the owner's or operator's independent certified public accountant to the owner or operator is required. The special report shall be based upon an agreed upon procedures engagement in accordance with professional auditing standards and shall describe the procedures performed in comparing the data in the chief financial officer's letter derived from the independently audited, year-end financial statements for the latest fiscal year with the amounts in such financial statements, the findings of the comparison, and the reasons for any difference.

(4) The owner or operator may obtain a one-time extension of the time allowed for submission of the documents specified in paragraph (f)(3) of this section if the fiscal year of the owner or operator ends during the 90 days prior to the effective date of these regulations and if the year-end financial statements for that fiscal year will be audited by an independent certified

public accountant. The extension will end no later than 90 days after the end of the owner's or operator's fiscal year. To obtain the extension, the owner's or operator's chief financial officer must send, by the effective date of these regulations, a letter to the Regional Administrator of each Region in which the owner's or operator's facilities to be covered by the financial test are located. This letter from the chief financial officer must:

(i) Request the extension;

(ii) Certify that he has grounds to believe that the owner or operator meets the criteria of the financial test;

(iii) Specify for each facility to be covered by the test the EPA Identification Number, name, address, the amount of liability coverage and, when applicable, current closure and post-closure cost estimates to be covered by the test;

(iv) Specify the date ending the owner's or operator's last complete fiscal year before the effective date of these regulations;

(v) Specify the date, no later than 90 days after the end of such fiscal year, when he will submit the documents specified in paragraph (f)(3) of this section; and

(vi) Certify that the year-end financial statements of the owner or operator for such fiscal year will be audited by an independent certified public accountant.

(5) After the initial submission of items specified in paragraph (f)(3) of this section, the owner or operator must send updated information to the Regional Administrator within 90 days after the close of each succeeding fiscal year. This information must consist of all three items specified in paragraph (f)(3) of this section.

(6) If the owner or operator no longer meets the requirements of paragraph (f)(1) of this section, he must obtain insurance, a letter of credit, a surety bond, a trust fund, or a guarantee for the entire amount of required liability coverage as specified in this section. Evidence of liability coverage must be submitted to the Regional Administrator within 90 days after the end of the fiscal year for which the year-end financial data show that the owner or

operator no longer meets the test requirements.

(7) The Regional Administrator may disallow use of this test on the basis of qualifications in the opinion expressed by the independent certified public accountant in his report on examination of the owner's or operator's financial statements (see paragraph (f)(3)(ii) of this section). An adverse opinion or a disclaimer of opinion will be cause for disallowance. The Regional Administrator will evaluate other qualifications on an individual basis. The owner or operator must provide evidence of insurance for the entire amount of required liability coverage as specified in this section within 30 days after notification of disallowance.

(g) *Guarantee for liability coverage.* (1) Subject to paragraph (g)(2) of this section, an owner or operator may meet the requirements of this section by obtaining a written guarantee, hereinafter referred to as "guarantee." The guarantor must be the direct or higher-tier parent corporation of the owner or operator, a firm whose parent corporation is also the parent corporation of the owner or operator, or a firm with a "substantial business relationship" with the owner or operator. The guarantor must meet the requirements for owners or operators in paragraphs (f)(1) through (f)(6) of this section. The wording of the guarantee must be identical to the wording specified in § 261.151(g)(2). A certified copy of the guarantee must accompany the items sent to the Regional Administrator as specified in paragraph (f)(3) of this section. One of these items must be the letter from the guarantor's chief financial officer. If the guarantor's parent corporation is also the parent corporation of the owner or operator, this letter must describe the value received in consideration of the guarantee. If the guarantor is a firm with a "substantial business relationship" with the owner or operator, this letter must describe this "substantial business relationship" and the value received in consideration of the guarantee.

(i) If the owner or operator fails to satisfy a judgment based on a determination of liability for bodily injury or property damage to third parties caused by sudden or nonsudden accidental occurrences (or both as the case may be), arising from the operation of facilities covered by this corporate guarantee, or fails to pay an amount agreed to in settlement of claims arising from or alleged to arise from such injury or damage, the guarantor will do so up to the limits of coverage.

(ii) [Reserved]

(2)(i) In the case of corporations incorporated in the United States, a guarantee may be used to satisfy the requirements of this section only if the Attorneys General or Insurance Commissioners of:

(A) The State in which the guarantor is incorporated; and

(B) Each State in which a facility covered by the guarantee is located have submitted a written statement to EPA that a guarantee executed as described in this section and § 264.151(g)(2) is a legally valid and enforceable obligation in that State.

(ii) In the case of corporations incorporated outside the United States, a guarantee may be used to satisfy the requirements of this section only if:

(A) The non-U.S. corporation has identified a registered agent for service of process in each State in which a facility covered by the guarantee is located and in the State in which it has its principal place of business; and if

(B) The Attorney General or Insurance Commissioner of each State in which a facility covered by the guarantee is located and the State in which the guarantor corporation has its principal place of business, has submitted a written statement to EPA that a guarantee executed as described in this section and § 261.151(h)(2) is a legally valid and enforceable obligation in that State.

(h) *Letter of credit for liability coverage.* (1) An owner or operator may satisfy the requirements of this section by obtaining an irrevocable standby letter of credit that conforms to the requirements of this paragraph and submitting a copy of the letter of credit to the Regional Administrator.

(2) The financial institution issuing the letter of credit must be an entity that has the authority to issue letters of credit and whose letter of credit operations are regulated and examined by a Federal or State agency.

(3) The wording of the letter of credit must be identical to the wording specified in §261.151(j).

(4) An owner or operator who uses a letter of credit to satisfy the requirements of this section may also establish a standby trust fund. Under the terms of such a letter of credit, all amounts paid pursuant to a draft by the trustee of the standby trust will be deposited by the issuing institution into the standby trust in accordance with instructions from the trustee. The trustee of the standby trust fund must be an entity which has the authority to act as a trustee and whose trust operations are regulated and examined by a Federal or State agency.

(5) The wording of the standby trust fund must be identical to the wording specified in §261.151(m).

(i) *Surety bond for liability coverage.* (1) An owner or operator may satisfy the requirements of this section by obtaining a surety bond that conforms to the requirements of this paragraph and submitting a copy of the bond to the Regional Administrator.

(2) The surety company issuing the bond must be among those listed as acceptable sureties on Federal bonds in the most recent Circular 570 of the U.S. Department of the Treasury.

(3) The wording of the surety bond must be identical to the wording specified in §261.151(k) of this chapter.

(4) A surety bond may be used to satisfy the requirements of this section only if the Attorneys General or Insurance Commissioners of:

(i) The State in which the surety is incorporated; and

(ii) Each State in which a facility covered by the surety bond is located have submitted a written statement to EPA that a surety bond executed as described in this section and §261.151(k) is a legally valid and enforceable obligation in that State.

(j) *Trust fund for liability coverage.* (1) An owner or operator may satisfy the requirements of this section by establishing a trust fund that conforms to the requirements of this paragraph and submitting an originally signed duplicate of the trust agreement to the Regional Administrator.

(2) The trustee must be an entity which has the authority to act as a

trustee and whose trust operations are regulated and examined by a Federal or State agency.

(3) The trust fund for liability coverage must be funded for the full amount of the liability coverage to be provided by the trust fund before it may be relied upon to satisfy the requirements of this section. If at any time after the trust fund is created the amount of funds in the trust fund is reduced below the full amount of the liability coverage to be provided, the owner or operator, by the anniversary date of the establishment of the Fund, must either add sufficient funds to the trust fund to cause its value to equal the full amount of liability coverage to be provided, or obtain other financial assurance as specified in this section to cover the difference. For purposes of this paragraph, "the full amount of the liability coverage to be provided" means the amount of coverage for sudden and/or nonsudden occurrences required to be provided by the owner or operator by this section, less the amount of financial assurance for liability coverage that is being provided by other financial assurance mechanisms being used to demonstrate financial assurance by the owner or operator.

(4) The wording of the trust fund must be identical to the wording specified in §261.151(l).

## §261.148 Incapacity of owners or operators, guarantors, or financial institutions.

(a) An owner or operator must notify the Regional Administrator by certified mail of the commencement of a voluntary or involuntary proceeding under Title 11 (Bankruptcy), U.S. Code, naming the owner or operator as debtor, within 10 days after commencement of the proceeding. A guarantor of a corporate guarantee as specified in §261.143(e) must make such a notification if he is named as debtor, as required under the terms of the corporate guarantee.

(b) An owner or operator who fulfills the requirements of §261.143 or §261.147 by obtaining a trust fund, surety bond, letter of credit, or insurance policy will be deemed to be without the required financial assurance or liability coverage in the event of bankruptcy of the trustee or issuing institution, or a suspension or revocation of the authority of the trustee institution to act as trustee or of the institution issuing the surety bond, letter of credit, or insurance policy to issue such instruments. The owner or operator must establish other financial assurance or liability coverage within 60 days after such an event.

## § 261.149  Use of State-required mechanisms.

(a) For a reclamation or intermediate facility located in a State where EPA is administering the requirements of this subpart but where the State has regulations that include requirements for financial assurance of closure or liability coverage, an owner or operator may use State-required financial mechanisms to meet the requirements of § 261.143 or § 261.147 if the Regional Administrator determines that the State mechanisms are at least equivalent to the financial mechanisms specified in this subpart. The Regional Administrator will evaluate the equivalency of the mechanisms principally in terms of certainty of the availability of: Funds for the required closure activities or liability coverage; and the amount of funds that will be made available. The Regional Administrator may also consider other factors as he deems appropriate. The owner or operator must submit to the Regional Administrator evidence of the establishment of the mechanism together with a letter requesting that the State-required mechanism be considered acceptable for meeting the requirements of this subpart. The submission must include the following information: The facility's EPA Identification Number (if available), name, and address, and the amount of funds for closure or liability coverage assured by the mechanism. The Regional Administrator will notify the owner or operator of his determination regarding the mechanism's acceptability in lieu of financial mechanisms specified in this subpart. The Regional Administrator may require the owner or operator to submit additional information as is deemed necessary to make this determination. Pending this determination, the owner

or operator will be deemed to be in compliance with the requirements of § 261.143 or § 261.147, as applicable.

(b) If a State-required mechanism is found acceptable as specified in paragraph (a) of this section except for the amount of funds available, the owner or operator may satisfy the requirements of this subpart by increasing the funds available through the State-required mechanism or using additional financial mechanisms as specified in this subpart. The amount of funds available through the State and Federal mechanisms must at least equal the amount required by this subpart.

## § 261.150  State assumption of responsibility.

(a) If a State either assumes legal responsibility for an owner's or operator's compliance with the closure or liability requirements of this part or assures that funds will be available from State sources to cover those requirements, the owner or operator will be in compliance with the requirements of § 261.143 or § 261.147 if the Regional Administrator determines that the State's assumption of responsibility is at least equivalent to the financial mechanisms specified in this subpart. The Regional Administrator will evaluate the equivalency of State guarantees principally in terms of: Certainty of the availability of funds for the required closure activities or liability coverage; and the amount of funds that will be made available. The Regional Administrator may also consider other factors as he deems appropriate. The owner or operator must submit to the Regional Administrator a letter from the State describing the nature of the State's assumption of responsibility together with a letter from the owner or operator requesting that the State's assumption of responsibility be considered acceptable for meeting the requirements of this subpart. The letter from the State must include, or have attached to it, the following information: The facility's EPA Identification Number (if available), name, and address, and the amount of funds for closure or liability coverage that are guaranteed by the State. The Regional Administrator will notify the owner or operator of his determination regarding the acceptability of the State's guarantee in lieu of financial mechanisms specified in this subpart. The Regional Administrator may require the owner or operator to submit additional information as is deemed necessary to make this determination. Pending this determination, the owner or operator will be deemed to be in compliance with the requirements of § 265.143 or § 265.147, as applicable.

(b) If a State's assumption of responsibility is found acceptable as specified in paragraph (a) of this section except for the amount of funds available, the owner or operator may satisfy the requirements of this subpart by use of both the State's assurance and additional financial mechanisms as specified in this subpart. The amount of funds available through the State and Federal mechanisms must at least equal the amount required by this subpart.

## § 261.151  Wording of the instruments.

(a)(1) A trust agreement for a trust fund, as specified in § 261.143(a) must be worded as follows, except that instructions in brackets are to be replaced with the relevant information and the brackets deleted:

TRUST AGREEMENT

Trust Agreement, the "Agreement," entered into as of [date] by and between [name of the owner or operator], a [name of State] [insert "corporation," "partnership," "association," or "proprietorship"], the "Grantor," and [name of corporate trustee], [insert "incorporated in the State of _____" or "a national bank"], the "Trustee."

Whereas, the United States Environmental Protection Agency, "EPA," an agency of the United States Government, has established certain regulations applicable to the Grantor, requiring that an owner or operator of a facility regulated under parts 264, or 265, or satisfying the conditions of the exclusion under § 261.4(a)(24) shall provide assurance that funds will be available if needed for care of the facility under 40 CFR parts 264 or 265, subparts G, as applicable ,

Whereas, the Grantor has elected to establish a trust to provide all or part of such financial assurance for the facilities identified herein,

Whereas, the Grantor, acting through its duly authorized officers, has selected the Trustee to be the trustee under this agreement, and the Trustee is willing to act as trustee,

Now, Therefore, the Grantor and the Trustee agree as follows:

Section 1. Definitions. As used in this Agreement:

(a) The term "Grantor" means the owner or operator who enters into this Agreement and any successors or assigns of the Grantor.

(b) The term "Trustee" means the Trustee who enters into this Agreement and any successor Trustee.

Section 2. Identification of Facilities and Cost Estimates. This Agreement pertains to the facilities and cost estimates identified on attached Schedule A [on Schedule A, for each facility list the EPA Identification Number (if available), name, address, and the current cost estimates, or portions thereof, for which financial assurance is demonstrated by this Agreement].

Section 3. Establishment of Fund. The Grantor and the Trustee hereby establish a trust fund, the "Fund," for the benefit of EPA in the event that the hazardous secondary materials of the grantor no longer meet the conditions of the exclusion under § 261.4(a)(24). The Grantor and the Trustee intend that no third party have access to the Fund except as herein provided. The Fund is established initially as consisting of the property, which is acceptable to the Trustee, described in Schedule B attached hereto. Such property and any other property subsequently transferred to the Trustee is referred to as the Fund, together with all earnings and profits thereon, less any payments or distributions made by the Trustee pursuant to this Agreement. The Fund shall be held by the Trustee, IN TRUST, as hereinafter provided. The Trustee shall not be responsible nor shall it undertake any responsibility for the amount or adequacy of, nor any duty to collect from the Grantor, any payments necessary to discharge any liabilities of the Grantor established by EPA.

Section 4. Payments from the Fund. The Trustee shall make payments from the Fund as the EPA Regional Administrator shall direct, in writing, to provide for the payment of the costs of the performance of activities required under subpart G of 40 CFR parts 264 or 265 for the facilities covered by this Agreement. The Trustee shall reimburse the Grantor or other persons as specified by the EPA Regional Administrator from the Fund for expenditures for such activities in such amounts as the beneficiary shall direct in writing. In addition, the Trustee shall refund to the Grantor such amounts as the EPA Regional Administrator specifies in writing. Upon refund, such funds shall no longer constitute part of the Fund as defined herein.

Section 5. Payments Comprising the Fund. Payments made to the Trustee for the Fund shall consist of cash or securities acceptable to the Trustee.

Section 6. Trustee Management. The Trustee shall invest and reinvest the principal and income of the Fund and keep the Fund invested as a single fund, without distinction between principal and income, in accordance with general investment policies and guidelines which the Grantor may communicate in writing to the Trustee from time to time, subject, however, to the provisions of this section. In investing, reinvesting, exchanging, selling, and managing the Fund, the Trustee shall discharge his duties with respect to the trust fund solely in the interest of the beneficiary and with the care, skill, prudence, and diligence under the circumstances then prevailing which persons of prudence, acting in a like capacity and familiar with such matters, would use in the conduct of an enterprise of a like character and with like aims; except that:

(i) Securities or other obligations of the Grantor, or any other owner or operator of the facilities, or any of their affiliates as defined in the Investment Company Act of 1940, as amended, 15 U.S.C. 80a–2.(a), shall not be acquired or held, unless they are securities or other obligations of the Federal or a State government;

(ii) The Trustee is authorized to invest the Fund in time or demand deposits of the Trustee, to the extent insured by an agency of the Federal or State government; and

(iii) The Trustee is authorized to hold cash awaiting investment or distribution uninvested for a reasonable time and without liability for the payment of interest thereon.

Section 7. Commingling and Investment. The Trustee is expressly authorized in its discretion:

(a) To transfer from time to time any or all of the assets of the Fund to any common, commingled, or collective trust fund created by the Trustee in which the Fund is eligible to participate, subject to all of the provisions thereof, to be commingled with the assets of other trusts participating therein; and

(b) To purchase shares in any investment company registered under the Investment Company Act of 1940, 15 U.S.C. 80a–1 *et seq.*, including one which may be created, managed, underwritten, or to which investment advice is rendered or the shares of which are sold by the Trustee. The Trustee may vote such shares in its discretion.

Section 8. Express Powers of Trustee. Without in any way limiting the powers and discretions conferred upon the Trustee by the other provisions of this Agreement or by law, the Trustee is expressly authorized and empowered:

(a) To sell, exchange, convey, transfer, or otherwise dispose of any property held by it, by public or private sale. No person dealing with the Trustee shall be bound to see to the application of the purchase money or to inquire into the validity or expediency of any such sale or other disposition;

(b) To make, execute, acknowledge, and deliver any and all documents of transfer and conveyance and any and all other instruments that may be necessary or appropriate to carry out the powers herein granted;

(c) To register any securities held in the Fund in its own name or in the name of a nominee and to hold any security in bearer form or in book entry, or to combine certificates representing such securities with certificates of the same issue held by the Trustee in other fiduciary capacities, or to deposit or arrange for the deposit of such securities in a qualified central depositary even though, when so deposited, such securities may be merged and held in bulk in the name of the nominee of such depositary with other securities deposited therein by another person, or to deposit or arrange for the deposit of any securities issued by the United States Government, or any agency or instrumentality thereof, with a Federal Reserve bank, but the books and records of the Trustee shall at all times show that all such securities are part of the Fund;

(d) To deposit any cash in the Fund in interest-bearing accounts maintained or savings certificates issued by the Trustee, in its separate corporate capacity, or in any other banking institution affiliated with the Trustee, to the extent insured by an agency of the Federal or State government; and

(e) To compromise or otherwise adjust all claims in favor of or against the Fund.

Section 9. Taxes and Expenses. All taxes of any kind that may be assessed or levied against or in respect of the Fund and all brokerage commissions incurred by the Fund shall be paid from the Fund. All other expenses incurred by the Trustee in connection with the administration of this Trust, including fees for legal services rendered to the Trustee, the compensation of the Trustee to the extent not paid directly by the Grantor, and all other proper charges and disbursements of the Trustee shall be paid from the Fund.

Section 10. Annual Valuation. The Trustee shall annually, at least 30 days prior to the anniversary date of establishment of the Fund, furnish to the Grantor and to the appropriate EPA Regional Administrator a statement confirming the value of the Trust. Any securities in the Fund shall be valued at market value as of no more than 60 days prior to the anniversary date of establishment of the Fund. The failure of the Grantor to object in writing to the Trustee within 90 days after the statement has been furnished to the Grantor and the EPA Regional Administrator shall constitute a conclusively binding assent by the Grantor, barring the Grantor from asserting any claim or liability against the Trustee with respect to matters disclosed in the statement.

Section 11. Advice of Counsel. The Trustee may from time to time consult with counsel, who may be counsel to the Grantor, with respect to any question arising as to the construction of this Agreement or any action to be taken hereunder. The Trustee shall be fully protected, to the extent permitted by law, in acting upon the advice of counsel.

Section 12. Trustee Compensation. The Trustee shall be entitled to reasonable compensation for its services as agreed upon in writing from time to time with the Grantor.

Section 13. Successor Trustee. The Trustee may resign or the Grantor may replace the Trustee, but such resignation or replacement shall not be effective until the Grantor has appointed a successor trustee and this successor accepts the appointment. The successor trustee shall have the same powers and duties as those conferred upon the Trustee hereunder. Upon the successor trustee's acceptance of the appointment, the Trustee shall assign, transfer, and pay over to the successor trustee the funds and properties then constituting the Fund. If for any reason the Grantor cannot or does not act in the event of the resignation of the Trustee, the Trustee may apply to a court of competent jurisdiction for the appointment of a successor trustee or for instructions. The successor trustee shall specify the date on which it assumes administration of the trust in a writing sent to the Grantor, the EPA Regional Administrator, and the present Trustee by certified mail 10 days before such change becomes effective. Any expenses incurred by the Trustee as a result of any of the acts contemplated by this Section shall be paid as provided in Section 9.

Section 14. Instructions to the Trustee. All orders, requests, and instructions by the Grantor to the Trustee shall be in writing, signed by such persons as are designated in the attached Exhibit A or such other designees as the Grantor may designate by amendment to Exhibit A. The Trustee shall be fully protected in acting without inquiry in accordance with the Grantor's orders, requests, and instructions. All orders, requests, and instructions by the EPA Regional Administrator to the Trustee shall be in writing, signed by the EPA Regional Administrators of the Regions in which the facilities are located, or their designees, and the Trustee shall act and shall be fully protected in acting in accordance with such orders, requests, and instructions. The Trustee shall have the right to assume, in the absence of written notice to the contrary, that no event constituting a change or a termination of the authority of any person to act on behalf of the Grantor or EPA hereunder has occurred. The Trustee shall have no duty to act in the absence of such orders, requests, and instructions from the Grantor and/or EPA, except as provided for herein.

Section 15. Amendment of Agreement. This Agreement may be amended by an instrument in writing executed by the Grantor, the Trustee, and the appropriate EPA Regional Administrator, or by the Trustee and the appropriate EPA Regional Administrator if the Grantor ceases to exist.

Section 16. Irrevocability and Termination. Subject to the right of the parties to amend this Agreement as provided in Section 16, this Trust shall be irrevocable and shall continue until terminated at the written agreement of the Grantor, the Trustee, and the EPA Regional Administrator, or by the Trustee and the EPA Regional Administrator, if the Grantor ceases to exist. Upon termination of the Trust, all remaining trust property, less final trust administration expenses, shall be delivered to the Grantor.

Section 17. Immunity and Indemnification. The Trustee shall not incur personal liability of any nature in connection with any act or omission, made in good faith, in the administration of this Trust, or in carrying out any directions by the Grantor or the EPA Regional Administrator issued in accordance with this Agreement. The Trustee shall be indemnified and saved harmless by the Grantor or from the Trust Fund, or both, from and against any personal liability to which the Trustee may be subjected by reason of any act or conduct in its official capacity, including all expenses reasonably incurred in its defense in the event the Grantor fails to provide such defense.

Section 18. Choice of Law. This Agreement shall be administered, construed, and enforced according to the laws of the State of [insert name of State].

Section 19. Interpretation. As used in this Agreement, words in the singular include the plural and words in the plural include the singular. The descriptive headings for each Section of this Agreement shall not affect the interpretation or the legal efficacy of this Agreement.

In Witness Whereof the parties have caused this Agreement to be executed by their respective officers duly authorized and their corporate seals to be hereunto affixed and attested as of the date first above written: The parties below certify that the wording of this Agreement is identical to the wording specified in 40 CFR 261.151(a)(1) as such regulations were constituted on the date first above written.

[Signature of Grantor]
[Title]

Attest:
[Title]
[Seal]
[Signature of Trustee]

Attest:
[Title]
[Seal]

133

(2) The following is an example of the certification of acknowledgment which must accompany the trust agreement for a trust fund as specified in § 261.143(a) of this chapter. State requirements may differ on the proper content of this acknowledgment.

State of _____
County of _____
On this [date], before me personally came [owner or operator] to me known, who, being by me duly sworn, did depose and say that she/he resides at [address], that she/he is [title] of [corporation], the corporation described in and which executed the above instrument; that she/he knows the seal of said corporation; that the seal affixed to such instrument is such corporate seal; that it was so affixed by order of the Board of Directors of said corporation, and that she/he signed her/his name thereto by like order.

[Signature of Notary Public]

(b) A surety bond guaranteeing payment into a trust fund, as specified in § 261.143(b) of this chapter, must be worded as follows, except that instructions in brackets are to be replaced with the relevant information and the brackets deleted:

FINANCIAL GUARANTEE BOND

Date bond executed:
Effective date:
Principal: [legal name and business address of owner or operator]
Type of Organization: [insert "individual," "joint venture," "partnership," or "corporation"]
State of incorporation: _____
Surety(ies): [name(s) and business address(es)]
EPA Identification Number, name, address and amount(s) for each facility guaranteed by this bond:
Total penal sum of bond: $ _____
Surety's bond number: _____
Know All Persons By These Presents, That we, the Principal and Surety(ies) are firmly bound to the U.S. EPA in the event that the hazardous secondary materials at the reclamation or intermediate facility listed below no longer meet the conditions of the exclusion under 40 CFR 261.4(a)(24), in the above penal sum for the payment of which we bind ourselves, our heirs, executors, administrators, successors, and assigns jointly and severally; provided that, where the Surety(ies) are corporations acting as co-sureties, we, the Sureties, bind ourselves in such sum "jointly and severally" only for the purpose of allowing a joint action or actions against any or all of us, and for all other purposes each Surety binds itself, jointly and severally with the Principal, for the payment of such sum only as is set forth opposite the name of such Surety, but if no limit of liability is indicated, the limit of liability shall be the full amount of the penal sum.

Whereas said Principal is required, under the Resource Conservation and Recovery Act as amended (RCRA), to have a permit or interim status in order to own or operate each facility identified above, or to meet conditions under 40 CFR sections 261.4(a)(24), and

Whereas said Principal is required to provide financial assurance as a condition of permit or interim status or as a condition of an exclusion under 40 CFR sections 261.4(a)(24) and

Whereas said Principal shall establish a standby trust fund as is required when a surety bond is used to provide such financial assurance;

Now, Therefore, the conditions of the obligation are such that if the Principal shall faithfully, before the beginning of final closure of each facility identified above, fund the standby trust fund in the amount(s) identified above for the facility,

Or, if the Principal shall satisfy all the conditions established for exclusion of hazardous secondary materials from coverage as solid waste under 40 CFR sections 261.4(a)(24),

Or, if the Principal shall fund the standby trust fund in such amount(s) within 15 days after a final order to begin closure is issued by an EPA Regional Administrator or a U.S. district court or other court of competent jurisdiction,

Or, if the Principal shall provide alternate financial assurance, as specified in subpart H of 40 CFR part 261, as applicable, and obtain the EPA Regional Administrator's written approval of such assurance, within 90 days after the date notice of cancellation is received by both the Principal and the EPA Regional Administrator(s) from the Surety(ies), then this obligation shall be null and void; otherwise it is to remain in full force and effect.

The Surety(ies) shall become liable on this bond obligation only when the Principal has failed to fulfill the conditions described above. Upon notification by an EPA Regional Administrator that the Principal has failed to perform as guaranteed by this bond, the Surety(ies) shall place funds in the amount guaranteed for the facility(ies) into the standby trust fund as directed by the EPA Regional Administrator.

The liability of the Surety(ies) shall not be discharged by any payment or succession of payments hereunder, unless and until such payment or payments shall amount in the aggregate to the penal sum of the bond, but in no event shall the obligation of the Surety(ies) hereunder exceed the amount of said penal sum.

The Surety(ies) may cancel the bond by sending notice of cancellation by certified mail to the Principal and to the EPA Regional Administrator(s) for the Region(s) in which the facility(ies) is (are) located, provided, however, that cancellation shall not occur during the 120 days beginning on the date of receipt of the notice of cancellation by both the Principal and the EPA Regional Administrator(s), as evidenced by the return receipts.

The Principal may terminate this bond by sending written notice to the Surety(ies), provided, however, that no such notice shall become effective until the Surety(ies) receive(s) written authorization for termination of the bond by the EPA Regional Administrator(s) of the EPA Region(s) in which the bonded facility(ies) is (are) located.

[The following paragraph is an optional rider that may be included but is not required.]

Principal and Surety(ies) hereby agree to adjust the penal sum of the bond yearly so that it guarantees a new amount, provided that the penal sum does not increase by more than 20 percent in any one year, and no decrease in the penal sum takes place without the written permission of the EPA Regional Administrator(s).

In Witness Whereof, the Principal and Surety(ies) have executed this Financial Guarantee Bond and have affixed their seals on the date set forth above.

The persons whose signatures appear below hereby certify that they are authorized to execute this surety bond on behalf of the Principal and Surety(ies) and that the wording of this surety bond is identical to the wording specified in 40 CFR 261.151(b) as such regulations were constituted on the date this bond was executed.

PRINCIPAL

[Signature(s)]

[Name(s)]

[Title(s)]

[Corporate seal]

CORPORATE SURETY(IES)

[Name and address]
State of incorporation:
Liability limit:
$
[Signature(s)]
[Name(s) and title(s)]
[Corporate seal]
[For every co-surety, provide signature(s), corporate seal, and other information in the same manner as for Surety above.]
Bond premium: $

(c) A letter of credit, as specified in §261.143(c) of this chapter, must be worded as follows, except that instructions in brackets are to be replaced with the relevant information and the brackets deleted:

Irrevocable Standby Letter of Credit

Regional Administrator(s)

Region(s) _____

U.S. Environmental Protection Agency

Dear Sir or Madam: We hereby establish our Irrevocable Standby Letter of Credit No._____ in your favor, in the event that the hazardous secondary materials at the covered reclamation or intermediary facility(ies) no longer meet the conditions of the exclusion under 40 CFR 261.4(a)(24), at the request and for the account of [owner's or operator's name and address] up to the aggregate amount of [in words] U.S. dollars $_____, available upon presentation of

(1) your sight draft, bearing reference to this letter of credit No.___, and

(2) your signed statement reading as follows: "I certify that the amount of the draft is payable pursuant to regulations issued under authority of the Resource Conservation and Recovery Act of 1976 as amended."

This letter of credit is effective as of [date] and shall expire on [date at least 1 year later], but such expiration date shall be automatically extended for a period of [at least 1 year] on [date] and on each successive expiration date, unless, at least 120 days before the current expiration date, we notify both you and [owner's or operator's name] by certified mail that we have decided not to extend this letter of credit beyond the current expiration date. In the event you are so notified, any unused portion of the credit shall be available upon presentation of your sight draft for 120 days after the date of receipt by both you and [owner's or operator's name], as shown on the signed return receipts.

Whenever this letter of credit is drawn on under and in compliance with the terms of this credit, we shall duly honor such draft upon presentation to us, and we shall deposit the amount of the draft directly into the standby trust fund of [owner's or operator's name] in accordance with your instructions.

We certify that the wording of this letter of credit is identical to the wording specified in 40 CFR 261.151(c) as such regulations were constituted on the date shown immediately below.

[Signature(s) and title(s) of official(s) of issuing institution] [Date]

This credit is subject to [insert "the most recent edition of the Uniform Customs and Practice for Documentary Credits, published and copyrighted by the International Chamber of Commerce," or "the Uniform Commercial Code"].

(d) A certificate of insurance, as specified in § 261.143(e) of this chapter, must be worded as follows, except that instructions in brackets are to be replaced with the relevant information and the brackets deleted:

Certificate of Insurance

Name and Address of Insurer (herein called the "Insurer"):

_____

Name and Address of Insured (herein called the "Insured"):

_____

Facilities Covered: [List for each facility: The EPA Identification Number (if any issued), name, address, and the amount of insurance for all facilities covered, which must total the face amount shown below.

Face Amount:

_____

Policy Number: _____

Effective Date:

_____

The Insurer hereby certifies that it has issued to the Insured the policy of insurance identified above to provide financial assurance so that in accordance with applicable regulations all hazardous secondary materials can be removed from the facility or any unit at the facility and the facility or any unit at the facility can be decontaminated at the facilities identified above. The Insurer further warrants that such policy conforms in all respects with the requirements of 40 CFR 261.143(d) as applicable and as such regulations were constituted on the date shown immediately below. It is agreed that any provision of the policy inconsistent with such regulations is hereby amended to eliminate such inconsistency.

Whenever requested by the EPA Regional Administrator(s) of the U.S. Environmental Protection Agency, the Insurer agrees to furnish to the EPA Regional Administrator(s) a duplicate original of the policy listed above, including all endorsements thereon.

I hereby certify that the wording of this certificate is identical to the wording specified in 40 CFR 261.151(d) such regulations were constituted on the date shown immediately below.

[Authorized signature for Insurer]

[Name of person signing]

[Title of person signing]

Signature of witness or notary: _____

[Date]

(e) A letter from the chief financial officer, as specified in § 261.143(e) of this chapter, must be worded as follows, except that instructions in brackets are to be replaced with the relevant information and the brackets deleted:

Letter From Chief Financial Officer

[Address to Regional Administrator of every Region in which facilities for which financial responsibility is to be demonstrated through the financial test are located].

I am the chief financial officer of [name and address of firm]. This letter is in support of this firm's use of the financial test to demonstrate financial assurance, as specified in subpart H of 40 CFR part 261.

[Fill out the following nine paragraphs regarding facilities and associated cost estimates. If your firm has no facilities that belong in a particular paragraph, write "None" in the space indicated. For each facility, include its EPA Identification Number (if any issued), name, address, and current cost estimates.]

1. This firm is the owner or operator of the following facilities for which financial assurance is demonstrated through the financial test specified in subpart H of 40 CFR 261. The current cost estimates covered by the test are shown for each facility: _____.

2. This firm guarantees, through the guarantee specified in subpart H of 40 CFR part 261, the following facilities owned or operated by the guaranteed party. The current cost estimates so guaranteed are shown for each facility: _____. The firm identified above is [insert one or more: (1) The direct or higher-tier parent corporation of the owner or operator; (2) owned by the same parent corporation as the parent corporation of the owner or operator, and receiving the following value in consideration of this guarantee_____, or (3) engaged in the following substantial business relationship with the owner or operator _____, and receiving the following value in consideration of this guarantee_____]. [Attach a written description of the business relationship or a copy of the contract establishing such relationship to this letter.]

3. In States where EPA is not administering the financial requirements of subpart H of 40 CFR part 261, this firm, as owner or operator or guarantor, is demonstrating financial assurance for the following facilities through the use of a test equivalent or substantially equivalent to the financial test specified in subpart H of 40 CFR part 261. The current cost estimates covered by such a test are shown for each facility:_____.

4. This firm is the owner or operator of the following hazardous secondary materials management facilities for which financial

assurance is not demonstrated either to EPA or a State through the financial test or any other financial assurance mechanism specified in subpart H of 40 CFR part 261 or equivalent or substantially equivalent State mechanisms. The current cost estimates not covered by such financial assurance are shown for each facility:_____.

5. This firm is the owner or operator of the following UIC facilities for which financial assurance for plugging and abandonment is required under part 144. The current closure cost estimates as required by 40 CFR 144.62 are shown for each facility:_____.

6. This firm is the owner or operator of the following facilities for which financial assurance for closure or post-closure care is demonstrated through the financial test specified in subpart H of 40 CFR parts 264 and 265. The current closure and/or post-closure cost estimates covered by the test are shown for each facility:_____.

7. This firm guarantees, through the guarantee specified in subpart H of 40 CFR parts 264 and 265, the closure or post-closure care of the following facilities owned or operated by the guaranteed party. The current cost estimates for the closure or post-closure care so guaranteed are shown for each facility:_____. The firm identified above is [insert one or more: (1) The direct or higher-tier parent corporation of the owner or operator; (2) owned by the same parent corporation as the parent corporation of the owner or operator, and receiving the following value in consideration of this guarantee _____; or (3) engaged in the following substantial business relationship with the owner or operator _____, and receiving the following value in consideration of this guarantee _____]. [Attach a written description of the business relationship or a copy of the contract establishing such relationship to this letter].

8. In States where EPA is not administering the financial requirements of subpart H of 40 CFR part 264 or 265, this firm, as owner or operator or guarantor, is demonstrating financial assurance for the closure or post-closure care of the following facilities through the use of a test equivalent or substantially equivalent to the financial test specified in subpart H of 40 CFR parts 264 and 265. The current closure and/or post-closure cost estimates covered by such a test are shown for each facility:____.

9. This firm is the owner or operator of the following hazardous waste management facilities for which financial assurance for closure or, if a disposal facility, post-closure care, is not demonstrated either to EPA or a State through the financial test or any other financial assurance mechanism specified in subpart H of 40 CFR parts 264 and 265 or equivalent or substantially equivalent State mechanisms. The current closure and/or post-closure cost estimates not covered by

such financial assurance are shown for each facility: ____.

This firm [insert "is required" or "is not required"] to file a Form 10K with the Securities and Exchange Commission (SEC) for the latest fiscal year.

The fiscal year of this firm ends on [month, day]. The figures for the following items marked with an asterisk are derived from this firm's independently audited, year-end financial statements for the latest completed fiscal year, ended [date].

[Fill in Alternative I if the criteria of paragraph (e)(1)(i) of §261.143 of this chapter are used. Fill in Alternative II if the criteria of paragraph (e)(1)(ii) of §261.143(e) of this chapter are used.]

### Alternative I

1. Sum of current cost estimates [total of all cost estimates shown in the nine paragraphs above] $____
*2. Total liabilities [if any portion of the cost estimates is included in total liabilities, you may deduct the amount of that portion from this line and add that amount to lines 3 and 4] $____
*3. Tangible net worth $_____
*4. Net worth $_____ -
*5. Current assets $_____
*6. Current liabilities $_____
7. Net working capital [line 5 minus line 6] $_____
*8. The sum of net income plus depreciation, depletion, and amortization $_____
*9. Total assets in U.S. (required only if less than 90% of firm's assets are located in the U.S.) $_____ -
10. Is line 3 at least $10 million? (Yes/No) _____
11. Is line 3 at least 6 times line 1? (Yes/No) _____ -
12. Is line 7 at least 6 times line 1? (Yes/No) _____
*13. Are at least 90% of firm's assets located in the U.S.? If not, complete line 14 (Yes/No) _____
14. Is line 9 at least 6 times line 1? (Yes/No) _____ -
15. Is line 2 divided by line 4 less than 2.0? (Yes/No) _____ -
16. Is line 8 divided by line 2 greater than 0.1? (Yes/No) _____ -
17. Is line 5 divided by line 6 greater than 1.5? (Yes/No) _____ -

### Alternative II

1. Sum of current cost estimates [total of all cost estimates shown in the eight paragraphs above] $_____ -
2. Current bond rating of most recent issuance of this firm and name of rating service _____ -
3. Date of issuance of bond _____ -
4. Date of maturity of bond _____ -

*5. Tangible net worth [if any portion of the cost estimates is included in "total liabilities" on your firm's financial statements, you may add the amount of that portion to this line] $_____-

*6. Total assets in U.S. (required only if less than 90% of firm's assets are located in the U.S.) $_____

7. Is line 5 at least $10 million? (Yes/No)
_____

8. Is line 5 at least 6 times line 1? (Yes/No)
_____

*9. Are at least 90% of firm's assets located in the U.S.? If not, complete line 10 (Yes/No)
_____

10. Is line 6 at least 6 times line 1? (Yes/No)
_____-

I hereby certify that the wording of this letter is identical to the wording specified in 40 CFR 261.151(e) as such regulations were constituted on the date shown immediately below.
[Signature] _____
[Name] _____
[Title] _____

[Date]

(f) A letter from the chief financial officer, as specified in Sec. 261.147(f) of this chapter, must be worded as follows, except that instructions in brackets are to be replaced with the relevant information and the brackets deleted.

Letter From Chief Financial Officer

[Address to Regional Administrator of every Region in which facilities for which financial responsibility is to be demonstrated through the financial test are located].

I am the chief financial officer of [firm's name and address]. This letter is in support of the use of the financial test to demonstrate financial responsibility for liability coverage under § 261.147[insert "and costs assured § 261.143(e)" if applicable] as specified in subpart H of 40 CFR part 261.

[Fill out the following paragraphs regarding facilities and liability coverage. If there are no facilities that belong in a particular paragraph, write "None" in the space indicated. For each facility, include its EPA Identification Number (if any issued), name, and address].

The firm identified above is the owner or operator of the following facilities for which liability coverage for [insert "sudden" or "nonsudden" or "both sudden and nonsudden"] accidental occurrences is being demonstrated through the financial test specified in subpart H of 40 CFR part 261:_____

The firm identified above guarantees, through the guarantee specified in subpart H of 40 CFR part 261, liability coverage for [insert "sudden" or "nonsudden" or "both sud-

den and nonsudden"] accidental occurrences at the following facilities owned or operated by the following:_____-. The firm identified above is [insert one or more: (1) The direct or higher-tier parent corporation of the owner or operator; (2) owned by the same parent corporation as the parent corporation of the owner or operator, and receiving the following value in consideration of this guarantee -_____; or (3) engaged in the following substantial business relationship with the owner or operator _____-, and receiving the following value in consideration of this guarantee _____-]. [Attach a written description of the business relationship or a copy of the contract establishing such relationship to this letter.]

The firm identified above is the owner or operator of the following facilities for which liability coverage for [insert "sudden" or "nonsudden" or "both sudden and nonsudden"] accidental occurrences is being demonstrated through the financial test specified in subpart H of 40 CFR parts 264 and 265:

The firm identified above guarantees, through the guarantee specified in subpart H of 40 CFR parts 264 and 265, liability coverage for [insert "sudden" or "nonsudden" or "both sudden and nonsudden"] accidental occurrences at the following facilities owned or operated by the following: ___. The firm identified above is [insert one or more: (1) The direct or higher-tier parent corporation of the owner or operator; (2) owned by the same parent corporation as the parent corporation of the owner or operator, and receiving the following value in consideration of this guarantee ___; or (3) engaged in the following substantial business relationship with the owner or operator ___, and receiving the following value in consideration of this guarantee ___]. [Attach a written description of the business relationship or a copy of the contract establishing such relationship to this letter.]

[If you are using the financial test to demonstrate coverage of both liability and costs assured under § 261.143(e) or closure or post-closure care costs under 40 CFR 264.143, 264.145, 265.143 or 265.145, fill in the following nine paragraphs regarding facilities and associated cost estimates. If there are no facilities that belong in a particular paragraph, write "None" in the space indicated. For each facility, include its EPA identification number (if any issued), name, address, and current cost estimates.]

1. This firm is the owner or operator of the following facilities for which financial assurance is demonstrated through the financial test specified in subpart H of 40 CFR 261. The current cost estimates covered by the test are shown for each facility:_____.

2. This firm guarantees, through the guarantee specified in subpart H of 40 CFR part

261, the following facilities owned or operated by the guaranteed party. The current cost estimates so guaranteed are shown for each facility:_____. The firm identified above is [insert one or more: (1) The direct or higher-tier parent corporation of the owner or operator; (2) owned by the same parent corporation as the parent corporation of the owner or operator, and receiving the following value in consideration of this guarantee_____, or (3) engaged in the following substantial business relationship with the owner or operator _____, and receiving the following value in consideration of this guarantee_____]. [Attach a written description of the business relationship or a copy of the contract establishing such relationship to this letter].

3. In States where EPA is not administering the financial requirements of subpart H of 40 CFR part 261, this firm, as owner or operator or guarantor, is demonstrating financial assurance for the following facilities through the use of a test equivalent or substantially equivalent to the financial test specified in subpart H of 40 CFR part 261. The current cost estimates covered by such a test are shown for each facility:_____.

4. This firm is the owner or operator of the following hazardous secondary materials management facilities for which financial assurance is not demonstrated either to EPA or a State through the financial test or any other financial assurance mechanism specified in subpart H of 40 CFR part 261 or equivalent or substantially equivalent State mechanisms. The current cost estimates not covered by such financial assurance are shown for each facility:_____.

5. This firm is the owner or operator of the following UIC facilities for which financial assurance for plugging and abandonment is required under part 144. The current closure cost estimates as required by 40 CFR 144.62 are shown for each facility:_____.

6. This firm is the owner or operator of the following facilities for which financial assurance for closure or post-closure care is demonstrated through the financial test specified in subpart H of 40 CFR parts 264 and 265. The current closure and/or post-closure cost estimates covered by the test are shown for each facility:_____.

7. This firm guarantees, through the guarantee specified in subpart H of 40 CFR parts 264 and 265, the closure or post-closure care of the following facilities owned or operated by the guaranteed party. The current cost estimates for the closure or post-closure care so guaranteed are shown for each facility:_____. The firm identified above is [insert one or more: (1) The direct or higher-tier parent corporation of the owner or operator; (2) owned by the same parent corporation as the parent corporation of the owner or operator, and receiving the following value in consideration of this guarantee _____; or

(3) engaged in the following substantial business relationship with the owner or operator _____, and receiving the following value in consideration of this guarantee _____].

[Attach a written description of the business relationship or a copy of the contract establishing such relationship to this letter].

8. In States where EPA is not administering the financial requirements of subpart H of 40 CFR part 264 or 265, this firm, as owner or operator or guarantor, is demonstrating financial assurance for the closure or post-closure care of the following facilities through the use of a test equivalent or substantially equivalent to the financial test specified in subpart H of 40 CFR parts 264 and 265. The current closure and/or post-closure cost estimates covered by such a test are shown for each facility:_____.

9. This firm is the owner or operator of the following hazardous waste management facilities for which financial assurance for closure or, if a disposal facility, post-closure care, is not demonstrated either to EPA or a State through the financial test or any other financial assurance mechanism specified in subpart H of 40 CFR parts 264 and 265 or equivalent or substantially equivalent State mechanisms. The current closure and/or post-closure cost estimates not covered by such financial assurance are shown for each facility:_____.

This firm [insert "is required" or "is not required"] to file a Form 10K with the Securities and Exchange Commission (SEC) for the latest fiscal year.

The fiscal year of this firm ends on [month, day]. The figures for the following items marked with an asterisk are derived from this firm's independently audited, year-end financial statements for the latest completed fiscal year, ended [date].

PART A. LIABILITY COVERAGE FOR ACCIDENTAL OCCURRENCES

[Fill in Alternative I if the criteria of paragraph (f)(1)(i) of Sec. 261.147 are used. Fill in Alternative II if the criteria of paragraph (f)(1)(ii) of Sec. 261.147 are used.]

ALTERNATIVE I

1. Amount of annual aggregate liability coverage to be demonstrated $_____.
*2. Current assets $_____.
*3. Current liabilities $_____.
4. Net working capital (line 2 minus line 3) $_____.
*5. Tangible net worth $_____.
*6. If less than 90% of assets are located in the U.S., give total U.S. assets $_____.
7. Is line 5 at least $10 million? (Yes/No)_____.
8. Is line 4 at least 6 times line 1? (Yes/No)_____.
9. Is line 5 at least 6 times line 1? (Yes/No)_____.

139

*10. Are at least 90% of assets located in the U.S.? (Yes/No) _____ . If not, complete line 11.

11. Is line 6 at least 6 times line 1? (Yes/No) _____ .

### ALTERNATIVE II

1. Amount of annual aggregate liability coverage to be demonstrated $ _____ -.

2. Current bond rating of most recent issuance and name of rating service _____ -.

3. Date of issuance of bond _____ .

4. Date of maturity of bond _____ .

*5. Tangible net worth $ _____ -.

*6. Total assets in U.S. (required only if less than 90% of assets are located in the U.S.) $ _____ -.

7. Is line 5 at least $10 million? (Yes/No) _____ -.

8. Is line 5 at least 6 times line 1? _____

9. Are at least 90% of assets located in the U.S.? If not, complete line 10. (Yes/No) ____ .

10. Is line 6 at least 6 times line 1? _____

[Fill in part B if you are using the financial test to demonstrate assurance of both liability coverage and costs assured under § 261.143(e) or closure or post-closure care costs under 40 CFR 264.143, 264.145, 265.143 or 265.145.]

### PART B. FACILITY CARE AND LIABILITY COVERAGE

[Fill in Alternative I if the criteria of paragraphs (e)(1)(i) of Sec. 261.143 and (f)(1)(i) of Sec. 261.147 are used. Fill in Alternative II if the criteria of paragraphs (e)(1)(ii) of Sec. 261.143 and (f)(1)(ii) of Sec. 261.147 are used.]

### ALTERNATIVE I

1. Sum of current cost estimates (total of all cost estimates listed above) $ _____ -

2. Amount of annual aggregate liability coverage to be demonstrated $ _____ -

3. Sum of lines 1 and 2 $ _____

*4. Total liabilities (if any portion of your cost estimates is included in your total liabilities, you may deduct that portion from this line and add that amount to lines 5 and 6) $ _____ -

*5. Tangible net worth $ _____

*6. Net worth $ _____ -

*7. Current assets $ _____

*8. Current liabilities $ _____

9. Net working capital (line 7 minus line 8) $ _____

*10. The sum of net income plus depreciation, depletion, and amortization $ _____ -

*11. Total assets in U.S. (required only if less than 90% of assets are located in the U.S.) $ _____

12. Is line 5 at least $10 million? (Yes/No)

13. Is line 5 at least 6 times line 3? (Yes/No)

14. Is line 9 at least 6 times line 3? (Yes/No)

*15. Are at least 90% of assets located in the U.S.? (Yes/No) If not, complete line 16.

16. Is line 11 at least 6 times line 3? (Yes/No)

17. Is line 4 divided by line 6 less than 2.0? (Yes/No)

18. Is line 10 divided by line 4 greater than 0.1? (Yes/No)

19. Is line 7 divided by line 8 greater than 1.5? (Yes/No)

### ALTERNATIVE II

1. Sum of current cost estimates (total of all cost estimates listed above) $ _____

2. Amount of annual aggregate liability coverage to be demonstrated $ _____ -

3. Sum of lines 1 and 2 $ _____

4. Current bond rating of most recent issuance and name of rating service _____ -

5. Date of issuance of bond _____ —

6. Date of maturity of bond _____ —

*7. Tangible net worth (if any portion of the cost estimates is included in "total liabilities" on your financial statements you may add that portion to this line) $ _____ -

*8. Total assets in the U.S. (required only if less than 90% of assets are located in the U.S.) $ _____ -

9. Is line 7 at least $10 million? (Yes/No)

10. Is line 7 at least 6 times line 3? (Yes/No)

*11. Are at least 90% of assets located in the U.S.? (Yes/No) If not complete line 12.

12. Is line 8 at least 6 times line 3? (Yes/No)

I hereby certify that the wording of this letter is identical to the wording specified in 40 CFR 261.151(f) as such regulations were constituted on the date shown immediately below.

[Signature] _____

[Name] _____

[Title] _____

[Date] _____

(g)(1) A corporate guarantee, as specified in § 261.143(e) of this chapter, must be worded as follows, except that instructions in brackets are to be replaced with the relevant information and the brackets deleted:

### CORPORATE GUARANTEE FOR FACILITY CARE

Guarantee made this [date] by [name of guaranteeing entity], a business corporation organized under the laws of the State of [insert name of State], herein referred to as guarantor. This guarantee is made on behalf of the [owner or operator] of [business address], which is [one of the following: "our subsidiary"; "a subsidiary of [name and address of common parent corporation], of which guarantor is a subsidiary"; or "an entity with which guarantor has a substantial business relationship, as defined in 40 CFR

264.141(h) and 265.141(h)'' to the United States Environmental Protection Agency (EPA).

RECITALS

1. Guarantor meets or exceeds the financial test criteria and agrees to comply with the reporting requirements for guarantors as specified in 40 CFR 261.143(e).

2. [Owner or operator] owns or operates the following facility(ies) covered by this guarantee: [List for each facility: EPA Identification Number (if any issued), name, and address.

3. "Closure plans" as used below refer to the plans maintained as required by subpart H of 40 CFR part 261 for the care of facilities as identified above.

4. For value received from [owner or operator], guarantor guarantees that in the event of a determination by the Regional Administrator that the hazardous secondary materials at the owner or operator's facility covered by this guarantee do not meet the conditions of the exclusion under §261.4(a)(24), the guarantor will dispose of any hazardous secondary material as hazardous waste, and close the facility in accordance with closure requirements found in parts 264 or 265 of this chapter, as applicable, or establish a trust fund as specified in §261.143(a) in the name of the owner or operator in the amount of the current cost estimate.

5. Guarantor agrees that if, at the end of any fiscal year before termination of this guarantee, the guarantor fails to meet the financial test criteria, guarantor shall send within 90 days, by certified mail, notice to the EPA Regional Administrator(s) for the Region(s) in which the facility(ies) is(are) located and to [owner or operator] that he intends to provide alternate financial assurance as specified in subpart H of 40 CFR part 261, as applicable, in the name of [owner or operator]. Within 120 days after the end of such fiscal year, the guarantor shall establish such financial assurance unless [owner or operator] has done so.

6. The guarantor agrees to notify the EPA Regional Administrator by certified mail, of a voluntary or involuntary proceeding under Title 11 (Bankruptcy), U.S. Code, naming guarantor as debtor, within 10 days after commencement of the proceeding.

7. Guarantor agrees that within 30 days after being notified by an EPA Regional Administrator of a determination that guarantor no longer meets the financial test criteria or that he is disallowed from continuing as a guarantor, he shall establish alternate financial assurance as specified in of 40 CFR parts 264, 265, or subpart H of 40 CFR part 261, as applicable, in the name of [owner or operator] unless [owner or operator] has done so.

8. Guarantor agrees to remain bound under this guarantee notwithstanding any or all of the following: amendment or modification of the closure plan, the extension or reduction of the time of performance, or any other modification or alteration of an obligation of the owner or operator pursuant to 40 CFR parts 264, 265, or Subpart H of 40 CFR part 261.

9. Guarantor agrees to remain bound under this guarantee for as long as [owner or operator] must comply with the applicable financial assurance requirements of 40 CFR parts 264 and 265 or the financial assurance condition of 40 CFR 261.4(a)(24)(vi)(F) for the above-listed facilities, except as provided in paragraph 10 of this agreement.

10. [Insert the following language if the guarantor is (a) a direct or higher-tier corporate parent, or (b) a firm whose parent corporation is also the parent corporation of the owner or operator):

Guarantor may terminate this guarantee by sending notice by certified mail to the EPA Regional Administrator(s) for the Region(s) in which the facility(ies) is(are) located and to [owner or operator], provided that this guarantee may not be terminated unless and until [the owner or operator] obtains, and the EPA Regional Administrator(s) approve(s), alternate coverage complying with 40 CFR 261.143.

[Insert the following language if the guarantor is a firm qualifying as a guarantor due to its "substantial business relationship" with the owner or operator]

Guarantor may terminate this guarantee 120 days following the receipt of notification, through certified mail, by the EPA Regional Administrator(s) for the Region(s) in which the facility(ies) is(are) located and by [the owner or operator].

11. Guarantor agrees that if [owner or operator] fails to provide alternate financial assurance as specified in 40 CFR parts 264, 265, or subpart H of 40 CFR 261, as applicable, and obtain written approval of such assurance from the EPA Regional Administrator(s) within 90 days after a notice of cancellation by the guarantor is received by an EPA Regional Administrator from guarantor, guarantor shall provide such alternate financial assurance in the name of [owner or operator].

12. Guarantor expressly waives notice of acceptance of this guarantee by the EPA or by [owner or operator]. Guarantor also expressly waives notice of amendments or modifications of the closure plan and of amendments or modifications of the applicable requirements of 40 CFR parts 264, 265, or subpart H of 40 CFR 261.

I hereby certify that the wording of this guarantee is identical to the wording specified in 40 CFR 261.151(g)(1) as such regulations were constituted on the date first above written.

Effective date: _____

[Name of guarantor] _____

[Authorized signature for guarantor] _____

141

[Name of person signing] _____
[Title of person signing] _____
Signature of witness or notary: _____

(2) A guarantee, as specified in Sec. 261.147(g) of this chapter, must be worded as follows, except that instructions in brackets are to be replaced with the relevant information and the brackets deleted:

GUARANTEE FOR LIABILITY COVERAGE

Guarantee made this [date] by [name of guaranteeing entity], a business corporation organized under the laws of [if incorporated within the United States insert "the State of _____-" and insert name of State; if incorporated outside the United States insert the name of the country in which incorporated, the principal place of business within the United States, and the name and address of the registered agent in the State of the principal place of business], herein referred to as guarantor. This guarantee is made on behalf of [owner or operator] of [business address], which is one of the following: "our subsidiary;" "a subsidiary of [name and address of common parent corporation], of which guarantor is a subsidiary;" or "an entity with which guarantor has a substantial business relationship, as defined in 40 CFR [either 264.141(h) or 265.141(h)]", to any and all third parties who have sustained or may sustain bodily injury or property damage caused by [sudden and/or nonsudden] accidental occurrences arising from operation of the facility(ies) covered by this guarantee.

RECITALS

1. Guarantor meets or exceeds the financial test criteria and agrees to comply with the reporting requirements for guarantors as specified in 40 CFR 261.147(g).

2. [Owner or operator] owns or operates the following facility(ies) covered by this guarantee: [List for each facility: EPA identification number (if any issued), name, and address; and if guarantor is incorporated outside the United States list the name and address of the guarantor's registered agent in each State.] This corporate guarantee satisfies RCRA third-party liability requirements for [insert "sudden" or "nonsudden" or "both sudden and nonsudden"] accidental occurrences in above-named owner or operator facilities for coverage in the amount of [insert dollar amount] for each occurrence and [insert dollar amount] annual aggregate.

3. For value received from [owner or operator], guarantor guarantees to any and all third parties who have sustained or may sustain bodily injury or property damage caused by [sudden and/or nonsudden] accidental occurrences arising from operations of the facility(ies) covered by this guarantee that in the event that [owner or operator] fails to satisfy a judgment or award based on a de-

termination of liability for bodily injury or property damage to third parties caused by [sudden and/or nonsudden] accidental occurrences, arising from the operation of the above-named facilities, or fails to pay an amount agreed to in settlement of a claim arising from or alleged to arise from such injury or damage, the guarantor will satisfy such judgment(s), award(s) or settlement agreement(s) up to the limits of coverage identified above.

4. Such obligation does not apply to any of the following:

(a) Bodily injury or property damage for which [insert owner or operator] is obligated to pay damages by reason of the assumption of liability in a contract or agreement. This exclusion does not apply to liability for damages that [insert owner or operator] would be obligated to pay in the absence of the contract or agreement.

(b) Any obligation of [insert owner or operator] under a workers' compensation, disability benefits, or unemployment compensation law or any similar law.

(c) Bodily injury to:

(1) An employee of [insert owner or operator] arising from, and in the course of, employment by [insert owner or operator]; or

(2) The spouse, child, parent, brother, or sister of that employee as a consequence of, or arising from, and in the course of employment by [insert owner or operator]. This exclusion applies:

(A) Whether [insert owner or operator] may be liable as an employer or in any other capacity; and

(B) To any obligation to share damages with or repay another person who must pay damages because of the injury to persons identified in paragraphs (1) and (2).

(d) Bodily injury or property damage arising out of the ownership, maintenance, use, or entrustment to others of any aircraft, motor vehicle or watercraft.

(e) Property damage to:

(1) Any property owned, rented, or occupied by [insert owner or operator];

(2) Premises that are sold, given away or abandoned by [insert owner or operator] if the property damage arises out of any part of those premises;

(3) Property loaned to [insert owner or operator];

(4) Personal property in the care, custody or control of [insert owner or operator];

(5) That particular part of real property on which [insert owner or operator] or any contractors or subcontractors working directly or indirectly on behalf of [insert owner or operator] are performing operations, if the property damage arises out of these operations.

5. Guarantor agrees that if, at the end of any fiscal year before termination of this guarantee, the guarantor fails to meet the financial test criteria, guarantor shall send

within 90 days, by certified mail, notice to the EPA Regional Administrator[s] for the Region[s] in which the facility[ies] is[are] located and to [owner or operator] that he intends to provide alternate liability coverage as specified in 40 CFR 261.147, as applicable, in the name of [owner or operator]. Within 120 days after the end of such fiscal year, the guarantor shall establish such liability coverage unless [owner or operator] has done so.

6. The guarantor agrees to notify the EPA Regional Administrator by certified mail of a voluntary or involuntary proceeding under title 11 (Bankruptcy), U.S. Code, naming guarantor as debtor, within 10 days after commencement of the proceeding. Guarantor agrees that within 30 days after being notified by an EPA Regional Administrator of a determination that guarantor no longer meets the financial test criteria or that he is disallowed from continuing as a guarantor, he shall establish alternate liability coverage as specified in 40 CFR 261.147 in the name of [owner or operator], unless [owner or operator] has done so.

7. Guarantor reserves the right to modify this agreement to take into account amendment or modification of the liability requirements set by 40 CFR 261.147, provided that such modification shall become effective only if a Regional Administrator does not disapprove the modification within 30 days of receipt of notification of the modification.

8. Guarantor agrees to remain bound under this guarantee for so long as [owner or operator] must comply with the applicable requirements of 40 CFR 261.147 for the above-listed facility(ies), except as provided in paragraph 10 of this agreement.

9. [Insert the following language if the guarantor is (a) a direct or higher-tier corporate parent, or (b) a firm whose parent corporation is also the parent corporation of the owner or operator]:

10. Guarantor may terminate this guarantee by sending notice by certified mail to the EPA Regional Administrator(s) for the Region(s) in which the facility(ies) is(are) located and to [owner or operator], provided that this guarantee may not be terminated unless and until [the owner or operator] obtains, and the EPA Regional Administrator(s) approve(s), alternate liability coverage complying with 40 CFR 261.147.

[Insert the following language if the guarantor is a firm qualifying as a guarantor due to its "substantial business relationship" with the owner or operator]:

Guarantor may terminate this guarantee 120 days following receipt of notification, through certified mail, by the EPA Regional Administrator(s) for the Region(s) in which the facility(ies) is(are) located and by [the owner or operator].

11. Guarantor hereby expressly waives notice of acceptance of this guarantee by any party.

12. Guarantor agrees that this guarantee is in addition to and does not affect any other responsibility or liability of the guarantor with respect to the covered facilities.

13. The Guarantor shall satisfy a third-party liability claim only on receipt of one of the following documents:

(a) Certification from the Principal and the third-party claimant(s) that the liability claim should be paid. The certification must be worded as follows, except that instructions in brackets are to be replaced with the relevant information and the brackets deleted:

CERTIFICATION OF VALID CLAIM

The undersigned, as parties [insert Principal] and [insert name and address of third-party claimant(s)], hereby certify that the claim of bodily injury and/or property damage caused by a [sudden or nonsudden] accidental occurrence arising from operating [Principal's] facility should be paid in the amount of $ .
[Signatures] _____
Principal _____
(Notary) Date _____
[Signatures] _____
Claimant(s) _____
(Notary) Date _____

(b) A valid final court order establishing a judgment against the Principal for bodily injury or property damage caused by sudden or nonsudden accidental occurrences arising from the operation of the Principal's facility or group of facilities.

14. In the event of combination of this guarantee with another mechanism to meet liability requirements, this guarantee will be considered [insert "primary" or "excess"] coverage.

I hereby certify that the wording of the guarantee is identical to the wording specified in 40 CFR 261.151(g)(2) as such regulations were constituted on the date shown immediately below.
Effective date: _____
[Name of guarantor] _____
[Authorized signature for guarantor] _____
[Name of person signing] _____
[Title of person signing] _____
Signature of witness or notary: _____

(h) A hazardous waste facility liability endorsement as required § 261.147 must be worded as follows, except that instructions in brackets are to be replaced with the relevant information and the brackets deleted:

HAZARDOUS SECONDARY MATERIAL RECLAMATION/INTERMEDIATE FACILITY LIABILITY ENDORSEMENT

1. This endorsement certifies that the policy to which the endorsement is attached provides liability insurance covering bodily injury and property damage in connection

with the insured's obligation to demonstrate financial responsibility under 40 CFR 261.147. The coverage applies at [list EPA Identification Number (if any issued), name, and address for each facility] for [insert "sudden accidental occurrences," "nonsudden accidental occurrences," or "sudden and nonsudden accidental occurrences"; if coverage is for multiple facilities and the coverage is different for different facilities, indicate which facilities are insured for sudden accidental occurrences, which are insured for nonsudden accidental occurrences, and which are insured for both]. The limits of liability are [insert the dollar amount of the "each occurrence" and "annual aggregate" limits of the Insurer's liability], exclusive of legal defense costs.

2. The insurance afforded with respect to such occurrences is subject to all of the terms and conditions of the policy; provided, however, that any provisions of the policy inconsistent with subsections (a) through (e) of this Paragraph 2 are hereby amended to conform with subsections (a) through (e):

(a) Bankruptcy or insolvency of the insured shall not relieve the Insurer of its obligations under the policy to which this endorsement is attached.

(b) The Insurer is liable for the payment of amounts within any deductible applicable to the policy, with a right of reimbursement by the insured for any such payment made by the Insurer. This provision does not apply with respect to that amount of any deductible for which coverage is demonstrated as specified in 40 CFR 261.147(f).

(c) Whenever requested by a Regional Administrator of the U.S. Environmental Protection Agency (EPA), the Insurer agrees to furnish to the Regional Administrator a signed duplicate original of the policy and all endorsements.

(d) Cancellation of this endorsement, whether by the Insurer, the insured, a parent corporation providing insurance coverage for its subsidiary, or by a firm having an insurable interest in and obtaining liability insurance on behalf of the owner or operator of the facility, will be effective only upon written notice and only after the expiration of 60 days after a copy of such written notice is received by the Regional Administrator(s) of the EPA Region(s) in which the facility(ies) is(are) located.

(e) Any other termination of this endorsement will be effective only upon written notice and only after the expiration of thirty (30) days after a copy of such written notice is received by the Regional Administrator(s) of the EPA Region(s) in which the facility(ies) is (are) located.

Attached to and forming part of policy No. ____ issued by [name of Insurer], herein called the Insurer, of [address of Insurer] to [name of insured] of [address] this _____ day of _____, 19___.

The effective date of said policy is _____ day of _____, 19___.

I hereby certify that the wording of this endorsement is identical to the wording specified in 40 CFR 261.151(h) as such regulation was constituted on the date first above written, and that the Insurer is licensed to transact the business of insurance, or eligible to provide insurance as an excess or surplus lines insurer, in one or more States.

[Signature of Authorized Representative of Insurer]

[Type name]

[Title], Authorized Representative of [name of Insurer]

[Address of Representative]

(i) A certificate of liability insurance as required in § 261.147 must be worded as follows, except that the instructions in brackets are to be replaced with the relevant information and the brackets deleted:

HAZARDOUS SECONDARY MATERIAL RECLAMATION/INTERMEDIATE FACILITY CERTIFICATE OF LIABILITY INSURANCE

1. [Name of Insurer], (the "Insurer"), of [address of Insurer] hereby certifies that it has issued liability insurance covering bodily injury and property damage to [name of insured], (the "insured"), of [address of insured] in connection with the insured's obligation to demonstrate financial responsibility under 40 CFR parts 264, 265, and the financial assurance condition of 40 CFR 261.4(a)(24)(vi)(F). The coverage applies at [list EPA Identification Number (if any issued), name, and address for each facility] for [insert "sudden accidental occurrences," "nonsudden accidental occurrences," or "sudden and nonsudden accidental occurrences"; if coverage is for multiple facilities and the coverage is different for different facilities, indicate which facilities are insured for sudden accidental occurrences, which are insured for nonsudden accidental occurrences, and which are insured for both]. The limits of liability are [insert the dollar amount of the "each occurrence" and "annual aggregate" limits of the Insurer's liability], exclusive of legal defense costs. The coverage is provided under policy number, issued on [date]. The effective date of said policy is [date].

2. The Insurer further certifies the following with respect to the insurance described in Paragraph 1:

(a) Bankruptcy or insolvency of the insured shall not relieve the Insurer of its obligations under the policy.

(b) The Insurer is liable for the payment of amounts within any deductible applicable to the policy, with a right of reimbursement by the insured for any such payment made by the Insurer. This provision does not apply with respect to that amount of any deductible for which coverage is demonstrated as specified in 40 CFR 261.147.

(c) Whenever requested by a Regional Administrator of the U.S. Environmental Protection Agency (EPA), the Insurer agrees to furnish to the Regional Administrator a signed duplicate original of the policy and all endorsements.

(d) Cancellation of the insurance, whether by the insurer, the insured, a parent corporation providing insurance coverage for its subsidiary, or by a firm having an insurable interest in and obtaining liability insurance on behalf of the owner or operator of the hazardous waste management facility, will be effective only upon written notice and only after the expiration of 60 days after a copy of such written notice is received by the Regional Administrator(s) of the EPA Region(s) in which the facility(ies) is(are) located.

(e) Any other termination of the insurance will be effective only upon written notice and only after the expiration of thirty (30) days after a copy of such written notice is received by the Regional Administrator(s) of the EPA Region(s) in which the facility(ies) is (are) located.

I hereby certify that the wording of this instrument is identical to the wording specified in 40 CFR 261.151(i) as such regulation was constituted on the date first above written, and that the Insurer is licensed to transact the business of insurance, or eligible to provide insurance as an excess or surplus lines insurer, in one or more States.

[Signature of authorized representative of Insurer]

[Type name]

[Title], Authorized Representative of [name of Insurer]

[Address of Representative]

(j) A letter of credit, as specified in §261.147(h) of this chapter, must be worded as follows, except that instructions in brackets are to be replaced with the relevant information and the brackets deleted:

IRREVOCABLE STANDBY LETTER OF CREDIT

Name and Address of Issuing Institution _____
Regional Administrator(s) _____
Region(s) _____
U.S. Environmental Protection Agency _____

Dear Sir or Madam: We hereby establish our Irrevocable Standby Letter of Credit No. _____ in the favor of ["any and all third-party liability claimants" or insert name of trustee of the standby trust fund], at the request and for the account of [owner or operator's name and address] for third-party liability awards or settlements up to [in words] U.S. dollars $_____ ----- per occurrence and the annual aggregate amount of [in words] U.S. dollars $__ —, for sudden accidental occurrences and/or for third-party liability awards or settlements up to the amount of [in words] U.S. dollars $_____ -- --- per occurrence, and the annual aggregate amount of [in words] U.S. dollars $_____ -- ---, for nonsudden accidental occurrences available upon presentation of a sight draft bearing reference to this letter of credit No. _____ -----, and [insert the following language if the letter of credit is being used without a standby trust fund: (1) a signed certificate reading as follows:

CERTIFICATE OF VALID CLAIM

The undersigned, as parties [insert principal] and [insert name and address of third party claimant(s)], hereby certify that the claim of bodily injury and/or property damage caused by a [sudden or nonsudden] accidental occurrence arising from operations of [principal's] facility should be paid in the amount of $[ ]. We hereby certify that the claim does not apply to any of the following:

(a) Bodily injury or property damage for which [insert principal] is obligated to pay damages by reason of the assumption of liability in a contract or agreement. This exclusion does not apply to liability for damages that [insert principal] would be obligated to pay in the absence of the contract or agreement.

(b) Any obligation of [insert principal] under a workers' compensation, disability benefits, or unemployment compensation law or any similar law.

(c) Bodily injury to:

(1) An employee of [insert principal] arising from, and in the course of, employment by [insert principal]; or

(2) The spouse, child, parent, brother or sister of that employee as a consequence of, or arising from, and in the course of employment by [insert principal].

This exclusion applies:

(A) Whether [insert principal] may be liable as an employer or in any other capacity; and

(B) To any obligation to share damages with or repay another person who must pay damages because of the injury to persons identified in paragraphs (1) and (2).

(d) Bodily injury or property damage arising out of the ownership, maintenance, use, or entrustment to others of any aircraft, motor vehicle or watercraft.

(e) Property damage to:

(1) Any property owned, rented, or occupied by [insert principal];

(2) Premises that are sold, given away or abandoned by [insert principal] if the property damage arises out of any part of those premises;

(3) Property loaned to [insert principal];

(4) Personal property in the care, custody or control of [insert principal];

(5) That particular part of real property on which [insert principal] or any contractors

145

or subcontractors working directly or indirectly on behalf of [insert principal] are performing operations, if the property damage arises out of these operations.

[Signatures] _____
Grantor _____
[Signatures] _____
Claimant(s)

or (2) a valid final court order establishing a judgment against the Grantor for bodily injury or property damage caused by sudden or nonsudden accidental occurrences arising from the operation of the Grantor's facility or group of facilities.]

This letter of credit is effective as of [date] and shall expire on [date at least one year later], but such expiration date shall be automatically extended for a period of [at least one year] on [date and on each successive expiration date, unless, at least 120 days before the current expiration date, we notify you, the USEPA Regional Administrator for Region [Region], and [owner's or operator's name] by certified mail that we have decided not to extend this letter of credit beyond the current expiration date.

Whenever this letter of credit is drawn on under and in compliance with the terms of this credit, we shall duly honor such draft upon presentation to us.

[Insert the following language if a standby trust fund is not being used: "In the event that this letter of credit is used in combination with another mechanism for liability coverage, this letter of credit shall be considered [insert "primary" or "excess" coverage]."

We certify that the wording of this letter of credit is identical to the wording specified in 40 CFR 261.151(j) as such regulations were constituted on the date shown immediately below. [Signature(s) and title(s) of official(s) of issuing institution] [Date].

This credit is subject to [insert "the most recent edition of the Uniform Customs and Practice for Documentary Credits, published and copyrighted by the International Chamber of Commerce," or "the Uniform Commercial Code"].

(k) A surety bond, as specified in Sec. 261.147(i) of this chapter, must be worded as follows: except that instructions in brackets are to be replaced with the relevant information and the brackets deleted:

PAYMENT BOND

Surety Bond No. [Insert number]

Parties [Insert name and address of owner or operator], Principal, incorporated in [Insert State of incorporation] of [Insert city and State of principal place of business] and [Insert name and address of surety company(ies)], Surety Company(ies), of [Insert surety(ies) place of business].

EPA Identification Number (if any issued), name, and address for each facility guaranteed by this bond: ____

_____

_____

Nonsudden

Sudden accidental

accidental

occurrences

occurrences

_____

_____

| Penal Sum Per Occurrence ................... | [insert amount] .... [insert amount] |
| Annual Aggregate ................................ | [insert amount] .... [insert amount] |

_____

Purpose: This is an agreement between the Surety(ies) and the Principal under which the Surety(ies), its(their) successors and assignees, agree to be responsible for the payment of claims against the Principal for bodily injury and/or property damage to third parties caused by ["sudden" and/or "nonsudden"] accidental occurrences arising from operations of the facility or group of facilities in the sums prescribed herein; subject to the governing provisions and the following conditions.

Governing Provisions:

(1) Section 3004 of the Resource Conservation and Recovery Act of 1976, as amended.

(2) Rules and regulations of the U.S. Environmental Protection Agency (EPA), particularly 40 CFR parts 264, 265, and Subpart H of 40 CFR part 261 (if applicable).

(3) Rules and regulations of the governing State agency (if applicable) [insert citation].

Conditions:

(1) The Principal is subject to the applicable governing provisions that require the Principal to have and maintain liability coverage for bodily injury and property damage to third parties caused by ["sudden" and/or "nonsudden"] accidental occurrences arising from operations of the facility or group of facilities. Such obligation does not apply to any of the following:

(a) Bodily injury or property damage for which [insert Principal] is obligated to pay damages by reason of the assumption of liability in a contract or agreement. This exclusion does not apply to liability for damages that [insert Principal] would be obligated to pay in the absence of the contract or agreement.

(b) Any obligation of [insert Principal] under a workers' compensation, disability benefits, or unemployment compensation law or similar law.

(c) Bodily injury to:

(1) An employee of [insert Principal] arising from, and in the course of, employment by [insert principal]; or

(2) The spouse, child, parent, brother or sister of that employee as a consequence of, or arising from, and in the course of employment by [insert Principal]. This exclusion applies:

(A) Whether [insert Principal] may be liable as an employer or in any other capacity; and

(B) To any obligation to share damages with or repay another person who must pay damages because of the injury to persons identified in paragraphs (1) and (2).

(d) Bodily injury or property damage arising out of the ownership, maintenance, use, or entrustment to others of any aircraft, motor vehicle or watercraft.

(e) Property damage to:

(1) Any property owned, rented, or occupied by [insert Principal];

(2) Premises that are sold, given away or abandoned by [insert Principal] if the property damage arises out of any part of those premises;

(3) Property loaned to [insert Principal];

(4) Personal property in the care, custody or control of [insert Principal];

(5) That particular part of real property on which [insert Principal] or any contractors or subcontractors working directly or indirectly on behalf of [insert Principal] are performing operations, if the property damage arises out of these operations.

(2) This bond assures that the Principal will satisfy valid third party liability claims, as described in condition 1.

(3) If the Principal fails to satisfy a valid third party liability claim, as described above, the Surety(ies) becomes liable on this bond obligation.

(4) The Surety(ies) shall satisfy a third party liability claim only upon the receipt of one of the following documents:

(a) Certification from the Principal and the third party claimant(s) that the liability claim should be paid. The certification must be worded as follows, except that instructions in brackets are to be replaced with the relevant information and the brackets deleted:

CERTIFICATION OF VALID CLAIM

The undersigned, as parties [insert name of Principal] and [insert name and address of third party claimant(s)], hereby certify that the claim of bodily injury and/or property damage caused by a [sudden or nonsudden] accidental occurrence arising from operating [Principal's] facility should be paid in the amount of $[ ].

[Signature]
Principal
[Notary] Date
[Signature(s)]
Claimant(s)

[Notary] Date

or (b) A valid final court order establishing a judgment against the Principal for bodily injury or property damage caused by sudden or nonsudden accidental occurrences arising from the operation of the Principal's facility or group of facilities.

(5) In the event of combination of this bond with another mechanism for liability coverage, this bond will be considered [insert "primary" or "excess"] coverage.

(6) The liability of the Surety(ies) shall not be discharged by any payment or succession of payments hereunder, unless and until such payment or payments shall amount in the aggregate to the penal sum of the bond. In no event shall the obligation of the Surety(ies) hereunder exceed the amount of said annual aggregate penal sum, provided that the Surety(ies) furnish(es) notice to the Regional Administrator forthwith of all claims filed and payments made by the Surety(ies) under this bond.

(7) The Surety(ies) may cancel the bond by sending notice of cancellation by certified mail to the Principal and the USEPA Regional Administrator for Region [Region ], provided, however, that cancellation shall not occur during the 120 days beginning on the date of receipt of the notice of cancellation by the Principal and the Regional Administrator, as evidenced by the return receipt.

(8) The Principal may terminate this bond by sending written notice to the Surety(ies) and to the EPA Regional Administrator(s) of the EPA Region(s) in which the bonded facility(ies) is (are) located.

(9) The Surety(ies) hereby waive(s) notification of amendments to applicable laws, statutes, rules and regulations and agree(s) that no such amendment shall in any way alleviate its (their) obligation on this bond.

(10) This bond is effective from [insert date] (12:01 a.m., standard time, at the address of the Principal as stated herein) and shall continue in force until terminated as described above.

In Witness Whereof, the Principal and Surety(ies) have executed this Bond and have affixed their seals on the date set forth above.

The persons whose signatures appear below hereby certify that they are authorized to execute this surety bond on behalf of the

Principal and Surety(ies) and that the wording of this surety bond is identical to the wording specified in 40 CFR 261.151(k), as such regulations were constituted on the date this bond was executed.

PRINCIPAL

[Signature(s)]
[Name(s)]
[Title(s)]
[Corporate Seal]

CORPORATE SURETY[IES]

[Name and address]
State of incorporation: _____
Liability Limit: $ _____
[Signature(s)]
[Name(s) and title(s)]
[Corporate seal]
[For every co-surety, provide signature(s), corporate seal, and other information in the same manner as for Surety above.]
Bond premium: $ _____

(l)(1) A trust agreement, as specified in § 261.147(j) of this chapter, must be worded as follows, except that instructions in brackets are to be replaced with the relevant information and the brackets deleted:

Trust Agreement

Trust Agreement, the "Agreement," entered into as of [date] by and between [name of the owner or operator] a [name of State] [insert "corporation," "partnership," "association," or "proprietorship"], the "Grantor," and [name of corporate trustee], [insert, "incorporated in the State of _____" or "a national bank"], the "trustee."

Whereas, the United States Environmental Protection Agency, "EPA," an agency of the United States Government, has established certain regulations applicable to the Grantor, requiring that an owner or operator must demonstrate financial responsibility for bodily injury and property damage to third parties caused by sudden accidental and/or nonsudden accidental occurrences arising from operations of the facility or group of facilities.

Whereas, the Grantor has elected to establish a trust to assure all or part of such financial responsibility for the facilities identified herein.

Whereas, the Grantor, acting through its duly authorized officers, has selected the Trustee to be the trustee under this agreement, and the Trustee is willing to act as trustee.

Now, therefore, the Grantor and the Trustee agree as follows:

Section 1. Definitions. As used in this Agreement:

(a) The term "Grantor" means the owner or operator who enters into this Agreement and any successors or assigns of the Grantor.

(b) The term "Trustee" means the Trustee who enters into this Agreement and any successor Trustee.

Section 2. Identification of Facilities. This agreement pertains to the facilities identified on attached schedule A [on schedule A, for each facility list the EPA Identification Number (if any issued), name, and address of the facility(ies) and the amount of liability coverage, or portions thereof, if more than one instrument affords combined coverage as demonstrated by this Agreement].

Section 3. Establishment of Fund. The Grantor and the Trustee hereby establish a trust fund, hereinafter the "Fund," for the benefit of any and all third parties injured or damaged by [sudden and/or nonsudden] accidental occurrences arising from operation of the facility(ies) covered by this guarantee, in the amounts of _____-[up to $1 million] per occurrence and [up to $2 million] annual aggregate for sudden accidental occurrences and _____ [up to $3 million] per occurrence and _____-[up to $6 million] annual aggregate for nonsudden occurrences, except that the Fund is not established for the benefit of third parties for the following:

(a) Bodily injury or property damage for which [insert Grantor] is obligated to pay damages by reason of the assumption of liability in a contract or agreement. This exclusion does not apply to liability for damages that [insert Grantor] would be obligated to pay in the absence of the contract or agreement.

(b) Any obligation of [insert Grantor] under a workers' compensation, disability benefits, or unemployment compensation law or any similar law.

(c) Bodily injury to:

(1) An employee of [insert Grantor] arising from, and in the course of, employment by [insert Grantor]; or

(2) The spouse, child, parent, brother or sister of that employee as a consequence of, or arising from, and in the course of employment by [insert Grantor]. This exclusion applies:

(A) Whether [insert Grantor] may be liable as an employer or in any other capacity; and

(B) To any obligation to share damages with or repay another person who must pay damages because of the injury to persons identified in paragraphs (1) and (2).

(d) Bodily injury or property damage arising out of the ownership, maintenance, use, or entrustment to others of any aircraft, motor vehicle or watercraft.

(e) Property damage to:

(1) Any property owned, rented, or occupied by [insert Grantor];

(2) Premises that are sold, given away or abandoned by [insert Grantor] if the property damage arises out of any part of those premises;

(3) Property loaned to [insert Grantor];

(4) Personal property in the care, custody or control of [insert Grantor];

(5) That particular part of real property on which [insert Grantor] or any contractors or subcontractors working directly or indirectly on behalf of [insert Grantor] are performing operations, if the property damage arises out of these operations.

In the event of combination with another mechanism for liability coverage, the Fund shall be considered [insert "primary" or "excess"] coverage.

The Fund is established initially as consisting of the property, which is acceptable to the Trustee, described in Schedule B attached hereto. Such property and any other property subsequently transferred to the Trustee is referred to as the Fund, together with all earnings and profits thereon, less any payments or distributions made by the Trustee pursuant to this Agreement. The Fund shall be held by the Trustee, IN TRUST, as hereinafter provided. The Trustee shall not be responsible nor shall it undertake any responsibility for the amount or adequacy of, nor any duty to collect from the Grantor, any payments necessary to discharge any liabilities of the Grantor established by EPA.

Section 4. Payment for Bodily Injury or Property Damage. The Trustee shall satisfy a third party liability claim by making payments from the Fund only upon receipt of one of the following documents:

(a) Certification from the Grantor and the third party claimant(s) that the liability claim should be paid. The certification must be worded as follows, except that instructions in brackets are to be replaced with the relevant information and the brackets deleted:

Certification of Valid Claim

The undersigned, as parties [insert Grantor] and [insert name and address of third party claimant(s)], hereby certify that the claim of bodily injury and/or property damage caused by a [sudden or nonsudden] accidental occurrence arising from operating [Grantor's] facility or group of facilities should be paid in the amount of $[ ].

[Signatures]

Grantor

[Signatures]

Claimant(s)

(b) A valid final court order establishing a judgment against the Grantor for bodily injury or property damage caused by sudden or nonsudden accidental occurrences arising from the operation of the Grantor's facility or group of facilities.

Section 5. Payments Comprising the Fund. Payments made to the Trustee for the Fund shall consist of cash or securities acceptable to the Trustee.

Section 6. Trustee Management. The Trustee shall invest and reinvest the principal and income, in accordance with general investment policies and guidelines which the Grantor may communicate in writing to the Trustee from time to time, subject, however, to the provisions of this section. In investing, reinvesting, exchanging, selling, and managing the Fund, the Trustee shall discharge his duties with respect to the trust fund solely in the interest of the beneficiary and with the care, skill, prudence, and diligence under the circumstance then prevailing which persons of prudence, acting in a like capacity and familiar with such matters, would use in the conduct of an enterprise of a like character and with like aims; except that:

(i) Securities or other obligations of the Grantor, or any other owner or operator of the facilities, or any of their affiliates as defined in the Investment Company Act of 1940, as amended, 15 U.S.C. 80a–2.(a), shall not be acquired or held unless they are securities or other obligations of the Federal or a State government;

(ii) The Trustee is authorized to invest the Fund in time or demand deposits of the Trustee, to the extent insured by an agency of the Federal or State government; and

(iii) The Trustee is authorized to hold cash awaiting investment or distribution uninvested for a reasonable time and without liability for the payment of interest thereon.

Section 7. Commingling and Investment. The Trustee is expressly authorized in its discretion:

(a) To transfer from time to time any or all of the assets of the Fund to any common commingled, or collective trust fund created by the Trustee in which the fund is eligible to participate, subject to all of the provisions thereof, to be commingled with the assets of other trusts participating therein; and

(b) To purchase shares in any investment company registered under the Investment Company Act of 1940, 15 U.S.C. 81a–1 et seq., including one which may be created, managed, underwritten, or to which investment advice is rendered or the shares of which are sold by the Trustee. The Trustee may vote such shares in its discretion.

Section 8. Express Powers of Trustee. Without in any way limiting the powers and discretions conferred upon the Trustee by the other provisions of this Agreement or by law, the Trustee is expressly authorized and empowered:

149

(a) To sell, exchange, convey, transfer, or otherwise dispose of any property held by it, by public or private sale. No person dealing with the Trustee shall be bound to see to the application of the purchase money or to inquire into the validity or expediency of any such sale or other disposition;

(b) To make, execute, acknowledge, and deliver any and all documents of transfer and conveyance and any and all other instruments that may be necessary or appropriate to carry out the powers herein granted;

(c) To register any securities held in the Fund in its own name or in the name of a nominee and to hold any security in bearer form or in book entry, or to combine certificates representing such securities with certificates of the same issue held by the Trustee in other fiduciary capacities, or to deposit or arrange for the deposit of such securities in a qualified central depository even though, when so deposited, such securities may be merged and held in bulk in the name of the nominee of such depository with other securities deposited therein by another person, or to deposit or arrange for the deposit of any securities issued by the United States Government, or any agency or instrumentality thereof, with a Federal Reserve bank, but the books and records of the Trustee shall at all times show that all such securities are part of the Fund;

(d) To deposit any cash in the Fund in interest-bearing accounts maintained or savings certificates issued by the Trustee, in its separate corporate capacity, or in any other banking institution affiliated with the Trustee, to the extent insured by an agency of the Federal or State government; and

(e) To compromise or otherwise adjust all claims in favor of or against the Fund.

Section 9. Taxes and Expenses. All taxes of any kind that may be assessed or levied against or in respect of the Fund and all brokerage commissions incurred by the Fund shall be paid from the Fund. All other expenses incurred by the Trustee in connection with the administration of this Trust, including fees for legal services rendered to the Trustee, the compensation of the Trustee to the extent not paid directly by the Grantor, and all other proper charges and disbursements of the Trustee shall be paid from the Fund.

Section 10. Annual Valuations. The Trustee shall annually, at least 30 days prior to the anniversary date of establishment of the Fund, furnish to the Grantor and to the appropriate EPA Regional Administrator a statement confirming the value of the Trust. Any securities in the Fund shall be valued at market value as of no more than 60 days prior to the anniversary date of establishment of the Fund. The failure of the Grantor to object in writing to the Trustee within 90 days after the statement has been furnished to the Grantor and the EPA Regional Administrator shall constitute a conclusively binding assent by the Grantor barring the Grantor from asserting any claim or liability against the Trustee with respect to matters disclosed in the statement.

Section 11. Advice of Counsel. The Trustee may from time to time consult with counsel, who may be counsel to the Grantor with respect to any question arising as to the construction of this Agreement or any action to be taken hereunder. The Trustee shall be fully protected, to the extent permitted by law, in acting upon the advice of counsel.

Section 12. Trustee Compensation. The Trustee shall be entitled to reasonable compensation for its services as agreed upon in writing from time to time with the Grantor.

Section 13. Successor Trustee. The Trustee may resign or the Grantor may replace the Trustee, but such resignation or replacement shall not be effective until the Grantor has appointed a successor trustee and this successor accepts the appointment. The successor trustee shall have the same powers and duties as those conferred upon the Trustee hereunder. Upon the successor trustee's acceptance of the appointment, the Trustee shall assign, transfer, and pay over to the successor trustee the funds and properties then constituting the Fund. If for any reason the Grantor cannot or does not act in the event of the resignation of the Trustee, the Trustee may apply to a court of competent jurisdiction for the appointment of a successor trustee or for instructions. The successor trustee shall specify the date on which it assumes administration of the trust in a writing sent to the Grantor, the EPA Regional Administrator, and the present Trustee by certified mail 10 days before such change becomes effective. Any expenses incurred by the Trustee as a result of any of the acts contemplated by this section shall be paid as provided in Section 9.

Section 14. Instructions to the Trustee. All orders, requests, and instructions by the Grantor to the Trustee shall be in writing, signed by such persons as are designated in the attached Exhibit A or such other designees as the Grantor may designate by amendments to Exhibit A. The Trustee shall be fully protected in acting without inquiry in accordance with the Grantor's orders, requests, and instructions. All orders, requests, and instructions by the EPA Regional Administrator to the Trustee shall be in writing, signed by the EPA Regional Administrators of the Regions in which the facilities are located, or their designees, and the Trustee shall act and shall be fully protected in acting in accordance with such orders, requests, and instructions. The Trustee shall have the right to assume, in the absence of written notice to the contrary, that no event constituting a change or a termination of the authority of any person to act on behalf of the Grantor or EPA hereunder

150

has occurred. The Trustee shall have no duty to act in the absence of such orders, requests, and instructions from the Grantor and/or EPA, except as provided for herein.

Section 15. Notice of Nonpayment. If a payment for bodily injury or property damage is made under Section 4 of this trust, the Trustee shall notify the Grantor of such payment and the amount(s) thereof within five (5) working days. The Grantor shall, on or before the anniversary date of the establishment of the Fund following such notice, either make payments to the Trustee in amounts sufficient to cause the trust to return to its value immediately prior to the payment of claims under Section 4, or shall provide written proof to the Trustee that other financial assurance for liability coverage has been obtained equaling the amount necessary to return the trust to its value prior to the payment of claims. If the Grantor does not either make payments to the Trustee or provide the Trustee with such proof, the Trustee shall within 10 working days after the anniversary date of the establishment of the Fund provide a written notice of nonpayment to the EPA Regional Administrator.

Section 16. Amendment of Agreement. This Agreement may be amended by an instrument in writing executed by the Grantor, the Trustee, and the appropriate EPA Regional Administrator, or by the Trustee and the appropriate EPA Regional Administrator if the Grantor ceases to exist.

Section 17. Irrevocability and Termination. Subject to the right of the parties to amend this Agreement as provided in Section 16, this Trust shall be irrevocable and shall continue until terminated at the written agreement of the Grantor, the Trustee, and the EPA Regional Administrator, or by the Trustee and the EPA Regional Administrator, if the Grantor ceases to exist. Upon termination of the Trust, all remaining trust property, less final trust administration expenses, shall be delivered to the Grantor.

The Regional Administrator will agree to termination of the Trust when the owner or operator substitutes alternate financial assurance as specified in this section.

Section 18. Immunity and Indemnification. The Trustee shall not incur personal liability of any nature in connection with any act or omission, made in good faith, in the administration of this Trust, or in carrying out any directions by the Grantor or the EPA Regional Administrator issued in accordance with this Agreement. The Trustee shall be indemnified and saved harmless by the Grantor or from the Trust Fund, or both, from and against any personal liability to which the Trustee may be subjected by reason of any act or conduct in its official capacity, including all expenses reasonably incurred in its defense in the event the Grantor fails to provide such defense.

Section 19. Choice of Law. This Agreement shall be administered, construed, and enforced according to the laws of the State of [enter name of State].

Section 20. Interpretation. As used in this Agreement, words in the singular include the plural and words in the plural include the singular. The descriptive headings for each section of this Agreement shall not affect the interpretation or the legal efficacy of this Agreement.

In Witness Whereof the parties have caused this Agreement to be executed by their respective officers duly authorized and their corporate seals to be hereunto affixed and attested as of the date first above written. The parties below certify that the wording of this Agreement is identical to the wording specified in 40 CFR 261.151(l) as such regulations were constituted on the date first above written.

[Signature of Grantor]

[Title]

Attest:

[Title]

[Seal]

[Signature of Trustee]

Attest:

[Title]

[Seal]

(2) The following is an example of the certification of acknowledgement which must accompany the trust agreement for a trust fund as specified in Sec. 261.147(j) of this chapter. State requirements may differ on the proper State of _____

County of _____

On this [date], before me personally came [owner or operator] to me known, who, being by me duly sworn, did depose and say that she/he resides at [address], that she/he is [title] of [corporation], the corporation described in and which executed the above instrument; that she/he knows the seal of said corporation; that the seal affixed to such instrument is such corporate seal; that it was so affixed by order of the Board of Directors of said corporation, and that she/he signed her/ his name thereto by like order.

[Signature of Notary Public]

(m)(1) A standby trust agreement, as specified in §261.147(h) of this chapter, must be worded as follows, except that instructions in brackets are to be replaced with the relevant information and the brackets deleted:

### Standby Trust Agreement

Trust Agreement, the "Agreement," entered into as of [date] by and between [name of the owner or operator] a [name of a State] [insert "corporation," "partnership," "association," or "proprietorship"], the "Grantor," and [name of corporate trustee], [insert, "incorporated in the State of _____" or "a national bank"], the "trustee."

Whereas the United States Environmental Protection Agency, "EPA," an agency of the United States Government, has established certain regulations applicable to the Grantor, requiring that an owner or operator must demonstrate financial responsibility for bodily injury and property damage to third parties caused by sudden accidental and/or nonsudden accidental occurrences arising from operations of the facility or group of facilities.

Whereas, the Grantor has elected to establish a standby trust into which the proceeds from a letter of credit may be deposited to assure all or part of such financial responsibility for the facilities identified herein.

Whereas, the Grantor, acting through its duly authorized officers, has selected the Trustee to be the trustee under this agreement, and the Trustee is willing to act as trustee.

Now, therefore, the Grantor and the Trustee agree as follows:

Section 1. Definitions. As used in this Agreement:

(a) The term Grantor means the owner or operator who enters into this Agreement and any successors or assigns of the Grantor.

(b) The term Trustee means the Trustee who enters into this Agreement and any successor Trustee.

Section 2. Identification of Facilities. This Agreement pertains to the facilities identified on attached schedule A [on schedule A, for each facility list the EPA Identification Number (if any issued), name, and address of the facility(ies) and the amount of liability coverage, or portions thereof, if more than one instrument affords combined coverage as demonstrated by this Agreement].

Section 3. Establishment of Fund. The Grantor and the Trustee hereby establish a standby trust fund, hereafter the "Fund," for the benefit of any and all third parties injured or damaged by [sudden and/or nonsudden] accidental occurrences arising from operation of the facility(ies) covered by this guarantee, in the amounts of _____ -[up to $1 million] per occurrence and _____ -[up to $2 million] annual aggregate for sudden accidental occurrences and _____ -[up to $3 million] per occurrence and _____ -[up to $6 million] annual aggregate for nonsudden occurrences, except that the Fund is not established for the benefit of third parties for the following:

(a) Bodily injury or property damage for which [insert Grantor] is obligated to pay damages by reason of the assumption of liability in a contract or agreement. This exclusion does not apply to liability for damages that [insert Grantor] would be obligated to pay in the absence of the contract or agreement.

(b) Any obligation of [insert Grantor] under a workers' compensation, disability benefits, or unemployment compensation law or any similar law.

(c) Bodily injury to:

(1) An employee of [insert Grantor] arising from, and in the course of, employment by [insert Grantor]; or

(2) The spouse, child, parent, brother or sister of that employee as a consequence of, or arising from, and in the course of employment by [insert Grantor].

This exclusion applies:

(A) Whether [insert Grantor] may be liable as an employer or in any other capacity; and

(B) To any obligation to share damages with or repay another person who must pay damages because of the injury to persons identified in paragraphs (1) and (2).

(d) Bodily injury or property damage arising out of the ownership, maintenance, use, or entrustment to others of any aircraft, motor vehicle or watercraft.

(e) Property damage to:

(1) Any property owned, rented, or occupied by [insert Grantor];

(2) Premises that are sold, given away or abandoned by [insert Grantor] if the property damage arises out of any part of those premises;

(3) Property loaned by [insert Grantor];

(4) Personal property in the care, custody or control of [insert Grantor];

(5) That particular part of real property on which [insert Grantor] or any contractors or subcontractors working directly or indirectly on behalf of [insert Grantor] are performing operations, if the property damage arises out of these operations.

In the event of combination with another mechanism for liability coverage, the Fund shall be considered [insert "primary" or "excess"] coverage.

The Fund is established initially as consisting of the proceeds of the letter of credit deposited into the Fund. Such proceeds and any other property subsequently transferred to the Trustee is referred to as the Fund, together with all earnings and profits thereon, less any payments or distributions made by the Trustee pursuant to this Agreement. The Fund shall be held by the Trustee, IN TRUST, as hereinafter provided. The Trustee shall not be responsible nor shall it undertake any responsibility for the amount or adequacy of, nor any duty to collect from the Grantor, any payments necessary to discharge any liabilities of the Grantor established by EPA.

Section 4. Payment for Bodily Injury or Property Damage. The Trustee shall satisfy a third party liability claim by drawing on the letter of credit described in Schedule B and by making payments from the Fund only upon receipt of one of the following documents:

(a) Certification from the Grantor and the third party claimant(s) that the liability claim should be paid. The certification must be worded as follows, except that instructions in brackets are to be replaced with the relevant information and the brackets deleted:

#### Certification of Valid Claim

The undersigned, as parties [insert Grantor] and [insert name and address of third party claimant(s)], hereby certify that the claim of bodily injury and/or property damage caused by a [sudden or nonsudden] accidental occurrence arising from operating [Grantor's] facility should be paid in the amount of $[ ]

[Signature] _____

Grantor

[Signatures] _____

Claimant(s)

(b) A valid final court order establishing a judgment against the Grantor for bodily injury or property damage caused by sudden or nonsudden accidental occurrences arising from the operation of the Grantor's facility or group of facilities.

Section 5. Payments Comprising the Fund. Payments made to the Trustee for the Fund shall consist of the proceeds from the letter of credit drawn upon by the Trustee in accordance with the requirements of 40 CFR 261.151(k) and Section 4 of this Agreement.

Section 6. Trustee Management. The Trustee shall invest and reinvest the principal and income, in accordance with general investment policies and guidelines which the Grantor may communicate in writing to the Trustee from time to time, subject, however, to the provisions of this Section. In investing, reinvesting, exchanging, selling, and managing the Fund, the Trustee shall discharge his duties with respect to the trust fund solely in the interest of the beneficiary and with the care, skill, prudence, and diligence under the circumstances then prevailing which persons of prudence, acting in a like capacity and familiar with such matters, would use in the conduct of an enterprise of a like character and with like aims; except that:

(i) Securities or other obligations of the Grantor, or any other owner or operator of the facilities, or any of their affiliates as defined in the Investment Company Act of 1940, as amended, 15 U.S.C. 80a–2(a), shall not be acquired or held, unless they are securities or other obligations of the Federal or a State government;

(ii) The Trustee is authorized to invest the Fund in time or demand deposits of the Trustee, to the extent insured by an agency of the Federal or a State government; and

(iii) The Trustee is authorized to hold cash awaiting investment or distribution uninvested for a reasonable time and without liability for the payment of interest thereon.

Section 7. Commingling and Investment. The Trustee is expressly authorized in its discretion:

(a) To transfer from time to time any or all of the assets of the Fund to any common, commingled, or collective trust fund created by the Trustee in which the Fund is eligible to participate, subject to all of the provisions thereof, to be commingled with the assets of other trusts participating therein; and

(b) To purchase shares in any investment company registered under the Investment Company Act of 1940, 15 U.S.C. 80a–1 et seq., including one which may be created, managed, underwritten, or to which investment advice is rendered or the shares of which are sold by the Trustee. The Trustee may vote such shares in its discretion.

Section 8. Express Powers of Trustee. Without in any way limiting the powers and discretions conferred upon the Trustee by the other provisions of this Agreement or by law, the Trustee is expressly authorized and empowered:

(a) To sell, exchange, convey, transfer, or otherwise dispose of any property held by it, by public or private sale. No person dealing with the Trustee shall be bound to see to the application of the purchase money or to inquire into the validity or expediency of any such sale or other disposition;

(b) To make, execute, acknowledge, and deliver any and all documents of transfer and conveyance and any and all other instruments that may be necessary or appropriate to carry out the powers herein granted;

(c) To register any securities held in the Fund in its own name or in the name of a nominee and to hold any security in bearer form or in book entry, or to combine certificates representing such securities with certificates of the same issue held by the Trustee in other fiduciary capacities, or to deposit or arrange for the deposit of such securities in a qualified central depositary even though, when so deposited, such securities may be merged and held in bulk in the name of the nominee of such depositary with other securities deposited therein by another person, or to deposit or arrange for the deposit of any securities issued by the United States Government, or any agency or instrumentality thereof, with a Federal Reserve Bank, but the books and records of the Trustee shall at all times show that all such securities are part of the Fund;

(d) To deposit any cash in the Fund in interest-bearing accounts maintained or savings certificates issued by the Trustee, in its separate corporate capacity, or in any other banking institution affiliated with the Trustee, to the extent insured by an agency of the Federal or State government; and

(e) To compromise or otherwise adjust all claims in favor of or against the Fund.

Section 9. Taxes and Expenses. All taxes of any kind that may be assessed or levied against or in respect of the Fund and all brokerage commissions incurred by the Fund shall be paid from the Fund. All other expenses incurred by the Trustee in connection with the administration of this Trust, including fees for legal services rendered to the Trustee, the compensation of the Trustee to the extent not paid directly by the Grantor, and all other proper charges and disbursements to the Trustee shall be paid from the Fund.

Section 10. Advice of Counsel. The Trustee may from time to time consult with counsel, who may be counsel to the Grantor, with respect to any question arising as to the construction of this Agreement or any action to be taken hereunder. The Trustee shall be fully protected, to the extent permitted by law, in acting upon the advice of counsel.

Section 11. Trustee Compensation. The Trustee shall be entitled to reasonable compensation for its services as agreed upon in writing from time to time with the Grantor.

Section 12. Successor Trustee. The Trustee may resign or the Grantor may replace the Trustee, but such resignation or replacement shall not be effective until the Grantor has appointed a successor trustee and this successor accepts the appointment. The successor trustee shall have the same powers and duties as those conferred upon the Trustee hereunder. Upon the successor trustee's acceptance of the appointment, the Trustee shall assign, transfer, and pay over to the successor trustee the funds and properties then constituting the Fund. If for any reason the Grantor cannot or does not act in the event of the resignation of the Trustee, the Trustee may apply to a court of competent jurisdiction for the appointment of a successor trustee or for instructions. The successor trustee shall specify the date on which it assumes administration of the trust in a writing sent to the Grantor, the EPA Regional Administrator and the present Trustee by certified mail 10 days before such change becomes effective. Any expenses incurred by the Trustee as a result of any of the acts contemplated by this Section shall be paid as provided in Section 9.

Section 13. Instructions to the Trustee. All orders, requests, certifications of valid claims, and instructions to the Trustee shall be in writing, signed by such persons as are designated in the attached Exhibit A or such other designees as the Grantor may designate by amendments to Exhibit A. The Trustee shall be fully protected in acting without inquiry in accordance with the Grantor's orders, requests, and instructions. The Trustee shall have the right to assume, in the absence of written notice to the contrary, that no event constituting a change or a termination of the authority of any person to act on behalf of the Grantor or the EPA Regional Administrator hereunder has occurred. The Trustee shall have no duty to act in the absence of such orders, requests, and instructions from the Grantor and/or EPA, except as provided for herein.

Section 14. Amendment of Agreement. This Agreement may be amended by an instrument in writing executed by the Grantor, the Trustee, and the EPA Regional Administrator, or by the Trustee and the EPA Regional Administrator if the Grantor ceases to exist.

Section 15. Irrevocability and Termination. Subject to the right of the parties to amend this Agreement as provided in Section 14, this Trust shall be irrevocable and shall continue until terminated at the written agreement of the Grantor, the Trustee, and the EPA Regional Administrator, or by the Trustee and the EPA Regional Administrator, if the Grantor ceases to exist. Upon termination of the Trust, all remaining trust property, less final trust administration expenses, shall be paid to the Grantor.

The Regional Administrator will agree to termination of the Trust when the owner or operator substitutes alternative financial assurance as specified in this section.

Section 16. Immunity and indemnification. The Trustee shall not incur personal liability of any nature in connection with any act or omission, made in good faith, in the administration of this Trust, or in carrying out any directions by the Grantor and the EPA Regional Administrator issued in accordance with this Agreement. The Trustee shall be indemnified and saved harmless by the Grantor or from the Trust Fund, or both, from and against any personal liability to which the Trustee may be subjected by reason of any act or conduct in its official capacity, including all expenses reasonably incurred in its defense in the event the Grantor fails to provide such defense.

Section 17. Choice of Law. This Agreement shall be administered, construed, and enforced according to the laws of the State of [enter name of State].

Section 18. Interpretation. As used in this Agreement, words in the singular include the plural and words in the plural include the singular. The descriptive headings for each Section of this Agreement shall not affect the interpretation of the legal efficacy of this Agreement.

In Witness Whereof the parties have caused this Agreement to be executed by their respective officers duly authorized and their corporate seals to be hereunto affixed and attested as of the date first above written. The parties below certify that the wording of this Agreement is identical to the wording specified in 40 CFR 261.151(m) as such regulations were constituted on the date first above written.

[Signature of Grantor]

[Title]

Attest:

[Title]

[Seal]

[Signature of Trustee]

Attest:

[Title]

[Seal]

(2) The following is an example of the certification of acknowledgement which must accompany the trust agreement for a standby trust fund as specified in section 261.147(h) of this chapter. State requirements may differ on the proper content of this acknowledgement.

State of _____

County of _____

On this [date], before me personally came [owner or operator] to me known, who, being by me duly sworn, did depose and say that she/he resides at [address], that she/he is [title] of [corporation], the corporation described in and which executed the above instrument; that she/he knows the seal of said corporation; that the seal affixed to such instrument is such corporate seal; that it was so affixed by order of the Board of Directors of said corporation, and that she/he signed her/ his name thereto by like order.

[Signature of Notary Public]

APPENDIX I TO PART 261—
REPRESENTATIVE SAMPLING METHODS

The methods and equipment used for sampling waste materials will vary with the form and consistency of the waste materials to be sampled. Samples collected using the sampling protocols listed below, for sampling waste with properties similar to the indicated materials, will be considered by the Agency to be representative of the waste.

Extremely viscous liquid—ASTM Standard D140–70 Crushed or powdered material—ASTM Standard D346–75 Soil or rock-like material—ASTM Standard D420–69 Soil-like material—ASTM Standard D1452–65 Fly Ash-like material—ASTM Standard D2234–76 [ASTM Standards are available

from ASTM, 1916 Race St., Philadelphia, PA 19103]

Containerized liquid waste—"COLIWASA."

Liquid waste in pits, ponds, lagoons, and similar reservoirs—"Pond Sampler."

This manual also contains additional information on application of these protocols.

[45 FR 33119, May 19, 1980, as amended at 70 FR 34562, June 14, 2005]

APPENDIX II TO PART 261 [RESERVED]

APPENDIX III TO PART 261 [RESERVED]

APPENDIX IV TO PART 261 [RESERVED FOR RADIOACTIVE WASTE TEST METHODS]

APPENDIX V TO PART 261 [RESERVED FOR INFECTIOUS WASTE TREATMENT SPECIFICATIONS]

APPENDIX VI TO PART 261 [RESERVED FOR ETIOLOGIC AGENTS]

APPENDIX VII TO PART 261—BASIS FOR LISTING HAZARDOUS WASTE

| EPA hazardous waste No. | Hazardous constituents for which listed |
|---|---|
| F001 ..... | Tetrachloroethylene, methylene chloride trichloroethylene, 1,1,1-trichloroethane, carbon tetrachloride, chlorinated fluorocarbons. |
| F002 ..... | Tetrachloroethylene, methylene chloride, trichloroethylene, 1,1,1-trichloroethane, 1,1,2-trichloroethane, chlorobenzene, 1,1,2-trichloro-1,2,2-trifluoroethane, ortho-dichlorobenzene, trichlorofluoromethane. |
| F003 ..... | N.A. |
| F004 ..... | Cresols and cresylic acid, nitrobenzene. |
| F005 ..... | Toluene, methyl ethyl ketone, carbon disulfide, isobutanol, pyridine, 2-ethoxyethanol, benzene, 2-nitropropane. |
| F006 ..... | Cadmium, hexavalent chromium, nickel, cyanide (complexed). |
| F007 ..... | Cyanide (salts). |
| F008 ..... | Cyanide (salts). |
| F009 ..... | Cyanide (salts). |
| F010 ..... | Cyanide (salts). |
| F011 ..... | Cyanide (salts). |
| F012 ..... | Cyanide (complexed). |
| F019 ..... | Hexavalent chromium, cyanide (complexed). |
| F020 ..... | Tetra- and pentachlorodibenzo-p- dioxins; tetra and pentachlorodi-benzofurans; tri- and tetrachlorophenols and their chlorophenoxy derivative acids, esters, ethers, amine and other salts. |
| F021 ..... | Penta- and hexachlorodibenzo-p- dioxins; penta- and hexachlorodibenzofurans; pentachlorophenol and its derivatives. |
| F022 ..... | Tetra-, penta-, and hexachlorodibenzo-p-dioxins; tetra-, penta-, and hexachlorodibenzofurans. |
| F023 ..... | Tetra-, and pentachlorodibenzo-p-dioxins; tetra- and pentachlorodibenzofurans; tri- and tetrachlorophenols and their chlorophenoxy derivative acids, esters, ethers, amine and other salts. |

| EPA hazardous waste No. | Hazardous constituents for which listed |
|---|---|
| F024 ..... | Chloromethane, dichloromethane, trichloromethane, carbon tetrachloride, chloroethylene, 1,1-dichloroethane, 1,2-dichloroethane, trans-1-2-dichloroethylene, 1,1-dichloroethylene, 1,1,1-trichloroethane, 1,1,2-trichloroethane, trichloroethylene, 1,1,1,2-tetra-chloroethane, 1,1,2,2-tetrachloroethane, tetrachloroethylene, pentachloroethane, hexachloroethane, allyl chloride (3-chloropropene), dichloropropane, dichloropropene, 2-chloro-1,3-butadiene, hexachloro-1,3-butadiene, hexachlorocyclopentadiene, hexachlorocyclohexane, benzene, chlorbenzene, dichlorobenzenes, 1,2,4-trichlorobenzene, tetrachlorobenzene, pentachlorobenzene, hexachlorobenzene, toluene, naphthalene. |
| F025 ..... | Chloromethane; Dichloromethane; Trichloromethane; Carbon tetrachloride; Chloroethylene; 1,1-Dichloroethane; 1,2-Dichloroethane; trans-1,2-Dichloroethylene; 1,1-Dichloroethylene; 1,1,1-Trichloroethane; 1,1,2-Trichloroethane; Trichloroethylene; 1,1,1,2-Tetrachloroethane; 1,1,2,2-Tetrachloroethane; Tetrachloroethylene; Pentachloroethane; Hexachloroethane; Allyl chloride (3-Chloropropene); Dichloropropane; Dichloropropene; 2-Chloro-1,3-butadiene; Hexachloro-1,3-butadiene; Hexachlorocyclopentadiene; Benzene; Chlorobenzene; Dichlorobenzene; 1,2,4-Trichlorobenzene; Tetrachlorobenzene; Pentachlorobenzene; Hexachlorobenzene; Toluene; Naphthalene. |
| F026 ..... | Tetra-, penta-, and hexachlorodibenzo-*p*-dioxins; tetra-, penta-, and hexachlorodibenzofurans. |
| F027 ..... | Tetra-, penta-, and hexachlorodibenzo-*p*- dioxins; tetra-, penta-, and hexachlorodibenzofurans; tri-, tetra-, and pentachlorophenols and their chlorophenoxy derivative acids, esters, ethers, amine and other salts. |
| F028 ..... | Tetra-, penta-, and hexachlorodibenzo-*p*- dioxins; tetra-, penta-, and hexachlorodibenzofurans; tri-, tetra-, and pentachlorophenols and their chlorophenoxy derivative acids, esters, ethers, amine and other salts. |
| F032 ..... | Benz(a)anthracene, benzo(a)pyrene, dibenz(a,h)anthracene, indeno(1,2,3-cd)pyrene, pentachlorophenol, arsenic, chromium, tetra-, penta-, hexa-, heptachlorodibenzo-p-dioxins, tetra-, penta-, hexa-, heptachlorodibenzofurans. |
| F034 ..... | Benz(a)anthracene, benzo(k)fluoranthene, benzo(a)pyrene, dibenz(a,h)anthracene, indeno(1,2,3-cd)pyrene, naphthalene, arsenic, chromium. |
| F035 ..... | Arsenic, chromium, lead. |
| F037 ..... | Benzene, benzo(a)pyrene, chrysene, lead, chromium. |
| F038 ..... | Benzene, benzo(a)pyrene, chrysene, lead, chromium. |
| F039 ..... | All constituents for which treatment standards are specified for multi-source leachate (wastewaters and nonwastewaters) under 40 CFR 268.43, Table CCW. |

| EPA hazardous waste No. | Hazardous constituents for which listed |
|---|---|
| K001 ..... | Pentachlorophenol, phenol, 2-chlorophenol, p-chloro-m-cresol, 2,4-dimethylphenyl, 2,4-dinitrophenol, trichlorophenols, tetrachlorophenols, 2,4-dinitrophenol, creosote, chrysene, naphthalene, fluoranthene, benzo(b)fluoranthene, benzo(a)pyrene, indeno(1,2,3-cd)pyrene, benz(a)anthracene, dibenz(a)anthracene, acenaphthalene. |
| K002 ..... | Hexavalent chromium, lead |
| K003 ..... | Hexavalent chromium, lead. |
| K004 ..... | Hexavalent chromium. |
| K005 ..... | Hexavalent chromium, lead. |
| K006 ..... | Hexavalent chromium. |
| K007 ..... | Cyanide (complexed), hexavalent chromium. |
| K008 ..... | Hexavalent chromium. |
| K009 ..... | Chloroform, formaldehyde, methylene chloride, methyl chloride, paraldehyde, formic acid. |
| K010 ..... | Chloroform, formaldehyde, methylene chloride, methyl chloride, paraldehyde, formic acid, chloroacetaldehyde. |
| K011 ..... | Acrylonitrile, acetonitrile, hydrocyanic acid. |
| K013 ..... | Hydrocyanic acid, acrylonitrile, acetonitrile. |
| K014 ..... | Acetonitrile, acrylamide. |
| K015 ..... | Benzyl chloride, chlorobenzene, toluene, benzotrichloride. |
| K016 ..... | Hexachlorobenzene, hexachlorobutadiene, carbon tetrachloride, hexachloroethane, perchloroethylene. |
| K017 ..... | Epichlorohydrin, chloroethers [bis(chloromethyl) ether and bis (2-chloroethyl) ethers], trichloropropane, dichloropropanols. |
| K018 ..... | 1,2-dichloroethane, trichloroethylene, hexachlorobutadiene, hexachlorobenzene. |
| K019 ..... | Ethylene dichloride, 1,1,1-trichloroethane, 1,1,2-trichloroethane, tetrachloroethanes (1,1,2,2-tetrachloroethane and 1,1,1,2-tetrachloroethane), trichloroethylene, tetrachloroethylene, carbon tetrachloride, chloroform, vinyl chloride, vinylidene chloride. |
| K020 ..... | Ethylene dichloride, 1,1,1-trichloroethane, 1,1,2-trichloroethane, tetrachloroethanes (1,1,2,2-tetrachloroethane and 1,1,1,2-tetrachloroethane), trichloroethylene, tetrachloroethylene, carbon tetrachloride, chloroform, vinyl chloride, vinylidene chloride. |
| K021 ..... | Antimony, carbon tetrachloride, chloroform. |
| K022 ..... | Phenol, tars (polycyclic aromatic hydrocarbons). |
| K023 ..... | Phthalic anhydride, maleic anhydride. |
| K024 ..... | Phthalic anhydride, 1,4-naphthoquinone. |
| K025 ..... | Meta-dinitrobenzene, 2,4-dinitrotoluene. |
| K026 ..... | Paraldehyde, pyridines, 2-picoline. |
| K027 ..... | Toluene diisocyanate, toluene-2, 4-diamine. |
| K028 ..... | 1,1,1-trichloroethane, vinyl chloride. |
| K029 ..... | 1,2-dichloroethane, 1,1,1-trichloroethane, vinyl chloride, vinylidene chloride, chloroform. |
| K030 ..... | Hexachlorobenzene, hexachlorobutadiene, hexachloroethane, 1,1,1,2-tetrachloroethane, 1,1,2,2-tetrachloroethane, ethylene dichloride. |
| K031 ..... | Arsenic. |
| K032 ..... | Hexachlorocyclopentadiene. |
| K033 ..... | Hexachlorocyclopentadiene. |
| K034 ..... | Hexachlorocyclopentadiene. |
| K035 ..... | Creosote, chrysene, naphthalene, fluoranthene benzo(b) fluoranthene, benzo(a)pyrene, indeno(1,2,3-cd) pyrene, benzo(a)anthracene, dibenzo(a)anthracene, acenaphthalene. |
| K036 ..... | Toluene, phosphorodithioic and phosphorothioic acid esters. |
| K037 ..... | Toluene, phosphorodithioic and phosphorothioic acid esters. |

| EPA hazardous waste No. | Hazardous constituents for which listed |
|---|---|
| K038 ..... | Phorate, formaldehyde, phosphorodithioic and phosphorothioic acid esters. |
| K039 ..... | Phosphorodithioic and phosphorothioic acid esters. |
| K040 ..... | Phorate, formaldehyde, phosphorodithioic and phosphorothioic acid esters. |
| K041 ..... | Toxaphene. |
| K042 ..... | Hexachlorobenzene, ortho-dichlorobenzene. |
| K043 ..... | 2,4-dichlorophenol, 2,6-dichlorophenol, 2,4,6-trichlorophenol. |
| K044 ..... | N.A. |
| K045 ..... | N.A. |
| K046 ..... | Lead. |
| K047 ..... | N.A. |
| K048 ..... | Hexavalent chromium, lead. |
| K049 ..... | Hexavalent chromium, lead. |
| K050 ..... | Hexavalent chromium. |
| K051 ..... | Hexavalent chromium, lead. |
| K052 ..... | Lead. |
| K060 ..... | Cyanide, napthalene, phenolic compounds, arsenic. |
| K061 ..... | Hexavalent chromium, lead, cadmium. |
| K062 ..... | Hexavalent chromium, lead. |
| K069 ..... | Hexavalent chromium, lead, cadmium. |
| K071 ..... | Mercury. |
| K073 ..... | Chloroform, carbon tetrachloride, hexachloroethane, trichloroethane, tetrachloroethylene, dichloroethylene, 1,1,2,2-tetrachloroethane. |
| K083 ..... | Aniline, diphenylamine, nitrobenzene, phenylenediamine. |
| K084 ..... | Arsenic. |
| K085 ..... | Benzene, dichlorobenzenes, trichlorobenzenes, tetrachlorobenzenes, pentachlorobenzenes, hexachlorobenzene, benzyl chloride. |
| K086 ..... | Lead, hexavalent chromium. |
| K087 ..... | Phenol, naphthalene. |
| K088 ..... | Cyanide (complexes). |
| K093 ..... | Phthalic anhydride, maleic anhydride. |
| K094 ..... | Phthalic anhydride. |
| K095 ..... | 1,1,2-trichloroethane, 1,1,1,2-tetrachloroethane, 1,1,2,2-tetrachloroethane. |
| K096 ..... | 1,2-dichloroethane, 1,1,1-trichloroethane, 1,1,2-trichloroethane. |
| K097 ..... | Chlordane, heptachlor. |
| K098 ..... | Toxaphene. |
| K099 ..... | 2,4-dichlorophenol, 2,4,6-trichlorophenol. |
| K100 ..... | Hexavalent chromium, lead, cadmium. |
| K101 ..... | Arsenic. |
| K102 ..... | Arsenic. |
| K103 ..... | Aniline, nitrobenzene, phenylenediamine. |
| K104 ..... | Aniline, benzene, diphenylamine, nitrobenzene, phenylenediamine. |
| K105 ..... | Benzene, monochlorobenzene, dichlorobenzenes, 2,4,6-trichlorophenol. |
| K106 ..... | Mercury. |
| K107 ..... | 1,1-Dimethylhydrazine (UDMH). |
| K108 ..... | 1,1-Dimethylhydrazine (UDMH). |
| K109 ..... | 1,1-Dimethylhydrazine (UDMH). |
| K110 ..... | 1,1-Dimethylhydrazine (UDMH). |
| K111 ..... | 2,4-Dinitrotoluene. |
| K112 ..... | 2,4-Toluenediamine, o-toluidine, p-toluidine, aniline. |
| K113 ..... | 2,4-Toluenediamine, o-toluidine, p-toluidine, aniline. |
| K114 ..... | 2,4-Toluenediamine, o-toluidine, p-toluidine. |
| K115 ..... | 2,4-Toluenediamine. |
| K116 ..... | Carbon tetrachloride, tetrachloroethylene, chloroform, phosgene. |
| K117 ..... | Ethylene dibromide. |
| K118 ..... | Ethylene dibromide. |
| K123 ..... | Ethylene thiourea. |

| EPA hazardous waste No. | Hazardous constituents for which listed |
|---|---|
| K124 ..... | Ethylene thiourea. |
| K125 ..... | Ethylene thiourea. |
| K126 ..... | Ethylene thiourea. |
| K131 ..... | Dimethyl sulfate, methyl bromide. |
| K132 ..... | Methyl bromide. |
| K136 ..... | Ethylene dibromide. |
| K141 ..... | Benzene, benz(a)anthracene, benzo(a)pyrene, benzo(b)fluoranthene, benzo(k)fluoranthene, dibenz(a,h)anthracene, indeno(1,2,3-cd)pyrene. |
| K142 ..... | Benzene, benz(a)anthracene, benzo(a)pyrene, benzo(b)fluoranthene, benzo(k)fluoranthene, dibenz(a,h)anthracene, indeno(1,2,3-cd)pyrene. |
| K143 ..... | Benzene, benz(a)anthracene, benzo(b)fluoranthene, benzo(k)fluoranthene. |
| K144 ..... | Benzene, benz(a)anthracene, benzo(a)pyrene, benzo(b)fluoranthene, benzo(k)fluoranthene, dibenz(a,h)anthracene. |
| K145 ..... | Benzene, benz(a)anthracene, benzo(a)pyrene, dibenz(a,h)anthracene, naphthalene. |
| K147 ..... | Benzene, benz(a)anthracene, benzo(a)pyrene, benzo(b)fluoranthene, benzo(k)fluoranthene, dibenz(a,h)anthracene, indeno(1,2,3-cd)pyrene. |
| K148 ..... | Benz(a)anthracene, benzo(a)pyrene, benzo(b)fluoranthene, benzo(k)fluoranthene, dibenz(a,h)anthracene, indeno(1,2,3-cd)pyrene. |
| K149 ..... | Benzotrichloride, benzyl chloride, chloroform, chloromethane, chlorobenzene, 1,4-dichlorobenzene, hexachlorobenzene, pentachlorobenzene, 1,2,4,5-tetrachlorobenzene, toluene. |
| K150 ..... | Carbon tetrachloride, chloroform, chloromethane, 1,4-dichlorobenzene, hexachlorobenzene, pentachlorobenzene, 1,2,4,5-tetrachlorobenzene, 1,1,2,2-tetrachloroethane, tetrachloroethylene, 1,2,4-trichlorobenzene. |
| K151 ..... | Benzene, carbon tetrachloride, chloroform, hexachlorobenzene, pentachlorobenzene, toluene, 1,2,4,5-tetrachlorobenzene, tetrachloroethylene. |
| K156 ..... | Benomyl, carbaryl, carbendazim, carbofuran, carbosulfan, formaldehyde, methylene chloride, triethylamine. |
| K157 ..... | Carbon tetrachloride, formaldehyde, methyl chloride, methylene chloride, pyridine, triethylamine. |
| K158 ..... | Benomyl, carbendazim, carbofuran, carbosulfan, chloroform, methylene chloride. |
| K159 ..... | Benzene, butylate, eptc, molinate, pebulate, vernolate. |
| K161 ..... | Antimony, arsenic, metam-sodium, ziram |
| K169 ..... | Benzene. |
| K170 ..... | Benzo(a)pyrene, dibenz(a,h)anthracene, benzo (a) anthracene, benzo (b)fluoranthene, benzo(k)fluoranthene, 3-methylcholanthrene, 7, 12-dimethylbenz(a)anthracene. |
| K171 ..... | Benzene, arsenic. |
| K172 ..... | Benzene, arsenic. |
| K174 ..... | 1,2,3,4,6,7,8-Heptachlorodibenzo-p-dioxin (1,2,3,4,6,7,8-HpCDD), 1,2,3,4,6,7,8-Heptachlorodibenzofuran (1,2,3,4,6,7,8-HpCDF), 1,2,3,4,7,8,9-Heptachlorodibenzofuran (1,2,3,6,7,8,9-HpCDF), HxCDDs (All Hexachlorodibenzo-p-dioxins), HxCDFs (All Hexachlorodibenzofurans), PeCDDs (All Pentachlorodibenzo-p-dioxins), OCDD (1,2,3,4,6,7,8,9-Octachlorodibenzo-p-dioxin, OCDF (1,2,3,4,6,7,8,9-Octachlorodibenzofuran), PeCDFs (All Pentachlorodibenzofurans), TCDDs (All tetrachlorodi-benzo-p-dioxins), TCDFs (All tetrachlorodibenzofurans). |
| K175 ..... | Mercury |

157

| EPA haz-ardous waste No. | Hazardous constituents for which listed |
|---|---|
| K176 ..... | Arsenic, Lead. |
| K177 ..... | Antimony. |
| K178 ..... | Thallium. |
| K181 ..... | Aniline, o-anisidine, 4-chloroaniline, p-cresidine, 2,4-dimethylaniline, 1,2-phenylenediamine, 1,3-phenylenediamine. |

N.A.—Waste is hazardous because it fails the test for the characteristic of ignitability, corrosivity, or reactivity.

[46 FR 4619, Jan. 16, 1981]

EDITORIAL NOTE: For FEDERAL REGISTER citations affecting Appendix VII, part 261, see the List of CFR Sections Affected, which appears in the Finding Aids section of the printed volume and on GPO Access.

APPENDIX VIII TO PART 261—HAZARDOUS CONSTITUENTS

| Common name | Chemical abstracts name | Chemical abstracts No. | Hazardous waste No. |
|---|---|---|---|
| A2213 ............................................ | Ethanimidothioic acid, 2- (dimethylamino) -N-hydroxy-2-oxo-, methyl ester. | 30558–43–1 | U394 |
| Acetonitrile ................................... | Same ............................................... | 75–05–8 | U003 |
| Acetophenone ............................... | Ethanone, 1-phenyl- ..................... | 98–86–2 | U004 |
| 2-Acetylaminefluarone .................. | Acetamide, N-9H-fluoren-2-yl- ........ | 53–96–3 | U005 |
| Acetyl chloride ............................. | Same ............................................... | 75–36–5 | U006 |
| 1-Acetyl-2-thiourea ...................... | Acetamide, N-(aminothioxomethyl)- ... | 591–08–2 | P002 |
| Acrolein ........................................ | 2-Propenal ..................................... | 107–02–8 | P003 |
| Acrylamide .................................... | 2-Propenamide .............................. | 79–06–1 | U007 |
| Acrylonitrile .................................. | 2-Propenenitrile ............................ | 107–13–1 | U009 |
| Aflatoxins ..................................... | Same ............................................... | 1402–68–2 | .................. |
| Aldicarb ........................................ | Propanal, 2-methyl-2-(methylthio)-, O-[(methylamino)carbonyl]oxime. | 116–06–3 | P070 |
| Aldicarb sulfone ........................... | Propanal, 2-methyl-2- (methylsulfonyl) -, O-[(methylamino) carbonyl] oxime. | 1646–88–4 | P203 |
| Aldrin ............................................ | 1,4,5,8- Dimethanonaphthalene, 1,2,3,4,10,10-10-hexachloro-1,4,4a,5,8,8a-hexahydro-, (1alpha,4alpha,4abeta,5alpha,8alpha, 8abeta)-. | 309–00–2 | P004 |
| Allyl alcohol ................................. | 2-Propen-1-ol ................................ | 107–18–6 | P005 |
| Allyl chloride ................................ | 1-Propane, 3-chloro ...................... | 107–05–1 | .................. |
| Aluminum phosphide ..................... | Same ............................................... | 20859–73–8 | P006 |
| 4-Aminobiphenyl ........................... | [1,1'-Biphenyl]-4-amine .................. | 92–67–1 | .................. |
| 5-(Aminomethyl)-3-isoxazolol ........ | 3(2H)-Isoxazolone, 5-(aminomethyl)- ... | 2763–96–4 | P007 |
| 4-Aminopyridine ............................ | 4-Pyridinamine .............................. | 504–24–5 | P008 |
| Amitrole ........................................ | 1H-1,2,4-Triazol-3-amine ............... | 61–82–5 | U011 |
| Ammonium vanadate ...................... | Vanadic acid, ammonium salt .......... | 7803–55–6 | P119 |
| Aniline .......................................... | Benzenamine .................................. | 62–53–3 | U012 |
| o-Anisidine (2-methoxyaniline) ....... | Benzenamine, 2-Methoxy- .............. | 90–04–0 | .................. |
| Antimony ....................................... | Same ............................................... | 7440–36–0 | .................. |
| Antimony compounds, N.O.S.[1] ...... | | .................. | .................. |
| Aramite ......................................... | Sulfurous acid, 2-chloroethyl 2-[4-(1,1-dimethylethyl)phenoxy]-1-methylethyl ester. | 140–57–8 | .................. |
| Arsenic ......................................... | Same ............................................... | 7440–38–2 | .................. |
| Arsenic compounds, N.O.S.[1] ........ | | .................. | .................. |
| Arsenic acid ................................. | Arsenic acid $H_3 AsO_4$ ............... | 7778–39–4 | P010 |
| Arsenic pentoxide ......................... | Arsenic oxide $As_2 O_5$ ............... | 1303–28–2 | P011 |
| Arsenic trioxide ............................ | Arsenic oxide $As_2 O_3$ ............... | 1327–53–3 | P012 |
| Auramine ....................................... | Benzenamine, 4,4'-carbonimidoylbis[N,N-dimethyl. | 492–80–8 | U014 |
| Azaserine ...................................... | L-Serine, diazoacetate (ester) ......... | 115–02–6 | U015 |
| Barban .......................................... | Carbamic acid, (3-chlorophenyl) -, 4-chloro-2-butynyl ester. | 101–27–9 | U280 |
| Barium ........................................... | Same ............................................... | 7440–39–3 | .................. |
| Barium compounds, N.O.S.[1] ......... | | .................. | .................. |
| Barium cyanide .............................. | Same ............................................... | 542–62–1 | P013 |
| Bendiocarb .................................... | 1,3-Benzodioxol-4-ol, 2,2-dimethyl-, methyl carbamate. | 22781–23–3 | U278 |
| Bendiocarb phenol ........................ | 1,3-Benzodioxol-4-ol, 2,2-dimethyl-, ... | 22961–82–6 | U364 |
| Benomyl ........................................ | Carbamic acid, [1- [(butylamino) carbonyl]-1H-benzimidazol-2-yl] -, methyl ester. | 17804–35–2 | U271 |
| Benz[c]acridine ............................. | Same ............................................... | 225–51–4 | U016 |
| Benz[a]anthracene ........................ | Same ............................................... | 56–55–3 | U018 |
| Benzal chloride ............................. | Benzene, (dichloromethyl)- ............. | 98–87–3 | U017 |

| Common name | Chemical abstracts name | Chemical abstracts No. | Hazardous waste No. |
|---|---|---|---|
| Benzene | Same | 71–43–2 | U019 |
| Benzenearsonic acid | Arsonic acid, phenyl- | 98–05–5 | |
| Benzidine | [1,1′-Biphenyl]-4,4′-diamine | 92–87–5 | U021 |
| Benzo[b]fluoranthene | Benz[e]acephenanthrylene | 205–99–2 | |
| Benzo[j]fluoranthene | Same | 205–82–3 | |
| Benzo(k)fluoranthene | Same | 207–08–9 | |
| Benzo[a]pyrene | Same | 50–32–8 | U022 |
| p-Benzoquinone | 2,5-Cyclohexadiene-1,4-dione | 106–51–4 | U197 |
| Benzotrichloride | Benzene, (trichloromethyl)- | 98–07–7 | U023 |
| Benzyl chloride | Benzene, (chloromethyl)- | 100–44–7 | P028 |
| Beryllium powder | Same | 7440–41–7 | P015 |
| Beryllium compounds, N.O.S.[1] | | | |
| Bis(pentamethylene)-thiuram tetrasulfide | Piperidine, 1,1′-(tetrathiodicarbonothioyl)-bis- | 120–54–7 | |
| Bromoacetone | 2-Propanone, 1-bromo- | 598–31–2 | P017 |
| Bromoform | Methane, tribromo- | 75–25–2 | U225 |
| 4-Bromophenyl phenyl ether | Benzene, 1-bromo-4-phenoxy- | 101–55–3 | U030 |
| Brucine | Strychnidin-10-one, 2,3-dimethoxy- | 357–57–3 | P018 |
| Butyl benzyl phthalate | 1,2-Benzenedicarboxylic acid, butyl phenylmethyl ester. | 85–68–7 | |
| Butylate | Carbamothioic acid, bis(2-methylpropyl)-, S-ethyl ester. | 2008–41–5 | |
| Cacodylic acid | Arsinic acid, dimethyl- | 75–60–5 | U136 |
| Cadmium | Same | 7440–43–9 | |
| Cadmium compounds, N.O.S.[1] | | | |
| Calcium chromate | Chromic acid H₂ CrO₄, calcium salt | 13765–19–0 | U032 |
| Calcium cyanide | Calcium cyanide Ca(CN)₂ | 592–01–8 | P021 |
| Carbaryl | 1-Naphthalenol, methylcarbamate | 63–25–2 | U279 |
| Carbendazim | Carbamic acid, 1H-benzimidazol-2-yl, methyl ester. | 10605–21–7 | U372 |
| Carbofuran | 7-Benzofuranol, 2,3-dihydro-2,2-dimethyl-, methylcarbamate. | 1563–66–2 | P127 |
| Carbofuran phenol | 7-Benzofuranol, 2,3-dihydro-2,2-dimethyl- | 1563–38–8 | U367 |
| Carbon disulfide | Same | 75–15–0 | P022 |
| Carbon oxyfluoride | Carbonic difluoride | 353–50–4 | U033 |
| Carbon tetrachloride | Methane, tetrachloro- | 56–23–5 | U211 |
| Carbosulfan | Carbamic acid, [(dibutylamino) thio] methyl-, 2,3-dihydro-2,2-dimethyl-7-benzofuranyl ester. | 55285–14–8 | P189 |
| Chloral | Acetaldehyde, trichloro- | 75–87–6 | U034 |
| Chlorambucil | Benzenebutanoic acid, 4-[bis(2-chloroethyl)amino]-. | 305–03–3 | U035 |
| Chlordane | 4,7-Methano-1H-indene, 1,2,4,5,6,7,8,8-octachloro-2,3,3a,4,7,7a-hexahydro-. | 57–74–9 | U036 |
| Chlordane (alpha and gamma isomers) | | | U036 |
| Chlorinated benzenes, N.O.S.[1] | | | |
| Chlorinated ethane, N.O.S.[1] | | | |
| Chlorinated fluorocarbons, N.O.S.[1] | | | |
| Chlorinated naphthalene, N.O.S.[1] | | | |
| Chlorinated phenol, N.O.S.[1] | | | |
| Chlornaphazin | Naphthalenamine, N,N′-bis(2-chloroethyl)- | 494–03–1 | U026 |
| Chloroacetaldehyde | Acetaldehyde, chloro- | 107–20–0 | P023 |
| Chloroalkyl ethers, N.O.S.[1] | | | |
| p-Chloroaniline | Benzenamine, 4-chloro- | 106–47–8 | P024 |
| Chlorobenzene | Benzene, chloro- | 108–90–7 | U037 |
| Chlorobenzilate | Benzeneacetic acid, 4-chloro-alpha-(4-chlorophenyl)-alpha-hydroxy-, ethyl ester. | 510–15–6 | U038 |
| p-Chloro-m-cresol | Phenol, 4-chloro-3-methyl- | 59–50–7 | U039 |
| 2-Chloroethyl vinyl ether | Ethene, (2-chloroethoxy)- | 110–75–8 | U042 |
| Chloroform | Methane, trichloro- | 67–66–3 | U044 |
| Chloromethyl methyl ether | Methane, chloromethoxy- | 107–30–2 | U046 |
| beta-Chloronaphthalene | Naphthalene, 2-chloro- | 91–58–7 | U047 |
| o-Chlorophenol | Phenol, 2-chloro- | 95–57–8 | U048 |
| 1-(o-Chlorophenyl)thiourea | Thiourea, (2-chlorophenyl)- | 5344–82–1 | P026 |
| Chloroprene | 1,3-Butadiene, 2-chloro- | 126–99–8 | |
| 3-Chloropropionitrile | Propanenitrile, 3-chloro- | 542–76–7 | P027 |
| Chromium | Same | 7440–47–3 | |
| Chromium compounds, N.O.S.[1] | | | |
| Chrysene | Same | 218–01–9 | U050 |
| Citrus red No. 2 | 2-Naphthalenol, 1-[(2,5-dimethoxyphenyl)azo]-. | 6358–53–8 | |
| Coal tar creosote | Same | 8007–45–2 | |
| Copper cyanide | Copper cyanide CuCN | 544–92–3 | P029 |
| Copper dimethyldithiocarbamate | Copper, bis(dimethylcarbamodithioato-S,S′)-, | 137–29–1 | |

| Common name | Chemical abstracts name | Chemical abstracts No. | Hazardous waste No. |
|---|---|---|---|
| Creosote | Same | | U051 |
| p-Cresidine | 2-Methoxy-5-methylbenzenamine | 120–71–8 | |
| Cresol (Cresylic acid) | Phenol, methyl- | 1319–77–3 | U052 |
| Crotonaldehyde | 2-Butenal | 4170–30–3 | U053 |
| m-Cumenyl methylcarbamate | Phenol, 3-(methylethyl)-, methyl carbamate | 64–00–6 | P202 |
| Cyanides (soluble salts and complexes) N.O.S.[1]. | | | P030 |
| Cyanogen | Ethanedinitrile | 460–19–5 | P031 |
| Cyanogen bromide | Cyanogen bromide (CN)Br | 506–68–3 | U246 |
| Cyanogen chloride | Cyanogen chloride (CN)Cl | 506–77–4 | P033 |
| Cycasin | beta-D-Glucopyranoside, (methyl-ONN-azoxy)methyl. | 14901–08–7 | |
| Cycloate | Carbamothioic acid, cyclohexylethyl-, S-ethyl ester. | 1134–23–2 | |
| 2-Cyclohexyl-4,6-dinitrophenol | Phenol, 2-cyclohexyl-4,6-dinitro- | 131–89–5 | P034 |
| Cyclophosphamide | 2H-1,3,2-Oxazaphosphorin-2-amine, N,N-bis(2-chloroethyl)tetrahydro-, 2-oxide. | 50–18–0 | U058 |
| 2,4-D | Acetic acid, (2,4-dichlorophenoxy)- | 94–75–7 | U240 |
| 2,4-D, salts, esters | | | U240 |
| Daunomycin | 5,12-Naphthacenedione, 8-acetyl-10-[(3-amino-2,3,6-trideoxy-alpha-L-lyxo-hexopyranosyl)oxy]-7,8,9,10-tetrahydro-6,8,11-trihydroxy-1-methoxy-, (8S-cis)-. | 20830–81–3 | U059 |
| Dazomet | 2H–1,3,5-thiadiazine-2-thione, tetrahydro-3,5-dimethyl. | 533–74–4 | |
| DDD | Benzene, 1,1′-(2,2-dichloroethylidene)bis[4-chloro-. | 72–54–8 | U060 |
| DDE | Benzene, 1,1′-(dichloroethenylidene)bis[4-chloro-. | 72–55–9 | |
| DDT | Benzene, 1,1′-(2,2,2-trichloroethylidene)bis[4-chloro-. | 50–29–3 | U061 |
| Diallate | Carbamothioic acid, bis(1-methylethyl)-, S-(2,3-dichloro-2-propenyl) ester. | 2303–16–4 | U062 |
| Dibenz[a,h]acridine | Same | 226–36–8 | |
| Dibenz[a,j]acridine | Same | 224–42–0 | |
| Dibenz[a,h]anthracene | Same | 53–70–3 | U063 |
| 7H-Dibenzo[c,g]carbazole | Same | 194–59–2 | |
| Dibenzo[a,e]pyrene | Naphtho[1,2,3,4-def]chrysene | 192–65–4 | |
| Dibenzo[a,h]pyrene | Dibenzo[b,def]chrysene | 189–64–0 | |
| Dibenzo[a,i]pyrene | Benzo[rst]pentaphene | 189–55–9 | U064 |
| 1,2-Dibromo-3-chloropropane | Propane, 1,2-dibromo-3-chloro- | 96–12–8 | U066 |
| Dibutyl phthalate | 1,2-Benzenedicarboxylic acid, dibutyl ester | 84–74–2 | U069 |
| o-Dichlorobenzene | Benzene, 1,2-dichloro- | 95–50–1 | U070 |
| m-Dichlorobenzene | Benzene, 1,3-dichloro- | 541–73–1 | U071 |
| p-Dichlorobenzene | Benzene, 1,4-dichloro- | 106–46–7 | U072 |
| Dichlorobenzene, N.O.S.[1] | Benzene, dichloro- | 25321–22–6 | |
| 3,3′-Dichlorobenzidine | [1,1′-Biphenyl]-4,4′-diamine, 3,3′-dichloro- | 91–94–1 | U073 |
| 1,4-Dichloro-2-butene | 2-Butene, 1,4-dichloro- | 764–41–0 | U074 |
| Dichlordifluoromethane | Methane, dichlorodifluoro- | 75–71–8 | U075 |
| Dichloroethylene, N.O.S.[1] | Dichloroethylene | 25323–30–2 | |
| 1,1-Dichloroethylene | Ethene, 1,1-dichloro- | 75–35–4 | U078 |
| 1,2-Dichloroethylene | Ethene, 1,2-dichloro-, (E)- | 156–60–5 | U079 |
| Dichloroethyl ether | Ethane, 1,1′oxybis[2-chloro- | 111–44–4 | U025 |
| Dichloroisopropyl ether | Propane, 2,2′-oxybis[2-chloro- | 108–60–1 | U027 |
| Dichloromethoxy ethane | Ethane, 1,1′-[methylenebis(oxy)]bis[2-chloro- | 111–91–1 | U024 |
| Dichloromethyl ether | Methane, oxybis[chloro- | 542-88-1 | P016 |
| 2,4-Dichlorophenol | Phenol, 2,4-dichloro- | 120–83–2 | U081 |
| 2,6-Dichlorophenol | Phenol, 2,6-dichloro- | 87–65–0 | U082 |
| Dichlorophenylarsine | Arsonous dichloride, phenyl- | 696–28–6 | P036 |
| Dichloropropane, N.O.S.[1] | Propane, dichloro- | 26638–19–7 | |
| Dichloropropanol, N.O.S.[1] | Propanol, dichloro- | 26545–73–3 | |
| Dichloropropene, N.O.S.[1] | 1-Propene, dichloro- | 26952–23–8 | |
| 1,3-Dichloropropene | 1-Propene, 1,3-dichloro- | 542–75–6 | U084 |
| Dieldrin | 2,7:3,6-Dimethanonaphth[2,3-b]oxirene, 3,4,5,6,9,9-hexachloro-1a,2,2a,3,6,6a,7,7a-octahydro-, (1aalpha,2beta,2aalpha,3beta,6beta,6aalpha,7beta,7aalpha)-. | 60–57–1 | P037 |
| 1,2:3,4-Diepoxybutane | 2,2′-Bioxirane | 1464–53–5 | U085 |
| Diethylarsine | Arsine, diethyl- | 692–42–2 | P038 |
| Diethylene glycol, dicarbamate | Ethanol, 2,2′-oxybis-, dicarbamate | 5952–26–1 | U395 |
| 1,4-Diethyleneoxide | 1,4-Dioxane | 123–91–1 | U108 |
| Diethylhexyl phthalate | 1,2-Benzenedicarboxylic acid, bis(2-ethylhexyl) ester. | 117–81–7 | U028 |

160

| Common name | Chemical abstracts name | Chemical abstracts No. | Hazardous waste No. |
|---|---|---|---|
| N,N′-Diethylhydrazine .................................. | Hydrazine, 1,2-diethyl- .............................. | 1615–80–1 | U086 |
| O,O-Diethyl S-methyl dithiophosphate ........... | Phosphorodithioic acid, O,O-diethyl S-methyl ester. | 3288–58–2 | U087 |
| Diethyl-p-nitrophenyl phosphate .................... | Phosphoric acid, diethyl 4-nitrophenyl ester | 311–45–5 | P041 |
| Diethyl phthalate ........................................ | 1,2-Benzenedicarboxylic acid, diethyl ester .. | 84–66–2 | U088 |
| O,O-Diethyl O-pyrazinyl phosphoro- thioate ..... | Phosphorothioic acid, O,O-diethyl O-pyrazinyl ester. | 297–97–2 | P040 |
| Diethylstilbesterol ....................................... | Phenol, 4,4′-(1,2-diethyl-1,2-ethenediyl)bis-, (E)-. | 56–53–1 | U089 |
| Dihydrosafrole ........................................... | 1,3-Benzodioxole, 5-propyl- ....................... | 94–58–6 | U090 |
| Diisopropylfluorophosphate (DFP) ................ | Phosphorofluoridic acid, bis(1-methylethyl) ester. | 55–91–4 | P043 |
| Dimethoate ............................................... | Phosphorodithioic acid, O,O-dimethyl S-[2-(methylamino)-2-oxoethyl] ester. | 60–51–5 | P044 |
| 3,3′-Dimethoxybenzidine ............................. | [1,1′-Biphenyl]-4,4′-diamine, 3,3′-dimethoxy- | 119–90–4 | U091 |
| p-Dimethylaminoazobenzene ........................ | Benzenamine, N,N-dimethyl-4-(phenylazo)- | 60–11–7 | U093 |
| 2,4-Dimethylaniline (2,4-xylidine) ................. | Benzenamine, 2,4-dimethyl- ....................... | 95–68–1 | .................... |
| 7,12-Dimethylbenz[a]anthracene .................. | Benz[a]anthracene, 7,12-dimethyl- .............. | 57–97–6 | U094 |
| 3,3′-Dimethylbenzidine ............................... | [1,1′-Biphenyl]-4,4′-diamine, 3,3′-dimethyl- ... | 119–93–7 | U095 |
| Dimethylcarbamoyl chloride ......................... | Carbamic chloride, dimethyl- ...................... | 79–44–7 | U097 |
| 1,1-Dimethylhydrazine ................................ | Hydrazine, 1,1-dimethyl- ............................ | 57–14–7 | U098 |
| 1,2-Dimethylhydrazine ................................ | Hydrazine, 1,2-dimethyl- ............................ | 540–73–8 | U099 |
| alpha,alpha-Dimethylphenethylamine ............ | Benzeneethanamine, alpha,alpha-dimethyl- ... | 122–09–8 | P046 |
| 2,4-Dimethylphenol .................................... | Phenol, 2,4-dimethyl- ................................ | 105–67–9 | U101 |
| Dimethyl phthalate ..................................... | 1,2-Benzenedicarboxylic acid, dimethyl ester | 131–11–3 | U102 |
| Dimethyl sulfate ........................................ | Sulfuric acid, dimethyl ester ....................... | 77–78–1 | U103 |
| Dimetilan .................................................. | Carbamic acid, dimethyl-, 1-[(dimethylamino) carbonyl]-5-methyl-1H-pyrazol-3-yl ester. | 644–64–4 | P191 |
| Dinitrobenzene, N.O.S. [1] ........................... | Benzene, dinitro- ...................................... | 25154–54–5 | .................... |
| 4,6-Dinitro-o-cresol ................................... | Phenol, 2-methyl-4,6-dinitro- ..................... | 534–52–1 | P047 |
| 4,6-Dinitro-o-cresol salts ........................... | .............................................................. | | P047 |
| 2,4-Dinitrophenol ...................................... | Phenol, 2,4-dinitro- ................................... | 51–28–5 | P048 |
| 2,4-Dinitrotoluene ..................................... | Benzene, 1-methyl-2,4-dinitro- ................... | 121–14–2 | U105 |
| 2,6-Dinitrotoluene ..................................... | Benzene, 2-methyl-1,3-dinitro- ................... | 606–20–2 | U106 |
| Dinoseb .................................................. | Phenol, 2-(1-methylpropyl)-4,6-dinitro- ........ | 88–85–7 | P020 |
| Di-n-octyl phthalate ................................... | 1,2-Benzenedicarboxylic acid, dioctyl ester .. | 117–84–0 | U017 |
| Diphenylamine .......................................... | Benzenamine, N-phenyl- ............................ | 122–39–4 | .................... |
| 1,2-Diphenylhydrazine ............................... | Hydrazine, 1,2-diphenyl- ............................ | 122–66–7 | U109 |
| Di-n-propylnitrosamine ............................... | 1-Propanamine, N-nitroso-N-propyl- ............ | 621–64–7 | U111 |
| Disulfiram ................................................ | Thioperoxydicarbonic diamide, tetraethyl ..... | 97–77–8 | .................... |
| Disulfoton ................................................ | Phosphorodithioic acid, O,O-diethyl S-[2-(ethylthio)ethyl] ester. | 298–04–4 | P039 |
| Dithiobiuret .............................................. | Thioimidodicarbonic diamide [(H$_2$ N)C(S)]$_2$ NH. | 541–53–7 | P049 |
| Endosulfan .............................................. | 6,9-Methano-2,4,3-benzodioxathiepin, 6,7,8,9,10,10-hexachloro-1,5,5a,6,9,9a-hexahydro-, 3-oxide. | 115–29–7 | P050 |
| Endothall ................................................ | 7-Oxabicyclo[2.2.1]heptane-2,3-dicarboxylic acid. | 145–73–3 | P088 |
| Endrin .................................................... | 2,7:3,6-Dimethanonaphth[2,3-b]oxirene, 3,4,5,6,9,9-hexachloro-1a,2,2a,3,6,6a,7,7a-octahydro-, (1aalpha,2beta,2abeta,3alpha,6alpha, 6abeta,7beta,7aalpha)-. | 72–20–8 | P051 |
| Endrin metabolites .................................... | .............................................................. | .................... | P051 |
| Epichlorohydrin ......................................... | Oxirane, (chloromethyl)- ............................ | 106–89–8 | U041 |
| Epinephrine ............................................. | 1,2-Benzenediol, 4-[1-hydroxy-2-(methylamino)ethyl]-, (R)-. | 51–43–4 | P042 |
| EPTC ..................................................... | Carbamothioic acid, dipropyl-, S-ethyl ester | 759–94–4 | .................... |
| Ethyl carbamate (urethane) ........................ | Carbamic acid, ethyl ester ......................... | 51–79–6 | U238 |
| Ethyl cyanide .......................................... | Propanenitrile .......................................... | 107–12–0 | P101 |
| Ethyl Ziram ............................................. | Zinc, bis(diethylcarbamodithioato-S,S′)- ....... | 14324–55–1 | .................... |
| Ethylenebisdithiocarbamic acid ................... | Carbamodithioic acid, 1,2-ethanediylbis- ...... | 111–54–6 | U114 |
| Ethylenebisdithiocarbamic acid, salts and esters. | .............................................................. | .................... | U114 |
| Ethylene dibromide ................................... | Ethane, 1,2-dibromo- ............................... | 106–93–4 | U067 |
| Ethylene dichloride ................................... | Ethane, 1,2-dichloro- ............................... | 107–06–2 | U077 |
| Ethylene glycol monoethyl ether ................. | Ethanol, 2-ethoxy- ................................... | 110–80–5 | U359 |
| Ethyleneimine .......................................... | Aziridine ................................................. | 151–56–4 | P054 |
| Ethylene oxide ......................................... | Oxirane .................................................. | 75–21–8 | U115 |
| Ethylenethiourea ...................................... | 2-Imidazolidinethione ............................... | 96–45–7 | U116 |
| Ethylidene dichloride ................................. | Ethane, 1,1-dichloro- ............................... | 75–34–3 | U076 |

| Common name | Chemical abstracts name | Chemical abstracts No. | Hazardous waste No. |
|---|---|---|---|
| Ethyl methacrylate | 2-Propenoic acid, 2-methyl-, ethyl ester | 97-63-2 | U118 |
| Ethyl methanesulfonate | Methanesulfonic acid, ethyl ester | 62-50-0 | U119 |
| Famphur | Phosphorothioic acid, O-[4-[(dimethylamino)sulfonyl]phenyl] O,O-dimethyl ester. | 52-85-7 | P097 |
| Ferbam | Iron, tris(dimethylcarbamodithioato-S,S')-, | 14484-64-1 | |
| Fluoranthene | Same | 206-44-0 | U120 |
| Fluorine | Same | 7782-41-4 | P056 |
| Fluoroacetamide | Acetamide, 2-fluoro- | 640-19-7 | P057 |
| Fluoroacetic acid, sodium salt | Acetic acid, fluoro-, sodium salt | 62-74-8 | P058 |
| Formaldehyde | Same | 50-00-0 | U122 |
| Formetanate hydrochloride | Methanimidamide, N,N-dimethyl-N'-[3-[[(methylamino) carbonyl]oxy]phenyl]-, monohydrochloride. | 23422-53-9 | P198 |
| Formic acid | Same | 64-18-6 | U123 |
| Formparanate | Methanimidamide, N,N-dimethyl-N'-[2-methyl-4-[[(methylamino) carbonyl]oxy]phenyl]-. | 17702-57-7 | P197 |
| Glycidylaldehyde | Oxiranecarboxyaldehyde | 765-34-4 | U126 |
| Halomethanes, N.O.S. [1] | | | |
| Heptachlor | 4,7-Methano-1H-indene, 1,4,5,6,7,8,8-heptachloro-3a,4,7,7a-tetrahydro-. | 76-44-8 | P059 |
| Heptachlor epoxide | 2,5-Methano-2H-indeno[1,2-b]oxirene, 2,3,4,5,6,7,7-heptachloro-1a,1b,5,5a,6,6a-hexa- hydro-, (1aalpha,1bbeta,2alpha,5alpha, 5abeta,6beta,6aalpha)-. | 1024-57-3 | |
| Heptachlor epoxide (alpha, beta, and gamma isomers). | | | |
| Heptachlorodibenzofurans | | | |
| Heptachlorodibenzo-p-dioxins | | | |
| Hexachlorobenzene | Benzene, hexachloro- | 118-74-1 | U127 |
| Hexachlorobutadiene | 1,3-Butadiene, 1,1,2,3,4,4-hexachloro- | 87-68-3 | U128 |
| Hexachlorocyclopentadiene | 1,3-Cyclopentadiene, 1,2,3,4,5,5-hexachloro- | 77-47-4 | U130 |
| Hexachlorodibenzo-p-dioxins | | | |
| Hexachlorodibenzofurans | | | |
| Hexachloroethane | Ethane, hexachloro- | 67-72-1 | U131 |
| Hexachlorophene | Phenol, 2,2'-methylenebis[3,4,6-trichloro- | 70-30-4 | U132 |
| Hexachloropropene | 1-Propene, 1,1,2,3,3,3-hexachloro- | 1888-71-7 | U243 |
| Hexaethyl tetraphosphate | Tetraphosphoric acid, hexaethyl ester | 757-58-4 | P062 |
| Hydrazine | Same | 302-01-2 | U133 |
| Hydrogen cyanide | Hydrocyanic acid | 74-90-8 | P063 |
| Hydrogen fluoride | Hydrofluoric acid | 7664-39-3 | U134 |
| Hydrogen sulfide | Hydrogen sulfide H$_2$S | 7783-06-4 | U135 |
| Indeno[1,2,3-cd]pyrene | Same | 193-39-5 | U137 |
| 3-Iodo-2-propynyl n-butylcarbamate | Carbamic acid, butyl-, 3-iodo-2-propynyl ester. | 55406-53-6 | |
| Isobutyl alcohol | 1-Propanol, 2-methyl- | 78-83-1 | U140 |
| Isodrin | 1,4,5,8-Dimethanonaphthalene, 1,2,3,4,10,10-hexachloro-1,4,4a,5,8,8a-hexahydro-, (1alpha,4alpha,4abeta,5beta, 8beta,8abeta)-. | 465-73-6 | P060 |
| Isolan | Carbamic acid, dimethyl-, 3-methyl-1-(1-methylethyl)-1H-pyrazol-5-yl ester. | 119-38-0 | P192 |
| Isosafrole | 1,3-Benzodioxole, 5-(1-propenyl)- | 120-58-1 | U141 |
| Kepone | 1,3,4-Metheno-2H-cyclobuta[cd]pentalen-2-one, 1,1a,3,3a,4,5,5,5a,5b,6-decachlorooctahydro-. | 143-50-0 | U142 |
| Lasiocarpine | 2-Butenoic acid, 2-methyl-,7-[[2,3-dihydroxy-2-(1-methoxyethyl)-3-methyl-1-oxobutoxy]methyl]-2,3,5,7a-tetrahydro-1H-pyrrolizin-1-yl ester, [1S-[1alpha(Z),7(2S*,3R*),7aalpha]]- | 303-34-4 | U143 |
| Lead | Same | 7439-92-1 | |
| Lead compounds, N.O.S. [1] | | | |
| Lead acetate | Acetic acid, lead(2+) salt | 301-04-2 | U144 |
| Lead phosphate | Phosphoric acid, lead(2+) salt (2:3) | 7446-27-7 | U145 |
| Lead subacetate | Lead, bis(acetato-O)tetrahydroxytri- | 1335-32-6 | U146 |
| Lindane | Cyclohexane, 1,2,3,4,5,6-hexachloro-, (1alpha,2alpha,3beta,4alpha, 5alpha,6beta)-. | 58-89-9 | U129 |
| Maleic anhydride | 2,5-Furandione | 108-31-6 | U147 |
| Maleic hydrazide | 3,6-Pyridazinedione, 1,2-dihydro- | 123-33-1 | U148 |

| Common name | Chemical abstracts name | Chemical abstracts No. | Hazardous waste No. |
|---|---|---|---|
| Malononitrile | Propanedinitrile | 109–77–3 | U149 |
| Manganese dimethyldithiocarbamate | Manganese, bis(dimethylcarbamodithioato-S,S')-,. | 15339–36–3 | P196 |
| Melphalan | L-Phenylalanine, 4-[bis(2-chloroethyl)aminol]-. | 148–82–3 | . U150 |
| Mercury | Same | 7439–97–6 | U151 |
| Mercury compounds, N.O.S. [1] | | | |
| Mercury fulminate | Fulminic acid, mercury(2+) salt | 628–86–4 | P065 |
| Metam Sodium | Carbamodithioic acid, methyl-, monosodium salt. | 137–42–8 | |
| Methacrylonitrile | 2-Propenenitrile, 2-methyl- | 126–98–7 | U152 |
| Methapyrilene | 1,2-Ethanediamine, N,N-dimethyl-N'-2-pyridinyl-N'-(2-thienylmethyl)-. | 91–80–5 | U155 |
| Methiocarb | Phenol, (3,5-dimethyl-4-(methylthio)-, methylcarbamate. | 2032–65–7 | P199 |
| Methomyl | Ethanimidothioic acid, N-[[(methylamino)carbonyl]oxy]-, methyl ester. | 16752–77–5 | P066 |
| Methoxychlor | Benzene, 1,1'-(2,2,2-trichloroethylidene)bis[4-methoxy-. | 72–43–5 | U247 |
| Methyl bromide | Methane, bromo- | 74–83–9 | U029 |
| Methyl chloride | Methane, chloro- | 74–87–3 | U045 |
| Methyl chlorocarbonate | Carbonochloridic acid, methyl ester | 79–22–1 | U156 |
| Methyl chloroform | Ethane, 1,1,1-trichloro- | 71–55–6 | U226 |
| 3-Methylcholanthrene | Benz[j]aceanthrylene, 1,2-dihydro-3-methyl- | 56–49–5 | U157 |
| 4,4'-Methylenebis(2-chloroaniline) | Benzenamine, 4,4'-methylenebis[2-chloro- ... | 101–14–4 | U158 |
| Methylene bromide | Methane, dibromo- | 74–95–3 | U068 |
| Methylene chloride | Methane, dichloro- | 75–09–2 | U080 |
| Methyl ethyl ketone (MEK) | 2-Butanone | 78–93–3 | U159 |
| Methyl ethyl ketone peroxide | 2-Butanone, peroxide | 1338–23–4 | U160 |
| Methyl hydrazine | Hydrazine, methyl- | 60–34–4 | P068 |
| Methyl iodide | Methane, iodo- | 74–88–4 | U138 |
| Methyl isocyanate | Methane, isocyanato- | 624–83–9 | P064 |
| 2-Methyllactonitrile | Propanenitrile, 2-hydroxy-2-methyl- | 75–86–5 | P069 |
| Methyl methacrylate | 2-Propenoic acid, 2-methyl-, methyl ester .... | 80–62–6 | U162 |
| Methyl methanesulfonate | Methanesulfonic acid, methyl ester | 66–27–3 | |
| Methyl parathion | Phosphorothioic acid, O,O-dimethyl O-(4-nitrophenyl) ester. | 298–00–0 | P071 |
| Methylthiouracil | 4(1H)-Pyrimidinone, 2,3-dihydro-6-methyl-2-thioxo-. | 56–04–2 | U164 |
| Metolcarb | Carbamic acid, methyl-, 3-methylphenyl ester. | 1129–41–5 | P190 |
| Mexacarbate | Phenol, 4-(dimethylamino)-3,5-dimethyl-, methylcarbamate (ester). | 315–18–4 | P128 |
| Mitomycin C | Azirino[2',3':3,4]pyrrolo[1,2-a]indole-4,7-dione, 6-amino-8-[[(aminocarbonyl)oxy]methyl]-1,1a,2,8,8a,8b-hexahydro-8a-methoxy-5-methyl-, [1aS-(1aalpha,8beta,8aalpha,8balpha)]-. | 50–07–7 | U010 |
| MNNG | Guanidine, N-methyl-N'-nitro-N-nitroso- | 70–25–7 | U163 |
| Molinate | 1H-Azepine-1-carbothioic acid, hexahydro-, S-ethyl ester. | 2212–67–1 | |
| Mustard gas | Ethane, 1,1'-thiobis[2-chloro- | 505–60–2 | |
| Naphthalone | Same | 91–20–3 | U165 |
| 1,4-Naphthoquinone | 1,4-Naphthalenedione | 130–15–4 | U166 |
| alpha-Naphthylamine | 1-Naphthalenamine | 134–32–7 | U167 |
| beta-Naphthylamine | 2-Naphthalenamine | 91–59–8 | U168 |
| alpha-Naphthylthiourea | Thiourea, 1-naphthalenyl- | 86–88–4 | P072 |
| Nickel | Same | 7440–02–0 | |
| Nickel compounds, N.O.S. [1] | | | |
| Nickel carbonyl | Nickel carbonyl Ni(CO)₄, (T-4)- | 13463–39–3 | P073 |
| Nickel cyanide | Nickel cyanide Ni(CN)₂ | 557–19–7 | P074 |
| Nicotine | Pyridine, 3-(1-methyl-2-pyrrolidinyl)-, (S)- ..... | 54–11–5 | P075 |
| Nicotine salts | | | P075 |
| Nitric oxide | Nitrogen oxide NO | 10102–43–9 | P076 |
| p-Nitroaniline | Benzenamine, 4-nitro- | 100–01–6 | P077 |
| Nitrobenzene | Benzene, nitro- | 98–95–3 | U169 |
| Nitrogen dioxide | Nitrogen oxide NO₂ | 10102–44–0 | P078 |
| Nitrogen mustard | Ethanamine, 2-chloro-N-(2-chloroethyl)-N-methyl-. | 51–75–2 | |
| Nitrogen mustard, hydrochloride salt | | | |
| Nitrogen mustard N-oxide | Ethanamine, 2-chloro-N-(2-chloroethyl)-N-methyl-, N-oxide. | 126–85–2 | |

| Common name | Chemical abstracts name | Chemical abstracts No. | Hazardous waste No. |
|---|---|---|---|
| Nitrogen mustard, N-oxide, hydro- chloride salt. | | | |
| Nitroglycerin | 1,2,3-Propanetriol, trinitrate | 55–63–0 | P081 |
| p-Nitrophenol | Phenol, 4-nitro- | 100–02–7 | U170 |
| 2-Nitropropane | Propane, 2-nitro- | 79–46–9 | U171 |
| Nitrosamines, N.O.S. [1] | | 35576–91–1 | |
| N-Nitrosodi-n-butylamine | 1-Butanamine, N-butyl-N-nitroso- | 924–16–3 | U172 |
| N-Nitrosodiethanolamine | Ethanol, 2,2'-(nitrosoimino)bis- | 1116–54–7 | U173 |
| N-Nitrosodiethylamine | Ethanamine, N-ethyl-N-nitroso- | 55–18–5 | U174 |
| N-Nitrosodimethylamine | Methanamine, N-methyl-N-nitroso- | 62–75–9 | P082 |
| N-Nitroso-N-ethylurea | Urea, N-ethyl-N-nitroso- | 759–73–9 | U176 |
| N-Nitrosomethylethylamine | Ethanamine, N-methyl-N-nitroso- | 10595–95–6 | |
| N-Nitroso-N-methylurea | Urea, N-methyl-N-nitroso- | 684–93–5 | U177 |
| N-Nitroso-N-methylurethane | Carbamic acid, methylnitroso-, ethyl ester | 615–53–2 | U178 |
| N-Nitrosomethylvinylamine | Vinylamine, N-methyl-N-nitroso- | 4549–40–0 | P084 |
| N-Nitrosomorpholine | Morpholine, 4-nitroso- | 59–89–2 | |
| N-Nitrosonornicotine | Pyridine, 3-(1-nitroso-2-pyrrolidinyl)-, (S)- | 16543–55–8 | |
| N-Nitrosopiperidine | Piperidine, 1-nitroso- | 100–75–4 | U179 |
| N-Nitrosopyrrolidine | Pyrrolidine, 1-nitroso- | 930–55–2 | U180 |
| N-Nitrososarcosine | Glycine, N-methyl-N-nitroso- | 13256–22–9 | |
| 5-Nitro-o-toluidine | Benzenamine, 2-methyl-5-nitro- | 99–55–8 | U181 |
| Octachlorodibenzo-p-dioxin (OCDD) | 1,2,3,4,6,7,8,9-Octachlorodibenzo-p-dioxin | 3268–87–9 | |
| Octachlorodibenzofuran (OCDF) | 1,2,3,4,6,7,8,9-Octachlorodibenofuran | 39001–02–0 | |
| Octamethylpyrophosphoramide | Diphosphoramide, octamethyl- | 152–16–9 | P085 |
| Osmium tetroxide | Osmium oxide OsO₄, (T-4)- | 20816–12–0 | P087 |
| Oxamyl | Ethanimidothioc acid, 2-(dimethylamino)-N-[[(methylamino)carbonyl]oxy]-2-oxo-, methyl ester. | 23135–22–0 | P194 |
| Paraldehyde | 1,3,5-Trioxane, 2,4,6-trimethyl- | 123–63–7 | U182 |
| Parathion | Phosphorothioic acid, O,O-diethyl O-(4-nitrophenyl) ester. | 56–38–2 | P089 |
| Pebulate | Carbamothioic acid, butylethyl-, S-propyl ester. | 1114–71–2 | |
| Pentachlorobenzene | Benzene, pentachloro- | 608–93–5 | U183 |
| Pentachlorodibenzo-p-dioxins | | | |
| Pentachlorodibenzofurans | | | |
| Pentachloroethane | Ethane, pentachloro- | 76–01–7 | U184 |
| Pentachloronitrobenzene (PCNB) | Benzene, pentachloronitro- | 82–68–8 | U185 |
| Pentachlorophenol | Phenol, pentachloro- | 87–86–5 | See F027 |
| Phenacetin | Acetamide, N-(4-ethoxyphenyl)- | 62–44–2 | U187 |
| Phenol | Same | 108–95–2 | U188 |
| 1,2-Phenylenediamine | 1,2-Benzenediamine | 95–54–5 | |
| 1,3-Phenylenediamine | 1,3-Benzenediamine | 108–45–2 | |
| Phenylenediamine | Benzenediamine | 25265–76–3 | |
| Phenylmercury acetate | Mercury, (acetato-O)phenyl- | 62–38–4 | P092 |
| Phenylthiourea | Thiourea, phenyl- | 103–85–5 | P093 |
| Phosgene | Carbonic dichloride | 75–44–5 | P095 |
| Phosphine | Same | 7803–51–2 | P096 |
| Phorate | Phosphorodithioic acid, O,O-diethyl S-[(ethylthio)methyl] ester. | 298–02–2 | P094 |
| Phthalic acid esters, N.O.S. [1] | | | |
| Phthalic anhydride | 1,3-Isobenzofurandione | 85–44–9 | U190 |
| Physostigmine | Pyrrolo[2,3-b]indol-5-01, 1,2,3,3a,8,8a-hexahydro-1,3a,8-trimethyl-, methylcarbamate (ester), (3aS-cis)-. | 57–47–6 | P204 |
| Physostigmine salicylate | Benzoic acid, 2-hydroxy-, compd. with (3aS-cis) −1,2,3,3a,8,8a-hexahydro-1,3a,8-trimethylpyrrolo [2,3-b]indol-5-yl methylcarbamate ester (1:1). | 57–64–7 | P188 |
| 2-Picoline | Pyridine, 2-methyl- | 109–06–8 | U191 |
| Polychlorinated biphenyls, N.O.S. [1] | | | |
| Potassium cyanide | Potassium cyanide K(CN) | 151–50–8 | P098 |
| Potassium dimethyldithiocarbamate | Carbamodithioic acid, dimethyl, potassium salt. | 128–03–0 | |
| Potassium n-hydroxymethyl-n-methyl-dithiocarbamate. | Carbamodithioic acid, (hydroxymethyl)methyl-, monopotassium salt. | 51026–28–9 | |
| Potassium n-methyldithiocarbamate | Carbamodithioic acid, methyl-monopotassium salt. | 137–41–7 | |
| Potassium pentachlorophenate | Pentachlorophenol, potassium salt | 7778736 | None |
| Potassium silver cyanide | Argentate(1-), bis(cyano-C)-, potassium | 506–61–6 | P099 |
| Promecarb | Phenol, 3-methyl-5-(1-methylethyl)-, methyl carbamate. | 2631–37–0 | P201 |

| Common name | Chemical abstracts name | Chemical abstracts No. | Hazardous waste No. |
|---|---|---|---|
| Pronamide | Benzamide, 3,5-dichloro-N-(1,1-dimethyl-2-propynyl)-. | 23950–58–5 | U192 |
| 1,3-Propane sultone | 1,2-Oxathiolane, 2,2-dioxide | 1120–71–4 | U193 |
| n-Propylamine | 1-Propanamine | 107–10–8 | U194 |
| Propargyl alcohol | 2-Propyn-1-ol | 107–19–7 | P102 |
| Propham | Carbamic acid, phenyl-, 1-methylethyl ester | 122–42–9 | U373 |
| Propoxur | Phenol, 2-(1-methylethoxy)-, methylcarbamate. | 114–26–1 | U411 |
| Propylene dichloride | Propane, 1,2-dichloro- | 78–87–5 | U083 |
| 1,2-Propylenimine | Aziridine, 2-methyl- | 75–55–8 | P067 |
| Propylthiouracil | 4(1H)-Pyrimidinone, 2,3-dihydro-6-propyl-2-thioxo-. | 51–52–5 | .............. |
| Prosulfocarb | Carbamothioic acid, dipropyl-, S-(phenylmethyl) ester. | 52888–80–9 | U387 |
| Pyridine | Same | 110–86–1 | U196 |
| Reserpine | Yohimban-16-carboxylic acid, 11,17-dimethoxy-18-[(3,4,5-trimethoxybenzoyl)oxy]-smethyl ester, (3beta,16beta,17alpha,18beta,20alpha)-. | 50–55–5 | U200 |
| Resorcinol | 1,3-Benzenediol | 108–46–3 | U201 |
| Saccharin | 1,2-Benzisothiazol-3(2H)-one, 1,1-dioxide .... | 81–07–2 | U202 |
| Saccharin salts | | .............. | U202 |
| Safrole | 1,3-Benzodioxole, 5-(2-propenyl)- | 94–59–7 | U203 |
| Selenium | Same | 7782–49–2 | .............. |
| Selenium compounds, N.O.S. [1] | | .............. | .............. |
| Selenium dioxide | Selenious acid | 7783–00–8 | U204 |
| Selenium sulfide | Selenium sulfide SeS$_2$ | 7488–56–4 | U205 |
| Selenium, tetrakis(dimethyl-dithiocarbamate) | Carbamodithioic acid, dimethyl-, tetraanhydrosulfide with orthothioselenious acid. | 144–34–3 | .............. |
| Selenourea | Same | 630–10–4 | P103 |
| Silver | Same | 7440–22–4 | .............. |
| Silver compounds, N.O.S. [1] | | .............. | .............. |
| Silver cyanide | Silver cyanide Ag(CN) | 506–64–9 | P104 |
| Silvex (2,4,5-TP) | Propanoic acid, 2-(2,4,5-trichlorophenoxy)- .. | 93–72–1 | See F027 |
| Sodium cyanide | Sodium cyanide Na(CN) | 143–33–9 | P106 |
| Sodium dibutyldithiocarbamate | Carbamodithioic acid, dibutyl, sodium salt .... | 136–30–1 | .............. |
| Sodium diethyldithiocarbamate | Carbamodithioic acid, diethyl-, sodium salt .. | 148–18–5 | .............. |
| Sodium dimethyldithiocarbamate | Carbamodithioic acid, dimethyl-, sodium salt | 128–04–1 | .............. |
| Sodium pentachlorophenate | Pentachlorophenol, sodium salt | 131522 | None |
| Streptozotocin | D-Glucose, 2-deoxy-2-[[(methylnitrosoamino)carbonyl]amino]-. | 18883–66–4 | U206 |
| Strychnine | Strychnidin-10-one | 57–24–9 | P108 |
| Strychnine salts | | .............. | P108 |
| Sulfallate | Carbamodithioic acid, diethyl-, 2-chloro-2-propenyl ester. | 95–06–7 | .............. |
| TCDD | Dibenzo[b,e][1,4]dioxin, 2,3,7,8-tetrachloro- | 1746–01–6 | .............. |
| Tetrabutylthiuram disulfide | Thioperoxydicarbonic diamide, tetrabutyl ...... | 1634–02–2 | .............. |
| 1,2,4,5-Tetrachlorobenzene | Benzene, 1,2,4,5-tetrachloro- | 95–94–3 | U207 |
| Tetrachlorodibenzo-p-dioxins | | .............. | .............. |
| Tetrachlorodibenzofurans | | .............. | .............. |
| Tetrachloroethane, N.O.S. [1] | Ethane, tetrachloro-, N.O.S. | 25322–20–7 | .............. |
| 1,1,1,2-Tetrachloroethane | Ethane, 1,1,1,2-tetrachloro- | 630–20–6 | U208 |
| 1,1,2,2-Tetrachloroethane | Ethane, 1,1,2,2-tetrachloro- | 79–34–5 | U209 |
| Tetrachloroethylene | Ethene, tetrachloro- | 127–18–4 | U210 |
| 2,3,4,6-Tetrachlorophenol | Phenol, 2,3,4,6-tetrachloro- | 58–90–2 | See F027 |
| 2,3,4,6-tetrachlorophenol, potassium salt | same | 53535276 | None |
| 2,3,4,6-tetrachlorophenol, sodium salt ........ | same | 25567559 | None |
| Tetraethyldithiopyrophosphate | Thiodiphosphoric acid, tetraethyl ester | 3689–24–5 | P109 |
| Tetraethyl lead | Plumbane, tetraethyl- | 78–00–2 | P110 |
| Tetraethyl pyrophosphate | Diphosphoric acid, tetraethyl ester | 107–49–3 | P111 |
| Tetramethylthiuram monosulfide | Bis(dimethylthiocarbamoyl) sulfide | 97–74–5 | .............. |
| Tetranitromethane | Methane, tetranitro- | 509–14–8 | P112 |
| Thallium | Same | 7440–28–0 | .............. |
| Thallium compounds, N.O.S. [1] | | .............. | .............. |
| Thallic oxide | Thallium oxide Tl$_2$ O$_3$ | 1314–32–5 | P113 |
| Thallium(I) acetate | Acetic acid, thallium(1+) salt | 563–68–8 | U214 |
| Thallium(I) carbonate | Carbonic acid, dithallium(1+) salt | 6533–73–9 | U215 |
| Thallium(I) chloride | Thallium chloride TlCl | 7791–12–0 | U216 |
| Thallium(I) nitrate | Nitric acid, thallium(1+) salt | 10102–45–1 | U217 |
| Thallium selenite | Selenious acid, dithallium(1+) salt | 12039–52–0 | P114 |
| Thallium(I) sulfate | Sulfuric acid, dithallium(1+) salt | 7446–18–6 | P115 |
| Thioacetamide | Ethanethioamide | 62–55–5 | U218 |

165

| Common name | Chemical abstracts name | Chemical abstracts No. | Hazardous waste No. |
|---|---|---|---|
| Thiodicarb | Ethanimidothioic acid, N,N′-[thiobis [(methylimino) carbonyloxy]] bis-, dimethyl ester. | 59669–26–0 | U410 |
| Thiofanox | 2-Butanone, 3,3-dimethyl-1-(methylthio)-, 0-[(methylamino)carbonyl] oxime. | 39196–18–4 | P045 |
| Thiomethanol | Methanethiol | 74–93–1 | U153 |
| Thiophanate-methyl | Carbamic acid, [1,2-phyenylenebis (iminocarbonothioyl)] bis-, dimethyl ester. | 23564–05–8 | U409 |
| Thiophenol | Benzenethiol | 108–98–5 | P014 |
| Thiosemicarbazide | Hydrazinecarbothioamide | 79–19–6 | P116 |
| Thiourea | Same | 62–56–6 | U219 |
| Thiram | Thioperoxydicarbonic diamide [(H$_2$ N)C(S)]$_2$ S$_2$, tetramethyl-. | 137-26-8 | U244 |
| Tirpate | 1,3-Dithiolane-2-carboxaldehyde, 2,4-dimethyl-, O-[(methylamino) carbonyl] oxime. | 26419–73–8 | P185 |
| Toluene | Benzene, methyl- | 108–88–3 | U220 |
| Toluenediamine | Benzenediamine, ar-methyl- | 25376–45–8 | U221 |
| Toluene-2,4-diamine | 1,3-Benzenediamine, 4-methyl- | 95–80–7 | |
| Toluene-2,6-diamine | 1,3-Benzenediamine, 2-methyl- | 823–40–5 | |
| Toluene-3,4-diamine | 1,2-Benzenediamine, 4-methyl- | 496–72–0 | |
| Toluene diisocyanate | Benzene, 1,3-diisocyanatomethyl- | 26471–62–5 | U223 |
| o-Toluidine | Benzenamine, 2-methyl- | 95–53–4 | U328 |
| o-Toluidine hydrochloride | Benzenamine, 2-methyl-, hydrochloride | 636–21–5 | U222 |
| p-Toluidine | Benzenamine, 4-methyl- | 106–49–0 | U353 |
| Toxaphene | Same | 8001–35–2 | P123 |
| Triallate | Carbamothioic acid, bis(1-methylethyl)-, S-(2,3,3-trichloro-2-propenyl) ester. | 2303–17–5 | U389 |
| 1,2,4-Trichlorobenzene | Benzene, 1,2,4-trichloro- | 120–82–1 | |
| 1,1,2-Trichloroethane | Ethane, 1,1,2-trichloro- | 79–00–5 | U227 |
| Trichloroethylene | Ethene, trichloro- | 79–01–6 | U228 |
| Trichloromethanethiol | Methanethiol, trichloro- | 75–70–7 | P118 |
| Trichloromonofluoromethane | Methane, trichlorofluoro- | 75–69–4 | U121 |
| 2,4,5-Trichlorophenol | Phenol, 2,4,5-trichloro- | 95–95–4 | See F027 |
| 2,4,6-Trichlorophenol | Phenol, 2,4,6-trichloro- | 88–06–2 | See F027 |
| 2,4,5-T | Acetic acid, (2,4,5-trichlorophenoxy)- | 93–76–5 | See F027 |
| Trichloropropane, N.O.S.[1] | | 25735–29–9 | |
| 1,2,3-Trichloropropane | Propane, 1,2,3-trichloro- | 96–18–4 | |
| Triethylamine | Ethanamine, N,N-diethyl- | 121–44–8 | U404 |
| O,O,O-Triethyl phosphorothioate | Phosphorothioic acid, O,O,O-triethyl ester | 126–68–1 | |
| 1,3,5-Trinitrobenzene | Benzene, 1,3,5-trinitro- | 99–35–4 | U234 |
| Tris(1-aziridinyl)phosphine sulfide | Aziridine, 1,1′,1″-phosphinothioylidynetris- | 52–24–4 | |
| Tris(2,3-dibromopropyl) phosphate | 1-Propanol, 2,3-dibromo-, phosphate (3:1) | 126–72–7 | U235 |
| Trypan blue | 2,7-Naphthalenedisulfonic acid, 3,3′-[(3,3′-dimethyl[1,1′-biphenyl]-4,4′-diyl)bis(azo)]-bis[5-amino-4-hydroxy-, tetrasodium salt. | 72–57–1 | U236 |
| Uracil mustard | 2,4-(1H,3H)-Pyrimidinedione, 5-[bis(2-chloroethyl)amino]-. | 66–75–1 | U237 |
| Vanadium pentoxide | Vanadium oxide V$_2$ O$_5$ | 1314–62–1 | P120 |
| Vernolate | Carbamothioic acid, dipropyl-,S-propyl ester | 1929–77–7 | |
| Vinyl chloride | Ethene, chloro- | 75–01–4 | U043 |
| Warfarin | 2H-1-Benzopyran-2-one, 4-hydroxy-3-(3-oxo-1-phenylbutyl)-, when present at concentrations less than 0.3%. | 81–81–2 | U248 |
| Warfarin | 2H-1-Benzopyran-2-one, 4-hydroxy-3-(3-oxo-1-phenylbutyl)-, when present at concentrations greater than 0.3%. | 81–81–2 | P001 |
| Warfarin salts, when present at concentrations less than 0.3%. | | | U248 |
| Warfarin salts, when present at concentrations greater than 0.3%. | | | P001 |
| Zinc cyanide | Zinc cyanide Zn(CN)$_2$ | 557–21–1 | P121 |
| Zinc phosphide | Zinc phosphide Zn$_3$ P$_2$, when present at concentrations greater than 10%. | 1314–84–7 | P122 |
| Zinc phosphide | Zinc phosphide Zn$_3$ P$_2$, when present at concentrations of 10% or less. | 1314–84–7 | U249 |
| Ziram | Zinc, bis(dimethylcarbamodithioato-S,S′)-, (T-4)-. | 137–30–4 | P205 |

[1] The abbreviation N.O.S. (not otherwise specified) signifies those members of the general class not specifically listed by name in this appendix.

[53 FR 13388, Apr. 22, 1988, as amended at 53 FR 43881, Oct. 31, 1988; 54 FR 50978, Dec. 11, 1989; 55 FR 50483, Dec. 6, 1990; 56 FR 7568, Feb. 25, 1991; 59 FR 468, Jan. 4, 1994; 59 FR 31551, June 20, 1994; 60 FR 7853, Feb. 9, 1995; 60 FR 19165, Apr. 17, 1995; 62 FR 32977, June 17, 1997; 63 FR 24625, May 4, 1998; 65 FR 14475, Mar. 17, 2000; 65 FR 67127, Nov. 8, 2000; 70 FR 9177, Feb. 24, 2005; 71 FR 40271, July 14, 2006]

APPENDIX IX TO PART 261—WASTES EXCLUDED UNDER §§ 260.20 AND 260.22

TABLE 1—WASTES EXCLUDED FROM NON-SPECIFIC SOURCES

| Facility | Address | Waste description |
|---|---|---|
| Aluminum Company of America. | 750 Norcold Ave., Sidney, Ohio 45365. | Wastewater treatment plant (WWTP) sludges generated from the chemical conversion coating of aluminum (EPA Hazardous Waste No. F019) and WWTP sludges generated from electroplating operations (EPA Hazardous Waste No. F006) and stored in an on-site landfill. This is an exclusion for approximately 16,772 cubic yards of landfilled WWTP filter cake. This exclusion applies only if the waste filter cake remains in place or, if excavated, is disposed of in a Subtitle D landfill which is permitted, licensed, or registered by a state to manage industrial solid waste. This exclusion was published on April 6, 1999.<br>1. The constituent concentrations measured in the TCLP extract may not exceed the following levels (mg/L): Arsenic—5; Barium—100; Chromium—5; Cobalt—210; Copper—130; Nickel—70; Vanadium—30; Zinc—1000; Fluoride—400; Acetone—400; Methylene Chloride—0.5; Bis(2-ethylhexyl)phthalate—0.6.<br>2. (a) If, anytime after disposal of the delisted waste, Alcoa possesses or is otherwise made aware of any environmental data (including but not limited to leachate data or groundwater monitoring data) or any other data relevant to the delisted waste indicating that any constituent identified in Condition (1) is at a level in the leachate higher than the delisting level established in Condition (1), or is at a level in the ground water or soil higher than the health based level, then Alcoa must report such data, in writing, to the Regional Administrator within 10 days of first possessing or being made aware of that data.<br>(b) Based on the information described in paragraph (a) and any other information received from any source, the Regional Administrator will make a preliminary determination as to whether the reported information requires Agency action to protect human health or the environment. Further action may include suspending or revoking this exclusion, or other appropriate response necessary to protect human health and the environment.<br>(c) If the Regional Administrator determines that the reported information does require Agency action, the Regional Administrator will notify the facility in writing of the actions the Regional Administrator believes are necessary to protect human health and the environment. The notice shall include a statement of the proposed action and a statement providing the facility with an opportunity to present information as to why the proposed Agency action is not necessary or to suggest an alternative action. The facility shall have 10 days from the date of the Regional Administrator's notice to present such information.<br>(d) Following the receipt of information from the facility described in paragraph (c) or (if no information is presented under paragraph (c) the initial receipt of information described in paragraph (a), the Regional Administrator will issue a final written determination describing the Agency actions that are necessary to protect human health or the environment. Any required action described in the Regional Administrator's determination shall become effective immediately, unless the Regional Administrator provides otherwise. |
| Alumnitec, Inc. (formerly Profile Extrusion Co., formerly United Technologies Automotive, Inc.). | Jeffersonville, IN. | Dewatered wastewater treatment sludge (EPA Hazardous Waste No. F019) generated from the chemical conversion of aluminum after April 29, 1986. |
| American Metals Corporation. | Westlake, Ohio. | Wastewater treatment plant (WWTP) sludges from the chemical conversion coating (phosphating) of aluminum (EPA Hazardous Waste No. F019) and other solid wastes previously disposed in an on-site landfill. This is a one-time exclusion for 12,400 cubic yards of landfilled WWTP sludge. This exclusion is effective on January 15, 2002.<br>1. *Delisting Levels:*<br>(A) The constituent concentrations measured in the TCLP extract may not exceed the following levels (mg/L): antimony—1.52; arsenic—0.691; barium—100; beryllium—3.07; cadmium—1; chromium—5; cobalt—166; copper—67,300; lead—5; mercury—0.2; nickel—209; selenium—1; silver—5; thallium—0.65; tin—1,660; vanadium—156; and zinc—2,070.<br>(B) The total constituent concentrations in any sample may not exceed the following levels (mg/kg): arsenic—9,280; mercury—94; and polychlorinated biphenyls—0.265.<br>(C) Concentrations of dioxin and furan congeners cannot exceed values which would result in a cancer risk greater than or equal to $10^{-6}$ as predicted by the model. |

167

TABLE 1—WASTES EXCLUDED FROM NON-SPECIFIC SOURCES—Continued

| Facility | Address | Waste description |
|---|---|---|
| | | 2. *Verification Sampling*—USG shall collect six additional vertically composited samples of sludge from locations that compliment historical data and shall analyze the samples by TCLP for metals including antimony, arsenic, barium, beryllium, cadmium, chromium, lead, mercury, nickel, selenium, silver, thallium, tin, vanadium, and zinc. If the samples exceed the levels in Condition (1)(a), USG must notify EPA. The corresponding sludge and all sludge yet to be disposed remains hazardous until USG has demonstrated by additional sampling that all constituents of concern are below the levels set forth in condition 1. |
| | | 3. *Reopener Language*—(a) If, anytime after disposal of the delisted waste, USG possesses or is otherwise made aware of any data (including but not limited to leachate data or groundwater monitoring data) or any other data relevant to the delisted waste indicating that any constituent identified in Condition (1) is at a level higher than the delisting level established in Condition (1), or is at a level in the groundwater exceeding maximum allowable point of exposure concentration referenced by the model, then USG must report such data, in writing, to the Regional Administrator within 10 days of first possessing or being made aware of that data. |
| | | (b) Based on the information described in paragraph (a) and any other information received from any source, the Regional Administrator will make a preliminary determination as to whether the reported information requires Agency action to protect human health or the environment. Further action may include suspending, or revoking the exclusion, or other appropriate response necessary to protect human health and the environment. |
| | | (c) If the Regional Administrator determines that the reported information does require Agency action, the Regional Administrator will notify USG in writing of the actions the Regional Administrator believes are necessary to protect human health and the environment. The notice shall include a statement of the proposed action and a statement providing USG with an opportunity to present information as to why the proposed Agency action is not necessary or to suggest an alternative action. USG shall have 10 days from the date of the Regional Administrator's notice to present the information. |
| | | (d) If after 10 days USG presents no further information, the Regional Administrator will issue a final written determination describing the Agency actions that are necessary to protect human health or the environment. Any required action described in the Regional Administrator's determination shall become effective immediately, unless the Regional Administrator provides otherwise. |
| | | 4. *Notifications*—USG must provide a one-time written notification to any State Regulatory Agency to which or through which the waste described above will be transported for disposal at least 60 days prior to the commencement of such activities. Failure to provide such a notification will result in a violation of the delisting petition and a possible revocation of the decision. |
| American Steel Cord. | Scottsburg, IN | Wastewater treatment plant (WWTP) sludge from electroplating operations (EPA Hazardous Waste No. F006) generated at a maximum annual rate of 3,000 cubic yards per year, after January 26, 1999, and disposed of in a Subtitle D landfill. |
| | | 1. Verification Testing: American Steel Cord must implement an annual testing program to demonstrate, based on the analysis of a minimum of four representative samples, that the constituent concentrations measured in the TCLP extract of the waste are within specific levels. The constituent concentrations must not exceed the following levels (mg/l) which are back-calculated from the delisting health-based levels and a DAF of 68. Arsenic—3.4; Barium—100; Cadmium—.34; Chromium—5; Copper—88.4.; Lead—1.02; Mercury—.136; Nickel—6.8.; Selenium—1; Silver—5; Zinc—680; Cyanide—13.6; Acetone—272; Benzo butyl phthlate—476; Chloroform—.68; 1,4-Dichlorobenzene—.272; cis-1,2-Dichloroethone—27.2; Methylene chloride—.34; Naphthalene—68; Styrene—6.8; Tetrachloroethene—.34; Toluene—68; and Xylene—680. American Steel Cord must measure and record the pH of the waste using SW 846 method 9045 and must record all pH measurements performed in accordance with the TCLP. |
| | | 2. Changes in Operating Conditions: If American Steel Cord significantly changes the manufacturing or treatment process or the chemicals used in the manufacturing or treatment process, American Steel Cord may handle the WWTP filter press sludge generated from the new process under this exclusion only after the facility has demonstrated that the waste meets the levels set forth in paragraph 1 and that no new hazardous constituents listed in Appendix VIII of Part 261 have been introduced. |
| | | 3. Data Submittals: The data obtained through annual verification testing or compliance with paragraph 2 must be submitted to U.S. EPA Region 5, 77 W. Jackson Blvd., Chicago, IL 60604–3590, within 60 days of sampling. Records of operating conditions and analytical data must be compiled, summarized, and maintained on site for a minimum of five years and must be made available for inspection. All data must be accompanied by a signed copy of the certification statement in 260.22(I)(12). |

TABLE 1—WASTES EXCLUDED FROM NON-SPECIFIC SOURCES—Continued

| Facility | Address | Waste description |
|---|---|---|
| | | 4. (a) If, anytime after disposal of the delisted waste, American Steel Cord possesses or is otherwise made aware of any environmental data (including but not limited to leachate data or groundwater monitoring data) or any other data relevant to the delisted waste indicating that any constituent identified in Condition (1) is at a level in the leachate higher than the delisting level established in Condition (1), or is at a level in the ground water or soil higher than the health based level, then American Steel Cord must report such data, in writing, to the Regional Administrator within 10 days of first possessing or being made aware of that data. |
| | | (b) Based on the information described in paragraph (a) and any other information received from any source, the Regional Administrator will make a preliminary determination as to whether the reported information requires Agency action to protect human health or the environment. Further action may include suspending, or revoking the exclusion, or other appropriate response necessary to protect human health and the environment. |
| | | (c) If the Regional Administrator determines that the reported information does require Agency action, the Regional Administrator will notify the facility in writing of the actions the Regional Administrator believes are necessary to protect human health and the environment. The notice shall include a statement of the proposed action and a statement providing the facility with an opportunity to present information as to why the proposed Agency action is not necessary or to suggest an alternative action. The facility shall have 10 days from the date of the Regional Administrator's notice to present such information. |
| | | (d) Following the receipt of information from the facility described in paragraph (c) or (if no information is presented under paragraph (c) the initial receipt of information described in paragraph (a), the Regional Administrator will issue a final written determination describing the Agency actions that are necessary to protect human health or the environment. Any required action described in the Regional Administrator's determination shall become effective immediately, unless the Regional Administrator provides otherwise. |
| Ampex Recording Media Corporation. | Opelika, Alabama. | Solvent recovery residues in the powder or pellet form (EPA Hazardous Waste Nos. F003 and F005) generated from the recovery of spent solvents from the manufacture of tape recording media (generated at a maximum annual rate of 1,000 cubic yards in the powder or pellet form) after August 9, 1993. In order to confirm that the characteristics of the wastes do not change significantly, the facility must, on an annual basis, analyze a representative composite sample of the waste (in its final form) for the constituents listed in 40 CFR 261.24 using the method specified therein. The annual analytical results, including quality control information, must be compiled, certified according to 40 CFR 260.22(i)(12), maintained on-site for a minimum of five years, and made available for inspection upon request by any employee or representative of EPA or the State of Alabama. Failure to maintain the required records on-site will be considered by EPA, at its discretion, sufficient basis to revoke the exclusion to the extent directed by EPA. |

TABLE 1—WASTES EXCLUDED FROM NON-SPECIFIC SOURCES—Continued

| Facility | Address | Waste description |
|---|---|---|
| Aptus, Inc. ...... | Coffeyville, Kansas. | Kiln residue and spray dryer/baghouse residue (EPA Hazardous Waste No. F027) generated during the treatment of cancelled pesticides containing 2,4,5–T and Silvex and related materials by Aptus' incinerator at Coffeyville, Kansas after December 27, 1991, so long as:<br>(1) The incinerator is monitored continuously and is in compliance with operating permit conditions. Should the incinerator fail to comply with the permit conditions relevant to the mechanical operation of the incinerator, Aptus must test the residues generated during the run when the failure occurred according to the requirements of Conditions (2) through (4), regardless of whether or not the demonstration in Condition (5) has been made.<br>(2) A minimum of four grab samples must be taken from each hopper (or other container) of kiln residue generated during each 24-hour run; all grabs collected during a given 24-hour run must then be composited to form one composite sample. A minimum of four grab samples must also be taken from each hopper (or other container) of spray dryer/baghouse residue generated during each 24-hour run; all grabs collected during a given 24-hour run must then be composited to form one composite sample. Prior to the disposal of the residues from each 24-hour run, a TCLP leachate test must be performed on these composite samples and the leachate analyzed for the TC toxic metals, nickel, and cyanide. If arsenic, chromium, lead or silver TC leachate test results exceed 1.6 ppm, barium levels exceed 32 ppm, cadmium or selenium levels exceed 0.3 ppm, mercury levels exceed 0.07 ppm, nickel levels exceed 10 ppm, or cyanide levels exceed 6.5 ppm, the wastes must be retreated to achieve these levels or must be disposed in accordance with subtitle C of RCRA. Analyses must be performed according to appropriate methods. As applicable to the method-defined parameters of concern, analyses requiring use of SW–846 methods incorporated by reference in 40 CFR 260.11 must be used without substitution. As applicable, the SW–846 methods might include Methods 0010, 0011, 0020, 0023A, 0030, 0031, 0040, 0050, 0051, 0060, 0061, 1010A, 1020B, 1110A, 1310B, 1311, 1312, 1320, 1330A, 9010C, 9012B, 9040C, 9045D, 9060A, 9070A (uses EPA Method 1664, Rev. A), 9071B, and 9095B.<br>(3) Aptus must generate, prior to the disposal of the residues, verification data from each 24 hour run for each treatment residue (i.e., kiln residue, spray dryer/baghouse residue) to demonstrate that the maximum allowable treatment residue concentrations listed below are not exceeded. Samples must be collected as specified in Condition (2). Analyses must be performed according to appropriate methods. As applicable to the method-defined parameters of concern, analyses requiring the use of SW–846 methods incorporated by reference in 40 CFR 260.11 must be used without substitution. As applicable, the SW–846 methods might include Methods 0010, 0011, 0020, 0023A, 0030, 0031, 0040, 0050, 0051, 0060, 0061, 1010A, 1020B, 1110A, 1310B, 1311, 1312, 1320, 1330A, 9040C, 9045D, 9060A, 9070A (uses EPA Method 1664, Rev. A), 9071B, and 9095B. Any residues which exceed any of the levels listed below must be retreated or must be disposed of as hazardous. Kiln residue and spray dryer/baghouse residue must not exceed the following levels:<br>Aldrin—0.015 ppm, Benzene—9.7 ppm, Benzo(a)pyrene—0.43 ppm, Benzo(b)fluoranthene]—1.8 ppm, Chlordane—0.37 ppm, Chloroform—5.4 ppm, Chrysene—170 ppm, Dibenz(a,h)anthracene—0.083 ppm, 1.2-Dichloroethane—4.1 ppm, Dichloromethane—2.4 ppm, 2,4-Dichlorophenol—480 ppm, Dichlorvos—260 ppm, Disulfaton—23 ppm, Endosulfan I—310 ppm, Fluorene—120 ppm, Indeno(1,2,3,cd)-pyrene—330 ppm, Methyl parathion—210 ppm, Nitrosodiphenylamine—130 ppm, Phenanthrene—150 ppm, Polychlorinated biphenyls—0.31 ppm, Tetrachlorethylene—59 ppm, 2,4,5-TP (silvex)—110 ppm, 2,4,6-Trichlorophenol—3.9 ppm.<br>(4) Aptus must generate, prior to disposal of residues, verification data from each 24-hour run for each treatment residue (i.e., kiln residue, spray dryer/baghouse residue) to demonstrate that the residues do not contain tetra-, penta-, or hexachlorodibenzo-p-dioxins or furans at levels of regulatory concern. Samples must be collected as specified in Condition (2). The TCDD equivalent levels for the solid residues must be less than 5 ppt. Any residues with detected dioxins or furans in excess of this level must be retreated or must be disposed of as acutely hazardous. For tetra- and penta-chlorinated dioxin and furan homologs, the maximum practical quantitation limit must not exceed 15 ppt for the solid residues. For hexachlorinated dioxin and furan homologs, the maximum practical quantitation limit must not exceed 37 ppt for the solid residues.<br>(5) The test data from Conditions (1), (2), (3), and (4) must be kept on file by Aptus for inspection purposes and must be compiled, summarized, and submitted to the Director for the Materials Recovery and Waste Management Division, Office of Resource Conservation and Recovery, by certified mail on a monthly basis and when the treatment of the cancelled pesticides and related materials is concluded. The testing requirements for Conditions (2), (3), and (4) will continue until Aptus provides the Director with the results of four consecutive batch analyses for the petitioned wastes, none of which exceed the maximum allowable levels listed in these conditions and the Director notifies Aptus that the conditions have been lifted. All data submitted will be placed in the RCRA public docket. |
| Arco Building Products. | Sugarcreek, Ohio. | Dewatered wastewater treatment sludge (EPA Hazardous Waste No. F019) generated from the chemical conversion coating of aluminum after August 15, 1986. |
| Arco Chemical Co.. | Miami, FL ....... | Dewatered wastewater treatment sludge (EPA Hazardous Waste No. FO19) generated from the chemical conversion coating of aluminum after April 29, 1986. |

TABLE 1—WASTES EXCLUDED FROM NON-SPECIFIC SOURCES—Continued

| Facility | Address | Waste description |
|---|---|---|
| Arkansas Department of Pollution Control and Ecology. | Vertac Superfund site, Jacksonville, Arkansas. | Kiln ash, cyclone ash, and calcium chloride salts from incineration of residues (EPA Hazardous Waste No. F020 and F023) generated from the primary production of 2,4,5–T and 2,4–D after August 24, 1990. This one-time exclusion applies only to the incineration of the waste materials described in the petition, and it is conditional upon the data obtained from ADPC&E's full-scale incineration facility. To ensure that hazardous constituents are not present in the waste at levels of regulatory concern once the full-scale treatment facility is in operation, ADPC&E must implement a testing program for the petitioned waste. This testing program must meet the following conditions for the exclusion to be valid:<br>(1) *Testing:* Sample collection and analyses (including quality control (QC) procedures) must be performed according to appropriate methods. As applicable to the method-defined parameters of concern, analyses requiring the use of SW–846 methods incorporated by reference in 40 CFR 260.11 must be used without substitution. As applicable, the SW–846 methods might include Methods 0010, 0011, 0020, 0023A, 0030, 0031, 0040, 0050, 0051, 0060, 0061, 1010A, 1020B, 1110A, 1310B, 1311, 1312, 1320, 1330A, 9010C, 9012B, 9040C, 9045D, 9060A, 9070A (uses EPA Method 1664, Rev. A), 9071B, and 9095B.<br>(A) *Initial testing:* Representative grab samples must be taken from each drum and kiln ash and cyclone ash generated from each 24 hours of operation, and the grab samples composited to form one composite sample of ash for each 24-hour period. Representative grab samples must also be taken from each drum of calcium chloride salts generated from each 24 hours of operation and composited to form one composite sample of calcium chloride salts for each 24-hour period. The initial testing requirements must be fullfilled for the following wastes: (i) Incineration by-products generated prior to and during the incinerator's trial burn; (ii) incineration by-products from the treatment of 2,4–D wastes for one week (or 7 days if incineration is not on consecutive days) after completion of the trial burn; (iii) incineration by-products from the treatment of blended 2,4–D and 2,4, 5–T wastes for two weeks (or 14 days if incineration is not on consecutive days) after completion of the trial burn; and (iv) incineration by-products from the treatment of blended 2,4–D and 2,4,5–T wastes for one week (or 7 days if incineration is not on consecutive days) when the percentage of 2, 4, 5–T wastes exceeds the maximum percentage treated under Condition (1)(A)(iii). Prior to disposal of the residues from each 24-hour sampling period, the daily composite must be analyzed for all the constituents listed in Condition (3). ADPC&E must report the analytical test data, including quality control information, obtained during this initial period no later than 90 days after the start of the operation.<br>(B) *Subsequent testing:* Representative grab samples of each drum of kiln and cyclone ash generated from each week of operation must be composited to form one composite sample of ash for each weekly period. Representative grab samples of each drum of calcium chloride salts generated from each week of operation must also be composited to form one composite sample of calcium chloride salts for each weekly period.<br>Prior to disposal of the residues from each weekly sampling period, the weekly composites must be analyzed for all of the constituents listed in Condition (3). The analytical data, including quality control information, must be compiled and maintained on site for a minimum of three years. These data must be furnished upon request and made available for inspection by any employee or representative of EPA.<br>(2) *Waste holding:* The incineration residues that are generated must be stored as hazardous until the initial verification analyses or subsequent analyses are completed.<br>If the composite incineration residue samples (from either Condition (1)(A) or Condition (1)(B)) do not exceed any of the delisting levels set in Condition (3), the incineration residues corresponding to these samples may be managed and disposed of in accordance with all applicable solid waste regulations.<br>If any composite incineration residue sample exceeds any of the delisting levels set in Condition (3), the incineration residues generated during the time period corresponding to this sample must be retreated until they meet these levels (analyses must be repeated) or managed and disposed of in accordance with subtitle C of RCRA. Incineration residues which are generated but for which analysis is not complete or valid must be managed and disposed of in accordance with subtitle C of RCRA, until valid analyses demonstrate that the wastes meet the delisting levels.<br>(3) *Delisting levels:* If concentrations in one or more of the incineration residues for any of the hazardous constituents listed below exceed their respective maximum allowable concentrations also listed below, the batch of failing waste must either be re-treated until it meets these levels or managed and disposed of in accordance with subtitle C of RCRA.<br>(A) Inorganics (Leachable): Arsenic, 0.32 ppm; Barium, 6.3 ppm; Cadmium, 0.06 ppm; Chromium, 0.32 ppm; Cyanide, 4.4 ppm; Lead, 0.32 ppm; Mercury, 0.01 ppm; Nickel, 4.4 ppm; Selenium, 0.06 ppm; Silver, 0.32 ppm. Metal concentrations must be measured in the waste leachate as per 40 CFR 261.24. Cyanide extractions must be conducted using distilled water. |

TABLE 1—WASTES EXCLUDED FROM NON-SPECIFIC SOURCES—Continued

| Facility | Address | Waste description |
|---|---|---|
| | | (B) Organics: Benzene, 0.87 ppm; Benzo(a)anthracene, 0.10 ppm; Benzo(a)pyrene, 0.04 ppm; Benzo (b)fluoranthene, 0.16 ppm; Chlorobenzene, 152 ppm; o-Chlorophenol, 44 ppm; Chrysene, 15 ppm; 2, 4–D, 107 ppm; DDE, 1.0 ppm; Dibenz(a,h)anthracene, 0.007 ppm; 1, 4-Dichlorobenzene, 265 ppm; 1, 1-Dichloroethylene, 1.3 ppm; trans-1,2-Dichloroethylene, 37 ppm; Dichloromethane, 0.23 ppm; 2,4-Dichlorophenol, 43 ppm; Hexachlorobenzene, 0.26 ppm; Indeno (1,2,3-cd) pyrene, 30 ppm; Polychlorinated biphenyls, 12 ppm; 2,4,5–T, $1 \times 10^6$ ppm; 1,2,4,5-Tetrachlorobenzene, 56 ppm; Tetrachloroethylene, 3.4 ppm; Tri-chloroethylene, 1.1 ppm; 2,4,5-Trichlorophenol, 21,000 ppm; 2,4,6-Trichlorophenol, 0.35 ppm. |
| | | (C) *Chlorinated dioxins and furans:* 2,3,7,8-Tetrachlorodibenzo-p-dioxin equivalents, $4 \times 10^{-7}$ ppm. The petitioned by-product must be analyzed for the tetra-, penta-, hexa-, and heptachlorodibenzo-p-dioxins, and the tetra-, penta-, hexa-, and heptachlorodibenzofurans to determine the 2, 3, 7, 8-tetra-chlorodibenzo-p-dioxin equivalent concentration. The analysis must be conducted using a measurement system that achieves practical quantitation limits of 15 parts per trillion (ppt) for the tetra- and penta-homologs, and 37 ppt for the hexa- and hepta-homologs. |
| | | (4) *Termination of testing:* Due to the possible variability of the incinerator feeds, the testing requirements of Condition (1)(B) will continue indefinitely. |
| | | (5) *Data submittals:* Within one week of system start-up, ADPC&E must notify the Section Chief, Variances Section (see address below) when the full-scale incineration system is on-line and waste treatment has begun. The data obtained through Condition (1)(A) must be submitted to PSPD/OSW (5303W), U.S. EPA, 1200 Pennsylvania Ave., NW., Washington, DC 20460, within the time period specified. At the Section Chief's request, ADPC&E must submit analytical data obtained through Condition (1)(B) within the time period specified by the Section Chief. Failure to submit the required data obtained from Condition (1)(A) within the specified time period or to maintain the required records for the time specified in Condition (1)(B) (or to submit data within the time specified by the Section Chief) will be considered by the Agency, at its discretion, sufficient basis to revoke ADPC&E's exclusion to the extent directed by EPA. All data must be accompanied by the following certification statement: |
| | | "Under civil and criminal penalty of law for the making or submission of false or fraudulent statements or representations (pursuant to the applicable provisions of the Federal Code, which include, but may not be limited to, 18 U.S.C. 1001 and 42 U.S.C. 6928), I certify that the information contained in or accompanying this document is true, accurate and complete. As to the (those) identified section(s) of this document for which I cannot personally verify its (their) truth and accuracy, I certify as the company official having supervisory responsibility for the persons who, acting under my direct instructions, made the verification that this information is true, accurate and complete. In the event that any of this information is determined by EPA in its sole discretion to be false, inaccurate or incomplete, and upon conveyance of this fact to the company, I recognize and agree that this exclusion of wastes will be void as if it never had effect or to the extent directed by EPA and that the company will be liable for any actions taken in contravention of the company's RCRA and CERCLA obligations premised upon the company's reliance on the void exclusion." |
| AutoAlliance International Inc.. | Flat Rock, Michigan. | Wastewater treatment sludges, F019, that are generated by AutoAlliance International, Inc. (AAI) at Flat Rock, Michigan at a maximum annual rate of 2,000 cubic yards per year. The sludges must be disposed of in a lined landfill with leachate collection which is licensed, permitted, or otherwise authorized to accept the delisted wastewater treatment sludges in accordance with 40 CFR part 258. The exclusion becomes effective as of April 6, 2007. |
| | | (1) Delisting Levels: (A) The concentrations in a leachate extract of the waste measured in any sample must not exceed the following levels (mg/L): arsenic—0.3; cadmium—0.5; chromium—4.95; lead—5; nickel—90.5; selenium—1; tin—721; zinc—898; p-cresol—11.4; and formaldehyde—84.2. |
| | .................. | (B) The total concentration measured in any sample must not exceed the following levels (mg/kg): mercury—8.92; and formaldehyde—689. |
| | | (2) Quarterly Verification Testing: To verify that the waste does not exceed the specified delisting levels, AAI must collect and analyze one representative sample of the waste on a quarterly basis. Sample collection and analyses, including quality control procedures, must be performed using appropriate methods. SW–846 Method 1311 must be used for generation of the leachate extract used in the testing of the delisting levels if oil and grease comprise less than 1% of the waste. SW–846 Method 1330A must be used for generation of the leaching extract if oil and grease comprise 1% or more of the waste. SW–846 Method 9071B must be used for determination of oil and grease. SW–846 Methods 1311, 1330A, and 9071B are incorporated by reference in 40 CFR 260.11. |

172

TABLE 1—WASTES EXCLUDED FROM NON-SPECIFIC SOURCES—Continued

| Facility | Address | Waste description |
|---|---|---|
| | | (3) Changes in Operating Conditions: AAI must notify the EPA in writing if the manufacturing process, the chemicals used in the manufacturing process, the treatment process, or the chemicals used in the treatment process change significantly. AAI must handle wastes generated after the process change as hazardous until it has demonstrated that the wastes continue to meet the delisting levels and that no new hazardous constituents listed in Appendix VIII of part 261 have been introduced and it has received written approval from EPA. |
| | | (4) Data Submittals: AAI must submit the data obtained through verification testing or as required by other conditions of this rule to both U.S. EPA Region 5, 77 W. Jackson Blvd., Chicago, IL 60604 and MDEQ, Waste and Hazardous Materials Division, Hazardous Waste Section, at P.O. Box 30241, Lansing, Michigan 48909. The quarterly verification data and certification of proper disposal must be submitted annually upon the anniversary of the effective date of this exclusion. AAI must compile, summarize and maintain on site for a minimum of five years records of operating conditions and analytical data. AAI must make these records available for inspection. A signed copy of the certification statement in 40 CFR 260.22(i)(12) must accompany all data. |
| | | (5) Reopener Language: (A) If, anytime after disposal of the delisted waste AAI possesses or is otherwise made aware of any data (including but not limited to leachate data or groundwater monitoring data) relevant to the delisted waste indicating that any constituent is at a level in the leachate higher than the specified delisting level, or is in the groundwater at a concentration higher than the maximum allowable groundwater concentration in paragraph (e), then AAI must report such data, in writing, to the Regional Administrator within 10 days of first possessing or being made aware of that data. |
| | | (B) Based on the information described in paragraph (a) and any other information received from any source, the Regional Administrator will make a preliminary determination as to whether the reported information requires Agency action to protect human health or the environment. Further action may include suspending, or revoking the exclusion, or other appropriate response necessary to protect human health and the environment. |
| | | (C) If the Regional Administrator determines that the reported information does require Agency action, the Regional Administrator will inform AAI in writing of the actions the Regional Administrator believes are necessary to protect human health and the environment. The notice shall include a statement of the proposed action and a statement providing AAI with an opportunity to present information as to why the proposed Agency action is not necessary or to suggest an alternative action. AAI shall have 30 days from the date of the Regional Administrator's notice to present the information. |
| | | (D) If after 30 days AAI presents no further information, the Regional Administrator will issue a final written determination describing the Agency actions that are necessary to protect human health or the environment. Any required action described in the Regional Administrator's determination shall become effective immediately, unless the Regional Administrator provides otherwise. |
| | | (E) Maximum Allowable Groundwater Concentrations (μg/L): arsenic—5; cadmium—5; chromium—100; lead—15; nickel—750; selenium—50; tin—22,500; zinc—11,300; p-cresol—188; and formaldehyde—1,380. |
| BAE Systems, Inc,. | Sealy, TX ....... | Filter Cake (EPA Hazardous Waste Number F019) generated at a maximum rate of 1,200 cubic yards per calendar year after April 15, 2009. |
| | | For the exclusion to be valid, BAE must implement a verification testing program that meets the following Paragraphs: |
| | | (1) Delisting Levels: All concentrations for those constituents must not exceed the maximum allowable concentrations in mg/l specified in this paragraph. |
| | | Filter Cake Leachable Concentrations (mg/l): Acetone—3211; Arsenic—0.052; Barium—100; Bis(2-ethylhexyl)phthalate—103; Cadmium—0.561; Chloroform—0.4924; Chromium—5.0; Copper 149; Cyanide—19; Furans—3.57; Hexavalent Chromium—5.0; Lead—3.57; Lindane—0.4; Methyl Ethyl Ketone—200; Nickel—82.2; Selenium—1.0; 2,4,5–TP (Silvex)—1.0; 2,4–D—6.65; Tin—9001; Tetrachlorodibenzo-p-dioxin—249; Tetrachloroethylene—0.125685; Zinc—1240. |
| | | (2) Waste Holding and Handling: |
| | | (A) Waste classification as non-hazardous can not begin until compliance with the limits set in paragraph (1) for filter cake has occurred for two consecutive quarterly sampling events. |
| | | (B) If constituent levels in any sample taken by BAE exceed any of the delisting levels set in paragraph (1) for the filter cake, BAE must do the following: |
| | | (i) notify EPA in accordance with paragraph (6) and |
| | | (ii) manage and dispose the filter cake as hazardous waste generated under Subtitle C of RCRA. |
| | | (3) Testing Requirements: |
| | | Upon this exclusion becoming final, BAE may perform quarterly analytical testing by sampling and analyzing the filter cake as follows: |
| | | (A) Quarterly Testing: |
| | | (i) Collect two representative composite samples of the filter cake at quarterly intervals after EPA grants the final exclusion. The first composite samples may be taken at any time after EPA grants the final approval. Sampling must be performed in accordance with the sampling plan approved by EPA in support of the exclusion. |

173

TABLE 1—WASTES EXCLUDED FROM NON-SPECIFIC SOURCES—Continued

| Facility | Address | Waste description |
|---|---|---|
| | | (ii) Analyze the samples for all constituents listed in paragraph (1). Any composite sample taken that exceeds the delisting levels listed in paragraph (1) for the filter cake must be disposed as hazardous waste in accordance with the applicable hazardous waste requirements. |
| | | (iii) Within thirty (30) days after taking its first quarterly sample, BAE will report its first quarterly analytical test data to EPA. If levels of constituents measured in the samples of the filter cake do not exceed the levels set forth in paragraph (1) of this exclusion for two consecutive quarters, BAE can manage and dispose the non-hazardous filter cake according to all applicable solid waste regulations. |
| | | (B) Annual Testing: |
| | | (i) If BAE completes the quarterly testing specified in paragraph (3) above and no sample contains a constituent at a level which exceeds the limits set forth in paragraph (1), BAE may begin annual testing as follows: BAE must test two representative composite samples of the filter cake for all constituents listed in paragraph (1) at least once per calendar year. |
| | | (ii) The samples for the annual testing shall be a representative composite sample according to appropriate methods. As applicable to the method-defined parameters of concern, analyses requiring the use of SW–846 methods incorporated by reference in 40 CFR 260.11 must be used without substitution. As applicable, the SW–846 methods might include Methods 0010, 0011, 0020, 0023A, 0030, 0031, 0040, 0050, 0051, 0060, 0061, 1010A, 1020B,1110A, 1310B, 1311, 1312, 1320, 1330A, 9010C, 9012B, 9040C, 9045D, 9060A, 9070A (uses EPA Method 1664, Rev. A), 9071B, and 9095B. Methods must meet Performance Based Measurement System Criteria in which the Data Quality Objectives are to demonstrate that samples of the BAE filter cake are representative for all constituents listed in paragraph (1). |
| | | (iii) The samples for the annual testing taken for the second and subsequent annual testing events shall be taken within the same calendar month as the first annual sample taken. |
| | | (iv) The annual testing report should include the total amount of waste in cubic yards disposed during the calendar year. |
| | | (4) Changes in Operating Conditions: If BAE significantly changes the process described in its petition or starts any processes that generate(s) the waste that may or could affect the composition or type of waste generated (by illustration, but not limitation, changes in equipment or operating conditions of the treatment process), it must notify EPA in writing and it may no longer handle the wastes generated from the new process as non-hazardous until the wastes meet the delisting levels set in paragraph (1) and it has received written approval to do so from EPA. |
| | | BAE must submit a modification to the petition complete with full sampling and analysis for circumstances where the waste volume changes and/or additional waste codes are added to the waste stream. |
| | | (5) Data Submittals: |
| | | BAE must submit the information described below. If BAE fails to submit the required data within the specified time or maintain the required records on-site for the specified time, EPA, at its discretion, will consider this sufficient basis to reopen the exclusion as described in paragraph (6). BAE must: |
| | | (A) Submit the data obtained through paragraph (3) to the Chief, Corrective Action and Waste Minimization Section, Multimedia Planning and Permitting Division, U.S. Environmental Protection Agency Region 6, 1445 Ross Ave., Dallas, Texas 75202, within the time specified. All supporting data can be submitted on CD–ROM or some comparable electronic media. |
| | | (B) Compile records of analytical data from paragraph (3), summarized, and maintained on-site for a minimum of five years. |
| | | (C) Furnish these records and data when either EPA or the State of Texas requests them for inspection. |
| | | (D) Send along with all data a signed copy of the following certification statement, to attest to the truth and accuracy of the data submitted: |
| | | "Under civil and criminal penalty of law for the making or submission of false or fraudulent statements or representations (pursuant to the applicable provisions of the Federal Code, which include, but may not be limited to, 18 U.S.C. 1001 and 42 U.S.C. 6928), I certify that the information contained in or accompanying this document is true, accurate and complete. |
| | | As to the (those) identified section(s) of this document for which I cannot personally verify its (their) truth and accuracy, I certify as the company official having supervisory responsibility for the persons who, acting under my direct instructions, made the verification that this information is true, accurate and complete. |
| | | If any of this information is determined by EPA in its sole discretion to be false, inaccurate or incomplete, and upon conveyance of this fact to the company, I recognize and agree that this exclusion of waste will be void as if it never had effect or to the extent directed by EPA and that the company will be liable for any actions taken in contravention of the company's RCRA and CERCLA obligations premised upon the company's reliance on the void exclusion." |
| | | (6) Reopener |

TABLE 1—WASTES EXCLUDED FROM NON-SPECIFIC SOURCES—Continued

| Facility | Address | Waste description |
|---|---|---|
| | | (A) If, anytime after disposal of the delisted waste BAE possesses or is otherwise made aware of any environmental data (including but not limited to leachate data or ground water monitoring data) or any other data relevant to the delisted waste indicating that any constituent identified for the delisting verification testing is at level higher than the delisting level allowed by the Division Director in granting the petition, then the facility must report the data, in writing, to the Division Director within 10 days of first possessing or being made aware of that data. |
| | | (B) If either the quarterly or annual testing of the waste does not meet the delisting requirements in paragraph (1), BAE must report the data, in writing, to the Division Director within 10 days of first possessing or being made aware of that data. |
| | | (C) If BAE fails to submit the information described in paragraphs (5), (6)(A) or (6)(B) or if any other information is received from any source, the Division Director will make a preliminary determination as to whether the reported information requires EPA action to protect human health and/or the environment. Further action may include suspending, or revoking the exclusion, or other appropriate response necessary to protect human health and the environment. |
| | | (D) If the Division Director determines that the reported information requires action by EPA, the Division Director will notify the facility in writing of the actions the Division Director believes are necessary to protect human health and the environment. The notice shall include a statement of the proposed action and a statement providing the facility with an opportunity to present information as to why the proposed EPA action is not necessary. The facility shall have 10 days from the date of the Division Director's notice to present such information. |
| | | (E) Following the receipt of information from the facility described in paragraph (6)(D) or (if no information is presented under paragraph (6)(D)) the initial receipt of information described in paragraphs (5), (6)(A) or (6)(B), the Division Director will issue a final written determination describing EPA actions that are necessary to protect human health and/or the environment. Any required action described in the Division Director's determination shall become effective immediately, unless the Division Director provides otherwise. |
| | | (7) Notification Requirements |
| | | BAE Systems must do the following before transporting the delisted waste. Failure to provide this notification will result in a violation of the delisting petition and a possible revocation of the decision. |
| | | (A) Provide a one-time written notification to any state Regulatory Agency to which or through which it will transport the delisted waste described above for disposal, 60 days before beginning such activities. |
| | | (B) Update the one-time written notification if it ships the delisted waste into a different disposal facility. |
| | | (C) Failure to provide this notification will result in a violation of the delisting variance and possible revocation of the decision. |
| Bayer Material Science LLC. | Baytown, TX .. | Toluene Diisocyanate (TDI) Residue (EPA Hazardous Waste No. K027) generated at a maximum rate of 9,780 cubic yards per calendar year after March 12, 2009. |
| | | For the exclusion to be valid, Bayer must implement a verification testing program that meets the following Paragraphs: |
| | | (1) Delisting Levels: |
| | | All concentrations for those constituents must not exceed the maximum allowable concentrations in mg/l specified in this paragraph. |
| | | TDI Residue Leachable Concentrations (mg/l): Arsenic—0.10, Barium—36.0; Chloromethane—6.06; Chromium—2.27; Cobalt—13.6; Copper—25.9; Cyanide—3.08; Dichlorophenoxyacetic acid—1.08; Diethyl phthalate—1000.0; Endrin—0.02; Lead—0.702; Nickel—13.5; ortho-dichlorobenzene—9.72; Selenium—0.89; Tin—22.5; Vanadium—0.976; Zinc—197.0; 2,4-Toluenediamine—0.0459; Toluene Diisocyanate—0.039. |
| | | (2) Waste Holding and Handling: |
| | | (A) Bayer must manage the TDI residue in a manner to ensure that the residues are offloaded safely and opportunities for chemical self-reaction and expansion are minimized. The TDI residue must be handled to ensure that contact with water is minimized. |
| | | (B) Waste classification as non-hazardous cannot begin until compliance with the limits set in paragraph (1) for the TDI residue has occurred for two consecutive quarterly sampling events and the reports have been approved by EPA. |
| | | (C) If constituent levels in any sample taken by Bayer exceed any of the delisting levels set in paragraph (1) for the TDI residue, Bayer must do the following: |
| | | (i) notify EPA in accordance with paragraph (6) and |
| | | (ii) manage and dispose the TDI residue as hazardous waste generated under Subtitle C of RCRA. |
| | | (3) Testing Requirements: |
| | | Upon this exclusion becoming final, Bayer must perform quarterly analytical testing by sampling and analyzing the TDI residue as follows: |
| | | (A) Quarterly Testing: |
| | | (i) Collect two representative composite samples of the TDI residue at quarterly intervals after EPA grants the final exclusion. The first composite samples may be taken at any time after EPA grants the final approval. Sampling should be performed in accordance with the sampling plan approved by EPA in support of the exclusion. |

TABLE 1—WASTES EXCLUDED FROM NON-SPECIFIC SOURCES—Continued

| Facility | Address | Waste description |
|---|---|---|
| | | (ii) Analyze the samples for all constituents listed in paragraph (1). Any composite sample taken that exceeds the delisting levels listed in paragraph (1) for the TDI residue must be disposed as hazardous waste in accordance with the applicable hazardous waste requirements. |
| | | (iii) Within thirty (30) days after taking its first quarterly sample, Bayer will report its first quarterly analytical test data to EPA. If levels of constituents measured in the samples of the TDI residue do not exceed the levels set forth in paragraph (1) of this exclusion for two consecutive quarters, Bayer can manage and dispose the non-hazardous TDI residue according to all applicable solid waste regulations. |
| | | (B) Annual Testing: |
| | | (i) If Bayer completes the quarterly testing specified in paragraph (3) above and no sample contains a constituent at a level which exceeds the limits set forth in paragraph (1), Bayer can begin annual testing as follows: Bayer must test two representative composite samples of the TDI residue for all constituents listed in paragraph (1) at least once per calendar year. |
| | | (ii) The samples for the annual testing shall be a representative composite sample according to appropriate methods. As applicable to the method-defined parameters of concern, analyses requiring the use of SW–846 methods incorporated by reference in 40 CFR 260.11 must be used without substitution. As applicable, the SW–846 methods might include Methods 0010, 0011, 0020, 0023A, 0030, 0031, 0040, 0050, 0051, 0060, 0061, 1010A, 1020B, 1110A, 1310B, 1311, 1312, 1320, 1330A, 9010C, 9012B, 9040C, 9045D, 9060A, 9070A (uses EPA Method 1664, Rev. A), 9071B, and 9095B. Methods must meet Performance Based Measurement System Criteria in which the Data Quality Objectives are to demonstrate that samples of the Bayer spent carbon are representative for all constituents listed in paragraph (1). |
| | | (iii) The samples for the annual testing taken for the second and subsequent annual testing events shall be taken within the same calendar month as the first annual sample taken. |
| | | (iv) The annual testing report must include the total amount of waste in cubic yards disposed during the calendar year. |
| | | (4) Changes in Operating Conditions: |
| | | If Bayer significantly changes the process described in its petition or starts any process that generates the waste that may or could affect the composition or type of waste generated (by illustration, but not limitation, changes in equipment or operating conditions of the treatment process), it must notify EPA in writing and it may no longer handle the wastes generated from the new process as non-hazardous until the wastes meet the delisting levels set in paragraph (1) and it has received written approval to do so from EPA. |
| | | Bayer must submit a modification to the petition complete with full sampling and analysis for circumstances where the waste volume changes and/or additional waste codes are added to the waste stream. |
| | | (5) Data Submittals: |
| | | Bayer must submit the information described below. If Bayer fails to submit the required data within the specified time or maintain the required records on-site for the specified time, EPA, at its discretion, will consider this sufficient basis to reopen the exclusion as described in paragraph (6). Bayer must: |
| | | (A) Submit the data obtained through paragraph 3 to the Chief, Corrective Action and Waste Minimization Section, Multimedia Planning and Permitting Division, U.S. Environmental Protection Agency Region 6, 1445 Ross Ave., Dallas, Texas 75202, within the time specified. All supporting data can be submitted on CD–ROM or some comparable electronic media. |
| | | (B) Compile records of analytical data from paragraph (3), summarized, and maintained on-site for a minimum of five years. |
| | | (C) Furnish these records and data when either EPA or the State of Texas requests them for inspection. |
| | | (D) Send along with all data a signed copy of the following certification statement, to attest to the truth and accuracy of the data submitted. "Under civil and criminal penalty of law for the making or submission of false or fraudulent statements or representations (pursuant to the applicable provisions of the Federal Code, which include, but may not be limited to, 18 U.S.C. 1001 and 42 U.S.C. 6928), I certify that the information contained in or accompanying this document is true, accurate and complete. |
| | | As to the (those) identified section(s) of this document for which I cannot personally verify its (their) truth and accuracy, I certify as the company official having supervisory responsibility for the persons who, acting under my direct instructions, made the verification that this information is true, accurate and complete. |
| | | If any of this information is determined by EPA in its sole discretion to be false, inaccurate or incomplete, and upon conveyance of this fact to the company, I recognize and agree that this exclusion of waste will be void as if it never had effect or to the extent directed by EPA and that the company will be liable for any actions taken in contravention of the company's RCRA and CERCLA obligations premised upon the company's reliance on the void exclusion." |
| | | (6) Reopener: |

TABLE 1—WASTES EXCLUDED FROM NON-SPECIFIC SOURCES—Continued

| Facility | Address | Waste description |
|---|---|---|
| | | (A) If, anytime after disposal of the delisted waste Bayer possesses or is otherwise made aware of any environmental data (including but not limited to leachate data or ground water monitoring data) or any other data relevant to the delisted waste indicating that any constituent identified for the delisting verification testing is at a level higher than the delisting level allowed by EPA in granting the petition, then the facility must report the data, in writing, to EPA within 10 days of first possessing or being made aware of that data. |
| | | (B) If either the quarterly or annual testing of the waste does not meet the delisting requirements in paragraph 1, Bayer must report the data, in writing, to EPA within 10 days of first possessing or being made aware of that data. |
| | | (C) If Bayer fails to submit the information described in paragraphs (5), (6)(A) or (6)(B) or if any other information is received from any source, EPA will make a preliminary determination as to whether the reported information requires action to protect human health and/or the environment. Further action may include suspending, or revoking the exclusion, or other appropriate response necessary to protect human health and the environment. |
| | | (D) If EPA determines that the reported information requires action, EPA will notify the facility in writing of the actions it believes are necessary to protect human health and the environment. The notice shall include a statement of the proposed action and a statement providing the facility with an opportunity to present information explaining why the proposed EPA action is not necessary. The facility shall have 10 days from the date of EPA's notice to present such information. |
| | | (E) Following the receipt of information from the facility described in paragraph (6)(D) or (if no information is presented under paragraph (6)(D)) the initial receipt of information described in paragraphs (5), (6)(A) or (6)(B), EPA will issue a final written determination describing the actions that are necessary to protect human health and/or the environment. Any required action described in EPA's determination shall become effective immediately, unless EPA provides otherwise. |
| | | (7) Notification Requirements |
| | | Bayer must do the following before transporting the delisted waste. Failure to provide this notification will result in a violation of the delisting petition and a possible revocation of the decision. |
| | | (A) Provide a one-time written notification to any state Regulatory Agency to which or through which it will transport the delisted waste described above for disposal, 60 days before beginning such activities. |
| | | (B) Update the one-time written notification if it ships the delisted waste into a different disposal facility. |
| | | (C) Failure to provide this notification will result in a violation of the delisting variance and a possible revocation of the decision. |
| BBC Brown Boveri, Inc.. | Sanford, FL .... | Dewatered Wastewater treatment sludges (EPA Hazardous Waste No. F006) generated from electroplating operations after October 17, 1986. |
| Bekaert Corp .. | Dyersburg, TN | Dewatered wastewater treatment plant (WWTP) sludge (EPA Hazardous Waste Nos. F006) generated at a maximum rate of 1250 cubic yards per calendar year after May 27, 2004, and disposed in a Subtitle D landfill. |
| | | For the exclusion to be valid, Bekaert must implement a verification testing program that meets the following paragraphs: |
| | | (1) Delisting Levels: All leachable concentrations for those constituents must not exceed the maximum allowable concentrations in mg/l specified in this paragraph. Bekaert must use the leaching method specified at 40 CFR 261.24 to measure constituents in the waste leachate. |
| | | (A) Inorganic Constituents TCLP (mg/l): Cadmium—0.672; Chromium—5.0; Nickel—127; Zinc—1260.0. |
| | | (D) Organic Constituents TCLP (mg/l): Methyl ethyl ketone—200.0. |
| | | (2) Waste Holding and Handling: |
| | | (A) Bekaert must accumulate the hazardous waste dewatered WWTP sludge in accordance with the applicable regulations of 40 CFR 262.34 and continue to dispose of the dewatered WWTP sludge as hazardous waste. |
| | | (B) Once the first quarterly sampling and analyses event described in paragraph (3) is completed and valid analyses demonstrate that no constituent is present in the sample at a level which exceeds the delisting levels set in paragraph (1), Bekaert can manage and dispose of the dewatered WWTP sludge as nonhazardous according to all applicable solid waste regulations. |
| | | (C) If constituent levels in any sample taken by Bekaert exceed any of the delisting levels set in paragraph (1), Bekaert must do the following: (i) notify EPA in accordance with paragraph (7) and (ii) manage and dispose the dewatered WWTP sludge as hazardous waste generated under Subtitle C of RCRA. |
| | | (D) Quarterly Verification Testing Requirements: Upon this exclusion becoming final, Bekaert may begin the quarterly testing requirements of paragraph (3) on its dewatered WWTP sludge. |
| | | (3) Quarterly Testing Requirements: Upon this exclusion becoming final, Bekaert may perform quarterly analytical testing by sampling and analyzing the dewatered WWTP sludge as follows: |

TABLE 1—WASTES EXCLUDED FROM NON-SPECIFIC SOURCES—Continued

| Facility | Address | Waste description |
|---|---|---|
| | | (A)(i) Collect four representative composite samples of the hazardous waste dewatered WWTP sludge at quarterly (ninety (90) day) intervals after EPA grants the final exclusion. The first composite sample may be taken at any time after EPA grants the final approval. |
| | | (ii) Analyze the samples for all constituents listed in paragraph (1). Any roll-offs from which the composite sample is taken exceeding the delisting levels listed in paragraph (1) must be disposed as hazardous waste in a Subtitle C landfill. |
| | | (iii) Within forty-five (45) days after taking its first quarterly sample, Bekaert will report its first quarterly analytical test data to EPA. If levels of constituents measured in the sample of the dewatered WWTP sludge do not exceed the levels set forth in paragraph (1) of this exclusion, Bekaert can manage and dispose the nonhazardous dewatered WWTP sludge according to all applicable solid waste regulations. |
| | | (4) Annual Testing: |
| | | (A) If Bekaert completes the quarterly testing specified in paragraph (3) above and no sample contains a constituent with a level which exceeds the limits set forth in paragraph (1), Bekaert may begin annual testing as follows: Bekaert must test one representative composite sample of the dewatered WWTP sludge for all constituents listed in paragraph (1) at least once per calendar year. |
| | | (B) The sample for the annual testing shall be a representative composite sample for all constituents listed in paragraph (1). |
| | | (C) The sample for the annual testing taken for the second and subsequent annual testing events shall be taken within the same calendar month as the first annual sample taken. |
| | | (5) Changes in Operating Conditions: If Bekaert significantly changes the process described in its petition or starts any processes that generate(s) the waste that may or could affect the composition or type of waste generated as established under paragraph (1) (by illustration, but not limitation, changes in equipment or operating conditions of the treatment process), it must notify the EPA in writing; it may no longer handle the wastes generated from the new process as nonhazardous until the wastes meet the delisting levels set in paragraph (1) and it has received written approval to do so from the EPA. |
| | | (6) Data Submittals: Bekaert must submit the information described below. If Bekaert fails to submit the required data within the specified time or maintain the required records on-site for the specified time, the EPA, at its discretion, will consider this sufficient basis to reopen the exclusion as described in paragraph (7). Bekaert must: |
| | | (A) Submit the data obtained through paragraph (3) to the Chief, North Section, RCRA Enforcement and Compliance Branch, Waste Division, U. S. Environmental Protection Agency Region 4, 61 Forsyth Street, SW., Atlanta, Georgia, 30303, within the time specified. |
| | | (B) Compile records of analytical data from paragraph (3), summarized, and maintained on-site for a minimum of five years. |
| | | (C) Furnish these records and data when either the EPA or the State of Tennessee request them for inspection. |
| | | (D) Send along with all data a signed copy of the following certification statement, to attest to the truth and accuracy of the data submitted: |
| | | "Under civil and criminal penalty of law for the making or submission of false or fraudulent statements or representations (pursuant to the applicable provisions of the Federal Code, which include, but may not be limited to, 18 U.S.C. 1001 and 42 U.S.C. 6928), I certify that the information contained in or accompanying this document is true, accurate and complete. |
| | | As to the (those) identified section(s) of this document for which I cannot personally verify its (their) truth and accuracy, I certify as the company official having supervisory responsibility for the persons who, acting under my direct instructions, made the verification that this information is true, accurate and complete. If any of this information is determined by the EPA in its sole discretion to be false, inaccurate or incomplete, and upon conveyance of this fact to the company, I recognize and agree that this exclusion of waste will be void as if it never had effect or to the extent directed by the EPA and that the company will be liable for any actions taken in contravention of the company's RCRA and CERCLA obligations premised upon the company's reliance on the void exclusion." |
| | | (7) Reopener: |
| | | (A) If, anytime after disposal of the delisted waste Bekaert possesses or is otherwise made aware of any environmental data (including but not limited to leachate data or ground water monitoring data) or any other data relevant to the delisted waste indicating that any constituent identified for the delisting verification testing is at level higher than the delisting level allowed by the Regional Administrator or his delegate in granting the petition, then the facility must report the data, in writing, to the Regional Administrator or his delegate within ten (10) days of first possessing or being made aware of that data. |
| | | (B) If either the quarterly or annual testing of the waste does not meet the delisting requirements in paragraph (1), Bekaert must report the data, in writing, to the Regional Administrator or his delegate within ten (10) days of first possessing or being made aware of that data. |

178

TABLE 1—WASTES EXCLUDED FROM NON-SPECIFIC SOURCES—Continued

| Facility | Address | Waste description |
|---|---|---|
| | | (C) If Bekaert fails to submit the information described in paragraphs (5), (6)(A) or (6)(B) or if any other information is received from any source, the Regional Administrator or his delegate will make a preliminary determination as to whether the reported information requires the EPA action to protect human health or the environment. Further action may include suspending, or revoking the exclusion, or other appropriate response necessary to protect human health and the environment. |
| | | (D) If the Regional Administrator or his delegate determines that the reported information requires action the EPA, the Regional Administrator or his delegate will notify the facility in writing of the actions the Regional Administrator or his delegate believes are necessary to protect human health and the environment. The notification shall include a statement of the proposed action and a statement providing the facility with an opportunity to present information as to why the proposed the EPA action is not necessary. The facility shall have ten (10) days from the date of the Regional Administrator or his delegate's notice to present such information. |
| | | (E) Following the receipt of information from the facility described in paragraph (6)(D) or (if no information is presented under paragraph (6)(D)) the initial receipt of information described in paragraphs (5), (6)(A) or (6)(B), the Regional Administrator or his delegate will issue a final written determination describing the EPA actions that are necessary to protect human health or the environment. Any required action described in the Regional Administrator or his delegate's determination shall become effective immediately, unless the Regional Administrator or his delegate provides otherwise. |
| | | (8) Notification Requirements: Bekaert must do following before transporting the delisted waste: |
| | | (A) Provide a one-time written notification to any State Regulatory Agency to which or through which it will transport the delisted waste described above for disposal, sixty (60) days before beginning such activities. |
| | | (B) Update the one-time written notification if Bekaert ships the delisted waste into a different disposal facility. |
| | | (C) Failure to provide this notification will result in a violation of the delisting variance and a possible revocation of the decision. |
| Bethlehem Steel Corporation. | Sparrows Point, Maryland. | Stabilized filter cake (at a maximum annual rate of 1100 cubic yards) from the treatment of wastewater treatment sludges (EPA Hazardous Waste No. F006) generated from electroplating operations after [insert date of publication in FEDERAL REGISTER]. Bethlehem Steel (BSC) must implement a testing program that meets the following conditions for the exclusion to be valid: |
| | | (1) *Testing:* Sample collection and analyses (including quality control (QC) procedures) must be performed using appropriate methods. As applicable to the method-defined parameters of concern, analyses requiring the use of SW–846 methods incorporated by reference in 40 CFR 260.11 must be used without substitution. As applicable, the SW–846 methods might include Methods 0010, 0011, 0020, 0023A, 0030, 0031, 0040, 0050, 0051, 0060, 0061, 1010A, 1020B, 1110A, 1310B, 1311, 1312, 1320, 1330A, 9010C, 9012B, 9040C, 9045D, 9060A, 9070A (uses EPA Method 1664, Rev. A), 9071B, and 9095B. If EPA judges the stabilization process to be effective under the conditions used during the initial verification testing, BSC may replace the testing required in Condition (1)(A) with the testing required in Condition (1)(B). BSC must continue to test as specified in Condition (1)(A) until and unless notified by EPA in writing that testing in Condition (1)(A) may be replaced by Condition (1)(B) (to the extent directed by EPA). |
| | | (A) *Initial Verification Testing:* During at least the first eight weeks of operation of the full-scale treatment system, BSC must collect and analyze weekly composites representative of the stabilized waste. Weekly composites must be composed of representative grab samples collected from every batch during each week of stabilization. The composite samples must be collected and analyzed, prior to the disposal of the stabilized filter cake, for all constituents listed in Condition (3). BSC must report the analytical test data, including a record of the ratios of lime kiln dust and fly ash used and quality control information, obtained during this initial period no later than 60 days after the collection of the last composite of stabilized filter cake. |
| | | (B) *Subsequent Verification Testing:* Following written notification by EPA, BSC may substitute the testing condition in (1)(B) for (1)(A). BSC must collect and analyze at least one composite representative of the stabilized filter cake generated each month. Monthly composites must be comprised of representative samples collected from all batches that are stabilized in a one-month period. The monthly samples must be analyzed prior to the disposal of the stabilized filter cake for chromium, lead and nickel. BSC may, at its discretion, analyze composite samples more frequently to demonstrate that smaller batches of waste are non-hazardous. |
| | | (C) *Annual Verification Testing:* In order to confirm that the characteristics of the treated waste do not change significantly, BSC must, on an annual basis, analyze a representative composite sample of stabilized filter cake for all TC constituents listed in 40 CFR §261.24 using the method specified therein. This composite sample must represent the stabilized filter cake generated over one week. |

179

TABLE 1—WASTES EXCLUDED FROM NON-SPECIFIC SOURCES—Continued

| Facility | Address | Waste description |
|---|---|---|
| | | (2) *Waste Holding and Handling:* BSC must store, as hazardous, all stabilized filter cake generated until verification testing (as specified in Conditions (1)(A) and (1)(B)) is completed and valid analyses demonstrate that the delisting levels set forth in Condition (3) are met. If the levels of hazardous constituents measured in the samples of stabilized filter cake generated are below all the levels set forth in Condition (3), then the stabilized filter cake is non-hazardous and may be managed and disposed of in accordance with all applicable solid waste regulations. If hazardous constituent levels in any weekly or monthly composite sample equal or exceed any of the delisting levels set in Condition (3), the stabilized filter cake generated during the time period corresponding to this sample must be retreated until it is below these levels or managed and disposed of in accordance with Subtitle C of RCRA. |
| | | (3) *Delisting Levels:* All concentrations must be measured in the waste leachate by the method specified in 40 CFR §261.24. The leachable concentrations for the constituents must be below the following levels (ppm): arsenic—4.8; barium—100; cadmium—0.48; chromium—5.0; lead—1.4; mercury—0.19; nickel—9.6; selenium—1.0; silver—5.0. |
| | | (4) *Changes in Operating Conditions:* After completing the initial verification test period in Condition (1)(A), if BSC decides to significantly change the stabilization process (e.g., stabilization reagents) developed under Condition (1), then BSC must notify EPA in writing prior to instituting the change. After written approval by EPA, BSC may manage waste generated from the changed process as non-hazardous under this exclusion, provided the other conditions of this exclusion are fulfilled. |
| | | (5) *Data Submittals:* Two weeks prior to system start-up, BSC must notify in writing (see address below) when stabilization of the dewatered filter cake will begin. The data obtained through Condition (1)(A) must be submitted to Waste and Chemicals Management Division (Mail Code 3HW11), U.S. EPA Region III, 1650 Arch St., Philadelphia, PA 19103 within the time period specified. The analytical data, including quality control information and records of ratios of lime kiln dust and fly ash used, must be compiled and maintained on site for a minimum of five years. These data must be furnished upon request and made available for inspection by EPA or the State of Maryland. Failure to submit the required data within the specified time period or maintain the required records on site for the specified time will be considered by the Agency, at its discretion, sufficient basis to revoke the exclusion to the extent directed by EPA. All data must be accompanied by a signed copy of the following certification statement to attest to the truth and accuracy of the data submitted: |
| | | "Under civil and criminal penalty of law for the making or submission of false or fraudulent statements or representations (pursuant to the applicable provisions of the Federal Code, which include, but may not be limited to, 18 U.S.C §1001 and 42 U.S.C §6928), I certify that the information contained in or accompanying this document is true, accurate and complete. |
| | | As to the (those) identified section(s) of this document for which I cannot personally verify its (their) truth and accuracy, I certify as the company official having supervisory responsibility for the persons who, acting under my direct instructions, made the verification that this information is true, accurate and complete. |
| | | In the event that any of this information is determined by EPA in its sole discretion to be false, inaccurate or incomplete, and upon conveyance of this fact to the company, I recognize and agree that this exclusion of waste will be void as if it never had effect or to the extent directed by EPA and that the company will be liable for any actions taken in contravention of the company's RCRA and CERCLA obligations premised upon the company's reliance on the void exclusion." |
| BMW Manufacturing Co., LLC. | Greer, South Carolina. | Wastewater treatment sludge (EPA Hazardous Waste No. F019) that BMW Manufacturing Corporation (BMW) generates by treating wastewater from automobile assembly plant located on Highway 101 South in Greer, South Carolina. This is a conditional exclusion for up to 2,850 cubic yards of waste (hereinafter referred to as "BMW Sludge") that will be generated each year and disposed in a Subtitle D landfill after August 31, 2005. With prior approval by the EPA, following a public comment period, BMW may also beneficially reuse the sludge. BMW must demonstrate that the following conditions are met for the exclusion to be valid. |
| | | (1) Delisting Levels: All leachable concentrations for these metals and cyanide must not exceed the following levels (ppm): Barium-100; Cadmium-1; Chromium-5; Cyanide-33.6, Lead-5; and Nickel-70.3. These metal and cyanide concentrations must be measured in the waste leachate obtained by the method specified in 40 CFR 261.24, except that for cyanide, deionized water must be the leaching medium. Cyanide concentrations in waste or leachate must be measured by the method specified in 40 CFR 268.40, Note 7. |

TABLE 1—WASTES EXCLUDED FROM NON-SPECIFIC SOURCES—Continued

| Facility | Address | Waste description |
|---|---|---|
| | | (2) Annual Verification Testing Requirements: Sample collection and analyses, including quality control procedures, must be performed using appropriate methods. As applicable to the method-defined parameters of concern, analyses requiring the use of SW–846 methods incorporated by reference in 40 CFR 260.11 must be used without substitution. As applicable, the SW–846 methods might include Methods 0010, 0011, 0020, 0023A, 0030, 0031, 0040, 0050, 0051, 0060, 0061, 1010A, 1020B, 1110A, 1310B, 1311, 1312, 1320, 1330A, 9010C, 9012B, 9040C, 9045D, 9060A, 9070A, (uses EPA Method 1664, Rev. A), 9071B, and 9095B. Methods must meet Performance Based Measurement System Criteria in which the Data Quality Objectives are to demonstrate that representative samples of the BMW Sludge meet the delisting levels in Condition (1). (A) Annual Verification Testing: BMW must implement an annual testing program to demonstrate that constituent concentrations measured in the TCLP extract do not exceed the delisting levels established in Condition (1). |
| | | (3) Waste Holding and Handling: BMW must hold sludge containers utilized for verification sampling until composite sample results are obtained. If the levels of constituents measured in the composite samples of BMW Sludge do not exceed the levels set forth in Condition (1), then the BMW Sludge is non-hazardous and must be managed in accordance with all applicable solid waste regulations. If constituent levels in a composite sample exceed any of the delisting levels set forth in Condition (1), the batch of BMW Sludge generated during the time period corresponding to this sample must be managed and disposed of in accordance with Subtitle C of RCRA. |
| | | (4) Changes in Operating Conditions: BMW must notify EPA in writing when significant changes in the manufacturing or wastewater treatment processes are implemented. EPA will determine whether these changes will result in additional constituents of concern. If so, EPA will notify BMW in writing that the BMW Sludge must be managed as hazardous waste F019 until BMW has demonstrated that the wastes meet the delisting levels set forth in Condition (1) and any levels established by EPA for the additional constituents of concern, and BMW has received written approval from EPA. If EPA determines that the changes do not result in additional constituents of concern, EPA will notify BMW, in writing, that BMW must verify that the BMW Sludge continues to meet Condition (1) delisting levels. |
| | | (5) Data Retention: Records of analytical data from Condition (2) must be compiled, summarized, and maintained by BMW for a minimum of three years, and must be furnished upon request by EPA or the State of South Carolina, and made available for inspection. Failure to maintain the required records for the specified time will be considered by EPA, at its discretion, sufficient basis to revoke the exclusion to the extent directed by EPA. All data must be accompanied by a signed copy of the certification statement in 40 CFR 260.22(i)(12). |
| | | (6) Reopener Language: (A) If, at any time after disposal of the delisted waste, BMW possesses or is otherwise made aware of any environmental data (including but not limited to leachate data or groundwater monitoring data) or any other data relevant to the delisted waste indicating that any constituent identified in the delisting verification testing is at a level higher than the delisting level allowed by EPA in granting the petition, BMW must report the data, in writing, to EPA and South Carolina within 10 days of first possessing or being made aware of that data. (B) If the testing of the waste, as required by Condition (2)(A), does not meet the delisting requirements of Condition (1), BMW must report the data, in writing, to EPA and South Carolina within 10 days of first possessing or being made aware of that data. (C) Based on the information described in paragraphs (6)(A) or (6)(B) and any other information received from any source, EPA will make a preliminary determination as to whether the reported information requires that EPA take action to protect human health or the environment. Further action may include suspending or revoking the exclusion, or other appropriate response necessary to protect human health and the environment. (D) If EPA determines that the reported information does require Agency action, EPA will notify the facility in writing of the action believed necessary to protect human health and the environment. The notice shall include a statement of the proposed action and a statement providing BMW with an opportunity to present information as to why the proposed action is not necessary. BMW shall have 10 days from the date of EPA's notice to present such information. (E) Following the receipt of information from BMW, as described in paragraph (6)(D), or if no such information is received within 10 days, EPA will issue a final written determination describing the Agency actions that are necessary to protect human health or the environment, given the information received in accordance with paragraphs (6)(A) or (6)(B). Any required action described in EPA's determination shall become effective immediately, unless EPA provides otherwise. |
| | | (7) Notification Requirements: BMW must provide a one-time written notification to any State Regulatory Agency in a State to which or through which the delisted waste described above will be transported, at least 60 days prior to the commencement of such activities. Failure to provide such a notification will result in a violation of the delisting conditions and a possible revocation of the decision to delist. |
| Boeing Commercial Airplane Co.. | Auburn, Washington. | Residually contaminated soils in an inactive sludge pile containment area on March 27, 1990, previously used to store wastewater treatment sludges generated from electroplating operations (EPA Hazardous Waste No. F006). |

TABLE 1—WASTES EXCLUDED FROM NON-SPECIFIC SOURCES—Continued

| Facility | Address | Waste description |
|---|---|---|
| Bommer Industries Inc.. | Landrum, SC | Wastewater treatment sludges (EPA Hazardous Waste No. F006) generated from their electroplating operations and contained in evaporation ponds #1 and #2 on August 12, 1987. |
| BWX] Technologies. | Lynchburg, VA | Wastewater treatment sludge from electroplating operations (EPA Hazardous Waste No. F006) generated at a maximum annual rate of 500 cubic yards per year, after January 14, 2000, and disposed of in a Subtitle D landfill. BWX Technologies must meet the following conditions for the exclusion to be valid: |
| | | (1) Delisting Levels: All leachable concentrations for the following constituents measure using the SW–846 method 1311 (the TCLP) must not exceed the following levels (mg/l). (a) Inorganic constituents—Antimony-0.6; Arsenic-5.0; Barium-100; Beryllium-0.4; Cadmium-0.5; Chromium-5.0; Cobalt-210; Copper-130; Lead-1.5; Mercury-0.2; Nickel-70; Silver-5.0; Thallium-0.2; Tin-2100; Zinc-1000; Fluoride-400. (b) Organic constituents—Acetone-400; Methylene Chloride-0.5. |
| | | (2) Verification testing schedule: BWX Technologies must analyze a representative sample of the filter cake from the pickle acid treatment system on an annual, calendar year basis using methods with appropriate detection levels and quality control procedures. If the level of any constituent measured in the sample of filter cake exceeds the levels set forth in Paragraph 1, then the waste is hazardous and must be managed in accordance with Subtitle C of RCRA. Data from the annual verification testing must be submitted to EPA within 60 days of the sampling event. |
| | | (3) Changes in Operating Conditions: If BWX Technologies significantly changes the manufacturing or treatment process described in the petition, or the chemicals used in the manufacturing or treatment process, BWX Technologies may not manage the filter cake generated from the new process under this exclusion until it has met the following conditions: (a) BWX Technologies must demonstrate that the waste meets the delisting levels set forth in Paragraph 1; (b) it must demonstrate that no new hazardous constituents listed in appendix VIII of part 261 have been introduced into the manufacturing or treatment process: and (c) it must obtain prior written approval from EPA to manage the waste under this exclusion. |
| | | (4) Data Submittals: The data obtained under Paragraphs 2 and 3 must be submitted to The Waste and Chemicals Management Division, U.S. EPA Region III, 1650 Arch Street, Philadelphia, PA 19103. Records of operating conditions and analytical data must be compiled, summarized, and maintained on site for a minimum of five years and must be furnished upon request by EPA or the Commonwealth of Virginia, and made available for inspection. Failure to submit the required data within the specified time period or to maintain the required records on site for the specified time period will be considered by EPA, at its discretion, sufficient basis to revoke the exclusion to the extent determined necessary by EPA. All data must be accompanied by a signed copy of the certification statement set forth in 40 CFR 260.22(i)(12) to attest to the truth and accuracy of the data submitted. |
| | | (5) Reopener: |
| | | (a) If BWX Technologies discovers that a condition at the facility or an assumption related to the disposal of the excluded waste that was modeled or predicted in the petition does not occur as modeled or predicted, then BWX Technologies must report any information relevant to that condition, in writing, to the Regional Administrator or his delegate within 10 days of discovering that condition. |
| | | (b) Upon receiving information described in paragraph (a) of this section, regardless of its source, the Regional Administrator or his delegate will determine whether the reported condition requires further action. Further action may include repealing the exclusion, modifying the exclusion, or other appropriate response necessary to protect human health and the environment. |
| | | (6) Notification Requirements: BWX Technologies must provide a one-time written notification to any State Regulatory Agency to which or through which the delisted waste described above will be transported for disposal at least 60 days prior to the commencement of such activities. Failure to provide such a notification will be deemed to be a violation of this exclusion and may result in a revocation of the decision. |
| Capitol Products Corp.. | Harrisburg, PA | Dewatered wastewater treatment sludges (EPA Hazardous Waste No. FO19) generated from the chemical conversion coating of aluminum after September 12, 1986. |
| Capitol Products Corporation. | Kentland, IN ... | Dewatered wastewater treatment sludges (EPA Hazardous Waste No. FO19) generated from the chemical conversion coating of aluminum after November 17, 1986. |
| Care Free Aluminum Products, Inc.. | Charlotte, Michigan. | Wastewater treatment sludge (EPA Hazardous Waste No. F019) generated from the chemical conversion coating of aluminum (generated at a maximum annual rate of 100 cubic yards), after August 21, 1992. In order to confirm that the characteristics of the waste do not change significantly, the facility must, on an annual basis, analyze a representative composite sample for the constituents listed in § 261.24 using the method specified therein. The annual analytical results, including quality control information, must be compiled, certified according to § 260.22(i)(12), maintained on-site for a minimum of five years, and made available for inspection upon request by any employee or representative of EPA or the State of Michigan. Failure to maintain the required records on-site will be considered by EPA, at its discretion, sufficient basis to revoke the exclusion to the extent directed by EPA. |

TABLE 1—WASTES EXCLUDED FROM NON-SPECIFIC SOURCES—Continued

| Facility | Address | Waste description |
|---|---|---|
| Chamberlian-Featherlite, Inc.. | Hot Springs, AR. | Dewatered wastewater treatment sludges (EPA Hazardous Waste No. F019) generated from the chemical conversion coating of aluminum after July 16, 1986. |
| Chrysler Group LLC at the Old Carco LLC Sterling Heights Assembly Plant. | Sterling Heights, Michigan. | Wastewater treatment sludges, F019, that are generated at Old Carco LLC's Sterling Heights Assembly Plant, (SHAP), Sterling Heights, Michigan by Chrysler Group LLC at a maximum annual rate of 3,000 cubic yards per year. The sludges must be disposed of in a lined landfill with leachate collection which is licensed, permitted, or otherwise authorized to accept the delisted wastewater treatment sludges in accordance with 40 CFR part 258. The exclusion becomes effective as of November 6, 2009.<br><br>1. *Delisting Levels:* The concentrations in a leachate extract of the waste measured in any sample must not exceed the following levels (mg/L): arsenic—0.22; nickel—67.8; benzene—0.057; hexachlorobenzene—0.0000724; naphthalene—0.00822; and pentachlorophenol—0.00607.<br><br>2. *Quarterly Verification Testing:* To verify that the waste does not exceed the specified delisting levels, Chrysler Group LLC or Old Carco LLC must collect and analyze one representative sample of the waste on a quarterly basis. Sample collection and analyses, including quality control procedures, must be performed using appropriate methods. SW–846 Method 1311 must be used for generation of the leachate extract used in the testing of the delisting levels if oil and grease comprise less than 1% of the waste. SW–846 Method 1330A must be used for generation of the leaching extract if oil and grease comprise 1% or more of the waste. SW–846 Method 9071B must be used for determination of oil and grease. SW–846 Methods 1311, 1330A, and 9071B are incorporated by reference in 40 CFR 260.11.<br><br>3. *Changes in Operating Conditions:* Chrysler Group LLC or Old Carco LLC must notify the EPA in writing if the manufacturing process, the chemicals used in the manufacturing process, the treatment process, or the chemicals used in the treatment process change significantly. Chrysler Group LLC or Old Carco LLC must handle wastes generated after the process change as hazardous until it has demonstrated that the wastes continue to meet the delisting levels and that no new hazardous constituents listed in Appendix VIII of part 261 have been introduced and it has received written approval from EPA.<br><br>4. *Data Submittals:* Chrysler Group LLC or Old Carco LLC must submit the data obtained through verification testing or as required by other conditions of this rule to both U.S. EPA Region 5, 77 W. Jackson Blvd., Chicago, IL 60604 and MDEQ, Waste and Hazardous Materials Division, Hazardous Waste Section, at P.O. Box 30241, Lansing, Michigan 48909. The quarterly verification data and certification of proper disposal must be submitted annually upon the anniversary of the effective date of this exclusion. Chrysler Group LLC or Old Carco LLC must compile, summarize, and maintain on site for a minimum of five years records of operating conditions and analytical data. Chrysler Group LLC or Old Carco LLC must make these records available for inspection. A signed copy of the certification statement in 40 CFR 260.22(i)(12) must accompany all data.<br><br>5. *Reopener Language*—(a) If, anytime after disposal of the delisted waste Chrysler Group LLC or Old Carco LLC possesses or is otherwise made aware of any data (including but not limited to leachate data or groundwater monitoring data) relevant to the delisted waste indicating that any constituent is at a level in the leachate higher than the specified delisting level, or is in the groundwater at a concentration higher than the maximum allowable groundwater concentration in paragraph (e), then Chrysler Group LLC or Old Carco LLC must report such data, in writing, to the Regional Administrator within 10 days of first possessing or being made aware of that data.<br><br>(b) Based on the information described in paragraph (a) and any other information received from any source, the Regional Administrator will make a preliminary determination as to whether the reported information requires Agency action to protect human health or the environment. Further action may include suspending, or revoking the exclusion, or other appropriate response necessary to protect human health and the environment.<br><br>(c) If the Regional Administrator determines that the reported information does require Agency action, the Regional Administrator will inform Chrysler Group LLC or Old Carco LLC in writing of the actions the Regional Administrator believes are necessary to protect human health and the environment. The notice shall include a statement of the proposed action and a statement providing Chrysler Group LLC or Old Carco LLC with an opportunity to present information as to why the proposed Agency action is not necessary or to suggest an alternative action. Chrysler Group LLC or Old Carco LLC shall have 30 days from the date of the Regional Administrator's notice to present the information.<br><br>(d) If after 30 days Chrysler Group LLC or Old Carco LLC presents no further information, the Regional Administrator will issue a final written determination describing the Agency actions that are necessary to protect human health or the environment. Any required action described in the Regional Administrator's determination shall become effective immediately, unless the Regional Administrator provides otherwise.<br><br>(e) Maximum Allowable Groundwater Concentrations (µg/L): arsenic—4.87; nickel—750; benzene—2.5; hexachlorobenzene—0.00168; naphthalene—245; and pentachlorophenol—0.071. |

TABLE 1—WASTES EXCLUDED FROM NON-SPECIFIC SOURCES—Continued

| Facility | Address | Waste description |
|---|---|---|
| Cincinnati Metropolitan Sewer District. | Cincinnati, OH | Sluiced bottom ash (approximately 25,000 cubic yards) contained in the South Lagoon, on September 13, 1985 which contains EPA Hazardous Waste Nos. F001, F002, F003, F004, and F005. |
| Clay Equipment Corporation. | Cedar Falls, Iowa. | Dewatered wastewater treatment sludges (EPA Hazardous Waste No. F006) and spent cyanide bath solutions (EPA Hazardous Waste No. F009) generated from electroplating operations and disposed of in an on-site surface impoundment. This is a onetime exclusion. This exclusion was published on August 1, 1989. |
| Continental Can Co.. | Olympia, WA | Dewatered wastewater treatment sludges (DPA Hazardous Waste No. FO19) generated from the chemical conversion coating of aluminum after September 12, 1986. |
| Cooper Crouse-Hinds. | Amarillo, TX ... | Wastewater Treatment Sludge (Hazardous Waste No. F006) generated at a maximum annual rate of 816 cubic yards per calendar year after April 15, 2009 and disposed in Subtitle D Landfill. |
|  |  | For the exclusion to be valid, Cooper Crouse-Hinds must implement a verification testing program that meets the following Paragraphs: |
|  |  | (1) Delisting Levels: All concentrations for those constituents must not exceed the maximum allowable concentrations in mg/l specified in this paragraph. |
|  |  | WWTP Sludge Leachable Concentrations (mg/l): |
|  |  | (i) Inorganic Constituents: |
|  |  | Arsenic-0.0759; Barium-100; Cadmium-0.819; Copper-216; Iron-1.24; Manganese-145; Nickel-119; Zinc-18. |
|  |  | (ii) Organic Constituents: |
|  |  | Benzene-0.5. |
|  |  | (2) Waste Holding and Handling: |
|  |  | (A) Waste classification as non-hazardous can not begin until compliance with the limits set in paragraph (1) for WWTP sludge has occurred for two consecutive quarterly sampling events. |
|  |  | (B) If constituent levels in any sample taken by Cooper Crouse-Hinds exceed any of the delisting levels set in paragraph (1) for the WWTP sludge, Cooper Crouse-Hinds must do the following: |
|  |  | (i) Notify EPA in accordance with paragraph (6) and |
|  |  | (ii) Manage and dispose WWTP sludge as hazardous waste generated under Subtitle C of RCRA. |
|  |  | (3) Testing Requirements: |
|  |  | Upon this exclusion becoming final, Cooper Crouse-Hinds may perform quarterly analytical testing by sampling and analyzing the WWTP sludge as follows: |
|  |  | (A) Quarterly Testing: |
|  |  | (i) Collect two representative composite samples of the sludge at quarterly intervals after EPA grants the final exclusion. The first composite samples may be taken at any time after EPA grants the final approval. Sampling must be performed in accordance with the sampling plan approved by EPA in support of the exclusion. |
|  |  | (ii) Analyze the samples for all constituents listed in paragraph (1). Any composite sample taken that exceeds the delisting levels listed in paragraph (1) for the sludge must be disposed as hazardous waste in accordance with the applicable hazardous waste requirements. |
|  |  | (iii) Within thirty (30) days after taking its first quarterly sample, Cooper Crouse-Hinds will report its first quarterly analytical test data to EPA. If levels of constituents measured in the samples of the sludge do not exceed the levels set forth in paragraph (1) of this exclusion for two consecutive quarters, Cooper Crouse-Hinds can manage and dispose the non-hazardous WWTP sludge according to all applicable solid waste regulations. |
|  |  | (B) Annual Testing: |
|  |  | (i) If Cooper Crouse-Hinds completes the quarterly testing specified in paragraph (3) above and no sample contains a constituent at a level which exceeds the limits set forth in paragraph (1), Cooper Crouse-Hinds may begin annual testing as follows: Cooper Crouse-Hinds must test two representative composite samples of the WWTP sludge for all constituents listed in paragraph (1) at least once per calendar year. |
|  |  | (ii) The samples for the annual testing shall be a representative composite sample according to appropriate methods. As applicable to the method-defined parameters of concern, analyses requiring the use of SW–846 methods incorporated by reference in 40 CFR 260.11 must be used without substitution. As applicable, the SW–846 methods might include Methods 0010, 0011, 0020, 0023A, 0030, 0031, 0040, 0050, 0051, 0060, 0061, 1010A, 1020B,1110A, 1310B, 1311, 1312, 1320, 1330A, 9010C, 9012B, 9040C, 9045D, 9060A, 9070A (uses EPA Method 1664, Rev. A), 9071B, and 9095B. Methods must meet Performance Based Measurement System Criteria in which the Data Quality Objectives are to demonstrate that samples of the WWTP sludge is representative for all constituents listed in paragraph (1). |
|  |  | (iii) The samples for the annual testing taken for the second and subsequent annual testing events shall be taken within the same calendar month as the first annual sample taken. |
|  |  | (iv) The annual testing report should include the total amount of delisted waste in cubic yards disposed as non-hazardous waste during the calendar year. |

184

TABLE 1—WASTES EXCLUDED FROM NON-SPECIFIC SOURCES—Continued

| Facility | Address | Waste description |
|---|---|---|
| | | (4) Changes in Operating Conditions: If Cooper Crouse-Hinds significantly changes the process described in its petition or starts any processes that generate(s) the waste that may or could affect the composition or type of waste generated (by illustration, but not limitation, changes in equipment or operating conditions of the treatment process), it must notify EPA in writing and it may no longer handle the wastes generated from the new process as non-hazardous until the wastes meet the delisting levels set in paragraph (1) and it has received written approval to do so from EPA. |
| | | Cooper Crouse-Hinds must submit a modification to the petition, complete with full sampling and analysis, for circumstances where the waste volume changes and/or additional waste codes are added to the waste stream, if it wishes to dispose of the material as non-hazardous. |
| | | (5) Data Submittals: |
| | | Cooper Crouse-Hinds must submit the information described below. If Cooper Crouse-Hinds fails to submit the required data within the specified time or maintain the required records on-site for the specified time, EPA, at its discretion, will consider this sufficient basis to reopen the exclusion as described in paragraph (6). Cooper Crouse-Hinds must: |
| | | (A) Submit the data obtained through paragraph (3) to the Chief, Corrective Action and Waste Minimization Section, Multimedia Planning and Permitting Division, U. S. Environmental Protection Agency Region 6, 1445 Ross Ave., Dallas, Texas, 75202, within the time specified. All supporting data can be submitted on CD–ROM or comparable electronic media. |
| | | (B) Compile records of analytical data from paragraph (3), summarized, and maintained on-site for a minimum of five years. |
| | | (C) Furnish these records and data when either EPA or the State of Texas requests them for inspection. |
| | | (D) Send along with all data a signed copy of the following certification statement, to attest to the truth and accuracy of the data submitted: |
| | | "Under civil and criminal penalty of law for the making or submission of false or fraudulent statements or representations (pursuant to the applicable provisions of the Federal Code, which include, but may not be limited to, 18 U.S.C. 1001 and 42 U.S.C. 6928), I certify that the information contained in or accompanying this document is true, accurate and complete. |
| | | "As to the (those) identified section(s) of this document for which I cannot personally verify its (their) truth and accuracy, I certify as the company official having supervisory responsibility for the persons who, acting under my direct instructions, made the verification that this information is true, accurate and complete. |
| | | "If any of this information is determined by EPA in its sole discretion to be false, inaccurate or incomplete, and upon conveyance of this fact to the company, I recognize and agree that this exclusion of waste will be void as if it never had effect or to the extent directed by EPA and that the company will be liable for any actions taken in contravention of the company's RCRA and CERCLA obligations premised upon the company's reliance on the void exclusion." |
| | | (6) Re-opener: |
| | | (A) If, anytime after disposal of the delisted waste Cooper Crouse-Hinds possesses or is otherwise made aware of any environmental data (including but not limited to leachate data or ground water monitoring data) or any other data relevant to the delisted waste indicating that any constituent identified for the delisting verification testing is at level higher than the delisting level allowed by the Division Director in granting the petition, then the facility must report the data, in writing, to the Division Director within 10 days of first possessing or being made aware of that data. |
| | | (B) If either the quarterly or annual testing of the waste does not meet the delisting requirements in paragraph (1), Cooper Crouse-Hinds must report the data, in writing, to the Division Director within 10 days of first possessing or being made aware of that data. |
| | | (C) If Cooper Crouse-Hinds fails to submit the information described in paragraphs (5), (6)(A) or (6)(B) or if any other information is received from any source, the Division Director will make a preliminary determination as to whether the reported information requires EPA action to protect human health and/or the environment. Further action may include suspending, or revoking the exclusion, or other appropriate response necessary to protect human health and the environment. |
| | | (D) If the Division Director determines that the reported information requires action by EPA, the Division Director will notify the facility in writing of the actions the Division Director believes are necessary to protect human health and the environment. The notice shall include a statement of the proposed action and a statement providing the facility with an opportunity to present information as to why the proposed EPA action is not necessary. The facility shall have 10 days from the date of the Division Director's notice to present such information. |
| | | (E) Following the receipt of information from the facility described in paragraph (6)(D) or (if no information is presented under paragraph (6)(D)) the initial receipt of information described in paragraphs (5), (6)(A) or (6)(B), the Division Director will issue a final written determination describing EPA actions that are necessary to protect human health and/or the environment. Any required action described in the Division Director's determination shall become effective immediately, unless the Division Director provides otherwise. |

TABLE 1—WASTES EXCLUDED FROM NON-SPECIFIC SOURCES—Continued

| Facility | Address | Waste description |
|---|---|---|
| | | (7) Notification Requirements: |
| | | Cooper Crouse-Hinds must do the following before transporting the delisted waste. Failure to provide this notification will result in a violation of the delisting petition and a possible revocation of the decision. |
| | | (A) Provide a one-time written notification to any state Regulatory Agency to which or through which it will transport the delisted waste described above for disposal, 60 days before beginning such activities. |
| | | (B) Update the one-time written notification if it ships the delisted waste into a different disposal facility. |
| | | (C) Failure to provide this notification will result in a violation of the delisting variance and a possible revocation of the decision. |
| DaimlerChrysler Corporation. | Jefferson North Assembly Plant, Detroit, Michigan. | Waste water treatment plant sludge, F019, that is generated by DaimlerChrysler Corporation at the Jefferson North Assembly Plant (DCC-JNAP) at a maximum annual rate of 2,000 cubic yards per year. The sludge must be disposed of in a lined landfill with leachate collection, which is licensed, permitted, or otherwise authorized to accept the delisted wastewater treatment sludge in accordance with 40 CFR part 258. The exclusion becomes effective as of February 26, 2004. |
| | | 1. *Delisting Levels:* (A) The concentrations in a TCLP extract of the waste measured in any sample may not exceed the following levels (mg/L): Antimony—0.659; Arsenic—0.3; Cadmium—0.48; Chromium—4.95; Lead—5; Nickel—90.5; Selenium—1; Thallium—0.282; Tin—721; Zinc—898; Acetone—228; p-Cresol—11.4; Formaldehyde—84.2; and Methylene chloride—0.288. (B) The total concentrations measured in any sample may not exceed the following levels (mg/kg): Mercury—8.92; and Formaldehyde—689. (C) The sum of the ratios of the TCLP concentrations to the delisting levels for nickel and either thallium or cadmium shall not exceed 1.0. |
| | | 2. *Quarterly Verification Testing:* To verify that the waste does not exceed the specified delisting levels, DCC-JNAP must collect and analyze one representative sample of the waste on a quarterly basis. |
| | | 3. *Changes in Operating Conditions:* DCC-JNAP must notify the EPA in writing if the manufacturing process, the chemicals used in the manufacturing process, the treatment process, or the chemicals used in the treatment process significantly change. DCC-JNAP must handle wastes generated after the process change as hazardous until it has demonstrated that the wastes continue to meet the delisting levels and that no new hazardous constituents listed in appendix VIII of part 261 have been introduced and it has received written approval from EPA. |
| | | 4. *Data Submittals:* DCC-JNAP must submit the data obtained through verification testing or as required by other conditions of this rule to both U.S. EPA Region 5, Waste Management Branch (DW–8J), 77 W. Jackson Blvd., Chicago, IL 60604 and MDEQ, Waste Management Division, Hazardous Waste Program Section, at P.O. Box 30241, Lansing, Michigan 48909. The quarterly verification data and certification of proper disposal must be submitted annually upon the anniversary of the effective date of this exclusion. The facility must compile, summarize, and maintain on site for a minimum of five years records of operating conditions and analytical data. The facility must make these records available for inspection. All data must be accompanied by a signed copy of the certification statement in 40 CFR 260.22(i)(12). |
| | | 5. *Reopener Language*—(a) If, anytime after disposal of the delisted waste, DCC-JNAP possesses or is otherwise made aware of any data (including but not limited to leachate data or groundwater monitoring data) relevant to the delisted waste indicating that any constituent is at a level in the leachate higher than the specified delisting level, or is in the groundwater at a concentration higher than the maximum allowable groundwater concentration in paragraph (e), then DCC-JNAP must report such data, in writing, to the Regional Administrator within 10 days of first possessing or being made aware of that data. |
| | | (b) Based on the information described in paragraph (a) and any other information received from any source, the Regional Administrator will make a preliminary determination as to whether the reported information requires Agency action to protect human health or the environment. Further action may include suspending, or revoking the exclusion, or other appropriate response necessary to protect human health and the environment. |
| | | (c) If the Regional Administrator determines that the reported information does require Agency action, the Regional Administrator will notify DCC-JNAP in writing of the actions the Regional Administrator believes are necessary to protect human health and the environment. The notice shall include a statement of the proposed action and a statement providing DCC-JNAP with an opportunity to present information as to why the proposed Agency action is not necessary or to suggest an alternative action. DCC-JNAP shall have 30 days from the date of the Regional Administrator's notice to present the information. |
| | | (d) If after 30 days the facility presents no further information, the Regional Administrator will issue a final written determination describing the Agency actions that are necessary to protect human health or the environment. Any required action described in the Regional Administrator's determination shall become effective immediately, unless the Regional Administrator provides otherwise. |

TABLE 1—WASTES EXCLUDED FROM NON-SPECIFIC SOURCES—Continued

| Facility | Address | Waste description |
|---|---|---|
| | | (e) *Maximum Allowable Groundwater Concentrations* (µg/L): Antimony—6; Arsenic—4.87; Cadmium—5; Chromium—100; Lead—15; Nickel—750; Selenium—50; Thallium—2; Tin—22,500; Zinc—11,300; acetone—3,750; p-Cresol—188; Formaldehyde—1,380; and Methylene chloride—5. |
| Dover Corp., Norris Div.. DuraTherm, Incorporated. | Tulsa, OK ...... San Leon, Texas. | Dewatered wastewater treatment sludge (EPA Hazardous Waste No. FO06) generated from their electroplating operations after April 29, 1986. |
| | | Desorber solids, (at a maximum generation of 20,000 cubic yards per calendar year) generated by DuraTherm using the thermal desorption treatment process, (EPA Hazardous Waste No. F037 and F038) and that is disposed of in subtitle D landfills after April 24, 2000. |
| | | For the exclusion to be valid, DuraTherm must implement a testing program that meets the following Paragraphs: |
| | | (1) *Delisting Levels:* All leachable concentrations for those constituents must not exceed the following levels (ppm). The petitioner must use an acceptable leaching method, for example SW–846, Method 1311 to measure constituents in the waste leachate. |
| | | Desorber solids (i) Inorganic Constituents Arsenic—1.35; Antimony—0.162; Barium—54.0; Beryllium—0.108; Cadmium—0.135; Chromium—0.6; Lead—0.405; Nickel—2.7; Selenium—1.0; Silver—5.0; Vanadium—5.4; Zinc—270. |
| | | (ii) Organic Constituents Anthracene—0.28; Benzene—0.135; Benzo(a) anthracene—0.059; Benzo(b)fluoranthene—0.11; Benzo(a)pyrene—0.061; Bis-ethylhexylphthalate—0.28; Carbon Disulfide—3.8; Chlorobenzene—0.057; Chrysene—0.059; o,m,p Cresols—54; Dibenzo (a,h) anthracene—0.055; 2,4 Dimethyl phenol—18.9; Dioctyl phthalate—0.017; Ethylbenzene—0.057; Fluoranthene—0.068; Fluorene—0.059; Naphthalene—0.059; Phenanthrene—0.059; Phenol—6.2; Pyrene—0.067; Styrene—2.7; Trichloroethylene—0.054; Toluene—0.08; Xylene—0.032 |
| | | (2) *Waste Holding and Handling:* (A) DuraTherm must store the desorber solids as described in its RCRA permit, or continue to dispose of as hazardous all desorber solids generated, until they have completed verification testing described in Paragraph (3)(A) and (B), as appropriate, and valid analyses show that paragraph (1) is satisfied. |
| | | (B) In order to isolate wastes that have been processed in the unit prior to one of the waste codes to be delisted, DuraTherm must designate the first batch of F037, F038, K048, K049, K050, or K051 wastes as hazardous. Subsequent batches of these wastes which satisfy paragraph (1) are eligible for delisting if they meet the criteria in paragraph (1) and no additional constituents (other than those of the delisted waste streams) from the previously processed wastes are detected. |
| | | (C) Levels of constituents measured in the samples of the desorber solids that do not exceed the levels set forth in Paragraph (1) are nonhazardous. DuraTherm can manage and dispose the nonhazardous desorber solids according to all applicable solid waste regulations. |
| | | (D) If constituent levels in a sample exceed any of the delisting levels set in Paragraph (1), DuraTherm must retreat or stabilize the batches of waste used to generate the representative sample until it meets the levels in paragraph (1). DuraTherm must repeat the analyses of the treated waste. |
| | | (E) If the facility has not treated the waste, DuraTherm must manage and dispose the waste generated under subtitle C of RCRA. |
| | | (3) *Verification Testing Requirements:* DuraTherm must perform sample collection and analyses, including quality control procedures, using appropriate methods. As applicable to the method-defined constituents of concern, analyses requiring the use of SW–846 methods incorporated by reference in 40 CFR 260.11 must be used without substitution. As applicable, the SW–846 methods might include Methods 0010, 0011, 0020, 0023A, 0030, 0031, 0040, 0050, 0051, 0060, 0061, 1010A, 1020B, 1110A, 1310B, 1311, 1312, 1320, 1330A, 9010C, 9012B, 9040C, 9045D, 9060A, 9070A (uses EPA Method 1664, Rev. A), 9071B, and 9095B. If EPA judges the process to be effective under the operating conditions used during the initial verification testing, DuraTherm may replace the testing required in Paragraph (3)(A) with the testing required in Paragraph (3)(B). DuraTherm must continue to test as specified in Paragraph (3)(A) until and unless notified by EPA in writing that testing in Paragraph (3)(A) may be replaced by Paragraph (3)(B). |
| | | (A) *Initial Verification Testing:* After EPA grants the final exclusion, DuraTherm must do the following: |
| | | (i) Collect and analyze composites of the desorber solids. |
| | | (ii) Make two composites of representative grab samples collected. |
| | | (iii) Analyze the waste, before disposal, for all of the constituents listed in Paragraph 1. |
| | | (iv) Sixty (60) days after this exclusion becomes final, report the operational and analytical test data, including quality control information. |
| | | (v) Submit the test plan for conducting the multiple pH leaching procedure to EPA for approval at least 10 days before conducting the analysis. |
| | | (vi) Conduct a multiple pH leaching procedure on 10 samples collected during the sixty-day test period. |
| | | (vii) The ten samples should include both non-stabilized and stabilized residual solids. If none of the samples collected during the sixty-day test period need to be stabilized, DuraTherm should provide multiple pH data on the first sample of stabilized wastes generated. |

187

Table 1—Wastes Excluded From Non-Specific Sources—Continued

| Facility | Address | Waste description |
|---|---|---|
| | | (vii) Perform the toxicity characteristic leaching procedure using three different pH extraction fluids to simulate disposal under three conditions and submit the results within 60 days of completion. Simulate an acidic landfill environment, basic landfill environment, and a landfill environment similar to the pH of the waste. |
| | | (B) *Subsequent Verification Testing:* Following written notification by EPA, DuraTherm may substitute the testing conditions in (3)(B) for (3)(A)(i). DuraTherm must continue to monitor operating conditions, and analyze representative samples each quarter of operation during the first year of waste generation. The samples must represent the waste generated in one quarter. DuraTherm must run the multiple pH procedure on these waste samples. |
| | | (C) *Termination of Organic Testing:* (i) DuraTherm must continue testing as required under Paragraph (3)(B) for organic constituents in Paragraph (1)(A)(ii), until the analytical results submitted under Paragraph (3)(B) show a minimum of two consecutive samples below the delisting levels in Paragraph (1)(A)(i), DuraTherm may then request that EPA stop quarterly organic testing. After EPA notifies DuraTherm in writing, the company may end quarterly organic testing. |
| | | (ii) Following cancellation of the quarterly testing, DuraTherm must continue to test a representative composite sample for all constituents listed in Paragraph (1) annually (by twelve months after final exclusion). |
| | | (4) *Changes in Operating Conditions:* If DuraTherm significantly changes the process described in its petition or starts any processes that generate(s) the waste that may or could affect the composition or type of waste generated as established under Paragraph (1) (by illustration, but not limitation, changes in equipment or operating conditions of the treatment process), they must notify EPA in writing; they may no longer handle the wastes generated from the new process as nonhazardous until the wastes meet the delisting levels set in Paragraph (1) and they have received written approval to do so from EPA. |
| | | (5) *Data Submittals:* DuraTherm must submit the information described below. If DuraTherm fails to submit the required data within the specified time or maintain the required records on-site for the specified time, EPA, at its discretion, will consider this sufficient basis to reopen the exclusion as described in Paragraph 6. DuraTherm must: |
| | | (A) Submit the data obtained through Paragraph 3 to Mr. William Gallagher, Chief, Region 6 Delisting Program, EPA, 1445 Ross Avenue, Dallas, Texas 75202-2733, Mail Code, (6PD-O) within the time specified. |
| | | (B) Compile records of operating conditions and analytical data from Paragraph (3), summarized, and maintained on-site for a minimum of five years. |
| | | (C) Furnish these records and data when EPA or the State of Texas request them for inspection. |
| | | (D) Send along with all data a signed copy of the following certification statement, to attest to the truth and accuracy of the data submitted: |
| | | Under civil and criminal penalty of law for the making or submission of false or fraudulent statements or representations (pursuant to the applicable provisions of the Federal Code, which include, but may not be limited to, 18 U.S.C. 1001 and 42 U.S.C. 6928), I certify that the information contained in or accompanying this document is true, accurate and complete. |
| | | As to the (those) identified section(s) of this document for which I cannot personally verify its (their) truth and accuracy, I certify as the company official having supervisory responsibility for the persons who, acting under my direct instructions, made the verification that this information is true, accurate and complete. |
| | | If any of this information is determined by EPA in its sole discretion to be false, inaccurate or incomplete, and upon conveyance of this fact to the company, I recognize and agree that this exclusion of waste will be void as if it never had effect or to the extent directed by EPA and that the company will be liable for any actions taken in contravention of the company's RCRA and CERCLA obligations premised upon the company's reliance on the void exclusion. |
| | | (6) *Reopener Language:* (A) If, anytime after disposal of the delisted waste, DuraTherm possesses or is otherwise made aware of any environmental data (including but not limited to leachate data or groundwater monitoring data) or any other data relevant to the delisted waste indicating that any constituent identified for the delisting verification testing is at level higher than the delisting level allowed by the Regional Administrator or his delegate in granting the petition, then the facility must report the data, in writing, to the Regional Administrator or his delegate within 10 days of first possessing or being made aware of that data. |
| | | (B) If the annual testing of the waste does not meet the delisting requirements in Paragraph 1, DuraTherm must report the data, in writing, to the Regional Administrator or his delegate within 10 days of first possessing or being made aware of that data. |
| | | (C) If DuraTherm fails to submit the information described in paragraphs (5),(6)(A) or (6)(B) or if any other information is received from any source, the Regional Administrator or his delegate will make a preliminary determination as to whether the reported information requires Agency action to protect human health or the environment. Further action may include suspending, or revoking the exclusion, or other appropriate response necessary to protect human health and the environment. |

TABLE 1—WASTES EXCLUDED FROM NON-SPECIFIC SOURCES—Continued

| Facility | Address | Waste description |
|---|---|---|
| | | (D) If the Regional Administrator or his delegate determines that the reported information does require Agency action, the Regional Administrator or his delegate will notify the facility in writing of the actions the Regional Administrator or his delegate believes are necessary to protect human health and the environment. The notice shall include a statement of the proposed action and a statement providing the facility with an opportunity to present information as to why the proposed Agency action is not necessary. The facility shall have 10 days from the date of the Regional Administrator or his delegate's notice to present such information. |
| | | (E) Following the receipt of information from the facility described in paragraph (6)(D) or (if no information is presented under paragraph (6)(D)) the initial receipt of information described in paragraphs (5), (6)(A) or (6)(B), the Regional Administrator or his delegate will issue a final written determination describing the Agency actions that are necessary to protect human health or the environment. Any required action described in the Regional Administrator or his delegate's determination shall become effective immediately, unless the Regional Administrator or his delegate provides otherwise. |
| | | (7) *Notification Requirements:* DuraTherm must do following before transporting the delisted waste: Failure to provide this notification will result in a violation of the delisting petition and a possible revocation of the decision. |
| | | (A) Provide a one-time written notification to any State Regulatory Agency to which or through which they will transport the delisted waste described above for disposal, 60 days before beginning such activities. |
| | | (B) Update the one-time written notification if they ship the delisted waste into a different disposal facility. |
| Eastman Chemical Company. | Longview, Texas. | Wastewater treatment sludge, (at a maximum generation of 82,100 cubic yards per calendar year) generated by Eastman (EPA Hazardous Waste Nos. F001, F002, F003, F005 generated at Eastman when disposed of in a Subtitle D landfill. |
| | | Eastman must implement a testing program that meets the following conditions for the exclusion to be valid: |
| | | (1) *Delisting Levels:* All concentrations for the following constituents must not exceed the following levels (mg/l). For the wastewater treatment sludge constituents must be measured in the waste leachate by the method specified in 40 CFR 261.24. Wastewater treatment sludge: |
| | | (i) Inorganic Constituents: Antimony-0.0515; Barium-7.30; Cobalt-2.25; Chromium-5.0; Lead-5.0; Mercury-0.0015; Nickel-2.83; Selenium-0.22; Silver-0.384; Vanadium-2.11; Zinc-28.0 |
| | | (ii) Organic Constituents: Acenaphthene-1.25; Acetone—7.13; bis(2-ethylhexylphthalate—0.28; 2-butanone—42.8; Chloroform—0.0099; Fluorene—0.55; Methanol-35.7; Methylene Chloride—0.486; naphthalene-0.0321. |
| | | (2) *Waste Holding and Handling:* If the concentrations of the sludge exceed the levels provided in Condition 1, then the sludge must be treated in the Fluidized Bed Incinerator (FBI) and meet the requirements of that September 25, 1996 delisting exclusion to be non-hazardous (as FBI ash). If the sludge meets the delisting levels provided in Condition 1, then it's non-hazardous (as sludge). If the waste water treatment sludge is not managed in the manner above, Eastman must manage it in accordance with applicable RCRA Subtitle C requirements. If the levels of constituents measured in the samples of the waste water treatment sludge do not exceed the levels set forth in Condition (1), then the waste is non-hazardous and may be managed and disposed of in accordance with all applicable solid waste regulations. During the verification period, Eastman must manage the waste in the FBI incinerator prior to disposal. |
| | | (3) *Verification Testing Requirements:* Eastman must perform sample collection and analyses, including quality control procedures, using appropriate methods. As applicable to the method-defined parameters of concern, analyses requiring the use of SW–846 methods incorporated by reference in 40 CFR 260.11 must be used without substitution. As applicable, the SW–846 methods might include Methods 0010, 0011, 0020, 0023A, 0030, 0031, 0040, 0050, 0051, 0060, 0061, 1010A, 1020B, 1110A, 1310B, 1311, 1312, 1320, 1330A, 9010C, 9012B, 9040C, 9045D, 9060A, 9070A (uses EPA Method 1664, Rev. A), 9071B, and 9095B. After completion of the initial verification period, Eastman may replace the testing required in Condition (3)(A) with the testing required in Condition (3)(B). Eastman must continue to test as specified in Condition (3)(A) until and unless notified by EPA in writing that testing in Condition (3)(A) may be replaced by Condition (3)(B). |
| | | (A) *Initial Verification Testing:* At quarterly intervals for one year after the final exclusion is granted, Eastman must collect and analyze composites of the wastewater treatment sludge for constituents listed in Condition (1). |
| | | (B) *Subsequent Verification Testing:* Following termination of the quarterly testing, Eastman must continue to test a representative composite sample for all constituents listed in Condition (1) on an annual basis (no later than twelve months after the final exclusion). |

TABLE 1—WASTES EXCLUDED FROM NON-SPECIFIC SOURCES—Continued

| Facility | Address | Waste description |
|---|---|---|
| | | (4) *Changes in Operating Conditions.* If Eastman significantly changes the process which generate(s) the waste(s) and which may or could affect the composition or type of waste(s) generated as established under Condition (1) (by illustration, but not limitation, change in equipment or operating conditions of the treatment process or generation of volumes in excess 82,100 cubic yards of waste annually), Eastman must (A) notify the EPA in writing of the change and (B) may no longer handle or manage the waste generated from the new process as nonhazardous until Eastman has demonstrated through testing the waste meets the delisting levels set in Condition (1) and (C) Eastman has received written approval to begin managing the wastes as non-hazardous from EPA. |
| | | (5) *Data Submittals.* Eastman must submit or maintain, as applicable, the information described below. If Eastman fails to submit the required data within the specified time or maintain the required records on-site for the specified time, EPA, at its discretion, will consider this sufficient basis to reopen the exclusion as described in Condition (6). Eastman must: |
| | | (A) Submit the data obtained through Condition (3) to Mr. William Gallagher, Chief, Region 6 Delisting Program, EPA, 1445 Ross Avenue, Dallas, Texas 75202–2733, Mail Code, (6PD-O) within the time specified. |
| | | (B) Compile records of operating conditions and analytical data from Condition (3), summarized, and maintained on-site for a minimum of five years. |
| | | (C) Furnish these records and data when EPA or the State of Texas request them for inspection. |
| | | (D) Send along with all data a signed copy of the following certification statement, to attest to the truth and accuracy of the data submitted: |
| | | (i) Under civil and criminal penalty of law for the making or submission of false or fraudulent statements or representations (pursuant to the applicable provisions of the Federal Code, which include, but may not be limited to, 18 U.S.C. 1001 and 42 U.S.C. 6928), I certify that the information contained in or accompanying this document is true, accurate and complete. |
| | | (ii) As to the (those) identified section(s) of this document for which I cannot personally verify its (their) truth and accuracy, I certify as the company official having supervisory responsibility for the persons who, acting under my direct instructions, made the verification that this information is true, accurate and complete. |
| | | (iii) If any of this information is determined by EPA in its sole discretion to be false, inaccurate or incomplete, and upon conveyance of this fact to the company, I recognize and agree that this exclusion of waste will be void as if it never had effect or to the extent directed by EPA and that the company will be liable for any actions taken in contravention of the company's RCRA and CERCLA obligations premised upon the company's reliance on the void exclusion. |
| | | (6) *Reopener Language:* |
| | | (A) If, anytime after disposal of the delisted waste, Eastman possesses or is otherwise made aware of any environmental data (including but not limited to leachate data or groundwater monitoring data) or any other data relevant to the delisted waste indicating that any constituent identified for the delisting verification testing is at level higher than the delisting level allowed by the Regional Administrator or his delegate in granting the petition, then the facility must report the data, in writing, to the Regional Administrator or his delegate within 10 days of first possessing or being made aware of that data. |
| | | (B) If the annual testing of the waste does not meet the delisting requirements in Condition (1), Eastman must report the data, in writing, to the Regional Administrator or his delegate within 10 days of first possessing or being made aware of that data. |
| | | (C) If Eastman fails to submit the information described in Conditions (5),(6)(A) or (6)(B) or if any other information is received from any source, the Regional Administrator or his delegate will make a preliminary determination as to whether the reported information requires Agency action to protect human health or the environment. Further action may include suspending, or revoking the exclusion, or other appropriate response necessary to protect human health and the environment. |
| | | (D) If the Regional Administrator or his delegate determines that the reported information does require Agency action, the Regional Administrator or his delegate will notify the facility in writing of the actions the Regional Administrator or his delegate believes are necessary to protect human health and the environment. The notice shall include a statement of the proposed action and a statement providing the facility with an opportunity to present information as to why the proposed Agency action is not necessary. The facility shall have 10 days from the date of the Regional Administrator or his delegate's notice to present such information. |
| | | (E) Following the receipt of information from the facility described in Condition (6)(D) or (if no information is presented under Condition (6)(D)) the initial receipt of information described in Conditions (5), (6)(A) or (6)(B), the Regional Administrator or his delegate will issue a final written determination describing the Agency actions that are necessary to protect human health or the environment. Any required action described in the Regional Administrator or his delegate's determination shall become effective immediately, unless the Regional Administrator or his delegate provides otherwise. |

TABLE 1—WASTES EXCLUDED FROM NON-SPECIFIC SOURCES—Continued

| Facility | Address | Waste description |
|---|---|---|
| | | (7) *Notification Requirements.* Eastman must do following before transporting the delisted waste off-site: Failure to provide this notification will result in a violation of the delisting petition and a possible revocation of the exclusion. |
| | | (A) Provide a one-time written notification to any State Regulatory Agency to which or through which they will transport the delisted waste described above for disposal, 60 days before beginning such activities. |
| | | (B) Update the one-time written notification if they ship the delisted waste into a different disposal facility. |
| Eli Lilly and Company. | Clinton, Indiana. | Incinerator scrubber liquids, entering and contained in their onsite surface impoundment, and solids settling from these liquids originating from the burning of spent solvents (EPA Hazardous Waste Nos. F002, F003, and F005) contained in their onsite surface impoundment and solids retention area on August 18, 1988 and any new incinerator scubber liquids and settled solids generated in the surface impoundment and disposed of in the retention are after August 12, 1988. |
| Envirite of Illinois (formerly Envirite Corporation). | Harvey, Illinois | See waste description under Envirite of Pennsylvania. |
| Envirite of Ohio (formerly Envirite Corporation). | Canton, Ohio | See waste description under Envirite of Pennsylvania. |
| Envirite of Pennsylvania (formerly Envirite Corporation). | York, Pennsylvania. | Dewatered wastewater sludges (EPA Hazardous Waste No .F006) generated from electroplating operations; spent cyanide plating solutions (EPA Hazardous Waste No. F007) generated from electroplating operations; plating bath residues from the bottom of plating baths (EPA Hazardous Waste No. F008) generated from electroplating operations where cyanides are used in the process; spent stripping and cleaning bath solutions (EPA Hazardous Waste No. F009) generated from electroplating operations where cyanides are used in the process; spent cyanide solutions from salt bath pot cleaning (EPA Hazardous Waste No. F011) generated from metal heat treating operations; quenching wastewater treatment sludges (EPA Hazardous Waste No. F012) generated from metal heat treating where cyanides are used in the process; wastewater treatment sludges (EPA Hazardous Waste No. F019) generated from the chemical conversion coating of aluminum after November 14, 1986. To ensure that hazardous constituents are not present in the waste at levels of regulatory concern, the facility must implement a contingency testing program for the petitioned waste. This testing program must meet the following conditions for the exclusions to be valid: |
| | | (1) Each batch of treatment residue must be representatively sampled and tested using the EP Toxicity test for barium, cadmium, chromium, lead, selenium, silver, mercury, and nickel. If the extract concentrations for chromium, lead, arsenic, and silver exceed 0.315 ppm; barium levels exceed 6.3 ppm; cadmium and selenium exceed 0.063 ppm; mercury exceeds 0.0126 ppm; or nickel levels exceed 2.205 ppm; the waste must be re-treated or managed and disposed as a hazardous waste under 40 CFR Parts 262 to 265 and the permitting standards of 40 CFR Part 270. |
| | | (2) Each batch of treatment residue must be tested for leachable cyanide. If the leachable cyanide levels (using the EP Toxicity test without acetic acid adjustment) exceed 1.26 ppm, the waste must be re-treated or managed and disposed as a hazardous waste under 40 CFR Parts 262 to 265 and the permitting standards of 40 CFR Part 270. |
| | | (3) Each batch of waste must be tested for the total content of specific organic toxicanto. If the total content of anthraocne exceeds 76.8 ppm, 1,2-diphenyl hydrazine exceeds 0.001 ppm, methylene chloride exceeds 8.18 ppm, methyl ethyl ketone exceeds 326 ppm, n-nitrosodiphenylamine exceeds 11.9 ppm, phenol exceeds 1,566 ppm, tetrachloroethylene exceeds 0.188 ppm, or trichloroethylene exceeds 0.592 ppm, the waste must be managed and disposed as a hazardous waste under 40 CFR Parts 262 to 265 and the permitting standards of 40 CFR Part 270. |
| | | (4) A grab sample must be collected from each batch to form one monthly composite sample which must be tested using GC/MS analysis for the compounds listed in #3, above, as well as the remaining organics on the priority pollutant list. (See 47 FR 52309, November 19, 1982, for a list of the priority pollutants.) |
| | | (5) The data from conditions 1–4 must be kept on file at the facility for inspection purposes and must be compiled, summarized, and submitted to the Administrator by certified mail semi-annually. The Agency will review this information and if needed will propose to modify or withdraw the exclusion. The organics testing described in conditions 3 and 4, above, are not required until six months from the date of promulgation. The Agency's decision to conditionally exclude the treatment residue generated from the wastewater treatment systems at these facilities applies only to the wastewater and solids treatment systems as they presently exist as described in the delisting petition. The exclusion does not apply to the proposed process additions described in the petition as recovery including crystallization, electrolytic metals recovery, evaporative recovery, and ion exchange. |

191

TABLE 1—WASTES EXCLUDED FROM NON-SPECIFIC SOURCES—Continued

| Facility | Address | Waste description |
|---|---|---|
| EPA's Mobile Incineration System. | Denney Farm Site; McDowell, MO. | Process wastewater, rotary kiln ash, CHEAF media, and other solids (except spent activated carbon) (EPA Hazardous Waste Nos. F020, F022, F023, F026, F027, and F028) generated during the field demonstration of EPA's Mobile Incinerator at the Denney Farm Site in McDowell, Missouri, after July 25, 1985, so long as: (1) The incinerator is functioning properly; (2) a grab sample is taken from each tank of wastewater generated and the EP leachate values do not exceed 0.03 ppm for mercury, 0.14 ppm for selenium, and 0.68 ppm for chromium; and (3) a grab sample is taken from each drum of soil or ash generated and a core sample is collected from each CHEAF roll generated and the EP leachate values of daily composites do not exceed 0.044 ppm in ash or CHEAF media for mercury or 0.22 ppm in ash or CHEAF media for selenium. |
| Falconer Glass Indust., Inc.. | Falconer, NY | Wastewater treatment sludges from the filter press and magnetic drum separator (EPA Hazardous Waste No. F006) generated from electroplating operations after July 16, 1986. |
| Florida Production Engineering Company. | Daytona Beach, Florida. | This is a one-time exclusion. Wastewater treatment sludges (EPA Hazardous Waste No. F006) generated from electroplating operations and contained in four on-site trenches on January 23, 1987. |
| Ford Motor Company, Dearborn Truck Assembly Plant. | Dearborn, Michigan. | Wastewater treatment plant sludge, F019, that is generated by Ford Motor Company at the Dearborn Truck Assembly Plant at a maximum annual rate of 2,000 cubic yards per year. The sludge must be disposed of in a lined landfill with leachate collection which is licensed, permitted, or otherwise authorized to accept the delisted wastewater treatment sludge in accordance with 40 CFR part 258. The exclusion becomes effective as of April 25, 2005.<br>1. *Delisting Levels:* (A) The concentrations in a TCLP extract of the waste measured in any sample may not exceed the following levels (mg/L): antimony—0.7; arsenic—0.3; barium—100; cadmium—0.5; chromium—5; lead—5; nickel—90; selenium—1; thallium—0.3; zinc—900; p-cresol—11; di-n-octyl phthlate—0.11; formaldehyde—80; and pentachlorophenol—0.009. (B) The total concentration measured in any sample may not exceed the following levels (mg/kg): mercury—9; and formaldehyde—700.<br>2. *Quarterly Verification Testing:* To verify that the waste does not exceed the specified delisting levels, Dearborn Truck Assembly Plant must collect and analyze one representative sample of the waste on a quarterly basis.<br>3. *Changes in Operating Conditions:* Dearborn Truck Assembly Plant must notify the EPA in writing if the manufacturing process, the chemicals used in the manufacturing process, the treatment process, or the chemicals used in the treatment process change significantly. Dearborn Truck Assembly Plant must handle wastes generated after the process change as hazardous until it has demonstrated that the wastes continue to meet the delisting levels and that no new hazardous constituents listed in appendix VIII of part 261 have been introduced and it has received written approval from EPA.<br>4. *Data Submittals:* Dearborn Truck Assembly Plant [Redln Off] must submit the data obtained through verification testing or as required by other conditions of this rule to both U.S. EPA Region 5, Waste Management Branch (DW–8J), 77 W. Jackson Blvd., Chicago, IL 60604 and MDEQ, Waste Management Division, Hazardous Waste Program Section, at P.O. Box 30241, Lansing, Michigan 48909. The quarterly verification data and certification of proper disposal must be submitted annually upon the anniversary of the effective date of this exclusion. Dearborn Truck Assembly Plant must compile, summarize and maintain on site for a minimum of five years records of operating conditions and analytical data. Dearborn Truck Assembly Plant must make these records available for inspection. All data must be accompanied by a signed copy of the certification statement in 40 CFR 260.22(i)(12).<br>5. *Reopener Language*—(a) If, anytime after disposal of the delisted waste, Dearborn Truck Assembly Plant possesses or is otherwise made aware of any data (including but not limited to leachate data or groundwater monitoring data) relevant to the delisted waste indicating that any constituent is at a level in the leachate higher than the specified delisting level, or is in the groundwater at a concentration higher than the maximum allowable groundwater concentration in paragraph (e), then Dearborn Truck Assembly Plant must report such data, in writing, to the Regional Administrator within 10 days of first possessing or being made aware of that data.<br>(b) Based on the information described in paragraph (a) and any other information received from any source, the Regional Administrator will make a preliminary determination as to whether the reported information requires Agency action to protect human health or the environment. Further action may include suspending, or revoking the exclusion, or other appropriate response necessary to protect human health and the environment.<br>(c) If the Regional Administrator determines that the reported information does require Agency action, the Regional Administrator will notify Dearborn Truck Assembly Plant in writing of the actions the Regional Administrator believes are necessary to protect human health and the environment. The notice shall include a statement of the proposed action and a statement providing Dearborn Truck Assembly Plant with an opportunity to present information as to why the proposed Agency action is not necessary or to suggest an alternative action. Dearborn Truck Assembly Plant shall have 30 days from the date of the Regional Administrator's notice to present the information. |

TABLE 1—WASTES EXCLUDED FROM NON-SPECIFIC SOURCES—Continued

| Facility | Address | Waste description |
|---|---|---|
| | | (d) If after 30 days the Dearborn Truck Assembly Plant presents no further information, the Regional Administrator will issue a final written determination describing the Agency actions that are necessary to protect human health or the environment. Any required action described in the Regional Administrator's determination shall become effective immediately, unless the Regional Administrator provides otherwise. |
| | | (e) Maximum Allowable Groundwater Concentrations (µg/L): antimony—6; arsenic—5; barium—2,000; cadmium—5; chromium—100; lead—15; nickel—800; selenium—50; thallium—2; tin—20,000; zinc—11,000; p-Cresol—200; Di-n-octyl phthlate—1.3; Formaldehyde—1,400; and Pentachlorophenol—0.15. |
| Ford Motor Company, Kansas City Assembly Plant. | Claycomo, Missouri. | Wastewater treatment sludge, F019, that is generated at the Ford Motor Company (Ford) Kansas City Assembly Plant (KCAP) at a maximum annual rate of 2,000 cubic yards per year. The sludge must be disposed of in a lined landfill with leachate collection, which is licensed, permitted, or otherwise authorized to accept the delisted wastewater treatment sludge in accordance with 40 CFR part 258. The exclusion becomes effective as of June 6, 2007. |
| | | 1. Delisting Levels: (a) The concentrations in a TCLP extract of the waste measured in any sample may not equal or exceed the following levels (mg/L): barium—100; chromium—5; mercury—0.155; nickel—90; thallium—0.282; zinc—898; cyanides—11.5; ethyl benzene—42.6; toluene—60.8; total xylenes—18.9; bis(2-ethylhexyl) phthalate—0.365; p-cresol—11.4; 2,4-dinitrotoluene—0.13; formaldehyde—343; and napthalene—.728; |
| | | (b) The total concentrations measured in any sample may not exceed the following levels (mg/kg): chromium 760000; mercury—10.4; thallium—116000; 2,4-dinitrotoluene—100000; and formaldehyde—6880. |
| | | 2. Quarterly Verification Testing: To verify that the waste does not exceed the specified delisting levels, Ford must collect and analyze one representative sample of KCAP's sludge on a quarterly basis. |
| | | 3. Changes in Operating Conditions: Ford must notify the EPA in writing if the manufacturing process, the chemicals used in the manufacturing process, the treatment process, or the chemicals used in the treatment process at KCAP significantly change. Ford must handle wastes generated at KCAP after the process change as hazardous until it has demonstrated that the waste continues to meet the delisting levels and that no new hazardous constituents listed in appendix VIII of part 261 have been introduced and Ford has received written approval from EPA for the changes. |
| | | 4. Data Submittals: Ford must submit the data obtained through verification testing at KCAP or as required by other conditions of this rule to EPA Region 7, Air, RCRA and Toxics Division, 901 N. 5th, Kansas City, Kansas 66101. The quarterly verification data and certification of proper disposal must be submitted annually upon the anniversary of the effective date of this exclusion. Ford must compile, summarize, and maintain at KCAP records of operating conditions and analytical data for a minimum of five years. Ford must make these records available for inspection. All data must be accompanied by a signed copy of the certification statement in 40 CFR 260.22(i)(12). |
| | | 5. Reopener Language—(a) If, anytime after disposal of the delisted waste, Ford possesses or is otherwise made aware of any data (including but not limited to leachate data or groundwater monitoring data) relevant to the delisted waste at KCAP indicating that any constituent is at a level in the leachate higher than the specified delisting level, or is in the groundwater at a concentration higher than the maximum allowable groundwater concentration in paragraph (e), then Ford must report such data in writing to the Regional Administrator within 10 days of first possessing or being made aware of that data. |
| | | (b) Based on the information described in paragraph (a) and any other information received from any source, the Regional Administrator will make a preliminary determination as to whether the reported information requires Agency action to protect human health or the environment. Further action may include suspending, or revoking the exclusion, or other appropriate response necessary to protect human health and the environment. |
| | | (c) If the Regional Administrator determines that the reported information does require Agency action, the Regional Administrator will notify Ford in writing of the actions the Regional Administrator believes are necessary to protect human health and the environment. The notice shall include a statement of the proposed action and a statement providing Ford with an opportunity to present information as to why the proposed Agency action is not necessary or to suggest an alternative action. Ford shall have 30 days from the date of the Regional Administrator's notice to present the information. |
| | | (d) If after 30 days Ford presents no further information, the Regional Administrator will issue a final written determination describing the Agency actions that are necessary to protect human health or the environment. Any required action described in the Regional Administrator's determination shall become effective immediately, unless the Regional Administrator provides otherwise. |

TABLE 1—WASTES EXCLUDED FROM NON-SPECIFIC SOURCES—Continued

| Facility | Address | Waste description |
|---|---|---|
| Ford Motor Company, Michigan Truck Plant and Wayne Integrated Stamping and Assembly Plant.. | Wayne, Michigan. | Waste water treatment plant sludge, F019, that is generated by Ford Motor Company at the Wayne Integrated Stamping and Assembly Plant from wastewaters from both the Wayne Integrated Stamping and Assembly Plant and the Michigan Truck Plant, Wayne, Michigan at a maximum annual rate of 2,000 cubic yards per year. The sludge must be disposed of in a lined landfill with leachate collection, which is licensed, permitted, or otherwise authorized to accept the delisted wastewater treatment sludge in accordance with 40 CFR part 258. The exclusion becomes effective as of July 30, 2003. |
| | | 1. Delisting Levels: (A) The TCLP concentrations measured in any sample may not exceed the following levels (mg/L): Antimony—0.659; Arsenic—0.3; Cadmium—0.48; Chromium—4.95; Lead—5; Nickel—90.5; Selenium—1; Thallium—0.282; Tin—721; Zinc—898; p-Cresol—11.4; and Formaldehyde—84.2. (B) The total concentrations measured in any sample may not exceed the following levels (mg/kg): Mercury—8.92; and Formaldehyde—689. (C) The sum of the ratios of the TCLP concentrations to the delisting levels for nickel and thallium and for nickel and cadmium shall not exceed 1.0. |
| | | 2. Quarterly Verification Testing: To verify that the waste does not exceed the specified delisting levels, the facility must collect and analyze one waste sample on a quarterly basis. |
| | | 3. Changes in Operating Conditions: The facility must notify the EPA in writing if the manufacturing process, the chemicals used in the manufacturing process, the treatment process, or the chemicals used in the treatment process significantly change. The facility must handle wastes generated after the process change as hazardous until it has demonstrated that the wastes continue to meet the delisting levels and that no new hazardous constituents listed in appendix VIII of part 261 have been introduced and it has received written approval from EPA. |
| | | 4. Data Submittals: The facility must submit the data obtained through verification testing or as required by other conditions of this rule to both U.S. EPA Region 5, Waste Management Branch (DW–8J), 77 W. Jackson Blvd., Chicago, IL 60604 and MDEQ, Waste Management Division, Hazardous Waste Program Section, at P.O. Box 30241, Lansing, Michigan 48909. The quarterly verification data and certification of proper disposal must be submitted annually upon the anniversary of the effective date of this exclusion. The facility must compile, summarize, and maintain on site for a minimum of five years records of operating conditions and analytical data. The facility must make these records available for inspection. All data must be accompanied by a signed copy of the certification statement in 40 CFR 260.22(i)(12). |
| | | 5. Reopener Language—(a) If, anytime after disposal of the delisted waste, the facility possesses or is otherwise made aware of any data (including but not limited to leachate data or groundwater monitoring data) relevant to the delisted waste indicating that any constituent is at a level in the leachate higher than the specified delisting level, or is in the groundwater at a concentration higher than the maximum allowable groundwater concentration in paragraph (e), then the facility must report such data, in writing, to the Regional Administrator within 10 days of first possessing or being made aware of that data. |
| | | (b) Based on the information described in paragraph (a) and any other information received from any source, the Regional Administrator will make a preliminary determination as to whether the reported information requires Agency action to protect human health or the environment. Further action may include suspending, or revoking the exclusion, or other appropriate response necessary to protect human health and the environment. |
| | | (c) If the Regional Administrator determines that the reported information does require Agency action, the Regional Administrator will notify the facility in writing of the actions the Regional Administrator believes are necessary to protect human health and the environment. The notice shall include a statement of the proposed action and a statement providing the facility with an opportunity to present information as to why the proposed Agency action is not necessary or to suggest an alternative action. The facility shall have 30 days from the date of the Regional Administrator's notice to present the information. |
| | | (d) If after 30 days the facility presents no further information, the Regional Administrator will issue a final written determination describing the Agency actions that are necessary to protect human health or the environment. Any required action described in the Regional Administrator's determination shall become effective immediately, unless the Regional Administrator provides otherwise. |
| | | (e) Maximum Allowable Groundwater Concentrations (ug/L): Antimony—6; Arsenic—4.87; Cadmium—5; Chromium—100; Lead—15; Nickel—750; Selenium—50; Thallium—2; Tin—22,500; Zinc—11,300; p-Cresol—188; and Formaldehyde—1,380. |
| Ford Motor Company, Wixom Assembly Plant. | Wixom, Michigan. | Waste water treatment plant sludge, F019, that is generated by Ford Motor Company at the Wixom Assembly Plant, Wixom, Michigan at a maximum annual rate of 2,000 cubic yards per year. The sludge must be disposed of in a lined landfill with leachate collection, which is licensed, permitted, or otherwise authorized to accept the delisted wastewater treatment sludge in accordance with 40 CFR Part 258. The exclusion becomes effective as of July 30, 2003. The conditions in paragraphs (2) through (5) for Ford Motor Company—Michigan Truck Plant and Wayne Integrated Stamping Plant—Wayne, Michigan also apply. |

TABLE 1—WASTES EXCLUDED FROM NON-SPECIFIC SOURCES—Continued

| Facility | Address | Waste description |
|---|---|---|
| | | Delisting Levels: (A) The TCLP concentrations measured in any sample may not exceed the following levels (mg/L): Antimony—0.659; Arsenic—0.3; Cadmium—0.48; Chromium—4.95; Lead—5; Nickel—90.5; Selenium—1; Thallium—0.282; Tin—721; Zinc—898; p-Cresol—11.4; and Formaldehyde—84.2. (B) The total concentrations measured in any sample may not exceed the following levels (mg/kg): Mercury—8.92; and Formaldehyde—689. (C) The sum of the ratios of the TCLP concentrations to the delisting levels for nickel and thallium and for nickel and cadmium shall not exceed 1.0. |
| GE's Former RCA del Caribe. | Barceloneta, PR. | Wastewater treatment plant (WWTP) sludges from chemical etching operation (EPA Hazardous Waste No. F006) and contaminated soil mixed with sludge. This is a one-time exclusion for a range of 5,000 to 15,000 cubic yards of WWTP sludge on condition of disposal in a Subtitle D landfill. This exclusion was published on February 1, 2007. 1. Reopener Language—(a) If, anytime after disposal of the delisted waste, GE discovers that any condition or assumption related to the characterization of the excluded waste which was used in the evaluation of the petition or that was predicted through modeling is not as reported in the petition, then GE must report any information relevant to that condition or assumption, in writing, to the Director of the Division of Environmental Planning and Protection in Region 2 within 10 days of first of discovering that information. (b) Upon receiving information described in paragraph (a) of this section, regardless of its source, the Director will determine whether the reported condition requires further action. Further action may include repealing the exclusion, modifying the exclusion, or other appropriate action deemed necessary to protect human health or the environment. |
| | | 2. Notifications—GE must provide a one-time written notification to any State or Commonwealth Regulatory Agency in any State or Commonwealth to which or through which the waste described above will be transported for disposal at least 60 days prior to the commencement of such activities. Failure to provide such a notification will result in a violation of the waste exclusion and a possible revocation of the decision. |
| General Electric Company. | Shreveport Louisiana. | Wastewater treatment sludges (EPA Hazardous Waste No. F006) generated from electroplating operations and contained in four on-site treatment ponds on August 12, 1987. |
| General Motors | Arlington, TX .. | Wastewater Treatment Sludge (WWTP) (EPA Hazardous Waste No. F019) generated at a maximum rate of 3,000 cubic yards per calendar year after January 3, 2007 and disposed in a Subtitle D landfill. |
| | | For the exclusion to be valid, GM-Arlington must implement a verification testing program that meets the following paragraphs: |
| | | (1) *Delisting Levels:* All leachable concentrations for those constituents must not exceed the following levels (mg/l for TCLP). |
| | | (i) Inorganic Constituents: Barium-100; Cadmium-0.36; Chromium-5 (3.71) ; Cobalt-18.02; Lead-5; Nickel-67.8; Silver-5; Tin-540; Zinc-673. |
| | | (ii) Organic Constituents: Acetone-171; Ethylbenzene-31.9; N-Butyl Alcohol-171; Toluene-45.6; Bis(2-Ethylhexyl) Phthalate-0.27; p-Cresol-8.55; Naphthalene-3.11. |
| | | (2) *Waste Management:* (A) GM-Arlington must manage as hazardous all WWTP sludge generated, until it has completed initial verification testing described in paragraph (3)(A) and (B), as appropriate, and valid analyses show that paragraph (1) is satisfied. |
| | | (B) Levels of constituents measured in the samples of the WWTP sludge that do not exceed the levels set forth in paragraph (1) are non-hazardous. GM-Arlington can manage and dispose of the non-hazardous WWTP sludge according to all applicable solid waste regulations. |
| | | (C) If constituent levels in a sample exceed any of the delisting levels set in paragraph (1), GM-Arlington can collect one additional sample and perform expedited analyses to verify if the constituent exceeds the delisting level. If this sample confirms the exceedance, GM-Arlington must, from that point forward, treat the waste as hazardous until it is demonstrated that the waste again meets the levels in paragraph (1). GM-Arlington must manage and dispose of the waste generated under Subtitle C of RCRA from the time it becomes aware of any exceedance. |
| | | (D) Upon completion of the Verification Testing described in paragraph 3(A) and (B), as appropriate, and the transmittal of the results to EPA, and if the testing results meet the requirements of paragraph (1), GM-Arlington may proceed to manage its WWTP sludge as non-hazardous waste. If subsequent Verification Testing indicates an exceedance of the Delisting Levels in paragraph (1), GM-Arlington must manage the WWTP sludge as a hazardous waste until two consecutive quarterly testing samples show levels below the Delisting Levels in paragraph (1). |
| | | (3) *Verification Testing Requirements:* GM-Arlington must perform sample collection and analyses, including quality control procedures, according to appropriate methods such as those found in SW–846 or other reliable sources (with the exception of analyses requiring the use of SW–846 methods incorporated by reference in 40 CFR 260.11, which must be used without substitution) for all constituents listed in paragraph (1). If EPA judges the process to be effective under the operating conditions used during the initial verification testing, GM-Arlington may replace the testing required in paragraph (3)(A) with the testing required in paragraph (3)(B). GM-Arlington Plant must continue to test as specified in paragraph (3)(A) until and unless notified by EPA in writing that testing in paragraph (3)(A) may be replaced by paragraph (3)(B). |
| | | (A) *Initial Verification Testing:* After EPA grants the final exclusion, GM-Arlington must do the following: |

TABLE 1—WASTES EXCLUDED FROM NON-SPECIFIC SOURCES—Continued

| Facility | Address | Waste description |
|---|---|---|
| | | (i) Within 30 days of this exclusion becoming final, collect two (2) samples, before disposal, of the WWTP sludge.<br>(ii) The samples are to be analyzed and compared against the Delisting Levels in paragraph (1).<br>(iii) Within 60 days of the exclusion becoming final, GM-Arlington must report to EPA the initial verification analytical test data for the WWTP sludge, including analytical quality control information for the first thirty (30) days of operation after this exclusion becomes final.<br>If levels of constituents measured in these samples of the WWTP sludge do not exceed the levels set forth in paragraph (1), GM-Arlington can manage and dispose of the WWTP sludge according to all applicable solid waste regulations.<br>(B) *Subsequent Verification Testing:* Following written notification by EPA, GM-Arlington may substitute the testing conditions in paragraph (3)(B) for paragraph (3)(A). GM-Arlington must continue to monitor operating conditions, and analyze two representative samples of the WWTP sludge for the next three quarters of operation during the first year of waste generation. The samples must represent the waste generated during the quarter. Quarterly reports are due to EPA, thirty days after the samples are taken.<br>After the first year of analytical sampling, verification sampling can be performed on a single annual sample of the WWTP sludge. The results are to be compared to the delisting levels in paragraph (1).<br>(C) *Termination of Testing:*<br>(i) After the first year of quarterly testing, if the delisting levels in paragraph (1) are being met, GM-Arlington may then request that EPA not require quarterly testing.<br>(ii) Following cancellation of the quarterly testing by EPA letter, GM-Arlington must continue to test one representative sample for all constituents listed in paragraph (1) annually. Results must be provided to EPA within 30 days of the testing.<br>(4) *Changes in Operating Conditions:* If GM-Arlington significantly changes the process described in its petition or starts any process that generates the waste that may or could significantly affect the composition or type of waste generated as established under paragraph (1) (by illustration, but not limitation, changes in equipment or operating conditions of the treatment process), it must notify EPA in writing; it may no longer handle the wastes generated from the new process as nonhazardous until the wastes meet the delisting levels set in paragraph (1) and it has received written approval to do so from EPA.<br>(5) *Data Submittals:* GM-Arlington must submit the information described below. If GM-Arlington fails to submit the required data within the specified time or maintain the required records on-site for the specified time, EPA, at its discretion, will consider this sufficient basis to reopen the exclusion as described in paragraph 6. GM-Arlington must:<br>(A) Submit the data obtained through paragraph (3) to the Section Chief, Region 6 Corrective Action and Waste Minimization Section, EPA, 1445 Ross Avenue, Dallas, Texas 75202–2733, Mail Code, (6PD–C) within the time specified.<br>(B) Compile records of operating conditions and analytical data from paragraph (3), summarized, and maintained on-site for a minimum of five years.<br>(C) Furnish these records and data when EPA or the State of Texas requests them for inspection.<br>(D) Send along with all data a signed copy of the following certification statement, to attest to the truth and accuracy of the data submitted:<br>"Under civil and criminal penalty of law for the making or submission of false or fraudulent statements or representations (pursuant to the applicable provisions of the Federal Code, which include, but may not be limited to, 18 U.S.C. 1001 and 42 U.S.C. 6928), I certify that the information contained in or accompanying this document is true, accurate and complete.<br>As to the (those) identified section(s) of this document for which I cannot personally verify its (their) truth and accuracy, I certify as the company official having supervisory responsibility for the persons who, acting under my direct instructions, made the verification that this information is true, accurate and complete.<br>If any of this information is determined by EPA in its sole discretion to be false, inaccurate or incomplete, and upon conveyance of this fact to the company, I recognize and agree that this exclusion of waste will be void as if it never had effect or to the extent directed by EPA and that the company will be liable for any actions taken in contravention of the company's RCRA and CERCLA obligations premised upon the company's reliance on the void exclusion."<br>(6) *Re-opener;*<br>(A) If, anytime after disposal of the delisted waste, GM-Arlington possesses or is otherwise made aware of any environmental data (including but not limited to leachate data or groundwater monitoring data) or any other data relevant to the delisted waste indicating that any constituent identified for the delisting verification testing is at a level higher than the delisting level allowed by EPA in granting the petition, then the facility must report the data, in writing, to EPA within 10 days of first possessing or being made aware of that data.<br>(B) If either the quarterly or annual testing of the waste does not meet the delisting requirements in paragraph 1, GM-Arlington must report the data, in writing, to EPA within 10 days of first possessing or being made aware of that data. |

TABLE 1—WASTES EXCLUDED FROM NON-SPECIFIC SOURCES—Continued

| Facility | Address | Waste description |
|---|---|---|
| | | (C) If GM-Arlington fails to submit the information described in paragraphs (5), (6)(A) or (6)(B) or if any other information is received from any source, EPA will make a preliminary determination as to whether the reported information requires action to protect human health and/or the environment. Further action may include suspending, or revoking the exclusion, or other appropriate response necessary to protect human health and the environment. |
| | | (D) If EPA determines that the reported information requires action, EPA will notify the facility in writing of the actions it believes are necessary to protect human health and the environment. The notice shall include a statement of the proposed action and a statement providing the facility with an opportunity to present information explaining why the proposed EPA action is not necessary. The facility shall have 10 days from the date of EPA's notice to present such information. |
| | | (E) Following the receipt of information from the facility described in paragraph (6)(D) or (if no information is presented under paragraph (6)(D)) the initial receipt of information described in paragraphs (5), (6)(A) or (6)(B), EPA will issue a final written determination describing the actions that are necessary to protect human health and/or the environment. Any required action described in EPA's determination shall become effective immediately, unless EPA provides otherwise. |
| | | (7) *Notification Requirements:* GM-Arlington must do the following before transporting the delisted waste. Failure to provide this notification will result in a violation of the delisting petition and a possible revocation of the decision. |
| | | (A) Provide a one-time written notification to any state Regulatory Agency to which or through which it will transport the delisted waste described above for disposal, 60 days before beginning such activities. |
| | | (B) Update the one-time written notification if it ships the delisted waste into a different disposal facility. |
| | | (C) Failure to provide this notification will result in a violation of the delisting variance and a possible revocation of the decision. |
| General Motors Corporation. | Lake Orion, Michigan. | Wastewater treatment plant (WWTP) sludge from the chemical conversion coating (phosphate coating) of aluminum (EPA Hazardous Waste No. F019) generated at a maximum annual rate of 1,500 tons per year (or 1,500 cubic yards per year), after October 24, 1997 and disposed of in a Subtitle D landfill. |
| | | 1. *Verification Testing:* GM must implement an annual testing program to demonstrate, based on the analysis of a minimum of four representative samples, that the constituent concentrations measured in the TCLP (or OWEP, where appropriate) extract of the waste are within specific levels. The constituent concentrations must not exceed the following levels (mg/l) which are back-calculated from the delisting health-based levels and a DAF of 90: Arsenic—4.5; Cobalt—189; Copper—126; Nickel—63; Vanadium—18; 1,2-Dichloroethane—0.45; Ethylbenzene—63; 4-Methylphenol—16.2; Naphthalene—90; Phenol—1800; and Xylene—900. The constituent concentrations must also be less than the following levels (mg/l) which are the toxicity characteristic levels: Barium—100.0; and Chromium (total)—5.0. |
| | | 2. *Changes in Operating Conditions:* If GM significantly changes the manufacturing or treatment process or the chemicals used in the manufacturing or treatment process, GM may handle the WWTP filter press sludge generated from the new process under this exclusion after the facility has demonstrated that the waste meets the levels set forth in paragraph 1 and that no new hazardous constituents listed in Appendix VIII of Part 261 have been introduced. |
| | | 3. *Data Submittals:* The data obtained through annual verification testing or paragraph 2 must be submitted to U.S. EPA Region 5, 77 W. Jackson Blvd., Chicago, IL 60604–3590, within 60 days of sampling. Records of operating conditions and analytical data must be compiled, summarized, and maintained on site for a minimum of five years and must be made available for inspection. All data must be accompanied by a signed copy of the certification statement in 260.22(I)(12). |
| General Motors Corporation Assembly Plant | Lordstown, Ohio. | Waste water treatment plant sludge, F019, that is generated at General Motors Corporation's Lordstown Assembly Plant at a maximum annual rate of 2,000 cubic yards per year. The sludge must be disposed of in a Subtitle D landfill which is licensed, permitted, or otherwise authorized by a state to accept the delisted wastewater treatment sludge. The exclusion becomes effective as of October 12, 2004. |
| | | 1. Delisting Levels: (A) The constituent concentrations measured in the TCLP extract may not exceed the following levels (mg/L): antimony—0.66; arsenic—0.30; chromium—5; lead—5; mercury—0.15; nickel—90; selenium—1; silver—5; thallium—0.28; tin—720; zinc—900; fluoride—130; p-cresol—11; formaldehyde—84; and methylene chloride—0.29 (B) The total constituent concentration measured in any sample of the waste may not exceed the following levels (mg/kg): chromium—4,100 ; formaldehyde—700; and mercury—10. (C) Maximum allowable groundwater concentrations (µg/L) are as follows: antimony—6; arsenic—4.88; chromium—100; lead—15; mercury—2; nickel—750; selenium—50; silver—188; thallium—2; tin—22,500; zinc—11,300; fluoride—4,000; p-cresol—188; formaldehyde—1,390; and methylene chloride—5. |
| | | 2. Quarterly Verification Testing: To verify that the waste does not exceed the specified delisting levels, GM must collect and analyze one waste sample on a quarterly basis using methods with appropriate detection levels and elements of quality control. |

197

TABLE 1—WASTES EXCLUDED FROM NON-SPECIFIC SOURCES—Continued

| Facility | Address | Waste description |
|---|---|---|
| | | 3. Changes in Operating Conditions: The facility must notify the EPA in writing if the manufacturing process, the chemicals used in the manufacturing process, the treatment process, or the chemicals used in the treatment process significantly change. GM must handle wastes generated after the process change as hazardous until it has demonstrated that the wastes continue to meet the delisting levels and that no new hazardous constituents listed in appendix VIII of part 261 have been introduced and it has received written approval from EPA. |
| | | 4. Data Submittals: The facility must submit the data obtained through verification testing or as required by other conditions of this rule to U.S. EPA Region 5, Waste Management Branch, RCRA Delisting Program (DW–8J), 77 W. Jackson Blvd., Chicago, IL 60604. The quarterly verification data and certification of proper disposal must be submitted annually upon the anniversary of the effective date of this exclusion. The facility must compile, summarize, and maintain on site for a minimum of five years records of operating conditions and analytical data. The facility must make these records available for inspection. All data must be accompanied by a signed copy of the certification statement in 40 CFR 260.22(i)(12). |
| | | 5. Reopener Language: (A) If, anytime after disposal of the delisted waste, GM possesses or is otherwise made aware of any data (including but not limited to leachate data or groundwater monitoring data) relevant to the delisted waste indicating that any constituent is at a level in the leachate higher than the specified delisting level, or is in the groundwater at a concentration higher than the maximum allowable groundwater concentration in paragraph (1), then GM must report such data, in writing, to the Regional Administrator within 10 days of first possessing or being made aware of that data. (B) Based on the information described in paragraph (A) and any other information received from any source, the Regional Administrator will make a preliminary determination as to whether the reported information requires Agency action to protect human health or the environment. Further action may include suspending, or revoking the exclusion, or other appropriate response necessary to protect human health and the environment. (C) If the Regional Administrator determines that the reported information does require Agency action, the Regional Administrator will notify the facility in writing of the actions the Regional Administrator believes are necessary to protect human health and the environment. The notice shall include a statement of the proposed action and a statement providing GM with an opportunity to present information as to why the proposed Agency action is not necessary or to suggest an alternative action. GM shall have 30 days from the date of the Regional Administrator's notice to present the information. (D) If after 30 days GM presents no further information, the Regional Administrator will issue a final written determination describing the Agency actions that are necessary to protect human health or the environment. Any required action described in the Regional Administrator's determination shall become effective immediately, unless the Regional Administrator provides otherwise. |
| General Motors Corp., Fisher Body Division. | Elyria, OH ...... | The residue generated from the use of the Chemfix® treatment process on sludge (EPA Hazardous Waste No. F006) generated from electroplating operations and contained in three on-site surface impoundments on November 14, 1986. To assure that stabilization occurs, the following conditions apply to this exclusion:<br>(1) Mixing ratios shall be monitored continuously to assure consistent treatment.<br>(2) One grab sample of the treated waste shall be taken each hour as it is pumped to the holding area (cell) from each trailer unit. At the end of each production day, the grab samples from the individual trailer units will be composited and the EP toxicity test will be run on each composite sample. If lead or total chromium concentrations exceed 0.315 ppm or if nickel exceeds 2.17 ppm, in the EP extract, the waste will be removed and retreated or disposed of as a hazardous waste.<br>(3) The treated waste shall be pumped into bermed cells which are constructed to assure that the treated waste is identifiable and retrievable (*i.e.*, the material can be removed and either disposed of as a hazardous waste or retreated if conditions 1 or 2 are not met).<br>Failure to satisfy any of these conditions would render the exclusion void. This is a one-time exclusion, applicable only to the residue generated from the use of the Chemfix® treatment process on the sludge currently contained in the three on-site surface impoundments. |
| General Motors Corporation, Flint Truck. | Flint, Michigan | Waste water treatment plant sludge, F019, that is generated by General Motors Corporation at Flint Truck, Flint, Michigan at a maximum annual rate of 3,000 cubic yards per year. The sludge must be disposed of in a lined landfill with leachate collection, which is licensed, permitted, or otherwise authorized to accept the delisted wastewater treatment sludge in accordance with 40 CFR part 258. The exclusion becomes effective as of July 30, 2003. The conditions in paragraphs (2) through (5) for Ford Motor Company—Michigan Truck Plant and Wayne Integrated Stamping Plant—Wayne, Michigan also apply.<br>Delisting Levels: (A) The TCLP concentrations measured in any sample may not exceed the following levels (mg/L): Antimony—0.494; Arsenic—0.224; Cadmium—0.36; Chromium—3.71; Lead—5; Nickel—67.8; Selenium—1; Thallium—0.211; Tin—540; Zinc—673; p-Cresol—8.55; and Formaldehyde—63. (B) The total concentrations measured in any sample may not exceed the following levels (mg/kg): Mercury—6.34; and Formaldehyde—535. (C) The sum of the ratios of the TCLP concentration to the delisting level for nickel and thallium and for nickel and cadmium shall not exceed 1.0. |

TABLE 1—WASTES EXCLUDED FROM NON-SPECIFIC SOURCES—Continued

| Facility | Address | Waste description |
|---|---|---|
| General Motors Corporation, Hamtramck. | Detroit, Michigan. | Waste water treatment plant sludge, F019, that is generated by General Motors Corporation at Hamtramck, Detroit, Michigan at a maximum annual rate of 3,000 cubic yards per year. The sludge must be disposed of in a lined landfill with leachate collection, which is licensed, permitted, or otherwise authorized to accept the delisted wastewater treatment sludge in accordance with 40 CFR part 258. The exclusion becomes effective as of July 30, 2003. The conditions in paragraphs (2) through (5) for Ford Motor Company—Michigan Truck Plant and Wayne Integrated Stamping Plant—Wayne, Michigan also apply. A maximum allowable groundwater concentration of 3,750 µg/L for n-butyl alcohol is added to paragraph (5)(e). |
| | | Delisting Levels: (A) The TCLP concentrations measured in any sample may not exceed the following levels (mg/L): Antimony—0.494; Arsenic—0.224; Cadmium—0.36; Chromium—3.71; Lead—5; Nickel—67.8; Selenium—1; Thallium—0.211; Tin—540; Zinc—673; p-Cresol—8.55; Formaldehyde—63; and n-Butyl alcohol—171. (B) The total concentrations measured in any sample may not exceed the following levels (mg/kg): Mercury—6.34; and Formaldehyde—535. (C) The sum of the ratios of the TCLP concentration to the delisting level for nickel and thallium and for nickel and cadmium shall not exceed 1.0. |
| General Motors Corporation, Janesville Truck Assembly Plant | Janesville, Wisconsin. | Wastewater treatment sludge, F019, that is generated at the General Motors Corporation (GM) Janesville Truck Assembly Plant (JTAP) at a maximum annual rate of 3,000 cubic yards per year. The sludge must be disposed of in a lined landfill with leachate collection, which is licensed, permitted, or otherwise authorized to accept the delisted wastewater treatment sludge in accordance with 40 CFR part 258. The exclusion becomes effective as of January 24, 2006. |
| | | 1. Delisting Levels: (A) The concentrations in a TCLP extract of the waste measured in any sample may not exceed the following levels (mg/L): antimony—0.49; arsenic—0.22; cadmium—0.36; chromium—3.7; lead—5; nickel—68; selenium—1; thallium—0.21; tin—540; zinc—670; p-cresol—8.5; and formaldehyde—43. (B) The total concentrations measured in any sample may not exceed the following levels (mg/kg): chromium—5,300; mercury—7; and formaldehyde—540. |
| | | 2. Quarterly Verification Testing: To verify that the waste does not exceed the specified delisting levels, GM must collect and analyze one representative sample of JTAP's sludge on a quarterly basis. |
| | | 3. Changes in Operating Conditions: GM must notify the EPA in writing if the manufacturing process, the chemicals used in the manufacturing process, the treatment process, or the chemicals used in the treatment process at JTAP significantly change. GM must handle wastes generated at JTAP after the process change as hazardous until it has demonstrated that the waste continues to meet the delisting levels and that no new hazardous constituents listed in appendix VIII of part 261 have been introduced and GM has received written approval from EPA. |
| | | 4. Data Submittals: GM must submit the data obtained through verification testing at JTAP or as required by other conditions of this rule to EPA Region 5, Waste Management Branch (DW–8J), 77 W. Jackson Blvd., Chicago, IL 60604. The quarterly verification data and certification of proper disposal must be submitted annually upon the anniversary of the effective date of this exclusion. GM must compile, summarize, and maintain at JTAP records of operating conditions and analytical data for a minimum of five years. GM must make these records available for inspection. All data must be accompanied by a signed copy of the certification statement in 40 CFR 260.22(i)(12). |
| | | 5. Reopener Language—(a) If, anytime after disposal of the delisted waste, GM possesses or is otherwise made aware of any data (including but not limited to leachate data or groundwater monitoring data) relevant to the delisted waste at JTAP indicating that any constituent is at a level in the leachate higher than the specified delisting level, or is in the groundwater at a concentration higher than the maximum allowable groundwater concentration in paragraph (e), then GM must report such data in writing to the Regional Administrator within 10 days of first possessing or being made aware of that data. |
| | | (b) Based on the information described in paragraph (a) and any other information received from any source, the Regional Administrator will make a preliminary determination as to whether the reported information requires Agency action to protect human health or the environment. Further action may include suspending, or revoking the exclusion, or other appropriate response necessary to protect human health and the environment. |
| | | (c) If the Regional Administrator determines that the reported information does require Agency action, the Regional Administrator will notify GM in writing of the actions the Regional Administrator believes are necessary to protect human health and the environment. The notice shall include a statement of the proposed action and a statement providing GM with an opportunity to present information as to why the proposed Agency action is not necessary or to suggest an alternative action. GM shall have 30 days from the date of the Regional Administrator's notice to present the information. |
| | | (d) If after 30 days GM presents no further information, the Regional Administrator will issue a final written determination describing the Agency actions that are necessary to protect human health or the environment. Any required action described in the Regional Administrator's determination shall become effective immediately, unless the Regional Administrator provides otherwise. |

TABLE 1—WASTES EXCLUDED FROM NON-SPECIFIC SOURCES—Continued

| Facility | Address | Waste description |
|---|---|---|
| General Motors Corporation. Lansing Car Assembly—Body Plant. | Lansing, Michigan. | (e) Maximum Allowable Groundwater Concentrations (mg/L):; antimony—0.006; arsenic—0.005; cadmium—0.005; chromium—0.1; lead—0.015; nickel—0.750; selenium—0.050; tin—23; zinc—11; p-Cresol—0.190; and formaldehyde—0.950. Wastewater treatment plant (WWTP) sludge from the chemical conversion coating (phosphate coating) of aluminum (EPA Hazardous Waste No. F019) generated at a maximum annual rate of 1,250 cubic yards per year and disposed of in a Subtitle D landfill, after May 16, 2000. |
| | | 1. Delisting Levels: |
| | | (A) The constituent concentrations measured in the TCLP extract may not exceed the following levels (mg/L): Antimony—0.576; Arsenic—4.8; Barium—100; Beryllium—0.384; Cadmium—0.48; Chromium (total)—5; Cobalt—201.6; Copper—124.8; Lead—1.44; Mercury—0.192; Nickel—67.2; Selenium—1; Silver—5; Thallium—0.192; Tin—2016; Vanadium—28.8; Zinc—960; Cyanide—19.2; Fluoride—384; Acetone—336; m,p—Cresol—19.2; 1,1—Dichloroethane—0.0864; Ethylbenzene—67.2; Formaldehyde—672; Phenol—1920; Toluene—96; 1,1,1—Trichloroethane—19.2; Xylene—960. |
| | | (B) The total concentration of formaldehyde in the waste may not exceed 2100 mg/kg. |
| | | (C) Analysis for determining reactivity from sulfide must be added to verification testing when an EPA-approved method becomes available. |
| | | 2. Verification Testing: GM must implement an annual testing program to demonstrate that the constituent concentrations measured in the TCLP extract (or OWEP, where appropriate) of the waste do not exceed the delisting levels established in Condition (1). |
| | | 3. Changes in Operating Conditions: If GM significantly changes the manufacturing or treatment process or the chemicals used in the manufacturing or treatment process, GM must notify the EPA of the changes in writing. GM must handle wastes generated after the process change as hazardous until GM has demonstrated that the wastes meet the delisting levels set forth in Condition (1), that no new hazardous constituents listed in Appendix VIII of Part 261 have been introduced, and GM has received written approval from EPA. |
| | | 4. Data Submittals: GM must submit the data obtained through annual verification testing or as required by other conditions of this rule to U.S. EPA Region 5, 77 W. Jackson Blvd. (DW–8J), Chicago, IL 60604, within 60 days of sampling. GM must compile, summarize, and maintain on site for a minimum of five years records of operating conditions and analytical data. GM must make these records available for inspection. All data must be accompanied by a signed copy of the certification statement in 40 CFR 260.22(i)(12). |
| | | 5. Reopener Language—(a) If, anytime after disposal of the delisted waste, GM possesses or is otherwise made aware of any environmental data (including but not limited to leachate data or groundwater monitoring data) or any other data relevant to the delisted waste indicating that any constituent identified in Condition (1) is at a level in the leachate higher than the delisting level established in Condition (1), or is at a level in the ground water or soil higher than the level predicted by the CML model, then GM must notify the Regional Administrator in writing within 10 days and must report the data within 45 days of first possessing or being made aware of that data. |
| | | (b) Based on the information described in paragraph (a) and any other information received from any source, the Regional Administrator will make a preliminary determination as to whether the reported information requires Agency action to protect human health or the environment. Further action may include suspending, or revoking the exclusion, or other appropriate response necessary to protect human health and the environment. |
| | | (c) If the Regional Administrator determines that the reported information does require Agency action, the Regional Administrator will notify GM in writing of the actions the Regional Administrator believes are necessary to protect human health and the environment. The notice shall include a statement of the proposed action and a statement providing GM with an opportunity to present information as to why the proposed Agency action is not necessary or to suggest an alternative action. GM shall have 10 days from the date of the Regional Administrator's notice to present the information. |
| | | (d) If after 10 days GM presents no further information, the Regional Administrator will issue a final written determination describing the Agency actions that are necessary to protect human health or the environment. Any required action described in the Regional Administrator's determination shall become effective immediately, unless the Regional Administrator provides otherwise. |
| General Motors Corporation, Pontiac East. | Pontiac, Michigan. | Waste water treatment plant sludge, F019, that is generated by General Motors Corporation at Pontiac East, Pontiac, Michigan at a maximum annual rate of 3,000 cubic yards per year. The sludge must be disposed of in a lined landfill with leachate collection, which is licensed, permitted, or otherwise authorized to accept the delisted wastewater treatment sludge in accordance with 40 CFR part 258. The exclusion becomes effective as of July 30, 2003. The conditions in paragraphs (2) through (5) for Ford Motor Company—Michigan Truck Plant and Wayne Integrated Stamping Plant—Wayne, Michigan also apply. |

Environmental Protection Agency

# Environmental Protection Agency

Pt. 261, App. IX

TABLE 1—WASTES EXCLUDED FROM NON-SPECIFIC SOURCES—Continued

| Facility | Address | Waste description |
|---|---|---|
| Geological Reclamation Operations and Waste Systems, Inc. | Morrisville, Pennsylvania. | Delisting Levels: (A) The TCLP concentrations measured in any sample may not exceed the following levels (mg/L): Antimony—0.494; Arsenic—0.224; Cadmium—0.36; Chromium—3.71; Lead—5; Nickel—67.8; Selenium—1; Thallium—0.211; Tin—540; Zinc—673; p-Cresol—8.55; and Formaldehyde—63. (B) The total concentrations measured in any sample may not exceed the following levels (mg/kg): Mercury—6.34; and Formaldehyde—535. (C) The sum of the ratios of the TCLP concentrations to the delisting levels for nickel and thallium and for nickel and cadmium shall not exceed 1.0. Wastewater treatment sludge filter cake from the treatment of EPA Hazardous Waste No. F039, generated at a maximum annual rate of 2000 cubic yards, after December 4, 2001, and disposed of in a Subtitle D landfill. The exclusion covers the filter cake resulting from the treatment of hazardous waste leachate derived from only "old" GROWS and non-hazardous leachate derived from only non-hazardous waste sources. The exclusion does not address the waste disposed of in the "old" GROWS' Landfill or the grit generated during the removal of heavy solids from the landfill leachate. To ensure that hazardous constituents are not present in the filter cake at levels of regulatory concern, GROWS must implement a testing program for the petitioned waste. This testing program must meet the conditions listed below in order for the exclusion to be valid:<br>(1) *Testing:* Sample collection and analyses, including quality control (QC) procedures, must be performed using appropriate methods. As applicable to the method-defined parameters of concern, analyses requiring the use of SW–846 methods incorporated by reference in 40 CFR 260.11 must be used without substitution. As applicable, the SW–846 methods might include Methods 0010, 0011, 0020, 0023A, 0030, 0031, 0040, 0050, 0051, 0060, 0061, 1010A, 1020B, 1110A, 1310B, 1311, 1312, 1320, 1330A, 9010C, 9012B, 9040C, 9045D, 9060A, 9070A (uses EPA Method 1664, Rev. A), 9071B, and 9095B.<br>(A) *Sample Collection:* Each batch of waste generated over a four-week period must be collected in containers with a maximum capacity of 20-cubic yards. At the end of the four-week period, each container must be divided into four quadrants and a single, full-depth core sample shall be collected from each quadrant. All of the full-depth core samples then must be composited under laboratory conditions to produce one representative composite sample for the four-week period.<br>(B) *Sample Analysis:* Each four-week composite sample must be analyzed for all of the constituents listed in Condition (3). The analytical data, including quality control information, must be submitted to The Waste and Chemicals Management Division, U.S. EPA Region III, 1650 Arch Street, Philadelphia, PA 19103, and the Pennsylvania Department of Environmental Protection, Bureau of Land Recycling and Waste Management, Rachel Carson State Office Building, 400 Market Street, 14th Floor, Harrisburg, PA 17105. Data from the annual verification testing must be compiled and submitted to EPA and the Pennsylvania Department of Environmental Protection within sixty (60) days from the end of the calendar year. All data must be accompanied by a signed copy of the statement set forth in 40 CFR 260.22(i)(12) to certify to the truth and accuracy of the data submitted. Records of operating conditions and analytical data must be compiled, summarized, and maintained on-site for a minimum of three years and must be furnished upon request by any employee or representative of EPA or the Pennsylvania Department of Environmental Protection, and made available for inspection.<br>(2) *Waste Holding:* The dewatered filter cake must be stored as hazardous until the verification analyses are completed. If the four-week composite sample does not exceed any of the delisting levels set forth in Condition (3), the filter cake corresponding to this sample may be managed and disposed of in accordance with all applicable solid waste regulations. If the four-week composite sample exceeds any of the delisting levels set forth in Condition (3), the filter cake waste generated during the time period corresponding to the four-week composite sample must be retreated until it meets these levels (analyses must be repeated) or managed and disposed of in accordance with Subtitle C of RCRA. Filter cake which is generated but for which analyses are not complete or valid must be managed and disposed of in accordance with Subtitle C of RCRA, until valid analyses demonstrate that the waste meets the delisting levels.<br>(3) *Delisting Levels:* If the concentrations in the four-week composite sample of the filter cake waste for any of the hazardous constituents listed below exceed their respective maximum allowable concentrations (mg/l or mg/kg) also listed below, the four-week batch of failing filter cake waste must either be retreated until it meets these levels or managed and disposed of in accordance with Subtitle C of RCRA. GROWS has the option of determining whether the filter cake waste exceeds the maximum allowable concentrations for the organic constituents by either performing the analysis on a TCLP leachate of the waste or performing total constituent analysis on the waste, and then comparing the results to the corresponding maximum allowable concentration level. |
| | | (A) Inorganics<br><br>Constituent:<br>Arsenic ........................................................................ | Maximum Allowable Leachate Conc. (mg/l)<br><br><br>3.00e-01 |

201

| | |
|---|---|
| Barium | 2.34e+01 |
| Cadmium | 1.80e-01 |
| Chromium | 5.00e+00 |
| Lead | 5.00e+00 |
| Mercury | 7.70e-02 |
| Nickel | 9.05e+00 |
| Selenium | 6.97e-01 |
| Silver | 1.23e+00 |
| Cyanide | 4.33e+00 |

Cyanide extractions must be conducted using distilled water in place of the leaching media specified in the TCLP procedure.

| (B) Organics | Maximum allowable leachate conc. (mg/l) | Maximum allowable total conc. (mg/kg) |
|---|---|---|
| Constituent: | | |
| Acetone | 2.28e+01 | 4.56e+02 |
| Acetonitrile | 3.92e+00 | 7.84e+01 |
| Acetophenone | 2.28e+01 | 4.56e+02 |
| Acrolein | 1.53e+03 | 3.06e+04 |
| Acrylonitrile | 7.80e-03 | 1.56e-01 |
| Aldrin | 5.81e-06 | 1.16e-04 |
| Aniline | 7.39e-01 | 1.48e+01 |
| Anthracene | 8.00e+00 | 1.60e+02 |
| Benz(a)anthracene | 1.93e-04 | 3.86e-03 |
| Benzene | 1.45e-01 | 2.90e+00 |
| Benzo(a)pyrene | 1.18e-05 | 2.36e-04 |
| Benzo(b)fluoranthene | 1.07e-04 | 2.14e-03 |
| Benzo(k)fluoranthene | 1.49e-03 | 2.98e-02 |
| Bis(2-chloroethyl)ether | 3.19e-02 | 6.38e-01 |
| Bis(2-ethylhexyl)phthalate | 8.96e-02 | 1.79e+00 |
| Bromodichloromethane | 6.80e-02 | 1.36e+00 |
| Bromoform (Tribromomethane) | 5.33e-01 | 1.07e+01 |
| Butyl-4,6-dinitrophenol, 2-sec-(Dinoseb) | 2.28e-01 | 4.56e+00 |
| Butylbenzylphthalate | 9.29e+00 | 1.86e+02 |
| Carbon disulfide | 2.28e+01 | 4.56e+02 |
| Carbon tetrachloride | 4.50e-02 | 9.00e-01 |
| Chlordane | 5.11e-04 | 1.02e-02 |
| Chloro-3-methylphenol 4- | 2.97e+02 | 5.94e+03 |
| Chloroaniline, p- | 9.14e-01 | 1.83e+01 |
| Chlorobenzene | 6.08e+00 | 1.22e+02 |
| Chlorobenzilate | 4.85e-02 | 9.70e-01 |
| Chlorodibromomethane | 5.02e-02 | 1.00e+00 |
| Chloroform | 7.79e-02 | 1.56e+00 |
| Chlorophenol, 2- | 1.14e+00 | 2.28e+01 |
| Chrysene | 2.04e-02 | 4.08e-01 |
| Cresol | 1.14e+00 | 2.28e+01 |
| DDD | 5.83e-04 | 1.17e-02 |
| DDE | 1.37e-04 | 2.74e-03 |
| DDT | 2.57e-04 | 5.14e-03 |
| Dibenz(a,h)anthracene | 5.59e-06 | 1.12e-04 |
| Dibromo-3-chloropropane, 1,2- | 3.51e-03 | 7.02e-02 |
| Dichlorobenzene 1,3- | 9.35e+00 | 1.87e+02 |
| Dichlorobenzene, 1,2- | 1.25e+01 | 2.50e+02 |
| Dichlorobenzene, 1,4- | 1.39e-01 | 2.78e+00 |
| Dichlorobenzidine, 3,3'- | 9.36e-03 | 1.87e-01 |
| Dichlorodifluoromethane | 4.57e+01 | 9.14e+02 |
| Dichloroethane, 1,1- | 1.20e+00 | 2.40e+01 |
| Dichloroethane, 1,2- | 2.57e-03 | 5.14e-02 |
| Dichloroethylene, 1,1- | 7.02e-03 | 1.40e-01 |
| Dichloroethylene, trans-1,2- | 4.57e+00 | 9.14e+01 |
| Dichlorophenol, 2,4- | 6.85e-01 | 1.37e+01 |
| Dichlorophenoxyacetic acid, 2,4-(2,4-D) | 2.28e+00 | 4.56e+01 |
| Dichloropropane, 1,2- | 1.14e-01 | 2.28e+00 |
| Dichloropropene, 1,3- | 2.34e-02 | 4.68e-01 |
| Dieldrin | 6.23e+01 | 1.25e+03 |
| Diethyl phthalate | 2.21e+02 | 4.42e+03 |
| Dimethoate | 6.01e+01 | 1.20e+03 |
| Dimethyl phthalate | 1.20e+02 | 2.40e+03 |
| Dimethylbenz(a)anthracene, 7,12- | 1.55e-06 | 3.10e-05 |
| Dimethylphenol, 2,4- | 4.57e+00 | 9.14e+01 |
| Di-n-butyl phthalate | 5.29e+00 | 1.06e+02 |
| Dinitrobenzene, 1,3- | 2.28e-02 | 4.56e-01 |
| Dinitromethylphenol, 4,6-,2- | 2.16e-02 | 4.32e-01 |

| | | |
|---|---|---|
| Dinitrophenol, 2,4- | 4.57e-01 | 9.14e+00 |
| Dinitrotoluene, 2,6- | 6.54e-03 | 1.31e-01 |
| Di-n-octyl phthalate | 1.12e-02 | 2.24e-01 |
| Dioxane, 1,4- | 3.83e-01 | 7.66e+00 |
| Diphenylamine | 3.76e+00 | 7.52e+01 |
| Disulfoton | 3.80e+02 | 7.60e+03 |
| Endosulfan | 1.37e+00 | 2.74e+01 |
| Endrin | 2.00e-02 | 4.00e-01 |
| Ethylbenzene | 1.66e+01 | 3.32e+02 |
| Ethylene Dibromide | 4.13e-03 | 8.26e-02 |
| Fluoranthene | 5.16e-01 | 1.03e+01 |
| Fluorene | 1.78e+00 | 3.56e+01 |
| Heptachlor | 8.00e-03 | 1.60e-01 |
| Heptachlor epoxide | 8.00e-03 | 1.60e-01 |
| Hexachloro-1,3-butadiene | 9.61e-03 | 1.92e-01 |
| Hexachlorobenzene | 9.67e-05 | 1.93e-03 |
| Hexachlorocyclohexane, gamma-(Lindane) | 4.00e-01 | 8.00e+00 |
| Hexachlorocyclopentadiene | 1.66e+04 | 3.32e+05 |
| Hexachloroethane | 1.76e-01 | 3.52e+00 |
| Hexachlorophene | 3.13e-04 | 6.26e-03 |
| Indeno(1,2,3-cd) pyrene | 6.04e-05 | 1.21e-03 |
| Isobutyl alcohol | 6.85e+01 | 1.37e+03 |
| Isophorone | 4.44e+00 | 8.88e+01 |
| Methacrylonitrile | 2.28e-02 | 4.56e-01 |
| Methoxychlor | 1.00e+01 | 2.00e+02 |
| Methyl bromide (Bromomethane) | 1.28e+02 | 2.56e+03 |
| Methyl chloride (Chloromethane) | 1.80e-01 | 3.60e+00 |
| Methyl ethyl ketone | 1.37e+02 | 2.74e+03 |
| Methyl isobutyl ketone | 1.83e+01 | 3.66e+02 |
| Methyl methacrylate | 1.03e+03 | 2.06e+04 |
| Methyl parathion | 1.27e+02 | 2.54e+03 |
| Methylene chloride | 2.88e-01 | 5.76e+00 |
| Naphthalene | 1.50e+00 | 3.00e+01 |
| Nitrobenzene | 1.14e-01 | 2.28e+00 |
| Nitrosodiethylamine | 2.81e-05 | 5.62e-04 |
| Nitrosodimethylamine | 8.26e-05 | 1.65e-03 |
| Nitrosodi-n-butylamine | 7.80e-04 | 1.56e-02 |
| N-Nitrosodi-n-propylamine | 6.02e-04 | 1.20e-02 |
| N-Nitrosodiphenylamine | 8.60e-01 | 1.72e+01 |
| N-Nitrosopyrrolidine | 2.01e-03 | 4.02e-02 |
| Pentachlorobenzene | 1.15e-02 | 2.30e-01 |
| Pentachloronitrobenzene (PCNB) | 5.00e-03 | 1.00e-01 |
| Pentachlorophenol | 4.10e-03 | 8.20e-02 |
| Phenanthrene | 2.09e-01 | 4.18e+00 |
| Phenol | 1.37e+02 | 2.74e+03 |
| Polychlorinated biphenyls | 3.00e-05 | 6.00e-04 |
| Pronamide | 1.71e+01 | 3.42e+02 |
| Pyrene | 3.96e-01 | 7.92e+00 |
| Pyridine | 2.28e-01 | 4.56e+00 |
| Styrene | 6.08e+00 | 1.22e+02 |
| Tetrachlorobenzene, 1,2,4,5- | 9.43e-03 | 1.89e-01 |
| Tetrachloroethane, 1,1,2,2- | 4.39e-01 | 8.78e+00 |
| Tetrachloroethylene | 8.55e-02 | 1.71e+00 |
| Tetrachlorophenol, 2,3,4,6- | 1.81e+00 | 3.62e+01 |
| Tetraethyl dithiopyrophosphate (Sulfotep) | 3.01e+05 | 6.02e+06 |
| Toluene | 4.57e+01 | 9.14e+02 |
| Toxaphene | 5.00e-01 | 1.00e+01 |
| Trichlorobenzene, 1,2,4- | 7.24e-01 | 1.45e+01 |
| Trichloroethane, 1,1,1- | 7.60e+00 | 1.52e+02 |
| Trichloroethane, 1,1,2- | 7.80e-02 | 1.56e+00 |
| Trichloroethylene | 3.04e-01 | 6.08e+00 |
| Trichlorofluoromethane | 6.85e+01 | 1.37e+03 |
| Trichlorophenol, 2,4,5- | 9.16e+00 | 1.83e+02 |
| Trichlorophenol, 2,4,6- | 2.76e-01 | 5.52e+00 |
| Trichlorophenoxyacetic acid, 2,4,5-(245–T) | 2.28e+00 | 4.56e+01 |
| Trichlorophenoxypropionic acid, 2,4,5-(Silvex) | 1.00e+00 | 2.00e+01 |
| Trichloropropane, 1,2,3- | 7.69e-04 | 1.54e-02 |
| Trinitrobenzene, sym- | 6.49e+00 | 1.30e+02 |
| Vinyl chloride | 2.34e-03 | 4.68e-02 |
| Xylenes (total) | 3.20e+02 | 6.40e+03 |

TABLE 1—WASTES EXCLUDED FROM NON-SPECIFIC SOURCES

| Facility | Address | Waste description |
|---|---|---|
| | | (4) *Changes in Operating Conditions:* If GROWS significantly changes the treatment process or the chemicals used in the treatment process, GROWS may not manage the treatment sludge filter cake generated from the new process under this exclusion until it has met the following conditions: (a) GROWS must demonstrate that the waste meets the delisting levels set forth in Paragraph 3; (b) it must demonstrate that no new hazardous constituents listed in Appendix VIII of Part 261 have been introduced into the manufacturing or treatment process: and (c) it must obtain prior written approval from EPA and the Pennsylvania Department of Environmental Protection to manage the waste under this exclusion. |
| | | (5) *Reopener:* |
| | | (a) If GROWS discovers that a condition at the facility or an assumption related to the disposal of the excluded waste that was modeled or predicted in the petition does not occur as modeled or predicted, then GROWS must report any information relevant to that condition, in writing, to the Regional Administrator or his delegate and to the Pennsylvania Department of Environmental Protection within 10 days of discovering that condition. |
| | | (b) Upon receiving information described in paragraph (a) of this section, regardless of its source, the Regional Administrator or his delegate and the Pennsylvania Department of Environmental Protection will determine whether the reported condition requires further action. Further action may include repealing the exclusion, modifying the exclusion, or other appropriate response necessary to protect human health and the environment. |
| Goodyear Tire and Rubber Co. | Randleman, NC. | Dewatered wastewater treatment sludges (EPA Hazardous Waste No. F006) generated from electroplating operations. |
| Gould, Inc. ...... | McConnelsville, OH. | Wastewater treatment sludge (EPA Hazardous Waste No. F006) generated from electroplating operations after November 27, 1985. |
| Hoechst Celanese Corporation. | Bucks, Alabama. | Distillation bottoms generated (at a maximum annual rate of 31,500 cubic yards) from the production of sodium hydrosulfite (EPA Hazardous Waste No. F003). This exclusion was published on July 17, 1990. This exclusion does not include the waste contained in Hoechst Celanese's on-site surface impoundment. |
| Hoechst Celanese Corporation. | Leeds, South Carolina. | Distillation bottoms generated (at a maximum annual rate of 38,500 cubic yards) from the production of sodium hydrosulfite (EPA Hazardous Waste No. F003). This exclusion was published on July 17, 1990. |
| Hanover Wire Cloth Division. | Hanover, Pennsylvania. | Dewatered filter cake (EPA Hazardous Waste No. F006) generated from electroplating operations after August 15, 1986. |
| Holston Army Ammunition Plant. | Kingsport, Tennessee. | Dewatered wastewater treatment sludges (EPA Hazardous Waste Nos. F003, F005, and K044) generated from the manufacturing and processing of explosives and containing spent non-halogenated solvents after November 14, 1986. |
| Imperial Clevite | Salem, IN ....... | Solid resin cakes containing EPA Hazardous Waste No. F002 generated after August 27, 1985, from solvent recovery operations. |
| Indiana Steel & Wire Corporation (formerly General Cable Co.). | Munci, IN ....... | Dewatered wastewater treatment sludges (EPA Hazardous Waste Nos. F006 and K062) generated from electroplating operations and steel finishing operations after October 24, 1986. This exclusion does not apply to sludges in any on-site impoundments as of this date. |
| International Minerals and Chemical Corporation. | Terre Haute, Indiana. | Spent non-halogenated solvents and still bottoms (EPA Hazardous Waste No. F003) generated from the recovery of n-butyl alchohol after August 15, 1986. |
| Kawneer Company, Incorporated. | Springdale, Arkansas. | Wastewater treatment filter press sludge (EPA Hazardous Waste No. F019) generated (at a maximum annual rate of 26 cubic yards) from the chemical conversion coating of aluminum. This exclusion was published on November 13, 1990. |
| Kay-Fries, Inc. | Stoney Point, NY. | Biological aeration lagoon sludge and filter press sludge generated after September 21, 1984, which contain EPA Hazardous Waste Nos. F003 and F005 as well as that disposed of in a holding lagoon as of September 21, 1984. |
| Keymark Corp. | Fonda, NY ..... | Wastewater treatment sludge (EPA Hazardous Waste No. F019) generated from chemical conversion coating of aluminum after November 27, 1985. |
| Keymark Corp. | Fonda, NY ..... | Wastewater treatment sludges (EPA Hazardous Waste No. F019) generated from the chemical conversion coating of aluminum and contained in an on-site impoundment on August 12, 1987. This is a one-time exclusion. |
| Lawrence Berkeley National Laboratory. | Berkeley, California. | Treated ignitable and spent halogenated and non-halogenated solvent mixed waste (D001, F002, F003, and F005), and bubbler water on silica gel generated during treatment at the National Tritium Labeling Facility (NTLF) of the Lawrence Berkeley National Laboratory (LBNL). This is a one-time exclusion for 200 U.S. gallons of treatment residues that will be disposed of in a Nuclear Regulatory Commission (NRC) licensed or Department of Energy (DOE) approved low-level radioactive waste disposal facility, after August 7, 2003. |
| | | (1) Waste Management: The treated waste residue and bubbler water on silica gel must be managed in accordance with DOE or NRC requirements prior to and during disposal. |

TABLE 1—WASTES EXCLUDED FROM NON-SPECIFIC SOURCES—Continued

| Facility | Address | Waste description |
|---|---|---|
| | | (2) Reopener Language: (A) If, anytime after disposal of the delisted waste, LBNL possesses or is otherwise made aware of any data (including but not limited to leachate data or groundwater monitoring data) relevant to the delisted waste indicating that any organic constituent from the waste is detected in the leachate or the groundwater, then LBNL must report such data, in writing, to the Regional Administrator within 10 days of first possessing or being made aware of that data. |
| | | (B) Based on the information described in paragraph (2)(A) and any other information received from any source, the Regional Administrator will make a preliminary determination as to whether the reported information requires Agency action to protect human health or the environment. Further action may include suspending, or revoking the exclusion, or other appropriate response necessary to protect human health and the environment. |
| | | (C) If the Regional Administrator determines that the reported information does require Agency action, the Regional Administrator will notify LBNL in writing of the actions the Regional Administrator believes are necessary to protect human health and the environment. The notice shall include a statement of the proposed action and a statement providing LBNL with an opportunity to present information as to why the proposed Agency action is not necessary or to suggest an alternative action. LBNL shall have 30 days from the date of the Regional Administrator's notice to present the information. (D) If after 30 days LBNL presents no further information, the Regional Administrator will issue a final written determination describing the Agency actions that are necessary to protect human health or the environment. Any required action described in the Regional Administrator's determination shall become effective immediately, unless the Regional Administrator provides otherwise. |
| | | (3) Notification Requirements: LBNL must do the following before transporting the delisted waste off-site:(A) Provide a one-time written notification to any State Regulatory Agency to which or through which they will transport the delisted waste described above for disposal, 60 days before beginning such activities. (B) Update the one-time written notification if LBNL ships the delisted waste to a different disposal facility. Failure to provide this notification will result in a violation of the delisting petition and a possible revocation of the exclusion. |
| Lederle Laboratories. | Pearl River, NY. | Spent non-halogenated solvents and still bottoms (EPA Hazardous Waste Nos. F003 and F005) generated from the recovery of the following solvents: Xylene, acetone, ethyl acetate, ethyl ether, methyl isobutyl ketone, n-butyl alcohol, cyclohexanone, methanol, toluene, and pyridine after August 2, 1988. Excusion applies to primary and secondary filter press sludges and compost soils generated from these sludges. |
| Lincoln Plating Company. | Lincoln, NE .... | Wastewater treatment sludges (EPA Hazardous Waste No. F006) generated from electroplating operations after November 17, 1986. |
| Lockheed Martin Aeronautics Company. | Fort Worth, TX | Sludge (EPA Hazardous Waste Number F019) generated at a maximum rate of 90 cubic yards per calendar year after October 9, 2008. |
| | | For the exclusion to be valid, Lockheed Martin Aeronautics Company must implement a verification testing program that meets the following Paragraphs: |
| | | (1) Delisting Levels: All concentrations for those constituents must not exceed the maximum allowable concentrations in mg/l specified in this paragraph. |
| | | Sludge Leachable Concentrations (mg/l): Antimony—8.45; Arsenic—0.657; Barium—100.0; Cadmium—1.00; Chromium—5.0; Chromium, Hexavalent—5.0; Cobalt—1040; Copper—1810; Cyanide—240; Lead—5.0; Mercury—0.20; Nickel—1040; Selenium—1.0; Silver—5.0; Vanadium—51.5; Zinc—15800; Acetone—40600; Acetonitrile—766; Carbon Disulfide—4400; Ethylbenzene—846; Methyl Ethyl Ketone—200.0; Methyl Isobutyl Ketone—3610; Methylene Chloride—6.16; Toluene—1180; Xylenes—745. |
| | | (2) Waste Holding and Handling: |
| | | (A) Waste classification as non-hazardous can not begin until compliance with the limits set in paragraph (1) for sludge has occurred for two consecutive quarterly sampling events. |
| | | (B) If constituent levels in any sample taken by Lockheed Martin Aeronautics Company exceed any of the delisting levels set in paragraph (1) for the sludge, Lockheed Martin Aeronautics Company must do the following: |
| | | (i) notify EPA in accordance with paragraph (6) and |
| | | (ii) manage and dispose the sludge as hazardous waste generated under Subtitle C of RCRA. |
| | | (3) Testing Requirements: |
| | | Upon this exclusion becoming final, Lockheed Martin Aeronautics Company may perform quarterly analytical testing by sampling and analyzing the sludge as follows: |
| | | (A) Quarterly Testing: |
| | | (i) Collect two representative composite samples of the sludge at quarterly intervals after EPA grants the final exclusion. The first composite samples may be taken at any time after EPA grants the final approval. Sampling should be performed in accordance with the sampling plan approved by EPA in support of the exclusion. |
| | | (ii) Analyze the samples for all constituents listed in paragraph (1). Any composite sample taken that exceeds the delisting levels listed in paragraph (1) for the sludge must be disposed as hazardous waste in accordance with the applicable hazardous waste requirements. |

TABLE 1—WASTES EXCLUDED FROM NON-SPECIFIC SOURCES—Continued

| Facility | Address | Waste description |
|---|---|---|
| | | (iii) Within thirty (30) days after taking each quarterly sample, Lockheed Martin Aeronautics Company will report its quarterly analytical test data to EPA. If levels of constituents measured in the samples of the sludge do not exceed the levels set forth in paragraph (1) of this exclusion for two consecutive quarters or sampling events, Lockheed Martin Aeronautics Company can manage and dispose the non-hazardous sludge according to all applicable solid waste regulations. |
| | | (B) Annual Testing: |
| | | (i) If Lockheed Martin Aeronautics Company completes the quarterly testing specified in paragraph (3) above and no sample contains a constituent at a level which exceeds the limits set forth in paragraph (1), Lockheed Martin Aeronautics Company may begin annual testing as follows: Lockheed Martin Aeronautics Company must test two representative composite samples of the sludge for all constituents listed in paragraph (1) at least once per calendar year. |
| | | (ii) The samples for the annual testing shall be a representative composite sample according to appropriate methods. As applicable to the method-defined parameters of concern, analyses requiring the use of SW–846 methods incorporated by reference in 40 CFR 260.11 must be used without substitution. As applicable, the SW–846 methods might include Methods 0010, 0011, 0020, 0023A, 0030, 0031, 0040, 0050, 0051, 0060, 0061, 1010A, 1020B, 1110A, 1310B, 1311, 1312, 1320, 1330A, 9010C, 9012B, 9040C, 9045D, 9060A, 9070A (uses EPA Method 1664, Rev. A), 9071B, and 9095B. Methods must meet Performance Based Measurement System Criteria in which the Data Quality Objectives are to demonstrate that samples of the Lockheed Martin Aeronautics Company sludge are representative for all constituents listed in paragraph (1). |
| | | (iii) The samples for the annual testing taken for the second and subsequent annual testing events shall be taken within the same calendar month as the first annual sample taken. |
| | | (iv) The annual testing report should include the total amount of waste in cubic yards disposed during the calendar year. |
| | | (4) Changes in Operating Conditions: If Lockheed Martin Aeronautics Company significantly changes the process described in its petition or starts any processes that generate(s) the waste that may or could affect the composition or type of waste generated (by illustration, but not limitation, changes in equipment or operating conditions of the treatment process), it must notify EPA in writing and it may no longer handle the wastes generated from the new process as non-hazardous until the wastes meet the delisting levels set in paragraph (1) and it has received written approval to do so from EPA. |
| | | Lockheed Martin Aeronautics Company must submit a modification to the petition complete with full sampling and analysis for circumstances where the waste volume changes and/or additional waste codes are added to the waste stream. |
| | | (5) Data Submittals: |
| | | Lockheed Martin Aeronautics Company must submit the information described below. If Lockheed Martin Aeronautics Company fails to submit the required data within the specified time or maintain the required records on-site for the specified time, EPA, at its discretion, will consider this sufficient basis to reopen the exclusion as described in paragraph (6). Lockheed Martin Aeronautics Company must: |
| | | (A) Submit the data obtained through paragraph (3) to the Chief, Corrective Action and Waste Minimization Section, Multimedia Planning and Permitting Division, U.S. Environmental Protection Agency Region 6, 1445 Ross Ave., Dallas, Texas, 75202, within the time specified. All supporting data can be submitted on CD-ROM or some comparable electronic media. |
| | | (B) Compile records of analytical data from paragraph (3), summarized, and maintained on-site for a minimum of five years. |
| | | (C) Furnish these records and data when either EPA or the State of Texas requests them for inspection. |
| | | (D) Send along with all data a signed copy of the following certification statement, to attest to the truth and accuracy of the data submitted: |
| | | "Under civil and criminal penalty of law for the making or submission of false or fraudulent statements or representations (pursuant to the applicable provisions of the Federal Code, which include, but may not be limited to, 18 U.S.C. 1001 and 42 U.S.C. 6928), I certify that the information contained in or accompanying this document is true, accurate and complete. |
| | | As to the (those) identified section(s) of this document for which I cannot personally verify its (their) truth and accuracy, I certify as the company official having supervisory responsibility for the persons who, acting under my direct instructions, made the verification that this information is true, accurate and complete. |
| | | If any of this information is determined by EPA in its sole discretion to be false, inaccurate or incomplete, and upon conveyance of this fact to the company, I recognize and agree that this exclusion of waste will be void as if it never had effect or to the extent directed by EPA and that the company will be liable for any actions taken in contravention of the company's RCRA and CERCLA obligations premised upon the company's reliance on the void exclusion." |
| | | (6) Reopener: |

TABLE 1—WASTES EXCLUDED FROM NON-SPECIFIC SOURCES—Continued

| Facility | Address | Waste description |
|---|---|---|
| | | (A) If, anytime after disposal of the delisted waste Lockheed Martin Aeronautics Company possesses or is otherwise made aware of any environmental data (including but not limited to leachate data or ground water monitoring data) or any other data relevant to the delisted waste indicating that any constituent identified for the delisting verification testing is at level higher than the delisting level allowed by the Division Director in granting the petition, then the facility must report the data, in writing, to the Division Director within 10 days of first possessing or being made aware of that data. |
| | | (B) If either the quarterly or annual testing of the waste does not meet the delisting requirements in paragraph 1, Lockheed Martin Aeronautics Company must report the data, in writing, to the Division Director within 10 days of first possessing or being made aware of that data. |
| | | (C) If Lockheed Martin Aeronautics Company fails to submit the information described in paragraphs (5), (6)(A) or (6)(B) or if any other information is received from any source, the Division Director will make a preliminary determination as to whether the reported information requires EPA action to protect human health and/or the environment. Further action may include suspending, or revoking the exclusion, or other appropriate response necessary to protect human health and the environment. |
| | | (D) If the Division Director determines that the reported information requires action by EPA, the Division Director will notify the facility in writing of the actions the Division Director believes are necessary to protect human health and the environment. The notice shall include a statement of the proposed action and a statement providing the facility with an opportunity to present information as to why the proposed EPA action is not necessary. The facility shall have 10 days from the date of the Division Director's notice to present such information. |
| | | (E) Following the receipt of information from the facility described in paragraph (6)(D) or (if no information is presented under paragraph (6)(D)) the initial receipt of information described in paragraphs (5), (6)(A) or (6)(B), the Division Director will issue a final written determination describing EPA actions that are necessary to protect human health and/or the environment. Any required action described in the Division Director's determination shall become effective immediately, unless the Division Director provides otherwise. |
| | | (7) *Notification Requirements:* Lockheed Martin Aeronautics Company must do the following before transporting the delisted waste. Failure to provide this notification will result in a violation of the delisting petition and a possible revocation of the decision. |
| | | (A) Provide a one-time written notification to any state Regulatory Agency to which or through which it will transport the delisted waste described above for disposal, 60 days before beginning such activities. |
| | | (B) Update one-time written notification, if it ships the delisted waste into a different disposal facility. |
| | | (C) Failure to provide this notification will result in a violation of the delisting variance and a possible revocation of the decision. |
| Loxcreen Company, Inc.. | Hayti, MO ...... | Dewatered wastewater treatment sludges (EPA Hazardous Waste No. F019) generated from the chemical conversion coating of aluminum after July 16, 1986. |
| MAHLE, Inc. ... | Morristown, Tennessee. | Wastewater treatment sludge filter cake (EPA Hazardous Waste No. F019) generated from the chemical conversion coating of aluminum (generated at a maximum annual rate of 33 cubic yards), after August 21, 1992. In order to confirm that the characteristics of the waste do not change significantly, the facility must, on an annual basis sample and test for the constituents listed in 40 CFR 261.24 using the method specified therein. The annual analytical results (including quality control information) must be compiled, certified according to 40 CFR 260.22(i)(12), maintained on-site for a minimum of five years, and made available for inspection upon request by representatives of EPA or the State of Tennessee. Failure to maintain the required records on-site will be considered by EPA, at its discretion, sufficient basis to revoke the exclusion to the extent directed by EPA. |
| Marquette Electronics Incorporated. | Milwaukee, Wisconsin. | Wastewater treatment sludge (EPA Hazardous Waste No. F006) generated from electroplating operations. This exclusion was published on April 20, 1989. |
| Martin Marietta Aerospace. | Ocala, Florida | Dewatered wastewater treatment sludges (EPA Hazardous Waste No. F006) generated from electroplating operations after January 23, 1987. |
| Mason Chamberlain, Incorporated. | Bay St. Louis, Mississippi. | Wastewater treatment sludge filter cake (EPA Hazardous Waste No. F019) generated (at a maximum annual rate of 1,262 cubic yards) from the chemical conversion coating of aluminum. This exclusion was published on October 27, 1989. |
| Maytag Company. | Newton, IA ..... | Wastewater treatment sludges (EPA Hazardous Waste No. F006) generated from electroplating operations and wastewater treatment sludges (EPA Hazardous Waste No. F019) generated from the chemical conversion coating of aluminum November 17, 1986. |
| McDonnell Douglas Corporation. | Tulsa, Oklahoma. | Stabilized wastewater treatment sludges from surface impoundments previously closed as a landfill (at a maximum generation of 85,000 cubic yards on a one-time basis). EPA Hazardous Waste No. F019, F002, F003, and F005 generated at U.S. Air Force Plant No. 3, Tulsa, Oklahoma and is disposed of in Subtitle D landfills after February 26, 1999. McDonnell Douglas must implement a testing program that meets the following conditions for the exclusion to be valid: |

TABLE 1—WASTES EXCLUDED FROM NON-SPECIFIC SOURCES—Continued

| Facility | Address | Waste description |
|---|---|---|
| | | (1) *Delisting Levels:* All leachable concentrations for the constituents in Conditions (1)(A) and (1)(B) in the approximately 5,000 cubic yards of combined stabilization materials and excavated sludges from the bottom portion of the northwest lagoon of the surface impoundments which are closed as a landfill must not exceed the following levels (ppm) after the stabilization process is completed in accordance with Condition (3). Constituents must be measured in the waste leachate by the method specified in 40 CFR 261.24. Cyanide extractions must be conducted using distilled water in the place of the leaching media per 40 CFR 261.24. Constituents in Condition (1)(C) must be measured as the total concentrations in the waste(ppm). |
| | | (A) Inorganic Constituents (leachate) |
| | | Antimony-0.336; Cadmium-0.280; Chromium (total)-5.0; Lead-0.84; Cyanide-11.2; |
| | | (B) Organic Constituents (leachate) |
| | | Benzene-0.28; trans-1,2-Dichloroethene-5.6; Tetrachloroethylene-0.280; Trichloroethylene-0.280 |
| | | (C) Organic Constituents (total analysis). |
| | | Benzene-10.; Ethylbenzene-10.; Toluene-30.; Xylenes-30.; trans-1,2-Dichloroethene-30.; Tetrachloroethylene-6.0; Trichloroethylene-6.0. |
| | | McDonnell Douglas Corporation shall control volatile emissions from the stabilization process by collection of the volatile chemicals as they are emitted from the waste but before release to the ambient air. and the facility shall use dust control measures. These two controls must be adequate to protect human health and the environment. |
| | | The approximately 80,000 cubic yards of previously stabilized waste in the upper northwest lagoon, entire northeast lagoon, and entire south lagoon of the surface impoundments which were closed as a landfill requires no verification testing. |
| | | (2) *Waste Holding and Handling:* McDonnell Douglas must store as hazardous all stabilized waste from the bottom portion of the northwest lagoon area of the closed landfill as generated until verification testing as specified in Condition (3), is completed and valid analyses demonstrate that Condition (1) is satisfied. If the levels of constituents measured in the samples of the stabilized waste do not exceed the levels set forth in Condition (1), then the waste is nonhazardous and may be managed and disposed of in a Subtitle D landfill in accordance with all applicable solid waste regulations. If constituent levels in a sample exceed any of the delisting levels set in Condition (1), the waste generated during the time period corresponding to this sample must be restabilized until delisting levels are met or managed and disposed of in accordance with Subtitle C of RCRA. |
| | | (3) *Verification Testing Requirements:* Sample collection and analyses, including quality control procedures, must be performed using appropriate methods. As applicable to the method-defined parameters of concern, analyses requiring the use of SW-846 methods incorporated by reference in 40 CFR 260.11 must be used without substitution. As applicable, the SW-846 methods might include Methods 0010, 0011, 0020, 0023A, 0030, 0031, 0040, 0050, 0051, 0060, 0061, 1010A, 1020B, 1110A, 1310B, 1311, 1312, 1320, 1330A, 9010C, 9012B, 9040C, 9045D, 9060A, 9070A (uses EPA Method 1664, Rev. A), 9071B, and 9095B. McDonnell Douglas must stabilize the previously unstabilized waste from the bottom portion of the northwest lagoon of the surface impoundment (which was closed as a landfill) using fly ash, kiln dust or similar accepted materials in batches of 500 cubic yards or less. McDonnell Douglas must analyze one composite sample from each batch of 500 cubic yards or less. A minimum of four grab samples must be taken from each waste pile (or other designated holding area) of stabilized waste generated from each batch run. Each composited batch sample must be analyzed, prior to disposal of the waste in the batch represented by that sample, for constituents listed in Condition (1). There are no verification testing requirements for the stabilized wastes in the upper portions of the northwest lagoon, the entire northeast lagoon, and the entire south lagoon of the surface impoundments which were closed as a landfill. |
| | | (4) *Changes in Operating Conditions:* If McDonnell Douglas significantly changes the stabilization process established under Condition (3) (e.g., use of new stabilization agents), McDonnell Douglas must notify the Agency in writing. After written approval by EPA, McDonnell Douglas may handle the wastes generated as non-hazardous, if the wastes meet the delisting levels set in Condition (1). |
| | | (5) *Data Submittals:* Records of operating conditions and analytical data from Condition (3) must be compiled, summarized, and maintained on site for a minimum of five years. These records and data must be furnished upon request by EPA, or the State of Oklahoma, or both, and made available for inspection. Failure to submit the required data within the specified time period or maintain the required records on site for the specified time will be considered by EPA, at its discretion, sufficient basis to revoke the exclusion to the extent directed by EPA. All data must be accompanied by a signed copy of the following certification statement to attest to the truth and accuracy of the data submitted: |
| | | Under civil and criminal penalty of law for the making or submission of false or fraudulent statements or representations (pursuant to the applicable provisions of the Federal Code, which include, but may not be limited to, 18 U.S.C. § 1001 and 42 U.S.C. § 6928), I certify that the information contained in or accompanying this document is true, accurate and complete. |

TABLE 1—WASTES EXCLUDED FROM NON-SPECIFIC SOURCES—Continued

| Facility | Address | Waste description |
|---|---|---|
| | | As to the (those) identified section(s) of this document for which I cannot personally verify its (their) truth and accuracy, I certify as the company official having supervisory responsibility for the persons who, acting under my direct instructions, made the verification that this information is true, accurate and complete. |
| | | In the event that any of this information is determined by EPA in its sole discretion to be false, inaccurate or incomplete, and upon conveyance of this fact to the company, I recognize and agree that this exclusion of waste will be void as if it never had effect or to the extent directed by EPA and that the company will be liable for any actions taken in contravention of the company's RCRA and CERCLA obligations premised upon the company's reliance on the void exclusion. |
| | | (6) *Reopener Language* |
| | | (a) If McDonnell Douglas discovers that a condition at the facility or an assumption related to the disposal of the excluded waste that was modeled or predicted in the petition does not occur as modeled or predicted, then McDonnell Douglas must report any information relevant to that condition, in writing, to the Regional Administrator or his delegate within 10 days of discovering that condition. |
| | | (b) Upon receiving information described in paragraph (a) from any source, the Regional Administrator or his delegate will determine whether the reported condition requires further action. Further action may include revoking the exclusion, modifying the exclusion, or other appropriate response necessary to protect human health and the environment. |
| | | (7) *Notification Requirements:* McDonnell Douglas must provide a one-time written notification to any State Regulatory Agency to which or through which the delisted waste described above will be transported for disposal at least 60 days prior to the commencement of such activity. The one-time written notification must be updated if the delisted waste is shipped to a different disposal facility. Failure to provide such a notification will result in a violation of the delisting petition and a possible revocation of the decision. |
| Merck & Company, Incorporated. | Elkton, Virginia | One-time exclusion for fly ash (EPA Hazardous Waste No. F002) from the incineration of wastewater treatment sludge generated from pharmaceutical production processes and stored in an on-site fly ash lagoon. This exclusion was published on May 12, 1989. |
| Metropolitan Sewer District of Greater Cincinnati. | Cincinnati, OH | Sluiced bottom ash sludge (approximately 25,000 cubic yards), contained in the North Lagoon, on September 21, 1984, which contains EPA Hazardous Wastes Nos. F001, F002, F003, F004, and F005. |
| Michelin Tire Corp.. | Sandy Springs, South Carolina. | Dewatered wastewater treatment sludge (EPA Hazardous Wastes No. F006) generated from electroplating operations after November 14, 1986. |
| Monroe Auto Equipment. | Paragould, AR | Wastewater treatment sludge (EPA Hazardous Waste No. F006) generated from electroplating operations after vacuum filtration after November 27, 1985. This exclusion does not apply to the sludge contained in the on-site impoundment. |
| Nissan North America, Inc.. | Smyrna, Tennessee. | Wastewater treatment sludge (EPA Hazardous Waste No. F019) that Nissan North American, Inc. (Nissan) generates by treating wastewater from automobile assembly plant located on 983 Nissan Drive in Smyrna, Tennessee. This is a conditional exclusion for up to 3,500 cubic yards of waste (hereinafter referred to as "Nissan Sludge") that will be generated each year and disposed in a Subtitle D landfill after February 27, 2006. Nissan must continue to demonstrate that the following conditions are met for the exclusion to be valid. |
| | | (1) *Delisting Levels:* All leachable concentrations for these metals, cyanide, and organic constituents must not exceed the following levels (ppm): Barium-100.0; Cadmium-0.422; Chromium-5.0; Cyanide-7.73, Lead-5.0; and Nickel-60.7; Bis-(2-ethylhexyl) phthalate-0.601; Di-n-octyl phthalate-0.0752; and 4-Methylphenol-7.66. These concentrations must be measured in the waste leachate obtained by the method specified in 40 CFR 261.24, except that for cyanide, deionized water must be the leaching medium. Cyanide concentrations in waste or leachate must be measured by the method specified in 40 CFR 268.40, Note 7. |
| | | (2) *Verification Testing Requirements:* Sample collection and analyses, including quality control procedures, must be performed using appropriate methods. As applicable to the method-defined parameters of concern, analyses requiring the use of SW-846 methods incorporated by reference in 40 CFR 260.11 must be used without substitution. As applicable, the SW-846 methods might include Methods 0010, 0011, 0020, 0023A, 0030, 0031, 0040, 0050, 0051, 0060, 0061, 1010A, 1020B, 1110A, 1310B, 1311, 1312, 1320, 1330A, 9010C, 9012B, 9040C, 9045D, 9060A, 9070A, (uses EPA Method 1664, Rev. A), 9071B, and 9095B. Methods must meet Performance Based Measurement System Criteria in which the Data Quality Objectives are to demonstrate that representative samples of the Nissan Sludge meet the delisting levels in Condition (1). Nissan must perform an annual testing program to demonstrate that constituent concentrations measured in the TCLP extract do not exceed the delisting levels established in Condition (1). |

TABLE 1—WASTES EXCLUDED FROM NON-SPECIFIC SOURCES—Continued

| Facility | Address | Waste description |
|----------|---------|-------------------|
| | | (3) *Waste Holding and Handling:* Nissan must hold sludge containers utilized for verification sampling until composite sample results are obtained. If the levels of constituents measured in Nissan's annual testing program do not exceed the levels set forth in Condition (1), then the Nissan Sludge is non-hazardous and must be managed in accordance with all applicable solid waste regulations. If constituent levels in a composite sample exceed any of the delisting levels set forth in Condition (1), the batch of Nissan Sludge generated during the time period corresponding to this sample must be managed and disposed of in accordance with Subtitle C of RCRA. |
| | | (4) *Changes in Operating Conditions:* Nissan must notify EPA in writing when significant changes in the manufacturing or wastewater treatment processes are implemented. EPA will determine whether these changes will result in additional constituents of concern. If so, EPA will notify Nissan in writing that the Nissan Sludge must be managed as hazardous waste F019 until Nissan has demonstrated that the wastes meet the delisting levels set forth in Condition (1) and any levels established by EPA for the additional constituents of concern, and Nissan has received written approval from EPA. If EPA determines that the changes do not result in additional constituents of concern, EPA will notify Nissan, in writing, that Nissan must verify that the Nissan Sludge continues to meet Condition (1) delisting levels. |
| | | (5) *Data Submittals:* Data obtained in accordance with Condition (2) must be submitted to Narindar M. Kumar, Chief, RCRA Enforcement and Compliance Branch, Mail Code: 4WD–RCRA, U.S. EPA, Region 4, Sam Nunn Atlanta Federal Center, 61 Forsyth Street, SW., Atlanta, Georgia 30303. The submission is due no later than 60 days after taking each annual verification samples in accordance with delisting Conditions (1) through (7). Records of analytical data from Condition (2) must be compiled, summarized, and maintained by Nissan for a minimum of three years, and must be furnished upon request by EPA or the State of Tennessee, and made available for inspection. Failure to submit the required data within the specified time period or maintain the required records for the specified time will be considered by EPA, at its discretion, sufficient basis to revoke the exclusion to the extent directed by EPA. All data must be accompanied by a signed copy of the certification statement in 40 CFR 260.22(i)(12). |
| | | (6) *Reopener Language:* (A) If, at any time after disposal of the delisted waste, Nissan possesses or is otherwise made aware of any environmental data (including but not limited to leachate data or groundwater monitoring data) or any other data relevant to the delisted waste indicating that any constituent identified in the delisting verification testing is at a level higher than the delisting level allowed by EPA in granting the petition, Nissan must report the data, in writing, to EPA and Tennessee within 10 days of first possessing or being made aware of that data. (B) If the testing of the waste, as required by Condition (2), does not meet the delisting requirements of Condition (1), Nissan must report the data, in writing, to EPA and Tennessee within 10 days of first possessing or being made aware of that data. (C) Based on the information described in paragraphs (6)(A) or (6)(B) and any other information received from any source, EPA will make a preliminary determination as to whether the reported information requires that EPA take action to protect human health or the environment. Further action may include suspending or revoking the exclusion, or other appropriate response necessary to protect human health and the environment. (D) If EPA determines that the reported information does require Agency action, EPA will notify the facility in writing of the action believed necessary to protect human health and the environment. The notice shall include a statement of the proposed action and a statement providing Nissan with an opportunity to present information as to why the proposed action is not necessary. Nissan shall have 10 days from the date of EPA's notice to present such information. (E) Following the receipt of information from Nissan, as described in paragraph (6)(D), or if no such information is received within 10 days, EPA will issue a final written determination describing the Agency actions that are necessary to protect human health or the environment, given the information received in accordance with paragraphs (6)(A) or (6)(B). Any required action described in EPA's determination shall become effective immediately, unless EPA provides otherwise. |
| | | (7) *Notification Requirements:* Nissan must provide a one-time written notification to any State Regulatory Agency in a State to which or through which the delisted waste described above will be transported, at least 60 days prior to the commencement of such activities. Failure to provide such a notification will result in a violation of the delisting conditions and a possible revocation of the decision to delist. |
| North American Philips Consumer Electronics Corporation. | Greenville, Tennessee. | Wastewater treatment sludges (EPA Hazardous Waste No. F006) generated from electroplating operations. This exclusion was published on April 20, 1989. |
| Occidental Chemical. | Ingleside, Texas. | Limestone Sludge, (at a maximum generation 1,114 cubic yards per calender year) Rockbox Residue, (at a maximum generation of 1,000 cubic yards per calender year) generated by Occidental Chemical using the wastewater treatment process to treat the Rockbox Residue and the Limestone Sludge (EPA Hazardous Waste No. F025, F001, F003, and F005) generated at Occidental Chemical. |

TABLE 1—WASTES EXCLUDED FROM NON-SPECIFIC SOURCES—Continued

| Facility | Address | Waste description |
|---|---|---|
|  |  | Occidental Chemical must implement a testing program that meets the following conditions for the exclusion to be valid:<br>(1) *Delisting Levels:* All concentrations for the following constituents must not exceed the following levels (ppm). The Rockbox Residue and the Limestone Sludge, must be measured in the waste leachate by the method specified in 40 CFR Part 261.24.<br>(A) Rockbox Residue<br>(i) Inorganic Constituents: Barium-100; Chromium-5; Copper-130; Lead-1.5; Selenium-1; Tin-2100; Vanadium-30; Zinc-1,000<br>(ii) Organic Constituents: Acetone-400; Bromodichloromethane-0.14; Bromoform-1.0; Chlorodibromomethane-0.1; Chloroform-1.0; Dichloromethane-1.0; Ethylbenzene-7,000; 2,3,7,8-TCDD Equivalent-0.00000006<br>(B) Limestone Sludge<br>(i) Inorganic Constituents: Antimony-0.6; Arsenic-5; Barium-100; Beryllium-0.4; Chromium-5; Cobalt-210; Copper-130; Lead-1.5; Nickel-70; Selenium-5; Silver-5; Vanadium-30; Zinc-1,000<br>(ii) Organic Constituents Acetone-400; Bromoform-1.0; Chlorodibromomethane-0.1; Dichloromethane-1.0; Diethyl phthalate-3,000, Ethylbenzene-7,000; 1,1,1-Trichloroethane-20; Toluene-700; Trichlorofluoromethane-1,000, Xylene-10,000, 2,3,7,8-TCDD Equivalent-0.00000006;<br>(2) *Waste Holding and Handling:* Occidental Chemical must store in accordance with its RCRA permit, or continue to dispose of as hazardous waste all Rockbox Residue and the Limestone Sludge generated until the verification testing described in Condition (3)(B), as appropriate, is completed and valid analyses demonstrate that condition (3) is satisfied. If the levels of constituents measured in the samples of the Rockbox Residue and the Limestone Sludge do not exceed the levels set forth in Condition (1), then the waste is nonhazardous and may be managed and disposed of in accordance with all applicable solid waste regulations. If constituent levels in a sample exceed any of the delisting levels waste generated during the time period corresponding to this sample must be managed and disposed of in accordance with Subtitle C of RCRA.<br>(3) *Verification Testing Requirements:* Sample collection and analyses, including quality control procedures, must be performed using appropriate methods. As applicable to the method-defined parameters of concern, any analyses requiring use of SW–846 methods incorporated by reference in 40 CFR 260.11 must use those methods without substitution. As applicable, the SW–846 methods might include Methods 0010, 0011, 0020, 0023A, 0030, 0031, 0040, 0050, 0051, 0060, 0061, 1010A, 1020B, 1110A, 1310B, 1311, 1312, 1320, 1330A, 9010C, 9012B, 9040C, 9045D, 9060A, 9070A (uses EPA Method 1664, Rev. A), 9071B, and 9095B. If EPA judges the incineration process to be effective under the operating conditions used during the initial verification testing, Occidental Chemical may replace the testing required in Condition (3)(A) with the testing required in Condition (3)(B). Occidental Chemical must continue to test as specified in Condition (3)(A) until and unless notified by EPA in writing that testing in Condition (3)(A) may be replaced by Condition (3)(B).<br>(A) *Initial Verification Testing:* (i) During the first 40 operating days of the Incinerator Offgas Treatment System after the final exclusion is granted, Occidental Chemical must collect and analyze composites of the Limestone Sludge. Daily composites must be representative grab samples collected every 6 hours during each unit operating cycle. The two wastes must be analyzed, prior to disposal, for all of the constituents listed in Paragraph 1. The waste must also be analyzed for pH. Occidental Chemical must report the operational and analytical test data, including quality control information, obtained during this initial period no later than 90 days after the generation of the two wastes.<br>(ii) When the Rockbox unit is decommissioned for cleanout, after the final exclusion is granted, Occidental Chemical must collect and analyze composites of the Rockbox Residue. Two composites must be composed of representative grab samples collected from the Rockbox unit. The waste must be analyzed, prior to disposal, for all of the constituents listed in Paragraph 1. The waste must be analyzed for pH. No later than 90 days after the Rockbox is decommissioned for cleanout the first two times after this exclusion becomes final, Occidental Chemical must report the operational and analytical test data, including quality control information.<br>(B) *Subsequent Verification Testing:* Following written notification by EPA, Occidental Chemical may substitute the testing conditions in (3)(B) for (3)(A)(i). Occidental Chemical must continue to monitor operating conditions, analyze samples representative of each quarter of operation during the first year of waste generation. The samples must represent the waste generated over one quarter. (This provision does not apply to the Rockbox Residue.)<br>(C) *Termination of Organic Testing for the Limestone Sludge:* Occidental Chemical must continue testing as required under Condition (3)(B) for organic constituents specified under Condition (3)(B) for organic constituents specified in Condition (1)(A)(ii) and (1)(B)(ii) until the analyses submitted under Condition (3)(B) show a minimum of two consecutive quarterly samples below the delisting levels in Condition (1)(A)(ii) and (1)(B)(ii), Occidental Chemical may then request that quarterly organic testing be terminated. After EPA notifies Occidental Chemical in writing it may terminate quarterly organic testing. Following termination of the quarterly testing, Occidental Chemical must continue to test a representative composite sample for all constituents listed in Condition (1) on an annual basis (no later than twelve months after exclusion). |

TABLE 1—WASTES EXCLUDED FROM NON-SPECIFIC SOURCES—Continued

| Facility | Address | Waste description |
|---|---|---|
| | | (4) *Changes in Operating Conditions:* If Occidental Chemical significantly changes the process which generate(s) the waste(s) and which may or could affect the composition or type waste(s) generated as established under Condition (1) (by illustration, but not limitation, change in equipment or operating conditions of the treatment process), Occidental Chemical must notify the EPA in writing and may no longer handle the wastes generated from the new process or no longer discharges as nonhazardous until the wastes meet the delisting levels set Condition (1) and it has received written approval to do so from EPA. |
| | | (5) *Data Submittals:* The data obtained through Condition 3 must be submitted to Mr. William Gallagher, Chief, Region 6 Delisting Program, U.S. EPA, 1445 Ross Avenue, Dallas, Texas 75202–2733, Mail Code, (6PD-O) within the time period specified. Records of operating conditions and analytical data from Condition (1) must be compiled, summarized, and maintained on site for a minimum of five years. These records and data must be furnished upon request by EPA, or the State of Texas, and made available for inspection. Failure to submit the required data within the specified time period or maintain the required records on site for the specified time will be considered by EPA, at its discretion, sufficient basis to revoke the exclusion to the extent directed by EPA. All data must be accompanied by a signed copy of the following certification statement to attest to the truth and accuracy of the data submitted: |
| | | Under civil and criminal penalty of law for the making or submission of false or fraudulent statements or representations (pursuant to the applicable provisions of the Federal Code, which include, but may not be limited to, 18 U.S.C. §1001 and 42 U.S.C. §6928), I certify that the information contained in or accompanying this document is true, accurate and complete. |
| | | As to the (those) identified section(s) of this document for which I cannot personally verify its (their) truth and accuracy, I certify as the company official having supervisory responsibility for the persons who, acting under my direct instructions, made the verification that this information is true, accurate and complete. |
| | | In the event that any of this information is determined by EPA in its sole discretion to be false, inaccurate or incomplete, and upon conveyance of this fact to the company, I recognize and agree that this exclusion of waste will be void as if it never had effect or to the extent directed by EPA and that the company will be liable for any actions taken in contravention of the company's RCRA and CERCLA obligations premised upon the company's reliance on the void exclusion. |
| | | (6) *Reopener:* (a) If Occidental Chemical discovers that a condition at the facility or an assumption related to the disposal of the excluded waste that was modeled or predicted in the petition does not occur as modeled or predicted, then Occidental Chemical must report any information relevant to that condition, in writing, to the Director of the Multimedia Planning and Permitting Division or his delegate within 10 days of discovering that condition. (b) Upon receiving information described in paragraph (a) from any source, the Director or his delegate will determine whether the reported condition requires further action. Further action may include revoking the exclusion, modifying the exclusion, or other appropriate response necessary to protect human health and the environment. |
| | | (7) *Notification Requirements:* Occidental Chemical must provide a one-time written notification to any State Regulatory Agency to which or through which the delisted waste described above will be transported for disposal at least 60 days prior to the commencement of such activities. Failure to provide such a notification will result in a violation of the delisting petition and a possible revocation of the decision. |
| Philway Products, Incorporated. | Ashland, Ohio | Filter press sludge generated (at a maximum annual rate of 96 cubic yards) during the treatment of electroplating wastewaters using lime (EPA Hazardous Waste No. F006). This exclusion was published on October 26, 1990. |
| Plastene Supply Company. | Portageville, Missouri. | Dewatered wastewater treatment sludges (EPA Hazardous Waste No. F006) generated from electroplating operations after August 15, 1986. |
| POP Fasteners | Shelton, Connecticut. | Wastewater treatment sludge (EPA Hazardous Waste No. F006) generated from electroplating operations (at a maximum annual rate of 1,000 cubic yards) after September 19, 1994. In order to confirm that the characteristics of the waste do not change significantly, the facility must, on an annual basis, analyze a representative composite sample for the constituents listed in §261.24 using the method specified therein. The annual analytical results, including quality control information, must be compiled, certified according to §260.22(i)(12), maintained on site for a minimum of five years, and made available for inspection upon request by any employee or representative of EPA or the State of Connecticut. Failure to maintain the required records on site will be considered by EPA, at its discretion, sufficient basis to revoke the exclusion to the extent directed by EPA. |
| Professional Plating, Incorporated. | Brillion, Wisconsin. | Wastewater treatment sludges, F019, which are generated at the Professional Plating, Incorporated (PPI) Brillion facility at a maximum annual rate of 140 cubic yards per year. The sludge must be disposed of in a Subtitle D landfill which is licensed, permitted, or otherwise authorized by a State to accept the delisted wastewater treatment sludge. The exclusion becomes effective as of March 1, 2010. |
| | | 1. *Delisting Levels:* The constituent concentrations measured in a leachate extract may not exceed the following levels (mg/L): chromium—5, cobalt—10.4; manganese—815; and nickel—638. |

212

TABLE 1—WASTES EXCLUDED FROM NON-SPECIFIC SOURCES—Continued

| Facility | Address | Waste description |
|---|---|---|
| | | 2. *Annual Verification Testing:* To verify that the waste does not exceed the specified delisting levels, PPI must collect and analyze, annually, one waste sample for the constituents in Section 1. using methods with appropriate detection levels and elements of quality control. SW–846 Method 1311 must be used for generation of the leachate extract used in the testing of the delisting levels if oil and grease comprise less than 1% of the waste. SW–846 Method 1330A must be used for generation of the leaching extract if oil and grease comprise 1% or more of the waste. SW–846 Method 9071B must be used for determination of oil and grease. SW–846 Methods 1311, 1330A, and 9071B are incorporated by reference in 40 CFR 260.11. |
| | | 3. *Changes in Operating Conditions:* PPI must notify the EPA in writing if the manufacturing process, the chemicals used in the manufacturing process, the treatment process, or the chemicals used in the treatment process significantly change. PPI must handle wastes generated after the process change as hazardous until it has demonstrated that the wastes continue to meet the maximum allowable concentrations in Section 1. and that no new hazardous constituents listed in appendix VIII of part 261 have been introduced and it has received written approval from EPA. |
| | | 4. *Reopener Language*—(a) If, anytime after disposal of the delisted waste, PPI possesses or is otherwise made aware of any data (including but not limited to leachate data or groundwater monitoring data) relevant to the delisted waste indicating that any constituent is at a concentration in the waste or waste leachate higher than the maximum allowable concentrations in Section 1. above or is in the groundwater at a concentration higher than the maximum allowable groundwater concentrations in Paragraph (e), then PPI must report such data, in writing, to the Regional Administrator within 10 days of first possessing or being made aware of that data. |
| | | (b) Based on the information described in paragraph (a) and any other information received from any source, the Regional Administrator will make a preliminary determination as to whether the reported information requires Agency action to protect human health or the environment. Further action may include suspending, or revoking the exclusion, or other appropriate response necessary to protect human health and the environment. |
| | | (c) If the Regional Administrator determines that the reported information does require Agency action, the Regional Administrator will notify the facility in writing of the actions the Regional Administrator believes are necessary to protect human health and the environment. The notice shall include a statement of the proposed action and a statement providing PPI with an opportunity to present information as to why the proposed Agency action is not necessary or to suggest an alternative action. PPI shall have 30 days from the date of the Regional Administrator's notice to present the information. |
| | | (d) If after 30 days PPI presents no further information, the Regional Administrator will issue a final written determination describing the Agency actions that are necessary to protect human health or the environment. Any required action described in the Regional Administrator's determination shall become effective immediately, unless the Regional Administrator provides otherwise. |
| | | (e) Maximum allowable groundwater concentrations (mg/L) are as follows: chromium—0.1; cobalt—0.0113; manganese—0.9; and nickel—0.75. |
| Reynolds Metals Company. | Sheffield, AL .. | Dewatered wastewater treatment sludges (EPA Hazardous Waste No. F019) generated from the chemical conversion coating of aluminum after August 15, 1990. |
| Reynolds Metals Company. | Sheffield, AL .. | Wastewater treatment filter press sludge (EPA Hazardous Waste No. F019) generated (at a maximum annual rate of 3,840 cubic yards) from the chemical conversion coating of aluminum. This exclusion was published on July 17, 1990. |
| Rhodia ............ | Houston,Texas | Filter-cake Sludge, (at a maximum generation of 1,200 cubic yards per calendar year) generated by Rhodia using the SARU and AWT treatment process to treat the filter-cake sludge (EPA Hazardous Waste Nos. D001–D43, F001–F012, F019, F024, F025, F032, F034, F037–F039) generated at Rhodia. |
| | | Rhodia must implement a testing program that meets the following conditions for the exclusion to be valid: |
| | | (1) *Delisting Levels:* All concentrations for the following constituents must not exceed the following levels (mg/l). For the filter-cake constituents must be measured in the waste leachate by the method specified in 40 CFR 261.24. |
| | | (A) Filter-cake Sludge |
| | | (i) Inorganic Constituents: Antimony-1.15; Arsenic-1.40; Barium-21.00; Beryllium-1.22; Cadmium-0.11; Cobalt-189.00; Copper-90.00; Chromium-0.60; Lead-0.75; Mercury-0.025; Nickel-9.00; Selenium-4.50; Silver-0.14; Thallium-0.20; Vanadium-1.60; Zinc-4.30 |
| | | (ii) Organic Constituents: Chlorobenzene-Non Detect; Carbon Tetrachloride-Non Detect; Acetone-360; Chloroform-0.9 |
| | | (2) *Waste Holding and Handling:* Rhodia must store in accordance with its RCRA permit, or continue to dispose of as hazardous waste all Filter-cake Sludge until the verification testing described in Condition (3)(A), as appropriate, is completed and valid analyses demonstrate that condition (3) is satisfied. If the levels of constituents measured in the samples of the Filter-cake Sludge do not exceed the levels set forth in Condition (1), then the waste is nonhazardous and may be managed and disposed of in accordance with all applicable solid waste regulations. |

TABLE 1—WASTES EXCLUDED FROM NON-SPECIFIC SOURCES—Continued

| Facility | Address | Waste description |
|---|---|---|
| | | (3) *Verification Testing Requirements:* Rhodia must perform sample collection and analyses, including quality control procedures, using appropriate methods. As applicable to the method-defined parameters of concern, analyses requiring the use of SW–846 methods incorporated by reference in 40 CFR 260.11 must be used without substitution. As applicable, the SW–846 methods might include Methods 0010, 0011, 0020, 0023A, 0030, 0031, 0040, 0050, 0051, 0060, 0061, 1010A, 1020B, 1110A, 1310B, 1311, 1312, 1320, 1330A, 9010C, 9012B, 9040C, 9045D, 9060A, 9070A (uses EPA Method 1664, Rev. A), 9071B, and 9095B. If EPA judges the process to be effective under the operating conditions used during the initial verification testing, Rhodia may replace the testing required in Condition (3)(A) with the testing required in Condition (3)(B). Rhodia must continue to test as specified in Condition (3)(A) until and unless notified by EPA in writing that testing in Condition (3)(A) may be replaced by Condition (3)(B). <br><br>(A) *Initial Verification Testing:* At quarterly intervals for one year after the final exclusion is granted, Rhodia must collect and analyze composites of the filter-cake sludge. From Paragraph 1 TCLP must be run on all waste and any constituents for which total concentrations have been identified. Rhodia must conduct a multiple pH leaching procedure on samples collected during the quarterly intervals. Rhodia must perform the TCLP procedure using distilled water and three different pH extraction fluids to simulate disposal under three conditions. Simulate an acidic landfill environment, basic landfill environment and a landfill environment similar to the pH of the waste. Rhodia must report the operational and analytical test data, including quality control information, obtained during this initial period no later than 90 days after the generation of the waste. <br><br>(B) *Subsequent Verification Testing:* Following termination of the quarterly testing, Rhodia must continue to test a representative composite sample for all constituents listed in Condition (1) on an annual basis (no later than twelve months after the final exclusion). <br><br>(4) *Changes in Operating Conditions:* If Rhodia significantly changes the process which generate(s) the waste(s) and which may or could affect the composition or type waste(s) generated as established under Condition (1) (by illustration, but not limitation, change in equipment or operating conditions of the treatment process), or its NPDES permit is changed, revoked or not reissued, Rhodia must notify the EPA in writing and may no longer handle the waste generated from the new process or no longer discharge as nonhazardous until the waste meet the delisting levels set in Condition (1) and it has received written approval to do so from EPA. <br><br>(5) *Data Submittals:* Rhodia must submit the information described below. If Rhodia fails to submit the required data within the specified time or maintain the required records on-site for the specified time, EPA, at its discretion, will consider this sufficient basis to reopen the exclusion as described in Paragraph 6. Rhodia must: <br><br>(A) Submit the data obtained through Paragraph 3 to Mr. William Gallagher, Chief, Region 6 Delisting Program, EPA, 1445 Ross Avenue, Dallas, Texas 75202–2733, Mail Code, (6PD-O) within the time specified. <br><br>(B) Compile records of operating conditions and analytical data from Paragraph (3), summarized, and maintained on-site for a minimum of five years. <br><br>(C) Furnish these records and data when EPA or the State of Texas request them for inspection. <br><br>(D) Send along with all data a signed copy of the following certification statement, to attest to the truth and accuracy of the data submitted: <br><br>(i) Under civil and criminal penalty of law for the making or submission of false or fraudulent statements or representations (pursuant to the applicable provisions of the Federal Code, which include, but may not be limited to, 18 U.S.C. 1001 and 42 U.S.C. 6928), I certify that the information contained in or accompanying this document is true, accurate and complete. <br><br>(ii) As to the (those) identified section(s) of this document for which I cannot personally verify its (their) truth and accuracy, I certify as the company official having supervisory responsibility for the persons who, acting under my direct instructions, made the verification that this information is true, accurate and complete. <br><br>(iii) If any of this information is determined by EPA in its sole discretion to be false, inaccurate or incomplete, and upon conveyance of this fact to the company, I recognize and agree that this exclusion of waste will be void as if it never had effect or to the extent directed by EPA and that the company will be liable for any actions taken in contravention of the company's RCRA and CERCLA obligations premised upon the company's reliance on the void exclusion. <br><br>(6) *Reopener Language* <br><br>(A) If, anytime after disposal of the delisted waste, Rhodia possesses or is otherwise made aware of any environmental data (including but not limited to leachate data or groundwater monitoring data) or any other data relevant to the delisted waste indicating that any constituent identified for the delisting verification testing is at level higher than the delisting level allowed by the Regional Administrator or his delegate in granting the petition, then the facility must report the data, in writing, to the Regional Administrator or his delegate within 10 days of first possessing or being made aware of that data. <br><br>(B) If the annual testing of the waste does not meet the delisting requirements in Paragraph 1, Rhodia must report the data, in writing, to the Regional Administrator or his delegate within 10 days of first possessing or being made aware of that data. |

TABLE 1—WASTES EXCLUDED FROM NON-SPECIFIC SOURCES—Continued

| Facility | Address | Waste description |
|---|---|---|
| | | (C) If Rhodia fails to submit the information described in paragraphs (5), (6)(A) or (6)(B) or if any other information is received from any source, the Regional Administrator or his delegate will make a preliminary determination as to whether the reported information requires Agency action to protect human health or the environment. Further action may include suspending, or revoking the exclusion, or other appropriate response necessary to protect human health and the environment. |
| | | (D) If the Regional Administrator or his delegate determines that the reported information does require Agency action, the Regional Administrator or his delegate will notify the facility in writing of the actions the Regional Administrator or his delegate believes are necessary to protect human health and the environment. The notice shall include a statement of the proposed action and a statement providing the facility with an opportunity to present information as to why the proposed Agency action is not necessary. The facility shall have 10 days from the date of the Regional Administrator or his delegate's notice to present such information. |
| | | (E) Following the receipt of information from the facility described in paragraph (6)(D) or (if no information is presented under paragraph (6)(D)) the initial receipt of information described in paragraphs (5), (6)(A) or (6)(B), the Regional Administrator or his delegate will issue a final written determination describing the Agency actions that are necessary to protect human health or the environment. Any required action described in the Regional Administrator or his delegate's determination shall become effective immediately, unless the Regional Administrator or his delegate provides otherwise. |
| | | (7) *Notification Requirements:* Rhodia must do following before transporting the delisted waste: Failure to provide this notification will result in a violation of the delisting petition and a possible revocation of the decision. |
| | | (A) Provide a one-time written notification to any State Regulatory Agency to which or through which they will transport the delisted waste described above for disposal, 60 days before beginning such activities. |
| | | (B) Update the one-time written notification if they ship the delisted waste into a different disposal facility. |
| Saturn Corporation. | Spring Hill, Tennessee. | Dewatered wastewater treatment plant (WWTP) sludge (EPA Hazardous Waste No. F019) generated at a maximum rate of 3,000 cubic yards per calendar year. The sludge must be disposed in a lined, Subtitle D landfill with leachate collection that is licensed, permitted, or otherwise authorized to accept the delisted WWTP sludge in accordance with 40 CFR part 258. The exclusion becomes effective on December 23, 2005. |
| | | For the exclusion to be valid, Saturn must implement a verification testing program that meets the following conditions: |
| | | 1. Delisting Levels: The constituent concentrations in an extract of the waste must not exceed the following maximum allowable concentrations in mg/l: antimony—0.494; arsenic—0.224; total chromium—3.71; lead—5.0; nickel—68; thallium—0.211; and zinc—673. Sample collection and analyses, including quality control procedures, must be performed using appropriate methods. As applicable to the method-defined parameters of concern, analyses requiring the use of SW–846 methods incorporated by reference in 40 CFR 260.11 must be used without substitution. As applicable, the SW–846 methods might include Methods 0010, 0011, 0020, 0023A, 0030, 0031, 0040, 0050, 0051, 0060, 0061, 1010A, 1020B, 1110A, 1310B, 1311, 1312, 1320, 1330A, 9010C, 9012B, 9040C, 9045D, 9060A, 9070A, (uses EPA Method 1664, Rev. A), 9071B, and 9095B. Methods must meet Performance Based Measurement System Criteria in which the Data Quality Objectives are to demonstrate that representative samples of Saturn's sludge meet the delisting levels in this condition. |
| | | 2. Waste Holding and Handling: |
| | | (a) Saturn must accumulate the hazardous waste dewatered WWTP sludge in accordance with the applicable regulations of 40 CFR 262.34 and continue to dispose of the dewatered WWTP sludge as hazardous waste until the results of the first quarterly verification testing are available. |
| | | (b) After the first quarterly verification sampling event described in Condition (3) has been completed and the laboratory data demonstrates that no constituent is present in the sample at a level which exceeds the delisting levels set in Condition (1), Saturn can manage and dispose of the dewatered WWTP sludge as nonhazardous according to all applicable solid waste regulations. |
| | | (c) If constituent levels in any sample taken by Saturn exceed any of the delisting levels set in Condition (1), Saturn must do the following: |
| | | (i) Notify EPA in accordance with Condition (7) and |
| | | (ii) Manage and dispose the dewatered WWTP sludge as hazardous waste generated under Subtitle C of RCRA. |
| | | 3. Quarterly Testing Requirements: Upon this exclusion becoming final, Saturn may perform quarterly analytical testing by sampling and analyzing the dewatered WWTP sludge as follows: |
| | | (i) Collect one representative composite sample (consisting of four grab samples) of the hazardous waste dewatered WWTP sludge at any time after EPA grants the final delisting. In addition, collect the second, third, and fourth quarterly samples at approximately ninety (90)-day intervals after EPA grants the final exclusion. |

215

TABLE 1—WASTES EXCLUDED FROM NON-SPECIFIC SOURCES—Continued

| Facility | Address | Waste description |
|---|---|---|
| | | (ii) Analyze the samples for all constituents listed in Condition (1). Any roll-offs from which the composite sample is taken exceeding the delisting levels listed in Condition (1) must be disposed as hazardous waste in a Subtitle C landfill. |
| | | (iii) Within forty-five (45) days after taking its first quarterly sample, Saturn will report its first quarterly analytical test data to EPA and will include the certification statement required in condition (6). If levels of constituents measured in the sample of the dewatered WWTP sludge do not exceed the levels set forth in Condition (1) of this exclusion, Saturn can manage and dispose the nonhazardous dewatered WWTP sludge according to all applicable solid waste regulations. |
| | | 4. Annual Verification Testing: |
| | | (i) If Saturn completes the quarterly testing specified in Condition (3) above, and no sample contains a constituent with a level which exceeds the limits set forth in Condition (1), Saturn may begin annual verification testing on an annual basis. Saturn must collect and analyze one sample of the WWTP sludge on an annual basis as follows: Saturn must test one representative composite sample of the dewatered WWTP sludge for all constituents listed in Condition (1) at least once per calendar year. |
| | | (ii) The sample collected for annual verification testing shall be a representative composite sample consisting of four grab samples that will be collected in accordance with the appropriate methods described in Condition (1). |
| | | (iii) The sample for the annual testing for the second and subsequent annual testing events shall be collected within the same calendar month as the first annual verification sample. Saturn will report the results of the annual verification testing to EPA on an annual basis and will include the certification statement required by Condition (6). |
| | | 5. Changes in Operating Conditions: Saturn must notify EPA in writing when significant changes in the manufacturing or wastewater treatment processes are implemented. EPA will determine whether these changes will result in additional constituents of concern. If so, EPA will notify Saturn in writing that Saturn's sludge must be managed as hazardous waste F019 until Saturn has demonstrated that the wastes meet the delisting levels set forth in Condition (1) and any levels established by EPA for the additional constituents of concern, and Saturn has received written approval from EPA. If EPA determines that the changes do not result in additional constituents of concern, EPA will notify Saturn, in writing, that Saturn must verify that Saturn's sludge continues to meet Condition (1) delisting levels. |
| | | 6. Data Submittals: Saturn must submit data obtained through verification testing at Saturn or as required by other conditions of this rule to: Chief, North Section, RCRA Enforcement and Compliance Branch, Waste Management Division, U.S. Environmental Protection Agency Region 4, Sam Nunn Atlanta Federal Center, 61 Forsyth Street SW., Atlanta, Georgia 30303. If Saturn fails to submit the required data within the specified time or maintain the required records on-site for the specified time, the EPA, at its discretion, will consider this sufficient basis to re-open the exclusion as described in Condition (7). Saturn must: |
| | | (A) Submit the data obtained through Condition (3) within the time specified. The quarterly verification data must be submitted to EPA in accordance with Condition (3). The annual verification data and certification statement of proper disposal must be submitted to EPA annually upon the anniversary of the effective date of this exclusion. All data must be accompanied by a signed copy of the certification statement in 40 CFR 260.22(i)(12). |
| | | (B) Compile, Summarize, and Maintain Records: Saturn must compile, summarize, and maintain at Saturn records of operating conditions and analytical data records of analytical data from Condition (3), summarized, and maintained on-site for a minimum of five years. Saturn must furnish these records and data when either the EPA or the State of Tennessee requests them for inspection. |
| | | (C) Send along with all data a signed copy of the following certification statement, to attest to the truth and accuracy of the data submitted: "I certify under penalty of law that I have personally examined and am familiar with the information submitted in this demonstration and all attached documents, and that, based on my inquiry of those individuals immediately responsible for getting the information, I believe that the submitted information is true, accurate, and complete. I am aware that there are significant penalties for sending false information, including the possibility of fine and imprisonment." |
| | | 7. Reopener. |
| | | (A) If, at any time after disposal of the delisted waste, Saturn possesses or is otherwise made aware of any data (including but not limited to leachate data or groundwater monitoring data) relevant to the delisted WWTP sludge at Saturn indicating that any constituent is at a level in the leachate higher than the specified delisting level or TCLP regulatory level, then Saturn must report the data, in writing, to the Regional Administrator within ten (10) days of first possessing or being made aware of that data. |
| | | (B) Based upon the information described in Paragraph (A) and any other information received from any source, the EPA Regional Administrator will make a preliminary determination as to whether the reported information requires EPA action to protect human health or the environment. Further action may include suspending, or revoking the exclusion, or other appropriate response necessary to protect human health and the environment. |

TABLE 1—WASTES EXCLUDED FROM NON-SPECIFIC SOURCES—Continued

| Facility | Address | Waste description |
|----------|---------|-------------------|
| | | (C) If the Regional Administrator determines that the reported information does require EPA action, the Regional Administrator will notify Saturn in writing of the actions the Regional Administrator believes are necessary to protect human health and the environment. The notification shall include a statement of the proposed action and a statement providing Saturn with an opportunity to present information as to why the proposed EPA action is not necessary. Saturn shall have ten (10) days from the date of the Regional Administrator's notice to present the information. |
| | | (D) Following the receipt of information from Saturn, or if Saturn presents no further information after 10 days, the Regional Administrator will issue a final written determination describing the EPA actions that are necessary to protect human health or the environment. Any required action described in the Regional Administrator's determination shall become effective immediately, unless the Regional Administrator provides otherwise. |
| | | 8. Notification Requirements: Before transporting the delisted waste, Saturn must provide a one-time written notification to any State Regulatory Agency to which or through which it will transport the delisted WWTP sludge for disposal. The notification will be updated if Saturn transports the delisted WWTP sludge to a different disposal facility. Failure to provide this notification will result in a violation of the delisting variance and a possible revocation of the decision. |
| Savannah River Site (SRS). | Aiken, South Carolina. | Vitrified waste (EPA Hazardous Waste Nos. F006 and F028) that the United States Department of Energy Savannah River Operations Office (DOE-SR) generated by treating the following waste streams from the M-Area of the Savannah River Site (SRS) in Aiken, South Carolina, as designated in the SRS Site Treatment Plan: W-004, Plating Line Sludge from Supernate Treatment; W-995, Mark 15 Filter Cake; W-029, Sludge Treatability Samples (glass and cementitious); W-031, Uranium/Chromium Solution; W-037, High Nickel Plating Line Sludge; W-038, Plating Line Sump Material; W-039, Nickel Plating Line Solution; W-048, Soils from Spill Remediation and Sampling Programs; W-054, Uranium/Lead Solution; W-082, Soils from Chemicals, Metals, and Pesticides Pits Excavation; and Dilute Effluent Treatment Facility (DETF) Filtercake (no Site Treatment Plan code). This is a one-time exclusion for 538 cubic yards of waste (hereinafter referred to as "DOE-SR Vitrified Waste") that was generated from 1996 through 1999 and 0.12 cubic yard of cementitious treatability samples (hereinafter referred to as "CTS") generated from 1988 through 1991 (EPA Hazardous Waste No. F006). The one-time exclusion for these wastes is contingent on their being disposed in a low-level radioactive waste landfill, in accordance with the Atomic Energy Act, after [insert date of final rule.] DOE-SR has demonstrated that concentrations of toxic constituents in the DOE-SR Vitrified Waste and CTS do not exceed the following levels:<br><br>(1) *TCLP Concentrations:* All leachable concentrations for these metals did not exceed the Land Disposal Restrictions (LDR) Universal Treatment Standards (UTS): (mg/l TCLP): Arsenic—5.0; Barium—21; Beryllium—1.22; Cadmium—0.11; Chromium—0.60; Lead—0.75; Nickel—11; and Silver—0.14. In addition, none of the metals in the DOE-SR Vitrified Waste exceeded the allowable delisting levels of the EPA, Region 6 Delisting Risk Assessment Software (DRAS): (mg/l TCLP): Arsenic—0.0649; Barium—100.0; Beryllium—0.40; Cadmium—1.0; Chromium—5.0; Lead—5.0; Nickel—10.0; and Silver—5.0. These metal concentrations were measured in the waste leachate obtained by the method specified in 40 CFR 261.24.<br><br>*Total Concentrations in Unextracted Waste:* The total concentrations in the DOE-SR Vitrified Waste, not the waste leachate, did not exceed the following levels (mg/kg): Arsenic—10; Barium—200; Beryllium—10; Cadmium—10; Chromium—500; Lead—200; Nickel—10,000; Silver—20; Acetonitrile—1.0, which is below the LDR UTS of 38 mg/kg; and Fluoride—1.0<br><br>(2) *Data Records:* Records of analytical data for the petitioned waste must be maintained by DOE-SR for a minimum of three years, and must be furnished upon request by EPA or the State of South Carolina, and made available for inspection. Failure to maintain the required records for the specified time will be considered by EPA, at its discretion, sufficient basis to revoke the exclusion to the extent directed by EPA. All data must be maintained with a signed copy of the certification statement in 40 CFR 260.22(i)(12). |

TABLE 1—WASTES EXCLUDED FROM NON-SPECIFIC SOURCES—Continued

| Facility | Address | Waste description |
|---|---|---|
| | | (3) *Reopener Language:* (A) If, at any time after disposal of the delisted waste, DOE-SR possesses or is otherwise made aware of any environmental data (including but not limited to leachate data or groundwater monitoring data) or any other data relevant to the delisted waste indicating that any constituent is identified at a level higher than the delisting level allowed by EPA in granting the petition, DOE-SR must report the data, in writing, to EPA within 10 days of first possessing or being made aware of that data. (B) Based on the information described in paragraph (3)(A) and any other information received from any source, EPA will make a preliminary determination as to whether the reported information requires that EPA take action to protect human health or the environment. Further action may include suspending or revoking the exclusion, or other appropriate response necessary to protect human health and the environment. (C) If EPA determines that the reported information does require Agency action, EPA will notify the facility in writing of the action believed necessary to protect human health and the environment. The notice shall include a statement of the proposed action and a statement providing DOE-SR with an opportunity to present information as to why the proposed action is not necessary. DOE-SR shall have 10 days from the date of EPA's notice to present such information.(E) Following the receipt of information from DOE-SR, as described in paragraph (3)(D), or if no such information is received within 10 days, EPA will issue a final written determination describing the Agency actions that are necessary to protect human health or the environment, given the information received in accordance with paragraphs (3)(A) or (3)(B). Any required action described in EPA's determination shall become effective immediately, unless EPA provides otherwise. (4) *Notification Requirements:* DOE-SR must provide a one-time written notification to any State Regulatory Agency in a State to which or through which the delisted waste described above will be transported, at least 60 days prior to the commencement of such activities. Failure to provide such a notification will result in a violation of the delisting conditions and a possible revocation of the decision to delist. |
| Siegel-Robert, Inc.. | St. Louis, MO | Wastewater treatment sludge (EPA Hazardous Waste No. F006) generated from electroplating operations after November 27, 1985. |
| Shell Oil Company. | Deer Park, TX | North Pond Sludge (EPA Hazardous Waste No. F037) generated one time at a volume of 15,000 cubic yards August 23, 2005 and disposed in a Subtitle D landfill. This is a one time exclusion and applies to 15,000 cubic yards of North Pond Sludge. (1) Reopener: (A) If, anytime after disposal of the delisted waste, Shell possesses or is otherwise made aware of any environmental data (including but not limited to leachate data or ground water monitoring data) or any other data relevant to the delisted waste indicating that any constituent identified for the delisting verification testing is at level higher than the delisting level allowed by the Division Director in granting the petition, then the facility must report the data, in writing, to the Division Director within 10 days of first possessing or being made aware of that data. (B) If Shell fails to submit the information described in paragraph (A) or if any other information is received from any source, the Division Director will make a preliminary determination as to whether the reported information requires EPA action to protect human health or the environment. Further action may include suspending, or revoking the exclusion, or other appropriate response necessary to protect human health and the environment. (C) If the Division Director determines that the reported information does require EPA action, the Division Director will notify the facility in writing of the actions the Division Director believes are necessary to protect human health and the environment. The notice shall include a statement of the proposed action and a statement providing the facility with an opportunity to present information as to why the proposed EPA action is not necessary. The facility shall have 10 days from the date of the Division Director's notice to present such information. (D) Following the receipt of information from the facility described in paragraph (C) or if no information is presented under paragraph (C), the Division Director will issue a final written determination describing the actions that are necessary to protect human health or the environment. Any required action described in the Division Director's determination shall become effective immediately, unless the Division Director provides otherwise. (2) Notification Requirements: Shell must do the following before transporting the delisted waste: Failure to provide this notification will result in a violation of the delisting petition and a possible revocation of the decision. (A) Provide a one-time written notification to any state regulatory agency to which or through which they will transport the delisted waste described above for disposal, 60 days before beginning such activities. (B) Update the one-time written notification, if they ship the delisted waste to a different disposal facility. (C) Failure to provide this notification will result in a violation of the delisting variance and a possible revocation of the decision. |

TABLE 1—WASTES EXCLUDED FROM NON-SPECIFIC SOURCES—Continued

| Facility | Address | Waste description |
|---|---|---|
| Shell Oil Company. | Deer Park, TX | Multi-source landfill leachate (EPA Hazardous Waste No. F039) generated at a maximum annual rate of 3.36 million gallons (16,619 cu. yards) per calendar year after August 23, 2005 and disposed of in accordance with the TPDES permit. |

The delisting levels set do not relieve Shell Oil Company of its duty to comply with the limits set in its TPDES permit. For the exclusion to be valid, Shell Oil Company must implement a verification testing program that meets the following paragraphs:

(1) Delisting Levels: All total concentrations for those constituents must not exceed the following levels (mg/l). The petitioner must analyze the aqueous waste on a total basis to measure constituents in the multi-source landfill leachate.

Multi-source landfill leachate (i) Inorganic Constituents Antimony-0.0204; Arsenic-0.385; Barium-2.92; Copper-418.00; Chromium-5.0; Cobalt-2.25; Nickel-1.13; Selenium-0.0863; Thallium-0.005; Vanadium-0.838

(ii) Organic Constituents Acetone-1.46; Acetophenone-1.58; Benzene-0.0222; p-Cresol-0.0788; Bis(2-ethylhexyl)phthlate-15800.00; Dichloroethane, 1,2–0.0803; Ethylbenzene-4.51; Fluorene-1.87; Napthalene-1.05; Phenol-9.46; Phenanthrene-1.36; Pyridine-0.0146; 2,3,7,8-TCDD equivalents as TEQ–0.0000926; Toluene-4.43; Trichloropropane-0.000574; Xylenes (total)-97.60

(2) Waste Management:

(A) Shell Oil Company must manage as hazardous all multi-source landfill leachate generated, until it has completed initial verification testing described in paragraph (3)(A) and (B), as appropriate, and valid analyses show that paragraph (1) is satisfied.

(B) Levels of constituents measured in the samples of the multi-source landfill leachate that do not exceed the levels set forth in paragraph (1) are non-hazardous. Shell Oil Company can manage and dispose of the non-hazardous multi-source landfill leachate according to all applicable solid waste regulations.

(C) If constituent levels in a sample exceed any of the delisting levels set in paragraph (1), Shell Oil Company can collect one additional sample and perform expedited analyses to verify if the constituent exceeds the delisting level. If this sample confirms the exceedance, Shell Oil Company must, from that point forward, treat the waste as hazardous until it is demonstrated that the waste again meets the levels in paragraph (1).

(D) If the facility has not treated the waste, Shell Oil Company must manage and dispose of the waste generated under Subtitle C of RCRA from the time that it becomes aware of any exceedance.

(E) Upon completion of the Verification Testing described in paragraph 3(A) and (B) as appropriate and the transmittal of the results to EPA, and if the testing results meet the requirements of paragraph (1), Shell Oil Company may proceed to manage its multi-source landfill leachate as non-hazardous waste. If Subsequent Verification Testing indicates an exceedance of the delisting levels in paragraph (1), Shell Oil Company must manage the multi-source landfill leachate as a hazardous waste until two consecutive quarterly testing samples show levels below the delisting levels in Table I.

(3) Verification Testing Requirements: Shell Oil Company must perform sample collection and analyses, including quality control procedures, using appropriate methods. As applicable to the method-defined parameters of concern, analyses requiring the use of SW–846 methods incorporated by reference in 40 CFR 260.11 must be used without substitution. As applicable, the SW–846 methods might include Methods 0010, 0011, 0020, 0023A, 0030, 0031, 0040, 0050, 0051, 0060, 0061, 1010A, 1020B, 1110A, 1310B, 1311, 1312, 1320, 1330A, 9010C, 9012B, 9040C, 9045D, 9060A, 9070A (uses EPA Method 1664, Rev. A), 9071B, and 9095B. Methods used must meet Performance Based Measurement System Criteria in which the Data Quality Objectives demonstrate that representative samples of the Shell-Deer Park multi-source landfill leachate are collected and meet the delisting levels in paragraph (1).

(A) Initial Verification Testing: After EPA grants the final exclusion, Shell Oil Company must do the following:

(i) Within 60 days of this exclusions becoming final, collect four samples, before disposal, of the multi-source landfill leachate.

(ii) The samples are to be analyzed and compared against the delisting levels in paragraph (1).

(iii) Within sixty (60) days after this exclusion becomes final, Shell Oil Company will report initial verification analytical test data for the multi-source landfill leachate, including analytical quality control information for the first thirty (30) days of operation after this exclusion becomes final. If levels of constituents measured in the samples of the multi-source landfill leachate that do not exceed the levels set forth in paragraph (1) are also non-hazardous in two consecutive quarters after the first thirty (30) days of operation after this exclusion become effective, Shell Oil Company can manage and dispose of the multi-source landfill leachate according to all applicable solid waste regulations.

TABLE 1—WASTES EXCLUDED FROM NON-SPECIFIC SOURCES—Continued

| Facility | Address | Waste description |
|---|---|---|
| | | (B) Subsequent Verification Testing: Following written notification by EPA, Shell Oil Company may substitute the testing conditions in (3)(B) for (3)(A). Shell Oil Company must continue to monitor operating conditions, and analyze one representative sample of the multi-source landfill leachate for each quarter of operation during the first year of waste generation. The sample must represent the waste generated during the quarter. After the first year of analytical sampling verification sampling can be performed on a single annual sample of the multi-source landfill leachate. The results are to be compared to the delisting levels in paragraph (1). |
| | | (C) Termination of Testing: |
| | | (i) After the first year of quarterly testing, if the delisting levels in paragraph (1) are being met, Shell Oil Company may then request that EPA not require quarterly testing. After EPA notifies Shell Oil Company in writing, the company may end quarterly testing. |
| | | (ii) Following cancellation of the quarterly testing, Shell Oil Company must continue to test a representative sample for all constituents listed in paragraph (1) annually. |
| | | (4) Changes in Operating Conditions: If Shell Oil Company significantly changes the process described in its petition or starts any processes that generate(s) the waste that may or could significantly affect the composition or type of waste generated as established under paragraph (1) (by illustration, but not limitation, changes in equipment or operating conditions of the treatment process), it must notify EPA in writing; it may no longer handle the wastes generated from the new process as nonhazardous until the wastes meet the delisting levels set in paragraph (1) and it has received written approval to do so from EPA. |
| | | (5) Data Submittals: Shell Oil Company must submit the information described below. If Shell Oil Company fails to submit the required data within the specified time or maintain the required records on-site for the specified time, EPA, at its discretion, will consider this sufficient basis to reopen the exclusion as described in paragraph 6. Shell Oil Company must: |
| | | (A) Submit the data obtained through paragraph 3 to the Section Chief, Region 6 Corrective Action and Waste Minimization Section, EPA, 1445 Ross Avenue, Dallas, Texas 75202–2733, Mail Code, (6PD–C) within the time specified. |
| | | (B) Compile records of operating conditions and analytical data from paragraph (3), summarized, and maintained on-site for a minimum of five years. |
| | | (C) Furnish these records and data when EPA or the state of Texas request them for inspection. |
| | | (D) Send along with all data a signed copy of the following certification statement, to attest to the truth and accuracy of the data submitted: |
| | | Under civil and criminal penalty of law for the making or submission of false or fraudulent statements or representations (pursuant to the applicable provisions of the Federal Code, which include, but may not be limited to, 18 U.S.C. 1001 and 42 U.S.C. 6928), I certify that the information contained in or accompanying this document is true, accurate and complete. |
| | | As to the (those) identified section(s) of this document for which I cannot personally verify its (their) truth and accuracy, I certify as the company official having supervisory responsibility for the persons who, acting under my direct instructions, made the verification that this information is true, accurate and complete. |
| | | If any of this information is determined by EPA in its sole discretion to be false, inaccurate or incomplete, and upon conveyance of this fact to the company, I recognize and agree that this exclusion of waste will be void as if it never had effect or to the extent directed by EPA and that the company will be liable for any actions taken in contravention of the company's RCRA and CERCLA obligations premised upon the company's reliance on the void exclusion. |
| | | (6) Reopener: |
| | | (A) If, anytime after disposal of the delisted waste, Shell Oil Company possesses or is otherwise made aware of any environmental data (including but not limited to leachate data or groundwater monitoring data) or any other data relevant to the delisted waste indicating that any constituent identified for the delisting verification testing is at a level higher than the delisting level allowed by the Division Director in granting the petition, then the facility must report the data, in writing, to the Division Director within 10 days of first possessing or being made aware of that data. |
| | | (B) If the annual testing of the waste does not meet the delisting requirements in paragraph 1, Shell Oil Company must report the data, in writing, to the Division Director within 10 days of first possessing or being made aware of that data. |
| | | (C) If Shell Oil Company fails to submit the information described in paragraphs (5),(6)(A) or (6)(B) or if any other information is received from any source, the Division Director will make a preliminary determination as to whether the reported information requires EPA action to protect human health and/or the environment. Further action may include suspending, or revoking the exclusion, or other appropriate response necessary to protect human health and the environment. |
| | | (D) If the Division Director determines that the reported information does require action, he will notify the facility in writing of the actions the Division Director believes are necessary to protect human health and the environment. The notice shall include a statement of the proposed action and a statement providing the facility with an opportunity to present information as to why the proposed action by EPA is not necessary. The facility shall have 10 days from the date of the Division Director's notice to present such information. |

TABLE 1—WASTES EXCLUDED FROM NON-SPECIFIC SOURCES—Continued

| Facility | Address | Waste description |
|---|---|---|
| | | (E) Following the receipt of information from the facility described in paragraph (6)(D) or if no information is presented under paragraph (6)(D), the Division Director will issue a final written determination describing the actions that are necessary to protect human health and/or the environment. Any required action described in the Division Director's determination shall become effective immediately, unless the Division Director provides otherwise. |
| | | (7) Notification Requirements: Shell Oil Company must do the following before transporting the delisted waste. Failure to provide this notification will result in a violation of the delisting petition and a possible revocation of the decision. |
| | | (A) Provide a one-time written notification to any state regulatory agency to which or through which it will transport the delisted waste described above for disposal, 60 days before beginning such activities. |
| | | (B) Update the one-time written notification if it ships the delisted waste into a different disposal facility. |
| | | (C) Failure to provide this notification will result in a violation of the delisting exclusion and a possible revocation of the decision. |
| Southeastern Public Service Authority (SPSA) and Onyx Environmental Service (Onyx). | Suffolk, Virginia. | Combustion ash generated from the burning of spent solvent methyl ethyl ketone (Hazardous Waste Number F005) and disposed in a Subtitle D landfill. This is a one-time exclusion for 1410 cubic yards of ash and is effective after September 11, 2003. |
| | | (1) *Reopener Language* (a) If SPSA and/or Onyx discovers that any condition or assumption related to the characterization of the excluded waste which was used in the evaluation of the petition or that was predicted through modeling is not as reported in the petition, then SPSA and/or Onyx must report any information relevant to that condition or assumption, in writing, to the Regional Administrator and the Virginia Department of Environmental Quality within 10 calendar days of discovering that information. |
| | | (b) Upon receiving information described in paragraph (a) of this section, regardless of its source, the Regional Administrator will determine whether the reported condition requires further action. Further action may include repealing the exclusion, modifying the exclusion, or other appropriate action deemed necessary to protect human health or the environment. |
| | | (2) *Notification Requirements* In the event that the delisted waste is transported off-site for disposal, SPSA/Onyx must provide a one-time written notification to any State Regulatory Agency to which or through which the delisted waste described above will be transported at least sixty (60) calendar days prior to the commencement of such activities. Failure to provide such notification will be deemed to be a violation of this exclusion and may result in revocation of the decision and other enforcement action. |
| Square D Company. | Oxford, Ohio .. | Dewatered filter press sludge (EPA Hazardous Waste No. F006) generated from electroplating operations after August 15, 1986. |
| Syntex Agribusiness. | Springfield, MO. | Kiln ash, cyclone ash, separator sludge, and filtered wastewater (except spent activiated carbon) (EPA Hazardous Waste No. F020 generated during the treatment of wastewater treatment sludge by the EPA's Mobile Incineration System at the Denney Farm Site in McDowell, Missouri after June 2, 1988, so long as: |
| | | (1) The incinerator is monitored continuously and is in compliance with operating permit conditions. Should the incinerator fail to comply with the permit conditions relevant to the mechanical operation of the incinerator, Syntex must test the residues generated during the run when the failure occurred according to the requirements of Conditions (2) through (6), regardless of whether or not the demonstration in Condition (7) has been made. |
| | | (2) Four grab samples of wastewater must be composited from the volume of filtered wastewater collected after each eight hour run and, prior to disposal the composite samples must be analyzed for the EP toxic metals, nickel, and cyanide. If arsenic, chromium, lead, and silver EP leachate test results exceed 0.61 ppm; barium levels exceed 12 ppm; cadmium and selenium levels exceed 0.12 ppm; mercury levels exceed 0.02 ppm; nickel levels exceed 6.1 ppm; or cyanide levels exceed 2.4 ppm, the wastewater must be retreated to achieve these levels or must be disposed in accordance with all applicable hazardous waste regulations. Analyses must be performed using appropriate methods. As applicable to the method- defined parameters of concern, analyses requiring the use of SW–846 methods incorporated by reference in 40 CFR 260.11 must be used without substitution. As applicable, the SW–846 methods might include Methods 0010, 0011, 0020, 0023A, 0030, 0031, 0040, 0050, 0051, 0060, 0061, 1010A, 1020B, 1110A, 1310B, 1311, 1312, 1320, 1330A, 9010C, 9012B, 9040C, 9045D, 9060A, 9070A (uses EPA Method 1664, Rev. A), 9071B, and 9095B. |

TABLE 1—WASTES EXCLUDED FROM NON-SPECIFIC SOURCES—Continued

| Facility | Address | Waste description |
|---|---|---|
| | | (3) One grab sample must be taken from each drum of kiln and cyclone ash generated during each eight-hour run; all grabs collected during a given eight-hour run must then be composited to form one composite sample. A composite sample of four grab samples of the separator sludge must be collected at the end of each eight-hour run. Prior to the disposal of the residues from each eight-hour run, an EP leachate test must be performed on these composite samples and the leachate analyzed for the EP toxic metals, nickel, and cyanide (using a distilled water extraction for the cyanide extraction) to demonstrate that the following maximum allowable treatment residue concentrations listed below are not exceeded. Analyses must be performed using appropriate methods. As applicable to the method-defined parameters of concern, analyses requiring the use of SW–846 methods incorporated by reference in 40 CFR 260.11 must be used without substitution. As applicable, the SW–846 methods might include Methods 0010, 0011, 0020, 0023A, 0030, 0031, 0040, 0050, 0051, 0060, 0061, 1010A, 1020B, 1110A, 1310B, 1311, 1312, 1320, 1330A, 9010C, 9012B, 9040C, 9045D, 9060A, 9070A (uses EPA Method 1664, Rev. A), 9071B, and 9095B. Any residues which exceed any of the levels listed below must be retreated to achieve these levels or must be disposed in accordance with all applicable hazardous waste regulations. |
| | | Maximum Allowable Solids Treatment Residue EP Leachate Concentrations (mg/L) |
| | | Arsenic—1.6, Barium—32, Cadmium—0.32, Chromium—1.6, Lead—1.6, Mercury—0.065, Nickel—16, Selenium—0.32, Silver—1.6, Cyanide—6.5. |
| | | (4) If Syntex stabilizes any of the kiln and cyclone ash or separator sludge, a Portland cement-type stabilization process must be used and Syntex must collect a composite sample of four grab samples from each batch of stabilized waste. An MEP leachate test must be performed on these composite samples and the leachate analyzed for the EP toxic metals, nickel, and cyanide (using a distilled water extraction for the cyanide leachate analysis) to demonstrate that the maximum allowable treatment residue concentrations listed in condition (3) are not exceeded during any run of the MEP extraction. Analyses must be performed using appropriate methods. As applicable to the method-defined parameters of concern, analyses requiring the use of SW–846 methods incorporated by reference in 40 CFR 260.11 must be used without substitution. As applicable, the SW–846 methods might include Methods 0010, 0011, 0020, 0023A, 0030, 0031, 0040, 0050, 0051, 0060, 0061, 1010A, 1020B, 1110A, 1310B, 1311, 1312, 1320, 1330A, 9010C, 9012B, 9040C 9045D, 9060A, 9070A (uses EPA Method 1664, Rev. A), 9071B, and 9095B. Any residues which exceed any of the levels listed in Condition (3) must be retreated to achieve these levels or must be disposed in accordance with all applicable hazardous waste regulations. (If the residues are stabilized, the analyses required in this condition supercede the analyses required in Condition (3).) |
| | | (5) Syntex must generate, prior to disposal of residues, verification data from each eight hour run from each treatment residue (i.e., kiln and cyclone ash, separator sludge, and filtered wastewater) to demonstrate that the maximum allowable treatment residue concentrations listed below are not exceeded. Samples must be collected as specified in Conditions (2) and (3). Analyses must be performed using appropriate methods. As applicable to the method-defined parameters of concern, analyses requiring the use of SW–846 methods incorporated by reference in 40 CFR 260.11 must be used without substitution. As applicable, the SW–846 methods might include Methods 0010, 0011, 0020, 0023A, 0030, 0031, 0040, 0050, 0051, 0060, 0061, 1010A, 1020B, 1110A, 1310B, 1311, 1312, 1320, 1330A, 9010C, 9012B, 9040C, 9045D, 9060A, 9070A (uses EPA Method 1664, Rev. A), 9071B, and 9095B. Any solid or liquid residues which exceed any of the levels listed below must be retreated to achieve these levels or must be disposed in accordance with Subtitle C of RCRA. Maximum Allowable Wastewater Concentrations (ppm): |
| | | Benz(a)anthracene—$1 \times 10^{-4}$, Benzo(a)pyrene—$4 \times 10^{-5}$, Benzo(b)fluoranthene—$2 \times 10^{-4}$, Chloroform—0.07, Chrysene—0.002, Dibenz(a,h)anthracene—$9 \times 10^{-6}$, 1,2-Dichloroethane—0.06, Dichloromethane—0.06, Indeno(1,2,3-cd)pyrene—0.002, Polychlorinated biphenyls—$1 \times 10^{-4}$, 1,2,4,5-Tetrachlorobenzene—0.13, 2,3,4,6-Tetrachlorophenol—12, Toluene—120, Trichloroethylene—0.04, 2,4,5-Trichlorophenol—49, 2,4,6-Trichlorophenol—0.02, Maximum Allowable Solid Treatment Residue. |
| | | Concentrations (ppm); Benz(a)anthracene—1.1, Benzo(a)pyrene—0.43, benzo(b)fluoranthene—1.8, Chloroform—5.4, Chrysene—170, Dibenz(a,h)anthracene—0.083, Dichloromethane—2.4, 1,2-Dichloroethane—4.1, Indeno(1,2,3-cd)pyrene—330, Polychlorinated biphenyls—0.31, 1,2,4,5-Tetrachlorobenzene—720, Trichloroethylene—6.6, 2,4,6-Trichlorophenol—3.9. |

TABLE 1—WASTES EXCLUDED FROM NON-SPECIFIC SOURCES—Continued

| Facility | Address | Waste description |
|---|---|---|
| | | (6) Syntex must generate, prior to disposal of residues, verification data from each eight-hour run for each treatment residue (i.e., kiln and cyclone ash, separator sludge, and filtered wastewater) to demonstrate that the residues do not contain tetra-, penta-, or hexachlorodibenzo-p-dioxins or furans at levels of regulatory concern. Samples must be collected as specified in Conditions (2) and (3). The TCDD equivalent levels for wastewaters must be less than 2 ppq and less than 5 ppt for the solid treatment residues. Any residues with detected dioxins or furans in excess of these levels must be retreated or must be disposed as acutely hazardous. For this analysis, Syntex must use appropriate methods. For tetra- and pentachloronated dioxin and furan homologs, the maximum practical quantitation limit must not exceed 15 ppt for solids and 120 ppq for wastewaters. For hexachlorinated homologs, the maximum practical quantitation limit must not exceed 37 ppt for solids and 300 ppq for wastewaters. |
| | | (7)(A) The test data from Conditions (1), (2), (3), (4), (5) and (6) must be kept on file by Syntex for inspection purposes and must be compiled, summarized, and submitted to the Section Chief, Variances Section, PSPD/OSW (WH–563), US EPA, 1200 Pennsylvania Ave., NW., Washington, DC 20460 by certified mail on a monthly basis and when the treatment of the lagoon sludge is concluded. All data submitted will be placed in the RCRA docket. |
| | | (B) The testing requirements for Conditions (2), (3), (4), (5), and (6) will continue until Syntex provides the Section Chief, Variances Section, with the results of four consecutive batch analyses for the petitioned wastes, none of which exceed the maximum allowable treatment residue concentrations listed in these conditions and the Section Chief, Variances Section, notifies Syntex that the conditions have been lifted. |
| | | (8) Syntex must provide a signed copy of the following certification statement when submitting data in response to the conditions listed above: "Under civil and criminal penalty of law for the making or submission of false or fraudulent statements or representations, I certify that the information contained in or accompanying this document is true, accurate, and complete. As to the (those) identified section(s) of this document for which I cannot personally verify its (their) accuracy, I certify as the company official having supervisory responsibility for the persons who, acting under my direct instructions, made the verification that this information is true, accurate and complete." |
| SR of Tennessee.  Tenneco Automotive. | Ripley, TN ...... Paragould, AR | Dewatered wastewater treatment sludges (EPA Hazardous Waste No. F006) generated from the copper, nickel, and chromium electroplating of plastic parts after November 17, 1986. Stabilized sludge from electroplating operations, excavated from the Finch Road Landfill and currently stored in containment cells by Tenneco (EPA Hazardous Waste Nos. F006). This is a one-time exclusion for 1,800 cubic yards of stabilized sludge when it is disposed of in a Subtitle D landfill. This exclusion was published on August 9, 2001. |
| | | (1) *Reopener Language:* |
| | | (A) If, anytime after disposal of the delisted waste, Tenneco possesses or is otherwise made aware of any environmental data (including but not limited to leachate data or groundwater monitoring data) or any other data relevant to the delisted waste indicating that any constituent identified for the delisting verification testing is at level higher than the delisting level allowed by the Regional Administrator or his delegate in granting the petition, then the facility must report the data, in writing, to the Regional Administrator or his delegate within 10 days of first possessing or being made aware of that data. |
| | | (B) If Tenneco fails to submit the information described in (2)(A) or if any other information is received from any source, the Regional Administrator or his delegate will make a preliminary determination as to whether the reported information requires Agency action to protect human health or the environment. Further action may include suspending, or revoking the exclusion, or other appropriate response necessary to protect human health and the environment. |
| | | (C) If the Regional Administrator or his delegate determines the reported information does require Agency action, the Regional Administrator or his delegate will notify the facility in writing of the actions the Regional Administrator or his delegate believes are necessary to protect human health and the environment. The notice shall include a statement of the proposed action and a statement providing the facility with an opportunity to present information as to why the proposed Agency action is not necessary. The facility shall have 10 days from the date of the Regional Administrator or his delegate's notice to present such information. |
| | | (D) Following the receipt of information from the facility described in (1)(C) or (if no information is presented under (1)(C)) the initial receipt of information described in (1)(A), the Regional Administrator or his delegate will issue a final written determination describing the Agency actions that are necessary to protect human health or the environment. Any required action described in the Regional Administrator or his delegate's determination shall become effective immediately, unless the Regional Administrator or his delegate provides otherwise. |
| | | (2) *Notification Requirements:* |
| | | Tenneco must do following before transporting the delisted waste off-site: Failure to provide this notification will result in a violation of the delisting petition and a possible revocation of the exclusion. |

223

TABLE 1—WASTES EXCLUDED FROM NON-SPECIFIC SOURCES—Continued

| Facility | Address | Waste description |
|---|---|---|
| | | (A) Provide a one-time written notification to any State Regulatory Agency to which or through which they will transport the delisted waste described above for disposal, 60 days before beginning such activities. |
| | | (B) Update the one-time written notification if Tenneco ships the delisted waste to a different disposal facility. |
| Tennessee Electroplating. | Ripley, Tennessee. | Dewatered wastewater treatment sludges (EPA Hazardous Waste Nos. F006) generated from electroplating operations after November 17, 1986. To ensure chromium levels do not exceed the regulatory standards there must be continuous batch testing of the filter press sludge for chromium for 45 days after the exclusion is granted. Each batch of treatment residue must be representatively sampled and tested using the EP toxicity test for chromium. This data must be kept on file at the facility for inspection purposes. If the extract levels exceed 0.922 ppm of chromium the waste must be managed and disposed of as hazardous. If these conditions are not met, the exclusion does not apply. This exclusion does not apply to sludges in any on-site impoundments as of this date. |
| Tennessee Electroplating. | Ripley, TN ...... | Wastewater treatment sludge (EPA Hazardous Waste No. F006) generated from electroplating operations and contained in an on-site surface impoundment (maximum volume of 6,300 cubic yards). This is a one-time exclusion. This exclusion was published on April 8, 1991. |
| Texas Eastman | Longview, Texas. | Incinerator ash (at a maximum generation of 7,000 cubic yards per calendar year) generated from the incineration of sludge from the wastewater treatment plant (EPA Hazardous Waste No. D001, D003, D018, D019, D021, D022, D027, D028, D029, D030, D032, D033, D034, D035, D036, D038, D039, D040, F001, F002, F003, F005, and that is disposed of in Subtitle D landfills after September 25, 1996. Texas Eastman must implement a testing program that meets the following conditions for the petition to be valid: |
| | | 1. *Delisting Levels:* All leachable concentrations for those metals must not exceed the following levels (mg/l). Metal concentrations must be measured in the waste leachate by the method specified in 40 CFR §261.24. |
| | | (A) Inorganic Constituents |
| | | Antimony—0.27; Arsenic—2.25; Barium—90.0; Beryllium—0.0009; Cadmium—0.225; Chromium—4.5; Cobalt—94.5; Copper—58.5; Lead—0.675; Mercury—0.045; Nickel—4.5; Selenium—1.0; Silver—5.0; Thallium—0.135; Tin—945.0; Vanadium—13.5; Zinc—450.0 |
| | | (B) Organic Constituents |
| | | Acenaphthene—90.0; Acetone—180.0; Benzene—0.135; Benzo(a)anthracene—0.00347; Benzo(a)pyrene—0.00045; Benzo(b) fluoranthene—0.00320; Bis(2 ethylhexyl) phthalate—0.27; Butylbenzyl phthalate—315.0; Chloroform—0.45; Chlorobenzene—31.5; Carbon Disulfide—180.0; Chrysene—0.1215; 1,2–Dichlorobenzene—135.0; 1,4–Dichlorobenzene—0.18; Di-n-butyl phthalate—180.0; Di-n-octyl phthalate—35.0; 1,4 Dioxane—0.36; Ethyl Acetate—1350.0; Ethyl Ether—315.0; Ethylbenzene—180.0; Flouranthene—45.0; Fluorene—45.0; 1–Butanol—180.0; Methyl Ethyl Ketone—200.0; Methylene Chloride—0.45; Methyl Isobutyl Ketone—90.0; Naphthalene—45.0; Pyrene—45.0; Toluene—315.0; Xylenes—3150.0 |
| | | 2. *Waste Holding and Handling:* Texas Eastman must store in accordance with its RCRA permit, or continue to dispose of as hazardous all FBI ash generated until the Initial and Subsequent Verification Testing described in Paragraph 4 and 5 below is completed and valid analyses demonstrate that all Verification Testing Conditions are satisfied. After completion of Initial and Subsequent Verification Testing, if the levels of constituents measured in the samples of the FBI ash do not exceed the levels set forth in Paragraph 1 above, and written notification is given by EPA, then the waste is non-hazardous and may be managed and disposed of in accordance with all applicable solid waste regulations. |
| | | 3. *Verification Testing Requirements:* Sample collection and analyses, including quality control procedures, must be performed using appropriate methods. As applicable to the method-defined parameters of concern, analyses requiring the use of SW–846 methods incorporated by reference in 40 CFR 260.11 must be used without substitution. As applicable, the SW–846 methods might include Methods 0010, 0011, 0020, 0023A, 0030, 0031, 0040, 0050, 0051, 0060, 0061, 1010A, 1020B, 1110A, 1310B, 1311, 1312, 1320, 1330A, 9010C, 9012B, 9040C, 9045D, 9060A, 9070A (uses EPA Method 1664, Rev. A), 9071B, and 9095B. If EPA judges the incineration process to be effective under the operating conditions used during the initial verification testing described in Condition (4) Texas Eastman may replace the testing required in Condition (4) with the testing required in Condition (5) below. Texas Eastman must, however, continue to test as specified in Condition (4) until notified by EPA in writing that testing in Condition (4) may be replaced by the testing described in Condition (5). |
| | | 4. *Initial Verification Testing:* During the first 40 operating days of the FBI incinerator after the final exclusion is granted, Texas Eastman must collect and analyze daily composites of the FBI ash. Daily composites must be composed of representative grab samples collected every 6 hours during each 24-hour FBI operating cycle. The FBI ash must be analyzed, prior to disposal of the ash, for all constituents listed in Paragraph 1. Texas Eastman must report the operational and analytical test data, including quality control information, obtained during this initial period no later than 90 days after receipt of the validated analytical results. |

TABLE 1—WASTES EXCLUDED FROM NON-SPECIFIC SOURCES—Continued

| Facility | Address | Waste description |
|---|---|---|
| | | 5. *Subsequent Verification Testing:* Following the completion of the Initial Verification Testing, Texas Eastman may request to monitor operating conditions and analyze samples representative of each quarter of operation during the first year of ash generation. The samples must represent the untreated ash generated over one quarter. Following written notification from EPA, Texas Eastman may begin the quarterly testing described in this Paragraph. |
| | | 6. *Termination of Organic Testing:* Texas Eastman must continue testing as required under Paragraph 5 for organic constituents specified in Paragraph 1 until the analyses submitted under Paragraph 5 show a minimum of two consecutive quarterly samples below the delisting levels in Paragraph 1. Texas Eastman may then request that quarterly organic testing be terminated. After EPA notifies Texas Eastman in writing it may terminate quarterly organic testing. |
| | | 7. *Annual Testing:* Following termination of quarterly testing under either Paragraphs 5 or 6, Texas Eastman must continue to test a representative composite sample for all constituents listed in Paragraph 1 (including organics) on an annual basis (no later than twelve months after the date that the final exclusion is effective). |
| | | 8. *Changes in Operating Conditions:* If Texas Eastman significantly changes the incineration process described in its petition or implements any new manufacturing or production process(es) which generate(s) the ash and which may or could affect the composition or type of waste generated established under Paragraph 3 (by illustration {but not limitation}, use of stabilization reagents or operating conditions of the fluidized bed incinerator), Texas Eastman must notify the EPA in writing and may no longer handle the wastes generated from the new process as non-hazardous until the wastes meet the delisting levels set in Paragraph 1 and it has received written approval to do so from EPA. |
| | | 9. *Data Submittals:* The data obtained through Paragraph 3 must be submitted to Mr. William Gallagher, Chief, Region 6 Delisting Program, U.S. EPA, 1445 Ross Avenue, Dallas, Texas 75202–2733, Mail Code, (6PD-O) within the time period specified. Records of operating conditions and analytical data from Paragraph 3 must be compiled, summarized, and maintained on site for a minimum of five years. These records and data must be furnished upon request by EPA, or the State of Texas, and made available for inspection. Failure to submit the required data within the specified time period or maintain the required records on site for the specified time will be considered by EPA, at its discretion, sufficient basis to revoke the exclusion to the extent directed by EPA. All data must be accompanied by a signed copy of the following certification statement to attest to the truth and accuracy of the data submitted: |
| | | Under civil and criminal penalty of law for the making or submission of false or fraudulent statements or representations (pursuant to the applicable provisions of the Federal Code, which include, but may not be limited to, 18 USC 1001 and 42 USC 6928), I certify that the information contained in or accompanying this document is true, accurate and complete. |
| | | As to the (those) identified section(s) of this document for which I cannot personally verify its (their) truth and accuracy, I certify as the company official having supervisory responsibility for the persons who, acting under my direct instructions, made the verification that this information is true, accurate and complete. |
| | | In the event that any of this information is determined by EPA in its sole discretion to be false, inaccurate or incomplete, and upon conveyance of this fact to the company, I recognize and agree that this exclusion of waste will be void as if it never had effect or to the extent directed by EPA and that the company will be liable for any actions taken in contravention of the company's RCRA and CERCLA obligations premised upon the company's reliance on the void exclusion. |
| | | 10. *Notification Requirements:* Texas Eastman must provide a one-time written notification to any State Regulatory Agency to which or through which the delisted waste described above will be transported for disposal at least 60 days prior to the commencement of such activities. Failure to provide such a notification will result in a violation of the delisting petition and a possible revocation of the decision. |
| Trigen/Cinergy-USFOS of Lansing LLC at General Motors Corporation, Lansing Grand River. | Lansing, Michigan. | Waste water treatment plant sludge, F019, that is generated at General Motors Corporation's Lansing Grand River (GM-Grand River) facility by Trigen/Cinergy-USFOS of Lansing LLC exclusively from wastewaters from GM-Grand River, Lansing, Michigan at a maximum annual rate of 2,000 cubic yards per year. The sludge must be disposed of in a lined landfill with leachate collection, which is licensed, permitted, or otherwise authorized to accept the delisted wastewater treatment sludge in accordance with 40 CFR Part 258. The exclusion becomes effective as of July 30, 2003. The conditions in paragraphs (2) through (5) for Ford Motor Company—Michigan Truck Plant and Wayne Integrated Stamping Plant—Wayne, Michigan also apply. |
| | | Delisting Levels: (A) The TCLP concentrations measured in any sample may not exceed the following levels (mg/L): Antimony—0.659; Arsenic—0.3; Cadmium—0.48; Chromium—4.95; Lead—5; Nickel—90.5; Selenium—1; Thallium—0.282; Tin—721; Zinc—898; p-Cresol—11.4; and Formaldehyde—84.2. (B) The total concentrations measured in any sample may not exceed the following levels (mg/kg): Mercury—8.92; and Formaldehyde—689. (C) The sum of the ratios of the TCLP concentrations to the delisting levels for nickel and thallium and for nickel and cadmium shall not exceed 1.0. |

225

TABLE 1—WASTES EXCLUDED FROM NON-SPECIFIC SOURCES—Continued

| Facility | Address | Waste description |
|---|---|---|
| Tyco Printed Circuit Group, Melbourne Division. | Melbourne, Florida. | Wastewater treatment sludge (EPA Hazardous Waste No. F006) that Tyco Printed Circuit Group, Melbourne Division (Tyco) generates by treating wastewater from its circuit board manufacturing plant located on John Rodes Blvd. in Melbourne, Florida. This is a conditional exclusion for up to 590 cubic yards of waste (hereinafter referred to as "Tyco Sludge") that will be generated each year and disposed in a Subtitle D landfill or shipped to a smelter for metal recovery after May 14, 2001. Tyco must demonstrate that the following conditions are met for the exclusion to be valid. (Please see Condition (8) for certification and recordkeeping requirements that must be met in order for the exclusion to be valid for waste that is sent to a smelter for metal recovery.) |
| | | (1) *Verification Testing Requirements:* Sample collection and analyses, including quality control procedures must be performed using appropriate methods. As applicable to the method-defined parameters of concern, analyses requiring the use of SW–846 methods incorporated by reference in 40 CDFR 260.11 must be used without substitution. As applicable, the SW–846 methods might include Methods 0010, 0011, 0020, 0023A, 0030, 0031, 0040, 0050, 0051, 0060, 0061, 1010A, 1020B, 1110A, 1310B, 1311, 1312, 1320, 1330A, 9010C, 9012B, 9040C, 9045D, 9060A, 9070A (uses EPA Method 1664, Rev. A), 9071B, and 9095B. Methods must meet Performance Based Measurement System Criteria in which the Data Quality Objectives are to demonstrate that representative samples of the Tyco Sludge meet the delisting levels in Condition (3). |
| | | (A) *Initial Verification Testing:* Tyco must collect and analyze a representative sample of every batch, for eight sequential batches of Tyco sludge generated in its wastewater treatment system after May 14, 2001. A batch is the Tyco Sludge generated during one day of wastewater treatment. Tyco must analyze for the constituents listed in Condition (3). A minimum of four composite samples must be collected as representative of each batch. Tyco must report analytical test data, including quality control information, no later than 60 days after generating the first batch of Tyco Sludge to be disposed in accordance with the delisting Conditions (1) through (7). |
| | | (B) *Subsequent Verification Testing:* If the initial verification testing in Condition (1)(A) is successful, i.e., delisting levels of condition (3) are met for all of the eight initial batches, Tyco must test a minimum of 5% of the Tyco Sludge generated each year. Tyco must collect and analyze at least one composite sample representative of that 5%. The composite must be made up of representative samples collected from each batch included in the 5%. Tyco may, at its discretion, analyze composite samples gathered more frequently to demonstrate that smaller batches of waste are non-hazardous. |
| | | (2) *Waste Holding and Handling:* Tyco must store as hazardous all Tyco Sludge generated until verification testing as specified in Condition (1)(A) or (1)(B), as appropriate, is completed and valid analyses demonstrate that Condition (3) is satisfied. If the levels of constituents measured in the samples of Tyco Sludge do not exceed the levels set forth in Condition (3), then the Tyco Sludge is non-hazardous and must be managed in accordance with all applicable solid waste regulations. If constituent levels in a sample exceed any of the delisting levels set forth in Condition (3), the batch of Tyco Sludge generated during the time period corresponding to this sample must be retreated until it meets the delisting levels set forth in Condition (3), or managed and disposed of in accordance with Subtitle C of RCRA. |
| | | (3) *Delisting Levels:* All leachable concentrations for these metals and cyanide must not exceed the following levels (ppm): Barium—100; Cadmium—0.5; Chromium—5.0; Cyanide—20, Lead—1.5; and Nickel—73. These metal and cyanide concentrations must be measured in the waste leachate obtained by the method specified in 40 CFR 261.24, except that for cyanide, deionized water must be the leaching medium. The total concentration of cyanide (total, not amenable) in the waste, not the waste leachate, must not exceed 200 mg/kg. Cyanide concentrations in waste or leachate must be measured by the method specified in 40 CFR 268.40, Note 7. The total concentrations of metals in the waste, not the waste leachate, must not exceed the following levels (ppm): Barium—2,000; Cadmium—500; Chromium—1,000; Lead—2,000; and Nickel—20,000. |
| | | (4) *Changes in Operating Conditions:* Tyco must notify EPA in writing when significant changes in the manufacturing or wastewater treatment processes are necessary (e.g., use of new chemicals not specified in the petition). EPA will determine whether these changes will result in additional constituents of concern. If so, EPA will notify Tyco in writing that the Tyco sludge must be managed as hazardous waste F006, pending receipt and evaluation of a new delisting petition. If EPA determines that the changes do not result in additional constituents of concern, EPA will notify Tyco, in writing, that Tyco must repeat Condition (1)(A) to verify that the Tyco Sludge continues to meet Condition (3) delisting levels. |

TABLE 1—WASTES EXCLUDED FROM NON-SPECIFIC SOURCES—Continued

| Facility | Address | Waste description |
|---|---|---|
| | | (5) *Data Submittals:* Data obtained in accordance with Condition (1)(A) must be submitted to Jewell Grubbs, Chief, RCRA Enforcement and Compliance Branch, Mail Code: 4WD-RCRA, U.S. EPA, Region 4, Sam Nunn Atlanta Federal Center, 61 Forsyth Street, Atlanta, Georgia 30303. This notification is due no later than 60 days after generating the first batch of Tyco Sludge to be disposed in accordance with delisting Conditions (1) through (7). Records of analytical data from Condition (1) must be compiled, summarized, and maintained by Tyco for a minimum of three years, and must be furnished upon request by EPA or the State of Florida, and made available for inspection. Failure to submit the required data within the specified time period or maintain the required records for the specified time will be considered by EPA, at its discretion, sufficient basis to revoke the exclusion to the extent directed by EPA. All data must be accompanied by a signed copy of the following certification statement to attest to the truth and accuracy of the data submitted: |
| | | Under civil and criminal penalty of law for the making or submission of false or fraudulent statements or representations (pursuant to the applicable provisions of the Federal Code, which include, but may not be limited to, 18 U.S.C. 1001 and 42 U.S.C. 6928), I certify that the information contained or accompanying this document is true, accurate and complete. |
| | | As to the (those) identified section(s) of this document for which I cannot personally verify its (their) truth and accuracy, I certify as the company official having supervisory responsibility for the persons who, acting under my direct instructions, made the verification that this information is true, accurate and complete. |
| | | In the event that any of this information is determined by EPA in its sole discretion to be false, inaccurate or incomplete, and upon conveyance of this fact to the company, I recognize and agree that this exclusion of waste will be void as if it never had effect or to the extent directed by EPA and that the company will be liable for any actions taken in contravention of the company's RCRA and CERCLA obligations premised upon the company's void exclusion. |
| | | (6) *Reopener Language:* (A) If, anytime after disposal or shipment to a smelter of the delisted waste, Tyco possesses or is otherwise made aware of any environmental data (including but not limited to leachate data or groundwater monitoring data) or any other data relevant to the delisted waste indicating that any constituent identified in the delisting verification testing is at a level higher than the delisting level allowed by EPA in granting the petition, Tyco must report the data, in writing, to EPA within 10 days of first possessing or being made aware of that data. (B) If the testing of the waste, as required by Condition (1)(B), does not meet the delisting requirements of Condition (3), Tyco must report the data, in writing, to EPA within 10 days of first possessing or being made aware of that data. (C) Based on the information described in paragraphs (6)(A) or (6)(B) and any other information received from any source, EPA will make a preliminary determination as to whether the reported information requires that EPA take action to protect human health or the environment. Further action may include suspending, or revoking the exclusion, or other appropriate response necessary to protect human health and the environment. (D) If EPA determines that the reported information does require Agency action, EPA will notify the facility in writing of the action believed necessary to protect human health and the environment. The notice shall include a statement of the proposed action and a statement providing Tyco with an opportunity to present information as to why the proposed action is not necessary. Tyco shall have 10 days from the date of EPA's notice to present such information. (E) Following the receipt of information from Tyco, as described in paragraph (6)(D) or if no such information is received within 10 days, EPA will issue a final written determination describing the Agency actions that are necessary to protect human health or the environment, given the information received in accordance with paragraphs (6)(A) or (6)(B). Any required action described in EPA's determination shall become effective immediately. |
| | | (7) *Notification Requirements:* Tyco must provide a one-time written notification to any State Regulatory Agency in a State to which or through which the delisted waste described above will be transported, at least 60 days prior to the commencement of such activities. Failure to provide such a notification will result in a violation of the delisting conditions and a possible revocation of the decision to delist. |

TABLE 1—WASTES EXCLUDED FROM NON-SPECIFIC SOURCES—Continued

| Facility | Address | Waste description |
|---|---|---|
| | | (8) *Recordkeeping and Certification Requirements for Waste to be Smelted for Metal Recovery:* Tyco must maintain in its facility files, and make available for inspection by EPA and the Florida Department of Environmental Protection (FDEP), records that include the name, address, telephone number, and contact person of each smelting facility used by Tyco for its delisted waste, quantities of waste shipped, analytical data for demonstrating that the delisting levels of Condition (3) are met, and a certification that the smelter(s) is(are) subject to regulatory controls on discharges to air, water, and land. The certification statement must be signed by a responsible official and contain the following language: Under civil and criminal penalty of law for the making or submission of false or fraudulent statements or representations (pursuant to the applicable provisions of the Federal Code, which include, but may not be limited to, 18 U.S.C. 1001 and 42 U.S.C. 6928), I certify that the smelter(s) used for Tyco's delisted waste is(are) subject to regulatory controls on discharges to air, water, and land. As the company official having supervisory responsibility for plant operations, I certify that to the best of my knowledge this information is true, accurate and complete. In the event that any of this information is determined by EPA in its sole discretion to be false, inaccurate or incomplete, and upon conveyance of this fact to the company, I recognize and agree that this exclusion of waste will be void as if it never had effect or to the extent directed by EPA and that the company will be liable for any actions taken in contravention of the company's RCRA and CERCLA obligations premised upon the company's void exclusion. |
| Universal Oil Products. | Decatur, Alabama. | Wastewater treatment sludges (EPA Hazardous Waste No. F006) generated from electroplating operations and contained in two on-site lagoons on August 15, 1986. This is a one-time exclusion. |
| U.S. EPA Combustion Research Facility. | Jefferson, Arkansas. | One-time exclusion for scrubber water (EPA Hazardous Waste No. F020) generated in 1985 from the incineration of Vertac still bottoms. This exclusion was published on June 28, 1989. |
| U.S. Nameplate Company, Inc.. | Mount Vernon, Iowa. | Retreated wastewater treatment sludges (EPA Hazardous Waste No. F006) previously generated from electroplating operations and currently contained in an on-site surface impoundment after September 28, 1988. This is a one-time exclusion for the reteated wastes only. This exclution does not relieve the waste unit from regulatory compliance under Subtitle C. |
| The Valero Refining Company—Tennessee, LLC. | Memphis, TN | Storm Water Basin sediment (EPA Hazardous Waste No. F037) generated one-time at a volume of 2,700 cubic yards March 10, 2010 and disposed in Subtitle D landfill. This is a one-time exclusion and applies to 2,700 cubic yards of Storm Water Basin sediment. |
| | | (1) Reopener. (A) If, anytime after disposal of the delisted waste, Valero possesses or is otherwise made aware of any environmental data (including but not limited to leachate data or ground water monitoring data) or any other data relevant to the delisted waste indicating that any constituent identified for the delisting verification testing is at level higher than the delisting level allowed by the Division Director in granting the petition, then the facility must report the data, in writing, to the Division Director within 10 days of first possessing or being made aware of that data. |
| | | (B) If Valero fails to submit the information described in paragraph (A) or if any other information is received from any source, the Division Director will make a preliminary determination as to whether the reported information requires EPA action to protect human health or the environment. Further action may include suspending, or revoking the exclusion, or other appropriate response necessary to protect human health and the environment. |
| | | (C) If the Division Director determines that the reported information does require EPA action, the Division Director will notify the facility in writing of the actions the Division Director believes are necessary to protect human health and the environment. The notice shall include a statement of the proposed action and a statement providing the facility with an opportunity to present information as to why the proposed EPA action is not necessary. The facility shall have 10 days from the date of the Division Director's notice to present such information. |
| | | (D) Following the receipt of information from the facility described in paragraph (C) or if no information is presented under paragraph initial receipt of information described in paragraphs (A) or (B), the Division Director will issue a final written determination describing EPA actions that are necessary to protect human health or the environment. Any required action described in the Division Director's determination shall become effective immediately, unless the Division Director provides otherwise. |
| | | (2) Notification Requirements: Valero must do the following before transporting the delisted waste: Failure to provide this notification will result in a violation of the delisting petition and a possible revocation of the decision. |
| | | (A) Provide a one-time written notification to any State Regulatory Agency to which or through which they will transport the delisted waste described above for disposal, 60 days before beginning such activities. |
| | | (B) Update the one-time written notification, if they ship the delisted waste to a different disposal facility. |
| | | (C) Failure to provide this notification will result in a violation of the delisting variance and a possible revocation of the decision. |

TABLE 1—WASTES EXCLUDED FROM NON-SPECIFIC SOURCES—Continued

| Facility | Address | Waste description |
|---|---|---|
| VAW of America Incorporated. | St. Augustine, Florida. | Wastewater treatment sludge filter cake (EPA Hazardous Waste No. F019) generated from the chemical conversion coating of aluminum. This exclusion was published on February 1, 1989. |
| Vermont American, Corp.. | Newark, OH ... | Wastewater treatment sludge (EPA Hazardous Waste No. F006) generated from electroplating operations after November 27, 1985. |
| Waterloo Industries. | Pocahontas, AR. | Wastewater treatment sludges (EPA Hazardous Waste No. F006) generated from electroplating operations after dewatering and held on-site on July 17, 1986 and any such sludge generated (after dewatering) after July 17, 1986. |
| Watervliet Arsenal. | Watervliet, NY | Wastewater treatment sludges (EPA Hazardous Waste No. F006) generated from electroplating operations after January 10, 1986. |
| Weirton Steel Corporation. | Weirton, West Virginia. | Wastewater treatment sludge (known as C&E sludge) containing EPA Hazardous Waste Numbers F007 and F008, subsequent to its excavation from the East Lagoon and the Figure 8 tanks for the purpose of transportation and disposal in a Subtitle D landfill after May 23, 2002. This is a one-time exclusion for a maximum volume of 18,000 cubic yards of C&E sludge. |
| | | (1) Reopener language. |
| | | (a) If Weirton discovers that any condition or assumption related to the characterization of the excluded waste which was used in the evaluation of the petition or that was predicted through modeling is not as reported in the petition, then Weirton must report any information relevant to that condition or assumption, in writing, to the Regional Administrator and the West Virginia Department of Environmental Protection within 10 calendar days of discovering that information. |
| | | (b) Upon receiving information described in paragraph (a) of this section, regardless of its source, the Regional Administrator and the West Virginia Department of Environmental Protection will determine whether the reported condition requires further action. Further action may include repealing the exclusion, modifying the exclusion, or other appropriate response necessary to protect human health or the environment. |
| | | (2) Notification Requirements. |
| | | Weirton must provide a one-time written notification to any State Regulatory Agency to which or through which the delisted waste described above will be transported for disposal at least 60 calendar days prior to the commencement of such activities. Failure to provide such notification will be deemed to be a violation of this exclusion and may result in revocation of the decision and other enforcement action. |
| William L. Bonnell Co.. | Newnan, Georgia. | Dewatered wastewater treatment sludges (EPA Hazardous Waste No. F019) generated from the chemical conversion coating of aluminum after November 14, 1986. This exclusion does not include sludges contained in Bonnell's on-site surface impoundments. |
| Windsor Plastics, Inc. | Evansville, IN | Spent non-halogenated solvents and still bottoms (EPA Hazardous Waste No. F003) generated from the recovery of acetone after November 17, 1986. |
| WRB Refining, LLC. | Borger, TX ..... | Thermal desorber residual solids (Hazardous Waste Nos. F037, F038, K048, K049, K050, and K051) generated at a maximum annual rate of 5,000 cubic yards per calendar year after September 29, 2009 and disposed in Subtitle D Landfill. |
| | | For the exclusion to be valid, WRB Refining LLC must implement a verification testing program that meets the following Paragraphs: |
| | | (1) Delisting Levels: All concentrations for those constituents must not exceed the maximum allowable concentrations in mg/l specified in this paragraph. |
| | | Thermal Desorber Residual Solid Leachable Concentrations (mg/l): Antimony–0.165; Arsenic–0.0129; Barium–54.8; Beryllium–0.119; Cadmium–0.139; Chromium–3.23; Chromium, Hexavalent–3.23; Cobalt–20.7; Copper–38.6; Cyanide–4.69; Lead–1.07; Mercury–0.104; Nickel–20.6; Selenium–1.0; Silver–5.0; Tin–3790.00; Vanadium–1.46; Zinc–320.0; Acenaphene–16.2; Anthracene–39.5; Benzene–0.117; Carbon Disulfide–86.0; 2-chlorophenol–4.41; Dibenzofuran–0.0226; 1,4-Dichlorobenzene–0.518; Ethylbenzene–16.5; Fluoranthene–3.75; Methylene Chloride–0.077; Naphthalene–0.0498; Phenol–264.0; Pyrene–6.78; Toluene–23.0; 1,2,4-Trichlorobenzene–1.51; Trichlorofluoromethane–23.5; Xylenes–14.6. |
| | | (2) Waste Holding and Handling: |
| | | (A) Waste classification as non-hazardous can not begin until compliance with the limits set in paragraph (1) for thermal desorber residual solids has occurred for two consecutive quarterly sampling events. |
| | | (B) If constituent levels in any sample taken by WRB Refining LLC exceed any of the delisting levels set in paragraph (1) for the thermal desorber residual solids, WRB Refining LLC must do the following: |
| | | (i) Notify EPA in accordance with paragraph (6) and |
| | | (ii) Manage and dispose the thermal desorber residual solids as hazardous waste generated under Subtitle C of RCRA. |
| | | (3) Testing Requirements: |
| | | Upon this exclusion becoming final, WRB Refining LLC may perform quarterly analytical testing by sampling and analyzing the desorber residual solids as follows: |
| | | (A) Quarterly Testing: |
| | | (i) Collect two representative composite samples of the sludge at quarterly intervals after EPA grants the final exclusion. The first composite samples may be taken at any time after EPA grants the final approval. Sampling should be performed in accordance with the sampling plan approved by EPA in support of the exclusion. |

TABLE 1—WASTES EXCLUDED FROM NON-SPECIFIC SOURCES—Continued

| Facility | Address | Waste description |
|---|---|---|
| | | (ii) Analyze the samples for all constituents listed in paragraph (1). Any composite sample taken that exceeds the delisting levels listed in paragraph (1) for the sludge must be disposed as hazardous waste in accordance with the applicable hazardous waste requirements. |
| | | (iii) Within thirty (30) days after taking its first quarterly sample, WRB Refining LLC will report its first quarterly analytical test data to EPA. If levels of constituents measured in the samples of the sludge do not exceed the levels set forth in paragraph (1) of this exclusion for two consecutive quarters, WRB Refining LLC can manage and dispose the non-hazardous thermal desorber residual solids according to all applicable solid waste regulations. |
| | | (B) Annual Testing: (i) If WRB Refining LLC completes the quarterly testing specified in paragraph (3) above and no sample contains a constituent at a level which exceeds the limits set forth in paragraph (1), WRB Refining LLC may begin annual testing as follows: WRB Refining LLC must test two representative composite samples of the thermal desorber residual solids for all constituents listed in paragraph (1) at least once per calendar year. |
| | | (ii) The samples for the annual testing shall be a representative composite sample according to appropriate methods. As applicable to the method-defined parameters of concern, analyses requiring the use of SW–846 methods incorporated by reference in 40 CFR 260.11 must be used without substitution. As applicable, the SW–846 methods might include Methods 0010, 0011, 0020, 0023A, 0030, 0031, 0040, 0050, 0051, 0060, 0061, 1010A, 1020B,1110A, 1310B, 1311, 1312, 1320, 1330A, 9010C, 9012B, 9040C, 9045D, 9060A, 9070A (uses EPA Method 1664, Rev. A), 9071B, and 9095B. Methods must meet Performance Based Measurement System Criteria in which the Data Quality Objectives are to demonstrate that samples of the WRB Refining thermal desorber residual solids are representative for all constituents listed in paragraph (1). |
| | | (iii) The samples for the annual testing taken for the second and subsequent annual testing events shall be taken within the same calendar month as the first annual sample taken. |
| | | (iv) The annual testing report should include the total amount of delisted waste in cubic yards disposed as non-hazardous waste during the calendar year. |
| | | (4) Changes in Operating Conditions: If WRB Refining LLC significantly changes the process described in its petition or starts any processes that generate(s) the waste that may or could affect the composition or type of waste generated (by illustration, but not limitation, changes in equipment or operating conditions of the treatment process), it must notify EPA in writing and it may no longer handle the wastes generated from the new process as non-hazardous until the wastes meet the delisting levels set in paragraph (1) and it has received written approval to do so from EPA. |
| | | WRB Refining LLC must submit a modification to the petition, complete with full sampling and analysis, for circumstances where the waste volume changes and/or additional waste codes are added to the waste stream, if it wishes to dispose of the material as non-hazardous. |
| | | (5) Data Submittals: |
| | | WRB Refining LLC must submit the information described below. If WRB Refining LLC fails to submit the required data within the specified time or maintain the required records on-site for the specified time, EPA, at its discretion, will consider this sufficient basis to reopen the exclusion as described in paragraph (6). WRB Refining LLC must: |
| | | (A) Submit the data obtained through paragraph (3) to the Chief, Corrective Action and Waste Minimization Section, Multimedia Planning and Permitting Division, U.S. Environmental Protection Agency Region 6, 1445 Ross Ave., Dallas, Texas, 75202, within the time specified. All supporting data can be submitted on CD–ROM or comparable electronic media. |
| | | (B) Compile records of analytical data from paragraph (3), summarized, and maintained on-site for a minimum of five years. |
| | | (C) Furnish these records and data when either EPA or the State of Texas requests them for inspection. |
| | | (D) Send along with all data a signed copy of the following certification statement, to attest to the truth and accuracy of the data submitted: |
| | | "Under civil and criminal penalty of law for the making or submission of false or fraudulent statements or representations (pursuant to the applicable provisions of the Federal Code, which include, but may not be limited to, 18 U.S.C. § 1001 and 42 U.S.C. § 6928), I certify that the information contained in or accompanying this document is true, accurate and complete. |
| | | As to the (those) identified section(s) of this document for which I cannot personally verify its (their) truth and accuracy, I certify as the company official having supervisory responsibility for the persons who, acting under my direct instructions, made the verification that this information is true, accurate and complete. |
| | | If any of this information is determined by EPA in its sole discretion to be false, inaccurate or incomplete, and upon conveyance of this fact to the company, I recognize and agree that this exclusion of waste will be void as if it never had effect or to the extent directed by EPA and that the company will be liable for any actions taken in contravention of the company's RCRA and CERCLA obligations premised upon the company's reliance on the void exclusion." |
| | | (6) Re-opener |

TABLE 1—WASTES EXCLUDED FROM NON-SPECIFIC SOURCES—Continued

| Facility | Address | Waste description |
|---|---|---|
| | | (A) If, anytime after disposal of the delisted waste WRB Refining LLC possesses or is otherwise made aware of any environmental data (including but not limited to leachate data or ground water monitoring data) or any other data relevant to the delisted waste indicating that any constituent identified for the delisting verification testing is at level higher than the delisting level allowed by the Division Director in granting the petition, then the facility must report the data, in writing, to the Division Director within 10 days of first possessing or being made aware of that data. |
| | | (B) If either the quarterly or annual testing of the waste does not meet the delisting requirements in paragraph 1, WRB Refining LLC must report the data, in writing, to the Division Director within 10 days of first possessing or being made aware of that data. |
| | | (C) If WRB Refining LLC fails to submit the information described in paragraphs (5), (6)(A) or (6)(B) or if any other information is received from any source, the Division Director will make a preliminary determination as to whether the reported information requires EPA action to protect human health and/or the environment. Further action may include suspending, or revoking the exclusion, or other appropriate response necessary to protect human health and the environment. |
| | | (D) If the Division Director determines that the reported information requires action by EPA, the Division Director will notify the facility in writing of the actions the Division Director believes are necessary to protect human health and the environment. The notice shall include a statement of the proposed action and a statement providing the facility with an opportunity to present information as to why the proposed EPA action is not necessary. The facility shall have 10 days from the date of the Division Director's notice to present such information. |
| | | (E) Following the receipt of information from the facility described in paragraph (6)(D) or (if no information is presented under paragraph (6)(D)) the initial receipt of information described in paragraphs (5), (6)(A) or (6)(B), the Division Director will issue a final written determination describing EPA actions that are necessary to protect human health and/or the environment. Any required action described in the Division Director's determination shall become effective immediately, unless the Division Director provides otherwise. |
| | | (7) Notification Requirements |
| | | WRB Refining LLC must do the following before transporting the delisted waste. Failure to provide this notification will result in a violation of the delisting petition and a possible revocation of the decision. |
| | | (A) Provide a one-time written notification to any state Regulatory Agency to which or through which it will transport the delisted waste described above for disposal, 60 days before beginning such activities. |
| | | (B) Update the one-time written notification if it ships the delisted waste into a different disposal facility. |
| | | (C) Failure to provide this notification will result in a violation of the delisting variance and a possible revocation of the decision. |

TABLE 2—WASTES EXCLUDED FROM SPECIFIC SOURCES

| Facility | Address | Waste description |
|---|---|---|
| American Chrome & Chemical. | Corpus Christi, Texas. | Dewatered sludge (the EPA Hazardous Waste No. K006) generated at a maximum generation of 1450 cubic yards per calendar year after September 21, 2004 and disposed in a Subtitle D landfill. ACC must implement a verification program that meets the following Paragraphs: |
| | | (1) Delisting Levels: All leachable constituent concentrations must not exceed the following levels (mg/l). The petitioner must use the method specified in 40 CFR 261.24 to measure constituents in the waste leachate. Dewatered wastewater sludge: Arsenic-0.0377; Barium-100.0; Chromium-5.0; Thallium-0.355; Zinc-1130.0. |
| | | (2) Waste Holding and Handling: |
| | | (A) ACC is a 90 day facility and does not have a RCRA permit, therefore, ACC must store the dewatered sludge following the requirements specified in 40 CFR 262.34, or continue to dispose of as hazardous all dewatered sludge generated, until they have completed verification testing described in Paragraph (3), as appropriate, and valid analyses show that paragraph (1) is satisfied. |
| | | (B) Levels of constituents measured in the samples of the dewatered sludge that do not exceed the levels set forth in Paragraph (1) are non-hazardous. ACC can manage and dispose the non-hazardous dewatered sludge according to all applicable solid waste regulations. |
| | | (C) If constituent levels in a sample exceed any of the delisting levels set in Paragraph (1), ACC must retreat the batches of waste used to generate the representative sample until it meets the levels. ACC must repeat the analyses of the treated waste. |
| | | (D) If the facility does not treat the waste or retreat it until it meets the delisting levels in Paragraph (1), ACC must manage and dispose the waste generated under Subtitle C of RCRA. |

TABLE 2—WASTES EXCLUDED FROM SPECIFIC SOURCES—Continued

| Facility | Address | Waste description |
|---|---|---|
| | | (E) The dewatered sludge must pass paint filter test as described in SW 846, Method 9095 or another appropriate method found in a reliable source before it is allowed to leave the facility. ACC must maintain a record of the actual volume of the dewatered sludge to be disposed of-site according to the requirements in Paragraph (5). |
| | | (3) Verification Testing Requirements: ACC must perform sample collection and analyses, including quality control procedures, according to appropriate methods such as those found in SW–846 or other reliable sources (with the exception of analyses requiring the use of SW–846 methods incorporated by reference in 40 CFR 260.11, which must be used without substitution. ACC must conduct verification testing each time it decides to evacuate the tank contents. Four (4) representative composite samples shall be collected from the dewatered sludge. ACC shall analyze the verification samples according to the constituent list specified in Paragraph (1) and submit the analytical results to EPA within 10 days of receiving the analytical results. If the EPA determines that the data collected under this Paragraph do not support the data provided for the petition, the exclusion will not cover the generated wastes. The EPA will notify ACC the decision in writing within two weeks of receiving this information. |
| | | (4) Changes in Operating Conditions: If ACC significantly changes the process described in its petition or starts any processes that may or could affect the composition or type of waste generated as established under Paragraph (1) (by illustration, but not limitation, changes in equipment or operating conditions of the treatment process), they must notify the EPA in writing; they may no longer handle the wastes generated from the new process as nonhazardous until the test results of the wastes meet the delisting levels set in Paragraph (1) and they have received written approval to do so from the EPA. |
| | | (5) Data Submittals: ACC must submit the information described below. If ACC fails to submit the required data within the specified time or maintain the required records on-site for the specified time, the EPA, at its discretion, will consider this sufficient basis to reopen the exclusion as described in Paragraph 6. ACC must: |
| | | (A) Submit the data obtained through Paragraph 3 to the Section Chief, Corrective Action and Waste Minimization Section, Environmental Protection Agency, 1445 Ross Avenue, Dallas, Texas 75202–2733, Mail Code, (6PD-C) within the time specified. |
| | | (B) Compile records of operating conditions and analytical data from Paragraph (3), summarized, and maintained on-site for a minimum of five years. |
| | | (C) Furnish these records and data when the EPA or the State of Texas request them for inspection. |
| | | (D) Send along with all data a signed copy of the following certification statement, to attest to the truth and accuracy of the data submitted: Under civil and criminal penalty of law for the making or submission of false or fraudulent statements or representations (pursuant to the applicable provisions of the Federal Code, which include, but may not be limited to, 18 U.S.C. 1001 and 42 U.S.C. 6928), I certify that the information contained in or accompanying this document is true, accurate and complete. As to the (those) identified section(s) of this document for which I cannot personally verify its (their) truth and accuracy, I certify as the company official having supervisory responsibility for the persons who, acting under my direct instructions, made the verification that this information is true, accurate and complete. If any of this information is determined by the EPA in its sole discretion to be false, inaccurate or incomplete, and upon conveyance of this fact to the company, I recognize and agree that this exclusion of waste will be void as if it never had effect or to the extent directed by the EPA and that the company will be liable for any actions taken in contravention of the company's RCRA and CERCLA obligations premised upon the company's reliance on the void exclusion. |
| | | (6) Reopener: |
| | | (A) If, anytime after disposal of the delisted waste, ACC possesses or is otherwise made aware of any environmental data (including but not limited to leachate data or ground water monitoring data) or any other data relevant to the delisted waste indicating that any constituent identified for the delisting verification testing is at level higher than the delisting level allowed by the Division Director in granting the petition, then the facility must report the data, in writing, to the Division Director within 10 days of first possessing or being made aware of that data. |
| | | (B) If the verification testing of the waste does not meet the delisting requirements in Paragraph 1, ACC must report the data, in writing, to the Division Director within 10 days of first possessing or being made aware of that data. |
| | | (C) If ACC fails to submit the information described in paragraphs (5),(6)(A) or (6)(B) or if any other information is received from any source, the Division Director will make a preliminary determination as to whether the reported information requires Agency action to protect human health or the environment. Further action may include suspending, or revoking the exclusion, or other appropriate response necessary to protect human health and the environment. |

TABLE 2—WASTES EXCLUDED FROM SPECIFIC SOURCES—Continued

| Facility | Address | Waste description |
|---|---|---|
| | | (D) If the Division Director determines that the reported information does require Agency action, the Division Director will notify the facility in writing of the actions the Division Director believes are necessary to protect human health and the environment. The notice shall include a statement of the proposed action and a statement providing the facility with an opportunity to present information as to why the proposed Agency action is not necessary. The facility shall have 10 days from the date of the Division Director's notice to present such information. |
| | | (E) Following the receipt of information from the facility described in paragraph (6)(D) or (if no information is presented under paragraph (6)(D)) the initial receipt of information described in paragraphs (5), (6)(A) or (6)(B), the Division Director will issue a final written determination describing the Agency actions that are necessary to protect human health or the environment. Any required action described in the Division Director's determination shall become effective immediately, unless the Division Director provides otherwise. |
| | | (7) Notification Requirements: ACC must do the following before transporting the delisted waste: Failure to provide this notification will result in a violation of the delisting petition and a possible revocation of the decision. |
| | | (A) Provide a one-time written notification to any State Regulatory Agency to which or through which they will transport the delisted waste described above for disposal, 60 days before beginning such activities. If ACC transports the excluded waste to or manages the waste in any state with delisting authorization, ACC must obtain delisting authorization from that state before it can manage the waste as nonhazardous in the state. |
| | | (B) Update the one-time written notification if they ship the delisted waste to a different disposal facility. |
| | | (C) Failure to provide the notification will result in a violation of the delisting variance and a possible revocation of the exclusion. |
| American Cyanamid. | Hannibal, Missouri. | Wastewater and sludge (EPA Hazardous Waste No. K038) generated from the washing and stripping of phorate production and contained in on-site lagoons on May 8, 1987, and such wastewater and sludge generated after May 8, 1987. |
| Amoco Oil Co. | Wood River, IL | 150 million gallons of DAF from petroleum refining contained in four surge ponds after treatment with the Chemifix® stabilization process. This waste contains EPA Hazardous Waste No. K048. This exclusion applies to the 150 million gallons of waste after chemical stabilization as long as the mixing ratios of the reagent with the waste are monitored continuously and do not vary outside of the limits presented in the demonstration samples; one grab sample is taken each hour from each treatment unit, composited, and EP toxicity tests performed on each sample. If the levels of lead or total chromium exceed 0.5 ppm in the EP extract, then the waste that was processed during the compositing period is considered hazardous; the treatment residue shall be pumped into bermed cells to ensure that the waste is identifiable in the event that removal is necessary. |
| Akzo Chemicals, Inc. (formerly Stauffer Chemical Company). | Axis, AL ......... | Brine purification muds generated from their chlor-alkali manufacturing operations (EPA Hazardous Waste No. K071) and disposed of in brine mud pond HWTF: 5 EP–201. |
| Bayer Material Science LLC. | Baytown, TX .. | Outfall 007 Treated Effluent (EPA Hazardous Waste Nos. K027, K104, K111, and K112) generated at a maximum rate of 18,071,150 cubic yards (5.475 billion gallons) per calendar year after July 25, 2005 as it exits the Outfall Tank and disposed in accordance with the TPDES permit. |
| | | The delisting levels set do not relieve Bayer of its duty to comply with the limits set in its TPDES permit. For the exclusion to be valid, Bayer must implement a verification testing program that meets the following Paragraphs: |
| | | (1) Delisting Levels: All concentrations for those constituents must not exceed the maximum allowable concentrations in mg/kg specified in this paragraph. |
| | | Outfall 007 Treated Effluent Total Concentrations (mg/kg): Antimony—0.0816; Arsenic—0.385, Barium—22.2; Chromium—153.0; Copper—3620.0; Cyanide—0.46; Mercury—0.0323; Nickel—11.3; Selenium—0.23; Thallium—0.0334; Vanadium—8.38; Zinc—112.0; Acetone—14.6; Acetophenone—15.8; Aniline—0.680; Benzene—0.0590; Bis (2-ethylhexyl)phthalate—1260.0; Bromodichloromethane—0.0719; Chloroform—0.077; Di-n-octyl phthalate—454.0; 2,4-Dinitrotoluene—0.00451; Diphenylamine—11.8; 1,4-Dioxane—1.76; Di-n-butyl phthalate—149.0; Fluoranthene—24.6; Methylene chloride—0.029; Methyl ethyl ketone—87.9; Nitrobenzene—0.0788; m-phenylenediamine—0.879; Pyrene—39.0; 1,1,1,2-Tetrachloroethane—0.703; o-Toluidine—0.0171; p-Toluidine—0.215; 2,4-Toluenediamine—0.00121. Toluene diisocyanate—0.001. |
| | | (2) Waste Holding and Handling: (A) Waste classification as non-hazardous can not begin until compliance with the limits set in paragraph (1) for the treated effluent has occurred for two consecutive quarterly sampling events and those reports have been approved by EPA. |
| | | The delisting for the treated effluent applies only during periods of TPDES compliance. |

TABLE 2—WASTES EXCLUDED FROM SPECIFIC SOURCES—Continued

| Facility | Address | Waste description |
|---|---|---|
| | | (B) If constituent levels in any sample taken by Bayer exceed any of the delisting levels set in paragraph (1) for the treated effluent, Bayer must do the following:<br>(i) notify EPA in accordance with paragraph (6) and<br>(ii) Manage and dispose the treated effluent as hazardous waste generated under Subtitle C of RCRA.<br>(iii) Routine inspection and regular maintenance of the effluent pipe line must occur to prevent spills and leaks of the treated effluent prior to discharge.<br>(3) Testing Requirements: Sample collection and analyses, including quality control procedures, must be performed using appropriate methods. As applicable to the method-defined parameters of concern, analyses requiring the use of SW–846 methods incorporated by reference in 40 CFR 260.11 must be used without substitution. As applicable, the SW–846 methods might include Methods 0010, 0011, 0020, 0023A, 0030, 0031, 0040, 0050, 0051, 0060, 0061, 1010A, 1020B, 1110A, 1310B, 1311, 1312, 1320, 1330A, 9010C, 9012B, 9040C, 9045D, 9060A, 9070A (uses EPA Method 1664, Rev. A), 9071B, and 9095B. Methods must meet Performance Based Measurement System Criteria in which the Data Quality Objectives are to demonstrate that representative samples of the Bayer treated effluent meet the delisting levels in paragraph (1).<br>(A) Quarterly Testing: Upon this exclusion becoming final, Bayer may perform quarterly analytical testing by sampling and analyzing the treated effluent as follows:<br>(i) Collect two representative composite samples of the treated effluent at quarterly intervals after EPA grants the final exclusion. The first composite samples may be taken at any time after EPA grants the final approval. Sampling should be performed in accordance with the sampling plan approved by EPA in support of the exclusion.<br>(ii) Analyze the samples for all constituents listed in paragraph (1). Any composite sample taken that exceeds the delisting levels listed in paragraph (1) for the treated effluent must be disposed of as hazardous waste in accordance with the applicable hazardous waste requirements in its TPDES discharge permit.<br>(iii) Within thirty (30) days after taking its first quarterly sample, Bayer will report its first quarterly analytical test data to EPA. If levels of constituents measured in the samples of the treated effluent do not exceed the levels set forth in paragraph (1) of this exclusion for two consecutive quarters, Bayer can manage and dispose the nonhazardous treated effluent according to all applicable solid waste regulations.<br>(B) Annual Testing:<br>(i) If Bayer completes the four (4) quarterly testing events specified in paragraph (3)(A) above and no sample contains a constituent with a level which exceeds the limits set forth in paragraph (1), Bayer may begin annual testing as follows: Bayer must test two representative composite samples of the treated effluent for all constituents listed in paragraph (1) at least once per calendar year.<br>(ii) The samples for the annual testing shall be a representative composite sample according to appropriate methods. As applicable to the method-defined parameters of concern, analyses requiring the use of SW–846 methods incorporated by reference in 40 CFR 260.11 must be used without substitution. As applicable, the SW–846 methods might include Methods 0010, 0011, 0020, 0023A, 0030, 0031, 0040, 0050, 0051, 0060, 0061, 1010A, 1020B, 1110A, 1310B, 1311, 1312, 1320, 1330A, 9010C, 9012B, 9040C, 9045D, 9060A, 9070A (uses EPA Method 1664, Rev. A), 9071B, and 9095B. Methods must meet Performance Based Measurement System Criteria in which the Data Quality Objectives are to demonstrate that representative samples of the Bayer treated effluent for all constituents listed in paragraph (1).<br>(iii) The samples for the annual testing taken for the second and subsequent annual testing events shall be taken within the same calendar month as the first annual sample taken.<br>(4) Changes in Operating Conditions: If Bayer significantly changes the process described in its petition or starts any processes that generate(s) the waste that may or could affect the composition or type of waste generated as established under paragraph (1) (by illustration, but not limitation, changes in equipment or operating conditions of the treatment process), it must notify EPA in writing; it may no longer handle the wastes generated from the new process as nonhazardous until the wastes meet the delisting levels set in paragraph (1) and it has received written approval to do so from EPA.<br>Bayer must submit a modification to the petition complete with full sampling and analysis for circumstances where the waste volume changes and/or additional waste codes are added to the waste stream. |

TABLE 2—WASTES EXCLUDED FROM SPECIFIC SOURCES—Continued

| Facility | Address | Waste description |
|---|---|---|
| | | (5) Data Submittals:<br>Bayer must submit the information described below. If Bayer fails to submit the required data within the specified time or maintain the required records on-site for the specified time, EPA, at its discretion, will consider this sufficient basis to reopen the exclusion as described in paragraph (6). Bayer must:<br>(i) Submit the data obtained through paragraph (3) to the Chief, Corrective Action and Waste Minimization Section, Multimedia Planning and Permitting Division, U.S. Environmental Protection Agency Region 6, 1445 Ross Ave., Dallas, Texas, 75202, within the time specified. All supporting data can be submitted on CD-ROM or some comparable electronic media.<br>(ii) Compile records of analytical data from paragraph (3), summarized, and maintained on-site for a minimum of five years.<br>(iii) Furnish these records and data when either EPA or the State of Texas request them for inspection.<br>(iv) Send along with all data a signed copy of the following certification statement, to attest to the truth and accuracy of the data submitted:<br>"Under civil and criminal penalty of law for the making or submission of false or fraudulent statements or representations (pursuant to the applicable provisions of the Federal Code, which include, but may not be limited to, 18 U.S.C. 1001 and 42 U.S.C. 6928), I certify that the information contained in or accompanying this document is true, accurate and complete.<br>As to the (those) identified section(s) of this document for which I cannot personally verify its (their) truth and accuracy, I certify as the company official having supervisory responsibility for the persons who, acting under my direct instructions, made the verification that this information is true, accurate and complete.<br>If any of this information is determined by EPA in its sole discretion to be false, inaccurate or incomplete, and upon conveyance of this fact to the company, I recognize and agree that this exclusion of waste will be void as if it never had effect or to the extent directed by EPA and that the company will be liable for any actions taken in contravention of the company's RCRA and CERCLA obligations premised upon the company's reliance on the void exclusion."<br>(6) Reopener:<br>(i) If, anytime after disposal of the delisted waste Bayer possesses or is otherwise made aware of any environmental data (including but not limited to leachate data or ground water monitoring data) or any other data relevant to the delisted waste indicating that any constituent identified for the delisting verification testing is at level higher than the delisting level allowed by the Division Director in granting the petition, then the facility must report the data, in writing, to the Division Director within 10 days of first possessing or being made aware of that data.<br>(ii) If either the quarterly or annual testing of the waste does not meet the delisting requirements in paragraph (1), Bayer must report the data, in writing, to the Division Director within 10 days of first possessing or being made aware of that data.<br>(iii) If Bayer fails to submit the information described in paragraphs (5), (6)(i) or (6)(ii) or if any other information is received from any source, the Division Director will make a preliminary determination as to whether the reported information requires EPA action to protect human health and/or the environment. Further action may include suspending, or revoking the exclusion, or other appropriate response necessary to protect human health and the environment.<br>(iv) If the Division Director determines that the reported information requires action by EPA, the Division Director will notify the facility in writing of the actions the Division Director believes are necessary to protect human health and the environment. The notice shall include a statement of the proposed action and a statement providing the facility with an opportunity to present information as to why the proposed EPA action is not necessary. The facility shall have 10 days from the date of the Division Director's notice to present such information.<br>(v) Following the receipt of information from the facility described in paragraph (6)(iv) or (if no information is presented under paragraph (6)(iv)) the initial receipt of information described in paragraphs (5), (6)(i) or (6)(ii), the Division Director will issue a final written determination describing EPA actions that are necessary to protect human health and/or the environment. Any required action described in the Division Director's determination shall become effective immediately, unless the Division Director provides otherwise. |
| Bayer Material Science LLC. | Baytown, TX .. | Spent Carbon (EPA Hazardous Waste Nos. K027, K104, K111, and K112) generated at a maximum rate of 7,728 cubic yards per calendar year after May 16, 2006.<br>For the exclusion to be valid, Bayer must implement a verification testing program that meets the following Paragraphs:<br>(1) Delisting Levels:<br>All concentrations for those constituents must not exceed the maximum allowable concentrations in mg/l specified in this paragraph. |

235

TABLE 2—WASTES EXCLUDED FROM SPECIFIC SOURCES—Continued

| Facility | Address | Waste description |
|---|---|---|
|  |  | Spent Carbon Leachable Concentrations (mg/l): Antimony–0.251; Arsenic–0.385, Barium–8.93; Beryllium–0.953; Cadmium–0.687; Chromium–5.0; Cobalt–2.75; Copper–128.0; Cyanide–1.65; Lead–5.0; Mercury–0.0294; Nickel–3.45; Selenium–0.266; Tin–2.75; Vanadium–2.58; Zinc–34.2; Aldrin–0.0000482; Acetophenone–87.1; Aniline–2.82; Benzene–0.554; Bis(2-ethylhexyl)phthalate–0.342; Benzyl alcohol–261; Butylbenzylphthalate–3.54; Chloroform–0.297; Di-n-octyl phthalate–0.00427; 2,4-Dinitrotoluene–0.0249; 2,6-Dinitrotoluene–0.0249 Diphenylamine–1.43; 1,4-Dioxane–14.6; Di-n-butylphthalate–2.02; Kepone–0.000373; 2-Nitrophenol–87.9; N-Nitrodiphenylamine–3.28; Phenol–52.2; 2,4-Toluenediamine–0.00502; Toluene diisocyanate–0.001. |

(2) Waste Holding and Handling:

(A) Waste classification as non-hazardous can not begin until compliance with the limits set in paragraph (1) for spent carbon has occurred for two consecutive quarterly sampling events and the reports have been approved by EPA.

(B) If constituent levels in any sample taken by Bayer exceed any of the delisting levels set in paragraph (1) for the spent carbon, Bayer must do the following:

(i) notify EPA in accordance with paragraph (6) and

(ii) manage and dispose the spent carbon as hazardous waste generated under Subtitle C of RCRA.

(3) Testing Requirements:

Upon this exclusion becoming final, Bayer must perform quarterly analytical testing by sampling and analyzing the spent carbon as follows:

(A) Quarterly Testing:

(i) Collect two representative composite samples of the spent carbon at quarterly intervals after EPA grants the final exclusion. The first composite samples may be taken at any time after EPA grants the final approval. Sampling should be performed in accordance with the sampling plan approved by EPA in support of the exclusion.

(ii) Analyze the samples for all constituents listed in paragraph (1). Any composite sample taken that exceeds the delisting levels listed in paragraph (1) for the spent carbon must be disposed as hazardous waste in accordance with the applicable hazardous waste requirements.

(iii) Within thirty (30) days after taking its first quarterly sample, Bayer will report its first quarterly analytical test data to EPA. If levels of constituents measured in the samples of the spent carbon do not exceed the levels set forth in paragraph (1) of this exclusion for two consecutive quarters, Bayer can manage and dispose the non-hazardous spent carbon according to all applicable solid waste regulations.

(B) Annual Testing:

(i) If Bayer completes the quarterly testing specified in paragraph (3) above and no sample contains a constituent at a level which exceeds the limits set forth in paragraph (1), Bayer can begin annual testing as follows: Bayer must test two representative composite samples of the spent carbon for all constituents listed in paragraph (1) at least once per calendar year.

(ii) The samples for the annual testing shall be a representative composite sample according to appropriate methods. As applicable to the method-defined parameters of concern, analyses requiring the use of SW–846 methods incorporated by reference in 40 CFR 260.11 must be used without substitution. As applicable, the SW–846 methods might include Methods 0010, 0011, 0020, 0023A, 0030, 0031, 0040, 0050, 0051, 0060, 0061, 1010A, 1020B, 1110A, 1310B, 1311, 1312, 1320, 1330A, 9010C, 9012B, 9040C, 9045D, 9060A, 9070A (uses EPA Method 1664, Rev. A), 9071B, and 9095B.

Methods must meet Performance Based Measurement System Criteria in which the Data Quality Objectives are to demonstrate that samples of the Bayer spent carbon are representative for all constituents listed in paragraph (1).

(iii) The samples for the annual testing taken for the second and subsequent annual testing events shall be taken within the same calendar month as the first annual sample taken.

(iv) The annual testing report must include the total amount of waste in cubic yards disposed during the calendar year.

(4) Changes in Operating Conditions:

If Bayer significantly changes the process described in its petition or starts any process that generates the waste that may or could affect the composition or type of waste generated (by illustration, but not limitation, changes in equipment or operating conditions of the treatment process), it must notify EPA in writing and it may no longer handle the wastes generated from the new process as non-hazardous until the wastes meet the delisting levels set in paragraph (1) and it has received written approval to do so from EPA.

Bayer must submit a modification to the petition complete with full sampling and analysis for circumstances where the waste volume changes and/or additional waste codes are added to the waste stream.

(5) Data Submittals:

Bayer must submit the information described below. If Bayer fails to submit the required data within the specified time or maintain the required records on-site for the specified time, EPA, at its discretion, will consider this sufficient basis to reopen the exclusion as described in paragraph (6). Bayer must:

TABLE 2—WASTES EXCLUDED FROM SPECIFIC SOURCES—Continued

| Facility | Address | Waste description |
|---|---|---|
| | | (A) Submit the data obtained through paragraph 3 to the Chief, Corrective Action and Waste Minimization Section, Multimedia Planning and Permitting Division, U. S. Environmental Protection Agency Region 6, 1445 Ross Ave., Dallas, Texas, 75202, within the time specified. All supporting data can be submitted on CD-ROM or some comparable electronic media. |
| | | (B) Compile records of analytical data from paragraph (3), summarized, and maintained on-site for a minimum of five years. |
| | | (C) Furnish these records and data when either EPA or the State of Texas requests them for inspection. |
| | | (D) Send along with all data a signed copy of the following certification statement, to attest to the truth and accuracy of the data submitted: |
| | | "Under civil and criminal penalty of law for the making or submission of false or fraudulent statements or representations (pursuant to the applicable provisions of the Federal Code, which include, but may not be limited to, 18 U.S.C. 1001 and 42 U.S.C. 6928), I certify that the information contained in or accompanying this document is true, accurate and complete. |
| | | As to the (those) identified section(s) of this document for which I cannot personally verify its (their) truth and accuracy, I certify as the company official having supervisory responsibility for the persons who, acting under my direct instructions, made the verification that this information is true, accurate and complete. |
| | | If any of this information is determined by EPA in its sole discretion to be false, inaccurate or incomplete, and upon conveyance of this fact to the company, I recognize and agree that this exclusion of waste will be void as if it never had effect or to the extent directed by EPA and that the company will be liable for any actions taken in contravention of the company's RCRA and CERCLA obligations premised upon the company's reliance on the void exclusion." |
| | | (6) Reopener: |
| | | (A) If, anytime after disposal of the delisted waste Bayer possesses or is otherwise made aware of any environmental data (including but not limited to leachate data or ground water monitoring data) or any other data relevant to the delisted waste indicating that any constituent identified for the delisting verification testing is at a level higher than the delisting level allowed by EPA in granting the petition, then the facility must report the data, in writing, to EPA within 10 days of first possessing or being made aware of that data. |
| | | (B) If either the quarterly or annual testing of the waste does not meet the delisting requirements in paragraph 1, Bayer must report the data, in writing, to EPA within 10 days of first possessing or being made aware of that data. |
| | | (C) If Bayer fails to submit the information described in paragraphs (5),(6)(A) or (6)(B) or if any other information is received from any source, EPA will make a preliminary determination as to whether the reported information requires action to protect human health and/or the environment. Further action may include suspending, or revoking the exclusion, or other appropriate response necessary to protect human health and the environment. |
| | | (D) If EPA determines that the reported information requires action, EPA will notify the facility in writing of the actions it believes are necessary to protect human health and the environment. The notice shall include a statement of the proposed action and a statement providing the facility with an opportunity to present information explaining why the proposed EPA action is not necessary. The facility shall have 10 days from the date of EPA's notice to present such information. |
| | | (E) Following the receipt of information from the facility described in paragraph (6)(D) or (if no information is presented under paragraph (6)(D)) the initial receipt of information described in paragraphs (5), (6)(A) or (6)(B), EPA will issue a final written determination describing the actions that are necessary to protect human health and/or the environment. Any required action described in EPA's determination shall become effective immediately, unless EPA provides otherwise. |
| Bekaert Steel Corporation. | Rogers, Arkansas. | Wastewater treatment sludge (EPA Hazardous Waste No. F006) generated from electroplating operations (at a maximum annual rate of 1250 cubic yards to be measured on a calendar year basis) after [insert publication date of the final rule]. In order to confirm that the characteristics of the waste do not change significantly, the facility must, on an annual basis, before July 1 of each year, analyze a representative composite sample for the constituents listed in § 261.24 as well as antimony, copper, nickel, and zinc using the method specified therein. The annual analytical results, including quality control information, must be compiled, certified according to § 260.22(i)(12) of this chapter, maintained on site for a minimum of five years, and made available for inspection upon request of any employee or representative of EPA or the State of Arkansas. Failure to maintain the required documents on site will be considered by EPA, at its discretion, sufficient basis to revoke the exclusion to the extent directed by EPA. |
| | | *Notification Requirements:* |
| | | Bekaert Steel Corporation must provide a one-time written notification to any State Regulatory Agency to which or through which the delisted waste described above will be transported for disposal at least 60 days prior to the commencement of such activities. Failure to provide such a notification will result in a violation of the delisting petition and a possible revocation of the decision. |

TABLE 2—WASTES EXCLUDED FROM SPECIFIC SOURCES—Continued

| Facility | Address | Waste description |
|---|---|---|
| Bethlehem Steel Corporation. | Lackawanna, New York. | Ammonia still lime sludge (EPA Hazardous Waste No. K060) and other solid waste generated from primary metal-making and coking operations. This is a one-time exclusion for 118,000 cubic yards of waste contained in the on-site landfill referred to as HWM–2. This exclusion was published on April 24, 1996. |
| Bethlehem Steel Corp.. | Steelton, PA .. | Uncured and cured chemically stabilized electric arc furnace dust/sludge (CSEAFD) treatment residue (K061) generated from the primary production of steel after May 22, 1989. This exclusion is conditioned upon the data obtained from Bethlehem's full-scale CSEAFD treatment facility because Bethlehem's original data were obtained from a laboratory-scale CSEAFD treatment process. To ensure that hazardous constituents are not present in the waste at levels of regulatory concern once the full-scale treatment facility is in operation, Bethlehem must implement a testing program for the petitioned waste. This testing program must meet the following conditions for the exclusion to be valid: |

(1) *Testing:*

(A) *Initial Testing:* During the first four weeks of operation of the full-scale treatment system, Bethlehem must collect representative grab samples of each treated batch of the CSEAFD and composite the grab samples daily. The daily composites, prior to disposal, must be analyzed for the EP leachate concentrations of all the EP toxic metals, nickel and cyanide (using distilled water in the cyanide extractions). Analyses must be performed using appropriate methods. As applicable to the method-defined parameters of concern, analyses requiring the use of SW–846 methods incorporated by reference in 40 CFR 260.11 must be used without substitution. As applicable, the SW–846 methods might include Methods 0010, 0011, 0020, 0023A, 0030, 0031, 0040, 0050, 0051, 0060, 0061, 1010A, 1020B, 1110A, 1310B, 1311, 1312, 1320, 1330A, 9010C, 9012B, 9040C, 9045D, 9060A, 9070A (uses EPA Method 1664, Rev. A), 9071B, and 9095B. Bethlehem must report the analytical test data obtained during this initial period no later than 90 days after the treatment of the first full-scale batch.

(B) *Subsequent Testing:* Bethlehem must collect representative grab samples from every treated batch of CSEAFD generated daily and composite all of the grab samples to produce a weekly composite sample. Bethlehem then must analyze each weekly composite sample for the EP leachate concentrations of all the EP toxic metals and nickel. Analyses must be performed using appropriate methods. As applicable to the method-defined parameters of concern, analyses requiring the use of SW–846 methods incorporated by reference in 40 CFR 260.11 must be used without substitution. As applicable, the SW–846 methods might include Methods 0010, 0011, 0020, 0023A, 0030, 0031, 0040, 0050, 0051, 0060, 0061, 1010A, 1020B, 1110A, 1310B, 1311, 1312, 1320, 1330A, 9010C, 9012B, 9040C, 9045D, 9060A, 9070A (uses EPA Method 1664, Rev. A), 9071B, and 9095B. The analytical data, including all quality control information, must be compiled and maintained on site for a minimum of three years. These data must be furnished upon request and made available for inspection by any employee or representative of EPA or the State of Pennsylvania.

(2) *Delisting Levels:* If the EP extract concentrations resulting from the testing in condition (1)(A) or (1)(B) for chromium, lead, arsenic, or silver exceeds 0.315 mg/l; for barium exceeds 6.3 mg/l; for cadmium or selenium exceed 0.063 mg/l; for mercury exooods 0.0126 mg/l; for nickel exceeds 3.15 mg/l; or for cyanide exceeds 4.42 mg/l, the waste must either be re-treated or managed and disposed in accordance with subtitle C of RCRA.

(3) *Data submittals:* Within one week of system start-up, Bethlehem must notify the Section Chief, Variances Section (see address below) when their full-scale stabilization system is on-line and waste treatment has begun. All data obtained through the initial testing condition (1)(A), must be submitted to PSPD/OSW (5303W), U.S. EPA, 1200 Pennsylvania Ave., NW., Washington, DC 20460 within the time period specified in condition (1)(A). At the Section Chief's request, Bethlehem must submit analytical data obtained through condition (1)(B) to the above address, within the time period specified by the Section Chief. Failure to submit the required data obtained from either condition (1)(A) or (1)(B) within the specified time periods will be considered by the Agency sufficient basis to revoke Bethlehem's exclusion to the extent directed by EPA. All data must be accompanied by the following certification statement:

"Under civil and criminal penalty of law for the making or submission of false or fraudulent statements or representations (pursuant to the applicable provisions of the Federal Code which include, but may not be limited to, 18 U.S.C. 6928), I certify that the information contained in or accompanying this document is true, accurate and complete.

"As to the (those) identified section(s) of this document for which I cannot personally verify its (their) truth and accuracy, I certify as the company official having supervisory responsibility for the persons who, acting under my direct instructions, made the verification that this information is true, accurate and complete.

"In the event that any of this information is determined by EPA in its sole discretion to be false, inaccurate or incomplete, and upon conveyance of this fact to the company, I recognize and agree that this exclusion of wastes will be void as if it never had effect or to the extent directed by EPA and that the company will be liable for any actions taken in contravention of the company's RCRA and CERCLA obligations premised upon the company's reliance on the void exclusion."

238

TABLE 2—WASTES EXCLUDED FROM SPECIFIC SOURCES—Continued

| Facility | Address | Waste description |
|---|---|---|
| Bethlehem Steel Corp.. | Johnstown, PA | Uncured and cured chemically stabilized electric arc furnace dust/sludge (CSEAFD) treatment residue (K061) generated from the primary production of steel after May 22, 1989. This exclusion is conditioned upon the data obtained from Bethlehem's full-scale CSEAFD treatment facility because Bethlehem's original data were obtained from a labortory-scale CSEAFD treatment process. To ensure that hazardous constituents are not present in the waste at levels of regulatory concern once the full-scale treatment facility is in operation, Bethlehem must implement a testing program for the petitioned waste. This testing program must meet the following conditions for the exclusion to be valid: |

(1) *Testing:*

(A) *Initial Testing:* During the first four weeks of operation of the full-scale treatment system, Bethlehem must collect representative grab samples of each treated batch of the CSEAFD and composite the grab samples daily. The daily composites, prior to disposal, must be analyzed for the EP leachate concentrations of all the EP toxic metals, nickel, and cyanide (using distilled water in the cyanide extractions). Analyses must be performed using appropriate methods. As applicable to the method-defined parameters of concern, analyses requiring the use of SW–846 methods incorporated by reference in 40 CFR 260.11 must be used without substitution. As applicable, the SW–846 methods might include Methods 0010, 0011, 0020, 0023A, 0030, 0031, 0040, 0050, 0051, 0060, 0061, 1010A, 1020B, 1110A, 1310B, 1311, 1312, 1320, 1330A, 9010C, 9012B, 9040C, 9045D, 9060A, 9070A (uses EPA Method 1664, Rev. A), 9071B, and 9095B. Bethlehem must report the analytical test data obtained during this initial period no later than 90 days after the treatment of the first full-scale batch.

(B) *Subsequent Testing:* Bethlehem must collect representative grab samples from every treated batch of CSEAFD generated daily and composite all of the grab samples to produce a weekly composite sample. Bethlehem then must analyze each weekly composite sample for the EP leachate concentrations of all the EP toxic metals and nickel. Analyses must be performed using appropriate methods. As applicable to the method-defined parameters of concern, analyses requiring the use of SW–846 methods incorporated by reference in 40 CFR 260.11 must be used without substitution. As applicable, the SW–846 methods might include Methods 0010, 0011, 0020, 0023A, 0030, 0031, 0040, 0050, 0051, 0060, 0061, 1010A, 1020B, 1110A, 1310B, 1311, 1312, 1320, 1330A, 9010C, 9012B, 9040C, 9045D, 9060A, 9070A (uses EPA Method 1664, Rev. A), and 9095B. The analytical data, including all quality control information, must be compiled and maintained on site for a minimum of three years. These data must be furnished upon request and made available for inspection by any employee or representative of EPA or the State of Pennsylvania.

(2) *Delisting Levels:* If the EP extract concentrations resulting from the testing in condition (1)(A) or (1)(B) for chromium, lead, arsenic, or silver exceed 0.315 mg/l; for barium exceeds 6.3 mg/l; for cadmium or selenium exceed 0.063 mg/l; for mercury exceeds 0.0126 mg/l; for nickel exceeds 3.15 mg/l; or for cyanide exceeds 4.42 mg/l, the waste must either be retreated until it meets these levels or managed and disposed in accordance with subtitle C of RCRA.

(3) *Data submittals:* Within one week of system start-up, Bethlehem must notify the Section Chief, Variances Section (see address below) when their full-scale stabilization system is on-line and waste treatment has begun. All data obtained through the initial testing condition (1)(A), must be submitted to the Section Chief, Variances Section, PSPD/OSW, (OS–343), U.S. EPA, 1200 Pennsylvania Ave., NW., Washington, DC 20406 within the time period specified in condition (1)(A). At the Section Chief's request, Bethlehem must submit analytical data obtained through condition (1)(B) to the above address, within the time period specified by the Section Chief. Failure to submit the required data obtained from either condition (1)(A) or (1)(B) within the specified time periods will be considered by the Agency sufficient basis to revoke Bethlehem's exclusion to the extent directed by EPA. All data must be accompanied by the following certification statement:

"Under civil and criminal penalty of law for the making or submission of false or fraudulent statements or representations (pursuant to the applicable provisions of the Federal Code which include, but may not be limited to, 18 U.S.C. 6928), I certify that the information contained in or accompanying this document is true, accurate and complete.

"As to the (those) identified section(s) of this document for which I cannot personally verify its (their) truth and accuracy, I certify as the company official having supervisory responsibility for the persons who, acting under my direct instructions, made the verification that this information is true, accurate and complete.

"In the event that any of this information is determined by EPA in its sole discretion to be false, inaccurate or incomplete, and upon conveyance of this fact to the company, I recognize and agree that this exclusion of wastes will be void as if it never had effect or to the extent directed by EPA and that the company will be liable for any actions taken in contravention of the company's RCRA and CERCLA obligations premised upon the company's reliance on the void exclusion."

TABLE 2—WASTES EXCLUDED FROM SPECIFIC SOURCES—Continued

| Facility | Address | Waste description |
|---|---|---|
| BF Goodrich Intermediates Company, Inc. | Calvert City, Kentucky. | Brine purification muds and saturator insolubles (EPA Hazardous Waste No. K071) after August 18, 1989. This exclusion is conditional upon the collection and submission of data obtained from BFG's full-scale treatment system because BFG's original data was based on data presented by another petitioner using an identical treatment process. To ensure that hazardous constituents are not present in the waste at levels of regulatory concern once the full-scale treatment facility is in operation, BFG must implement a testing program. All sampling and analyses (including quality control procedures) must be performed using appropriate methods. As applicable to the method-defined parameters of concern, analyses requiring the use of SW–846 methods incorporated by reference in 40 CFR 260.11 must be used without substitution. As applicable, the SW–846 methods might include Methods 0010, 0011, 0020, 0023A, 0030, 0031, 0040, 0050, 0051, 0060, 0061, 1010A, 1020B, 1110A, 1310B, 1311, 1312, 1320, 1330A, 9010C, 9012B, 9040C, 9045D, 9060A, 9070A (uses EPA Method 1664, Rev. A), 9071B, and 9095B. This testing program must meet the following conditions for the exclusion to be valid: |
| | | (1) Initial Testing: During the first four weeks of full-scale operation, BFG must do the following: |
| | | (A) Collect representative grab samples from every batch of the treated mercury brine purification muds and treated saturator insolubles on a daily basis and composite the grab samples to produce two separate daily composite samples (one of the treated mercury brine purification muds and one of the treated saturator insolubles). Prior to disposal of the treated batches, two daily composite samples must be analyzed for EP leachate concentration of mercury. BFG must report the analytical test data, including all quality control data, within 90 days after the treatment of the first full-scale batch. |
| | | (B) Collect representative grab samples from every batch of treated mercury brine purification muds and treated saturator insolubles on a daily basis and composite the grab samples to produce two separate weekly composite samples (one of the treated mercury brine muds and one of the treated saturator insolubles). Prior to disposal of the treated batches, two weekly composite samples must be analyzed for the EP leachate concentrations of all the EP toxic metals (except mercury), nickel, and cyanide (using distilled water in the cyanide extractions). BFG must report the analytical test data, including all quality control data, obtained during this initial period no later than 90 days after the treatment of the first full-scale batch. |
| | | (2) Subsequent Testing: After the first four weeks of full-scale operation, BFG must do the following: |
| | | (A) Continue to sample and test as described in condition (1)(A). BFG must compile and store on-site for a minimum of three years all analytical data and quality control data. These data must be furnished upon request and made available for inspection by any employee or representative of EPA or the State of Kentucky. |
| | | (B) Continue to sample and test as described in condition (1)(B). BFG must compile and store on-site for a minimum of three years all analytical data and quality control data. These data must be furnished upon request and made available for inspection by any employee or representative of EPA or the State of Kentucky. These testing requirements shall be terminated by EPA when the results of four consecutive weekly composite samples of both the treated mercury brine muds and treated saturator insolubles, obtained from either the initial testing or subsequent testing, show the maximum allowable levels in condition (3) are not exceeded and the Section Chief, Variances Section, notifies BFG that the requirements of this condition have been lifted. |
| | | (3) If, under condition (1) or (2), the EP leachate concentrations for chromium, lead, arsenic, or silver exceed 0.316 mg/l; for barium exceeds 6.31 mg/l; for cadmium or selenium exceed 0.063 mg/l; for mercury exceeds 0.0126 mg/l, for nickel exceeds 3.16 mg/l; or for cyanide exceeds 4.42 mg/l, the waste must either be retreated until it meets these levels or managed and disposed of in accordance with subtitle C of RCRA. |
| | | (4) Within one week of system start-up, BFG must notify the Section Chief, Variances Section (see address below) when the full-scale system is on-line and waste treatment has begun. All data obtained through condition (1) must be submitted to PSPD/OSW (5303W), U.S. EPA, 1200 Pennsylvania Ave., NW., Washington, DC 20460 within the time period specified in condition (1). At the Section Chief's request, BFG must submit any other analytical data obtained through condition (2) to the above address, within the time period specified by the Section Chief. Failure to submit the required data will be considered by the Agency sufficient basis to revoke BFG's exclusion to the extent directed by EPA. All data must be accompanied by the following certification statement: |
| | | "Under civil and criminal penalty of law for the making or submission of false or fraudulent statements or representations (pursuant to the applicable provisions of the Federal Code which include, but may not be limited to, 18 U.S.C. §6928), I certify that the information contained in or accompanying this document is true, accurate and complete. |
| | | As to the (those) identified section(s) of this document for which I cannot personally verify its (their) truth and accuracy, I certify as the company official having supervisory responsibility for the persons who, acting under my direct instructions, made the verification that this information is true, accurate and complete. |

TABLE 2—WASTES EXCLUDED FROM SPECIFIC SOURCES—Continued

| Facility | Address | Waste description |
|---|---|---|
| | | In the event that any of this information is determined by EPA in its sole discretion to be false, inaccurate or incomplete, and upon conveyance of this fact to the company, I recognize and agree that this exclusion of wastes will be void as if it never had effect or to the extent directed by EPA and that the company will be liable for any actions taken in contravention of the company's RCRA and CERCLA obligations premised upon the company's reliance on the void exclusion." |
| CF&I Steel Corporation. | Pueblo, Colorado. | Fully-cured chemically stabilized electric arc furnace dust/sludge (CSEAFD) treatment residue (EPA Hazardous Waste No. K061) generated from the primary production of steel after May 9, 1989. This exclusion is conditioned upon the data obtained from CF&I's full-scale CSEAFD treatment facility because CF&I's original data was obtained from a laboratory-scale CSEAFD treatment process. To ensure that hazardous constituents are not present in the waste at levels of regulatory concern once the full-scale treatment facility is in operation, CF&I must implement a testing program for the petitioned waste. This testing program must meet the following conditions for the exclusion to be vaild: |
| | | (1) *Testing:* |
| | | (A) *Initial Testing:* During the first four weeks of operation of the full-scale treatment system, CF&I must collect representative grab samples of each treated batch of the CSEAFD and composite the grab samples daily. The daily composites, prior to disposal, must be analyzed for the EP leachate concentrations of all the EP toxic metals, nickel, and cyanide (using distilled water in the cyanide extractions). Analyses must be performed using appropriate methods. As applicable to the method-defined parameters of concern, analyses requiring the use of SW–846 methods incorporated by reference in 40 CFR 260.11 must be used without substitution. As applicable, the SW–846 methods might include Methods 0010, 0011, 0020, 0023A, 0030, 0031, 0040, 0050, 0051, 0060, 0061, 1010A, 1020B, 1110A, 1310B, 1311, 1312, 1320, 1330A, 9010C, 9012B, 9040C, 9045D, 9060A, 9070A (uses EPA Method 1664, Rev. A), 9071B, and 9095B. CF&I must report the analytical test data obtained during this initial period no later than 90 days after the treatment of the first full-scale batch. |
| | | (B) *Subsequent Testing:* CF&I must collect representative grab samples from every treated batch of CSEAFD generated daily and composite all of the grab samples to produce a weekly composite sample. CF&I then must analyze each weekly composite sample for the EP leachate concentrations of all of the EP toxic metals and nickel. Analyses must be performed using appropriate methods. As applicable to the method-defined parameters of concern, analyses requiring the use of SW–846 methods incorporated by reference in 40 CFR 260.11 must be used without substitution. As applicable, the SW–846 methods might include Methods 0010, 0011, 0020, 0023A, 0030, 0031, 0040, 0050, 0051, 0060, 0061, 1010A, 1020B, 1110A, 1310B, 1311, 1312, 1320, 1330A, 9010C, 9012B, 9040C, 9045D, 9060A, 9070A (uses EPA Method 1664, Rev. A), 9071B, and 9095B. The analytical data, including all quality control information, must be compiled and maintained on site for a minimum of three years. These data must be furnished upon request and made available for inspection by any employee or representative of EPA or the State of Colorado. |
| | | (2) *Delisting levels:* If the EP extract concentrations determined in conditions (1)(A) or (1)(B) for chromium, lead, arsenic, or silver exceed 0.315 mg/l; for barium exceeds 6.3 mg/l; for cadmium or selenium exceed 0.063 mg/l; for mercury exceeds 0.0126 mg/l; for nickel exceeds 3.15 mg/l; or for cyanide exceeds 4.42 mg/l, the waste must either be re-treated or managed and disposed in accordance with Subtitle C of RCRA. |
| | | (3) *Data submittals:* Within one week of system start-up, CF&I must notify the Section Chief, Variances Section (see address below) when their full-scale stabilization system is on-line and waste treatment has begun. All data obtained through the initial testing condition (1)(A), must be submitted to the Section Chief, Variances Section, PSPD/OSW, (OS–343), U.S. EPA, 1200 Pennsylvania Ave., NW., Washington, DC 20460 within the time period specified in condition (1)(A). At the Section Chief's request, CF&I must submit analytical data obtained through condition (1)(B) to the above address, within the time period specified by the Section Chief. Failure to submit the required data obtained from either condition (1)(A) or (1)(B) within the specified time periods will be considered by the Agency sufficient basis to revoke CF&I's exclusion to the extent directed by EPA. All data must be accompanied by the following certification statement: "Under civil and criminal penalty of law for the making of submission of false or fraudulent statements or representations (pursuant to the applicable provisions of the Federal Code which include, but may not be limited to, 18 U.S.C. 6928), I certify that the information contained in or accompanying this document is true, accurate and complete. As to the (those) identified section(s) of this document for which I cannot personally verify its (their) truth and accuracy, I certify as the company official having supervisory responsibility for the persons who, acting under my direct instructions, made the verification that this information is true, accurate and complete. In the event that any of this information is determined by EPA in its sole discretion to be false, inaccurate or incomplete, and upon conveyance of this fact to the company, I recognize and agree that this exclusion of wastes will be void as if it never had effect or to the extent directed by EPA and that the company will be liable for any actions taken in contravention of the company's RCRA and CERCLA obligations premised upon the company's reliance on the void exclusion." |

TABLE 2—WASTES EXCLUDED FROM SPECIFIC SOURCES—Continued

| Facility | Address | Waste description |
|---|---|---|
| Chaparral Steel Midlothian, L.P. | Midlothian, Texas. | Leachate from Landfill No. 3, storm water from the baghouse area, and other K061 wastewaters which have been pumped to tank storage (at a maximum generation of 2500 cubic yards or 500,000 gallons per calender year) (EPA Hazardous Waste No. K061) generated at Chaparral Steel Midlothian, L.P., Midlothian, Texas, and is managed as nonhazardous solid waste after February 23, 2000. |

Chaparral Steel must implement a testing program that meets the following conditions for the exclusion to be valid:

(1) *Delisting Levels:* All concentrations for the constituent total lead in the approximately 2,500 cubic yards (500,000 gallons) per calender year of raw leachate from Landfill No. 3, storm water from the baghouse area, and other K061 wastewaters that is transferred from the storage tank to nonhazardous management must not exceed 0.69 mg/l (ppm). Constituents must be measured in the waste by appropriate methods. As applicable to the method-defined parameters of concern, analyses requiring the use of SW–846 methods incorporated by reference in 40 CFR 260.11 must be used without substitution. As applicable, the SW–846 methods might include Methods 0010, 0020, 0023A, 0030, 0031, 0040, 0050, 0051, 0060, 0061, 1010A, 1020B, 1110A, 1310B, 1311, 1312, 1320, 1330A, 9010C, 9012B, 9040C, 9045D, 9060A, 9070A (uses EPA Method 1664, Rev. A), 9071B, and 9095B.

(2) *Waste Holding and Handling:* Chaparral Steel must store as hazardous all leachate waste from Landfill No. 3, storm water from the bag house area, and other K061 wastewaters until verification testing as specified in Condition (3), is completed and valid analyses demonstrate that condition (1) is satisfied. If the levels of constituents measured in the samples of the waste do not exceed the levels set forth in Condition (1), then the waste is nonhazardous and may be managed and disposed of in accordance with all applicable solid waste regulations. If constituent levels in a sample exceed the delisting levels set in Condition (1), the waste volume corresponding to this sample must be treated until delisting levels are met or returned to the original storage tank. Treatment is designated as precipitation, flocculation, and filtering in a wastewater treatment system to remove metals from the wastewater. Treatment residuals precipitated will be designated as a hazardous waste. If the delisting level cannot be met, then the waste must be managed and disposed of in accordance with subtitle C of RCRA.

(3) *Verification Testing Requirements:* Sample collection and analyses, including quality control procedures, must be performed using appropriate methods. As applicable to the method-defined parameters of concern, analyses requiring the use of SW–846 methods incorporated by reference in 40 CFR 260.11 must be used without substitution. As applicable, the SW–846 methods might include Methods 0010, 0011, 0020, 0023A, 0030, 0031, 0040, 0050, 0051, 0060, 0061, 1010A, 1020B, 1110A, 1310B, 1311, 1312, 1320, 1330A, 9010C, 9012B, 9040C, 9045D, 9060A, 9070A (uses EPA Method 1664, Rev. A), 9071B, and 9095B. Chaparral Steel must analyze one composite sample from each batch of untreated wastewater transferred from the hazardous waste storage tank to non-hazardous waste management. Each composited batch sample must be analyzed, prior to non-hazardous management of the waste in the batch, for the constituent lead as listed in Condition (1). Chaparral may treat the waste as specified in Condition (2). If EPA judges the treatment process to be effective during the operating conditions used during the initial verification testing, Chaparral Steel may replace the testing requirement in Condition (3)(A) with the testing requirement in Condition (3)(B). Chaparral must continue to test as specified in (3)(A) until and unless notified by EPA or designated authority that testing in Condition (3)(A) may be replaced by Condition (3)(B).

(A) Initial Verification Testing: Representative composite samples from the first eight (8) full-scale treated batches of wastewater from the K061 leachate/wastewater storage tank must be analyzed for the constituent lead as listed in Condition (1), Chaparral must report to EPA the operational and analytical test data, including quality control information, obtained from these initial full scale treatment batches within 90 days of the eighth treatment batch.

(B) Subsequent Verification Testing: Following notification by EPA, Chaparral Steel may substitute the testing conditions in (3)(B) for (3)(A). Chaparral Steel must analyze representative composite samples from the treated full scale batches on an annual basis. If delisting levels for any constituent listed in Condition (1) are exceeded in the annual sample, Chaparral must reinstitute complete testing as required in Condition (3)(A). As stated in Condition (3) Chaparral must continue to test all batches of untreated waste to determine if delisting criteria are met before managing the wastewater from the K061 tank as nonhazardous.

(4) *Changes in Operating Conditions:* If Chaparral Steel significantly changes the treatment process established under Condition (3) (e.g., use of new treatment agents), Chaparral Steel must notify the Agency in writing. After written approval by EPA, Chaparral Steel may handle the wastes generated as non-hazardous, if the wastes meet the delisting levels set in Condition (1).

TABLE 2—WASTES EXCLUDED FROM SPECIFIC SOURCES—Continued

| Facility | Address | Waste description |
|---|---|---|
| | | (5) *Data Submittals:* Records of operating conditions and analytical data from Condition (3) must be compiled, summarized, and maintained on site for a minimum of five years. These records and data must be furnished upon request by EPA, or the State of Texas, or both, and be made available for inspection. Failure to submit the required data within the specified time period or maintain the required records on site for the specified time will be considered by EPA, at its discretion, sufficient basis to reopen the exclusion as described in Paragraph (6). All data must be accompanied by a signed copy of the following certification statement to attest to the truth and accuracy of the data submitted: |
| | | Under civil and criminal penalty of law for the making or submission of false or fraudulent statements or representations (pursuant to the applicable provisions of the Federal Code, which include, but may not be limited to, 18 U.S.C. 1001 and 42 U.S.C. 6928), I certify that the information contained in or accompanying this document is true, accurate and complete. |
| | | As to the (those) identified section(s) of this document for which I cannot personally verify its (their) truth and accuracy, I certify as the company official having supervisory responsibility for the persons who, acting under my direct instructions, made the verification that this information is true, accurate and complete. |
| | | In the event that any of this information is determined by EPA in its sole discretion to be false, inaccurate or incomplete, and upon conveyance of this fact to the company, I recognize and agree that this exclusion of waste will be void as if it never had effect or to the extent directed by EPA and that the company will be liable for any actions taken in contravention of the company's RCRA and CERCLA obligations premised upon the company's reliance on the void exclusion. |
| | | (6) *Reopener Language* |
| | | (A) If, anytime after disposal of the delisted waste, Chaparral Steel possesses or is otherwise made aware of any environmental data (including but not limited to leachate data or groundwater monitoring data) or any other data relevant to the delisted waste indicating that any constituent identified for the delisting verification testing is at level higher than the delisting level allowed by the Regional Administrator or his delegate in granting the petition, then the facility must report the data, in writing, to the Regional Administrator or his delegate within 10 days of first possessing or being made aware of that data. |
| | | (B) Based on the information described in paragraphs (5), or (6)(A) and any other information received from any source, the Regional Administrator or his delegate will make a preliminary determination as to whether the reported information requires Agency action to protect human health or the environment. Further action may include suspending, or revoking the exclusion, or other appropriate response necessary to protect human health and the environment. |
| | | (C) If the Regional Administrator or his delegate determines that the reported information does require Agency action, the Regional Administrator or his delegate will notify the facility in writing of the actions the Regional Administrator or his delegate believes are necessary to protect human health and the environment. The notice shall include a statement of the proposed action and a statement providing the facility with an opportunity to present information as to why the proposed Agency action is not necessary. The facility shall have 10 days from the date of the Regional Administrator or delegate's notice to present such information. |
| | | (D) Following the receipt of information from the facility described in paragraph (6)(C) or (if no information is presented under paragraph (6)(C)) the initial receipt of information described in paragraph (5) or (6)(A), the Regional Administrator or his delegate will issue a final written determination describing the Agency actions that are necessary to protect human health or the environment. Any required action described in the Regional Administrator or delegate's determination shall become effective immediately, unless the Regional Administrator or his delegate provides otherwise. |
| | | (7) *Notification Requirements:* Chaparral Steel must provide a one-time written notification to any State Regulatory Agency to which or through which the delisted waste described above will be transported for disposal at least 60 days prior to the commencement of such activity. The one-time written notification must be updated if the delisted waste is shipped to a different disposal facility. Failure to provide such a notification will result in a violation of the delisting petition and a possible revocation of the decision. |
| Conversion Systems, Inc. | Horsham, Pennsylvania. | Chemically Stabilized Electric Arc Furnace Dust (CSEAFD) that is generated by Conversion Systems, Inc. (CSI) (using the Super Detox™ treatment process as modified by CSI to treat EAFD (EPA Hazardous Waste No. K061)) at the following sites and that is disposed of in Subtitle D landfills: |
| | | Northwestern Steel, Sterling, Illinois after June 13, 1995. |
| | | CSI must implement a testing program for each site that meets the following conditions for the exclusion to be valid: |

TABLE 2—WASTES EXCLUDED FROM SPECIFIC SOURCES—Continued

| Facility | Address | Waste description |
|---|---|---|
| | | (1) *Verification Testing Requirements:* Sample collection and analyses, including quality control procedures, must be performed using appropriate methods. As applicable to the method-defined parameters of concern, analyses requiring the use of SW–846 methods incorporated by reference in 40 CFR 260.11 must be used without substitution. As applicable, the SW–846 methods might include Methods 0010, 0011, 0020, 0023A, 0030, 0031, 0040, 0050, 0051, 0060, 0061, 1010A, 1020B, 1110A, 1310B, 1311, 1312, 1320, 1330A, 9010C, 9012B, 9040C, 9045D, 9060A, 9070A (uses EPA Method 1664, Rev. A), 9071B, and 9095B. |
| | | (A) *Initial Verification Testing:* During the first 20 operating days of full-scale operation of a newly constructed Super Detox™ treatment facility, CSI must analyze a minimum of four (4) composite samples of CSEAFD representative of the full 20-day period. Composites must be comprised of representative samples collected from every batch generated. The CSEAFD samples must be analyzed for the constituents listed in Condition (3). CSI must report the operational and analytical test data, including quality control information, obtained during this initial period no later than 60 days after the generation of the first batch of CSEAFD. |
| | | (B) *Addition of New Super Detox™ Treatment Facilities to Exclusion:* If the Agency's review of the data obtained during initial verification testing indicates that the CSEAFD generated by a specific Super Detox™ treatment facility consistently meets the delisting levels specified in Condition (3), the Agency will publish a notice adding to this exclusion the location of the new Super Detox™ treatment facility and the name of the steel mill contracting CSI's services. If the Agency's review of the data obtained during initial verification testing indicates that the CSEAFD generated by a specific Super Detox™ treatment facility fails to consistently meet the conditions of the exclusion, the Agency will not publish the notice adding the new facility. |
| | | (C) *Subsequent Verification Testing:* For the Sterling, Illinois facility and any new facility subsequently added to CSI's conditional multiple-site exclusion, CSI must collect and analyze at least one composite sample of CSEAFD each month. The composite samples must be composed of representative samples collected from all batches treated in each month. These monthly representative samples must be analyzed, prior to the disposal of the CSEAFD, for the constituents listed in Condition (3). CSI may, at its discretion, analyze composite samples gathered more frequently to demonstrate that smaller batches of waste are nonhazardous. |
| | | (2) *Waste Holding and Handling:* CSI must store as hazardous all CSEAFD generated until verification testing as specified in Conditions (1)(A) and (1)(C), as appropriate, is completed and valid analyses demonstrate that Condition (3) is satisfied. If the levels of constituents measured in the samples of CSEAFD do not exceed the levels set forth in Condition (3), then the CSEAFD is non-hazardous and may be disposed of in Subtitle D landfills. If constituent levels in a sample exceed any of the delisting levels set in Condition (3), the CSEAFD generated during the time period corresponding to this sample must be retreated until it meets these levels, or managed and disposed of in accordance with Subtitle C of RCRA. CSEAFD generated by a new CSI treatment facility must be managed as a hazardous waste prior to the addition of the name and location of the facility to the exclusion. After addition of the new facility to the exclusion, CSEAFD generated during the verification testing in Condition (1)(A) is also non-hazardous, if the delisting levels in Condition (3) are satisfied. |
| | | (3) *Delisting Levels:* All leachable concentrations for those metals must not exceed the following levels (ppm): Antimony—0.06; arsenic—0.50; barium—7.6; beryllium—0.010; cadmium—0.050; chromium—0.33; lead—0.15; mercury—0.009; nickel—1; selenium—0.16; silver—0.30; thallium—0.020; vanadium—2; and zinc—70. Metal concentrations must be measured in the waste leachate by the method specified in 40 CFR 261.24. |
| | | (4) *Changes in Operating Conditions:* After initiating subsequent testing as described in Condition (1)(C), if CSI significantly changes the stabilization process established under Condition (1) (e.g., use of new stabilization reagents), CSI must notify the Agency in writing. After written approval by EPA, CSI may handle CSEAFD wastes generated from the new process as non-hazardous, if the wastes meet the delisting levels set in Condition (3). |
| | | (5) *Data Submittals:* At least one month prior to operation of a new Super Detox™ treatment facility, CSI must notify, in writing, the Chief of the Waste Identification Branch (see address below) when the Super Detox™ treatment facility is scheduled to be on-line. The data obtained through Condition (1)(A) must be submitted to the Branch Chief of the Waste Identification Branch, OSW (Mail Code 5304), U.S. EPA, 1200 Pennsylvania Ave., NW., Washington, DC 20460 within the time period specified. Records of operating conditions and analytical data from Condition (1) must be compiled, summarized, and maintained on site for a minimum of five years. These records and data must be furnished upon request by EPA, or the State in which the CSI facility is located, and made available for inspection. Failure to submit the required data within the specified time period or maintain the required records on site for the specified time will be considered by EPA, at its discretion, sufficient basis to revoke the exclusion to the extent directed by EPA. All data must be accompanied by a signed copy of the following certification statement to attest to the truth and accuracy of the data submitted: |

244

TABLE 2—WASTES EXCLUDED FROM SPECIFIC SOURCES—Continued

| Facility | Address | Waste description |
|---|---|---|
| | | Under civil and criminal penalty of law for the making or submission of false or fraudulent statements or representations (pursuant to the applicable provisions of the Federal Code, which include, but may not be limited to, 18 U.S.C. 1001 and 42 U.S.C. 6928), I certify that the information contained in or accompanying this document is true, accurate and complete. |
| | | As to the (those) identified section(s) of this document for which I cannot personally verify its (their) truth and accuracy, I certify as the company official having supervisory responsibility for the persons who, acting under my direct instructions, made the verification that this information is true, accurate and complete. |
| | | In the event that any of this information is determined by EPA in its sole discretion to be false, inaccurate or incomplete, and upon conveyance of this fact to the company, I recognize and agree that this exclusion of waste will be void as if it never had effect or to the extent directed by EPA and that the company will be liable for any actions taken in contravention of the company's RCRA and CERCLA obligations premised upon the company's reliance on the void exclusion. |
| Conversion Systems, Inc. | Willow Grove, PA. | Chemically Stabilized Electric Arc Furnace Dust (CSEAFD) that is generated by Conversion Systems Inc. (CSI) using the Super Detox™ process as modified by CSI to treat EAFD (EPA Hazardous Waste No. K061) at the following sites and that is disposed of in Subtitle C landfills: |
| | | Northwestern Steel, Sterling, Illinois after June 13, 1995. |
| | | Structural Metals, Inc. treated at U.S. Ecology, Robstown, Texas after September 23, 2008. |
| | | (1) Verification Testing Requirements: Sample collection and analyses, including quality control procedures must be performed using appropriate methods. As applicable to the method-defined parameters of concern, analyses requiring the use of SW–846 methods incorporated by reference in 40 CFR 260.11 must be used without substitution. As applicable, the SW–846 methods might include Methods 0010, 0011, 0020, 0023A, 0030, 0031, 0040, 0050, 0051, 0060, 0061, 1010A, 1020B, 1110A, 1310B, 1311, 1312, 1320, 1330A, 9010C, 9012B, 9040C, 9045D, 9060A, 9070A (uses EPA Method 1664, Rev. A), 9071B, and 9095B. |
| | | (A) *Initial Verification Testing:* During the first 20 operating days of full scale operation of a newly constructed Super Detox™ treatment facility, CSI must analyze a minimum of four (4) composite samples of CSEAFD representative of the full 20-day period. Composites must be comprised of representative samples collected from every batch generated. The CSEAFD samples must be analyzed for the constituents listed in Condition (3). CSI must report the operational and analytical test data, including quality control information, obtained during this initial period no later than 60 days after the generation of the first batch of CSEAFD. |
| | | (B) *Addition of New Super Detox™ Treatment Facilities to Exclusion:* If the Agency's review of the data obtained during initial verification testing indicates that the CSEAFD generated by a specific Super Detox™ treatment facility consistently meets the delisting levels specified in Condition (3), the Agency will publish a notice adding to this exclusion the location of the new Super Detox™ treatment facility and the name of the steel mill contracting CSI's services. If the Agency's review of the data obtained during initial verification testing indicates that the CSEAFD generated by a specific Super Detox™ treatment facility fails to consistently meet the conditions of this exclusion, the Agency will not publish the notice adding the new facility. |
| | | (C) Subsequent Verification Testing: For the Sterling, Illinois facility and any new facility subsequently added to CSI's conditional multiple-site exclusion, CSI must collect and analyze at least one composite sample of CSEAFD each month. The composite samples must be composed of representative samples collected from all batches treated in each month. The composite samples must be composed representative samples collected from all batches treated in each month. These monthly representative samples must be analyzed, prior to disposal of the CSEAFD, for the constituents listed in Condition (3). CSI may, at its discretion, analyze composite samples gathered more frequently to demonstrate that smaller batches of waste are non-hazardous. |
| | | (2) *Waste Holding and Handling:* CSI must store as hazardous all CSEAFD generated until verification testing as specified in Conditions (1)(A) and (1)(C), as appropriate, is completed and valid analyses demonstrate that Condition (3) is satisfied. If the levels of constituents measured in the samples of CSEAFD do not exceed the levels set forth in Condition (3), then the CSEAFD is non-hazardous and may be managed and disposed of in Subtitle D landfills. If constituent levels in a sample exceed any of the delisting levels set in Condition (3), the CSEAFD·generated during the time period corresponding to this sample must be retreated until it meets these levels, or managed and disposed of in accordance with Subtitle C of RCRA. CSEAFD generated by a new CSI treatment facility must be managed as a hazardous waste prior to the addition of the name and location of the facility to the exclusion. After addition of the new facility to the exclusion, CSEAFD generated during the verification testing in Condition (1)(A) is also non-hazardous, if the delisting levels in Condition (3) are satisfied. |

TABLE 2—WASTES EXCLUDED FROM SPECIFIC SOURCES—Continued

| Facility | Address | Waste description |
|---|---|---|
| | | (3) *Delisting Levels:* All leachable constituents for those metals must not exceed the following levels (ppm): Antimony-0.06; Arsenic-0.50; Barium-7.6; Beryllium-0.010; Cadmium-0.050; Chromium-0.33; Lead-0.15; Mercury-0.009; Nickel-1.00; Selenium-0.16; Silver-0.30; Thallium-0.020; Vanadium-2.0; Zinc-70. Metal concentrations must be measured in the waste leachate by the method specified in 40 CFR 261.24. |
| | | (4) *Changes in Operating Conditions:* After initiating subsequent testing described in Condition (1)(C), if CSI significantly changes the stabilization process established under Condition (1) (e.g., use of new stabilization reagents), CSI must notify the Agency in writing. After written approval by EPA, CSI may handle CSEAFD generated from the new process as non-hazardous, if the wastes meet the delisting levels set in Condition (3). |
| | | (5) *Data Submittals:* CSI must submit the information described below. If CSI fails to submit the required data within the specified time or maintain the required records on-site for the specified time, EPA, at its discretion, will consider this sufficient basis to reopen the exclusion as described in paragraph (6). CSI must: |
| | | (A) At least one month prior to operation of a new Super Detox™ treatment facility, CSI must notify, in writing, the EPA Regional Administrator or his designee, when the new Super Detox™ treatment facility is scheduled to be on-line. The data obtained through paragraph 1(A) must be submitted to the Regional Administrator or his designee within the time period specified. All supporting data can be submitted on CD-ROM or some comparable electronic media. |
| | | (B) CSI shall submit and receive EPA approval of the Quality Assurance Project Plan for data collection for each new facility added to this exclusion prior to conducting sampling events in paragraph 1(A). |
| | | (C) Compile records of analytical data from paragraph (3), summarized, and maintained on-site for a minimum of five years. |
| | | (D) Furnish these records and data when either EPA or the State agency requests them for inspection. |
| | | (E) Send along with all data a signed copy of the following certification statement, to attest to the truth and accuracy of the data submitted. "Under civil and criminal penalty of law for the making or submission of false or fraudulent statements or representations (pursuant to the applicable provisions of the Federal Code, which include, but may not be limited to, 18 U.S.C. 1001 and 42 U.S.C. 6928), I certify that the information contained in or accompanying this document is true, accurate and complete. |
| | | As to the (those) identified section(s) of this document for which I cannot personally verify its (their) truth and accuracy, I certify as the company official having supervisory responsibility for the persons who, acting under my direct instructions, made the verification that this information is true, accurate and complete. If any of this information is determined by EPA in its sole discretion to be false, inaccurate or incomplete, and upon conveyance of this fact to the company, I recognize and agree that this exclusion of waste will be void as if it never had effect or to the extent directed by EPA and that the company will be liable for any actions taken in contravention of the company's RCRA and CERCLA obligations premised upon the company's reliance on the void exclusion." |
| | | (6) Reopener: (A) If, anytime after disposal of the delisted waste CSI, the treatment facility, or the steel mill possess or is otherwise made aware of any data (including but not limited to leachate data or ground water monitoring data) relevant to the delisted waste indicating that any constituent identified for the delisting verification testing is at a level higher than the delisting level allowed by EPA in granting the petition, then the facility must report the data, in writing, to EPA within 10 days of first possessing or being made aware of that data. |
| | | (B) If subsequent verification testing of the waste as required by paragraph 1(C) does not meet the delisting requirements in paragraph 3 and the waste is subsequently managed as non-hazardous waste, CSI must report the data, in writing, to EPA within 10 days of first possessing or being made aware of that data. |
| | | (C) If CSI fails to submit the information described in paragraphs (5), (6)(A) or (6)(B) or if any other information is received from any source, EPA will make a preliminary determination as to whether the reported information requires action to protect human health and/or the environment. Further action may include suspending, or revoking the exclusion, or other appropriate response necessary to protect human health and the environment. |
| | | (D) If EPA determines that the reported information requires action, EPA will notify the facility in writing of the actions it believes are necessary to protect human health and the environment. The notice shall include a statement of the proposed action and a statement providing the facility with an opportunity to present information explaining why the proposed EPA action is not necessary. The facility shall have 10 days from the date of EPA's notice to present such information. |
| | | (E) Following the receipt of information from the facility described in paragraph (6)(D) or (if no information is presented under paragraph (6)(D)) the initial receipt of information described in paragraphs (5), (6)(A) or (6)(B), EPA will issue a final written determination describing the actions that are necessary to protect human health and/or the environment. Any required action described in EPA's determination shall become effective immediately, unless EPA provides otherwise. |

TABLE 2—WASTES EXCLUDED FROM SPECIFIC SOURCES—Continued

| Facility | Address | Waste description |
|---|---|---|
| | | (7) Notification Requirements: CSI or the treatment facility must do the following before transporting the delisted waste. Failure to provide this notification will result in a violation of the delisting petition and a possible revocation of the decision. |
| | | (A) Provide a one-time written notification to any state Regulatory Agency to which or through which it will transport the delisted waste described above for disposal, 60 days before beginning such activities. |
| | | (B) Update the one-time written notification if it ships the delisted waste into a different disposal facility. |
| | | (C) Failure to provide this notification will result in a violation of the delisting exclusion and a possible revocation of the decision. |
| DuraTherm, Incorporated. | San Leon, Texas. | Desorber Solids, (at a maximum generation of 20,000 cubic yards per calendar year) generated by DuraTherm using the treatment process to treat the Desorber solids, (EPA Hazardous Waste No. K048, K049, K050, and K051 and disposed of in a subtitle D landfill. DuraTherm must implement the testing program found in Table 1. Wastes Excluded From Non-Specific Sources, for the petition to be valid. |
| Eastman Chemical Company. | Longview, Texas. | Wastewater treatment sludge, (at a maximum generation of 82,100 cubic yards per calendar year) (EPA Hazardous Waste Nos. K009, K010) generated at Eastman. Eastman must implement the testing program described in Table 1. Waste Excluded From Non-Specific Sources for the petition to be valid. |
| Envirite of Illinois (formerly Envirite Corporation). | Harvey, Illinois | See waste description under Envirite of Pennsylvania. |
| Envirite of Ohio (formerly Envirite Corporation). | Canton, Ohio | See waste description under Envirite of Pennsylvania. |
| Envirite of Pennsylvania (formerly Envirite Corporation). | York, Pennsylvania. | Spent pickle liquor (EPA Hazardous Waste No. K062) generated from steel finishing operations of facilities within the iron and steel industry (SIC Codes 331 and 332); wastewater treatment sludge (EPA Hazardous Waste No. K002) generated from the production of chrome yellow and orange pigments; wastewater treatment sludge (EPA Hazardous Waste No. K003) generated from the production of molybdate orange pigments; wastewater treatment sludge (EPA Hazardous Waste No. K004) generated from the production of zinc yellow pigments; wastewater treatment sludge (EPA Hazardous Waste K005) generated from the production of chrome green pigments; wastewater treatment sludge (EPA Hazardous Waste No. K006) generated from the production of chrome oxide green pigments (anhydrous and hydrated); wastewater treatment sludge (EPA Hazardous Waste No. K007) generated from the production of iron blue pigments; oven residues (EPA Hazardous Waste No. K008) generated from the production of chrome oxide green pigments after November 14, 1986. To ensure that hazardous constituents are not present in the waste at levels of regulatory concern, the facility must implement a contingency testing program for the petitioned wastes. This testing program must meet the following conditions for the exclusions to be valid: |
| | | (1) Each batch of treatment residue must be representatively sampled and tested using the EP Toxicity test for arsenic, barium, cadmium, chromium, lead, selenium, silver, mercury, and nickel. If the extract concentrations for chromium, lead, arsenic, and silver exceed 0.315 ppm; barium levels exceed 6.3 ppm; cadmium and selenium exceed 0.063 ppm; mercury exceeds 0.0126 ppm; or nickel levels exceed 2.205 ppm, the waste must be re-treated or managed and disposed as a hazardous waste under 40 CFR Parts 262 to 265 and the permitting standards of 40 CFR Part 270. |
| | | (2) Each batch of treatment residue (formerly must be tested for leachable cyanide. If the leachable cyanide levels Corporation) (using the EP Toxicity test without acetic acid adjustment) exceed 1.26 ppm, the waste must be re-treated or managed and disposed as a hazardous waste under 40 CFR Parts 262 to 265 and the permitting standards of 40 CFR Part 270. |
| | | (3) Each batch of waste must be tested for the total content of specific organic toxicants. If the total content of anthracene exceeds 76.8 ppm, 1.2-diphenyl hydrazine exceeds 0.001 ppm, methylene chloride exceeds 8.18 ppm, methyl ethyl ketone exceeds 326 ppm, n-nitrosodiphenylamine exceeds 11.9 ppm, phenol exceeds 1,566 ppm, tetrachloroethylene exceeds 0.188 ppm, or trichloroethylene exceeds 0.592 ppm, the waste must be managed and disposed as a hazardous waste under 40 CFR Parts 262 to 265 and the permitting standards of 40 CFR Part 27 0. |
| | | (4) A grab sample must be collected from each batch to form one monthly composite sample which must be tested using GC/MS analysis for the compounds listed in #3, above, as well as the remaining organics on the priority pollutant list. (See 47 FR 52309, November 19, 1982, for a list of the priority pollutants.) |

TABLE 2—WASTES EXCLUDED FROM SPECIFIC SOURCES—Continued

| Facility | Address | Waste description |
|----------|---------|-------------------|
| | | (5) The data from conditions 1–4 must be kept on file at the facility for inspection purposes and must be compiled, summarized, and submitted to the Administrator by certified mail semi-annually. The Agency will review this information and if needed will propose to modify or withdraw the exclusion. The organics testing described in conditions 3 and 4, above, is not required until six months from the date of promulgation. The Agency's decision to conditionally exclude the treatment residue generated from the wastewater treatment systems at these facilities applies only to the wastewater and solids treatment systems as they presently exist as described in the delisting petition. The exclusion does not apply to the proposed process additions described in the petition as recovery, including crystallization, electrolytic metals recovery, evaporative recovery, and ion exchange. |
| ERCO Worldwide (USA) Inc. (formerly Vulcan Materials Company). | Port Edwards, Wisconsin. | Brine purification muds (EPA Hazardous Waste No. K071) generated from the mercury cell process in chlorine production, where separately purified brine is not used after November 17, 1986. To assure that mercury levels in this waste are maintained at acceptable levels, the following conditions apply to this exclusion: Each batch of treated brine clarifier muds and saturator insolubles must be tested (by the extraction procedure) prior to disposal and the leachate concentration of mercury must be less than or equal to 0.0129 ppm. If the waste does not meet this requirement, then it must be re-treated or disposed of as hazardous. This exclusion does not apply to wastes for which either of these conditions is not satisfied. |
| Giant Refining Company, Inc. | Bloomfield, New Mexico. | Waste generated during the excavation of soils from two wastewater treatment impoundments (referred to as the South and North Oily Water Ponds) used to contain water outflow from an API separator (EPA Hazardous Waste No. K051). This is a one-time exclusion for approximately 2,000 cubic yards of stockpiled waste. This exclusion was published on September 3, 1996. |
| | | Notification Requirements: Giant Refining Company must provide a one-time written notification to any State Regulatory Agency to which or through which the delisted waste described above will be transported for disposal at least 60 days prior to the commencement of such activities. Failure to provide such a notification will result in a violation of the delisting petition and a possible revocation of the decision. |
| Heritage Environmental Services, LLC., at the Nucor Steel facility. | Crawfordsville, Indiana. | Electric arc furnace dust (EAFD) that has been generated by Nucor Steel at its Crawfordsville, Indiana facility and treated on site by Heritage Environmental Services, LLC (Heritage) at a maximum annual rate of 30,000 cubic yards per year and disposed of in a Subtitle D landfill which has groundwater monitoring, after January 15, 2002. |
| | | (1) *Delisting Levels:* |
| | | (A) The constituent concentrations measured in either of the extracts specified in Paragraph (2) may not exceed the following levels (mg/L): Antimony—0.206; Arsenic—0.0936; Barium—55.7; Beryllium—0.416; Cadmium—0.15; Chromium (total)—1.55; Lead—5.0; Mercury—0.149; Nickel—28.30; Selenium—0.58; Silver—3.84; Thallium—0.088; Vanadium—21.1; Zinc—280.0. |
| | | (B) Total mercury may not exceed 1 mg/kg. |
| | | (2) *Verification Testing:* On a monthly basis, Heritage or Nucor must analyze two samples of the waste using the TCLP, SW-846 Method 1311, with an extraction fluid of pH 12 ±0.05 standard units and for the mercury determinative analysis of the leachate using an appropriate method. The constituent concentrations measured must be less than the delisting levels established in Paragraph (1). |
| | | (3) *Changes in Operating Conditions:* If Nucor significantly changes the manufacturing process or chemicals used in the manufacturing process or Heritage significantly changes the treatment process or the chemicals used in the treatment process, Heritage or Nucor must notify the EPA of the changes in writing. Heritage and Nucor must handle wastes generated after the process change as hazardous until Heritage or Nucor has demonstrated that the wastes continue to meet the delisting levels set forth in Paragraph (1) and that no new hazardous constituents listed in Appendix VIII of Part 261 have been introduced and Heritage and Nucor have received written approval from EPA. |
| | | (4) *Data Submittals:* Heritage must submit the data obtained through monthly verification testing or as required by other conditions of this rule to U.S. EPA Region 5, Waste Management Branch (DW–8J), 77 W. Jackson Blvd., Chicago, IL 60604 by February 1 of each calendar year for the prior calendar year. Heritage or Nucor must compile, summarize, and maintain on site for a minimum of five years records of operating conditions and analytical data. Heritage or Nucor must make these records available for inspection. All data must be accompanied by a signed copy of the certification statement in 40 CFR 260.22(i)(12). |
| | | (5) *Reopener Language*—(A) If, anytime after disposal of the delisted waste, Heritage or Nucor possesses or is otherwise made aware of any data (including but not limited to leachate data or groundwater monitoring data) relevant to the delisted waste indicating that any constituent identified in Paragraph (1) is at a level in the leachate higher than the delisting level established in Paragraph (1), or is at a level in the groundwater higher than the maximum allowable point of exposure concentration predicted by the CMTP model, then Heritage or Nucor must report such data, in writing, to the Regional Administrator within 10 days of first possessing or being made aware of that data. |

TABLE 2—WASTES EXCLUDED FROM SPECIFIC SOURCES—Continued

| Facility | Address | Waste description |
|---|---|---|
| | | (B) Based on the information described in paragraph (5)(A) and any other information received from any source, the Regional Administrator will make a preliminary determination as to whether the reported information requires Agency action to protect human health or the environment. Further action may include suspending, or revoking the exclusion, or other appropriate response necessary to protect human health and the environment. |
| | | (C) If the Regional Administrator determines that the reported information does require Agency action, the Regional Administrator will notify Heritage and Nucor in writing of the actions the Regional Administrator believes are necessary to protect human health and the environment. The notice shall include a statement of the proposed action and a statement providing Heritage and Nucor with an opportunity to present information as to why the proposed Agency action is not necessary or to suggest an alternative action. Heritage and Nucor shall have 30 days from the date of the Regional Administrator's notice to present the information. |
| | | (D) If after 30 days Heritage or Nucor presents no further information, the Regional Administrator will issue a final written determination describing the Agency actions that are necessary to protect human health or the environment. Any required action described in the Regional Administrator's determination shall become effective immediately, unless the Regional Administrator provides otherwise. |
| LCP Chemical | Orrington, ME | Brine purification muds and wastewater treatment sludges generated after August 27, 1985 from their chlor-alkali manufacturing operations (EPA Hazardous Waste Nos. K071 and K106) that have been batch tested for mercury using the EP toxicity procedures and have been found to contain less than 0.05 ppm mercury in the EP extract. Brine purification muds and wastewater treatment sludges that exceed this level will be considered a hazardous waste. |
| Marathon Oil Co. | Texas City, Texas. | Residual solids (at a maximum annual generation rate of 1,000 cubic yards) generated from the thermal desorption treatment and, where necessary, stabilization of wastewater treatment plant API/DAF filter cake (EPA Hazardous Waste Nos. K048 and K051), after [insert date of publication]. Marathon must implement a testing program that meets the following conditions for the exclusion to be valid: |
| | | (1) *Testing:* Sample collection and analyses (including quality control (QC) procedures) must be performed using appropriate methods. As applicable to the method-defined parameters of concern, analyses requiring the use of SW–846 methods incorporated by reference in 40 CFR 260.11 must be used without substitution. As applicable, the SW–846 methods might include Methods 0010, 0011, 0020, 0023A, 0030, 0031, 0040, 0050, 0051, 0060, 0061, 1010A, 1020B, 1110A, 1310B, 1311, 1312, 1320, 1330A, 9010C, 9012B, 9040C, 9045D, 9060A, 9070A (uses EPA Method 1664, Rev. A), 9071B, and 9095B. If EPA judges the treatment process to be effective under the operating conditions used during the initial verification testing, Marathon may replace the testing required in Condition (1)(A) with the testing required in Condition (1)(B). Marathon must continue to test as specified in Condition (1)(A), including testing for organics in Conditions (3)(B) and (3)(C), until and unless notified by EPA in writing that testing in Condition (1)(A) may be replaced by Condition (1)(B), or that testing for organics may be terminated as described in (1)(C) (to the extent directed by EPA). |
| | | (A) *Initial Verification Testing:* During at least the first 40 operating days of full-scale operation of the thermal desorption unit, Marathon must monitor the operating conditions and analyze 5-day composites of residual solids. 5-day composites must be composed of representative grab samples collected from every batch during each 5-day period of operation. The samples must be analyzed prior to disposal of the residual solids for constituents listed in Condition (3). Marathon must report the operational and analytical test data, including quality control information, obtained during this initial period no later than 90 days after the treatment of the first full-scale batch. |
| | | (B) *Subsequent Verification Testing:* Following notification by EPA, Marathon may substitute the testing conditions in (1)(B) for (1)(A). Marathon must continue to monitor operating conditions, and analyze samples representative of each month of operation. The samples must be composed of representative grab samples collected during at least the first five days of operation of each month. These monthly representative samples must be analyzed for the constituents listed in Condition (3) prior to the disposal of the residual solids. Marathon may, at its discretion, analyze composite samples gathered more frequently to demonstrate that smaller batches of waste are nonhazardous. |
| | | (C) *Termination of Organic Testing:* Marathon must continue testing as required under Condition (1)(B) for organic constituents specified in Conditions (3)(B) and (3)(C) until the analyses submitted under Condition (1)(B) show a minimum of four consecutive monthly representative samples with levels of specific constituents significantly below the delisting levels in Conditions (3)(B) and (3)(C), and EPA notifies Marathon in writing that monthly testing for specific organic constituents may be terminated. Following termination of monthly testing, Marathon must continue to test a representative 5-day composite sample for all constituents listed in Conditions (3)(B) and (3)(C) on an annual basis. If delisting levels for any constituents listed in Conditions (3)(B) and (3)(C) are exceeded in the annual sample, Marathon must reinstitute complete testing as required in Condition (1)(B). |

TABLE 2—WASTES EXCLUDED FROM SPECIFIC SOURCES—Continued

| Facility | Address | Waste description |
|---|---|---|
| | | (2) *Waste Holding and Handling:* Marathon must store as hazardous all residual solids generated until verification testing (as specified in Conditions (1)(A) and (1)(B)) is completed and valid analysis demonstrates that Condition (3) is satisfied. If the levels of hazardous constituents in the samples of residual solids are below all of the levels set forth in Condition (3), then the residual solids are non-hazardous and may be managed and disposed of in accordance with all applicable solid waste regulations. If hazardous constituent levels in any 5-day composite or other representative sample equal or exceed any of the delisting levels set in Condition (3), the residual solids generated during the corresponding time period must be retreated and/or stabilized as allowed below, until the residual solids meet these levels, or managed and disposed of in accordance with Subtitle C of RCRA. |
| | | If the residual solids contain leachable inorganic concentrations at or above the delisting levels set forth in Condition (3)(A), then Marathon may stabilize the material with Type 1 portland cement as demonstrated in the petition to immobilize the metals. Following stabilization, Marathon must repeat analyses in Condition (3)(A) prior to disposal. |
| | | (3) *Delisting Levels:* Leachable concentrations in Conditions (3)(A) and (3)(B) must be measured in the waste leachate by the method specified in 40 CFR 261.24. The indicator parameters in Condition (3)(C) must be measured as the total concentration in the waste. Concentrations must be less than the following levels (ppm): |
| | | (A) *Inorganic Constituents:* antimony-0.6; arsenic, chromium, or silver-5.0; barium-100.0; beryllium-0.4; cadmium-0.5; lead-1.5; mercury-0.2; nickel-10.0; selenium-1.0; vanadium-20.0. |
| | | (B) *Organic Constituents:* acenaphthene-200; benzene-0.5; benzo(a)anthracene-0.01; benzo(a)pyrene-0.02; benzo(b)fluoranthene-0.02; chrysene-0.02; ethyl benzene-70; fluoranthene-100; fluorene-100; naphthalene-100; pyrene-100; toluene-100. |
| | | (C) *Indicator Parameters:* 1-methyl naphthalene-3; benzo(a)pyrene-3. |
| | | (4) *Changes in Operating Conditions:* After completing the initial verification test period in Condition (1)(A), if Marathon significantly changes the operating conditions established under Condition (1), Marathon must notify the Agency in writing. After written approval by EPA, Marathon must re-institute the testing required in Condition (1)(A) for a minimum of four 5-day operating periods. Marathon must report the operations and test data, required by Condition (1)(A), including quality control data, obtained during this period no later than 60 days after the changes take place. Following written notification by EPA, Marathon may replace testing Condition (1)(A) with (1)(B). Marathon must fulfill all other requirements in Condition (1), as appropriate. |
| | | (5) *Data Submittals:* At least two weeks prior to system start-up, Marathon must notify in writing the Section Chief Delisting Section (see address below) when the thermal desorption and stabilization units will be on-line and waste treatment will begin. The data obtained through Condition (1)(A) must be submitted to HWID/OSW (5304W) (OS–333), U.S. EPA, 1200 Pennsylvania Ave., NW., Washington, DC 20460 within the time period specified. Records of operating conditions and analytical data from Condition (1) must be compiled, summarized, and maintained on site for a minimum of five years. These records and data must be furnished upon request by EPA or the State of Texas and made available for inspection. Failure to submit the required data within the specified time period or maintain the required records on site for the specified time will be considered by EPA, at its discretion, sufficient basis to revoke the exclusion to the extent directed by EPA. All data must be accompanied by a signed copy of the following certification statement to attest to the truth and accuracy of the data submitted: |
| | | "Under civil and criminal penalty of law for the making or submission of false or fraudulent statements or representations (pursuant to the applicable provisions of the Federal Code, which include, but may not be limited to, 18 U.S.C. 1001 and 42 U.S.C 6928), I certify that the information contained in or accompanying this document is true, accurate, and complete. |
| | | As to the (those) identified sections(s) of this document for which I cannot personally verify its (their) truth and accuracy, I certify as the company official having supervisory responsibility for the persons who, acting under my direct instructions, made the verification that this information is true, accurate, and complete. |
| | | In the event that any of this information is determined by EPA in its sole discretion to be false, inaccurate, or incomplete, and upon conveyance of this fact to the company, I recognize and agree that this exclusion of waste will be void as if it never had effect or to the extent directed by EPA and that the company will be liable for any actions taken in contravention of the company's RCRA and CERCLA obligations premised upon the company's reliance on the void exclusion." |
| Mearl Corp ...... | Peekskill, NY | Wastewater treatment sludge (EPA Hazardous Waste Nos. K006 and K007) generated from the production of chrome oxide green and iron blue pigments after November 27, 1985. |
| Monsanto Industrial Chemicals Company. | Sauget, Illinois | Brine purification muds (EPA Hazardous Waste No. K071) generated from the mercury cell process in chlorine production, where separately prepurified brine is not used after August 15, 1986. |

TABLE 2—WASTES EXCLUDED FROM SPECIFIC SOURCES—Continued

| Facility | Address | Waste description |
|---|---|---|
| Occidental Chemical. | Ingleside, Texas. | Limestone Sludge, (at a maximum generation of 1,114 cubic yards per calendar year) Rockbox Residue, (at a maximum generation of 1,000 cubic yards per calendar year) generated by Occidental Chemical using the wastewater treatment process to treat the Rockbox Residue and the Limestone Sludge (EPA Hazardous Waste No. K019, K020). Occidental Chemical must implement a testing program that meets conditions found in Table 1. Wastes Excluded From Non-Specific Sources from the petition to be valid. |
| Occidental Chemical Corp., Muscle Shoals Plant. | Sheffield, Alabama. | Retorted wastewater treatment sludge from the mercury cell process in chlorine production (EPA Hazardous Plant Waste No. K106) after September 19, 1989. This exclusion is conditional upon the submission of data obtained from Occidental's full-scale retort treatment system because Occidental's original data were based on a pilot-scale retort system. To ensure that hazardous constituents are not present in the waste at levels of regulatory concern once the full-scale treatment facility is in operation, Occidental must implement a testing program. All sampling and analyses (including quality control procedures) must be performed using appropriate methods. As applicable to the method-defined parameters of concern, analyses requiring the use of SW–846 methods incorporated by reference in 40 CFR 260.11 must be used without substitution. As applicable, the SW–846 methods might include Methods 0010, 0011, 0020, 0023A, 0030, 0031, 0040, 0050, 0051, 0060, 0061, 1010A, 1020B, 1110A, 1310B, 1311, 1312, 1320, 1330A, 9010C, 9012B, 9040C, 9045D, 9060A, 9070A (uses EPA Method 1664, Rev. A), 9071B, and 9095B. This testing program must meet the following conditions for the exclusion to be valid: |
|  |  | (1) Initial Testing—During the first four weeks of full-scale retort operation, Occidental must do the following: |
|  |  | (A) Collect representative grab samples from every batch of retorted material and composite the grab samples to produce a weekly composite sample. The weekly composite samples, prior to disposal or recycling, must be analyzed for the EP leachate concentrations of all the EP toxic metals (except mercury), nickel, and cyanide (using distilled water in the cyanide extractions). Occidental must report the analytical test data, including all quality control data, obtained during this initial period no later than 90 days after the treatment of the first full-scale batch. |
|  |  | (B) Collect representative grab samples of every batch of retorted material prior to its disposal or recycling and analyze the sample for EP leachate concentration of mercury. Occidental must report the analytical test data, including all quality control data, within 90 days after the treatment of the first full-scale batch. |
|  |  | (2) Subsequent Testing—After the first four weeks of full-scale retort operation, Occidental must do the following: |
|  |  | (A) Continue to sample and test as described in condition (1)(A). Occidental must compile and store on-site for a minimum of three years all analytical data and quality control data. These data must be furnished upon request and made available for inspection by any employee or representative of EPA or the State of Alabama. These testing requirements shall be terminated by EPA when the results of four consecutive weekly composite samples of the petitioned waste, obtained from either the initial testing or subsequent testing show the maximum allowable levels in condition (3) are not exceeded and the Section Chief, Variances Section, notifies Occidental that the requirements of this condition have been lifted. |
|  |  | (B) Continue to sample and test for mercury as described in condition (1)(B). Occidental must compile and store on-site for a minimum of three years all analytical data and quality control data. These data must be furnished upon request and made available for inspection by any employee or representative of EPA or the State of Alabama. These testing requirements shall remain in effect until Occidental provides EPA with analytical and quality control data for thirty consecutive batches of retorted material, collected as described in condition (1)(B), demonstrating that the EP leachable levels of mercury are below the maximum allowable level in condition (3) and the Section Chief, Variances Section, notifies Occidental that the testing in condition (2)(B) may be replaced with (2)(C). |
|  |  | (C) [If the conditions in (2)(B) are satisfied, the testing requirements for mercury in (2)(B) shall be replaced with the following condition]. Collect representative grab samples from every batch of retorted material on a daily basis and composite the grab samples to produce a weekly composite sample. Occidental must analyze each weekly composite sample prior to its disposal or recycling for the EP leachate concentration of mercury. Occidental must compile and store on-site for a minimum of three years all analytical data and quality control data. These data must be furnished upon request and made available for inspection by any employee or representative of EPA or the State of Alabama. |
|  |  | (3) If, under condition (1) or (2), the EP leachate concentrations for chromium, lead, arsenic, or silver exceed 1.616 mg/l; for barium exceeds 32.3 mg/l; for cadmium or selenium exceed 0.323 mg/l; for mercury exceeds 0.065 mg/l, for nickel exceeds 16.15 mg/l; or for cyanide exceeds 22.61 mg/l, the waste must either be retreated until it meets these levels or managed and disposed of in accordance with subtitle C of RCRA. |

TABLE 2—WASTES EXCLUDED FROM SPECIFIC SOURCES—Continued

| Facility | Address | Waste description |
|---|---|---|
| | | (4) Within one week of system start-up, Occidental must notify the Section Chief, Variances Section (see address below) when the full-scale retort system is on-line and waste treatment has begun. All data obtained through condition (1) must be submitted to PSPD/OSW (5303W), U.S. EPA, 1200 Pennsylvania Ave., NW., Washington, DC 20460 within the time period specified in condition (1). At the Section Chief's request, Occidental must submit any other analytical data obtained through condition (2) to the above address, within the time period specified by the Section Chief. Failure to submit the required data will be considered by the Agency sufficient basis to revoke Occidental's exclusion to the extent directed by EPA. All data must be accompanied by the following certification statement: |
| | | "Under civil and criminal penalty of law for the making or submission of false or fraudulent statements or representations (pursuant to the applicable provisions of the Federal Code which include, but may not be limited to, 18 U.S.C. 6928), I certify that the information contained in or accompanying this document is true, accurate and complete. |
| | | As to the (those) identified section(s) of this document for which I cannot personally verify its (their) truth and accuracy, I certify as the company official having supervisory responsibility for the persons who, acting under my direct instructions, made the verification that this information is true, accurate and complete. |
| | | In the event that any of this information is determined by EPA in its sole discretion to be false, inaccurate or incomplete, and upon conveyance of this fact to the company, I recognize and agree that this exclusion of wastes will be void as if it never had effect or to the extent directed by EPA and that the company will be liable for any actions taken in contravention of the company's RCRA and CERCLA obligations premised upon the company's reliance on the void exclusion." |
| Occidental Chemical Corporation. | Delaware City, Delaware. | Sodium chloride treatment muds (NaCl-TM), sodium chloride saturator cleanings (NaCl-SC), and potassium chloride treatment muds (KCl-TM) (all classified as EPA Hazardous Waste No. K071) generated at a maximum combined rate (for all three wastes) of 1,018 tons per year. This exclusion was published on April 29, 1991 and is conditioned upon the collection of data from Occidental's full-scale brine treatment system because Occidental's request for exclusion was based on data from a laboratory-scale brine treatment process. To ensure that hazardous constituents are not present in the waste at levels of regulatory concern once the full-scale treatment system is in operation, Occidental must implement a testing program for the petitioned waste. All sampling and analyses (including quality control (QC) procedures) must be performed using appropriate methods. As applicable to the method-defined parameters of concern, analyses requiring the use of SW–846 methods incorporated by reference in 40 CFR 260.11 must be used without substitution. As applicable, the SW–846 methods might include Methods 0010, 0011, 0020, 0023A, 0030, 0031, 0040, 0050, 0051, 0060, 0061, 1010A, 1020B, 1110A, 1310B, 1311, 1312, 1320, 1330A, 9010C, 9012B, 9040C, 9045D, 9060A, 9070A (uses EPA Method 1664, Rev. A), 9071B, and 9095B. This testing program must meet the following conditions for the exclusion to be valid: |
| | | (1) *Initial Testing:* During the first four weeks of full-scale treatment system operation, Occidental must do the following: |
| | | (A) Collect representative grab samples from each batch of the three treated wastestreams (sodium chloride saturator cleanings (NaCl-SC), sodium chloride treatment muds (NaCl-TM)) and potassium chloride treatment muds (KCl-TM)) on an as generated basis and composite the samples to produce three separate weekly composite samples (of each type of K071 waste). The three weekly composite samples, prior to disposal, must be analyzed for the EP leachate concentrations of all the EP toxic metals (except mercury), nickel, and cyanide (using distilled water in the cyanide extractions). Occidental must report the waste volumes produced and the analytical test data, including all quality control data, obtained during this initial period, no later than 90 days after the treatment of the first full-scale batch. |
| | | (B) Collect representative grab samples of each batch of the three treated wastestreams (NaCl-SC, NACl-TM and KCl-TM) and composite the grab samples to produce three separate daily composite samples (of each type of K071 waste) on an as generated basis. The three daily composite samples, prior to disposal, must be analyzed for the EP leachate concentration of mercury. Occidental must report the waste volumes produced and the analytical test data, including all quality control data, obtained during this initial period, no later than 90 days after the treatment of the first full-scale batch. |
| | | (2) *Subsequent Testing:* After the first four weeks of full-scale treatment operations, Occidental must do the following; all sampling and analyses (including quality control procedures) must be performed using appropriate methods, and as applicable to the method-defined parameters of concern, analyses requiring the use of SW–846 methods incorporated by reference in 40 CFR 260.11 must be used without substitution. As applicable, the SW–846 methods might include Methods 0010, 0011, 0020, 0023A, 0030, 0031, 0040, 0050, 0051, 0060, 0061, 1010A, 1020B, 1110A, 1310B, 1311, 1312, 1320, 1330A, 9010C, 9012B, 9040C, 9045D, 9060A, 9070A (uses EPA Method 1664, Rev. A), 9071B, and 9095B: |

TABLE 2—WASTES EXCLUDED FROM SPECIFIC SOURCES—Continued

| Facility | Address | Waste description |
|---|---|---|
| | | (A) Continue to sample and test as described in condition (1)(A). Occidental must compile and store on-site for a minimum of three years the records of waste volumes produced and all analytical data and quality control data. These data must be furnished upon request and made available for inspection by any employee or representative of EPA or the State of Delaware. These testing requirements shall be terminated by EPA when the results of four consecutive weekly composite samples of the petitioned waste, obtained from either the initial testing or subsequent testing, show the maximum allowable levels in condition (3) are not exceeded and the Section Chief, Variances Section, notifies Occidental that the requirements of this condition have been lifted. |
| | | (B) Continue to sample and test for mercury as described in condition (1)(B). Occidental must compile and store on-site for a minimum of three years the records of waste volumes produced and all analytical data and quality control data. These data must be furnished upon request and made available for inspection by any employee or representative of EPA or the State of Delaware. These testing requirements shall be terminated and replaced with the requirements of condition (2)(C) if Occidental provides EPA with analytical and quality control data for thirty consecutive batches of treated material, collected as described in condition (1)(B), demonstrating that the EP leachable level of mercury in condition (3) is not exceeded (in all three treated wastes), and the Section Chief, Variances Section, notifies Occidental that the testing in condition (2)(B) may be replaced with (2)(C). |
| | | (C) [If the conditions in (2)(B) are satisfied, the testing requirements for mercury in (2)(B) shall be replaced with the following condition.] Collect representative grab samples from each batch of the three treated wastestreams (NaCl-SC, NaCl-TM and KCl-TM) on an as generated basis and composite the grab samples to produce three separate weekly composite samples (of each type of K071 waste). The three weekly composite samples, prior to disposal, must be analyzed for the EP leachate concentration of mercury. Occidental must compile and store on-site for a minimum of three years the records of waste volumes produced and all analytical data and quality control data. These data must be furnished upon request and made available for inspection by any employee or representative of EPA or the State of Delaware. |
| | | (3) If, under conditions (1) or (2), the EP leachate concentrations for chromium, lead, arsenic, or silver exceed 0.77 mg/l; for barium exceeds 15.5 mg/l; for cadmium or selenium exceed 0.16 mg/l; for mercury exceeds 0.031 mg/l, or for nickel or total cyanide exceed 10.9 mg/l, the waste must either be retreated or managed and disposed of in accordance with all applicable hazardous waste regulations. |
| | | (4) Within one week of system start-up, Occidental must notify the Section Chief, Variances Section (see address below) when the full-scale system is on-line and waste treatment has begun. All data obtained through condition (1) must be submitted to the Section Chief, Variances Section, PSPD/OSW, (OS–333), U.S. EPA, 1200 Pennsylvania Ave., NW., Washington, DC 20460 within the time period required in condition (1). At the Section Chief's request, Occidental must submit any other analytical data obtained through conditions (1) and (2) to the above address within the time period specified by the Section Chief. Failure to submit the required data will be considered by the Agency sufficient basis to revoke Occidental's exclusion to the extent directed by EPA. All data (either submitted to EPA or maintained at the site) must be accompanied by the following statement: |
| | | "Under civil and criminal penalty of law for the making or submission of false or fraudulent statements or representations (pursuant to the applicable provisions of the Federal Code, which include, but may not be limited to 18 U.S.C. 1001 and 42 U.S.C. 6926), I certify that the information contained in or accompanying this document is true, accurate and complete. |
| | | As to the (those) identified section(s) of this document for which I cannot personally verify its (their) truth and accuracy, I certify as the company official having supervisory responsibility for the persons who, acting under my direct instructions, made the verification that this information is true, accurate and complete. |
| | | In the event that any of this information is determined by EPA in its sole discretion to be false, inaccurate or incomplete, and upon conveyance of this fact to the company, I recognize and agree that this exclusion of wastes will be void as if it never had effect or to the extent directed by EPA and that the company will be liable for any actions taken in contravention of the company's RCRA and CERCLA obligations premised upon the company's reliance on the void exclusion." |
| Olin Corporation. | Charleston, TN. | Sodium chloride purification muds and potassium chloride purification muds (both classified as EPA Hazardous Waste No. K071) that have been batch tested using EPA's Toxicity Characteristic Leaching Procedure and have been found to contain less than 0.05 ppm mercury. Purification muds that have been found to contain less than 0.05 ppm mercury will be disposed in Olin's on-site non-hazardous waste landfill or another Subtitle D landfill. Purification muds that exceed this level will be considered a hazardous waste. |
| Ormet Primary Aluminum Corporation. | Hannibal, OH | Vitrified spent potliner (VSP), K088, that is generated by Ormet Primary Aluminum Corporation in Hannibal (Ormet), Ohio at a maximum annual rate of 8,500 cubic yards per year and disposed of in a Subtitle D landfill, licensed, permitted, or registered by a state. The exclusion becomes effective as of July 25, 2002. |

TABLE 2—WASTES EXCLUDED FROM SPECIFIC SOURCES—Continued

| Facility | Address | Waste description |
|---|---|---|
| | | 1. *Delisting Levels:* (A) The constituent concentrations measured in any of the extracts specified in paragraph (2) may not exceed the following levels (mg/L): Antimony—0.235; Arsenic—0.107; Barium—63.5; Beryllium—0.474; Cadmium—0.171; Chromium (total)—1.76; Lead—5; Mercury—0.17; Nickel—32.2; Selenium—0.661; Silver—4.38; Thallium—0.1; Tin—257; Vanadium—24.1; Zinc—320; Cyanide—4.11. (B) Land disposal restrictions (LDR) treatment standards for K088 must also be met before the VSP can be land disposed. Ormet must comply with any future LDR treatment standards promulgated under 40 CFR 268.40 for K088. |
| | | 2. *Verification Testing:* (A) On a quarterly basis, Ormet must collect two samples of the waste and analyze them for the constituents listed in paragraph (1) using the methodologies specified in an EPA-approved sampling plan specifying (a) the TCLP method, and (b) the TCLP procedure with an extraction fluid of 0.1 Normal sodium hydroxide solution. The constituent concentrations measured in the extract must be less than the delisting levels established in paragraph (1). Ormet must also comply with LDR treatment standards in accordance with 40 CFR 268.40. (B) If the quarterly testing of the waste does not meet the delisting levels set forth in paragraph (1), Ormet must notify the Agency in writing in accordance with paragraph (5). The exclusion will be suspended and the waste managed as hazardous until Ormet has received written approval for the exclusion from the Agency. Ormet may provide sampling results that support the continuation of the delisting exclusion. |
| | | 3. *Changes in Operating Conditions:* If Ormet significantly changes the manufacturing process, the treatment process, or the chemicals used, Ormet must notify the EPA of the changes in writing. Ormet must handle wastes generated after the process change as hazardous until Ormet has demonstrated that the wastes continue to meet the delisting levels set forth in paragraph (1) and that no new hazardous constituents listed in Appendix VIII of part 261 have been introduced and Ormet has received written approval from EPA. |
| | | 4. *Data Submittals:* Ormet must submit the data obtained through quarterly verification testing or as required by other conditions of this rule to U.S. EPA Region 5, Waste Management Branch (DW–8J), 77 W. Jackson Blvd., Chicago, IL 60604 by February 1 of each calendar year for the prior calendar year. Ormet must compile, summarize, and maintain on site for a minimum of five years records of operating conditions and analytical data. Ormet must make these records available for inspection. All data must be accompanied by a signed copy of the certification statement in 40 CFR 260.22(i)(12). |
| | | 5. *Reopener Language*—(a) If, anytime after disposal of the delisted waste, Ormet possesses or is otherwise made aware of any data (including but not limited to leachate data or groundwater monitoring data) relevant to the delisted waste indicating that any constituent identified in paragraph (1) is at a level in the leachate higher than the delisting level established in paragraph (1), or is at a level in the groundwater higher than the point of exposure groundwater levels referenced by the model, then Ormet must report such data, in writing, to the Regional Administrator within 10 days of first possessing or being made aware of that data. |
| | | (b) Based on the information described in paragraph (5)(a) or any other information received from any source, the Regional Administrator will make a preliminary determination as to whether the reported information requires Agency action to protect human health or the environment. Further action may include suspending, or revoking the exclusion, or other appropriate response necessary to protect human health and the environment. |
| | | (c) If the Regional Administrator determines that the information does require Agency action, the Regional Administrator will notify Ormet in writing of the actions the Regional Administrator believes are necessary to protect human health and the environment. The notice shall include a statement of the proposed action and a statement providing Ormet with an opportunity to present information as to why the proposed Agency action is not necessary or to suggest an alternative action. Ormet shall have 30 days from the date of the Regional Administrator's notice to present the information. |
| | | (d) If after 30 days Ormet presents no further information, the Regional Administrator will issue a final written determination describing the Agency actions that are necessary to protect human health or the environment. Any required action described in the Regional Administrator's determination shall become effective immediately, unless the Regional Administrator provides otherwise. |
| Oxy Vinyls ...... | Deer Park, Texas. | Rockbox Residue, (at a maximum generation of 1,000 cubic yards per calender year) generated by Oxy Vinyls using the wastewater treatment process to treat the Rockbox Residue (EPA Hazardous Waste No. K017, K019, and K020). |
| | | Oxy Vinyls must implement a testing program that meets the following conditions for the exclusion to be valid: |
| | | (1) *Delisting Levels:* All concentrations for the following constituents must not exceed the following levels (ppm). The Rockbox Residue must be measured in the waste leachate by the method specified in 40 CFR 261.24. |
| | | (A) Rockbox Residue: |
| | | (i) Inorganic Constituents: Barium—200; Chromium—5.0; Copper—130; Lead+1.5; Tin—2,100; Vanadium—30; Zinc—1,000 |
| | | (ii) Organic Constituents: Acetone—400; Dichloromethane—1.0; Dimethylphthalate—4,000; Xylene—10,000; 2,3,7,8-TCDD Equivalent—0.00000006 |

TABLE 2—WASTES EXCLUDED FROM SPECIFIC SOURCES—Continued

| Facility | Address | Waste description |
|---|---|---|
| | | (2) *Waste Holding and Handling:* Oxy Vinyls must store in accordance with its RCRA permit, or continue to dispose of as hazardous waste all Rockbox Residue generated until the verification testing described in Condition (3)(B), as appropriate, is completed and valid analyses demonstrate that condition (3) is satisfied. If the levels of constituents measured in the samples of the Rockbox Residue do not exceed the levels set forth in Condition (1), then the waste is nonhazardous and may be managed and disposed of in accordance with all applicable solid waste regulations. If constituent levels in a sample exceed any of the delisting levels set in Condition 1, waste generated during the time period corresponding to this sample must be managed and disposed of in accordance with subtitle C of RCRA. |
| | | (3) *Verification Testing Requirements:* Sample collection and analyses, including quality control procedures, must be performed using appropriate methods. As applicable to the method-defined parameters of concern, analyses requiring the use of SW–846 methods incorporated by reference in 40 CFR 260.11 must be used without substitution. As applicable, the SW–846 methods might include Methods 0010, 0011, 0020, 0023A, 0030, 0031, 0040, 0050, 0051, 0060, 0061, 1010A, 1020B, 1110A, 1310B, 1311, 1312, 1320, 1330A, 9010C, 9012B, 9040C, 9045D, 9060A, 9070A (uses EPA Method 1664, Rev. A), 9071B, and 9095B. If EPA judges the incineration process to be effective under the operating conditions used during the initial verification testing, OxyVinyls may replace the testing required in Condition (3)(A) with the testing required in Condition (3)(B). OxyVinyls must continue to test as specified in Condition (3)(A) until and unless notified by EPA in writing that testing in Condition (3)(A) may be replaced by Condition (3)(B). |
| | | (A) *Initial Verification Testing:* (i) When the Rockbox unit is decommissioned for clean out, after the final exclusion is granted, Oxy Vinyls must collect and analyze composites of the Rockbox Residue. Two composites must be composed of representative grab samples collected from the Rockbox unit. The waste must be analyzed, prior to disposal, for all of the constituents listed in Condition 1. No later than 90 days after the Rockbox unit is decommissioned for clean out the first two times after this exclusion becomes final, Oxy Vinyls must report the operational and analytical test data, including quality control information. |
| | | (B) *Subsequent Verification Testing:* Following written notification by EPA, Oxy Vinyls may substitute the testing conditions in (3)(B) for (3)(A)(i). Oxy Vinyls must continue to monitor operating conditions, analyze samples representative of each cleanout of the Rockbox of operation during the first year of waste generation. |
| | | (C) *Termination of Organic Testing for the Rockbox Residue:* Oxy Vinyls must continue testing as required under Condition (3)(B) for organic constituents specified under Condition (3)(B) for organic constituents specified in Condition (1)(A)(ii) until the analyses submitted under Condition (3)(B) show a minimum of two consecutive annual samples below the delisting levels in Condition (1)(A)(ii), Oxy Vinyls may then request that annual organic testing be terminated. Following termination of the quarterly testing, Oxy Vinyls must continue to test a representative composite sample for all constituents listed in Condition (1) on an annual basis (no later than twelve months after exclusion). |
| | | (4) *Changes in Operating Conditions:* If Oxy Vinyls significantly changes the process which generate(s) the waste(s) and which may or could affect the composition or type waste(s) generated as established under Condition (1) (by illustration, but not limitation, change in equipment or operating conditions of the treatment process), Oxy Vinyls must notify the EPA in writing and may no longer handle the wastes generated from the new process or no longer discharges as nonhazardous until the wastes meet the delisting levels set Condition (1) and it has received written approval to do so from EPA. |
| | | (5) *Data Submittals:* The data obtained through Condition 3 must be submitted to Mr. William Gallagher, Chief, Region 6 Delisting Program, U.S. EPA, 1445 Ross Avenue, Dallas, Texas 75202–2733, Mail Code, (6PD-O) within the time period specified. Records of operating conditions and analytical data from Condition (1) must be compiled, summarized, and maintained on site for a minimum of five years. These records and data must be furnished upon request by EPA, or the State of Texas, and made available for inspection. Failure to submit the required data within the specified time period or maintain the required records on site for the specified time will be considered by EPA, at its discretion, sufficient basis to revoke the exclusion to the extent directed by EPA. All data must be accompanied by a signed copy of the following certification statement to attest to the truth and accuracy of the data submitted:<br><br>Under civil and criminal penalty of law for the making or submission of false or fraudulent statements or representations (pursuant to the applicable provisions of the Federal Code, which include, but may not be limited to, 18 U.S.C. 1001 and 42 U.S.C. 6928), I certify that the information contained in or accompanying this document is true, accurate and complete.<br><br>As to the (those) identified section(s) of this document for which I cannot personally verify its (their) truth and accuracy, I certify as the company official having supervisory responsibility for the persons who, acting under my direct instructions, made the verification that this information is true, accurate and complete. |

TABLE 2—WASTES EXCLUDED FROM SPECIFIC SOURCES—Continued

| Facility | Address | Waste description |
|---|---|---|
| | | In the event that any of this information is determined by EPA in its sole discretion to be false, inaccurate or incomplete, and upon conveyance of this fact to the company, I recognize and agree that this exclusion of waste will be void as if it never had effect or to the extent directed by EPA and that the company will be liable for any actions taken in contravention of the company's RCRA and CERCLA obligations premised upon the company's reliance on the void exclusion. |
| | | (6) *Reopener Language:* |
| | | (A) If, anytime after disposal of the delisted waste, Oxy Vinyls possesses or is otherwise made aware of any environmental data (including but not limited to leachate data or groundwater monitoring data) or any other data relevant to the delisted waste indicating that any constituent identified for the delisting verification testing is at level higher than the delisting level allowed by the Director in granting the petition, then the facility must report the data, in writing, to the Director within 10 days of first possessing or being made aware of that data. |
| | | (B) If the annual testing of the waste does not meet the delisting requirements in Paragraph 1, Oxy Vinyls must report the data, in writing, to the Director within 10 days of first possessing or being made aware of that data. |
| | | (C) Based on the information described in paragraphs (A) or (B) and any other information received from any source, the Director will make a preliminary determination as to whether the reported information requires Agency action to protect human health or the environment. Further action may include suspending, or revoking the exclusion, or other appropriate response necessary to protect human health and the environment. |
| | | (D) If the Director determines that the reported information does require Agency action, the Director will notify the facility in writing of the actions the Director believes are necessary to protect human health and the environment. The notice shall include a statement of the proposed action and a statement providing the facility with an opportunity to present information as to why the proposed Agency action is not necessary. The facility shall have 10 days from the date of the Director's notice to present such information. |
| | | (E) Following the receipt of information from the facility described in paragraph (D) or (if no information is presented under paragraph (D)) the initial receipt of information described in paragraphs (A) or (B), the Director will issue a final written determination describing the Agency actions that are necessary to protect human health or the environment. Any required action described in the Director's determination shall become effective immediately, unless the Director provides otherwise. |
| | | (7) *Notification Requirements:* Oxy Vinyls must provide a one-time written notification to any State Regulatory Agency to which or through which the delisted waste described above will be transported for disposal at least 60 days prior to the commencement of such activities. Failure to provide such a notification will result in a violation of the delisting petition and a possible revocation of the decision. |
| OxyVinyls, L.P. | Deer Park, TX | Incinerator Offgas Scrubber Water (EPA Hazardous Waste Nos. K017, K019 and K020) generated at a maximum annual rate of 919,990 cubic yards per calendar year after April 22, 2004, and disposed in accordance with the TPDES permit. For the exclusion to be valid, OxyVinyls must implement a testing program that meets the following Paragraphs: |
| | | (1) *Delisting Levels:* All total concentrations for those constituents must not exceed the following levels (mg/kg) in the incinerator offgas scrubber water. Incinerator offgas treatment scrubber water (i) Inorganic Constituents Antimony—0.0204; Arsenic—0.385; Barium—2.92; Beryllium—0.166; Cadmium—0.0225; Chromium—5.0; Cobalt—13.14; Copper—418.00; Lead—5.0; Nickel—1.13; Mercury—0.0111; Vanadium—0.838; Zinc—2.61 (ii) Organic Constituents Acetone—1.46; Bromoform—0.481; Bromomethane—8.2; Bromodichloromethane—0.0719; Chloroform—0.683; Dibromochloromethane—0.057; Iodomethane—0.19; Methylene Chloride—0.029; 2,3,7,8—TCDD equivalents as TEQ—0.0000926 |
| | | (2) *Waste Management:* (A) OxyVinyls must manage as hazardous all incinerator offgas treatment scrubber water generated, until it has completed initial verification testing described in Paragraphs (3)(A) and (B), as appropriate, and valid analyses show that paragraph (1) is satisfied. |
| | | (B) Levels of constituents measured in the samples of the incinerator offgas treatment scrubber water that do not exceed the levels set forth in Paragraph (1) are non-hazardous. OxyVinyls can manage and dispose the non-hazardous incinerator offgas treatment scrubber water according to all applicable solid waste regulations. |
| | | (C) If constituent levels in a sample exceed any of the delisting levels set in Paragraph (1), OxyVinyls must collect one additional sample and perform the expedited analyses to confirm if the constituent exceeds the delisting level. If this sample confirms the exceedance, OxyVinyls must, from that point forward, treat the waste as hazardous until it is demonstrated that the waste again meets the levels set in Paragraph (1). OxyVinyls must notify EPA of the exceedance and resampling analytical results prior to disposing of the waste. |
| | | (D) If the waste exceeds the levels in paragraph (1) OxyVinyls must manage and dispose of the waste generated under Subtitle C of RCRA from the time that it becomes aware of any exceedance. |

TABLE 2—WASTES EXCLUDED FROM SPECIFIC SOURCES—Continued

| Facility | Address | Waste description |
|---|---|---|
| | | (E) Upon completion of the Verification Testing described in Paragraphs 3(A) and (B) as appropriate and the transmittal of the results to EPA, and if the testing results meet the requirements of Paragraph (1), OxyVinyls may proceed to manage its incinerator offgas treatment scrubber water as non-hazardous waste. If subsequent verification testing indicates an exceedance of the Delisting Levels in Paragraph (1), OxyVinyls must manage the incinerator offgas treatment scrubber water as a hazardous waste until two consecutive quarterly testing samples show levels below the Delisting Levels. |
| | | (3) *Verification Testing Requirements:* OxyVinyls must perform sample collection and analyses, including quality control procedures, using appropriate methods. As applicable to the method-defined parameters of concern, analyses requiring the use of SW–846 methods incorporated by reference in 40 CFR 260.11 must be used without substitution. As applicable, the SW–846 methods might include Methods 0010, 0011, 0020, 0023A, 0030, 0031, 0040, 0050, 0051, 0060, 0061, 1010A, 1020B, 1110A, 1310B, 1311, 1312, 1320, 1330A, 9010C, 9012B, 9040C, 9045D, 9060A, 9070A (uses EPA Method 1664, Rev. A), 9071B, and 9095B. If EPA judges the process to be effective under the operating conditions used during the initial verification testing, OxyVinyls may replace the testing required in Paragraph (3)(A) with the testing required in Paragraph (3)(B). OxyVinyls must continue to test as specified in Paragraph (3)(A) until and unless notified by EPA in writing that testing in Paragraph (3)(A) may be replaced by Paragraph (3)(B). |
| | | (A) *Initial Verification Testing:* After EPA grants the final exclusion, OxyVinyls must do the following: (i) Within 60 days of this exclusion becoming final, collect four samples, before disposal, of the incinerator offgas treatment scrubber water. (ii) The samples are to be analyzed and compared against the delisting levels in Paragraph (1) (iii). Within sixty (60) days after the exclusion becomes final, OxyVinyls will report initial verification analytical test data, including analytical quality control information for the first sixty (30) days of operation after this exclusion becomes final of the incinerator offgas treatment scrubber water. If levels of constituents measured in the samples of the incinerator offgas treatment scrubber water that do not exceed the levels set forth in Paragraph (1) and are also non-hazardous in two consecutive quarters after the first thirty (30) days of operation after this exclusion, OxyVinyls can manage and dispose of the incinerator offgas treatment scrubber water according to all applicable solid water regulations after reporting the analytical results to EPA. |
| | | (B) *Subsequent Verification Testing:* Following written notification by EPA, OxyVinyls may substitute the testing conditions in Paragraph (3)(B) for (3)(A). OxyVinyls must continue to monitor operating conditions, and analyze representative samples of each quarter of operation during the first year of waste generation. The samples must represent the waste generated during the quarter. After the first year of analytical sampling verification sampling can be performed on a single annual composite sample of the incinerator offgas treatment scrubber water. The results are to be compared to the delisting levels in Condition (1). |
| | | (C) *Termination of Testing:* (i) After the first year of quarterly testing, if the Delisting Levels in Paragraph (1) are being met, OxyVinyls may then request that EPA stop requiring quarterly testing. After EPA notifies OxyVinyls in writing, the company may end quarterly testing. (ii) Following cancellation of the quarterly testing, OxyVinyls must continue to test a representative sample for all constituents listed in Paragraph (1) annually. |
| | | (4) *Changes in Operating Conditions:* If OxyVinyls significantly changes the process described in its petition or starts any processes that generate(s) the waste that may or could significantly affect the composition or type of waste generated as established under Paragraph (1) (by illustration, but not limitation, changes in equipment or operating conditions of the treatment process), it must notify EPA in writing; OxyVinyls may no longer handle the wastes generated from the new process as nonhazardous until the wastes meet the delisting levels set in Paragraph (1) and it has received written approval to do so from EPA. |
| | | (5) *Data Submittals:* OxyVinyls must submit the information described below. If OxyVinyls fails to submit the required data within the specified time or maintain the required records on-site for the specified time, EPA, at its discretion, will consider this sufficient basis to reopen the exclusion as described in Paragraph 6. OxyVinyls must: |
| | | (A) Submit the data obtained through Paragraph 3 to the Section Chief, EPA Region 6 Corrective Action and Waste Minimization Section, 1445 Ross Avenue, Dallas, Texas 75202–2733, Mail Code, (6PD–C) within the time specified. |
| | | (B) Compile records of operating conditions and analytical data from Paragraph (3), summarized, and maintained on-site for a minimum of five years. |
| | | (C) Finish these records and data when EPA or the State of Texas request them for inspection. |

TABLE 2—WASTES EXCLUDED FROM SPECIFIC SOURCES—Continued

| Facility | Address | Waste description |
|---|---|---|
| | | (D) Send along with all data a signed copy of the following certification statement, to attest to the truth and accuracy of the data submitted: Under civil and criminal penalty of law for the making or submission of false or fraudulent statements or representations (pursuant to the applicable provisions of the Federal Code, which include, but may not be limited to, 18 U.S.C. 1001 and 42 U.S.C. 6928), I certify that the information contained in or accompanying this document is true, accurate and complete. As to the (those) identified section(s) of this document for which I cannot personally verify its (their) truth and accuracy, I certify as the company official having supervisory responsibility for the persons who, acting under my direct instructions, made the verification that this information is true, accurate and complete. If any of this information is determined by EPA in its sole discretion to be false, inaccurate or incomplete, and upon conveyance of this fact to the company, I recognize and agree that this exclusion of waste will be void as if its never had effect or to the extent directed by EPA and that the company will be liable for any actions taken in contravention of the company's RCRA and CERCLA obligations premised upon the company's reliance on the void exclusion. |
| | | (6) *Reopener:* (A) If, anytime after disposal of the delisted waste OxyVinyls possesses or is otherwise made aware of any environmental data (including but not limited to leachate data or groundwater monitoring data) or any other data relevant to the delisted waste indicating that any constituent identified for the delisting verification testing is at a level higher than the delisting level allowed by the Regional Administrator or his delegate in granting the petition, then the facility must report the data, in writing, to the Regional Administrator or his delegate within 10 days of first possessing or being made aware of that data. |
| | | (B) If the annual testing of the waste does not meet the delisting requirements in Paragraph 1, OxyVinyls must report the data, in writing, to the Regional Administrator or his delegate within 10 days of first possessing or being made aware of that data. |
| | | (C) If OxyVinyls fails to submit the information described in paragraphs (5), (6)(A) or (6)(B) or if any other information is received from any source, the Regional Administrator or his delegate will make a preliminary determination as to whether the reported information requires EPA action to protect human health or the environment. Further action may include suspending, or revoking the exclusion, or other appropriate response necessary to protect human health and environment. |
| | | (D) If the Regional Administrator or his delegate determines that the reported information does require action by EPA's Regional Administrator or his delegate will notify the facility in writing of the actions the Regional Administrator or his delegate believes are necessary to protect human health and the environment. The notice shall include a statement of the proposed action and a statement providing the facility with an opportunity to present information as to why the proposed EPA action is not necessary. The facility shall have 10 days from the date of the Regional Administrator or his delegate's notice to present such information. |
| | | (E) Following the receipt of information from the facility described in paragraph (6)(D) or (of no information is presented under paragraph (6)(D)) the initial receipt of information described in paragraphs (5), (6)(A) or (6)(B), the Regional Administrator or his delegate will issue a final written determination describing EPA actions that are necessary to protect human health or the environment. Any require action described in the Regional Administrator or his delegate's determination shall become effective immediately, unless the Regional Administrator or his delegate provides otherwise. |
| | | (7) *Notification Requirements:* OxyVinyls must do the following before transporting the delisted waste. Failure to provide this notification will result in a violation of the delisting petition and a possible revocation of the decision. |
| | | (A) Provide a one-time written notification to any State Regulatory Agency to which or through which it will transport the delisted waste described above for disposal, 60 days before beginning such activities. |
| | | (B) Update the one-time written notification if it ships the delisted waste into a different disposal facility. |
| | | (C) Failure to provide this notification will result in a violation of the delisting variance and a possible revocation of the decision. |
| Perox, Incorporated. | Sharon, Pennsylvania. | Iron oxide (EPA Hazardous Waste No. K062) generated (at a maximum annual rate of 4800 cubic yards) from a spent hydrochloric acid pickle liquor regeneration plant for spent pickle liquor generated from steel finishing operations. This exclusion was published on November 13, 1990. |
| Pioneer Chlor Alkai Company, Inc. (formerly Stauffer Chemical Company). | St. Gabriel, LA | Brine purification muds, which have been washed and vacuum filtered, generated after August 27, 1985 from their chlor-alkali manufacturing operations (EPA Hazardous Waste No. K071) that have been batch tested for mercury using the EP toxicity procedure and have been found to contain less than 0.05 ppm in mercury in the EP extract. Brine purification muds that exceed this level will be considered a hazardous waste. |

TABLE 2—WASTES EXCLUDED FROM SPECIFIC SOURCES—Continued

| Facility | Address | Waste description |
|---|---|---|
| POP Fasteners | Shelton, Connecticut. | Wastewater treatment sludge (EPA Hazardous Waste No. F006) generated from electroplating operations (at a maximum annual rate of 300 cubic yards) after December 7, 1992. In order to confirm that the characteristics of the waste do not change significantly, the facility must, on an annual basis, analyze a representative composite sample for the constituents listed in §261.24 using the method specified therein. The annual analytical results, including quality control information, must be compiled, certified according to §260.22(i)(12) of this chapter, maintained on site for a minimum of five years, and made available for inspection upon request by any employee or representative of EPA or the State of Connecticut. Failure to maintain the required records on site will be considered by EPA, at its discretion, sufficient basis to revoke the exclusion to the extent directed by EPA. |
| Rhodia ............ | Houston, Texas. | Filter-cake Sludge, (at a maximum generation of 1,200 cubic yards per calendar year) generated by Rhodia using the SARU and AWT treatment process to treat the filter-cake sludge (EPA Hazardous Waste Nos. K002–004, K006–K011, K013–K052, K060–K062, K064–K066, K069, K071, K073, K083–K088, K090–K091, K093–K118, K123–K126, K131–K133, K136, K141–K145, K147–K151, K156–K161) generated at Rhodia. Rhodia must implement the testing program described in Table 1. Waste Excluded From Non-Specific Sources for the petition to be valid. |
| Roanoke Electric Steel Corp. | Roanoke, VA | Fully-cured chemically stabilized electric arc furnace dust/sludge (CSEAFD) treatment residue (EPA Hazardous Waste No. K061) generated from the primary production of steel after March 22, 1989. This exclusion is conditioned upon the data obtained from Roanoke's full-scale CSEAFD treatment facility because Roanoke's original data were obtained from a laboratory-scale CSEAFD treatment process. To ensure that hazardous constituents are not present in the waste at levels of regulatory concern once the full-scale treatment facility is in operation, Roanoke must implement a testing program for the petitioned waste.<br>This testing program must meet the following conditions for the exclusion to be valid:<br>(1) *Testing:*<br>(A) *Initial Testing:* During the first four weeks of operation of the full-scale treatment system, Roanoke must collect representative grab samples of each treated batch of the CSEAFD and composite the grab samples daily. The daily composites, prior to disposal, must be analyzed for the EP leachate concentrations of all the EP toxic metals, nickel and cyanide (using distilled water in the cyanide extractions). Analyses must be performed using appropriate methods. As applicable to the method-defined parameters of concern, analyses requiring the use of SW–846 methods incorporated by reference in 40 CFR 260.11 must be used without substitution. As applicable, the SW–846 methods might include Methods 0010, 0011, 0020, 0023A, 0030, 0031, 0040, 0050, 0051, 0060, 0061, 1010A, 1020B, 1110A, 1310B, 1311, 1312, 1320, 1330A, 9010C, 9012B, 9040C, 9045D, 9060A, 9070A (uses EPA Method 1664, Rev. A), 9071B, and 9095B. Roanoke must report the analytical test data obtained during this initial period no later than 90 days after the treatment of the first full-scale batch. |

TABLE 2—WASTES EXCLUDED FROM SPECIFIC SOURCES—Continued

| Facility | Address | Waste description |
|---|---|---|
| | | (B) *Subsequent Testing:* Roanoke must collect representative grab samples from every treated batch of CSEAFD generated daily and composite all of the grab samples to produce a weekly composite sample. Roanoke then must analyze each weekly composite sample for all of the EP toxic metals and nickel. Analyses must be performed using appropriate methods. As applicable to the method-defined parameters of concern, analyses requiring the use of SW-846 methods incorporated by reference in 40 CFR 260.11 must be used without substitution. As applicable, the SW-846 methods might include Methods 0010, 0011, 0020, 0023A, 0030, 0031, 0040, 0050, 0051,0060,0061, 1010A, 1020B, 1110A, 1310B, 1311, 1312, 1320, 1330A, 9010C, 9012B, 9040C, 9045D, 9060A, 9070A (uses EPA Method 1664, Rev. A), 9071B, and 9095B. The analytical data, including all quality control information, must be compiled and maintained on site for a minimum of three years. These data must be furnished upon request and made available for inspection for any employee or representative of EPA or the State of Virginia. |
| | | (2) *Delisting levels:* If the EP extract concentrations for chromium, lead, arsenic, or silver exceed 0.315 mg/l; for barium exceeds 6.3 mg/l; for cadmium or selenium exceed 0.063 mg/l; for mercury exceeds 0.0126 mg/l, for nickel exceeds 3.15 mg/l, or for cyanide exceeds 1.26 mg/l, the waste must either be re-treated or managed and disposed in accordance with subtitle C of RCRA. |
| | | (3) *Data submittals: Within one week of system start-up, Roanoke must notify the Section Chief, Variances Section (see address below) when their full-scale stabilization system in on-line and waste treatment has begun. All data obtained through the initial testing condition (1)(A), must be submitted to the Section Chief, Variances Section, PSPD/OSW, (OS–343), U.S. EPA, 1200 Pennsylvania Ave., NW., Washington, DC 20460 within the time period specified in condition (1)(A). Failure to submit the required data or keep the required records will be considered by the Agency, at its discretion, sufficient basis to revoke Roanoke's exclusion. All data must be accompanied by the following certification statement:* "Under civil and criminal penalty of law for the making or submission of false or fraudulent statements or representations (pursuant to the applicable provisions of the Federal Code which include, but may not be limited to, 18 USC 6928), I certify that the information contained in or accompanying this document is true, accurate and complete. As to the (those) identified section(s) of this document for which I cannot personally verify its (their) truth and accuracy, I certify as the company official having supervisory responsibility for the persons who, acting under my direct instructions, made the verification that this information is true, accurate and complete. In the event that any of this information is determined by EPA in its sole discretion to be false, inaccurate or incomplete, and upon conveyance of this fact to the company, I recognize and agree that this exclusion of wastes will be void as if it never had effect or to the extent directed by EPA and that the company will be liable for any actions taken in contravention of the company's RCRA and CERCLA obligations premised upon the company's reliance on the void exclusion." |
| Texas Eastman | Longview, Texas. | Incinerator ash (at a maximum generation of 7,000 cubic yards per calendar year) generated from the incineration of sludge from the wastewater treatment plant (EPA Hazardous Waste No. K009 and K010, and that is disposed of in Subtitle D landfills after September 25, 1996. Texas Eastman must implement a testing program that meets conditions found in Table 1. Wastes Excluded From Non-Specific Sources for the petition to be valid. |
| United States Department of Energy (Energy). | Richland, Washington. | Treated effluents bearing the waste numbers identified below, from the 200 Area Effluent Treatment Facility (ETF) located at the Hanford Facility, at a maximum generation rate of 210 million liters per year, subject to Conditions 1–7: This conditional exclusion applies to Environmental Protection Agency (EPA) Hazardous Waste Nos. F001, F002, F003, F004, F005, and F039. This exclusion also applies to EPA Hazardous Waste Nos. F006–F012, F019 and F027 provided that the as-generated waste streams bearing these waste numbers prior to treatment in the 200 Area ETF is in the form of dilute wastewater containing a maximum of 1.0 weight percent of any hazardous constituent. In addition, this conditional exclusion applies to all other U- and P-listed waste numbers that meet the following criteria: The U/P listed substance has a treatment standard for wastewater forms of F039 multi-source leachate under 40 CFR 268.40,"Treatment Standards for Hazardous Wastes"; and the as-generated waste stream prior to treatment in the 200 Area ETF is in the form of dilute wastewater containing a maximum of 1.0 weight percent of any hazardous constituent. This exclusion shall apply at the point of discharge from the 200 Area ETF verification tanks after satisfaction of Conditions 1–7. |
| | | Conditions: |
| | | (1) Waste Influent Characterization and Processing Strategy Preparation |
| | | (a) Prior to treatment of any waste stream in the 200 Area ETF, Energy must: |
| | | (i) Complete sufficient characterization of the waste stream to demonstrate that the waste stream is within the treatability envelope of 200 Area ETF as specified in Tables C–1 and C–2 of the delisting petition dated November 29, 2001. Results of the waste stream characterization and the treatability evaluation must be in writing and placed in the facility operating record, along with a copy of the November 29, 2001 petition. Waste stream characterization may be carried out in whole or in part using the waste analysis procedures in the Hanford Facility RCRA Permit, WA7 89000 8967; |

TABLE 2—WASTES EXCLUDED FROM SPECIFIC SOURCES—Continued

| Facility | Address | Waste description |
|---|---|---|
| | | (ii) Prepare a written waste processing strategy specific to the waste stream, based on the ETF process model documented in the November 29, 2001 petition. For waste processing strategies applicable to waste streams for which inorganic envelope data is provided in Table C–2 of the November 29, 2001 petition, Energy shall use envelope data specific to that waste stream, if available. Otherwise, Energy shall use the minimum envelope in Table C–2. |
| | | (b) Energy may modify the 200 Area ETF treatability envelope specified in Tables C–1 and C–2 of the November 29, 2001 delisting petition to reflect changes in treatment technology or operating practices upon written approval of the Regional Administrator. Requests for modification shall be accompanied by an engineering report detailing the basis for a modified treatment envelope. Data supporting modified envelopes must be based on at least four influent waste stream characterization data points and corresponding treated effluent verification sample data points for wastes managed under a particular waste processing strategy. Treatment efficiencies must be calculated based on a comparison of upper 95 percent confidence level constituent concentrations. Upon written EPA approval of the engineering report, the associated inorganic treatment efficiency data may be used in lieu of those in Tables C–1 and C–2 for purposes of condition (1)(a)(i). |
| | | (c) Energy shall conduct all 200 Area ETF treatment operations for a particular waste stream according to the written waste processing strategy, as may be modified by Condition 3(b)(i). |
| | | (d) The following definitions apply: |
| | | (i) A waste stream is defined as all wastewater received by the 200 Area ETF that meet the 200 Area ETF waste acceptance criteria as defined by the Hanford Facility RCRA Permit, WA7 89000 8967 and are managed under the same 200 Area ETF waste processing strategy. |
| | | (ii) A waste processing strategy is defined as a specific 200 Area ETF unit operation configuration, primary operating parameters and expected maximum influent total dissolved solids (TDS) and total organic carbon (TOC). Each waste processing strategy shall require monitoring and recording of treated effluent conductivity for purposes of Condition (2)(b)(i)(E), and for monitoring and recording of primary operating parameters as necessary to demonstrate that 200 Area ETF operations are in accordance with the associated waste processing strategy. |
| | | (iii) Primary operating parameters are defined as ultraviolet oxidation (UV/OX) peroxide addition rate, reverse osmosis reject ratio, and processing flow rate as measured at the 200 Area ETF surge tank outlet. |
| | | (iv) Key unit operations are defined as filtration, UV/OX, reverse osmosis, ion exchange, and secondary waste treatment. |
| | | (2) Testing. Energy shall perform verification testing of treated effluents according to Conditions (a), (b), and (c) below. |
| | | (a) No later than 45 days after the effective date of this rule, or such other time as may be approved of in advance and in writing by EPA, Energy shall submit to EPA a report proposing required data quality parameters and data acceptance criteria (parameter values) for sampling and analysis which may be conducted pursuant to the requirements of this rule. This report shall explicitly consider verification sampling and analysis for purposes of demonstrating compliance with exclusion limits in Condition 5, as well as any sampling and analysis which may be required pursuant to Conditions (1)(a)(i) and (1)(d)(ii). This report shall contain a detailed justification for the proposed data quality parameters and data acceptance criteria. Following review and approval of this report, the proposed data quality parameters and data acceptance criteria shall become enforceable conditions of this exclusion. Pending EPA approval of this report, Energy may demonstrate compliance with sampling and analysis requirements of this rule through application of methods appearing in EPA Publication SW–846 or equivalent methods. Energy shall maintain a written sampling and analysis plan, including QA/QC requirements and procedures, based upon these enforceable data quality parameters and data acceptance criteria in the facility operating record, and shall conduct all sampling and analysis conducted pursuant to this rule according to this written plan. Records of all sampling and analysis, including quality assurance QA/QC information, shall be placed in the facility operating record. As applicable to the method-defined parameters of concern, analyses requiring the use of SW–846 methods incorporated by reference in 40 CFR 260.11 must be used without substitution. As applicable, the SW–846 methods might include Methods 0010, 0011, 0020, 0023A, 0030, 0031, 0040, 0050, 0051, 0060, 0061, 1010A, 1020B, 1110A, 1310B, 1311, 1312, 1320, 1330A, 9010C, 9012B, 9040C, 9045D, 9060A, 9070A (uses EPA Method 1664, Rev. A), 9071B, and 9095B. |
| | | (b) Initial verification testing. |
| | | (i) Verification sampling shall consist of a representative sample of one filled effluent discharge tank, analyzed for all constituents in Condition (5), and for conductivity for purposes of establishing a conductivity baseline with respect to Condition (2)(b)(i)(E). Verification sampling shall be required under each of the following conditions: |
| | | (A) Any new or modified waste strategy; |
| | | (B) Influent wastewater total dissolved solids or total organic carbon concentration increases by an order of magnitude or more above values established in the waste processing strategy; |

TABLE 2—WASTES EXCLUDED FROM SPECIFIC SOURCES—Continued

| Facility | Address | Waste description |
|---|---|---|
| | | (C) Changes in primary operating parameters; |
| | | (D) Changes in influent flow rate outside a range of 150 to 570 liters per minute; |
| | | (E) Increase greater than a factor of ten (10) in treated effluent conductivity (conductivity changes indicate changes in dissolved ionic constituents, which in turn are a good indicator of 200 Area ETF treatment efficiency). |
| | | (F) Any failure of initial verification required by this condition, or subsequent verification required by Condition (2)(c). |
| | | (ii) Treated effluents shall be managed according to Condition 3. Once Condition (3)(a) is satisfied, subsequent verification testing shall be performed according to Condition (2)(c). |
| | | (c) Subsequent Verification: Following successful initial verification associated with a specific waste processing strategy, Energy must continue to monitor primary operating parameters, and collect and analyze representative samples from every fifteenth (15th) verification tank filled with 200 Area ETF effluents processed according to the associated waste processing strategy. These representative samples must be analyzed prior to disposal of 200 Area ETF effluents for all constituents in Condition (5). Treated effluent from tanks sampled according to this condition must be managed according to Condition (3). |
| | | (3) Waste Holding and Handling: Energy must store as hazardous waste all 200 Area ETF effluents subject to verification testing in Condition (2)(b) and (2)(c), that is, until valid analyses demonstrate Condition (5) is satisfied. |
| | | (a) If the levels of hazardous constituents in the samples of 200 Area ETF effluent are equal to or below the levels set forth in Condition (5), the 200 Area ETF effluents are not listed as hazardous wastes provided they are disposed of in the State Authorized Land Disposal Site (SALDS) (except as provided pursuant to Condition (7)) according to applicable requirements and permits. Subsequent treated effluent batches shall be subject to verification requirements of Condition (2)(c). |
| | | (b) If hazardous constituent levels in any representative sample collected from a verification tank exceed any of the delisting levels set in Condition (5), Energy must: |
| | | (i) Review waste characterization data, and review and change accordingly the waste processing strategy as necessary to ensure subsequent batches of treated effluent do not exceed delisting criteria; |
| | | (ii) Retreat the contents of the failing verification tank; |
| | | (iii) Perform verification testing on the retreated effluent. If constituent concentrations are at or below delisting levels in Condition (5), the treated effluent are not listed hazardous waste provided they are disposed at SALDS according to applicable requirements and permits (except as provided pursuant to Condition (7)), otherwise repeat the requirements of Condition (3)(b). |
| | | (iv) Perform initial verification sampling according to Condition (2)(b) on the next treated effluent tank once testing required by Condition (3)(b)(iii) demonstrates compliance with delisting requirements. |
| | | (4) Re-opener Language |
| | | (a) If, anytime before, during, or after treatment of waste in the 200 Area ETF, Energy possesses or is otherwise made aware of any data (including but not limited to groundwater monitoring data, as well as data concerning the accuracy of site conditions or the validity of assumptions upon which the November 29, 2001 petition was based) relevant to the delisted waste indicating that the treated effluent no longer meets delisting criteria (excluding record keeping and data submissions required by Condition (6)), or that groundwater affected by discharge of the treated effluent exhibits hazardous constituent concentrations above health-based limits, Energy must report such data, in writing, to the Regional Administrator within 10 days of first possessing or being made aware of that data. |
| | | (b) Energy shall provide written notification to the Regional Administrator no less than 180 days prior to any planned or proposed substantial modifications to the 200 Area ETF, exclusive of routine maintenance activities, that could affect waste processing strategies or primary operating parameters. This condition shall specifically include, but not be limited to, changes that do or would require Class II or III modification to the Hanford Facility RCRA Permit WA7 89000 8967 (in the case of permittee-initiated modifications) or equivalent modifications in the case of agency-initiated permit modifications operations. Energy may request a modification to the 180-day notification requirement of this condition in the instance of agency-initiated permit modifications for purposes of ensuring coordination with permitting activities. |
| | | (c) Based on the information described in paragraph (4)(a) or (4)(b) or any other relevant information received from any source, the Regional Administrator will make a preliminary determination as to whether the reported information requires Agency action to protect human health or the environment. Further action could include suspending or revoking the exclusion, or other appropriate response necessary to protect human health and the environment. |
| | | (5) Delisting Levels: All total constituent concentrations in treated effluents managed under this exclusion must be equal to or less than the following levels, expressed as mg/L: |
| | | Inorganic Constituents |
| | | Ammonia—6.0 |
| | | Barium—1.6 |
| | | Beryllium—$4.5 \times 10^{-2}$ |

TABLE 2—WASTES EXCLUDED FROM SPECIFIC SOURCES—Continued

| Facility | Address | Waste description |
|---|---|---|
| | | Nickel—$4.5 \times 10^{-1}$<br>Silver—$1.1 \times 10^{-1}$<br>Vanadium—$1.6 \times 10^{-1}$<br>Zinc—6.8<br>Arsenic—$1.5 \times 10^{-2}$<br>Cadmium—$1.1 \times 10^{-2}$<br>Chromium—$6.8 \times 10^{-2}$<br>Lead—$9.0 \times 10^{-2}$<br>Mercury—$6.8 \times 10^{-3}$<br>Selenium—$1.1 \times 10^{-1}$<br>Fluoride—1.2<br>Cyanides—$4.8 \times 10^{-1}$<br><br>Organic Constituents:<br><br>Cresol—1.2<br>2,4,6 Trichlorophenol—$3.6 \times 10^{-1}$<br>Benzene—$6.0 \times 10^{-2}$<br>Chrysene—$5.6 \times 10^{-1}$<br>Hexachlorobenzne—$2.0 \times 10^{-3}$<br>Hexachlorocyclopentadiene—$1.8 \times 10^{-1}$<br>Dichloroisopropyl ether<br>[Bis(2-Chloroisopropyl) either]—$6.0 \times 10^{-2}$<br>Di-n-octylphthalate—$4.8 \times 10^{-1}$<br>1-Butanol—2.4<br>Isophorone—4.2<br>Diphenylamine—$5.6 \times 10^{-1}$<br>p-Chloroaniline—$1.2 \times 10^{-1}$<br>Acetonitrile—1.2<br>Carbazole—$1.8 \times 10^{-1}$<br>N-Nitrosodimethylamine—$2.0 \times 10^{-2}$<br>Pyridine—$2.4 \times 10^{-2}$<br>Lindane [gamma-BHC]—$3.0 \times 10^{-3}$<br>Arochlor [total of Arochlors 1016, 1221, 1232, 1242, 1248, 1254, 1260]—$5.0 \times 10^{-4}$<br>Carbon tetrachloride—$1.8 \times 10^{-2}$<br>Tetrahydrofuran—$5.6 \times 10^{-1}$<br>Acetone—2.4<br>Carbon disulfide—2.3<br>Tributyl phosphate—$1.2 \times 10^{-1}$<br>(6) Recordkeeping and Data Submittals.<br>(a) Energy shall maintain records of all waste characterization, and waste processing strategies required by Condition (1), and verification sampling data, including QA/QC results, in the facility operating record for a period of no less than three (3) years. However, this period is automatically extended during the course of any unresolved enforcement action regarding the 200 Area ETF or as requested by EPA.<br>(b) No less than thirty (30) days after receipt of verification data indicating a failure to meet delisting criteria of Condition (5), Energy shall notify the Regional Administrator. This notification shall include a summary of waste characterization data for the associated influent, verification data, and any corrective actions taken according to Condition (3)(b)(i).<br>(c) Records required by Condition (6)(a) must be furnished on request by EPA or the State of Washington and made available for inspection. All data must be accompanied by a signed copy of the following certification statement to attest to the truth and accuracy of the data submitted:<br>"Under civil and criminal penalty of law for the making or submission of false or fraudulent statements or representations (pursuant to the applicable provisions of the Federal Code, which include, but may not be limited to, 18 U.S.C. 1001 and 42 U.S.C. 6928). I certify that the information contained in or accompanying this document is true, accurate, and complete.<br>As to the (those) identified section(s) of the document for which I cannot personally verify its (their) truth and accuracy, I certify as the official having supervisory responsibility of the persons who, acting under my direct instructions, made the verification that this information is true, accurate, and complete.<br>In the event that any of this information is determined by EPA in its sole discretion to be false, inaccurate, or incomplete, and upon conveyance of this fact to Energy, I recognize and agree that this exclusion of waste will be void as if it never had effect to the extent directed by EPA and that the Energy will be liable for Energy's reliance on the void exclusion." |

TABLE 2—WASTES EXCLUDED FROM SPECIFIC SOURCES—Continued

| Facility | Address | Waste description |
|---|---|---|
| | | (7) *Treated Effluent Disposal Requirements.* Energy may at any time propose alternate reuse practices for treated effluent managed under terms of this exclusion in lieu of disposal at the SALDS. Such proposals must be in writing to the Regional Administrator, and demonstrate that the risks and potential human health or environmental exposures from alternate treated effluent disposal or reuse practices do not warrant retaining the waste as a hazardous waste. Upon written approval by EPA of such a proposal, non-hazardous treated effluents may be managed according to the proposed alternate practices in lieu of the SALDS disposal requirement in paragraph (3)(a). The effect of such approved proposals shall be explicitly limited to approving alternate disposal practices in lieu of the requirements in paragraph (3)(a) to dispose of treated effluent in SALDS. |
| USX Steel Corporation, USS Division, Southworks Plant, Gary Works. | Chicago, Illinois. | Fully-cured chemically stabilized electric arc furnace dust/sludge (CSEAFD) treatment residue (EPA Hazardous Waste No. K061) generated from the primary production of steel after April 29, 1991. This exclusion (for 35,000 tons of CSEAFD per year) is conditioned upon the data obtained from USX's full-scale CSEAFD treatment facility. To ensure that hazardous constituents are not present in the waste at levels of regulatory concern once the full-scale treatment facility is in operation, USX must implement a testing program for the petitioned waste. This testing program must meet the following conditions for the exclusion to be valid: |
| | | (1) *Testing:* Sample collection and analyses (including quality control (QC) procedures) must be performed using appropriate methods. As applicable to the method-defined parameters of concern, analyses requiring the use of SW–846 methods incorporated by reference in 40 CFR 260.11 must be used without substitution. As applicable, the SW–846 methods might include Methods 0010, 0011, 0020, 0023A, 0030, 0031, 0040, 0050, 0051, 0060, 0061,1010A, 1020B, 1110A, 1310B, 1311, 1312, 1320, 1330A, 9010C, 9012B, 9040C, 9045D, 9060A, 9070A (uses EPA Method 1664, Rev. A), 9071B, and 9095B. |
| | | (A) *Initial Testing:* During the first four weeks of operation of the full-scale treatment system, USX must collect representative grab samples of each treated batch of the CSEAFD and composite the grab samples daily. The daily composites, prior to disposal, must be analyzed for the EP leachate concentrations of all the EP toxic metals, nickel, and cyanide (using distilled water in the cyanide extractions). USX must report the analytical test data, including quality control information, obtained during this initial period no later than 90 days after the treatment of the first full-scale batch. |
| | | (B) *Subsequent Testing:* USX must collect representative grab samples from every treated batch of CSEAFD generated daily and composite all of the grab samples to produce a weekly composite sample. USX then must analyze each weekly composite sample for all of the EP toxic metals, and nickel. The analytical data, including quality control information, must be compiled and maintained on site for a minimum of three years. These data must be furnished upon request and made available for inspection by any employee or representative of EPA or the State of Illinois. |
| | | (2) *Delisting levels:* If the EP extract concentrations for chromium, lead, arsenic, or silver exceed 0.315 mg/l; for barium exceeds 6.3 mg/l; for cadmium or selenium exceed 0.063 mg/l; for mercury exceeds 0.0126 mg/l; for nickel exceeds 3.15 mg/l; or for cyanide exceeds 4.42 mg/l, the waste must either be re-treated until it meets these levels or managed and disposed in accordance with subtitle C of RCRA. |
| | | (3) *Data submittals:* Within one week of system start-up USX must notify the Section Chief, Delisting Section (see address below) when their full-scale stabilization system is on-line and waste treatment has begun. The data obtained through condition (1)(A) must be submitted to the Section Chief, Delisting Section, CAD/OSW (OS–333), U.S. EPA, 1200 Pennsylvania Ave., NW., Washington, DC 20460 within the time period specified. At the Section Chief's request, USX must submit any other analytical data obtained through conditions (1)(A) or (1)(B) within the time period specified by the Section Chief. Failure to submit the required data obtained from conditions (1)(A) or (1)(B) within the specified time period or maintain the required records for the specified time will be considered by the Agency, at its discretion, sufficient basis to revoke USX's exclusion to the extent directed by EPA. All data must be accompanied by the following certification statement: "Under civil and criminal penalty of law for the making or submission of false or fraudulent statements or representations (pursuant to the applicable provisions of the Federal Code which include, but may not be limited to, 18 U.S.C. §6928), I certify that the information contained in or accompanying this document is true, accurate and complete. As to the (those) identified section(s) of this document for which I cannot personally verify its (their) truth and accuracy, I certify as the company official having supervisory responsibility for the persons who, acting under my direct instructions, made the verification that this information is true, accurate and complete. In the event that any of this information is determined by EPA in its sole discretion to be false, inaccurate or incomplete, and upon conveyance of this fact to the company, I recognize and agree that this exclusion of wastes will be void as if it never had effect or to the extent directed by EPA and that the company will be liable for any actions taken in contravention of the company's RCRA and CERCLA obligations premised upon the company's reliance on the void exclusion." |

TABLE 3—WASTES EXCLUDED FROM COMMERCIAL CHEMICAL PRODUCTS, OFF-SPECIFICATION
SPECIES, CONTAINER RESIDUES, AND SOIL RESIDUES THEREOF

| Facility | Address | Waste description |
|---|---|---|
| Eastman Chemical Company. | Longview, Texas. | Wastewater treatment sludge, (at a maximum generation of 82,100 cubic yards per calendar year) generated by Eastman (EPA Hazardous Waste Nos. U001, U002, U028, U031, U069, U088, U112, U115, U117, U122, U140, U147, U154, U159, U161, U220, U226, U239, U359). Eastman must implement the testing program described in Table 1. Waste Excluded From Non-Specific Sources for the petition to be valid. |
| Rhodia ............ | Houston, Texas. | Filter-cake Sludge, (at a maximum generation of 1,200 cubic yards per calendar year) generated by Rhodia using the SARU and AWT treatment process to treat the filter-cake sludge (EPA Hazardous Waste Nos. P001–P024, P026-P031, P033–P034, P036–P051, P054, P056-P060, P062–P078, P081–P082, P084–P085, P087–P089, P092–P116, P118–P123, P127-P128, P185, P188–P192, P194, P196–P199, P201–P205, U001–U012, U014–U039, U041-U053, U055–U064, U066–U099, U101–U103, U105–U138, U140–U174, U176–U194, U196-U197, U200–U211, U213–U223, U225–U228, U234–U240, U243–U244, U246–U249, U271, U277–U280, U328, U353, U359, U364–U367, U372–U373, U375–U379, U381–U396, U400-U404, U407, U409–U411) generated at Rhodia. Rhodia must implement the testing program described in Table 1. Waste Excluded From Non-Specific Sources for the petition to be valid. |
| Texas Eastman | Longview, Texas. | Incinerator ash (at a maximum generation of 7,000 cubic yards per calendar year) generated from the incineration of sludge from the wastewater treatment plant (EPA Hazardous Waste No. U001, U002, U003, U019, U028, U031, U037, U044, U056, U069, U070, U107, U108, U112, U113, U115, U117, U122, U140, U147, U151, U154, U159, U161, U169, U190, U196, U211, U213, U226, U239, and U359, and that is disposed of in Subtitle D landfills after September 25, 1996. Texas Eastman must implement the testing program described in Table 1. Wastes Excluded From Non-Specific Sources for the petition to be valid. |
| Union Carbide Corp. | Taft, LA ......... | Contaminated soil (approximately 11,000 cubic yards), which contains acrolein in concentrations of less than 9 ppm. |

[49 FR 37070, Sept. 21, 1984]

EDITORIAL NOTE: For FEDERAL REGISTER citations affecting appendix IX of part 261, see the List of CFR Sections Affected, which appears in the Finding Aids section of the printed volume and on GPO Access.

# PART 262—STANDARDS APPLICABLE TO GENERATORS OF HAZARDOUS WASTE

## Subpart A—General

Sec. .
262.10   Purpose, scope, and applicability.
262.11   Hazardous waste determination.
262.12   EPA identification numbers.

## Subpart B—The Manifest

262.20   General requirements.
262.21   Manifest tracking numbers, manifest printing, and obtaining manifests.
262.22   Number of copies.
262.23   Use of the manifest.
262.27   Waste minimization certification.

## Subpart C—Pre-Transport Requirements

262.30   Packaging.
262.31   Labeling.
262.32   Marking.
262.33   Placarding.
262.34   Accumulation time.

## Subpart D—Recordkeeping and Reporting

262.40   Recordkeeping.

262.41   Biennial report.
262.42   Exception reporting.
262.43   Additional reporting.
262.44   Special requirements for generators of between 100 and 1000 kg/mo.

## Subpart E—Exports of Hazardous Waste

262.50   Applicability.
262.51   Definitions.
262.52   General requirements.
262.53   Notification of intent to export.
262.54   Special manifest requirements.
262.55   Exception reports.
262.56   Annual reports.
262.57   Recordkeeping.
262.58   International agreements.

## Subpart F—Imports of Hazardous Waste

262.60   Imports of hazardous waste.

## Subpart G—Farmers

262.70   Farmers.

## Subpart H—Transfrontier shipments of hazardous waste for recovery within the OECD

262.80   Applicability.

AUTHORITY: 42 U.S.C 6906, 6912, 6922–6925, 6937, and 6938.

SOURCE: 45 FR 33142, May 19, 1980, unless otherwise noted.

## Subpart A—General

### § 262.10   Purpose, scope, and applicability.

(a) These regulations establish standards for generators of hazardous waste.

(b) 40 CFR 261.5(c) and (d) must be used to determine the applicability of provisions of this part that are dependent on calculations of the quantity of hazardous waste generated per month.

(c) A generator who treats, stores, or disposes of hazardous waste on-site must only comply with the following sections of this part with respect to that waste: Section 262.11 for determining whether or not he has a hazardous waste, § 262.12 for obtaining an EPA identification number, § 262.34 for accumulation of hazardous waste, § 262.40 (c) and (d) for recordkeeping, § 262.43 for additional reporting, and if applicable, § 262.70 for farmers.

(d) Any person who exports or imports hazardous waste subject to the Federal manifesting requirements of part 262, or subject to the universal waste management standards of 40 CFR Part 273, or subject to State requirements analogous to 40 CFR Part 273, to or from the countries listed in § 262.58(a)(1) for recovery must comply with subpart H of this part.

(e) Any person who imports hazardous waste into the United States must comply with the standards applicable to generators established in this part.

(f) A farmer who generates waste pesticides which are hazardous waste and who complies with all of the requirements of §262.70 is not required to comply with other standards in this part or 40 CFR parts 270, 264, 265, 267, or 268 with respect to such pesticides.

(g) A person who generates a hazardous waste as defined by 40 CFR part 261 is subject to the compliance requirements and penalties prescribed in section 3008 of the Act if he does not comply with the requirements of this part.

(h) An owner or operator who initiates a shipment of hazardous waste from a treatment, storage, or disposal facility must comply with the generator standards established in this part.

(i) Persons responding to an explosives or munitions emergency in accordance with 40 CFR 264.1(g)(8)(i)(D) or (iv) or 265.1(c)(11)(i)(D) or (iv), and 270.1(c)(3)(i)(D) or (iii) are not required to comply with the standards of this part.

(j)(1) Universities that are participating in the Laboratory XL project are the University of Massachusetts Boston in Boston, Massachusetts, Boston College in Chestnut Hill, Massachusetts, and the University of Vermont in Burlington, Vermont ("Universities"). The Universities generate laboratory wastes (as defined in §262.102), some of which will be hazardous wastes. As long as the Universities comply with all the requirements of subpart J of this part the Universities' laboratories that are participating in the University Laboratories XL Project as identified in Table 1 of this section, are not subject to the provisions of §§262.11, 262.34(c), 40 CFR parts 264 and 265, 267, and the permit requirements of 40 CFR part 270 with respect to said laboratory wastes.

TABLE 1—LABORATORY XL PROJECT PARTICIPANT INFORMATION

| Institution | Approx. number of labs | Departments participating | Location of current hazardous waste accumulation areas |
|---|---|---|---|
| Boston College, Chestnut Hill, MA. | 120 | Chemistry, Biology, Geology, Physics, Psychology. | Merkert Chemistry Building, 2609 Beacon St., Boston, MA, Higgins Building, 140 Commonwealth Ave., Chestnut Hill, MA. |
| University of Massachusetts Boston, Boston, MA. | 150 | Chemistry, Biology, Psychology, Anthropology, Geology and Earth Sciences, and Environmental, Coastal and Ocean Sciences. | Science Building (Bldg. #080); McCormack Building (Bldg. #020); and Wheatley Building (Bldg. #010), 100 Morrissey Blvd., Boston, MA. |
| University of Vermont, Burlington, VT. | 400 | Colleges of: Agriculture and Life Sciences, Arts and Sciences, Medicine, and Engineering and Mathematics; and Schools of: Nursing, Allied Heath Sciences, and Natural Resources. | Given Bunker, 89 Beaumont Ave., Burlington, VT. |

(2) Each University shall have the right to change its respective departments or the on-site location of its hazardous waste accumulation areas listed in Table 1 of this section upon written notice to the Regional Administrator for EPA-Region I and the appropriate state agency. Such written notice will be provided at least ten days prior to the effective date of any such changes.

(k) Generators in the Commonwealth of Massachusetts may comply with the State regulations regarding Class A recyclable materials in 310 C.M.R. 30.200, when authorized by the EPA under 40 CFR part 271, with respect to those recyclable materials and matters covered by the authorization, instead of complying with the hazardous waste accumulation requirements of §262.34, the reporting requirements of §262.41, the storage facility operator requirements of 40 CFR parts 264, 265 and 267, and the permitting requirements of 40 CFR part 270. Such generators must also comply with any other applicable requirements, including any applicable authorized State regulations governing hazardous wastes not being recycled and any applicable Federal requirements which are being directly implemented by the EPA within Massachusetts pursuant to the Hazardous and Solid Waste Amendments of 1984.

(l) The laboratories owned by an eligible academic entity that chooses to be subject to the requirements of Subpart K of this part are not subject to (for purposes of this paragraph, the terms "laboratory" and "eligible academic entity" shall have the meaning as defined in § 262.200 of Subpart K of this part).:

(1) The requirements of § 262.11 or § 262.34(c), for large quantity generators and small quantity generators, except as provided in Subpart K, and

(2) The conditions of § 261.5(b), for conditionally exempt small quantity generators, except as provided in Subpart K.

NOTE 1: The provisions of § 262.34 are applicable to the on-site accumulation of hazardous waste by generators. Therefore, the provisions of § 262.34 only apply to owners or operators who are shipping hazardous waste which they generated at that facility.

NOTE 2: A generator who treats, stores, or disposes of hazardous waste on-site must comply with the applicable standards and permit requirements set forth in 40 CFR parts 264, 265, 266, 268, and 270.

[45 FR 33142, May 19, 1980, as amended at 45 FR 86970, Dec. 31, 1980; 47 FR 1251, Jan. 11, 1982; 48 FR 14294, Apr. 1, 1983; 53 FR 27164, July 19, 1988; 56 FR 3877, Jan. 31, 1991; 60 FR 25541, May 11, 1995; 61 FR 16309, Apr. 12, 1996; 62 FR 6651, Feb. 12, 1997; 64 FR 52392, Sept. 28, 1999; 69 FR 11813, Mar. 12, 2004; 73 FR 72954, Dec. 1, 2008; 75 FR 13003, Mar. 18, 2010]

EFFECTIVE DATE NOTE: At 75 FR 1253, Jan. 8, 2010, § 262.10 was amended by revising paragraph (d), effective July 7, 2010. For the convenience of the user, the revised text is set forth as follows:

§ 262.10  Purpose, scope, and applicability.

*       *       *       *       *

(d) Any person who exports or imports wastes that are considered hazardous under U.S. national procedures to or from the countries listed in § 262.58(a)(1) for recovery must comply with subpart H of this part. A waste is considered hazardous under U.S. national procedures if the waste meets the Federal definition of hazardous waste in 40 CFR 261.3 and is subject to either the Federal RCRA manifesting requirements at 40 CFR part 262, subpart B, the universal waste management standards of 40 CFR part 273, State requirements analogous to 40 CFR part 273, the export requirements in the spent lead-acid battery management standards of 40 CFR part 266, subpart G, or State requirements analogous to the export requirements in 40 CFR part 266, subpart G.

*       *       *       *       *

§ 262.11  Hazardous waste determination.

A person who generates a solid waste, as defined in 40 CFR 261.2, must determine if that waste is a hazardous waste using the following method:

(a) He should first determine if the waste is excluded from regulation under 40 CFR 261.4.

(b) He must then determine if the waste is listed as a hazardous waste in subpart D of 40 CFR part 261.

NOTE: Even if the waste is listed, the generator still has an opportunity under 40 CFR 260.22 to demonstrate to the Administrator that the waste from his particular facility or operation is not a hazardous waste.

(c) For purposes of compliance with 40 CFR part 268, or if the waste is not listed in subpart D of 40 CFR part 261, the generator must then determine whether the waste is identified in subpart C of 40 CFR part 261 by either:

(1) Testing the waste according to the methods set forth in subpart C of 40 CFR part 261, or according to an equivalent method approved by the Administrator under 40 CFR 260.21; or

(2) Applying knowledge of the hazard characteristic of the waste in light of the materials or the processes used.

(d) If the waste is determined to be hazardous, the generator must refer to parts 261, 264, 265, 266, 267, 268, and 273 of this chapter for possible exclusions or restrictions pertaining to management of the specific waste.

[45 FR 33142, May 19, 1980, as amended at 45 FR 76624, Nov. 19, 1980; 51 FR 40637, Nov. 7, 1986; 55 FR 22684, June 1, 1990; 56 FR 3877, Jan. 31, 1991; 60 FR 25541, May 11, 1995; 75 FR 13004, Mar. 18, 2010]

§ 262.12  EPA identification numbers.

(a) A generator must not treat, store, dispose of, transport, or offer for transportation, hazardous waste without having received an EPA identification number from the Administrator.

(b) A generator who has not received an EPA identification number may obtain one by applying to the Administrator using EPA form 8700–12. Upon receiving the request the Administrator will assign an EPA identification number to the generator.

(c) A generator must not offer his hazardous waste to transporters or to treatment, storage, or disposal facilities that have not received an EPA identification number.

## Subpart B—The Manifest

### § 262.20  General requirements.

(a)(1) A generator who transports, or offers for transport a hazardous waste for offsite treatment, storage, or disposal, or a treatment, storage, and disposal facility who offers for transport a rejected hazardous waste load, must prepare a Manifest (OMB Control number 2050–0039) on EPA Form 8700–22, and, if necessary, EPA Form 8700–22A, according to the instructions included in the appendix to this part.

(2) The revised manifest form and procedures in 40 CFR 260.10, 261.7, 262.20, 262.21, 262.27, 262.32, 262.34, 262.54, 262.60, and the appendix to part 262, shall not apply until September 5, 2006. The manifest form and procedures in 40 CFR 260.10, 261.7, 262.20, 262.21, 262.32, 262.34, 262.54, 262.60, and the Appendix to part 262, contained in the 40 CFR, parts 260 to 265, edition revised as of July 1, 2004, shall be applicable until September 5, 2006.

(b) A generator must designate on the manifest one facility which is permitted to handle the waste described on the manifest.

(c) A generator may also designate on the manifest one alternate facility which is permitted to handle his waste in the event an emergency prevents delivery of the waste to the primary designated facility.

(d) If the transporter is unable to deliver the hazardous waste to the designated facility or the alternate facility, the generator must either designate another facility or instruct the transporter to return the waste.

(e) The requirements of this subpart do not apply to hazardous waste produced by generators of greater than 100 kg but less than 1000 kg in a calendar month where:

(1) The waste is reclaimed under a contractual agreement pursuant to which:

(i) The type of waste and frequency of shipments are specified in the agreement;

(ii) The vehicle used to transport the waste to the recycling facility and to deliver regenerated material back to the generator is owned and operated by the reclaimer of the waste; and

(2) The generator maintains a copy of the reclamation agreement in his files for a period of at least three years after termination or expiration of the agreement.

(f) The requirements of this subpart and §262.32(b) do not apply to the transport of hazardous wastes on a public or private right-of-way within or along the border of contiguous property under the control of the same person, even if such contiguous property is divided by a public or private right-of-way. Notwithstanding 40 CFR 263.10(a), the generator or transporter must comply with the requirements for transporters set forth in 40 CFR 263.30 and 263.31 in the event of a discharge of hazardous waste on a public or private right-of-way.

[45 FR 33142, May 19, 1980, as amended at 49 FR 10500, Mar. 20, 1984; 51 FR 10175, Mar. 24, 1986; 53 FR 45090, Nov. 8, 1988; 62 FR 6651, Feb. 12, 1997; 70 FR 10815, Mar. 4, 2005; 70 FR 35037, June 16, 2005]

### § 262.21  Manifest tracking numbers, manifest printing, and obtaining manifests.

(a)(1) A registrant may not print, or have printed, the manifest for use of distribution unless it has received approval from the EPA Director of the Office of Resource Conservation and Recovery to do so under paragraphs (c) and (e) of this section.

(2) The approved registrant is responsible for ensuring that the organizations identified in its application are in compliance with the procedures of its approved application and the requirements of this section. The registrant is responsible for assigning manifest tracking numbers to its manifests.

(b) A registrant must submit an initial application to the EPA Director of the Office of Resource Conservation and Recovery that contains the following information:

(1) Name and mailing address of registrant;

(2) Name, telephone number and email address of contact person;

(3) Brief description of registrant's government or business activity;

(4) EPA identification number of the registrant, if applicable;

(5) Description of the scope of the operations that the registrant plans to undertake in printing, distributing, and using its manifests, including:

(i) A description of the printing operation. The description should include an explanation of whether the registrant intends to print its manifests in-house (*i.e.*, using its own printing establishments) or through a separate (*i.e.*, unaffiliated) printing company. If the registrant intends to use a separate printing company to print the manifest on its behalf, the application must identify this printing company and discuss how the registrant will oversee the company. If this includes the use of intermediaries (e.g., prime and subcontractor relationships), the role of each must be discussed. The application must provide the name and mailing address of each company. It also must provide the name and telephone number of the contact person at each company.

(ii) A description of how the registrant will ensure that its organization and unaffiliated companies, if any, comply with the requirements of this section. The application must discuss how the registrant will ensure that a unique manifest tracking number will be pre-printed on each manifest. The application must describe the internal control procedures to be followed by the registrant and unaffiliated companies to ensure that numbers are tightly controlled and remain unique. In particular, the application must describe how the registrant will assign manifest tracking numbers to its manifests. If computer systems or other infrastructure will be used to maintain, track, or assign numbers, these should be indicated. The application must also indicate how the printer will pre-print a unique number on each form (e.g., crash or press numbering). The applica-

tion also must explain the other quality procedures to be followed by each establishment and printing company to ensure that all required print specifications are consistently achieved and that printing violations are identified and corrected at the earliest practicable time.

(iii) An indication of whether the registrant intends to use the manifests for its own business operations or to distribute the manifests to a separate company or to the general public (e.g., for purchase).

(6) A brief description of the qualifications of the company that will print the manifest. The registrant may use readily available information to do so (e.g., corporate brochures, product samples, customer references, documentation of ISO certification), so long as such information pertains to the establishments or company being proposed to print the manifest.

(7) Proposed unique three-letter manifest tracking number suffix. If the registrant is approved to print the manifest, the registrant must use this suffix to pre-print a unique manifest tracking number on each manifest.

(8) A signed certification by a duly authorized employee of the registrant that the organizations and companies in its application will comply with the procedures of its approved application and the requirements of this section and that it will notify the EPA Director of the Office of Resource Conservation and Recovery of any duplicated manifest tracking numbers on manifests that have been used or distributed to other parties as soon as this becomes known.

(c) EPA will review the application submitted under paragraph (b) of this section and either approve it or request additional information or modification before approving it.

(d)(1) Upon EPA approval of the application under paragraph (c) of this section, EPA will provide the registrant an electronic file of the manifest, continuation sheet, and manifest instructions and ask the registrant to submit three fully assembled manifests and continuation sheet samples, except as noted in paragraph (d)(3) of this section. The registrant's samples must meet all of the specifications in paragraph (f) of this section and be printed by the company that will print the manifest as identified in the application approved under paragraph (c) of this section.

(2) The registrant must submit a description of the manifest samples as follows:

(i) Paper type (*i.e.*, manufacturer and grade of the manifest paper);

(ii) Paper weight of each copy;

(iii) Ink color of the manifest's instructions. If screening of the ink was used, the registrant must indicate the extent of the screening; and

(iv) Method of binding the copies.

(3) The registrant need not submit samples of the continuation sheet if it will print its continuation sheet using the same paper type, paper weight of each copy, ink color of the instructions, and binding method as its manifest form samples.

(e) EPA will evaluate the forms and either approve the registrant to print them as proposed or request additional information or modification to them before approval. EPA will notify the registrant of its decision by mail. The registrant cannot use or distribute its forms until EPA approves them. An approved registrant must print the manifest and continuation sheet according to its application approved under paragraph (c) of this section and the manifest specifications in paragraph (f) of this section. It also must print the forms according to the paper type, paper weight, ink color of the manifest instructions and binding method of its approved forms.

(f) Paper manifests and continuation sheets must be printed according to the following specifications:

(1) The manifest and continuation sheet must be printed with the exact format and appearance as EPA Forms 8700–22 and 8700–22A, respectively. However, information required to complete the manifest may be pre-printed on the manifest form.

(2) A unique manifest tracking number assigned in accordance with a numbering system approved by EPA must be pre-printed in Item 4 of the manifest. The tracking number must consist of a unique three-letter suffix following nine digits.

(3) The manifest and continuation sheet must be printed on 8½ × 11-inch white paper, excluding common stubs (e.g., top- or side-bound stubs). The paper must be durable enough to withstand normal use.

(4) The manifest and continuation sheet must be printed in black ink that can be legibly photocopied, scanned, and faxed, except that the marginal words indicating copy distribution must be in red ink.

(5) The manifest and continuation sheet must be printed as six-copy forms. Copy-to-copy registration must be exact within ⅟₃₂nd of an inch. Handwritten and typed impressions on the form must be legible on all six copies. Copies must be bound together by one or more common stubs that reasonably ensure that they will not become detached inadvertently during normal use.

(6) Each copy of the manifest and continuation sheet must indicate how the copy must be distributed, as follows:

(i) Page 1 (top copy): "Designated facility to destination State (if required)".

(ii) Page 2: "Designated facility to generator State (if required)".

(iii) Page 3: "Designated facility to generator".

(iv) Page 4: "Designated facility's copy".

(v) Page 5: "Transporter's copy".

(vi) Page 6 (bottom copy): "Generator's initial copy".

(7) The instructions in the appendix to 40 CFR part 262 must appear legibly on the back of the copies of the manifest and continuation sheet as provided in this paragraph (f). The instructions must not be visible through the front of the copies when photocopied or faxed.

(i) Manifest Form 8700–22.

(A) The "Instructions for Generators" on Copy 6;

(B) The "Instructions for International Shipment Block" and "Instructions for Transporters" on Copy 5; and

(C) The "Instructions for Treatment, Storage, and Disposal Facilities" on Copy 4.

(ii) Manifest Form 8700–22A.

(A) The "Instructions for Generators" on Copy 6;

(B) The "Instructions for Transporters" on Copy 5; and

(C) The "Instructions for Treatment, Storage, and Disposal Facilities" on Copy 4.

(g)(1) A generator may use manifests printed by any source so long as the source of the printed form has received approval from EPA to print the manifest under paragraphs (c) and (e) of this section. A registered source may be a:

(i) State agency;

(ii) Commercial printer;

(iii) Hazardous waste generator, transporter or TSDF; or

(iv) Hazardous waste broker or other preparer who prepares or arranges shipments of hazardous waste for transportation.

(2) A generator must determine whether the generator state or the consignment state for a shipment regulates any additional wastes (beyond those regulated Federally) as hazardous wastes under these states' authorized programs. Generators also must determine whether the consignment state or generator state requires the generator to submit any copies of the manifest to these states. In cases where the generator must supply copies to either the generator's state or the consignment state, the generator is responsible for supplying legible photocopies of the manifest to these states.

(h)(1) If an approved registrant would like to update any of the information provided in its application approved under paragraph (c) of this section (e.g., to update a company phone number or name of contact person), the registrant must revise the application and submit it to the EPA Director of the Office of Resource Conservation and Recovery, along with an indication or explanation of the update, as soon as practicable after the change occurs. The Agency either will approve or deny the revision. If the Agency denies the revision, it will explain the reasons for the denial, and it will contact the registrant and request further modification before approval.

(2) If the registrant would like a new tracking number suffix, the registrant must submit a proposed suffix to the EPA Director of the Office of Resource Conservation and Recovery, along with the reason for requesting it. The Agency will either approve the suffix or deny the suffix and provide an explanation why it is not acceptable.

(3) If a registrant would like to change the paper type, paper weight, ink color of the manifest instructions, or binding method of its manifest or continuation sheet subsequent to approval under paragraph (e) of this section, then the registrant must submit three samples of the revised form for EPA review and approval. If the approved registrant would like to use a new printer, the registrant must submit three manifest samples printed by the new printer, along with a brief description of the printer's qualifications to print the manifest. EPA will evaluate the manifests and either approve the registrant to print the forms as proposed or request additional information or modification to them before approval. EPA will notify the registrant of its decision by mail. The registrant cannot use or distribute its revised forms until EPA approves them.

(i) If, subsequent to its approval under paragraph (e) of this section, a registrant typesets its manifest or continuation sheet instead of using the electronic file of the forms provided by EPA, it must submit three samples of the manifest or continuation sheet to the registry for approval. EPA will evaluate the manifests or continuation sheets and either approve the registrant to print them as proposed or request additional information or modification to them before approval. EPA will notify the registrant of its decision by mail. The registrant cannot use or distribute its typeset forms until EPA approves them.

(j) EPA may exempt a registrant from the requirement to submit form samples under paragraph (d) or (h)(3) of this section if the Agency is persuaded that a separate review of the registrant's forms would serve little purpose in informing an approval decision (e.g., a registrant certifies that it will print the manifest using the same paper type, paper weight, ink color of the instructions and binding method of the form samples approved for some other registrant). A registrant may request an exemption from EPA by indicating why an exemption is warranted.

(k) An approved registrant must notify EPA by phone or email as soon as it becomes aware that it has duplicated tracking numbers on any manifests that have been used or distributed to other parties.

(l) If, subsequent to approval of a registrant under paragraph (e) of this section, EPA becomes aware that the approved paper type, paper weight, ink color of the instructions, or binding method of the registrant's form is unsatisfactory, EPA will contact the registrant and require modifications to the form.

(m)(1) EPA may suspend and, if necessary, revoke printing privileges if we find that the registrant:

(i) Has used or distributed forms that deviate from its approved form samples in regard to paper weight, paper type, ink color of the instructions, or binding method; or

(ii) Exhibits a continuing pattern of behavior in using or distributing manifests that contain duplicate manifest tracking numbers.

(2) EPA will send a warning letter to the registrant that specifies the date by which it must come into compliance with the requirements. If the registrant does not come in compliance by the specified date, EPA will send a second letter notifying the registrant that EPA has suspended or revoked its printing privileges. An approved registrant must provide information on its printing activities to EPA if requested.

[70 FR 10815, Mar. 4, 2005, as amended at 74 FR 30230, June 25, 2009]

## §262.22 Number of copies.

The manifest consists of at least the number of copies which will provide the generator, each transporter, and the owner or operator of the designated facility with one copy each for their records and another copy to be returned to the generator.

## §262.23 Use of the manifest.

(a) The generator must:

(1) Sign the manifest certification by hand; and

(2) Obtain the handwritten signature of the initial transporter and date of acceptance on the manifest; and

(3) Retain one copy, in accordance with §262.40(a).

(b) The generator must give the transporter the remaining copies of the manifest.

(c) For shipments of hazardous waste within the United States solely by water (bulk shipments only), the generator must send three copies of the manifest dated and signed in accordance with this section to the owner or operator of the designated facility or the last water (bulk shipment) transporter to handle the waste in the United States if exported by water. Copies of the manifest are not required for each transporter.

(d) For rail shipments of hazardous waste within the United States which originate at the site of generation, the generator must send at least three copies of the manifest dated and signed in accordance with this section to:

(1) The next non-rail transporter, if any; or

(2) The designated facility if transported solely by rail; or

(3) The last rail transporter to handle the waste in the United States if exported by rail.

(e) For shipments of hazardous waste to a designated facility in an authorized State which has not yet obtained authorization to regulate that particular waste as hazardous, the generator must assure that the designated facility agrees to sign and return the manifest to the generator, and that any out-of-state transporter signs and forwards the manifest to the designated facility.

NOTE: See §263.20(e) and (f) for special provisions for rail or water (bulk shipment) transporters.

(f) For rejected shipments of hazardous waste or container residues contained in non-empty containers that are returned to the generator by the designated facility (following the procedures of 40 CFR 264.72(f) or 265.72(f)), the generator must:

(1) Sign either:

(i) Item 20 of the new manifest if a new manifest is used for the returned shipment; or

273

(ii) Item 18c of the original manifest if the original manifest is used for the returned shipment;

(2) Provide the transporter a copy of the manifest;

(3) Within 30 days of delivery of the rejected shipment or container residues contained in non-empty containers, send a copy of the manifest to the designated facility that returned the shipment to the generator; and

(4) Retain at the generator's site a copy of each manifest for at least three years from the date of delivery.

[45 FR 33142, May 19, 1980, as amended at 45 FR 86973, Dec. 31, 1980; 55 FR 2354, Jan. 23, 1990; 75 FR 13004, Mar. 18, 2010]

### § 262.27 Waste minimization certification.

A generator who initiates a shipment of hazardous waste must certify to one of the following statements in Item 15 of the uniform hazardous waste manifest:

(a) "I am a large quantity generator. I have a program in place to reduce the volume and toxicity of waste generated to the degree I have determined to be economically practicable and I have selected the practicable method of treatment, storage, or disposal currently available to me which minimizes the present and future threat to human health and the environment;" or

(b) "I am a small quantity generator. I have made a good faith effort to minimize my waste generation and select the best waste management method that is available to me and that I can afford."

[70 FR 10817, Mar. 4, 2005]

## Subpart C—Pre-Transport Requirements

### § 262.30 Packaging.

Before transporting hazardous waste or offering hazardous waste for transportation off-site, a generator must package the waste in accordance with the applicable Department of Transportation regulations on packaging under 49 CFR parts 173, 178, and 179.

### § 262.31 Labeling.

Before transporting or offering hazardous waste for transportation off-site, a generator must label each package in accordance with the applicable Department of Transportation regulations on hazardous materials under 49 CFR part 172.

### § 262.32 Marking.

(a) Before transporting or offering hazardous waste for transportation off-site, a generator must mark each package of hazardous waste in accordance with the applicable Department of Transportation regulations on hazardous materials under 49 CFR part 172;

(b) Before transporting hazardous waste or offering hazardous waste for transportation off-site, a generator must mark each container of 119 gallons or less used in such transportation with the following words and information in accordance with the requirements of 49 CFR 172.304:

HAZARDOUS WASTE—Federal Law Prohibits Improper Disposal. If found, contact the nearest police or public safety authority or the U.S. Environmental Protection Agency.

Generator's Name and Address _____.
Generator's EPA Identification Number _____.

Manifest Tracking Number _____.

[45 FR 33142, May 19, 1980, as amended at 70 FR 10817, Mar. 4, 2005]

### § 262.33 Placarding.

Before transporting hazardous waste or offering hazardous waste for transportation off-site, a generator must placard or offer the initial transporter the appropriate placards according to Department of Transportation regulations for hazardous materials under 49 CFR part 172, subpart F.

[70 FR 35037, June 16, 2005]

### § 262.34 Accumulation time.

(a) Except as provided in paragraphs (d), (e), and (f) of this section, a generator may accumulate hazardous waste on-site for 90 days or less without a permit or without having interim status, provided that:

(1) The waste is placed:

(i) In containers and the generator complies with the applicable requirements of subparts I, AA, BB, and CC of 40 CFR part 265; and/or

(ii) In tanks and the generator complies with the applicable requirements of subparts J, AA, BB, and CC of 40 CFR part 265 except §§265.197(c) and 265.200; and/or

(iii) On drip pads and the generator complies with subpart W of 40 CFR part 265 and maintains the following records at the facility:

(A) A description of procedures that will be followed to ensure that all wastes are removed from the drip pad and associated collection system at least once every 90 days; and

(B) Documentation of each waste removal, including the quantity of waste removed from the drip pad and the sump or collection system and the date and time of removal; and/or

(iv) In containment buildings and the generator complies with subpart DD of 40 CFR part 265, has placed its professional engineer certification that the building complies with the design standards specified in 40 CFR 265.1101 in the facility's operating record no later than 60 days after the date of initial operation of the unit. After February 18, 1993, PE certification will be required prior to operation of the unit. The owner or operator shall maintain the following records at the facility:

(A) A written description of procedures to ensure that each waste volume remains in the unit for no more than 90 days, a written description of the waste generation and management practices for the facility showing that they are consistent with respecting the 90 day limit, and documentation that the procedures are complied with; or

(B) Documentation that the unit is emptied at least once every 90 days.

In addition, such a generator is exempt from all the requirements in subparts G and H of 40 CFR part 265, except for §§265.111 and 265.114.

(2) The date upon which each period of accumulation begins is clearly marked and visible for inspection on each container;

(3) While being accumulated on-site, each container and tank is labeled or marked clearly with the words, "Hazardous Waste"; and

(4) The generator complies with the requirements for owners or operators in subparts C and D in 40 CFR part 265, with §265.16, and with all applicable requirements under 40 CFR part 268.

(b) A generator of 1,000 kilograms or greater of hazardous waste in a calendar month, or greater than 1 kg of acute hazardous waste listed in §§261.31 or 261.33(e) in a calendar month, who accumulates hazardous waste or acute hazardous waste for more than 90 days is an operator of a storage facility and is subject to the requirements of 40 CFR parts 264, 265, and 267 and the permit requirements of 40 CFR part 270 unless he has been granted an extension to the 90-day period. Such extension may be granted by EPA if hazardous wastes must remain on-site for longer than 90 days due to unforeseen, temporary, and uncontrollable circumstances. An extension of up to 30 days may be granted at the discretion of the Regional Administrator on a case-by-case basis.

(c)(1) A generator may accumulate as much as 55 gallons of hazardous waste or one quart of acutely hazardous waste listed in §261.31 or §261.33(e) in containers at or near any point of generation where wastes initially accumulate which is under the control of the operator of the process generating the waste, without a permit or interim status and without complying with paragraph (a) or (d) of this section provided he:

(i) Complies with §§265.171, 265.172, and 265.173(a) of this chapter; and

(ii) Marks his containers either with the words "Hazardous Waste" or with other words that identify the contents of the containers.

(2) A generator who accumulates either hazardous waste or acutely hazardous waste listed in §261.31 or §261.33(e) in excess of the amounts listed in paragraph (c)(1) of this section at or near any point of generation must, with respect to that amount of excess waste, comply within three days with paragraph (a) of this section or other applicable provisions of this chapter. During the three day period the generator must continue to comply with paragraphs (c)(1)(i) and (ii) of this section. The generator must mark the container holding the excess accumulation of hazardous waste with the date the excess amount began accumulating.

(d) A generator who generates greater than 100 kilograms but less than 1000 kilograms of hazardous waste in a calendar month may accumulate hazardous waste on-site for 180 days or less without a permit or without having interim status provided that:

(1) The quantity of waste accumulated on-site never exceeds 6000 kilograms;

(2) The generator complies with the requirements of subpart I of part 265 of this chapter, except for §§ 265.176 and 265.178;

(3) The generator complies with the requirements of § 265.201 in subpart J of part 265;

(4) The generator complies with the requirements of paragraphs (a)(2) and (a)(3) of this section, the requirements of subpart C of part 265, with all applicable requirements under 40 CFR part 268; and

(5) The generator complies with the following requirements:

(i) At all times there must be at least one employee either on the premises or on call (*i.e.*, available to respond to an emergency by reaching the facility within a short period of time) with the responsibility for coordinating all emergency response measures specified in paragraph (d)(5)(iv) of this section. This employee is the emergency coordinator.

(ii) The generator must post the following information next to the telephone:

(A) The name and telephone number of the emergency coordinator;

(B) Location of fire extinguishers and spill control material, and, if present, fire alarm; and

(C) The telephone number of the fire department, unless the facility has a direct alarm.

(iii) The generator must ensure that all employees are thoroughly familiar with proper waste handling and emergency procedures, relevant to their responsibilities during normal facility operations and emergencies;

(iv) The emergency coordinator or his designee must respond to any emergencies that arise. The applicable responses are as follows:

(A) In the event of a fire, call the fire department or attempt to extinguish it using a fire extinguisher;

(B) In the event of a spill, contain the flow of hazardous waste to the extent possible, and as soon as is practicable, clean up the hazardous waste and any contaminated materials or soil;

(C) In the event of a fire, explosion, or other release which could threaten human health outside the facility or when the generator has knowledge that a spill has reached surface water, the generator must immediately notify the National Response Center (using their 24-hour toll free number 800/424–8802). The report must include the following information:

(*1*) The name, address, and U.S. EPA Identification Number of the generator;

(*2*) Date, time, and type of incident (e.g., spill or fire);

(*3*) Quantity and type of hazardous waste involved in the incident;

(*4*) Extent of injuries, if any; and

(*5*) Estimated quantity and disposition of recovered materials, if any.

(e) A generator who generates greater than 100 kilograms but less than 1000 kilograms of hazardous waste in a calendar month and who must transport his waste, or offer his waste for transportation, over a distance of 200 miles or more for off-site treatment, storage or disposal may accumulate hazardous waste on-site for 270 days or less without a permit or without having interim status provided that he complies with the requirements of paragraph (d) of this section.

(f) A generator who generates greater than 100 kilograms but less than 1000 kilograms of hazardous waste in a calendar month and who accumulates hazardous waste in quantities exceeding 6000 kg or accumulates hazardous waste for more than 180 days (or for more than 270 days if he must transport his waste, or offer his waste for transportation, over a distance of 200 miles or more) is an operator of a storage facility and is subject to the requirements of 40 CFR parts 264, 265 and 267, and the permit requirements of 40 CFR part 270 unless he has been granted an extension to the 180-day (or 270-day if applicable) period. Such extension may be granted by EPA if hazardous wastes must remain on-site for longer than 180 days (or 270 days if applicable) due to unforeseen, temporary, and uncontrollable circumstances. An extension of up to 30 days may be granted at the discretion of the Regional Administrator on a case-by-case basis.

(g) A generator who generates 1,000 kilograms or greater of hazardous waste per calendar month who also generates wastewater treatment sludges from electroplating operations that meet the listing description for the RCRA hazardous waste code F006, may accumulate F006 waste on-site for more than 90 days, but not more than 180 days without a permit or without having interim status provided that:

(1) The generator has implemented pollution prevention practices that reduce the amount of any hazardous substances, pollutants or contaminants entering F006 or otherwise released to the environment prior to its recycling;

(2) The F006 waste is legitimately recycled through metals recovery;

(3) No more than 20,000 kilograms of F006 waste is accumulated on-site at any one time; and

(4) The F006 waste is managed in accordance with the following:

(i) The F006 waste is placed:

(A) In containers and the generator complies with the applicable requirements of subparts I, AA, BB, and CC of 40 CFR part 265; and/or

(B) In tanks and the generator complies with the applicable requirements of subparts J, AA, BB, and CC of 40 CFR part 265, except §§ 265.197(c) and 265.200; and/or

(C) In containment buildings and the generator complies with subpart DD of 40 CFR part 265, and has placed its professional engineer certification that the building complies with the design standards specified in 40 CFR 265.1101 in the facility's operating record prior to operation of the unit. The owner or operator must maintain the following records at the facility:

(1) A written description of procedures to ensure that the F006 waste remains in the unit for no more than 180 days, a written description of the waste generation and management practices for the facility showing that they are consistent with the 180-day limit, and documentation that the generator is complying with the procedures; or

(2) Documentation that the unit is emptied at least once every 180 days.

(ii) In addition, such a generator is exempt from all the requirements in subparts G and H of 40 CFR part 265, except for §§ 265.111 and 265.114.

(iii) The date upon which each period of accumulation begins is clearly marked and visible for inspection on each container;

(iv) While being accumulated on-site, each container and tank is labeled or marked clearly with the words, "Hazardous Waste;" and

(v) The generator complies with the requirements for owners or operators in subparts C and D in 40 CFR part 265, with 40 CFR 265.16, and with 40 CFR 268.7(a)(5).

(h) A generator who generates 1,000 kilograms or greater of hazardous waste per calendar month who also generates wastewater treatment sludges from electroplating operations that meet the listing description for the RCRA hazardous waste code F006, and who must transport this waste, or offer this waste for transportation, over a distance of 200 miles or more for off-site metals recovery, may accumulate F006 waste on-site for more than 90 days, but not more than 270 days without a permit or without having interim status if the generator complies with the requirements of paragraphs (g)(1) through (g)(4) of this section.

(i) A generator accumulating F006 in accordance with paragraphs (g) and (h) of this section who accumulates F006 waste on-site for more than 180 days (or for more than 270 days if the generator must transport this waste, or offer this waste for transportation, over a distance of 200 miles or more), or who accumulates more than 20,000 kilograms of F006 waste on-site is an operator of a storage facility and is subject to the requirements of 40 CFR parts 264, 265 and 267, and the permit requirements of 40 CFR part 270 unless the generator has been granted an extension to the 180-day (or 270-day if applicable) period or an exception to the 20,000 kilogram accumulation limit. Such extensions and exceptions may be granted by EPA if F006 waste must remain on-site for longer than 180 days (or 270 days if applicable) or if more than 20,000 kilograms of F006 waste must remain on-site due to unforeseen, temporary, and uncontrollable circumstances. An extension of up to 30 days or an exception to the accumulation limit may be granted at the discretion of the Regional Administrator on a case-by-case basis.

(j) A member of the Performance Track Program who generates 1000 kg or greater of hazardous waste per month (or one kilogram or more of acute hazardous waste) may accumulate hazardous waste on-site without a permit or interim status for an extended period of time, provided that:

(1) The generator accumulates the hazardous waste for no more than 180 days, or for no more than 270 days if the generator must transport the waste (or offer the waste for transport) more than 200 miles from the generating facility; and

(2) The generator first notifies the Regional Administrator and the Director of the authorized State in writing of its intent to begin accumulation of hazardous waste for extended time periods under the provisions of this section. Such advance notice must include:

(i) Name and EPA ID number of the facility, and specification of when the facility will begin accumulation of hazardous wastes for extended periods of time in accordance with this section; and

(ii) A description of the types of hazardous wastes that will be accumulated for extended periods of time, and the units that will be used for such extended accumulation; and

(iii) A Statement that the facility has made all changes to its operations, procedures, including emergency preparedness procedures, and equipment, including equipment needed for emergency preparedness, that will be necessary to accommodate extended time periods for accumulating hazardous wastes; and

(iv) If the generator intends to accumulate hazardous wastes on-site for up to 270 days, a certification that a facility that is permitted (or operating under interim status) under part 270 of this chapter to receive these wastes is not available within 200 miles of the generating facility; and

(3) The waste is managed in:

(i) Containers, in accordance with the applicable requirements of subparts I, AA, BB, and CC of 40 CFR part 265 and 40 CFR 264.175; or

(ii) Tanks, in accordance with the applicable requirements of subparts J, AA, BB, and CC of 40 CFR part 265, except for §§ 265.197(c) and 265.200; or

(iii) Drip pads, in accordance with subpart W of 40 CFR part 265; or

(iv) Containment buildings, in accordance with subpart DD of 40 CFR part 265; and

(4) The quantity of hazardous waste that is accumulated for extended time periods at the facility does not exceed 30,000 kg; and

(5) The generator maintains the following records at the facility for each unit used for extended accumulation times:

(i) A written description of procedures to ensure that each waste volume remains in the unit for no more than 180 days (or 270 days, as applicable), a description of the waste generation and management practices at the facility showing that they are consistent with the extended accumulation time limit, and documentation that the procedures are complied with; or

(ii) Documentation that the unit is emptied at least once every 180 days (or 270 days, if applicable); and

(6) Each container or tank that is used for extended accumulation time periods is labeled or marked clearly with the words "Hazardous Waste," and for each container the date upon which each period of accumulation begins is clearly marked and visible for inspection; and

(7) The generator complies with the requirements for owners and operators in 40 CFR part 265, with § 265.16, and with § 268.7(a)(5). In addition, such a generator is exempt from all the requirements in subparts G and H of part 265, except for §§ 265.111 and 265.114; and

(8) The generator has implemented pollution prevention practices that reduce the amount of any hazardous substances, pollutants, or contaminants released to the environment prior to its recycling, treatment, or disposal; and

(9) The generator includes the following with its Performance Track Annual Performance Report, which must

be submitted to the Regional Administrator and the Director of the authorized State:

(i) Information on the total quantity of each hazardous waste generated at the facility that has been managed in the previous year according to extended accumulation time periods; and

(ii) Information for the previous year on the number of off-site shipments of hazardous wastes generated at the facility, the types and locations of destination facilities, how the wastes were managed at the destination facilities (e.g., recycling, treatment, storage, or disposal), and what changes in on-site or off-site waste management practices have occurred as a result of extended accumulation times or other pollution prevention provisions of this section; and

(iii) Information for the previous year on any hazardous waste spills or accidents occurring at extended accumulation units at the facility, or during off-site transport of accumulated wastes; and

(iv) If the generator intends to accumulate hazardous wastes on-site for up to 270 days, a certification that a facility that is permitted (or operating under interim status) under part 270 of this chapter to receive these wastes is not available within 200 miles of the generating facility; and

(k) If hazardous wastes must remain on-site at a Performance Track member facility for longer than 180 days (or 270 days, if applicable) due to unforseen, temporary, and uncontrollable circumstances, an extension to the extended accumulation time period of up to 30 days may be granted at the discretion of the Regional Administrator on a case-by-case basis.

(1) If a generator who is a member of the Performance Track Program withdraws from the Performance Track Program, or if the Regional Administrator terminates a generator's membership, the generator must return to compliance with all otherwise applicable hazardous waste regulations as soon as possible, but no later than six months after the date of withdrawal or termination.

(m) A generator who sends a shipment of hazardous waste to a designated facility with the understanding that the designated facility can accept and manage the waste and later receives that shipment back as a rejected load or residue in accordance with manifest discrepancy provisions of §264.72 or §265.72 of this chapter may accumulate the returned waste on-site in accordance with paragraphs (a) and (b) or (d), (e) and (f) of this section, depending on the amount of hazardous waste on-site in that calendar month. Upon receipt of the returned shipment, the generator must:

(1) Sign Item 18c of the manifest, if the transporter returned the shipment using the original manifest; or

(2) Sign Item 20 of the manifest, if the transporter returned the shipment using a new manifest.

[47 FR 1251, Jan. 11, 1982, as amended at 48 FR 14294, Apr. 1, 1983; 49 FR 49571, Dec. 20, 1984; 51 FR 10175, Mar. 24, 1986; 51 FR 25472, July 14, 1986; 55 FR 22684, June 1, 1990; 55 FR 50483, Dec. 6, 1990; 56 FR 3877, Jan. 31, 1991; 56 FR 30195, July 1, 1991; 57 FR 37264, Aug. 18, 1992; 59 FR 62926, Dec. 6, 1994; 61 FR 4911, Feb. 9, 1996; 61 FR 59950, Nov. 25, 1996; 64 FR 3388, Jan. 21, 1999; 64 FR 25414, May 11, 1999; 64 FR 56471, Oct. 20, 1999; 65 FR 12397, Mar. 8, 2000; 69 FR 21753, Apr. 22, 2004; 69 FR 62224, Oct. 25, 2004; 70 FR 10817, Mar. 4, 2005; 71 FR 40271, July 14, 2006; 75 FR 13004, Mar. 18, 2010]

## Subpart D—Recordkeeping and Reporting

### §262.40  Recordkeeping.

(a) A generator must keep a copy of each manifest signed in accordance with §262.23(a) for three years or until he receives a signed copy from the designated facility which received the waste. This signed copy must be retained as a record for at least three years from the date the waste was accepted by the initial transporter.

(b) A generator must keep a copy of each Biennial Report and Exception Report for a period of at least three years from the due date of the report.

(c) A generator must keep records of any test results, waste analyses, or other determinations made in accordance with §262.11 for at least three years from the date that the waste was last sent to on-site or off-site treatment, storage, or disposal.

(d) The periods or retention referred to in this section are extended automatically during the course of any unresolved enforcement action regarding the regulated activity or as requested by the Administrator.

[45 FR 33142, May 19, 1980, as amended at 48 FR 3981, Jan. 28, 1983]

### § 262.41  Biennial report.

(a) A generator who ships any hazardous waste off-site to a treatment, storage or disposal facility within the United States must prepare and submit a single copy of a Biennial Report to the Regional Administrator by March 1 of each even numbered year. The Biennial Report must be submitted on EPA Form 8700–13A, must cover generator activities during the previous year, and must include the following information:

(1) The EPA identification number, name, and address of the generator;

(2) The calendar year covered by the report;

(3) The EPA identification number, name, and address for each off-site treatment, storage, or disposal facility in the United States to which waste was shipped during the year;

(4) The name and EPA identification number of each transporter used during the reporting year for shipments to a treatment, storage or disposal facility within the United States;

(5) A description, EPA hazardous waste number (from 40 CFR part 261, subpart C or D), DOT hazard class, and quantity of each hazardous waste shipped off-site for shipments to a treatment, storage or disposal facility within the United States. This information must be listed by EPA identification number of each such off-site facility to which waste was shipped.

(6) A description of the efforts undertaken during the year to reduce the volume and toxicity of waste generated.

(7) A description of the changes in volume and toxicity of waste actually achieved during the year in comparison to previous years to the extent such information is available for years prior to 1984.

(8) The certification signed by the generator or authorized representative.

(b) Any generator who treats, stores, or disposes of hazardous waste on-site must submit a biennial report covering those wastes in accordance with the provisions of 40 CFR parts 270, 264, 265, 266, and 267. Reporting for exports of hazardous waste is not required on the Biennial Report form. A separate annual report requirement is set forth at 40 CFR 262.56.

[48 FR 3981, Jan. 28, 1983, as amended at 48 FR 14294, Apr. 1, 1983; 50 FR 28746, July 15, 1985; 51 FR 28682, Aug. 8, 1986; 75 FR 13005, Mar. 18, 2010]

### § 262.42  Exception reporting.

(a)(1) A generator of 1,000 kilograms or greater of hazardous waste in a calendar month, or greater than 1 kg of acute hazardous waste listed in § 261.31 or § 261.33(e) in a calendar month, who does not receive a copy of the manifest with the handwritten signature of the owner or operator of the designated facility within 35 days of the date the waste was accepted by the initial transporter must contact the transporter and/or the owner or operator of the designated facility to determine the status of the hazardous waste.

(2) A generator of 1,000 kilograms or greater of hazardous waste in a calendar month, or greater than 1 kg of acute hazardous waste listed in § 261.31 or § 261.33(e) in a calendar month, must submit an Exception Report to the EPA Regional Administrator for the Region in which the generator is located if he has not received a copy of the manifest with the handwritten signature of the owner or operator of the designated facility within 45 days of the date the waste was accepted by the initial transporter. The Exception Report must include:

(i) A legible copy of the manifest for which the generator does not have confirmation of delivery;

(ii) A cover letter signed by the generator or his authorized representative explaining the efforts taken to locate the hazardous waste and the results of those efforts.

(b) A generator of greater than 100 kilograms but less than 1000 kilograms of hazardous waste in a calendar month who does not receive a copy of the manifest with the handwritten signature of the owner or operator of the designated facility within 60 days of the date the waste was accepted by the initial transporter must submit a legible copy of the manifest, with some indication that the generator has not received confirmation of delivery, to the EPA Regional Administrator for the Region in which the generator is located.

NOTE: The submission to EPA need only be a handwritten or typed note on the manifest itself, or on an attached sheet of paper, stating that the return copy was not received.

(c) For rejected shipments of hazardous waste or container residues contained in non-empty containers that are forwarded to an alternate facility by a designated facility using a new manifest (following the procedures of 40 CFR 264.72(e)(1) through (6) or 40 CFR 265.72(e)(1) through (6)), the generator must comply with the requirements of paragraph (a) or (b) of this section, as applicable, for the shipment forwarding the material from the designated facility to the alternate facility instead of for the shipment from the generator to the designated facility. For purposes of paragraph (a) or (b) of this section for a shipment forwarding such waste to an alternate facility by a designated facility:

(1) The copy of the manifest received by the generator must have the handwritten signature of the owner or operator of the alternate facility in place of the signature of the owner or operator of the designated facility, and

(2) The 35/45/60-day timeframes begin the date the waste was accepted by the initial transporter forwarding the hazardous waste shipment from the designated facility to the alternate facility.

[52 FR 35898, Sept. 23, 1987, as amended at 75 FR 13005, Mar. 18, 2010]

**§262.43 Additional reporting.**

The Administrator, as he deems necessary under sections 2002(a) and 3002(6) of the Act, may require generators to furnish additional reports concerning the quantities and disposition of wastes identified or listed in 40 CFR part 261.

**§262.44 Special requirements for generators of between 100 and 1000 kg/mo.**

A generator of greater than 100 kilograms but less than 1000 kilograms of hazardous waste in a calendar month is subject only to the following requirements in this subpart:

(a) Section 262.40(a), (c), and (d), recordkeeping;

(b) Section 262.42(b), exception reporting; and

(c) Section 262.43, additional reporting.

[52 FR 35899, Sept. 23, 1987]

## Subpart E—Exports of Hazardous Waste

SOURCE: 51 FR 28682, Aug. 8, 1986, unless otherwise noted.

**§262.50 Applicability.**

This subpart establishes requirements applicable to exports of hazardous waste. Except to the extent §262.58 provides otherwise, a primary exporter of hazardous waste must comply with the special requirements of this subpart and a transporter transporting hazardous waste for export must comply with applicable requirements of part 263. Section 262.58 sets forth the requirements of international agreements between the United States and receiving countries which establish different notice, export, and enforcement procedures for the transportation, treatment, storage and disposal of hazardous waste for shipments between the United States and those countries.

**§262.51 Definitions.**

In addition to the definitions set forth at 40 CFR 260.10, the following definitions apply to this subpart:

*Consignee* means the ultimate treatment, storage or disposal facility in a receiving country to which the hazardous waste will be sent.

*EPA Acknowledgement of Consent* means the cable sent to EPA from the U.S. Embassy in a receiving country that acknowledges the written consent of the receiving country to accept the hazardous waste and describes the terms and conditions of the receiving country's consent to the shipment.

*Primary Exporter* means any person who is required to originate the manifest for a shipment of hazardous waste in accordance with 40 CFR part 262, subpart B, or equivalent State provision, which specifies a treatment, storage, or disposal facility in a receiving country as the facility to which the hazardous waste will be sent and any intermediary arranging for the export.

*Receiving country* means a foreign country to which a hazardous waste is sent for the purpose of treatment, storage or disposal (except short-term storage incidental to transportation).

*Transit country* means any foreign country, other than a receiving country, through which a hazardous waste is transported.

[53 FR 27164, July 19, 1988]

### § 262.52   General requirements.

Exports of hazardous waste are prohibited except in compliance with the applicable requirements of this subpart and part 263. Exports of hazardous waste are prohibited unless:

(a) Notification in accordance with § 262.53 has been provided;

(b) The receiving country has consented to accept the hazardous waste;

(c) A copy of the EPA Acknowledgment of Consent to the shipment accompanies the hazardous waste shipment and, unless exported by rail, is attached to the manifest (or shipping paper for exports by water (bulk shipment));

(d) The hazardous waste shipment conforms to the terms of the receiving country's written consent as reflected in the EPA Acknowledgment of Consent.

### § 262.53   Notification of intent to export.

(a) A primary exporter of hazardous waste must notify EPA of an intended export before such waste is scheduled to leave the United States. A complete notification should be submitted sixty (60) days before the initial shipment is intended to be shipped off site. This notification may cover export activities extending over a twelve (12) month or lesser period. The notification must be in writing, signed by the primary ex-

porter, and include the following information:

(1) Name, mailing address, telephone number and EPA ID number of the primary exporter;

(2) By consignee, for each hazardous waste type:

(i) A description of the hazardous waste and the EPA hazardous waste number (from 40 CFR part 261, subparts C and D), U.S. DOT proper shipping name, hazard class and ID number (UN/NA) for each hazardous waste as identified in 49 CFR parts 171 through 177;

(ii) The estimated frequency or rate at which such waste is to be exported and the period of time over which such waste is to be exported;

(iii) The estimated total quantity of the hazardous waste in units as specified in the instructions to the Uniform Hazardous Waste Manifest Form (8700–22);

(iv) All points of entry to and departure from each foreign country through which the hazardous waste will pass;

(v) A description of the means by which each shipment of the hazardous waste will be transported (e.g., mode of transportation vehicle (air, highway, rail, water, etc.), type(s) of container (drums, boxes, tanks, etc.));

(vi) A description of the manner in which the hazardous waste will be treated, stored or disposed of in the receiving country (e.g., land or ocean incineration, other land disposal, ocean dumping, recycling);

(vii) The name and site address of the consignee and any alternate consignee; and

(viii) The name of any transit countries through which the hazardous waste will be sent and a description of the approximate length of time the hazardous waste will remain in such country and the nature of its handling while there;

(b) Notifications submitted by mail should be sent to the following mailing address: Office of Enforcement and Compliance Assurance, Office of Federal Activities, International Compliance Assurance Division (2254A), Environmental Protection Agency, 1200 Pennsylvania Ave., NW., Washington, DC 20460. Hand-delivered notifications should be sent to: Office of Enforcement and Compliance Assurance, Office of Federal Activities, International Compliance Assurance Division, Environmental Protection Agency, Ariel Rios Bldg., Room 6144, 12th St. and Pennsylvania Ave., NW., Washington, DC 20004. In both cases, the following shall be prominently displayed on the front of the envelope: "Attention: Notification of Intent to Export.".

(c) Except for changes to the telephone number in paragraph (a)(1) of this section, changes to paragraph (a)(2)(v) of this section and decreases in the quantity indicated pursuant to paragraph (a)(2)(iii) of this section when the conditions specified on the original notification change (including any exceedance of the estimate of the quantity of hazardous waste specified in the original notification), the primary exporter must provide EPA with a written renotification of the change. The shipment cannot take place until consent of the receiving country to the changes (except for changes to paragraph (a)(2)(viii) of this section and in the ports of entry to and departure from transit countries pursuant to paragraph (a)(2)(iv) of this section) has been obtained and the primary exporter receives an EPA Acknowledgment of Consent reflecting the receiving country's consent to the changes.

(d) Upon request by EPA, a primary exporter shall furnish to EPA any additional information which a receiving country requests in order to respond to a notification.

(e) In conjunction with the Department of State, EPA will provide a complete notification to the receiving country and any transit countries. A notification is complete when EPA receives a notification which EPA determines satisfies the requirements of paragraph (a) of this section. Where a claim of confidentiality is asserted with respect to any notification information required by paragraph (a) of this section, EPA may find the notification not complete until any such claim is resolved in accordance with 40 CFR 260.2.

(f) Where the receiving country consents to the receipt of the hazardous waste, EPA will forward an EPA Acknowledgment of Consent to the primary exporter for purposes of §262.54(h). Where the receiving country

objects to receipt of the hazardous waste or withdraws a prior consent, EPA will notify the primary exporter in writing. EPA will also notify the primary exporter of any responses from transit countries.

[51 FR 28682, Aug. 8, 1986, as amended at 56 FR 43705, Sept. 4, 1991; 61 FR 16309, Apr. 12, 1996; 71 FR 40271, July 14, 2006]

## §262.54 Special manifest requirements.

A primary exporter must comply with the manifest requirements of 40 CFR 262.20 through 262.23 except that:

(a) In lieu of the name, site address and EPA ID number of the designated permitted facility, the primary exporter must enter the name and site address of the consignee;

(b) In lieu of the name, site address and EPA ID number of a permitted alternate facility, the primary exporter may enter the name and site address of any alternate consignee.

(c) In the International Shipments block, the primary exporter must check the export box and enter the point of exit (city and State) from the United States.

(d) The following statement must be added to the end of the first sentence of the certification set forth in Item 16 of the Uniform Hazardous Waste Manifest Form: "and conforms to the terms of the attached EPA Acknowledgment of Consent";

(e) The primary exporter may obtain the manifest from any source that is registered with the U.S. EPA as a supplier of manifests (e.g., states, waste handlers, and/or commercial forms printers).

(f) The primary exporter must require the consignee to confirm in writing the delivery of the hazardous waste to that facility and to describe any significant discrepancies (as defined in 40 CFR 264.72(a)) between the manifest and the shipment. A copy of the manifest signed by such facility may be used to confirm delivery of the hazardous waste.

(g) In lieu of the requirements of §262.20(d), where a shipment cannot be delivered for any reason to the designated or alternate consignee, the primary exporter must:

(1) Renotify EPA of a change in the conditions of the original notification to allow shipment to a new consignee in accordance with § 262.53(c) and obtain an EPA Acknowledgment of Consent prior to delivery; or

(2) Instruct the transporter to return the waste to the primary exporter in the United States or designate another facility within the United States; and

(3) Instruct the transporter to revise the manifest in accordance with the primary exporter's instructions.

(h) The primary exporter must attach a copy of the EPA Acknowledgment of Consent to the shipment to the manifest which must accompany the hazardous waste shipment. For exports by rail or water (bulk shipment), the primary exporter must provide the transporter with an EPA Acknowledgment of Consent which must accompany the hazardous waste but which need not be attached to the manifest except that for exports by water (bulk shipment) the primary exporter must attach the copy of the EPA Acknowledgment of Consent to the shipping paper.

(i) The primary exporter shall provide the transporter with an additional copy of the manifest for delivery to the U.S. Customs official at the point the hazardous waste leaves the United States in accordance with § 263.20(g)(4).

[51 FR 28682, Aug. 8, 1986, as amended at 70 FR 10818, Mar. 4, 2005]

## § 262.55   Exception reports.

In lieu of the requirements of § 262.42, a primary exporter must file an exception report with the Administrator if:

(a) He has not received a copy of the manifest signed by the transporter stating the date and place of departure from the United States within forty-five (45) days from the date it was accepted by the initial transporter;

(b) Within ninety (90) days from the date the waste was accepted by the initial transporter, the primary exporter has not received written confirmation from the consignee that the hazardous waste was received;

(c) The waste is returned to the United States.

EFFECTIVE DATE NOTE: At 75 FR 1253, Jan. 8, 2010, § 262.55 was amended by revising the introductory text, effective July 7, 2010. For the convenience of the user, the revised text is set forth as follows:

## § 262.55   Exception reports.

In lieu of the requirements of § 262.42, a primary exporter must file an exception report with the Office of Enforcement and Compliance Assurance, Office of Federal Activities, International Compliance Assurance Division (2254A), Environmental Protection Agency, 1200 Pennsylvania Avenue, NW., Washington, DC 20460, if any of the following occurs:

\*      \*      \*      \*      \*

## § 262.56   Annual reports.

(a) Primary exporters of hazardous waste shall file with the Administrator no later than March 1 of each year, a report summarizing the types, quantities, frequency, and ultimate destination of all hazardous waste exported during the previous calendar year. Such reports shall include the following:

(1) The EPA identification number, name, and mailing and site address of the exporter;

(2) The calendar year covered by the report;

(3) The name and site address of each consignee;

(4) By consignee, for each hazardous waste exported, a description of the hazardous waste, the EPA hazardous waste number (from 40 CFR part 261, subpart C or D), DOT hazard class, the name and US EPA ID number (where applicable) for each transporter used, the total amount of waste shipped and number of shipments pursuant to each notification;

(5) Except for hazardous waste produced by exporters of greater than 100 kg but less than 1000 kg in a calendar month, unless provided pursuant to § 262.41, in even numbered years:

(i) A description of the efforts undertaken during the year to reduce the volume and toxicity of waste generated; and

(ii) A description of the changes in volume and toxicity of waste actually achieved during the year in comparison to previous years to the extent such information is available for years prior to 1984.

(6) A certification signed by the primary exporter which states:

I certify under penalty of law that I have personally examined and am familiar with the information submitted in this and all attached documents, and that based on my inquiry of those individuals immediately responsible for obtaining the information, I believe that the submitted information is true, accurate, and complete. I am aware that there are significant penalties for submitting false information including the possibility of fine and imprisonment.

(b) Annual reports submitted by mail should be sent to the following mailing address: Office of Enforcement and Compliance Assurance, Office of Federal Activities, International Compliance Assurance Division (2254A), Environmental Protection Agency, 1200 Pennsylvania Ave., NW., Washington, DC 20460. Hand-delivered reports should be sent to: Office of Enforcement and Compliance Assurance, Office of Federal Activities, International Compliance Assurance Division, Environmental Protection Agency, Ariel Rios Bldg., Room 6144, 12th St. and Pennsylvania Ave., NW., Washington, DC 20004.

[51 FR 28682, Aug. 8, 1986, as amended at 56 FR 43705, Sept. 4, 1991; 61 FR 16309, Apr. 12, 1996; 71 FR 40271, July 14, 2006]

## §262.57 Recordkeeping.

(a) For all exports a primary exporter must:

(1) Keep a copy of each notification of intent to export for a period of at least three years from the date the hazardous waste was accepted by the initial transporter;

(2) Keep a copy of each EPA Acknowledgment of Consent for a period of at least three years from the date the hazardous waste was accepted by the initial transporter;

(3) Keep a copy of each confirmation of delivery of the hazardous waste from the consignee for at least three years from the date the hazardous waste was accepted by the initial transporter; and

(4) Keep a copy of each annual report for a period of at least three years from the due date of the report.

(b) The periods of retention referred to in this section are extended automatically during the course of any unresolved enforcement action regarding the regulated activity or as requested by the Administrator.

## §262.58 International agreements.

(a) Any person who exports or imports hazardous waste subject to Federal manifest requirements of Part 262, or subject to the universal waste management standards of 40 CFR Part 273, or subject to State requirements analogous to 40 CFR Part 273, to or from designated member countries of the Organization for Economic Cooperation and Development (OECD) as defined in paragraph (a)(1) of this section for purposes of recovery is subject to Subpart H of this part. The requirements of Subparts E and F do not apply.

(1) For the purposes of subpart H, the designated OECD Member countries consist of Australia, Austria, Belgium, the Czech Republic, Denmark, Finland, France, Germany, Greece, Hungary, Iceland, Ireland, Italy, Japan, Luxembourg, the Netherlands, New Zealand, Norway, Poland, Portugal, the Slovak Republic, South Korea, Spain, Sweden, Switzerland, Turkey, the United Kingdom, and the United States.

(2) For the purposes of this Subpart, Canada and Mexico are considered OECD member countries only for the purpose of transit.

(b) Any person who exports hazardous waste to or imports hazardous waste from: a designated OECD member country for purposes other than recovery (e.g., incineration, disposal), Mexico (for any purpose), or Canada (for any purpose) remains subject to the requirements of subparts E and F of this part.

[61 FR 16310, Apr. 12, 1996, as amended at 71 FR 40271, July 14, 2006]

EFFECTIVE DATE NOTE: At 75 FR 1253, Jan. 8, 2010, §262.58 was revised, effective July 7, 2010. For the convenience of the user, the revised text is set forth as follows:

## §262.58 International agreements.

(a) Any person who exports or imports wastes that are considered hazardous under U.S. national procedures to or from designated Member countries of the Organization for Economic Cooperation and Development (OECD) as defined in paragraph (a)(1) of this section for purposes of recovery is subject to subpart H of this part. The requirements of subparts E and F of this part do not apply to such exports and imports. A waste is considered hazardous under U.S. national procedures if the waste meets the Federal definition of hazardous waste in 40 CFR

261.3 and is subject to either the Federal RCRA manifesting requirements at 40 CFR part 262, subpart B, the universal waste management standards of 40 CFR part 273, State requirements analogous to 40 CFR part 273, the export requirements in the spent lead-acid battery management standards of 40 CFR part 266, subpart G, or State requirements analogous to the export requirements in 40 CFR part 266, subpart G.

(1) For the purposes of subpart H, the designated OECD Member countries consist of Australia, Austria, Belgium, the Czech Republic, Denmark, Finland, France, Germany, Greece, Hungary, Iceland, Ireland, Italy, Japan, Luxembourg, the Netherlands, New Zealand, Norway, Poland, Portugal, the Republic of Korea, the Slovak Republic, Spain, Sweden, Switzerland, Turkey, the United Kingdom, and the United States.

(2) For the purposes of subpart H of this part, Canada and Mexico are considered OECD Member countries only for the purpose of transit.

(b) Any person who exports hazardous waste to or imports hazardous waste from: A designated OECD Member country for purposes other than recovery (*e.g.*, incineration, disposal), Mexico (for any purpose), or Canada (for any purpose) remains subject to the requirements of subparts E and F of this part, and is not subject to the requirements of subpart H of this part.

## Subpart F—Imports of Hazardous Waste

### § 262.60   Imports of hazardous waste.

(a) Any person who imports hazardous waste from a foreign country into the United States must comply with the requirements of this part and the special requirements of this subpart.

(b) When importing hazardous waste, a person must meet all the requirements of § 262.20 for the manifest except that:

(1) In place of the generator's name, address and EPA identification number, the name and address of the foreign generator and the importer's name, address and EPA identification number must be used.

(2) In place of the generator's signature on the certification statement, the U.S. importer or his agent must sign and date the certification and obtain the signature of the initial transporter.

(c) A person who imports hazardous waste may obtain the manifest form from any source that is registered with the U.S. EPA as a supplier of manifests (*e.g.*, states, waste handlers, and/or commercial forms printers).

(d) In the International Shipments block, the importer must check the import box and enter the point of entry (city and State) into the United States.

(e) The importer must provide the transporter with an additional copy of the manifest to be submitted by the receiving facility to U.S. EPA in accordance with § 264.71(a)(3) and § 265.71(a)(3) of this chapter.

[51 FR 28685, Aug. 8, 1986, as amended at 70 FR 10818, Mar. 4, 2005; 75 FR 13005, Mar. 18, 2010]

## Subpart G—Farmers

### § 262.70   Farmers.

A farmer disposing of waste pesticides from his own use which are hazardous wastes is not required to comply with the standards in this part or other standards in 40 CFR parts 264, 265, 268, or 270 for those wastes provided he triple rinses each emptied pesticide container in accordance with § 261.7(b)(3) and disposes of the pesticide residues on his own farm in a manner consistent with the disposal instructions on the pesticide label.

[53 FR 27165, July 19, 1988, as amended at 71 FR 40271, July 14, 2006]

## Subpart H—Transfrontier Shipments of Hazardous Waste for Recovery within the OECD

EFFECTIVE DATE NOTE: At 75 FR 1253, Jan. 8, 2010, subpart H was revised, effective July 7, 2010. The new subpart appears after the text of this subpart.

SOURCE: 61 FR 16310, Apr. 12, 1996, unless otherwise noted.

### § 262.80   Applicability.

(a) The requirements of this subpart apply to imports and exports of wastes that are considered hazardous under U.S. national procedures and are destined for recovery operations in the countries listed in § 262.58(a)(1). A waste is considered hazardous under U.S. national procedures if it meets the Federal definition of hazardous waste in 40 CFR 261.3 and it is subject to either the Federal manifesting requirements at 40

CFR Part 262, Subpart B, to the universal waste management standards of 40 CFR Part 273, or to State requirements analogous to 40 CFR Part 273.

(b) Any person (notifier, consignee, or recovery facility operator) who mixes two or more wastes (including hazardous and non-hazardous wastes) or otherwise subjects two or more wastes (including hazardous and non-hazardous wastes) to physical or chemical transformation operations, and thereby creates a new hazardous waste, becomes a generator and assumes all subsequent generator duties under RCRA and any notifier duties, if applicable, under this subpart.

## §262.81 Definitions.

The following definitions apply to this subpart.

(a) *Competent authorities* means the regulatory authorities of concerned countries having jurisdiction over transfrontier movements of wastes destined for recovery operations.

(b) *Concerned countries* means the exporting and importing OECD member countries and any OECD member countries of transit.

(c) *Consignee* means the person to whom possession or other form of legal control of the waste is assigned at the time the waste is received in the importing country.

(d) *Country of transit* means any designated OECD country in §262.58(a)(1) and (a)(2) other than the exporting or importing country across which a transfrontier movement of wastes is planned or takes place.

(e) *Exporting country* means any designated OECD member country in §262.58(a)(1) from which a transfrontier movement of wastes is planned or has commenced.

(f) *Importing country* means any designated OECD country in §262.58(a)(1) to which a transfrontier movement of wastes is planned or takes place for the purpose of submitting the wastes to recovery operations therein.

(g) *Notifier* means the person under the jurisdiction of the exporting country who has, or will have at the time the planned transfrontier movement commences, possession or other forms of legal control of the wastes and who proposes their transfrontier movement for the ultimate purpose of submitting them to recovery operations. When the United States (U.S.) is the exporting country, notifier is interpreted to mean a person domiciled in the U.S.

(h) *OECD area* means all land or marine areas under the national jurisdiction of any designated OECD member country in §262.58. When the regulations refer to shipments to or from an OECD country, this means OECD area.

(i) *Recognized trader* means a person who, with appropriate authorization of concerned countries, acts in the role of principal to purchase and subsequently sell wastes; this person has legal control of such wastes from time of purchase to time of sale; such a person may act to arrange and facilitate transfrontier movements of wastes destined for recovery operations.

(j) *Recovery facility* means an entity which, under applicable domestic law, is operating or is authorized to operate in the importing country to receive wastes and to perform recovery operations on them.

(k) *Recovery operations* means activities leading to resource recovery, recycling, reclamation, direct re-use or alternative uses as listed in Table 2.B of the Annex of OECD Council Decision C(88)90(Final) of 27 May 1988, (available from the Environmental Protection Agency, RCRA Docket, EPA/DC, EPA West, Room B102, 1301 Constitution Ave., NW., Washington, DC 20460 (Docket # F–94–IEHF-FFFFF) and the Organisation for Economic Co-operation and Development, Environment Directorate, 2 rue Andre Pascal, 75775 Paris Cedex 16, France) which include:

R1 Use as a fuel (other than in direct incineration) or other means to generate energy

R2 Solvent reclamation/regeneration

R3 Recycling/reclamation of organic substances which are not used as solvents

R4 Recycling/reclamation of metals and metal compounds

R5 Recycling/reclamation of other inorganic materials

R6 Regeneration of acids or bases

R7 Recovery of components used for pollution control

R8 Recovery of components from catalysts

R9 Used oil re-refining or other reuses of previously used oil

R10 Land treatment resulting in benefit to agriculture or ecological improvement

R11 Uses of residual materials obtained from any of the operations numbered R1–R10

R12 Exchange of wastes for submission to any of the operations numbered R1–R11

6R13 Accumulation of material intended for any operation in Table 2.B

(1) *Transfrontier movement* means any shipment of wastes destined for recovery operations from an area under the national jurisdiction of one OECD member country to an area under the national jurisdiction of another OECD member country.

[61 FR 16310, Apr. 12, 1996, as amended at 71 FR 40271, July 14, 2006]

### § 262.82   General conditions.

(a) *Scope.* The level of control for exports and imports of waste is indicated by assignment of the waste to a green, amber, or red list and by U.S. national procedures as defined in § 262.80(a). The green, amber, and red lists are incorporated by reference in § 262.89 (e).

(1) Wastes on the green list are subject to existing controls normally applied to commercial transactions, except as provided below:

(i) Green-list wastes that are considered hazardous under U.S. national procedures are subject to amber-list controls.

(ii) Green-list wastes that are sufficiently contaminated or mixed with amber-list wastes, such that the waste or waste mixture is considered hazardous under U.S. national procedures, are subject to amber-list controls.

(iii) Green-list wastes that are sufficiently contaminated or mixed with other wastes subject to red-list controls such that the waste or waste mixture is considered hazardous under U.S. national procedures must be handled in accordance with the red-list controls.

(2) Wastes on the amber list that are considered hazardous under U.S. national procedures as defined in § 262.80(a) are subject to the amber-list controls of this Subpart.

(i) If amber-list wastes are sufficiently contaminated or mixed with other wastes subject to red-list controls such that the waste or waste mixture is considered hazardous under U.S. national procedures, the wastes must be handled in accordance with the red-list controls.

(ii) [Reserved]

(3) Wastes on the red list that are considered hazardous under U.S. national procedures as defined in § 262.80(a) are subject to the red-list controls of this subpart.

NOTE TO PARAGRAPH (a)(3): Some wastes on the amber or red lists are not listed or otherwise identified as hazardous under RCRA (*e.g.,* polychlorinated biphenyls) and therefore are not subject to the amber- or red-list controls of this subpart. Regardless of the status of the waste under RCRA, however, other Federal environmental statutes (e.g., the Toxic Substances Control Act) may restrict certain waste imports or exports. Such restrictions continue to apply without regard to this Subpart.

(4) Wastes not yet assigned to a list are eligible for transfrontier movements, as follows:

(i) If such wastes are considered hazardous under U.S. national procedures as defined in § 262.80(a), these wastes are subject to the red-list controls; or

(ii) If such wastes are not considered hazardous under U.S. national procedures as defined in § 262.80(a), such wastes may move as though they appeared on the green list.

(b) *General conditions applicable to transfrontier movements of hazardous waste.* (1) The waste must be destined for recovery operations at a facility that, under applicable domestic law, is operating or is authorized to operate in the importing country;

(2) The transfrontier movement must be in compliance with applicable international transport agreements; and

NOTE TO PARAGRAPH (b)(2): These international agreements include, but are not limited to, the Chicago Convention (1944), ADR (1957), ADNR (1970), MARPOL Convention (1973/1978), SOLAS Convention (1974), IMDG Code (1985), COTIF (1985), and RID (1985).

(3) Any transit of waste through a non-OECD member country must be conducted in compliance with all applicable international and national laws and regulations.

(c) *Provisions relating to re-export for recovery to a third country.* (1) Re-export of wastes subject to the amber-list control system from the U.S., as the importing country, to a third country listed in § 262.58(a)(1) may occur only after a notifier in the U.S. provides notification to and obtains consent of the competent authorities in the third country, the original exporting country, and new transit countries. The notification must comply with the notice and consent procedures in §262.83 for all concerned countries and the original exporting country. The competent authorities of the original exporting country as well as the competent authorities of all other concerned countries have 30 days to object to the proposed movement.

(i) The 30-day period begins once the competent authorities of both the initial exporting country and new importing country issue Acknowledgements of Receipt of the notification.

(ii) The transfrontier movement may commence if no objection has been lodged after the 30-day period has passed or immediately after written consent is received from all relevant OECD importing and transit countries.

(2) Re-export of waste subject to the red-list control system from the original importing country to a third country listed in §262.58(a)(1) may occur only following notification of the competent authorities of the third country, the original exporting country, and new transit countries by a notifier in the original importing country in accordance with §262.83. The transfrontier movement may not proceed until receipt by the original importing country of written consent from the competent authorities of the third country, the original exporting country, and new transit countries.

(3) In the case of re-export of amber or red-list wastes to a country other than those in §262.58(a)(1), notification to and consent of the competent authorities of the original OECD member country of export and any OECD member countries of transit is required as specified in paragraphs (c)(1) and (c)(2) of this section in addition to compliance with all international agreements and arrangements to which the first importing OECD member country is a party and all applicable regulatory requirements for exports from the first importing country.

[61 FR 16310, Apr. 12, 1996, as amended at 71 FR 40271, July 14, 2006]

## §262.83 Notification and consent.

(a) *Applicability.* Consent must be obtained from the competent authorities of the relevant OECD importing and transit countries prior to exporting hazardous waste destined for recovery operations subject to this Subpart. Hazardous wastes subject to amber-list controls are subject to the requirements of paragraph (b) of this section; hazardous wastes subject to red-list controls are subject to the requirements of paragraph (c) of this section; and wastes not identified on any list are subject to the requirements of paragraph (d) of this section.

(b) *Amber-list wastes.* The export from the U.S. of hazardous wastes as described in §262.80(a) that appear on the amber list is prohibited unless the notification and consent requirements of paragraph (b)(1) or paragraph (b)(2) of this section are met.

(1) Transactions requiring specific consent:

(i) *Notification.* At least 45 days prior to commencement of the transfrontier movement, the notifier must provide written notification in English of the proposed transfrontier movement to the Office of Enforcement and Compliance Assurance, Office of Federal Activities, International Compliance Assurance Division (2254A), Environmental Protection Agency, 1200 Pennsylvania Ave., NW., Washington, DC 20460, with the words "Attention: OECD Export Notification" prominently displayed on the envelope. This notification must include all of the information identified in paragraph (e) of this section. In cases where wastes having similar physical and chemical characteristics, the same United Nations classification, and the same RCRA waste codes are to be sent periodically to the same recovery facility by the same notifier, the notifier may submit one notification of intent to export these wastes in multiple shipments during a period of up to one year.

(ii) *Tacit consent.* If no objection has been lodged by any concerned country (*i.e.*, exporting, importing, or transit countries) to a notification provided pursuant to paragraph (b)(1)(i) of this section within 30 days after the date of issuance of the Acknowledgment of Receipt of notification by the competent authority of the importing country, the transfrontier movement may commence. Tacit consent expires one calendar year after the close of the 30 day period; renotification and renewal of all consents is required for exports after that date.

(iii) *Written consent.* If the competent authorities of all the relevant OECD importing and transit countries provide written consent in a period less than 30 days, the transfrontier movement may commence immediately after all necessary consents are received. Written consent expires for each relevant OECD importing and transit country one calendar year after the date of that country's consent unless otherwise specified; renotification and renewal of each expired consent is required for exports after that date.

(2) Shipments to facilities pre-approved by the competent authorities of the importing countries to accept specific wastes for recovery:

(i) The notifier must provide EPA the information identified in paragraph (e) of this section in English, at least 10 days in advance of commencing shipment to a pre-approved facility. The notification should indicate that the recovery facility is pre-approved, and may apply to a single specific shipment or to multiple shipments as described in paragraph (b)(1)(i) of this section. This information must be sent to the Office of Enforcement and Compliance Assurance, Office of Federal Activities, International Compliance Assurance Division (2254A), Environmental Protection Agency, 1200 Pennsylvania Ave., NW., Washington, DC 20460, with the words "Attention: OECD Export Notification—Pre-approved Facility" prominently displayed on the envelope.

(ii) Shipments may commence after the notification required in paragraph (b)(1)(i) of this section has been received by the competent authorities of all concerned countries, unless the notifier has received information indicating that the competent authorities of one or more concerned countries objects to the shipment.

(c) *Red-list wastes.* The export from the U.S. of hazardous wastes as described in § 262.80(a) that appear on the red list is prohibited unless notice is given pursuant to paragraph (b)(1)(i) of

this section and the notifier receives *written* consent from the importing country and any transit countries prior to commencement of the transfrontier movement.

(d) *Unlisted wastes.* Wastes not assigned to the green, amber, or red list that are considered hazardous under U.S. national procedures as defined in § 262.80(a) are subject to the notification and consent requirements established for red-list wastes in accordance with paragraph (c) of this section. Unlisted wastes that are not considered hazardous under U.S. national procedures as defined in § 262.80(a) are not subject to amber or red controls when exported or imported.

(e) *Notification information.* Notifications submitted under this section must include:

(1) Serial number or other accepted identifier of the notification form;

(2) Notifier name and EPA identification number (if applicable), address, and telephone and telefax numbers;

(3) Importing recovery facility name, address, telephone and telefax numbers, and technologies employed;

(4) Consignee name (if not the owner or operator of the recovery facility) address, and telephone and telefax numbers; whether the consignee will engage in waste exchange or storage prior to delivering the waste to the final recovery facility and identification of recovery operations to be employed at the final recovery facility;

(5) Intended transporters and/or their agents;

(6) Country of export and relevant competent authority, and point of departure;

(7) Countries of transit and relevant competent authorities and points of entry and departure;

(8) Country of import and relevant competent authority, and point of entry;

(9) Statement of whether the notification is a single notification or a general notification. If general, include period of validity requested;

(10) Date foreseen for commencement of transfrontier movement;

(11) Designation of waste type(s) from the appropriate list (amber or red and waste list code), descriptions of each waste type, estimated total quantity of

each, RCRA waste code, and United Nations number for each waste type; and

(12) Certification/Declaration signed by the notifier that states:

I certify that the above information is complete and correct to the best of my knowledge. I also certify that legally-enforceable written contractual obligations have been entered into, and that any applicable insurance or other financial guarantees are or shall be in force covering the transfrontier movement.

Name: _____
Signature: _____
Date: _____

NOTE TO PARAGRAPH (e)(12): The U.S. does not currently require financial assurance; however, U.S. exporters may be asked by other governments to provide and certify to such assurance as a condition of obtaining consent to a proposed movement.

[61 FR 16310, Apr. 12, 1996, as amended at 71 FR 40271, July 14, 2006]

### §262.84  Tracking document.

(a) All U.S. parties subject to the contract provisions of §262.85 must ensure that a tracking document meeting the conditions of §262.84(b) accompanies each transfrontier shipment of wastes subject to amber-list or red-list controls from the initiation of the shipment until it reaches the final recovery facility, including cases in which the waste is stored and/or exchanged by the consignee prior to shipment to the final recovery facility, except as provided in §§262.84(a)(1) and (2).

(1) For shipments of hazardous waste within the U.S. solely by water (bulk shipments only) the generator must forward the tracking document with the manifest to the last water (bulk shipment) transporter to handle the waste in the U.S. if exported by water, (in accordance with the manifest routing procedures at §262.23(c)).

(2) For rail shipments of hazardous waste within the U.S. which originate at the site of generation, the generator must forward the tracking document with the manifest (in accordance with the routing procedures for the manifest in §262.23(d)) to the next non-rail transporter, if any, or the last rail transporter to handle the waste in the U.S. if exported by rail.

(b) The tracking document must include all information required under §262.83 (for notification), and the following:

(1) Date shipment commenced.

(2) Name (if not notifier), address, and telephone and telefax numbers of primary exporter.

(3) Company name and EPA ID number of all transporters.

(4) Identification (license, registered name or registration number) of means of transport, including types of packaging.

(5) Any special precautions to be taken by transporters.

(6) Certification/declaration signed by notifier that no objection to the shipment has been lodged as follows:

I certify that the above information is complete and correct to the best of my knowledge. I also certify that legally-enforceable written contractual obligations have been entered into, that any applicable insurance or other financial guarantees are or shall be in force covering the transfrontier movement, and that:
1. All necessary consents have been received; OR
2. The shipment is directed at a recovery facility within the OECD area and no objection has been received from any of the concerned countries within the 30 day tacit consent period; OR
3. The shipment is directed at a recovery facility pre-authorized for that type of waste within the OECD area; such an authorization has not been revoked, and no objection has been received from any of the concerned countries.

(delete sentences that are not applicable)
Name: _____
Signature: _____
Date: _____

(7) Appropriate signatures for each custody transfer (e.g. transporter, consignee, and owner or operator of the recovery facility).

(c) Notifiers also must comply with the special manifest requirements of 40 CFR 262.54(a), (b), (c), (e), and (i) and consignees must comply with the import requirements of 40 CFR part 262, subpart F.

(d) Each U.S. person that has physical custody of the waste from the time the movement commences until it arrives at the recovery facility must sign the tracking document (e.g. transporter, consignee, and owner or operator of the recovery facility).

(e) Within three working days of the receipt of imports subject to this Subpart, the owner or operator of the U.S. recovery facility must send signed copies of the tracking document to the notifier, to the Office of Enforcement and Compliance Assurance, Office of Federal Activities, International Compliance Assurance Division (2254A), Environmental Protection Agency, 1200 Pennsylvania Ave., NW., Washington, DC 20460, and to the competent authorities of the exporting and transit countries.

[61 FR 16310, Apr. 12, 1996, as amended at 71 FR 40271, July 14, 2006]

## § 262.85  Contracts.

(a) Transfrontier movements of hazardous wastes subject to amber or red control procedures are prohibited unless they occur under the terms of a valid written contract, chain of contracts, or equivalent arrangements (when the movement occurs between parties controlled by the same corporate or legal entity). Such contracts or equivalent arrangements must be executed by the notifier and the owner or operator of the recovery facility, and must specify responsibilities for each. Contracts or equivalent arrangements are valid for the purposes of this section only if persons assuming obligations under the contracts or equivalent arrangements have appropriate legal status to conduct the operations specified in the contract or equivalent arrangement.

(b) Contracts or equivalent arrangements must specify the name and EPA ID number, where available, of:

(1) The generator of each type of waste;

(2) Each person who will have physical custody of the wastes;

(3) Each person who will have legal control of the wastes; and

(4) The recovery facility.

(c) Contracts or equivalent arrangements must specify which party to the contract will assume responsibility for alternate management of the wastes if its disposition cannot be carried out as described in the notification of intent to export. In such cases, contracts must specify that:

(1) The person having actual possession or physical control over the wastes will immediately inform the notifier and the competent authorities of the exporting and importing countries and, if the wastes are located in a country of transit, the competent authorities of that country; and

(2) The person specified in the contract will assume responsibility for the adequate management of the wastes in compliance with applicable laws and regulations including, if necessary, arranging their return to the original country of export.

(d) Contracts must specify that the consignee will provide the notification required in § 262.82(c) prior to re-export of controlled wastes to a third country.

(e) Contracts or equivalent arrangements must include provisions for financial guarantees, if required by the competent authorities of any concerned country, in accordance with applicable national or international law requirements.

NOTE TO PARAGRAPH (e): Financial guarantees so required are intended to provide for alternate recycling, disposal or other means of sound management of the wastes in cases where arrangements for the shipment and the recovery operations cannot be carried out as foreseen. The U.S. does not require such financial guarantees at this time; however, some OECD countries do. It is the responsibility of the notifier to ascertain and comply with such requirements; in some cases, transporters or consignees may refuse to enter into the necessary contracts absent specific references or certifications to financial guarantees.

(f) Contracts or equivalent arrangements must contain provisions requiring each contracting party to comply with all applicable requirements of this subpart.

(g) Upon request by EPA, U.S. notifiers, consignees, or recovery facilities must submit to EPA copies of contracts, chain of contracts, or equivalent arrangements (when the movement occurs between parties controlled by the same corporate or legal entity). Information contained in the contracts or equivalent arrangements for which a claim of confidentiality is asserted accordance with 40 CFR 2.203(b) will be treated as confidential and will be disclosed by EPA only as provided in 40 CFR 260.2.

NOTE TO PARAGRAPH (g): Although the U.S. does not require routine submission of contracts at this time, OECD Council Decision C(92)39/FINAL allows members to impose such requirements. When other OECD countries require submission of partial or complete copies of the contract as a condition to granting consent to proposed movements, EPA will request the required information; absent submission of such information, some OECD countries may deny consent for the proposed movement.

## §262.86 Provisions relating to recognized traders.

(a) A recognized trader who takes physical custody of a waste and conducts recovery operations (including storage prior to recovery) is acting as the owner or operator of a recovery facility and must be so authorized in accordance with all applicable Federal laws.

(b) A recognized trader acting as a notifier or consignee for transfrontier shipments of waste must comply with all the requirements of this Subpart associated with being a notifier or consignee.

## §262.87 Reporting and recordkeeping.

(a) *Annual reports.* For all waste movements subject to this Subpart, persons (e.g., notifiers, recognized traders) who meet the definition of primary exporter in §262.51 shall file an annual report with the Office of Enforcement and Compliance Assurance, Office of Federal Activities, International Compliance Assurance Division (2254A), Environmental Protection Agency, 1200 Pennsylvania Ave., NW., Washington, DC 20460, no later than March 1 of each year summarizing the types, quantities, frequency, and ultimate destination of all such hazardous waste exported during the previous calendar year. (If the primary exporter is required to file an annual report for waste exports that are not covered under this Subpart, he may include all export information in one report provided the following information on exports of waste destined for recovery within the designated OECD member countries is contained in a separate section). Such reports shall include the following:

(1) The EPA identification number, name, and mailing and site address of the notifier filing the report;

(2) The calendar year covered by the report;

(3) The name and site address of each final recovery facility;

(4) By final recovery facility, for each hazardous waste exported, a description of the hazardous waste, the EPA hazardous waste number (from 40 CFR part 261, subpart C or D), designation of waste type(s) from OECD waste list and applicable waste code from the OECD lists, DOT hazard class, the name and U.S. EPA identification number (where applicable) for each transporter used, the total amount of hazardous waste shipped pursuant to this Subpart, and number of shipments pursuant to each notification;

(5) In even numbered years, for each hazardous waste exported, except for hazardous waste produced by exporters of greater than 100 kg but less than 1000 kg in a calendar month, and except for hazardous waste for which information was already provided pursuant to §262.41:

(i) A description of the efforts undertaken during the year to reduce the volume and toxicity of waste generated; and

(ii) A description of the changes in volume and toxicity of the waste actually achieved during the year in comparison to previous years to the extent such information is available for years prior to 1984; and

(6) A certification signed by the person acting as primary exporter that states:

I certify under penalty of law that I have personally examined and am familiar with the information submitted in this and all attached documents, and that based on my inquiry of those individuals immediately responsible for obtaining the information, I believe that the submitted information is true, accurate, and complete. I am aware that there are significant penalties for submitting false information including the possibility of fine and imprisonment.

(b) *Exception reports.* Any person who meets the definition of primary exporter in §262.51 must file an exception report in lieu of the requirements of

§ 262.42 with the Administrator if any of the following occurs:

(1) He has not received a copy of the tracking documentation signed by the transporter stating point of departure of the waste from the United States, within forty-five (45) days from the date it was accepted by the initial transporter;

(2) Within ninety (90) days from the date the waste was accepted by the initial transporter, the notifier has not received written confirmation from the recovery facility that the hazardous waste was received;

(3) The waste is returned to the United States.

(c) *Recordkeeping.* (1) Persons who meet the definition of primary exporter in § 262.51 shall keep the following records: § 262.89

(i) A copy of each notification of intent to export and all written consents obtained from the competent authorities of concerned countries for a period of at least three years from the date the hazardous waste was accepted by the initial transporter;

(ii) A copy of each annual report for a period of at least three years from the due date of the report; and

(iii) A copy of any exception reports and a copy of each confirmation of delivery (*i.e.*, tracking documentation) sent by the recovery facility to the notifier for at least three years from the date the hazardous waste was accepted by the initial transporter or received by the recovery facility, whichever is applicable.

(2) The periods of retention referred to in this section are extended automatically during the course of any unresolved enforcement action regarding the regulated activity or as requested by the Administrator.

[61 FR 16310, Apr. 12, 1996, as amended at 71 FR 40272, July 14, 2006]

## § 262.88 Pre-approval for U.S. Recovery Facilities. [Reserved]

## § 262.89 OECD Waste Lists.

(a) *General.* For the purposes of this Subpart, a waste is considered hazardous under U.S. national procedures, and hence subject to this Subpart, if the waste:

(1) Meets the Federal definition of hazardous waste in 40 CFR 261.3; and

(2) Is subject to either the Federal RCRA manifesting requirements at 40 CFR part 262, subpart B, to the universal waste management standards of 40 CFR part 273, or to State requirements analogous to 40 CFR part 273.

(b) If a waste is hazardous under paragraph (a) of this section and it appears on the amber or red list, it is subject to amber- or red-list requirements respectively;

(c) If a waste is hazardous under paragraph (a) of this section and it does not appear on either amber or red lists, it is subject to red-list requirements.

(d) The appropriate control procedures for hazardous wastes and hazardous waste mixtures are addressed in § 262.82.

(e) The OECD Green List of Wastes (revised May 1994), Amber List of Wastes and Red List of Wastes (both revised May 1993) as set forth in Appendix 3, Appendix 4 and Appendix 5, respectively, to the OECD Council Decision C(92)39/FINAL (Concerning the Control of Transfrontier Movements of Wastes Destined for Recovery Operations) are incorporated by reference. These incorporations by reference were approved by the Director of the Federal Register in accordance with 5 U.S.C. 552(a) and 1 CFR part 51 on July 11, 1996. These materials are incorporated as they exist on the date of the approval and a notice of any change in these materials will be published in the FEDERAL REGISTER. The materials are available for inspection at: the U.S. Environmental Protection Agency, RCRA Information Center (RIC), 1235 Jefferson-Davis Highway, first floor, Arlington, VA 22203 (Docket # F-94-IEHF-FFFFF) or at the National Archives and Records Administration (NARA), and may be obtained from the Organisation for Economic Co-operation and Development, Environment Direcorate, 2 rue Andre Pascal, 75775 Paris Cedex 16, France. For information on the availability of this material at NARA, call 202-741-6030, or go to: *http://www.archives.gov/ federal_register/*

*code_of_federal_regulations/
ibr_locations.html.*

[61 FR 16310, Apr. 12, 1996, as amended at 69
FR 18803, Apr. 9, 2004]

EFFECTIVE DATE NOTE: At 75 FR 1253, Jan.
8, 2010, subpart H was revised, effective July
7, 2010. For the convenience of the user, the
revised text is set forth as follows:

## Subpart H—Transboundary Movements of Hazardous Waste for Recovery Within the OECD

Sec.
262.80  Applicability.
262.81  Definitions.
262.82  General conditions.
262.83  Notification and consent.
262.84  Movement document.
262.85  Contracts.
262.86  Provisions relating to recognized traders.
262.87  Reporting and recordkeeping.
262.88  Pre-approval for U.S. recovery facilities [Reserved]
262.89  OECD waste lists.

## Subpart H—Transboundary Movements of Hazardous Waste for Recovery Within the OECD

### §262.80  Applicability.

(a) The requirements of this subpart apply
to imports and exports of wastes that are
considered hazardous under U.S. national
procedures and are destined for recovery operations in the countries listed in
§262.58(a)(1). A waste is considered hazardous
under U.S. national procedures if the waste:

(1) Meets the Federal definition of hazardous waste in 40 CFR 261.3; and

(2) Is subject to either the Federal RCRA
manifesting requirements at 40 CFR part 262,
subpart B, the universal waste management
standards of 40 CFR part 273, State requirements analogous to 40 CFR part 273, the export requirements in the spent lead-acid battery management standards of 40 CFR part
266, subpart G, or State requirements analogous to the export requirements in 40 CFR
part 266, subpart G.

(b) Any person (exporter, importer, or recovery facility operator) who mixes two or
more wastes (including hazardous and non-
hazardous wastes) or otherwise subjects two
or more wastes (including hazardous and
non-hazardous wastes) to physical or chemical transformation operations, and thereby
creates a new hazardous waste, becomes a
generator and assumes all subsequent generator duties under RCRA and any exporter duties, if applicable, under this subpart.

### §262.81  Definitions.

The following definitions apply to this subpart.

*Competent authority* means the regulatory
authority or authorities of concerned countries having jurisdiction over transboundary
movements of wastes destined for recovery
operations.

*Countries concerned* means the OECD Member countries of export or import and any
OECD Member countries of transit.

*Country of export* means any designated
OECD Member country listed in §262.58(a)(1)
from which a transboundary movement of
hazardous wastes is planned to be initiated
or is initiated.

*Country of import* means any designated
OECD Member country listed in §262.58(a)(1)
to which a transboundary movement of hazardous wastes is planned or takes place for
the purpose of submitting the wastes to recovery operations therein.

*Country of transit* means any designated
OECD Member country listed in §262.58(a)(1)
and (a)(2) other than the country of export or
country of import across which a
transboundary movement of hazardous
wastes is planned or takes place.

*Exporter* means the person under the jurisdiction of the country of export who has, or
will have at the time the planned
transboundary movement commences, possession or other forms of legal control of the
wastes and who proposes transboundary
movement of the hazardous wastes for the
ultimate purpose of submitting them to recovery operations. When the United States
(U.S.) is the country of export, *exporter* is interpreted to mean a person domiciled in the
United States.

*Importer* means the person to whom possession or other form of legal control of the
waste is assigned at the time the waste is received in the country of import.

*OECD area* means all land or marine areas
under the national jurisdiction of any OECD
Member country listed in §262.58. When the
regulations refer to shipments to or from an
OECD Member country, this means OECD
area.

*OECD* means the Organization for Economic Cooperation and Development.

*Recognized trader* means a person who, with
appropriate authorization of countries concerned, acts in the role of principal to purchase and subsequently sell wastes; this person has legal control of such wastes from
time of purchase to time of sale; such a person may act to arrange and facilitate
transboundary movements of wastes destined for recovery operations.

*Recovery facility* means a facility which,
under applicable domestic law, is operating
or is authorized to operate in the country of
import to receive wastes and to perform recovery operations on them.

*Recovery operations* means activities leading to resource recovery, recycling, reclamation, direct re-use or alternative uses, which include:

R1   Use as a fuel (other than in direct incineration) or other means to generate energy.

R2   Solvent reclamation/regeneration.

R3   Recycling/reclamation of organic substances which are not used as solvents.

R4   Recycling/reclamation of metals and metal compounds.

R5   Recycling/reclamation of other inorganic materials.

R6   Regeneration of acids or bases.

R7   Recovery of components used for pollution abatement.

R8   Recovery of components used from catalysts.

R9   Used oil re-refining or other reuses of previously used oil.

R10   Land treatment resulting in benefit to agriculture or ecological improvement.

R11   Uses of residual materials obtained from any of the operations numbered R1–R10.

R12   Exchange of wastes for submission to any of the operations numbered R1–R11.

R13   Accumulation of material intended for any operation numbered R1–R12.

*Transboundary movement* means any movement of wastes from an area under the national jurisdiction of one OECD Member country to an area under the national jurisdiction of another OECD Member country.

### § 262.82   General conditions.

(a) *Scope.* The level of control for exports and imports of waste is indicated by assignment of the waste to either a list of wastes subject to the Green control procedures or a list of wastes subject to the Amber control procedures and by the national procedures of the United States, as defined in § 262.80(a). The OECD Green and Amber lists are incorporated by reference in § 262.89(d).

(1) Listed wastes subject to the Green control procedures.

(i) Green wastes that are not considered hazardous under U.S. national procedures as defined in § 262.80(a) are subject to existing controls normally applied to commercial transactions.

(ii) Green wastes that are considered hazardous under U.S. national procedures as defined in § 262.80(a) are subject to the Amber control procedures set forth in this subpart.

(2) Listed wastes subject to the Amber control procedures.

(i) Amber wastes that are considered hazardous under U.S. national procedures as defined in § 262.80(a) are subject to the Amber control procedures set forth in this subpart.

(ii) Amber wastes that are considered hazardous under U.S. national procedures as defined in § 262.80(a), are subject to the Amber control procedures in the United States, even if they are imported to or exported from a designated OECD Member country listed in § 262.58(a)(1) that does not consider the waste to be hazardous. In such an event, the responsibilities of the Amber control procedures shift as provided:

(A) For U.S. exports, the United States shall issue an acknowledgement of receipt and assume other responsibilities of the competent authority of the country of import.

(B) For U.S. imports, the U.S. recovery facility/importer and the United States shall assume the obligations associated with the Amber control procedures that normally apply to the exporter and country of export, respectively.

(iii) Amber wastes that are not considered hazardous under U.S. national procedures as defined in § 262.80(a), but are considered hazardous by an OECD Member country are subject to the Amber control procedures in the OECD Member country that considers the waste hazardous. All responsibilities of the U.S. importer/exporter shift to the importer/exporter of the OECD Member country that considers the waste hazardous unless the parties make other arrangements through contracts.

NOTE TO PARAGRAPH (A)(2): Some wastes subject to the Amber control procedures are not listed or otherwise identified as hazardous under RCRA, and therefore are not subject to the Amber control procedures of this subpart. Regardless of the status of the waste under RCRA, however, other Federal environmental statutes (*e.g.*, the Toxic Substances Control Act) restrict certain waste imports or exports. Such restrictions continue to apply with regard to this subpart.

(3) Procedures for mixtures of wastes.

(i) A Green waste that is mixed with one or more other Green wastes such that the resulting mixture is not considered hazardous under U.S. national procedures as defined in § 262.80(a) shall be subject to the Green control procedures, provided the composition of this mixture does not impair its environmentally sound recovery.

NOTE TO PARAGRAPH (A)(3)(I): The regulated community should note that some OECD Member countries may require, by domestic law, that mixtures of different Green wastes be subject to the Amber control procedures.

(ii) A Green waste that is mixed with one or more Amber wastes, in any amount, *de minimis* or otherwise, or a mixture of two or more Amber wastes, such that the resulting waste mixture is considered hazardous under U.S. national procedures as defined in § 262.80(a) are subject to the Amber control procedures, provided the composition of this mixture does not impair its environmentally sound recovery.

NOTE TO PARAGRAPH (A)(3)(II): The regulated community should note that some

OECD Member countries may require, by domestic law, that a mixture of a Green waste and more than a *de minimis* amount of an Amber waste or a mixture of two or more Amber wastes be subject to the Amber control procedures.

(4) Wastes not yet assigned to an OECD waste list are eligible for transboundary movements, as follows:

(i) If such wastes are considered hazardous under U.S. national procedures as defined in §262.80(a), such wastes are subject to the Amber control procedures.

(ii) If such wastes are not considered hazardous under U.S. national procedures as defined in §262.80(a), such wastes are subject to the Green control procedures.

(b) *General conditions applicable to transboundary movements of hazardous waste:*
(1) The waste must be destined for recovery operations at a facility that, under applicable domestic law, is operating or is authorized to operate in the importing country;

(2) The transboundary movement must be in compliance with applicable international transport agreements; and

NOTE TO PARAGRAPH (B)(2): These international agreements include, but are not limited to, the Chicago Convention (1944), ADR (1957), ADNR (1970), MARPOL Convention (1973/1978), SOLAS Convention (1974), IMDG Code (1985), COTIF (1985), and RID (1985).

(3) Any transit of waste through a non-OECD Member country must be conducted in compliance with all applicable international and national laws and regulations.

(c) *Provisions relating to re-export for recovery to a third country:* (1) Re-export of wastes subject to the Amber control procedures from the United States, as the country of import, to a third country listed in §262.58(a)(1) may occur only after an exporter in the United States provides notification to and obtains consent from the competent authorities in the third country, the original country of export, and any transit countries. The notification must comply with the notice and consent procedures in §262.83 for all countries concerned and the original country of export. The competent authorities of the original country of export, as well as the competent authorities of all other countries concerned have thirty (30) days to object to the proposed movement.

(i) The thirty (30) day period begins once the competent authorities of both the initial country of export and new country of import issue Acknowledgements of Receipt of the notification.

(ii) The transboundary movement may commence if no objection has been lodged after the thirty (30) day period has passed or immediately after written consent is received from all relevant OECD importing and transit countries.

(2) In the case of re-export of Amber wastes to a country other than those listed in §262.58(a)(1), notification to and consent of the competent authorities of the original OECD Member country of export and any OECD Member countries of transit is required as specified in paragraph (c)(1) of this section, in addition to compliance with all international agreements and arrangements to which the first importing OECD Member country is a party and all applicable regulatory requirements for exports from the first country of import.

(d) *Duty to return or re-export wastes subject to the Amber control procedures.* When a transboundary movement of wastes subject to the Amber control procedures cannot be completed in accordance with the terms of the contract or the consent(s) and alternative arrangements cannot be made to recover the waste in an environmentally sound manner in the country of import, the waste must be returned to the country of export or re-exported to a third country. The provisions of paragraph (c) of this section apply to any shipments to be re-exported to a third country. The following provisions apply to shipments to be returned to the country of export as appropriate:

(1) Return from the United States to the country of export: The U.S. importer must inform EPA at the specified address in §262.83(b)(1)(i) of the need to return the shipment. EPA will then inform the competent authorities of the countries of export and transit, citing the reason(s) for returning the waste. The U.S. importer must complete the return within ninety (90) days from the time EPA informs the country of export of the need to return the waste, unless informed in writing by EPA of another timeframe agreed to by the concerned Member countries. If the return shipment will cross any transit country, the return shipment may only occur after EPA provides notification to and obtains consent from the competent authority of the country of transit, and provides a copy of that consent to the U.S. importer.

(2) Return from the country of import to the United States: The U.S. exporter must provide for the return of the hazardous waste shipment within ninety (90) days from the time the country of import informs EPA of the need to return the waste or such other period of time as the concerned Member countries agree. The U.S. exporter must submit an exception report to EPA in accordance with §262.87(b).

(e) *Duty to return wastes subject to the Amber control procedures from a country of transit.* When a transboundary movement of wastes subject to the Amber control procedures does not comply with the requirements of the notification and movement documents or otherwise constitutes illegal shipment, and if alternative arrangements cannot be made to recover these wastes in an environmentally sound manner, the waste must be returned to the country of export. The following provisions apply as appropriate:

(1) Return from the United States (as country of transit) to the country of export: The U.S. transporter must inform EPA at the specified address in § 262.83(b)(1)(i) of the need to return the shipment. EPA will then inform the competent authority of the country of export, citing the reason(s) for returning the waste. The U.S. transporter must complete the return within ninety (90) days from the time EPA informs the country of export of the need to return the waste, unless informed in writing by EPA of another timeframe agreed to by the concerned Member countries.

(2) Return from the country of transit to the United States (as country of export): The U.S. exporter must provide for the return of the hazardous waste shipment within ninety (90) days from the time the competent authority of the country of transit informs EPA of the need to return the waste or such other period of time as the concerned Member countries agree. The U.S. exporter must submit an exception report to EPA in accordance with § 262.87(b).

(f) *Requirements for wastes destined for and received by R12 and R13 facilities.* The transboundary movement of wastes destined for R12 and R13 operations must comply with all Amber control procedures for notification and consent as set forth in § 262.83 and for the movement document as set forth in § 262.84. Additional responsibilities of R12/R13 facilities include:

(1) Indicating in the notification document the foreseen recovery facility or facilities where the subsequent R1–R11 recovery operation takes place or may take place.

(2) Within three (3) days of the receipt of the wastes by the R12/R13 recovery facility or facilities, the facility(ies) shall return a signed copy of the movement document to the exporter and to the competent authorities of the countries of export and import. The facility(ies) shall retain the original of the movement document for three (3) years.

(3) As soon as possible, but no later than thirty (30) days after the completion of the R12/R13 recovery operation and no later than one (1) calendar year following the receipt of the waste, the R12 or R13 facility(ies) shall send a certificate of recovery to the foreign exporter and to the competent authority of the country of export and to the Office of Enforcement and Compliance Assurance, Office of Federal Activities, International Compliance Assurance Division (2254A), Environmental Protection Agency, 1200 Pennsylvania Avenue, NW. Washington, DC 20460, by mail, e-mail without digital signature followed by mail, or fax followed by mail.

(4) When an R12/R13 recovery facility delivers wastes for recovery to an R1–R11 recovery facility located in the country of import, it shall obtain as soon as possible, but no later than one (1) calendar year following delivery of the waste, a certification from the R1–R11 facility that recovery of the wastes at that facility has been completed. The R12/R13 facility must promptly transmit the applicable certification to the competent authorities of the countries of import and export, identifying the transboundary movements to which the certification pertain.

(5) When an R12/R13 recovery facility delivers wastes for recovery to an R1–R11 recovery facility located:

(i) In the initial country of export, Amber control procedures apply, including a new notification;

(ii) In a third country other than the initial country of export, Amber control procedures apply, with the additional provision that the competent authority of the initial country of export shall also be notified of the transboundary movement.

(g) *Laboratory analysis exemption.* The transboundary movement of an Amber waste is exempt from the Amber control procedures if it is in certain quantities and destined for laboratory analysis to assess its physical or chemical characteristics, or to determine its suitability for recovery operations. The quantity of such waste shall be determined by the minimum quantity reasonably needed to perform the analysis in each particular case adequately, but in no case exceed twenty-five kilograms (25 kg). Waste destined for laboratory analysis must still be appropriately packaged and labeled.

§ 262.83   **Notification and consent.**

(a) *Applicability.* Consent must be obtained from the competent authorities of the relevant OECD countries of import and transit prior to exporting hazardous waste destined for recovery operations subject to this subpart. Hazardous wastes subject to the Amber control procedures are subject to the requirements of paragraph (b) of this section; and wastes not identified on any list are subject to the requirements of paragraph (c) of this section.

(b) *Amber wastes.* Exports of hazardous wastes from the United States as described in § 262.80(a) that are subject to the Amber control procedures are prohibited unless the notification and consent requirements of paragraph (b)(1) or paragraph (b)(2) of this section are met.

(1) Transactions requiring specific consent:

(i) *Notification.* At least forty-five (45) days prior to commencement of each transboundary movement, the exporter must provide written notification in English of the proposed transboundary movement to the Office of Enforcement and Compliance Assurance, Office of Federal Activities, International Compliance Assurance Division (2254A), Environmental Protection Agency, 1200 Pennsylvania Avenue, NW., Washington,

DC 20460, with the words "Attention: OECD Export Notification" prominently displayed on the envelope. This notification must include all of the information identified in paragraph (d) of this section. In cases where wastes having similar physical and chemical characteristics, the same United Nations classification, the same RCRA waste codes, and are to be sent periodically to the same recovery facility by the same exporter, the exporter may submit one general notification of intent to export these wastes in multiple shipments during a period of up to one (1) year. Even when a general notification is used for multiple shipments, each shipment still must be accompanied by its own movement document pursuant to §262.84.

(ii) *Tacit consent.* If no objection has been lodged by any countries concerned (*i.e.*, exporting, importing, or transit) to a notification provided pursuant to paragraph (b)(1)(i) of this section within thirty (30) days after the date of issuance of the Acknowledgement of Receipt of notification by the competent authority of the country of import, the transboundary movement may commence. Tacit consent expires one (1) calendar year after the close of the thirty (30) day period; renotification and renewal of all consents is required for exports after that date.

(iii) *Written consent.* If the competent authorities of all the relevant OECD importing and transit countries provide written consent in a period less than thirty (30) days, the transboundary movement may commence immediately after all necessary consents are received. Written consent expires for each relevant OECD importing and transit country one (1) calendar year after the date of that country's consent unless otherwise specified; renotification and renewal of each expired consent is required for exports after that date.

(2) Transboundary movements to facilities pre-approved by the competent authorities of the importing countries to accept specific wastes for recovery:

(i) *Notification.* The exporter must provide EPA a notification that contains all the information identified in paragraph (d) of this section in English, at least ten (10) days in advance of commencing shipment to a pre-approved facility. The notification must indicate that the recovery facility is pre-approved, and may apply to a single specific shipment or to multiple shipments as described in paragraph (b)(1)(i) of this section. This information must be sent to the Office of Enforcement and Compliance Assurance, Office of Federal Activities, International Compliance Assurance Division (2254A), Environmental Protection Agency, 1200 Pennsylvania Avenue, NW., Washington, DC 20460, with the words "OECD Export Notification—Pre-approved Facility" prominently displayed on the envelope. General notifications that cover multiple shipments as de-

scribed in paragraph (b)(1)(i) of this section may cover a period of up to three (3) years. Even when a general notification is used for multiple shipments, each shipment still must be accompanied by its own movement document pursuant to §262.84.

(ii) Exports to pre-approved facilities may take place after the elapse of seven (7) working days from the issuance of an Acknowledgement of Receipt of the notification by the competent authority of the country of import unless the exporter has received information indicating that the competent authority of any countries concerned objects to the shipment.

(c) *Wastes not covered in the OECD Green and Amber lists.* Wastes destined for recovery operations, that have not been assigned to the OECD Green and Amber lists, incorporated by reference in §262.89(d), but which are considered hazardous under U.S. national procedures as defined in §262.80(a), are subject to the notification and consent requirements established for the Amber control procedures in accordance with paragraph (b) of this section. Wastes destined for recovery operations, that have not been assigned to the OECD Green and Amber lists incorporated by reference in §262.89(d), and are not considered hazardous under U.S. national procedures as defined by §262.80(a) are subject to the Green control procedures.

(d) *Notifications submitted under this section must include the information specified in paragraphs (d)(1) through (d)(14) of this section:* (1) Serial number or other accepted identifier of the notification document;

(2) Exporter name and EPA identification number (if applicable), address, telephone, fax numbers, and e-mail address;

(3) Importing recovery facility name, address, telephone, fax numbers, e-mail address, and technologies employed;

(4) Importer name (if not the owner or operator of the recovery facility), address, telephone, fax numbers, and e-mail address; whether the importer will engage in waste exchange recovery operation R12 or waste accumulation recovery operation R13 prior to delivering the waste to the final recovery facility and identification of recovery operations to be employed at the final recovery facility;

(5) Intended transporter(s) and/or their agent(s); address, telephone, fax, and e-mail address;

(6) Country of export and relevant competent authority, and point of departure;

(7) Countries of transit and relevant competent authorities and points of entry and departure;

(8) Country of import and relevant competent authority, and point of entry;

(9) Statement of whether the notification is a single notification or a general notification. If general, include period of validity requested;

(10) Date(s) foreseen for commencement of transboundary movement(s);

(11) Means of transport envisaged;

(12) Designation of waste type(s) from the appropriate OECD list incorporated by reference in § 262.89(d), description(s) of each waste type, estimated total quantity of each, RCRA waste code, and the United Nations number for each waste type;

(13) Specification of the recovery operation(s) as defined in § 262.81.

(14) Certification/Declaration signed by the exporter that states:

I certify that the above information is complete and correct to the best of my knowledge. I also certify that legally-enforceable written contractual obligations have been entered into, and that any applicable insurance or other financial guarantees are or shall be in force covering the transboundary movement.
Name: _____
Signature: _____
Date: _____

NOTE TO PARAGRAPH (D)(14): The United States does not currently require financial assurance for these waste shipments. However, U.S. exporters may be asked by other governments to provide and certify to such assurance as a condition of obtaining consent to a proposed movement.

(e) *Certificate of Recovery.* As soon as possible, but no later than thirty (30) days after the completion of recovery and no later than one (1) calendar year following receipt of the waste, the U.S. recovery facility shall send a certificate of recovery to the exporter and to the competent authorities of the countries of export and import by mail, e-mail without a digital signature followed by mail, or fax followed by mail. The certificate of recovery shall include a signed, written and dated statement that affirms that the waste materials were recovered in the manner agreed to by the parties to the contract required under § 262.85.

## § 262.84  Movement document.

(a) All U.S. parties subject to the contract provisions of § 262.85 must ensure that a movement document meeting the conditions of paragraph (b) of this section accompanies each transboundary movement of wastes subject to the Amber control procedures from the initiation of the shipment until it reaches the final recovery facility, including cases in which the waste is stored and/or sorted by the importer prior to shipment to the final recovery facility, except as provided in paragraphs (a)(1) and (2) of this section.

(1) For shipments of hazardous waste within the United States solely by water (bulk shipments only), the generator must forward the movement document with the manifest to the last water (bulk shipment) transporter to handle the waste in the United States if exported by water, (in accordance with the manifest routing procedures at § 262.23(c)).

(2) For rail shipments of hazardous waste within the United States which originate at the site of generation, the generator must forward the movement document with the manifest (in accordance with the routing procedures for the manifest in § 262.23(d)) to the next non-rail transporter, if any, or the last rail transporter to handle the waste in the United States if exported by rail.

(b) The movement document must include all information required under § 262.83 (for notification), as well as the following paragraphs (b)(1) through (b)(7) of this section:

(1) Date movement commenced;

(2) Name (if not exporter), address, telephone, fax numbers, and e-mail of primary exporter;

(3) Company name and EPA ID number of all transporters;

(4) Identification (license, registered name or registration number) of means of transport, including types of packaging envisaged;

(5) Any special precautions to be taken by transporter(s);

(6) Certification/declaration signed by the exporter that no objection to the shipment has been lodged, as follows:

I certify that the above information is complete and correct to the best of my knowledge. I also certify that legally-enforceable written contractual obligations have been entered into, that any applicable insurance or other financial guarantees are or shall be in force covering the transboundary movement, and that:

1. All necessary consents have been received; OR

2. The shipment is directed to a recovery facility within the OECD area and no objection has been received from any of the countries concerned within the thirty (30) day tacit consent period; OR

3. The shipment is directed to a recovery facility pre-approved for that type of waste within the OECD area; such an authorization has not been revoked, and no objection has been received from any of the countries concerned.

(Delete sentences that are not applicable)
Name: _____
Signature: _____
Date: _____

(7) Appropriate signatures for each custody transfer (*e.g.*, transporter, importer, and owner or operator of the recovery facility).

(c) Exporters also must comply with the special manifest requirements of 40 CFR 262.54(a), (b), (c), (e), and (i) and importers must comply with the import requirements of 40 CFR part 262, subpart F.

(d) Each U.S. person that has physical custody of the waste from the time the movement commences until it arrives at the recovery facility must sign the movement document (*e.g.*, transporter, importer, and owner or operator of the recovery facility).

(e) Within three (3) working days of the receipt of imports subject to this subpart, the owner or operator of the U.S. recovery facility must send signed copies of the movement document to the exporter, to the Office of Enforcement and Compliance Assurance, Office of Federal Activities, International Compliance Assurance Division (2254A), Environmental Protection Agency, 1200 Pennsylvania Avenue, NW., Washington, DC 20460, and to the competent authorities of the countries of export and transit. If the concerned U.S. recovery facility is a R12/R13 recovery facility as defined under §262.81, the facility shall retain the original of the movement document for three (3) years.

## §262.85 Contracts.

(a) Transboundary movements of hazardous wastes subject to the Amber control procedures are prohibited unless they occur under the terms of a valid written contract, chain of contracts, or equivalent arrangements (when the movement occurs between parties controlled by the same corporate or legal entity). Such contracts or equivalent arrangements must be executed by the exporter and the owner or operator of the recovery facility, and must specify responsibilities for each. Contracts or equivalent arrangements are valid for the purposes of this section only if persons assuming obligations under the contracts or equivalent arrangements have appropriate legal status to conduct the operations specified in the contract or equivalent arrangements.

(b) Contracts or equivalent arrangements must specify the name and EPA ID number, where available, of paragraph (b)(1) through (b)(4) of this section:

(1) The generator of each type of waste;

(2) Each person who will have physical custody of the wastes;

(3) Each person who will have legal control of the wastes; and

(4) The recovery facility.

(c) Contracts or equivalent arrangements must specify which party to the contract will assume responsibility for alternate management of the wastes if their disposition cannot be carried out as described in the notification of intent to export. In such cases, contracts must specify that:

(1) The person having actual possession or physical control over the wastes will immediately inform the exporter and the competent authorities of the countries of export and import and, if the wastes are located in a country of transit, the competent authorities of that country; and

(2) The person specified in the contract will assume responsibility for the adequate management of the wastes in compliance with applicable laws and regulations including, if necessary, arranging the return of wastes and, as the case may be, shall provide the notification for re-export.

(d) Contracts must specify that the importer will provide the notification required in §262.82(c) prior to the re-export of controlled wastes to a third country.

(e) Contracts or equivalent arrangements must include provisions for financial guarantees, if required by the competent authorities of any countries concerned, in accordance with applicable national or international law requirements.

NOTE TO PARAGRAPH (E): Financial guarantees so required are intended to provide for alternate recycling, disposal or other means of sound management of the wastes in cases where arrangements for the shipment and the recovery operations cannot be carried out as foreseen. The United States does not require such financial guarantees at this time; however, some OECD Member countries do. It is the responsibility of the exporter to ascertain and comply with such requirements; in some cases, transporters or importers may refuse to enter into the necessary contracts absent specific references or certifications to financial guarantees.

(f) Contracts or equivalent arrangements must contain provisions requiring each contracting party to comply with all applicable requirements of this subpart.

(g) Upon request by EPA, U.S. exporters, importers, or recovery facilities must submit to EPA copies of contracts, chain of contracts, or equivalent arrangements (when the movement occurs between parties controlled by the same corporate or legal entity). Information contained in the contracts or equivalent arrangements for which a claim of confidentiality is asserted in accordance with 40 CFR 2.203(b) will be treated as confidential and will be disclosed by EPA only as provided in 40 CFR 260.2.

NOTE TO PARAGRAPH (G): Although the United States does not require routine submission of contracts at this time, the OECD Decision allows Member countries to impose such requirements. When other OECD Member countries require submission of partial or complete copies of the contract as a condition to granting consent to proposed movements, EPA will request the required information; absent submission of such information, some OECD Member countries may deny consent for the proposed movement.

## §262.86 Provisions relating to recognized traders.

(a) A recognized trader who takes physical custody of a waste and conducts recovery operations (including storage prior to recovery) is acting as the owner or operator of a recovery facility and must be so authorized in accordance with all applicable Federal laws.

(b) A recognized trader acting as an exporter or importer for transboundary shipments of waste must comply with all the requirements of this subpart associated with being an exporter or importer.

### § 262.87  Reporting and recordkeeping.

(a) *Annual reports.* For all waste movements subject to this subpart, persons (*e.g.*, exporters, recognized traders) who meet the definition of primary exporter in § 262.51 or who initiate the movement documentation under § 262.84 shall file an annual report with the Office of Enforcement and Compliance Assurance, Office of Federal Activities, International Compliance Assurance Division (2254A), Environmental Protection Agency, 1200 Pennsylvania Avenue, NW., Washington, DC 20460, no later than March 1 of each year summarizing the types, quantities, frequency, and ultimate destination of all such hazardous waste exported during the previous calendar year. (If the primary exporter or the person who initiates the movement document under § 262.84 is required to file an annual report for waste exports that are not covered under this subpart, he may include all export information in one report provided the following information on exports of waste destined for recovery within the designated OECD Member countries is contained in a separate section.) Such reports shall include all of the following paragraphs (a)(1) through (a)(6) of this section specified as follows:

(1) The EPA identification number, name, and mailing and site address of the exporter filing the report;

(2) The calendar year covered by the report;

(3) The name and site address of each final recovery facility;

(4) By final recovery facility, for each hazardous waste exported, a description of the hazardous waste, the EPA hazardous waste number (from 40 CFR part 261, subpart C or D), designation of waste type(s) and applicable waste code(s) from the appropriate OECD waste list incorporated by reference in § 262.89(d), DOT hazard class, the name and U.S. EPA identification number (where applicable) for each transporter used, the total amount of hazardous waste shipped pursuant to this subpart, and number of shipments pursuant to each notification;

(5) In even numbered years, for each hazardous waste exported, except for hazardous waste produced by exporters of greater than 100kg but less than 1,000kg in a calendar month, and except for hazardous waste for which information was already provided pursuant to § 262.41:

(i) A description of the efforts undertaken during the year to reduce the volume and toxicity of the waste generated; and

(ii) A description of the changes in volume and toxicity of the waste actually achieved during the year in comparison to previous years to the extent such information is available for years prior to 1984; and

(6) A certification signed by the person acting as primary exporter or initiator of the movement document under § 262.84 that states:

I certify under penalty of law that I have personally examined and am familiar with the information submitted in this and all attached documents, and that based on my inquiry of those individuals immediately responsible for obtaining the information, I believe that the submitted information is true, accurate, and complete. I am aware that there are significant penalties for submitting false information including the possibility of fine and imprisonment.

(b) *Exception reports.* Any person who meets the definition of primary exporter in § 262.51 or who initiates the movement document under § 262.84 must file an exception report in lieu of the requirements of § 262.42 (if applicable) with the Office of Enforcement and Compliance Assurance, Office of Federal Activities, International Compliance Assurance Division (2254A), Environmental Protection Agency, 1200 Pennsylvania Avenue, NW., Washington, DC 20460, if any of the following occurs:

(1) He has not received a copy of the RCRA hazardous waste manifest (if applicable) signed by the transporter identifying the point of departure of the waste from the United States, within forty-five (45) days from the date it was accepted by the initial transporter;

(2) Within ninety (90) days from the date the waste was accepted by the initial transporter, the exporter has not received written confirmation from the recovery facility that the hazardous waste was received;

(3) The waste is returned to the United States.

(c) *Recordkeeping.* (1) Persons who meet the definition of primary exporter in § 262.51 or who initiate the movement document under § 262.84 shall keep the following records in paragraphs (c)(1)(i) through (c)(1)(iv) of this section:

(i) A copy of each notification of intent to export and all written consents obtained from the competent authorities of countries concerned for a period of at least three (3) years from the date the hazardous waste was accepted by the initial transporter;

(ii) A copy of each annual report for a period of at least three (3) years from the due date of the report;

(iii) A copy of any exception reports and a copy of each confirmation of delivery (*i.e.*, movement document) sent by the recovery facility to the exporter for at least three (3) years from the date the hazardous waste was accepted by the initial transporter or received by the recovery facility, whichever is applicable; and

(iv) A copy of each certificate of recovery sent by the recovery facility to the exporter for at least three (3) years from the date that the recovery facility completed processing the waste shipment.

(2) The periods of retention referred to in this section are extended automatically during the course of any unresolved enforcement action regarding the regulated activity or as requested by the Administrator.

**§262.88  Pre-approval for U.S. recovery facilities [Reserved]**

**§262.89  OECD waste lists.**

(a) *General.* For the purposes of this subpart, a waste is considered hazardous under U.S. national procedures, and hence subject to this subpart, if the waste:

(1) Meets the Federal definition of hazardous waste in 40 CFR 261.3; and

(2) Is subject to either the Federal RCRA manifesting requirements at 40 CFR part 262, subpart B, the universal waste management standards of 40 CFR part 273, State requirements analogous to 40 CFR part 273, the export requirements in the spent lead-acid battery management standards of 40 CFR part 266, subpart G, or State requirements analogous to the export requirements in 40 CFR part 266, subpart G.

(b) If a waste is hazardous under paragraph (a) of this section, it is subject to the Amber control procedures, regardless of whether it appears in Appendix 4 of the OECD Decision, as defined in §262.81.

(c) The appropriate control procedures for hazardous wastes and hazardous waste mixtures are addressed in §262.82.

(d) The OECD waste lists, as set forth in Annex B ("Green List") and Annex C ("Amber List") (collectively "OECD waste lists") of the 2009 "Guidance Manual for the Implementation of Council Decision C(2001)107/FINAL, as Amended, on the Control of Transboundary Movements of Wastes Destined for Recovery Operations," are incorporated by reference. This incorporation by reference was approved by the Director of the Federal Register in accordance with 5 U.S.C. 552(a) and 1 CFR part 51. This material is incorporated as it exists on the date of the approval and a notice of any change in these materials will be published in the FEDERAL REGISTER. The materials are available for inspection at: the U.S. Environmental Protection Agency, Docket Center Public Reading Room, EPA West, Room 3334, 1301 Constitution Avenue NW., Washington, DC 20004 (Docket # EPA–HQ–RCRA–2005–0018) or at the National Archives and Records Ad-

ministration (NARA), and may be obtained from the Organization for Economic Cooperation and Development, Environment Directorate, 2 rue André Pascal, F–75775 Paris Cedex 16, France. For information on the availability of this material at NARA, call 202–741–6030, or go to: *http://www.archives.gov/federal-register/cfr/ibr-locations.html.* To contact the EPA Docket Center Public Reading Room, call (202) 566–1744. To contact the OECD, call +33 (0) 1 45 24 81 67.

## Subpart I—New York State Public Utilities

SOURCE: 64 FR 37636, July 12, 1999, unless otherwise noted.

**§262.90  Project XL for Public Utilities in New York State.**

(a) The following definitions apply to this section:

(1) A *Utility* is any company that operates wholesale and/or retail oil and gas pipelines, or any company that provides electric power or telephone service and is regulated by New York State's Public Service Commission or the New York Power Authority.

(2) A *right-of-way* is a fixed, integrated network of aboveground or underground conveyances, including land structures, fixed equipment, and other appurtenances, controlled or owned by a Utility, and used for the purpose of conveying its products or services to customers.

(3) A *remote location* is a location in New York State within a Utility's right-of-way network that is not permanently staffed.

(4) A *Utility's central collection facility (UCCF)* is a Utility-owned facility within the Utility's right-of-way network to which hazardous waste, generated by the Utility at remote locations within the same right-of-way network, is brought.

(b) A UCCF designated pursuant to paragraph (e) of this section may consolidate hazardous waste (with the exception of mixed waste) generated by that Utility at its remote locations (and at that UCCF) for up to 90 days without a permit or without having interim status, provided that:

(1) The Utility complies with all applicable requirements for generators in 40 CFR part 262 (except §262.34 (d)

through (f)) for hazardous waste generated at its remote locations and at the UCCF, including the manifest and pretransport requirements for all shipments greater than 100 kilograms sent from a remote location to a UCCF.

(2) The Utility transports the hazardous waste from the remote location to a UCCF immediately after collection of all hazardous waste at the remote location is complete or when the staff collecting the hazardous waste leave the remote location, whichever comes first.

(3) The Utility complies with all applicable requirements for transporters in 40 CFR part 263 for each shipment of hazardous waste greater than 100 kilograms which is sent from remote location to the UCCF, and all applicable Department of Transportation requirements.

(4)(i) The Utility complies with 40 CFR 262.34 (a) through (c), regardless of the total quantity of hazardous waste generated or consolidated at the UCCF per calendar month;

(ii) The Utility complies with 40 CFR 264.178; and

(iii) Secondary containment is provided for all liquid hazardous waste consolidated in containers if:

(A) The UCCF is consolidating 8,800 gallons or more of liquid hazardous waste, or

(B) The UCCF is consolidating 185 gallons or more of liquid hazardous waste and is located in an area designated by New York State that overlays a sole-source aquifer.

(5) The Utility submits a biennial report in accordance with 40 CFR 262.41 including all hazardous waste shipped from remote locations to the UCCF. This UCCF biennial report may be submitted in lieu of submitting a biennial report for each remote location. However, for hazardous waste generated at a particular remote location that exceeds 1000 kg per calendar month and that is not sent to the UCCF, the Utility must submit a separate biennial report.

(6) Waste generated at a remote location that is not sent to a UCCF is managed according to the requirements of parts 260 through 270 of this chapter.

(7) The Utility maintains records at the UCCF in accordance with all the recordkeeping requirements set forth in subpart D of 40 CFR part 262, including 40 CFR 262.40, and maintains records on any PCB test results for hazardous wastes brought to the facility from remote locations.

(8) The UCCF obtains an EPA identification number.

(9) The UCCF receives hazardous waste only from its remote location.

(10) The Utility reinvests at least one-third of the direct savings described in paragraph (h) of this section in one or more environmentally beneficial projects, such as remediation or pollution prevention, that are over and above existing legal requirements and that have not been initiated prior to the Utility's receipt of approval to consolidate hazardous waste pursuant to this section.

(c) Utilities seeking to have UCCFs designated under paragraph (e) of this section must comply with the following requirements:

(1) Any New York State Utility seeking approval to consolidate hazardous waste under this section must notify local governments and communities of the Utility's intent to designate specific UCCFs.

(2) In carrying out paragraph (c)(1) of this section, the Utility must solicit public comment. In soliciting public comment, the Utility must use the notice method set forth in paragraph (c)(2)(i) of this section, as well as at least two of the methods set forth in paragraphs (c)(2)(ii) through (vii) of this section. Each Utility must also notify by mail all parties who commented on the proposed rule for this XL project.

(i) A public notice in a newspaper of general circulation within the area in which each proposed UCCF is located;

(ii) A radio announcement in each affected community during peak listening hours;

(iii) Mailings to all citizens within a five-mile radius of proposed UCCF;

(iv) Well-publicized community meetings;

(v) Presentations to the local community board;

(vi) Placement of copies of this section and the Final Project Agreement that explains the regulatory relief outlined in this section in the local library nearest the proposed UCCF, and inclusion of the name and address of the library in the newspaper notice; and

(vii) Placement of copies of this section and the Final Project Agreement that explains the regulatory relief outlined in this section on the Utility's web site, and inclusion of the web site's address in the newspaper notice.

(3) All outreach efforts made under paragraph (c)(2) of this section shall be prepared in English (and any other language spoken by a large number of persons in the community of concern) and at a minimum shall include the following information:

(i) A brief description of the XL project, the intended new use of the facility, and a request for comments on the proposed UCCF.

(ii) The name, if any, and address of the proposed UCCF and its current status under the RCRA Subtitle C program.

(iii) The intended duration of use of the UCCF under the requirements of this section.

(iv) Names, addresses, and telephone numbers of contact persons, representing the Utility, to whom questions or comments may be directed.

(v) Notification of when the comment period of no less than 30 days will close.

(4) Prior to the solicitation of public comment pursuant to paragraph (c)(2) of this section, the Utility must submit copies of each notice, announcement or mailing directly to local governments and to EPA.

(5) At the close of the comment period, the Utility shall prepare a Responsiveness Package containing a summary of public outreach efforts, all comments and questions received as a result of its outreach efforts, and the Utility's written responses to all comments and questions. The Utility shall provide copies of its Responsiveness Package to any citizens that participated in the public notice process, local governments and EPA.

(d) Upon completion of the public notice procedures described in paragraph (c) of this section, the Utility must provide written notice to EPA of its intent to participate. The Notice of Intent must contain the following information:

(1) The name of the Utility, corporate address, and corporate mailing address, if different.

(2) The name, mailing address, and telephone number of a corporate-level contact person to whom communications and inquiries may be directed. This contact person may be changed by written notification to EPA.

(3) A list of the names, addresses, and EPA identification numbers, if applicable, of all Utility-owned facilities in New York State that are proposed UCCFs and the names and telephone numbers of a designated contact person at each facility.

(4) A summary of public outreach efforts undertaken pursuant to paragraph (c) of this section.

(5) A commitment that one-third of the direct cost savings outlined in paragraph (h) of this section due to project participation will be reinvested in one or more environmentally beneficial projects which are over and above existing legal requirements and which have not been initiated prior to the Utility's receipt of approval to consolidate hazardous waste pursuant to this section.

(6) An acknowledgment that the signatory is personally familiar with the terms and conditions of this section and has the authority to obligate and does obligate the Utility to comply with all such terms and conditions. The Utility shall comply with the signatory requirements set forth in 40 CFR 270.11(a)(1).

(e) The procedures for designating UCCFs are as follows:

(1) Subject to paragraphs (e)(2) through (5) of this section, the Utility and specified UCCF shall receive approval to comply with the requirements set forth in paragraph (b) of this section upon the receipt of written acknowledgment from EPA that the Notice of Intent described in paragraph (d) of this section has been received and found to be complete and in compliance with all the requirements set forth in paragraph (d) of this section. This acknowledgment will state whether the UCCF has been designated under this section and any additional limitations which have been placed on the UCCF.

(2) Based on information provided and comments received during the public notice and comment period, EPA shall prepare a response to the comments received. The response to comments shall be attached to the acknowledgment described in paragraph (e)(1). Both the acknowledgment and the response to comments shall be sent to all persons who commented on the designation of the UCCF(s) that are the subject of the acknowledgment.

(3) Based on information provided and comments received during or after the public notice and comment period, designated UCCFs may be rejected for the proposed use, or, if EPA determines that acceptance for the proposed use under the conditions of paragraph (b) of this section may not fully protect human health and the environment based on the Utility's compliance history or other appropriate factors, the acknowledgment may impose conditions in addition to those in paragraph (b) of this section.

(4) If EPA determines that a site-specific informational public meeting is warranted prior to determining the acceptability of a designated UCCF, the acknowledgment will so state.

(5) Subsequent to any public meeting, EPA may reject or prohibit UCCFs from participating in this project based on information provided or comments received during or after the public notice process or based on a determination that acceptance for the proposed use under the conditions of paragraph (b) of this section may not fully protect human health and the environment based on the Utility's compliance history or other appropriate factors.

(f) At any time, a Utility may add or remove UCCF designations by complying with the following requirements:

(1) A Utility may notify EPA of its intent to designate additional UCCFs. Such a notification shall be submitted to, and processed by, EPA, in the manner indicated in paragraphs (d) and (e) of this section.

(2) To have one or more additional UCCFs designated, the Utility must comply with paragraph (c) of this section.

(3) A Utility can discontinue use of a facility as a UCCF by notifying EPA in writing.

(g) Each Utility that receives approval to consolidate hazardous waste pursuant to this section shall submit an Annual Progress Report with the following information for the preceding year:

(1) The number of remote locations statewide for which hazardous waste was handled in accordance with paragraph (b) of this section.

(2) The total tonnage of each type of hazardous waste handled by each UCCF.

(3) The number of remote locations statewide from which 1,000 kilograms or more of hazardous waste were collected per calendar month.

(4) The number of remote locations statewide from which between 100 and 1,000 kilograms of hazardous waste were collected per calendar month.

(5) An estimate of the monetary value, on a Utility-wide basis, of the direct savings realized by participation in this project. Direct savings at a minimum include those outlined in paragraph (h) of this section.

(6) Descriptions of the environmental compliance, remediation, or pollution prevention projects or activities into which the savings, described in paragraph (h) of this section, have been reinvested, with an estimate of the savings reinvested in each. Any such projects must consist of activities that are over and above existing legal requirements and that have not been initiated prior to the Utility's receipt of approval to consolidate hazardous waste pursuant to this section.

(7) The addresses and EPA identification numbers for all facilities that served as UCCFs for hazardous waste from remote locations.

(h) Utilities that receive approval to consolidate hazardous waste pursuant to this section must assess the direct savings realized as a result. Cost estimates shall include direct savings based on relief from any regulatory requirements, which the facility expects to be relieved from due to compliance with the provisions of this section including, but not limited to, the following:

(1) Database management for each remote location as an individual generator;

(2) Biennial Report preparation costs; and/or

(3) Cost savings realized from consolidation of waste for economical shipment (including no longer shipping waste directly to a TSD from remote locations).

(i) If any UCCF or Utility that receives approval under this section fails to comply with any of the requirements of this section, EPA may terminate or suspend the UCCF's or Utility's participation. EPA will provide a UCCF or Utility with 15 days written notice of its intent to terminate or suspend participation. During this period, the UCCF will have the opportunity to come back into compliance or provide a written explanation as to why it was not in compliance with the terms of this section and how it will come back into compliance. If EPA then issues a written notice terminating or suspending participation, the Utility must take immediate action to come into compliance with all otherwise applicable federal requirements. EPA may also take enforcement action against a Utility for non-compliance with the provisions of this section.

(j) This section will expire on May 24, 2011.

[64 FR 37636, July 12, 1999, as amended at 70 FR 29913, May 24, 2005; 71 FR 40272, July 14, 2006]

## Subpart J—University Laboratories XL Project—Laboratory Environmental Management Standard

Source: 64 FR 53292, Sept. 28, 1999, unless otherwise noted.

### §262.100  To what organizations does this subpart apply?

This subpart applies to an organization that meets all three of the following conditions:

(a) It is one of the three following academic institutions: The University of Massachusetts Boston in Boston, Massachusetts, Boston College in Chestnut Hill, Massachusetts, or the University of Vermont in Burlington, Vermont ("Universities"); and

(b) It is a laboratory at one of the Universities (identified pursuant to §262.105(c)(2)(ii)) where laboratory scale activities, as defined in §262.102, result in laboratory waste; and

(c) It complies with all the requirements of this subpart.

### §262.101  What is in this subpart?

This subpart provides a framework for a new management system for wastes that are generated in University laboratories. This framework is called the Laboratory Environmental Management Standard. The standard includes some specific definitions that apply to the University laboratories. It contains specific requirements for how to handle laboratory waste that are called Minimum Performance Criteria. The standard identifies the requirements for developing and implementing an environmental management plan. It outlines the responsibilities of the management staff of each participating university. Finally, the standard identifies requirements for training people who will work in the laboratories or manage laboratory waste. This Subpart contains requirements for RCRA solid and hazardous waste determination, and circumstances for termination and expiration of this pilot.

### §262.102  What special definitions are included in this subpart?

For purposes of this subpart, the following definitions apply:

*Acutely Hazardous Laboratory Waste* means a laboratory waste, defined in the Environmental Management Plan as posing significant potential hazards to human health or the environment and which must include RCRA "P" wastes, and may include particularly hazardous substances as designated in a University's Chemical Hygiene Plan under OSHA, or Extremely Hazardous Substances under the Emergency Planning and Community Right to Know Act.

*Emergency* means any occurrence such as, but not limited to, equipment failure, rupture of containers or failure of control equipment which results in the potential uncontrolled release of a

307

hazardous chemical into the environment and which requires agency or fire department notification and/or reporting.

*Environmental Management Plan* (EMP) means a written program developed and implemented by the university which sets forth standards and procedures, responsibilities, pollution control equipment, performance criteria, resources and work practices that both protect human health and the environment from the hazards presented by laboratory wastes within a laboratory and between a laboratory and the hazardous waste accumulation area, and satisfies the plan requirements defined elsewhere in this Subpart. Certain requirements of this plan are satisfied through the use of the Chemical Hygiene Plan (see, 29 CFR 1910.1450), or equivalent, and other relevant plans, including a waste minimization plan. The elements of the Environmental Management Plan must be easily accessible, but may be integrated into existing plans, incorporated as an attachment, or developed as a separate document.

*Environmental Objective* means an overall environmental goal of the organization which is verifiable.

*Environmental Performance* means results of the data collected pursuant to implementation of the Environmental Management Plan as measured against policy, objectives and targets.

*Environmental Target* means an environmental performance requirement of the organization which is quantifiable, where practicable, verifiable and designed to be achieved within a specified time frame.

*Hazardous Chemical* means any chemical which is a physical hazard or a health hazard. A physical hazard means a chemical for which there is scientifically valid evidence that it is a combustible liquid, a compressed gas, explosive, flammable, an organic peroxide, an oxidizer, pyrophoric, unstable (reactive) or water-reactive. A health hazard means a chemical for which there is statistically significant evidence based on at least one study conducted in accordance with established scientific principles that acute or chronic health effects may occur in exposed employees. The term "health hazard" includes chemicals which are carcinogens, toxic or highly toxic agents, reproductive toxins, irritants, corrosives, sensitizers, hepatotoxins, nephrotoxins, neurotoxins, agents which act on the hematopoietic system and agents which damage the lungs, skin, eyes or mucous membranes.

*Hazardous Chemical of Concern* means a chemical that the organization has identified as having the potential to be of significant risk to human health or the environment if not managed in accordance with procedures or practices defined by the organization.

*Hazardous Waste Accumulation Area* means the on-site area at a University where the University will make a solid and hazardous waste determination with respect to laboratory wastes.

*In-Line Waste Collection* means a system for the automatic collection of laboratory waste which is directly connected to or part of a laboratory scale activity and which is constructed or operated in a manner which prevents the release of any laboratory waste therein into the environment during collection.

*Laboratory* means, for the purpose of this Subpart, an area within a facility where the laboratory use of hazardous chemicals occurs. It is a workplace where relatively small quantities of hazardous chemicals are used on a non-production basis. The physical extent of individual laboratories within an organization will be defined by the Environmental Management Plan. A laboratory may include more than a single room if the rooms are in the same building and under the common supervision of a laboratory supervisor.

*Laboratory Clean-Out* means an evaluation of the chemical inventory of a laboratory as a result of laboratory renovation, relocation or a change in laboratory supervision that may result in the transfer of laboratory wastes to the hazardous waste accumulation area.

*Laboratory Environmental Management Standard* means the provisions of this Subpart and includes the requirements for preparation of Environmental Management Plans and the inclusion of Minimum Performance Criteria within each Environmental Management Plan.

*Laboratory Scale* means work with substances in which containers used for reactions, transfers and other handling of substances are designed to be safely and easily manipulated by one person. "Laboratory Scale" excludes those workplaces whose function is to produce commercial quantities of chemicals.

*Laboratory Waste* means a hazardous chemical that results from laboratory scale activities and includes the following: excess or unused hazardous chemicals that may or may not be re-used outside their laboratory of origin; hazardous chemicals determined to be RCRA hazardous waste as defined in 40 CFR Part 261; and hazardous chemicals that will be determined not to be RCRA hazardous waste pursuant to §262.106.

*Laboratory Worker* means a person who is assigned to handle hazardous chemicals in the laboratory and may include researchers, students or technicians.

*Legal and Other Requirements* means requirements imposed by, or as a result of, governmental permits, governmental laws and regulations, judicial and administrative enforcement orders, non-governmental legally enforceable contracts, research grants and agreements, certification specifications, formal voluntary commitments and organizational policies and standards.

*Senior Management* means senior personnel with overall responsibility, authority and accountability for managing laboratory activities within the organization.

*Universities* means the following academic institutions; University of Vermont, Boston College, and the University of Massachusetts Boston, which are participants in this Laboratory XL project and which are subject to the requirements set forth in this Subpart J.

## §262.103 What is the scope of the laboratory environmental management standard?

The Laboratory Environmental Management Standard will not affect or supersede any legal requirements other than those described in §262.10(j). The requirements that continue to apply include, but are not limited to, OSHA, Fire Codes, wastewater permit limitations, emergency response notification provisions, or other legal requirements applicable to University laboratories.

## §262.104 What are the minimum performance criteria?

The Minimum Performance Criteria that each University must meet in managing its Laboratory Waste are:

(a) Each University must label all laboratory waste with the general hazard class and either the words "laboratory waste" or with the chemical name of the contents. If the container is too small to hold a label, the label must be placed on a secondary container.

(b) Each University may temporarily hold up to 55 gallons of laboratory waste or one quart of acutely hazardous laboratory waste, or weight equivalent, in each laboratory, but upon reaching these thresholds, each University must mark that laboratory waste with the date when this threshold requirement was met (by dating the container(s) or secondary container(s)).

(c) Each university must remove all of the dated laboratory waste from the laboratory for delivery to a location identified in paragraph (i) of this section within 30 days of reaching the threshold amount identified in paragraph (b) of this section.

(d) In no event shall the excess laboratory waste that a laboratory temporarily holds before dated laboratory waste is removed exceed an additional 55 gallons of laboratory waste (or one additional quart of acutely hazardous laboratory waste). No more than 110 gallons of laboratory waste total (or no more than two quarts of acutely hazardous laboratory waste total) may be temporarily held in a laboratory at any one time. Excess laboratory waste must be dated and removed in accordance with the requirements of paragraphs (b) and (c) of this section.

(e) Containers of laboratory wastes must be:

(1) Closed at all times except when wastes are being added to (including during in-line waste collection) or removed from the container;

(2) Maintained in good condition and stored in the laboratory in a manner to avoid leaks;

(3) Compatible with their contents to avoid reactions between the waste and its container; and must be made of, or lined with, materials which are compatible with the laboratory wastes to be temporarily held in the laboratory so that the container is not impaired; and

(4) Inspected regularly (at least annually) to ensure that they meet requirements for container management.

(f) The management of laboratory waste must not result in the release of hazardous constituents into the land, air and water where such release is prohibited under federal law.

(g) The requirements for emergency response are:

(1) Each University must post notification procedures, location of emergency response equipment to be used by laboratory workers and evacuation procedures;

(2) Emergency response equipment and procedures for emergency response must be appropriate to the hazards in the laboratory such that hazards to human health and the environment will be minimized in the event of an emergency;

(3) In the event of a fire, explosion or other release of laboratory waste which could threaten human health or the environment, the laboratory worker must follow the notification procedures under paragraph (g)(1) of this section.

(h) Each University must investigate, document, and take actions to correct and prevent future incidents of hazardous chemical spills, exposures and other incidents that trigger a reportable emergency or that require reporting under paragraph (g) of this section.

(i) Each University may only transfer laboratory wastes from a laboratory:

(1) Directly to an on-site designated hazardous waste accumulation area. Notwithstanding 40 CFR 263.10(a), each University must comply with requirements for transporters set forth in 40 CFR 263.30 and 263.31 in the event of a discharge of laboratory waste en route from a laboratory to an on-site hazardous waste accumulation area; or

(2) To a treatment, storage or disposal (TSD) facility permitted to handle the waste under 40 CFR part 270 or in interim status under 40 CFR parts 265 and 270 (or authorized to handle the

waste by a state with a hazardous waste management program approved under 40 CFR part 271) if it is determined in the laboratory by the individuals identified in § 262.105(b)(3) to be responsible for waste management decisions that the waste is a hazardous waste and that it is prudent to transfer it directly to a treatment, storage, and disposal facility rather than an on-site accumulation area.

(j) Each University must ensure that laboratory workers receive training and are provided with information so that they can implement and comply with these Minimum Performance Criteria.

§ 262.105  What must be included in the laboratory environmental management plan?

(a) Each University must include specific measures it will take to protect human health and the environment from hazards associated with the management of laboratory wastes and from the reuse, recycling or disposal of such materials outside the laboratory.

(b) Each University must write, implement and comply with an Environmental Management Plan that includes the following:

(1) The specific procedures to assure compliance with each of the Minimum Performance Criteria set forth in § 262.104.

(2) An environmental policy, or environmental, health and safety policy, signed by the University's senior management, which must include commitments to regulatory compliance, waste minimization, risk reduction and continual improvement of the environmental management system.

(3) A description of roles and responsibilities for the implementation and maintenance of the Laboratory Environmental Management Plan.

(4) A system for identifying and tracking legal and other requirements applicable to laboratory waste, including the procedures for providing updates to laboratory supervisors.

(5) Criteria for the identification of physical and chemical hazards and the control measures to reduce the potential for releases of laboratory wastes to

the environment, including engineering controls, the use of personal protective equipment and hygiene practices, containment strategies and other control measures.

(6) A pollution prevention plan, including, but not limited to, roles and responsibilities, training, pollution prevention activities, and performance review.

(7) A system for conducting and updating annual surveys of hazardous chemicals of concern and procedures for identifying acutely hazardous laboratory waste.

(8) The procedures for conducting laboratory clean-outs with regard to the safe management and disposal of laboratory wastes.

(9) The criteria that laboratory workers must comply with for managing, containing and labeling laboratory wastes, including: an evaluation of the need for and the use of any special containers or labeling circumstances, and the use of laboratory wastes secondary containers including packaging, bottles, or test tube racks.

(10) The procedures relevant to the safe and timely removal of laboratory wastes from the laboratory.

(11) The emergency preparedness and response procedures to be implemented for laboratory waste.

(12) Provisions for information dissemination and training, provided for in paragraph (d) of this section.

(13) The procedures for the development and approval of changes to the Environmental Management Plan.

(14) The procedures and work practices for safely transferring or moving laboratory wastes from a laboratory to a location identified in § 262.104(i).

(15) The procedures for regularly inspecting a laboratory to assess conformance with the requirements of the Environmental Management Plan.

(16) The procedures for the identification of environmental management plan noncompliance, and the assignment of responsibility, timelines and corrective actions to prevent their reoccurrence.

(17) The record keeping requirements to document conformance with this Plan.

(c) Organizational responsibilities for each university. Each University must:

(1) Develop and oversee implementation of its Laboratory Environmental Management Plan.

(2) Identify the following:

(i) Annual environmental objectives and targets;

(ii) Those laboratories covered by the requirements of the Laboratory Environmental Management Plan.

(3) Assign roles and responsibilities for the effective implementation of the Environmental Management Plan.

(4) Determine whether laboratory wastes are solid wastes under RCRA and, if so, whether they are hazardous.

(5) Develop, implement, and maintain:

(i) Policies, procedures and practices governing its compliance with the Environmental Management Plan and applicable federal and state hazardous waste regulations.

(ii) Procedures to monitor and measure relevant conformance and environmental performance data for the purpose of supporting continual improvement of the Environmental Management Plan.

(iii) Policies and procedures for managing environmental documents and records applicable to this Environmental Management Standard.

(6) Ensure that:

(i) Its Environmental Management Plan is available to laboratory workers, vendors, employee representatives, visitors, on-site contractors, and upon request, to governmental representatives.

(ii) Personnel designated by each University to handle laboratory wastes and RCRA hazardous waste receive appropriate training.

(iii) The Environmental Management Plan is reviewed at least annually by senior management to ensure its continuing suitability, adequacy and effectiveness. The reviews may include, but not be limited to, a consideration of monitoring and measuring information, Laboratory Environmental Management Standard performance data, assessment and audit results and other relevant information and data.

(d) What are the Information and Training Requirements for Each University?

(1) Each University must ensure that laboratory workers receive training

and are provided with the information to understand and implement the elements of each University's Environmental Management Plan that are relevant to the laboratory workers' responsibilities.

(2) When must each University ensure that laboratory workers receive training and information?

(i) Each University must provide the information to each laboratory worker when he/she is first assigned to a work area where laboratory wastes may be generated.

(ii) Each University must ensure that each laboratory worker has had training within six months of when he/she is first assigned to a work area where laboratory wastes may be generated. Each University must retrain a laboratory worker when a laboratory waste poses a new or unique hazard for which the laboratory worker has not received prior training and as frequently as needed to maintain knowledge of the procedures of the Environmental Management Plan.

(3) Each University must provide an outline of training and specify who is to receive training in its Environmental Management Plan.

(4) Each University must ensure that laboratory workers are informed of:

(i) The contents of this Subpart and the Laboratory Environmental Management Plan(s) for the laboratory(ies) in which they will be performing work;

(ii) The location and availability of the Environmental Management Plan;

(iii) Emergency response measures applicable to laboratories;

(iv) Signs and indicators of a hazardous substance release;

(v) The location and availability of known reference materials relevant to implementation of the Environmental Management Plan; and

(vi) Environmental training requirements applicable to laboratory workers.

(5) Each University must ensure that Laboratory workers have received training in:

(i) Methods and observations that may be used to detect the presence or release of a hazardous substance;

(ii) The chemical and physical hazards associated with laboratory wastes in their work area;

(iii) The relevant measures a laboratory worker can take to protect human health and the environment; and

(iv) Details of the Environmental Management Plan sufficient to ensure they manage laboratory waste in accordance with the requirements of this Subpart.

(6) Requirements pertaining to Laboratory visitors:

(i) Laboratory visitors, such as on-site contractors or environmental vendors, that require information and training under this standard must be identified in the Environmental Management Plan.

(ii) Laboratory visitors identified in the Environmental Management Plan must be informed of the existence and location of the Environmental Management Plan.

(iii) Laboratory visitors identified in the Environmental Management Plan must be informed of relevant policies, procedures or work practices to ensure compliance with the requirements of the Environmental Management Plan.

(7) Each University must define methods of providing objective evidence and records of training and information dissemination in its Environmental Management Plan.

§ 262.106 When must a hazardous waste determination be made?

(a) For laboratory waste sent from a laboratory to an on-site hazardous waste accumulation area, each University must evaluate the laboratory wastes to determine whether they are solid wastes under RCRA and, if so, determine pursuant to § 262.11 (a) through (d) whether they are hazardous wastes, as soon as the laboratory wastes reach the University's Hazardous Waste Accumulation area(s). At this point each University must determine whether the laboratory waste will be reused or whether it must be managed as RCRA solid or hazardous waste.

(b) For laboratory waste that will be sent from a laboratory to a TSD facility permitted to handle the waste, each University must evaluate such laboratory wastes to determine whether they are solid wastes under RCRA and, if so, determine pursuant to § 262.11 (a) through (d) whether they are hazardous wastes, prior to the 30-day deadline for removing dated laboratory waste from the laboratory.

(c) Laboratory waste that is determined to be hazardous waste is no longer subject to the provisions of this subpart and must be managed in accordance with all applicable provisions of 40 CFR Parts 260 through 270.

**§262.107 Under what circumstances will a university's participation in this environmental management standard pilot be terminated?**

(a) EPA retains the right to terminate a University's participation in this Laboratory XL project if the University:

(1) Is in non-compliance with the Minimum Performance Criteria in §262.104; or

(2) Has actual environmental management practices in the laboratory that do not conform to its Environmental Management Plan; or

(3) Is in non-compliance with the Hazardous Waste Determination requirements of §262.106.

(b) In the event of termination, EPA will provide the University with 15 days written notice of its intent to terminate. During this period, which commences upon receipt of the notice, the University will have the opportunity to come back into compliance with the Minimum Performance Criteria, its Environmental Management Plan, or the requirements for making a hazardous waste determination at §262.106 or to provide a written explanation as to why it was not in compliance and how it intends to return to compliance. If, upon review of the University's written explanation, EPA then reissues a written notice terminating the University from this XL Project, the provisions of paragraph (c) of this section will immediately apply and the University shall have 90 days to come into compliance with the applicable RCRA requirements deferred by §262.10(j). During the 90-day transition period, the provisions of this subpart shall continue to apply to the University.

(c) If a University withdraws from this XL project, or receives a notice of termination pursuant to this section, it must submit to EPA and the state a schedule for returning to full compliance with RCRA requirements at the laboratory level. The schedule must show how the University will return to full compliance with RCRA within 90 days from the date of the notice of termination or withdrawal.

**§262.108 When will this subpart expire?**

This subpart will expire on April 15, 2009.

[71 FR 35550, June 21, 2006]

## Subpart K—Alternative Requirements for Hazardous Waste Determination and Accumulation of Unwanted Material for Laboratories Owned by Eligible Academic Entities

Source: 73 72954, Dec. 1, 2008, unless otherwise noted.

**§262.200 Definitions for this subpart.**

The following definitions apply to this subpart:

*Central accumulation area* means an on-site hazardous waste accumulation area subject to either §262.34(a) (or 262.34(j) and (k) for Performance Track members) of this part (large quantity generators); or §262.34(d)–(f) of this part (small quantity generators). A central accumulation area at an eligible academic entity that chooses to be subject to this subpart must also comply with §262.211 when accumulating unwanted material and/or hazardous waste.

*College/University* means a private or public, post-secondary, degree-granting, academic institution, that is accredited by an accrediting agency listed annually by the U.S. Department of Education.

*Eligible academic entity* means a college or university, or a non-profit research institute that is owned by or has a formal written affiliation agreement with a college or university, or a teaching hospital that is owned by or has a formal written affiliation agreement with a college or university.

*Formal written affiliation agreement* for a non-profit research institute means a written document that establishes a relationship between institutions for the purposes of research and/or education and is signed by authorized representatives, as defined by § 260.10, from each institution. A relationship on a project-by-project or grant-by-grant basis is not considered a formal written affiliation agreement. A *formal written affiliation agreement* for a teaching hospital means a master affiliation agreement and program letter of agreement, as defined by the Accreditation Council for Graduate Medical Education, with an accredited medical program or medical school.

*Laboratory* means an area owned by an eligible academic entity where relatively small quantities of chemicals and other substances are used on a non-production basis for teaching or research (or diagnostic purposes at a teaching hospital) and are stored and used in containers that are easily manipulated by one person. Photo laboratories, art studios, and field laboratories are considered laboratories. Areas such as chemical stockrooms and preparatory laboratories that provide a support function to teaching or research laboratories (or diagnostic laboratories at teaching hospitals) are also considered laboratories.

*Laboratory clean-out* means an evaluation of the inventory of chemicals and other materials in a laboratory that are no longer needed or that have expired and the subsequent removal of those chemicals or other unwanted materials from the laboratory. A clean-out may occur for several reasons. It may be on a routine basis (e.g., at the end of a semester or academic year) or as a result of a renovation, relocation, or change in laboratory supervisor/occupant. A regularly scheduled removal of unwanted material as required by § 262.208 does not qualify as a laboratory clean-out.

*Laboratory worker* means a person who handles chemicals and/or unwanted material in a laboratory and may include, but is not limited to, faculty, staff, post-doctoral fellows, interns, researchers, technicians, supervisors/managers, and principal investigators. A person does not need to be paid or otherwise compensated for his/her work in the laboratory to be considered a laboratory worker. Undergraduate and graduate students in a supervised classroom setting are not laboratory workers.

*Non-profit research institute* means an organization that conducts research as its primary function and files as a non-profit organization under the tax code of 26 U.S.C. 501(c)(3).

*Reactive acutely hazardous unwanted material* means an unwanted material that is one of the acutely hazardous commercial chemical products listed in § 261.33(e) for reactivity.

*Teaching hospital* means a hospital that trains students to become physicians, nurses or other health or laboratory personnel.

*Trained professional* means a person who has completed the applicable RCRA training requirements of § 265.16 for large quantity generators, or is knowledgeable about normal operations and emergencies in accordance with § 262.34(d)(5)(iii) for small quantity generators and conditionally exempt small quantity generators. A trained professional may be an employee of the eligible academic entity or may be a contractor or vendor who meets the requisite training requirements.

*Unwanted material* means any chemical, mixtures of chemicals, products of experiments or other material from a laboratory that is no longer needed, wanted or usable in the laboratory and that is destined for hazardous waste determination by a trained professional. Unwanted materials include reactive acutely hazardous unwanted materials and materials that may eventually be determined not to be solid waste pursuant to § 261.2, or a hazardous waste pursuant to § 261.3. If an eligible academic entity elects to use another equally effective term in lieu of "unwanted material," as allowed by § 262.206(a)(1)(i), the equally effective term has the same meaning and is subject to the same requirements as "unwanted material" under this subpart.

*Working container* means a small container (*i.e.*, two gallons or less) that is in use at a laboratory bench, hood, or other work station, to collect unwanted material from a laboratory experiment or procedure.

## §262.201 Applicability of this subpart.

(a) Large quantity generators and small quantity generators. This subpart provides alternative requirements to the requirements in §§262.11 and 262.34(c) for the hazardous waste determination and accumulation of hazardous waste in laboratories owned by eligible academic entities that choose to be subject to this subpart, provided that they complete the notification requirements of §262.203.

(b) Conditionally exempt small quantity generators. This subpart provides alternative requirements to the conditional exemption in §261.5(b) for the accumulation of hazardous waste in laboratories owned by eligible academic entities that choose to be subject to this subpart, provided that they complete the notification requirements of §262.203.

## §262.202 This subpart is optional.

(a) Large quantity generators and small quantity generators: Eligible academic entities have the option of complying with this subpart with respect to its laboratories, as an alternative to complying with the requirements of §§262.11 and 262.34(c).

(b) Conditionally exempt small quantity generators. Eligible academic entities have the option of complying with this subpart with respect to its laboratories, as an alternative to complying with the conditional exemption of §261.5(b).

## §262.203 How an eligible academic entity indicates it will be subject to the requirements of this subpart.

(a) An eligible academic entity must notify the appropriate EPA Regional Administrator in writing, using the RCRA Subtitle C Site Identification Form (EPA Form 8700–12), that it is electing to be subject to the requirements of this subpart for all the laboratories owned by the eligible academic entity under the same EPA Identification Number. An eligible academic entity that is a conditionally exempt small quantity generator and does not have an EPA Identification Number must notify that it is electing to be subject to the requirements of this subpart for all the laboratories owned by the eligible academic entity

that are on-site, as defined by §260.10. An eligible academic entity must submit a separate notification (Site Identification Form) for each EPA Identification Number (or site, for conditionally exempt small quantity generators) that is electing to be subject to the requirements of this subpart, and must submit the Site Identification Form before it begins operating under this subpart.

(b) When submitting the Site Identification Form, the eligible academic entity must, at a minimum, fill out the following fields on the form:

(1) Reason for Submittal.

(2) Site EPA Identification Number (except for conditionally exempt small quantity generators).

(3) Site Name.

(4) Site Location Information.

(5) Site Land Type.

(6) North American Industry Classification System (NAICS) Code(s) for the Site.

(7) Site Mailing Address.

(8) Site Contact Person.

(9) Operator and Legal Owner of the Site.

(10) Type of Regulated Waste Activity.

(11) Certification.

(c) An eligible academic entity must keep a copy of the notification on file at the eligible academic entity for as long as its laboratories are subject to this subpart.

(d) A teaching hospital that is not owned by a college or university must keep a copy of its formal written affiliation agreement with a college or university on file at the teaching hospital for as long as its laboratories are subject to this subpart.

(e) A non-profit research institute that is not owned by a college or university must keep a copy of its formal written affiliation agreement with a college or university on file at the non-profit research institute for as long as its laboratories are subject to this subpart.

## §262.204 How an eligible academic entity indicates it will withdraw from the requirements of this subpart.

(a) An eligible academic entity must notify the appropriate EPA Regional Administrator in writing, using the

RCRA Subtitle C Site Identification Form (EPA Form 8700–12), that it is electing to no longer be subject to the requirements of this subpart for all the laboratories owned by the eligible academic entity under the same EPA Identification Number and that it will comply with the requirements of §§ 262.11 and 262.34(c) for small quantity generators and large quantity generators. An eligible academic entity that is a conditionally exempt small quantity generator and does not have an EPA Identification Number must notify that it is withdrawing from the requirements of this subpart for all the laboratories owned by the eligible academic entity that are on-site and that it will comply with the conditional exemption in § 261.5(b). An eligible academic entity must submit a separate notification (Site Identification Form) for each EPA Identification Number (or site, for conditionally exempt small quantity generators) that is withdrawing from the requirements of this subpart and must submit the Site Identification Form before it begins operating under the requirements of §§ 262.11 and 262.34(c) for small quantity generators and large quantity generators, or § 261.5(b) for conditionally exempt small quantity generators.

(b) When submitting the Site Identification Form, the eligible academic entity must, at a minimum, fill out the following fields on the form:

(1) Reason for Submittal.

(2) Site EPA Identification Number (except for conditionally exempt small quantity generators).

(3) Site Name.

(4) Site Location Information.

(5) Site Land Type.

(6) North American Industry Classification System (NAICS) Code(s) for the Site.

(7) Site Mailing Address.

(8) Site Contact Person.

(9) Operator and Legal Owner of the Site.

(10) Type of Regulated Waste Activity.

(11) Certification.

(c) An eligible academic entity must keep a copy of the withdrawal notice on file at the eligible academic entity for three years from the date of the notification.

**§ 262.205 Summary of the requirements of this subpart.**

An eligible academic entity that chooses to be subject to this subpart is not required to have interim status or a RCRA Part B permit for the accumulation of unwanted material and hazardous waste in its laboratories, provided the laboratories comply with the provisions of this subpart and the eligible academic entity has a Laboratory Management Plan (LMP) in accordance with § 262.214 that describes how the laboratories owned by the eligible academic entity will comply with the requirements of this subpart.

**§ 262.206 Labeling and management standards for containers of unwanted material in the laboratory.**

An eligible academic entity must manage containers of unwanted material while in the laboratory in accordance with the requirements in this section.

(a) Labeling: Label unwanted material as follows:

(1) The following information must be affixed or attached to the container:

(i) The words "unwanted material" or another equally effective term that is to be used consistently by the eligible academic entity and that is identified in Part I of the Laboratory Management Plan, and

(ii) Sufficient information to alert emergency responders to the contents of the container. Examples of information that would be sufficient to alert emergency responders to the contents of the container include, but are not limited to:

(A) The name of the chemical(s),

(B) The type or class of chemical, such as organic solvents or halogenated organic solvents.

(2) The following information may be affixed or attached to the container, but must at a minimum be associated with the container:

(i) The date that the unwanted material first began accumulating in the container, and

(ii) Information sufficient to allow a trained professional to properly identify whether an unwanted material is a solid and hazardous waste and to assign the proper hazardous waste code(s), pursuant to §262.11. Examples of information that would allow a trained professional to properly identify whether an unwanted material is a solid or hazardous waste include, but are not limited to:

(A) The name and/or description of the chemical contents or composition of the unwanted material, or, if known, the product of the chemical reaction,

(B) Whether the unwanted material has been used or is unused,

(C) A description of the manner in which the chemical was produced or processed, if applicable.

(b) Management of Containers in the Laboratory: An eligible academic entity must properly manage containers of unwanted material in the laboratory to assure safe storage of the unwanted material, to prevent leaks, spills, emissions to the air, adverse chemical reactions, and dangerous situations that may result in harm to human health or the environment. Proper container management must include the following:

(1) Containers are maintained and kept in good condition and damaged containers are replaced, overpacked, or repaired, and

(2) Containers are compatible with their contents to avoid reactions between the contents and the container; and are made of, or lined with, material that is compatible with the unwanted material so that the container's integrity is not impaired, and

(3) Containers must be kept closed at all times, except:

(i) When adding, removing or consolidating unwanted material, or

(ii) A working container may be open until the end of the procedure or work shift, or until it is full, whichever comes first, at which time the working container must either be closed or the contents emptied into a separate container that is then closed, or

(iii) When venting of a container is necessary.

(A) For the proper operation of laboratory equipment, such as with in-line collection of unwanted materials from high performance liquid chromatographs, or

(B) To prevent dangerous situations, such as build-up of extreme pressure.

## §262.207 Training.

An eligible academic entity must provide training to all individuals working in a laboratory at the eligible academic entity, as follows:

(a) Training for laboratory workers and students must be commensurate with their duties so they understand the requirements in this subpart and can implement them.

(b) An eligible academic entity can provide training for laboratory workers and students in a variety of ways, including, but not limited to:

(1) Instruction by the professor or laboratory manager before or during an experiment; or

(2) Formal classroom training; or

(3) Electronic/written training; or

(4) On-the-job training; or

(5) Written or oral exams.

(c) An eligible academic entity that is a large quantity generator must maintain documentation for the durations specified in §265.16(e) demonstrating training for all laboratory workers that is sufficient to determine whether laboratory workers have been trained. Examples of documentation demonstrating training can include, but are not limited to, the following:

(1) Sign-in/attendance sheet(s) for training session(s); or

(2) Syllabus for training session; or

(3) Certificate of training completion; or

(4) Test results.

(d) A trained professional must:

(1) Accompany the transfer of unwanted material and hazardous waste when the unwanted material and hazardous waste is removed from the laboratory, and

(2) Make the hazardous waste determination, pursuant to §262.11, for unwanted material.

## §262.208 Removing containers of unwanted material from the laboratory.

(a) Removing containers of unwanted material on a regular schedule. An eligible academic entity must either:

(1) Remove all containers of unwanted material from each laboratory on a regular interval, not to exceed 6 months; or

(2) Remove containers of unwanted material from each laboratory within 6

months of each container's accumulation start date.

(b) The eligible academic entity must specify in Part I of its Laboratory Management Plan whether it will comply with paragraph (a)(1) or (a)(2) of this section for the regular removal of unwanted material from its laboratories.

(c) The eligible academic entity must specify in Part II of its Laboratory Management Plan how it will comply with paragraph (a)(1) or (a)(2) of this section and develop a schedule for regular removals of unwanted material from its laboratories.

(d) Removing containers of unwanted material when volumes are exceeded.

(1) If a laboratory accumulates a total volume of unwanted material (including reactive acutely hazardous unwanted material) in excess of 55 gallons before the regularly scheduled removal, the eligible academic entity must ensure that all containers of unwanted material in the laboratory (including reactive acutely hazardous unwanted material):

(i) Are marked on the label that is associated with the container (or on the label that is affixed or attached to the container, if that is preferred) with the date that 55 gallons is exceeded; and

(ii) Are removed from the laboratory within 10 calendar days of the date that 55 gallons was exceeded, or at the next regularly scheduled removal, whichever comes first.

(2) If a laboratory accumulates more than 1 quart of reactive acutely hazardous unwanted material before the regularly scheduled removal, then the eligible academic entity must ensure that all containers of reactive acutely hazardous unwanted material:

(i) Are marked on the label that is associated with the container (or on the label that is affixed or attached to the container, if that is preferred) with the date that 1 quart is exceeded; and

(ii) Are removed from the laboratory within 10 calendar days of the date that 1 quart was exceeded, or at the next regularly scheduled removal, whichever comes first.

§ 262.209 Where and when to make the hazardous waste determination and where to send containers of unwanted material upon removal from the laboratory.

(a) Large quantity generators and small quantity generators—an eligible academic entity must ensure that a trained professional makes a hazardous waste determination, pursuant to § 262.11, for unwanted material in any of the following areas:

(1) In the laboratory before the unwanted material is removed from the laboratory, in accordance with § 262.210;

(2) Within 4 calendar days of arriving at an on-site central accumulation area, in accordance with § 262.211; and

(3) Within 4 calendar days of arriving at an on-site interim status or permitted treatment, storage or disposal facility, in accordance with § 262.212.

(b) Conditionally exempt small quantity generators—an eligible academic entity must ensure that a trained professional makes a hazardous waste determination, pursuant to § 262.11, for unwanted material in the laboratory before the unwanted material is removed from the laboratory, in accordance with § 262.210.

§ 262.210 Making the hazardous waste determination in the laboratory before the unwanted material is removed from the laboratory.

If an eligible academic entity makes the hazardous waste determination, pursuant to § 262.11, for unwanted material in the laboratory, it must comply with the following:

(a) A trained professional must make the hazardous waste determination, pursuant to § 262.11, before the unwanted material is removed from the laboratory.

(b) If an unwanted material is a hazardous waste, the eligible academic entity must:

(1) Write the words "hazardous waste" on the container label that is affixed or attached to the container, before the hazardous waste may be removed from the laboratory; and

(2) Write the appropriate hazardous waste code(s) on the label that is associated with the container (or on the label that is affixed or attached to the container, if that is preferred) before

the hazardous waste is transported off-site.

(3) Count the hazardous waste toward the eligible academic entity's generator status, pursuant to §261.5(c) and (d), in the calendar month that the hazardous waste determination was made.

(c) A trained professional must accompany all hazardous waste that is transferred from the laboratory(ies) to an on-site central accumulation area or on-site interim status or permitted treatment, storage or disposal facility.

(d) When hazardous waste is removed from the laboratory:

(1) Large quantity generators and small quantity generators must ensure it is taken directly from the laboratory(ies) to an on-site central accumulation area, or on-site interim status or permitted treatment, storage or disposal facility, or transported off-site.

(2) Conditionally exempt small quantity generators must ensure it is taken directly from the laboratory(ies) to any of the types of facilities listed in §261.5(f)(3) for acute hazardous waste, or §261.5(g)(3) for hazardous waste.

(e) An unwanted material that is a hazardous waste is subject to all applicable hazardous waste regulations when it is removed from the laboratory.

## §262.211 Making the hazardous waste determination at an on-site central accumulation area.

If an eligible academic entity makes the hazardous waste determination, pursuant to §262.11, for unwanted material at an on-site central accumulation area, it must comply with the following:

(a) A trained professional must accompany all unwanted material that is transferred from the laboratory(ies) to an on-site central accumulation area.

(b) All unwanted material removed from the laboratory(ies) must be taken directly from the laboratory(ies) to the on-site central accumulation area.

(c) The unwanted material becomes subject to the generator accumulation regulations of §262.34(a) (or §262.34(j) and (k) for Performance Track members) for large quantity generators or §262.34(d)–(f) for small quantity generators as soon as it arrives in the central

accumulation area, except for the "hazardous waste" labeling requirements of §262.34(a)(3) (or §262.34(j)(6) for Performance Track members).

(d) A trained professional must determine, pursuant to §262.11, if the unwanted material is a hazardous waste within 4 calendar days of the unwanted materials' arrival at the on-site central accumulation area.

(e) If the unwanted material is a hazardous waste, the eligible academic entity must:

(1) Write the words "hazardous waste" on the container label that is affixed or attached to the container, within 4 calendar days of arriving at the on-site central accumulation area and before the hazardous waste may be removed from the on-site central accumulation area, and

(2) Write the appropriate hazardous waste code(s) on the container label that is associated with the container (or on the label that is affixed or attached to the container, if that is preferred) before the hazardous waste may be treated or disposed of on-site or transported off-site, and

(3) Count the hazardous waste toward the eligible academic entity's generator status, pursuant to §261.5(c) and (d) in the calendar month that the hazardous waste determination was made, and

(4) Manage the hazardous waste according to all applicable hazardous waste regulations.

## §262.212 Making the hazardous waste determination at an on-site interim status or permitted treatment, storage or disposal facility.

If an eligible academic entity makes the hazardous waste determination, pursuant to §262.11, for unwanted material at an on-site interim status or permitted treatment, storage or disposal facility, it must comply with the following:

(a) A trained professional must accompany all unwanted material that is transferred from the laboratory(ies) to an on-site interim status or permitted treatment, storage or disposal facility.

(b) All unwanted material removed from the laboratory(ies) must be taken directly from the laboratory(ies) to the on-site interim status or permitted treatment, storage or disposal facility.

(c) The unwanted material becomes subject to the terms of the eligible academic entity's hazardous waste permit or interim status as soon as it arrives in the on-site treatment, storage or disposal facility.

(d) A trained professional must determine, pursuant to § 262.11, if the unwanted material is a hazardous waste within 4 calendar days of the unwanted materials' arrival at an on-site interim status or permitted treatment, storage or disposal facility.

(e) If the unwanted material is a hazardous waste, the eligible academic entity must:

(1) Write the words "hazardous waste" on the container label that is affixed or attached to the container (or on the label that is affixed or attached to the container, if that is preferred) within 4 calendar days of arriving at the on-site interim status or permitted treatment, storage or disposal facility and before the hazardous waste may be removed from the on-site interim status or permitted treatment, storage or disposal facility, and

(2) Write the appropriate hazardous waste code(s) on the container label that is associated with the container (or on the label that is affixed or attached to the container, if that is preferred) before the hazardous waste may be treated or disposed on-site or transported off-site, and

(3) Count the hazardous waste toward the eligible academic entity's generator status, pursuant to § 261.5(c) and (d) in the calendar month that the hazardous waste determination was made, and

(4) Manage the hazardous waste according to all applicable hazardous waste regulations.

## § 262.213   Laboratory clean-outs.

(a) One time per 12 month period for each laboratory, an eligible academic entity may opt to conduct a laboratory clean-out that is subject to all the applicable requirements of this subpart, except that:

(1) If the volume of unwanted material in the laboratory exceeds 55 gallons (or 1 quart of reactive acutely hazardous unwanted material), the eligible academic entity is not required to remove all unwanted materials from the laboratory within 10 calendar days of exceeding 55 gallons (or 1 quart of reactive acutely hazardous unwanted material), as required by § 262.208. Instead, the eligible academic entity must remove all unwanted materials from the laboratory within 30 calendar days from the start of the laboratory clean-out; and

(2) For the purposes of on-site accumulation, an eligible academic entity is not required to count a hazardous waste that is an unused commercial chemical product (listed in 40 CFR part 261, subpart D or exhibiting one or more characteristics in 40 CFR part 261, subpart C) generated solely during the laboratory clean-out toward its hazardous waste generator status, pursuant to § 261.5(c) and (d). An unwanted material that is generated prior to the beginning of the laboratory clean-out and is still in the laboratory at the time the laboratory clean-out commences must be counted toward hazardous waste generator status, pursuant to § 261.5(c) and (d), if it is determined to be hazardous waste; and

(3) For the purposes of off-site management, an eligible academic entity must count all its hazardous waste, regardless of whether the hazardous waste was counted toward generator status under paragraph (a)(2) of this section, and if it generates more than 1 kg/month of acute hazardous waste or more than 100 kg/month of hazardous waste (i.e., the conditionally exempt small quantity generator limits of § 261.5), the hazardous waste is subject to all applicable hazardous waste regulations when it is transported off-site; and

(4) An eligible academic entity must document the activities of the laboratory clean-out. The documentation must, at a minimum, identify the laboratory being cleaned out, the date the laboratory clean-out begins and ends, and the volume of hazardous waste generated during the laboratory clean-out. The eligible academic entity must maintain the records for a period of three years from the date the clean-out ends; and

(b) For all other laboratory clean-outs conducted during the same 12-

month period, an eligible academic entity is subject to all the applicable requirements of this subpart, including, but not limited to:

(1) The requirement to remove all unwanted materials from the laboratory within 10 calendar days of exceeding 55 gallons (or 1 quart of reactive acutely hazardous unwanted material), as required by §262.208; and

(2) The requirement to count all hazardous waste, including unused hazardous waste, generated during the laboratory clean-out toward its hazardous waste generator status, pursuant to §261.5(c) and (d).

**§262.214  Laboratory management plan.**

An eligible academic entity must develop and retain a written Laboratory Management Plan, or revise an existing written plan. The Laboratory Management Plan is a site-specific document that describes how the eligible academic entity will manage unwanted materials in compliance with this subpart. An eligible academic entity may write one Laboratory Management Plan for all the laboratories owned by the eligible academic entity that have opted into this subpart, even if the laboratories are located at sites with different EPA Identification Numbers. The Laboratory Management Plan must contain two parts with a total of nine elements identified in paragraphs (a) and (b) of this section. In Part I of its Laboratory Management Plan, an eligible academic entity must describe its procedures for each of the elements listed in paragraph (a) of this section. An eligible academic entity must implement and comply with the specific provisions that it develops to address the elements in Part I of the Laboratory Management Plan. In Part II of its Laboratory Management Plan, an eligible academic entity must describe its best management practices for each of the elements listed in paragraph (b) of this section. The specific actions taken by an eligible academic entity to implement each element in Part II of its Laboratory Management Plan may vary from the procedures described in the eligible academic entity's Laboratory Management Plan, without constituting a violation of this subpart.

An eligible academic entity may include additional elements and best management practices in Part II of its Laboratory Management Plan if it chooses.

(a) The eligible academic entity must implement and comply with the specific provisions of Part I of its Laboratory Management Plan. In Part I of its Laboratory Management Plan, an eligible academic entity must:

(1) Describe procedures for container labeling in accordance with §262.206(a), including:

(i) Identifying whether the eligible academic entity will use the term "unwanted material" on the containers in the laboratory. If not, identify an equally effective term that will be used in lieu of "unwanted material" and consistently by the eligible academic entity. The equally effective term, if used, has the same meaning and is subject to the same requirements as "unwanted material."

(ii) Identifying the manner in which information that is "associated with the container" will be imparted.

(2) Identify whether the eligible academic entity will comply with §262.208(a)(1) or (a)(2) for regularly scheduled removals of unwanted material from the laboratory.

(b) In Part II of its Laboratory Management Plan, an eligible academic entity must:

(1) Describe its intended best practices for container labeling and management, including how the eligible academic entity will manage containers used for in-line collection of unwanted materials, such as with high performance liquid chromatographs and other laboratory equipment (see the required standards at §262.206).

(2) Describe its intended best practices for providing training for laboratory workers and students commensurate with their duties (see the required standards at §262.207(a)).

(3) Describe its intended best practices for providing training to ensure safe on-site transfers of unwanted material and hazardous waste by trained professionals (see the required standards at §262.207(d)(1)).

(4) Describe its intended best practices for removing unwanted material from the laboratory, including:

(i) For regularly scheduled removals—Develop a regular schedule for identifying and removing unwanted materials from its laboratories (see the required standards at § 262.208(a)(1) and (a)(2)).

(ii) For removals when maximum volumes are exceeded:

(A) Describe its intended best practices for removing unwanted materials from the laboratory within 10 calendar days when unwanted materials have exceeded their maximum volumes (see the required standards at § 262.208(d)).

(B) Describe its intended best practices for communicating that unwanted materials have exceeded their maximum volumes.

(5) Describe its intended best practices for making hazardous waste determinations, including specifying the duties of the individuals involved in the process (see the required standards at § 262.11 and §§ 262.209 through 262.212).

(6) Describe its intended best practices for laboratory clean-outs, if the eligible academic entity plans to use the incentives for laboratory clean-outs provided in § 262.213, including:

(i) Procedures for conducting laboratory clean-outs (see the required standards at § 262.213(a)(1) through (3)); and

(ii) Procedures for documenting laboratory clean-outs (see the required standards at § 262.213(a)(4)).

(7) Describe its intended best practices for emergency prevention, including:

(i) Procedures for emergency prevention, notification, and response, appropriate to the hazards in the laboratory; and

(ii) A list of chemicals that the eligible academic entity has, or is likely to have, that become more dangerous when they exceed their expiration date and/or as they degrade; and

(iii) Procedures to safely dispose of chemicals that become more dangerous when they exceed their expiration date and/or as they degrade; and

(iv) Procedures for the timely characterization of unknown chemicals.

(c) An eligible academic entity must make its Laboratory Management Plan available to laboratory workers, students, or any others at the eligible academic entity who request it.

(d) An eligible academic entity must review and revise its Laboratory Management Plan, as needed.

## § 262.215 Unwanted material that is not solid or hazardous waste.

(a) If an unwanted material does not meet the definition of solid waste in § 261.2, it is no longer subject to this subpart or to the RCRA hazardous waste regulations.

(b) If an unwanted material does not meet the definition of hazardous waste in § 261.3, it is no longer subject to this subpart or to the RCRA hazardous waste regulations, but must be managed in compliance with any other applicable regulations and/or conditions.

## § 262.216 Non-laboratory hazardous waste generated at an eligible academic entity.

An eligible academic entity that generates hazardous waste outside of a laboratory is not eligible to manage that hazardous waste under this subpart; and

(a) Remains subject to the generator requirements of §§ 262.11 and 262.34(c) for large quantity generators and small quantity generators (if the hazardous waste is managed in a satellite accumulation area), and all other applicable generator requirements of 40 CFR part 262, with respect to that hazardous waste; or

(b) Remains subject to the conditional exemption of § 261.5(b) for conditionally exempt small quantity generators, with respect to that hazardous waste.

APPENDIX TO PART 262—UNIFORM HAZARDOUS WASTE MANIFEST AND INSTRUCTIONS (EPA FORMS 8700–22 AND 8700–22A AND THEIR INSTRUCTIONS)

U.S. EPA FORM 8700–22

Read all instructions before completing this form.

1. This form has been designed for use on a 12-pitch (elite) typewriter which is also compatible with standard computer printers; a firm point pen may also be used—press down hard.

2. Federal regulations require generators and transporters of hazardous waste and owners or operators of hazardous waste treatment, storage, and disposal facilities to complete this form (FORM 8700–22) and, if

necessary, the continuation sheet (FORM 8700–22A) for both inter- and intrastate transportation of hazardous waste.

---

Please print or type. (Form designed for use on elite (12-pitch) typewriter.)                                    Form Approved. OMB No. 2050-0039

| UNIFORM HAZARDOUS WASTE MANIFEST | 1. Generator ID Number | 2. Page 1 of | 3. Emergency Response Phone | 4. Manifest Tracking Number |
|---|---|---|---|---|

**5. Generator's Name and Mailing Address**          Generator's Site Address (if different than mailing address)

Generator's Phone:

**6. Transporter 1 Company Name**                                                    U.S. EPA ID Number

**7. Transporter 2 Company Name**                                                    U.S. EPA ID Number

**8. Designated Facility Name and Site Address**                                     U.S. EPA ID Number

Facility's Phone:

| 9a. HM | 9b. U.S. DOT Description (including Proper Shipping Name, Hazard Class, ID Number, and Packing Group (if any)) | 10. Containers | | 11. Total Quantity | 12. Unit Wt./Vol. | 13. Waste Codes |
|---|---|---|---|---|---|---|
| | | No. | Type | | | |
| | 1. | | | | | |
| | 2. | | | | | |
| | 3. | | | | | |
| | 4. | | | | | |

**14. Special Handling Instructions and Additional Information**

**15. GENERATOR'S/OFFEROR'S CERTIFICATION:** I hereby declare that the contents of this consignment are fully and accurately described above by the proper shipping name, and are classified, packaged, marked and labeled/placarded, and are in all respects in proper condition for transport according to applicable international and national governmental regulations. If export shipment and I am the Primary Exporter, I certify that the contents of this consignment conform to the terms of the attached EPA Acknowledgment of Consent.
I certify that the waste minimization statement identified in 40 CFR 262.27(a) (if I am a large quantity generator) or (b) (if I am a small quantity generator) is true.

| Generator's/Offeror's Printed/Typed Name | Signature | Month | Day | Year |
|---|---|---|---|---|

| 16. International Shipments | ☐ Import to U.S. | ☐ Export from U.S. | Port of entry/exit: _____ |
|---|---|---|---|

Transporter signature (for exports only):                    Date leaving U.S.:

**17. Transporter Acknowledgment of Receipt of Materials**

| Transporter 1 Printed/Typed Name | Signature | Month | Day | Year |
|---|---|---|---|---|
| Transporter 2 Printed/Typed Name | Signature | Month | Day | Year |

**18. Discrepancy**

18a. Discrepancy Indication Space    ☐ Quantity    ☐ Type    ☐ Residue    ☐ Partial Rejection    ☐ Full Rejection

Manifest Reference Number:

| 18b. Alternate Facility (or Generator) | U.S. EPA ID Number |
|---|---|

Facility's Phone:

| 18c. Signature of Alternate Facility (or Generator) | | Month | Day | Year |
|---|---|---|---|---|

**19. Hazardous Waste Report Management Method Codes (i.e., codes for hazardous waste treatment, disposal, and recycling systems)**

| 1. | 2. | 3. | 4. |
|---|---|---|---|

**20. Designated Facility Owner or Operator:** Certification of receipt of hazardous materials covered by the manifest except as noted in Item 18a

| Printed/Typed Name | Signature | Month | Day | Year |
|---|---|---|---|---|

EPA Form 8700-22 (Rev. 3-05) Previous editions are obsolete.          **DESIGNATED FACILITY TO DESTINATION STATE (IF REQUIRED)**

*(Left margin labels: GENERATOR, TRANSPORTER INT'L, DESIGNATED FACILITY)*

---

### MANIFEST 8700–22

The following statement must be included with each Uniform Hazardous Waste Manifest, either on the form, in the instructions to the form, or accompanying the form:

Public reporting burden for this collection of information is estimated to average: 30

minutes for generators, 10 minutes for transporters, and 25 minutes for owners or operators of treatment, storage, and disposal facilities. This includes time for reviewing instructions, gathering data, completing, reviewing and transmitting the form. Any correspondence regarding the PRA burden statement for the manifest must be sent to the Director of the Collection Strategies Division in EPA's Office of Information Collection at the following address: U.S. Environmental Protection Agency (2822T), 1200 Pennsylvania Ave., NW., Washington, DC 20460. Do not send the completed form to this address.

### I. INSTRUCTIONS FOR GENERATORS

#### MANIFEST 8700-22

The following statement must be included with each Uniform Hazardous Waste Manifest, either on the form, in the instructions to the form, or accompanying the form:

Public reporting burden for this collection of information is estimated to average: 30 minutes for generators, 10 minutes for transporters, and 25 minutes for owners or operators of treatment, storage, and disposal facilities. This includes time for reviewing instructions, gathering data, completing, reviewing and transmitting · the form. Send comments regarding the burden estimate, including suggestions for reducing this burden, to: Chief, Information Policy Branch (2136), U.S. Environmental Protection Agency, Ariel Rios Building; 1200 Pennsylvania Ave., NW., Washington, DC 20460; and to the Office of Information and Regulatory Affairs, Office of Management and Budget, Washington, DC 20503.

### I. INSTRUCTIONS FOR GENERATORS

#### Item 1. Generator's U.S. EPA Identification Number

Enter the generator's U.S. EPA twelve digit identification number, or the State generator identification number if the generator site does not have an EPA identification number.

#### Item 2. Page 1 of __

Enter the total number of pages used to complete this Manifest (i.e., the first page (EPA Form 8700-22) plus the number of Continuation Sheets (EPA Form 8700-22A), if any).

#### Item 3. Emergency Response Phone Number

Enter a phone number for which emergency response information can be obtained in the event of an incident during transportation. The emergency response phone number must:

1. Be the number of the generator or the number of an agency or organization who is capable of and accepts responsibility for providing detailed information about the shipment;

2. Reach a phone that is monitored 24 hours a day at all times the waste is in transportation (including transportation related storage); and

3. Reach someone who is either knowledgeable of the hazardous waste being shipped and has comprehensive emergency response and spill cleanup/incident mitigation information for the material being shipped or has immediate access to a person who has that knowledge and information about the shipment.

NOTE: Emergency Response phone number information should only be entered in Item 3 when there is one phone number that applies to all the waste materials described in Item 9b. If a situation (e.g., consolidated shipments) arises where more than one Emergency Response phone number applies to the various wastes listed on the· manifest, the phone numbers associated with each specific material should be entered after its description in Item 9b.

#### Item 4. Manifest Tracking Number

This unique tracking number must be preprinted on the manifest by the forms printer.

#### Item 5. Generator's Mailing Address, Phone Number and Site Address

Enter the name of the generator, the mailing address to which the completed manifest signed by the designated facility should be mailed, and the generator's telephone number. Note, the telephone number (including area code) should be the normal business number for the generator, or the number where the generator or his authorized agent may be reached to provide instructions in the event the designated and/or alternate (if any) facility rejects some or all of the shipment. Also enter the physical site address from which the shipment originates only if this address is different than the mailing address.

#### Item 6. Transporter 1 Company Name, and U.S. EPA ID Number

Enter the company name and U.S. EPA ID number of the first transporter who will transport the waste. Vehicle or driver information may not be entered here.

#### Item 7. Transporter 2 Company Name and U.S. EPA ID Number

If applicable, enter the company name and U.S. EPA ID number of the second transporter who will transport the waste. Vehicle or driver information may not be entered here.

If more than two transporters are needed, use a Continuation Sheet(s) (EPA Form 8700–22A).

*Item 8. Designated Facility Name, Site Address, and U.S. EPA ID Number*

Enter the company name and site address of the facility designated to receive the waste listed on this manifest. Also enter the facility's phone number and the U.S. EPA twelve digit identification number of the facility.

*Item 9. U.S. DOT Description (Including Proper Shipping Name, Hazard Class or Division, Identification Number, and Packing Group)*

*Item 9a.* If the wastes identified in Item 9b consist of both hazardous and nonhazardous materials, then identify the hazardous materials by entering an "X" in this Item next to the corresponding hazardous material identified in Item 9b.

If applicable, enter the name of the person accepting the waste on behalf of the second transporter. That person must acknowledge acceptance of the waste described on the manifest by signing and entering the date of receipt.

*Item 9b.* Enter the U.S. DOT Proper Shipping Name, Hazard Class or Division, Identification Number (UN/NA) and Packing Group for each waste as identified in 49 CFR 172. Include technical name(s) and reportable quantity references, if applicable.

NOTE: If additional space is needed for waste descriptions, enter these additional descriptions in Item 27 on the Continuation Sheet (EPA Form 8700–22A). Also, if more than one Emergency Response phone number applies to the various wastes described in either Item 9b or Item 27, enter applicable Emergency Response phone numbers immediately following the shipping descriptions for those Items.

*Item 10. Containers (Number and Type)*

Enter the number of containers for each waste and the appropriate abbreviation from Table I (below) for the type of container.

TABLE I—TYPES OF CONTAINERS

BA = Burlap, cloth, paper, or plastic bags.
CF = Fiber or plastic boxes, cartons, cases.
CM = Metal boxes, cartons, cases (including roll-offs).
CW = Wooden boxes, cartons, cases.
CY = Cylinders.
DF = Fiberboard or plastic drums, barrels, kegs.
DM = Metal drums, barrels, kegs.
DT = Dump truck.
DW = Wooden drums, barrels, kegs.
HG = Hopper or gondola cars.
TC = Tank cars.
TP = Portable tanks.
TT = Cargo tanks (tank trucks).

*Item 11. Total Quantity*

Enter, in designated boxes, the total quantity of waste. Round partial units to the nearest whole unit, and *do not* enter decimals or fractions. To the extent practical, report quantities using appropriate units of measure that will allow you to report quantities with precision. Waste quantities entered should be based on actual measurements or reasonably accurate estimates of actual quantities shipped. Container capacities are not acceptable as estimates.

*Item 12. Units of Measure (Weight/Volume)*

Enter, in designated boxes, the appropriate abbreviation from Table II (below) for the unit of measure.

TABLE II—UNITS OF MEASURE

G = Gallons (liquids only).
K = Kilograms.
L = Liters (liquids only).
M = Metric Tons (1000 kilograms).
N = Cubic Meters.
P = Pounds.
T = Tons (2000 pounds).
Y = Cubic Yards.
**Note:** Tons, Metric Tons, Cubic Meters, and Cubic Yards should only be reported in connection with very large bulk shipments, such as rail cars, tank trucks, or barges.

*Item 13. Waste Codes*

Enter up to six federal and state waste codes to describe each waste stream identified in Item 9b. State waste codes that are not redundant with federal codes must be entered here, in addition to the federal waste codes which are most representative of the properties of the waste.

*Item 14. Special Handling Instructions and Additional Information.*

1. Generators may enter any special handling or shipment-specific information necessary for the proper management or tracking of the materials under the generator's or other handler's business processes, such as waste profile numbers, container codes, bar codes, or response guide numbers. Generators also may use this space to enter additional descriptive information about their shipped materials, such as chemical names, constituent percentages, physical state, or specific gravity of wastes identified with volume units in Item 12.

2. This space may be used to record limited types of federally required information for which there is no specific space provided on the manifest, including any alternate facility designations; the manifest tracking number of the original manifest for rejected wastes and residues that are re-shipped under a second manifest; and the specification of PCB waste descriptions and PCB out-of-service dates required under 40 CFR

761.207. Generators, however, cannot be required to enter information in this space to meet state regulatory requirements.

*Item 15. Generator's/Offeror's Certifications*

1. The generator must read, sign, and date the waste minimization certification statement. In signing the waste minimization certification statement, those generators who have not been exempted by statute or regulation from the duty to make a waste minimization certification under section 3002(b) of RCRA are also certifying that they have complied with the waste minimization requirements. The Generator's Certification also contains the required attestation that the shipment has been properly prepared and is in proper condition for transportation (the shipper's certification). The content of the shipper's certification statement is as follows: "I hereby declare that the contents of this consignment are fully and accurately described above by the proper shipping name, and are classified, packaged, marked, and labeled/placarded, and are in all respects in proper condition for transport by highway according to applicable international and national governmental regulations. If export shipment and I am the Primary Exporter, I certify that the contents of this consignment conform to the terms of the attached EPA Acknowledgment of Consent." When a party other than the generator prepares the shipment for transportation, this party may also sign the shipper's certification statement as the offeror of the shipment.

2. Generator or Offeror personnel may preprint the words, "On behalf of" in the signature block or may hand write this statement in the signature block prior to signing the generator/offeror certification, to indicate that the individual signs as the employee or agent of the named principal.

NOTE: All of the above information except the handwritten signature required in Item 15 may be pre-printed.

II. INSTRUCTIONS FOR INTERNATIONAL SHIPMENT BLOCK

*Item 16. International Shipments*

For export shipments, the primary exporter must check the export box, and enter the point of exit (city and state) from the United States. For import shipments, the importer must check the import box and enter the point of entry (city and state) into the United States. For exports, the transporter must sign and date the manifest to indicate the day the shipment left the United States. Transporters of hazardous waste shipments must deliver a copy of the manifest to the U.S. Customs when exporting the waste across U.S. borders.

III. INSTRUCTIONS FOR TRANSPORTERS

*Item 17. Transporters' Acknowledgments of Receipt*

Enter the name of the person accepting the waste on behalf of the first transporter. That person must acknowledge acceptance of the waste described on the manifest by signing and entering the date of receipt. Only one signature per transportation company is required. Signatures are not required to track the movement of wastes in and out of transfer facilities, unless there is a change of custody between transporters.

If applicable, enter the name of the person accepting the waste on behalf of the second transporter. That person must acknowledge acceptance of the waste described on the manifest by signing and entering the date of receipt.

NOTE: Transporters carrying imports, who are acting as importers, may have responsibilities to enter information in the International Shipments Block. Transporters carrying exports may also have responsibilities to enter information in the International Shipments Block. See above instructions for Item 16.

IV. INSTRUCTIONS FOR OWNERS AND OPERATORS OF TREATMENT, STORAGE, AND DISPOSAL FACILITIES

*Item 18. Discrepancy*

Item 18a. Discrepancy Indication Space

1. The authorized representative of the designated (or alternate) facility's owner or operator must note in this space any discrepancies between the waste described on the Manifest and the waste actually received at the facility. Manifest discrepancies are: significant differences (as defined by §§ 264.72(b) and 265.72(b)) between the quantity or type of hazardous waste designated on the manifest or shipping paper, and the quantity and type of hazardous waste a facility actually receives, rejected wastes, which may be a full or partial shipment of hazardous waste that the TSDF cannot accept, or container residues, which are residues that exceed the quantity limits for "empty" containers set forth in 40 CFR 261.7(b).

2. For rejected loads and residues (40 CFR 264.72(d), (e), and (f), or 40 CFR 265.72(d), (e), or (f)), check the appropriate box if the shipment is a rejected load (*i.e.*, rejected by the designated and/or alternate facility and is sent to an alternate facility or returned to the generator) or a regulated residue that cannot be removed from a container. Enter the reason for the rejection or the inability to remove the residue and a description of the waste. Also, reference the manifest tracking number for any additional manifests being used to track the rejected waste

or residue shipment on the original manifest. Indicate the original manifest tracking number in Item 14, the Special Handling Block and Additional Information Block of the additional manifests.

3. Owners or operators of facilities located in unauthorized States (*i.e.*, states in which the U.S. EPA administers the hazardous waste management program) who cannot resolve significant differences in quantity or type within 15 days of receiving the waste must submit to their Regional Administrator a letter with a copy of the Manifest at issue describing the discrepancy and attempts to reconcile it (40 CFR 264.72(c) and 265.72(c)).

4. Owners or operators of facilities located in authorized States (*i.e.*, those States that have received authorization from the U.S. EPA to administer the hazardous waste management program) should contact their State agency for information on where to report discrepancies involving "significant differences" to state officials.

Item 18b. Alternate Facility (or Generator) for Receipt of Full Load Rejections

Enter the name, address, phone number, and EPA Identification Number of the Alternate Facility which the rejecting TSDF has designated, after consulting with the generator, to receive a fully rejected waste shipment. In the event that a fully rejected shipment is being returned to the generator, the rejecting TSDF may enter the generator's site information in this space. This field is not to be used to forward partially rejected loads or residue waste shipments.

Item 18c. Alternate Facility (or Generator) Signature

The authorized representative of the alternate facility (or the generator in the event of a returned shipment) must sign and date this field of the form to acknowledge receipt of the fully rejected wastes or residues identified by the initial TSDF.

*Item 19. Hazardous Waste Report Management Method Codes*

Enter the most appropriate Hazardous Waste Report Management Method code for each waste listed in Item 9. The Hazardous Waste Report Management Method code is to be entered by the first treatment, storage, or disposal facility (TSDF) that receives the waste and is the code that best describes the way in which the waste is to be managed when received by the TSDF.

*Item 20. Designated Facility Owner or Operator Certification of Receipt (Except As Noted in Item 18a)*

Enter the name of the person receiving the waste on behalf of the owner or operator of the facility. That person must acknowledge receipt or rejection of the waste described on the Manifest by signing and entering the date of receipt or rejection where indicated. Since the Facility Certification acknowledges receipt of the waste except as noted in the Discrepancy Space in Item 18a, the certification should be signed for both waste receipt and waste rejection, with the rejection being noted and described in the space provided in Item 18a. Fully rejected wastes may be forwarded or returned using Item 18b after consultation with the generator. Enter the name of the person accepting the waste on behalf of the owner or operator of the alternate facility or the original generator. That person must acknowledge receipt or rejection of the waste described on the Manifest by signing and entering the date they received or rejected the waste in Item 18c. Partially rejected wastes and residues must be re-shipped under a new manifest, to be initiated and signed by the rejecting TSDF as offeror of the shipment.

MANIFEST CONTINUATION SHEET

Please print or type. (Form designed for use on elite (12-pitch) typewriter.)

Form Approved. OMB No. 2050-0039

| UNIFORM HAZARDOUS WASTE MANIFEST (Continuation Sheet) | 21. Generator ID Number | | 22. Page | 23. Manifest Tracking Number |
|---|---|---|---|---|

24. Generator's Name

25. Transporter _____ Company Name — U.S. EPA ID Number

26. Transporter _____ Company Name — U.S. EPA ID Number

| 27a. HM | 27b. U.S. DOT Description (including Proper Shipping Name, Hazard Class, ID Number, and Packing Group (if any)) | 28. Containers No. | Type | 29. Total Quantity | 30. Unit Wt./Vol. | 31. Waste Codes |
|---|---|---|---|---|---|---|

**GENERATOR**

32. Special Handling Instructions and Additional Information

**TRANSPORTER**

| 33. Transporter _____ Acknowledgment of Receipt of Materials | | | |
|---|---|---|---|
| Printed/Typed Name | Signature | | Month Day Year |

| 34. Transporter _____ Acknowledgment of Receipt of Materials | | | |
|---|---|---|---|
| Printed/Typed Name | Signature | | Month Day Year |

**DESIGNATED FACILITY**

35. Discrepancy

36. Hazardous Waste Report Management Method Codes (i.e., codes for hazardous waste treatment, disposal, and recycling systems)

EPA Form 8700-22A (Rev. 3-05) Previous editions are obsolete.

DESIGNATED FACILITY TO DESTINATION STATE (IF REQUIRED)

INSTRUCTIONS—CONTINUATION SHEET, U.S. EPA FORM 8700–22A

Read all instructions before completing this form. This form has been designed for use on a 12-pitch (elite) typewriter; a firm point pen may also be used—press down hard.

This form must be used as a continuation sheet to U.S. EPA Form 8700–22 if:

• More than two transporters are to be used to transport the waste; or

• More space is required for the U.S. DOT descriptions and related information in Item 9 of U.S. EPA Form 8700–22.

Federal regulations require generators and transporters of hazardous waste and owners or operators of hazardous waste treatment, storage, or disposal facilities to use the uniform hazardous waste manifest (EPA Form 8700–22) and, if necessary, this continuation sheet (EPA Form 8700–22A) for both interstate and intrastate transportation.

### Item 21. Generator's ID Number

Enter the generator's U.S. EPA twelve digit identification number or, the State generator identification number if the generator site does not have an EPA identification number.

### Item 22. Page ___—

Enter the page number of this Continuation Sheet.

### Item 23. Manifest Tracking Number

Enter the Manifest Tracking number from Item 4 of the Manifest form to which this continuation sheet is attached.

### Item 24. Generator's Name—

Enter the generator's name as it appears in Item 5 on the first page of the Manifest.

### Item 25. Transporter—Company Name

If additional transporters are used to transport the waste described on this Manifest, enter the company name of each additional transporter in the order in which they will transport the waste. Enter after the word "Transporter" the order of the transporter. For example, Transporter 3 Company Name. Also enter the U.S. EPA twelve digit identification number of the transporter described in Item 25.

### Item 26. Transporter—Company Name

If additional transporters are used to transport the waste described on this Manifest, enter the company name of each additional transporter in the order in which they will transport the waste. Enter after the word "Transporter" the order of the transporter. For example, Transporter 4 Company Name. Each Continuation Sheet can record the names of two additional transporters. Also enter the U.S. EPA twelve digit identification number of the transporter named in Item 26.

### Item 27. U.S. D.O.T. Description Including Proper Shipping Name, Hazardous Class, and ID Number (UN/NA)

For each row enter a sequential number under Item 27b that corresponds to the order of waste codes from one continuation sheet to the next, to reflect the total number of wastes being shipped. Refer to instructions for Item 9 of the manifest for the information to be entered.

### Item 28. Containers (No. And Type)

Refer to the instructions for Item 10 of the manifest for information to be entered.

### Item 29. Total Quantity

Refer to the instructions for Item 11 of the manifest form.

### Item 30. Units of Measure (Weight/Volume)

Refer to the instructions for Item 12 of the manifest form.

### Item 31. Waste Codes

Refer to the instructions for Item 13 of the manifest form.

### Item 32. Special Handling Instructions and Additional Information

Refer to the instructions for Item 14 of the manifest form.

### TRANSPORTERS

### Item 33. Transporter—Acknowledgment of Receipt of Materials

Enter the same number of the Transporter as identified in Item 25. Enter also the name of the person accepting the waste on behalf of the Transporter (Company Name) identified in Item 25. That person must acknowledge acceptance of the waste described on the Manifest by signing and entering the date of receipt.

### Item 34. Transporter—Acknowledgment of Receipt of Materials

Enter the same number of the Transporter as identified in Item 26. Enter also the name of the person accepting the waste on behalf of the Transporter (Company Name) identified in Item 26. That person must acknowledge acceptance of the waste described on the Manifest by signing and entering the date of receipt.

### OWNER AND OPERATORS OF TREATMENT, STORAGE, OR DISPOSAL FACILITIES

### Item 35. Discrepancy Indication Space

Refer to Item 18. This space may be used to more fully describe information on discrepancies identified in Item 18a of the manifest form.

### Item 36. Hazardous Waste Report Management Method Codes

For each field here, enter the sequential number that corresponds to the waste materials described under Item 27, and enter the appropriate process code that describes how the materials will be processed when received. If additional continuation sheets are attached, continue numbering the waste materials and process code fields sequentially, and enter on each sheet the process codes corresponding to the waste materials identified on that sheet.

[45 FR 33142, May 19, 1980, as amended at 70 FR 10818, Mar. 4, 2005]

# PART 263—STANDARDS APPLICABLE TO TRANSPORTERS OF HAZARDOUS WASTE

## Subpart A—General

Sec.
263.10   Scope.
263.11   EPA identification number.
263.12   Transfer facility requirements.

## Subpart B—Compliance With the Manifest System and Recordkeeping

263.20   The manifest system.
263.21   Compliance with the manifest.
263.22   Recordkeeping.

## Subpart C—Hazardous Waste Discharges

263.30   Immediate action.
263.31   Discharge clean up.

AUTHORITY: 42 U.S.C. 6906, 6912, 6922–6925, 6937, and 6938.

SOURCE: 45 FR 33151, May 19, 1980, unless otherwise noted.

## Subpart A—General

### § 263.10   Scope.

(a) These regulations establish standards which apply to persons transporting hazardous waste within the United States if the transportation requires a manifest under 40 CFR part 262.

NOTE: The regulations set forth in parts 262 and 263 establish the responsibilities of generators and transporters of hazardous waste in the handling, transportation, and management of that waste. In these regulations, EPA has expressly adopted certain regulations of the Department of Transportation (DOT) governing the transportation of hazardous materials. These regulations concern, among other things, labeling, marking, placarding, using proper containers, and reporting discharges. EPA has expressly adopted these regulations in order to satisfy its statutory obligation to promulgate regulations which are necessary to protect human health and the environment in the transportation of hazardous waste. EPA's adoption of these DOT regulations ensures consistency with the requirements of DOT and thus avoids the establishment of duplicative or conflicting requirements with respect to

these matters. These EPA regulations which apply to both interstate and intrastate transportation of hazardous waste are enforceable by EPA.

DOT has revised its hazardous materials transportation regulations in order to encompass the transportation of hazardous waste and to regulate intrastate, as well as interstate, transportation of hazardous waste. Transporters of hazardous waste are cautioned that DOT's regulations are fully applicable to their activities and enforceable by DOT. These DOT regulations are codified in title 49, Code of Federal Regulations, subchapter C.

EPA and DOT worked together to develop standards for transporters of hazardous waste in order to avoid conflicting requirements. Except for transporters of bulk shipments of hazardous waste by water, a transporter who meets all applicable requirements of 49 CFR parts 171 through 179 and the requirements of 40 CFR 263.11 and 263.31 will be deemed in compliance with this part. Regardless of DOT's action, EPA retains its authority to enforce these regulations.

(b) These regulations do not apply to on-site transportation of hazardous waste by generators or by owners or operators of permitted hazardous waste management facilities.

(c) A transporter of hazardous waste must also comply with 40 CFR part 262, Standards Applicable to Generators of Hazardous Waste, if he:

(1) Transports hazardous waste into the United States from abroad; or

(2) Mixes hazardous wastes of different DOT shipping descriptions by placing them into a single container.

(d) A transporter of hazardous waste subject to the Federal manifesting requirements of 40 CFR part 262, or subject to the waste management standards of 40 CFR part 273, or subject to State requirements analogous to 40 CFR part 273, that is being imported from or exported to any of the countries listed in 40 CFR 262.58(a)(1) for purposes of recovery is subject to this Subpart and to all other relevant requirements of subpart H of 40 CFR part 262, including, but not limited to, 40 CFR 262.84 for tracking documents.

(e) The regulations in this part do not apply to transportation during an explosives or munitions emergency response, conducted in accordance with 40 CFR 264.1(g)(8)(i)(D) or (iv) or 265.1(c)(11)(i)(D) or (iv), and 270.1(c)(3)(i)(D) or (iii).

(f) Section 266.203 of this chapter identifies how the requirements of this part apply to military munitions classified as solid waste under 40 CFR 266.202.

[45 FR 33151, May 19, 1980, as amended at 45 FR 86968, Dec. 31, 1980; 61 FR 16314, Apr. 12, 1996; 62 FR 6651, Feb. 12, 1997]

EFFECTIVE DATE NOTE: At 75 FR 1259, Jan. 8, 2010, §263.10 was amended by revising paragraph (d), effective July 7, 2010. For the convenience of the user, the revised text is set forth as follows:

§263.10  Scope.

\*        \*        \*        \*        \*

(d) A transporter of hazardous waste subject to the Federal manifesting requirements of 40 CFR part 262, or subject to the waste management standards of 40 CFR part 273, or subject to State requirements analogous to 40 CFR part 273, that is being imported from or exported to any of the countries listed in 40 CFR 262.58(a)(1) for purposes of recovery is subject to this Subpart and to all other relevant requirements of subpart H of 40 CFR part 262, including, but not limited to, 40 CFR 262.84 for movement documents.

\*        \*        \*        \*        \*

§263.11   EPA identification number.

(a) A transporter must not transport hazardous wastes without having received an EPA identification number from the Administrator.

(b) A transporter who has not received an EPA identification number may obtain one by applying to the Administrator using EPA Form 8700–12. Upon receiving the request, the Administrator will assign an EPA identification number to the transporter.

§263.12   Transfer facility requirements.

A transporter who stores manifested shipments of hazardous waste in containers meeting the requirements of §262.30 at a transfer facility for a period of ten days or less is not subject to regulation under parts 270, 264, 265, 267, and 268 of this chapter with respect to the storage of those wastes.

[75 FR 13005, Mar. 18, 2010]

## Subpart B—Compliance With the Manifest System and Record-keeping

§263.20   The manifest system.

(a)(1) *Manifest requirement.* A transporter may not accept hazardous waste from a generator unless the transporter is also provided with a manifest signed in accordance with the requirements of §262.23.

(2) *Exports.* In the case of exports other than those subject to subpart H of 40 CFR part 262, a transporter may not accept such waste from a primary exporter or other person if he knows the shipment does not conform to the EPA Acknowledgment of Consent; and unless, in addition to a manifest signed by the generator as provided in this section, the transporter shall also be provided with an EPA Acknowledgment of Consent which, except for shipments by rail, is attached to the manifest (or shipping paper for exports by water (bulk shipment)). For exports of hazardous waste subject to the requirements of subpart H of 40 CFR part 262, a transporter may not accept hazardous waste without a tracking document that includes all information required by 40 CFR 262.84.

(3) *Compliance Date for Form Revisions.* The revised Manifest form and procedures in 40 CFR 260.10, 261.7, 263.20, and 263.21, shall not apply until September 5, 2006. The Manifest form and procedures in 40 CFR 260.10, 261.7, 263.20, and 263.21, contained in the 40 CFR, parts 260 to 265, edition revised as of July 1, 2004, shall be applicable until September 5, 2006.

(b) Before transporting the hazardous waste, the transporter must sign and date the manifest acknowledging acceptance of the hazardous waste from the generator. The transporter must return a signed copy to the generator before leaving the generator's property.

(c) The transporter must ensure that the manifest accompanies the hazardous waste. In the case of exports, the transporter must ensure that a copy of the EPA Acknowledgment of Consent also accompanies the hazardous waste.

(d) A transporter who delivers a hazardous waste to another transporter or to the designated facility must:

(1) Obtain the date of delivery and the handwritten signature of that transporter or of the owner or operator of the designated facility on the manifest; and

(2) Retain one copy of the manifest in accordance with § 263.22; and

(3) Give the remaining copies of the manifest to the accepting transporter or designated facility.

(e) The requirements of paragraphs (c), (d) and (f) of this section do not apply to water (bulk shipment) transporters if:

(1) The hazardous waste is delivered by water (bulk shipment) to the designated facility; and

(2) A shipping paper containing all the information required on the manifest (excluding the EPA identification numbers, generator certification, and signatures) and, for exports, an EPA Acknowledgment of Consent accompanies the hazardous waste; and

(3) The delivering transporter obtains the date of delivery and handwritten signature of the owner or operator of the designated facility on either the manifest or the shipping paper; and

(4) The person delivering the hazardous waste to the initial water (bulk shipment) transporter obtains the date of delivery and signature of the water (bulk shipment) transporter on the manifest and forwards it to the designated facility; and

(5) A copy of the shipping paper or manifest is retained by each water (bulk shipment) transporter in accordance with § 263.22.

(f) For shipments involving rail transportation, the requirements of paragraphs (c), (d) and (e) do not apply and the following requirements do apply:

(1) When accepting hazardous waste from a non-rail transporter, the initial rail transporter must:

(i) Sign and date the manifest acknowledging acceptance of the hazardous waste;

(ii) Return a signed copy of the manifest to the non-rail transporter;

(iii) Forward at least three copies of the manifest to:

(A) The next non-rail transporter, if any; or,

(B) The designated facility, if the shipment is delivered to that facility by rail; or

(C) The last rail transporter designated to handle the waste in the United States;

(iv) Retain one copy of the manifest and rail shipping paper in accordance with § 263.22.

(2) Rail transporters must ensure that a shipping paper containing all the information required on the manifest (excluding the EPA identification numbers, generator certification, and signatures) and, for exports an EPA Acknowledgment of Consent accompanies the hazardous waste at all times.

NOTE: Intermediate rail transporters are not required to sign either the manifest or shipping paper.

(3) When delivering hazardous waste to the designated facility, a rail transporter must:

(i) Obtain the date of delivery and handwritten signature of the owner or operator of the designated facility on the manifest or the shipping paper (if the manifest has not been received by the facility); and

(ii) Retain a copy of the manifest or signed shipping paper in accordance with § 263.22.

(4) When delivering hazardous waste to a non-rail transporter a rail transporter must:

(i) Obtain the date of delivery and the handwritten signature of the next non-rail transporter on the manifest; and

(ii) Retain a copy of the manifest in accordance with § 263.22.

(5) Before accepting hazardous waste from a rail transporter, a non-rail transporter must sign and date the manifest and provide a copy to the rail transporter.

(g) Transporters who transport hazardous waste out of the United States must:

(1) Sign and date the manifest in the International Shipments block to indicate the date that the shipment left the United States;

(2) Retain one copy in accordance with § 263.22(d);

(3) Return a signed copy of the manifest to the generator; and

(4) Give a copy of the manifest to a U.S. Customs official at the point of departure from the United States.

(h) A transporter transporting hazardous waste from a generator who generates greater than 100 kilograms but less than 1000 kilograms of hazardous waste in a calendar month need not comply with the requirements of this section or those of §263.22 provided that:

(1) The waste is being transported pursuant to a reclamation agreement as provided for in §262.20(e);

(2) The transporter records, on a log or shipping paper, the following information for each shipment:

(i) The name, address, and U.S. EPA Identification Number of the generator of the waste;

(ii) The quantity of waste accepted;

(iii) All DOT-required shipping information;

(iv) The date the waste is accepted; and

(3) The transporter carries this record when transporting waste to the reclamation facility; and

(4) The transporter retains these records for a period of at least three years after termination or expiration of the agreement.

[45 FR 33151, May 19, 1980, as amended at 45 FR 86973, Dec. 31, 1980; 51 FR 10176, Mar. 24, 1986; 51 FR 28685, Aug. 8, 1986; 61 FR 16315, Apr. 12, 1996; 70 FR 10821, Mar. 4, 2005]

## §263.21 Compliance with the manifest.

(a) The transporter must deliver the entire quantity of hazardous waste which he has accepted from a generator or a transporter to:

(1) The designated facility listed on the manifest; or

(2) The alternate designated facility, if the hazardous waste cannot be delivered to the designated facility because an emergency prevents delivery; or

(3) The next designated transporter; or

(4) The place outside the United States designated by the generator.

(b)(1) If the hazardous waste cannot be delivered in accordance with paragraph (a) of this section because of an emergency condition other than rejection of the waste by the designated facility, then the transporter must contact the generator for further directions and must revise the manifest according to the generator's instructions.

(2) If hazardous waste is rejected by the designated facility while the transporter is on the facility's premises, then the transporter must obtain the following:

(i) For a partial load rejection or for regulated quantities of container residues, a copy of the original manifest that includes the facility's date and signature, and the Manifest Tracking Number of the new manifest that will accompany the shipment, and a description of the partial rejection or container residue in the discrepancy block of the original manifest. The transporter must retain a copy of this manifest in accordance with §263.22, and give the remaining copies of the original manifest to the rejecting designated facility. If the transporter is forwarding the rejected part of the shipment or a regulated container residue to an alternate facility or returning it to the generator, the transporter must obtain a new manifest to accompany the shipment, and the new manifest must include all of the information required in 40 CFR 264.72(e)(1) through (6) or (f)(1) through (6) or 40 CFR 265.72(e)(1) through (6) or (f)(1) through (6).

(ii) For a full load rejection that will be taken back by the transporter, a copy of the original manifest that includes the rejecting facility's signature and date attesting to the rejection, the description of the rejection in the discrepancy block of the manifest, and the name, address, phone number, and Identification Number for the alternate facility or generator to whom the shipment must be delivered. The transporter must retain a copy of the manifest in accordance with §263.22, and give a copy of the manifest containing this information to the rejecting designated facility. If the original manifest is not used, then the transporter must obtain a new manifest for the shipment and comply with 40 CFR 264.72(e)(1) through (6) or 40 CFR 265.72(e)(1) through (6).

[45 FR 33151, May 19, 1980, as amended at 70 FR 10821, Mar. 2005]

## § 263.22  Recordkeeping.

(a) A transporter of hazardous waste must keep a copy of the manifest signed by the generator, himself, and the next designated transporter or the owner or operator of the designated facility for a period of three years from the date the hazardous waste was accepted by the initial transporter.

(b) For shipments delivered to the designated facility by water (bulk shipment), each water (bulk shipment) transporter must retain a copy of the shipping paper containing all the information required in § 263.20(e)(2) for a period of three years from the date the hazardous waste was accepted by the initial transporter.

(c) For shipments of hazardous waste by rail within the United States:

(1) The initial rail transporter must keep a copy of the manifest and shipping paper with all the information required in § 263.20(f)(2) for a period of three years from the date the hazardous waste was accepted by the initial transporter; and

(2) The final rail transporter must keep a copy of the signed manifest (or the shipping paper if signed by the designated facility in lieu of the manifest) for a period of three years from the date the hazardous waste was accepted by the initial transporter.

NOTE: Intermediate rail transporters are not required to keep records pursuant to these regulations.

(d) A transporter who transports hazardous waste out of the United States must keep a copy of the manifest indicating that the hazardous waste left the United States for a period of three years from the date the hazardous waste was accepted by the initial transporter.

(e) The periods of retention referred to in this Section are extended automatically during the course of any unresolved enforcement action regarding the regulated activity or as requested by the Administrator.

[45 FR 33151, May 19, 1980, as amended at 45 FR 86973, Dec. 31, 1980]

## Subpart C—Hazardous Waste Discharges

## § 263.30  Immediate action.

(a) In the event of a discharge of hazardous waste during transportation, the transporter must take appropriate immediate action to protect human health and the environment (e.g., notify local authorities, dike the discharge area).

(b) If a discharge of hazardous waste occurs during transportation and an official (State or local government or a Federal Agency) acting within the scope of his official responsibilities determines that immediate removal of the waste is necessary to protect human health or the environment, that official may authorize the removal of the waste by transporters who do not have EPA identification numbers and without the preparation of a manifest.

(c) An air, rail, highway, or water transporter who has discharged hazardous waste must:

(1) Give notice, if required by 49 CFR 171.15, to the National Response Center (800–424–8802 or 202–426–2675); and

(2) Report in writing as required by 49 CFR 171.16 to the Director, Office of Hazardous Materials Regulations, Materials Transportation Bureau, Department of Transportation, Washington, DC 20590.

(d) A water (bulk shipment) transporter who has discharged hazardous waste must give the same notice as required by 33 CFR 153.203 for oil and hazardous substances.

## § 263.31  Discharge clean up.

A transporter must clean up any hazardous waste discharge that occurs during transportation or take such action as may be required or approved by Federal, State, or local officials so that the hazardous waste discharge no longer presents a hazard to human health or the environment.

# PART 264—STANDARDS FOR OWNERS AND OPERATORS OF HAZARDOUS WASTE TREATMENT, STORAGE, AND DISPOSAL FACILITIES

AUTHORITY: 42 U.S.C. 6905, 6912(a), 6924, and 6925.

SOURCE: 45 FR 33221, May 19, 1980, unless otherwise noted.

## Subpart A—General

### § 264.1  Purpose, scope and applicability.

(a) The purpose of this part is to establish minimum national standards which define the acceptable management of hazardous waste.

(b) The standards in this part apply to owners and operators of all facilities which treat, store, or dispose of hazardous waste, except as specifically provided otherwise in this part or part 261 of this chapter.

(c) The requirements of this part apply to a person disposing of hazardous waste by means of ocean disposal subject to a permit issued under the Marine Protection, Research, and Sanctuaries Act only to the extent they are included in a RCRA permit by rule granted to such a person under part 270 of this chapter.

[Comment: These part 264 regulations do apply to the treatment or storage of hazardous waste before it is loaded onto an ocean vessel for incineration or disposal at sea.]

(d) The requirements of this part apply to a person disposing of hazardous waste by means of underground injection subject to a permit issued under an Underground Injection Control (UIC) program approved or promulgated under the Safe Drinking Water Act only to the extent they are required by § 144.14 of this chapter.

[Comment: These part 264 regulations do apply to the above-ground treatment or storage of hazardous waste before it is injected underground.]

(e) The requirements of this part apply to the owner or operator of a POTW which treats, stores, or disposes of hazardous waste only to the extent they are included in a RCRA permit by rule granted to such a person under part 270 of this chapter.

(f) The requirements of this part do not apply to a person who treats, stores, or disposes of hazardous waste in a State with a RCRA hazardous waste program authorized under subpart A of part 271 of this chapter, or in a State authorized under subpart B of part 271 of this chapter for the component or components of Phase II interim authorization which correspond to the person's treatment, storage or disposal processes; except that this part will apply:

(1) As stated in paragraph (d) of this section, if the authorized State RCRA program does not cover disposal of hazardous waste by means of underground injection; and

(2) To a person who treats, stores or disposes of hazardous waste in a State authorized under subpart A of part 271 of this chapter, at a facility which was not covered by standards under this part when the State obtained authorization, and for which EPA promulgates standards under this part after the State is authorized. This paragraph will only apply until the State is authorized to permit such facilities under subpart A of part 271 of this chapter.

(3) To a person who treats, stores, or disposes of hazardous waste in a State which is authorized under subpart A or B of part 271 of this chapter if the State has not been authorized to carry out the requirements and prohibitions applicable to the treatment, storage, or disposal of hazardous waste at his facility which are imposed pursuant to the Hazardous and Solid Waste Amendments of 1984. The requirements and prohibitions that are applicable until a State receives authorization to carry them out include all Federal program requirements identified in § 271.1(j).

(g) The requirements of this part do not apply to:

(1) The owner or operator of a facility permitted, licensed, or registered by a State to manage municipal or industrial solid waste, if the only hazardous waste the facility treats, stores, or disposes of is excluded from regulation under this part by § 261.5 of this chapter;

(2) The owner or operator of a facility managing recyclable materials described in § 261.6 (a)(2), (3), and (4) of this chapter (except to the extent they are referred to in part 279 or subparts C, F, G, or H of part 266 of this chapter).

(3) A generator accumulating waste on-site in compliance with § 262.34 of this chapter;

(4) A farmer disposing of waste pesticides from his own use in compliance with § 262.70 of this chapter; or

(5) The owner or operator of a totally enclosed treatment facility, as defined in §260.10.

(6) The owner or operator of an elementary neutralization unit or a wastewater treatment unit as defined in §260.10 of this chapter, provided that if the owner or operator is diluting hazardous ignitable (D001) wastes (other than the D001 High TOC Subcategory defined in §268.40 of this chapter, Table Treatment Standards for Hazardous Wastes), or reactive (D003) waste, to remove the characteristic before land disposal, the owner/operator must comply with the requirements set out in §264.17(b).

(7) [Reserved]

(8)(i) Except as provided in paragraph (g)(8)(ii) of this section, a person engaged in treatment or containment activities during immediate response to any of the following situations:

(A) A discharge of a hazardous waste;

(B) An imminent and substantial threat of a discharge of hazardous waste;

(C) A discharge of a material which, when discharged, becomes a hazardous waste.

(D) An immediate threat to human health, public safety, property, or the environment, from the known or suspected presence of military munitions, other explosive material, or an explosive device, as determined by an explosive or munitions emergency response specialist as defined in 40 CFR 260.10.

(ii) An owner or operator of a facility otherwise regulated by this part must comply with all applicable requirements of subparts C and D.

(iii) Any person who is covered by paragraph (g)(8)(i) of this section and who continues or initiates hazardous waste treatment or containment activities after the immediate response is over is subject to all applicable requirements of this part and parts 122 through 124 of this chapter for those activities.

(iv) In the case of an explosives or munitions emergency response, if a Federal, State, Tribal or local official acting within the scope of his or her official responsibilities, or an explosives or munitions emergency response specialist, determines that immediate removal of the material or waste is nec-

essary to protect human health or the environment, that official or specialist may authorize the removal of the material or waste by transporters who do not have EPA identification numbers and without the preparation of a manifest. In the case of emergencies involving military munitions, the responding military emergency response specialist's organizational unit must retain records for three years identifying the dates of the response, the responsible persons responding, the type and description of material addressed, and its disposition.

(9) A transporter storing manifested shipments of hazardous waste in containers meeting the requirements of 40 CFR 262.30 at a transfer facility for a period of ten days or less.

(10) The addition of absorbent material to waste in a container (as defined in §260.10 of this chapter) or the addition of waste to absorbent material in a container, provided that these actions occur at the time waste is first placed in the container; and §§264.17(b), 264.171, and 264.172 are complied with.

(11) Universal waste handlers and universal waste transporters (as defined in 40 CFR 260.10) handling the wastes listed below. These handlers are subject to regulation under 40 CFR part 273, when handling the below listed universal wastes.

(i) Batteries as described in 40 CFR 273.2;

(ii) Pesticides as described in §273.3 of this chapter;

(iii) Mercury-containing equipment as described in §273.4 of this chapter; and

(iv) Lamps as described in §273.5 of this chapter.

(12) A New York State Utility central collection facility consolidating hazardous waste in accordance with 40 CFR 262.90.

(h) The requirements of this part apply to owners or operators of all facilities which treat, store, or dispose of hazardous wastes referred to in part 268.

(i) Section 266.205 of this chapter identifies when the requirements of this part apply to the storage of military munitions classified as solid waste under §266.202 of this chapter. The treatment and disposal of hazardous waste military munitions are subject to the applicable permitting, procedural, and technical standards in 40 CFR parts 260 through 270.

(j) The requirements of subparts B, C, and D of this part and § 264.101 do not apply to remediation waste management sites. (However, some remediation waste management sites may be a part of a facility that is subject to a traditional RCRA permit because the facility is also treating, storing or disposing of hazardous wastes that are not remediation wastes. In these cases, Subparts B, C, and D of this part, and § 264.101 do apply to the facility subject to the traditional RCRA permit.) Instead of the requirements of subparts B, C, and D of this part, owners or operators of remediation waste management sites must:

(1) Obtain an EPA identification number by applying to the Administrator using EPA Form 8700–12;

(2) Obtain a detailed chemical and physical analysis of a representative sample of the hazardous remediation wastes to be managed at the site. At a minimum, the analysis must contain all of the information which must be known to treat, store or dispose of the waste according to this part and part 268 of this chapter, and must be kept accurate and up to date;

(3) Prevent people who are unaware of the danger from entering, and minimize the possibility for unauthorized people or livestock to enter onto the active portion of the remediation waste management site, unless the owner or operator can demonstrate to the Director that:

(i) Physical contact with the waste, structures, or equipment within the active portion of the remediation waste management site will not injure people or livestock who may enter the active portion of the remediation waste management site; and

(ii) Disturbance of the waste or equipment by people or livestock who enter onto the active portion of the remediation waste management site, will not cause a violation of the requirements of this part;

(4) Inspect the remediation waste management site for malfunctions, deterioration, operator errors, and discharges that may be causing, or may

lead to, a release of hazardous waste constituents to the environment, or a threat to human health. The owner or operator must conduct these inspections often enough to identify problems in time to correct them before they harm human health or the environment, and must remedy the problem before it leads to a human health or environmental hazard. Where a hazard is imminent or has already occurred, the owner/operator must take remedial action immediately;

(5) Provide personnel with classroom or on-the-job training on how to perform their duties in a way that ensures the remediation waste management site complies with the requirements of this part, and on how to respond effectively to emergencies;

(6) Take precautions to prevent accidental ignition or reaction of ignitable or reactive waste, and prevent threats to human health and the environment from ignitable, reactive and incompatible waste;

(7) For remediation waste management sites subject to regulation under subparts I through O and subpart X of this part, the owner/operator must design, construct, operate, and maintain a unit within a 100-year floodplain to prevent washout of any hazardous waste by a 100-year flood, unless the owner/operator can meet the demonstration of § 264.18(b);

(8) Not place any non-containerized or bulk liquid hazardous waste in any salt dome formation, salt bed formation, underground mine or cave;

(9) Develop and maintain a construction quality assurance program for all surface impoundments, waste piles and landfill units that are required to comply with §§ 264.221(c) and (d), 264.251(c) and (d), and 264.301(c) and (d) at the remediation waste management site, according to the requirements of § 264.19;

(10) Develop and maintain procedures to prevent accidents and a contingency and emergency plan to control accidents that occur. These procedures must address proper design, construction, maintenance, and operation of remediation waste management units at the site. The goal of the plan must be to minimize the possibility of, and the hazards from a fire, explosion, or any

unplanned sudden or non-sudden release of hazardous waste or hazardous waste constituents to air, soil, or surface water that could threaten human health or the environment. The plan must explain specifically how to treat, store and dispose of the hazardous remediation waste in question, and must be implemented immediately whenever a fire, explosion, or release of hazardous waste or hazardous waste constituents which could threaten human health or the environment;

(11) Designate at least one employee, either on the facility premises or on call (that is, available to respond to an emergency by reaching the facility quickly), to coordinate all emergency response measures. This emergency coordinator must be thoroughly familiar with all aspects of the facility's contingency plan, all operations and activities at the facility, the location and characteristics of waste handled, the location of all records within the facility, and the facility layout. In addition, this person must have the authority to commit the resources needed to carry out the contingency plan;

(12) Develop, maintain and implement a plan to meet the requirements in paragraphs (j)(2) through (j)(6) and (j)(9) through (j)(10) of this section; and

(13) Maintain records documenting compliance with paragraphs (j)(1) through (j)(12) of this section.

[45 FR 33221, May 19, 1980]

EDITORIAL NOTE: For FEDERAL REGISTER citations affecting §264.1, see the List of CFR Sections Affected, which appears in the Finding Aids section of the printed volume and on GPO Access.

### §264.2 [Reserved]

### §264.3 Relationship to interim status standards.

A facility owner or operator who has fully complied with the requirements for interim status—as defined in section 3005(e) of RCRA and regulations under §270.70 of this chapter—must comply with the regulations specified in part 265 of this chapter in lieu of the regulations in this part, until final administrative disposition of his permit application is made, except as provided under 40 CFR part 264 subpart S.

[*Comment:* As stated in section 3005(a) of RCRA, after the effective date of regulations under that section, i.e., parts 270 and 124 of this chapter, the treatment, storage, or disposal of hazardous waste is prohibited except in accordance with a permit. Section 3005(e) of RCRA provides for the continued operation of an existing facility which meets certain conditions until final administrative disposition of the owner's or operator's permit application is made.]

[45 FR 33221, May 19, 1980, as amended at 48 FR 14294, Apr. 1, 1983; 58 FR 8683, Feb. 16, 1993]

### §264.4 Imminent hazard action.

Notwithstanding any other provisions of these regulations, enforcement actions may be brought pursuant to section 7003 of RCRA.

[45 FR 33221, May 19, 1980, as amended at 71 FR 40272, July 14, 2006]

## Subpart B—General Facility Standards

### §264.10 Applicability.

(a) The regulations in this subpart apply to owners and operators of all hazardous waste facilities, except as provided in §264.1 and in paragraph (b) of this section.

(b) Section 264.18(b) applies only to facilities subject to regulation under subparts I through O and subpart X of this part.

[46 FR 2848, Jan. 12, 1981, as amended at 52 FR 46963, Dec. 10, 1987]

### §264.11 Identification number.

Every facility owner or operator must apply to EPA for an EPA identification number in accordance with the EPA notification procedures (45 FR 12746).

[45 FR 33221, May 19, 1980, as amended at 50 FR 4514, Jan. 31, 1985]

### §264.12 Required notices.

(a)(1) The owner or operator of a facility that has arranged to receive hazardous waste from a foreign source must notify the Regional Administrator in writing at least four weeks in advance of the date the waste is expected to arrive at the facility. Notice of subsequent shipments of the same waste from the same foreign source is not required.

(2) The owner or operator of a recovery facility that has arranged to receive hazardous waste subject to 40 CFR part 262, subpart H must provide a copy of the tracking document bearing all required signatures to the notifier, to the Office of Enforcement and Compliance Assurance, Office of Compliance, Enforcement Planning, Targeting and Data Division (2222A), Environmental Protection Agency, 1200 Pennsylvania Ave., NW., Washington, DC 20460; and to the competent authorities of all other concerned countries within three working days of receipt of the shipment. The original of the signed tracking document must be maintained at the facility for at least three years.

(b) The owner or operator of a facility that receives hazardous waste from an off-site source (except where the owner or operator is also the generator) must inform the generator in writing that he has the appropriate permit(s) for, and will accept, the waste the generator is shipping. The owner or operator must keep a copy of this written notice as part of the operating record.

(c) Before transferring ownership or operation of a facility during its operating life, or of a disposal facility during the post-closure care period, the owner or operator must notify the new owner or operator in writing of the requirements of this part and part 270 of this chapter.

[*Comment:* An owner's or operator's failure to notify the new owner or operator of the requirements of this part in no way relieves the new owner or operator of his obligation to comply with all applicable requirements.]

[45 FR 33221, May 19, 1980, as amended at 48 FR 14294, Apr. 1, 1983; 50 FR 4514, Jan. 31, 1985; 61 FR 16315, Apr. 12, 1996]

EFFECTIVE DATE NOTE: At 75 FR 1260, Jan. 8, 2010, § 264.12 was amended by revising paragraph (a)(2), effective July 7, 2010. For the convenience of the user, the revised text is set forth as follows:

§ 264.12   Required notices.

(a) * * *

(2) The owner or operator of a recovery facility that has arranged to receive hazardous waste subject to 40 CFR part 262, subpart H must provide a copy of the movement document bearing all required signatures to the foreign exporter; to the Office of Enforcement and Compliance Assurance, Office of

Federal Activities, International Compliance Assurance Division (2254A), Environmental Protection Agency, 1200 Pennsylvania Avenue, NW., Washington, DC 20460; and to the competent authorities of all other countries concerned within three (3) working days of receipt of the shipment. The original of the signed movement document must be maintained at the facility for at least three (3) years. In addition, such owner or operator shall, as soon as possible, but no later than thirty (30) days after the completion of recovery and no later than one (1) calendar year following the receipt of the hazardous waste, send a certificate of recovery to the foreign exporter and to the competent authority of the country of export and to EPA's Office of Enforcement and Compliance Assurance at the above address by mail, e-mail without a digital signature followed by mail, or fax followed by mail.

\*     \*     \*     \*     \*

§ 264.13   General waste analysis.

(a)(1) Before an owner or operator treats, stores, or disposes of any hazardous wastes, or nonhazardous wastes if applicable under § 264.113(d), he must obtain a detailed chemical and physical analysis of a representative sample of the wastes. At a minimum, the analysis must contain all the information which must be known to treat, store, or dispose of the waste in accordance with this part and part 268 of this chapter.

(2) The analysis may include data developed under part 261 of this chapter, and existing published or documented data on the hazardous waste or on hazardous waste generated from similar processes.

[*1:* For example, the facility's records of analyses performed on the waste before the effective date of these regulations, or studies conducted on hazardous waste generated from processes similar to that which generated the waste to be managed at the facility, may be included in the data base required to comply with paragraph (a)(1) of this section. The owner or operator of an off-site facility may arrange for the generator of the hazardous waste to supply part of the information required by paragraph (a)(1) of this section, except as otherwise specified in 40 CFR 268.7 (b) and (c). If the generator does not supply the information, and the owner or operator chooses to accept a hazardous waste, the owner or operator is responsible for obtaining the information required to comply with this section.]

(3) The analysis must be repeated as necessary to ensure that it is accurate and up to date. At a minimum, the analysis must be repeated:

(i) When the owner or operator is notified, or has reason to believe, that the process or operation generating the hazardous wastes, or non-hazardous wastes if applicable under §264.113(d), has changed; and

(ii) For off-site facilities, when the results of the inspection required in paragraph (a)(4) of this section indicate that the hazardous waste received at the facility does not match the waste designated on the accompanying manifest or shipping paper.

(4) The owner or operator of an off-site facility must inspect and, if necessary, analyze each hazardous waste movement received at the facility to determine whether it matches the identity of the waste specified on the accompanying manifest or shipping paper.

(b) The owner or operator must develop and follow a written waste analysis plan which describes the procedures which he will carry out to comply with paragraph (a) of this section. He must keep this plan at the facility. At a minimum, the plan must specify:

(1) The parameters for which each hazardous waste, or non-hazardous waste if applicable under §264.113(d), will be analyzed and the rationale for the selection of these parameters (i.e., how analysis for these parameters will provide sufficient information on the waste's properties to comply with paragraph (a) of this section);

(2) The test methods which will be used to test for these parameters;

(3) The sampling method which will be used to obtain a representative sample of the waste to be analyzed. A representative sample may be obtained using either:

(i) One of the sampling methods described in appendix I of part 261 of this chapter; or

(ii) An equivalent sampling method.

[*Comment:* See §260.21 of this chapter for related discussion.]

(4) The frequency with which the initial analysis of the waste will be reviewed or repeated to ensure that the analysis is accurate and up to date; and

(5) For off-site facilities, the waste analyses that hazardous waste generators have agreed to supply.

(6) Where applicable, the methods that will be used to meet the additional waste analysis requirements for specific waste management methods as specified in §§264.17, 264.314, 264.341, 264.1034(d), 264.1063(d), 264.1083, and 268.7 of this chapter.

(7) For surface impoundments exempted from land disposal restrictions under §268.4(a), the procedures and schedules for:

(i) The sampling of impoundment contents;

(ii) The analysis of test data; and,

(iii) The annual removal of residues which are not delisted under §268.22 of this chapter or which exhibit a characteristic of hazardous waste and either:

(A) Do not meet applicable treatment standards of part 268, subpart D; or

(B) Where no treatment standards have been established:

(*1*) Such residues are prohibited from land disposal under §268.32 or RCRA section 3004(d); or

(*2*) Such residues are prohibited from land disposal under §268.33(f).

(8) For owners and operators seeking an exemption to the air emission standards of subpart CC in accordance with §264.1082—

(i) If direct measurement is used for the waste determination, the procedures and schedules for waste sampling and analysis, and the results of the analysis of test data to verify the exemption.

(ii) If knowledge of the waste is used for the waste determination, any information prepared by the facility owner or operator or by the generator of the hazardous waste, if the waste is received from off-site, that is used as the basis for knowledge of the waste.

(c) For off-site facilities, the waste analysis plan required in paragraph (b) of this section must also specify the procedures which will be used to inspect and, if necessary, analyze each movement of hazardous waste received at the facility to ensure that it matches the identity of the waste designated on the accompanying manifest or shipping paper. At a minimum, the plan must describe:

(1) The procedures which will be used to determine the identity of each movement of waste managed at the facility; and

(2) The sampling method which will be used to obtain a representative sample of the waste to be identified, if the identification method includes sampling.

(3) The procedures that the owner or operator of an off-site landfill receiving containerized hazardous waste will use to determine whether a hazardous waste generator or treater has added a biodegradable sorbent to the waste in the container.

[*Comment:* Part 270 of this chapter requires that the waste analysis plan be submitted with part B of the permit application.]

[45 FR 33221, May 19, 1980, as amended at 46 FR 2848, Jan. 12, 1981; 50 FR 4514, Jan. 31, 1985; 51 FR 40637, Nov. 7, 1986; 53 FR 31211, Aug. 17, 1988; 54 FR 33394, Aug. 14, 1989; 55 FR 22685, June 1, 1990; 55 FR 25494, June 21, 1990; 57 FR 8088, Mar. 6, 1992; 57 FR 54460, Nov. 18, 1992; 59 FR 62926, Dec. 6, 1994; 61 FR 4911, Feb. 9, 1996; 71 FR 40272, July 14, 2006]

## § 264.14  Security.

(a) The owner or operator must prevent the unknowing entry, and minimize the possibility for the unauthorized entry, of persons or livestock onto the active portion of his facility, *unless* he can demonstrate to the Regional Administrator that:

(1) Physical contact with the waste, structures, or equipment within the active portion of the facility will not injure unknowing or unauthorized persons or livestock which may enter the active portion of a facility; and

(2) Disturbance of the waste or equipment, by the unknowing or unauthorized entry of persons or livestock onto the active portion of a facility, will not cause a violation of the requirements of this part.

[*Comment:* Part 270 of this chapter requires that an owner or operator who wishes to make the demonstration referred to above must do so with part B of the permit application.]

(b) Unless the owner or operator has made a successful demonstration under paragraphs (a) (1) and (2) of this section, a facility must have:

(1) A 24-hour surveillance system (e.g., television monitoring or surveil-

lance by guards or facility personnel) which continuously monitors and controls entry onto the active portion of the facility; or

(2)(i) An artificial or natural barrier (e.g., a fence in good repair or a fence combined with a cliff), which completely surrounds the active portion of the facility; and

(ii) A means to control entry, at all times, through the gates or other entrances to the active portion of the facility (e.g., an attendant, television monitors, locked entrance, or controlled roadway access to the facility).

[*Comment:* The requirements of paragraph (b) of this section are satisfied if the facility or plant within which the active portion is located itself has a surveillance system, or a barrier and a means to control entry, which complies with the requirements of paragraph (b) (1) or (2) of this section.]

(c) Unless the owner or operator has made a successful demonstration under paragraphs (a) (1) and (2) of this section, a sign with the legend, "Danger—Unauthorized Personnel Keep Out", must be posted at each entrance to the active portion of a facility, and at other locations, in sufficient numbers to be seen from any approach to this active portion. The legend must be written in English and in any other language predominant in the area surrounding the facility (e.g., facilities in counties bordering the Canadian province of Quebec must post signs in French; facilities in counties bordering Mexico must post signs in Spanish), and must be legible from a distance of at least 25 feet. Existing signs with a legend other than "Danger—Unauthorized Personnel Keep Out" may be used if the legend on the sign indicates that only authorized personnel are allowed to enter the active portion, and that entry onto the active portion can be dangerous.

[*Comment:* See § 264.117(b) for discussion of security requirements at disposal facilities during the post-closure care period.]

[45 FR 33221, May 19, 1980, as amended at 46 FR 2848, Jan. 12, 1981; 48 FR 14294, Apr. 1, 1983; 50 FR 4514, Jan. 31, 1985]

## § 264.15  General inspection requirements.

(a) The owner or operator must inspect his facility for malfunctions and deterioration, operator errors, and discharges which may be causing—or may lead to—(1) release of hazardous waste constituents to the environment or (2) a threat to human health. The owner or operator must conduct these inspections often enough to identify problems in time to correct them before they harm human health or the environment.

(b)(1) The owner or operator must develop and follow a written schedule for inspecting monitoring equipment, safety and emergency equipment, security devices, and operating and structural equipment (such as dikes and sump pumps) that are important to preventing, detecting, or responding to environmental or human health hazards.

(2) He must keep this schedule at the facility.

(3) The schedule must identify the types of problems (e.g., malfunctions or deterioration) which are to be looked for during the inspection (e.g., inoperative sump pump, leaking fitting, eroding dike, etc.).

(4) The frequency of inspection may vary for the items on the schedule. However, the frequency should be based on the rate of deterioration of the equipment and the probability of an environmental or human health incident if the deterioration, malfunction, or operator error goes undetected between inspections. Areas subject to spills, such as loading and unloading areas, must be inspected daily when in use, except for Performance Track member facilities, that must inspect at least once each month, upon approval by the Director, as described in paragraph (b)(5) of this section. At a minimum, the inspection schedule must include the items and frequencies called for in §§264.174, 264.193, 264.195, 264.226, 264.254, 264.278, 264.303, 264.347, 264.602, 264.1033, 264.1052, 264.1053, 264.1058, and 264.1083 through 264.1089 of this part, where applicable.

[*Comment:* Part 270 of this chapter requires the inspection schedule to be submitted with part B of the permit application. EPA will evaluate the schedule along with the rest of the application to ensure that it adequately protects human health and the environment. As part of this review, EPA may modify or amend the schedule as may be necessary.]

(5) Performance Track member facilities that choose to reduce their inspection frequency must:

(i) Submit a request for a Class I permit modification with prior approval to the Director. The modification request must identify the facility as a member of the National Environmental Performance Track Program and identify the management units for reduced inspections and the proposed frequency of inspections. The modification request must also specify, in writing, that the reduced inspection frequency will apply for as long as the facility is a Performance Track member facility, and that within seven calendar days of ceasing to be a Performance Track member, the facility will revert to the non-Performance Track inspection frequency. Inspections must be conducted at least once each month.

(ii) Within 60 days, the Director will notify the Performance Track member facility, in writing, if the request is approved, denied, or if an extension to the 60-day deadline is needed. This notice must be placed in the facility's operating record. The Performance Track member facility should consider the application approved if the Director does not: deny the application; or notify the Performance Track member facility of an extension to the 60-day deadline. In these situations, the Performance Track member facility must adhere to the revised inspection schedule outlined in its request for a Class 1 permit modification and keep a copy of the application in the facility's operating record.

(iii) Any Performance Track member facility that discontinues their membership or is terminated from the program must immediately notify the Director of their change in status. The facility must place in its operating record a dated copy of this notification and revert back to the non-Performance Track inspection frequencies within seven calendar days.

(c) The owner or operator must remedy any deterioration or malfunction of equipment or structures which the inspection reveals on a schedule which ensures that the problem does not lead to an environmental or human health hazard. Where a hazard is imminent or has already occurred, remedial action must be taken immediately.

(d) The owner or operator must record inspections in an inspection log or summary. He must keep these records for at least three years from the date of inspection. At a minimum, these records must include the date and time of the inspection, the name of the inspector, a notation of the observations made, and the date and nature of any repairs or other remedial actions.

[45 FR 33221, May 19, 1980, as amended at 48 FR 14294, Apr. 1, 1983; 50 FR 4514, Jan. 31, 1985; 57 FR 3486, Jan. 29, 1992; 59 FR 62926, Dec. 6, 1994; 62 FR 64656, Dec. 8, 1997; 71 FR 16903, Apr. 4, 2006]

## § 264.16  Personnel training.

(a)(1) Facility personnel must successfully complete a program of classroom instruction or on-the-job training that teaches them to perform their duties in a way that ensures the facility's compliance with the requirements of this part. The owner or operator must ensure that this program includes all the elements described in the document required under paragraph (d)(3) of this section.

[*Comment:* Part 270 of this chapter requires that owners and operators submit with part B of the RCRA permit application, an outline of the training program used (or to be used) at the facility and a brief description of how the training program is designed to meet actual job tasks.]

(2) This program must be directed by a person trained in hazardous waste management procedures, and must include instruction which teaches facility personnel hazardous waste management procedures (including contingency plan implementation) relevant to the positions in which they are employed.

(3) At a minimum, the training program must be designed to ensure that facility personnel are able to respond effectively to emergencies by familiarizing them with emergency procedures, emergency equipment, and emergency systems, including, where applicable:

(i) Procedures for using, inspecting, repairing, and replacing facility emergency and monitoring equipment;

(ii) Key parameters for automatic waste feed cut-off systems;

(iii) Communications or alarm systems;

(iv) Response to fires or explosions;

(v) Response to ground-water contamination incidents; and

(vi) Shutdown of operations.

(4) For facility employees that receive emergency response training pursuant to Occupational Safety and Health Administration (OSHA) regulations 29 CFR 1910.120(p)(8) and 1910.120(q), the facility is not required to provide separate emergency response training pursuant to this section, provided that the overall facility training meets all the requirements of this section.

(b) Facility personnel must successfully complete the program required in paragraph (a) of this section within six months after the effective date of these regulations or six months after the date of their employment or assignment to a facility, or to a new position at a facility, whichever is later. Employees hired after the effective date of these regulations must not work in unsupervised positions until they have completed the training requirements of paragraph (a) of this section.

(c) Facility personnel must take part in an annual review of the initial training required in paragraph (a) of this section.

(d) The owner or operator must maintain the following documents and records at the facility:

(1) The job title for each position at the facility related to hazardous waste management, and the name of the employee filling each job;

(2) A written job description for each position listed under paragraph (d)(1) of this section. This description may be consistent in its degree of specificity with descriptions for other similar positions in the same company location or bargaining unit, but must include the requisite skill, education, or other qualifications, and duties of employees assigned to each position;

(3) A written description of the type and amount of both introductory and continuing training that will be given to each person filling a position listed under paragraph (d)(1) of this section;

(4) Records that document that the training or job experience required under paragraphs (a), (b), and (c) of this

section has been given to, and completed by, facility personnel.

(e) Training records on current personnel must be kept until closure of the facility; training records on former employees must be kept for at least three years from the date the employee last worked at the facility. Personnel training records may accompany personnel transferred within the same company.

[45 FR 33221, May 19, 1980, as amended at 46 FR 2848, Jan. 12, 1981; 48 FR 14294, Apr. 1, 1983; 50 FR 4514, Jan. 31, 1985; 71 FR 16903, Apr. 4, 2006]

## §264.17 General requirements for ignitable, reactive, or incompatible wastes.

(a) The owner or operator must take precautions to prevent accidental ignition or reaction of ignitable or reactive waste. This waste must be separated and protected from sources of ignition or reaction including but not limited to: open flames, smoking, cutting and welding, hot surfaces, frictional heat, sparks (static, electrical, or mechanical), spontaneous ignition (e.g., from heat-producing chemical reactions), and radiant heat. While ignitable or reactive waste is being handled, the owner or operator must confine smoking and open flame to specially designated locations. "No Smoking" signs must be conspicuously placed wherever there is a hazard from ignitable or reactive waste.

(b) Where specifically required by other sections of this part, the owner or operator of a facility that treats, stores or disposes ignitable or reactive waste, or mixes incompatible waste or incompatible wastes and other materials, must take precautions to prevent reactions which:

(1) Generate extreme heat or pressure, fire or explosions, or violent reactions;

(2) Produce uncontrolled toxic mists, fumes, dusts, or gases in sufficient quantities to threaten human health or the environment;

(3) Produce uncontrolled flammable fumes or gases in sufficient quantities to pose a risk of fire or explosions;

(4) Damage the structural integrity of the device or facility;

(5) Through other like means threaten human health or the environment.

(c) When required to comply with paragraph (a) or (b) of this section, the owner or operator must document that compliance. This documentation may be based on references to published scientific or engineering literature, data from trial tests (e.g., bench scale or pilot scale tests), waste analyses (as specified in §264.13), or the results of the treatment of similar wastes by similar treatment processes and under similar operating conditions.

[46 FR 2848, Jan. 12, 1981, as amended at 50 FR 4514, Jan. 31, 1985; 71 FR 40272, July 14, 2006]

## §264.18 Location standards.

(a) *Seismic considerations.* (1) Portions of new facilities where treatment, storage, or disposal of hazardous waste will be conducted must not be located within 61 meters (200 feet) of a fault which has had displacement in Holocene time.

(2) As used in paragraph (a)(1) of this section:

(i) "Fault" means a fracture along which rocks on one side have been displaced with respect to those on the other side.

(ii) "Displacement" means the relative movement of any two sides of a fault measured in any direction.

(iii) "Holocene" means the most recent epoch of the Quaternary period, extending from the end of the Pleistocene to the present.

[*Comment:* Procedures for demonstrating compliance with this standard in part B of the permit application are specified in §270.14(b)(11). Facilities which are located in political jurisdictions other than those listed in appendix VI of this part, are assumed to be in compliance with this requirement.]

(b) *Floodplains.* (1) A facility located in a 100-year floodplain must be designed, constructed, operated, and maintained to prevent washout or any hazardous waste by a 100-year flood, *unless* the owner or operator can demonstrate to the Regional Administrator's satisfaction that:

(i) Procedures are in effect which will cause the waste to be removed safely, before flood waters can reach the facility, to a location where the wastes will not be vulnerable to flood waters; or

(ii) For existing surface impoundments, waste piles, land treatment units, landfills, and miscellaneous units, no adverse effects on human health or the environment will result if washout occurs, considering:

(A) The volume and physical and chemical characteristics of the waste in the facility;

(B) The concentration of hazardous constituents that would potentially affect surface waters as a result of washout;

(C) The impact of such concentrations on the current or potential uses of and water quality standards established for the affected surface waters; and

(D) The impact of hazardous constituents on the sediments of affected surface waters or the soils of the 100-year floodplain that could result from washout.

[*Comment:* The location where wastes are moved must be a facility which is either permitted by EPA under part 270 of this chapter, authorized to manage hazardous waste by a State with a hazardous waste management program authorized under part 271 of this chapter, or in interim status under parts 270 and 265 of this chapter.]

(2) As used in paragraph (b)(1) of this section:

(i) "100-year floodplain" means any land area which is subject to a one percent or greater chance of flooding in any given year from any source.

(ii) "Washout" means the movement of hazardous waste from the active portion of the facility as a result of flooding.

(iii) "100-year flood" means a flood that has a one percent chance of being equalled or exceeded in any given year.

[*Comment:* (1) Requirements pertaining to other Federal laws which affect the location and permitting of facilities are found in § 270.3 of this chapter. For details relative to these laws, see EPA's manual for SEA (special environmental area) requirements for hazardous waste facility permits. Though EPA is responsible for complying with these requirements, applicants are advised to consider them in planning the location of a facility to help prevent subsequent project delays.]

(c) *Salt dome formations, salt bed formations, underground mines and caves.* The placement of any noncontainerized or bulk liquid hazardous waste in any salt dome formation, salt bed formation, underground mine or cave is prohibited, except for the Department of Energy Waste Isolation Pilot Project in New Mexico.

[46 FR 2848, Jan. 12, 1981, as amended at 47 FR 32350, July 26, 1982; 48 FR 14294, Apr. 1, 1983; 48 FR 30115, June 30, 1983; 50 FR 4514, Jan. 31, 1985; 50 FR 28746, July 15, 1985; 52 FR 46963, Dec. 10, 1987; 71 FR 40272, July 14, 2006]

## § 264.19  Construction quality assurance program.

(a) *CQA program.* (1) A construction quality assurance (CQA) program is required for all surface impoundment, waste pile, and landfill units that are required to comply with §§ 264.221 (c) and (d), 264.251 (c) and (d), and 264.301 (c) and (d). The program must ensure that the constructed unit meets or exceeds all design criteria and specifications in the permit. The program must be developed and implemented under the direction of a CQA officer who is a registered professional engineer.

(2) The CQA program must address the following physical components, where applicable:

(i) Foundations;

(ii) Dikes;

(iii) Low-permeability soil liners;

(iv) Geomembranes (flexible membrane liners);

(v) Leachate collection and removal systems and leak detection systems; and

(vi) Final cover systems.

(b) *Written CQA plan.* The owner or operator of units subject to the CQA program under paragraph (a) of this section must develop and implement a written CQA plan. The plan must identify steps that will be used to monitor and document the quality of materials and the condition and manner of their installation. The CQA plan must include:

(1) Identification of applicable units, and a description of how they will be constructed.

(2) Identification of key personnel in the development and implementation of the CQA plan, and CQA officer qualifications.

(3) A description of inspection and sampling activities for all unit components identified in paragraph (a)(2) of this section, including observations

and tests that will be used before, during, and after construction to ensure that the construction materials and the installed unit components meet the design specifications. The description must cover: Sampling size and locations; frequency of testing; data evaluation procedures; acceptance and rejection criteria for construction materials; plans for implementing corrective measures; and data or other information to be recorded and retained in the operating record under § 264.73.

(c) *Contents of program.* (1) The CQA program must include observations, inspections, tests, and measurements sufficient to ensure:

(i) Structural stability and integrity of all components of the unit identified in paragraph (a)(2) of this section;

(ii) Proper construction of all components of the liners, leachate collection and removal system, leak detection system, and final cover system, according to permit specifications and good engineering practices, and proper installation of all components (e.g., pipes) according to design specifications;

(iii) Conformity of all materials used with design and other material specifications under §§ 264.221, 264.251, and 264.301.

(2) The CQA program shall include test fills for compacted soil liners, using the same compaction methods as in the full scale unit, to ensure that the liners are constructed to meet the hydraulic conductivity requirements of §§ 264.221(c)(1)(i)(B), 264.251(c)(1)(i)(B), and 264.301(c)(1)(i)(B) in the field. Compliance with the hydraulic conductivity requirements must be verified by using in-situ testing on the constructed test fill. The Regional Administrator may accept an alternative demonstration, in lieu of a test fill, where data are sufficient to show that a constructed soil liner will meet the hydraulic conductivity requirements of §§ 264.221(c)(1)(i)(B), 264.251(c)(1)(i)(B), and 264.301(c)(1)(i)(B) in the field.

(d) *Certification.* Waste shall not be received in a unit subject to § 264.19 until the owner or operator has submitted to the Regional Administrator by certified mail or hand delivery a certification signed by the CQA officer that the approved CQA plan has been successfully carried out and that the unit meets the requirements of §§ 264.221 (c) or (d), 264.251 (c) or (d), or 264.301 (c) or (d); and the procedure in § 270.30(1)(2)(ii) of this chapter has been completed. Documentation supporting the CQA officer's certification must be furnished to the Regional Administrator upon request.

[57 FR 3486, Jan. 29, 1992]

## Subpart C—Preparedness and Prevention

### § 264.30  Applicability.

The regulations in this subpart apply to owners and operators of all hazardous waste facilities, except as § 264.1 provides otherwise.

### § 264.31  Design and operation of facility.

Facilities must be designed, constructed, maintained, and operated to minimize the possibility of a fire, explosion, or any unplanned sudden or non-sudden release of hazardous waste or hazardous waste constituents to air, soil, or surface water which could threaten human health or the environment.

### § 264.32  Required equipment.

All facilities must be equipped with the following, *unless* it can be demonstrated to the Regional Administrator that none of the hazards posed by waste handled at the facility could require a particular kind of equipment specified below:

(a) An internal communications or alarm system capable of providing immediate emergency instruction (voice or signal) to facility personnel;

(b) A device, such as a telephone (immediately available at the scene of operations) or a hand-held two-way radio, capable of summoning emergency assistance from local police departments, fire departments, or State or local emergency response teams;

(c) Portable fire extinguishers, fire control equipment (including special extinguishing equipment, such as that using foam, inert gas, or dry chemicals), spill control equipment, and decontamination equipment; and

(d) Water at adequate volume and pressure to supply water hose streams, or foam producing equipment, or automatic sprinklers, or water spray systems.

[*Comment:* Part 270 of this chapter requires that an owner or operator who wishes to make the demonstration referred to above must do so with part B of the permit application.]

[45 FR 33221, May 19, 1980, as amended at 48 FR 14294, Apr. 1, 1983]

### § 264.33 Testing and maintenance of equipment.

All facility communications or alarm systems, fire protection equipment, spill control equipment, and decontamination equipment, where required, must be tested and maintained as necessary to assure its proper operation in time of emergency.

### § 264.34 Access to communications or alarm system.

(a) Whenever hazardous waste is being poured, mixed, spread, or otherwise handled, all personnel involved in the operation must have immediate access to an internal alarm or emergency communication device, either directly or through visual or voice contact with another employee, *unless* the Regional Administrator has ruled that such a device is not required under § 264.32.

(b) If there is ever just one employee on the premises while the facility is operating, he must have immediate access to a device, such as a telephone (immediately available at the scene of operation) or a hand-held two-way radio, capable of summoning external emergency assistance, *unless* the Regional Administrator has ruled that such a device is not required under § 264.32.

### § 264.35 Required aisle space.

The owner or operator must maintain aisle space to allow the unobstructed movement of personnel, fire protection equipment, spill control equipment, and decontamination equipment to any area of facility operation in an emergency, *unless* it can be demonstrated to the Regional Administrator that aisle space is not needed for any of these purposes.

[*Comment:* Part 270 of this chapter requires that an owner or operator who wishes to make the demonstration referred to above must do so with part B of the permit application.]

[45 FR 33221, May 19, 1980, as amended at 48 FR 14294, Apr. 1, 1983]

### § 264.36 [Reserved]

### § 264.37 Arrangements with local authorities.

(a) The owner or operator must attempt to make the following arrangements, as appropriate for the type of waste handled at his facility and the potential need for the services of these organizations:

(1) Arrangements to familiarize police, fire departments, and emergency response teams with the layout of the facility, properties of hazardous waste handled at the facility and associated hazards, places where facility personnel would normally be working, entrances to and roads inside the facility, and possible evacuation routes;

(2) Where more than one police and fire department might respond to an emergency, agreements designating primary emergency authority to a specific police and a specific fire department, and agreements with any others to provide support to the primary emergency authority;

(3) Agreements with State emergency response teams, emergency response contractors, and equipment suppliers; and

(4) Arrangements to familiarize local hospitals with the properties of hazardous waste handled at the facility and the types of injuries or illnesses which could result from fires, explosions, or releases at the facility.

(b) Where State or local authorities decline to enter into such arrangements, the owner or operator must document the refusal in the operating record.

## Subpart D—Contingency Plan and Emergency Procedures

### § 264.50 Applicability.

The regulations in this subpart apply to owners and operators of all hazardous waste facilities, except as § 264.1 provides otherwise.

## §264.51 Purpose and implementation of contingency plan.

(a) Each owner or operator must have a contingency plan for his facility. The contingency plan must be designed to minimize hazards to human health or the environment from fires, explosions, or any unplanned sudden or non-sudden release of hazardous waste or hazardous waste constituents to air, soil, or surface water.

(b) The provisions of the plan must be carried out immediately whenever there is a fire, explosion, or release of hazardous waste or hazardous waste constituents which could threaten human health or the environment.

[45 FR 33221, May 19, 1980, as amended at 50 FR 4514, Jan. 31, 1985]

## §264.52 Content of contingency plan.

(a) The contingency plan must describe the actions facility personnel must take to comply with §§264.51 and 264.56 in response to fires, explosions, or any unplanned sudden or non-sudden release of hazardous waste or hazardous waste constituents to air, soil, or surface water at the facility.

(b) If the owner or operator has already prepared a Spill Prevention, Control, and Countermeasures (SPCC) Plan in accordance with part 112 of this chapter, or some other emergency or contingency plan, he need only amend that plan to incorporate hazardous waste management provisions that are sufficient to comply with the requirements of this part. The owner or operator may develop one contingency plan which meets all regulatory requirements. EPA recommends that the plan be based on the National Response Team's Integrated Contingency Plan Guidance ("One Plan"). When modifications are made to non-RCRA provisions in an integrated contingency plan, the changes do not trigger the need for a RCRA permit modification.

(c) The plan must describe arrangements agreed to by local police departments, fire departments, hospitals, contractors, and State and local emergency response teams to coordinate emergency services, pursuant to §264.37.

(d) The plan must list names, addresses, and phone numbers (office and home) of all persons qualified to act as emergency coordinator (see §264.55), and this list must be kept up to date. Where more than one person is listed, one must be named as primary emergency coordinator and others must be listed in the order in which they will assume responsibility as alternates. *For new facilities*, this information must be supplied to the Regional Administrator at the time of certification, rather than at the time of permit application.

(e) The plan must include a list of all emergency equipment at the facility (such as fire extinguishing systems, spill control equipment, communications and alarm systems (internal and external), and decontamination equipment), where this equipment is required. This list must be kept up to date. In addition, the plan must include the location and a physical description of each item on the list, and a brief outline of its capabilities.

(f) The plan must include an evacuation plan for facility personnel where there is a possibility that evacuation could be necessary. This plan must describe signal(s) to be used to begin evacuation, evacuation routes, and alternate evacuation routes (in cases where the primary routes could be blocked by releases of hazardous waste or fires).

[45 FR 33221, May 19, 1980, as amended at 46 FR 27480, May 20, 1981; 50 FR 4514, Jan. 31, 1985; 71 FR 16903, Apr. 4, 2006; 75 FR 13005, Mar. 18, 2010]

## §264.53 Copies of contingency plan.

A copy of the contingency plan and all revisions to the plan must be:

(a) Maintained at the facility; and

(b) Submitted to all local police departments, fire departments, hospitals, and State and local emergency response teams that may be called upon to provide emergency services.

[*Comment:* The contingency plan must be submitted to the Regional Administrator with Part B of the permit application under part 270, of this chapter and, after modification or approval, will become a condition of any permit issued.]

[45 FR 33221, May 19, 1980, as amended at 48 FR 30115, June 30, 1983; 50 FR 4514, Jan. 31, 1985]

## § 264.54 Amendment of contingency plan.

The contingency plan must be reviewed, and immediately amended, if necessary, whenever:

(a) The facility permit is revised;

(b) The plan fails in an emergency;

(c) The facility changes—in its design, construction, operation, maintenance, or other circumstances—in a way that materially increases the potential for fires, explosions, or releases of hazardous waste or hazardous waste constituents, or changes the response necessary in an emergency;

(d) The list of emergency coordinators changes; or

(e) The list of emergency equipment changes.

[45 FR 33221, May 19, 1980, as amended at 50 FR 4514, Jan. 31, 1985; 53 FR 37935, Sept. 28, 1988]

## § 264.55 Emergency coordinator.

At all times, there must be at least one employee either on the facility premises or on call (i.e., available to respond to an emergency by reaching the facility within a short period of time) with the responsibility for coordinating all emergency response measures. This emergency coordinator must be thoroughly familiar with all aspects of the facility's contingency plan, all operations and activities at the facility, the location and characteristics of waste handled, the location of all records within the facility, and the facility layout. In addition, this person must have the authority to commit the resources needed to carry out the contingency plan.

[*Comment:* The emergency coordinator's responsibilities are more fully spelled out in § 264.56. Applicable responsibilities for the emergency coordinator vary, depending on factors such as type and variety of waste(s) handled by the facility, and type and complexity of the facility.]

## § 264.56 Emergency procedures.

(a) Whenever there is an imminent or actual emergency situation, the emergency coordinator (or his designee when the emergency coordinator is on call) must immediately:

(1) Activate internal facility alarms or communication systems, where applicable, to notify all facility personnel; and

(2) Notify appropriate State or local agencies with designated response roles if their help is needed.

(b) Whenever there is a release, fire, or explosion, the emergency coordinator must immediately identify the character, exact source, amount, and areal extent of any released materials. He may do this by observation or review of facility records or manifests, and, if necessary, by chemical analysis.

(c) Concurrently, the emergency coordinator must assess possible hazards to human health or the environment that may result from the release, fire, or explosion. This assessment must consider both direct and indirect effects of the release, fire, or explosion (e.g., the effects of any toxic, irritating, or asphyxiating gases that are generated, or the effects of any hazardous surface water run-off from water or chemical agents used to control fire and heat-induced explosions).

(d) If the emergency coordinator determines that the facility has had a release, fire, or explosion which could threaten human health, or the environment, outside the facility, he must report his findings as follows:

(1) If his assessment indicates that evacuation of local areas may be advisable, he must immediately notify appropriate local authorities. He must be available to help appropriate officials decide whether local areas should be evacuated; and

(2) He must immediately notify either the government official designated as the on-scene coordinator for that geographical area, or the National Response Center (using their 24-hour toll free number 800/424–8802). The report must include:

(i) Name and telephone number of reporter;

(ii) Name and address of facility;

(iii) Time and type of incident (e.g., release, fire);

(iv) Name and quantity of material(s) involved, to the extent known;

(v) The extent of injuries, if any; and

(vi) The possible hazards to human health, or the environment, outside the facility.

(e) During an emergency, the emergency coordinator must take all reasonable measures necessary to ensure that fires, explosions, and releases do not occur, recur, or spread to other hazardous waste at the facility. These measures must include, where applicable, stopping processes and operations, collecting and containing release waste, and removing or isolating containers.

(f) If the facility stops operations in response to a fire, explosion, or release, the emergency coordinator must monitor for leaks, pressure buildup, gas generation, or ruptures in valves, pipes, or other equipment, wherever this is appropriate.

(g) Immediately after an emergency, the emergency coordinator must provide for treating, storing, or disposing of recovered waste, contaminated soil or surface water, or any other material that results from a release, fire, or explosion at the facility.

[*Comment:* Unless the owner or operator can demonstrate, in accordance with §261.3(c) or (d) of this chapter, that the recovered material is not a hazardous waste, the owner or operator becomes a generator of hazardous waste and must manage it in accordance with all applicable requirements of parts 262, 263, and 264 of this chapter.]

(h) The emergency coordinator must ensure that, in the affected area(s) of the facility:

(1) No waste that may be incompatible with the released material is treated, stored, or disposed of until cleanup procedures are completed; and

(2) All emergency equipment listed in the contingency plan is cleaned and fit for its intended use before operations are resumed.

(i) The owner or operator must note in the operating record the time, date, and details of any incident that requires implementing the contingency plan. Within 15 days after the incident, he must submit a written report on the incident to the Regional Administrator. The report must include:

(1) Name, address, and telephone number of the owner or operator;

(2) Name, address, and telephone number of the facility;

(3) Date, time, and type of incident (e.g., fire, explosion);

(4) Name and quantity of material(s) involved;

(5) The extent of injuries, if any;

(6) An assessment of actual or potential hazards to human health or the environment, where this is applicable; and

(7) Estimated quantity and disposition of recovered material that resulted from the incident.

[45 FR 33221, May 19, 1980, as amended at 50 FR 4514, Jan. 31, 1985; 71 FR 16903, Apr. 4, 2006; 75 FR 13005, Mar. 18, 2010]

## Subpart E—Manifest System, Recordkeeping, and Reporting

### §264.70  Applicability.

(a) The regulations in this subpart apply to owners and operators of both on-site and off-site facilities, except as §264.1 provides otherwise. Sections 264.71, 264.72, and 264.76 do not apply to owners and operators of on-site facilities that do not receive any hazardous waste from off-site sources, nor to owners and operators of off-site facilities with respect to waste military munitions exempted from manifest requirements under 40 CFR 266.203(a). Section 264.73(b) only applies to permittees who treat, store, or dispose of hazardous wastes on-site where such wastes were generated.

(b) The revised Manifest form and procedures in 40 CFR 260.10, 261.7, 264.70, 264.71. 264.72, and 264.76, shall not apply until September 5, 2006. The Manifest form and procedures in 40 CFR 260.10, 261.7, 264.70, 264.71. 264.72, and 264.76, contained in the 40 CFR, parts 260 to 265, edition revised as of July 1, 2004, shall be applicable until September 5, 2006.

[70 FR 10821, Mar. 4, 2005]

### §264.71  Use of manifest system.

(a)(1) If a facility receives hazardous waste accompanied by a manifest, the owner, operator or his/her agent must sign and date the manifest as indicated in paragraph (a)(2) of this section to certify that the hazardous waste covered by the manifest was received, that the hazardous waste was received except as noted in the discrepancy space of the manifest, or that the hazardous waste was rejected as noted in the manifest discrepancy space.

(2) If a facility receives a hazardous waste shipment accompanied by a manifest, the owner, operator or his agent must:

(i) Sign and date, by hand, each copy of the manifest;

(ii) Note any discrepancies (as defined in § 264.72(a)) on each copy of the manifest;

(iii) Immediately give the transporter at least one copy of the manifest;

(iv) Within 30 days of delivery, send a copy of the manifest to the generator; and

(v) Retain at the facility a copy of each manifest for at least three years from the date of delivery.

(3) If a facility receives hazardous waste imported from a foreign source, the receiving facility must mail a copy of the manifest to the following address within 30 days of delivery: International Compliance Assurance Division, OFA/OECA (2254A), U.S. Environmental Protection Agency, Ariel Rios Building, 1200 Pennsylvania Avenue, NW., Washington, DC 20460.

(b) If a facility receives, from a rail or water (bulk shipment) transporter, hazardous waste which is accompanied by a shipping paper containing all the information required on the manifest (excluding the EPA identification numbers, generator's certification, and signatures), the owner or operator, or his agent, must:

(1) Sign and date each copy of the manifest or shipping paper (if the manifest has not been received) to certify that the hazardous waste covered by the manifest or shipping paper was received;

(2) Note any significant discrepancies (as defined in § 264.72(a)) in the manifest or shipping paper (if the manifest has not been received) on each copy of the manifest or shipping paper.

[Comment: The Agency does not intend that the owner or operator of a facility whose procedures under § 264.13(c) include waste analysis must perform that analysis before signing the shipping paper and giving it to the transporter. Section 264.72(b), however, requires reporting an unreconciled discrepancy discovered during later analysis.]

(3) Immediately give the rail or water (bulk shipment) transporter at least one copy of the manifest or shipping paper (if the manifest has not been received);

(4) Within 30 days after the delivery, send a copy of the signed and dated manifest or a signed and dated copy of the shipping paper (if the manifest has not been received within 30 days after delivery) to the generator; and

[Comment: Section 262.23(c) of this chapter requires the generator to send three copies of the manifest to the facility when hazardous waste is sent by rail or water (bulk shipment).]

(5) Retain at the facility a copy of the manifest and shipping paper (if signed in lieu of the manifest at the time of delivery) for at least three years from the date of delivery.

(c) Whenever a shipment of hazardous waste is initiated from a facility, the owner or operator of that facility must comply with the requirements of part 262 of this chapter.

[Comment: The provisions of § 262.34 are applicable to the on-site accumulation of hazardous wastes by generators. Therefore, the provisions of § 262.34 only apply to owners or operators who are shipping hazardous waste which they generated at that facility.]

(d) Within three working days of the receipt of a shipment subject to 40 CFR part 262, subpart H, the owner or operator of the facility must provide a copy of the tracking document bearing all required signatures to the notifier, to the Office of Enforcement and Compliance Assurance, Office of Compliance, Enforcement Planning, Targeting and Data Division (2222A), Environmental Protection Agency, 1200 Pennsylvania Ave., NW., Washington, DC 20460, and to competent authorities of all other concerned countries. The original copy of the tracking document must be maintained at the facility for at least three years from the date of signature.

(e) A facility must determine whether the consignment state for a shipment regulates any additional wastes (beyond those regulated Federally) as hazardous wastes under its state hazardous waste program. Facilities must also determine whether the consignment state or generator state requires the facility to submit any copies of the manifest to these states.

[45 FR 33221, May 19, 1980, as amended at 45 FR 86970, 86974, Dec. 31, 1980; 61 FR 16315, Apr. 12, 1996; 70 FR 10821, Mar. 4, 2005]

EFFECTIVE DATE NOTE: At 75 FR 1260, Jan. 8, 2010, §264.71 was amended by revising paragraphs (a)(3) and (d), effective July 7, 2010. For the convenience of the user, the revised text is set forth as follows:

**§264.71   Use of manifest system.**

(a) * * *

(3) If a facility receives hazardous waste imported from a foreign source, the receiving facility must mail a copy of the manifest and documentation confirming EPA's consent to the import of hazardous waste to the following address within thirty (30) days of delivery: Office of Enforcement and Compliance Assurance, Office of Federal Activities, International Compliance Assurance Division (2254A), Environmental Protection Agency, 1200 Pennsylvania Avenue, NW., Washington, DC 20460.

\*         \*         \*         \*         \*

(d) Within three (3) working days of the receipt of a shipment subject to 40 CFR part 262, subpart H, the owner or operator of a facility must provide a copy of the movement document bearing all required signatures to the exporter, to the Office of Enforcement and Compliance Assurance, Office of Federal Activities, International Compliance Assurance Division (2254A), Environmental Protection Agency, 1200 Pennsylvania Avenue, NW., Washington, DC 20460, and to competent authorities of all other concerned countries. The original copy of the movement document must be maintained at the facility for at least three (3) years from the date of signature.

\*         \*         \*         \*         \*

**§264.72   Manifest discrepancies.**

(a) Manifest discrepancies are:

(1) Significant differences (as defined by paragraph (b) of this section) between the quantity or type of hazardous waste designated on the manifest or shipping paper, and the quantity and type of hazardous waste a facility actually receives;

(2) Rejected wastes, which may be a full or partial shipment of hazardous waste that the TSDF cannot accept; or

(3) Container residues, which are residues that exceed the quantity limits for "empty" containers set forth in 40 CFR 261.7(b).

(b) Significant differences in quantity are: For bulk waste, variations greater than 10 percent in weight; for batch waste, any variation in piece count, such as a discrepancy of one drum in a truckload. Significant differences in type are obvious differences which can be discovered by inspection or waste analysis, such as waste solvent substituted for waste acid, or toxic constituents not reported on the manifest or shipping paper.

(c) Upon discovering a significant difference in quantity or type, the owner or operator must attempt to reconcile the discrepancy with the waste generator or transporter (*e.g.*, with telephone conversations). If the discrepancy is not resolved within 15 days after receiving the waste, the owner or operator must immediately submit to the Regional Administrator a letter describing the discrepancy and attempts to reconcile it, and a copy of the manifest or shipping paper at issue.

(d)(1) Upon rejecting waste or identifying a container residue that exceeds the quantity limits for "empty" containers set forth in 40 CFR 261.7(b), the facility must consult with the generator prior to forwarding the waste to another facility that can manage the waste. If it is impossible to locate an alternative facility that can receive the waste, the facility may return the rejected waste or residue to the generator. The facility must send the waste to the alternative facility or to the generator within 60 days of the rejection or the container residue identification.

(2) While the facility is making arrangements for forwarding rejected wastes or residues to another facility under this section, it must ensure that either the delivering transporter retains custody of the waste, or, the facility must provide for secure, temporary custody of the waste, pending delivery of the waste to the first transporter designated on the manifest prepared under paragraph (e) or (f) of this section.

(e) Except as provided in paragraph (e)(7) of this section, for full or partial load rejections and residues that are to be sent off-site to an alternate facility, the facility is required to prepare a new manifest in accordance with

§ 262.20(a) of this chapter and the following instructions:

(1) Write the generator's U.S. EPA ID number in Item 1 of the new manifest. Write the generator's name and mailing address in Item 5 of the new manifest. If the mailing address is different from the generator's site address, then write the generator's site address in the designated space for Item 5.

(2) Write the name of the alternate designated facility and the facility's U.S. EPA ID number in the designated facility block (Item 8) of the new manifest.

(3) Copy the manifest tracking number found in Item 4 of the old manifest to the Special Handling and Additional Information Block of the new manifest, and indicate that the shipment is a residue or rejected waste from the previous shipment.

(4) Copy the manifest tracking number found in Item 4 of the new manifest to the manifest reference number line in the Discrepancy Block of the old manifest (Item 18a).

(5) Write the DOT description for the rejected load or the residue in Item 9 (U.S. DOT Description) of the new manifest and write the container types, quantity, and volume(s) of waste.

(6) Sign the Generator's/Offeror's Certification to certify, as the offeror of the shipment, that the waste has been properly packaged, marked and labeled and is in proper condition for transportation, and mail a signed copy of the manifest to the generator identified in Item 5 of the new manifest.

(7) For full load rejections that are made while the transporter remains present at the facility, the facility may forward the rejected shipment to the alternate facility by completing Item 18b of the original manifest and supplying the information on the next destination facility in the Alternate Facility space. The facility must retain a copy of this manifest for its records, and then give the remaining copies of the manifest to the transporter to accompany the shipment. If the original manifest is not used, then the facility must use a new manifest and comply with paragraphs (e)(1), (2), (3), (4), (5), and (6) of this section.

(f) Except as provided in paragraph (f)(7) of this section, for rejected wastes and residues that must be sent back to the generator, the facility is required to prepare a new manifest in accordance with § 262.20(a) of this chapter and the following instructions:

(1) Write the facility's U.S. EPA ID number in Item 1 of the new manifest. Write the facility's name and mailing address in Item 5 of the new manifest. If the mailing address is different from the facility's site address, then write the facility's site address in the designated space for Item 5 of the new manifest.

(2) Write the name of the initial generator and the generator's U.S. EPA ID number in the designated facility block (Item 8) of the new manifest.

(3) Copy the manifest tracking number found in Item 4 of the old manifest to the Special Handling and Additional Information Block of the new manifest, and indicate that the shipment is a residue or rejected waste from the previous shipment.

(4) Copy the manifest tracking number found in Item 4 of the new manifest to the manifest reference number line in the Discrepancy Block of the old manifest (Item 18a).

(5) Write the DOT description for the rejected load or the residue in Item 9 (U.S. DOT Description) of the new manifest and write the container types, quantity, and volume(s) of waste.

(6) Sign the Generator's/Offeror's Certification to certify, as offeror of the shipment, that the waste has been properly packaged, marked and labeled and is in proper condition for transportation.

(7) For full load rejections that are made while the transporter remains at the facility, the facility may return the shipment to the generator with the original manifest by completing Item 18a and 18b of the manifest and supplying the generator's information in the Alternate Facility space. The facility must retain a copy for its records and then give the remaining copies of the manifest to the transporter to accompany the shipment. If the original manifest is not used, then the facility must use a new manifest and comply with paragraphs (f)(1), (2), (3), (4), (5), (6), and (8) of this section.

(8) For full or partial load rejections and container residues contained in

non-empty containers that are returned to the generator, the facility must also comply with the exception reporting requirements in §262.42(a).

(g) If a facility rejects a waste or identifies a container residue that exceeds the quantity limits for "empty" containers set forth in 40 CFR 261.7(b) after it has signed, dated, and returned a copy of the manifest to the delivering transporter or to the generator, the facility must amend its copy of the manifest to indicate the rejected wastes or residues in the discrepancy space of the amended manifest. The facility must also copy the manifest tracking number from Item 4 of the new manifest to the Discrepancy space of the amended manifest, and must re-sign and date the manifest to certify to the information as amended. The facility must retain the amended manifest for at least three years from the date of amendment, and must within 30 days, send a copy of the amended manifest to the transporter and generator that received copies prior to their being amended.

[70 FR 10822, Mar. 4, 2005, as amended at 70 FR 35041, June 16, 2005; 75 FR 13005, Mar. 18, 2010]

## §264.73 Operating record.

(a) The owner or operator must keep a written operating record at his facility.

(b) The following information must be recorded, as it becomes available, and maintained in the operating record for three years unless noted as follows:

(1) A description and the quantity of each hazardous waste received, and the method(s) and date(s) of its treatment, storage, or disposal at the facility as required by appendix I of this part. This information must be maintained in the operating record until closure of the facility;

(2) The location of each hazardous waste within the facility and the quantity at each location. For disposal facilities, the location and quantity of each hazardous waste must be recorded on a map or diagram that shows each cell or disposal area. For all facilities, this information must include cross-references to manifest document numbers if the waste was accompanied by a manifest. This information must be

maintained in the operating record until closure of the facility.

[*Comment:* See §264.119 for related requirements.]

(3) Records and results of waste analyses and waste determinations performed as specified in §§264.13, 264.17, 264.314, 264.341, 264.1034, 264.1063, 264.1083, 268.4(a), and 268.7 of this chapter.

(4) Summary reports and details of all incidents that require implementing the contingency plan as specified in §264.56(j);

(5) Records and results of inspections as required by §264.15(d) (except these data need be kept only three years);

(6) Monitoring, testing or analytical data, and corrective action where required by subpart F of this part and §§264.19, 264.191, 264.193, 264.195, 264.222, 264.223, 264.226, 264.252–264.254, 264.276, 264.278, 264.280, 264.302–264.304, 264.309, 264.602, 264.1034(c)–264.1034(f), 264.1035, 264.1063(d)–264.1063(i), 264.1064, and 264.1082 through 264.1090 of this part. Maintain in the operating record for three years, except for records and results pertaining to ground-water monitoring and cleanup which must be maintained in the operating record until closure of the facility.

(7) For off-site facilities, notices to generators as specified in §264.12(b); and

(8) All closure cost estimates under §264.142, and for disposal facilities, all post-closure cost estimates under §264.144 of this part. This information must be maintained in the operating record until closure of the facility.

(9) A certification by the permittee no less often than annually, that the permittee has a program in place to reduce the volume and toxicity of hazardous waste that he generates to the degree determined by the permittee to be economically practicable; and the proposed method of treatment, storage or disposal is that practicable method currently available to the permittee which minimizes the present and future threat to human health and the environment.

(10) Records of the quantities and date of placement for each shipment of hazardous waste placed in land disposal

units under an extension to the effective date of any land disposal restriction granted pursuant to § 268.5 of this chapter, a petition pursuant to § 268.6 of this chapter, or a certification under § 268.8 of this chapter, and the applicable notice required by a generator under § 268.7(a) of this chapter. This information must be maintained in the operating record until closure of the facility.

(11) For an off-site treatment facility, a copy of the notice, and the certification and demonstration, if applicable, required by the generator or the owner or operator under § 268.7 or § 268.8;

(12) For an on-site treatment facility, the information contained in the notice (except the manifest number), and the certification and demonstration if applicable, required by the generator or the owner or operator under § 268.7 or § 268.8;

(13) For an off-site land disposal facility, a copy of the notice, and the certification and demonstration if applicable, required by the generator or the owner or operator of a treatment facility under §§ 268.7 and 268.8, whichever is applicable; and

(14) For an on-site land disposal facility, the information contained in the notice required by the generator or owner or operator of a treatment facility under § 268.7, except for the manifest number, and the certification and demonstration if applicable, required under § 268.8, whichever is applicable.

(15) For an off-site storage facility, a copy of the notice, and the certification and demonstration if applicable, required by the generator or the owner or operator under § 268.7 or § 268.8; and

(16) For an on-site storage facility, the information contained in the notice (except the manifest number), and the certification and demonstration if applicable, required by the generator or the owner or operator under § 268.7 or § 268.8.

(17) Any records required under § 264.1(j)(13).

(18) Monitoring, testing or analytical data where required by § 264.347 must be maintained in the operating record for five years.

(19) Certifications as required by § 264.196(f) must be maintained in the operating record until closure of the facility.

[45 FR 33221, May 19, 1980]

EDITORIAL NOTE: For FEDERAL REGISTER citations affecting § 264.73, see the List of CFR Sections Affected, which appears in the Finding Aids section of the printed volume and on GPO Access.

## § 264.74 Availability, retention, and disposition of records.

(a) All records, including plans, required under this part must be furnished upon request, and made available at all reasonable times for inspection, by any officer, employee, or representative of EPA who is duly designated by the Administrator.

(b) The retention period for all records required under this part is extended automatically during the course of any unresolved enforcement action regarding the facility or as requested by the Administrator.

(c) A copy of records of waste disposal locations and quantities under § 264.73(b)(2) must be submitted to the Regional Administrator and local land authority upon closure of the facility.

## § 264.75 Biennial report.

The owner or operator must prepare and submit a single copy of a biennial report to the Regional Administrator by March 1 of each even numbered year. The biennial report must be submitted on EPA form 8700-13B. The report must cover facility activities during the previous calendar year and must include:

(a) The EPA identification number, name, and address of the facility;

(b) The calendar year covered by the report;

(c) For off-site facilities, the EPA identification number of each hazardous waste generator from which the facility received a hazardous waste during the year; for imported shipments, the report must give the name and address of the foreign generator;

(d) A description and the quantity of each hazardous waste the facility received during the year. For off-site facilities, this information must be listed by EPA identification number of each generator;

(e) The method of treatment, storage, or disposal for each hazardous waste;

(f) [Reserved]

(g) The most recent closure cost estimate under § 264.142, and, for disposal facilities, the most recent post-closure cost estimate under § 264.144; and

(h) For generators who treat, store, or dispose of hazardous waste on-site, a description of the efforts undertaken during the year to reduce the volume and toxicity of waste generated.

(i) For generators who treat, store, or dispose of hazardous waste on-site, a description of the changes in volume and toxicity of waste actually achieved during the year in comparison to previous years to the extent such information is available for the years prior to 1984.

(j) The certification signed by the owner or operator of the facility or his authorized representative.

[45 FR 33221, May 19, 1980, as amended at 46 FR 2849, Jan. 12, 1981; 48 FR 3982, Jan. 28, 1983; 50 FR 4514, Jan. 31, 1985; 51 FR 28556, Aug. 8, 1986]

§ 264.76 **Unmanifested waste report.**

(a) If a facility accepts for treatment, storage, or disposal any hazardous waste from an off-site source without an accompanying manifest, or without an accompanying shipping paper as described by § 263.20(e) of this chapter, and if the waste is not excluded from the manifest requirement by this chapter, then the owner or operator must prepare and submit a letter to the Regional Administrator within 15 days after receiving the waste. The unmanifested waste report must contain the following information:

(1) The EPA identification number, name and address of the facility;

(2) The date the facility received the waste;

(3) The EPA identification number, name and address of the generator and the transporter, if available;

(4) A description and the quantity of each unmanifested hazardous waste the facility received;

(5) The method of treatment, storage, or disposal for each hazardous waste;

(6) The certification signed by the owner or operator of the facility or his authorized representative; and,

(7) A brief explanation of why the waste was unmanifested, if known.

(b) [Reserved]

[70 FR 10823, Mar. 4, 2005]

§ 264.77 **Additional reports.**

In addition to submitting the biennial reports and unmanifested waste reports described in §§ 264.75 and 264.76, the owner or operator must also report to the Regional Administrator:

(a) Releases, fires, and explosions as specified in § 264.56(j);

(b) Facility closures specified in § 264.115; and

(c) As otherwise required by subparts F, K through N, AA, BB, and CC of this part.

[46 FR 2849, Jan. 12, 1981, as amended at 47 FR 32350, July 26, 1982; 48 FR 3982, Jan. 28, 1983; 55 FR 25494, June 21, 1990; 59 FR 62926, Dec. 6, 1994]

## Subpart F—Releases From Solid Waste Management Units

SOURCE: 47 FR 32350, July 26, 1982, unless otherwise noted.

§ 264.90 **Applicability.**

(a)(1) Except as provided in paragraph (b) of this section, the regulations in this subpart apply to owners or operators of facilities that treat, store or dispose of hazardous waste. The owner or operator must satisfy the requirements identified in paragraph (a)(2) of this section for all wastes (or constituents thereof) contained in solid waste management units at the facility, regardless of the time at which waste was placed in such units.

(2) All solid waste management units must comply with the requirements in § 264.101. A surface impoundment, waste pile, and land treatment unit or landfill that receives hazardous waste after July 26, 1982 (hereinafter referred to as a "regulated unit") must comply with the requirements of §§ 264.91 through 264.100 in lieu of § 264.101 for purposes of detecting, characterizing and responding to releases to the uppermost aquifer. The financial responsibility requirements of § 264.101 apply to regulated units.

(b) The owner or operator's regulated unit or units are not subject to regulation for releases into the uppermost aquifer under this subpart if:

359

(1) The owner or operator is exempted under § 264.1; or

(2) He operates a unit which the Regional Administrator finds:

(i) Is an engineered structure,

(ii) Does not receive or contain liquid waste or waste containing free liquids,

(iii) Is designed and operated to exclude liquid, precipitation, and other run-on and run-off,

(iv) Has both inner and outer layers of containment enclosing the waste,

(v) Has a leak detection system built into each containment layer,

(vi) The owner or operator will provide continuing operation and maintenance of these leak detection systems during the active life of the unit and the closure and post-closure care periods, and

(vii) To a reasonable degree of certainty, will not allow hazardous constituents to migrate beyond the outer containment layer prior to the end of the post-closure care period.

(3) The Regional Administrator finds, pursuant to § 264.280(d), that the treatment zone of a land treatment unit that qualifies as a regulated unit does not contain levels of hazardous constituents that are above background levels of those constituents by an amount that is statistically significant, and if an unsaturated zone monitoring program meeting the requirements of § 264.278 has not shown a statistically significant increase in hazardous constituents below the treatment zone during the operating life of the unit. An exemption under this paragraph can only relieve an owner or operator of responsibility to meet the requirements of this subpart during the post-closure care period; or

(4) The Regional Administrator finds that there is no potential for migration of liquid from a regulated unit to the uppermost aquifer during the active life of the regulated unit (including the closure period) and the post-closure care period specified under § 264.117. This demonstration must be certified by a qualified geologist or geotechnical engineer. In order to provide an adequate margin of safety in the prediction of potential migration of liquid, the owner or operator must base any predictions made under this paragraph on assumptions that maximize the rate of liquid migration.

(5) He designs and operates a pile in compliance with § 264.250(c).

(c) The regulations under this subpart apply during the active life of the regulated unit (including the closure period). After closure of the regulated unit, the regulations in this subpart:

(1) Do not apply if all waste, waste residues, contaminated containment system components, and contaminated subsoils are removed or decontaminated at closure;

(2) Apply during the post-closure care period under § 264.117 if the owner or operator is conducting a detection monitoring program under § 264.98; or

(3) Apply during the compliance period under § 264.96 if the owner or operator is conducting a compliance monitoring program under § 264.99 or a corrective action program under § 264.100.

(d) Regulations in this subpart may apply to miscellaneous units when necessary to comply with §§ 264.601 through 264.603.

(e) The regulations of this subpart apply to all owners and operators subject to the requirements of 40 CFR 270.1(c)(7), when the Agency issues either a post-closure permit or an enforceable document (as defined in 40 CFR 270.1(c)(7)) at the facility. When the Agency issues an enforceable document, references in this subpart to "in the permit" mean "in the enforceable document."

(f) The Regional Administrator may replace all or part of the requirements of §§ 264.91 through 264.100 applying to a regulated unit with alternative requirements for groundwater monitoring and corrective action for releases to groundwater set out in the permit (or in an enforceable document) (as defined in 40 CFR 270.1(c)(7)) where the Regional Administrator determines that:

(1) The regulated unit is situated among solid waste management units (or areas of concern), a release has occurred, and both the regulated unit and one or more solid waste management unit(s) (or areas of concern) are likely to have contributed to the release; and

(2) It is not necessary to apply the groundwater monitoring and corrective action requirements of §§ 264.91 through

264.100 because alternative requirements will protect human health and the environment.

[47 FR 32350, July 26, 1982, as amended at 50 FR 28746, July 15, 1985; 52 FR 46963, Dec. 10, 1987; 63 FR 56733, Oct. 22, 1998]

§ 264.91 Required programs.

(a) Owners and operators subject to this subpart must conduct a monitoring and response program as follows:

(1) Whenever hazardous constituents under § 264.93 from a regulated unit are detected at a compliance point under § 264.95, the owner or operator must institute a compliance monitoring program under § 264.99. Detected is defined as statistically significant evidence of contamination as described in § 264.98(f);

(2) Whenever the ground-water protection standard under § 264.92 is exceeded, the owner or operator must institute a corrective action program under § 264.100. Exceeded is defined as statistically significant evidence of increased contamination as described in § 264.99(d);

(3) Whenever hazardous constituents under § 264.93 from a regulated unit exceed concentration limits under § 264.94 in ground water between the compliance point under § 264.95 and the downgradient facility property boundary, the owner or operator must institute a corrective action program under § 264.100; or

(4) In all other cases, the owner or operator must institute a detection monitoring program under § 264.98.

(b) The Regional Administrator will specify in the facility permit the specific elements of the monitoring and response program. The Regional Administrator may include one or more of the programs identified in paragraph (a) of this section in the facility permit as may be necessary to protect human health and the environment and will specify the circumstances under which each of the programs will be required. In deciding whether to require the owner or operator to be prepared to institute a particular program, the Regional Administrator will consider the potential adverse effects on human health and the environment that might occur before final administrative action on a permit modification application to incorporate such a program could be taken.

[47 FR 32350, July 26, 1982, as amended at 53 FR 39728, Oct. 11, 1988]

§ 264.92 Ground-water protection standard.

The owner or operator must comply with conditions specified in the facility permit that are designed to ensure that hazardous constituents under § 264.93 detected in the ground water from a regulated unit do not exceed the concentration limits under § 264.94 in the uppermost aquifer underlying the waste management area beyond the point of compliance under § 264.95 during the compliance period under § 264.96. The Regional Administrator will establish this ground-water protection standard in the facility permit when hazardous constituents have been detected in the ground water.

[53 FR 39728, Oct. 11, 1988]

§ 264.93 Hazardous constituents.

(a) The Regional Administrator will specify in the facility permit the hazardous constituents to which the ground-water protection standard of § 264.92 applies. Hazardous constituents are constituents identified in appendix VIII of part 261 of this chapter that have been detected in ground water in the uppermost aquifer underlying a regulated unit and that are reasonably expected to be in or derived from waste contained in a regulated unit, unless the Regional Administrator has excluded them under paragraph (b) of this section.

(b) The Regional Administrator will exclude an appendix VIII constituent from the list of hazardous constituents specified in the facility permit if he finds that the constituent is not capable of posing a substantial present or potential hazard to human health or the environment. In deciding whether to grant an exemption, the Regional Administrator will consider the following:

(1) Potential adverse effects on ground-water quality, considering:

(i) The physical and chemical characteristics of the waste in the regulated unit, including its potential for migration;

(ii) The hydrogeological characteristics of the facility and surrounding land;

(iii) The quantity of ground water and the direction of ground-water flow;

(iv) The proximity and withdrawal rates of ground-water users;

(v) The current and future uses of ground water in the area;

(vi) The existing quality of ground water, including other sources of contamination and their cumulative impact on the ground-water quality;

(vii) The potential for health risks caused by human exposure to waste constituents;

(viii) The potential damage to wildlife, crops, vegetation, and physical structures caused by exposure to waste constituents;

(ix) The persistence and permanence of the potential adverse effects; and

(2) Potential adverse effects on hydraulically-connected surface water quality, considering:

(i) The volume and physical and chemical characteristics of the waste in the regulated unit;

(ii) The hydrogeological characteristics of the facility and surrounding land;

(iii) The quantity and quality of ground water, and the direction of ground-water flow;

(iv) The patterns of rainfall in the region;

(v) The proximity of the regulated unit to surface waters;

(vi) The current and future uses of surface waters in the area and any water quality standards established for those surface waters;

(vii) The existing quality of surface water, including other sources of contamination and the cumulative impact on surface-water quality;

(viii) The potential for health risks caused by human exposure to waste constituents;

(ix) The potential damage to wildlife, crops, vegetation, and physical structures caused by exposure to waste constituents; and

(x) The persistence and permanence of the potential adverse effects.

(c) In making any determination under paragraph (b) of this section

about the use of ground water in the area around the facility, the Regional Administrator will consider any identification of underground sources of drinking water and exempted aquifers made under § 144.8 of this chapter.

[47 FR 32350, July 26, 1982, as amended at 48 FR 14294, Apr. 1, 1983]

§ 264.94   Concentration limits.

(a) The Regional Administrator will specify in the facility permit concentration limits in the ground water for hazardous constituents established under § 264.93. The concentration of a hazardous constituent:

(1) Must not exceed the background level of that constituent in the ground water at the time that limit is specified in the permit; or

(2) For any of the constituents listed in Table 1, must not exceed the respective value given in that table if the background level of the constituent is below the value given in Table 1; or

TABLE 1—MAXIMUM CONCENTRATION OF CONSTITUENTS FOR GROUND-WATER PROTECTION

| Constituent | Maximum concentration [1] |
|---|---|
| Arsenic | 0.05 |
| Barium | 1.0 |
| Cadmium | 0.01 |
| Chromium | 0.05 |
| Lead | 0.05 |
| Mercury | 0.002 |
| Selenium | 0.01 |
| Silver | 0.05 |
| Endrin (1,2,3,4,10,10-hexachloro-1,7-epoxy 1,4,4a,5,6,7,8,9a-octahydro-1, 4-endo, endo-5,8-dimethano naphthalene) | 0.0002 |
| Lindane (1,2,3,4,5,6-hexachlorocyclohexane, gamma isomer) | 0.004 |
| Methoxychlor (1,1,1-Trichloro-2,2-bis (p-methoxyphenylethane) | 0.1 |
| Toxaphene ($C_{10}H_{10}Cl_6$, Technical chlorinated camphene, 67–69 percent chlorine) | 0.005 |
| 2,4-D (2,4-Dichlorophenoxyacetic acid) | 0.1 |
| 2,4,5-TP Silvex (2,4,5-Trichlorophenoxypropionic acid) | 0.01 |

[1] Milligrams per liter.

(3) Must not exceed an alternate limit established by the Regional Administrator under paragraph (b) of this section.

(b) The Regional Administrator will establish an alternate concentration limit for a hazardous constituent if he finds that the constituent will not pose a substantial present or potential hazard to human health or the environment as long as the alternate concentration limit is not exceeded. In establishing alternate concentration limits, the Regional Administrator will consider the following factors:

(1) Potential adverse effects on ground-water quality, considering:

(i) The physical and chemical characteristics of the waste in the regulated unit, including its potential for migration;

(ii) The hydrogeological characteristics of the facility and surrounding land;

(iii) The quantity of ground water and the direction of ground-water flow;

(iv) The proximity and withdrawal rates of ground-water users;

(v) The current and future uses of ground water in the area;

(vi) The existing quality of ground water, including other sources of contamination and their cumulative impact on the ground-water quality;

(vii) The potential for health risks caused by human exposure to waste constituents;

(viii) The potential damage to wildlife, crops, vegetation, and physical structures caused by exposure to waste constituents;

(ix) The persistence and permanence of the potential adverse effects; and

(2) Potential adverse effects on hydraulically-connected surface-water quality, considering:

(i) The volume and physical and chemical characteristics of the waste in the regulated unit;

(ii) The hydrogeological characteristics of the facility and surrounding land;

(iii) The quantity and quality of ground water, and the direction of ground-water flow;

(iv) The patterns of rainfall in the region;

(v) The proximity of the regulated unit to surface waters;

(vi) The current and future uses of surface waters in the area and any water quality standards established for those surface waters;

(vii) The existing quality of surface water, including other sources of contamination and the cumulative impact on surface water quality;

(viii) The potential for health risks caused by human exposure to waste constituents;

(ix) The potential damage to wildlife, crops, vegetation, and physical structures caused by exposure to waste constituents; and

(x) The persistence and permanence of the potential adverse effects.

(c) In making any determination under paragraph (b) of this section about the use of ground water in the area around the facility the Regional Administrator will consider any identification of underground sources of drinking water and exempted aquifers made under §144.8 of this chapter.

[47 FR 32350, July 26, 1982, as amended at 48 FR 14294, Apr. 1, 1983]

§264.95 Point of compliance.

(a) The Regional Administrator will specify in the facility permit the point of compliance at which the ground-water protection standard of §264.92 applies and at which monitoring must be conducted. The point of compliance is a vertical surface located at the hydraulically downgradient limit of the waste management area that extends down into the uppermost aquifer underlying the regulated units.

(b) The waste management area is the limit projected in the horizontal plane of the area on which waste will be placed during the active life of a regulated unit.

(1) The waste management area includes horizontal space taken up by any liner, dike, or other barrier designed to contain waste in a regulated unit.

(2) If the facility contains more than one regulated unit, the waste management area is described by an imaginary line circumscribing the several regulated units.

§264.96 Compliance period.

(a) The Regional Administrator will specify in the facility permit the compliance period during which the ground-water protection standard of §264.92 applies. The compliance period is the number of years equal to the active life of the waste management area (including any waste management activity prior to permitting, and the closure period.)

(b) The compliance period begins when the owner or operator initiates a compliance monitoring program meeting the requirements of § 264.99.

(c) If the owner or operator is engaged in a corrective action program at the end of the compliance period specified in paragraph (a) of this section, the compliance period is extended until the owner or operator can demonstrate that the ground-water protection standard of § 264.92 has not been exceeded for a period of three consecutive years.

§ 264.97 General ground-water monitoring requirements.

The owner or operator must comply with the following requirements for any ground-water monitoring program developed to satisfy § 264.98, § 264.99, or § 264.100:

(a) The ground-water monitoring system must consist of a sufficient number of wells, installed at appropriate locations and depths to yield ground-water samples from the uppermost aquifer that:

(1) Represent the quality of background ground water that has not been affected by leakage from a regulated unit;

(i) A determination of background ground-water quality may include sampling of wells that are not hydraulically upgradient of the waste management area where:

(A) Hydrogeologic conditions do not allow the owner or operator to determine what wells are hydraulically upgradient; and

(B) Sampling at other wells will provide an indication of background ground-water quality that is representative or more representative than that provided by the upgradient wells; and

(2) Represent the quality of ground water passing the point of compliance.

(3) Allow for the detection of contamination when hazardous waste or hazardous constituents have migrated from the waste management area to the uppermost aquifer.

(b) If a facility contains more than one regulated unit, separate ground-water monitoring systems are not required for each regulated unit provided that provisions for sampling the ground water in the uppermost aquifer

will enable detection and measurement at the compliance point of hazardous constituents from the regulated units that have entered the ground water in the uppermost aquifer.

(c) All monitoring wells must be cased in a manner that maintains the integrity of the monitoring-well bore hole. This casing must be screened or perforated and packed with gravel or sand, where necessary, to enable collection of ground-water samples. The annular space (i.e., the space between the bore hole and well casing) above the sampling depth must be sealed to prevent contamination of samples and the ground water.

(d) The ground-water monitoring program must include consistent sampling and analysis procedures that are designed to ensure monitoring results that provide a reliable indication of ground-water quality below the waste management area. At a minimum the program must include procedures and techniques for:

(1) Sample collection;

(2) Sample preservation and shipment;

(3) Analytical procedures; and

(4) Chain of custody control.

(e) The ground-water monitoring program must include sampling and analytical methods that are appropriate for ground-water sampling and that accurately measure hazardous constituents in ground-water samples.

(f) The ground-water monitoring program must include a determination of the ground-water surface elevation each time ground water is sampled.

(g) In detection monitoring or where appropriate in compliance monitoring, data on each hazardous constituent specified in the permit will be collected from background wells and wells at the compliance point(s). The number and kinds of samples collected to establish background shall be appropriate for the form of statistical test employed, following generally accepted statistical principles. The sample size shall be as large as necessary to ensure with reasonable confidence that a contaminant release to ground water from a facility will be detected. The owner or operator will determine an appropriate sampling procedure and interval for each hazardous constituent listed in the facility permit which shall be specified in the unit permit upon approval by the Regional Administrator. This sampling procedure shall be:

(1) A sequence of at least four samples, taken at an interval that assures, to the greatest extent technically feasible, that an independent sample is obtained, by reference to the uppermost aquifer's effective porosity, hydraulic conductivity, and hydraulic gradient, and the fate and transport characteristics of the potential contaminants, or

(2) an alternate sampling procedure proposed by the owner or operator and approved by the Regional Administrator.

(h) The owner or operator will specify one of the following statistical methods to be used in evaluating groundwater monitoring data for each hazardous constituent which, upon approval by the Regional Administrator, will be specified in the unit permit. The statistical test chosen shall be conducted separately for each hazardous constituent in each well. Where practical quantification limits (pql's) are used in any of the following statistical procedures to comply with §264.97(i)(5), the pql must be proposed by the owner or operator and approved by the Regional Administrator. Use of any of the following statistical methods must be protective of human health and the environment and must comply with the performance standards outlined in paragraph (i) of this section.

(1) A parametric analysis of variance (ANOVA) followed by multiple comparisons procedures to identify statistically significant evidence of contamination. The method must include estimation and testing of the contrasts between each compliance well's mean and the background mean levels for each constituent.

(2) An analysis of variance (ANOVA) based on ranks followed by multiple comparisons procedures to identify statistically significant evidence of contamination. The method must include estimation and testing of the contrasts between each compliance well's median and the background median levels for each constituent.

(3) A tolerance or prediction interval procedure in which an interval for each constituent is established from the distribution of the background data, and the level of each constituent in each compliance well is compared to the upper tolerance or prediction limit.

(4) A control chart approach that gives control limits for each constituent.

(5) Another statistical test method submitted by the owner or operator and approved by the Regional Administrator.

(i) Any statistical method chosen under §264.97(h) for specification in the unit permit shall comply with the following performance standards, as appropriate:

(1) The statistical method used to evaluate ground-water monitoring data shall be appropriate for the distribution of chemical parameters or hazardous constituents. If the distribution of the chemical parameters or hazardous constituents is shown by the owner or operator to be inappropriate for a normal theory test, then the data should be transformed or a distribution-free theory test should be used. If the distributions for the constituents differ, more than one statistical method may be needed.

(2) If an individual well comparison procedure is used to compare an individual compliance well constituent concentration with background constituent concentrations or a ground-water protection standard, the test shall be done at a Type I error level no less than 0.01 for each testing period. If a multiple comparisons procedure is used, the Type I experimentwise error rate for each testing period shall be no less than 0.05; however, the Type I error of no less than 0.01 for individual well comparisons must be maintained. This performance standard does not apply to tolerance intervals, prediction intervals or control charts.

(3) If a control chart approach is used to evaluate ground-water monitoring data, the specific type of control chart and its associated parameter values shall be proposed by the owner or operator and approved by the Regional Administrator if he or she finds it to be protective of human health and the environment.

(4) If a tolerance interval or a prediction interval is used to evaluate

groundwater monitoring data, the levels of confidence and, for tolerance intervals, the percentage of the population that the interval must contain, shall be proposed by the owner or operator and approved by the Regional Administrator if he or she finds these parameters to be protective of human health and the environment. These parameters will be determined after considering the number of samples in the background data base, the data distribution, and the range of the concentration values for each constituent of concern.

(5) The statistical method shall account for data below the limit of detection with one or more statistical procedures that are protective of human health and the environment. Any practical quantification limit (pql) approved by the Regional Administrator under § 264.97(h) that is used in the statistical method shall be the lowest concentration level that can be reliably achieved within specified limits of precision and accuracy during routine laboratory operating conditions that are available to the facility.

(6) If necessary, the statistical method shall include procedures to control or correct for seasonal and spatial variability as well as temporal correlation in the data.

(j) Ground-water monitoring data collected in accordance with paragraph (g) of this section including actual levels of constituents must be maintained in the facility operating record. The Regional Administrator will specify in the permit when the data must be submitted for review.

[47 FR 32350, July 26, 1982, as amended at 50 FR 4514, Jan. 31, 1985; 53 FR 39728, Oct. 11, 1988; 71 FR 40272, July 14, 2006]

## § 264.98 Detection monitoring program.

An owner or operator required to establish a detection monitoring program under this subpart must, at a minimum, discharge the following responsibilities:

(a) The owner or operator must monitor for indicator parameters (e.g., specific conductance, total organic carbon, or total organic halogen), waste constituents, or reaction products that provide a reliable indication of the presence of hazardous constituents in ground water. The Regional Administrator will specify the parameters or constituents to be monitored in the facility permit, after considering the following factors:

(1) The types, quantities, and concentrations of constituents in wastes managed at the regulated unit;

(2) The mobility, stability, and persistence of waste constituents or their reaction products in the unsaturated zone beneath the waste management area;

(3) The detectability of indicator parameters, waste constituents, and reaction products in ground water; and

(4) The concentrations or values and coefficients of variation of proposed monitoring parameters or constituents in the ground-water background.

(b) The owner or operator must install a ground-water monitoring system at the compliance point as specified under § 264.95. The ground-water monitoring system must comply with § 264.97(a)(2), (b), and (c).

(c) The owner or operator must conduct a ground-water monitoring program for each chemical parameter and hazardous constituent specified in the permit pursuant to paragraph (a) of this section in accordance with § 264.97(g). The owner or operator must maintain a record of ground-water analytical data as measured and in a form necessary for the determination of statistical significance under § 264.97(h).

(d) The Regional Administrator will specify the frequencies for collecting samples and conducting statistical tests to determine whether there is statistically significant evidence of contamination for any parameter or hazardous constituent specified in the permit conditions under paragraph (a) of this section in accordance with § 264.97(g).

(e) The owner or operator must determine the ground-water flow rate and direction in the uppermost aquifer at least annually.

(f) The owner or operator must determine whether there is statistically significant evidence of contamination for any chemical parameter of hazardous constituent specified in the permit pursuant to paragraph (a) of this section at a frequency specified under paragraph (d) of this section.

(1) In determining whether statistically significant evidence of contamination exists, the owner or operator must use the method(s) specified in the permit under §264.97(h). These method(s) must compare data collected at the compliance point(s) to the background ground-water quality data.

(2) The owner or operator must determine whether there is statistically significant evidence of contamination at each monitoring well as the compliance point within a reasonable period of time after completion of sampling. The Regional Administrator will specify in the facility permit what period of time is reasonable, after considering the complexity of the statistical test and the availability of laboratory facilities to perform the analysis of ground-water samples.

(g) If the owner or operator determines pursuant to paragraph (f) of this section that there is statistically significant evidence of contamination for chemical parameters or hazardous constituents specified pursuant to paragraph (a) of this section at any monitoring well at the compliance point, he or she must:

(1) Notify the Regional Administrator of this finding in writing within seven days. The notification must indicate what chemical parameters or hazardous constituents have shown statistically significant evidence of contamination;

(2) Immediately sample the ground water in all monitoring wells and determine whether constituents in the list of appendix IX of this part are present, and if so, in what concentration. However, the Regional Administrator, on a discretionary basis, may allow sampling for a site-specific subset of constituents from the Appendix IX list of this part and other representative/related waste constituents.

(3) For any appendix IX compounds found in the analysis pursuant to paragraph (g)(2) of this section, the owner or operator may resample within one month or at an alternative site-specific schedule approved by the Administrator and repeat the analysis for those compounds detected. If the results of the second analysis confirm the initial

results, then these constituents will form the basis for compliance monitoring. If the owner or operator does not resample for the compounds in paragraph (g)(2) of this section, the hazardous constituents found during this initial appendix IX analysis will form the basis for compliance monitoring.

(4) Within 90 days, submit to the Regional Administrator an application for a permit modification to establish a compliance monitoring program meeting the requirements of §264.99. The application must include the following information:

(i) An identification of the concentration of any appendix IX constituent detected in the ground water at each monitoring well at the compliance point;

(ii) Any proposed changes to the ground-water monitoring system at the facility necessary to meet the requirements of §264.99;

(iii) Any proposed additions or changes to the monitoring frequency, sampling and analysis procedures or methods, or statistical methods used at the facility necessary to meet the requirements of §264.99;

(iv) For each hazardous constituent detected at the compliance point, a proposed concentration limit under §264.94(a) (1) or (2), or a notice of intent to seek an alternate concentration limit under §264.94(b); and

(5) Within 180 days, submit to the Regional Administrator:

(i) All data necessary to justify an alternate concentration limit sought under §264.94(b); and

(ii) An engineering feasibility plan for a corrective action program necessary to meet the requirement of §264.100, unless:

(A) All hazardous constituents identified under paragraph (g)(2) of this section are listed in Table 1 of §264.94 and their concentrations do not exceed the respective values given in that Table; or

(B) The owner or operator has sought an alternate concentration limit under §264.94(b) for every hazardous constituent identified under paragraph (g)(2) of this section.

(6) If the owner or operator determines, pursuant to paragraph (f) of this

section, that there is a statistically significant difference for chemical parameters or hazardous constituents specified pursuant to paragraph (a) of this section at any monitoring well at the compliance point, he or she may demonstrate that a source other than a regulated unit caused the contamination or that the detection is an artifact caused by an error in sampling, analysis, or statistical evaluation or natural variation in the ground water. The owner operator may make a demonstration under this paragraph in addition to, or in lieu of, submitting a permit modification application under paragraph (g)(4) of this section; however, the owner or operator is not relieved of the requirement to submit a permit modification application within the time specified in paragraph (g)(4) of this section unless the demonstration made under this paragraph successfully shows that a source other than a regulated unit caused the increase, or that the increase resulted from error in sampling, analysis, or evaluation. In making a demonstration under this paragraph, the owner or operator must:

(i) Notify the Regional Administrator in writing within seven days of determining statistically significant evidence of contamination at the compliance point that he intends to make a demonstration under this paragraph;

(ii) Within 90 days, submit a report to the Regional Administrator which demonstrates that a source other than a regulated unit caused the contamination or that the contamination resulted from error in sampling, analysis, or evaluation;

(iii) Within 90 days, submit to the Regional Administrator an application for a permit modification to make any appropriate changes to the detection monitoring program facility; and

(iv) Continue to monitor in accordance with the detection monitoring program established under this section.

(h) If the owner or operator determines that the detection monitoring program no longer satisfies the requirements of this section, he or she must, within 90 days, submit an application for a permit modification to

make any appropriate changes to the program.

[47 FR 32350, July 26, 1982, as amended at 50 FR 4514, Jan. 31, 1985; 52 FR 25946, July 9, 1987; 53 FR 39729, Oct. 11, 1988; 71 FR 16904, Apr. 4, 2006; 71 FR 40272, July 14, 2006]

### § 264.99  Compliance monitoring program.

An owner or operator required to establish a compliance monitoring program under this subpart must, at a minimum, discharge the following responsibilities:

(a) The owner or operator must monitor the ground water to determine whether regulated units are in compliance with the ground-water protection standard under § 264.92. The Regional Administrator will specify the ground-water protection standard in the facility permit, including:

(1) A list of the hazardous constituents identified under § 264.93;

(2) Concentration limits under § 264.94 for each of those hazardous constituents;

(3) The compliance point under § 264.95; and

(4) The compliance period under § 264.96.

(b) The owner or operator must install a ground-water monitoring system at the compliance point as specified under § 264.95. The ground-water monitoring system must comply with § 264.97(a)(2), (b), and (c).

(c) The Regional Administrator will specify the sampling procedures and statistical methods appropriate for the constituents and the facility, consistent with § 264.97 (g) and (h).

(1) The owner or operator must conduct a sampling program for each chemical parameter or hazardous constituent in accordance with § 264.97(g).

(2) The owner or operator must record ground-water analytical data as measured and in form necessary for the determination of statistical significance under § 264.97(h) for the compliance period of the facility.

(d) The owner or operator must determine whether there is statistically significant evidence of increased contamination for any chemical parameter or hazardous constituent specified in the permit, pursuant to paragraph

(a) of this section, at a frequency specified under paragraph (f) under this section.

(1) In determining whether statistically significant evidence of increased contamination exists, the owner or operator must use the method(s) specified in the permit under §264.97(h). The methods(s) must compare data collected at the compliance point(s) to a concentration limit developed in accordance with §264.94.

(2) The owner or operator must determine whether there is statistically significant evidence of increased contamination at each monitoring well at the compliance point within a reasonable time period after completion of sampling. The Regional Administrator will specify that time period in the facility permit, after considering the complexity of the statistical test and the availability of laboratory facilities to perform the analysis of ground-water samples.

(e) The owner or operator must determine the ground-water flow rate and direction in the uppermost aquifer at least annually.

(f) The Regional Administrator will specify the frequencies for collecting samples and conducting statistical tests to determine statistically significant evidence of increased contamination in accordance with §264.97(g).

(g) Annually, the owner or operator must determine whether additional hazardous constituents from Appendix IX of this part, which could possibly be present but are not on the detection monitoring list in the permit, are actually present in the uppermost aquifer and, if so, at what concentration, pursuant to procedures in §264.98(f). To accomplish this, the owner or operator must consult with the Regional Administrator to determine on a case-by-case basis: which sample collection event during the year will involve enhanced sampling; the number of monitoring wells at the compliance point to undergo enhanced sampling; the number of samples to be collected from each of these monitoring wells; and, the specific constituents from Appendix IX of this part for which these samples must be analyzed. If the enhanced sampling event indicates that Appendix IX constituents are present in the ground

water that are not already identified in the permit as monitoring constituents, the owner or operator may resample within one month or at an alternative site-specific schedule approved by the Regional Administrator, and repeat the analysis. If the second analysis confirms the presence of new constituents, the owner or operator must report the concentration of these additional constituents to the Regional Administrator within seven days after the completion of the second analysis and add them to the monitoring list. If the owner or operator chooses not to resample, then he or she must report the concentrations of these additional constituents to the Regional Administrator within seven days after completion of the initial analysis, and add them to the monitoring list.

(h) If the owner or operator determines pursuant to paragraph (d) of this section that any concentration limits under §264.94 are being exceeded at any monitoring well at the point of compliance he or she must:

(1) Notify the Regional Administrator of this finding in writing within seven days. The notification must indicate what concentration limits have been exceeded.

(2) Submit to the Regional Administrator an application for a permit modification to establish a corrective action program meeting the requirements of §264.100 within 180 days, or within 90 days if an engineering feasibility study has been previously submitted to the Regional Administrator under §264.98(g)(5). The application must at a minimum include the following information:

(i) A detailed description of corrective actions that will achieve compliance with the ground-water protection standard specified in the permit under paragraph (a) of this section; and

(ii) A plan for a ground-water monitoring program that will demonstrate the effectiveness of the corrective action. Such a ground-water monitoring program may be based on a compliance monitoring program developed to meet the requirements of this section.

(i) If the owner or operator determines, pursuant to paragraph (d) of this section, that the ground-water concentration limits under this section are being exceeded at any monitoring well at the point of compliance, he or she may demonstrate that a source other than a regulated unit caused the contamination or that the detection is an artifact caused by an error in sampling, analysis, or statistical evaluation or natural variation in the ground water. In making a demonstration under this paragraph, the owner or operator must:

(1) Notify the Regional Administrator in writing within seven days that he intends to make a demonstration under this paragraph;

(2) Within 90 days, submit a report to the Regional Administrator which demonstrates that a source other than a regulated unit caused the standard to be exceeded or that the apparent noncompliance with the standards resulted from error in sampling, analysis, or evaluation;

(3) Within 90 days, submit to the Regional Administrator an application for a permit modification to make any appropriate changes to the compliance monitoring program at the facility; and

(4) Continue to monitor in accord with the compliance monitoring program established under this section.

(j) If the owner or operator determines that the compliance monitoring program no longer satisfies the requirements of this section, he must, within 90 days, submit an application for a permit modification to make any appropriate changes to the program.

[47 FR 32350, July 26, 1982, as amended at 50 FR 4514, Jan. 31, 1985; 52 FR 25946, July 9, 1987; 53 FR 39730, Oct. 11, 1988; 71 FR 16904, Apr. 4, 2006; 71 FR 40272, July 14, 2006]

§ 264.100  Corrective action program.

An owner or operator required to establish a corrective action program under this subpart must, at a minimum, discharge the following responsibilities:

(a) The owner or operator must take corrective action to ensure that regulated units are in compliance with the ground-water protection standard under § 264.92. The Regional Administrator will specify the ground-water protection standard in the facility permit, including:

(1) A list of the hazardous constituents identified under § 264.93;

(2) Concentration limits under § 264.94 for each of those hazardous constituents;

(3) The compliance point under § 264.95; and

(4) The compliance period under § 264.96.

(b) The owner or operator must implement a corrective action program that prevents hazardous constituents from exceeding their respective concentration limits at the compliance point by removing the hazardous waste constituents or treating them in place. The permit will specify the specific measures that will be taken.

(c) The owner or operator must begin corrective action within a reasonable time period after the ground-water protection standard is exceeded. The Regional Administrator will specify that time period in the facility permit. If a facility permit includes a corrective action program in addition to a compliance monitoring program, the permit will specify when the corrective action will begin and such a requirement will operate in lieu of § 264.99(i)(2).

(d) In conjunction with a corrective action program, the owner or operator must establish and implement a ground-water monitoring program to demonstrate the effectiveness of the corrective action program. Such a monitoring program may be based on the requirements for a compliance monitoring program under § 264.99 and must be as effective as that program in determining compliance with the ground-water protection standard under § 264.92 and in determining the success of a corrective action program under paragraph (e) of this section, where appropriate.

(e) In addition to the other requirements of this section, the owner or operator must conduct a corrective action program to remove or treat in place any hazardous constituents under § 264.93 that exceed concentration limits under § 264.94 in groundwater:

(1) Between the compliance point under § 264.95 and the downgradient property boundary; and

(2) Beyond the facility boundary, where necessary to protect human

health and the environment, unless the owner or operator demonstrates to the satisfaction of the Regional Administrator that, despite the owner's or operator's best efforts, the owner or operator was unable to obtain the necessary permission to undertake such action. The owner/operator is not relieved of all responsibility to clean up a release that has migrated beyond the facility boundary where off-site access is denied. On-site measures to address such releases will be determined on a case-by-case basis.

(3) Corrective action measures under this paragraph must be initiated and completed within a reasonable period of time considering the extent of contamination.

(4) Corrective action measures under this paragraph may be terminated once the concentration of hazardous constituents under §264.93 is reduced to levels below their respective concentration limits under §264.94.

(f) The owner or operator must continue corrective action measures during the compliance period to the extent necessary to ensure that the ground-water protection standard is not exceeded. If the owner or operator is conducting corrective action at the end of the compliance period, he must continue that corrective action for as long as necessary to achieve compliance with the ground-water protection standard. The owner or operator may terminate corrective action measures taken beyond the period equal to the active life of the waste management area (including the closure period) if he can demonstrate, based on data from the ground-water monitoring program under paragraph (d) of this section, that the ground-water protection standard of §264.92 has not been exceeded for a period of three consecutive years.

(g) The owner or operator must report in writing to the Regional Administrator on the effectiveness of the corrective action program. The owner or operator must submit these reports annually.

(h) If the owner or operator determines that the corrective action program no longer satisfies the requirements of this section, he must, within 90 days, submit an application for a permit modification to make any appropriate changes to the program.

[47 FR 32350, July 26, 1985, as amended at 50 FR 4514, Jan. 31, 1985; 52 FR 45798, Dec. 1, 1987; 71 FR 16904, Apr. 4, 2006]

### §264.101 Corrective action for solid waste management units.

(a) The owner or operator of a facility seeking a permit for the treatment, storage or disposal of hazardous waste must institute corrective action as necessary to protect human health and the environment for all releases of hazardous waste or constituents from any solid waste management unit at the facility, regardless of the time at which waste was placed in such unit.

(b) Corrective action will be specified in the permit in accordance with this section and subpart S of this part. The permit will contain schedules of compliance for such corrective action (where such corrective action cannot be completed prior to issuance of the permit) and assurances of financial responsibility for completing such corrective action.

(c) The owner or operator must implement corrective actions beyond the facility property boundary, where necessary to protect human health and the environment, unless the owner or operator demonstrates to the satisfaction of the Regional Administrator that, despite the owner's or operator's best efforts, the owner or operator was unable to obtain the necessary permission to undertake such actions. The owner/operator is not relieved of all responsibility to clean up a release that has migrated beyond the facility boundary where off-site access is denied. On-site measures to address such releases will be determined on a case-by-case basis. Assurances of financial responsibility for such corrective action must be provided.

(d) This section does not apply to remediation waste management sites unless they are part of a facility subject to a permit for treating, storing or disposing of hazardous wastes that are not remediation wastes.

[50 FR 28747, July 15, 1985, as amended at 52 FR 45798, Dec. 1, 1987; 58 FR 8683, Feb. 16, 1993; 63 FR 65938, Nov. 30, 1998; 71 FR 40272, July 14, 2006]

## Subpart G—Closure and Post-Closure

SOURCE: 51 FR 16444, May 2, 1986, unless otherwise noted.

### § 264.110 Applicability.

Except as § 264.1 provides otherwise:

(a) Sections 264.111 through 264.115 (which concern closure) apply to the owners and operators of all hazardous waste management facilities; and

(b) Sections 264.116 through 264.120 (which concern post-closure care) apply to the owners and operators of:

(1) All hazardous waste disposal facilities;

(2) Waste piles and surface impoundments from which the owner or operator intends to remove the wastes at closure to the extent that these sections are made applicable to such facilities in § 264.228 or § 264.258;

(3) Tank systems that are required under § 264.197 to meet the requirements for landfills; and

(4) Containment buildings that are required under § 264.1102 to meet the requirement for landfills.

(c) The Regional Administrator may replace all or part of the requirements of this subpart (and the unit-specific standards referenced in § 264.111(c) applying to a regulated unit), with alternative requirements set out in a permit or in an enforceable document (as defined in 40 CFR 270.1(c)(7)), where the Regional Administrator determines that:

(1) The regulated unit is situated among solid waste management units (or areas of concern), a release has occurred, and both the regulated unit and one or more solid waste management unit(s) (or areas of concern) are likely to have contributed to the release; and

(2) It is not necessary to apply the closure requirements of this subpart (and those referenced herein) because the alternative requirements will protect human health and the environment and will satisfy the closure performance standard of § 264.111 (a) and (b).

[51 FR 16444, May 2, 1986, as amended at 51 FR 25472, July 14, 1986; 57 FR 37264, Aug. 18, 1992; 63 FR 56733, Oct. 22, 1998]

### § 264.111 Closure performance standard.

The owner or operator must close the facility in a manner that:

(a) Minimizes the need for further maintenance; and

(b) Controls, minimizes or eliminates, to the extent necessary to protect human health and the environment, post-closure escape of hazardous waste, hazardous constituents, leachate, contaminated run-off, or hazardous waste decomposition products to the ground or surface waters or to the atmosphere; and

(c) Complies with the closure requirements of this part, including, but not limited to, the requirements of §§ 264.178, 264.197, 264.228, 264.258, 264.280, 264.310, 264.351, 264.601 through 264.603, and 264.1102.

[51 FR 16444, May 2, 1986, as amended at 52 FR 46963, Dec. 10, 1987; 57 FR 37265, Aug. 18, 1992; 71 FR 40272, July 14, 2006]

### § 264.112 Closure plan; amendment of plan.

(a) *Written plan.* (1) The owner or operator of a hazardous waste management facility must have a written closure plan. In addition, certain surface impoundments and waste piles from which the owner or operator intends to remove or decontaminate the hazardous waste at partial or final closure are required by §§ 264.228(c)(1)(i) and 264.258(c)(1)(i) to have contingent closure plans. The plan must be submitted with the permit application, in accordance with § 270.14(b)(13) of this chapter, and approved by the Regional Administrator as part of the permit issuance procedures under part 124 of this chapter. In accordance with § 270.32 of this chapter, the approved closure plan will become a condition of any RCRA permit.

(2) The Director's approval of the plan must ensure that the approved closure plan is consistent with §§ 264.111 through 264.115 and the applicable requirements of subpart F of this part, §§ 264.178, 264.197, 264.228, 264.258, 264.280, 264.310, 264.351, 264.601, and 264.1102. Until final closure is completed and certified in accordance with § 264.115, a copy of the approved plan and all approved revisions must be furnished to

the Director upon request, including requests by mail.

(b) *Content of plan.* The plan must identify steps necessary to perform partial and/or final closure of the facility at any point during its active life. The closure plan must include, at least:

. (1) A description of how each hazardous waste management unit at the facility will be closed in accordance with §264.111;

(2) A description of how final closure of the facility will be conducted in accordance with §264.111. The description must identify the maximum extent of the operations which will be unclosed during the active life of the facility; and

(3) An estimate of the maximum inventory of hazardous wastes ever onsite over the active life of the facility and a detailed description of the methods to be used during partial closures and final closure, including, but not limited to, methods for removing, transporting, treating, storing, or disposing of all hazardous wastes, and identification of the type(s) of the offsite hazardous waste management units to be used, if applicable; and

(4) A detailed description of the steps needed to remove or decontaminate all hazardous waste residues and contaminated containment system components, equipment, structures, and soils during partial and final closure, including, but not limited to, procedures for cleaning equipment and removing contaminated soils, methods for sampling and testing surrounding soils, and criteria for determining the extent of decontamination required to satisfy the closure performance standard; and

(5) A detailed description of other activities necessary during the closure period to ensure that all partial closures and final closure satisfy the closure performance standards, including, but not limited to, ground-water monitoring, leachate collection, and run-on and run-off control; and

(6) A schedule for closure of each hazardous waste management unit and for final closure of the facility. The schedule must include, at a minimum, the total time required to close each hazardous waste management unit and the time required for intervening closure activities which will allow tracking of the progress of partial and final closure. (For example, in the case of a landfill unit, estimates of the time required to treat or dispose of all hazardous waste inventory and of the time required to place a final cover must be included.)

(7) For facilities that use trust funds to establish financial assurance under §264.143 or §264.145 and that are expected to close prior to the expiration of the permit, an estimate of the expected year of final closure.

(8) For facilities where the Regional Administrator has applied alternative requirements at a regulated unit under §§264.90(f), 264.110(c), and/or §264.140(d), either the alternative requirements applying to the regulated unit, or a reference to the enforceable document containing those alternative requirements.

(c) *Amendment of plan.* The owner or operator must submit a written notification of or request for a permit modification to authorize a change in operating plans, facility design, or the approved closure plan in accordance with the applicable procedures in parts 124 and 270. The written notification or request must include a copy of the amended closure plan for review or approval by the Regional Administrator.

(1) The owner or operator may submit a written notification or request to the Regional Administrator for a permit modification to amend the closure plan at any time prior to the notification of partial or final closure of the facility.

(2) The owner or operator must submit a written notification of or request for a permit modification to authorize a change in the approved closure plan whenever:

(i) Changes in operating plans or facility design affect the closure plan, or

(ii) There is a change in the expected year of closure, if applicable, or

(iii) In conducting partial or final closure activities, unexpected events require a modification of the approved closure plan.

(iv) The owner or operator requests the Regional Administrator to apply alternative requirements to a regulated unit under §§264.90(f), 264.110(c), and/or §264.140(d).

(3) The owner or operator must submit a written request for a permit modification including a copy of the amended closure plan for approval at least 60 days prior to the proposed change in facility design or operation, or no later than 60 days after an unexpected event has occurred which has affected the closure plan. If an unexpected event occurs during the partial or final closure period, the owner or operator must request a permit modification no later than 30 days after the unexpected event. An owner or operator of a surface impoundment or waste pile that intends to remove all hazardous waste at closure and is not otherwise required to prepare a contingent closure plan under § 264.228(c)(1)(i) or § 264.258(c)(1)(i), must submit an amended closure plan to the Regional Administrator no later than 60 days from the date that the owner or operator or Regional Administrator determines that the hazardous waste management unit must be closed as a landfill, subject to the requirements of § 264.310, or no later than 30 days from that date if the determination is made during partial or final closure. The Regional Administrator will approve, disapprove, or modify this amended plan in accordance with the procedures in parts 124 and 270. In accordance with § 270.32 of this chapter, the approved closure plan will become a condition of any RCRA permit issued.

(4) The Regional Administrator may request modifications to the plan under the conditions described in § 264.112(c)(2). The owner or operator must submit the modified plan within 60 days of the Regional Administrator's request, or within 30 days if the change in facility conditions occurs during partial or final closure. Any modifications requested by the Regional Administrator will be approved in accordance with the procedures in parts 124 and 270.

(d) *Notification of partial closure and final closure.* (1) The owner or operator must notify the Regional Administrator in writing at least 60 days prior to the date on which he expects to begin closure of a surface impoundment, waste pile, land treatment or landfill unit, or final closure of a facility with such a unit. The owner or operator must notify the Regional Administrator in writing at least 45 days prior to the date on which he expects to begin final closure of a facility with only treatment or storage tanks, container storage, or incinerator units to be closed. The owner or operator must notify the Regional Administrator in writing at least 45 days prior to the date on which he expects to begin partial or final closure of a boiler or industrial furnace, whichever is earlier.

(2) The date when he "expects to begin closure" must be either:

(i) No later than 30 days after the date on which any hazardous waste management unit receives the known final volume of hazardous wastes, or if there is a reasonable possibility that the hazardous waste management unit will receive additional hazardous wastes, no later than one year after the date on which the unit received the most recent volume of hazardous wastes. If the owner or operator of a hazardous waste management unit can demonstrate to the Regional Administrator that the hazardous waste management unit or facility has the capacity to receive additional hazardous wastes and he has taken all steps to prevent threats to human health and the environment, including compliance with all applicable permit requirements, the Regional Administrator may approve an extension to this one-year limit; or

(ii) For units meeting the requirements of § 264.113(d), no later than 30 days after the date on which the hazardous waste management unit receives the known final volume of non-hazardous wastes, or if there is a reasonable possibility that the hazardous waste management unit will receive additional non-hazardous wastes, no later than one year after the date on which the unit received the most recent volume of non-hazardous wastes. If the owner or operator can demonstrate to the Regional Administrator that the hazardous waste management unit has the capacity to receive additional non-hazardous wastes and he has taken, and will continue to take, all steps to prevent threats to human health and the environment, including compliance with all applicable permit requirements, the Regional Administrator may approve an extension to this one-year limit.

(3) If the facility's permit is terminated, or if the facility is otherwise ordered, by judicial decree or final order under section 3008 of RCRA, to cease receiving hazardous wastes or to close, then the requirements of this paragraph do not apply. However, the owner or operator must close the facility in accordance with the deadlines established in §264.113.

(e) *Removal of wastes and decontamination or dismantling of equipment.* Nothing in this section shall preclude the owner or operator from removing hazardous wastes and decontaminating or dismantling equipment in accordance with the approved partial or final closure plan at any time before or after notification of partial or final closure.

[51 FR 16444, May 2, 1986, as amended at 52 FR 46963, Dec. 10, 1987; 53 FR 37935, Sept. 28, 1988; 54 FR 33394, Aug. 14, 1989; 56 FR 7207, Feb. 21, 1991; 57 FR 37265, Aug. 18, 1992; 63 FR 56733, Oct. 22, 1998; 71 FR 40272, July 14, 2006]

### §264.113 Closure; time allowed for closure.

(a) Within 90 days after receiving the final volume of hazardous wastes, or the final volume of non-hazardous wastes if the owner or operator complies with all applicable requirements in paragraphs (d) and (e) of this section, at a hazardous waste management unit or facility, the owner or operator must treat, remove from the unit or facility, or dispose of on-site, all hazardous wastes in accordance with the approved closure plan. The Regional Administrator may approve a longer period if the owner or operator complies with all applicable requirements for requesting a modification to the permit and demonstrates that:

(1)(i) The activities required to comply with this paragraph will, of necessity, take longer than 90 days to complete; or

(ii)(A) The hazardous waste management unit or facility has the capacity to receive additional hazardous wastes, or has the capacity to receive non-hazardous wastes if the owner or operator complies with paragraphs (d) and (e) of this section; and

(B) There is a reasonable likelihood that he or another person will recommence operation of the hazardous waste management unit or the facility within one year; and

(C) Closure of the hazardous waste management unit or facility would be incompatible with continued operation of the site; and

(2) He has taken and will continue to take all steps to prevent threats to human health and the environment, including compliance with all applicable permit requirements.

(b) The owner or operator must complete partial and final closure activities in accordance with the approved closure plan and within 180 days after receiving the final volume of hazardous wastes, or the final volume of non-hazardous wastes if the owner or operator complies with all applicable requirements in paragraphs (d) and (e) of this section, at the hazardous waste management unit or facility. The Regional Administrator may approve an extension to the closure period if the owner or operator complies with all applicable requirements for requesting a modification to the permit and demonstrates that:

(1)(i) The partial or final closure activities will, of necessity, take longer than 180 days to complete; or

(ii)(A) The hazardous waste management unit or facility has the capacity to receive additional hazardous wastes, or has the capacity to receive non-hazardous wastes if the owner or operator complies with paragraphs (d) and (e) of this section; and

(B) There is reasonable likelihood that he or another person will recommence operation of the hazardous waste management unit or the facility within one year; and

(C) Closure of the hazardous waste management unit or facility would be incompatible with continued operation of the site; and

(2) He has taken and will continue to take all steps to prevent threats to human health and the environment from the unclosed but not operating hazardous waste management unit or facility, including compliance with all applicable permit requirements.

(c) The demonstrations referred to in paragraphs (a)(1) and (b)(1) of this section must be made as follows:

(1) The demonstrations in paragraph (a)(1) of this section must be made at least 30 days prior to the expiration of the 90-day period in paragraph (a) of this section; and

(2) The demonstration in paragraph (b)(1) of this section must be made at least 30 days prior to the expiration of the 180-day period in paragraph (b) of this section, unless the owner or operator is otherwise subject to the deadlines in paragraph (d) of this section.

(d) The Regional Administrator may allow an owner or operator to receive only non-hazardous wastes in a landfill, land treatment, or surface impoundment unit after the final receipt of hazardous wastes at that unit if:

(1) The owner or operator requests a permit modification in compliance with all applicable requirements in parts 270 and 124 of this title and in the permit modification request demonstrates that:

(i) The unit has the existing design capacity as indicated on the part A application to receive non-hazardous wastes; and

(ii) There is a reasonable likelihood that the owner or operator or another person will receive non-hazardous wastes in the unit within one year after the final receipt of hazardous wastes; and

(iii) The non-hazardous wastes will not be incompatible with any remaining wastes in the unit, or with the facility design and operating requirements of the unit or facility under this part; and

(iv) Closure of the hazardous waste management unit would be incompatible with continued operation of the unit or facility; and

(v) The owner or operator is operating and will continue to operate in compliance with all applicable permit requirements; and

(2) The request to modify the permit includes an amended waste analysis plan, ground-water monitoring and response program, human exposure assessment required under RCRA section 3019, and closure and post-closure plans, and updated cost estimates and demonstrations of financial assurance for closure and post-closure care as necessary and appropriate, to reflect any changes due to the presence of hazardous constituents in the non-hazardous wastes, and changes in closure activities, including the expected year of closure if applicable under § 264.112(b)(7), as a result of the receipt of non-hazardous wastes following the final receipt of hazardous wastes; and

(3) The request to modify the permit includes revisions, as necessary and appropriate, to affected conditions of the permit to account for the receipt of non-hazardous wastes following receipt of the final volume of hazardous wastes; and

(4) The request to modify the permit and the demonstrations referred to in paragraphs (d)(1) and (d)(2) of this section are submitted to the Regional Administrator no later than 120 days prior to the date on which the owner or operator of the facility receives the known final volume of hazardous wastes at the unit, or no later than 90 days after the effective date of this rule in the state in which the unit is located, whichever is later.

(e) In addition to the requirements in paragraph (d) of this section, an owner or operator of a hazardous waste surface impoundment that is not in compliance with the liner and leachate collection system requirements in 42 U.S.C. 3004(o)(1) and 3005(j)(1) or 42 U.S.C. 3004(o) (2) or (3) or 3005(j) (2), (3), (4) or (13) must:

(1) Submit with the request to modify the permit:

(i) A contingent corrective measures plan, unless a corrective action plan has already been submitted under § 264.99; and

(ii) A plan for removing hazardous wastes in compliance with paragraph (e)(2) of this section; and

(2) Remove all hazardous wastes from the unit by removing all hazardous liquids, and removing all hazardous sludges to the extent practicable without impairing the integrity of the liner(s), if any.

(3) Removal of hazardous wastes must be completed no later than 90 days after the final receipt of hazardous wastes. The Regional Administrator may approve an extension to this deadline if the owner or operator

demonstrates that the removal of hazardous wastes will, of necessity, take longer than the allotted period to complete and that an extension will not pose a threat to human health and the environment.

(4) If a release that is a statistically significant increase (or decrease in the case of pH) over background values for detection monitoring parameters or constituents specified in the permit or that exceeds the facility's ground-water protection standard at the point of compliance, if applicable, is detected in accordance with the requirements in subpart F of this part, the owner or operator of the unit:

(i) Must implement corrective measures in accordance with the approved contingent corrective measures plan required by paragraph (e)(1) of this section no later than one year after detection of the release, or approval of the contingent corrective measures plan, whichever is later;

(ii) May continue to receive wastes at the unit following detection of the release only if the approved corrective measures plan includes a demonstration that continued receipt of wastes will not impede corrective action; and

(iii) May be required by the Regional Administrator to implement corrective measures in less than one year or to cease the receipt of wastes until corrective measures have been implemented if necessary to protect human health and the environment.

(5) During the period of corrective action, the owner or operator shall provide annual reports to the Regional Administrator describing the progress of the corrective action program, compile all ground-water monitoring data, and evaluate the effect of the continued receipt of non-hazardous wastes on the effectiveness of the corrective action.

(6) The Regional Administrator may require the owner or operator to commence closure of the unit if the owner or operator fails to implement corrective action measures in accordance with the approved contingent corrective measures plan within one year as required in paragraph (e)(4) of this section, or fails to make substantial progress in implementing corrective action and achieving the facility's ground-water protection standard or background levels if the facility has not yet established a ground-water protection standard.

(7) If the owner or operator fails to implement corrective measures as required in paragraph (e)(4) of this section, or if the Regional Administrator determines that substantial progress has not been made pursuant to paragraph (e)(6) of this section he shall:

(i) Notify the owner or operator in writing that the owner or operator must begin closure in accordance with the deadlines in paragraphs (a) and (b) of this section and provide a detailed statement of reasons for this determination, and

(ii) Provide the owner or operator and the public, through a newspaper notice, the opportunity to submit written comments on the decision no later than 20 days after the date of the notice.

(iii) If the Regional Administrator receives no written comments, the decision will become final five days after the close of the comment period. The Regional Administrator will notify the owner or operator that the decision is final, and that a revised closure plan, if necessary, must be submitted within 15 days of the final notice and that closure must begin in accordance with the deadlines in paragraphs (a) and (b) of this section.

(iv) If the Regional Administrator receives written comments on the decision, he shall make a final decision within 30 days after the end of the comment period, and provide the owner or operator in writing and the public through a newspaper notice, a detailed statement of reasons for the final decision. If the Regional Administrator determines that substantial progress has not been made, closure must be initiated in accordance with the deadlines in paragraphs (a) and (b) of this section.

(v) The final determinations made by the Regional Administrator under paragraphs (e)(7) (iii) and (iv) of this section are not subject to administrative appeal.

[51 FR 16444, May 2, 1986, as amended at 54 FR 33394, Aug. 14, 1989; 71 FR 16904, Apr. 4, 2006]

## § 264.114 Disposal or decontamination of equipment, structures and soils.

During the partial and final closure periods, all contaminated equipment, structures and soils must be properly disposed of or decontaminated unless otherwise specified in §§ 264.197, 264.228, 264.258, 264.280 or § 264.310. By removing any hazardous wastes or hazardous constituents during partial and final closure, the owner or operator may become a generator of hazardous waste and must handle that waste in accordance with all applicable requirements of part 262 of this chapter.

[51 FR 16444, May 2, 1986, as amended at 52 FR 46963, Dec. 10, 1987; 53 FR 34086, Sept. 2, 1988]

## § 264.115 Certification of closure.

Within 60 days of completion of closure of each hazardous waste surface impoundment, waste pile, land treatment, and landfill unit, and within 60 days of the completion of final closure, the owner or operator must submit to the Regional Administrator, by registered mail, a certification that the hazardous waste management unit or facility, as applicable, has been closed in accordance with the specifications in the approved closure plan. The certification must be signed by the owner or operator and by a qualified Professional Engineer. Documentation supporting the Professional Engineer's certification must be furnished to the Regional Administrator upon request until he releases the owner or operator from the financial assurance requirements for closure under § 264.143(i).

[71 FR 16904, Apr. 4, 2006, as amended at 71 FR 40272, July 14, 2006]

## § 264.116 Survey plat.

No later than the submission of the certification of closure of each hazardous waste disposal unit, the owner or operator must submit to the local zoning authority, or the authority with jurisdiction over local land use, and to the Regional Administrator, a survey plat indicating the location and dimensions of landfill cells or other hazardous waste disposal units with respect to permanently surveyed benchmarks. This plat must be prepared and certified by a professional land surveyor. The plat filed with the local zoning authority, or the authority with jurisdiction over local land use, must contain a note, prominently displayed, which states the owner's or operator's obligation to restrict disturbance of the hazardous waste disposal unit in accordance with the applicable subpart G regulations.

[51 FR 16444, May 2, 1986, as amended at 71 FR 40272, July 14, 2006]

## § 264.117 Post-closure care and use of property.

(a)(1) Post-closure care for each hazardous waste management unit subject to the requirements of §§ 264.117 through 264.120 must begin after completion of closure of the unit and continue for 30 years after that date and must consist of at least the following:

(i) Monitoring and reporting in accordance with the requirements of subparts F, K, L, M, N, and X of this part; and

(ii) Maintenance and monitoring of waste containment systems in accordance with the requirements of subparts F, K, L, M, N, and X of this part.

(2) Any time preceding partial closure of a hazardous waste management unit subject to post-closure care requirements or final closure, or any time during the post-closure period for a particular unit, the Regional Administrator may, in accordance with the permit modification procedures in parts 124 and 270:

(i) Shorten the post-closure care period applicable to the hazardous waste management unit, or facility, if all disposal units have been closed, if he finds that the reduced period is sufficient to protect human health and the environment (e.g., leachate or ground-water monitoring results, characteristics of the hazardous wastes, application of advanced technology, or alternative disposal, treatment, or re-use techniques indicate that the hazardous waste management unit or facility is secure); or

(ii) Extend the post-closure care period applicable to the hazardous waste management unit or facility if he finds that the extended period is necessary to protect human health and the environment (e.g., leachate or ground-water monitoring results indicate a potential for migration of hazardous wastes at levels which may be harmful to human health and the environment).

(b) The Regional Administrator may require, at partial and final closure, continuation of any of the security requirements of §264.14 during part or all of the post-closure period when:

(1) Hazardous wastes may remain exposed after completion of partial or final closure; or

(2) Access by the public or domestic livestock may pose a hazard to human health.

(c) Post-closure use of property on or in which hazardous wastes remain after partial or final closure must never be allowed to disturb the integrity of the final cover, liner(s), or any other components of the containment system, or the function of the facility's monitoring systems, unless the Regional Administrator finds that the disturbance:

(1) Is necessary to the proposed use of the property, and will not increase the potential hazard to human health or the environment; or

(2) Is necessary to reduce a threat to human health or the environment.

(d) All post-closure care activities must be in accordance with the provisions of the approved post-closure plan as specified in §264.118.

[51 FR 16444, May 2, 1986, as amended at 52 FR 46963, Dec. 10, 1987]

### §264.118 Post-closure plan; amendment of plan.

(a) *Written Plan.* The owner or operator of a hazardous waste disposal unit must have a written post-closure plan. In addition, certain surface impoundments and waste piles from which the owner or operator intends to remove or decontaminate the hazardous wastes at partial or final closure are required by §§264.228(c)(1)(ii) and 264.258(c)(1)(ii) to have contingent post-closure plans. Owners or operators of surface impoundments and waste piles not otherwise required to prepare contingent post-closure plans under §§264.228(c)(1)(ii) and 264.258(c)(1)(ii) must submit a post-closure plan to the Regional Administrator within 90 days from the date that the owner or operator or Regional administrator determines that the hazardous waste management unit must be closed as a landfill, subject to the requirements of §§264.117 through 264.120. The plan must be submitted with the permit application, in accordance with §270.14(b)(13) of this chapter, and approved by the Regional Administrator as part of the permit issuance procedures under part 124 of this chapter. In accordance with §270.32 of this chapter, the approved post-closure plan will become a condition of any RCRA permit issued.

(b) For each hazardous waste management unit subject to the requirements of this section, the post-closure plan must identify the activities that will be carried on after closure of each disposal unit and the frequency of these activities, and include at least:

(1) A description of the planned monitoring activities and frequencies at which they will be performed to comply with subparts F, K, L, M, N, and X of this part during the post-closure care period; and

(2) A description of the planned maintenance activities, and frequencies at which they will be performed, to ensure:

(i) The integrity of the cap and final cover or other containment systems in accordance with the requirements of subparts F, K, L, M, N, and X of this part; and

(ii) The function of the monitoring equipment in accordance with the requirements of subparts F, K, L, M, N, and X of this part; and

(3) The name, address, and phone number of the person or office to contact about the hazardous waste disposal unit or facility during the post-closure care period.

(4) For facilities where the Regional Administrator has applied alternative requirements at a regulated unit under §§264.90(f), 264.110(c), and/or §§264.140(d), either the alternative requirements that apply to the regulated unit, or a reference to the enforceable document containing those requirements.

(c) Until final closure of the facility, a copy of the approved post-closure plan must be furnished to the Regional Administrator upon request, including request by mail. After final closure has been certified, the person or office specified in §264.118(b)(3) must keep the approved post-closure plan during the remainder of the post-closure period.

(d) *Amendment of plan.* The owner or operator must submit a written notification of or request for a permit modification to authorize a change in the approved post-closure plan in accordance with the applicable requirements in parts 124 and 270. The written notification or request must include a copy of the amended post-closure plan for review or approval by the Regional Administrator.

(1) The owner or operator may submit a written notification or request to the Regional Administrator for a permit modification to amend the post-closure plan at any time during the active life of the facility or during the post-closure care period.

(2) The owner or operator must submit a written notification of or request for a permit modification to authorize a change in the approved post-closure plan whenever:

(i) Changes in operating plans or facility design affect the approved post-closure plan, or

(ii) There is a change in the expected year of final closure, if applicable, or

(iii) Events which occur during the active life of the facility, including partial and final closures, affect the approved post-closure plan.

(iv) The owner or operator requests the Regional Administrator to apply alternative requirements to a regulated unit under §§ 264.90(f), 264.110(c), and/or § 264.140(d).

(3) The owner or operator must submit a written request for a permit modification at least 60 days prior to the proposed change in facility design or operation, or no later than 60 days after an unexpected event has occurred which has affected the post-closure plan. An owner or operator of a surface impoundment or waste pile that intends to remove all hazardous waste at closure and is not otherwise required to submit a contingent post-closure plan under §§ 264.228(c)(1)(ii) and 264.258(c)(1)(ii) must submit a post-closure plan to the Regional Administrator no later than 90 days after the date that the owner or operator or Regional Administrator determines that the hazardous waste management unit must be closed as a landfill, subject to

the requirements of § 264.310. The Regional Administrator will approve, disapprove or modify this plan in accordance with the procedures in parts 124 and 270. In accordance with § 270.32 of this chapter, the approved post-closure plan will become a permit condition.

(4) The Regional Administrator may request modifications to the plan under the conditions described in § 264.118(d)(2). The owner or operator must submit the modified plan no later than 60 days after the Regional Administrator's request, or no later than 90 days if the unit is a surface impoundment or waste pile not previously required to prepare a contingent post-closure plan. Any modifications requested by the Regional Administrator will be approved, disapproved, or modified in accordance with the procedures in parts 124 and 270.

[51 FR 16444, May 2, 1986, as amended at 52 FR 46964, Dec. 10, 1987; 53 FR 37935, Sept. 28, 1988; 63 FR 56733, Oct. 22, 1998; 71 FR 40272, July 14, 2006]

## § 264.119   Post-closure notices.

(a) No later than 60 days after certification of closure of each hazardous waste disposal unit, the owner or operator must submit to the local zoning authority, or the authority with jurisdiction over local land use, and to the Regional Administrator a record of the type, location, and quantity of hazardous wastes disposed of within each cell or other disposal unit of the facility. For hazardous wastes disposed of before January 12, 1981, the owner or operator must identify the type, location, and quantity of the hazardous wastes to the best of his knowledge and in accordance with any records he has kept.

(b) Within 60 days of certification of closure of the first hazardous waste disposal unit and within 60 days of certification of closure of the last hazardous waste disposal unit, the owner or operator must:

(1) Record, in accordance with State law, a notation on the deed to the facility property—or on some other instrument which is normally examined during title search—that will in perpetuity notify any potential purchaser of the property that:

(i) The land has been used to manage hazardous wastes; and

(ii) Its use is restricted under 40 CFR part 264, subpart G regulations; and

(iii) The survey plat and record of the type, location, and quantity of hazardous wastes disposed of within each cell or other hazardous waste disposal unit of the facility required by §§264.116 and 264.119(a) have been filed with the local zoning authority or the authority with jurisdiction over local land use and with the Regional Administrator; and

(2) Submit a certification, signed by the owner or operator, that he has recorded the notation specified in paragraph (b)(1) of this section, including a copy of the document in which the notation has been placed, to the Regional Administrator.

(c) If the owner or operator or any subsequent owner or operator of the land upon which a hazardous waste disposal unit is located wishes to remove hazardous wastes and hazardous waste residues, the liner, if any, or contaminated soils, he must request a modification to the post-closure permit in accordance with the applicable requirements in parts 124 and 270. The owner or operator must demonstrate that the removal of hazardous wastes will satisfy the criteria of §264.117(c). By removing hazardous waste, the owner or operator may become a generator of hazardous waste and must manage it in accordance with all applicable requirements of this chapter. If he is granted a permit modification or otherwise granted approval to conduct such removal activities, the owner or operator may request that the Regional Administrator approve either:

(1) The removal of the notation on the deed to the facility property or other instrument normally examined during title search; or

(2) The addition of a notation to the deed or instrument indicating the removal of the hazardous waste.

[51 FR 16444, May 2, 1986, as amended at 71 FR 40272, July 14, 2006]

### §264.120  Certification of completion of post-closure care.

No later than 60 days after completion of the established post-closure care period for each hazardous waste disposal unit, the owner or operator must submit to the Regional Administrator, by registered mail, a certification that the post-closure care period for the hazardous waste disposal unit was performed in accordance with the specifications in the approved post-closure plan. The certification must be signed by the owner or operator and a qualified Professional Engineer. Documentation supporting the Professional Engineer's certification must be furnished to the Regional Administrator upon request until he releases the owner or operator from the financial assurance requirements for post-closure care under §264.145(i).

[71 FR 16904, Apr. 4, 2006]

## Subpart H—Financial Requirements

SOURCE: 47 FR 15047, Apr. 7, 1982, unless otherwise noted.

### §264.140  Applicability.

(a) The requirements of §§264.142, 264.143, and 264.147 through 264.151 apply to owners and operators of all hazardous waste facilities, except as provided otherwise in this section or in §264.1.

(b) The requirements of §§264.144 and 264.145 apply only to owners and operators of:

(1) Disposal facilities;

(2) Piles, and surface impoundments from which the owner or operator intends to remove the wastes at closure, to the extent that these sections are made applicable to such facilities in §§264.228 and 264.258;

(3) Tank systems that are required under §264.197 to meet the requirements for landfills; and

(4) Containment buildings that are required under §264.1102 to meet the requirements for landfills.

(c) States and the Federal government are exempt from the requirements of this subpart.

(d) The Regional Administrator may replace all or part of the requirements of this subpart applying to a regulated unit with alternative requirements for financial assurance set out in the permit or in an enforceable document (as defined in 40 CFR 270.1(c)(7)), where the Regional Administrator:

(1) Prescribes alternative requirements for the regulated unit under § 264.90(f) and/or § 264.110(c); and

(2) Determines that it is not necessary to apply the requirements of this subpart because the alternative financial assurance requirements will protect human health and the environment.

[47 FR 15047, Apr. 7, 1982, as amended at 47 FR 32357, July 26, 1982; 51 FR 25472, July 14, 1986; 57 FR 37265, Aug. 18, 1992; 63 FR 56733, Oct. 22, 1998; 71 FR 40272, July 14, 2006]

### § 264.141 Definitions of terms as used in this subpart.

(a) *Closure plan* means the plan for closure prepared in accordance with the requirements of § 264.112.

(b) *Current closure cost estimate* means the most recent of the estimates prepared in accordance with § 264.142 (a), (b), and (c).

(c) *Current post-closure cost estimate* means the most recent of the estimates prepared in accordance with § 264.144 (a), (b), and (c).

(d) *Parent corporation* means a corporation which directly owns at least 50 percent of the voting stock of the corporation which is the facility owner or operator; the latter corporation is deemed a "subsidiary" of the parent corporation.

(e) *Post-closure plan* means the plan for post-closure care prepared in accordance with the requirements of §§ 264.117 through 264.120.

(f) The following terms are used in the specifications for the financial tests for closure, post-closure care, and liability coverage. The definitions are intended to assist in the understanding of these regulations and are not intended to limit the meanings of terms in a way that conflicts with generally accepted accounting practices.

*Assets* means all existing and all probable future economic benefits obtained or controlled by a particular entity.

*Current assets* means cash or other assets or resources commonly identified as those which are reasonably expected to be realized in cash or sold or consumed during the normal operating cycle of the business.

*Current liabilities* means obligations whose liquidation is reasonably expected to require the use of existing resources properly classifiable as current assets or the creation of other current liabilities.

*Current plugging and abandonment cost estimate* means the most recent of the estimates prepared in accordance with § 144.62(a), (b), and (c) of this title.

*Independently audited* refers to an audit performed by an independent certified public accountant in accordance with generally accepted auditing standards.

*Liabilities* means probable future sacrifices of economic benefits arising from present obligations to transfer assets or provide services to other entities in the future as a result of past transactions or events.

*Net working capital* means current assets minus current liabilities.

*Net worth* means total assets minus total liabilities and is equivalent to owner's equity.

*Tangible net worth* means the tangible assets that remain after deducting liabilities; such assets would not include intangibles such as goodwill and rights to patents or royalties.

(g) In the liability insurance requirements the terms *bodily injury* and *property damage* shall have the meanings given these terms by applicable State law. However, these terms do not include those liabilities which, consistent with standard industry practices, are excluded from coverage in liability policies for bodily injury and property damage. The Agency intends the meanings of other terms used in the liability insurance requirements to be consistent with their common meanings within the insurance industry. The definitions given below of several of the terms are intended to assist in the understanding of these regulations and are not intended to limit their meanings in a way that conflicts with general insurance industry usage.

*Accidental occurrence* means an accident, including continuous or repeated exposure to conditions, which results in bodily injury or property damage neither expected nor intended from the standpoint of the insured.

*Legal defense costs* means any expenses that an insurer incurs in defending against claims of third parties brought under the terms and conditions of an insurance policy.

*Nonsudden accidental occurrence* means an occurrence which takes place over time and involves continuous or repeated exposure.

*Sudden accidental occurrence* means an occurrence which is not continuous or repeated in nature.

(h) *Substantial business relationship* means the extent of a business relationship necessary under applicable State law to make a guarantee contract issued incident to that relationship valid and enforceable. A "substantial business relationship" must arise from a pattern of recent or ongoing business transactions, in addition to the guarantee itself, such that a currently existing business relationship between the guarantor and the owner or operator is demonstrated to the satisfaction of the applicable EPA Regional Administrator.

[47 FR 16554, Apr. 16, 1982, as amended at 51 FR 16447, May 2, 1986; 53 FR 33950, Sept. 1, 1988]

### §264.142  Cost estimate for closure.

(a) The owner or operator must have a detailed written estimate, in current dollars, of the cost of closing the facility in accordance with the requirements in §§264.111 through 264.115 and applicable closure requirements in §§264.178, 264.197, 264.228, 264.258, 264.280, 264.310, 264.351, 264.601 through 264.603, and 264.1102.

(1) The estimate must equal the cost of final closure at the point in the facility's active life when the extent and manner of its operation would make closure the most expensive, as indicated by its closure plan (see §264.112(b)); and

(2) The closure cost estimate must be based on the costs to the owner or operator of hiring a third party to close the facility. A third party is a party who is neither a parent nor a subsidiary of the owner or operator. (See definition of parent corporation in §264.141(d).) The owner or operator may use costs for on-site disposal if he can demonstrate that on-site disposal capacity will exist at all times over the life of the facility.

(3) The closure cost estimate may not incorporate any salvage value that may be realized with the sale of hazardous wastes, or non-hazardous wastes if applicable under §264.113(d), facility structures or equipment, land, or other assets associated with the facility at the time of partial or final closure.

(4) The owner or operator may not incorporate a zero cost for hazardous wastes, or non-hazardous wastes if applicable under §264.113(d), that might have economic value.

(b) During the active life of the facility, the owner or operator must adjust the closure cost estimate for inflation within 60 days prior to the anniversary date of the establishment of the financial instrument(s) used to comply with §264.143. For owners and operators using the financial test or corporate guarantee, the closure cost estimate must be updated for inflation within 30 days after the close of the firm's fiscal year and before submission of updated information to the Regional Administrator as specified in §264.143(f)(3). The adjustment may be made by recalculating the maximum costs of closure in current dollars, or by using an inflation factor derived from the most recent Implicit Price Deflator for Gross National Product published by the U.S. Department of Commerce in its *Survey of Current Business*, as specified in paragraphs (b)(1) and (2) of this section. The inflation factor is the result of dividing the latest published annual Deflator by the Deflator for the previous year.

(1) The first adjustment is made by multiplying the closure cost estimate by the inflation factor. The result is the adjusted closure cost estimate.

(2) Subsequent adjustments are made by multiplying the latest adjusted closure cost estimate by the latest inflation factor.

(c) During the active life of the facility, the owner or operator must revise the closure cost estimate no later than 30 days after the Regional Administrator has approved the request to modify the closure plan, if the change in the closure plan increases the cost of closure. The revised closure cost estimate must be adjusted for inflation as specified in §264.142(b).

(d) The owner or operator must keep the following at the facility during the operating life of the facility: The latest closure cost estimate prepared in accordance with § 264.142 (a) and (c) and, when this estimate has been adjusted in accordance with § 264.142(b), the latest adjusted closure cost estimate.

[47 FR 15047, Apr. 7, 1982, as amended at 50 FR 4514, Jan. 31, 1985; 51 FR 16447, May 2, 1986; 52 FR 46964, Dec. 10, 1987; 54 FR 33395, Aug. 14, 1989; 57 FR 37265, Aug. 18, 1992; 71 FR 40272, July 14, 2006]

### § 264.143  Financial assurance for closure.

An owner or operator of each facility must establish financial assurance for closure of the facility. He must choose from the options as specified in paragraphs (a) through (f) of this section.

(a) *Closure trust fund.* (1) An owner or operator may satisfy the requirements of this section by establishing a closure trust fund which conforms to the requirements of this paragraph and submitting an originally signed duplicate of the trust agreement to the Regional Administrator. An owner or operator of a new facility must submit the originally signed duplicate of the trust agreement to the Regional Administrator at least 60 days before the date on which hazardous waste is first received for treatment, storage, or disposal. The trustee must be an entity which has the authority to act as a trustee and whose trust operations are regulated and examined by a Federal or State agency.

(2) The wording of the trust agreement must be identical to the wording specified in § 264.151(a)(1), and the trust agreement must be accompanied by a formal certification of acknowledgment (for example, see § 264.151(a)(2)). Schedule A of the trust agreement must be updated within 60 days after a change in the amount of the current closure cost estimate covered by the agreement.

(3) Payments into the trust fund must be made annually by the owner or operator over the term of the initial RCRA permit or over the remaining operating life of the facility as estimated in the closure plan, whichever period is shorter; this period is hereafter referred to as the "pay-in period." The

payments into the closure trust fund must be made as follows:

(i) For a new facility, the first payment must be made before the initial receipt of hazardous waste for treatment, storage, or disposal. A receipt from the trustee for this payment must be submitted by the owner or operator to the Regional Administrator before this initial receipt of hazardous waste. The first payment must be at least equal to the current closure cost estimate, except as provided in § 264.143(g), divided by the number of years in the pay-in period. Subsequent payments must be made no later than 30 days after each anniversary date of the first payment. The amount of each subsequent payment must be determined by this formula:

$$\text{Next payment} = \frac{CE - CV}{Y}$$

where CE is the current closure cost estimate, CV is the current value of the trust fund, and Y is the number of years remaining in the pay-in period.

(ii) If an owner or operator establishes a trust fund as specified in § 265.143(a) of this chapter, and the value of that trust fund is less than the current closure cost estimate when a permit is awarded for the facility, the amount of the current closure cost estimate still to be paid into the trust fund must be paid in over the pay-in period as defined in paragraph (a)(3) of this section. Payments must continue to be made no later than 30 days after each anniversary date of the first payment made pursuant to part 265 of this chapter. The amount of each payment must be determined by this formula:

$$\text{Next payment} = \frac{CE - CV}{Y}$$

where CE is the current closure cost estimate, CV is the current value of the trust fund, and Y is the number of years remaining in the pay-in period.

(4) The owner or operator may accelerate payments into the trust fund or he may deposit the full amount of the current closure cost estimate at the time the fund is established. However, he must maintain the value of the fund at no less than the value that the fund

would have if annual payments were made as specified in paragraph (a)(3) of this section.

(5) If the owner or operator establishes a closure trust fund after having used one or more alternate mechanisms specified in this section or in §265.143 of this chapter, his first payment must be in at least the amount that the fund would contain if the trust fund were established initially and annual payments made according to specifications of this paragraph and §265.143(a) of this chapter, as applicable.

(6) After the pay-in period is completed, whenever the current closure cost estimate changes, the owner or operator must compare the new estimate with the trustee's most recent annual valuation of the trust fund. If the value of the fund is less than the amount of the new estimate, the owner or operator, within 60 days after the change in the cost estimate, must either deposit an amount into the fund so that its value after this deposit at least equals the amount of the current closure cost estimate, or obtain other financial assurance as specified in this section to cover the difference.

(7) If the value of the trust fund is greater than the total amount of the current closure cost estimate, the owner or operator may submit a written request to the Regional Administrator for release of the amount in excess of the current closure cost estimate.

(8) If an owner or operator substitutes other financial assurance as specified in this section for all or part of the trust fund, he may submit a written request to the Regional Administrator for release of the amount in excess of the current closure cost estimate covered by the trust fund.

(9) Within 60 days after receiving a request from the owner or operator for release of funds as specified in paragraph (a) (7) or (8) of this section, the Regional Administrator will instruct the trustee to release to the owner or operator such funds as the Regional Administrator specifies in writing.

(10) After beginning partial or final closure, an owner or operator or another person authorized to conduct partial or final closure may request reimbursements for partial or final closure expenditures by submitting itemized bills to the Regional Administrator. The owner or operator may request reimbursements for partial closure only if sufficient funds are remaining in the trust fund to cover the maximum costs of closing the facility over its remaining operating life. Within 60 days after receiving bills for partial or final closure activities, the Regional Administrator will instruct the trustee to make reimbursements in those amounts as the Regional Administrator specifies in writing, if the Regional Administrator determines that the partial or final closure expenditures are in accordance with the approved closure plan, or otherwise justified. If the Regional Administrator has reason to believe that the maximum cost of closure over the remaining life of the facility will be significantly greater than the value of the trust fund, he may withhold reimbursements of such amounts as he deems prudent until he determines, in accordance with §264.143(i) that the owner or operator is no longer required to maintain financial assurance for final closure of the facility. If the Regional Administrator does not instruct the trustee to make such reimbursements, he will provide the owner or operator with a detailed written statement of reasons.

(11) The Regional Administrator will agree to termination of the trust when:

(i) An owner or operator substitutes alternate financial assurance as specified in this section; or

(ii) The Regional Administrator releases the owner or operator from the requirements of this section in accordance with §264.143(i).

(b) *Surety bond guaranteeing payment into a closure trust fund.* (1) An owner or operator may satisfy the requirements of this section by obtaining a surety bond which conforms to the requirements of this paragraph and submitting the bond to the Regional Administrator. An owner or operator of a new facility must submit the bond to the Regional Administrator at least 60 days before the date on which hazardous waste is first received for treatment, storage, or disposal. The bond must be effective before this initial receipt of hazardous waste. The surety company issuing the bond must, at a minimum, be among those listed as acceptable sureties on Federal bonds in Circular 570 of the U.S. Department of the Treasury.

(2) The wording of the surety bond must be identical to the wording specified in § 264.151(b).

(3) The owner or operator who uses a surety bond to satisfy the requirements of this section must also establish a standby trust fund. Under the terms of the bond, all payments made thereunder will be deposited by the surety directly into the standby trust fund in accordance with instructions from the Regional Administrator. This standby trust fund must meet the requirements specified in § 264.143(a), except that:

(i) An originally signed duplicate of the trust agreement must be submitted to the Regional Administrator with the surety bond; and

(ii) Until the standby trust fund is funded pursuant to the requirements of this section, the following are not required by these regulations:

(A) Payments into the trust fund as specified in § 264.143(a);

(B) Updating of Schedule A of the trust agreement (see § 264.151(a)) to show current closure cost estimates;

(C) Annual valuations as required by the trust agreement; and

(D) Notices of nonpayment as required by the trust agreement.

(4) The bond must guarantee that the owner or operator will:

(i) Fund the standby trust fund in an amount equal to the penal sum of the bond before the beginning of final closure of the facility; or

(ii) Fund the standby trust fund in an amount equal to the penal sum within 15 days after an administrative order to begin final closure issued by the Regional Administrator becomes final, or within 15 days after an order to begin final closure is issued by a U.S. district court or other court of competent jurisdiction; or

(iii) Provide alternate financial assurance as specified in this section, and obtain the Regional Administrator's written approval of the assurance provided, within 90 days after receipt by both the owner or operator and the Regional Administrator of a notice of cancellation of the bond from the surety.

(5) Under the terms of the bond, the surety will become liable on the bond obligation when the owner or operator fails to perform as guaranteed by the bond.

(6) The penal sum of the bond must be in an amount at least equal to the current closure cost estimate, except as provided in § 264.143(g).

(7) Whenever the current closure cost estimate increases to an amount greater than the penal sum, the owner or operator, within 60 days after the increase, must either cause the penal sum to be increased to an amount at least equal to the current closure cost estimate and submit evidence of such increase to the Regional Administrator, or obtain other financial assurance as specified in this section to cover the increase. Whenever the current closure cost estimate decreases, the penal sum may be reduced to the amount of the current closure cost estimate following written approval by the Regional Administrator.

(8) Under the terms of the bond, the surety may cancel the bond by sending notice of cancellation by certified mail to the owner or operator and to the Regional Administrator. Cancellation may not occur, however, during the 120 days beginning on the date of receipt of the notice of cancellation by both the owner or operator and the Regional Administrator, as evidenced by the return receipts.

(9) The owner or operator may cancel the bond if the Regional Administrator has given prior written consent based on his receipt of evidence of alternate financial assurance as specified in this section.

(c) *Surety bond guaranteeing performance of closure.* (1) An owner or operator may satisfy the requirements of this section by obtaining a surety bond which conforms to the requirements of this paragraph and submitting the bond to the Regional Administrator. An owner or operator of a new facility must submit the bond to the Regional Administrator at least 60 days before the date on which hazardous waste is first received for treatment, storage, or disposal. The bond must be effective before this initial receipt of hazardous waste. The surety company issuing the bond must, at a minimum, be among those listed as acceptable sureties on Federal bonds in Circular 570 of the U.S. Department of the Treasury.

(2) The wording of the surety bond must be identical to the wording specified in §264.151(c).

(3) The owner or operator who uses a surety bond to satisfy the requirements of this section must also establish a standby trust fund. Under the terms of the bond, all payments made thereunder will be deposited by the surety directly into the standby trust fund in accordance with instructions from the Regional Administrator. This standby trust must meet the requirements specified in §264.143(a), except that:

(i) An originally signed duplicate of the trust agreement must be submitted to the Regional Administrator with the surety bond; and

(ii) Unless the standby trust fund is funded pursuant to the requirements of this section, the following are not required by these regulations:

(A) Payments into the trust fund as specified in §264.143(a);

(B) Updating of Schedule A of the trust agreement (see §264.151(a)) to show current closure cost estimates;

(C) Annual valuations as required by the trust agreement; and

(D) Notices of nonpayment as required by the trust agreement.

(4) The bond must guarantee that the owner or operator will:

(i) Perform final closure in accordance with the closure plan and other requirements of the permit for the facility whenever required to do so; or

(ii) Provide alternate financial assurance as specified in this section, and obtain the Regional Administrator's written approval of the assurance provided, within 90 days after receipt by both the owner or operator and the Regional Administrator of a notice of cancellation of the bond from the surety.

(5) Under the terms of the bond, the surety will become liable on the bond obligation when the owner or operator fails to perform as guaranteed by the bond. Following a final administrative determination pursuant to section 3008 of RCRA that the owner or operator has failed to perform final closure in accordance with the approved closure plan and other permit requirements when required to do so, under the terms of the bond the surety will perform final closure as guaranteed by the bond or will deposit the amount of the penal sum into the standby trust fund.

(6) The penal sum of the bond must be in an amount at least equal to the current closure cost estimate.

(7) Whenever the current closure cost estimate increases to an amount greater than the penal sum, the owner or operator, within 60 days after the increase, must either cause the penal sum to be increased to an amount at least equal to the current closure cost estimate and submit evidence of such increase to the Regional Administrator, or obtain other financial assurance as specified in this section. Whenever the current closure cost estimate decreases, the penal sum may be reduced to the amount of the current closure cost estimate following written approval by the Regional Administrator.

(8) Under the terms of the bond, the surety may cancel the bond by sending notice of cancellation by certified mail to the owner or operator and to the Regional Administrator. Cancellation may not occur, however, during the 120 days beginning on the date of receipt of the notice of cancellation by both the owner or operator and the Regional Administrator, as evidenced by the return receipts.

(9) The owner or operator may cancel the bond if the Regional Administrator has given prior written consent. The Regional Administrator will provide such written consent when:

(i) An owner or operator substitutes alternate financial assurance as specified in this section; or

(ii) The Regional Administrator releases the owner or operator from the requirements of this section in accordance with §264.143(i).

(10) The surety will not be liable for deficiencies in the performance of closure by the owner or operator after the Regional Administrator releases the owner or operator from the requirements of this section in accordance with §264.143(i).

(d) *Closure letter of credit.* (1) An owner or operator may satisfy the requirements of this section by obtaining an irrevocable standby letter of credit which conforms to the requirements of this paragraph and submitting the letter to the Regional Administrator. An owner or operator of a new facility must submit the letter of credit to the Regional Administrator at least 60 days before the date on which hazardous waste is first received for treatment, storage, or disposal. The letter of credit must be effective before this initial receipt of hazardous waste. The issuing institution must be an entity which has the authority to issue letters of credit and whose letter-of-credit operations are regulated and examined by a Federal or State agency.

(2) The wording of the letter of credit must be identical to the wording specified in § 264.151(d).

(3) An owner or operator who uses a letter of credit to satisfy the requirements of this section must also establish a standby trust fund. Under the terms of the letter of credit, all amounts paid pursuant to a draft by the Regional Administrator will be deposited by the issuing institution directly into the standby trust fund in accordance with instructions from the Regional Administrator. This standby trust fund must meet the requirements of the trust fund specified in § 264.143(a), except that:

(i) An originally signed duplicate of the trust agreement must be submitted to the Regional Administrator with the letter of credit; and

(ii) Unless the standby trust fund is funded pursuant to the requirements of this section, the following are not required by these regulations:

(A) Payments into the trust fund as specified in § 264.143(a);

(B) Updating of Schedule A of the trust agreement (see § 264.151(a)) to show current closure cost estimates;

(C) Annual valuations as required by the trust agreement; and

(D) Notices of nonpayment as required by the trust agreement.

(4) The letter of credit must be accompanied by a letter from the owner or operator referring to the letter of credit by number, issuing institution, and date, and providing the following information: the EPA Identification Number, name, and address of the facility, and the amount of funds assured for closure of the facility by the letter of credit.

(5) The letter of credit must be irrevocable and issued for a period of at least 1 year. The letter of credit must provide that the expiration date will be automatically extended for a period of at least 1 year unless, at least 120 days before the current expiration date, the issuing institution notifies both the owner or operator and the Regional Administrator by certified mail of a decision not to extend the expiration date. Under the terms of the letter of credit, the 120 days will begin on the date when both the owner or operator and the Regional Administrator have received the notice, as evidenced by the return receipts.

(6) The letter of credit must be issued in an amount at least equal to the current closure cost estimate, except as provided in § 264.143(g).

(7) Whenever the current closure cost estimate increases to an amount greater than the amount of the credit, the owner or operator, within 60 days after the increase, must either cause the amount of the credit to be increased so that it at least equals the current closure cost estimate and submit evidence of such increase to the Regional Administrator, or obtain other financial assurance as specified in this section to cover the increase. Whenever the current closure cost estimate decreases, the amount of the credit may be reduced to the amount of the current closure cost estimate following written approval by the Regional Administrator.

(8) Following a final administrative determination pursuant to section 3008 of RCRA that the owner or operator has failed to perform final closure in accordance with the closure plan and other permit requirements when required to do so, the Regional Administrator may draw on the letter of credit.

(9) If the owner or operator does not establish alternate financial assurance as specified in this section and obtain written approval of such alternate assurance from the Regional Administrator within 90 days after receipt by

both the owner or operator and the Regional Administrator of a notice from issuing institution that it has decided not to extend the letter of credit beyond the current expiration date, the Regional Administrator will draw on the letter of credit. The Regional Administrator may delay the drawing if the issuing institution grants an extension of the term of the credit. During the last 30 days of any such extension the Regional Administrator will draw on the letter of credit if the owner or operator has failed to provide alternate financial assurance as specified in this section and obtain written approval of such assurance from the Regional Administrator.

(10) The Regional Administrator will return the letter of credit to the issuing institution for termination when:

(i) An owner or operator substitutes alternate financial assurance as specified in this section; or

(ii) The Regional Administrator releases the owner or operator from the requirements of this section in accordance with §264.143(i).

(e) *Closure insurance.* (1) An owner or operator may satisfy the requirements of this section by obtaining closure insurance which conforms to the requirements of this paragraph and submitting a certificate of such insurance to the Regional Administrator. An owner or operator of a new facility must submit the certificate of insurance to the Regional Administrator at least 60 days before the date on which hazardous waste is first received for treatment, storage, or disposal. The insurance must be effective before this initial receipt of hazardous waste. At a minimum, the insurer must be licensed to transact the business of insurance, or eligible to provide insurance as an excess or surplus lines insurer, in one or more States.

(2) The wording of the certificate of insurance must be identical to the wording specified in §264.151(e).

(3) The closure insurance policy must be issued for a face amount at least equal to the current closure cost estimate, except as provided in §264.143(g). The term "face amount" means the total amount the insurer is obligated to pay under the policy. Actual payments by the insurer will not change the face amount, although the insurer's future liability will be lowered by the amount of the payments.

(4) The closure insurance policy must guarantee that funds will be available to close the facility whenever final closure occurs. The policy must also guarantee that once final closure begins, the insurer will be responsible for paying out funds, up to an amount equal to the face amount of the policy, upon the direction of the Regional Administrator, to such party or parties as the Regional Administrator specifies.

(5) After beginning partial or final closure, an owner or operator or any other person authorized to conduct closure may request reimbursements for closure expenditures by submitting itemized bills to the Regional Administrator. The owner or operator may request reimbursements for partial closure only if the remaining value of the policy is sufficient to cover the maximum costs of closing the facility over its remaining operating life. Within 60 days after receiving bills for closure activities, the Regional Administrator will instruct the insurer to make reimbursements in such amounts as the Regional Administrator specifies in writing, if the Regional Administrator determines that the partial or final closure expenditures are in accordance with the approved closure plan or otherwise justified. If the Regional Administrator has reason to believe that the maximum cost of closure over the remaining life of the facility will be significantly greater than the face amount of the policy, he may withhold reimbursements of such amounts as he deems prudent until he determines, in accordance with §264.143(i), that the owner or operator is no longer required to maintain financial assurance for final closure of the facility. If the Regional Administrator does not instruct the insurer to make such reimbursements, he will provide the owner or operator with a detailed written statement of reasons.

(6) The owner or operator must maintain the policy in full force and effect until the Regional Administrator consents to termination of the policy by the owner or operator as specified in

paragraph (e)(10) of this section. Failure to pay the premium, without substitution of alternate financial assurance as specified in this section, will constitute a significant violation of these regulations, warranting such remedy as the Regional Administrator deems necessary. Such violation will be deemed to begin upon receipt by the Regional Administrator of a notice of future cancellation, termination, or failure to renew due to nonpayment of the premium, rather than upon the date of expiration.

(7) Each policy must contain a provision allowing assignment of the policy to a successor owner or operator. Such assignment may be conditional upon consent of the insurer, provided such consent is not unreasonably refused.

(8) The policy must provide that the insurer may not cancel, terminate, or fail to renew the policy except for failure to pay the premium. The automatic renewal of the policy must, at a minimum, provide the insured with the option of renewal at the face amount of the expiring policy. If there is a failure to pay the premium, the insurer may elect to cancel, terminate, or fail to renew the policy by sending notice by certified mail to the owner or operator and the Regional Administrator. Cancellation, termination, or failure to renew may not occur, however, during the 120 days beginning with the date of receipt of the notice by both the Regional Administrator and the owner or operator, as evidenced by the return receipts. Cancellation, termination, or failure to renew may not occur and the policy will remain in full force and effect in the event that on or before the date of expiration:

(i) The Regional Administrator deems the facility abandoned; or

(ii) The permit is terminated or revoked or a new permit is denied; or

(iii) Closure is ordered by the Regional Administrator or a U.S. district court or other court of competent jurisdiction; or

(iv) The owner or operator is named as debtor in a voluntary or involuntary proceeding under Title 11 (Bankruptcy), U.S. Code; or

(v) The premium due is paid.

(9) Whenever the current closure cost estimate increases to an amount greater than the face amount of the policy, the owner or operator, within 60 days after the increase, must either cause the face amount to be increased to an amount at least equal to the current closure cost estimate and submit evidence of such increase to the Regional Administrator, or obtain other financial assurance as specified in this section to cover the increase. Whenever the current closure cost estimate decreases, the face amount may be reduced to the amount of the current closure cost estimate following written approval by the Regional Administrator.

(10) The Regional Administrator will give written consent to the owner or operator that he may terminate the insurance policy when:

(i) An owner or operator substitutes alternate financial assurance as specified in this section; or

(ii) The Regional Administrator releases the owner or operator from the requirements of this section in accordance with § 264.143(i).

(f) *Financial test and corporate guarantee for closure.* (1) An owner or operator may satisfy the requirements of this section by demonstrating that he passes a financial test as specified in this paragraph. To pass this test the owner or operator must meet the criteria of either paragraph (f)(1)(i) or (ii) of this section:

(i) The owner or operator must have:

(A) Two of the following three ratios: a ratio of total liabilities to net worth less than 2.0; a ratio of the sum of net income plus depreciation, depletion, and amortization to total liabilities greater than 0.1; and a ratio of current assets to current liabilities greater than 1.5; and

(B) Net working capital and tangible net worth each at least six times the sum of the current closure and postclosure cost estimates and the current plugging and abandonment cost estimates; and

(C) Tangible net worth of at least $10 million; and

(D) Assets located in the United States amounting to at least 90 percent of total assets or at least six times the sum of the current closure and postclosure cost estimates and the current plugging and abandonment cost estimates.

(ii) The owner or operator must have:

(A) A current rating for his most recent bond issuance of AAA, AA, A, or BBB as issued by Standard and Poor's or Aaa, Aa, A, or Baa as issued by Moody's; and

(B) Tangible net worth at least six times the sum of the current closure and post-closure cost estimates and the current plugging and abandonment cost estimates; and

(C) Tangible net worth of at least $10 million; and

(D) Assets located in the United States amounting to at least 90 percent of total assets or at least six times the sum of the current closure and post-closure cost estimates and the current plugging and abandonment cost estimates.

(2) The phrase "current closure and post-closure cost estimates" as used in paragraph (f)(1) of this section refers to the cost estimates required to be shown in paragraphs 1–4 of the letter from the owner's or operator's chief financial officer (§264.151(f)). The phrase "current plugging and abandonment cost estimates" as used in paragraph (f)(1) of this section refers to the cost estimates required to be shown in paragraphs 1–4 of the letter from the owner's or operator's chief financial officer (§144.70(f) of this title).

(3) To demonstrate that he meets this test, the owner or operator must submit the following items to the Regional Administrator:

(i) A letter signed by the owner's or operator's chief financial officer and worded as specified in §264.151(f); and

(ii) A copy of the independent certified public accountant's report on examination of the owner's or operator's financial statements for the latest completed fiscal year; and

(iii) A special report from the owner's or operator's independent certified public accountant to the owner or operator stating that:

(A) He has compared the data which the letter from the chief financial officer specifies as having been derived from the independently audited, year-end financial statements for the latest fiscal year with the amounts in such financial statements; and

(B) In connection with that procedure, no matters came to his attention which caused him to believe that the specified data should be adjusted.

(4) An owner or operator of a new facility must submit the items specified in paragraph (f)(3) of this section to the Regional Administrator at least 60 days before the date on which hazardous waste is first received for treatment, storage, or disposal.

(5) After the initial submission of items specified in paragraph (f)(3) of this section, the owner or operator must send updated information to the Regional Administrator within 90 days after the close of each succeeding fiscal year. This information must consist of all three items specified in paragraph (f)(3) of this section.

(6) If the owner or operator no longer meets the requirements of paragraph (f)(1) of this section, he must send notice to the Regional Administrator of intent to establish alternate financial assurance as specified in this section. The notice must be sent by certified mail within 90 days after the end of the fiscal year for which the year-end financial data show that the owner or operator no longer meets the requirements. The owner or operator must provide the alternate financial assurance within 120 days after the end of such fiscal year.

(7) The Regional Administrator may, based on a reasonable belief that the owner or operator may no longer meet the requirements of paragraph (f)(1) of this section, require reports of financial condition at any time from the owner or operator in addition to those specified in paragraph (f)(3) of this section. If the Regional Administrator finds, on the basis of such reports or other information, that the owner or operator no longer meets the requirements of paragraph (f)(1) of this section, the owner or operator must provide alternate financial assurance as specified in this section within 30 days after notification of such a finding.

(8) The Regional Administrator may disallow use of this test on the basis of qualifications in the opinion expressed by the independent certified public accountant in his report on examination of the owner's or operator's financial statements (see paragraph (f)(3)(ii) of

this section). An adverse opinion or a disclaimer of opinion will be cause for disallowance. The Regional Administrator will evaluate other qualifications on an individual basis. The owner or operator must provide alternate financial assurance as specified in this section within 30 days after notification of the disallowance.

(9) The owner or operator is no longer required to submit the items specified in paragraph (f)(3) of this section when:

(i) An owner or operator substitutes alternate financial assurance as specified in this section; or

(ii) The Regional Administrator releases the owner or operator from the requirements of this section in accordance with § 264.143(i).

(10) An owner or operator may meet the requirements of this section by obtaining a written guarantee. The guarantor must be the direct or higher-tier parent corporation of the owner or operator, a firm whose parent corporation is also the parent corporation of the owner or operator, or a firm with a "substantial business relationship" with the owner or operator. The guarantor must meet the requirements for owners or operators in paragraphs (f)(1) through (8) of this section and must comply with the terms of the guarantee. The wording of the guarantee must be identical to the wording specified in § 264.151(h). The certified copy of the guarantee must accompany the items sent to the Regional Administrator as specified in paragraph (f)(3) of this section. One of these items must be the letter from the guarantor's chief financial officer. If the guarantor's parent corporation is also the parent corporation of the owner or operator, the letter must describe the value received in consideration of the guarantee. If the guarantor is a firm with a "substantial business relationship" with the owner or operator, this letter must describe this "substantial business relationship" and the value received in consideration of the guarantee. The terms of the guarantee must provide that:

(i) If the owner or operator fails to perform final closure of a facility covered by the corporate guarantee in accordance with the closure plan and other permit requirements whenever

required to do so, the guarantor will do so or establish a trust fund as specified in § 264.143(a) in the name of the owner or operator.

(ii) The corporate guarantee will remain in force unless the guarantor sends notice of cancellation by certified mail to the owner or operator and to the Regional Administrator. Cancellation may not occur, however, during the 120 days beginning on the date of receipt of the notice of cancellation by both the owner or operator and the Regional Administrator, as evidenced by the return receipts.

(iii) If the owner or operator fails to provide alternate financial assurance as specified in this section and obtain the written approval of such alternate assurance from the Regional Administrator within 90 days after receipt by both the owner or operator and the Regional Administrator of a notice of cancellation of the corporate guarantee from the guarantor, the guarantor will provide such alternative financial assurance in the name of the owner or operator.

(g) *Use of multiple financial mechanisms.* An owner or operator may satisfy the requirements of this section by establishing more than one financial mechanism per facility. These mechanisms are limited to trust funds, surety bonds guaranteeing payment into a trust fund, letters of credit, and insurance. The mechanisms must be as specified in paragraphs (a), (b), (d), and (e), respectively, of this section, except that it is the combination of mechanisms, rather than the single mechanism, which must provide financial assurance for an amount at least equal to the current closure cost estimate. If an owner or operator uses a trust fund in combination with a surety bond or a letter of credit, he may use the trust fund as the standby trust fund for the other mechanisms. A single standby trust fund may be established for two or more mechanisms. The Regional Administrator may use any or all of the mechanisms to provide for closure of the facility.

(h) *Use of a financial mechanism for multiple facilities.* An owner or operator may use a financial assurance mechanism specified in this section to meet the requirements of this section for more than one facility. Evidence of financial assurance submitted to the Regional Administrator must include a list showing, for each facility, the EPA Identification Number, name, address, and the amount of funds for closure assured by the mechanism. If the facilities covered by the mechanism are in more than one Region, identical evidence of financial assurance must be submitted to and maintained with the Regional Administrators of all such Regions. The amount of funds available through the mechanism must be no less than the sum of funds that would be available if a separate mechanism had been established and maintained for each facility. In directing funds available through the mechanism for closure of any of the facilities covered by the mechanism, the Regional Administrator may direct only the amount of funds designated for that facility, unless the owner or operator agrees to the use of additional funds available under the mechanism.

(i) *Release of the owner or operator from the requirements of this section.* Within 60 days after receiving certifications from the owner or operator and a qualified Professional Engineer that final closure has been completed in accordance with the approved closure plan, the Regional Administrator will notify the owner or operator in writing that he is no longer required by this section to maintain financial assurance for final closure of the facility, unless the Regional Administrator has reason to believe that final closure has not been in accordance with the approved closure plan. The Regional Administrator shall provide the owner or operator a detailed written statement of any such reason to believe that closure has not been in accordance with the approved closure plan.

[47 FR 15047, Apr. 7, 1982, as amended at 51 FR 16448, May 2, 1986; 57 FR 42835, Sept. 16, 1992; 71 FR 16905, Apr. 4, 2006; 71 FR 40272, July 14, 2006]

§264.144 **Cost estimate for post-closure care.**

(a) The owner or operator of a disposal surface impoundment, disposal miscellaneous unit, land treatment unit, or landfill unit, or of a surface impoundment or waste pile required

under §§264.228 and 264.258 to prepare a contingent closure and post-closure plan, must have a detailed written estimate, in current dollars, of the annual cost of post-closure monitoring and maintenance of the ·facility in accordance with the applicable post-closure regulations in §§264.117 through 264.120, 264.228, 264.258, 264.280, 264.310, and 264.603.

(1) The post-closure cost estimate must be based on the costs to the owner or operator of hiring a third party to conduct post-closure care activities. A third party is a party who is neither a parent nor a subsidiary of the owner or operator. (See definition of parent corporation in §264.141(d).)

(2) The post-closure cost estimate is calculated by multiplying the annual post-closure cost estimate by the number of years of post-closure care required under §264.117.

(b) During the active life of the facility, the owner or operator must adjust the post-closure cost estimate for inflation within 60 days prior to the anniversary date of the establishment of the financial instrument(s) used to comply with §264.145. For owners or operators using the financial test or corporate guarantee, the post-closure cost estimate must be updated for inflation within 30 days after the close of the firm's fiscal year and before the submission of updated information to the Regional Administrator as specified in §264.145(f)(5). The adjustment may be made by recalculating the post-closure cost estimate in current dollars or by using an inflation factor derived from the most recent Implicit Price Deflator for Gross National Product published by the U.S. Department of Commerce in its *Survey of Current Business* as specified in §264.145(b)(1) and (2). The inflation factor is the result of dividing the latest published annual Deflator by the Deflator for the previous year.

(1) The first adjustment is made by multiplying the post-closure cost estimate by the inflation factor. The result is the adjusted post-closure cost estimate.

(2) Subsequent adjustments are made by multiplying the latest adjusted post-closure cost estimate by the latest inflation factor.

(c) During the active life of the facility, the owner or operator must revise the post-closure cost estimate within 30 days after the Regional Administrator has approved the request to modify the post-closure plan, if the change in the post-closure plan increases the cost of post-closure care. The revised post-closure cost estimate must be adjusted for inflation as specified in § 264.144(b).

(d) The owner or operator must keep the following at the facility during the operating life of the facility: The latest post-closure cost estimate prepared in accordance with § 264.144 (a) and (c) and, when this estimate has been adjusted in accordance with § 264.144(b), the latest adjusted post-closure cost estimate.

[47 FR 15047, Apr. 7, 1982, as amended at 47 FR 32357, July 26, 1982; 50 FR 4514, Jan. 31, 1985; 51 FR 16449, May 2, 1986; 52 FR 46964, Dec. 10, 1987]

## § 264.145   Financial assurance for post-closure care.

The owner or operator of a hazardous waste management unit subject to the requirements of § 264.144 must establish financial assurance for post-closure care in accordance with the approved post-closure plan for the facility 60 days prior to the initial receipt of hazardous waste or the effective date of the regulation, whichever is later. He must choose from the following options:

(a) *Post-closure trust fund.* (1) An owner or operator may satisfy the requirements of this section by establishing a post-closure trust fund which conforms to the requirements of this paragraph and submitting an originally signed duplicate of the trust agreement to the Regional Administrator. An owner or operator of a new facility must submit the originally signed duplicate of the trust agreement to the Regional Administrator at least 60 days before the date on which hazardous waste is first received for disposal. The trustee must be an entity which has the authority to act as a trustee and whose trust operations are regulated and examined by a Federal or State agency.

(2) The wording of the trust agreement must be identical to the wording specified in § 264.151(a)(1), and the trust agreement must be accompanied by a formal certification of acknowledgment (for example, see § 264.151(a)(2)). Schedule A of the trust agreement must be updated within 60 days after a change in the amount of the current post-closure cost estimate covered by the agreement.

(3) Payments into the trust fund must be made annually by the owner or operator over the term of the initial RCRA permit or over the remaining operating life of the facility as estimated in the closure plan, whichever period is shorter; this period is hereafter referred to as the "pay-in period." The payments into the post-closure trust fund must be made as follows:

(i) For a new facility, the first payment must be made before the initial receipt of hazardous waste for disposal. A receipt from the trustee for this payment must be submitted by the owner or operator to the Regional Administrator before this initial receipt of hazardous waste. The first payment must be at least equal to the current post-closure cost estimate, except as provided in § 264.145(g), divided by the number of years in the pay-in period. Subsequent payments must be made no later than 30 days after each anniversary date of the first payment. The amount of each subsequent payment must be determined by this formula:

$$\text{Next payment} = \frac{CE - CV}{Y}$$

where CE is the current post-closure cost estimate, CV is the current value of the trust fund, and Y is the number of years remaining in the pay-in period.

(ii) If an owner or operator establishes a trust fund as specified in § 265.145(a) of this chapter, and the value of that trust fund is less than the current post-closure cost estimate when a permit is awarded for the facility, the amount of the current post-closure cost estimate still to be paid into the fund must be paid in over the pay-in period as defined in paragraph (a)(3) of this section. Payments must continue to be made no later than 30 days after each anniversary date of the first payment made pursuant to Part 265 of this chapter. The amount of each payment must be determined by this formula:

$$\text{Next payment} = \frac{CE - CV}{Y}$$

where CE is the current post-closure cost estimate, CV is the current value of the trust fund, and Y is the number of years remaining in the pay-in period.

(4) The owner or operator may accelerate payments into the trust fund or he may deposit the full amount of the current post-closure cost estimate at the time the fund is established. However, he must maintain the value of the fund at no less than the value that the fund would have if annual payments were made as specified in paragraph (a)(3) of this section.

(5) If the owner or operator establishes a post-closure trust fund after having used one or more alternate mechanisms specified in this section or in §265.145 of this chapter, his first payment must be in at least the amount that the fund would contain if the trust fund were established initially and annual payments made according to specifications of this paragraph and §265.145(a) of this chapter, as applicable.

(6) After the pay-in period is completed, whenever the current post-closure cost estimate changes during the operating life of the facility, the owner or operator must compare the new estimate with the trustee's most recent annual valuation of the trust fund. If the value of the fund is less than the amount of the new estimate, the owner or operator, within 60 days after the change in the cost estimate, must either deposit an amount into the fund so that its value after this deposit at least equals the amount of the current post-closure cost estimate, or obtain other financial assurance as specified in this section to cover the difference.

(7) During the operating life of the facility, if the value of the trust fund is greater than the total amount of the current post-closure cost estimate, the owner or operator may submit a written request to the Regional Administrator for release of the amount in excess of the current post-closure cost estimate.

(8) If an owner or operator substitutes other financial assurance as specified in this section for all or part of the trust fund, he may submit a written request to the Regional Administrator for release of the amount in excess of the current post-closure cost estimate covered by the trust fund.

(9) Within 60 days after receiving a request from the owner or operator for release of funds as specified in paragraph (a) (7) or (8) of this section, the Regional Administrator will instruct the trustee to release to the owner or operator such funds as the Regional Administrator specifies in writing.

(10) During the period of post-closure care, the Regional Administrator may approve a release of funds if the owner or operator demonstrates to the Regional Administrator that the value of the trust fund exceeds the remaining cost of post-closure care.

(11) An owner or operator or any other person authorized to conduct post-closure care may request reimbursements for post-closure care expenditures by submitting itemized bills to the Regional Administrator. Within 60 days after receiving bills for post-closure care activities, the Regional Administrator will instruct the trustee to make reimbursements in those amounts as the Regional Administrator specifies in writing, if the Regional Administrator determines that the post-closure care expenditures are in accordance with the approved post-closure plan or otherwise justified. If the Regional Administrator does not instruct the trustee to make such reimbursements, he will provide the owner or operator with a detailed written statement of reasons.

(12) The Regional Administrator will agree to termination of the trust when:

(i) An owner or operator substitutes alternate financial assurance as specified in this section; or

(ii) The Regional Administrator releases the owner or operator from the requirements of this section in accordance with §264.145(i).

(b) *Surety bond guaranteeing payment into a post-closure trust fund.* (1) An owner or operator may satisfy the requirements of this section by obtaining a surety bond which conforms to the requirements of this paragraph and submitting the bond to the Regional Administrator. An owner or operator of a new facility must submit the bond to the Regional Administrator at least 60 days before the date on which hazardous waste is first received for disposal. The bond must be effective before this initial receipt of hazardous waste. The surety company issuing the bond must, at a minimum, be among those listed as acceptable sureties on Federal bonds in Circular 570 of the U.S. Department of the Treasury.

(2) The wording of the surety bond must be identical to the wording specified in § 264.151(b).

(3) The owner or operator who uses a surety bond to satisfy the requirements of this section must also establish a standby trust fund. Under the terms of the bond, all payments made thereunder will be deposited by the surety directly into the standby trust fund in accordance with instructions from the Regional Administrator. This standby trust fund must meet the requirements specified in § 264.145(a), except that:

(i) An originally signed duplicate of the trust agreement must be submitted to the Regional Administrator with the surety bond; and

(ii) Until the standby trust fund is funded pursuant to the requirements of this section, the following are not required by these regulations:

(A) Payments into the trust fund as specified in § 264.145(a);

(B) Updating of Schedule A of the trust agreement (see § 264.151(a)) to show current post-closure cost estimates;

(C) Annual valuations as required by the trust agreement; and

(D) Notices of nonpayment as required by the trust agreement.

(4) The bond must guarantee that the owner or operator will:

(i) Fund the standby trust fund in an amount equal to the penal sum of the bond before the beginning of final closure of the facility; or

(ii) Fund the standby trust fund in an amount equal to the penal sum within 15 days after an administrative order to begin final closure issued by the Regional Administrator becomes final, or within 15 days after an order to begin final closure is issued by a U.S. district

court or other court of competent jurisdiction; or

(iii) Provide alternate financial assurance as specified in this section, and obtain the Regional Administrator's written approval of the assurance provided, within 90 days after receipt by both the owner or operator and the Regional Administrator of a notice of cancellation of the bond from the surety.

(5) Under the terms of the bond, the surety will become liable on the bond obligation when the owner or operator fails to perform as guaranteed by the bond.

(6) The penal sum of the bond must be in an amount at least equal to the current post-closure cost estimate, except as provided in § 264.145(g).

(7) Whenever the current post-closure cost estimate increases to an amount greater than the penal sum, the owner or operator, within 60 days after the increase, must either cause the penal sum to be increased to an amount at least equal to the current post-closure cost estimate and submit evidence of such increase to the Regional Administrator, or obtain other financial assurance as specified in this section to cover the increase. Whenever the current post-closure cost estimate decreases, the penal sum may be reduced to the amount of the current post-closure cost estimate following written approval by the Regional Administrator.

(8) Under the terms of the bond, the surety may cancel the bond by sending notice of cancellation by certified mail to the owner or operator and to the Regional Administrator. Cancellation may not occur, however, during the 120 days beginning on the date of receipt of the notice of cancellation by both the owner or operator and the Regional Administrator, as evidenced by the return receipts.

(9) The owner or operator may cancel the bond if the Regional Administrator has given prior written consent based on his receipt of evidence of alternate financial assurance as specified in this section.

(c) *Surety bond guaranteeing performance of post-closure care.* (1) An owner or operator may satisfy the requirements of this section by obtaining a surety bond which conforms to the requirements of this paragraph and submitting the bond to the Regional Administrator. An owner or operator of a new facility must submit the bond to the Regional Administrator at least 60 days before the date on which hazardous waste is first received for disposal. The bond must be effective before this initial receipt of hazardous waste. The surety company issuing the bond must, at a minimum, be among those listed as acceptable sureties on Federal bonds in Circular 570 of the U.S. Department of the Treasury.

(2) The wording of the surety bond must be identical to the wording specified in §264.151(c).

(3) The owner or operator who uses a surety bond to satisfy the requirements of this section must also establish a standby trust fund. Under the terms of the bond, all payments made thereunder will be deposited by the surety directly into the standby trust fund in accordance with instructions from the Regional Administrator. This standby trust fund must meet the requirements specified in §264.145(a), except that:

(i) An originally signed duplicate of the trust agreement must be submitted to the Regional Administrator with the surety bond; and

(ii) Unless the standby trust fund is funded pursuant to the requirements of this section, the following are not required by these regulations:

(A) Payments into the trust fund as specified in §264.145(a);

(B) Updating of Schedule A of the trust agreement (see §264.151(a)) to show current post-closure cost estimates;

(C) Annual valuations as required by the trust agreement; and

(D) Notices of nonpayment as required by the trust agreement.

(4) The bond must guarantee that the owner or operator will:

(i) Perform post-closure care in accordance with the post-closure plan and other requirements of the permit for the facility; or

(ii) Provide alternate financial assurance as specified in this section, and obtain the Regional Administrator's written approval of the assurance provided, within 90 days of receipt by both the owner or operator and the Regional Administrator of a notice of cancellation of the bond from the surety.

(5) Under the terms of the bond, the surety will become liable on the bond obligation when the owner or operator fails to perform as guaranteed by the bond. Following a final administrative determination pursuant to section 3008 of RCRA that the owner or operator has failed to perform post-closure care in accordance with the approved post-closure plan and other permit requirements, under the terms of the bond the surety will perform post-closure care in accordance with the post-closure plan and other permit requirements or will deposit the amount of the penal sum into the standby trust fund.

(6) The penal sum of the bond must be in an amount at least equal to the current post-closure cost estimate.

(7) Whenever the current post-closure cost estimate increases to an amount greater than the penal sum during the operating life of the facility, the owner or operator, within 60 days after the increase, must either cause the penal sum to be increased to an amount at least equal to the current post-closure cost estimate and submit evidence of such increase to the Regional Administrator, or obtain other financial assurance as specified in this section. Whenever the current post-closure cost estimate decreases during the operating life of the facility, the penal sum may be reduced to the amount of the current post-closure cost estimate following written approval by the Regional Administrator.

(8) During the period of post-closure care, the Regional Administrator may approve a decrease in the penal sum if the owner or operator demonstrates to the Regional Administrator that the amount exceeds the remaining cost of post-closure care.

(9) Under the terms of the bond, the surety may cancel the bond by sending notice of cancellation by certified mail to the owner or operator and to the Regional Administrator. Cancellation may not occur, however, during the 120 days beginning on the date of receipt of the notice of cancellation by both the owner or operator and the Regional Administrator, as evidenced by the return receipts.

(10) The owner or operator may cancel the bond if the Regional Administrator has given prior written consent. The Regional Administrator will provide such written consent when:

(i) An owner or operator substitutes alternate financial assurance as specified in this section; or

(ii) The Regional Administrator releases the owner or operator from the requirements of this section in accordance with § 264.145(i).

(11) The surety will not be liable for deficiencies in the performance of post-closure care by the owner or operator after the Regional Administrator releases the owner or operator from the requirements of this section in accordance with § 264.145(i).

(d) *Post-closure letter of credit.* (1) An owner or operator may satisfy the requirements of this section by obtaining an irrevocable standby letter of credit which conforms to the requirements of this paragraph and submitting the letter to the Regional Administrator. An owner or operator of a new facility must submit the letter of credit to the Regional Administrator at least 60 days before the date on which hazardous waste is first received for disposal. The letter of credit must be effective before this initial receipt of hazardous waste. The issuing institution must be an entity which has the authority to issue letters of credit and whose letter-of-credit operations are regulated and examined by a Federal or State agency.

(2) The wording of the letter of credit must be identical to the wording specified in § 264.151(d).

(3) An owner or operator who uses a letter of credit to satisfy the requirements of this section must also establish a standby trust fund. Under the terms of the letter of credit, all amounts paid pursuant to a draft by the Regional Administrator will be deposited by the issuing institution directly into the standby trust fund in accordance with instructions from the Regional Administrator. This standby trust fund must meet the requirements of the trust fund specified in § 264.145(a), except that:

(i) An originally signed duplicate of the trust agreement must be submitted to the Regional Administrator with the letter of credit; and

(ii) Unless the standby trust fund is funded pursuant to the requirements of this section, the following are not required by these regulations:

(A) Payments into the trust fund as specified in § 264.145(a);

(B) Updating of Schedule A of the trust agreement (see § 264.151(a)) to show current post-closure cost estimates;

(C) Annual valuations as required by the trust agreement; and

(D) Notices of nonpayment as required by the trust agreement.

(4) The letter of credit must be accompanied by a letter from the owner or operator referring to the letter of credit by number, issuing institution, and date, and providing the following information: the EPA Identification Number, name, and address of the facility, and the amount of funds assured for post-closure care of the facility by the letter of credit.

(5) The letter of credit must be irrevocable and issued for a period of at least 1 year. The letter of credit must provide that the expiration date will be automatically extended for a period of at least 1 year unless, at least 120 days before the current expiration date, the issuing institution notifies both the owner or operator and the Regional Administrator by certified mail of a decision not to extend the expiration date. Under the terms of the letter of credit, the 120 days will begin on the date when both the owner or operator and the Regional Administrator have received the notice, as evidenced by the return receipts.

(6) The letter of credit must be issued in an amount at least equal to the current post-closure cost estimate, except as provided in § 264.145(g).

(7) Whenever the current post-closure cost estimate increases to an amount greater than the amount of the credit during the operating life of the facility, the owner or operator, within 60 days after the increase, must either cause the amount of the credit to be increased so that it at least equals the current post-closure cost estimate and submit evidence of such increase to the Regional Administrator, or obtain other financial assurance as specified

in this section to cover the increase. Whenever the current post-closure cost estimate decreases during the operating life of the facility, the amount of the credit may be reduced to the amount of the current post-closure cost estimate following written approval by the Regional Administrator.

(8) During the period of post-closure care, the Regional Administrator may approve a decrease in the amount of the letter of credit if the owner or operator demonstrates to the Regional Administrator that the amount exceeds the remaining cost of post-closure care.

(9) Following a final administrative determination pursuant to section 3008 of RCRA that the owner or operator has failed to perform post-closure care in accordance with the approved post-closure plan and other permit requirements, the Regional Administrator may draw on the letter of credit.

(10) If the owner or operator does not establish alternate financial assurance as specified in this section and obtain written approval of such alternate assurance from the Regional Administrator within 90 days after receipt by both the owner or operator and the Regional Administrator of a notice from the issuing institution that it has decided not to extend the letter of credit beyond the current expiration date, the Regional Administrator will draw on the letter of credit. The Regional Administrator may delay the drawing if the issuing institution grants an extension of the term of the credit. During the last 30 days of any such extension the Regional Administrator will draw on the letter of credit if the owner or operator has failed to provide alternate financial assurance as specified in this section and obtain written approval of such assurance from the Regional Administrator.

(11) The Regional Administrator will return the letter of credit to the issuing institution for termination when:

(i) An owner or operator substitutes alternate financial assurance as specified in this section; or

(ii) The Regional Administrator releases the owner or operator from the requirements of this section in accordance with §264.145(i).

(e) *Post-closure insurance.* (1) An owner or operator may satisfy the requirements of this section by obtaining post-closure insurance which conforms to the requirements of this paragraph and submitting a certificate of such insurance to the Regional Administrator. An owner or operator of a new facility must submit the certificate of insurance to the Regional Administrator at least 60 days before the date on which hazardous waste is first received for disposal. The insurance must be effective before this initial receipt of hazardous waste. At a minimum, the insurer must be licensed to transact the business of insurance, or eligible to provide insurance as an excess or surplus lines insurer, in one or more States.

(2) The wording of the certificate of insurance must be identical to the wording specified in §264.151(e).

(3) The post-closure insurance policy must be issued for a face amount at least equal to the current post-closure cost estimate, except as provided in §264.145(g). The term "face amount" means the total amount the insurer is obligated to pay under the policy. Actual payments by the insurer will not change the face amount, although the insurer's future liability will be lowered by the amount of the payments.

(4) The post-closure insurance policy must guarantee that funds will be available to provide post-closure care of the facility whenever the post-closure period begins. The policy must also guarantee that once post-closure care begins, the insurer will be responsible for paying out funds, up to an amount equal to the face amount of the policy, upon the direction of the Regional Administrator, to such party or parties as the Regional Administrator specifies.

(5) An owner or operator or any other person authorized to conduct post-closure care may request reimbursements for post-closure care expenditures by submitting itemized bills to the Regional Administrator. Within 60 days after receiving bills for post-closure care activities, the Regional Administrator will instruct the insurer to make reimbursements in those

amounts as the Regional Administrator specifies in writing, if the Regional Administrator determines that the post-closure care expenditures are in accordance with the approved post-closure plan or otherwise justified. If the Regional Administrator does not instruct the insurer to make such reimbursements, he will provide the owner or operator with a detailed written statement of reasons.

(6) The owner or operator must maintain the policy in full force and effect until the Regional Administrator consents to termination of the policy by the owner or operator as specified in paragraph (e)(11) of this section. Failure to pay the premium, without substitution of alternate financial assurance as specified in this section, will constitute a significant violation of these regulations, warranting such remedy as the Regional Administrator deems necessary. Such violation will be deemed to begin upon receipt by the Regional Administrator of a notice of future cancellation, termination, or failure to renew due to nonpayment of the premium, rather than upon the date of expiration.

(7) Each policy must contain a provision allowing assignment of the policy to a successor owner or operator. Such assignment may be conditional upon consent of the insurer, provided such consent is not unreasonably refused.

(8) The policy must provide that the insurer may not cancel, terminate, or fail to renew the policy except for failure to pay the premium. The automatic renewal of the policy must, at a minimum, provide the insured with the option of renewal at the face amount of the expiring policy. If there is a failure to pay the premium, the insurer may elect to cancel, terminate, or fail to renew the policy by sending notice by certified mail to the owner or operator and the Regional Administrator. Cancellation, termination, or failure to renew may not occur, however, during the 120 days beginning with the date of receipt of the notice by both the Regional Administrator and the owner or operator, as evidenced by the return receipts. Cancellation, termination, or failure to renew may not occur and the policy will remain in full force and effect in the event that on or before the date of expiration:

(i) The Regional Administrator deems the facility abandoned; or

(ii) The permit is terminated or revoked or a new permit is denied; or

(iii) Closure is ordered by the Regional Administrator or a U.S. district court or other court of competent jurisdiction; or

(iv) The owner or operator is named as debtor in a voluntary or involuntary proceeding under Title 11 (Bankruptcy), U.S. Code; or

(v) The premium due is paid.

(9) Whenever the current post-closure cost estimate increases to an amount greater than the face amount of the policy during the operating life of the facility, the owner or operator, within 60 days after the increase, must either cause the face amount to be increased to an amount at least equal to the current post-closure cost estimate and submit evidence of such increase to the Regional Administrator, or obtain other financial assurance as specified in this section to cover the increase. Whenever the current post-closure cost estimate decreases during the operating life of the facility, the face amount may be reduced to the amount of the current post-closure cost estimate following written approval by the Regional Administrator.

(10) Commencing on the date that liability to make payments pursuant to the policy accrues, the insurer will thereafter annually increase the face amount of the policy. Such increase must be equivalent to the face amount of the policy, less any payments made, multiplied by an amount equivalent to 85 percent of the most recent investment rate or of the equivalent coupon-issue yield announced by the U.S. Treasury for 26-week Treasury securities.

(11) The Regional Administrator will give written consent to the owner or operator that he may terminate the insurance policy when:

(i) An owner or operator substitutes alternate financial assurance as specified in this section; or

(ii) The Regional Administrator releases the owner or operator from the requirements of this section in accordance with § 264.145(i).

(f) *Financial test and corporate guarantee for post-closure care.* (1) An owner or operator may satisfy the requirements of this section by demonstrating that he passes a financial test as specified in this paragraph. To pass this test the owner or operator must meet the criteria of either paragraph (f)(1)(i) or (ii) of this section:

(i) The owner or operator must have:

(A) Two of the following three ratios: a ratio of total liabilities to net worth less than 2.0; a ratio of the sum of net income plus depreciation, depletion, and amortization to total liabilities greater than 0.1; and a ratio of current assets to current liabilities greater than 1.5; and

(B) Net working capital and tangible net worth each at least six times the sum of the current closure and post-closure cost estimates and the current plugging and abandonment cost estimates; and

(C) Tangible net worth of at least $10 million; and

(D) Assets in the United States amounting to at least 90 percent of his total assets or at least six times the sum of the current closure and post-closure cost estimates and the current plugging and abandonment cost estimates.

(ii) The owner or operator must have:

(A) A current rating for his most recent bond issuance of AAA, AA, A, or BBB as issued by Standard and Poor's or Aaa, Aa, A or Baa as issued by Moody's; and

(B) Tangible net worth at least six times the sum of the current closure and post-closure cost estimates and the current plugging and abandonment cost estimates; and

(C) Tangible net worth of at least $10 million; and

(D) Assets located in the United States amounting to at least 90 percent of his total assets or at least six times the sum of the current closure and post-closure cost estimates and the current plugging and abandonment cost estimates.

(2) The phrase "current closure and post-closure cost estimates" as used in paragraph (f)(1) of this section refers to the cost estimates required to be shown in paragraphs 1–4 of the letter from the owner's or operator's chief financial officer (§264.151(f)). The phrase "current plugging and abandonment cost estimates" as used in paragraph (f)(1) of this section refers to the cost estimates required to be shown in paragraphs 1–4 of the letter from the owner's or operator's chief financial officer (§144.70(f) of this title).

(3) To demonstrate that he meets this test, the owner or operator must submit the following items to the Regional Administrator:

(i) A letter signed by the owner's or operator's chief financial officer and worded as specified in §264.151(f); and

(ii) A copy of the independent certified public accountant's report on examination of the owner's or operator's financial statements for the latest completed fiscal year; and

(iii) A special report from the owner's or operator's independent certified public accountant to the owner or operator stating that:

(A) He has compared the data which the letter from the chief financial officer specifies as having been derived from the independently audited, year-end financial statements for the latest fiscal year with the amounts in such financial statements; and

(B) In connection with that procedure, no matters came to his attention which caused him to believe that the specified data should be adjusted.

(4) An owner or operator of a new facility must submit the items specified in paragraph (f)(3) of this section to the Regional Administrator at least 60 days before the date on which hazardous waste is first received for disposal.

(5) After the initial submission of items specified in paragraph (f)(3) of this section, the owner or operator must send updated information to the Regional Administrator within 90 days after the close of each succeeding fiscal year. This information must consist of all three items specified in paragraph (f)(3) of this section.

(6) If the owner or operator no longer meets the requirements of paragraph (f)(1) of this section, he must send notice to the Regional Administrator of intent to establish alternate financial assurance as specified in this section. The notice must be sent by certified mail within 90 days after the end of the

fiscal year for which the year-end financial data show that the owner or operator no longer meets the requirements. The owner or operator must provide the alternate financial assurance within 120 days after the end of such fiscal year.

(7) The Regional Administrator may, based on a reasonable belief that the owner or operator may no longer meet the requirements of paragraph (f)(1) of this section, require reports of financial condition at any time from the owner or operator in addition to those specified in paragraph (f)(3) of this section. If the Regional Administrator finds, on the basis of such reports or other information, that the owner or operator no longer meets the requirements of paragraph (f)(1) of this section, the owner or operator must provide alternate financial assurance as specified in this section within 30 days after notification of such a finding.

(8) The Regional Administrator may disallow use of this test on the basis of qualifications in the opinion expressed by the independent certified public accountant in his report on examination of the owner's or operator's financial statements (see paragraph (f)(3)(ii) of this section). An adverse opinion or a disclaimer of opinion will be cause for disallowance. The Regional Administrator will evaluate other qualifications on an individual basis. The owner or operator must provide alternate financial assurance as specified in this section within 30 days after notification of the disallowance.

(9) During the period of post-closure care, the Regional Administrator may approve a decrease in the current post-closure cost estimate for which this test demonstrates financial assurance if the owner or operator demonstrates to the Regional Administrator that the amount of the cost estimate exceeds the remaining cost of post-closure care.

(10) The owner or operator is no longer required to submit the items specified in paragraph (f)(3) of this section when:

(i) An owner or operator substitutes alternate financial assurance as specified in this section; or

(ii) The Regional Administrator releases the owner or operator from the requirements of this section in accordance with § 264.145(i).

(11) An owner or operator may meet the requirements of this section by obtaining a written guarantee. The guarantor must be the direct or higher-tier parent corporation of the owner or operator, a firm whose parent corporation is also the parent corporation of the owner or operator, or a firm with a "substantial business relationship" with the owner or operator. The guarantor must meet the requirements for owners or operators in paragraphs (f)(1) through (9) of this section and must comply with the terms of the guarantee. The wording of the guarantee must be identical to the wording specified in § 264.151(h). A certified copy of the guarantee must accompany the items sent to the Regional Administrator as specified in paragraph (f)(3) of this section. One of these items must be the letter from the guarantor's chief financial officer. If the guarantor's parent corporation is also the parent corporation of the owner or operator, the letter must describe the value received in consideration of the guarantee. If the guarantor is a firm with a "substantial business relationship" with the owner or operator, this letter must describe this "substantial business relationship" and the value received in consideration of the guarantee. The terms of the guarantee must provide that:

(i) If the owner or operator fails to perform post-closure care of a facility covered by the corporate guarantee in accordance with the post-closure plan and other permit requirements whenever required to do so, the guarantor will do so or establish a trust fund as specified in § 264.145(a) in the name of the owner or operator.

(ii) The corporate guarantee will remain in force unless the guarantor sends notice of cancellation by certified mail to the owner or operator and to the Regional Administrator. Cancellation may not occur, however, during the 120 days beginning on the date of receipt of the notice of cancellation by both the owner or operator and the Regional Administrator, as evidenced by the return receipts.

(iii) If the owner or operator fails to provide alternate financial assurance as specified in this section and obtain the written approval of such alternate assurance from the Regional Administrator within 90 days after receipt by both the owner or operator and the Regional Administrator of a notice of cancellation of the corporate guarantee from the guarantor, the guarantor will provide such alternate financial assurance in the name of the owner or operator.

(g) *Use of multiple financial mechanisms.* An owner or operator may satisfy the requirements of this section by establishing more than one financial mechanism per facility. These mechanisms are limited to trust funds, surety bonds guaranteeing payment into a trust fund, letters of credit, and insurance. The mechanisms must be as specified in paragraphs (a), (b), (d), and (e), respectively, of this section, except that it is the combination of mechanisms, rather than the single mechanism, which must provide financial assurance for an amount at least equal to the current post-closure cost estimate. If an owner or operator uses a trust fund in combination with a surety bond or a letter of credit, he may use the trust fund as the standby trust fund for the other mechanisms. A single standby trust fund may be established for two or more mechanisms. The Regional Administrator may use any or all of the mechanisms to provide for post-closure care of the facility.

(h) *Use of a financial mechanism for multiple facilities.* An owner or operator may use a financial assurance mechanism specified in this section to meet the requirements of this section for more than one facility. Evidence of financial assurance submitted to the Regional Administrator must include a list showing, for each facility, the EPA Identification Number, name, address, and the amount of funds for post-closure care assured by the mechanism. If the facilities covered by the mechanism are in more than one Region, identical evidence of financial assurance must be submitted to and maintained with the Regional Administrators of all such Regions. The amount of funds available through the mechanism must be no less than the sum of funds that would be available if a separate mechanism had been established and

maintained for each facility. In directing funds available through the mechanism for post-closure care of any of the facilities covered by the mechanism, the Regional Administrator may direct only the amount of funds designated for that facility, unless the owner or operator agrees to the use of additional funds available under the mechanism.

(i) *Release of the owner or operator from the requirements of this section.* Within 60 days after receiving certifications from the owner or operator and a qualified Professional Engineer that the post-closure care period has been completed for a hazardous waste disposal unit in accordance with the approved plan, the Regional Administrator will notify the owner or operator that he is no longer required to maintain financial assurance for post-closure of that unit, unless the Regional Administrator has reason to believe that post-closure care has not been in accordance with the approved post-closure plan. The Regional Administrator shall provide the owner or operator a detailed written statement of any such reason to believe that post-closure care has not been in accordance with the approved post-closure plan.

[47 FR 15047, Apr. 7, 1982, as amended at 51 FR 16449, May 2, 1986; 57 FR 42836, Sept. 16, 1992; 71 FR 16905, Apr. 4, 2006; 71 FR 40272, July 14, 2006]

### §264.146 Use of a mechanism for financial assurance of both closure and post-closure care.

An owner or operator may satisfy the requirements for financial assurance for both closure and post-closure care for one or more facilities by using a trust fund, surety bond, letter of credit, insurance, financial test, or corporate guarantee that meets the specifications for the mechanism in both §§264.143 and 264.145. The amount of funds available through the mechanism must be no less than the sum of funds that would be available if a separate mechanism had been established and maintained for financial assurance of closure and of post-closure care.

### §264.147 Liability requirements.

(a) *Coverage for sudden accidental occurrences.* An owner or operator of a hazardous waste treatment, storage, or disposal facility, or a group of such facilities, must demonstrate financial responsibility for bodily injury and property damage to third parties caused by sudden accidental occurrences arising from operations of the facility or group of facilities. The owner or operator must have and maintain liability coverage for sudden accidental occurrences in the amount of at least $1 million per occurrence with an annual aggregate of at least $2 million, exclusive of legal defense costs. This liability coverage may be demonstrated as specified in paragraphs (a) (1), (2), (3), (4), (5), or (6) of this section:

(1) An owner or operator may demonstrate the required liability coverage by having liability insurance as specified in this paragraph.

(i) Each insurance policy must be amended by attachment of the Hazardous Waste Facility Liability Endorsement or evidenced by a Certificate of Liability Insurance. The wording of the endorsement must be identical to the wording specified in § 264.151(i). The wording of the certificate of insurance must be identical to the wording specified in § 264.151(j). The owner or operator must submit a signed duplicate original of the endorsement or the certificate of insurance to the Regional Administrator, or Regional Administrators if the facilities are located in more than one Region. If requested by a Regional Administrator, the owner or operator must provide a signed duplicate original of the insurance policy. An owner or operator of a new facility must submit the signed duplicate original of the Hazardous Waste Facility Liability Endorsement or the Certificate of Liability Insurance to the Regional Administrator at least 60 days before the date on which hazardous waste is first received for treatment, storage, or disposal. The insurance must be effective before this initial receipt of hazardous waste.

(ii) Each insurance policy must be issued by an insurer which, at a minimum, is licensed to transact the business of insurance, or eligible to provide insurance as an excess or surplus lines insurer, in one or more States.

(2) An owner or operator may meet the requirements of this section by

passing a financial test or using the guarantee for liability coverage as specified in paragraphs (f) and (g) of this section.

(3) An owner or operator may meet the requirements of this section by obtaining a letter of credit for liability coverage as specified in paragraph (h) of this section.

(4) An owner or operator may meet the requirements of this section by obtaining a surety bond for liability coverage as specified in paragraph (i) of this section.

(5) An owner or operator may meet the requirements of this section by obtaining a trust fund for liability coverage as specified in paragraph (j) of this section.

(6) An owner or operator may demonstrate the required liability coverage through the use of combinations of insurance, financial test, guarantee, letter of credit, surety bond, and trust fund, except that the owner or operator may not combine a financial test covering part of the liability coverage requirement with a guarantee unless the financial statement of the owner or operator is not consolidated with the financial statement of the guarantor. The amounts of coverage demonstrated must total at least the minimum amounts required by this section. If the owner or operator demonstrates the required coverage through the use of a combination of financial assurances under this paragraph, the owner or operator shall specify at least one such assurance as "primary" coverage and shall specify other assurance as "excess" coverage.

(7) An owner or operator shall notify the Regional Administrator in writing within 30 days whenever:

(i) A claim results in a reduction in the amount of financial assurance for liability coverage provided by a financial instrument authorized in paragraphs (a)(1) through (a)(6) of this section; or

(ii) A Certification of Valid Claim for bodily injury or property damages caused by a sudden or non-sudden accidental occurrence arising from the operation of a hazardous waste treatment, storage, or disposal facility is entered between the owner or operator and third-party claimant for liability coverage under paragraphs (a)(1) through (a)(6) of this section; or

(iii) A final court order establishing a judgment for bodily injury or property damage caused by a sudden or non-sudden accidental occurrence arising from the operation of a hazardous waste treatment, storage, or disposal facility is issued against the owner or operator or an instrument that is providing financial assurance for liability coverage under paragraphs (a)(1) through (a)(6) of this section.

(b) *Coverage for nonsudden accidental occurrences.* An owner or operator of a surface impoundment, landfill, land treatment facility, or disposal miscellaneous unit that is used to manage hazardous waste, or a group of such facilities, must demonstrate financial responsibility for bodily injury and property damage to third parties caused by nonsudden accidental occurrences arising from operations of the facility or group of facilities. The owner or operator must have and maintain liability coverage for nonsudden accidental occurrences in the amount of at least $3 million per occurrence with an annual aggregate of at least $6 million, exclusive of legal defense costs. An owner or operator who must meet the requirements of this section may combine the required per-occurrence coverage levels for sudden and nonsudden accidental occurrences into a single per-occurrence level, and combine the required annual aggregate coverage levels for sudden and nonsudden accidental occurrences into a single annual aggregate level. Owners or operators who combine coverage levels for sudden and nonsudden accidental occurrences must maintain liability coverage in the amount of at least $4 million per occurrence and $8 million annual aggregate. This liability coverage may be demonstrated as specified in paragraphs (b) (1), (2), (3), (4), (5), or (6), of this section:

(1) An owner or operator may demonstrate the required liability coverage by having liability insurance as specified in this paragraph.

(i) Each insurance policy must be amended by attachment of the Hazardous Waste Facility Liability Endorsement or evidenced by a Certificate of Liability Insurance. The word-

ing of the endorsement must be identical to the wording specified in §264.151(i). The wording of the certificate of insurance must be identical to the wording specified in §264.151(j). The owner or operator must submit a signed duplicate original of the endorsement or the certificate of insurance to the Regional Administrator, or Regional Administrators if the facilities are located in more than one Region. If requested by a Regional Administrator, the owner or operator must provide a signed duplicate original of the insurance policy. An owner or operator of a new facility must submit the signed duplicate original of the Hazardous Waste Facility Liability Endorsement or the Certificate of Liability Insurance to the Regional Administrator at least 60 days before the date on which hazardous waste is first received for treatment, storage, or disposal. The insurance must be effective before this initial receipt of hazardous waste.

(ii) Each insurance policy must be issued by an insurer which, at a minimum, is licensed to transact the business of insurance, or eligible to provide insurance as an excess or surplus lines insurer, in one or more States.

(2) An owner or operator may meet the requirements of this section by passing a financial test or using the guarantee for liability coverage as specified in paragraphs (f) and (g) of this section.

(3) An owner or operator may meet the requirements of this section by obtaining a letter of credit for liability coverage as specified in paragraph (h) of this section.

(4) An owner or operator may meet the requirements of this section by obtaining a surety bond for liability coverage as specified in paragraph (i) of this section.

(5) An owner or operator may meet the requirements of this section by obtaining a trust fund for liability coverage as specified in paragraph (j) of this section.

(6) An owner or operator may demonstrate the required liability coverage through the use of combinations of insurance, financial test, guarantee, letter of credit, surety bond, and trust fund, except that the owner or operator

may not combine a financial test covering part of the liability coverage requirement with a guarantee unless the financial statement of the owner or operator is not consolidated with the financial statement of the guarantor. The amounts of coverage demonstrated must total at least the minimum amount required by this section. If the owner or operator demonstrates the required coverage through the use of a combination of financial assurances under this paragraph, the owner or operator shall specify at least one such assurance as "primary" coverage and shall specify other assurance as "excess" coverage.

(7) An owner or operator shall notify the Regional Administrator in writing within 30 days whenever:

(i) A Claim results in a reduction in the amount of financial assurance for liability coverage provided by a financial instrument authorized in paragraphs (b)(1) through (b)(6) of this section; or

(ii) A Certification of Valid Claim for bodily injury or property damages caused by a sudden or non-sudden accidental occurrence arising from the operation of a hazardous waste treatment, storage, or disposal facility is entered between the owner or operator and third-party claimant for liability coverage under paragraphs (b)(1) through (b)(6) of this section; or

(iii) A final court order establishing a judgment for bodily injury or property damage caused by a sudden or non-sudden accidental occurrence arising from the operation of a hazardous waste treatment, storage, or disposal facility is issued against the owner or operator or an instrument that is providing financial assurance for liability coverage under paragraphs (b)(1) through (b)(6) of this section.

(c) *Request for variance.* If an owner or operator can demonstrate to the satisfaction of the Regional Administrator that the levels of financial responsibility required by paragraph (a) or (b) of this section are not consistent with the degree and duration of risk associated with treatment, storage, or disposal at the facility or group of facilities, the owner or operator may obtain a variance from the Regional Administrator. The request for a variance must be submitted to the Regional Administrator as part of the application under § 270.14 of this chapter for a facility that does not have a permit, or pursuant to the procedures for permit modification under § 124.5 of this chapter for a facility that has a permit. If granted, the variance will take the form of an adjusted level of required liability coverage, such level to be based on the Regional Administrator's assessment of the degree and duration of risk associated with the ownership or operation of the facility or group of facilities. The Regional Administrator may require an owner or operator who requests a variance to provide such technical and engineering information as is deemed necessary by the Regional Administrator to determine a level of financial responsibility other than that required by paragraph (a) of this section. Any request for a variance for a permitted facility will be treated as a request for a permit modification under §§ 270.41(a)(5) and 124.5 of this chapter.

(d) *Adjustments by the Regional Administrator.* If the Regional Administrator determines that the levels of financial responsibility required by paragraph (a) or (b) of this section are not consistent with the degree and duration of risk associated with treatment, storage, or disposal at the facility or group of facilities, the Regional Administrator may adjust the level of financial responsibility required under paragraph (a) or (b) of this section as may be necessary to protect human health and the environment. This adjusted level will be based on the Regional Administrator's assessment of the degree and duration of risk associated with the ownership or operation of the facility or group of facilities. In addition, if the Regional Administrator determines that there is a significant risk to human health and the environment from nonsudden accidental occurrences resulting from the operations of a facility that is not a surface impoundment, landfill, or land treatment facility, he may require that an owner or operator of the facility comply with paragraph (b) of this section. An owner or operator must furnish to the Regional Administrator, within a reasonable time, any information which the

Regional Administrator requests to determine whether cause exists for such adjustments of level or type of coverage. Any adjustment of the level or type of coverage for a facility that has a permit will be treated as a permit modification under §§270.41(a)(5) and 124.5 of this chapter.

(e) *Period of coverage.* Within 60 days after receiving certifications from the owner or operator and a qualified Professional Engineer that final closure has been completed in accordance with the approved closure plan, the Regional Administrator will notify the owner or operator in writing that he is no longer required by this section to maintain liability coverage for that facility, unless the Regional Administrator has reason to believe that closure has not been in accordance with the approved closure plan.

(f) *Financial test for liability coverage.* (1) An owner or operator may satisfy the requirements of this section by demonstrating that he passes a financial test as specified in this paragraph. To pass this test the owner or operator must meet the criteria of paragraph (f)(1)(i) or (ii):

(i) The owner or operator must have:

(A) Net working capital and tangible net worth each at least six times the amount of liability coverage to be demonstrated by this test; and

(B) Tangible net worth of at least $10 million; and

(C) Assets in the United States amounting to either: (*1*) At least 90 percent of his total assets; or (*2*) at least six times the amount of liability coverage to be demonstrated by this test.

(ii) The owner or operator must have:

(A) A current rating for his most recent bond issuance of AAA, AA, A, or BBB as issued by Standard and Poor's, or Aaa, Aa, A, or Baa as issued by Moody's; and

(B) Tangible net worth of at least $10 million; and

(C) Tangible net worth at least six times the amount of liability coverage to be demonstrated by this test; and

(D) Assets in the United States amounting to either: (*1*) At least 90 percent of his total assets; or (*2*) at least six times the amount of liability coverage to be demonstrated by this test.

(2) The phrase "amount of liability coverage" as used in paragraph (f)(1) of this section refers to the annual aggregate amounts for which coverage is required under paragraphs (a) and (b) of this section.

(3) To demonstrate that he meets this test, the owner or operator must submit the following three items to the Regional Administrator:

(i) A letter signed by the owner's or operator's chief financial officer and worded as specified in §264.151(g). If an owner or operator is using the financial test to demonstrate both assurance for closure or post-closure care, as specified by §§264.143(f), 264.145(f), 265.143(e), and 265.145(e), and liability coverage, he must submit the letter specified in §264.151(g) to cover both forms of financial responsibility; a separate letter as specified in §264.151(f) is not required.

(ii) A copy of the independent certified public accountant's report on examination of the owner's or operator's financial statements for the latest completed fiscal year.

(iii) A special report from the owner's or operator's independent certified public accountant to the owner or operator stating that:

(A) He has compared the data which the letter from the chief financial officer specifies as having been derived from the independently audited, year-end financial statements for the latest fiscal year with the amounts in such financial statements; and

(B) In connection with that procedure, no matters came to his attention which caused him to believe that the specified data should be adjusted.

(4) An owner or operator of a new facility must submit the items specified in paragraph (f)(3) of this section to the Regional Administrator at least 60 days before the date on which hazardous waste is first received for treatment, storage, or disposal.

(5) After the initial submission of items specified in paragraph (f)(3) of this section, the owner or operator must send updated information to the Regional Administrator within 90 days after the close of each succeeding fiscal year. This information must consist of all three items specified in paragraph (f)(3) of this section.

(6) If the owner or operator no longer meets the requirements of paragraph (f)(1) of this section, he must obtain insurance, a letter of credit, a surety bond, a trust fund, or a guarantee for the entire amount of required liability coverage as specified in this section. Evidence of liability coverage must be submitted to the Regional Administrator within 90 days after the end of the fiscal year for which the year-end financial data show that the owner or operator no longer meets the test requirements.

(7) The Regional Administrator may disallow use of this test on the basis of qualifications in the opinion expressed by the independent certified public accountant in his report on examination of the owner's or operator's financial statements (see paragraph (f)(3)(ii) of this section). An adverse opinion or a disclaimer of opinion will be cause for disallowance. The Regional Administrator will evaluate other qualifications on an individual basis. The owner or operator must provide evidence of insurance for the entire amount of required liability coverage as specified in this section within 30 days after notification of disallowance.

(g) *Guarantee for liability coverage.* (1) Subject to paragraph (g)(2) of this section, an owner or operator may meet the requirements of this section by obtaining a written guarantee, hereinafter referred to as "guarantee." The guarantor must be the direct or higher-tier parent corporation of the owner or operator, a firm whose parent corporation is also the parent corporation of the owner or operator, or a firm with a "substantial business relationship" with the owner or operator. The guarantor must meet the requirements for owners or operators in paragraphs (f)(1) through (f)(6) of this section. The wording of the guarantee must be identical to the wording specified in § 264.151(h)(2) of this part. A certified copy of the guarantee must accompany the items sent to the Regional Administrator as specified in paragraph (f)(3) of this section. One of these items must be the letter from the guarantor's chief financial officer. If the guarantor's parent corporation is also the parent corporation of the owner or operator, this letter must describe the value received

in consideration of the guarantee. If the guarantor is a firm with a "substantial business relationship" with the owner or operator, this letter must describe this "substantial business relationship" and the value received in consideration of the guarantee.

(i) If the owner or operator fails to satisfy a judgment based on a determination of liability for bodily injury or property damage to third parties caused by sudden or nonsudden accidental occurrences (or both as the case may be), arising from the operation of facilities covered by this corporate guarantee, or fails to pay an amount agreed to in settlement of claims arising from or alleged to arise from such injury or damage, the guarantor will do so up to the limits of coverage.

(ii) [Reserved]

(2)(i) In the case of corporations incorporated in the United States, a guarantee may be used to satisfy the requirements of this section only if the Attorneys General or Insurance Commissioners of (A) the State in which the guarantor is incorporated, and (B) each State in which a facility covered by the guarantee is located have submitted a written statement to EPA that a guarantee executed as described in this section and § 264.151(h)(2) is a legally valid and enforceable obligation in that State.

(ii) In the case of corporations incorporated outside the United States, a guarantee may be used to satisfy the requirements of this section only if (A) the non-U.S. corporation has identified a registered agent for service of process in each State in which a facility covered by the guarantee is located and in the State in which it has its principal place of business, and (B) the Attorney General or Insurance Commissioner of each State in which a facility covered by the guarantee is located and the State in which the guarantor corporation has its principal place of business, has submitted a written statement to EPA that a guarantee executed as described in this section and § 264.151(h)(2) is a legally valid and enforceable obligation in that State.

(h) *Letter of credit for liability coverage.* (1) An owner or operator may satisfy the requirements of this section by obtaining an irrevocable standby letter of credit that conforms to the requirements of this paragraph and submitting a copy of the letter of credit to the Regional Administrator.

(2) The financial institution issuing the letter of credit must be an entity that has the authority to issue letters of credit and whose letter of credit operations are regulated and examined by a Federal or State agency.

(3) The wording of the letter of credit must be identical to the wording specified in §264.151(k) of this part.

(4) An owner or operator who uses a letter of credit to satisfy the requirements of this section may also establish a standby trust fund. Under the terms of such a letter of credit, all amounts paid pursuant to a draft by the trustee of the standby trust will be deposited by the issuing institution into the standby trust in accordance with instructions from the trustee. The trustee of the standby trust fund must be an entity which has the authority to act as a trustee and whose trust operations are regulated and examined by a Federal or State agency.

(5) The wording of the standby trust fund must be identical to the wording specified in §264.151(n).

(i) *Surety bond for liability coverage.* (1) An owner or operator may satisfy the requirements of this section by obtaining a surety bond that conforms to the requirements of this paragraph and submitting a copy of the bond to the Regional Administrator.

(2) The surety company issuing the bond must be among those listed as acceptable sureties on Federal bonds in the most recent Circular 570 of the U.S. Department of the Treasury.

(3) The wording of the surety bond must be identical to the wording specified in §264.151(l) of this part.

(4) A surety bond may be used to satisfy the requirements of this section only if the Attorneys General or Insurance Commissioners of (i) the State in which the surety is incorporated, and (ii) each State in which a facility covered by the surety bond is located have submitted a written statement to EPA that a surety bond executed as described in this section and §264.151(l) of this part is a legally valid and enforceable obligation in that State.

(j) *Trust fund for liability coverage.* (1) An owner or operator may satisfy the requirements of this section by establishing a trust fund that conforms to the requirements of this paragraph and submitting an originally signed duplicate of the trust agreement to the Regional Administrator.

(2) The trustee must be an entity which has the authority to act as a trustee and whose trust operations are regulated and examined by a Federal or State agency.

(3) The trust fund for liability coverage must be funded for the full amount of the liability coverage to be provided by the trust fund before it may be relied upon to satisfy the requirements of this section. If at any time after the trust fund is created the amount of funds in the trust fund is reduced below the full amount of the liability coverage to be provided, the owner or operator, by the anniversary date of the establishment of the fund, must either add sufficient funds to the trust fund to cause its value to equal the full amount of liability coverage to be provided, or obtain other financial assurance as specified in this section to cover the difference. For purposes of this paragraph, "the full amount of the liability coverage to be provided" means the amount of coverage for sudden and/or nonsudden occurrences required to be provided by the owner or operator by this section, less the amount of financial assurance for liability coverage that is being provided by other financial assurance mechanisms being used to demonstrate financial assurance by the owner or operator.

(4) The wording of the trust fund must be identical to the wording specified in §264.151(m) of this part.

(k) Notwithstanding any other provision of this part, an owner or operator using liability insurance to satisfy the requirements of this section may use, until October 16, 1982, a Hazardous Waste Facility Liability Endorsement or Certificate of Liability Insurance that does not certify that the insurer is licensed to transact the business of insurance, or eligible as an excess or surplus lines insurer, in one or more States.

[47 FR 16554, Apr. 16, 1982, as amended at 47 FR 28627, July 1, 1982; 48 FR 30115, June 30, 1983; 51 FR 16450, May 2, 1986; 51 FR 25354, July 11, 1986; 52 FR 44320, Nov. 18, 1987; 52 FR 46964, Dec. 10, 1987; 53 FR 33950, Sept. 1, 1988; 56 FR 30200, July 1, 1991; 57 FR 42836, Sept. 16, 1992; 71 FR 16905, Apr. 4, 2006; 71 FR 40272, July 14, 2006]

**§ 264.148  Incapacity of owners or operators, guarantors, or financial institutions.**

(a) An owner or operator must notify the Regional Administrator by certified mail of the commencement of a voluntary or involuntary proceeding under Title 11 (Bankruptcy), U.S. Code, naming the owner or operator as debtor, within 10 days after commencement of the proceeding. A guarantor of a corporate guarantee as specified in §§ 264.143(f) and 264.145(f) must make such a notification if he is named as debtor, as required under the terms of the corporate guarantee (§ 264.151(h)).

(b) An owner or operator who fulfills the requirements of § 264.143, § 264.145, or § 264.147 by obtaining a trust fund, surety bond, letter of credit, or insurance policy will be deemed to be without the required financial assurance or liability coverage in the event of bankruptcy of the trustee or issuing institution, or a suspension or revocation of the authority of the trustee institution to act as trustee or of the institution issuing the surety bond, letter of credit, or insurance policy to issue such instruments. The owner or operator must establish other financial assurance or liability coverage within 60 days after such an event.

**§ 264.149  Use of State-required mechanisms.**

(a) For a facility located in a State where EPA is administering the requirements of this subpart but where the State has hazardous waste regulations that include requirements for financial assurance of closure or post-closure care or liability coverage, an owner or operator may use State-required financial mechanisms to meet the requirements of § 264.143, § 264.145, or § 264.147, if the Regional Adminis-

trator determines that the State mechanisms are at least equivalent to the financial mechanism specified in this subpart. The Regional Administrator will evaluate the equivalency of the mechanisms principally in terms of (1) certainty of the availability of funds for the required closure or post-closure care activities or liability coverage and (2) the amount of funds that will be made available. The Regional Administrator may also consider other factors as he deems appropriate. The owner or operator must submit to the Regional Administrator evidence of the establishment of the mechanism together with a letter requesting that the State-required mechanism be considered acceptable for meeting the requirements of this subpart. The submission must include the following information: The facility's EPA Identification Number, name, and address, and the amount of funds for closure or post-closure care or liability coverage assured by the mechanism. The Regional Administrator will notify the owner or operator of his determination regarding the mechanism's acceptability in lieu of financial mechanisms specified in this subpart. The Regional Administrator may require the owner or operator to submit additional information as is deemed necessary to make this determination. Pending this determination, the owner or operator will be deemed to be in compliance with the requirements of § 264.143, § 264.145, or § 264.147, as applicable.

(b) If a State-required mechanism is found acceptable as specified in paragraph (a) of this section except for the amount of funds available, the owner or operator may satisfy the requirements of this subpart by increasing the funds available through the State-required mechanism or using additional financial mechanisms as specified in this subpart. The amount of funds available through the State and Federal mechanisms must at least equal the amount required by this subpart.

**§ 264.150  State assumption of responsibility.**

(a) If a State either assumes legal responsibility for an owner's or operator's compliance with the closure, post-closure care, or liability requirements

410

of this part or assures that funds will be available from State sources to cover those requirements, the owner or operator will be in compliance with the requirements of §264.143, §264.145, or §264.147 if the Regional Administrator determines that the State's assumption of responsibility is at least equivalent to the financial mechanisms specified in this subpart. The Regional Administrator will evaluate the equivalency of State guarantees principally in terms of (1) certainty of the availability of funds for the required closure or post-closure care activities or liability coverage and (2) the amount of funds that will be made available. The Regional Administrator may also consider other factors as he deems appropriate. The owner or operator must submit to the Regional Administrator a letter from the State describing the nature of the State's assumption of responsibility together with a letter from the owner or operator requesting that the State's assumption of responsibility be considered acceptable for meeting the requirements of this subpart. The letter from the State must include, or have attached to it, the following information: the facility's EPA Identification Number, name, and address, and the amount of funds for closure or post-closure care or liability coverage that are guaranteed by the State. The Regional Administrator will notify the owner or operator of his determination regarding the acceptability of the State's guarantee in lieu of financial mechanisms specified in this subpart. The Regional Administrator may require the owner or operator to submit additional information as is deemed necessary to make this determination. Pending this determination, the owner or operator will be deemed to be in compliance with the requirements of §264.143, §264.145, or §264.147, as applicable.

(b) If a State's assumption of responsibility is found acceptable as specified in paragraph (a) of this section except for the amount of funds available, the owner or operator may satisfy the requirements of this subpart by use of both the State's assurance *and* additional financial mechanisms as specified in this subpart. The amount of funds available through the State and

Federal mechanisms must at least equal the amount required by this subpart.

### §264.151 Wording of the instruments.

(a)(1) A trust agreement for a trust fund, as specified in §264.143(a) or §264.145(a) or §265.143(a) or §265.145(a) of this chapter, must be worded as follows, except that instructions in brackets are to be replaced with the relevant information and the brackets deleted:

TRUST AGREEMENT

Trust Agreement, the "Agreement," entered into as of [date] by and between [name of the owner or operator], a [name of State] [insert "corporation," "partnership," "association," or "proprietorship"], the "Grantor," and [name of corporate trustee], [insert "incorporated in the State of _____" or "a national bank"], the "Trustee."

Whereas, the United States Environmental Protection Agency, "EPA," an agency of the United States Government, has established certain regulations applicable to the Grantor, requiring that an owner or operator of a hazardous waste management facility shall provide assurance that funds will be available when needed for closure and/or post-closure care of the facility,

Whereas, the Grantor has elected to establish a trust to provide all or part of such financial assurance for the facilities identified herein,

Whereas, the Grantor, acting through its duly authorized officers, has selected the Trustee to be the trustee under this agreement, and the Trustee is willing to act as trustee,

Now, Therefore, the Grantor and the Trustee agree as follows:

*Section 1. Definitions.* As used in this Agreement:

(a) The term "Grantor" means the owner or operator who enters into this Agreement and any successors or assigns of the Grantor.

(b) The term "Trustee" means the Trustee who enters into this Agreement and any successor Trustee.

*Section 2. Identification of Facilities and Cost Estimates.* This Agreement pertains to the facilities and cost estimates identified on attached Schedule A [on Schedule A, for each facility list the EPA Identification Number, name, address, and the current closure and/or post-closure cost estimates, or portions thereof, for which financial assurance is demonstrated by this Agreement].

*Section 3. Establishment of Fund.* The Grantor and the Trustee hereby establish a trust fund, the "Fund," for the benefit of EPA. The Grantor and the Trustee intend that no third party have access to the Fund except as herein provided. The Fund is established initially as consisting of the property, which is acceptable to the Trustee, described in Schedule B attached hereto. Such property and any other property subsequently transferred to the Trustee is referred to as the Fund, together with all earnings and profits thereon, less any payments or distributions made by the Trustee pursuant to this Agreement. The Fund shall be held by the Trustee, IN TRUST, as hereinafter provided. The Trustee shall not be responsible nor shall it undertake any responsibility for the amount or adequacy of, nor any duty to collect from the Grantor, any payments necessary to discharge any liabilities of the Grantor established by EPA.

*Section 4. Payment for Closure and Post-Closure Care.* The Trustee shall make payments from the Fund as the EPA Regional Administrator shall direct, in writing, to provide for the payment of the costs of closure and/or post-closure care of the facilities covered by this Agreement. The Trustee shall reimburse the Grantor or other persons as specified by the EPA Regional Administrator from the Fund for closure and post-closure expenditures in such amounts as the EPA Regional Administrator shall direct in writing. In addition, the Trustee shall refund to the Grantor such amounts as the EPA Regional Administrator specifies in writing. Upon refund, such funds shall no longer constitute part of the Fund as defined herein.

*Section 5. Payments Comprising the Fund.* Payments made to the Trustee for the Fund shall consist of cash or securities acceptable to the Trustee.

*Section 6. Trustee Management.* The Trustee shall invest and reinvest the principal and income of the Fund and keep the Fund invested as a single fund, without distinction between principal and income, in accordance with general investment policies and guidelines which the Grantor may communicate in writing to the Trustee from time to time, subject, however, to the provisions of this section. In investing, reinvesting, exchanging, selling, and managing the Fund, the Trustee shall discharge his duties with respect to the trust fund solely in the interest of the beneficiary and with the care, skill, prudence, and diligence under the circumstances then prevailing which persons of prudence, acting in a like capacity and familiar with such matters, would use in the conduct of an enterprise of a like character and with like aims; *except that:*

(i) Securities or other obligations of the Grantor, or any other owner or operator of the facilities, or any of their affiliates as defined in the Investment Company Act of 1940, as amended, 15 U.S.C. 80a-2.(a), shall not be acquired or held, unless they are securities or other obligations of the Federal or a State government;

(ii) The Trustee is authorized to invest the Fund in time or demand deposits of the Trustee, to the extent insured by an agency of the Federal or State government; and

(iii) The Trustee is authorized to hold cash awaiting investment or distribution uninvested for a reasonable time and without liability for the payment of interest thereon.

*Section 7. Commingling and Investment.* The Trustee is expressly authorized in its discretion:

(a) To transfer from time to time any or all of the assets of the Fund to any common, commingled, or collective trust fund created by the Trustee in which the Fund is eligible to participate, subject to all of the provisions thereof, to be commingled with the assets of other trusts participating therein; and

(b) To purchase shares in any investment company registered under the Investment Company Act of 1940, 15 U.S.C. 80a-1 et seq., including one which may be created, managed, underwritten, or to which investment advice is rendered or the shares of which are sold by the Trustee. The Trustee may vote such shares in its discretion.

*Section 8. Express Powers of Trustee.* Without in any way limiting the powers and discretions conferred upon the Trustee by the other provisions of this Agreement or by law, the Trustee is expressly authorized and empowered:

(a) To sell, exchange, convey, transfer, or otherwise dispose of any property held by it, by public or private sale. No person dealing with the Trustee shall be bound to see to the application of the purchase money or to inquire into the validity or expediency of any such sale or other disposition;

(b) To make, execute, acknowledge, and deliver any and all documents of transfer and conveyance and any and all other instruments that may be necessary or appropriate to carry out the powers herein granted;

(c) To register any securities held in the Fund in its own name or in the name of a nominee and to hold any security in bearer form or in book entry, or to combine certificates representing such securities with certificates of the same issue held by the Trustee in other fiduciary capacities, or to deposit or arrange for the deposit of such securities in a qualified central depositary even though, when so deposited, such securities may be merged and held in bulk in the name of the nominee of such depositary with other securities deposited therein by another person, or to deposit or arrange for the deposit of any securities issued by the United States Government, or any agency or instrumentality thereof, with a Federal Reserve bank, but the books and records of the Trustee shall at all times show that all such securities are part of the Fund;

(d) To deposit any cash in the Fund in interest-bearing accounts maintained or savings certificates issued by the Trustee, in its separate corporate capacity, or in any other banking institution affiliated with the Trustee, to the extent insured by an agency of the Federal or State government; and

(e) To compromise or otherwise adjust all claims in favor of or against the Fund.

*Section 9. Taxes and Expenses.* All taxes of any kind that may be assessed or levied against or in respect of the Fund and all brokerage commissions incurred by the Fund shall be paid from the Fund. All other expenses incurred by the Trustee in connection with the administration of this Trust, including fees for legal services rendered to the Trustee, the compensation of the Trustee to the extent not paid directly by the Grantor, and all other proper charges and disbursements of the Trustee shall be paid from the Fund.

*Section 10. Annual Valuation.* The Trustee shall annually, at least 30 days prior to the anniversary date of establishment of the Fund, furnish to the Grantor and to the appropriate EPA Regional Administrator a statement confirming the value of the Trust. Any securities in the Fund shall be valued at market value as of no more than 60 days prior to the anniversary date of establishment of the Fund. The failure of the Grantor to object in writing to the Trustee within 90 days after the statement has been furnished to the Grantor and the EPA Regional Administrator shall constitute a conclusively binding assent by the Grantor, barring the Grantor from asserting any claim or liability against the Trustee with respect to matters disclosed in the statement.

*Section 11. Advice of Counsel.* The Trustee may from time to time consult with counsel, who may be counsel to the Grantor, with respect to any question arising as to the construction of this Agreement or any action to be taken hereunder. The Trustee shall be fully protected, to the extent permitted by law, in acting upon the advice of counsel.

*Section 12. Trustee Compensation.* The Trustee shall be entitled to reasonable compensation for its services as agreed upon in writing from time to time with the Grantor.

*Section 13. Successor Trustee.* The Trustee may resign or the Grantor may replace the Trustee, but such resignation or replacement shall not be effective until the Grantor has appointed a successor trustee and this successor accepts the appointment. The successor trustee shall have the same powers and duties as those conferred upon the Trustee hereunder. Upon the successor trustee's acceptance of the appointment, the Trustee shall assign, transfer, and pay over to the successor trustee the funds and properties then constituting the Fund. If for any reason the Grantor cannot or does not act in the event of the resignation of the Trustee, the Trustee may apply to a court of competent jurisdiction for the appointment of a successor trustee or for instructions. The successor trustee shall specify the date on which it assumes administration of the trust in a writing sent to the Grantor, the EPA Regional Administrator, and the present Trustee by certified mail 10 days before such change becomes effective. Any expenses incurred by the Trustee as a result of any of the acts contemplated by this Section shall be paid as provided in Section 9.

*Section 14. Instructions to the Trustee.* All orders, requests, and instructions by the Grantor to the Trustee shall be in writing, signed by such persons as are designated in the attached Exhibit A or such other designees as the Grantor may designate by amendment to Exhibit A. The Trustee shall be fully protected in acting without inquiry in accordance with the Grantor's orders, requests, and instructions. All orders, requests, and instructions by the EPA Regional Administrator to the Trustee shall be in writing, signed by the EPA Regional Administrators of the Regions in which the facilities are located, or their designees, and the Trustee shall act and shall be fully protected in acting in accordance with such orders, requests, and instructions. The Trustee shall have the right to assume, in the absence of written notice to the contrary, that no event constituting a change or a termination of the authority of any person to act on behalf of the Grantor or EPA hereunder has occurred. The Trustee shall have no duty to act in the absence of such orders, requests, and instructions from the Grantor and/or EPA, except as provided for herein.

*Section 15. Notice of Nonpayment.* The Trustee shall notify the Grantor and the appropriate EPA Regional Administrator, by certified mail within 10 days following the expiration of the 30-day period after the anniversary of the establishment of the Trust, if no payment is received from the Grantor during that period. After the pay-in period is completed, the Trustee shall not be required to send a notice of nonpayment.

*Section 16. Amendment of Agreement.* This Agreement may be amended by an instrument in writing executed by the Grantor, the Trustee, and the appropriate EPA Regional Administrator, or by the Trustee and the appropriate EPA Regional Administrator if the Grantor ceases to exist.

*Section 17. Irrevocability and Termination.* Subject to the right of the parties to amend this Agreement as provided in Section 16, this Trust shall be irrevocable and shall continue until terminated at the written agreement of the Grantor, the Trustee, and the EPA Regional Administrator, or by the Trustee and the EPA Regional Administrator, if the Grantor ceases to exist. Upon termination of the Trust, all remaining trust property, less final trust administration expenses, shall be delivered to the Grantor.

*Section 18. Immunity and Indemnification.* The Trustee shall not incur personal liability of any nature in connection with any act or omission, made in good faith, in the administration of this Trust, or in carrying out any directions by the Grantor or the EPA Regional Administrator issued in accordance with this Agreement. The Trustee shall be indemnified and saved harmless by the Grantor or from the Trust Fund, or both, from and against any personal liability to which the Trustee may be subjected by reason of any act or conduct in its official capacity, including all expenses reasonably incurred in its defense in the event the Grantor fails to provide such defense.

*Section 19. Choice of Law.* This Agreement shall be administered, construed, and enforced according to the laws of the State of [insert name of State].

*Section 20. Interpretation.* As used in this Agreement, words in the singular include the plural and words in the plural include the singular. The descriptive headings for each Section of this Agreement shall not affect the interpretation or the legal efficacy of this Agreement.

In Witness Whereof the parties have caused this Agreement to be executed by their respective officers duly authorized and their corporate seals to be hereunto affixed and attested as of the date first above written: The parties below certify that the wording of this Agreement is identical to the wording specified in 40 CFR 264.151(a)(1) as such regulations were constituted on the date first above written.

> [Signature of Grantor]
> [Title]

Attest:
> [Title]
> [Seal]
> [Signature of Trustee]

Attest:
> [Title]
> [Seal]

(2) The following is an example of the certification of acknowledgment which must accompany the trust agreement for a trust fund as specified in §§ 264.143(a) and 264.145(a) or §§ 265.143(a) or 265.145(a) of this chapter. State requirements may differ on the proper content of this acknowledgment.

State of _____
County of _____

On this [date], before me personally came [owner or operator] to me known, who, being by me duly sworn, did depose and say that she/he resides at [address], that she/he is [title] of [corporation], the corporation described in and which executed the above instrument; that she/he knows the seal of said corporation; that the seal affixed to such instrument is such corporate seal; that it was

so affixed by order of the Board of Directors of said corporation, and that she/he signed her/his name thereto by like order.

[Signature of Notary Public]

(b) A surety bond guaranteeing payment into a trust fund, as specified in § 264.143(b) or § 264.145(b) or § 265.143(b) or § 265.145(b) of this chapter, must be worded as follows, except that instructions in brackets are to be replaced with the relevant information and the brackets deleted:

FINANCIAL GUARANTEE BOND

Date bond executed:

Effective date:

Principal: [legal name and business address of owner or operator]

Type of Organization: [insert "individual," "joint venture," "partnership," or "corporation"]

State of incorporation: _____

Surety(ies): [name(s) and business address(es)]

EPA Identification Number, name, address and closure and/or post-closure amount(s) for each facility guaranteed by this bond [indicate closure and post-closure amounts separately]: _____

Total penal sum of bond: $ _____

Surety's bond number: _____

Know All Persons By These Presents, That we, the Principal and Surety(ies) hereto are firmly bound to the U.S. Environmental Protection Agency (hereinafter called EPA), in the above penal sum for the payment of which we bind ourselves, our heirs, executors, administrators, successors, and assigns jointly and severally; provided that, where the Surety(ies) are corporations acting as co-sureties, we, the Sureties, bind ourselves in such sum "jointly and severally" only for the purpose of allowing a joint action or actions against any or all of us, and for all other purposes each Surety binds itself, jointly and severally with the Principal, for the payment of such sum only as is set forth opposite the name of such Surety, but if no limit of liability is indicated, the limit of liability shall be the full amount of the penal sum.

Whereas said Principal is required, under the Resource Conservation and Recovery Act as amended (RCRA), to have a permit or interim status in order to own or operate each hazardous waste management facility identified above, and

Whereas said Principal is required to provide financial assurance for closure, or closure and post-closure care, as a condition of the permit or interim status, and

414

Whereas said Principal shall establish a standby trust fund as is required when a surety bond is used to provide such financial assurance;

Now, Therefore, the conditions of the obligation are such that if the Principal shall faithfully, before the beginning of final closure of each facility identified above, fund the standby trust fund in the amount(s) identified above for the facility,

Or, if the Principal shall fund the standby trust fund in such amount(s) within 15 days after a final order to begin closure is issued by an EPA Regional Administrator or a U.S. district court or other court of competent jurisdiction,

Or, if the Principal shall provide alternate financial assurance, as specified in subpart H of 40 CFR part 264 or 265, as applicable, and obtain the EPA Regional Administrator's written approval of such assurance, within 90 days after the date notice of cancellation is received by the Principal and the EPA Regional Administrator(s) from the Surety(ies), then this obligation shall be null and void; otherwise it is to remain in full force and effect.

The Surety(ies) shall become liable on this bond obligation only when the Principal has failed to fulfill the conditions described above. Upon notification by an EPA Regional Administrator that the Principal has failed to perform as guaranteed by this bond, the Surety(ies) shall place funds in the amount guaranteed for the facility(ies) into the standby trust fund as directed by the EPA Regional Administrator.

The liability of the Surety(ies) shall not be discharged by any payment or succession of payments hereunder, unless and until such payment or payments shall amount in the aggregate to the penal sum of the bond, but in no event shall the obligation of the Surety(ies) hereunder exceed the amount of said penal sum.

The Surety(ies) may cancel the bond by sending notice of cancellation by certified mail to the Principal and to the EPA Regional Administrator(s) for the Region(s) in which the facility(ies) is (are) located, provided, however, that cancellation shall not occur during the 120 days beginning on the date of receipt of the notice of cancellation by both the Principal and the EPA Regional Administrator(s), as evidenced by the return receipts.

The Principal may terminate this bond by sending written notice to the Surety(ies), provided, however, that no such notice shall become effective until the Surety(ies) receive(s) written authorization for termination of the bond by the EPA Regional Administrator(s) of the EPA Region(s) in which the bonded facility(ies) is (are) located.

[The following paragraph is an optional rider that may be included but is not required.]

Principal and Surety(ies) hereby agree to adjust the penal sum of the bond yearly so that it guarantees a new closure and/or post-closure amount, provided that the penal sum does not increase by more than 20 percent in any one year, and no decrease in the penal sum takes place without the written permission of the EPA Regional Administrator(s).

In Witness Whereof, the Principal and Surety(ies) have executed this Financial Guarantee Bond and have affixed their seals on the date set forth above.

The persons whose signatures appear below hereby certify that they are authorized to execute this surety bond on behalf of the Principal and Surety(ies) and that the wording of this surety bond is identical to the wording specified in 40 CFR 264.151(b) as such regulations were constituted on the date this bond was executed.

Principal

[Signature(s)] _____

[Name(s)] _____

[Title(s)] _____

[Corporate seal] _____

Corporate Surety(ies)

[Name and address]
State of incorporation: _____

Liability limit: $ _____

[Signature(s)]
[Name(s) and title(s)]
[Corporate seal]
[For every co-surety, provide signature(s), corporate seal, and other information in the same manner as for Surety above.]
Bond premium: $ _____

(c) A surety bond guaranteeing performance of closure and/or post-closure care, as specified in § 264.143(c) or § 264.145(c), must be worded as follows, except that the instructions in brackets are to be replaced with the relevant information and the brackets deleted:

PERFORMANCE BOND

Date bond executed: _____

Effective date: _____

Principal: [legal name and business address of owner or operator]

Type of organization: [insert "individual," "joint venture," "partnership," or "corporation"]

State of incorporation: _____

Surety(ies): [name(s) and business address(es)] _____

EPA Identification Number, name, address, and closure and/or post-closure amount(s) for each facility guaranteed by this bond [indicate closure and post-closure amounts separately]: _____

Total penal sum of bond: $ _____
Surety's bond number: _____
Know All Persons By These Presents, That we, the Principal and Surety(ies) hereto are firmly bound to the U.S. Environmental Protection Agency (hereinafter called EPA), in the above penal sum for the payment of which we bind ourselves, our heirs, executors, administrators, successors, and assigns jointly and severally; provided that, where the Surety(ies) are corporations acting as co-sureties, we, the Sureties, bind ourselves in such sum "jointly and severally" only for the purpose of allowing a joint action or actions against any or all of us, and for all other purposes each Surety binds itself, jointly and severally with the Principal, for the payment of such sum only as is set forth opposite the name of such Surety, but if no limit of liability is indicated, the limit of liability shall be the full amount of the penal sum.

Whereas said Principal is required, under the Resource Conservation and Recovery Act as amended (RCRA), to have a permit in order to own or operate each hazardous waste management facility identified above, and

Whereas said Principal is required to provide financial assurance for closure, or closure and post-closure care, as a condition of the permit, and

Whereas said Principal shall establish a standby trust fund as is required when a surety bond is used to provide such financial assurance;

Now, Therefore, the conditions of this obligation are such that if the Principal shall faithfully perform closure, whenever required to do so, of each facility for which this bond guarantees closure, in accordance with the closure plan and other requirements of the permit as such plan and permit may be amended, pursuant to all applicable laws, statutes, rules, and regulations, as such laws, statutes, rules, and regulations may be amended,

And, if the Principal shall faithfully perform post-closure care of each facility for which this bond guarantees post-closure care, in accordance with the post-closure plan and other requirements of the permit, as such plan and permit may be amended, pursuant to all applicable laws, statutes, rules, and regulations, as such laws, statutes, rules, and regulations may be amended,

Or, if the Principal shall provide alternate financial assurance as specified in subpart H of 40 CFR part 264, and obtain the EPA Regional Administrator's written approval of such assurance, within 90 days after the date notice of cancellation is received by both the Principal and the EPA Regional Administrator(s) from the Surety(ies), then this obligation shall be null and void, otherwise it is to remain in full force and effect.

The Surety(ies) shall become liable on this bond obligation only when the Principal has failed to fulfill the conditions described above.

Upon notification by an EPA Regional Administrator that the Principal has been found in violation of the closure requirements of 40 CFR part 264, for a facility for which this bond guarantees performance of closure, the Surety(ies) shall either perform closure in accordance with the closure plan and other permit requirements or place the closure amount guaranteed for the facility into the standby trust fund as directed by the EPA Regional Administrator.

Upon notification by an EPA Regional Administrator that the Principal has been found in violation of the post-closure requirements of 40 CFR part 264 for a facility for which this bond guarantees performance of post-closure care, the Surety(ies) shall either perform post-closure care in accordance with the post-closure plan and other permit requirements or place the post-closure amount guaranteed for the facility into the standby trust fund as directed by the EPA Regional Administrator.

Upon notification by an EPA Regional Administrator that the Principal has failed to provide alternate financial assurance as specified in subpart H of 40 CFR part 264, and obtain written approval of such assurance from the EPA Regional Administrator(s) during the 90 days following receipt by both the Principal and the EPA Regional Administrator(s) of a notice of cancellation of the bond, the Surety(ies) shall place funds in the amount guaranteed for the facility(ies) into the standby trust fund as directed by the EPA Regional Administrator.

The surety(ies) hereby waive(s) notification of amendments to closure plans, permits, applicable laws, statutes, rules, and regulations and agrees that no such amendment shall in any way alleviate its (their) obligation on this bond.

The liability of the Surety(ies) shall not be discharged by any payment or succession of payments hereunder, unless and until such payment or payments shall amount in the aggregate to the penal sum of the bond, but in no event shall the obligation of the Surety(ies) hereunder exceed the amount of said penal sum.

The Surety(ies) may cancel the bond by sending notice of cancellation by certified mail to the owner or operator and to the EPA Regional Administrator(s) for the Region(s) in which the facility(ies) is (are) located, provided, however, that cancellation shall not occur during the 120 days beginning on the date of receipt of the notice of cancellation by both the Principal and the EPA Regional Administrator(s), as evidenced by the return receipts.

The principal may terminate this bond by sending written notice to the Surety(ies), provided, however, that no such notice shall become effective until the Surety(ies) receive(s) written authorization for termination of the bond by the EPA Regional Administrator(s) of the EPA Region(s) in which the bonded facility(ies) is (are) located.

[The following paragraph is an *optional* rider that may be included but is not required.]

Principal and Surety(ies) hereby agree to adjust the penal sum of the bond yearly so that it guarantees a new closure and/or postclosure amount, provided that the penal sum does not increase by more than 20 percent in any one year, and no decrease in the penal sum takes place without the written permission of the EPA Regional Administrator(s).

In Witness Whereof, The Principal and Surety(ies) have executed this Performance Bond and have affixed their seals on the date set forth above.

The persons whose signatures appear below hereby certify that they are authorized to execute this surety bond on behalf of the Principal and Surety(ies) and that the wording of this surety bond is identical to the wording specified in 40 CFR 264.151(c) as such regulation was constituted on the date this bond was executed.

Principal

[Signature(s)]

[Name(s)]

[Title(s)]

[Corporate seal]

Corporate Surety(ies)

[Name and address]

State of incorporation: _____

Liability limit: $ _____

[Signature(s)]

[Name(s) and title(s)]

[Corporate seal]

[For every co-surety, provide signature(s), corporate seal, and other information in the same manner as for Surety above.]

Bond premium: $ _____

(d) A letter of credit, as specified in §264.143(d) or §264.145(d) or §265.143(c) or §265.145(c) of this chapter, must be worded as follows, except that instructions in brackets are to be replaced with the relevant information and the brackets deleted:

IRREVOCABLE STANDBY LETTER OF CREDIT

Regional Administrator(s)

Region(s) _____

U.S. Environmental Protection Agency

Dear Sir or Madam: We hereby establish our Irrevocable Standby Letter of Credit No. _____ in your favor, at the request and for the account of [owner's or operator's name and address] up to the aggregate amount of [in words] U.S. dollars $_____ , available upon presentation [insert, if more than one Regional Administrator is a beneficiary, "by any one of you"] of

(1) your sight draft, bearing reference to this letter of credit No. _____ , and

(2) your signed statement reading as follows: "I certify that the amount of the draft is payable pursuant to regulations issued under authority of the Resource Conservation and Recovery Act of 1976 as amended."

This letter of credit is effective as of [date] and shall expire on [date at least 1 year later], but such expiration date shall be automatically extended for a period of [at least 1 year] on [date] and on each successive expiration date, unless, at least 120 days before the current expiration date, we notify both you and [owner's or operator's name] by certified mail that we have decided not to extend this letter of credit beyond the current expiration date. In the event you are so notified, any unused portion of the credit shall be available upon presentation of your sight draft for 120 days after the date of receipt by both you and [owner's or operator's name], as shown on the signed return receipts.

Whenever this letter of credit is drawn on under and in compliance with the terms of this credit, we shall duly honor such draft upon presentation to us, and we shall deposit the amount of the draft directly into the standby trust fund of [owner's or operator's name] in accordance with your instructions.

We certify that the wording of this letter of credit is identical to the wording specified in 40 CFR 264.151(d) as such regulations were constituted on the date shown immediately below.

[Signature(s) and title(s) of official(s) of issuing institution] [Date]

This credit is subject to [insert "the most recent edition of the Uniform Customs and Practice for Documentary Credits, published and copyrighted by the International Chamber of Commerce," or "the Uniform Commercial Code"].

(e) A certificate of insurance, as specified in §264.143(e) or §264.145(e) or §265.143(d) or §265.145(d) of this chapter, must be worded as follows, except that instructions in brackets are to be replaced with the relevant information and the brackets deleted:

CERTIFICATE OF INSURANCE FOR<br>CLOSURE OR POST-CLOSURE CARE

Name and Address of Insurer
(herein called the "Insurer"): _____
Name and Address of Insured
(herein called the "Insured"): _____

417

Facilities Covered: [List for each facility: The EPA Identification Number, name, address, and the amount of insurance for closure and/or the amount for post-closure care (these amounts for all facilities covered must total the face amount shown below).]

Face Amount:
Policy Number: _____
Effective Date: _____

The Insurer hereby certifies that it has issued to the Insured the policy of insurance identified above to provide financial assurance for [insert "closure" or "closure and post-closure care" or "post-closure care"] for the facilities identified above. The Insurer further warrants that such policy conforms in all respects with the requirements of 40 CFR 264.143(e), 264.145(e), 265.143(d), and 265.145(d), as applicable and as such regulations were constituted on the date shown immediately below. It is agreed that any provision of the policy inconsistent with such regulations is hereby amended to eliminate such inconsistency.

Whenever requested by the EPA Regional Administrator(s) of the U.S. Environmental Protection Agency, the Insurer agrees to furnish to the EPA Regional Administrator(s) a duplicate original of the policy listed above, including all endorsements thereon.

I hereby certify that the wording of this certificate is identical to the wording specified in 40 CFR 264.151(e) as such regulations were constituted on the date shown immediately below.

[Authorized signature for Insurer]
[Name of person signing]
[Title of person signing]
Signature of witness or notary: _____
[Date]

(f) A letter from the chief financial officer, as specified in § 264.143(f) or 264.145(f), or § 265.143(e) or 265.145(e) of this chapter, must be worded as follows, except that instructions in brackets are to be replaced with the relevant information and the brackets deleted:

LETTER FROM CHIEF FINANCIAL OFFICER

[Address to Regional Administrator of every Region in which facilities for which financial responsibility is to be demonstrated through the financial test are located].

I am the chief financial officer of [name and address of firm]. This letter is in support of this firm's use of the financial test to demonstrate financial assurance for closure and/or post-closure costs, as specified in subpart H of 40 CFR parts 264 and 265.

[Fill out the following five paragraphs regarding facilities and associated cost estimates. If your firm has no facilities that belong in a particular paragraph, write "None"

in the space indicated. For each facility, include its EPA Identification Number, name, address, and current closure and/or post-closure cost estimates. Identify each cost estimate as to whether it is for closure or post-closure care].

1. This firm is the owner or operator of the following facilities for which financial assurance for closure or post-closure care is demonstrated through the financial test specified in subpart H of 40 CFR parts 264 and 265. The current closure and/or post-closure cost estimates covered by the test are shown for each facility: _____.

2. This firm guarantees, through the guarantee specified in subpart H of 40 CFR parts 264 and 265, the closure or post-closure care of the following facilities owned or operated by the guaranteed party. The current cost estimates for the closure or post-closure care so guaranteed are shown for each facility: _____. The firm identified above is [insert one or more: (1) The direct or higher-tier parent corporation of the owner or operator; (2) owned by the same parent corporation as the parent corporation of the owner or operator, and receiving the following value in consideration of this guarantee _____; or (3) engaged in the following substantial business relationship with the owner or operator _____, and receiving the following value in consideration of this guarantee _____]. [Attach a written description of the business relationship or a copy of the contract establishing such relationship to this letter].

3. In States where EPA is not administering the financial requirements of subpart H of 40 CFR part 264 or 265, this firm, as owner or operator or guarantor, is demonstrating financial assurance for the closure or post-closure care of the following facilities through the use of a test equivalent or substantially equivalent to the financial test specified in subpart H of 40 CFR parts 264 and 265. The current closure and/or post-closure cost estimates covered by such a test are shown for each facility: _____.

4. This firm is the owner or operator of the following hazardous waste management facilities for which financial assurance for closure or, if a disposal facility, post-closure care, is not demonstrated either to EPA or a State through the financial test or any other financial assurance mechanism specified in subpart H of 40 CFR parts 264 and 265 or equivalent or substantially equivalent State mechanisms. The current closure and/or post-closure cost estimates not covered by such financial assurance are shown for each facility: _____.

5. This firm is the owner or operator of the following UIC facilities for which financial assurance for plugging and abandonment is required under part 144. The current closure cost estimates as required by 40 CFR 144.62 are shown for each facility: _____.

This firm [insert "is required" or "is not required"] to file a Form 10K with the Securities and Exchange Commission (SEC) for the latest fiscal year.

The fiscal year of this firm ends on [month, day]. The figures for the following items marked with an asterisk are derived from this firm's independently audited, year-end financial statements for the latest completed fiscal year, ended [date].

[Fill in Alternative I if the criteria of paragraph (f)(1)(i) of §264.143 or §264.145, or of paragraph (e)(1)(i) of §265.143 or §265.145 of this chapter are used. Fill in Alternative II if the criteria of paragraph (f)(1)(ii) of §264.143 or §264.145, or of paragraph (e)(1)(ii) of §265.143 or §265.145 of this chapter are used.]

ALTERNATIVE I

1. Sum of current closure and post-closure cost estimate [total of all cost estimates shown in the five paragraphs above] $_____

*2. Total liabilities [if any portion of the closure or post-closure cost estimates is included in total liabilities, you may deduct the amount of that portion from this line and add that amount to lines 3 and 4]$_____

*3. Tangible net worth $_____

*4. Net worth $_____

*5. Current assets $_____

*6. Current liabilities $_____

7. Net working capital [line 5 minus line 6] $_____

*8. The sum of net income plus depreciation, depletion, and amortization $_____

*9. Total assets in U.S. (required only if less than 90% of firm's assets are located in the U.S.) $_____

10. Is line 3 at least $10 million? (Yes/No) _____

11. Is line 3 at least 6 times line 1? (Yes/No) _____

12. Is line 7 at least 6 times line 1? (Yes/No) _____

*13. Are at least 90% of firm's assets located in the U.S.? If not, complete line 14 (Yes/No) _____

14. Is line 9 at least 6 times line 1? (Yes/No) _____

15. Is line 2 divided by line 4 less than 2.0? (Yes/No) _____

16. Is line 8 divided by line 2 greater than 0.1? (Yes/No) _____

17. Is line 5 divided by line 6 greater than 1.5? (Yes/No) _____

ALTERNATIVE II

1. Sum of current closure and post-closure cost estimates [total of all cost estimates shown in the five paragraphs above] $_____

2. Current bond rating of most recent issuance of this firm and name of rating service _____

3. Date of issuance of bond _____

4. Date of maturity of bond _____

*5. Tangible net worth [if any portion of the closure and post-closure cost estimates is included in "total liabilities" on your firm's financial statements, you may add the amount of that portion to this line] $_____

*6. Total assets in U.S. (required only if less than 90% of firm's assets are located in the U.S.) $_____

7. Is line 5 at least $10 million? (Yes/No) _____

8. Is line 5 at least 6 times line 1? (Yes/No) _____

*9. Are at least 90% of firm's assets located in the U.S.? If not, complete line 10 (Yes/No) _____

10. Is line 6 at least 6 times line 1? (Yes/No) _____

I hereby certify that the wording of this letter is identical to the wording specified in 40 CFR 264.151(f) as such regulations were constituted on the date shown immediately below.

[Signature] _____

[Name] _____

[Title] _____

[Date] _____

(g) A letter from the chief financial officer, as specified in §264.147(f) or §265.147(f) of this chapter, must be worded as follows, except that instructions in brackets are to be replaced with the relevant information and the brackets deleted.

LETTER FROM CHIEF FINANCIAL OFFICER

[Address to Regional Administrator of every Region in which facilities for which financial responsibility is to be demonstrated through the financial test are located].

I am the chief financial officer of [firm's name and address]. This letter is in support of the use of the financial test to demonstrate financial responsibility for liability coverage [insert "and closure and/or post-closure care" if applicable] as specified in subpart H of 40 CFR parts 264 and 265.

[Fill out the following paragraphs regarding facilities and liability coverage. If there are no facilities that belong in a particular paragraph, write "None" in the space indicated. For each facility, include its EPA Identification Number, name, and address].

The firm identified above is the owner or operator of the following facilities for which liability coverage for [insert "sudden" or "nonsudden" or "both sudden and nonsudden"] accidental occurrences is being demonstrated through the financial test specified in subpart H of 40 CFR parts 264 and 265:_____

The firm identified above guarantees, through the guarantee specified in subpart H of 40 CFR parts 264 and 265, liability coverage for [insert "sudden" or "nonsudden" or

"both sudden and nonsudden"] accidental occurrences at the following facilities owned or
operated by the following: _____. The firm
identified above is [insert one or more: (1)
The direct or higher-tier parent corporation
of the owner or operator; (2) owned by the
same parent corporation as the parent corporation of the owner or operator, and receiving the following value in consideration
of this guarantee _____ ; or (3) engaged in
the following substantial business relationship with the owner or operator _____, and
receiving the following value in consideration of this guarantee _____]. [Attach a
written description of the business relationship or a copy of the contract establishing
such relationship to this letter.]

[If you are using the financial test to demonstrate coverage of both liability and closure and post-closure care, fill in the following five paragraphs regarding facilities
and associated closure and post-closure cost
estimates. If there are no facilities that belong in a particular paragraph, write "None"
in the space indicated. For each facility, include its EPA identification number, name,
address, and current closure and/or post-closure cost estimates. Identify each cost estimate as to whether it is for closure or post-
closure care.]

1. The firm identified above owns or operates the following facilities for which financial assurance for closure or post-closure
care or liability coverage is demonstrated
through the financial test specified in subpart H of 40 CFR parts 264 and 265. The current closure and/or post-closure cost estimate covered by the test are shown for each
facility: _____.

2. The firm identified above guarantees,
through the guarantee specified in subpart H
of 40 CFR parts 264 and 265, the closure and
post-closure care or liability coverage of the
following facilities owned or operated by the
guaranteed party. The current cost estimates for closure or post-closure care so
guaranteed are shown for each facility:
_____.

3. In States where EPA is not administering the financial requirements of subpart H of 40 CFR parts 264 and 265, this firm
is demonstrating financial assurance for the
closure or post-closure care of the following
facilities through the use of a test equivalent
or substantially equivalent to the financial
test specified in subpart H or 40 CFR parts
264 and 265. The current closure or post-closure cost estimates covered by such a test
are shown for each facility: _____.

4. The firm identified above owns or operates the following hazardous waste management facilities for which financial assurance
for closure or, if a disposal facility, post-closure care, is not demonstrated either to EPA
or a State through the financial test or any
other financial assurance mechanisms specified in subpart H of 40 CFR parts 264 and 265

or equivalent or substantially equivalent
State mechanisms. The current closure and/
or post-closure cost estimates not covered by
such financial assurance are shown for each
facility: _____.

5. This firm is the owner or operator or
guarantor of the following UIC facilities for
which financial assurance for plugging and
abandonment is required under part 144 and
is assured through a financial test. The current closure cost estimates as required by 40
CFR 144.62 are shown for each facility: _____.

This firm [insert "is required" or "is not
required"] to file a Form 10K with the Securities and Exchange Commission (SEC) for
the latest fiscal year.

The fiscal year of this firm ends on
[month, day]. The figures for the following
items marked with an asterisk are derived
from this firm's independently audited, year-
end financial statements for the latest completed fiscal year, ended [date].

*Part A. Liability Coverage for Accidental
Occurrences*

[Fill in Alternative I if the criteria of paragraph (f)(1)(i) of § 264.147 or § 265.147 are used.
Fill in Alternative II if the criteria of paragraph (f)(1)(ii) of § 264.147 or § 265.147 are
used.]

ALTERNATIVE I

1. Amount of annual aggregate liability
coverage to be demonstrated $_____.
*2. Current assets $_____.
*3. Current liabilities $_____.
4. Net working capital (line 2 minus line 3)
$_____.
*5. Tangible net worth $_____.
*6. If less than 90% of assets are located in
the U.S., give total U.S. assets $_____.
7. Is line 5 at least $10 million? (Yes/No)
_____.
8. Is line 4 at least 6 times line 1? (Yes/No)
_____.
9. Is line 5 at least 6 times line 1? (Yes/No)
_____.
*10. Are at least 90% of assets located in
the U.S.? (Yes/No) _____. If not, complete
line 11.
11. Is line 6 at least 6 times line 1? (Yes/No)
_____.

ALTERNATIVE II

1. Amount of annual aggregate liability
coverage to be demonstrated $_____.
2. Current bond rating of most recent
issuance and name of rating service
_____.
3. Date of issuance of bond _____.
4. Date of maturity of bond
_____.
*5. Tangible net worth $_____.

*6. Total assets in U.S. (required only if less than 90% of assets are located in the U.S.) $_____.

7. Is line 5 at least $10 million? (Yes/No) _____.

8. Is line 5 at least 6 times line 1? _____.

9. Are at least 90% of assets located in the U.S.? If not, complete line 10. (Yes/No) _____.

10. Is line 6 at least 6 times line 1? _____.

[Fill in part B if you are using the financial test to demonstrate assurance of both liability coverage and closure or post-closure care.]

### Part B. Closure or Post-Closure Care and Liability Coverage

[Fill in Alternative I if the criteria of paragraphs (f)(1)(i) of §264.143 or §264.145 and (f)(1)(i) of §264.147 are used or if the criteria of paragraphs (e)(1)(i) of §265.143 or §265.145 and (f)(1)(i) of §265.147 are used. Fill in Alternative II if the criteria of paragraphs (f)(1)(ii) of §264.143 or §264.145 and (f)(1)(ii) of §264.147 are used or if the criteria of paragraphs (e)(1)(i) of §265.143 or §265.145 and (f)(1)(ii) of §265.147 are used.]

#### ALTERNATIVE I

1. Sum of current closure and post-closure cost estimates (total of all cost estimates listed above) $_____

2. Amount of annual aggregate liability coverage to be demonstrated $_____

3. Sum of lines 1 and 2 $_____

*4. Total liabilities (if any portion of your closure or post-closure cost estimates is included in your total liabilities, you may deduct that portion from this line and add that amount to lines 5 and 6) $_____

*5. Tangible net worth $_____

*6. Net worth $_____

*7. Current assets $_____

*8. Current liabilities $_____

9. Net working capital (line 7 minus line 8) $_____

*10. The sum of net income plus depreciation, depletion, and amortization $_____

*11. Total assets in U.S. (required only if less than 90% of assets are located in the U.S.) $_____

12. Is line 5 at least $10 million? (Yes/No)

13. Is line 5 at least 6 times line 3? (Yes/No)

14. Is line 9 at least 6 times line 3? (Yes/No)

*15. Are at least 90% of assets located in the U.S.? (Yes/No) If not, complete line 16.

16. Is line 11 at least 6 times line 3? (Yes/No)

17. Is line 4 divided by line 6 less than 2.0? (Yes/No)

18. Is line 10 divided by line 4 greater than 0.1? (Yes/No)

19. Is line 7 divided by line 8 greater than 1.5? (Yes/No)

#### ALTERNATIVE II

1. Sum of current closure and post-closure cost estimates (total of all cost estimates listed above) $_____

2. Amount of annual aggregate liability coverage to be demonstrated $_____

3. Sum of lines 1 and 2 $_____

4. Current bond rating of most recent issuance and name of rating service _____

5. Date of issuance of bond _____

6. Date of maturity of bond _____

*7. Tangible net worth (if any portion of the closure or post-closure cost estimates is included in "total liabilities" on your financial statements you may add that portion to this line) $_____

*8. Total assets in the U.S. (required only if less than 90% of assets are located in the U.S.) $_____

9. Is line 7 at least $10 million? (Yes/No)

10. Is line 7 at least 6 times line 3? (Yes/No)

*11. Are at least 90% of assets located in the U.S.? (Yes/No) If not complete line 12.

12. Is line 8 at least 6 times line 3? (Yes/No)

I hereby certify that the wording of this letter is identical to the wording specified in 40 CFR 264.151(g) as such regulations were constituted on the date shown immediately below.

[Signature] _____

[Name] _____

[Title] _____

[Date] _____

(h)(1) A corporate guarantee, as specified in §264.143(f) or §264.145(f), or §265.143(e) or §265.145(e) of this chapter, must be worded as follows, except that instructions in brackets are to be replaced with the relevant information and the brackets deleted:

##### CORPORATE GUARANTEE FOR CLOSURE OR POST-CLOSURE CARE

Guarantee made this [date] by [name of guaranteeing entity], a business corporation organized under the laws of the State of [insert name of State], herein referred to as guarantor. This guarantee is made on behalf of the [owner or operator] of [business address], which is [one of the following: "our subsidiary"; "a subsidiary of [name and address of common parent corporation], of which guarantor is a subsidiary"; or "an entity with which guarantor has a substantial business relationship, as defined in 40 CFR [either 264.141(h) or 265.141(h)]" to the United States Environmental Protection Agency (EPA).

##### RECITALS

1. Guarantor meets or exceeds the financial test criteria and agrees to comply with the

reporting requirements for guarantors as specified in 40 CFR 264.143(f), 264.145(f), 265.143(e), and 265.145(e).

2. [Owner or operator] owns or operates the following hazardous waste management facility(ies) covered by this guarantee: [List for each facility: EPA Identification Number, name, and address. Indicate for each whether guarantee is for closure, post-closure care, or both.]

3. "Closure plans" and "post-closure plans" as used below refer to the plans maintained as required by subpart G of 40 CFR parts 264 and 265 for the closure and post-closure care of facilities as identified above.

4. For value received from [owner or operator], guarantor guarantees to EPA that in the event that [owner or operator] fails to perform "closure," "post-closure care" or "closure and post-closure care"] of the above facility(ies) in accordance with the closure or post-closure plans and other permit or interim status requirements whenever required to do so, the guarantor shall do so or establish a trust fund as specified in subpart H of 40 CFR part 264 or 265, as applicable, in the name of [owner or operator] in the amount of the current closure or post-closure cost estimates as specified in subpart H of 40 CFR parts 264 and 265.

5. Guarantor agrees that if, at the end of any fiscal year before termination of this guarantee, the guarantor fails to meet the financial test criteria, guarantor shall send within 90 days, by certified mail, notice to the EPA Regional Administrator(s) for the Region(s) in which the facility(ies) is(are) located and to [owner or operator] that he intends to provide alternate financial assurance as specified in subpart H of 40 CFR part 264 or 265, as applicable, in the name of [owner or operator]. Within 120 days after the end of such fiscal year, the guarantor shall establish such financial assurance unless [owner or operator] has done so.

6. The guarantor agrees to notify the EPA Regional Administrator by certified mail, of a voluntary or involuntary proceeding under Title 11 (Bankruptcy), U.S. Code, naming guarantor as debtor, within 10 days after commencement of the proceeding.

7. Guarantor agrees that within 30 days after being notified by an EPA Regional Administrator of a determination that guarantor no longer meets the financial test criteria or that he is disallowed from continuing as a guarantor of closure or post-closure care, he shall establish alternate financial assurance as specified in subpart H of 40 CFR part 264 or 265, as applicable, in the name of [owner or operator] unless [owner or operator] has done so.

8. Guarantor agrees to remain bound under this guarantee notwithstanding any or all of the following: amendment or modification of the closure or post-closure plan, amendment or modification of the permit, the extension or reduction of the time of performance of closure or post-closure, or any other modification or alteration of an obligation of the owner or operator pursuant to 40 CFR part 264 or 265.

9. Guarantor agrees to remain bound under this guarantee for as long as [owner or operator] must comply with the applicable financial assurance requirements of subpart H of 40 CFR parts 264 and 265 for the above-listed facilities, except as provided in paragraph 10 of this agreement.

10. [Insert the following language if the guarantor is (a) a direct or higher-tier corporate parent, or (b) a firm whose parent corporation is also the parent corporation of the owner or operator]:

Guarantor may terminate this guarantee by sending notice by certified mail to the EPA Regional Administrator(s) for the Region(s) in which the facility(ies) is(are) located and to [owner or operator], provided that this guarantee may not be terminated unless and until [the owner or operator] obtains, and the EPA Regional Administrator(s) approve(s), alternate closure and/or post-closure care coverage complying with 40 CFR 264.143, 264.145, 265.143, and/or 265.145.

[Insert the following language if the guarantor is a firm qualifying as a guarantor due to its "substantial business relationship" with its owner or operator]

Guarantor may terminate this guarantee 120 days following the receipt of notification, through certified mail, by the EPA Regional Administrator(s) for the Region(s) in which the facility(ies) is(are) located and by [the owner or operator].

11. Guarantor agrees that if [owner or operator] fails to provide alternate financial assurance as specified in subpart H of 40 CFR part 264 or 265, as applicable, and obtain written approval of such assurance from the EPA Regional Administrator(s) within 90 days after a notice of cancellation by the guarantor is received by an EPA Regional Administrator from guarantor, guarantor shall provide such alternate financial assurance in the name of [owner or operator].

12. Guarantor expressly waives notice of acceptance of this guarantee by the EPA or by [owner or operator]. Guarantor also expressly waives notice of amendments or modifications of the closure and/or post-closure plan and of amendments or modifications of the facility permit(s).

I hereby certify that the wording of this guarantee is identical to the wording specified in 40 CFR 264.151(h) as such regulations were constituted on the date first above written.

Effective date: _____

[Name of guarantor] _____

[Authorized signature for guarantor] _____

[Name of person signing] _____

[Title of person signing] _____
Signature of witness or notary: _____

(2) A guarantee, as specified in §264.147(g) or §265.147(g) of this chapter, must be worded as follows, except that instructions in brackets are to be replaced with the relevant information and the brackets deleted:

GUARANTEE FOR LIABILITY COVERAGE

Guarantee made this [date] by [name of guaranteeing entity], a business corporation organized under the laws of [if incorporated within the United States insert "the State of _____" and insert name of State; if incorporated outside the United States insert the name of the country in which incorporated, the principal place of business within the United States, and the name and address of the registered agent in the State of the principal place of business], herein referred to as guarantor. This guarantee is made on behalf of [owner or operator] of [business address], which is one of the following: "our subsidiary;" "a subsidiary of [name and address of common parent corporation], of which guarantor is a subsidiary;" or "an entity with which guarantor has a substantial business relationship, as defined in 40 CFR [either 264.141(h) or 265.141(h)]", to any and all third parties who have sustained or may sustain bodily injury or property damage caused by [sudden and/or nonsudden] accidental occurrences arising from operation of the facility(ies) covered by this guarantee.

RECITALS

1. Guarantor meets or exceeds the financial test criteria and agrees to comply with the reporting requirements for guarantors as specified in 40 CFR 264.147(g) and 265.147(g).
2. [Owner or operator] owns or operates the following hazardous waste management facility(ies) covered by this guarantee: [List for each facility: EPA identification number, name, and address; and if guarantor is incorporated outside the United States list the name and address of the guarantor's registered agent in each State.] This corporate guarantee satisfies RCRA third-party liability requirements for [insert "sudden" or "nonsudden" or "both sudden and nonsudden"] accidental occurrences in above-named owner or operator facilities for coverage in the amount of [insert dollar amount] for each occurrence and [insert dollar amount] annual aggregate.
3. For value received from [owner or operator], guarantor guarantees to any and all third parties who have sustained or may sustain bodily injury or property damage caused by [sudden and/or nonsudden] accidental occurrences arising from operations of the facility(ies) covered by this guarantee that in the event that [owner or operator] fails to

satisfy a judgment or award based on a determination of liability for bodily injury or property damage to third parties caused by [sudden and/or nonsudden] accidental occurrences, arising from the operation of the above-named facilities, or fails to pay an amount agreed to in settlement of a claim arising from or alleged to arise from such injury or damage, the guarantor will satisfy such judgment(s), award(s) or settlement agreement(s) up to the limits of coverage identified above.
4. Such obligation does not apply to any of the following:
(a) Bodily injury or property damage for which [insert owner or operator] is obligated to pay damages by reason of the assumption of liability in a contract or agreement. This exclusion does not apply to liability for damages that [insert owner or operator] would be obligated to pay in the absence of the contract or agreement.
(b) Any obligation of [insert owner or operator] under a workers' compensation, disability benefits, or unemployment compensation law or any similar law.
(c) Bodily injury to:
(1) An employee of [insert owner or operator] arising from, and in the course of, employment by [insert owner or operator]; or
(2) The spouse, child, parent, brother, or sister of that employee as a consequence of, or arising from, and in the course of employment by [insert owner or operator]. This exclusion applies:
(A) Whether [insert owner or operator] may be liable as an employer or in any other capacity; and
(B) To any obligation to share damages with or repay another person who must pay damages because of the injury to persons identified in paragraphs (1) and (2).
(d) Bodily injury or property damage arising out of the ownership, maintenance, use, or entrustment to others of any aircraft, motor vehicle or watercraft.
(e) Property damage to:
(1) Any property owned, rented, or occupied by [insert owner or operator];
(2) Premises that are sold, given away or abandoned by [insert owner or operator] if the property damage arises out of any part of those premises;
(3) Property loaned to [insert owner or operator];
(4) Personal property in the care, custody or control of [insert owner or operator];
(5) That particular part of real property on which [insert owner or operator] or any contractors or subcontractors working directly or indirectly on behalf of [insert owner or operator] are performing operations, if the property damage arises out of these operations.
5. Guarantor agrees that if, at the end of any fiscal year before termination of this

guarantee, the guarantor fails to meet the financial test criteria, guarantor shall send within 90 days, by certified mail, notice to the EPA Regional Administrator[s] for the Region[s] in which the facility[ies] is[are] located and to [owner or operator] that he intends to provide alternate liability coverage as specified in 40 CFR 264.147 and 265.147, as applicable, in the name of [owner or operator]. Within 120 days after the end of such fiscal year, the guarantor shall establish such liability coverage unless [owner or operator] has done so.

6. The guarantor agrees to notify the EPA Regional Administrator by certified mail of a voluntary or involuntary proceeding under title 11 (Bankruptcy), U.S. Code, naming guarantor as debtor, within 10 days after commencement of the proceeding.

7. Guarantor agrees that within 30 days after being notified by an EPA Regional Administrator of a determination that guarantor no longer meets the financial test criteria or that he is disallowed from continuing as a guarantor, he shall establish alternate liability coverage as specified in 40 CFR 264.147 or 265.147 in the name of [owner or operator], unless [owner or operator] has done so.

8. Guarantor reserves the right to modify this agreement to take into account amendment or modification of the liability requirements set by 40 CFR 264.147 and 265.147, provided that such modification shall become effective only if a Regional Administrator does not disapprove the modification within 30 days of receipt of notification of the modification.

9. Guarantor agrees to remain bound under this guarantee for so long as [owner or operator] must comply with the applicable requirements of 40 CFR 264.147 and 265.147 for the above-listed facility(ies), except as provided in paragraph 10 of this agreement.

10. [Insert the following language if the guarantor is (a) a direct or higher-tier corporate parent, or (b) a firm whose parent corporation is also the parent corporation of the owner or operator]:

Guarantor may terminate this guarantee by sending notice by certified mail to the EPA Regional Administrator(s) for the Region(s) in which the facility(ies) is(are) located and to [owner or operator], provided that this guarantee may not be terminated unless and until [the owner or operator] obtains, and the EPA Regional Administrator(s) approve(s), alternate liability coverage complying with 40 CFR 264.147 and/or 265.147.

[Insert the following language if the guarantor is a firm qualifying as a guarantor due to its "substantial business relationship" with the owner or operator]:

Guarantor may terminate this guarantee 120 days following receipt of notification, through certified mail, by the EPA Regional

Administrator(s) for the Region(s) in which the facility(ies) is(are) located and by [the owner or operator].

11. Guarantor hereby expressly waives notice of acceptance of this guarantee by any party.

12. Guarantor agrees that this guarantee is in addition to and does not affect any other responsibility or liability of the guarantor with respect to the covered facilities.

13. The Guarantor shall satisfy a third-party liability claim only on receipt of one of the following documents:

(a) Certification from the Principal and the third-party claimant(s) that the liability claim should be paid. The certification must be worded as follows, except that instructions in brackets are to be replaced with the relevant information and the brackets deleted:

CERTIFICATION OF VALID CLAIM

The undersigned, as parties [insert Principal] and [insert name and address of third-party claimant(s)], hereby certify that the claim of bodily injury and/or property damage caused by a [sudden or nonsudden] accidental occurrence arising from operating [Principal's] hazardous waste treatment, storage, or disposal facility should be paid in the amount of $ .

[Signatures] _____

Principal _____

(Notary) Date _____

[Signatures] _____

Claimant(s) _____

(Notary) Date _____

(b) A valid final court order establishing a judgment against the Principal for bodily injury or property damage caused by sudden or nonsudden accidental occurrences arising from the operation of the Principal's facility or group of facilities.

14. In the event of combination of this guarantee with another mechanism to meet liability requirements, this guarantee will be considered [insert "primary" or "excess"] coverage.

I hereby certify that the wording of the guarantee is identical to the wording specified in 40 CFR 264.151(h)(2) as such regulations were constituted on the date shown immediately below.

Effective date: _____

[Name of guarantor] _____

[Authorized signature for guarantor] _____

[Name of person signing] _____

[Title of person signing] _____

Signature of witness or notary: _____

(i) A hazardous waste facility liability endorsement as required in § 264.147 or § 265.147 must be worded as follows, except that instructions in brackets are to be replaced with the relevant information and the brackets deleted:

HAZARDOUS WASTE FACILITY LIABILITY ENDORSEMENT

1. This endorsement certifies that the policy to which the endorsement is attached provides liability insurance covering bodily injury and property damage in connection with the insured's obligation to demonstrate financial responsibility under 40 CFR 264.147 or 265.147. The coverage applies at [list EPA Identification Number, name, and address for each facility] for [insert "sudden accidental occurrences," "nonsudden accidental occurrences," or "sudden and nonsudden accidental occurrences"; if coverage is for multiple facilities and the coverage is different for different facilities, indicate which facilities are insured for sudden accidental occurrences, which are insured for nonsudden accidental occurrences, and which are insured for both]. The limits of liability are [insert the dollar amount of the "each occurrence" and "annual aggregate" limits of the Insurer's liability], exclusive of legal defense costs.

2. The insurance afforded with respect to such occurrences is subject to all of the terms and conditions of the policy; provided, however, that any provisions of the policy inconsistent with subsections (a) through (e) of this Paragraph 2 are hereby amended to conform with subsections (a) through (e):

(a) Bankruptcy or insolvency of the insured shall not relieve the Insurer of its obligations under the policy to which this endorsement is attached.

(b) The Insurer is liable for the payment of amounts within any deductible applicable to the policy, with a right of reimbursement by the insured for any such payment made by the Insurer. This provision does not apply with respect to that amount of any deductible for which coverage is demonstrated as specified in 40 CFR 264.147(f) or 265.147(f).

(c) Whenever requested by a Regional Administrator of the U.S. Environmental Protection Agency (EPA), the Insurer agrees to furnish to the Regional Administrator a signed duplicate original of the policy and all endorsements.

(d) Cancellation of this endorsement, whether by the Insurer, the insured, a parent corporation providing insurance coverage for its subsidiary, or by a firm having an insurable interest in and obtaining liability insurance on behalf of the owner or operator of the hazardous waste management facility, will be effective only upon written notice and only after the expiration of 60 days after a copy of such written notice is received by the Regional Administrator(s) of the EPA Region(s) in which the facility(ies) is(are) located.

(e) Any other termination of this endorsement will be effective only upon written notice and only after the expiration of thirty (30) days after a copy of such written notice is received by the Regional Administrator(s) of the EPA Region(s) in which the facility(ies) is (are) located.

Attached to and forming part of policy No. _____ issued by [name of Insurer], herein called the Insurer, of [address of Insurer] to [name of insured] of [address] this _____ day of _____, 19___. The effective date of said policy is _____ day of _____, 19___.

I hereby certify that the wording of this endorsement is identical to the wording specified in 40 CFR 264.151(i) as such regulation was constituted on the date first above written, and that the Insurer is licensed to transact the business of insurance, or eligible to provide insurance as an excess or surplus lines insurer, in one or more States.

[Signature of Authorized Representative of Insurer]
[Type name]
[Title], Authorized Representative of [name of Insurer]
[Address of Representative]

(j) A certificate of liability insurance as required in §264.147 or §265.147 must be worded as follows, except that the instructions in brackets are to be replaced with the relevant information and the brackets deleted:

HAZARDOUS WASTE FACILITY CERTIFICATE OF LIABILITY INSURANCE

1. [Name of Insurer], (the "Insurer"), of [address of Insurer] hereby certifies that it has issued liability insurance covering bodily injury and property damage to [name of insured], (the "insured"), of [address of insured] in connection with the insured's obligation to demonstrate financial responsibility under 40 CFR 264.147 or 265.147. The coverage applies at [list EPA Identification Number, name, and address for each facility] for [insert "sudden accidental occurrences," "nonsudden accidental occurrences," or "sudden and nonsudden accidental occurrences"; if coverage is for multiple facilities and the coverage is different for different facilities, indicate which facilities are insured for sudden accidental occurrences, which are insured for nonsudden accidental occurrences, and which are insured for both]. The limits of liability are [insert the dollar amount of the "each occurrence" and "annual aggregate" limits of the Insurer's liability], exclusive of legal defense costs. The coverage is provided under policy number _____, issued on [date]. The effective date of said policy is [date].

2. The Insurer further certifies the following with respect to the insurance described in Paragraph 1:

(a) Bankruptcy or insolvency of the insured shall not relieve the Insurer of its obligations under the policy.

(b) The Insurer is liable for the payment of amounts within any deductible applicable to the policy, with a right of reimbursement by the insured for any such payment made by the Insurer. This provision does not apply with respect to that amount of any deductible for which coverage is demonstrated as specified in 40 CFR 264.147(f) or 265.147(f).

(c) Whenever requested by a Regional Administrator of the U.S. Environmental Protection Agency (EPA), the Insurer agrees to furnish to the Regional Administrator a signed duplicate original of the policy and all endorsements.

(d) Cancellation of the insurance, whether by the insurer, the insured, a parent corporation providing insurance coverage for its subsidiary, or by a firm having an insurable interest in and obtaining liability insurance on behalf of the owner or operator of the hazardous waste management facility, will be effective only upon written notice and only after the expiration of 60 days after a copy of such written notice is received by the Regional Administrator(s) of the EPA Region(s) in which the facility(ies) is(are) located.

(e) Any other termination of the insurance will be effective only upon written notice and only after the expiration of thirty (30) days after a copy of such written notice is received by the Regional Administrator(s) of the EPA Region(s) in which the facility(ies) is (are) located.

I hereby certify that the wording of this instrument is identical to the wording specified in 40 CFR 264.151(j) as such regulation was constituted on the date first above written, and that the Insurer is licensed to transact the business of insurance, or eligible to provide insurance as an excess or surplus lines insurer, in one or more States.

[Signature of authorized representative of Insurer]

[Type name]

[Title], Authorized Representative of [name of Insurer]

[Address of Representative]

(k) A letter of credit, as specified in § 264.147(h) or 265.147(h) of this chapter, must be worded as follows, except that instructions in brackets are to be replaced with the relevant information and the brackets deleted:

IRREVOCABLE STANDBY LETTER OF CREDIT

Name and Address of Issuing Institution ____

Regional Administrator(s) _____

Region(s) _____

U.S. Environmental Protection Agency ____

Dear Sir or Madam: We hereby establish our Irrevocable Standby Letter of Credit No. _____ in the favor of ["any and all third-party liability claimants" or insert name of trustee of the standby trust fund], at the request and for the account of [owner or operator's name and address] for third-party liability awards or settlements up to [in words] U.S. dollars $_____ per occurrence and the annual aggregate amount of [in words] U.S. dollars $_____, for sudden accidental occurrences and/or for third-party liability awards or settlements up to the amount of [in words] U.S. dollars $_____ per occurrence, and the annual aggregate amount of [in words] U.S. dollars $_____, for nonsudden accidental occurrences available upon presentation of a sight draft bearing reference to this letter of credit No. _____, and [insert the following language if the letter of credit is being used without a standby trust fund: (1) a signed certificate reading as follows:

CERTIFICATE OF VALID CLAIM

The undersigned, as parties [insert principal] and [insert name and address of third party claimant(s)], hereby certify that the claim of bodily injury and/or property damage caused by a [sudden or nonsudden] accidental occurrence arising from operations of [principal's] hazardous waste treatment, storage, or disposal facility should be paid in the amount of $[          ]. We hereby certify that the claim does not apply to any of the following:

(a) Bodily injury or property damage for which [insert principal] is obligated to pay damages by reason of the assumption of liability in a contract or agreement. This exclusion does not apply to liability for damages that [insert principal] would be obligated to pay in the absence of the contract or agreement.

(b) Any obligation of [insert principal] under a workers' compensation, disability benefits, or unemployment compensation law or any similar law.

(c) Bodily injury to:

(1) An employee of [insert principal] arising from, and in the course of, employment by [insert principal]; or

(2) The spouse, child, parent, brother or sister of that employee as a consequence of, or arising from, and in the course of employment by [insert principal].

This exclusion applies:

(A) Whether [insert principal] may be liable as an employer or in any other capacity; and

(B) To any obligation to share damages with or repay another person who must pay damages because of the injury to persons identified in paragraphs (1) and (2).

(d) Bodily injury or property damage arising out of the ownership, maintenance, use, or entrustment to others of any aircraft, motor vehicle or watercraft.

(e) Property damage to:

(1) Any property owned, rented, or occupied by [insert principal];

(2) Premises that are sold, given away or abandoned by [insert principal] if the property damage arises out of any part of those premises;

(3) Property loaned to [insert principal];

(4) Personal property in the care, custody or control of [insert principal];

(5) That particular part of real property on which [insert principal] or any contractors or subcontractors working directly or indirectly on behalf of [insert principal] are performing operations, if the property damage arises out of these operations.

[Signatures] _____

Grantor _____

[Signatures] _____

Claimant(s) _____

or (2) a valid final court order establishing a judgment against the Grantor for bodily injury or property damage caused by sudden or nonsudden accidental occurrences arising from the operation of the Grantor's facility or group of facilities.]

This letter of credit is effective as of [date] and shall expire on [date at least one year later], but such expiration date shall be automatically extended for a period of [at least one year] on [date and on each successive expiration date, unless, at least 120 days before the current expiration date, we notify you, the USEPA Regional Administrator for Region [Region #], and [owner's or operator's name] by certified mail that we have decided not to extend this letter of credit beyond the current expiration date.

Whenever this letter of credit is drawn on under and in compliance with the terms of this credit, we shall duly honor such draft upon presentation to us.

[Insert the following language if a standby trust fund is not being used: "In the event that this letter of credit is used in combination with another mechanism for liability coverage, this letter of credit shall be considered [insert "primary" or "excess" coverage]."

We certify that the wording of this letter of credit is identical to the wording specified in 40 CFR 264.151(k) as such regulations were constituted on the date shown immediately below. [Signature(s) and title(s) of official(s) of issuing institution] [Date].

This credit is subject to [insert "the most recent edition of the Uniform Customs and Practice for Documentary Credits, published and copyrighted by the International Chamber of Commerce," or "the Uniform Commercial Code"].

(1) A surety bond, as specified in §264.147(i) or §265.147(i) of this chapter,

must be worded as follows: except that instructions in brackets are to be replaced with the relevant information and the brackets deleted:

PAYMENT BOND

Surety Bond No. [Insert number]

Parties [Insert name and address of owner or operator], Principal, incorporated in [Insert State of incorporation] of [Insert city and State of principal place of business] and [Insert name and address of surety company(ies)], Surety Company(ies), of [Insert surety(ies) place of business].

EPA Identification Number, name, and address for each facility guaranteed by this bond: _____

| | Sudden accidental occurrences | Nonsudden accidental occurrences |
|---|---|---|
| Penal Sum Per Occurrence. | [insert amount] ..... | [insert amount] |
| Annual Aggregate | [insert amount] ..... | [insert amount] |

Purpose: This is an agreement between the Surety(ies) and the Principal under which the Surety(ies), its(their) successors and assignees, agree to be responsible for the payment of claims against the Principal for bodily injury and/or property damage to third parties caused by ["sudden" and/or "nonsudden"] accidental occurrences arising from operations of the facility or group of facilities in the sums prescribed herein; subject to the governing provisions and the following conditions.

Governing Provisions:

(1) Section 3004 of the Resource Conservation and Recovery Act of 1976, as amended.

(2) Rules and regulations of the U.S. Environmental Protection Agency (EPA), particularly 40 CFR ["§264.147" or "§265.147"] (if applicable).

(3) Rules and regulations of the governing State agency (if applicable) [insert citation].

Conditions:

(1) The Principal is subject to the applicable governing provisions that require the Principal to have and maintain liability coverage for bodily injury and property damage to third parties caused by ["sudden" and/or "nonsudden"] accidental occurrences arising from operations of the facility or group of facilities. Such obligation does not apply to any of the following:

(a) Bodily injury or property damage for which [insert principal] is obligated to pay damages by reason of the assumption of liability in a contract or agreement. This exclusion does not apply to liability for damages that [insert principal] would be obligated to pay in the absence of the contract or agreement.

(b) Any obligation of [insert principal] under a workers' compensation, disability

427

benefits, or unemployment compensation law or similar law:

(c) Bodily injury to:

(1) An employee of [insert principal] arising from, and in the course of, employment by [insert principal]; or

(2) The spouse, child, parent, brother or sister of that employee as a consequence of, or arising from, and in the course of employment by [insert principal]. This exclusion applies:

(A) Whether [insert principal] may be liable as an employer or in any other capacity; and

(B) To any obligation to share damages with or repay another person who must pay damages because of the injury to persons identified in paragraphs (1) and (2).

(d) Bodily injury or property damage arising out of the ownership, maintenance, use, or entrustment to others of any aircraft, motor vehicle or watercraft.

(e) Property damage to:

(1) Any property owned, rented, or occupied by [insert principal];

(2) Premises that are sold, given away or abandoned by [insert principal] if the property damage arises out of any part of those premises;

(3) Property loaned to [insert principal];

(4) Personal property in the care, custody or control of [insert principal];

(5) That particular part of real property on which [insert principal] or any contractors or subcontractors working directly or indirectly on behalf of [insert principal] are performing operations, if the property damage arises out of these operations.

(2) This bond assures that the Principal will satisfy valid third party liability claims, as described in condition 1.

(3) If the Principal fails to satisfy a valid third party liability claim, as described above, the Surety(ies) becomes liable on this bond obligation.

(4) The Surety(ies) shall satisfy a third party liability claim only upon the receipt of one of the following documents:

(a) Certification from the Principal and the third party claimant(s) that the liability claim should be paid. The certification must be worded as follows, except that instructions in brackets are to be replaced with the relevant information and the brackets deleted:

CERTIFICATION OF VALID CLAIM

The undersigned, as parties [insert name of Principal] and [insert name and address of third party claimant(s)], hereby certify that the claim of bodily injury and/or property damage caused by a [sudden or nonsudden] accidental occurrence arising from operating [Principal's] hazardous waste treatment, storage, or disposal facility should be paid in the amount of $[     ].

[Signature]
Principal

[Notary]      Date

[Signature(s)]
Claimant(s)

[Notary]      Date

or (b) A valid final court order establishing a judgment against the Principal for bodily injury or property damage caused by sudden or nonsudden accidental occurrences arising from the operation of the Principal's facility or group of facilities.

(5) In the event of combination of this bond with another mechanism for liability coverage, this bond will be considered [insert "primary" or "excess"] coverage.

(6) The liability of the Surety(ies) shall not be discharged by any payment or succession of payments hereunder, unless and until such payment or payments shall amount in the aggregate to the penal sum of the bond. In no event shall the obligation of the Surety(ies) hereunder exceed the amount of said annual aggregate penal sum, provided that the Surety(ies) furnish(es) notice to the Regional Administrator forthwith of all claims filed and payments made by the Surety(ies) under this bond.

(7) The Surety(ies) may cancel the bond by sending notice of cancellation by certified mail to the Principal and the USEPA Regional Administrator for Region [Region #], provided, however, that cancellation shall not occur during the 120 days beginning on the date of receipt of the notice of cancellation by the Principal and the Regional Administrator, as evidenced by the return receipt.

(8) The Principal may terminate this bond by sending written notice to the Surety(ies) and to the EPA Regional Administrator(s) of the EPA Region(s) in which the bonded facility(ies) is (are) located.

(9) The Surety(ies) hereby waive(s) notification of amendments to applicable laws, statutes, rules and regulations and agree(s) that no such amendment shall in any way alleviate its (their) obligation on this bond.

(10) This bond is effective from [insert date] (12:01 a.m., standard time, at the address of the Principal as stated herein) and shall continue in force until terminated as described above.

In Witness Whereof, the Principal and Surety(ies) have executed this Bond and have affixed their seals on the date set forth above.

The persons whose signatures appear below hereby certify that they are authorized to execute this surety bond on behalf of the Principal and Surety(ies) and that the wording of this surety bond is identical to the wording specified in 40 CFR 264.151(1), as such regulations were constituted on the date this bond was executed.

PRINCIPAL

[Signature(s)]
[Name(s)]
[Title(s)]
[Corporate Seal]

CORPORATE SURETY[IES]

[Name and address]
State of incorporation: _____
Liability Limit: $ _____
[Signature(s)]
[Name(s) and title(s)]
[Corporate seal]
[For every co-surety, provide signature(s), corporate seal, and other information in the same manner as for Surety above.]
Bond premium: $ _____

(m)(1) A trust agreement, as specified in §264.147(j) or §265.147(j) of this chapter, must be worded as follows, except that instructions in brackets are to be replaced with the relevant information and the brackets deleted:

TRUST AGREEMENT

Trust Agreement, the "Agreement," entered into as of [date] by and between [name of the owner or operator] a [name of State] [insert "corporation," "partnership," "association," or "proprietorship"], the "Grantor," and [name of corporate trustee], [insert, "incorporated in the State of _____" or "a national bank"], the "trustee."

Whereas, the United States Environmental Protection Agency, "EPA," an agency of the United States Government, has established certain regulations applicable to the Grantor, requiring that an owner or operator of a hazardous waste management facility or group of facilities must demonstrate financial responsibility for bodily injury and property damage to third parties caused by sudden accidental and/or nonsudden accidental occurrences arising from operations of the facility or group of facilities.

Whereas, the Grantor has elected to establish a trust to assure all or part of such financial responsibility for the facilities identified herein.

Whereas, the Grantor, acting through its duly authorized officers, has selected the Trustee to be the trustee under this agreement, and the Trustee is willing to act as trustee.

Now, therefore, the Grantor and the Trustee agree as follows:

Section 1. *Definitions.* As used in this Agreement:

(a) The term "Grantor" means the owner or operator who enters into this Agreement and any successors or assigns of the Grantor.

(b) The term "Trustee" means the Trustee who enters into this Agreement and any successor Trustee.

Section 2. *Identification of Facilities.* This agreement pertains to the facilities identified on attached schedule A [on schedule A, for each facility list the EPA Identification Number, name, and address of the facility(ies) and the amount of liability coverage, or portions thereof, if more than one instrument affords combined coverage as demonstrated by this Agreement].

Section 3. *Establishment of Fund.* The Grantor and the Trustee hereby establish a trust fund, hereinafter the "Fund," for the benefit of any and all third parties injured or damaged by [sudden and/or nonsudden] accidental occurrences arising from operation of the facility(ies) covered by this guarantee, in the amounts of _____ [up to $1 million] per occurrence and _____ [up to $2 million] annual aggregate for sudden accidental occurrences and _____ [up to $3 million] per occurrence and _____ [up to $6 million] annual aggregate for nonsudden occurrences, except that the Fund is not established for the benefit of third parties for the following:

(a) Bodily injury or property damage for which [insert Grantor] is obligated to pay damages by reason of the assumption of liability in a contract or agreement. This exclusion does not apply to liability for damages that [insert Grantor] would be obligated to pay in the absence of the contract or agreement.

(b) Any obligation of [insert Grantor] under a workers' compensation, disability benefits, or unemployment compensation law or any similar law.

(c) Bodily injury to:

(1) An employee of [insert Grantor] arising from, and in the course of, employment by [insert Grantor]; or

(2) The spouse, child, parent, brother or sister of that employee as a consequence of, or arising from, and in the course of employment by [insert Grantor].

This exclusion applies:

(A) Whether [insert Grantor] may be liable as an employer or in any other capacity; and

(B) To any obligation to share damages with or repay another person who must pay damages because of the injury to persons identified in paragraphs (1) and (2).

(d) Bodily injury or property damage arising out of the ownership, maintenance, use, or entrustment to others of any aircraft, motor vehicle or watercraft.

(e) Property damage to:

(1) Any property owned, rented, or occupied by [insert Grantor];

(2) Premises that are sold, given away or abandoned by [insert Grantor] if the property damage arises out of any part of those premises;

(3) Property loaned to [insert Grantor];

(4) Personal property in the care, custody or control of [insert Grantor];

429

(5) That particular part of real property on which [insert Grantor] or any contractors or subcontractors working directly or indirectly on behalf of [insert Grantor] are performing operations, if the property damage arises out of these operations.

In the event of combination with another mechanism for liability coverage, the fund shall be considered [insert "primary" or "excess"] coverage.

The Fund is established initially as consisting of the property, which is acceptable to the Trustee, described in Schedule B attached hereto. Such property and any other property subsequently transferred to the Trustee is referred to as the Fund, together with all earnings and profits thereon, less any payments or distributions made by the Trustee pursuant to this Agreement. The Fund shall be held by the Trustee, IN TRUST, as hereinafter provided. The Trustee shall not be responsible nor shall it undertake any responsibility for the amount or adequacy of, nor any duty to collect from the Grantor, any payments necessary to discharge any liabilities of the Grantor established by EPA.

Section 4. *Payment for Bodily Injury or Property Damage.* The Trustee shall satisfy a third party liability claim by making payments from the Fund only upon receipt of one of the following documents:

(a) Certification from the Grantor and the third party claimant(s) that the liability claim should be paid. The certification must be worded as follows, except that instructions in brackets are to be replaced with the relevant information and the brackets deleted:

CERTIFICATION OF VALID CLAIM

The undersigned, as parties [insert Grantor] and [insert name and address of third party claimant(s)], hereby certify that the claim of bodily injury and/or property damage caused by a [sudden or nonsudden] accidental occurrence arising from operating [Grantor's] hazardous waste treatment, storage, or disposal facility should be paid in the amount of $[      ].

[Signatures]
Grantor

[Signatures]
Claimant(s)

(b) A valid final court order establishing a judgment against the Grantor for bodily injury or property damage caused by sudden or nonsudden accidental occurrences arising from the operation of the Grantor's facility or group of facilities.

Section 5. *Payments Comprising the Fund.* Payments made to the Trustee for the Fund shall consist of cash or securities acceptable to the Trustee.

Section 6. *Trustee Management.* The Trustee shall invest and reinvest the principal and income, in accordance with general investment policies and guidelines which the Grantor may communicate in writing to the Trustee from time to time, subject, however, to the provisions of this section. In investing, reinvesting, exchanging, selling, and managing the Fund, the Trustee shall discharge his duties with respect to the trust fund solely in the interest of the beneficiary and with the care, skill, prudence, and diligence under the circumstance then prevailing which persons of prudence, acting in a like capacity and familiar with such matters, would use in the conduct of an enterprise of a like character and with like aims; *except that:*

(i) Securities or other obligations of the Grantor, or any other owner or operator of the facilities, or any of their affiliates as defined in the Investment Company Act of 1940, as amended, 15 U.S.C. 80a–2.(a), shall not be acquired or *held* unless they are securities or other obligations of the Federal or a State government;

(ii) The Trustee is authorized to invest the Fund in time or demand deposits of the Trustee, to the extent insured by an agency of the Federal or State government; and

(iii) The Trustee is authorized to hold cash awaiting investment or distribution uninvested for a reasonable time and without liability for the payment of interest thereon.

Section 7. *Commingling and Investment.* The Trustee is expressly authorized in its discretion:

(a) To transfer from time to time any or all of the assets of the Fund to any common commingled, or collective trust fund created by the Trustee in which the fund is eligible to participate, subject to all of the provisions thereof, to be commingled with the assets of other trusts participating therein; and

(b) To purchase shares in any investment company registered under the Investment Company Act of 1940, 15 U.S.C. 81a–1 et seq., including one which may be created, managed, underwritten, or to which investment advice is rendered or the shares of which are sold by the Trustee. The Trustee may vote such shares in its discretion.

Section 8. *Express Powers of Trustee.* Without in any way limiting the powers and discretions conferred upon the Trustee by the other provisions of this Agreement or by law, the Trustee is expressly authorized and empowered:

(a) To sell, exchange, convey, transfer, or otherwise dispose of any property held by it, by public or private sale. No person dealing with the Trustee shall be bound to see to the application of the purchase money or to inquire into the validity or expediency of any such sale or other disposition;

(b) To make, execute, acknowledge, and deliver any and all documents of transfer and conveyance and any and all other instruments that may be necessary or appropriate to carry out the powers herein granted;

(c) To register any securities held in the Fund in its own name or in the name of a nominee and to hold any security in bearer form or in book entry, or to combine certificates representing such securities with certificates of the same issue held by the Trustee in other fiduciary capacities, or to deposit or arrange for the deposit of such securities in a qualified central depository even though, when so deposited, such securities may be merged and held in bulk in the name of the nominee of such depository with other securities deposited therein by another person, or to deposit or arrange for the deposit of any securities issued by the United States Government, or any agency or instrumentality thereof, with a Federal Reserve bank, but the books and records of the Trustee shall at all times show that all such securities are part of the Fund;

(d) To deposit any cash in the Fund in interest-bearing accounts maintained or savings certificates issued by the Trustee, in its separate corporate capacity, or in any other banking institution affiliated with the Trustee, to the extent insured by an agency of the Federal or State government; and

(e) To compromise or otherwise adjust all claims in favor of or against the Fund.

Section 9. *Taxes and Expenses.* All taxes of any kind that may be assessed or levied against or in respect of the Fund and all brokerage commissions incurred by the Fund shall be paid from the Fund. All other expenses incurred by the Trustee in connection with the administration of this Trust, including fees for legal services rendered to the Trustee, the compensation of the Trustee to the extent not paid directly by the Grantor, and all other proper charges and disbursements of the Trustee shall be paid from the Fund.

Section 10. *Annual Valuations.* The Trustee shall annually, at least 30 days prior to the anniversary date of establishment of the Fund, furnish to the Grantor and to the appropriate EPA Regional Administrator a statement confirming the value of the Trust. Any securities in the Fund shall be valued at market value as of no more than 60 days prior to the anniversary date of establishment of the Fund. The failure of the Grantor to object in writing to the Trustee within 90 days after the statement has been furnished to the Grantor and the EPA Regional Administrator shall constitute a conclusively binding assent by the Grantor barring the Grantor from asserting any claim or liability against the Trustee with respect to matters disclosed in the statement.

Section 11. *Advice of Counsel.* The Trustee may from time to time consult with counsel, who may be counsel to the Grantor with respect to any question arising as to the construction of this Agreement or any action to be taken hereunder. The Trustee shall be fully protected, to the extent permitted by law, in acting upon the advice of counsel.

Section 12. *Trustee Compensation.* The Trustee shall be entitled to reasonable compensation for its services as agreed upon in writing from time to time with the Grantor.

Section 13. *Successor Trustee.* The Trustee may resign or the Grantor may replace the Trustee, but such resignation or replacement shall not be effective until the Grantor has appointed a successor trustee and this successor accepts the appointment. The successor trustee shall have the same powers and duties as those conferred upon the Trustee hereunder. Upon the successor trustee's acceptance of the appointment, the Trustee shall assign, transfer, and pay over to the successor trustee the funds and properties then constituting the Fund. If for any reason the Grantor cannot or does not act in the event of the resignation of the Trustee, the Trustee may apply to a court of competent jurisdiction for the appointment of a successor trustee or for instructions. The successor trustee shall specify the date on which it assumes administration of the trust in a writing sent to the Grantor, the EPA Regional Administrator, and the present Trustee by certified mail 10 days before such change becomes effective. Any expenses incurred by the Trustee as a result of any of the acts contemplated by this section shall be paid as provided in Section 9.

Section 14. *Instructions to the Trustee.* All orders, requests, and instructions by the Grantor to the Trustee shall be in writing, signed by such persons as are designated in the attached Exhibit A or such other designees as the Grantor may designate by amendments to Exhibit A. The Trustee shall be fully protected in acting without inquiry in accordance with the Grantor's orders, requests, and instructions. All orders, requests, and instructions by the EPA Regional Administrator to the Trustee shall be in writing, signed by the EPA Regional Administrators of the Regions in which the facilities are located, or their designees, and the Trustee shall act and shall be fully protected in acting in accordance with such orders, requests, and instructions. The Trustee shall have the right to assume, in the absence of written notice to the contrary, that no event constituting a change or a termination of the authority of any person to act on behalf of the Grantor or EPA hereunder has occurred. The Trustee shall have no duty to act in the absence of such orders, requests, and instructions from the Grantor and/or EPA, except as provided for herein.

Section 15. *Notice of Nonpayment.* If a payment for bodily injury or property damage is made under Section 4 of this trust, the Trustee shall notify the Grantor of such payment and the amount(s) thereof within five

(5) working days. The Grantor shall, on or before the anniversary date of the establishment of the Fund following such notice, either make payments to the Trustee in amounts sufficient to cause the trust to return to its value immediately prior to the payment of claims under Section 4, or shall provide written proof to the Trustee that other financial assurance for liability coverage has been obtained equalling the amount necessary to return the trust to its value prior to the payment of claims. If the Grantor does not either make payments to the Trustee or provide the Trustee with such proof, the Trustee shall within 10 working days after the anniversary date of the establishment of the Fund provide a written notice of nonpayment to the EPA Regional Administrator.

Section 16. *Amendment of Agreement.* This Agreement may be amended by an instrument in writing executed by the Grantor, the Trustee, and the appropriate EPA Regional Administrator, or by the Trustee and the appropriate EPA Regional Administrator if the Grantor ceases to exist.

Section 17. *Irrevocability and Termination.* Subject to the right of the parties to amend this Agreement as provided in Section 16, this Trust shall be irrevocable and shall continue until terminated at the written agreement of the Grantor, the Trustee, and the EPA Regional Administrator, or by the Trustee and the EPA Regional Administrator, if the Grantor ceases to exist. Upon termination of the Trust, all remaining trust property, less final trust administration expenses, shall be delivered to the Grantor.

The Regional Administrator will agree to termination of the Trust when the owner or operator substitutes alternate financial assurance as specified in this section.

Section 18. *Immunity and Indemnification.* The Trustee shall not incur personal liability of any nature in connection with any act or omission, made in good faith, in the administration of this Trust, or in carrying out any directions by the Grantor or the EPA Regional Administrator issued in accordance with this Agreement. The Trustee shall be indemnified and saved harmless by the Grantor or from the Trust Fund, or both, from and against any personal liability to which the Trustee may be subjected by reason of any act or conduct in its official capacity, including all expenses reasonably incurred in its defense in the event the Grantor fails to provide such defense.

Section 19. *Choice of Law.* This Agreement shall be administered, construed, and enforced according to the laws of the State of [enter name of State].

Section 20. *Interpretation.* As used in this Agreement, words in the singular include the plural and words in the plural include the singular. The descriptive headings for each section of this Agreement shall not affect the interpretation or the legal efficacy of this Agreement.

In Witness Whereof the parties have caused this Agreement to be executed by their respective officers duly authorized and their corporate seals to be hereunto affixed and attested as of the date first above written. The parties below certify that the wording of this Agreement is identical to the wording specified in 40 CFR 264.151(m) as such regulations were constituted on the date first above written.

[Signature of Grantor]
[Title]
Attest:
[Title]
[Seal]

[Signature of Trustee]
Attest:
[Title]
[Seal]

(2) The following is an example of the certification of acknowledgement which must accompany the trust agreement for a trust fund as specified in §§ 264.147(j) or 265.147(j) of this chapter. State requirements may differ on the proper content of this acknowledgement.

State of
County of
On this [date], before me personally came [owner or operator] to me known, who, being by me duly sworn, did depose and say that she/he resides at [address], that she/he is [title] of [corporation], the corporation described in and which executed the above instrument; that she/he knows the seal of said corporation; that the seal affixed to such instrument is such corporate seal; that it was so affixed by order of the Board of Directors of said corporation, and that she/he signed her/his name thereto by like order.

[Signature of Notary Public]

(n)(1) A standby trust agreement, as specified in § 264.147(h) or 265.147(h) of this chapter, must be worded as follows, except that instructions in brackets are to be replaced with the relevant information and the brackets deleted:

STANDBY TRUST AGREEMENT

Trust Agreement, the "Agreement," entered into as of [date] by and between [name of the owner or operator] a [name of a State] [insert "corporation," "partnership," "association," or "proprietorship"], the "Grantor," and [name of corporate trustee], [insert, "incorporated in the State of _____" or "a national bank"], the "trustee."

Whereas the United States Environmental Protection Agency, "EPA," an agency of the United States Government, has established certain regulations applicable to the Grantor, requiring that an owner or operator of a hazardous waste management facility or group of facilities must demonstrate financial responsibility for bodily injury and property damage to third parties caused by sudden accidental and/or nonsudden accidental occurrences arising from operations of the facility or group of facilities.

Whereas, the Grantor has elected to establish a standby trust into which the proceeds from a letter of credit may be deposited to assure all or part of such financial responsibility for the facilities identified herein.

Whereas, the Grantor, acting through its duly authorized officers, has selected the Trustee to be the trustee under this agreement, and the Trustee is willing to act as trustee.

Now, therefore, the Grantor and the Trustee agree as follows:

Section 1. *Definitions.* As used in this Agreement:

(a) The term *Grantor* means the owner or operator who enters into this Agreement and any successors or assigns of the Grantor.

(b) The term *Trustee* means the Trustee who enters into this Agreement and any successor Trustee.

Section 2. *Identification of Facilities.* This agreement pertains to the facilities identified on attached schedule A [on schedule A, for each facility list the EPA Identification Number, name, and address of the facility(ies) and the amount of liability coverage, or portions thereof, if more than one instrument affords combined coverage as demonstrated by this Agreement].

Section 3. *Establishment of Fund.* The Grantor and the Trustee hereby establish a standby trust fund, hereafter the "Fund," for the benefit of any and all third parties injured or damaged by [sudden and/or nonsudden] accidental occurrences arising from operation of the facility(ies) covered by this guarantee, in the amounts of _____ [up to $1 million] per occurrence and [up to $2 million] annual aggregate for sudden accidental occurrences and [up to $3 million] per occurrence and _____ [up to $6 million] annual aggregate for nonsudden occurrences, except that the Fund is not established for the benefit of third parties for the following:

(a) Bodily injury or property damage for which [insert Grantor] is obligated to pay damages by reason of the assumption of liability in a contract or agreement. This exclusion does not apply to liability for damages that [insert Grantor] would be obligated to pay in the absence of the contract or agreement.

(b) Any obligation of [insert Grantor] under a workers' compensation, disability benefits, or unemployment compensation law or any similar law.

(c) Bodily injury to:

(1) An employee of [insert Grantor] arising from , and in the course of, employment by [insert Grantor]; or

(2) The spouse, child, parent, brother or sister of that employee as a consequence of, or arising from, and in the course of employment by [insert Grantor].

This exclusion applies:

(A) Whether [insert Grantor] may be liable as an employer or in any other capacity; and

(B) To any obligation to share damages with or repay another person who must pay damages because of the injury to persons identified in paragraphs (1) and (2).

(d) Bodily injury or property damage arising out of the ownership, maintenance, use, or entrustment to others of any aircraft, motor vehicle or watercraft.

(e) Property damage to:

(1) Any property owned, rented, or occupied by [insert Grantor];

(2) Premises that are sold, given away or abandoned by [insert Grantor] if the property damage arises out of any part of those premises;

(3) Property loaned by [insert Grantor];

(4) Personal property in the care, custody or control of [insert Grantor];

(5) That particular part of real property on which [insert Grantor] or any contractors or subcontractors working directly or indirectly on behalf of [insert Grantor] are performing operations, if the property damage arises out of these operations.

In the event of combination with another mechanism for liability coverage, the fund shall be considered [insert "primary" or "excess"] coverage.

The Fund is established initially as consisting of the proceeds of the letter of credit deposited into the Fund. Such proceeds and any other property subsequently transferred to the Trustee is referred to as the Fund, together with all earnings and profits thereon, less any payments or distributions made by the Trustee pursuant to this Agreement. The Fund shall be held by the Trustee, IN TRUST, as hereinafter provided. The Trustee shall not be responsible nor shall it undertake any responsibility for the amount or adequacy of, nor any duty to collect from the Grantor, any payments necessary to discharge any liabilities of the Grantor established by EPA.

Section 4. *Payment for Bodily Injury or Property Damage.* The Trustee shall satisfy a third party liability claim by drawing on the letter of credit described in Schedule B and by making payments from the Fund only upon receipt of one of the following documents:

(a) Certification from the Grantor and the third party claimant(s) that the liability claim should be paid. The certification must

433

be worded as follows, except that instructions in brackets are to be replaced with the relevant information and the brackets deleted:

CERTIFICATION OF VALID CLAIM

The undersigned, as parties [insert Grantor] and [insert name and address of third party claimant(s)], hereby certify that the claim of bodily injury and/or property damage caused by a [sudden or nonsudden] accidental occurrence arising from operating [Grantor's] hazardous waste treatment, storage, or disposal facility should be paid in the amount of $[        ].

[Signature] _____

Grantor _____

[Signatures] _____

Claimant(s) _____

(b) A valid final court order establishing a judgment against the Grantor for bodily injury or property damage caused by sudden or nonsudden accidental occurrences arising from the operation of the Grantor's facility or group of facilities.

Section 5. *Payments Comprising the Fund.* Payments made to the Trustee for the Fund shall consist of the proceeds from the letter of credit drawn upon by the Trustee in accordance with the requirements of 40 CFR 264.151(k) and Section 4 of this Agreement.

Section 6. *Trustee Management.* The Trustee shall invest and reinvest the principal and income, in accordance with general investment policies and guidelines which the Grantor may communicate in writing to the Trustee from time to time, subject, however, to the provisions of this Section. In investing, reinvesting, exchanging, selling, and managing the Fund, the Trustee shall discharge his duties with respect to the trust fund solely in the interest of the beneficiary and with the care, skill, prudence, and diligence under the circumstances then prevailing which persons of prudence, acting in a like capacity and familiar with such matters, would use in the conduct of an enterprise of a like character and with like aims; except that:

(i) Securities or other obligations of the Grantor, or any other owner or operator of the facilities, or any of their affiliates as defined in the Investment Company Act of 1940, as amended, 15 U.S.C. 80a–2(a), shall not be acquired or held, unless they are securities or other obligations of the Federal or a State government;

(ii) The Trustee is authorized to invest the Fund in time or demand deposits of the Trustee, to the extent insured by an agency of the Federal or a State government; and

(iii) The Trustee is authorized to hold cash awaiting investment or distribution uninvested for a reasonable time and without liability for the payment of interest thereon.

Section 7. *Commingling and Investment.* The Trustee is expressly authorized in its discretion:

(a) To transfer from time to time any or all of the assets of the Fund to any common, commingled, or collective trust fund created by the Trustee in which the Fund is eligible to participate, subject to all of the provisions thereof, to be commingled with the assets of other trusts participating therein; and

(b) To purchase shares in any investment company registered under the Investment Company Act of 1940, 15 U.S.C. 80a–1 et seq., including one which may be created, managed, underwritten, or to which investment advice is rendered or the shares of which are sold by the Trustee. The Trustee may vote such shares in its discretion.

Section 8. *Express Powers of Trustee.* Without in any way limiting the powers and discretions conferred upon the Trustee by the other provisions of this Agreement or by law, the Trustee is expressly authorized and empowered:

(a) To sell, exchange, convey, transfer, or otherwise dispose of any property held by it, by public or private sale. No person dealing with the Trustee shall be bound to see to the application of the purchase money or to inquire into the validity or expediency of any such sale or other disposition;

(b) To make, execute, acknowledge, and deliver any and all documents of transfer and conveyance and any and all other instruments that may be necessary or appropriate to carry out the powers herein granted;

(c) To register any securities held in the Fund in its own name or in the name of a nominee and to hold any security in bearer form or in book entry, or to combine certificates representing such securities with certificates of the same issue held by the Trustee in other fiduciary capacities, or to deposit or arrange for the deposit of such securities in a qualified central depositary even though, when so deposited, such securities may be merged and held in bulk in the name of the nominee of such depositary with other securities deposited therein by another person, or to deposit or arrange for the deposit of any securities issued by the United States Government, or any agency or instrumentality thereof, with a Federal Reserve Bank, but the books and records of the Trustee shall at all times show that all such securities are part of the Fund;

(d) To deposit any cash in the Fund in interest-bearing accounts maintained or savings certificates issued by the Trustee, in its separate corporate capacity, or in any other banking institution affiliated with the Trustee, to the extent insured by an agency of the Federal or State government; and

(e) To compromise or otherwise adjust all claims in favor of or against the Fund.

434

Section 9. *Taxes and Expenses.* All taxes of any kind that may be assessed or levied against or in respect of the Fund and all brokerage commissions incurred by the Fund shall be paid from the Fund. All other expenses incurred by the Trustee in connection with the administration of this Trust, including fees for legal services rendered to the Trustee, the compensation of the Trustee to the extent not paid directly by the Grantor, and all other proper charges and disbursements to the Trustee shall be paid from the Fund.

Section 10. *Advice of Counsel.* The Trustee may from time to time consult with counsel, who may be counsel to the Grantor, with respect to any question arising as to the construction of this Agreement or any action to be taken hereunder. The Trustee shall be fully protected, to the extent permitted by law, in acting upon the advice of counsel.

Section 11. *Trustee Compensation.* The Trustee shall be entitled to reasonable compensation for its services as agreed upon in writing from time to time with the Grantor.

Section 12. *Successor Trustee.* The Trustee may resign or the Grantor may replace the Trustee, but such resignation or replacement shall not be effective until the Grantor has appointed a successor trustee and this successor accepts the appointment. The successor trustee shall have the same powers and duties as those conferred upon the Trustee hereunder. Upon the successor trustee's acceptance of the appointment, the Trustee shall assign, transfer, and pay over to the successor trustee the funds and properties then constituting the Fund. If for any reason the Grantor cannot or does not act in the event of the resignation of the Trustee, the Trustee may apply to a court of competent jurisdiction for the appointment of a successor trustee or for instructions. The successor trustee shall specify the date on which it assumes administration of the trust in a writing sent to the Grantor, the EPA Regional Administrator and the present Trustee by certified mail 10 days before such change becomes effective. Any expenses incurred by the Trustee as a result of any of the acts contemplated by this Section shall be paid as provided in Section 9.

Section 13. *Instructions to the Trustee.* All orders, requests, certifications of valid claims, and instructions to the Trustee shall be in writing, signed by such persons as are designated in the attached Exhibit A or such other designees as the Grantor may designate by amendments to Exhibit A. The Trustee shall be fully protected in acting without inquiry in accordance with the Grantor's orders, requests, and instructions. The Trustee shall have the right to assume, in the absence of written notice to the contrary, that no event constituting a change or a termination of the authority of any person to act on behalf of the Grantor or the EPA Regional Administrator hereunder has occurred. The Trustee shall have no duty to act in the absence of such orders, requests, and instructions from the Grantor and/or EPA, except as provided for herein.

Section 14. *Amendment of Agreement.* This Agreement may be amended by an instrument in writing executed by the Grantor, the Trustee, and the EPA Regional Administrator, or by the Trustee and the EPA Regional Administrator if the Grantor ceases to exist.

Section 15. *Irrevocability and Termination.* Subject to the right of the parties to amend this Agreement as provided in Section 14, this Trust shall be irrevocable and shall continue until terminated at the written agreement of the Grantor, the Trustee, and the EPA Regional Administrator, or by the Trustee and the EPA Regional Administrator, if the Grantor ceases to exist. Upon termination of the Trust, all remaining trust property, less final trust administration expenses, shall be paid to the Grantor.

The Regional Administrator will agree to termination of the Trust when the owner or operator substitutes alternative financial assurance as specified in this section.

Section 16. *Immunity and indemnification.* The Trustee shall not incur personal liability of any nature in connection with any act or omission, made in good faith, in the administration of this Trust, or in carrying out any directions by the Grantor and the EPA Regional Administrator issued in accordance with this Agreement. The Trustee shall be indemnified and saved harmless by the Grantor or from the Trust Fund, or both, from and against any personal liability to which the Trustee may be subjected by reason of any act or conduct in its official capacity, including all expenses reasonably incurred in its defense in the event the Grantor fails to provide such defense.

Section 17. *Choice of Law.* This Agreement shall be administered, construed, and enforced according to the laws of the State of [enter name of State].

Section 18. *Interpretation.* As used in this Agreement, words in the singular include the plural and words in the plural include the singular. The descriptive headings for each Section of this Agreement shall not affect the interpretation of the legal efficacy of this Agreement.

In Witness Whereof the parties have caused this Agreement to be executed by their respective officers duly authorized and their corporate seals to be hereunto affixed and attested as of the date first above written. The parties below certify that the wording of this Agreement is identical to the wording specified in 40 CFR 264.151(n) as such regulations were constituted on the date first above written.

_____

[Signature of Grantor]

435

[Title]
Attest:
[Title]
[Seal]

_____

[Signature of Trustee]

Attest:
[Title]
[Seal]

(2) The following is an example of the certification of acknowledgement which must accompany the trust agreement for a standby trust fund as specified in section 264.147(h) or 265.147(h) of this chapter. State requirements may differ on the proper content of this acknowledgement.

State of _____

County of _____

On this [date], before me personally came [owner or operator] to me known, who, being by me duly sworn, did depose and say that she/he resides at [address], that she/he is [title] of [corporation], the corporation described in and which executed the above instrument; that she/he knows the seal of said corporation; that the seal affixed to such instrument is such corporate seal; that it was so affixed by order of the Board of Directors of said corporation, and that she/he signed her/his name thereto by like order.

_____

[Signature of Notary Public]

[47 FR 15059, Apr. 7, 1982, as amended at 47 FR 16556, Apr. 16, 1982; 47 FR 17989, Apr. 27, 1982; 47 FR 19995, May 10, 1982; 47 FR 28627, July 1, 1982; 51 FR 16450, May 2, 1986; 51 FR 25354, July 11, 1986; 52 FR 44320, Nov. 18, 1987; 53 FR 33952, Sept. 1, 1988; 57 FR 42836, Sept. 16, 1992; 59 FR 29960, June 10, 1994; 71 FR 40272, July 14, 2006]

## Subpart I—Use and Management of Containers

SOURCE: 46 FR 2866, Jan. 12, 1981, unless otherwise noted.

### § 264.170 Applicability.

The regulations in this subpart apply to owners and operators of all hazardous waste facilities that store containers of hazardous waste, except as § 264.1 provides otherwise.

[Comment: Under § 261.7 and § 261.33(c), if a hazardous waste is emptied from a container the residue remaining in the container is not considered a hazardous waste if the container is "empty" as defined in § 261.7. In that event, management of the container is exempt from the requirements of this subpart.]

### § 264.171 Condition of containers.

If a container holding hazardous waste is not in good condition (e.g., severe rusting, apparent structural defects) or if it begins to leak, the owner or operator must transfer the hazardous waste from this container to a container that is in good condition or manage the waste in some other way that complies with the requirements of this part.

### § 264.172 Compatibility of waste with containers.

The owner or operator must use a container made of or lined with materials which will not react with, and are otherwise compatible with, the hazardous waste to be stored, so that the ability of the container to contain the waste is not impaired.

### § 264.173 Management of containers.

(a) A container holding hazardous waste must always be closed during storage, except when it is necessary to add or remove waste.

(b) A container holding hazardous waste must not be opened, handled, or stored in a manner which may rupture the container or cause it to leak.

[Comment: Reuse of containers in transportation is governed by U.S. Department of Transportation regulations including those set forth in 49 CFR 173.28.]

### § 264.174 Inspections.

At least weekly, the owner or operator must inspect areas where containers are stored, except for Performance Track member facilities, that may conduct inspections at least once each month, upon approval by the Director. To apply for reduced inspection frequencies, the Performance Track member facility must follow the procedures identified in § 264.15(b)(5) of this part. The owner or operator must look for leaking containers and for deterioration of containers and the containment system caused by corrosion or other factors.

[Comment: See §§ 264.15(c) and 264.171 for remedial action required if deterioration or leaks are detected.]

[71 FR 16905, Apr. 4, 2006]

## §264.175  Containment.

(a) Container storage areas must have a containment system that is designed and operated in accordance with paragraph (b) of this section, except as otherwise provided by paragraph (c) of this section.

(b) A containment system must be designed and operated as follows:

(1) A base must underlie the containers which is free of cracks or gaps and is sufficiently impervious to contain leaks, spills, and accumulated precipitation until the collected material is detected and removed;

(2) The base must be sloped or the containment system must be otherwise designed and operated to drain and remove liquids resulting from leaks, spills, or precipitation, unless the containers are elevated or are otherwise protected from contact with accumulated liquids;

(3) The containment system must have sufficient capacity to contain 10% of the volume of containers or the volume of the largest container, whichever is greater. Containers that do not contain free liquids need not be considered in this determination;

(4) Run-on into the containment system must be prevented unless the collection system has sufficient excess capacity in addition to that required in paragraph (b)(3) of this section to contain any run-on which might enter the system; and

(5) Spilled or leaked waste and accumulated precipitation must be removed from the sump or collection area in as timely a manner as is necessary to prevent overflow of the collection system.

[*Comment:* If the collected material is a hazardous waste under part 261 of this Chapter, it must be managed as a hazardous waste in accordance with all applicable requirements of parts 262 through 266 of this chapter. If the collected material is discharged through a point source to waters of the United States, it is subject to the requirements of section 402 of the Clean Water Act, as amended.]

(c) Storage areas that store containers holding only wastes that do not contain free liquids need not have a containment system defined by paragraph (b) of this section, except as provided by paragraph (d) of this section or provided that:

(1) The storage area is sloped or is otherwise designed and operated to drain and remove liquid resulting from precipitation, or

(2) The containers are elevated or are otherwise protected from contact with accumulated liquid.

(d) Storage areas that store containers holding the wastes listed below that do not contain free liquids must have a containment system defined by paragraph (b) of this section:

(1) FO20, FO21, FO22, FO23, FO26, and FO27.

(2) [Reserved]

[46 FR 55112, Nov. 6, 1981, as amended at 50 FR 2003, Jan. 14, 1985; 71 FR 40273, July 14, 2006]

## §264.176  Special requirements for ignitable or reactive waste.

Containers holding ignitable or reactive waste must be located at least 15 meters (50 feet) from the facility's property line.

[*Comment:* See §264.17(a) for additional requirements.]

## §264.177  Special requirements for incompatible wastes.

(a) Incompatible wastes, or incompatible wastes and materials (see appendix V for examples), must not be placed in the same container, unless §264.17(b) is complied with.

(b) Hazardous waste must not be placed in an unwashed container that previously held an incompatible waste or material.

[*Comment:* As required by §264.13, the waste analysis plan must include analyses needed to comply with §264.177. Also, §264.17(c) requires wastes analyses, trial tests or other documentation to assure compliance with §264.17(b). As required by §264.73, the owner or operator must place the results of each waste analysis and trial test, and any documented information, in the operating record of the facility.]

(c) A storage container holding a hazardous waste that is incompatible with any waste or other materials stored nearby in other containers, piles, open tanks, or surface impoundments must be separated from the other materials or protected from them by means of a dike, berm, wall, or other device.

[*Comment:* The purpose of this section is to prevent fires, explosions, gaseous emission, leaching, or other discharge of hazardous waste or hazardous waste constituents which could result from the mixing of incompatible wastes or materials if containers break or leak.]

### § 264.178  Closure.

At closure, all hazardous waste and hazardous waste residues must be removed from the containment system. Remaining containers, liners, bases, and soil containing or contaminated with hazardous waste or hazardous waste residues must be decontaminated or removed.

[*Comment:* At closure, as throughout the operating period, unless the owner or operator can demonstrate in accordance with § 261.3(d) of this chapter that the solid waste removed from the containment system is not a hazardous waste, the owner or operator becomes a generator of hazardous waste and must manage it in accordance with all applicable requirements of parts 262 through 266 of this chapter].

### § 264.179  Air emission standards.

The owner or operator shall manage all hazardous waste placed in a container in accordance with the applicable requirements of subparts AA, BB, and CC of this part.

[61 FR 59950, Nov. 25, 1996]

## Subpart J—Tank Systems

SOURCE: 51 FR 25472, July 14, 1986, unless otherwise noted.

### § 264.190  Applicability.

The requirements of this subpart apply to owners and operators of facilities that use tank systems for storing or treating hazardous waste except as otherwise provided in paragraphs (a), (b), and (c) of this section or in § 264.1 of this part.

(a) Tank systems that are used to store or treat hazardous waste which contains no free liquids and are situated inside a building with an impermeable floor are exempted from the requirements in § 264.193. To demonstrate the absence or presence of free liquids in the stored/treated waste, the following test must be used: Method 9095B (Paint Filter Liquids Test) as described in "Test Methods for Evaluating Solid Waste, Physical/Chemical Methods,"

EPA Publication SW–846, as incorporated by reference in § 260.11 of this chapter.

(b) Tank systems, including sumps, as defined in § 260.10, that serve as part of a secondary containment system to collect or contain releases of hazardous wastes are exempted from the requirements in § 264.193(a).

(c) Tanks, sumps, and other such collection devices or systems used in conjunction with drip pads, as defined in § 260.10 of this chapter and regulated under 40 CFR part 264 subpart W, must meet the requirements of this subpart.

[51 FR 25472, July 14, 1986; 51 FR 29430, Aug. 15, 1986, as amended at 53 FR 34086, Sept. 2, 1988; 55 FR 50484, Dec. 6, 1990; 58 FR 46050, Aug. 31, 1993; 70 FR 34581, June 14, 2005]

### § 264.191  Assessment of existing tank system's integrity.

(a) For each existing tank system that does not have secondary containment meeting the requirements of § 264.193, the owner or operator must determine that the tank system is not leaking or is unfit for use. Except as provided in paragraph (c) of this section, the owner or operator must obtain and keep on file at the facility a written assessment reviewed and certified by a qualified Professional Engineer, in accordance with § 270.11(d) of this chapter, that attests to the tank system's integrity by January 12, 1988.

(b) This assessment must determine that the tank system is adequately designed and has sufficient structural strength and compatibility with the waste(s) to be stored or treated, to ensure that it will not collapse, rupture, or fail. At a minimum, this assessment must consider the following:

(1) Design standard(s), if available, according to which the tank and ancillary equipment were constructed;

(2) Hazardous characteristics of the waste(s) that have been and will be handled;

(3) Existing corrosion protection measures;

(4) Documented age of the tank system, if available (otherwise, an estimate of the age); and

(5) Results of a leak test, internal inspection, or other tank integrity examination such that:

438

(i) For non-enterable underground tanks, the assessment must include a leak test that is capable of taking into account the effects of temperature variations, tank end deflection, vapor pockets, and high water table effects, and

(ii) For other than non-enterable underground tanks and for ancillary equipment, this assessment must include either a leak test, as described above, or other integrity examination that is certified by a qualified Professional Engineer in accordance with §270.11(d) of this chapter, that addresses cracks, leaks, corrosion, and erosion.

[NOTE: The practices described in the American Petroleum Institute (API) Publication, Guide for Inspection of Refinery Equipment, Chapter XIII, "Atmospheric and Low-Pressure Storage Tanks," 4th edition, 1981, may be used, where applicable, as guidelines in conducting other than a leak test.]

(c) Tank systems that store or treat materials that become hazardous wastes subsequent to July 14, 1986, must conduct this assessment within 12 months after the date that the waste becomes a hazardous waste.

(d) If, as a result of the assessment conducted in accordance with paragraph (a), a tank system is found to be leaking or unfit for use, the owner or operator must comply with the requirements of §264.196.

[51 FR 25472, July 14, 1986; 51 FR 29430, Aug. 15, 1986, as amended at 71 FR 16905, Apr. 4, 2006]

## §264.192 Design and installation of new tank systems or components.

(a) Owners or operators of new tank systems or components must obtain and submit to the Regional Administrator, at time of submittal of part B information, a written assessment, reviewed and certified by a qualified Professional Engineer, in accordance with §270.11(d) of this chapter, attesting that the tank system has sufficient structural integrity and is acceptable for the storing and treating of hazardous waste. The assessment must show that the foundation, structural support, seams, connections, and pressure controls (if applicable) are adequately designed and that the tank system has sufficient structural strength, compatibility with the waste(s) to be stored or treated, and corrosion protection to ensure that it will not collapse, rupture, or fail. This assessment, which will be used by the Regional Administrator to review and approve or disapprove the acceptability of the tank system design, must include, at a minimum, the following information:

(1) Design standard(s) according to which tank(s) and/or the ancillary equipment are constructed;

(2) Hazardous characteristics of the waste(s) to be handled;

(3) For new tank systems or components in which the external shell of a metal tank or any external metal component of the tank system will be in contact with the soil or with water, a determination by a corrosion expert of:

(i) Factors affecting the potential for corrosion, including but not limited to:

(A) Soil moisture content;

(B) Soil pH;

(C) Soil sulfides level;

(D) Soil resistivity;

(E) Structure to soil potential;

(F) Influence of nearby underground metal structures (e.g., piping);

(G) Existence of stray electric current;

(H) Existing corrosion-protection measures (e.g., coating, cathodic protection), and

(ii) The type and degree of external corrosion protection that are needed to ensure the integrity of the tank system during the use of the tank system or component, consisting of one or more of the following:

(A) Corrosion-resistant materials of construction such as special alloys, fiberglass reinforced plastic, etc.;

(B) Corrosion-resistant coating (such as epoxy, fiberglass, etc.) with cathodic protection (e.g., impressed current or sacrificial anodes); and

(C) Electrical isolation devices such as insulating joints, flanges, etc.

[NOTE: The practices described in the National Association of Corrosion Engineers (NACE) standard, "Recommended Practice (RP-02-85)—Control of External Corrosion on Metallic Buried, Partially Buried, or Submerged Liquid Storage Systems," and the American Petroleum Institute (API) Publication 1632, "Cathodic Protection of Underground Petroleum Storage Tanks and Piping Systems," may be used, where applicable, as guidelines in providing corrosion protection for tank systems.]

(4) For underground tank system components that are likely to be adversely affected by vehicular traffic, a determination of design or operational measures that will protect the tank system against potential damage; and

(5) Design considerations to ensure that:

(i) Tank foundations will maintain the load of a full tank;

(ii) Tank systems will be anchored to prevent flotation or dislodgment where the tank system is placed in a saturated zone, or is located within a seismic fault zone subject to the standards of § 264.18(a); and

(iii) Tank systems will withstand the effects of frost heave.

(b) The owner or operator of a new tank system must ensure that proper handling procedures are adhered to in order to prevent damage to the system during installation. Prior to covering, enclosing, or placing a new tank system or component in use, an independent, qualified, installation inspector or a qualified Professional Engineer, either of whom is trained and experienced in the proper installation of tanks systems or components, must inspect the system for the presence of any of the following items:

(1) Weld breaks;

(2) Punctures;

(3) Scrapes of protective coatings;

(4) Cracks;

(5) Corrosion;

(6) Other structural damage or inadequate construction/installation.

All discrepancies must be remedied before the tank system is covered, enclosed, or placed in use.

(c) New tank systems or components that are placed underground and that are backfilled must be provided with a backfill material that is a noncorrosive, porous, homogeneous substance and that is installed so that the backfill is placed completely around the tank and compacted to ensure that the tank and piping are fully and uniformly supported.

(d) All new tanks and ancillary equipment must be tested for tightness prior to being covered, enclosed, or placed in use. If a tank system is found not to be tight, all repairs necessary to remedy the leak(s) in the system must be performed prior to the tank system being covered, enclosed, or placed into use.

(e) Ancillary equipment must be supported and protected against physical damage and excessive stress due to settlement, vibration, expansion, or contraction.

[NOTE: The piping system installation procedures described in American Petroleum Institute (API) Publication 1615 (November 1979), "Installation of Underground Petroleum Storage Systems," or ANSI Standard B31.3, "Petroleum Refinery Piping," and ANSI Standard B31.4 "Liquid Petroleum Transportation Piping System," may be used, where applicable, as guidelines for proper installation of piping systems.]

(f) The owner or operator must provide the type and degree of corrosion protection recommended by an independent corrosion expert, based on the information provided under paragraph (a)(3) of this section, or other corrosion protection if the Regional Administrator believes other corrosion protection is necessary to ensure the integrity of the tank system during use of the tank system. The installation of a corrosion protection system that is field fabricated must be supervised by an independent corrosion expert to ensure proper installation.

(g) The owner or operator must obtain and keep on file at the facility written statements by those persons required to certify the design of the tank system and supervise the installation of the tank system in accordance with the requirements of paragraphs (b) through (f) of this section, that attest that the tank system was properly designed and installed and that repairs, pursuant to paragraphs (b) and (d) of this section, were performed. These written statements must also include the certification statement as required in § 270.11(d) of this chapter.

[51 FR 25472, July 14, 1986; 51 FR 29430, Aug. 15, 1986, as amended at 71 FR 16905, Apr. 4, 2006]

§ 264.193 Containment and detection of releases.

(a) In order to prevent the release of hazardous waste or hazardous constituents to the environment, secondary containment that meets the requirements of this section must be provided (except as provided in paragraphs (f) and (g) of this section):

(1) For all new and existing tank systems or components, prior to their being put into service.

(2) For tank systems that store or treat materials that become hazardous wastes, within two years of the hazardous waste listing, or when the tank system has reached 15 years of age, whichever comes later.

(b) Secondary containment systems must be:

(1) Designed, installed, and operated to prevent any migration of wastes or accumulated liquid out of the system to the soil, ground water, or surface water at any time during the use of the tank system; and

(2) Capable of detecting and collecting releases and accumulated liquids until the collected material is removed.

(c) To meet the requirements of paragraph (b) of this section, secondary containment systems must be at a minimum:

(1) Constructed of or lined with materials that are compatible with the wastes(s) to be placed in the tank system and must have sufficient strength and thickness to prevent failure owing to pressure gradients (including static head and external hydrological forces), physical contact with the waste to which it is exposed, climatic conditions, and the stress of daily operation (including stresses from nearby vehicular traffic).

(2) Placed on a foundation or base capable of providing support to the secondary containment system, resistance to pressure gradients above and below the system, and capable of preventing failure due to settlement, compression, or uplift;

(3) Provided with a leak-detection system that is designed and operated so that it will detect the failure of either the primary or secondary containment structure or the presence of any release of hazardous waste or accumulated liquid in the secondary containment system within 24 hours, or at the earliest practicable time if the owner or operator can demonstrate to the Regional Administrator that existing detection technologies or site conditions will not allow detection of a release within 24 hours; and

(4) Sloped or otherwise designed or operated to drain and remove liquids resulting from leaks, spills, or precipitation. Spilled or leaked waste and accumulated precipitation must be removed from the secondary containment system within 24 hours, or in as timely a manner as is possible to prevent harm to human health and the environment, if the owner or operator can demonstrate to the Regional Administrator that removal of the released waste or accumulated precipitation cannot be accomplished within 24 hours.

[NOTE: If the collected material is a hazardous waste under part 261 of this chapter, it is subject to management as a hazardous waste in accordance with all applicable requirements of parts 262 through 265 of this chapter. If the collected material is discharged through a point source to waters of the United States, it is subject to the requirements of sections 301, 304, and 402 of the Clean Water Act, as amended. If discharged to a Publicly Owned Treatment Works (POTW), it is subject to the requirements of section 307 of the Clean Water Act, as amended. If the collected material is released to the environment, it may be subject to the reporting requirements of 40 CFR part 302.]

(d) Secondary containment for tanks must include one or more of the following devices:

(1) A liner (external to the tank);

(2) A vault;

(3) A double-walled tank; or

(4) An equivalent device as approved by the Regional Administrator.

(e) In addition to the requirements of paragraphs (b), (c), and (d) of this section, secondary containment systems must satisfy the following requirements:

(1) External liner systems must be:

(i) Designed or operated to contain 100 percent of the capacity of the largest tank within its boundary;

(ii) Designed or operated to prevent run-on or infiltration of precipitation into the secondary containment system unless the collection system has sufficient excess capacity to contain run-on or infiltration. Such additional capacity must be sufficient to contain precipitation from a 25-year, 24-hour rainfall event.

(iii) Free of cracks or gaps; and

(iv) Designed and installed to surround the tank completely and to cover all surrounding earth likely to come into contact with the waste if the waste is released from the tank(s) (i.e., capable of preventing lateral as well as vertical migration of the waste).

(2) Vault systems must be:

(i) Designed or operated to contain 100 percent of the capacity of the largest tank within its boundary;

(ii) Designed or operated to prevent run-on or infiltration of precipitation into the secondary containment system unless the collection system has sufficient excess capacity to contain run-on or infiltration. Such additional capacity must be sufficient to contain precipitation from a 25-year, 24-hour rainfall event;

(iii) Constructed with chemical-resistant water stops in place at all joints (if any);

(iv) Provided with an impermeable interior coating or lining that is compatible with the stored waste and that will prevent migration of waste into the concrete;

(v) Provided with a means to protect against the formation of and ignition of vapors within the vault, if the waste being stored or treated:

(A) Meets the definition of ignitable waste under § 261.21 of this chapter; or

(B) Meets the definition of reactive waste under § 261.23 of this chapter, and may form an ignitable or explosive vapor; and

(vi) Provided with an exterior moisture barrier or be otherwise designed or operated to prevent migration of moisture into the vault if the vault is subject to hydraulic pressure.

(3) Double-walled tanks must be:

(i) Designed as an integral structure (i.e., an inner tank completely enveloped within an outer shell) so that any release from the inner tank is contained by the outer shell;

(ii) Protected, if constructed of metal, from both corrosion of the primary tank interior and of the external surface of the outer shell; and

(iii) Provided with a built-in continuous leak detection system capable of detecting a release within 24 hours, or at the earliest practicable time, if the owner or operator can demonstrate to the Regional Administrator, and the Regional Administrator concludes, that the existing detection technology or site conditions would not allow detection of a release within 24 hours.

[NOTE: The provisions outlined in the Steel Tank Institute's (STI) "Standard for Dual Wall Underground Steel Storage Tanks" may be used as guidelines for aspects of the design of underground steel double-walled tanks.]

(f) Ancillary equipment must be provided with secondary containment (e.g., trench, jacketing, double-walled piping) that meets the requirements of paragraphs (b) and (c) of this section except for:

(1) Aboveground piping (exclusive of flanges, joints, valves, and other connections) that are visually inspected for leaks on a daily basis;

(2) Welded flanges, welded joints, and welded connections, that are visually inspected for leaks on a daily basis;

(3) Sealless or magnetic coupling pumps and sealless valves, that are visually inspected for leaks on a daily basis; and

(4) Pressurized aboveground piping systems with automatic shut-off devices (e.g., excess flow check valves, flow metering shutdown devices, loss of pressure actuated shut-off devices) that are visually inspected for leaks on a daily basis.

(g) The owner or operator may obtain a variance from the requirements of this section if the Regional Administrator finds, as a result of a demonstration by the owner or operator that alternative design and operating practices, together with location characteristics, will prevent the migration of any hazardous waste or hazardous constituents into the ground water; or surface water at least as effectively as secondary containment during the active life of the tank system *or* that in the event of a release that does migrate to ground water or surface water, no substantial present or potential hazard will be posed to human health or the environment. New underground tank systems may not, per a demonstration in accordance with paragraph (g)(2) of this section, be exempted from the secondary containment requirements of this section.

(1) In deciding whether to grant a variance based on a demonstration of equivalent protection of ground water and surface water, the Regional Administrator will consider:

(i) The nature and quantity of the wastes;

(ii) The proposed alternate design and operation;

(iii) The hydrogeologic setting of the facility, including the thickness of soils present between the tank system and ground water; and

(iv) All other factors that would influence the quality and mobility of the hazardous constituents and the potential for them to migrate to ground water or surface water.

(2) In deciding whether to grant a variance based on a demonstration of no substantial present or potential hazard, the Regional Administrator will consider:

(i) The potential adverse effects on ground water, surface water, and land quality taking into account:

(A) The physical and chemical characteristics of the waste in the tank system, including its potential for migration,

(B) The hydrogeological characteristics of the facility and surrounding land,

(C) The potential for health risks caused by human exposure to waste constituents,

(D) The potential for damage to wildlife, crops, vegetation, and physical structures caused by exposure to waste constituents, and

(E) The persistence and permanence of the potential adverse effects;

(ii) The potential adverse effects of a release on ground-water quality, taking into account:

(A) The quantity and quality of ground water and the direction of ground-water flow,

(B) The proximity and withdrawal rates of ground-water users,

(C) The current and future uses of ground water in the area, and

(D) The existing quality of ground water, including other sources of con-

tamination and their cumulative impact on the ground-water quality;

(iii) The potential adverse effects of a release on surface water quality, taking into account:

(A) The quantity and quality of ground water and the direction of ground-water flow,

(B) The patterns of rainfall in the region,

(C) The proximity of the tank system to surface waters,

(D) The current and future uses of surface waters in the area and any water quality standards established for those surface waters, and

(E) The existing quality of surface water, including other sources of contamination and the cumulative impact on surface-water quality; and

(iv) The potential adverse effects of a release on the land surrounding the tank system, taking into account:

(A) The patterns of rainfall in the region, and

(B) The current and future uses of the surrounding land.

(3) The owner or operator of a tank system, for which a variance from secondary containment had been granted in accordance with the requirements of paragraph (g)(1) of this section, at which a release of hazardous waste has occurred from the primary tank system but has not migrated beyond the zone of engineering control (as established in the variance), must:

(i) Comply with the requirements of §264.196, except paragraph (d), and

(ii) Decontaminate or remove contaminated soil to the extent necessary to:

(A) Enable the tank system for which the variance was granted to resume operation with the capability for the detection of releases at least equivalent to the capability it had prior to the release; and

(B) Prevent the migration of hazardous waste or hazardous constituents to ground water or surface water; and

(iii) If contaminated soil cannot be removed or decontaminated in accordance with paragraph (g)(3)(ii) of this section, comply with the requirement of §264.197(b).

(4) The owner or operator of a tank system, for which a variance from secondary containment had been granted

443

in accordance with the requirements of paragraph (g)(1) of this section, at which a release of hazardous waste has occurred from the primary tank system and has migrated beyond the zone of engineering control (as established in the variance), must:

(i) Comply with the requirements of § 264.196 (a), (b), (c), and (d); and

(ii) Prevent the migration of hazardous waste or hazardous constituents to ground water or surface water, if possible, and decontaminate or remove contaminated soil. If contaminated soil cannot be decontaminated or removed or if ground water has been contaminated, the owner or operator must comply with the requirements of § 264.197(b); and

(iii) If repairing, replacing, or reinstalling the tank system, provide secondary containment in accordance with the requirements of paragraphs (a) through (f) of this section or reapply for a variance from secondary containment and meet the requirements for new tank systems in § 264.192 if the tank system is replaced. The owner or operator must comply with these requirements even if contaminated soil can be decontaminated or removed and ground water or surface water has not been contaminated.

(h) The following procedures must be followed in order to request a variance from secondary containment:

(1) The Regional Administrator must be notified in writing by the owner or operator that he intends to conduct and submit a demonstration for a variance from secondary containment as allowed in paragraph (g) of this section according to the following schedule:

(i) For existing tank systems, at least 24 months prior to the date that secondary containment must be provided in accordance with paragraph (a) of this section.

(ii) For new tank systems, at least 30 days prior to entering into a contract for installation.

(2) As part of the notification, the owner or operator must also submit to the Regional Administrator a description of the steps necessary to conduct the demonstration and a timetable for completing each of the steps. The demonstration must address each of the

factors listed in paragraph (g)(1) or paragraph (g)(2) of this section;

(3) The demonstration for a variance must be completed within 180 days after notifying the Regional Administrator of an intent to conduct the demonstration; and

(4) If a variance is granted under this paragraph, the Regional Administrator will require the permittee to construct and operate the tank system in the manner that was demonstrated to meet the requirements for the variance.

(i) All tank systems, until such time as secondary containment that meets the requirements of this section is provided, must comply with the following:

(1) For non-enterable underground tanks, a leak test that meets the requirements of § 264.191(b)(5) or other tank integrity method, as approved or required by the Regional Administrator, must be conducted at least annually.

(2) For other than non-enterable underground tanks, the owner or operator must either conduct a leak test as in paragraph (i)(1) of this section or develop a schedule and procedure for an assessment of the overall condition of the tank system by a qualified Professional Engineer. The schedule and procedure must be adequate to detect obvious cracks, leaks, and corrosion or erosion that may lead to cracks and leaks. The owner or operator must remove the stored waste from the tank, if necessary, to allow the condition of all internal tank surfaces to be assessed. The frequency of these assessments must be based on the material of construction of the tank and its ancillary equipment, the age of the system, the type of corrosion or erosion protection used, the rate of corrosion or erosion observed during the previous inspection, and the characteristics of the waste being stored or treated.

(3) For ancillary equipment, a leak test or other integrity assessment as approved by the Regional Administrator must be conducted at least annually.

[NOTE: The practices described in the American Petroleum Institute (API) Publication Guide for Inspection of Refinery Equipment, Chapter XIII, "Atmospheric and Low-Pressure Storage Tanks," 4th edition,

444

1981, may be used, where applicable, as guidelines for assessing the overall condition of the tank system.]

(4) The owner or operator must maintain on file at the facility a record of the results of the assessments conducted in accordance with paragraphs (i)(1) through (i)(3) of this section.

(5) If a tank system or component is found to be leaking or unfit for use as a result of the leak test or assessment in paragraphs (i)(1) through (i)(3) of this section, the owner or operator must comply with the requirements of §264.196.

[51 FR 25472, July 14, 1986; 51 FR 29430, Aug. 15, 1986, as amended at 53 FR 34086, Sept. 2, 1988; 71 FR 16905, Apr. 4, 2006; 71 FR 40273, July 14, 2006]

§ **264.194 General operating requirements.**

(a) Hazardous wastes or treatment reagents must not be placed in a tank system if they could cause the tank, its ancillary equipment, or the containment system to rupture, leak, corrode, or otherwise fail.

(b) The owner or operator must use appropriate controls and practices to prevent spills and overflows from tank or containment systems. These include at a minimum:

(1) Spill prevention controls (e.g., check valves, dry disconnect couplings);

(2) Overfill prevention controls (e.g., level sensing devices, high level alarms, automatic feed cutoff, or bypass to a standby tank); and

(3) Maintenance of sufficient freeboard in uncovered tanks to prevent overtopping by wave or wind action or by precipitation.

(c) The owner or operator must comply with the requirements of §264.196 if a leak or spill occurs in the tank system.

§ **264.195 Inspections.**

(a) The owner or operator must develop and follow a schedule and procedure for inspecting overfill controls.

(b) The owner or operator must inspect at least once each operating day data gathered from monitoring and leak detection equipment (e.g., pressure or temperature gauges, monitoring wells) to ensure that the tank system is being operated according to its design.

[NOTE: Section 264.15(c) requires the owner or operator to remedy any deterioration or malfunction he finds. Section 264.196 requires the owner or operator to notify the Regional Administrator within 24 hours of confirming a leak. Also, 40 CFR part 302 may require the owner or operator to notify the National Response Center of a release.]

(c) In addition, except as noted under paragraph (d) of this section, the owner or operator must inspect at least once each operating day:

(1) Above ground portions of the tank system, if any, to detect corrosion or releases of waste.

(2) The construction materials and the area immediately surrounding the externally accessible portion of the tank system, including the secondary containment system (*e.g.*, dikes) to detect erosion or signs of releases of hazardous waste (*e.g.*, wet spots, dead vegetation).

(d) Owners or operators of tank systems that either use leak detection systems to alert facility personnel to leaks, or implement established workplace practices to ensure leaks are promptly identified, must inspect at least weekly those areas described in paragraphs (c)(1) and (c)(2) of this section. Use of the alternate inspection schedule must be documented in the facility's operating record. This documentation must include a description of the established workplace practices at the facility.

(e) Performance Track member facilities may inspect on a less frequent basis, upon approval by the Director, but must inspect at least once each month. To apply for a less than weekly inspection frequency, the Performance Track member facility must follow the procedures described in §264.15(b)(5).

(f) Ancillary equipment that is not provided with secondary containment, as described in §264.193(f)(1) through (4), must be inspected at least once each operating day.

(g) The owner or operator must inspect cathodic protection systems, if present, according to, at a minimum, the following schedule to ensure that they are functioning properly:

(1) The proper operation of the cathodic protection system must be confirmed within six months after initial installation and annually thereafter; and

(2) All sources of impressed current must be inspected and/or tested, as appropriate, at least bimonthly (i.e., every other month).

[NOTE: The practices described in the National Association of Corrosion Engineers (NACE) standard, "Recommended Practice (RP–02–85)—Control of External Corrosion on Metallic Buried, Partially Buried, or Submerged Liquid Storage Systems," and the American Petroleum Institute (API) Publication 1632, "Cathodic Protection of Underground Petroleum Storage Tanks and Piping Systems," may be used, where applicable, as guidelines in maintaining and inspecting cathodic protection systems.]

(h) The owner or operator must document in the operating record of the facility an inspection of those items in paragraphs (a) through (c) of this section.

[51 FR 25472, July 14, 1986, as amended at 71 FR 16906, Apr. 4, 2006]

## § 264.196 Response to leaks or spills and disposition of leaking or unfit-for-use tank systems.

A tank system or secondary containment system from which there has been a leak or spill, or which is unfit for use, must be removed from service immediately, and the owner or operator must satisfy the following requirements:

(a) *Cessation of use; prevent flow or addition of wastes.* The owner or operator must immediately stop the flow of hazardous waste into the tank system or secondary containment system and inspect the system to determine the cause of the release.

(b) *Removal of waste from tank system or secondary containment system.* (1) If the release was from the tank system, the owner/operator must, within 24 hours after detection of the leak or, if the owner/operator demonstrates that it is not possible, at the earliest practicable time, remove as much of the waste as is necessary to prevent further release of hazardous waste to the environment and to allow inspection and repair of the tank system to be performed.

(2) If the material released was to a secondary containment system, all released materials must be removed within 24 hours or in as timely a manner as is possible to prevent harm to human health and the environment.

(c) *Containment of visible releases to the environment.* The owner/operator must immediately conduct a visual inspection of the release and, based upon that inspection:

(1) Prevent further migration of the leak or spill to soils or surface water; and

(2) Remove, and properly dispose of, any visible contamination of the soil or surface water.

(d) *Notifications, reports.* (1) Any release to the environment, except as provided in paragraph (d)(2) of this section, must be reported to the Regional Administrator within 24 hours of its detection. If the release has been reported pursuant to 40 CFR part 302, that report will satisfy this requirement.

(2) A leak or spill of hazardous waste is exempted from the requirements of this paragraph if it is:

(i) Less than or equal to a quantity of one (1) pound, and

(ii) Immediately contained and cleaned up.

(3) Within 30 days of detection of a release to the environment, a report containing the following information must be submitted to the Regional Administrator:

(i) Likely route of migration of the release;

(ii) Characteristics of the surrounding soil (soil composition, geology, hydrogeology, climate);

(iii) Results of any monitoring or sampling conducted in connection with the release (if available). If sampling or monitoring data relating to the release are not available within 30 days, these data must be submitted to the Regional Administrator as soon as they become available.

(iv) Proximity to downgradient drinking water, surface water, and populated areas; and

(v) Description of response actions taken or planned.

(e) *Provision of secondary containment, repair, or closure.* (1) Unless the owner/operator satisfies the requirements of paragraphs (e)(2) through (4) of this section, the tank system must be closed in accordance with §264.197.

(2) If the cause of the release was a spill that has not damaged the integrity of the system, the owner/operator may return the system to service as soon as the released waste is removed and repairs, if necessary, are made.

(3) If the cause of the release was a leak from the primary tank system into the secondary containment system, the system must be repaired prior to returning the tank system to service.

(4) If the source of the release was a leak to the environment from a component of a tank system without secondary containment, the owner/operator must provide the component of the system from which the leak occurred with secondary containment that satisfies the requirements of §264.193 before it can be returned to service, unless the source of the leak is an aboveground portion of a tank system that can be inspected visually. If the source is an aboveground component that can be inspected visually, the component must be repaired and may be returned to service without secondary containment as long as the requirements of paragraph (f) of this section are satisfied. If a component is replaced to comply with the requirements of this subparagraph, that component must satisfy the requirements for new tank systems or components in §§264.192 and 264.193. Additionally, if a leak has occurred in any portion of a tank system component that is not readily accessible for visual inspection (e.g., the bottom of an inground or onground tank), the entire component must be provided with secondary containment in accordance with §264.193 prior to being returned to use.

(f) *Certification of major repairs.* If the owner/operator has repaired a tank system in accordance with paragraph (e) of this section, and the repair has been extensive (*e.g.*, installation of an internal liner; repair of a ruptured primary containment or secondary containment vessel), the tank system must not be returned to service unless the owner/operator has obtained a certification by a qualified Professional Engineer in accordance with §270.11(d) of this chapter that the repaired system is capable of handling hazardous wastes without release for the intended life of the system. This certification must be placed in the operating record and maintained until closure of the facility.

[NOTE: The Regional Administrator may, on the basis of any information received that there is or has been a release of hazardous waste or hazardous constituents into the environment, issue an order under RCRA section 3004(v), 3008(h), or 7003(a) requiring corrective action or such other response as deemed necessary to protect human health or the environment.]

[NOTE: See §264.15(c) for the requirements necessary to remedy a failure. Also, 40 CFR part 302 may require the owner or operator to notify the National Response Center of certain releases.]

[51 FR 25472, July 14, 1986; 51 FR 29430, Aug. 15, 1986, as amended at 53 FR 34086, Sept. 2, 1988; 71 FR 16906, Apr. 4, 2006]

## §264.197 Closure and post-closure care.

(a) At closure of a tank system, the owner or operator must remove or decontaminate all waste residues, contaminated containment system components (liners, etc.), contaminated soils, and structures and equipment contaminated with waste, and manage them as hazardous waste, unless §261.3(d) of this chapter applies. The closure plan, closure activities, cost estimates for closure, and financial responsibility for tank systems must meet all of the requirements specified in subparts G and H of this part.

(b) If the owner or operator demonstrates that not all contaminated soils can be practicably removed or decontaminated as required in paragraph (a) of this section, then the owner or operator must close the tank system and perform post-closure care in accordance with the closure and post-closure care requirements that apply to landfills (§264.310). In addition, for the purposes of closure, post-closure, and financial responsibility, such a tank system is then considered to be a landfill, and the owner or operator must meet all of the requirements for landfills specified in subparts G and H of this part.

(c) If an owner or operator has a tank system that does not have secondary containment that meets the requirements of § 264.193 (b) through (f) and has not been granted a variance from the secondary containment requirements in accordance with § 264.193(g), then:

(1) The closure plan for the tank system must include both a plan for complying with paragraph (a) of this section and a contingent plan for complying with paragraph (b) of this section.

(2) A contingent post-closure plan for complying with paragraph (b) of this section must be prepared and submitted as part of the permit application.

(3) The cost estimates calculated for closure and post-closure care must reflect the costs of complying with the contingent closure plan and the contingent post-closure plan, if those costs are greater than the costs of complying with the closure plan prepared for the expected closure under paragraph (a) of this section.

(4) Financial assurance must be based on the cost estimates in paragraph (c)(3) of this section.

(5) For the purposes of the contingent closure and post-closure plans, such a tank system is considered to be a landfill, and the contingent plans must meet all of the closure, post-closure, and financial responsibility requirements for landfills under subparts G and H of this part.

[51 FR 25472, July 14, 1986; 51 FR 29430, Aug. 15, 1986]

§ 264.198 Special requirements for ignitable or reactive wastes.

(a) Ignitable or reactive waste must not be placed in tank systems, unless:

(1) The waste is treated, rendered, or mixed before or immediately after placement in the tank system so that:

(i) The resulting waste, mixture, or dissolved material no longer meets the definition of ignitable or reactive waste under §§ 261.21 or 261.23 of this chapter, and

(ii) Section 264.17(b) is complied with; or

(2) The waste is stored or treated in such a way that it is protected from any material or conditions that may cause the waste to ignite or react; or

(3) The tank system is used solely for emergencies.

(b) The owner or operator of a facility where ignitable or reactive waste is stored or treated in a tank must comply with the requirements for the maintenance of protective distances between the waste management area and any public ways, streets, alleys, or an adjoining property line that can be built upon as required in Tables 2-1 through 2-6 of the National Fire Protection Association's "Flammable and Combustible Liquids Code," (1977 or 1981), (incorporated by reference, see § 260.11).

§ 264.199 Special requirements for incompatible wastes.

(a) Incompatible wastes, or incompatible wastes and materials, must not be placed in the same tank system, unless § 264.17(b) is complied with.

(b) Hazardous waste must not be placed in a tank system that has not been decontaminated and that previously held an incompatible waste or material, unless § 264.17(b) is complied with.

§ 264.200 Air emission standards.

The owner or operator shall manage all hazardous waste placed in a tank in accordance with the applicable requirements of subparts AA, BB, and CC of this part.

[61 FR 59950, Nov. 25, 1996]

## Subpart K—Surface Impoundments

SOURCE: 47 FR 32357, July 26, 1982, unless otherwise noted.

§ 264.220 Applicability.

The regulations in this subpart apply to owners and operators of facilities that use surface impoundments to treat, store, or dispose of hazardous waste except as § 264.1 provides otherwise.

§ 264.221 Design and operating requirements.

(a) Any surface impoundment that is not covered by paragraph (c) of this section or § 265.221 of this chapter must have a liner for all portions of the impoundment (except for existing portions of such impoundments). The liner must be designed, constructed, and installed to prevent any migration of wastes out of the impoundment to the adjacent subsurface soil or ground water or surface water at any time during the active life (including the closure period) of the impoundment. The liner may be constructed of materials that may allow wastes to migrate into the liner (but not into the adjacent subsurface soil or ground water or surface water) during the active life of the facility, provided that the impoundment is closed in accordance with §264.228(a)(1). For impoundments that will be closed in accordance with §264.228(a)(2), the liner must be constructed of materials that can prevent wastes from migrating into the liner during the active life of the facility. The liner must be:

(1) Constructed of materials that have appropriate chemical properties and sufficient strength and thickness to prevent failure due to pressure gradients (including static head and external hydrogeologic forces), physical contact with the waste or leachate to which they are exposed, climatic conditions, the stress of installation, and the stress of daily operation;

(2) Placed upon a foundation or base capable of providing support to the liner and resistance to pressure gradients above and below the liner to prevent failure of the liner due to settlement, compression, or uplift; and

(3) Installed to cover all surrounding earth likely to be in contact with the waste or leachate.

(b) The owner or operator will be exempted from the requirements of paragraph (a) of this section if the Regional Administrator finds, based on a demonstration by the owner or operator, that alternate design and operating practices, together with location characteristics, will prevent the migration of any hazardous constituents (see §264.93) into the ground water or surface water at any future time. In deciding whether to grant an exemption, the Regional Administrator will consider:

(1) The nature and quantity of the wastes;

(2) The proposed alternate design and operation;

(3) The hydrogeologic setting of the facility, including the attenuative capacity and thickness of the liners and soils present between the impoundment and ground water or surface water; and

(4) All other factors which would influence the quality and mobility of the leachate produced and the potential for it to migrate to ground water or surface water.

(c) The owner or operator of each new surface impoundment unit on which construction commences after January 29, 1992, each lateral expansion of a surface impoundment unit on which construction commences after July 29, 1992 and each replacement of an existing surface impoundment unit that is to commence reuse after July 29, 1992 must install two or more liners and a leachate collection and removal system between such liners. "Construction commences" is as defined in §260.10 of this chapter under "existing facility".

(1)(i) The *liner system* must include:

(A) A top liner designed and constructed of materials (e.g., a geomembrane) to prevent the migration of hazardous constituents into such liner during the active life and post-closure care period; and

(B) A composite bottom liner, consisting of at least two components. The upper component must be designed and constructed of materials (e.g., a geomembrane) to prevent the migration of hazardous constituents into this component during the active life and post-closure care period. The lower component must be designed and constructed of materials to minimize the migration of hazardous constituents if a breach in the upper component were to occur. The lower component must be constructed of at least 3 feet (91 cm) of compacted soil material with a hydraulic conductivity of no more than $1\times10^{-7}$ cm/sec.

(ii) The liners must comply with paragraphs (a) (1), (2), and (3) of this section.

(2) The *leachate collection and removal system* between the liners, and immediately above the bottom composite liner in the case of multiple leachate collection and removal systems, is also

a *leak detection system.* This leak detection system must be capable of detecting, collecting, and removing leaks of hazardous constituents at the earliest practicable time through all areas of the top liner likely to be exposed to waste or leachate during the active life and post-closure care period. The requirements for a leak detection system in this paragraph are satisfied by installation of a system that is, at a minimum:

(i) Constructed with a bottom slope of one percent or more;

(ii) Constructed of granular drainage materials with a hydraulic conductivity of $1\times10^{-1}$ cm/sec or more and a thickness of 12 inches (30.5 cm) or more; or constructed of synthetic or geonet drainage materials with a transmissivity of $3\times10^{-4}$ m²sec or more;

(iii) Constructed of materials that are chemically resistant to the waste managed in the surface impoundment and the leachate expected to be generated, and of sufficient strength and thickness to prevent collapse under the pressures exerted by overlying wastes and any waste cover materials or equipment used at the surface impoundment;

(iv) Designed and operated to minimize clogging during the active life and post-closure care period; and

(v) Constructed with sumps and liquid removal methods (e.g., pumps) of sufficient size to collect and remove liquids from the sump and prevent liquids from backing up into the drainage layer. Each unit must have its own sump(s). The design of each sump and removal system must provide a method for measuring and recording the volume of liquids present in the sump and of liquids removed.

(3) The owner or operator shall collect and remove pumpable liquids in the sumps to minimize the head on the bottom liner.

(4) The owner or operator of a leak detection system that is not located completely above the seasonal high water table must demonstrate that the operation of the leak detection system will not be adversely affected by the presence of ground water.

(d) The Regional Administrator may approve alternative design or operating practices to those specified in paragraph (c) of this section if the owner or operator demonstrates to the Regional Administrator that such design and operating practices, together with location characteristics:

(1) Will prevent the migration of any hazardous constituent into the ground water or surface water at least as effectively as the liners and leachate collection and removal system specified in paragraph (c) of this section; and

(2) Will allow detection of leaks of hazardous constituents through the top liner at least as effectively.

(e) The double liner requirement set forth in paragraph (c) of this section may be waived by the Regional Administrator for any monofill, if:

(1) The monofill contains only hazardous wastes from foundry furnace emission controls or metal casting molding sand, and such wastes do not contain constituents which would render the wastes hazardous for reasons other than the toxicity characteristic in § 261.24 of this chapter; and

(2)(i)(A) The monofill has at least one liner for which there is no evidence that such liner is leaking. For the purposes of this paragraph, the term "liner" means a liner designed, constructed, installed, and operated to prevent hazardous waste from passing into the liner at any time during the active life of the facility, or a liner designed, constructed, installed, and operated to prevent hazardous waste from migrating beyond the liner to adjacent subsurface soil, ground water, or surface water at any time during the active life of the facility. In the case of any surface impoundment which has been exempted from the requirements of paragraph (c) of this section on the basis of a liner designed, constructed, installed, and operated to prevent hazardous waste from passing beyond the liner, at the closure of such impoundment, the owner or operator must remove or decontaminate all waste residues, all contaminated liner material, and contaminated soil to the extent practicable. If all contaminated soil is not removed or decontaminated, the owner or operator of such impoundment will comply with appropriate post-closure requirements, including but not limited to ground-water monitoring and corrective action;

(B) The monofill is located more than one-quarter mile from an "underground source of drinking water" (as that term is defined in 40 CFR 270.2); and

(C) The monofill is in compliance with generally applicable ground-water monitoring requirements for facilities with permits under RCRA section 3005(c); or

(ii) The owner or operator demonstrates that the monofill is located, designed and operated so as to assure that there will be no migration of any hazardous constituent into ground water or surface water at any future time.

(f) The owner or operator of any replacement surface impoundment unit is exempt from paragraph (c) of this section if:

(1) The existing unit was constructed in compliance with the design standards of sections 3004 (o)(1)(A)(i) and (o)(5) of the Resource Conservation and Recovery Act; and

(2) There is no reason to believe that the liner is not functioning as designed.

(g) A surface impoundment must be designed, constructed, maintained, and operated to prevent overtopping resulting from normal or abnormal operations; overfilling; wind and wave action; rainfall; run-on; malfunctions of level controllers, alarms, and other equipment; and human error.

(h) A surface impoundment must have dikes that are designed, constructed, and maintained with sufficient structural integrity to prevent massive failure of the dikes. In ensuring structural integrity, it must not be presumed that the liner system will function without leakage during the active life of the unit.

(i) The Regional Administrator will specify in the permit all design and operating practices that are necessary to ensure that the requirements of this section are satisfied.

[47 FR 32357, July 26, 1982, as amended at 50 FR 4514, Jan. 31, 1985; 50 FR 28747, July 15, 1985; 57 FR 3487, Jan. 29, 1992; 71 FR 40273, July 14, 2006]

**§264.222  Action leakage rate.**

(a) The Regional Administrator shall approve an action leakage rate for surface impoundment units subject to §264.221 (c) or (d). The action leakage rate is the maximum design flow rate that the leak detection system (LDS) can remove without the fluid head on the bottom liner exceeding 1 foot. The action leakage rate must include an adequate safety margin to allow for uncertainties in the design (e.g., slope, hydraulic conductivity, thickness of drainage material), construction, operation, and location of the LDS, waste and leachate characteristics, likelihood and amounts of other sources of liquids in the LDS, and proposed response actions (e.g., the action leakage rate must consider decreases in the flow capacity of the system over time resulting from siltation and clogging, rib layover and creep of synthetic components of the system, overburden pressures, etc.).

(b) To determine if the action leakage rate has been exceeded, the owner or operator must convert the weekly or monthly flow rate from the monitoring data obtained under §264.226(d) to an average daily flow rate (gallons per acre per day) for each sump. Unless the Regional Administrator approves a different calculation, the average daily flow rate for each sump must be calculated weekly during the active life and closure period, and if the unit is closed in accordance with §264.228(b), monthly during the post-closure care period when monthly monitoring is required under §264.226(d).

[57 FR 3487, Jan. 29, 1992]

**§264.223  Response actions.**

(a) The owner or operator of surface impoundment units subject to §264.221 (c) or (d) must have an approved response action plan before receipt of waste. The response action plan must set forth the actions to be taken if the action leakage rate has been exceeded. At a minimum, the response action plan must describe the actions specified in paragraph (b) of this section.

(b) If the flow rate into the leak detection system exceeds the action leakage rate for any sump, the owner or operator must:

(1) Notify the Regional Administrator in writing of the exceedance within 7 days of the determination;

451

(2) Submit a preliminary written assessment to the Regional Administrator within 14 days of the determination, as to the amount of liquids, likely sources of liquids, possible location, size, and cause of any leaks, and short-term actions taken and planned;

(3) Determine to the extent practicable the location, size, and cause of any leak;

(4) Determine whether waste receipt should cease or be curtailed, whether any waste should be removed from the unit for inspection, repairs, or controls, and whether or not the unit should be closed;

(5) Determine any other short-term and longer-term actions to be taken to mitigate or stop any leaks; and

(6) Within 30 days after the notification that the action leakage rate has been exceeded, submit to the Regional Administrator the results of the analyses specified in paragraphs (b) (3), (4), and (5) of this section, the results of actions taken, and actions planned. Monthly thereafter, as long as the flow rate in the leak detection system exceeds the action leakage rate, the owner or operator must submit to the Regional Administrator a report summarizing the results of any remedial actions taken and actions planned.

(c) To make the leak and/or remediation determinations in paragraphs (b) (3), (4), and (5) of this section, the owner or operator must:

(1)(i) Assess the source of liquids and amounts of liquids by source,

(ii) Conduct a fingerprint, hazardous constituent, or other analyses of the liquids in the leak detection system to identify the source of liquids and possible location of any leaks, and the hazard and mobility of the liquid; and

(iii) Assess the seriousness of any leaks in terms of potential for escaping into the environment; or

(2) Document why such assessments are not needed.

[57 FR 3488, Jan. 29, 1992, as amended at 71 FR 40273, July 14, 2006]

## §§ 264.224–264.225   [Reserved]

## § 264.226   Monitoring and inspection.

(a) During construction and installation, liners (except in the case of existing portions of surface impoundments exempt from § 264.221(a)) and cover systems (e.g., membranes, sheets, or coatings) must be inspected for uniformity, damage, and imperfections (e.g., holes, cracks, thin spots, or foreign materials). Immediately after construction or installation:

(1) Synthetic liners and covers must be inspected to ensure tight seams and joints and the absence of tears, punctures, or blisters; and

(2) Soil-based and admixed liners and covers must be inspected for imperfections including lenses, cracks, channels, root holes, or other structural non-uniformities that may cause an increase in the permeability of the liner or cover.

(b) While a surface impoundment is in operation, it must be inspected weekly and after storms to detect evidence of any of the following:

(1) Deterioration, malfunctions, or improper operation of overtopping control systems;

(2) Sudden drops in the level of the impoundment's contents; and

(3) Severe erosion or other signs of deterioration in dikes or other containment devices.

(c) Prior to the issuance of a permit, and after any extended period of time (at least six months) during which the impoundment was not in service, the owner or operator must obtain a certification from a qualified engineer that the impoundment's dike, including that portion of any dike which provides freeboard, has structural integrity. The certification must establish, in particular, that the dike:

(1) Will withstand the stress of the pressure exerted by the types and amounts of wastes to be placed in the impoundment; and

(2) Will not fail due to scouring or piping, without dependence on any liner system included in the surface impoundment construction.

(d)(1) An owner or operator required to have a leak detection system under § 264.221 (c) or (d) must record the amount of liquids removed from each leak detection system sump at least once each week during the active life and closure period.

(2) After the final cover is installed, the amount of liquids removed from each leak detection system sump must

be recorded at least monthly. If the liquid level in the sump stays below the pump operating level for two consecutive months, the amount of liquids in the sumps must be recorded at least quarterly. If the liquid level in the sump stays below the pump operating level for two consecutive quarters, the amount of liquids in the sumps must be recorded at least semi-annually. If at any time during the post-closure care period the pump operating level is exceeded at units on quarterly or semi-annual recording schedules, the owner or operator must return to monthly recording of amounts of liquids removed from each sump until the liquid level again stays below the pump operating level for two consecutive months.

(3) "Pump operating level" is a liquid level proposed by the owner or operator and approved by the Regional Administrator based on pump activation level, sump dimensions, and level that avoids backup into the drainage layer and minimizes head in the sump.

[47 FR 32357, July 26, 1982, as amended at 50 FR 4514, Jan. 31, 1985; 50 FR 28748, July 15, 1985; 57 FR 3488, Jan. 29, 1992; 71 FR 40273, July 14, 2006]

### §264.227 Emergency repairs; contingency plans.

(a) A surface impoundment must be removed from service in accordance with paragraph (b) of this section when:

(1) The level of liquids in the impoundment suddenly drops and the drop is not known to be caused by changes in the flows into or out of the impoundment; or

(2) The dike leaks.

(b) When a surface impoundment must be removed from service as required by paragraph (a) of this section, the owner or operator must:

(1) Immediately shut off the flow or stop the addition of wastes into the impoundment;

(2) Immediately contain any surface leakage which has occurred or is occurring;

(3) Immediately stop the leak;

(4) Take any other necessary steps to stop or prevent catastrophic failure;

(5) If a leak cannot be stopped by any other means, empty the impoundment; and

(6) Notify the Regional Administrator of the problem in writing within seven days after detecting the problem.

(c) As part of the contingency plan required in subpart D of this part, the owner or operator must specify a procedure for complying with the requirements of paragraph (b) of this section.

(d) No surface impoundment that has been removed from service in accordance with the requirements of this section may be restored to service unless the portion of the impoundment which was failing is repaired and the following steps are taken:

(1) If the impoundment was removed from service as the result of actual or imminent dike failure, the dike's structural integrity must be recertified in accordance with §264.226(c).

(2) If the impoundment was removed from service as the result of a sudden drop in the liquid level, then:

(i) For any existing portion of the impoundment, a liner must be installed in compliance with §264.221(a); and

(ii) For any other portion of the impoundment, the repaired liner system must be certified by a qualified engineer as meeting the design specifications approved in the permit.

(e) A surface impoundment that has been removed from service in accordance with the requirements of this section and that is not being repaired must be closed in accordance with the provisions of §264.228.

[47 FR 32357, July 26, 1982, as amended at 50 FR 28748, July 15, 1985]

### §264.228 Closure and post-closure care.

(a) At closure, the owner or operator must:

(1) Remove or decontaminate all waste residues, contaminated containment system components (liners, etc.), contaminated subsoils, and structures and equipment contaminated with waste and leachate, and manage them as hazardous waste unless §261.3(d) of this chapter applies; or

(2)(i) Eliminate free liquids by removing liquid wastes or solidifying the remaining wastes and waste residues;

(ii) Stabilize remaining wastes to a bearing capacity sufficient to support final cover; and

(iii) Cover the surface impoundment with a final cover designed and constructed to:

(A) Provide long-term minimization of the migration of liquids through the closed impoundment;

(B) Function with minimum maintenance;

(C) Promote drainage and minimize erosion or abrasion of the final cover;

(D) Accommodate settling and subsidence so that the cover's integrity is maintained; and

(E) Have a permeability less than or equal to the permeability of any bottom liner system or natural subsoils present.

(b) If some waste residues or contaminated materials are left in place at final closure, the owner or operator must comply with all post-closure requirements contained in §§ 264.117 through 264.120, including maintenance and monitoring throughout the post-closure care period (specified in the permit under § 264.117). The owner or operator must:

(1) Maintain the integrity and effectiveness of the final cover, including making repairs to the cap as necessary to correct the effects of settling, subsidence, erosion, or other events;

(2) Maintain and monitor the leak detection system in accordance with §§ 264.221(c)(2)(iv) and (3) and 264.226(d), and comply with all other applicable leak detection system requirements of this part;

(3) Maintain and monitor the groundwater monitoring system and comply with all other applicable requirements of subpart F of this part; and

(4) Prevent run-on and run-off from eroding or otherwise damaging the final cover.

(c)(1) If an owner or operator plans to close a surface impoundment in accordance with paragraph (a)(1) of this section, and the impoundment does not comply with the liner requirements of § 264.221(a) and is not exempt from them in accordance with § 264.221(b), then:

(i) The closure plan for the impoundment under § 264.112 must include both a plan for complying with paragraph (a)(1) of this section and a contingent plan for complying with paragraph (a)(2) of this section in case not all con-

taminated subsoils can be practicably removed at closure; and

(ii) The owner or operator must prepare a contingent post-closure plan under § 264.118 for complying with paragraph (b) of this section in case not all contaminated subsoils can be practicably removed at closure.

(2) The cost estimates calculated under §§ 264.142 and 264.144 for closure and post-closure care of an impoundment subject to this paragraph must include the cost of complying with the contingent closure plan and the contingent post-closure plan, but are not required to include the cost of expected closure under paragraph (a)(1) of this section.

[47 FR 32357, July 26, 1982, as amended at 50 FR 28748, July 15, 1985; 57 FR 3488, Jan. 29, 1992]

§ 264.229 Special requirements for ignitable or reactive waste.

Ignitable or reactive waste must not be placed in a surface impoundment, unless the waste and impoundment satisfy all applicable requirements of 40 CFR part 268, and:

(a) The waste is treated, rendered, or mixed before or immediately after placement in the impoundment so that:

(1) The resulting waste, mixture, or dissolution of material no longer meets the definition of ignitable or reactive waste under § 261.21 or § 261.23 of this chapter; and

(2) Section 264.17(b) is complied with; or

(b) The waste is managed in such a way that it is protected from any material or conditions which may cause it to ignite or react; or

(c) The surface impoundment is used solely for emergencies.

[47 FR 32357, July 26, 1982, as amended at 55 FR 22685, June 1, 1990]

§ 264.230 Special requirements for incompatible wastes.

Incompatible wastes, or incompatible wastes and materials, (see appendix V of this part for examples) must not be placed in the same surface impoundment, unless § 264.17(b) is complied with.

**§ 264.231 Special requirements for hazardous wastes FO20, FO21, FO22, FO23, FO26, and FO27.**

(a) Hazardous Wastes FO20, FO21, FO22, FO23, FO26, and FO27 must not be placed in a surface impoundment unless the owner or operator operates the surface impoundment in accordance with a management plan for these wastes that is approved by the Regional Administrator pursuant to the standards set out in this paragraph, and in accord with all other applicable requirements of this part. The factors to be considered are:

(1) The volume, physical, and chemical characteristics of the wastes, including their potential to migrate through soil or to volatilize or escape into the atmosphere;

(2) The attenuative properties of underlying and surrounding soils or other materials;

(3) The mobilizing properties of other materials co-disposed with these wastes; and

(4) The effectiveness of additional treatment, design, or monitoring techniques.

(b) The Regional Administrator may determine that additional design, operating, and monitoring requirements are necessary for surface impoundments managing hazardous wastes FO20, FO21, FO22, FO23, FO26, and FO27 in order to reduce the possibility of migration of these wastes to ground water, surface water, or air so as to protect human health and the environment.

[50 FR 2004, Jan. 14, 1985]

**§ 264.232 Air emission standards.**

The owner or operator shall manage all hazardous waste placed in a surface impoundment in accordance with the applicable requirements of subparts BB and CC of this part.

[61 FR 59950, Nov. 25, 1996]

## Subpart L—Waste Piles

SOURCE: 47 FR 32359, July 26, 1982, unless otherwise noted.

**§ 264.250 Applicability.**

(a) The regulations in this subpart apply to owners and operators of facilities that store or treat hazardous waste in piles, except as § 264.1 provides otherwise.

(b) The regulations in this subpart do not apply to owners or operators of waste piles that are closed with wastes left in place. Such waste piles are subject to regulation under subpart N of this part (Landfills).

(c) The owner or operator of any waste pile that is inside or under a structure that provides protection from precipitation so that neither runoff nor leachate is generated is not subject to regulation under § 264.251 or under subpart F of this part, provided that:

(1) Liquids or materials containing free liquids are not placed in the pile;

(2) The pile is protected from surface water run-on by the structure or in some other manner;

(3) The pile is designed and operated to control dispersal of the waste by wind, where necessary, by means other than wetting; and

(4) The pile will not generate leachate through decomposition or other reactions.

**§ 264.251 Design and operating requirements.**

(a) A waste pile (except for an existing portion of a waste pile) must have:

(1) A liner that is designed, constructed, and installed to prevent any migration of wastes out of the pile into the adjacent subsurface soil or ground water or surface water at any time during the active life (including the closure period) of the waste pile. The liner may be constructed of materials that may allow waste to migrate into the liner itself (but not into the adjacent subsurface soil or ground water or surface water) during the active life of the facility. The liner must be:

(i) Constructed of materials that have appropriate chemical properties and sufficient strength and thickness to prevent failure due to pressure gradients (including static head and external hydrogeologic forces), physical contact with the waste or leachate to which they are exposed, climatic conditions, the stress of installation, and the stress of daily operation;

(ii) Placed upon a foundation or base capable of providing support to the

455

liner and resistance to pressure gradients above and below the liner to prevent failure of the liner due to settlement, compression, or uplift; and

(iii) Installed to cover all surrounding earth likely to be in contact with the waste or leachate; and

(2) A leachate collection and removal system immediately above the liner that is designed, constructed, maintained, and operated to collect and remove leachate from the pile. The Regional Administrator will specify design and operating conditions in the permit to ensure that the leachate depth over the liner does not exceed 30 cm (one foot). The leachate collection and removal system must be:

(i) Constructed of materials that are:

(A) Chemically resistant to the waste managed in the pile and the leachate expected to be generated; and

(B) Of sufficient strength and thickness to prevent collapse under the pressures exerted by overlaying wastes, waste cover materials, and by any equipment used at the pile; and

(ii) Designed and operated to function without clogging through the scheduled closure of the waste pile.

(b) The owner or operator will be exempted from the requirements of paragraph (a) of this section, if the Regional Administrator finds, based on a demonstration by the owner or operator, that alternate design and operating practices, together with location characteristics, will prevent the migration of any hazardous constituents (see § 264.93) into the ground water or surface water at any future time. In deciding whether to grant an exemption, the Regional Administrator will consider:

(1) The nature and quantity of the wastes;

(2) The proposed alternate design and operation;

(3) The hydrogeologic setting of the facility, including attenuative capacity and thickness of the liners and soils present between the pile and ground water or surface water; and

(4) All other factors which would influence the quality and mobility of the leachate produced and the potential for it to migrate to ground water or surface water.

(c) The owner or operator of each new waste pile unit, each lateral expansion

of a waste pile unit, and each replacement of an existing waste pile unit must install two or more liners and a leachate collection and removal system above and between such liners.

(1)(i) The liner system must include:

(A) A top liner designed and constructed of materials (e.g., a geomembrane) to prevent the migration of hazardous constituents into such liner during the active life and post-closure care period; and

(B) A composite bottom liner, consisting of at least two components. The upper component must be designed and constructed of materials (e.g., a geomembrane) to prevent the migration of hazardous constituents into this component during the active life and post-closure care period. The lower component must be designed and constructed of materials to minimize the migration of hazardous constituents if a breach in the upper component were to occur. The lower component must be constructed of at least 3 feet (91 cm) of compacted soil material with a hydraulic conductivity of no more than $1 \times 10^{-7}$ cm/sec.

(ii) The liners must comply with paragraphs (a)(1)(i), (ii), and (iii) of this section.

(2) The *leachate collection and removal system* immediately above the top liner must be designed, constructed, operated, and maintained to collect and remove leachate from the waste pile during the active life and post-closure care period. The Regional Administrator will specify design and operating conditions in the permit to ensure that the leachate depth over the liner does not exceed 30 cm (one foot). The leachate collection and removal system must comply with paragraphs (c)(3)(iii) and (iv) of this section.

(3) The *leachate collection and removal system* between the liners, and immediately above the bottom composite liner in the case of multiple leachate collection and removal systems, is also a *leak detection system*. This leak detection system must be capable of detecting, collecting, and removing leaks of hazardous constituents at the earliest practicable time through all areas of the top liner likely to be exposed to waste or leachate during the active life

and post-closure care period. The requirements for a leak detection system in this paragraph are satisfied by installation of a system that is, at a minimum:

(i) Constructed with a bottom slope of one percent or more;

(ii) Constructed of granular drainage materials with a hydraulic conductivity of $1\times10^{-2}$ cm/sec or more and a thickness of 12 inches (30.5 cm) or more; or constructed of synthetic or geonet drainage materials with a transmissivity of $3\times10^{-5}$ m²/sec or more;

(iii) Constructed of materials that are chemically resistant to the waste managed in the waste pile and the leachate expected to be generated, and of sufficient strength and thickness to prevent collapse under the pressures exerted by overlying wastes, waste cover materials, and equipment used at the waste pile;

(iv) Designed and operated to minimize clogging during the active life and post-closure care period; and

(v) Constructed with sumps and liquid removal methods (e.g., pumps) of sufficient size to collect and remove liquids from the sump and prevent liquids from backing up into the drainage layer. Each unit must have its own sump(s). The design of each sump and removal system must provide a method for measuring and recording the volume of liquids present in the sump and of liquids removed.

(4) The owner or operator shall collect and remove pumpable liquids in the leak detection system sumps to minimize the head on the bottom liner.

(5) The owner or operator of a leak detection system that is not located completely above the seasonal high water table must demonstrate that the operation of the leak detection system will not be adversely affected by the presence of ground water.

(d) The Regional Administrator may approve alternative design or operating practices to those specified in paragraph (c) of this section if the owner or operator demonstrates to the Regional Administrator that such design and operating practices, together with location characteristics:

(1) Will prevent the migration of any hazardous constituent into the ground

water or surface water at least as effectively as the liners and leachate collection and removal systems specified in paragraph (c) of this section; and

(2) Will allow detection of leaks of hazardous constituents through the top liner at least as effectively.

(e) Paragraph (c) of this section does not apply to monofills that are granted a waiver by the Regional Administrator in accordance with §264.221(e).

(f) The owner or operator of any replacement waste pile unit is exempt from paragraph (c) of this section if:

(1) The existing unit was constructed in compliance with the design standards of section 3004(o)(1)(A)(i) and (o)(5) of the Resource Conservation and Recovery Act; and

(2) There is no reason to believe that the liner is not functioning as designed.

(g) The owner or operator must design, construct, operate, and maintain a run-on control system capable of preventing flow onto the active portion of the pile during peak discharge from at least a 25-year storm.

(h) The owner or operator must design, construct, operate, and maintain a run-off management system to collect and control at least the water volume resulting from a 24-hour, 25-year storm.

(i) Collection and holding facilities (e.g., tanks or basins) associated with run-on and run-off control systems must be emptied or otherwise managed expeditiously after storms to maintain design capacity of the system.

(j) If the pile contains any particulate matter which may be subject to wind dispersal, the owner or operator must cover or otherwise manage the pile to control wind dispersal.

(k) The Regional Administrator will specify in the permit all design and operating practices that are necessary to ensure that the requirements of this section are satisfied.

[47 FR 32359, July 26, 1982, as amended at 50 FR 4514, Jan. 31, 1985; 57 FR 3488, Jan. 29, 1992; 71 FR 16906, Apr. 4, 2006; 71 FR 40273, July 14, 2006]

§264.252   **Action leakage rate.**

(a) The Regional Administrator shall approve an action leakage rate for waste pile units subject to §264.251(c) or (d). The action leakage rate is the maximum design flow rate that the leak detection system (LDS) can remove without the fluid head on the bottom liner exceeding 1 foot. The action leakage rate must include an adequate safety margin to allow for uncertainties in the design (e.g., slope, hydraulic conductivity, thickness of drainage material), construction, operation, and location of the LDS, waste and leachate characteristics, likelihood and amounts of other sources of liquids in the LDS, and proposed response actions (e.g., the action leakage rate must consider decreases in the flow capacity of the system over time resulting from siltation and clogging, rib layover and creep of synthetic components of the system, overburden pressures, etc.).

(b) To determine if the action leakage rate has been exceeded, the owner or operator must convert the weekly flow rate from the monitoring data obtained under § 264.254(c) to an average daily flow rate (gallons per acre per day) for each sump. Unless the Regional Administrator approves a different calculation, the average daily flow rate for each sump must be calculated weekly during the active life and closure period.

[57 FR 3489, Jan. 29, 1992, as amended at 71 FR 40273, July 14, 2006]

§ 264.253   Response actions.

(a) The owner or operator of waste pile units subject to § 264.251 (c) or (d) must have an approved response action plan before receipt of waste. The response action plan must set forth the actions to be taken if the action leakage rate has been exceeded. At a minimum, the response action plan must describe the actions specified in paragraph (b) of this section.

(b) If the flow rate into the leak detection system exceeds the action leakage rate for any sump, the owner or operator must:

(1) Notify the Regional Administrator in writing of the exceedance within 7 days of the determination;

(2) Submit a preliminary written assessment to the Regional Administrator within 14 days of the determination, as to the amount of liquids, likely sources of liquids, possible location, size, and cause of any leaks, and short-term actions taken and planned;

(3) Determine to the extent practicable the location, size, and cause of any leak;

(4) Determine whether waste receipt should cease or be curtailed, whether any waste should be removed from the unit for inspection, repairs, or controls, and whether or not the unit should be closed;

(5) Determine any other short-term and long-term actions to be taken to mitigate or stop any leaks; and

(6) Within 30 days after the notification that the action leakage rate has been exceeded, submit to the Regional Administrator the results of the analyses specified in paragraphs (b) (3), (4), and (5) of this section, the results of actions taken, and actions planned. Monthly thereafter, as long as the flow rate in the leak detection system exceeds the action leakage rate, the owner or operator must submit to the Regional Administrator a report summarizing the results of any remedial actions taken and actions planned.

(c) To make the leak and/or remediation determinations in paragraphs (b) (3), (4), and (5) of this section, the owner or operator must:

(1)(i) Assess the source of liquids and amounts of liquids by source,

(ii) Conduct a fingerprint, hazardous constituent, or other analyses of the liquids in the leak detection system to identify the source of liquids and possible location of any leaks, and the hazard and mobility of the liquid; and

(iii) Assess the seriousness of any leaks in terms of potential for escaping into the environment; or

(2) Document why such assessments are not needed.

[57 FR 3489, Jan. 29, 1992]

§ 264.254   Monitoring and inspection.

(a) During construction or installation, liners (except in the case of existing portions of piles exempt from § 264.251(a)) and cover systems (e.g., membranes, sheets, or coatings) must be inspected for uniformity, damage, and imperfections (e.g., holes, cracks, thin spots, or foreign materials). Immediately after construction or installation:

(1) Synthetic liners and covers must be inspected to ensure tight seams and joints and the absence of tears, punctures, or blisters; and

(2) Soil-based and admixed liners and covers must be inspected for imperfections including lenses, cracks, channels, root holes, or other structural non-uniformities that may cause an increase in the permeability of the liner or cover.

(b) While a waste pile is in operation, it must be inspected weekly and after storms to detect evidence of any of the following:

(1) Deterioration, malfunctions, or improper operation of run-on and run-off control systems;

(2) Proper functioning of wind dispersal control systems, where present; and

(3) The presence of leachate in and proper functioning of leachate collection and removal systems, where present.

(c) An owner or operator required to have a leak detection system under §264.251(c) must record the amount of liquids removed from each leak detection system sump at least once each week during the active life and closure period.

[47 FR 32359, July 26, 1982, as amended at 50 FR 4514, Jan. 31, 1985; 50 FR 28748, July 15, 1985; 57 FR 3489, Jan. 29, 1992]

### §264.255 [Reserved]

### §264.256 Special requirements for ignitable or reactive waste.

Ignitable or reactive waste must not be placed in a waste pile unless the waste and waste pile satisfy all applicable requirements of 40 CFR part 268, and:

(a) The waste is treated, rendered, or mixed before or immediately after placement in the pile so that:

(1) The resulting waste, mixture, or dissolution of material no longer meets the definition of ignitable or reactive waste under §261.21 or §261.23 of this chapter; and

(2) Section 264.17(b) is complied with; or

(b) The waste is managed in such a way that it is protected from any material or conditions which may cause it to ignite or react.

[47 FR 32359, July 26, 1982, as amended at 55 FR 22685, June 1, 1990]

### §264.257 Special requirements for incompatible wastes.

(a) Incompatible wastes, or incompatible wastes and materials, (see appendix V of this part for examples) must not be placed in the same pile, unless §264.17(b) is complied with.

(b) A pile of hazardous waste that is incompatible with any waste or other material stored nearby in containers, other piles, open tanks, or surface impoundments must be separated from the other materials, or protected from them by means of a dike, berm, wall, or other device.

(c) Hazardous waste must not be piled on the same base where incompatible wastes or materials were previously piled, unless the base has been decontaminated sufficiently to ensure compliance with §264.17(b).

### §264.258 Closure and post-closure care.

(a) At closure, the owner or operator must remove or decontaminate all waste residues, contaminated containment system components (liners, etc.), contaminated subsoils, and structures and equipment contaminated with waste and leachate, and manage them as hazardous waste unless §261.3(d) of this chapter applies.

(b) If, after removing or decontaminating all residues and making all reasonable efforts to effect removal or decontamination of contaminated components, subsoils, structures, and equipment as required in paragraph (a) of this section, the owner or operator finds that not all contaminated subsoils can be practically removed or decontaminated, he must close the facility and perform post-closure care in accordance with the closure and post-closure care requirements that apply to landfills (§264.310).

(c)(1) The owner or operator of a waste pile that does not comply with the liner requirements of §264.251(a)(1) and is not exempt from them in accordance with §264.250(c) or §264.251(b), must:

459

(i) Include in the closure plan for the pile under § 264.112 both a plan for complying with paragraph (a) of this section and a contingent plan for complying with paragraph (b) of this section in case not all contaminated subsoils can be practicably removed at closure; and

(ii) Prepare a contingent post-closure plan under § 264.118 for complying with paragraph (b) of this section in case not all contaminated subsoils can be practicably removed at closure.

(2) The cost estimates calculated under §§ 264.142 and 264.144 for closure and post-closure care of a pile subject to this paragraph must include the cost of complying with the contingent closure plan and the contingent post-closure plan, but are not required to include the cost of expected closure under paragraph (a) of this section.

§ 264.259 Special requirements for hazardous wastes FO20, FO21, FO22, FO23, FO26, and FO27.

(a) Hazardous Wastes FO20, FO21, FO22, FO23, FO26, and FO27 must not be placed in waste piles that are not enclosed (as defined in § 264.250(c)) unless the owner or operator operates the waste pile in accordance with a management plan for these wastes that is approved by the Regional Administrator pursuant to the standards set out in this paragraph, and in accord with all other applicable requirements of this part. The factors to be considered are:

(1) The volume, physical, and chemical characteristics of the wastes, including their potential to migrate through soil or to volatilize or escape into the atmosphere;

(2) The attenuative properties of underlying and surrounding soils or other materials;

(3) The mobilizing properties of other materials co-disposed with these wastes; and

(4) The effectiveness of additional treatment, design, or monitoring techniques.

(b) The Regional Administrator may determine that additional design, operating, and monitoring requirements are necessary for piles managing hazardous wastes FO20, FO21, FO22, FO23, FO26, and FO27 in order to reduce the possibility of migration of these wastes to ground water, surface water, or air so as to protect human health and the environment.

[50 FR 2004, Jan. 14, 1985, as amended at 71 FR 40273, July 14, 2006]

## Subpart M—Land Treatment

SOURCE: 47 FR 32361, July 26, 1982, unless otherwise noted.

§ 264.270 Applicability.

The regulations in this subpart apply to owners and operators of facilities that treat or dispose of hazardous waste in land treatment units, except as § 264.1 provides otherwise.

§ 264.271 Treatment program.

(a) An owner or operator subject to this subpart must establish a land treatment program that is designed to ensure that hazardous constituents placed in or on the treatment zone are degraded, transformed, or immobilized within the treatment zone. The Regional Administrator will specify in the facility permit the elements of the treatment program, including:

(1) The wastes that are capable of being treated at the unit based on a demonstration under § 264.272;

(2) Design measures and operating practices necessary to maximize the success of degradation, transformation, and immobilization processes in the treatment zone in accordance with § 264.273(a); and

(3) Unsaturated zone monitoring provisions meeting the requirements of § 264.278.

(b) The Regional Administrator will specify in the facility permit the hazardous constituents that must be degraded, transformed, or immobilized under this subpart. Hazardous constituents are constituents identified in appendix VIII of part 261 of this chapter that are reasonably expected to be in, or derived from, waste placed in or on the treatment zone.

(c) The Regional Administrator will specify the vertical and horizontal dimensions of the treatment zone in the facility permit. The treatment zone is the portion of the unsaturated zone below and including the land surface in which the owner or operator intends to maintain the conditions necessary for effective degradation, transformation, or immobilization of hazardous constituents. The maximum depth of the treatment zone must be:

(1) No more than 1.5 meters (5 feet) from the initial soil surface; and

(2) More than 1 meter (3 feet) above the seasonal high water table.

[47 FR 32361, July 26, 1982, as amended at 50 FR 4514, Jan. 31, 1985]

## §264.272 Treatment demonstration.

(a) For each waste that will be applied to the treatment zone, the owner or operator must demonstrate, prior to application of the waste, that hazardous constituents in the waste can be completely degraded, transformed, or immobilized in the treatment zone.

(b) In making this demonstration, the owner or operator may use field tests, laboratory analyses, available data, or, in the case of existing units, operating data. If the owner or operator intends to conduct field tests or laboratory analyses in order to make the demonstration required under paragraph (a) of this section, he must obtain a treatment or disposal permit under §270.63. The Regional Administrator will specify in this permit the testing, analytical, design, and operating requirements (including the duration of the tests and analyses, and, in the case of field tests, the horizontal and vertical dimensions of the treatment zone, monitoring procedures, closure and clean-up activities) necessary to meet the requirements in paragraph (c) of this section.

(c) Any field test or laboratory analysis conducted in order to make a demonstration under paragraph (a) of this section must:

(1) Accurately simulate the characteristics and operating conditions for the proposed land treatment unit including:

(i) The characteristics of the waste (including the presence of appendix VIII of part 261 of this chapter constituents);

(ii) The climate in the area;

(iii) The topography of the surrounding area;

(iv) The characteristics of the soil in the treatment zone (including depth); and

(v) The operating practices to be used at the unit.

(2) Be likely to show that hazardous constituents in the waste to be tested will be completely degraded, transformed, or immobilized in the treatment zone of the proposed land treatment unit; and

(3) Be conducted in a manner that protects human health and the environment considering:

(i) The characteristics of the waste to be tested;

(ii) The operating and monitoring measures taken during the course of the test;

(iii) The duration of the test;

(iv) The volume of waste used in the test;

(v) In the case of field tests, the potential for migration of hazardous constituents to ground water or surface water.

[47 FR 32361, July 26, 1982, as amended at 48 FR 14294, Apr. 1, 1983]

## §264.273 Design and operating requirements.

The Regional Administrator will specify in the facility permit how the owner or operator will design, construct, operate, and maintain the land treatment unit in compliance with this section.

(a) The owner or operator must design, construct, operate, and maintain the unit to maximize the degradation, transformation, and immobilization of hazardous constituents in the treatment zone. The owner or operator must design, construct, operate, and maintain the unit in accord with all design and operating conditions that were used in the treatment demonstration under §264.272. At a minimum, the Regional Administrator will specify the following in the facility permit:

(1) The rate and method of waste application to the treatment zone;

(2) Measures to control soil pH;

(3) Measures to enhance microbial or chemical reactions (*e.g.*, fertilization, tilling); and

(4) Measures to control the moisture content of the treatment zone.

(b) The owner or operator must design, construct, operate, and maintain the treatment zone to minimize run-off of hazardous constituents during the active life of the land treatment unit.

(c) The owner or operator must design, construct, operate, and maintain a run-on control system capable of preventing flow onto the treatment zone during peak discharge from at least a 25-year storm.

(d) The owner or operator must design, construct, operate, and maintain a run-off management system to collect and control at least the water volume resulting from a 24-hour, 25-year storm.

(e) Collection and holding facilities (e.g., tanks or basins) associated with run-on and run-off control systems must be emptied or otherwise managed expeditiously after storms to maintain the design capacity of the system.

(f) If the treatment zone contains particulate matter which may be subject to wind dispersal, the owner or operator must manage the unit to control wind dispersal.

(g) The owner or operator must inspect the unit weekly and after storms to detect evidence of:

(1) Deterioration, malfunctions, or improper operation of run-on and run-off control systems; and

(2) Improper functioning of wind dispersal control measures.

[47 FR 32361, July 26, 1982, as amended at 50 FR 4514, Jan. 31, 1985]

## §§ 264.274–264.275  [Reserved]

## § 264.276  Food-chain crops.

The Regional Administrator may allow the growth of food-chain crops in or on the treatment zone only if the owner or operator satisfies the conditions of this section. The Regional Administrator will specify in the facility permit the specific food-chain crops which may be grown.

(a)(1) The owner or operator must demonstrate that there is no substantial risk to human health caused by the growth of such crops in or on the treatment zone by demonstrating, prior to the planting of such crops, that hazardous constituents other than cadmium:

(i) Will not be transferred to the food or feed portions of the crop by plant uptake or direct contact, and will not otherwise be ingested by food-chain animals (e.g., by grazing); or

(ii) Will not occur in greater concentrations in or on the food or feed portions of crops grown on the treatment zone than in or on identical portions of the same crops grown on untreated soils under similar conditions in the same region.

(2) The owner or operator must make the demonstration required under this paragraph prior to the planting of crops at the facility for all constituents identified in appendix VIII of part 261 of this chapter that are reasonably expected to be in, or derived from, waste placed in or on the treatment zone.

(3) In making a demonstration under this paragraph, the owner or operator may use field tests, greenhouse studies, available data, or, in the case of existing units, operating data, and must:

(i) Base the demonstration on conditions similar to those present in the treatment zone, including soil characteristics (e.g., pH, cation exchange capacity), specific wastes, application rates, application methods, and crops to be grown; and

(ii) Describe the procedures used in conducting any tests, including the sample selection criteria, sample size, analytical methods, and statistical procedures.

(4) If the owner or operator intends to conduct field tests or greenhouse studies in order to make the demonstration required under this paragraph, he must obtain a permit for conducting such activities.

(b) The owner or operator must comply with the following conditions if cadmium is contained in wastes applied to the treatment zone:

(1)(i) The pH of the waste and soil mixture must be 6.5 or greater at the time of each waste application, except for waste containing cadmium at concentrations of 2 mg/kg (dry weight) or less;

(ii) The annual application of cadmium from waste must not exceed 0.5 kilograms per hectare (kg/ha) on land used for production of tobacco, leafy vegetables, or root crops grown for human consumption. For other food-chain crops, the annual cadmium application rate must not exceed:

| Time period | Annual Cd application rate (kilograms per hectare) |
|---|---|
| Present to June 30, 1984 | 2.0 |
| July 1, 1984 to December 31, 1986 | 1.25 |
| Beginning January 1, 1987 | 0.5 |

(iii) The cumulative application of cadmium from waste must not exceed 5 kg/ha if the waste and soil mixture has a pH of less than 6.5; and

(iv) If the waste and soil mixture has a pH of 6.5 or greater or is maintained at a pH of 6.5 or greater during crop growth, the cumulative application of cadmium from waste must not exceed: 5 kg/ha if soil cation exchange capacity (CEC) is less than 5 meq/100g; 10 kg/ha if soil CEC is 5–15 meq/100g; and 20 kg/ha if soil CEC is greater than 15 meq/100g; or

(2)(i) Animal feed must be the only food-chain crop produced;

(ii) The pH of the waste and soil mixture must be 6.5 or greater at the time of waste application or at the time the crop is planted, whichever occurs later, and this pH level must be maintained whenever food-chain crops are grown;

(iii) There must be an operating plan which demonstrates how the animal feed will be distributed to preclude ingestion by humans. The operating plan must describe the measures to be taken to safeguard against possible health hazards from cadmium entering the food chain, which may result from alternative land uses; and

(iv) Future property owners must be notified by a stipulation in the land record or property deed which states that the property has received waste at high cadmium application rates and that food-chain crops must not be grown except in compliance with paragraph (b)(2) of this section.

§264.277 [Reserved]

§264.278 Unsaturated zone monitoring.

An owner or operator subject to this subpart must establish an unsaturated

zone monitoring program to discharge the following responsibilities:

(a) The owner or operator must monitor the soil and soil-pore liquid to determine whether hazardous constituents migrate out of the treatment zone.

(1) The Regional Administrator will specify the hazardous constituents to be monitored in the facility permit. The hazardous constituents to be monitored are those specified under §264.271(b).

(2) The Regional Administrator may require monitoring for principal hazardous constituents (PHCs) in lieu of the constituents specified under §264.271(b). PHCs are hazardous constituents contained in the wastes to be applied at the unit that are the most difficult to treat, considering the combined effects of degradation, transformation, and immobilization. The Regional Administrator will establish PHCs if he finds, based on waste analyses, treatment demonstrations, or other data, that effective degradation, transformation, or immobilization of the PHCs will assure treatment at at least equivalent levels for the other hazardous constituents in the wastes.

(b) The owner or operator must install an unsaturated zone monitoring system that includes soil monitoring using soil cores and soil-pore liquid monitoring using devices such as lysimeters. The unsaturated zone monitoring system must consist of a sufficient number of sampling points at appropriate locations and depths to yield samples that:

(1) Represent the quality of background soil-pore liquid quality and the chemical make-up of soil that has not been affected by leakage from the treatment zone; and

(2) Indicate the quality of soil-pore liquid and the chemical make-up of the soil below the treatment zone.

(c) The owner or operator must establish a background value for each hazardous constituent to be monitored under paragraph (a) of this section. The permit will specify the background values for each constituent or specify the procedures to be used to calculate the background values.

(1) Background soil values may be based on a one-time sampling at a

463

background plot having characteristics similar to those of the treatment zone.

(2) Background soil-pore liquid values must be based on at least quarterly sampling for one year at a background plot having characteristics similar to those of the treatment zone.

(3) The owner or operator must express all background values in a form necessary for the determination of statistically significant increases under paragraph (f) of this section.

(4) In taking samples used in the determination of all background values, the owner or operator must use an unsaturated zone monitoring system that complies with paragraph (b)(1) of this section.

(d) The owner or operator must conduct soil monitoring and soil-pore liquid monitoring immediately below the treatment zone. The Regional Administrator will specify the frequency and timing of soil and soil-pore liquid monitoring in the facility permit after considering the frequency, timing, and rate of waste application, and the soil permeability. The owner or operator must express the results of soil and soil-pore liquid monitoring in a form necessary for the determination of statistically significant increases under paragraph (f) of this section.

(e) The owner or operator must use consistent sampling and analysis procedures that are designed to ensure sampling results that provide a reliable indication of soil-pore liquid quality and the chemical make-up of the soil below the treatment zone. At a minimum, the owner or operator must implement procedures and techniques for:

(1) Sample collection;

(2) Sample preservation and shipment;

(3) Analytical procedures; and

(4) Chain of custody control.

(f) The owner or operator must determine whether there is a statistically significant change over background values for any hazardous constituent to be monitored under paragraph (a) of this section below the treatment zone each time he conducts soil monitoring and soil-pore liquid monitoring under paragraph (d) of this section.

(1) In determining whether a statistically significant increase has occurred, the owner or operator must compare the value of each constituent, as determined under paragraph (d) of this section, to the background value for that constituent according to the statistical procedure specified in the facility permit under this paragraph.

(2) The owner or operator must determine whether there has been a statistically significant increase below the treatment zone within a reasonable time period after completion of sampling. The Regional Administrator will specify that time period in the facility permit after considering the complexity of the statistical test and the availability of laboratory facilities to perform the analysis of soil and soil-pore liquid samples.

(3) The owner or operator must determine whether there is a statistically significant increase below the treatment zone using a statistical procedure that provides reasonable confidence that migration from the treatment zone will be identified. The Regional Administrator will specify a statistical procedure in the facility permit that he finds:

(i) Is appropriate for the distribution of the data used to establish background values; and

(ii) Provides a reasonable balance between the probability of falsely identifying migration from the treatment zone and the probability of failing to identify real migration from the treatment zone.

(g) If the owner or operator determines, pursuant to paragraph (f) of this section, that there is a statistically significant increase of hazardous constituents below the treatment zone, he must:

(1) Notify the Regional Administrator of this finding in writing within seven days. The notification must indicate what constituents have shown statistically significant increases.

(2) Within 90 days, submit to the Regional Administrator an application for a permit modification to modify the operating practices at the facility in order to maximize the success of degradation, transformation, or immobilization processes in the treatment zone.

(h) If the owner or operator determines, pursuant to paragraph (f) of this section, that there is a statistically significant increase of hazardous constituents below the treatment zone, he may demonstrate that a source other than regulated units caused the increase or that the increase resulted from an error in sampling, analysis, or evaluation. While the owner or operator may make a demonstration under this paragraph in addition to, or in lieu of, submitting a permit modification application under paragraph (g)(2) of this section, he is not relieved of the requirement to submit a permit modification application within the time specified in paragraph (g)(2) of this section unless the demonstration made under this paragraph successfully shows that a source other than regulated units caused the increase or that the increase resulted from an error in sampling, analysis, or evaluation. In making a demonstration under this paragraph, the owner or operator must:

(1) Notify the Regional Administrator in writing within seven days of determining a statistically significant increase below the treatment zone that he intends to make a determination under this paragraph;

(2) Within 90 days, submit a report to the Regional Administrator demonstrating that a source other than the regulated units caused the increase or that the increase resulted from error in sampling, analysis, or evaluation;

(3) Within 90 days, submit to the Regional Administrator an application for a permit modification to make any appropriate changes to the unsaturated zone monitoring program at the facility; and

(4) Continue to monitor in accordance with the unsaturated zone monitoring program established under this section.

[47 FR 32361, July 26, 1982, as amended at 50 FR 4514, Jan. 31, 1985]

## § 264.279 Recordkeeping.

The owner or operator must include hazardous waste application dates and rates in the operating record required under § 264.73.

[47 FR 32361, July 26, 1982, as amended at 50 FR 4514, Jan. 31, 1985]

## § 264.280 Closure and post-closure care.

(a) During the closure period the owner or operator must:

(1) Continue all operations (including pH control) necessary to maximize degradation, transformation, or immobilization of hazardous constituents within the treatment zone as required under § 264.273(a), except to the extent such measures are inconsistent with paragraph (a)(8) of this section.

(2) Continue all operations in the treatment zone to minimize run-off of hazardous constituents as required under § 264.273(b);

(3) Maintain the run-on control system required under § 264.273(c);

(4) Maintain the run-off management system required under § 264.273(d);

(5) Control wind dispersal of hazardous waste if required under § 264.273(f);

(6) Continue to comply with any prohibitions or conditions concerning growth of food-chain crops under § 264.276;

(7) Continue unsaturated zone monitoring in compliance with § 264.278, except that soil-pore liquid monitoring may be terminated 90 days after the last application of waste to the treatment zone; and

(8) Establish a vegetative cover on the portion of the facility being closed at such time that the cover will not substantially impede degradation, transformation, or immobilization of hazardous constituents in the treatment zone. The vegetative cover must be capable of maintaining growth without extensive maintenance.

(b) For the purpose of complying with § 264.115 of this chapter, when closure is completed the owner or operator may submit to the Regional Administrator certification by an independent, qualified soil scientist, in lieu of a qualified Professional Engineer, that the facility has been closed in accordance with the specifications in the approved closure plan.

(c) During the post-closure care period the owner or operator must:

(1) Continue all operations (including pH control) necessary to enhance degradation and transformation and sustain immobilization of hazardous constituents in the treatment zone to the

extent that such measures are consistent with other post-closure care activities;

(2) Maintain a vegetative cover over closed portions of the facility;

(3) Maintain the run-on control system required under § 264.273(c);

(4) Maintain the run-off management system required under § 264.273(d);

(5) Control wind dispersal of hazardous waste if required under § 264.273(f);

(6) Continue to comply with any prohibitions or conditions concerning growth of food-chain crops under § 264.276; and

(7) Continue unsaturated zone monitoring in compliance with § 264.278, excect that soil-pore liquid monitoring may be terminated 90 days after the last application of waste to the treatment zone.

(d) The owner or operator is not subject to regulation under paragraphs (a)(8) and (c) of this section if the Regional Administrator finds that the level of hazardous constituents in the treatment zone soil does not exceed the background value of those constituents by an amount that is statistically significant when using the test specified in paragraph (d)(3) of this section. The owner or operator may submit such a demonstration to the Regional Administrator at any time during the closure or post-closure care periods. For the purposes of this paragraph:

(1) The owner or operator must establish background soil values and determine whether there is a statistically significant increase over those values for all hazardous constituents specified in the facility permit under § 264.271 (b).

(i) Background soil values may be based on a one-time sampling of a background plot having characteristics similar to those of the treatment zone.

(ii) The owner or operator must express background values and values for hazardous constituents in the treatment zone in a form necessary for the determination of statistically significant increases under paragraph (d)(3) of this section.

(2) In taking samples used in the determination of background and treatment zone values, the owner or operator must take samples at a sufficient number of sampling points and at appropriate locations and depths to yield samples that represent the chemical make-up of soil that has not been affected by leakage from the treatment zone and the soil within the treatment zone, respectively.

(3) In determining whether a statistically significant increase has occurred, the owner or operator must compare the value of each constituent in the treatment zone to the background value for that constituent using a statistical procedure that provides reasonable confidence that constituent presence in the treatment zone will be identified. The owner or operator must use a statistical procedure that:

(i) Is appropriate for the distribution of the data used to establish background values; and

(ii) Provides a reasonable balance between the probability of falsely identifying hazardous constituent presence in the treatment zone and the probability of failing to identify real presence in the treatment zone.

(e) The owner or operator is not subject to regulation under Subpart F of this chapter if the Regional Administrator finds that the owner or operator satisfies paragraph (d) of this section and if unsaturated zone monitoring under § 264.278 indicates that hazardous constituents have not migrated beyond the treatment zone during the active life of the land treatment unit.

[47 FR 32361, July 26, 1982, as amended at 71 FR 16906, Apr. 4, 2006; 71 FR 40273, July 14, 2006]

## § 264.281 Special requirements for ignitable or reactive waste.

The owner or operator must not apply ignitable or reactive waste to the treatment zone unless the waste and the treatment zone meet all applicable requirements of 40 CFR part 268, and:

(a) The waste is immediately incorporated into the soil so that:

(1) The resulting waste, mixture, or dissolution of material no longer meets the definition of ignitable or reactive waste under § 261.21 or § 261.23 of this chapter; and

(2) Section 264.17(b) is complied with; or

(b) The waste is managed in such a way that it is protected from any material or conditions which may cause it to ignite or react.

[47 FR 32361, July 26, 1982, as amended at 55 FR 22685, June 1, 1990]

### § 264.282 Special requirements for incompatible wastes.

The owner or operator must not place incompatible wastes, or incompatible wastes and materials (see appendix V of this part for examples), in or on the same treatment zone, unless § 264.17(b) is complied with.

### § 264.283 Special requirements for hazardous wastes FO20, FO21, FO22, FO23, FO26, and FO27.

(a) Hazardous Wastes FO20, FO21, FO22, FO23, FO26 and FO27 must not be placed in a land treatment unit unless the owner or operator operates the facility in accordance with a management plan for these wastes that is approved by the Regional Administrator pursuant to the standards set out in this paragraph, and in accord with all other applicable requirements of this part. The factors to be considered are:

(1) The volume, physical, and chemical characteristics of the wastes, including their potential to migrate through soil or to volatilize or escape into the atmosphere;

(2) The attenuative properties of underlying and surrounding soils or other materials;

(3) The mobilizing properties of other materials co-disposed with these wastes; and

(4) The effectiveness of additional treatment, design, or monitoring techniques.

(b) The Regional Administrator may determine that additional design, operating, and monitoring requirements are necessary for land treatment facilities managing hazardous wastes FO20, FO21, FO22, FO23, FO26, and FO27 in order to reduce the possibility of migration of these wastes to ground water, surface water, or air so as to protect human health and the environment.

[50 FR 2004, Jan. 14, 1985, as amended at 71 FR 40273, July 14, 2006]

## Subpart N—Landfills

SOURCE: 47 FR 32365, July 26, 1982, unless otherwise noted.

### § 264.300 Applicability.

The regulations in this subpart apply to owners and operators of facilities that dispose of hazardous waste in landfills, except as § 264.1 provides otherwise.

### § 264.301 Design and operating requirements.

(a) Any landfill that is not covered by paragraph (c) of this section or § 265.301(a) of this chapter must have a liner system for all portions of the landfill (except for existing portions of such landfill). The liner system must have:

(1) A liner that is designed, constructed, and installed to prevent any migration of wastes out of the landfill to the adjacent subsurface soil or ground water or surface water at anytime during the active life (including the closure period) of the landfill. The liner must be constructed of materials that prevent wastes from passing into the liner during the active life of the facility. The liner must be:

(i) Constructed of materials that have appropriate chemical properties and sufficient strength and thickness to prevent failure due to pressure gradients (including static head and external hydrogeologic forces), physical contact with the waste or leachate to which they are exposed, climatic conditions, the stress of installation, and the stress of daily operation;

(ii) Placed upon a foundation or base capable of providing support to the liner and resistance to pressure gradients above and below the liner to prevent failure of the liner due to settlement, compression, or uplift; and

(iii) Installed to cover all surrounding earth likely to be in contact with the waste or leachate; and

(2) A leachate collection and removal system immediately above the liner that is designed, constructed, maintained, and operated to collect and remove leachate from the landfill. The Regional Administrator will specify design and operating conditions in the permit to ensure that the leachate

depth over the liner does not exceed 30 cm (one foot). The leachate collection and removal system must be:

(i) Constructed of materials that are:

(A) Chemically resistant to the waste managed in the landfill and the leachate expected to be generated; and

(B) Of sufficient strength and thickness to prevent collapse under the pressures exerted by overlying wastes, waste cover materials, and by any equipment used at the landfill; and

(ii) Designed and operated to function without clogging through the scheduled closure of the landfill.

(b) The owner or operator will be exempted from the requirements of paragraph (a) of this section if the Regional Administrator finds, based on a demonstration by the owner or operator, that alternative design and operating practices, together with location characteristics, will prevent the migration of any hazardous constituents (see § 264.93) into the ground water or surface water at any future time. In deciding whether to grant an exemption, the Regional Administrator will consider:

(1) The nature and quantity of the wastes;

(2) The proposed alternate design and operation;

(3) The hydrogeologic setting of the facility, including the attenuative capacity and thickness of the liners and soils present between the landfill and ground water or surface water; and

(4) All other factors which would influence the quality and mobility of the leachate produced and the potential for it to migrate to ground water or surface water.

(c) The owner or operator of each new landfill unit on which construction commences after January 29, 1992, each lateral expansion of a landfill unit on which construction commences after July 29, 1992, and each replacement of an existing landfill unit that is to commence reuse after July 29, 1992 must install two or more liners and a leachate collection and removal system above and between such liners. "Construction commences" is as defined in § 260.10 of this chapter under "existing facility".

(1)(i) The *liner system* must include:

(A) A top liner designed and constructed of materials (e.g., a geomembrane) to prevent the migration of hazardous constituents into such liner during the active life and post-closure care period; and

(B) A composite bottom liner, consisting of at least two components. The upper component must be designed and constructed of materials (e.g., a geomembrane) to prevent the migration of hazardous constituents into this component during the active life and post-closure care period. The lower component must be designed and constructed of materials to minimize the migration of hazardous constituents if a breach in the upper component were to occur. The lower component must be constructed of at least 3 feet (91 cm) of compacted soil material with a hydraulic conductivity of no more than $1 \times 10^{-7}$ cm/sec.

(ii) The liners must comply with paragraphs (a)(1) (i), (ii), and (iii) of this section.

(2) The *leachate collection and removal system* immediately above the top liner must be designed, constructed, operated, and maintained to collect and remove leachate from the landfill during the active life and post-closure care period. The Regional Administrator will specify design and operating conditions in the permit to ensure that the leachate depth over the liner does not exceed 30 cm (one foot). The leachate collection and removal system must comply with paragraphs (c)(3) (iii) and (iv) of this section.

(3) The *leachate collection and removal system* between the liners, and immediately above the bottom composite liner in the case of multiple leachate collection and removal systems, is also a *leak detection system*. This leak detection system must be capable of detecting, collecting, and removing leaks of hazardous constituents at the earliest practicable time through all areas of the top liner likely to be exposed to waste or leachate during the active life and post-closure care period. The requirements for a leak detection system in this paragraph are satisfied by installation of a system that is, at a minimum:

(i) Constructed with a bottom slope of one percent or more;

(ii) Constructed of granular drainage materials with a hydraulic conductivity of $1 \times 10^{-2}$ cm/sec or more and a thickness of 12 inches (30.5 cm) or more; or constructed of synthetic or geonet drainage materials with a transmissivity of $3\times10^{-5}$ m$^2$/sec or more;

(iii) Constructed of materials that are chemically resistant to the waste managed in the landfill and the leachate expected to be generated, and of sufficient strength and thickness to prevent collapse under the pressures exerted by overlying wastes, waste cover materials, and equipment used at the landfill;

(iv) Designed and operated to minimize clogging during the active life and post-closure care period; and

(v) Constructed with sumps and liquid removal methods (e.g., pumps) of sufficient size to collect and remove liquids from the sump and prevent liquids from backing up into the drainage layer. Each unit must have its own sump(s). The design of each sump and removal system must provide a method for measuring and recording the volume of liquids present in the sump and of liquids removed.

(4) The owner or operator shall collect and remove pumpable liquids in the leak detection system sumps to minimize the head on the bottom liner.

(5) The owner or operator of a leak detection system that is not located completely above the seasonal high water table must demonstrate that the operation of the leak detection system will not be adversely affected by the presence of ground water.

(d) The Regional Administrator may approve alternative design or operating practices to those specified in paragraph (c) of this section if the owner or operator demonstrates to the Regional Administrator that such design and operating practices, together with location characteristics:

(1) Will prevent the migration of any hazardous constituent into the ground water or surface water at least as effectively as the liners and leachate collection and removal systems specified in paragraph (c) of this section; and

(2) Will allow detection of leaks of hazardous constituents through the top liner at least as effectively.

(e) The double liner requirement set forth in paragraph (c) of this section

may be waived by the Regional Administrator for any monofill, if:

(1) The monofill contains only hazardous wastes from foundry furnace emission controls or metal casting molding sand, and such wastes do not contain constituents which would render the wastes hazardous for reasons other than the Toxicity Characteristic in §261.24 of this chapter, with EPA Hazardous Waste Numbers D004 through D017; and

(2)(i)(A) The monofill has at least one liner for which there is no evidence that such liner is leaking;

(B) The monofill is located more than one-quarter mile from an "underground source of drinking water" (as that term is defined in 40 CFR 270.2); and

(C) The monofill is in compliance with generally applicable ground-water monitoring requirements for facilities with permits under RCRA 3005(c); or

(ii) The owner or operator demonstrates that the monofill is located, designed and operated so as to assure that there will be no migration of any hazardous constituent into ground water or surface water at any future time.

(f) The owner or operator of any replacement landfill unit is exempt from paragraph (c) of this section if:

(1) The existing unit was constructed in compliance with the design standards of section 3004(o)(1)(A)(i) and (o)(5) of the Resource Conservation and Recovery Act; and

(2) There is no reason to believe that the liner is not functioning as designed.

(g) The owner or operator must design, construct, operate, and maintain a run-on control system capable of preventing flow onto the active portion of the landfill during peak discharge from at least a 25-year storm.

(h) The owner or operator must design, construct, operate, and maintain a run-off management system to collect and control at least the water volume resulting from a 24-hour, 25-year storm.

(i) Collection and holding facilities (e.g., tanks or basins) associated with run-on and run-off control systems must be emptied or otherwise managed

469

expeditiously after storms to maintain design capacity of the system.

(j) If the landfill contains any particulate matter which may be subject to wind dispersal, the owner or operator must cover or otherwise manage the landfill to control wind dispersal.

(k) The Regional Administrator will specify in the permit all design and operating practices that are necessary to ensure that the requirements of this section are satisfied.

(l) Any permit under RCRA 3005(c) which is issued for a landfill located within the State of Alabama shall require the installation of two or more liners and a leachate collection system above and between such liners, notwithstanding any other provision of RCRA.

[47 FR 32365, July 26, 1982, as amended at 50 FR 4514, Jan. 31, 1985; 50 FR 28748, July 15, 1985; 55 FR 11875, Mar. 29, 1990; 57 FR 3489, Jan. 29, 1992; 71 FR 40273, July 14, 2006]

§ 264.302  Action leakage rate.

(a) The Regional Administrator shall approve an action leakage rate for landfill units subject to § 264.301(c) or (d). The action leakage rate is the maximum design flow rate that the leak detection system (LDS) can remove without the fluid head on the bottom liner exceeding 1 foot. The action leakage rate must include an adequate safety margin to allow for uncertainties in the design (e.g., slope, hydraulic conductivity, thickness of drainage material), construction, operation, and location of the LDS, waste and leachate characteristics, likelihood and amounts of other sources of liquids in the LDS, and proposed response actions (e.g., the action leakage rate must consider decreases in the flow capacity of the system over time resulting from siltation and clogging, rib layover and creep of synthetic components of the system, overburden pressures, etc.).

(b) To determine if the action leakage rate has been exceeded, the owner or operator must convert the weekly or monthly flow rate from the monitoring data obtained under § 264.303(c) to an average daily flow rate (gallons per acre per day) for each sump. Unless the Regional Administrator approves a different calculation, the average daily flow rate for each sump must be calculated weekly during the active life and closure period, and monthly during the post-closure care period when monthly monitoring is required under § 264.303(c).

[57 FR 3490, Jan. 29, 1992, as amended at 71 FR 40273, July 14, 2006]

§ 264.303  Monitoring and inspection.

(a) During construction or installation, liners (except in the case of existing portions of landfills exempt from § 264.301(a)) and cover systems (e.g., membranes, sheets, or coatings) must be inspected for uniformity, damage, and imperfections (e.g., holes, cracks, thin spots, or foreign materials). Immediately after construction or installation:

(1) Synthetic liners and covers must be inspected to ensure tight seams and joints and the absence of tears, punctures, or blisters; and

(2) Soil-based and admixed liners and covers must be inspected for imperfections including lenses, cracks, channels, root holes, or other structural non-uniformities that may cause an increase in the permeability of the liner or cover.

(b) While a landfill is in operation, it must be inspected weekly and after storms to detect evidence of any of the following:

(1) Deterioration, malfunctions, or improper operation of run-on and run-off control systems;

(2) Proper functioning of wind dispersal control systems, where present; and

(3) The presence of leachate in and proper functioning of leachate collection and removal systems, where present.

(c)(1) An owner or operator required to have a leak detection system under § 264.301(c) or (d) must record the amount of liquids removed from each leak detection system sump at least once each week during the active life and closure period.

(2) After the final cover is installed, the amount of liquids removed from each leak detection system sump must be recorded at least monthly. If the liquid level in the sump stays below the pump operating level for two consecutive months, the amount of liquids in the sumps must be recorded at least

quarterly. If the liquid level in the sump stays below the pump operating level for two consecutive quarters, the amount of liquids in the sumps must be recorded at least semi-annually. If at any time during the post-closure care period the pump operating level is exceeded at units on quarterly or semi-annual recording schedules, the owner or operator must return to monthly recording of amounts of liquids removed from each sump until the liquid level again stays below the pump operating level for two consecutive months.

(3) "Pump operating level" is a liquid level proposed by the owner or operator and approved by the Regional Administrator based on pump activation level, sump dimensions, and level that avoids backup into the drainage layer and minimizes head in the sump.

[47 FR 32365, July 26, 1982, as amended at 50 FR 28748, July 15, 1985; 57 FR 3490, Jan. 29, 1992]

## § 264.304  Response actions.

(a) The owner or operator of landfill units subject to §264.301(c) or (d) must have an approved response action plan before receipt of waste. The response action plan must set forth the actions to be taken if the action leakage rate has been exceeded. At a minimum, the response action plan must describe the actions specified in paragraph (b) of this section.

(b) If the flow rate into the leak detection system exceeds the action leakage rate for any sump, the owner or operator must:

(1) Notify the Regional Administrator in writing of the exceedance within 7 days of the determination;

(2) Submit a preliminary written assessment to the Regional Administrator within 14 days of the determination, as to the amount of liquids, likely sources of liquids, possible location, size, and cause of any leaks, and short-term actions taken and planned;

(3) Determine to the extent practicable the location, size, and cause of any leak;

(4) Determine whether waste receipt should cease or be curtailed, whether any waste should be removed from the unit for inspection, repairs, or controls, and whether or not the unit should be closed;

(5) Determine any other short-term and longer-term actions to be taken to mitigate or stop any leaks; and

(6) Within 30 days after the notification that the action leakage rate has been exceeded, submit to the Regional Administrator the results of the analyses specified in paragraphs (b)(3), (4), and (5) of this section, the results of actions taken, and actions planned. Monthly thereafter, as long as the flow rate in the leak detection system exceeds the action leakage rate, the owner or operator must submit to the Regional Administrator a report summarizing the results of any remedial actions taken and actions planned.

(c) To make the leak and/or remediation determinations in paragraphs (b)(3), (4), and (5) of this section, the owner or operator must:

(1)(i) Assess the source of liquids and amounts of liquids by source,

(ii) Conduct a fingerprint, hazardous constituent, or other analyses of the liquids in the leak detection system to identify the source of liquids and possible location of any leaks, and the hazard and mobility of the liquid; and

(iii) Assess the seriousness of any leaks in terms of potential for escaping into the environment; or

(2) Document why such assessments are not needed.

[57 FR 3491, Jan. 29, 1992, as amended at 71 FR 40273, July 14, 2006]

## §§ 264.305–264.308  [Reserved]

## § 264.309  Surveying and record-keeping.

The owner or operator of a landfill must maintain the following items in the operating record required under § 264.73:

(a) On a map, the exact location and dimensions, including depth, of each cell with respect to permanently surveyed benchmarks; and

(b) The contents of each cell and the approximate location of each hazardous waste type within each cell.

[47 FR 32365, July 26, 1982, as amended at 50 FR 4514, Jan. 31, 1985]

## § 264.310  Closure and post-closure care.

(a) At final closure of the landfill or upon closure of any cell, the owner or operator must cover the landfill or cell with a final cover designed and constructed to:

(1) Provide long-term minimization of migration of liquids through the closed landfill;

(2) Function with minimum maintenance;

(3) Promote drainage and minimize erosion or abrasion of the cover;

(4) Accommodate settling and subsidence so that the cover's integrity is maintained; and

(5) Have a permeability less than or equal to the permeability of any bottom liner system or natural subsoils present.

(b) After final closure, the owner or operator must comply with all post-closure requirements contained in §§ 264.117 through 264.120, including maintenance and monitoring throughout the post-closure care period (specified in the permit under § 264.117). The owner or operator must:

(1) Maintain the integrity and effectiveness of the final cover, including making repairs to the cap as necessary to correct the effects of settling, subsidence, erosion, or other events;

(2) Continue to operate the leachate collection and removal system until leachate is no longer detected;

(3) Maintain and monitor the leak detection system in accordance with §§ 264.301(c)(3)(iv) and (4) and 264.303(c), and comply with all other applicable leak detection system requirements of this part;

(4) Maintain and monitor the groundwater monitoring system and comply with all other applicable requirements of subpart F of this part;

(5) Prevent run-on and run-off from eroding or otherwise damaging the final cover; and

(6) Protect and maintain surveyed benchmarks used in complying with § 264.309.

[47 FR 32365, July 26, 1982, as amended at 50 FR 28748, July 15, 1985; 57 FR 3491, Jan. 29, 1992]

## § 264.311  [Reserved]

## § 264.312  Special requirements for ignitable or reactive waste.

(a) Except as provided in paragraph (b) of this section, and in § 264.316, ignitable or reactive waste must not be placed in a landfill, unless the waste and landfill meet all applicable requirements of part 268, and:

(1) The resulting waste, mixture, or dissolution of material no longer meets the definition of ignitable or reactive waste under § 261.21 or § 261.23 of this chapter; and

(2) Section 264.17(b) is complied with.

(b) Except for prohibited wastes which remain subject to treatment standards in subpart D of part 268, ignitable wastes in containers may be landfilled without meeting the requirements of paragraph (a) of this section, provided that the wastes are disposed of in such a way that they are protected from any material or conditions which may cause them to ignite. At a minimum, ignitable wastes must be disposed of in non-leaking containers which are carefully handled and placed so as to avoid heat, sparks, rupture, or any other condition that might cause ignition of the wastes; must be covered daily with soil or other non-combustible material to minimize the potential for ignition of the wastes; and must not be disposed of in cells that contain or will contain other wastes which may generate heat sufficient to cause ignition of the waste.

[47 FR 32365, July 26, 1982, as amended at 55 FR 22685, June 1, 1990]

## § 264.313  Special requirements for incompatible wastes.

Incompatible wastes, or incompatible wastes and materials, (see appendix V of this part for examples) must not be placed in the same landfill cell, unless § 264.17(b) is complied with.

## § 264.314  Special requirements for bulk and containerized liquids.

(a) The placement of bulk or non-containerized liquid hazardous waste or hazardous waste containing free liquids (whether or not sorbents have been added) in any landfill is prohibited.

(b) To demonstrate the absence or presence of free liquids in either a containerized or a bulk waste, the following test must be used: Method 9095B (Paint Filter Liquids Test) as described in "Test Methods for Evaluating Solid Waste, Physical/Chemical Methods,"

EPA Publication SW–846, as incorporated by reference in §260.11 of this chapter.

(c) Containers holding free liquids must not be placed in a landfill unless:

(1) All free-standing liquid:

(i) Has been removed by decanting, or other methods;

(ii) Has been mixed with sorbent or solidified so that free-standing liquid is no longer observed; or

(iii) Has been otherwise eliminated; or

(2) The container is very small, such as an ampule; or

(3) The container is designed to hold free liquids for use other than storage, such as a battery or capacitor; or

(4) The container is a lab pack as defined in §264.316 and is disposed of in accordance with §264.316.

(d) Sorbents used to treat free liquids to be disposed of in landfills must be nonbiodegradable. Nonbiodegradable sorbents are: materials listed or described in paragraph (d)(1) of this section; materials that pass one of the tests in paragraph (d)(2) of this section; or materials that are determined by EPA to be nonbiodegradable through the part 260 petition process.

(1) Nonbiodegradable sorbents. (i) Inorganic minerals, other inorganic materials, and elemental carbon (e.g., aluminosilicates, clays, smectites, Fuller's earth, bentonite, calcium bentonite, montmorillonite, calcined montmorillonite, kaolinite, micas (illite), vermiculites, zeolites; calcium carbonate (organic free limestone); oxides/hydroxides, alumina, lime, silica (sand), diatomaceous earth; perlite (volcanic glass); expanded volcanic rock; volcanic ash; cement kiln dust; fly ash; rice hull ash; activated charcoal/activated carbon); or

(ii) High molecular weight synthetic polymers (e.g., polyethylene, high density polyethylene (HDPE), polypropylene, polystyrene, polyurethane, polyacrylate, polynorborene, polyisobutylene, ground synthetic rubber, cross-linked allylstyrene and tertiary butyl copolymers). This does not include polymers derived from biological material or polymers specifically designed to be degradable; or

(iii) Mixtures of these nonbiodegradable materials.

(2) Tests for nonbiodegradable sorbents. (i) The sorbent material is determined to be nonbiodegradable under ASTM Method G21–70 (1984a)—Standard Practice for Determining Resistance of Synthetic Polymer Materials to Fungi; or

(ii) The sorbent material is determined to be nonbiodegradable under ASTM Method G22–76 (1984b)—Standard Practice for Determining Resistance of Plastics to Bacteria; or

(iii) The sorbent material is determined to be non-biodegradable under OECD test 301B: [$CO_2$ Evolution (Modified Sturm Test)].

(e) The placement of any liquid which is not a hazardous waste in a landfill is prohibited unless the owner or operator of such landfill demonstrates to the Regional Administrator, or the Regional Administrator determines that:

(1) The only reasonably available alternative to the placement in such landfill is placement in a landfill or unlined surface impoundment, whether or not permitted or operating under interim status, which contains, or may reasonably be anticipated to contain, hazardous waste; and

(2) Placement in such owner or operator's landfill will not present a risk of contamination of any "underground source of drinking water" (as that term is defined in 40 CFR 270.2.)

[47 FR 32365, July 26, 1982, as amended at 50 FR 18374, Apr. 30, 1985; 50 FR 28748, July 15, 1985; 57 FR 54460, Nov. 18, 1992; 58 FR 46050, Aug. 31, 1993; 60 FR 35705, July 11, 1995; 70 FR 34581, June 14, 2005; 71 FR 16906, Apr. 4, 2006; 71 FR 40273, July 14, 2006; 75 FR 13006, Mar. 18, 2010]

## §264.315 Special requirements for containers.

Unless they are very small, such as an ampule, containers must be either:

(a) At least 90 percent full when placed in the landfill; or

(b) Crushed, shredded, or similarly reduced in volume to the maximum practical extent before burial in the landfill.

## §264.316 Disposal of small containers of hazardous waste in overpacked drums (lab packs).

Small containers of hazardous waste in overpacked drums (lab packs) may be placed in a landfill if the following requirements are met:

(a) Hazardous waste must be packaged in non-leaking inside containers. The inside containers must be of a design and constructed of a material that will not react dangerously with, be decomposed by, or be ignited by the contained waste. Inside containers must be tightly and securely sealed. The inside containers must be of the size and type specified in the Department of Transportation (DOT) hazardous materials regulations (49 CFR parts 173, 178, and 179), if those regulations specify a particular inside container for the waste.

(b) The inside containers must be overpacked in an open head DOT-specification metal shipping container (49 CFR parts 178 and 179) of no more than 416-liter (110 gallon) capacity and surrounded by, at a minimum, a sufficient quantity of sorbent material, determined to be nonbiodegradable in accordance with § 264.314(d), to completely sorb all of the liquid contents of the inside containers. The metal outer container must be full after it has been packed with inside containers and sorbent material.

(c) The sorbent material used must not be capable of reacting dangerously with, being decomposed by, or being ignited by the contents of the inside containers, in accordance with § 264.17(b).

(d) Incompatible wastes, as defined in § 260.10 of this chapter, must not be placed in the same outside container.

(e) Reactive wastes, other than cyanide- or sulfide-bearing waste as defined in § 261.23(a)(5) of this chapter, must be treated or rendered non-reactive prior to packaging in accordance with paragraphs (a) through (d) of this section. Cyanide- and sulfide-bearing reactive waste may be packed in accordance with paragraphs (a) through (d) of this section without first being treated or rendered non-reactive.

(f) Such disposal is in compliance with the requirements of part 268. Persons who incinerate lab packs according to the requirements in 40 CFR 268.42(c)(1) may use fiber drums in place of metal outer containers. Such fiber drums must meet the DOT specifications in 49 CFR 173.12 and be over-packed according to the requirements in paragraph (b) of this section.

[47 FR 32365, July 26, 1982, as amended at 55 FR 22685, June 1, 1990; 57 FR 54460, Nov. 18, 1992; 75 FR 13006, Mar. 18, 2010]

§ 264.317  Special requirements for hazardous wastes FO20, FO21, FO22, FO23, FO26, and FO27.

(a) Hazardous Wastes FO20, FO21, FO22, FO23, FO26, and FO27 must not be placed in a landfill unless the owner or operator operates the landfill in accord with a management plan for these wastes that is approved by the Regional Administrator pursuant to the standards set out in this paragraph, and in accord with all other applicable requirements of this part. The factors to be considered are:

(1) The volume, physical, and chemical characteristics of the wastes, including their potential to migrate through the soil or to volatilize or escape into the atmosphere;

(2) The attenuative properties of underlying and surrounding soils or other materials;

(3) The mobilizing properties of other materials co-disposed with these wastes; and

(4) The effectiveness of additional treatment, design, or monitoring requirements.

(b) The Regional Administrator may determine that additional design, operating, and monitoring requirements are necessary for landfills managing hazardous wastes FO20, FO21, FO22, FO23, FO26, and FO27 in order to reduce the possibility of migration of these wastes to ground water, surface water, or air so as to protect human health and the environment.

[50 FR 2004, Jan. 14, 1985, as amended at 71 FR 40273, July 14, 2006]

## Subpart O—Incinerators

§ 264.340  Applicability.

(a) The regulations of this subpart apply to owners and operators of hazardous waste incinerators (as defined in § 260.10 of this chapter), except as § 264.1 provides otherwise.

(b) *Integration of the MACT standards.* (1) Except as provided by paragraphs (b)(2) through (b)(4) of this section, the standards of this part do not apply to a new hazardous waste incineration unit that becomes subject to RCRA permit requirements after October 12, 2005; or no longer apply when an owner or operator of an existing hazardous waste incineration unit demonstrates compliance with the maximum achievable control technology (MACT) requirements of part 63, subpart EEE, of this chapter by conducting a comprehensive performance test and submitting to the Administrator a Notification of Compliance under §§63.1207(j) and 63.1210(d) of this chapter documenting compliance with the requirements of part 63, subpart EEE, of this chapter. Nevertheless, even after this demonstration of compliance with the MACT standards, RCRA permit conditions that were based on the standards of this part will continue to be in effect until they are removed from the permit or the permit is terminated or revoked, unless the permit expressly provides otherwise.

(2) The MACT standards do not replace the closure requirements of §264.351 or the applicable requirements of subparts A through H, BB and CC of this part.

(3) The particulate matter standard of §264.343(c) remains in effect for incinerators that elect to comply with the alternative to the particulate matter standard under §§63.1206(b)(14) and 63.1219(e) of this chapter.

(4) The following requirements remain in effect for startup, shutdown, and malfunction events if you elect to comply with §270.235(a)(1)(i) of this chapter to minimize emissions of toxic compounds from these events:

(i) Section 264.345(a) requiring that an incinerator operate in accordance with operating requirements specified in the permit; and

(ii) Section 264.345(c) requiring compliance with the emission standards and operating requirements during startup and shutdown if hazardous waste is in the combustion chamber, except for particular hazardous wastes.

(c) After consideration of the waste analysis included with part B of the permit application, the Regional Administrator, in establishing the permit conditions, must exempt the applicant from all requirements of this subpart except §264.341 (Waste analysis) and §264.351 (Closure),

(1) If the Regional Administrator finds that the waste to be burned is:

(i) Listed as a hazardous waste in part 261, subpart D, of this chapter solely because it is ignitable (Hazard Code I), corrosive (Hazard Code C), or both; or

(ii) Listed as a hazardous waste in part 261, subpart D, of this chapter solely because it is reactive (Hazard Code R) for characteristics other than those listed in §261.23(a)(4) and (5), and will not be burned when other hazardous wastes are present in the combustion zone; or

(iii) A hazardous waste solely because it possesses the characteristic of ignitability, corrosivity, or both, as determined by the test for characteristics of hazardous wastes under part 261, subpart C, of this chapter; or

(iv) A hazardous waste solely because it possesses any of the reactivity characteristics described by §261.23(a)(1), (2), (3), (6), (7), and (8) of this chapter, and will not be burned when other hazardous wastes are present in the combustion zone; and

(2) If the waste analysis shows that the waste contains none of the hazardous constituents listed in part 261, appendix VIII, of this chapter, which would reasonably be expected to be in the waste.

(d) If the waste to be burned is one which is described by paragraphs (b)(1)(i), (ii), (iii), or (iv) of this section and contains insignificant concentrations of the hazardous constituents listed in part 261, appendix VIII, of this chapter, then the Regional Administrator may, in establishing permit conditions, exempt the applicant from all requirements of this subpart, except §264.341 (Waste analysis) and §264.351 (Closure), after consideration of the waste analysis included with part B of the permit application, unless the Regional Administrator finds that the waste will pose a threat to human health and the environment when burned in an incinerator.

(e) The owner or operator of an incinerator may conduct trial burns subject only to the requirements of §270.62 of this chapter (Short term and incinerator permits).

[46 FR 7678, Jan. 23, 1981, as amended at 47 FR 27532, June 24, 1982; 48 FR 14295, Apr. 1, 1983; 50 FR 665, Jan. 4, 1985; 50 FR 49203, Nov. 29, 1985; 56 FR 7207, Feb. 21, 1991; 64 FR 53074, Sept. 30, 1999; 66 FR 35106, July 3, 2001; 67 FR 6815, Feb. 13, 2002; 70 FR 59575, Oct. 12, 2005; 73 FR 18983, Apr. 8, 2008]

### § 264.341   Waste analysis.

(a) As a portion of the trial burn plan required by § 270.62 of this chapter, or with part B of the permit application, the owner or operator must have included an analysis of the waste feed sufficient to provide all information required by § 270.62(b) or § 270.19 of this chapter. Owners or operators of new hazardous waste incinerators must provide the information required by § 270.62(c) or § 270.19 of this chapter to the greatest extent possible.

(b) Throughout normal operation the owner or operator must conduct sufficient waste analysis to verify that waste feed to the incinerator is within the physical and chemical composition limits specified in his permit (under § 264.345(b)).

[46 FR 7678, Jan. 23, 1981, as amended at 47 FR 27532, June 24, 1982; 48 FR 14295, Apr. 1, 1983; 48 FR 30115, June 30, 1983; 50 FR 4514, Jan. 31, 1985]

### § 264.342   Principal organic hazardous constituents (POHCs).

(a) Principal Organic Hazardous Constituents (POHCs) in the waste feed must be treated to the extent required by the performance standard of § 264.343.

(b)(1) One or more POHCs will be specified in the facility's permit, from among those constituents listed in part 261, appendix VIII of this chapter, for each waste feed to be burned. This specification will be based on the degree of difficulty of incineration of the organic constituents in the waste and on their concentration or mass in the waste feed, considering the results of waste analyses and trial burns or alternative data submitted with part B of the facility's permit application. Organic constituents which represent the greatest degree of difficulty of incineration will be those most likely to be designated as POHCs. Constituents are more likely to be designated as POHCs if they are present in large quantities or concentrations in the waste.

(2) Trial POHCs will be designated for performance of trial burns in accordance with the procedure specified in § 270.62 of this chapter for obtaining trial burn permits.

[46 FR 7678, Jan. 23, 1981, as amended at 48 FR 14295, Apr. 1, 1983]

### § 264.343   Performance standards.

An incinerator burning hazardous waste must be designed, constructed, and maintained so that, when operated in accordance with operating requirements specified under § 264.345, it will meet the following performance standards:

(a)(1) Except as provided in paragraph (a)(2) of this section, an incinerator burning hazardous waste must achieve a destruction and removal efficiency (DRE) of 99.99% for each principal organic hazardous constituent (POHC) designated (under § 264.342) in its permit for each waste feed. DRE is determined for each POHC from the following equation:

$$DRE = \frac{\left(W_{in} - W_{out}\right)}{W_{in}} \times 100\%$$

where:

$W_{in}$=mass feed rate of one principal organic hazardous constituent (POHC) in the waste stream feeding the incinerator

and

$W_{out}$=mass emission rate of the same POHC present in exhaust emissions prior to release to the atmosphere.

(2) An incinerator burning hazardous wastes FO20, FO21, FO22, FO23, FO26, or FO27 must achieve a destruction and removal efficiency (DRE) of 99.9999% for each principal organic hazardous constituent (POHC) designated (under § 264.342) in its permit. This performance must be demonstrated on POHCs that are more difficult to incinerate than tetra-, penta-, and hexachlorodibenzo-p-dioxins and dibenzofurans. DRE is determined for each POHC from the equation in § 264.343(a)(1).

(b) An incinerator burning hazardous waste and producing stack emissions of more than 1.8 kilograms per hour (4 pounds per hour) of hydrogen chloride (HCl) must control HCl emissions such that the rate of emission is no greater than the larger of either 1.8 kilograms per hour or 1% of the HCl in the stack gas prior to entering any pollution control equipment.

(c) An incinerator burning hazardous waste must not emit particulate matter in excess of 180 milligrams per dry standard cubic meter (0.08 grains per dry standard cubic foot) when corrected for the amount of oxygen in the stack gas according to the formula:

$$P_c = P_m \times \frac{14}{21 - Y}$$

Where $P_c$ is the corrected concentration of particulate matter, $P_m$ is the measured concentration of particulate matter, and Y is the measured concentration of oxygen in the stack gas, using the Orsat method for oxygen analysis of dry flue gas, presented in part 60, appendix A (Method 3), of this chapter. This correction procedure is to be used by all hazardous waste incinerators except those operating under conditions of oxygen enrichment. For these facilities, the Regional Administrator will select an appropriate correction procedure, to be specified in the facility permit.

(d) For purposes of permit enforcement, compliance with the operating requirements specified in the permit (under §264.345) will be regarded as compliance with this section. However, evidence that compliance with those permit conditions is insufficient to ensure compliance with the performance requirements of this section may be "information" justifying modification, revocation, or reissuance of a permit under §270.41 of this chapter.

[46 FR 7678, Jan. 23, 1981, as amended at 47 FR 27532, June 24, 1982; 48 FR 14295, Apr. 1, 1983; 50 FR 2005, Jan. 14, 1985; 71 FR 16906, Apr. 4, 2006]

## §264.344 Hazardous waste incinerator permits.

(a) The owner or operator of a hazardous waste incinerator may burn only wastes specified in his permit and only under operating conditions specified for those wastes under §264.345, except:

(1) In approved trial burns under §270.62 of this chapter; or

(2) Under exemptions created by §264.340.

(b) Other hazardous wastes may be burned only after operating conditions have been specified in a new permit or a permit modification as applicable. Operating requirements for new wastes may be based on either trial burn results or alternative data included with part B of a permit application under §270.19 of this chapter.

(c) The permit for a new hazardous waste incinerator must establish appropriate conditions for each of the applicable requirements of this subpart, including but not limited to allowable waste feeds and operating conditions necessary to meet the requirements of §264.345, sufficient to comply with the following standards:

(1) For the period beginning with initial introduction of hazardous waste to the incinerator and ending with initiation of the trial burn, and only for the minimum time required to establish operating conditions required in paragraph (c)(2) of this section, not to exceed a duration of 720 hours operating time for treatment of hazardous waste, the operating requirements must be those most likely to ensure compliance with the performance standards of §264.343, based on the Regional Administrator's engineering judgment. The Regional Administrator may extend the duration of this period once for up to 720 additional hours when good cause for the extension is demonstrated by the applicant.

(2) For the duration of the trial burn, the operating requirements must be sufficient to demonstrate compliance with the performance standards of §264.343 and must be in accordance with the approved trial burn plan;

(3) For the period immediately following completion of the trial burn, and only for the minimum period sufficient to allow sample analysis, data computation, and submission of the trial burn results by the applicant, and review of the trial burn results and modification of the facility permit by the Regional Administrator, the operating requirements must be those most likely to ensure compliance with the performance standards of §264.343, based on the Regional Administrator's engineering judgement.

(4) For the remaining duration of the permit, the operating requirements

must be those demonstrated, in a trial burn or by alternative data specified in § 270.19(c) of this chapter, as sufficient to ensure compliance with the performance standards of § 264.343.

[46 FR 7678, Jan. 23, 1981, as amended at 47 FR 27532, June 24, 1982; 48 FR 14295, Apr. 1, 1983; 50 FR 4514, Jan. 31, 1985; 71 FR 40273, July 14, 2006]

### § 264.345  Operating requirements.

(a) An incinerator must be operated in accordance with operating requirements specified in the permit. These will be specified on a case-by-case basis as those demonstrated (in a trial burn or in alternative data as specified in § 264.344(b) and included with part B of a facility's permit application) to be sufficient to comply with the performance standards of § 264.343.

(b) Each set of operating requirements will specify the composition of the waste feed (including acceptable variations in the physical or chemical properties of the waste feed which will not affect compliance with the performance requirement of § 264.343) to which the operating requirements apply. For each such waste feed, the permit will specify acceptable operating limits including the following conditions:

(1) Carbon monoxide (CO) level in the stack exhaust gas;

(2) Waste feed rate;

(3) Combustion temperature;

(4) An appropriate indicator of combustion gas velocity;

(5) Allowable variations in incinerator system design or operating procedures; and

(6) Such other operating requirements as are necessary to ensure that the performance standards of § 264.343 are met.

(c) During start-up and shut-down of an incinerator, hazardous waste (except wastes exempted in accordance with § 264.340) must not be fed into the incinerator unless the incinerator is operating within the conditions of operation (temperature, air feed rate, etc.) specified in the permit.

(d) Fugitive emissions from the combustion zone must be controlled by:

(1) Keeping the combustion zone totally sealed against fugitive emissions; or

(2) Maintaining a combustion zone pressure lower than atmospheric pressure; or

(3) An alternate means of control demonstrated (with part B of the permit application) to provide fugitive emissions control equivalent to maintenance of combustion zone pressure lower than atmospheric pressure.

(e) An incinerator must be operated with a functioning system to automatically cut off waste feed to the incinerator when operating conditions deviate from limits established under paragraph (a) of this section.

(f) An incinerator must cease operation when changes in waste feed, incinerator design, or operating conditions exceed limits designated in its permit.

[46 FR 7678, Jan. 23, 1981, as amended at 47 FR 27532, June 24, 1982; 50 FR 4514, Jan. 31, 1985]

### § 264.346  [Reserved]

### § 264.347  Monitoring and inspections.

(a) The owner or operator must conduct, as a minimum, the following monitoring while incinerating hazardous waste:

(1) Combustion temperature, waste feed rate, and the indicator of combustion gas velocity specified in the facility permit must be monitored on a continuous basis.

(2) CO must be monitored on a continuous basis at a point in the incinerator downstream of the combustion zone and prior to release to the atmosphere.

(3) Upon request by the Regional Administrator, sampling and analysis of the waste and exhaust emissions must be conducted to verify that the operating requirements established in the permit achieve the performance standards of § 264.343.

(b) The incinerator and associated equipment (pumps, valves, conveyors, pipes, etc.) must be subjected to thorough visual inspection, at least daily, for leaks, spills, fugitive emissions, and signs of tampering.

(c) The emergency waste feed cutoff system and associated alarms must be tested at least weekly to verify operability, unless the applicant demonstrates to the Regional Administrator that weekly inspections will unduly restrict or upset operations and that less frequent inspection will be adequate. At a minimum, operational testing must be conducted at least monthly.

(d) This monitoring and inspection data must be recorded and the records must be placed in the operating record required by § 264.73 of this part and maintained in the operating record for five years.

[46 FR 7678, Jan. 23, 1981, as amended at 47 FR 27533, June 24, 1982; 50 FR 4514, Jan. 31, 1985; 71 FR 16907, Apr. 4, 2006]

## §§ 264.348–264.350 [Reserved]

### § 264.351 Closure.

At closure the owner or operator must remove all hazardous waste and hazardous waste residues (including, but not limited to, ash, scrubber waters, and scrubber sludges) from the incinerator site.

[*Comment:* At closure, as throughout the operating period, unless the owner or operator can demonstrate, in accordance with § 261.3(d) of this chapter, that the residue removed from the incinerator is not a hazardous waste, the owner or operator becomes a generator of hazardous waste and must manage it in accordance with applicable requirements of parts 262 through 266 of this chapter.]

[46 FR 7678, Jan. 23, 1981]

## Subparts P–R [Reserved]

## Subpart S—Special Provisions for Cleanup

### § 264.550 Applicability of Corrective Action Management Unit (CAMU) regulations.

(a) Except as provided in paragraph (b) of this section, CAMUs are subject to the requirements of § 264.552.

(b) CAMUs that were approved before April 22, 2002, or for which substantially complete applications (or equivalents) were submitted to the Agency on or before November 20, 2000, are subject to the requirements in § 264.551 for grandfathered CAMUs; CAMU waste, activities, and design will not be subject to the standards in § 264.552, so long as the waste, activities, and design remain within the general scope of the CAMU as approved.

[67 FR 3024, Jan. 22, 2002]

### § 264.551 Grandfathered Corrective Action Management Units (CAMUs).

(a) To implement remedies under § 264.101 or RCRA Section 3008(h), or to implement remedies at a permitted facility that is not subject to § 264.101, the Regional Administrator may designate an area at the facility as a corrective action management unit under the requirements in this section. Corrective action management unit means an area within a facility that is used only for managing remediation wastes for implementing corrective action or cleanup at the facility. A CAMU must be located within the contiguous property under the control of the owner or operator where the wastes to be managed in the CAMU originated. One or more CAMUs may be designated at a facility.

(1) Placement of remediation wastes into or within a CAMU does not constitute land disposal of hazardous wastes.

(2) Consolidation or placement of remediation wastes into or within a CAMU does not constitute creation of a unit subject to minimum technology requirements.

(b)(1) The Regional Administrator may designate a regulated unit (as defined in § 264.90(a)(2)) as a CAMU, or may incorporate a regulated unit into a CAMU, if:

(i) The regulated unit is closed or closing, meaning it has begun the closure process under § 264.113 or § 265.113; and

(ii) Inclusion of the regulated unit will enhance implementation of effective, protective and reliable remedial actions for the facility.

(2) The subpart F, G, and H requirements and the unit-specific requirements of part 264 or 265 that applied to that regulated unit will continue to apply to that portion of the CAMU after incorporation into the CAMU.

(c) The Regional Administrator shall designate a CAMU in accordance with the following:

(1) The CAMU shall facilitate the implementation of reliable, effective, protective, and cost-effective remedies;

(2) Waste management activities associated with the CAMU shall not create unacceptable risks to humans or to the environment resulting from exposure to hazardous wastes or hazardous constituents;

(3) The CAMU shall include uncontaminated areas of the facility, only if including such areas for the purpose of managing remediation waste is more protective than management of such wastes at contaminated areas of the facility;

(4) Areas within the CAMU, where wastes remain in place after closure of the CAMU, shall be managed and contained so as to minimize future releases, to the extent practicable;

(5) The CAMU shall expedite the timing of remedial activity implementation, when appropriate and practicable;

(6) The CAMU shall enable the use, when appropriate, of treatment technologies (including innovative technologies) to enhance the long-term effectiveness of remedial actions by reducing the toxicity, mobility, or volume of wastes that will remain in place after closure of the CAMU; and

(7) The CAMU shall, to the extent practicable, minimize the land area of the facility upon which wastes will remain in place after closure of the CAMU.

(d) The owner/operator shall provide sufficient information to enable the Regional Administrator to designate a CAMU in accordance with the criteria in § 264.552.

(e) The Regional Administrator shall specify, in the permit or order, requirements for CAMUs to include the following:

(1) The areal configuration of the CAMU.

(2) Requirements for remediation waste management to include the specification of applicable design, operation and closure requirements.

(3) Requirements for ground water monitoring that are sufficient to:

(i) Continue to detect and to characterize the nature, extent, concentration, direction, and movement of existing releases of hazardous constituents

in ground water from sources located within the CAMU; and

(ii) Detect and subsequently characterize releases of hazardous constituents to ground water that may occur from areas of the CAMU in which wastes will remain in place after closure of the CAMU.

(4) Closure and post-closure requirements.

(i) Closure of corrective action management units shall:

(A) Minimize the need for further maintenance; and

(B) Control, minimize, or eliminate, to the extent necessary to protect human health and the environment, for areas where wastes remain in place, post-closure escape of hazardous waste, hazardous constituents, leachate, contaminated runoff, or hazardous waste decomposition products to the ground, to surface waters, or to the atmosphere.

(ii) Requirements for closure of CAMUs shall include the following, as appropriate and as deemed necessary by the Regional Administrator for a given CAMU:

(A) Requirements for excavation, removal, treatment or containment of wastes;

(B) For areas in which wastes will remain after closure of the CAMU, requirements for capping of such areas; and

(C) Requirements for removal and decontamination of equipment, devices, and structures used in remediation waste management activities within the CAMU.

(iii) In establishing specific closure requirements for CAMUs under § 264.552(e), the Regional Administrator shall consider the following factors:

(A) CAMU characteristics;

(B) Volume of wastes which remain in place after closure;

(C) Potential for releases from the CAMU;

(D) Physical and chemical characteristics of the waste;

(E) Hydrological and other relevant environmental conditions at the facility which may influence the migration of any potential or actual releases; and

(F) Potential for exposure of humans and environmental receptors if releases were to occur from the CAMU.

(iv) Post-closure requirements as necessary to protect human health and the environment, to include, for areas where wastes will remain in place, monitoring and maintenance activities, and the frequency with which such activities shall be performed to ensure the integrity of any cap, final cover, or other containment system.

(f) The Regional Administrator shall document the rationale for designating CAMUs and shall make such documentation available to the public.

(g) Incorporation of a CAMU into an existing permit must be approved by the Regional Administrator according to the procedures for Agency-initiated permit modifications under §270.41 of this chapter, or according to the permit modification procedures of §270.42 of this chapter.

(h) The designation of a CAMU does not change EPA's existing authority to address clean-up levels, media-specific points of compliance to be applied to remediation at a facility, or other remedy selection decisions.

[58 FR 8683, Feb. 16, 1993, as amended at 63 FR 65939, Nov. 30, 1998. Redesignated and amended at 67 FR 3025, Jan. 22, 2002]

## §264.552 Corrective Action Management Units (CAMU).

(a) To implement remedies under §264.101 or RCRA Section 3008(h), or to implement remedies at a permitted facility that is not subject to §264.101, the Regional Administrator may designate an area at the facility as a corrective action management unit under the requirements in this section. Corrective action management unit means an area within a facility that is used only for managing CAMU-eligible wastes for implementing corrective action or cleanup at the facility. A CAMU must be located within the contiguous property under the control of the owner or operator where the wastes to be managed in the CAMU originated. One or more CAMUs may be designated at a facility.

(1) *CAMU-eligible waste* means:

(i) All solid and hazardous wastes, and all media (including ground water, surface water, soils, and sediments) and debris, that are managed for implementing cleanup. As-generated wastes (either hazardous or non-hazardous) from ongoing industrial operations at a site are not CAMU-eligible wastes.

(ii) Wastes that would otherwise meet the description in paragraph (a)(1)(i) of this section are not "CAMU-Eligible Wastes" where:

(A) The wastes are hazardous wastes found during cleanup in intact or substantially intact containers, tanks, or other non-land-based units found above ground, unless the wastes are first placed in the tanks, containers or non-land-based units as part of cleanup, or the containers or tanks are excavated during the course of cleanup; or

(B) The Regional Administrator exercises the discretion in paragraph (a)(2) of this section to prohibit the wastes from management in a CAMU.

(iii) Notwithstanding paragraph (a)(1)(i) of this section, where appropriate, as-generated non-hazardous waste may be placed in a CAMU where such waste is being used to facilitate treatment or the performance of the CAMU.

(2) The Regional Administrator may prohibit, where appropriate, the placement of waste in a CAMU where the Regional Administrator has or receives information that such wastes have not been managed in compliance with applicable land disposal treatment standards of part 268 of this chapter, or applicable unit design requirements of this part, or applicable unit design requirements of part 265 of this chapter, or that non-compliance with other applicable requirements of this chapter likely contributed to the release of the waste.

(3) *Prohibition against placing liquids in CAMUs.* (i) The placement of bulk or noncontainerized liquid hazardous waste or free liquids contained in hazardous waste (whether or not sorbents have been added) in any CAMU is prohibited except where placement of such wastes facilitates the remedy selected for the waste.

(ii) The requirements in §264.314(c) for placement of containers holding free liquids in landfills apply to placement in a CAMU except where placement facilitates the remedy selected for the waste.

(iii) The placement of any liquid which is not a hazardous waste in a

CAMU is prohibited unless such placement facilitates the remedy selected for the waste or a demonstration is made pursuant to § 264.314(e).

(iv) The absence or presence of free liquids in either a containerized or a bulk waste must be determined in accordance with § 264.314(b). Sorbents used to treat free liquids in CAMUs must meet the requirements of § 264.314(d).

(4) Placement of CAMU-eligible wastes into or within a CAMU does not constitute land disposal of hazardous wastes.

(5) Consolidation or placement of CAMU-eligible wastes into or within a CAMU does not constitute creation of a unit subject to minimum technology requirements.

(b)(1) The Regional Administrator may designate a regulated unit (as defined in § 264.90(a)(2)) as a CAMU, or may incorporate a regulated unit into a CAMU, if:

(i) The regulated unit is closed or closing, meaning it has begun the closure process under § 264.113 or § 265.113 of this chapter; and

(ii) Inclusion of the regulated unit will enhance implementation of effective, protective and reliable remedial actions for the facility.

(2) The subpart F, G, and H requirements and the unit-specific requirements of this part 264 or part 265 of this chapter that applied to the regulated unit will continue to apply to that portion of the CAMU after incorporation into the CAMU.

(c) The Regional Administrator shall designate a CAMU that will be used for storage and/or treatment only in accordance with paragraph (f) of this section. The Regional Administrator shall designate all other CAMUs in accordance with the following:

(1) The CAMU shall facilitate the implementation of reliable, effective, protective, and cost-effective remedies;

(2) Waste management activities associated with the CAMU shall not create unacceptable risks to humans or to the environment resulting from exposure to hazardous wastes or hazardous constituents;

(3) The CAMU shall include uncontaminated areas of the facility, only if including such areas for the

purpose of managing CAMU-eligible waste is more protective than management of such wastes at contaminated areas of the facility;

(4) Areas within the CAMU, where wastes remain in place after closure of the CAMU, shall be managed and contained so as to minimize future releases, to the extent practicable;

(5) The CAMU shall expedite the timing of remedial activity implementation, when appropriate and practicable;

(6) The CAMU shall enable the use, when appropriate, of treatment technologies (including innovative technologies) to enhance the long-term effectiveness of remedial actions by reducing the toxicity, mobility, or volume of wastes that will remain in place after closure of the CAMU; and

(7) The CAMU shall, to the extent practicable, minimize the land area of the facility upon which wastes will remain in place after closure of the CAMU.

(d) The owner/operator shall provide sufficient information to enable the Regional Administrator to designate a CAMU in accordance with the criteria in this section. This must include, unless not reasonably available, information on:

(1) The origin of the waste and how it was subsequently managed (including a description of the timing and circumstances surrounding the disposal and/or release);

(2) Whether the waste was listed or identified as hazardous at the time of disposal and/or release; and

(3) Whether the disposal and/or release of the waste occurred before or after the land disposal requirements of part 268 of this chapter were in effect for the waste listing or characteristic.

(e) The Regional Administrator shall specify, in the permit or order, requirements for CAMUs to include the following:

(1) The areal configuration of the CAMU.

(2) Except as provided in paragraph (g) of this section, requirements for CAMU-eligible waste management to include the specification of applicable design, operation, treatment and closure requirements.

(3) Minimum design requirements. CAMUs, except as provided in paragraph (f) of this section, into which wastes are placed must be designed in accordance with the following:

(i) Unless the Regional Administrator approves alternate requirements under paragraph (e)(3)(ii) of this section, CAMUs that consist of new, replacement, or laterally expanded units must include a composite liner and a leachate collection system that is designed and constructed to maintain less than a 30-cm depth of leachate over the liner. For purposes of this section, *composite liner* means a system consisting of two components; the upper component must consist of a minimum 30-mil flexible membrane liner (FML), and the lower component must consist of at least a two-foot layer of compacted soil with a hydraulic conductivity of no more than $1 \times 10 - 7$ cm/sec. FML components consisting of high density polyethylene (HDPE) must be at least 60 mil thick. The FML component must be installed in direct and uniform contact with the compacted soil component;

(ii) Alternate requirements. The Regional Administrator may approve alternate requirements if:

(A) The Regional Administrator finds that alternate design and operating practices, together with location characteristics, will prevent the migration of any hazardous constituents into the ground water or surface water at least as effectively as the liner and leachate collection systems in paragraph (e)(3)(i) of this section; or

(B) The CAMU is to be established in an area with existing significant levels of contamination, and the Regional Administrator finds that an alternative design, including a design that does not include a liner, would prevent migration from the unit that would exceed long-term remedial goals.

(4) Minimum treatment requirements: Unless the wastes will be placed in a CAMU for storage and/or treatment only in accordance with paragraph (f) of this section, CAMU-eligible wastes that, absent this section, would be subject to the treatment requirements of part 268 of this chapter, and that the Regional Administrator determines contain principal hazardous constituents must be treated to the standards specified in paragraph (e)(4)(iii) of this section.

(i) Principal hazardous constituents are those constituents that the Regional Administrator determines pose a risk to human health and the environment substantially higher than the cleanup levels or goals at the site.

(A) In general, the Regional Administrator will designate as principal hazardous constituents:

(1) Carcinogens that pose a potential direct risk from ingestion or inhalation at the site at or above $10^{-3}$; and

(2) Non-carcinogens that pose a potential direct risk from ingestion or inhalation at the site an order of magnitude or greater over their reference dose.

(B) The Regional Administrator will also designate constituents as principal hazardous constituents, where appropriate, when risks to human health and the environment posed by the potential migration of constituents in wastes to ground water are substantially higher than cleanup levels or goals at the site; when making such a designation, the Regional Administrator may consider such factors as constituent concentrations, and fate and transport characteristics under site conditions.

(C) The Regional Administrator may also designate other constituents as principal hazardous constituents that the Regional Administrator determines pose a risk to human health and the environment substantially higher than the cleanup levels or goals at the site.

(ii) In determining which constituents are "principal hazardous constituents," the Regional Administrator must consider all constituents which, absent this section, would be subject to the treatment requirements in 40 CFR part 268.

(iii) Waste that the Regional Administrator determines contains principal hazardous constituents must meet treatment standards determined in accordance with paragraph (e)(4)(iv) or (e)(4)(v) of this section.

(iv) Treatment standards for wastes placed in CAMUs.

(A) For non-metals, treatment must achieve 90 percent reduction in total

483

principal hazardous constituent concentrations, except as provided by paragraph (e)(4)(iv)(C) of this section.

(B) For metals, treatment must achieve 90 percent reduction in principal hazardous constituent concentrations as measured in leachate from the treated waste or media (tested according to the TCLP) or 90 percent reduction in total constituent concentrations (when a metal removal treatment technology is used), except as provided by paragraph (e)(4)(iv)(C) of this section.

(C) When treatment of any principal hazardous constituent to a 90 percent reduction standard would result in a concentration less than 10 times the Universal Treatment Standard for that constituent, treatment to achieve constituent concentrations less than 10 times the Universal Treatment Standard is not required. Universal Treatment Standards are identified in § 268.48 Table UTS of this chapter.

(D) For waste exhibiting the hazardous characteristic of ignitability, corrosivity or reactivity, the waste must also be treated to eliminate these characteristics.

(E) For debris, the debris must be treated in accordance with § 268.45 of this chapter, or by methods or to levels established under paragraphs (e)(4)(iv)(A) through (D) or paragraph (e)(4)(v) of this section, whichever the Regional Administrator determines is appropriate.

(F) Alternatives to TCLP. For metal bearing wastes for which metals removal treatment is not used, the Regional Administrator may specify a leaching test other than the TCLP (SW846 Method 1311, 40 CFR 260.11(c)(3)(v)) to measure treatment effectiveness, provided the Regional Administrator determines that an alternative leach testing protocol is appropriate for use, and that the alternative more accurately reflects conditions at the site that affect leaching.

(v) Adjusted standards. The Regional Administrator may adjust the treatment level or method in paragraph (e)(4)(iv) of this section to a higher or lower level, based on one or more of the following factors, as appropriate. The adjusted level or method must be pro-

tective of human health and the environment:

(A) The technical impracticability of treatment to the levels or by the methods in paragraph (e)(4)(iv) of this section;

(B) The levels or methods in paragraph (e)(4)(iv) of this section would result in concentrations of principal hazardous constituents (PHCs) that are significantly above or below cleanup standards applicable to the site (established either site-specifically, or promulgated under state or federal law);

(C) The views of the affected local community on the treatment levels or methods in paragraph (e)(4)(iv) of this section as applied at the site, and, for treatment levels, the treatment methods necessary to achieve these levels;

(D) The short-term risks presented by the on-site treatment method necessary to achieve the levels or treatment methods in paragraph (e)(4)(iv) of this section;

(E) The long-term protection offered by the engineering design of the CAMU and related engineering controls:

(1) Where the treatment standards in paragraph (e)(4)(iv) of this section are substantially met and the principal hazardous constituents in the waste or residuals are of very low mobility; or

(2) Where cost-effective treatment has been used and the CAMU meets the Subtitle C liner and leachate collection requirements for new land disposal units at § 264.301(c) and (d); or

(3) Where, after review of appropriate treatment technologies, the Regional Administrator determines that cost-effective treatment is not reasonably available, and the CAMU meets the Subtitle C liner and leachate collection requirements for new land disposal units at § 264.301(c) and (d); or

(4) Where cost-effective treatment has been used and the principal hazardous constituents in the treated wastes are of very low mobility; or

(5) Where, after review of appropriate treatment technologies, the Regional Administrator determines that cost-effective treatment is not reasonably available, the principal hazardous constituents in the wastes are of very low mobility, and either the CAMU meets or exceeds the liner standards for new, replacement, or laterally expanded

CAMUs in paragraphs (e)(3)(i) and (ii) of this section, or the CAMU provides substantially equivalent or greater protection.

(vi) The treatment required by the treatment standards must be completed prior to, or within a reasonable time after, placement in the CAMU.

(vii) For the purpose of determining whether wastes placed in CAMUs have met site-specific treatment standards, the Regional Administrator may, as appropriate, specify a subset of the principal hazardous constituents in the waste as analytical surrogates for determining whether treatment standards have been met for other principal hazardous constituents. This specification will be based on the degree of difficulty of treatment and analysis of constituents with similar treatment properties.

(5) Except as provided in paragraph (f) of this section, requirements for ground water monitoring and corrective action that are sufficient to:

(i) Continue to detect and to characterize the nature, extent, concentration, direction, and movement of existing releases of hazardous constituents in ground water from sources located within the CAMU; and

(ii) Detect and subsequently characterize releases of hazardous constituents to ground water that may occur from areas of the CAMU in which wastes will remain in place after closure of the CAMU; and

(iii) Require notification to the Regional Administrator and corrective action as necessary to protect human health and the environment for releases to ground water from the CAMU.

(6) Except as provided in paragraph (f) of this section, closure and post-closure requirements:

(i) Closure of corrective action management units shall:

(A) Minimize the need for further maintenance; and

(B) Control, minimize, or eliminate, to the extent necessary to protect human health and the environment, for areas where wastes remain in place, post-closure escape of hazardous wastes, hazardous constituents, leachate, contaminated runoff, or hazardous waste decomposition products to the ground, to surface waters, or to the atmosphere.

(ii) Requirements for closure of CAMUs shall include the following, as appropriate and as deemed necessary by the Regional Administrator for a given CAMU:

(A) Requirements for excavation, removal, treatment or containment of wastes; and

(B) Requirements for removal and decontamination of equipment, devices, and structures used in CAMU-eligible waste management activities within the CAMU.

(iii) In establishing specific closure requirements for CAMUs under paragraph (e) of this section, the Regional Administrator shall consider the following factors:

(A) CAMU characteristics;

(B) Volume of wastes which remain in place after closure;

(C) Potential for releases from the CAMU;

(D) Physical and chemical characteristics of the waste;

(E) Hydrogeological and other relevant environmental conditions at the facility which may influence the migration of any potential or actual releases; and

(F) Potential for exposure of humans and environmental receptors if releases were to occur from the CAMU.

(iv) Cap requirements:

(A) At final closure of the CAMU, for areas in which wastes will remain after closure of the CAMU, with constituent concentrations at or above remedial levels or goals applicable to the site, the owner or operator must cover the CAMU with a final cover designed and constructed to meet the following performance criteria, except as provided in paragraph (e)(6)(iv)(B) of this section:

(1) Provide long-term minimization of migration of liquids through the closed unit;

(2) Function with minimum maintenance;

(3) Promote drainage and minimize erosion or abrasion of the cover;

(4) Accommodate settling and subsidence so that the cover's integrity is maintained; and

(5) Have a permeability less than or equal to the permeability of any bottom liner system or natural subsoils present.

(B) The Regional Administrator may determine that modifications to paragraph (e)(6)(iv)(A) of this section are needed to facilitate treatment or the performance of the CAMU (e.g., to promote biodegradation).

(v) Post-closure requirements as necessary to protect human health and the environment, to include, for areas where wastes will remain in place, monitoring and maintenance activities, and the frequency with which such activities shall be performed to ensure the integrity of any cap, final cover, or other containment system.

(f) CAMUs used for storage and/or treatment only are CAMUs in which wastes will not remain after closure. Such CAMUs must be designated in accordance with all of the requirements of this section, except as follows.

(1) CAMUs that are used for storage and/or treatment only and that operate in accordance with the time limits established in the staging pile regulations at § 264.554(d)(1)(iii), (h), and (i) are subject to the requirements for staging piles at § 264.554(d)(1)(i) and (ii), § 264.554(d)(2), § 264.554(e) and (f), and § 264.554(j) and (k) in lieu of the performance standards and requirements for CAMUs in this section at paragraphs (c) and (e)(3) through (6).

(2) CAMUs that are used for storage and/or treatment only and that do not operate in accordance with the time limits established in the staging pile regulations at § 264.554(d)(1)(iii), (h), and (i):

(i) Must operate in accordance with a time limit, established by the Regional Administrator, that is no longer than necessary to achieve a timely remedy selected for the waste, and

(ii) Are subject to the requirements for staging piles at § 264.554(d)(1)(i) and (ii), § 264.554(d)(2), § 264.554(e) and (f), and § 264.554(j) and (k) in lieu of the performance standards and requirements for CAMUs in this section at paragraphs (c) and (e)(4) and (6).

(g) CAMUs into which wastes are placed where all wastes have constituent levels at or below remedial levels or goals applicable to the site do not have to comply with the requirements for liners at paragraph (e)(3)(i) of this section, caps at paragraph (e)(6)(iv) of this section, ground water monitoring requirements at paragraph (e)(5) of this section or, for treatment and/or storage-only CAMUs, the design standards at paragraph (f) of this section.

(h) The Regional Administrator shall provide public notice and a reasonable opportunity for public comment before designating a CAMU. Such notice shall include the rationale for any proposed adjustments under paragraph (e)(4)(iii) of this section to the treatment standards in paragraph (e)(4)(iv) of this section.

(i) Notwithstanding any other provision of this section, the Regional Administrator may impose additional requirements as necessary to protect human health and the environment.

(j) Incorporation of a CAMU into an existing permit must be approved by the Regional Administrator according to the procedures for Agency-initiated permit modifications under § 270.41 of this chapter, or according to the permit modification procedures of § 270.42 of this chapter.

(k) The designation of a CAMU does not change EPA's existing authority to address clean-up levels, media-specific points of compliance to be applied to remediation at a facility, or other remedy selection decisions.

[67 FR 3025, Jan. 22, 2002, as amended at 71 FR 40273, July 14, 2006; 75 FR 13006, Mar. 18, 2010]

## § 264.553  Temporary Units (TU).

(a) For temporary tanks and container storage areas used to treat or store hazardous remediation wastes during remedial activities required under § 264.101 or RCRA 3008(h), or at a permitted facility that is not subject to § 264.101, the Regional Administrator may designate a unit at the facility, as a temporary unit. A temporary unit must be located within the contiguous property under the control of the owner/operator where the wastes to be managed in the temporary unit originated. For temporary units, the Regional Administrator may replace the design, operating, or closure standard applicable to these units under this part 264 or part 265 of this chapter with alternative requirements which protect human health and the environment.

(b) Any temporary unit to which alternative requirements are applied in accordance with paragraph (a) of this section shall be:

(1) Located within the facility boundary; and

(2) Used only for treatment or storage of remediation wastes.

(c) In establishing standards to be applied to a temporary unit, the Regional Administrator shall consider the following factors:

(1) Length of time such unit will be in operation;

(2) Type of unit;

(3) Volumes of wastes to be managed;

(4) Physical and chemical characteristics of the wastes to be managed in the unit;

(5) Potential for releases from the unit;

(6) Hydrogeological and other relevant environmental conditions at the facility which may influence the migration of any potential releases; and

(7) Potential for exposure of humans and environmental receptors if releases were to occur from the unit.

(d) The Regional Administrator shall specify in the permit or order the length of time a temporary unit will be allowed to operate, to be no longer than a period of one year. The Regional Administrator shall also specify the design, operating, and closure requirements for the unit.

(e) The Regional Administrator may extend the operational period of a temporary unit once for no longer than a period of one year beyond that originally specified in the permit or order, if the Regional Administrator determines that:

(1) Continued operation of the unit will not pose a threat to human health and the environment; and

(2) Continued operation of the unit is necessary to ensure timely and efficient implementation of remedial actions at the facility.

(f) Incorporation of a temporary unit or a time extension for a temporary unit into an existing permit shall be:

(1) Approved in accordance with the procedures for Agency-initiated permit modifications under §270.41; or

(2) Requested by the owner/operator as a Class II modification according to the procedures under §270.42 of this chapter.

(g) The Regional Administrator shall document the rationale for designating a temporary unit and for granting time extensions for temporary units and shall make such documentation available to the public.

[58 FR 8683, Feb. 16, 1993, as amended at 63 FR 65939, Nov. 30, 1998; 71 FR 40273, July 14, 2006]

## §264.554 Staging piles.

This section is written in a special format to make it easier to understand the regulatory requirements. Like other Environmental Protection Agency (EPA) regulations, this establishes enforceable legal requirements. For this "I" and "you" refer to the owner/operator.

(a) *What is a staging pile?* A staging pile is an accumulation of solid, nonflowing remediation waste (as defined in §260.10 of this chapter) that is not a containment building and is used only during remedial operations for temporary storage at a facility. A staging pile must be located within the contiguous property under the control of the owner/operator where the wastes to be managed in the staging pile originated. Staging piles must be designated by the Director according to the requirements in this section.

(1) For the purposes of this section, storage includes mixing, sizing, blending, or other similar physical operations as long as they are intended to prepare the wastes for subsequent management or treatment.

(2) [Reserved]

(b) *When may I use a staging pile?* You may use a staging pile to store hazardous remediation waste (or remediation waste otherwise subject to land disposal restrictions) only if you follow the standards and design criteria the Director has designated for that staging pile. The Director must designate the staging pile in a permit or, at an interim status facility, in a closure plan or order (consistent with §270.72(a)(5) and (b)(5) of this chapter). The Director must establish conditions in the permit, closure plan, or order that comply with paragraphs (d) through (k) of this section.

(c) *What information must I provide to get a staging pile designated?* When seeking a staging pile designation, you must provide:

(1) Sufficient and accurate information to enable the Director to impose standards and design criteria for your staging pile according to paragraphs (d) through (k) of this section;

(2) Certification by a qualified Professional Engineer for technical data, such as design drawings and specifications, and engineering studies, unless the Director determines, based on information that you provide, that this certification is not necessary to ensure that a staging pile will protect human health and the environment; and

(3) Any additional information the Director determines is necessary to protect human health and the environment.

(d) *What performance criteria must a staging pile satisfy?* The Director must establish the standards and design criteria for the staging pile in the permit, closure plan, or order.

(1) The standards and design criteria must comply with the following:

(i) The staging pile must facilitate a reliable, effective and protective remedy;

(ii) The staging pile must be designed so as to prevent or minimize releases of hazardous wastes and hazardous constituents into the environment, and minimize or adequately control cross-media transfer, as necessary to protect human health and the environment (for example, through the use of liners, covers, run-off/run-on controls, as appropriate); and

(iii) The staging pile must not operate for more than two years, except when the Director grants an operating term extension under paragraph (i) of this section (entitled "May I receive an operating extension for a staging pile?"). You must measure the two-year limit, or other operating term specified by the Director in the permit, closure plan, or order, from the first time you place remediation waste into a staging pile. You must maintain a record of the date when you first placed remediation waste into the staging pile for the life of the permit, clo-

sure plan, or order, or for three years, whichever is longer.

(2) In setting the standards and design criteria, the Director must consider the following factors:

(i) Length of time the pile will be in operation;

(ii) Volumes of wastes you intend to store in the pile;

(iii) Physical and chemical characteristics of the wastes to be stored in the unit;

(iv) Potential for releases from the unit;

(v) Hydrogeological and other relevant environmental conditions at the facility that may influence the migration of any potential releases; and

(vi) Potential for human and environmental exposure to potential releases from the unit;

(e) *May a staging pile receive ignitable or reactive remediation waste?* You must not place ignitable or reactive remediation waste in a staging pile unless:

(1) You have treated, rendered or mixed the remediation waste before you placed it in the staging pile so that:

(i) The remediation waste no longer meets the definition of ignitable or reactive under § 261.21 or § 261.23 of this chapter; and

(ii) You have complied with § 264.17(b); or

(2) You manage the remediation waste to protect it from exposure to any material or condition that may cause it to ignite or react.

(f) *How do I handle incompatible remediation wastes in a staging pile?* The term "incompatible waste" is defined in § 260.10 of this chapter. You must comply with the following requirements for incompatible wastes in staging piles:

(1) You must not place incompatible remediation wastes in the same staging pile unless you have complied with § 264.17(b);

(2) If remediation waste in a staging pile is incompatible with any waste or material stored nearby in containers, other piles, open tanks or land disposal units (for example, surface impoundments), you must separate the incompatible materials, or protect them from one another by using a dike, berm, wall or other device; and

(3) You must not pile remediation waste on the same base where incompatible wastes or materials were previously piled, unless the base has been decontaminated sufficiently to comply with §264.17(b).

(g) *Are staging piles subject to Land Disposal Restrictions (LDR) and Minimum Technological Requirements (MTR)?* No. Placing hazardous remediation wastes into a staging pile does not constitute land disposal of hazardous wastes or create a unit that is subject to the minimum technological requirements of RCRA 3004(o).

(h) *How long may I operate a staging pile?* The Director may allow a staging pile to operate for up to two years after hazardous remediation waste is first placed into the pile. You must use a staging pile no longer than the length of time designated by the Director in the permit, closure plan, or order (the "operating term"), except as provided in paragraph (i) of this section.

(i) *May I receive an operating extension for a staging pile?* (1) The Director may grant one operating term extension of up to 180 days beyond the operating term limit contained in the permit, closure plan, or order (see paragraph (l) of this section for modification procedures). To justify to the Director the need for an extension, you must provide sufficient and accurate information to enable the Director to determine that continued operation of the staging pile:

(i) Will not pose a threat to human health and the environment; and

(ii) Is necessary to ensure timely and efficient implementation of remedial actions at the facility.

(2) The Director may, as a condition of the extension, specify further standards and design criteria in the permit, closure plan, or order, as necessary, to ensure protection of human health and the environment.

(j) *What is the closure requirement for a staging pile located in a previously contaminated area?* (1) Within 180 days after the operating term of the staging pile expires, you must close a staging pile located in a previously contaminated area of the site by removing or decontaminating all:

(i) Remediation waste;

(ii) Contaminated containment system components; and

(iii) Structures and equipment contaminated with waste and leachate.

(2) You must also decontaminate contaminated subsoils in a manner and according to a schedule that the Director determines will protect human health and the environment.

(3) The Director must include the above requirements in the permit, closure plan, or order in which the staging pile is designated.

(k) *What is the closure requirement for a staging pile located in an uncontaminated area?* (1) Within 180 days after the operating term of the staging pile expires, you must close a staging pile located in an uncontaminated area of the site according to §§264.258(a) and 264.111; or according to §§265.258(a) and 265.111 of this chapter.

(2) The Director must include the above requirement in the permit, closure plan, or order in which the staging pile is designated.

(l) *How may my existing permit (for example, RAP), closure plan, or order be modified to allow me to use a staging pile?* (1) To modify a permit, other than a RAP, to incorporate a staging pile or staging pile operating term extension, either:

(i) The Director must approve the modification under the procedures for Agency-initiated permit modifications in §270.41 of this chapter; or

- (ii) You must request a Class 2 modification under §270.42 of this chapter.

(2) To modify a RAP to incorporate a staging pile or staging pile operating term extension, you must comply with the RAP modification requirements under §§270.170 and 270.175 of this chapter.

(3) To modify a closure plan to incorporate a staging pile or staging pile operating term extension, you must follow the applicable requirements under §264.112(c) or §265.112(c) of this chapter.

(4) To modify an order to incorporate a staging pile or staging pile operating term extension, you must follow the terms of the order and the applicable provisions of §270.72(a)(5) or (b)(5) of this chapter.

(m) *Is information about the staging pile available to the public?* The Director must document the rationale for designating a staging pile or staging pile operating term extension and make this documentation available to the public.

[63 FR 65939, Nov. 30, 1998, as amended at 67 FR 3028, Jan. 22, 2002; 71 FR 16907, Apr. 4, 2006; 71 FR 40273, July 14, 2006]

## § 264.555 Disposal of CAMU-eligible wastes in permitted hazardous waste landfills.

(a) The Regional Administrator with regulatory oversight at the location where the cleanup is taking place may approve placement of CAMU-eligible wastes in hazardous waste landfills not located at the site from which the waste originated, without the wastes meeting the requirements of RCRA 40 CFR part 268, if the conditions in paragraphs (a)(1) through (3) of this section are met:

(1) The waste meets the definition of CAMU-eligible waste in § 264.552(a)(1) and (2).

(2) The Regional Administrator with regulatory oversight at the location where the cleanup is taking place identifies principal hazardous constitutes in such waste, in accordance with § 264.552(e)(4)(i) and (ii), and requires that such principal hazardous constituents are treated to any of the following standards specified for CAMU-eligible wastes:

(i) The treatment standards under § 264.552(e)(4)(iv); or

(ii) Treatment standards adjusted in accordance with § 264.552(e)(4)(v)(A), (C), (D) or (E)(1); or

(iii) Treatment standards adjusted in accordance with § 264.552(e)(4)(v)(E)(2), where treatment has been used and that treatment significantly reduces the toxicity or mobility of the principal hazardous constituents in the waste, minimizing the short-term and long-term threat posed by the waste, including the threat at the remediation site.

(3) The landfill receiving the CAMU-eligible waste must have a RCRA hazardous waste permit, meet the requirements for new landfills in Subpart N of this part, and be authorized to accept CAMU-eligible wastes; for the purposes of this requirement, "permit" does not include interim status.

(b) The person seeking approval shall provide sufficient information to enable the Regional Administrator with regulatory oversight at the location where the cleanup is taking place to approve placement of CAMU-eligible waste in accordance with paragraph (a) of this section. Information required by § 264.552(d)(1) through (3) for CAMU applications must be provided, unless not reasonably available.

(c) The Regional Administrator with regulatory oversight at the location where the cleanup is taking place shall provide public notice and a reasonable opportunity for public comment before approving CAMU eligible waste for placement in an off-site permitted hazardous waste landfill, consistent with the requirements for CAMU approval at § 264.552(h). The approval must be specific to a single remediation.

(d) Applicable hazardous waste management requirements in this part, including recordkeeping requirements to demonstrate compliance with treatment standards approved under this section, for CAMU-eligible waste must be incorporated into the receiving facility permit through permit issuance or a permit modification, providing notice and an opportunity for comment and a hearing. Notwithstanding 40 CFR 270.4(a), a landfill may not receive hazardous CAMU-eligible waste under this section unless its permit specifically authorizes receipt of such waste.

(e) For each remediation, CAMU-eligible waste may not be placed in an off-site landfill authorized to receive CAMU-eligible waste in accordance with paragraph (d) of this section until the following additional conditions have been met:

(1) The landfill owner/operator notifies the Regional Administrator responsible for oversight of the landfill and persons on the facility mailing list, maintained in accordance with 40 CFR 124.10(c)(1)(ix), of his or her intent to receive CAMU-eligible waste in accordance with this section; the notice must identify the source of the remediation waste, the principal hazardous constituents in the waste, and treatment requirements.

(2) Persons on the facility mailing list may provide comments, including objections to the receipt of the CAMU-

eligible waste, to the Regional Administrator within 15 days of notification.

(3) The Regional Administrator may object to the placement of the CAMU-eligible waste in the landfill within 30 days of notification; the Regional Administrator may extend the review period an additional 30 days because of public concerns or insufficient information.

(4) CAMU-eligible wastes may not be placed in the landfill until the Regional Administrator has notified the facility owner/operator that he or she does not object to its placement.

(5) If the Regional Administrator objects to the placement or does not notify the facility owner/operator that he or she has chosen not to object, the facility may not receive the waste, notwithstanding 40 CFR 270.4(a), until the objection has been resolved, or the owner/operator obtains a permit modification in accordance with the procedures of § 270.42 specifically authorizing receipt of the waste.

(6) As part of the permit issuance or permit modification process of paragraph (d) of this section, the Regional Administrator may modify, reduce, or eliminate the notification requirements of this paragraph as they apply to specific categories of CAMU-eligible waste, based on minimal risk.

(f) Generators of CAMU-eligible wastes sent off-site to a hazardous waste landfill under this section must comply with the requirements of 40 CFR 268.7(a)(4); off-site facilities treating CAMU-eligible wastes to comply with this section must comply with the requirements of § 268.7(b)(4), except that the certification must be with respect to the treatment requirements of paragraph (a)(2) of this section.

(g) For the purposes of this section only, the "design of the CAMU" in 40 CFR 264.552(e)(4)(v)(E) means design of the permitted Subtitle C landfill.

[67 FR 3028, Jan. 22, 2002, as amended at 71 FR 40274, July 14, 2006]

## Subparts T–V [Reserved]

## Subpart W—Drip Pads

SOURCE: 56 FR 30196, July 1, 1991, unless otherwise noted.

### § 264.570 Applicability.

(a) The requirements of this subpart apply to owners and operators of facilities that use new or existing drip pads to convey treated wood drippage, precipitation, and/or surface water run-off to an associated collection system. Existing drip pads are those constructed before December 6, 1990 and those for which the owner or operator has a design and has entered into binding financial or other agreements for construction prior to December 6, 1990. All other drip pads are new drip pads. The requirement at § 264.573(b)(3) to install a leak collection system applies only to those drip pads that are constructed after December 24, 1992 except for those constructed after December 24, 1992 for which the owner or operator has a design and has entered into binding financial or other agreements for construction prior to December 24, 1992.

(b) The owner or operator of any drip pad that is inside or under a structure that provides protection from precipitation so that neither run-off nor run-on is generated is not subject to regulation under § 264.573(e) or § 264.573(f), as appropriate.

(c) The requirements of this subpart are not applicable to the management of infrequent and incidental drippage in storage yards provided that:

(1) The owner or operator maintains and complies with a written contingency plan that describes how the owner or operator will respond immediately to the discharge of such infrequent and incidental drippage. At a minimum, the contingency plan must describe how the owner or operator will do the following:

(i) Clean up the drippage;

(ii) Document the cleanup of the drippage;

(iii) Retain documents regarding cleanup for three years; and

(iv) Manage the contaminated media in a manner consistent with Federal regulations.

[56 FR 30196, July 1, 1991, as amended at 57 FR 61502, Dec. 24, 1992]

### § 264.571 Assessment of existing drip pad integrity.

(a) For each existing drip pad as defined in § 264.570 of this subpart, the owner or operator must evaluate the drip pad and determine whether it meets all of the requirements of this subpart, except the requirements for liners and leak detection systems of § 264.573(b). No later than the effective date of this rule, the owner or operator must obtain and keep on file at the facility a written assessment of the drip pad, reviewed and certified by a qualified Professional Engineer that attests to the results of the evaluation. The assessment must be reviewed, updated and re-certified annually until all upgrades, repairs, or modifications necessary to achieve compliance with all the standards of § 264.573 are complete. The evaluation must document the extent to which the drip pad meets each of the design and operating standards of § 264.573, except the standards for liners and leak detection systems, specified in § 264.573(b).

(b) The owner or operator must develop a written plan for upgrading, repairing, and modifying the drip pad to meet the requirements of § 264.573(b) and submit the plan to the Regional Administrator no later than 2 years before the date that all repairs, upgrades, and modifications are complete. This written plan must describe all changes to be made to the drip pad in sufficient detail to document compliance with all the requirements of § 264.573. The plan must be reviewed and certified by a qualified Professional Engineer.

(c) Upon completion of all upgrades, repairs, and modifications, the owner or operator must submit to the Regional Administrator or state Director, the as-built drawings for the drip pad together with a certification by a qualified Professional Engineer attesting that the drip pad conforms to the drawings.

(d) If the drip pad is found to be leaking or unfit for use, the owner or operator must comply with the provisions of § 264.573 (m) of this subpart or close the drip pad in accordance with § 264.575 of this subpart.

[56 FR 30196, July 1, 1991, as amended at 57 FR 61503, Dec. 24, 1992; 71 FR 16907, Apr. 4, 2006]

## § 264.572 Design and installation of new drip pads.

Owners and operators of new drip pads must ensure that the pads are designed, installed, and operated in accordance with one of the following:

(a) all of the requirements of §§ 264.573 (except 264.573(a)(4)), 264.574 and 264.575 of this subpart, or

(b) all of the requirements of §§ 264.573 (except § 264.573(b)), 264.574 and 264.575 of this subpart.

[57 FR 61503, Dec. 24, 1992]

## § 264.573 Design and operating requirements.

(a) Drip pads must:

(1) Be constructed of non-earthen materials, excluding wood and non-structurally supported asphalt;

(2) Be sloped to free-drain treated wood drippage, rain and other waters, or solutions of drippage and water or other wastes to the associated collection system;

(3) Have a curb or berm around the perimeter;

(4)(i) Have a hydraulic conductivity of less than or equal to $1\times10^{-7}$ centimeters per second, e.g., existing concrete drip pads must be sealed, coated, or covered with a surface material with a hydraulic conductivity of less than or equal to $1\times10^{-7}$ centimeters per second such that the entire surface where drippage occurs or may run across is capable of containing such drippage and mixtures of drippage and precipitation, materials, or other wastes while being routed to an associated collection system. This surface material must be maintained free of cracks and gaps that could adversely affect its hydraulic conductivity, and the material must be chemically compatible with the preservatives that contact the drip pad. The requirements of this provision apply only to existing drip pads and those drip pads for which the owner or operator elects to comply with § 264.572(b) instead of § 264.572(a).

(ii) The owner or operator must obtain and keep on file at the facility a written assessment of the drip pad, reviewed and certified by a qualified Professional Engineer that attests to the results of the evaluation. The assessment must be reviewed, updated and recertified annually. The evaluation must document the extent to which the drip pad meets the design and operating standards of this section, except for paragraph (b) of this section.

(5) Be of sufficient structural strength and thickness to prevent failure due to physical contact, climatic conditions, the stress of daily operations, e.g., variable and moving loads such as vehicle traffic, movement of wood, etc.

[NOTE: EPA will generally consider applicable standards established by professional organizations generally recognized by the industry such as the American Concrete Institute (ACI) or the American Society of Testing and Materials (ASTM) in judging the structural integrity requirement of this paragraph.]

(b) If an owner/operator elects to comply with §264.572(a) instead of §264.572(b), the drip pad must have:

(1) A synthetic liner installed below the drip pad that is designed, constructed, and installed to prevent leakage from the drip pad into the adjacent subsurface soil or groundwater or surface water at any time during the active life (including the closure period) of the drip pad. The liner must be constructed of materials that will prevent waste from being absorbed into the liner and to prevent releases into the adjacent subsurface soil or groundwater or surface water during the active life of the facility. The liner must be:

(i) Constructed of materials that have appropriate chemical properties and sufficient strength and thickness to prevent failure due to pressure gradients (including static head and external hydrogeologic forces), physical contact with the waste or drip pad leakage to which they are exposed, climatic conditions, the stress of installation, and the stress of daily operation (including stresses from vehicular traffic on the drip pad);

(ii) Placed upon a foundation or base capable of providing support to the liner and resistance to pressure gradients above and below the liner to prevent failure of the liner due to settlement, compression or uplift; and

(iii) Installed to cover all surrounding earth that could come in contact with the waste or leakage; and

(2) A leakage detection system immediately above the liner that is designed, constructed, maintained and operated to detect leakage from the drip pad. The leakage detection system must be:

(i) Constructed of materials that are:

(A) Chemically resistant to the waste managed in the drip pad and the leakage that might be generated; and

(B) Of sufficient strength and thickness to prevent collapse under the pressures exerted by overlaying materials and by any equipment used at the drip pad;

(ii) Designed and operated to function without clogging through the scheduled closure of the drip pad; and

(iii) Designed so that it will detect the failure of the drip pad or the presence of a release of hazardous waste or accumulated liquid at the earliest practicable time.

(3) A leakage collection system immediately above the liner that is designed, constructed, maintained and operated to collect leakage from the drip pad such that it can be removed from below the drip pad. The date, time, and quantity of any leakage collected in this system and removed must be documented in the operating log.

(c) Drip pads must be maintained such that they remain free of cracks, gaps, corrosion, or other deterioration that could cause hazardous waste to be released from the drip pad.

[NOTE:See §264.573(m) for remedial action required if deterioration or leakage is detected.]

(d) The drip pad and associated collection system must be designed and operated to convey, drain, and collect liquid resulting from drippage or precipitation in order to prevent run-off.

(e) Unless protected by a structure, as described in §264.570(b) of this subpart, the owner or operator must design, construct, operate and maintain a run-on control system capable of preventing flow onto the drip pad during peak discharge from at least a 24-hour, 25-year storm, unless the system has sufficient excess capacity to contain any run-off that might enter the system.

(f) Unless protected by a structure or cover as described in §264.570(b) of this subpart, the owner or operator must design, construct, operate and maintain a run-off management system to collect and control at least the water volume resulting from a 24-hour, 25-year storm.

(g) The drip pad must be evaluated to determine that it meets the requirements of paragraphs (a) through (f) of this section and the owner or operator must obtain a statement from a qualified Professional Engineer certifying that the drip pad design meets the requirements of this section.

(h) Drippage and accumulated precipitation must be removed from the associated collection system as necessary to prevent overflow onto the drip pad.

(i) The drip pad surface must be cleaned thoroughly in a manner and frequency such that accumulated residues of hazardous waste or other materials are removed, with residues being properly managed as hazardous waste, so as to allow weekly inspections of the entire drip pad surface without interference or hindrance from accumulated residues of hazardous waste or other materials on the drip pad. The owner or operator must document the date and time of each cleaning and the cleaning procedure used in the facility's operating log. The owner/operator must determine if the residues are hazardous as per 40 CFR 262.11 and, if so, must manage them under parts 261–268, 270, and section 3010 of RCRA.

(j) Drip pads must be operated and maintained in a manner to minimize tracking of hazardous waste or hazardous waste constituents off the drip pad as a result of activities by personnel or equipment.

(k) After being removed from the treatment vessel, treated wood from pressure and non-pressure processes must be held on the drip pad until drippage has ceased. The owner or operator must maintain records sufficient to document that all treated wood is held on the pad following treatment in accordance with this requirement.

(l) Collection and holding units associated with run-on and run-off control systems must be emptied or otherwise managed as soon as possible after storms to maintain design capacity of the system.

(m) Throughout the active life of the drip pad and as specified in the permit, if the owner or operator detects a condition that may have caused or has caused a release of hazardous waste, the condition must be repaired within a reasonably prompt period of time following discovery, in accordance with the following procedures:

(1) Upon detection of a condition that may have caused or has caused a release of hazardous waste (e.g., upon detection of leakage in the leak detection system), the owner or operator must:

(i) Enter a record of the discovery in the facility operating log;

(ii) Immediately remove the portion of the drip pad affected by the condition from service;

(iii) Determine what steps must be taken to repair the drip pad and clean up any leakage from below the drip pad, and establish a schedule for accomplishing the repairs;

(iv) Within 24 hours after discovery of the condition, notify the Regional Administrator of the condition and, within 10 working days, provide written notice to the Regional Administrator with a description of the steps that will be taken to repair the drip pad and clean up any leakage, and the schedule for accomplishing this work.

(2) The Regional Administrator will review the information submitted, make a determination regarding whether the pad must be removed from service completely or partially until repairs and cleanup are complete and notify the owner or operator of the determination and the underlying rationale in writing.

(3) Upon completing all repairs and cleanup, the owner or operator must notify the Regional Administrator in writing and provide a certification signed by an independent, qualified registered professional engineer, that the repairs and cleanup have been completed according to the written plan submitted in accordance with paragraph (m)(1)(iv) of this section.

(n) Should a permit be necessary, the Regional Administrator will specify in the permit all design and operating practices that are necessary to ensure that the requirements of this section are satisfied.

(o) The owner or operator must maintain, as part of the facility operating log, documentation of past operating and waste handling practices. This must include identification of preservative formulations used in the past, a description of drippage management practices, and a description of treated wood storage and handling practices.

[56 FR 30196, July 1, 1991, as amended at 57 FR 5861, Feb. 18, 1992; 57 FR 61503, Dec. 24, 1992; 71 FR 16907, Apr. 4, 2006; 71 FR 40274, July 14, 2006]

### §264.574  Inspections.

(a) During construction or installation, liners and cover systems (*e.g.*, membranes, sheets, or coatings) must be inspected for uniformity, damage and imperfections (*e.g.*, holes, cracks, thin spots, or foreign materials). Immediately after construction or installation, liners must be inspected and certified as meeting the requirements in §264.573 of this subpart by a qualified Professional Engineer. This certification must be maintained at the facility as part of the facility operating record. After installation, liners and covers must be inspected to ensure tight seams and joints and the absence of tears, punctures, or blisters.

(b) While a drip pad is in operation, it must be inspected weekly and after storms to detect evidence of any of the following:

(1) Deterioration, malfunctions or improper operation of run-on and run-off control systems;

(2) The presence of leakage in and proper functioning of leak detection system.

(3) Deterioration or cracking of the drip pad surface.

NOTE: See §264.573(m) for remedial action required if deterioration or leakage is detected.

[56 FR 30196, July 1, 1991, as amended at 71 FR 16907, Apr. 4, 2006]

### §264.575  Closure.

(a) At closure, the owner or operator must remove or decontaminate all waste residues, contaminated containment system components (pad, liners, etc.), contaminated subsoils, and structures and equipment contaminated with waste and leakage, and manage them as hazardous waste.

(b) If, after removing or decontaminating all residues and making all reasonable efforts to effect removal or decontamination of contaminated components, subsoils, structures, and equipment as required in paragraph (a) of this section, the owner or operator finds that not all contaminated subsoils can be practically removed or decontaminated, he must close the facility and perform post-closure care in accordance with closure and post-closure care requirements that apply to landfills (§264.310). For permitted units, the requirement to have a permit continues throughout the post-closure period. In addition, for the purpose of closure, post-closure, and financial responsibility, such a drip pad is then considered to be landfill, and the owner or operator must meet all of the requirements for landfills specified in subparts G and H of this part.

(c)(1) The owner or operator of an existing drip pad, as defined in §264.570 of this subpart, that does not comply with the liner requirements of §264.573(b)(1) must:

(i) Include in the closure plan for the drip pad under §264.112 both a plan for complying with paragraph (a) of this section and a contingent plan for complying with paragraph (b) of this section in case not all contaminated subsoils can be practicably removed at closure; and

(ii) Prepare a contingent post-closure plan under §264.118 of this part for complying with paragraph (b) of this section in case not all contaminated subsoils can be practicably removed at closure.

(2) The cost estimates calculated under §§264.112 and 264.144 of this part for closure and post-closure care of a drip pad subject to this paragraph must include the cost of complying with the contingent closure plan and the contingent post-closure plan, but are not required to include the cost of expected closure under paragraph (a) of this section.

## Subpart X—Miscellaneous Units

Source: 52 FR 46964, Dec. 10, 1987, unless otherwise noted.

### § 264.600  Applicability.

The requirements in this subpart apply to owners and operators of facilities that treat, store, or dispose of hazardous waste in miscellaneous units, except as § 264.1 provide otherwise.

[52 FR 46964, Dec. 10, 1987, as amended at 71 FR 40274, July 14, 2006]

### § 264.601  Environmental performance standards.

A miscellaneous unit must be located, designed, constructed, operated, maintained, and closed in a manner that will ensure protection of human health and the environment. Permits for miscellaneous units are to contain such terms and provisions as necessary to protect human health and the environment, including, but not limited to, as appropriate, design and operating requirements, detection and monitoring requirements, and requirements for responses to releases of hazardous waste or hazardous constituents from the unit. Permit terms and provisions must include those requirements of subparts I through O and subparts AA through CC of this part, part 270, part 63 subpart EEE, and part 146 of this chapter that are appropriate for the miscellaneous unit being permitted. Protection of human health and the environment includes, but is not limited to:

(a) Prevention of any releases that may have adverse effects on human health or the environment due to migration of waste constituents in the ground water or subsurface environment, considering:

(1) The volume and physical and chemical characteristics of the waste in the unit, including its potential for migration through soil, liners, or other containing structures;

(2) The hydrologic and geologic characteristics of the unit and the surrounding area;

(3) The existing quality of ground water, including other sources of contamination and their cumulative impact on the ground water;

(4) The quantity and direction of ground-water flow;

(5) The proximity to and withdrawal rates of current and potential ground-water users;

(6) The patterns of land use in the region;

(7) The potential for deposition or migration of waste constituents into subsurface physical structures, and into the root zone of food-chain crops and other vegetation;

(8) The potential for health risks caused by human exposure to waste constituents; and

(9) The potential for damage to domestic animals, wildlife, crops, vegetation, and physical structures caused by exposure to waste constituents;

(b) Prevention of any releases that may have adverse effects on human health or the environment due to migration of waste constituents in surface water, or wetlands or on the soil surface considering:

(1) The volume and physical and chemical characteristics of the waste in the unit;

(2) The effectiveness and reliability of containing, confining, and collecting systems and structures in preventing migration;

(3) The hydrologic characteristics of the unit and the surrounding area, including the topography of the land around the unit;

(4) The patterns of precipitation in the region;

(5) The quantity, quality, and direction of ground-water flow;

(6) The proximity of the unit to surface waters;

(7) The current and potential uses of nearby surface waters and any water quality standards established for those surface waters;

(8) The existing quality of surface waters and surface soils, including other sources of contamination and their cumulative impact on surface waters and surface soils;

(9) The patterns of land use in the region;

(10) The potential for health risks caused by human exposure to waste constituents; and

(11) The potential for damage to domestic animals, wildlife, crops, vegetation, and physical structures caused by exposure to waste constituents.

(c) Prevention of any release that may have adverse effects on human health or the environment due to migration of waste constituents in the air, considering:

(1) The volume and physical and chemical characteristics of the waste in the unit, including its potential for the emission and dispersal of gases, aerosols and particulates;

(2) The effectiveness and reliability of systems and structures to reduce or prevent emissions of hazardous constituents to the air;

(3) The operating characteristics of the unit;

(4) The atmospheric, meteorologic, and topographic characteristics of the unit and the surrounding area;

(5) The existing quality of the air, including other sources of contamination and their cumulative impact on the air;

(6) The potential for health risks caused by human exposure to waste constituents; and

(7) The potential for damage to domestic animals, wildlife, crops, vegetation, and physical structures caused by exposure to waste constituents.

[59 FR 62927, Dec. 6, 1994, as amended at 64 FR 53074, Sept. 30, 1999; 71 FR 40274, July 14, 2006]

### § 264.602 Monitoring, analysis, inspection, response, reporting, and corrective action.

Monitoring, testing, analytical data, inspections, response, and reporting procedures and frequencies must ensure compliance with §§ 264.601, 264.15, 264.33, 264.75, 264.76, 264.77, and 264.101 as well as meet any additional requirements needed to protect human health and the environment as specified in the permit.

### § 264.603 Post-closure care.

A miscellaneous unit that is a disposal unit must be maintained in a manner that complies with § 264.601 during the post-closure care period. In addition, if a treatment or storage unit has contaminated soils or ground water that cannot be completely removed or decontaminated during closure, then that unit must also meet the requirements of § 264.601 during post-closure care. The post-closure plan under § 264.118 must specify the procedures that will be used to satisfy this requirement.

## Subparts Y–Z [Reserved]

## Subpart AA—Air Emission Standards for Process Vents

SOURCE: 55 FR 25494, June 21, 1990, unless otherwise noted.

### § 264.1030 Applicability.

(a) The regulations in this subpart apply to owners and operators of facilities that treat, store, or dispose of hazardous wastes (except as provided in § 264.1).

(b) Except for § 264.1034, paragraphs (d) and (e), this subpart applies to process vents associated with distillation, fractionation, thin-film evaporation, solvent extraction, or air or steam stripping operations that manage hazardous wastes with organic concentrations of at least 10 ppmw, if these operations are conducted in one of the following:

(1) A unit that is subject to the permitting requirements of 40 CFR part 270, or

(2) A unit (including a hazardous waste recycling unit) that is not exempt from permitting under the provisions of 40 CFR 262.34(a) (i.e., a hazardous waste recycling unit that is not a 90-day tank or container) and that is located at a hazardous waste management facility otherwise subject to the permitting requirements of 40 CFR part 270, or

(3) A unit that is exempt from permitting under the provisions of 40 CFR 262.34(a) (i.e., a "90-day" tank or container) and is not a recycling unit under the provisions of 40 CFR 261.6.

(c) For the owner and operator of a facility subject to this subpart and who received a final permit under RCRA section 3005 prior to December 6, 1996, the requirements of this subpart shall be incorporated into the permit when the permit is reissued in accordance with the requirements of 40 CFR 124.15

497

or reviewed in accordance with the requirements of 40 CFR 270.50(d). Until such date when the owner and operator receive a final permit incorporating the requirements of this subpart, the owner and operator are subject to the requirements of 40 CFR 265, subpart AA.

[NOTE: The requirements of §§ 264.1032 through 264.1036 apply to process vents on hazardous waste recycling units previously exempt under § 261.6(c)(1). Other exemptions under §§ 261.4, and 264.1(g) are not affected by these requirements.]

(d) The requirements of this subpart do not apply to the pharmaceutical manufacturing facility, commonly referred to as the Stonewall Plant, located at Route 340 South, Elkton, Virginia, provided that facility is operated in compliance with the requirements contained in a Clean Air Act permit issued pursuant to 40 CFR 52.2454. The requirements of this subpart shall apply to the facility upon termination of the Clean Air Act permit issued pursuant to 40 CFR 52.2454.

(e) The requirements of this subpart do not apply to the process vents at a facility where the facility owner or operator certifies that all of the process vents that would otherwise be subject to this subpart are equipped with and operating air emission controls in accordance with the process vent requirements of an applicable Clean Air Act regulation codified under 40 CFR part 60, part 61, or part 63. The documentation of compliance under regulations at 40 CFR part 60, part 61, or part 63 shall be kept with, or made readily available with, the facility operating record.

[55 FR 25494, June 21, 1990, as amended at 56 FR 19290, Apr. 26, 1991; 61 FR 59950, Nov. 25, 1996; 62 FR 52641, Oct. 8, 1997; 62 FR 64656, Dec. 8, 1997; 71 FR 40274, July 14, 2006]

## § 264.1031 Definitions.

As used in this subpart, all terms not defined herein shall have the meaning given them in the Act and parts 260–266.

*Air stripping operation* is a desorption operation employed to transfer one or more volatile components from a liquid mixture into a gas (air) either with or without the application of heat to the liquid. Packed towers, spray towers, and bubble-cap, sieve, or valve-type plate towers are among the process configurations used for contacting the air and a liquid.

*Bottoms receiver* means a container or tank used to receive and collect the heavier bottoms fractions of the distillation feed stream that remain in the liquid phase.

*Closed-vent system* means a system that is not open to the atmosphere and that is composed of piping, connections, and, if necessary, flow-inducing devices that transport gas or vapor from a piece or pieces of equipment to a control device.

*Condenser* means a heat-transfer device that reduces a thermodynamic fluid from its vapor phase to its liquid phase.

*Connector* means flanged, screwed, welded, or other joined fittings used to connect two pipelines or a pipeline and a piece of equipment. For the purposes of reporting and recordkeeping, connector means flanged fittings that are not covered by insulation or other materials that prevent location of the fittings.

*Continuous recorder* means a data-recording device recording an instantaneous data value at least once every 15 minutes.

*Control device* means an enclosed combustion device, vapor recovery system, or flare. Any device the primary function of which is the recovery or capture of solvents or other organics for use, reuse, or sale (e.g., a primary condenser on a solvent recovery unit) is not a control device.

*Control device shutdown* means the cessation of operation of a control device for any purpose.

*Distillate receiver* means a container or tank used to receive and collect liquid material (condensed) from the overhead condenser of a distillation unit and from which the condensed liquid is pumped to larger storage tanks or other process units.

*Distillation operation* means an operation, either batch or continuous, separating one or more feed stream(s) into two or more exit streams, each exit stream having component concentrations different from those in the feed stream(s). The separation is achieved by the redistribution of the components between the liquid and vapor phase as they approach equilibrium within the distillation unit.

*Double block and bleed system* means two block valves connected in series with a bleed valve or line that can vent the line between the two block valves.

*Equipment* means each valve, pump, compressor, pressure relief device, sampling connection system, open-ended valve or line, or flange or other connector, and any control devices or systems required by this subpart.

*Flame zone* means the portion of the combustion chamber in a boiler occupied by the flame envelope.

*Flow indicator* means a device that indicates whether gas flow is present in a vent stream.

*First attempt at repair* means to take rapid action for the purpose of stopping or reducing leakage of organic material to the atmosphere using best practices.

*Fractionation operation* means a distillation operation or method used to separate a mixture of several volatile components of different boiling points in successive stages, each stage removing from the mixture some proportion of one of the components.

*Hazardous waste management unit shutdown* means a work practice or operational procedure that stops operation of a hazardous waste management unit or part of a hazardous waste management unit. An unscheduled work practice or operational procedure that stops operation of a hazardous waste management unit or part of a hazardous waste management unit for less than 24 hours is not a hazardous waste management unit shutdown. The use of spare equipment and technically feasible bypassing of equipment without stopping operation are not hazardous waste management unit shutdowns.

*Hot well* means a container for collecting condensate as in a steam condenser serving a vacuum-jet or steam-jet ejector.

*In gas/vapor service* means that the piece of equipment contains or contacts a hazardous waste stream that is in the gaseous state at operating conditions.

*In heavy liquid service* means that the piece of equipment is not in gas/vapor service or in light liquid service.

*In light liquid service* means that the piece of equipment contains or contacts a waste stream where the vapor pressure of one or more of the organic components in the stream is greater than 0.3 kilopascals (kPa) at 20 °C, the total concentration of the pure organic components having a vapor pressure greater than 0.3 kilopascals (kPa) at 20 °C is equal to or greater than 20 percent by weight, and the fluid is a liquid at operating conditions.

*In situ sampling systems* means non-extractive samplers or in-line samplers.

*In vacuum service* means that equipment is operating at an internal pressure that is at least 5 kPa below ambient pressure.

*Malfunction* means any sudden failure of a control device or a hazardous waste management unit or failure of a hazardous waste management unit to operate in a normal or usual manner, so that organic emissions are increased.

*Open-ended valve or line* means any valve, except pressure relief valves, having one side of the valve seat in contact with hazardous waste and one side open to the atmosphere, either directly or through open piping.

*Pressure release* means the emission of materials resulting from the system pressure being greater than the set pressure of the pressure relief device.

*Process heater* means a device that transfers heat liberated by burning fuel to fluids contained in tubes, including all fluids except water that are heated to produce steam.

*Process vent* means any open-ended pipe or stack that is vented to the atmosphere either directly, through a vacuum-producing system, or through a tank (e.g., distillate receiver, condenser, bottoms receiver, surge control tank, separator tank, or hot well) associated with hazardous waste distillation, fractionation, thin-film evaporation, solvent extraction, or air or steam stripping operations.

*Repaired* means that equipment is adjusted, or otherwise altered, to eliminate a leak.

*Sampling connection system* means an assembly of equipment within a process or waste management unit used during periods of representative operation to take samples of the process or waste fluid. Equipment used to take non-routine grab samples is not considered a sampling connection system.

*Sensor* means a device that measures a physical quantity or the change in a physical quantity, such as temperature, pressure, flow rate, pH, or liquid level.

*Separator tank* means a device used for separation of two immiscible liquids.

*Solvent extraction operation* means an operation or method of separation in which a solid or solution is contacted with a liquid solvent (the two being mutually insoluble) to preferentially dissolve and transfer one or more components into the solvent.

*Startup* means the setting in operation of a hazardous waste management unit or control device for any purpose.

*Steam stripping operation* means a distillation operation in which vaporization of the volatile constituents of a liquid mixture takes place by the introduction of steam directly into the charge.

*Surge control tank* means a large-sized pipe or storage reservoir sufficient to contain the surging liquid discharge of the process tank to which it is connected.

*Thin-film evaporation operation* means a distillation operation that employs a heating surface consisting of a large diameter tube that may be either straight or tapered, horizontal or vertical. Liquid is spread on the tube wall by a rotating assembly of blades that maintain a close clearance from the wall or actually ride on the film of liquid on the wall.

*Vapor incinerator* means any enclosed combustion device that is used for destroying organic compounds and does not extract energy in the form of steam or process heat.

*Vented* means discharged through an opening, typically an open-ended pipe or stack, allowing the passage of a stream of liquids, gases, or fumes into the atmosphere. The passage of liquids, gases, or fumes is caused by mechanical means such as compressors or vacuum-producing systems or by process-related means such as evaporation produced by heating and not caused by tank loading and unloading (working losses) or by natural means such as diurnal temperature changes.

[55 FR 25494, June 21, 1990, as amended at 62 FR 64657, Dec. 8, 1997; 64 FR 3389, Jan. 21, 1999]

§ 264.1032   Standards: Process vents.

(a) The owner or operator of a facility with process vents associated with distillation, fractionation, thin-film evaporation, solvent extraction, or air or steam stripping operations managing hazardous wastes with organic concentrations of at least 10 ppmw shall either:

(1) Reduce total organic emissions from all affected process vents at the facility below 1.4 kg/h (3 lb/h) and 2.8 Mg/yr (3.1 tons/yr), or

(2) Reduce, by use of a control device, total organic emissions from all affected process vents at the facility by 95 weight percent.

(b) If the owner or operator installs a closed-vent system and control device to comply with the provisions of paragraph (a) of this section the closed-vent system and control device must meet the requirements of § 264.1033.

(c) Determinations of vent emissions and emission reductions or total organic compound concentrations achieved by add-on control devices may be based on engineering calculations or performance tests. If performance tests are used to determine vent emissions, emission reductions, or total organic compound concentrations achieved by add-on control devices, the performance tests must conform with the requirements of § 264.1034(c).

(d) When an owner or operator and the Regional Administrator do not agree on determinations of vent emissions and/or emission reductions or total organic compound concentrations achieved by add-on control devices based on engineering calculations, the procedures in § 264.1034(c) shall be used to resolve the disagreement.

§ 264.1033   Standards: Closed-vent systems and control devices.

(a)(1) Owners or operators of closed-vent systems and control devices used to comply with provisions of this part

500

shall comply with the provisions of this section.

(2)(i) The owner or operator of an existing facility who cannot install a closed-vent system and control device to comply with the provisions of this subpart on the effective date that the facility becomes subject to the provisions of this subpart must prepare an implementation schedule that includes dates by which the closed-vent system and control device will be installed and in operation. The controls must be installed as soon as possible, but the implementation schedule may allow up to 30 months after the effective date that the facility becomes subject to this subpart for installation and startup.

(ii) Any unit that begins operation after December 21, 1990, and is subject to the provisions of this subpart when operation begins, must comply with the rules immediately (i.e., must have control devices installed and operating on startup of the affected unit); the 30-month implementation schedule does not apply.

(iii) The owner or operator of any facility in existence on the effective date of a statutory or EPA regulatory amendment that renders the facility subject to this subpart shall comply with all requirements of this subpart as soon as practicable but no later than 30 months after the amendment's effective date. When control equipment required by this subpart can not be installed and begin operation by the effective date of the amendment, the facility owner or operator shall prepare an implementation schedule that includes the following information: Specific calendar dates for award of contracts or issuance of purchase orders for the control equipment, initiation of on-site installation of the control equipment, completion of the control equipment installation, and performance of any testing to demonstrate that the installed equipment meets the applicable standards of this subpart. The owner or operator shall enter the implementation schedule in the operating record or in a permanent, readily available file located at the facility.

(iv) Owners and operators of facilities and units that become newly subject to the requirements of this subpart after December 8, 1997, due to an action other than those described in paragraph (a)(2)(iii) of this section must comply with all applicable requirements immediately (i.e., must have control devices installed and operating on the date the facility or unit becomes subject to this subpart; the 30-month implementation schedule does not apply).

(b) A control device involving vapor recovery (e.g., a condenser or adsorber) shall be designed and operated to recover the organic vapors vented to it with an efficiency of 95 weight percent or greater unless the total organic emission limits of § 264.1032(a)(1) for all affected process vents can be attained at an efficiency less than 95 weight percent.

(c) An enclosed combustion device (e.g., a vapor incinerator, boiler, or process heater) shall be designed and operated to reduce the organic emissions vented to it by 95 weight percent or greater; to achieve a total organic compound concentration of 20 ppmv, expressed as the sum of the actual compounds, not carbon equivalents, on a dry basis corrected to 3 percent oxygen; or to provide a minimum residence time of 0.50 seconds at a minimum temperature of 760 °C. If a boiler or process heater is used as the control device, then the vent stream shall be introduced into the flame zone of the boiler or process heater.

(d)(1) A flare shall be designed for and operated with no visible emissions as determined by the methods specified in paragraph (e)(1) of this section, except for periods not to exceed a total of 5 minutes during any 2 consecutive hours.

(2) A flare shall be operated with a flame present at all times, as determined by the methods specified in paragraph (f)(2)(iii) of this section.

(3) A flare shall be used only if the net heating value of the gas being combusted is 11.2 MJ/scm (300 Btu/scf) or greater if the flare is steam-assisted or air-assisted; or if the net heating value of the gas being combusted is 7.45 MJ/scm (200 Btu/scf) or greater if the flare is nonassisted. The net heating value of the gas being combusted shall be determined by the methods specified in paragraph (e)(2) of this section.

(4)(i) A steam-assisted or nonassisted flare shall be designed for and operated with an exit velocity, as determined by the methods specified in paragraph (e)(3) of this section, less than 18.3 m/s (60 ft/s), except as provided in paragraphs (d)(4) (ii) and (iii) of this section.

(ii) A steam-assisted or nonassisted flare designed for and operated with an exit velocity, as determined by the methods specified in paragraph (e)(3) of this section, equal to or greater than 18.3 m/s (60 ft/s) but less than 122 m/s (400 ft/s) is allowed if the net heating value of the gas being combusted is greater than 37.3 MJ/scm (1,000 Btu/scf).

(iii) A steam-assisted or nonassisted flare designed for and operated with an exit velocity, as determined by the methods specified in paragraph (e)(3) of this section, less than the velocity, $V_{max}$, as determined by the method specified in paragraph (e)(4) of this section and less than 122 m/s (400 ft/s) is allowed.

(5) An air-assisted flare shall be designed and operated with an exit velocity less than the velocity, $V_{max}$, as determined by the method specified in paragraph (e)(5) of this section.

(6) A flare used to comply with this section shall be steam-assisted, air-assisted, or nonassisted.

(e)(1) Reference Method 22 in 40 CFR part 60 shall be used to determine the compliance of a flare with the visible emission provisions of this subpart. The observation period is 2 hours and shall be used according to Method 22.

(2) The net heating value of the gas being combusted in a flare shall be calculated using the following equation:

$$H_T = K \left[ \sum_{i=1}^{n} C_i H_i \right]$$

where:

$H_T$=Net heating value of the sample, MJ/scm; where the net enthalpy per mole of offgas is based on combustion at 25 °C and 760 mm Hg, but the standard temperature for determining the volume corresponding to 1 mol is 20 °C;

K=Constant, $1.74{\times}10^{-7}$ (1/ppm) (g mol/scm) (MJ/kcal) where standard temperature for (g mol/scm) is 20 °C;

$C_i$=Concentration of sample component i in ppm on a wet basis, as measured for organics by Reference Method 18 in 40 CFR

part 60 and measured for hydrogen and carbon monoxide by ASTM D 1946-82 (incorporated by reference as specified in § 260.11); and

$H_i$=Net heat of combustion of sample component i, kcal/9 mol at 25 °C and 760 mm Hg. The heats of combustion may be determined using ASTM D 2382-83 (incorporated by reference as specified in § 260.11) if published values are not available or cannot be calculated.

(3) The actual exit velocity of a flare shall be determined by dividing the volumetric flow rate (in units of standard temperature and pressure), as determined by Reference Methods 2, 2A, 2C, or 2D in 40 CFR part 60 as appropriate, by the unobstructed (free) cross-sectional area of the flare tip.

(4) The maximum allowed velocity in m/s, $V_{max}$, for a flare complying with paragraph (d)(4)(iii) of this section shall be determined by the following equation:

$$Log_{10}(V_{max})=(H_T+28.8)/31.7$$

where:

28.8=Constant,

31.7=Constant,

$H_T$=The net heating value as determined in paragraph (e)(2) of this section.

(5) The maximum allowed velocity in m/s, $V_{max}$, for an air-assisted flare shall be determined by the following equation:

$$V_{max}=8.706+0.7084 (H_T)$$

where:

8.706=Constant,

0.7084=Constant,

$H_T$=The net heating value as determined in paragraph (e)(2) of this section.

(f) The owner or operator shall monitor and inspect each control device required to comply with this section to ensure proper operation and maintenance of the control device by implementing the following requirements:

(1) Install, calibrate, maintain, and operate according to the manufacturer's specifications a flow indicator that provides a record of vent stream flow from each affected process vent to the control device at least once every hour. The flow indicator sensor shall be installed in the vent stream at the nearest feasible point to the control device inlet but before the point at which the vent streams are combined.

(2) Install, calibrate, maintain, and operate according to the manufacturer's specifications a device to continuously monitor control device operation as specified below:

(i) For a thermal vapor incinerator, a temperature monitoring device equipped with a continuous recorder. The device shall have an accuracy of ±1 percent of the temperature being monitored in °C or ±0.5 °C, whichever is greater. The temperature sensor shall be installed at a location in the combustion chamber downstream of the combustion zone.

(ii) For a catalytic vapor incinerator, a temperature monitoring device equipped with a continuous recorder. The device shall be capable of monitoring temperature at two locations and have an accuracy of ±1 percent of the temperature being monitored in °C or ±0.5 °C, whichever is greater. One temperature sensor shall be installed in the vent stream at the nearest feasible point to the catalyst bed inlet and a second temperature sensor shall be installed in the vent stream at the nearest feasible point to the catalyst bed outlet.

(iii) For a flare, a heat sensing monitoring device equipped with a continuous recorder that indicates the continuous ignition of the pilot flame.

(iv) For a boiler or process heater having a design heat input capacity less than 44 MW, a temperature monitoring device equipped with a continuous recorder. The device shall have an accuracy of ±1 percent of the temperature being monitored in °C or ±0.5 °C, whichever is greater. The temperature sensor shall be installed at a location in the furnace downstream of the combustion zone.

(v) For a boiler or process heater having a design heat input capacity greater than or equal to 44 MW, a monitoring device equipped with a continuous recorder to measure a parameter(s) that indicates good combustion operating practices are being used.

(vi) For a condenser, either:

(A) A monitoring device equipped with a continuous recorder to measure the concentration level of the organic compounds in the exhaust vent stream from the condenser, or

(B) A temperature monitoring device equipped with a continuous recorder. The device shall be capable of monitoring temperature with an accuracy of ±1 percent of the temperature being monitored in degrees Celsius ( °C) or ±0.5 °C, whichever is greater. The temperature sensor shall be installed at a location in the exhaust vent stream from the condenser exit (i.e., product side).

(vii) For a carbon adsorption system that regenerates the carbon bed directly in the control device such as a fixed-bed carbon adsorber, either:

(A) A monitoring device equipped with a continuous recorder to measure the concentration level of the organic compounds in the exhaust vent stream from the carbon bed, or

(B) A monitoring device equipped with a continuous recorder to measure a parameter that indicates the carbon bed is regenerated on a regular, predetermined time cycle.

(3) Inspect the readings from each monitoring device required by paragraphs (f)(1) and (2) of this section at least once each operating day to check control device operation and, if necessary, immediately implement the corrective measures necessary to ensure the control device operates in compliance with the requirements of this section.

(g) An owner or operator using a carbon adsorption system such as a fixed-bed carbon adsorber that regenerates the carbon bed directly onsite in the control device shall replace the existing carbon in the control device with fresh carbon at a regular, predetermined time interval that is no longer than the carbon service life established as a requirement of §264.1035(b)(4)(iii)(F).

(h) An owner or operator using a carbon adsorption system such as a carbon canister that does not regenerate the carbon bed directly onsite in the control device shall replace the existing carbon in the control device with fresh carbon on a regular basis by using one of the following procedures:

(1) Monitor the concentration level of the organic compounds in the exhaust vent stream from the carbon adsorption system on a regular schedule, and replace the existing carbon with fresh

carbon immediately when carbon breakthrough is indicated. The monitoring frequency shall be daily or at an interval no greater than 20 percent of the time required to consume the total carbon working capacity established as a requirement of § 264.1035(b)(4)(iii)(G), whichever is longer.

(2) Replace the existing carbon with fresh carbon at a regular, predetermined time interval that is less than the design carbon replacement interval established as a requirement of § 264.1035(b)(4)(iii)(G).

(i) An alternative operational or process parameter may be monitored if it can be demonstrated that another parameter will ensure that the control device is operated in conformance with these standards and the control device's design specifications.

(j) An owner or operator of an affected facility seeking to comply with the provisions of this part by using a control device other than a thermal vapor incinerator, catalytic vapor incinerator, flare, boiler, process heater, condenser, or carbon adsorption system is required to develop documentation including sufficient information to describe the control device operation and identify the process parameter or parameters that indicate proper operation and maintenance of the control device.

(k) A closed-vent system shall meet either of the following design requirements:

(1) A closed-vent system shall be designed to operate with no detectable emissions, as indicated by an instrument reading of less than 500 ppmv above background as determined by the procedure in § 264.1034(b) of this subpart, and by visual inspections; or

(2) A closed-vent system shall be designed to operate at a pressure below atmospheric pressure. The system shall be equipped with at least one pressure gauge or other pressure measurement device that can be read from a readily accessible location to verify that negative pressure is being maintained in the closed-vent system when the control device is operating.

(l) The owner or operator shall monitor and inspect each closed-vent system required to comply with this sec-

tion to ensure proper operation and maintenance of the closed-vent system by implementing the following requirements:

(1) Each closed-vent system that is used to comply with paragraph (k)(1) of this section shall be inspected and monitored in accordance with the following requirements:

(i) An initial leak detection monitoring of the closed-vent system shall be conducted by the owner or operator on or before the date that the system becomes subject to this section. The owner or operator shall monitor the closed-vent system components and connections using the procedures specified in § 264.1034(b) of this subpart to demonstrate that the closed-vent system operates with no detectable emissions, as indicated by an instrument reading of less than 500 ppmv above background.

(ii) After initial leak detection monitoring required in paragraph (l)(1)(i) of this section, the owner or operator shall inspect and monitor the closed-vent system as follows:

(A) Closed-vent system joints, seams, or other connections that are permanently or semi-permanently sealed (e.g., a welded joint between two sections of hard piping or a bolted and gasketed ducting flange) shall be visually inspected at least once per year to check for defects that could result in air pollutant emissions. The owner or operator shall monitor a component or connection using the procedures specified in § 264.1034(b) of this subpart to demonstrate that it operates with no detectable emissions following any time the component is repaired or replaced (e.g., a section of damaged hard piping is replaced with new hard piping) or the connection is unsealed (e.g., a flange is unbolted).

(B) Closed-vent system components or connections other than those specified in paragraph (l)(1)(ii)(A) of this section shall be monitored annually and at other times as requested by the Regional Administrator, except as provided for in paragraph (o) of this section, using the procedures specified in § 264.1034(b) of this subpart to demonstrate that the components or connections operate with no detectable emissions.

(iii) In the event that a defect or leak is detected, the owner or operator shall repair the defect or leak in accordance with the requirements of paragraph (l)(3) of this section.

(iv) The owner or operator shall maintain a record of the inspection and monitoring in accordance with the requirements specified in §264.1035 of this subpart.

(2) Each closed-vent system that is used to comply with paragraph (k)(2) of this section shall be inspected and monitored in accordance with the following requirements:

(i) The closed-vent system shall be visually inspected by the owner or operator to check for defects that could result in air pollutant emissions. Defects include, but are not limited to, visible cracks, holes, or gaps in ductwork or piping or loose connections.

(ii) The owner or operator shall perform an initial inspection of the closed-vent system on or before the date that the system becomes subject to this section. Thereafter, the owner or operator shall perform the inspections at least once every year.

(iii) In the event that a defect or leak is detected, the owner or operator shall repair the defect in accordance with the requirements of paragraph (l)(3) of this section.

(iv) The owner or operator shall maintain a record of the inspection and monitoring in accordance with the requirements specified in §264.1035 of this subpart.

(3) The owner or operator shall repair all detected defects as follows:

(i) Detectable emissions, as indicated by visual inspection, or by an instrument reading greater than 500 ppmv above background, shall be controlled as soon as practicable, but not later than 15 calendar days after the emission is detected, except as provided for in paragraph (l)(3)(iii) of this section.

(ii) A first attempt at repair shall be made no later than 5 calendar days after the emission is detected.

(iii) Delay of repair of a closed-vent system for which leaks have been detected is allowed if the repair is technically infeasible without a process unit shutdown, or if the owner or operator determines that emissions resulting from immediate repair would be greater than the fugitive emissions likely to result from delay of repair. Repair of such equipment shall be completed by the end of the next process unit shutdown.

(iv) The owner or operator shall maintain a record of the defect repair in accordance with the requirements specified in §264.1035 of this subpart.

(m) Closed-vent systems and control devices used to comply with provisions of this subpart shall be operated at all times when emissions may be vented to them.

(n) The owner or operator using a carbon adsorption system to control air pollutant emissions shall document that all carbon that is a hazardous waste and that is removed from the control device is managed in one of the following manners, regardless of the average volatile organic concentration of the carbon:

(1) Regenerated or reactivated in a thermal treatment unit that meets one of the following:

(i) The owner or operator of the unit has been issued a final permit under 40 CFR part 270 which implements the requirements of subpart X of this part; or

(ii) The unit is equipped with and operating air emission controls in accordance with the applicable requirements of subparts AA and CC of either this part or of 40 CFR part 265; or

(iii) The unit is equipped with and operating air emission controls in accordance with a national emission standard for hazardous air pollutants under 40 CFR part 61 or 40 CFR part 63.

(2) Incinerated in a hazardous waste incinerator for which the owner or operator either:

(i) Has been issued a final permit under 40 CFR part 270 which implements the requirements of subpart O of this part; or

(ii) Has designed and operates the incinerator in accordance with the interim status requirements of 40 CFR part 265, subpart O.

(3) Burned in a boiler or industrial furnace for which the owner or operator either:

(i) Has been issued a final permit under 40 CFR part 270 which implements the requirements of 40 CFR part 266, subpart H; or

(ii) Has designed and operates the boiler or industrial furnace in accordance with the interim status requirements of 40 CFR part 266, subpart H.

(o) Any components of a closed-vent system that are designated, as described in § 264.1035(c)(9) of this subpart, as unsafe to monitor are exempt from the requirements of paragraph (l)(1)(ii)(B) of this section if:

(1) The owner or operator of the closed-vent system determines that the components of the closed-vent system are unsafe to monitor because monitoring personnel would be exposed to an immediate danger as a consequence of complying with paragraph (l)(1)(ii)(B) of this section; and

(2) The owner or operator of the closed-vent system adheres to a written plan that requires monitoring the closed-vent system components using the procedure specified in paragraph (l)(1)(ii)(B) of this section as frequently as practicable during safe-to-monitor times.

[55 FR 25494, June 21, 1990, as amended at 56 FR 19290, Apr. 26, 1991; 59 FR 62927, Dec. 6, 1994; 61 FR 4911, Feb. 9, 1996; 61 FR 59950, Nov. 25, 1996; 62 FR 64657, Dec. 8, 1997; 71 FR 40274, July 14, 2006]

§ 264.1034  Test methods and procedures.

(a) Each owner or operator subject to the provisions of this subpart shall comply with the test methods and procedures requirements provided in this section.

(b) When a closed-vent system is tested for compliance with no detectable emissions, as required in § 264.1033(l) of this subpart, the test shall comply with the following requirements:

(1) Monitoring shall comply with Reference Method 21 in 40 CFR part 60.

(2) The detection instrument shall meet the performance criteria of Reference Method 21.

(3) The instrument shall be calibrated before use on each day of its use by the procedures specified in Reference Method 21.

(4) Calibration gases shall be:

(i) Zero air (less than 10 ppm of hydrocarbon in air).

(ii) A mixture of methane or n-hexane and air at a concentration of approximately, but less than, 10,000 ppm methane or n-hexane.

(5) The background level shall be determined as set forth in Reference Method 21.

(6) The instrument probe shall be traversed around all potential leak interfaces as close to the interface as possible as described in Reference Method 21.

(7) The arithmetic difference between the maximum concentration indicated by the instrument and the background level is compared with 500 ppm for determining compliance.

(c) Performance tests to determine compliance with § 264.1032(a) and with the total organic compound concentration limit of § 264.1033(c) shall comply with the following:

(1) Performance tests to determine total organic compound concentrations and mass flow rates entering and exiting control devices shall be conducted and data reduced in accordance with the following reference methods and calculation procedures:

(i) Method 2 in 40 CFR part 60 for velocity and volumetric flow rate.

(ii) Method 18 or Method 25A in 40 CFR part 60, appendix A, for organic content. If Method 25A is used, the organic HAP used as the calibration gas must be the single organic HAP representing the largest percent by volume of the emissions. The use of Method 25A is acceptable if the response from the high-level calibration gas is at least 20 times the standard deviation of the response from the zero calibration gas when the instrument is zeroed on the most sensitive scale.

(iii) Each performance test shall consist of three separate runs; each run conducted for at least 1 hour under the conditions that exist when the hazardous waste management unit is operating at the highest load or capacity level reasonably expected to occur. For the purpose of determining total organic compound concentrations and mass flow rates, the average of results of all runs shall apply. The average shall be computed on a time-weighted basis.

(iv) Total organic mass flow rates shall be determined by the following equation:

(A) For sources utilizing Method 18.

$$E_h = Q_{2sd}\left\{\sum_{i=1}^{n} C_i MW_i\right\}[0.0416]\left[10^{-6}\right]$$

Where:

$E_h$ = Total organic mass flow rate, kg/h;
$Q_{2sd}$ = Volumetric flow rate of gases entering or exiting control device, as determined by Method 2, dscm/h;
n = Number of organic compounds in the vent gas;
$C_i$ = Organic concentration in ppm, dry basis, of compound i in the vent gas, as determined by Method 18;
$MW_i$ = Molecular weight of organic compound i in the vent gas, kg/kg-mol;
0.0416 = Conversion factor for molar volume, kg-mol/m3 (@ 293 K and 760 mm Hg);
$10^{-6}$ = Conversion from ppm

(B) For sources utilizing Method 25A.

$$E_h = (Q)(C)(MW)(0.0416)(10^{-6})$$

Where:

$E_h$ = Total organic mass flow rate, kg/h;
Q = Volumetric flow rate of gases entering or exiting control device, as determined by Method 2, dscm/h;
C = Organic concentration in ppm, dry basis, as determined by Method 25A;
MW = Molecular weight of propane, 44;
0.0416 = Conversion factor for molar volume, kg-mol/m3 (@ 293 K and 760 mm Hg);
$10^{-6}$ = Conversion from ppm.

(v) The annual total organic emission rate shall be determined by the following equation:

$$E_A = (E_h)(H)$$

where:

$E_A$ = Total organic mass emission rate, kg/y;
$E_h$ = Total organic mass flow rate for the process vent, kg/h;
H = Total annual hours of operations for the affected unit, h.

(vi) Total organic emissions from all affected process vents at the facility shall be determined by summing the hourly total organic mass emission rates ($E_h$ as determined in paragraph (c)(1)(iv) of this section) and by summing the annual total organic mass emission rates ($E_A$, as determined in paragraph (c)(1)(v) of this section) for all affected process vents at the facility.

(2) The owner or operator shall record such process information as may be necessary to determine the conditions of the performance tests. Operations during periods of startup, shutdown, and malfunction shall not constitute representative conditions for the purpose of a performance test.

(3) The owner or operator of an affected facility shall provide, or cause to be provided, performance testing facilities as follows:

(i) Sampling ports adequate for the test methods specified in paragraph (c)(1) of this section.

(ii) Safe sampling platform(s).

(iii) Safe access to sampling platform(s).

(iv) Utilities for sampling and testing equipment.

(4) For the purpose of making compliance determinations, the time-weighted average of the results of the three runs shall apply. In the event that a sample is accidentally lost or conditions occur in which one of the three runs must be discontinued because of forced shutdown, failure of an irreplaceable portion of the sample train, extreme meteorological conditions, or other circumstances beyond the owner or operator's control, compliance may, upon the Regional Administrator's approval, be determined using the average of the results of the two other runs.

(d) To show that a process vent associated with a hazardous waste distillation, fractionation, thin-film evaporation, solvent extraction, or air or steam stripping operation is not subject to the requirements of this subpart, the owner or operator must make an initial determination that the time-weighted, annual average total organic concentration of the waste managed by the waste management unit is less than 10 ppmw using one of the following two methods:

(1) Direct measurement of the organic concentration of the waste using the following procedures:

(i) The owner or operator must take a minimum of four grab samples of waste for each waste stream managed in the affected unit under process conditions expected to cause the maximum waste organic concentration.

(ii) For waste generated onsite, the grab samples must be collected at a point before the waste is exposed to the atmosphere such as in an enclosed pipe or other closed system that is used to transfer the waste after generation to the first affected distillation, fractionation, thin-film evaporation, solvent

507

extraction, or air or steam stripping operation. For waste generated offsite, the grab samples must be collected at the inlet to the first waste management unit that receives the waste provided the waste has been transferred to the facility in a closed system such as a tank truck and the waste is not diluted or mixed with other waste.

(iii) Each sample shall be analyzed and the total organic concentration of the sample shall be computed using Method 9060A (incorporated by reference under 40 CFR 260.11) of "Test Methods for Evaluating Solid Waste, Physical/Chemical Methods," EPA Publication SW–846, or analyzed for its individual organic constituents.

(iv) The arithmetic mean of the results of the analyses of the four samples shall apply for each waste stream managed in the unit in determining the time-weighted, annual average total organic concentration of the waste. The time-weighted average is to be calculated using the annual quantity of each waste stream processed and the mean organic concentration of each waste stream managed in the unit.

(2) Using knowledge of the waste to determine that its total organic concentration is less than 10 ppmw. Documentation of the waste determination is required. Examples of documentation that shall be used to support a determination under this provision include production process information documenting that no organic compounds are used, information that the waste is generated by a process that is identical to a process at the same or another facility that has previously been demonstrated by direct measurement to generate a waste stream having a total organic content less than 10 ppmw, or prior speciation analysis results on the same waste stream where it can also be documented that no process changes have occurred since that analysis that could affect the waste total organic concentration.

(e) The determination that distillation, fractionation, thin-film evaporation, solvent extraction, or air or steam stripping operations manage hazardous wastes with time-weighted, annual average total organic concentrations less than 10 ppmw shall be made as follows:

(1) By the effective date that the facility becomes subject to the provisions of this subpart or by the date when the waste is first managed in a waste management unit, whichever is later, and

(2) For continuously generated waste, annually, or

(3) Whenever there is a change in the waste being managed or a change in the process that generates or treats the waste.

(f) When an owner or operator and the Regional Administrator do not agree on whether a distillation, fractionation, thin-film evaporation, solvent extraction, or air or steam stripping operation manages a hazardous waste with organic concentrations of at least 10 ppmw based on knowledge of the waste, the dispute may be resolved by using direct measurement as specified at paragraph (d)(1) of this section.

[55 FR 25494, June 21, 1990, as amended at 61 FR 59951, Nov. 25, 1996; 62 FR 32462, June 13, 1997; 70 FR 34581, June 14, 2005; 71 FR 40274, July 14, 2006]

## § 264.1035 Recordkeeping requirements.

(a)(1) Each owner or operator subject to the provisions of this subpart shall comply with the recordkeeping requirements of this section.

(2) An owner or operator of more than one hazardous waste management unit subject to the provisions of this subpart may comply with the recordkeeping requirements for these hazardous waste management units in one recordkeeping system if the system identifies each record by each hazardous waste management unit.

(b) Owners and operators must record the following information in the facility operating record:

(1) For facilities that comply with the provisions of § 264.1033(a)(2), an implementation schedule that includes dates by which the closed-vent system and control device will be installed and in operation. The schedule must also include a rationale of why the installation cannot be completed at an earlier date. The implementation schedule must be in the facility operating record by the effective date that the facility becomes subject to the provisions of this subpart.

(2) Up-to-date documentation of compliance with the process vent standards in §264.1032, including:

(i) Information and data identifying all affected process vents, annual throughput and operating hours of each affected unit, estimated emission rates for each affected vent and for the overall facility (i.e., the total emissions for all affected vents at the facility), and the approximate location within the facility of each affected unit (e.g., identify the hazardous waste management units on a facility plot plan).

(ii) Information and data supporting determinations of vent emissions and emission reductions achieved by add-on control devices based on engineering calculations or source tests. For the purpose of determining compliance, determinations of vent emissions and emission reductions must be made using operating parameter values (e.g., temperatures, flow rates, or vent stream organic compounds and concentrations) that represent the conditions that result in maximum organic emissions, such as when the waste management unit is operating at the highest load or capacity level reasonably expected to occur. If the owner or operator takes any action (e.g., managing a waste of different composition or increasing operating hours of affected waste management units) that would result in an increase in total organic emissions from affected process vents at the facility, then a new determination is required.

(3) Where an owner or operator chooses to use test data to determine the organic removal efficiency or total organic compound concentration achieved by the control device, a performance test plan. The test plan must include:

(i) A description of how it is determined that the planned test is going to be conducted when the hazardous waste management unit is operating at the highest load or capacity level reasonably expected to occur. This shall include the estimated or design flow rate and organic content of each vent stream and define the acceptable operating ranges of key process and control device parameters during the test program.

(ii) A detailed engineering description of the closed-vent system and control device including:

(A) Manufacturer's name and model number of control device.

(B) Type of control device.

(C) Dimensions of the control device.

(D) Capacity.

(E) Construction materials.

(iii) A detailed description of sampling and monitoring procedures, including sampling and monitoring locations in the system, the equipment to be used, sampling and monitoring frequency, and planned analytical procedures for sample analysis.

(4) Documentation of compliance with §264.1033 shall include the following information:

(i) A list of all information references and sources used in preparing the documentation.

(ii) Records, including the dates, of each compliance test required by §264.1033(k).

(iii) If engineering calculations are used, a design analysis, specifications, drawings, schematics, and piping and instrumentation diagrams based on the appropriate sections of "APTI Course 415: Control of Gaseous Emissions" (incorporated by reference as specified in §260.11) or other engineering texts acceptable to the Regional Administrator that present basic control device design information. Documentation provided by the control device manufacturer or vendor that describes the control device design in accordance with paragraphs (b)(4)(iii)(A) through (b)(4)(iii)(G) of this section may be used to comply with this requirement. The design analysis shall address the vent stream characteristics and control device operation parameters as specified below.

(A) For a thermal vapor incinerator, the design analysis shall consider the vent stream composition, constituent concentrations, and flow rate. The design analysis shall also establish the design minimum and average temperature in the combustion zone and the combustion zone residence time.

(B) For a catalytic vapor incinerator, the design analysis shall consider the vent stream composition, constituent concentrations, and flow rate. The design analysis shall also establish the design minimum and average temperatures across the catalyst bed inlet and outlet.

(C) For a boiler or process heater, the design analysis shall consider the vent stream composition, constituent concentrations, and flow rate. The design analysis shall also establish the design minimum and average flame zone temperatures, combustion zone residence time, and description of method and location where the vent stream is introduced into the combustion zone.

(D) For a flare, the design analysis shall consider the vent stream composition, constituent concentrations, and flow rate. The design analysis shall also consider the requirements specified in § 264.1033(d).

(E) For a condenser, the design analysis shall consider the vent stream composition, constituent concentrations, flow rate, relative humidity, and temperature. The design analysis shall also establish the design outlet organic compound concentration level, design average temperature of the condenser exhaust vent stream, and design average temperatures of the coolant fluid at the condenser inlet and outlet.

(F) For a carbon adsorption system such as a fixed-bed adsorber that regenerates the carbon bed directly onsite in the control device, the design analysis shall consider the vent stream composition, constituent concentrations, flow rate, relative humidity, and temperature. The design analysis shall also establish the design exhaust vent stream organic compound concentration level, number and capacity of carbon beds, type and working capacity of activated carbon used for carbon beds, design total steam flow over the period of each complete carbon bed regeneration cycle, duration of the carbon bed steaming and cooling/drying cycles, design carbon bed temperature after regeneration, design carbon bed regeneration time, and design service life of carbon.

(G) For a carbon adsorption system such as a carbon canister that does not regenerate the carbon bed directly onsite in the control device, the design analysis shall consider the vent stream composition, constituent concentrations, flow rate, relative humidity, and temperature. The design analysis shall

also establish the design outlet organic concentration level, capacity of carbon bed, type and working capacity of activated carbon used for carbon bed, and design carbon replacement interval based on the total carbon working capacity of the control device and source operating schedule.

(iv) A statement signed and dated by the owner or operator certifying that the operating parameters used in the design analysis reasonably represent the conditions that exist when the hazardous waste management unit is or would be operating at the highest load or capacity level reasonably expected to occur.

(v) A statement signed and dated by the owner or operator certifying that the control device is designed to operate at an efficiency of 95 percent or greater unless the total organic concentration limit of § 264.1032(a) is achieved at an efficiency less than 95 weight percent or the total organic emission limits of § 264.1032(a) for affected process vents at the facility can be attained by a control device involving vapor recovery at an efficiency less than 95 weight percent. A statement provided by the control device manufacturer or vendor certifying that the control equipment meets the design specifications may be used to comply with this requirement.

(vi) If performance tests are used to demonstrate compliance, all test results.

(c) Design documentation and monitoring, operating, and inspection information for each closed-vent system and control device required to comply with the provisions of this part shall be recorded and kept up-to-date in the facility operating record. The information shall include:

(1) Description and date of each modification that is made to the closed-vent system or control device design.

(2) Identification of operating parameter, description of monitoring device, and diagram of monitoring sensor location or locations used to comply with § 264.1033 (f)(1) and (f)(2).

(3) Monitoring, operating, and inspection information required by paragraphs (f) through (k) of § 264.1033.

(4) Date, time, and duration of each period that occurs while the control device is operating when any monitored parameter exceeds the value established in the control device design analysis as specified below:

(i) For a thermal vapor incinerator designed to operate with a minimum residence time of 0.50 second at a minimum temperature of 760 °C, period when the combustion temperature is below 760 °C.

(ii) For a thermal vapor incinerator designed to operate with an organic emission reduction efficiency of 95 weight percent or greater, period when the combustion zone temperature is more than 28 °C below the design average combustion zone temperature established as a requirement of paragraph (b)(4)(iii)(A) of this section.

(iii) For a catalytic vapor incinerator, period when:

(A) Temperature of the vent stream at the catalyst bed inlet is more than 28 °C below the average temperature of the inlet vent stream established as a requirement of paragraph (b)(4)(iii)(B) of this section, or

(B) Temperature difference across the catalyst bed is less than 80 percent of the design average temperature difference established as a requirement of paragraph (b)(4)(iii)(B) of this section.

(iv) For a boiler or process heater, period when:

(A) Flame zone temperature is more than 28 °C below the design average flame zone temperature established as a requirement of paragraph (b)(4)(iii)(C) of this section, or

(B) Position changes where the vent stream is introduced to the combustion zone from the location established as a requirement of paragraph (b)(4)(iii)(C) of this section.

(v) For a flare, period when the pilot flame is not ignited.

(vi) For a condenser that complies with §264.1033(f)(2)(vi)(A), period when the organic compound concentration level or readings of organic compounds in the exhaust vent stream from the condenser are more than 20 percent greater than the design outlet organic compound concentration level established as a requirement of paragraph (b)(4)(iii)(E) of this section.

(vii) For a condenser that complies with §264.1033(f)(2)(vi)(B), period when:

(A) Temperature of the exhaust vent stream from the condenser is more than 6 °C above the design average exhaust vent stream temperature established as a requirement of paragraph (b)(4)(iii)(E) of this section; or

(B) Temperature of the coolant fluid exiting the condenser is more than 6 °C above the design average coolant fluid temperature at the condenser outlet established as a requirement of paragraph (b)(4)(iii)(E) of this section.

(viii) For a carbon adsorption system such as a fixed-bed carbon adsorber that regenerates the carbon bed directly onsite in the control device and complies with §264.1033(f)(2)(vii)(A), period when the organic compound concentration level or readings of organic compounds in the exhaust vent stream from the carbon bed are more than 20 percent greater than the design exhaust vent stream organic compound concentration level established as a requirement of paragraph (b)(4)(iii)(F) of this section.

(ix) For a carbon adsorption system such as a fixed-bed carbon adsorber that regenerates the carbon bed directly onsite in the control device and complies with §264.1033(f)(2)(vii)(B), period when the vent stream continues to flow through the control device beyond the predetermined carbon bed regeneration time established as a requirement of paragraph (b)(4)(iii)(F) of this section.

(5) Explanation for each period recorded under paragraph (4) of the cause for control device operating parameter exceeding the design value and the measures implemented to correct the control device operation.

(6) For a carbon adsorption system operated subject to requirements specified in §264.1033(g) or §264.1033(h)(2), date when existing carbon in the control device is replaced with fresh carbon.

(7) For a carbon adsorption system operated subject to requirements specified in §264.1033(h)(1), a log that records:

(i) Date and time when control device is monitored for carbon breakthrough and the monitoring device reading.

(ii) Date when existing carbon in the control device is replaced with fresh carbon.

(8) Date of each control device start-up and shutdown.

(9) An owner or operator designating any components of a closed-vent system as unsafe to monitor pursuant to § 264.1033(o) of this subpart shall record in a log that is kept in the facility operating record the identification of closed-vent system components that are designated as unsafe to monitor in accordance with the requirements of § 264.1033(o) of this subpart, an explanation for each closed-vent system component stating why the closed-vent system component is unsafe to monitor, and the plan for monitoring each closed-vent system component.

(10) When each leak is detected as specified in § 264.1033(l) of this subpart, the following information shall be recorded:

(i) The instrument identification number, the closed-vent system component identification number, and the operator name, initials, or identification number.

(ii) The date the leak was detected and the date of first attempt to repair the leak.

(iii) The date of successful repair of the leak.

(iv) Maximum instrument reading measured by Method 21 of 40 CFR part 60, appendix A after it is successfully repaired or determined to be nonrepairable.

(v) "Repair delayed" and the reason for the delay if a leak is not repaired within 15 calendar days after discovery of the leak.

(A) The owner or operator may develop a written procedure that identifies the conditions that justify a delay of repair. In such cases, reasons for delay of repair may be documented by citing the relevant sections of the written procedure.

(B) If delay of repair was caused by depletion of stocked parts, there must be documentation that the spare parts were sufficiently stocked on-site before depletion and the reason for depletion.

(d) Records of the monitoring, operating, and inspection information required by paragraphs (c)(3) through (c)(10) of this section shall be maintained by the owner or operator for at least 3 years following the date of each occurrence, measurement, maintenance, corrective action, or record.

(e) For a control device other than a thermal vapor incinerator, catalytic vapor incinerator, flare, boiler, process heater, condenser, or carbon adsorption system, the Regional Administrator will specify the appropriate record-keeping requirements.

(f) Up-to-date information and data used to determine whether or not a process vent is subject to the requirements in § 264.1032 including supporting documentation as required by § 264.1034(d)(2) when application of the knowledge of the nature of the hazardous waste stream or the process by which it was produced is used, shall be recorded in a log that is kept in the facility operating record.

[55 FR 25494, June 21, 1990, as amended at 56 FR 19290, Apr. 26, 1991; 61 FR 59952, Nov. 25, 1996; 71 FR 40274, July 14, 2006]

## § 264.1036   Reporting requirements.

(a) A semiannual report shall be submitted by owners and operators subject to the requirements of this subpart to the Regional Administrator by dates specified by the Regional Administrator. The report shall include the following information:

(1) The Environmental Protection Agency identification number, name, and address of the facility.

(2) For each month during the semiannual reporting period, dates when the control device exceeded or operated outside of the design specifications as defined in § 264.1035(c)(4) and as indicated by the control device monitoring required by § 264.1033(f) and such exceedances were not corrected within 24 hours, or that a flare operated with visible emissions as defined in § 264.1033(d) and as determined by Method 22 monitoring, the duration and cause of each exceedance or visible emissions, and any corrective measures taken.

(b) If, during the semiannual reporting period, the control device does not exceed or operate outside of the design specifications as defined in § 264.1035(c)(4) for more than 24 hours or a flare does not operate with visible emissions as defined in § 264.1033(d), a report to the Regional Administrator is not required.

**§§ 264.1037–264.1049  [Reserved]**

## Subpart BB—Air Emission Standards for Equipment Leaks

SOURCE: 55 FR 25501, June 21, 1990, unless otherwise noted.

### § 264.1050  Applicability.

(a) The regulations in this subpart apply to owners and operators of facilities that treat, store, or dispose of hazardous wastes (except as provided in § 264.1).

(b) Except as provided in § 264.1064(k), this subpart applies to equipment that contains or contacts hazardous wastes with organic concentrations of at least 10 percent by weight that are managed in one of the following:

(1) A unit that is subject to the permitting requirements of 40 CFR part 270, or

(2) A unit (including a hazardous waste recycling unit) that is not exempt from permitting under the provisions of 40 CFR 262.34(a) (i.e., a hazardous waste recycling unit that is not a "90-day" tank or container) and that is located at a hazardous waste management facility otherwise subject to the permitting requirements of 40 CFR part 270, or

(3) A unit that is exempt from permitting under the provisions of 40 CFR 262.34(a) (i.e., a "90-day" tank or container) and is not a recycling unit under the provisions of 40 CFR 261.6.

(c) For the owner or operator of a facility subject to this subpart and who received a final permit under RCRA section 3005 prior to December 6, 1996, the requirements of this subpart shall be incorporated into the permit when the permit is reissued in accordance with the requirements of 40 CFR 124.15 or reviewed in accordance with the requirements of 40 CFR 270.50(d). Until such date when the owner or operator receives a final permit incorporating the requirements of this subpart, the owner or operator is subject to the requirements of 40 CFR part 265, subpart BB.

(d) Each piece of equipment to which this subpart applies shall be marked in such a manner that it can be distinguished readily from other pieces of equipment.

(e) Equipment that is in vacuum service is excluded from the requirements of § 264.1052 to § 264.1060 if it is identified as required in § 264.1064(g)(5).

(f) Equipment that contains or contacts hazardous waste with an organic concentration of at least 10 percent by weight for less than 300 hours per calendar year is excluded from the requirements of §§ 264.1052 through 264.1060 of this subpart if it is identified, as required in § 264.1064(g)(6) of this subpart.

(g) The requirements of this subpart do not apply to the pharmaceutical manufacturing facility, commonly referred to as the Stonewall Plant, located at Route 340 South, Elkton, Virginia, provided that facility is operated in compliance with the requirements contained in a Clean Air Act permit issued pursuant to 40 CFR 52.2454. The requirements of this subpart shall apply to the facility upon termination of the Clean Air Act permit issued pursuant to 40 CFR 52.2454.

(h) Purged coatings and solvents from surface coating operations subject to the national emission standards for hazardous air pollutants (NESHAP) for the surface coating of automobiles and light-duty trucks at 40 CFR part 63, subpart IIII, are not subject to the requirements of this subpart.

[NOTE: The requirements of §§ 264.1052 through 264.1065 apply to equipment associated with hazardous waste recycling units previously exempt under § 261.6(c)(1). Other exemptions under §§ 261.4, and 264.1(g) are not affected by these requirements.]

[55 FR 25501, June 21, 1990, as amended at 61 FR 59952, Nov. 25, 1996; 62 FR 52641, Oct. 8, 1997; 62 FR 64657, Dec. 8, 1997; 69 FR 22661, Apr. 26, 2004; 71 FR 40274, July 14, 2006]

### § 264.1051  Definitions.

As used in this subpart, all terms shall have the meaning given them in § 264.1031, the Act, and parts 260–266.

### § 264.1052  Standards: Pumps in light liquid service.

(a)(1) Each pump in light liquid service shall be monitored monthly to detect leaks by the methods specified in

§ 264.1063(b), except as provided in paragraphs (d), (e), and (f) of this section.

(2) Each pump in light liquid service shall be checked by visual inspection each calendar week for indications of liquids dripping from the pump seal.

(b)(1) If an instrument reading of 10,000 ppm or greater is measured, a leak is detected.

(2) If there are indications of liquids dripping from the pump seal, a leak is detected.

(c)(1) When a leak is detected, it shall be repaired as soon as practicable, but not later than 15 calendar days after it is detected, except as provided in § 264.1059.

(2) A first attempt at repair (e.g., tightening the packing gland) shall be made no later than 5 calendar days after each leak is detected.

(d) Each pump equipped with a dual mechanical seal system that includes a barrier fluid system is exempt from the requirements of paragraph (a) of this section, *provided* the following requirements are met:

(1) Each dual mechanical seal system must be:

(i) Operated with the barrier fluid at a pressure that is at all times greater than the pump stuffing box pressure, or

(ii) Equipped with a barrier fluid degassing reservoir that is connected by a closed-vent system to a control device that complies with the requirements of § 264.1060, or

(iii) Equipped with a system that purges the barrier fluid into a hazardous waste stream with no detectable emissions to the atmosphere.

(2) The barrier fluid system must not be a hazardous waste with organic concentrations 10 percent or greater by weight.

(3) Each barrier fluid system must be equipped with a sensor that will detect failure of the seal system, the barrier fluid system, or both.

(4) Each pump must be checked by visual inspection, each calendar week, for indications of liquids dripping from the pump seals.

(5)(i) Each sensor as described in paragraph (d)(3) of this section must be checked daily or be equipped with an audible alarm that must be checked monthly to ensure that it is functioning properly.

(ii) The owner or operator must determine, based on design considerations and operating experience, a criterion that indicates failure of the seal system, the barrier fluid system, or both.

(6)(i) If there are indications of liquids dripping from the pump seal or the sensor indicates failure of the seal system, the barrier fluid system, or both based on the criterion determined in paragraph (d)(5)(ii) of this section, a leak is detected.

(ii) When a leak is detected, it shall be repaired as soon as practicable, but not later than 15 calendar days after it is detected, except as provided in § 264.1059.

(iii) A first attempt at repair (e.g., relapping the seal) shall be made no later than 5 calendar days after each leak is detected.

(e) Any pump that is designated, as described in § 264.1064(g)(2), for no detectable emissions, as indicated by an instrument reading of less than 500 ppm above background, is exempt from the requirements of paragraphs (a), (c), and (d) of this section if the pump meets the following requirements:

(1) Must have no externally actuated shaft penetrating the pump housing.

(2) Must operate with no detectable emissions as indicated by an instrument reading of less than 500 ppm above background as measured by the methods specified in § 264.1063(c).

(3) Must be tested for compliance with paragraph (e)(2) of this section initially upon designation, annually, and at other times as requested by the Regional Administrator.

(f) If any pump is equipped with a closed-vent system capable of capturing and transporting any leakage from the seal or seals to a control device that complies with the requirements of § 264.1060, it is exempt from the requirements of paragraphs (a) through (e) of this section.

[55 FR 25501, June 21, 1990, as amended at 56 FR 19290, Apr. 26, 1991]

## § 264.1053   Standards: Compressors.

(a) Each compressor shall be equipped with a seal system that includes a barrier fluid system and that prevents leakage of total organic emissions to the atmosphere, except as provided in paragraphs (h) and (i) of this section.

(b) Each compressor seal system as required in paragraph (a) of this section shall be:

(1) Operated with the barrier fluid at a pressure that is at all times greater than the compressor stuffing box pressure, or

(2) Equipped with a barrier fluid system that is connected by a closed-vent system to a control device that complies with the requirements of §264.1060, or

(3) Equipped with a system that purges the barrier fluid into a hazardous waste stream with no detectable emissions to atmosphere.

(c) The barrier fluid must not be a hazardous waste with organic concentrations 10 percent or greater by weight.

(d) Each barrier fluid system as described in paragraphs (a) through (c) of this section shall be equipped with a sensor that will detect failure of the seal system, barrier fluid system, or both.

(e)(1) Each sensor as required in paragraph (d) of this section shall be checked daily or shall be equipped with an audible alarm that must be checked monthly to ensure that it is functioning properly unless the compressor is located within the boundary of an unmanned plant site, in which case the sensor must be checked daily.

(2) The owner or operator shall determine, based on design considerations and operating experience, a criterion that indicates failure of the seal system, the barrier fluid system, or both.

(f) If the sensor indicates failure of the seal system, the barrier fluid system, or both based on the criterion determined under paragraph (e)(2) of this section, a leak is detected.

(g)(1) When a leak is detected, it shall be repaired as soon as practicable, but not later than 15 calendar days after it is detected, except as provided in §264.1059.

(2) A first attempt at repair (e.g., tightening the packing gland) shall be made no later than 5 calendar days after each leak is detected.

(h) A compressor is exempt from the requirements of paragraphs (a) and (b) of this section if it is equipped with a closed-vent system capable of capturing and transporting any leakage from the seal to a control device that complies with the requirements of §264.1060, except as provided in paragraph (i) of this section.

(i) Any compressor that is designated, as described in §264.1064(g)(2), for no detectable emissions as indicated by an instrument reading of less than 500 ppm above background is exempt from the requirements of paragraphs (a) through (h) of this section if the compressor:

(1) Is determined to be operating with no detectable emissions, as indicated by an instrument reading of less than 500 ppm above background, as measured by the method specified in §264.1063(c).

(2) Is tested for compliance with paragraph (i)(1) of this section initially upon designation, annually, and at other times as requested by the Regional Administrator.

### §264.1054 Standards: Pressure relief devices in gas/vapor service.

(a) Except during pressure releases, each pressure relief device in gas/vapor service shall be operated with no detectable emissions, as indicated by an instrument reading of less than 500 ppm above background, as measured by the method specified in §264.1063(c).

(b)(1) After each pressure release, the pressure relief device shall be returned to a condition of no detectable emissions, as indicated by an instrument reading of less than 500 ppm above background, as soon as practicable, but no later than 5 calendar days after each pressure release, except as provided in §264.1059.

(2) No later than 5 calendar days after the pressure release, the pressure relief device shall be monitored to confirm the condition of no detectable emissions, as indicated by an instrument reading of less than 500 ppm above background, as measured by the method specified in §264.1063(c).

(c) Any pressure relief device that is equipped with a closed-vent system capable of capturing and transporting leakage from the pressure relief device to a control device as described in §264.1060 is exempt from the requirements of paragraphs (a) and (b) of this section.

## § 264.1055 Standards: Sampling connection systems.

(a) Each sampling connection system shall be equipped with a closed-purge, closed-loop, or closed-vent system. This system shall collect the sample purge for return to the process or for routing to the appropriate treatment system. Gases displaced during filling of the sample container are not required to be collected or captured.

(b) Each closed-purge, closed-loop, or closed-vent system as required in paragraph (a) of this section shall meet one of the following requirements:

(1) Return the purged process fluid directly to the process line;

(2) Collect and recycle the purged process fluid; or

(3) Be designed and operated to capture and transport all the purged process fluid to a waste management unit that complies with the applicable requirements of § 264.1084 through § 264.1086 of this subpart or a control device that complies with the requirements of § 264.1060 of this subpart.

(c) *In-situ* sampling systems and sampling systems without purges are exempt from the requirements of paragraphs (a) and (b) of this section.

[61 FR 59952, Nov. 25, 1996]

## § 264.1056 Standards: Open-ended valves or lines.

(a)(1) Each open-ended valve or line shall be equipped with a cap, blind flange, plug, or a second valve.

(2) The cap, blind flange, plug, or second valve shall seal the open end at all times except during operations requiring hazardous waste stream flow through the open-ended valve or line.

(b) Each open-ended valve or line equipped with a second valve shall be operated in a manner such that the valve on the hazardous waste stream end is closed before the second valve is closed.

(c) When a double block and bleed system is being used, the bleed valve or line may remain open during operations that require venting the line between the block valves but shall comply with paragraph (a) of this section at all other times.

## § 264.1057 Standards: Valves in gas/vapor service or in light liquid service.

(a) Each valve in gas/vapor or light liquid service shall be monitored monthly to detect leaks by the methods specified in §264.1063(b) and shall comply with paragraphs (b) through (e) of this section, except as provided in paragraphs (f), (g), and (h) of this section, and §§264.1061 and 264.1062.

(b) If an instrument reading of 10,000 ppm or greater is measured, a leak is detected.

(c)(1) Any valve for which a leak is not detected for two successive months may be monitored the first month of every succeeding quarter, beginning with the next quarter, until a leak is detected.

(2) If a leak is detected, the valve shall be monitored monthly until a leak is not detected for two successive months.

(d)(1) When a leak is detected, it shall be repaired as soon as practicable, but no later than 15 calendar days after the leak is detected, except as provided in § 264.1059.

(2) A first attempt at repair shall be made no later than 5 calendar days after each leak is detected.

(e) First attempts at repair include, but are not limited to, the following best practices where practicable:

(1) Tightening of bonnet bolts.

(2) Replacement of bonnet bolts.

(3) Tightening of packing gland nuts.

(4) Injection of lubricant into lubricated packing.

(f) Any valve that is designated, as described in § 264.1064(g)(2), for no detectable emissions, as indicated by an instrument reading of less than 500 ppm above background, is exempt from the requirements of paragraph (a) of this section if the valve:

(1) Has no external actuating mechanism in contact with the hazardous waste stream.

(2) Is operated with emissions less than 500 ppm above background as determined by the method specified in § 264.1063(c).

(3) Is tested for compliance with paragraph (f)(2) of this section initially

upon designation, annually, and at other times as requested by the Regional Administrator.

(g) Any valve that is designated, as described in § 264.1064(h)(1), as an unsafe-to-monitor valve is exempt from the requirements of paragraph (a) of this section if:

(1) The owner or operator of the valve determines that the valve is unsafe to monitor because monitoring personnel would be exposed to an immediate danger as a consequence of complying with paragraph (a) of this section.

(2) The owner or operator of the valve adheres to a written plan that requires monitoring of the valve as frequently as practicable during safe-to-monitor times.

(h) Any valve that is designated, as described in § 264.1064(h)(2), as a difficult-to-monitor valve is exempt from the requirements of paragraph (a) of this section if:

(1) The owner or operator of the valve determines that the valve cannot be monitored without elevating the monitoring personnel more than 2 meters above a support surface.

(2) The hazardous waste management unit within which the valve is located was in operation before June 21, 1990.

(3) The owner or operator of the valve follows a written plan that requires monitoring of the valve at least once per calendar year.

## § 264.1058 Standards: Pumps and valves in heavy liquid service, pressure relief devices in light liquid or heavy liquid service, and flanges and other connectors.

(a) Pumps and valves in heavy liquid service, pressure relief devices in light liquid or heavy liquid service, and flanges and other connectors shall be monitored within 5 days by the method specified in § 264.1063(b) if evidence of a potential leak is found by visual, audible, olfactory, or any other detection method.

(b) If an instrument reading of 10,000 ppm or greater is measured, a leak is detected.

(c)(1) When a leak is detected, it shall be repaired as soon as practicable, but not later than 15 calendar days after it is detected, except as provided in § 264.1059.

(2) The first attempt at repair shall be made no later than 5 calendar days after each leak is detected.

(d) First attempts at repair include, but are not limited to, the best practices described under § 264.1057(e).

(e) Any connector that is inaccessible or is ceramic or ceramic-lined (e.g., porcelain, glass, or glass-lined) is exempt from the monitoring requirements of paragraph (a) of this section and from the recordkeeping requirements of § 264.1064 of this subpart.

[55 FR 25501, June 21, 1990, as amended at 61 FR 59952, Nov. 25, 1996; 71 FR 40274, July 14, 2006]

## § 264.1059 Standards: Delay of repair.

(a) Delay of repair of equipment for which leaks have been detected will be allowed if the repair is technically infeasible without a hazardous waste management unit shutdown. In such a case, repair of this equipment shall occur before the end of the next hazardous waste management unit shutdown.

(b) Delay of repair of equipment for which leaks have been detected will be allowed for equipment that is isolated from the hazardous waste management unit and that does not continue to contain or contact hazardous waste with organic concentrations at least 10 percent by weight.

(c) Delay of repair for valves will be allowed if:

(1) The owner or operator determines that emissions of purged material resulting from immediate repair are greater than the emissions likely to result from delay of repair.

(2) When repair procedures are effected, the purged material is collected and destroyed or recovered in a control device complying with § 264.1060.

(d) Delay of repair for pumps will be allowed if:

(1) Repair requires the use of a dual mechanical seal system that includes a barrier fluid system.

(2) Repair is completed as soon as practicable, but not later than 6 months after the leak was detected.

(e) Delay of repair beyond a hazardous waste management unit shutdown will be allowed for a valve if valve assembly replacement is necessary during the hazardous waste management unit shutdown, valve assembly supplies have been depleted, and valve assembly supplies had been sufficiently stocked before the supplies were depleted. Delay of repair beyond the next hazardous waste management unit shutdown will not be allowed unless the next hazardous waste management unit shutdown occurs sooner than 6 months after the first hazardous waste management unit shutdown.

### § 264.1060 Standards: Closed-vent systems and control devices.

(a) Owners and operators of closed-vent systems and control devices subject to this subpart shall comply with the provisions of § 264.1033 of this part.

(b)(1) The owner or operator of an existing facility who cannot install a closed-vent system and control device to comply with the provisions of this subpart on the effective date that the facility becomes subject to the provisions of this subpart must prepare an implementation schedule that includes dates by which the closed-vent system and control device will be installed and in operation. The controls must be installed as soon as possible, but the implementation schedule may allow up to 30 months after the effective date that the facility becomes subject to this subpart for installation and startup.

(2) Any unit that begins operation after December 21, 1990, and is subject to the provisions of this subpart when operation begins, must comply with the rules immediately (i.e., must have control devices installed and operating on startup of the affected unit); the 30-month implementation schedule does not apply.

(3) The owner or operator of any facility in existence on the effective date of a statutory or EPA regulatory amendment that renders the facility subject to this subpart shall comply with all requirements of this subpart as soon as practicable but no later than 30 months after the amendment's effective date. When control equipment required by this subpart can not be installed and begin operation by the effective date of the amendment, the facility owner or operator shall prepare an implementation schedule that includes the following information: Specific calendar dates for award or contracts or issuance of purchase orders for the control equipment, initiation of on-site installation of the control equipment, completion of the control equipment installation, and performance of any testing to demonstrate that the installed equipment meets the applicable standards of this subpart. The owner or operator shall enter the implementation schedule in the operating record or in a permanent, readily available file located at the facility.

(4) Owners and operators of facilities and units that become newly subject to the requirements of this subpart after December 8, 1997, due to an action other than those described in paragraph (b)(3) of this section must comply with all applicable requirements immediately (i.e., must have control devices installed and operating on the date the facility or unit becomes subject to this subpart; the 30-month implementation schedule does not apply).

[62 FR 64657, Dec. 8, 1997]

### § 264.1061 Alternative standards for valves in gas/vapor service or in light liquid service: percentage of valves allowed to leak.

(a) An owner or operator subject to the requirements of § 264.1057 may elect to have all valves within a hazardous waste management unit comply with an alternative standard that allows no greater than 2 percent of the valves to leak.

(b) The following requirements shall be met if an owner or operator decides to comply with the alternative standard of allowing 2 percent of valves to leak:

(1) A performance test as specified in paragraph (c) of this section shall be conducted initially upon designation, annually, and at other times requested by the Regional Administrator.

(2) If a valve leak is detected, it shall be repaired in accordance with § 264.1057(d) and (e).

(c) Performance tests shall be conducted in the following manner:

(1) All valves subject to the requirements in § 264.1057 within the hazardous waste management unit shall be monitored within 1 week by the methods specified in § 264.1063(b).

(2) If an instrument reading of 10,000 ppm or greater is measured, a leak is detected.

(3) The leak percentage shall be determined by dividing the number of valves subject to the requirements in §264.1057 for which leaks are detected by the total number of valves subject to the requirements in §264.1057 within the hazardous waste management unit.

[55 FR 25501, June 21, 1990, as amended at 71 FR 16907, Apr. 4, 2006]

§264.1062 Alternative standards for valves in gas/vapor service or in light liquid service: skip period leak detection and repair.

(a) An owner or operator subject to the requirements of §264.1057 may elect for all valves within a hazardous waste management unit to comply with one of the alternative work practices specified in paragraphs (b) (2) and (3) of this section.

(b)(1) An owner or operator shall comply with the requirements for valves, as described in §264.1057, except as described in paragraphs (b)(2) and (b)(3) of this section.

(2) After two consecutive quarterly leak detection periods with the percentage of valves leaking equal to or less than 2 percent, an owner or operator may begin to skip one of the quarterly leak detection periods (i.e., monitor for leaks once every six months) for the valves subject to the requirements in §264.1057 of this subpart.

(3) After five consecutive quarterly leak detection periods with the percentage of valves leaking equal to or less than 2 percent, an owner or operator may begin to skip three of the quarterly leak detection periods (i.e., monitor for leaks once every year) for the valves subject to the requirements in §264.1057 of this subpart.

(4) If the percentage of valves leaking is greater than 2 percent, the owner or operator shall monitor monthly in compliance with the requirements in §264.1057, but may again elect to use this section after meeting the requirements of §264.1057(c)(1).

[55 FR 25501, June 21, 1990, as amended at 62 FR 64658, Dec. 8, 1997; 71 FR 16907, Apr. 4, 2006]

§264.1063 Test methods and procedures.

(a) Each owner or operator subject to the provisions of this subpart shall comply with the test methods and procedures requirements provided in this section.

(b) Leak detection monitoring, as required in §§264.1052–264.1062, shall comply with the following requirements:

(1) Monitoring shall comply with Reference Method 21 in 40 CFR part 60.

(2) The detection instrument shall meet the performance criteria of Reference Method 21.

(3) The instrument shall be calibrated before use on each day of its use by the procedures specified in Reference Method 21.

(4) Calibration gases shall be:

(i) Zero air (less than 10 ppm of hydrocarbon in air).

(ii) A mixture of methane or n-hexane and air at a concentration of approximately, but less than, 10,000 ppm methane or n-hexane.

(5) The instrument probe shall be traversed around all potential leak interfaces as close to the interface as possible as described in Reference Method 21.

(c) When equipment is tested for compliance with no detectable emissions, as required in §§264.1052(e), 264.1053(i), 264.1054, and 264.1057(f), the test shall comply with the following requirements:

(1) The requirements of paragraphs (b)(1) through (4) of this section shall apply.

(2) The background level shall be determined as set forth in Reference Method 21.

(3) The instrument probe shall be traversed around all potential leak interfaces as close to the interface as possible as described in Reference Method 21.

(4) The arithmetic difference between the maximum concentration indicated by the instrument and the background level is compared with 500 ppm for determining compliance.

(d) In accordance with the waste analysis plan required by §264.13(b), an owner or operator of a facility must determine, for each piece of equipment, whether the equipment contains or

519

contacts a hazardous waste with organic concentration that equals or exceeds 10 percent by weight using the following:

(1) Methods described in ASTM Methods D 2267–88, E 169–87, E 168–88, E 260–85 (incorporated by reference under § 260.11);

(2) Method 9060A (incorporated by reference under 40 CFR 260.11) of "Test Methods for Evaluating Solid Waste," EPA Publication SW–846, for computing total organic concentration of the sample, or analyzed for its individual organic constituents; or

(3) Application of the knowledge of the nature of the hazardous waste stream or the process by which it was produced. Documentation of a waste determination by knowledge is required. Examples of documentation that shall be used to support a determination under this provision include production process information documenting that no organic compounds are used, information that the waste is generated by a process that is identical to a process at the same or another facility that has previously been demonstrated by direct measurement to have a total organic content less than 10 percent, or prior speciation analysis results on the same waste stream where it can also be documented that no process changes have occurred since that analysis that could affect the waste total organic concentration.

(e) If an owner or operator determines that a piece of equipment contains or contacts a hazardous waste with organic concentrations at least 10 percent by weight, the determination can be revised only after following the procedures in paragraph (d)(1) or (d)(2) of this section.

(f) When an owner or operator and the Regional Administrator do not agree on whether a piece of equipment contains or contacts a hazardous waste with organic concentrations at least 10 percent by weight, the procedures in paragraph (d)(1) or (d)(2) of this section can be used to resolve the dispute.

(g) Samples used in determining the percent organic content shall be representative of the highest total organic content hazardous waste that is expected to be contained in or contact the equipment.

(h) To determine if pumps or valves are in light liquid service, the vapor pressures of constituents may be obtained from standard reference texts or may be determined by ASTM D–2879–86 (incorporated by reference under § 260.11).

(i) Performance tests to determine if a control device achieves 95 weight percent organic emission reduction shall comply with the procedures of § 264.1034(c)(1) through (c)(4).

[55 FR 25501, June 21, 1990, as amended at 62 FR 32462, June 13, 1997; 70 FR 34581, June 14, 2005]

§ 264.1064   Recordkeeping requirements.

(a)(1) Each owner or operator subject to the provisions of this subpart shall comply with the recordkeeping requirements of this section.

(2) An owner or operator of more than one hazardous waste management unit subject to the provisions of this subpart may comply with the recordkeeping requirements for these hazardous waste management units in one recordkeeping system if the system identifies each record by each hazardous waste management unit.

(b) Owners and operators must record the following information in the facility operating record:

(1) For each piece of equipment to which subpart BB of part 264 applies:

(i) Equipment identification number and hazardous waste management unit identification.

(ii) Approximate locations within the facility (e.g., identify the hazardous waste management unit on a facility plot plan).

(iii) Type of equipment (e.g., a pump or pipeline valve).

(iv) Percent-by-weight total organics in the hazardous waste stream at the equipment.

(v) Hazardous waste state at the equipment (e.g., gas/vapor or liquid).

(vi) Method of compliance with the standard (e.g., "monthly leak detection and repair" or "equipped with dual mechanical seals").

(2) For facilities that comply with the provisions of § 264.1033(a)(2), an implementation schedule as specified in § 264.1033(a)(2).

(3) Where an owner or operator chooses to use test data to demonstrate the organic removal efficiency or total organic compound concentration achieved by the control device, a performance test plan as specified in §264.1035(b)(3).

(4) Documentation of compliance with §264.1060, including the detailed design documentation or performance test results specified in §264.1035(b)(4).

(c) When each leak is detected as specified in §§264.1052, 264.1053, 264.1057, and 264.1058, the following requirements apply:

(1) A weatherproof and readily visible identification, marked with the equipment identification number, the date evidence of a potential leak was found in accordance with §264.1058(a), and the date the leak was detected, shall be attached to the leaking equipment.

(2) The identification on equipment, except on a valve, may be removed after it has been repaired.

(3) The identification on a valve may be removed after it has been monitored for 2 successive months as specified in §264.1057(c) and no leak has been detected during those 2 months.

(d) When each leak is detected as specified in §§264.1052, 264.1053, 264.1057, and 264.1058, the following information shall be recorded in an inspection log and shall be kept in the facility operating record:

(1) The instrument and operator identification numbers and the equipment identification number.

(2) The date evidence of a potential leak was found in accordance with §264.1058(a).

(3) The date the leak was detected and the dates of each attempt to repair the leak.

(4) Repair methods applied in each attempt to repair the leak.

(5) "Above 10,000" if the maximum instrument reading measured by the methods specified in §264.1063(b) after each repair attempt is equal to or greater than 10,000 ppm.

(6) "Repair delayed" and the reason for the delay if a leak is not repaired within 15 calendar days after discovery of the leak.

(7) Documentation supporting the delay of repair of a valve in compliance with §264.1059(c).

(8) The signature of the owner or operator (or designate) whose decision it was that repair could not be effected without a hazardous waste management unit shutdown.

(9) The expected date of successful repair of the leak if a leak is not repaired within 15 calendar days.

(10) The date of successful repair of the leak.

(e) Design documentation and monitoring, operating, and inspection information for each closed-vent system and control device required to comply with the provisions of §264.1060 shall be recorded and kept up-to-date in the facility operating record as specified in §264.1035(c). Design documentation is specified in §264.1035 (c)(1) and (c)(2) and monitoring, operating, and inspection information in §264.1035 (c)(3)–(c)(8).

(f) For a control device other than a thermal vapor incinerator, catalytic vapor incinerator, flare, boiler, process heater, condenser, or carbon adsorption system, the Regional Administrator will specify the appropriate recordkeeping requirements.

(g) The following information pertaining to all equipment subject to the requirements in §§264.1052 through 264.1060 shall be recorded in a log that is kept in the facility operating record:

(1) A list of identification numbers for equipment (except welded fittings) subject to the requirements of this subpart.

(2)(i) A list of identification numbers for equipment that the owner or operator elects to designate for no detectable emissions, as indicated by an instrument reading of less than 500 ppm above background, under the provisions of §§264.1052(e), 264.1053(i), and 264.1057(f).

(ii) The designation of this equipment as subject to the requirements of §§264.1052(e), 264.1053(i), or 264.1057(f) shall be signed by the owner or operator.

(3) A list of equipment identification numbers for pressure relief devices required to comply with §264.1054(a).

(4)(i) The dates of each compliance test required in §§264.1052(e), 264.1053(i), 264.1054, and 264.1057(f).

(ii) The background level measured during each compliance test.

(iii) The maximum instrument reading measured at the equipment during each compliance test.

(5) A list of identification numbers for equipment in vacuum service.

(6) Identification, either by list or location (area or group) of equipment that contains or contacts hazardous waste with an organic concentration of at least 10 percent by weight for less than 300 hours per calendar year.

(h) The following information pertaining to all valves subject to the requirements of § 264.1057 (g) and (h) shall be recorded in a log that is kept in the facility operating record:

(1) A list of identification numbers for valves that are designated as unsafe to monitor, an explanation for each valve stating why the valve is unsafe to monitor, and the plan for monitoring each valve.

(2) A list of identification numbers for valves that are designated as difficult to monitor, an explanation for each valve stating why the valve is difficult to monitor, and the planned schedule for monitoring each valve.

(i) The following information shall be recorded in the facility operating record for valves complying with § 264.1062:

(1) A schedule of monitoring.

(2) The percent of valves found leaking during each monitoring period.

(j) The following information shall be recorded in a log that is kept in the facility operating record:

(1) Criteria required in § 264.1052(d)(5)(ii) and § 264.1053(e)(2) and an explanation of the design criteria.

(2) Any changes to these criteria and the reasons for the changes.

(k) The following information shall be recorded in a log that is kept in the facility operating record for use in determining exemptions as provided in the applicability section of this subpart and other specific subparts:

(1) An analysis determining the design capacity of the hazardous waste management unit.

(2) A statement listing the hazardous waste influent to and effluent from each hazardous waste management unit subject to the requirements in §§ 264.1052 through 264.1060 and an analysis determining whether these hazardous wastes are heavy liquids.

(3) An up-to-date analysis and the supporting information and data used to determine whether or not equipment is subject to the requirements in §§ 264.1052 through 264.1060. The record shall include supporting documentation as required by § 264.1063(d)(3) when application of the knowledge of the nature of the hazardous waste stream or the process by which it was produced is used. If the owner or operator takes any action (e.g., changing the process that produced the waste) that could result in an increase in the total organic content of the waste contained in or contacted by equipment determined not to be subject to the requirements in §§ 264.1052 through 264.1060, then a new determination is required.

(l) Records of the equipment leak information required by paragraph (d) of this section and the operating information required by paragraph (e) of this section need be kept only 3 years.

(m) The owner or operator of a facility with equipment that is subject to this subpart and to regulations at 40 CFR part 60, part 61, or part 63 may elect to determine compliance with this subpart either by documentation pursuant to § 264.1064 of this subpart, or by documentation of compliance with the regulations at 40 CFR part 60, part 61, or part 63 pursuant to the relevant provisions of the regulations at 40 part 60, part 61, or part 63. The documentation of compliance under regulations at 40 CFR part 60, part 61, or part 63 shall be kept with or made readily available with the facility operating record.

[55 FR 25501, June 21, 1990, as amended at 61 FR 59952, Nov. 25, 1996; 62 FR 64658, Dec. 8, 1997; 71 FR 40274, July 14, 2006]

## § 264.1065  Reporting requirements.

(a) A semiannual report shall be submitted by owners and operators subject to the requirements of this subpart to the Regional Administrator by dates specified by the Regional Administrator. The report shall include the following information:

(1) The Environmental Protection Agency identification number, name, and address of the facility.

(2) For each month during the semiannual reporting period:

(i) The equipment identification number of each valve for which a leak was not repaired as required in §264.1057(d).

(ii) The equipment identification number of each pump for which a leak was not repaired as required in §264.1052 (c) and (d)(6).

(iii) The equipment identification number of each compressor for which a leak was not repaired as required in §264.1053(g).

(3) Dates of hazardous waste management unit shutdowns that occurred within the semiannual reporting period.

(4) For each month during the semiannual reporting period, dates when the control device installed as required by §264.1052, 264.1053, 264.1054, or 264.1055 exceeded or operated outside of the design specifications as defined in §264.1064(e) and as indicated by the control device monitoring required by §264.1060 and was not corrected within 24 hours, the duration and cause of each exceedance, and any corrective measures taken.

(b) If, during the semiannual reporting period, leaks from valves, pumps, and compressors are repaired as required in §§264.1057 (d), 264.1052 (c) and (d)(6), and 264.1053 (g), respectively, and the control device does not exceed or operate outside of the design specifications as defined in §264.1064(e) for more than 24 hours, a report to the Regional Administrator is not required.

**§§264.1066–264.1079  [Reserved]**

## Subpart CC—Air Emission Standards for Tanks, Surface Impoundments, and Containers

SOURCE: 59 FR 62927, Dec. 6, 1994, unless otherwise noted.

### §264.1080  Applicability.

(a) The requirements of this subpart apply to owners and operators of all facilities that treat, store, or dispose of hazardous waste in tanks, surface impoundments, or containers subject to either subpart I, J, or K of this part except as §264.1 and paragraph (b) of this section provide otherwise.

(b) The requirements of this subpart do not apply to the following waste management units at the facility:

(1) A waste management unit that holds hazardous waste placed in the unit before December 6, 1996, and in which no hazardous waste is added to the unit on or after December 6, 1996.

(2) A container that has a design capacity less than or equal to 0.1 m³.

(3) A tank in which an owner or operator has stopped adding hazardous waste and the owner or operator has begun implementing or completed closure pursuant to an approved closure plan.

(4) A surface impoundment in which an owner or operator has stopped adding hazardous waste (except to implement an approved closure plan) and the owner or operator has begun implementing or completed closure pursuant to an approved closure plan.

(5) A waste management unit that is used solely for on-site treatment or storage of hazardous waste that is placed in the unit as a result of implementing remedial activities required under the corrective action authorities of RCRA sections 3004(u), 3004(v), or 3008(h); CERCLA authorities; or similar Federal or State authorities.

(6) A waste management unit that is used solely for the management of radioactive mixed waste in accordance with all applicable regulations under the authority of the Atomic Energy Act and the Nuclear Waste Policy Act.

(7) A hazardous waste management unit that the owner or operator certifies is equipped with and operating air emission controls in accordance with the requirements of an applicable Clean Air Act regulation codified under 40 CFR part 60, part 61, or part 63. For the purpose of complying with this paragraph, a tank for which the air emission control includes an enclosure, as opposed to a cover, must be in compliance with the enclosure and control device requirements of §264.1084(i), except as provided in §264.1082(c)(5).

(8) A tank that has a process vent as defined in 40 CFR 264.1031.

(c) For the owner and operator of a facility subject to this subpart who received a final permit under RCRA section 3005 prior to December 6, 1996, the requirements of this subpart shall be incorporated into the permit when the permit is reissued in accordance with the requirements of 40 CFR 124.15 of this chapter or reviewed in accordance with the requirements of 40 CFR 270.50(d) of this chapter. Until such date when the permit is reissued in accordance with the requirements of 40 CFR 124.15 or reviewed in accordance with the requirements of 40 CFR 270.50(d), the owner and operator are subject to the requirements of 40 CFR part 265, subpart CC.

(d) The requirements of this subpart, except for the recordkeeping requirements specified in § 264.1089(i) of this subpart, are administratively stayed for a tank or a container used for the management of hazardous waste generated by organic peroxide manufacturing and its associated laboratory operations when the owner or operator of the unit meets all of the following conditions:

(1) The owner or operator identifies that the tank or container receives hazardous waste generated by an organic peroxide manufacturing process producing more than one functional family of organic peroxides or multiple organic peroxides within one functional family, that one or more of these organic peroxides could potentially undergo self-accelerating thermal decomposition at or below ambient temperatures, and that organic peroxides are the predominant products manufactured by the process. For the purpose of meeting the conditions of this paragraph, "organic peroxide" means an organic compound that contains the bivalent —O—O— structure and which may be considered to be a structural derivative of hydrogen peroxide where one or both of the hydrogen atoms has been replaced by an organic radical.

(2) The owner or operator prepares documentation, in accordance with the requirements of § 264.1089(i) of this subpart, explaining why an undue safety hazard would be created if air emission controls specified in §§ 264.1084 through 264.1087 of this subpart are installed and operated on the tanks and containers used at the facility to manage the hazardous waste generated by the organic peroxide manufacturing process or processes meeting the conditions of paragraph (d)(1) of this section.

(3) The owner or operator notifies the Regional Administrator in writing that

hazardous waste generated by an organic peroxide manufacturing process or processes meeting the conditions of paragraph (d)(1) of this section are managed at the facility in tanks or containers meeting the conditions of paragraph (d)(2) of this section. The notification shall state the name and address of the facility, and be signed and dated by an authorized representative of the facility owner or operator.

(e)(1) Except as provided in paragraph (e)(2) of this section, the requirements of this subpart do not apply to the pharmaceutical manufacturing facility, commonly referred to as the Stonewall Plant, located at Route 340 South, Elkton, Virginia, provided that facility is operated in compliance with the requirements contained in a Clean Air Act permit issued pursuant to 40 CFR 52.2454. The requirements of this subpart shall apply to the facility upon termination of the Clean Air Act permit issued pursuant to 40 CFR 52.2454.

(2) Notwithstanding paragraph (e)(1) of this section, any hazardous waste surface impoundment operated at the Stonewall Plant is subject to:

(i) The standards in § 264.1085 and all requirements related to hazardous waste surface impoundments that are referenced in or by § 264.1085, including the closed-vent system and control device requirements of § 264.1087 and the recordkeeping requirements of § 264.1089(c); and

(ii) The reporting requirements of § 264.1090 that are applicable to surface impoundments and/or to closed-vent systems and control devices associated with a surface impoundment.

(f) This section applies only to the facility commonly referred to as the OSi Specialties Plant, located on State Route 2, Sistersville, West Virginia ("Sistersville Plant").

(1)(i) Provided that the Sistersville Plant is in compliance with the requirements of paragraph (f)(2) of this section, the requirements referenced in paragraphs (f)(1)(iii) and (f)(1)(iv) of this section are temporarily deferred, as specified in paragraph (f)(3) of this section, with respect to the two hazardous waste surface impoundments at the Sistersville Plant. Beginning on the date that paragraph (f)(1)(ii) of this section is first implemented, the temporary deferral of this paragraph shall no longer be effective.

(ii)(A) In the event that a notice of revocation is issued pursuant to paragraph (f)(3)(iv) of this section, the requirements referenced in paragraphs (f)(1)(iii) and (f)(1)(iv) of this section are temporarily deferred, with respect to the two hazardous waste surface impoundments, provided that the Sistersville Plant is in compliance with the requirements of paragraphs (f)(2)(ii), (f)(2)(iii), (f)(2)(iv), (f)(2)(v), (f)(2)(vi) and (g) of this section, except as provided under paragraph (f)(1)(ii)(B) of this section. The temporary deferral of the previous sentence shall be effective beginning on the date the Sistersville Plant receives written notification of revocation, and continuing for a maximum period of 18 months from that date, provided that the Sistersville Plant is in compliance with the requirements of paragraphs (f)(2)(ii), (f)(2)(iii), (f)(2)(iv), (f)(2)(v), (f)(2)(vi) and (g) of this section at all times during that 18-month period. In no event shall the temporary deferral continue to be effective after the MON Compliance Date.

(B) In the event that a notification of revocation is issued pursuant to paragraph (f)(3)(iv) of this section as a result of the permanent removal of the capper unit from methyl capped polyether production service, the requirements referenced in paragraphs (f)(1)(iii) and (f)(1)(iv) of this section are temporarily deferred, with respect to the two hazardous waste surface impoundments, provided that the Sistersville Plant is in compliance with the requirements of paragraphs (f)(2)(vi), and (g) of this section. The temporary deferral of the previous sentence shall be effective beginning on the date the Sistersville Plant receives written notification of revocation, and continuing for a maximum period of 18 months from that date, provided that the Sistersville Plant is in compliance with the requirements of paragraphs (f)(2)(vi) and (g) of this section at all times during that 18-month period. In no event shall the temporary deferral continue to be effective after the MON Compliance Date.

(iii) The standards in §264.1085 of this part, and all requirements referenced in or by §264.1085 that otherwise would apply to the two hazardous waste surface impoundments, including the closed-vent system and control device requirements of §264.1087 of this part.

(iv) The reporting requirements of §264.1090 that are applicable to surface impoundments and/or to closed-vent systems and control devices associated with a surface impoundment.

(2) Notwithstanding the effective period and revocation provisions in paragraph (f)(3) of this section, the temporary deferral provided in paragraph (f)(1)(i) of this section is effective only if the Sistersville Plant meets the requirements of paragraph (f)(2) of this section.

(i) The Sistersville Plant shall install an air pollution control device on the polyether methyl capper unit ("capper unit"), implement a methanol recovery operation, and implement a waste minimization/pollution prevention ("WMPP") project. The installation and implementation of these requirements shall be conducted according to the schedule described in paragraphs (f)(2)(i) and (f)(2)(vi) of this section.

(A) The Sistersville Plant shall complete the initial start-up of a thermal incinerator on the capper unit's process vents from the first stage vacuum pump, from the flash pot and surge tank, and from the water stripper, no later than April 1, 1998.

(B) The Sistersville Plant shall provide to the EPA and the West Virginia Department of Environmental Protection, written notification of the actual date of initial start-up of the thermal incinerator, and commencement of the methanol recovery operation. The Sistersville Plant shall submit this written notification as soon as practicable, but in no event later than 15 days after such events.

(ii) The Sistersville Plant shall install and operate the capper unit process vent thermal incinerator according to the requirements of paragraphs (f)(2)(ii)(A) through (f)(2)(ii)(D) of this section.

(A) Capper unit process vent thermal incinerator.

(*1*) Except as provided under paragraph (f)(2)(ii)(D) of this section, the

Sistersville Plant shall operate the process vent thermal incinerator such that the incinerator reduces the total organic compounds ("TOC") from the process vent streams identified in paragraph (f)(2)(i)(A) of this section, by 98 weight-percent, or to a concentration of 20 parts per million by volume, on a dry basis, corrected to 3 percent oxygen, whichever is less stringent.

(i) Prior to conducting the initial performance test required under paragraph (f)(2)(ii)(B) of this section, the Sistersville Plant shall operate the thermal incinerator at or above a minimum temperature of 1600 Fahrenheit.

(ii) After the initial performance test required under paragraph (f)(2)(ii)(B) of this section, the Sistersville Plant shall operate the thermal incinerator at or above the minimum temperature established during that initial performance test.

(iii) The Sistersville Plant shall operate the process vent thermal incinerator at all times that the capper unit is being operated to manufacture product.

(2) The Sistersville Plant shall install, calibrate, and maintain all air pollution control and monitoring equipment described in paragraphs (f)(2)(i)(A) and (f)(2)(ii)(B)(3) of this section, according to the manufacturer's specifications, or other written procedures that provide adequate assurance that the equipment can reasonably be expected to control and monitor accurately, and in a manner consistent with good engineering practices during all periods when emissions are routed to the unit.

(B) The Sistersville Plant shall comply with the requirements of paragraphs (f)(2)(ii)(B)(1) through (f)(2)(ii)(B)(3) of this section for performance testing and monitoring of the capper unit process vent thermal incinerator.

(1) Within sixty (120) days after thermal incinerator initial start-up, the Sistersville Plant shall conduct a performance test to determine the minimum temperature at which compliance with the emission reduction requirement specified in paragraph (f)(4) of this section is achieved. This determination shall be made by measuring TOC minus methane and ethane, ac-

cording to the procedures specified in paragraph (f)(2)(ii)(B) of this section.

(2) The Sistersville Plant shall conduct the initial performance test in accordance with the standards set forth in paragraph (f)(4) of this section.

(3) Upon initial start-up, the Sistersville Plant shall install, calibrate, maintain and operate, according to manufacturer's specifications and in a manner consistent with good engineering practices, the monitoring equipment described in paragraphs (f)(2)(ii)(B)(3)(i) through (f)(2)(ii)(B)(3)(iii) of this section.

(i) A temperature monitoring device equipped with a continuous recorder. The temperature monitoring device shall be installed in the firebox or in the duct work immediately downstream of the firebox in a position before any substantial heat exchange is encountered.

(ii) A flow indicator that provides a record of vent stream flow to the incinerator at least once every fifteen minutes. The flow indicator shall be installed in the vent stream from the process vent at a point closest to the inlet of the incinerator.

(iii) If the closed-vent system includes bypass devices that could be used to divert the gas or vapor stream to the atmosphere before entering the control device, each bypass device shall be equipped with either a bypass flow indicator or a seal or locking device as specified in this paragraph. For the purpose of complying with this paragraph, low leg drains, high point bleeds, analyzer vents, open-ended valves or lines, spring-loaded pressure relief valves, and other fittings used for safety purposes are not considered to be bypass devices. If a bypass flow indicator is used to comply with this paragraph, the bypass flow indicator shall be installed at the inlet to the bypass line used to divert gases and vapors from the closed-vent system to the atmosphere at a point upstream of the control device inlet. If a seal or locking device (e.g. car-seal or lock-and-key configuration) is used to comply with this paragraph, the device shall be placed on the mechanism by which the bypass device position is controlled (e.g., valve handle, damper levels) when the bypass device is in the closed position such that the bypass device cannot be opened without breaking the seal or removing the lock. The Sistersville Plant shall visually inspect the seal or locking device at least once every month to verify that the bypass mechanism is maintained in the closed position.

(C) The Sistersville Plant shall keep on-site an up-to-date, readily accessible record of the information described in paragraphs (f)(2)(ii)(C)(*1*) through (f)(2)(ii)(C)(*4*) of this section.

(*1*) Data measured during the initial performance test regarding the firebox temperature of the incinerator and the percent reduction of TOC achieved by the incinerator, and/or such other information required in addition to or in lieu of that information by the WVDEP in its approval of equivalent test methods and procedures.

(*2*) Continuous records of the equipment operating procedures specified to be monitored under paragraph (f)(2)(ii)(B)(*3*) of this section, as well as records of periods of operation during which the firebox temperature falls below the minimum temperature established under paragraph (f)(2)(ii)(A)(*1*) of this section.

(*3*) Records of all periods during which the vent stream has no flow rate to the extent that the capper unit is being operated during such period.

(*4*) Records of all periods during which there is flow through a bypass device.

(D) The Sistersville Plant shall comply with the start-up, shutdown, maintenance and malfunction requirements contained in paragraphs (f)(2)(ii)(D)(*1*) through (f)(2)(ii)(D)(*6*) of this section, with respect to the capper unit process vent incinerator.

(*1*) The Sistersville Plant shall develop and implement a Start-up, Shutdown and Malfunction Plan as required by the provisions set forth in paragraph (f)(2)(ii)(D) of this section. The plan shall describe, in detail, procedures for operating and maintaining the thermal incinerator during periods of start-up, shutdown and malfunction, and a program of corrective action for malfunctions of the thermal incinerator.

(*2*) The plan shall include a detailed description of the actions the Sistersville Plant will take to perform the functions described in paragraphs (f)(2)(ii)(D)(*2*)(*i*) through (f)(2)(ii)(D)(*2*)(*iii*) of this section.

(*i*) Ensure that the thermal incinerator is operated in a manner consistent with good air pollution control practices.

(*ii*) Ensure that the Sistersville Plant is prepared to correct malfunctions as soon as practicable after their occurrence in order to minimize excess emissions.

(*iii*) Reduce the reporting requirements associated with periods of start-up, shutdown and malfunction.

(*3*) During periods of start-up, shutdown and malfunction, the Sistersville Plant shall maintain the process unit and the associated thermal incinerator in accordance with the procedures set forth in the plan.

(*4*) The plan shall contain record keeping requirements relating to periods of start-up, shutdown or malfunction, actions taken during such periods in conformance with the plan, and any failures to act in conformance with the plan during such periods.

(*5*) During periods of maintenance or malfunction of the thermal incinerator, the Sistersville Plant may continue to operate the capper unit, provided that operation of the capper unit without the thermal incinerator shall be limited to no more than 240 hours each calendar year.

(*6*) For the purposes of paragraph (f)(2)(iii)(D) of this section, the Sistersville Plant may use its operating procedures manual, or a plan developed for other reasons, provided that plan meets the requirements of paragraph (f)(2)(iii)(D) of this section for the start-up, shutdown and malfunction plan.

(iii) The Sistersville Plant shall operate the closed-vent system in accordance with the requirements of paragraphs (f)(2)(iii)(A) through (f)(2)(iii)(D) of this section.

(A) Closed-vent system.

(*1*) At all times when the process vent thermal incinerator is operating, the Sistersville Plant shall route the vent streams identified in paragraph (f)(2)(i) of this section from the capper unit to the thermal incinerator through a closed-vent system.

(2) The closed-vent system will be designed for and operated with no detectable emissions, as defined in paragraph (f)(6) of this section.

(B) The Sistersville Plant will comply with the performance standards set forth in paragraph (f)(2)(iii)(A)(*1*) of this section on and after the date on which the initial performance test referenced in paragraph (f)(2)(ii)(B) of this section is completed, but no later than sixty (60) days after the initial start-up date.

(C) The Sistersville Plant shall comply with the monitoring requirements of paragraphs (f)(2)(iii)(C)(*1*) through (f)(2)(iii)(C)(*3*) of this section, with respect to the closed-vent system.

(*1*) At the time of the performance test described in paragraph (f)(2)(ii)(B) of this section, the Sistersville Plant shall inspect the closed-vent system as specified in paragraph (f)(5) of this section.

(*2*) At the time of the performance test described in paragraph (f)(2)(ii)(B) of this section, and annually thereafter, the Sistersville Plant shall inspect the closed-vent system for visible, audible, or olfactory indications of leaks.

(*3*) If at any time a defect or leak is detected in the closed-vent system, the Sistersville Plant shall repair the defect or leak in accordance with the requirements of paragraphs (f)(2)(iii)(C)(*3*)(*i*) and (f)(2)(iii)(C)(*3*)(*ii*) of this section.

(*i*) The Sistersville Plant shall make first efforts at repair of the defect no later than five (5) calendar days after detection, and repair shall be completed as soon as possible but no later than forty-five (45) calendar days after detection.

(*ii*) The Sistersville Plant shall maintain a record of the defect repair in accordance with the requirements specified in paragraph (f)(2)(iii)(D) of this section.

(D) The Sistersville Plant shall keep on-site up-to-date, readily accessible records of the inspections and repairs required to be performed by paragraph (f)(2)(iii) of this section.

(iv) The Sistersville Plant shall operate the methanol recovery operation in accordance with paragraphs (f)(2)(iv)(A) through (f)(2)(iv)(C) of this section.

(A) The Sistersville Plant shall operate the condenser associated with the methanol recovery operation at all times during which the capper unit is being operated to manufacture product.

(B) The Sistersville Plant shall comply with the monitoring requirements described in paragraphs (f)(2)(B)(*1*) through (f)(2)(B)(*3*) of this section, with respect to the methanol recovery operation.

(*1*) The Sistersville Plant shall perform measurements necessary to determine the information described in paragraphs (f)(2)(iv)(B)(*1*)(*i*) and (f)(2)(iv)(B)(*1*)(*ii*) of this section to demonstrate the percentage recovery by weight of the methanol contained in the influent gas stream to the condenser.

(*i*) Information as is necessary to calculate the annual amount of methanol generated by operating the capper unit.

(*ii*) The annual amount of methanol recovered by the condenser associated with the methanol recovery operation.

(2) The Sistersville Plant shall install, calibrate, maintain and operate according to manufacturer specifications, a temperature monitoring device with a continuous recorder for the condenser associated with the methanol recovery operation, as an indicator that the condenser is operating.

(3) The Sistersville Plant shall record the dates and times during which the capper unit and the condenser are operating.

(C) The Sistersville Plant shall keep on-site up-to-date, readily-accessible records of the parameters specified to be monitored under paragraph (f)(2)(iv)(B) of this section.

(v) The Sistersville Plant shall comply with the requirements of paragraphs (f)(2)(v)(A) through (f)(2)(v)(C) of this section for the disposition of methanol collected by the methanol recovery operation.

(A) On an annual basis, the Sistersville Plant shall ensure that a minimum of 95% by weight of the methanol collected by the methanol recovery operation (also referred to as the "collected methanol") is utilized

528

for reuse, recovery, or thermal recovery/treatment. The Sistersville Plant may use the methanol on-site, or may transfer or sell the methanol for reuse, recovery, or thermal recovery/treatment at other facilities.

(1) Reuse. To the extent reuse of all of the collected methanol destined for reuse, recovery, or thermal recovery is not economically feasible, the Sistersville Plant shall ensure the residual portion is sent for recovery, as defined in paragraph (f)(6) of this section, except as provided in paragraph (f)(2)(v)(A)(2) of this section.

(2) Recovery. To the extent that reuse or recovery of all the collected methanol destined for reuse, recovery, or thermal recovery is not economically feasible, the Sistersville Plant shall ensure that the residual portion is sent for thermal recovery/treatment, as defined in paragraph (f)(6) of this section.

(3) The Sistersville Plant shall ensure that, on an annual basis, no more than 5% of the methanol recovery collected by the methanol recovery operation is subject to bio-treatment.

(4) In the event the Sistersville Plant receives written notification of revocation pursuant to paragraph (f)(3)(iv) of this section, the percent limitations set forth under paragraph (f)(2)(v)(A) of this section shall no longer be applicable, beginning on the date of receipt of written notification of revocation.

(B) The Sistersville Plant shall perform such measurements as are necessary to determine the pounds of collected methanol directed to reuse, recovery, thermal recovery/treatment and bio-treatment, respectively, on a monthly basis.

(C) The Sistersville Plant shall keep on-site up-to-date, readily accessible records of the amounts of collected methanol directed to reuse, recovery, thermal recovery/treatment and bio-treatment necessary for the measurements required under paragraph (f)(2)(iv)(B) of this section.

(vi) The Sistersville Plant shall perform a WMPP project in accordance with the requirements and schedules set forth in paragraphs (f)(2)(vi)(A) through (f)(2)(vi)(C) of this section.

(A) In performing the WMPP Project, the Sistersville Plant shall use a Study Team and an Advisory Committee as described in paragraphs (f)(2)(vi)(A)(1) through (f)(2)(vi)(A)(6) of this section.

(1) At a minimum, the multi-functional Study Team shall consist of Sistersville Plant personnel from appropriate plant departments (including both management and employees) and an independent contractor. The Sistersville Plant shall select a contractor that has experience and training in WMPP in the chemical manufacturing industry.

(2) The Sistersville Plant shall direct the Study Team such that the team performs the functions described in paragraphs (f)(2)(vi)(A)(2)(i) through (f)(2)(vi)(A)(2)(v) of this section.

(i) Review Sistersville Plant operations and waste streams.

(ii) Review prior WMPP efforts at the Sistersville Plant.

(iii) Develop criteria for the selection of waste streams to be evaluated for the WMPP Project.

(iv) Identify and prioritize the waste streams to be evaluated during the study phase of the WMPP Project, based on the criteria described in paragraph (f)(2)(vi)(A)(2)(iii) of this section.

(v) Perform the WMPP Study as required by paragraphs (f)(2)(vi)(A)(3) through (f)(2)(vi)(A)(5), paragraph (f)(2)(vi)(B), and paragraph (f)(2)(vi)(C) of this section.

(3)(i) The Sistersville Plant shall establish an Advisory Committee consisting of a representative from EPA, a representative from WVDEP, the Sistersville Plant Manager, the Sistersville Plant Director of Safety, Health and Environmental Affairs, and a stakeholder representative(s).

(ii) The Sistersville Plant shall select the stakeholder representative(s) by mutual agreement of EPA, WVDEP and the Sistersville Plant no later than 20 days after receiving from EPA and WVDEP the names of their respective committee members.

(4) The Sistersville Plant shall convene a meeting of the Advisory Committee no later than thirty days after selection of the stakeholder representatives, and shall convene meetings periodically thereafter as necessary for the Advisory Committee to perform its assigned functions. The Sistersville

Plant shall direct the Advisory Committee to perform the functions described in paragraphs (f)(2)(vi)(A)(*4*)(*i*) through (f)(2)(vi)(A)(*4*)(*iii*) of this section.

(*i*) Review and comment upon the Study Team's criteria for selection of waste streams, and the Study Team's identification and prioritization of the waste streams to be evaluated during the WMPP Project.

(*ii*) Review and comment upon the Study Team progress reports and the draft WMPP Study Report.

(*iii*) Periodically review the effectiveness of WMPP opportunities implemented as part of the WMPP Project, and, where appropriate, WMPP opportunities previously determined to be infeasible by the Sistersville Plant but which had potential for feasibility in the future.

(*5*) Beginning on January 15, 1998, and every ninety (90) days thereafter until submission of the final WMPP Study Report required by paragraph (f)(2)(vi)(C) of this section, the Sistersville Plant shall direct the Study Team to submit a progress report to the Advisory Committee detailing its efforts during the prior ninety (90) day period.

(B) The Sistersville Plant shall ensure that the WMPP Study and the WMPP Study Report meet the requirements of paragraphs (f)(2)(vi)(B)(*1*) through (f)(2)(vi)(B)(*3*) of this section.

(*1*) The WMPP Study shall consist of a technical, economic, and regulatory assessment of opportunities for source reduction and for environmentally sound recycling for waste streams identified by the Study Team.

(*2*) The WMPP Study shall evaluate the source, nature, and volume of the waste streams; describe all the WMPP opportunities identified by the Study Team; provide a feasibility screening to evaluate the technical and economical feasibility of each of the WMPP opportunities; identify any cross-media impacts or any anticipated transfers of risk associated with each feasible WMPP opportunity; and identify the projected economic savings and projected quantitative waste reduction estimates for each WMPP opportunity identified.

(*3*) No later than October 19, 1998, the Sistersville Plant shall prepare and submit to the members of the Advisory Committee a draft WMPP Study Report which, at a minimum, includes the results of the WMPP Study, identifies WMPP opportunities the Sistersville Plant determines to be feasible, discusses the basis for excluding other opportunities as not feasible, and makes recommendations as to whether the WMPP Study should be continued. The members of the Advisory Committee shall provide any comments to the Sistersville Plant within thirty (30) days of receiving the WMPP Study Report.

(C) Within thirty (30) days after receipt of comments from the members of the Advisory Committee, the Sistersville Plant shall submit to EPA and WVDEP a final WMPP Study Report which identifies those WMPP opportunities the Sistersville Plant determines to be feasible and includes an implementation schedule for each such WMPP opportunity. The Sistersville Plant shall make reasonable efforts to implement all feasible WMPP opportunities in accordance with the priorities identified in the implementation schedule.

(*1*) For purposes of this section, a WMPP opportunity is feasible if the Sistersville Plant considers it to be technically feasible (taking into account engineering and regulatory factors, product line specifications and customer needs) and economically practical (taking into account the full environmental costs and benefits associated with the WMPP opportunity and the company's internal requirements for approval of capital projects). For purposes of the WMPP Project, the Sistersville Plant shall use "An Introduction to Environmental Accounting as a Business Management Tool," (EPA 742/R–95/001) as one tool to identify the full environmental costs and benefits of each WMPP opportunity.

(*2*) In implementing each WMPP opportunity, the Sistersville Plant shall, after consulting with the other members of the Advisory Committee, develop appropriate protocols and methods for determining the information required by paragraphs (f)(2)(vi)(*2*)(*i*) through (f)(2)(vi)(*2*)(*iii*) of this section.

(*i*) The overall volume of wastes reduced.

(*ii*) The quantities of each constituent identified in paragraph (f)(8) of this section reduced in the wastes.

(*iii*) The economic benefits achieved.

(*3*) No requirements of paragraph (f)(2)(vi) of this section are intended to prevent or restrict the Sistersville Plant from evaluating and implementing any WMPP opportunities at the Sistersville Plant in the normal course of its operations or from implementing, prior to the completion of the WMPP Study, any WMPP opportunities identified by the Study Team.

(vii) The Sistersville Plant shall maintain on-site each record required by paragraph (f)(2) of this section, through the MON Compliance Date.

(viii) The Sistersville Plant shall comply with the reporting requirements of paragraphs (f)(2)(viii)(A) through (f)(2)(viii)(G) of this section.

(A) At least sixty days prior to conducting the initial performance test of the thermal incinerator, the Sistersville Plant shall submit to EPA and WVDEP copies of a notification of performance test, as described in 40 CFR 63.7(b). Following the initial performance test of the thermal incinerator, the Sistersville Plant shall submit to EPA and WVDEP copies of the performance test results that include the information relevant to initial performance tests of thermal incinerators contained in 40 CFR 63.7(g)(1), 40 CFR 63.117(a)(4)(i), and 40 CFR 63.117(a)(4)(ii).

(B) Beginning in 1999, on January 31 of each year, the Sistersville Plant shall submit a semiannual written report to the EPA and WVDEP, with respect to the preceding six month period ending on December 31, which contains the information described in paragraphs (f)(2)(viii)(B)(*1*) through (f)(2)(viii)(B)(*10*) of this section.

(*1*) Instances of operating below the minimum operating temperature established for the thermal incinerator under paragraph (f)(2)(ii)(A)(*1*) of this section which were not corrected within 24 hours of onset.

(*2*) Any periods during which the paper unit was being operated to manufacture product while the flow indicator the vent streams to the thermal incinerator showed no flow.

(*3*) Any periods during which the capper unit was being operated to manufacture product while the flow indicator for any bypass device on the closed vent system to the thermal incinerator showed flow.

(*4*) Information required to be reported during that six month period under the preconstruction permit issued under the state permitting program approved under subpart XX of 40 CFR Part 52—Approval and Promulgation of Implementation Plans for West Virginia.

(*5*) Any periods during which the capper unit was being operated to manufacture product while the condenser associated with the methanol recovery operation was not in operation.

(*6*) The amount (in pounds and by month) of methanol collected by the methanol recovery operation during the six month period.

(*7*) The amount (in pounds and by month) of collected methanol utilized for reuse, recovery, thermal recovery/treatment, or bio-treatment, respectively, during the six month period.

(*8*) The calculated amount (in pounds and by month) of methanol generated by operating the capper unit.

(*9*) The status of the WMPP Project, including the status of developing the WMPP Study Report.

(*10*) Beginning in the year after the Sistersville Plant submits the final WMPP Study Report required by paragraph (f)(2)(vi)(C) of this section, and continuing in each subsequent Semiannual Report required by paragraph (f)(2)(viii)(B) of this section, the Sistersville Plant shall report on the progress of the implementation of feasible WMPP opportunities identified in the WMPP Study Report. The Semiannual Report required by paragraph (f)(2)(viii)(B) of this section shall identify any cross-media impacts or impacts to worker safety or community health issues that have occurred as a result of implementation of the feasible WMPP opportunities.

(C) Beginning in 1999, on July 31 of each year, the Sistersville Plant shall provide an Annual Project Report to the EPA and WVDEP Project XL contacts containing the information required by paragraphs (f)(2)(viii)(C)(*1*) through (f)(2)(viii)(C)(*8*) of this section.

(*1*) The categories of information required to be submitted under paragraphs (f)(2)(viii)(B)(*1*) through (f)(2)(viii)(B)(*8*) of this section, for the preceding 12 month period ending on June 30.

(*2*) An updated Emissions Analysis for January through December of the preceding calendar year. The Sistersville Plant shall submit the updated Emissions Analysis in a form substantially equivalent to the previous Emissions Analysis prepared by the Sistersville Plant to support Project XL. The Emissions Analysis shall include a comparison of the volatile organic emissions associated with the capper unit process vents and the wastewater treatment system (using the EPA Water 8 model or other model agreed to by the Sistersville Plant, EPA and WVDEP) under Project XL with the expected emissions from those sources absent Project XL during that period.

(*3*) A discussion of the Sistersville Plant's performance in meeting the requirements of this section, specifically identifying any areas in which the Sistersville Plant either exceeded or failed to achieve any such standard.

(*4*) A description of any unanticipated problems in implementing the XL Project and any steps taken to resolve them.

(*5*) A WMPP Implementation Report that contains the information contained in paragraphs (f)(2)(viii)(C)(*5*)(*i*) through (viii)(C)(*5*)(*vi*) of this section.

(*i*) A summary of the WMPP opportunities selected for implementation.

(*ii*) A description of the WMPP opportunities initiated and/or completed.

(*iii*) Reductions in volume of waste generated and amounts of each constituent reduced in wastes including any constituents identified in paragraph (f)(8) of this section.

(*iv*) An economic benefits analysis.

(*v*) A summary of the results of the Advisory Committee's review of implemented WMPP opportunities.

(*vi*) A reevaluation of WMPP opportunities previously determined to be infeasible by the Sistersville Plant but which had potential for future feasibility.

(*6*) An assessment of the nature of, and the successes or problems associated with, the Sistersville Plant's interaction with the federal and state agencies under the Project.

(*7*) An update on stakeholder involvement efforts.

(*8*) An evaluation of the Project as implemented against the Project XL Criteria and the baseline scenario.

(D) The Sistersville Plant shall submit to the EPA and WVDEP Project XL contacts a written Final Project Report covering the period during which the temporary deferral was effective, as described in paragraph (f)(3) of this section.

(*1*) The Final Project Report shall contain the information required to be submitted for the Semiannual Report required under paragraph (f)(2)(viii)(B) of this section, and the Annual Project Report required under paragraph (f)(2)(viii)(C) of this section.

(*2*) The Sistersville Plant shall submit the Final Project Report to EPA and WVDEP no later than 180 days after the temporary deferral of paragraph (f)(1) of this section is revoked, or 180 days after the MON Compliance Date, whichever occurs first.

(E)(*1*) The Sistersville Plant shall retain on-site a complete copy of each of the report documents to be submitted to EPA and WVDEP in accordance with requirements under paragraph (f)(2) of this section. The Sistersville Plant shall retain this record until 180 days after the MON Compliance Date. The Sistersville Plant shall provide to stakeholders and interested parties a written notice of availability (to be mailed to all persons on the Project mailing list and to be provided to at least one local newspaper of general circulation) of each such document, and provide a copy of each document to any such person upon request, subject to the provisions of 40 CFR part 2.

(*2*) Any reports or other information submitted to EPA or WVDEP may be released to the public pursuant to the Federal Freedom of Information Act (42 U.S.C. 552 *et seq.*), subject to the provisions of 40 CFR part 2.

(F) The Sistersville Plant shall make all supporting monitoring results and records required under paragraph (f)(2) of this section available to EPA and WVDEP within a reasonable amount of time after receipt of a written request from those Agencies, subject to the provisions of 40 CFR part 2.

(G) Each report submitted by the Sistersville Plant under the requirements of paragraph (f)(2) of this section shall be certified by a Responsible Corporate Officer, as defined in 40 CFR 270.11(a)(1).

(H) For each report submitted in accordance with paragraph (f)(2) of this section, the Sistersville Plant shall send one copy each to the addresses in paragraphs (f)(2)(viii) (H)(1) through (H)(3) of this section.

(1) U.S. EPA Region 3, 1650 Arch Street, Philadelphia, PA 19103–2029, Attention Tad Radzinski, Mail Code 3WC11.

(2) U.S. EPA, 1200 Pennsylvania Ave., NW., Washington, DC 20460, Attention L. Nancy Birnbaum, Mail Code 2129.

(3) West Virginia Division of Environmental Protection, Office of Air Quality, 1558 Washington Street East, Charleston, WV 25311–2599, Attention John H. Johnston.

(3) Effective period and revocation of temporary deferral.

(i) The temporary deferral contained in this section is effective from April 1, 1998, and shall remain effective until the MON Compliance Date. The temporary deferral contained in this section may be revoked prior to the MON Compliance Date, as described in paragraph (f)(3)(iv) of this section.

(ii) On the MON Compliance Date, the temporary deferral contained in this section will no longer be effective.

(iii) The Sistersville Plant shall come into compliance with those requirements deferred by this section no later than the MON Compliance Date. No later than 18 months prior to the MON Compliance Date, the Sistersville Plant shall submit to EPA an implementation schedule that meets the requirements of paragraph (g)(1)(iii) of this section.

(iv) The temporary deferral contained in this section may be revoked for cause, as determined by EPA, prior to the MON Compliance Date. The Sistersville Plant may request EPA to revoke the temporary deferral con-

tained in this section at any time. The revocation shall be effective on the date that the Sistersville Plant receives written notification of revocation from EPA.

(v) Nothing in this section shall affect the provisions of the MON, as applicable to the Sistersville Plant.

(vi) Nothing in paragraph (f) or (g) of this section shall affect any regulatory requirements not referenced in paragraph (f)(1)(iii) or (f)(1)(iv) of this section, as applicable to the Sistersville Plant.

(4) The Sistersville Plant shall conduct the initial performance test required by paragraph (f)(2)(ii)(B) of this section using the procedures in paragraph (f)(4) of this section. The organic concentration and percent reduction shall be measured as TOC minus methane and ethane, according to the procedures specified in paragraph (f)(4) of this section.

(i) Method 1 or 1A of 40 CFR part 60, appendix A, as appropriate, shall be used for selection of the sampling sites.

(A) To determine compliance with the 98 percent reduction of TOC requirement of paragraph (f)(2)(ii)(A)(1) of this section, sampling sites shall be located at the inlet of the control device after the final product recovery device, and at the outlet of the control device.

(B) To determine compliance with the 20 parts per million by volume TOC limit in paragraph (f)(2)(ii)(A)(1) of this section, the sampling site shall be located at the outlet of the control device.

(ii) The gas volumetric flow rate shall be determined using Method 2, 2A, 2C, or 2D of 40 CFR part 60, appendix A, as appropriate.

(iii) To determine compliance with the 20 parts per million by volume TOC limit in paragraph (f)(2)(ii)(A)(1) of this section, the Sistersville Plant shall use Method 18 of 40 CFR part 60, appendix A to measure TOC minus methane and ethane. Alternatively, any other method or data that has been validated according to the applicable procedures in Method 301 of 40 CFR part 63, appendix A, may be used. The following procedures shall be used to calculate parts per million by volume concentration, corrected to 3 percent oxygen:

(A) The minimum sampling time for each run shall be 1 hour in which either an integrated sample or a minimum of four grab samples shall be taken. If grab sampling is used, then the samples shall be taken at approximately equal intervals in time, such as 15 minute intervals during the run.

(B) The concentration of TOC minus methane and ethane ($C_{TOC}$) shall be calculated as the sum of the concentrations of the individual components, and shall be computed for each run using the following equation:

$$C_{TOC} = \sum_{i=1}^{x} \frac{\left( \sum_{j=1}^{n} C_{ji} \right)}{x}$$

Where:

$_{TOC}$=Concentration of TOC (minus methane and ethane), dry basis, parts per million by volume.

$_{ji}$=Concentration of sample components j of sample i, dry basis, parts per million by volume.

n=Number of components in the sample.

x=Number of samples in the sample run.

(C) The concentration of TOC shall be corrected to 3 percent oxygen if a combustion device is the control device.

(1) The emission rate correction factor or excess air, integrated sampling and analysis procedures of Method 3B of 40 CFR part 60, appendix A shall be used to determine the oxygen concentration (%$O_{2d}$). The samples shall be taken during the same time that the TOC (minus methane or ethane) samples are taken.

(2) The concentration corrected to 3 percent oxygen ($C_c$) shall be computed using the following equation:

$$C_c = C_m \left( \frac{17.9}{20.9 \quad \%O_{2d}} \right)$$

Where:

$C_c$=Concentration of TOC corrected to 3 percent oxygen, dry basis, parts per million by volume.

$C_m$=Concentration of TOC (minus methane and ethane), dry basis, parts per million by volume.

%$O_{2d}$=Concentration of oxygen, dry basis, percent by volume.

(iv) To determine compliance with the 98 percent reduction requirement of paragraph (f)(2)(ii)(A)(1) of this section, the Sistersville Plant shall use Method 18 of 40 CFR part 60, appendix A; alternatively, any other method or data that has been validated according to the applicable procedures in Method 301 of 40 CFR part 63, appendix A may be used. The following procedures shall be used to calculate percent reduction efficiency:

(A) The minimum sampling time for each run shall be 1 hour in which either an integrated sample or a minimum of four grab samples shall be taken. If grab sampling is used, then the samples shall be taken at approximately equal intervals in time such as 15 minute intervals during the run.

(B) The mass rate of TOC minus methane and ethane ($E_i$, $E_o$) shall be computed. All organic compounds (minus methane and ethane) measured by Method 18 of 40 CFR part 60, appendix A are summed using the following equations:

$$E_i = K_2 \left( \sum_{j=1}^{n} C_{ij} M_{ij} \right) Q_i$$

$$E_o = K_2 \left( \sum_{j=1}^{n} C_{oj} M_{oj} \right) Q_o$$

Where:

$C_{ij}$, $C_{oj}$=Concentration of sample component j of the gas stream at the inlet and outlet of the control device, respectively, dry basis, parts per million by volume.

$E_i$, $E_o$=Mass rate of TOC (minus methane and ethane) at the inlet and outlet of the control device, respectively, dry basis, kilogram per hour.

$M_{ij}$, $M_{oj}$=Molecular weight of sample component j of the gas stream at the inlet and outlet of the control device, respectively, gram/gram-mole.

$Q_i$, $Q_o$=Flow rate of gas stream at the inlet and outlet of the control device, respectively, dry standard cubic meter per minute.

$K_2$=Constant, $2.494 \times 10^{-6}$ (parts per million)$^{-1}$ (gram-mole per standard cubic meter) (kilogram/gram) (minute/hour), where standard temperature (gram-mole per standard cubic meter) is 20 °C.

(C) The percent reduction in TOC (minus methane and ethane) shall be calculated as follows:

$$R = \frac{E_i E_o}{E_i} (100)$$

Where:

R=Control efficiency of control device, percent.

$E_i$=Mass rate of TOC (minus methane and ethane) at the inlet to the control device as calculated under paragraph (f)(4)(iv)(B) of this section, kilograms TOC per hour.

$E_o$=Mass rate of TOC (minus methane and ethane) at the outlet of the control device, as calculated under paragraph (f)(4)(iv)(B) of this section, kilograms TOC per hour.

(5) At the time of the initial performance test of the process vent thermal incinerator required under paragraph (f)(2)(ii)(B) of this section, the Sistersville Plant shall inspect each closed vent system according to the procedures specified in paragraphs (f)(5)(i) through (f)(5)(vi) of this section.

(i) The initial inspections shall be conducted in accordance with Method 21 of 40 CFR part 60, appendix A.

(ii) (A) Except as provided in paragraph (f)(5)(ii)(B) of this section, the detection instrument shall meet the performance criteria of Method 21 of 40 CFR part 60, appendix A, except the instrument response factor criteria in section 3.1.2(a) of Method 21 of 40 CFR part 60, appendix A shall be for the average composition of the process fluid not each individual volatile organic compound in the stream. For process streams that contain nitrogen, air, or other inerts which are not organic hazardous air pollutants or volatile organic compounds, the average stream response factor shall be calculated on an inert-free basis.

(B) If no instrument is available at the plant site that will meet the performance criteria specified in paragraph (f)(5)(ii)(A) of this section, the instrument readings may be adjusted by multiplying by the average response factor of the process fluid, calculated on an inert-free basis as described in paragraph (f)(5)(ii)(A) of this section.

(iii) The detection instrument shall be calibrated before use on each day of its use by the procedures specified in Method 21 of 40 CFR part 60, appendix A.

(iv) Calibration gases shall be as follows:

(A) Zero air (less than 10 parts per million hydrocarbon in air); and

(B) Mixtures of methane in air at a concentration less than 10,000 parts per million. A calibration gas other than methane in air may be used if the instrument does not respond to methane or if the instrument does not meet the performance criteria specified in paragraph (f)(5)(ii)(A) of this section. In such cases, the calibration gas may be a mixture of one or more of the compounds to be measured in air.

(v) The Sistersville Plant may elect to adjust or not adjust instrument readings for background. If the Sistersville Plant elects to not adjust readings for background, all such instrument readings shall be compared directly to the applicable leak definition to determine whether there is a leak. If the Sistersville Plant elects to adjust instrument readings for background, the Sistersville Plant shall measure background concentration using the procedures in 40 CFR 63.180(b) and (c). The Sistersville Plant shall subtract background reading from the maximum concentration indicated by the instrument.

(vi) The arithmetic difference between the maximum concentration indicated by the instrument and the background level shall be compared with 500 parts per million for determining compliance.

(6) Definitions of terms as used in paragraphs (f) and (g) of this section.

(i) Closed vent system is defined as a system that is not open to the atmosphere and that is composed of piping, connections and, if necessary, flow-inducing devices that transport gas or vapor from the capper unit process vent to the thermal incinerator.

(ii) No detectable emissions means an instrument reading of less than 500 parts per million by volume above background as determined by Method 21 in 40 CFR part 60.

(iii) Reuse includes the substitution of collected methanol (without reclamation subsequent to its collection) for virgin methanol as an ingredient (including uses as an intermediate) or as an effective substitute for a commercial product.

(iv) Recovery includes the substitution of collected methanol for virgin methanol as an ingredient (including uses as an intermediate) or as an effective substitute for a commercial product following reclamation of the methanol subsequent to its collection.

(v) Thermal recovery/treatment includes the use of collected methanol in fuels blending or as a feed to any combustion device to the extent permitted by Federal and state law.

(vi) Bio-treatment includes the treatment of the collected methanol through introduction into a biological treatment system, including the treatment of the collected methanol as a waste stream in an on-site or off-site wastewater treatment system. Introduction of the collected methanol to the on-site wastewater treatment system will be limited to points downstream of the surface impoundments, and will be consistent with the requirements of federal and state law.

(vii) Start-up shall have the meaning set forth at 40 CFR 63.2.

(viii) Flow indicator means a device which indicates whether gas flow is present in the vent stream, and, if required by the permit for the thermal incinerator, which measures the gas flow in that stream.

(ix) Continuous Recorder means a data recording device that records an instantaneous data value at least once every fifteen minutes.

(x) MON means the National Emission Standards for Hazardous Air Pollutants for the source category Miscellaneous Organic Chemical Production and Processes ("MON"), promulgated under the authority of Section 112 of the Clean Air Act.

(xi) MON Compliance Date means the date 3 years after the effective date of the National Emission Standards for Hazardous Air Pollutants for the source category Miscellaneous Organic Chemical Production and Processes ("MON").

(7) OSi Specialties, Incorporated, a subsidiary of Witco Corporation ("OSi"), may seek to transfer its rights and obligations under this section to a future owner of the Sistersville Plant in accordance with the requirements of paragraphs (f)(7)(i) through (f)(7)(iii) of this section.

(i) OSi will provide to EPA a written notice of any proposed transfer at least forty-five days prior to the effective date of any such transfer. The written notice will identify the proposed transferee.

(ii) The proposed transferee will provide to EPA a written request to assume the rights and obligations under this section at least forty-five days prior to the effective date of any such transfer. The written request will describe the transferee's financial and technical capability to assume the obligations under this section, and will include a statement of the transferee's intention to fully comply with the terms of this section and to sign the Final Project Agreement for this XL Project as an additional party.

(iii) Within thirty days of receipt of both the written notice and written request described in paragraphs (f)(7)(i) and (f)(7)(ii) of this section, EPA will determine, based on all relevant information, whether to approve a transfer of rights and obligations under this section from OSi to a different owner.

(8) The constituents to be identified by the Sistersville Plant pursuant to paragraphs (f)(2)(vi)(C)(2)(ii) and (f)(2)(viii)(C)(5)(iii) of this section are: 1 Naphthalenamine; 1,2,4 Trichlorobenzene; 1,1 Dichloroethylene; 1,1,1 Trichloroethane; 1,1,1,2 Tetrachloroethane; 1,1,2 Trichloro 1,2,2 Triflouroethane; 1,1,2 Trichloroethane; 1,1,2,2 Tetrachloroethane; 1,2 Dichlorobenzene; 1,2 Dichloroethane; 1,2 Dichloropropane; 1,2 Dichloropropanone; 1,2 Transdichloroethene; 1,2, Trans-Dichloroethene; 1,2,4,5 Tetrachlorobenzine; 1,3 Dichlorobenzene; 1,4 Dichloro 2 butene; 1,4 Dioxane; 2 Chlorophenol; 2 Cyclohexyl 4,6 dinitrophenol; 2 Methyl Pyridine; 2 Nitropropane; 2, 4-Di-nitrotoluene; Acetone; Acetonitrile; Acrylonitrile; Allyl Alcohol; Aniline; Antimony; Arsenic; Barium; Benzene; Benzotrichloride; Benzyl Chloride; Beryllium; Bis (2 ethyl Hexyl) Phthalate; Butyl Alcohol, n; Butyl Benzyl Phthalate; Cadmium; Carbon Disulfide; Carbon Tetrachloride; Chlorobenzene; Chloroform; Chloromethane; Chromium; Chrysene; Copper; Creosol; Creosol, m-; Creosol, o; Creosol, p; Cyanide; Cyclohexanone; Di-n-octyl phthalate; Dichlorodiflouromethane;

Diethyl Phthalate; Dihydrosafrole; Dimethylamine; Ethyl Acetate; Ethyl benzene; Ethyl Ether; Ethylene Glycol Ethyl Ether; Ethylene Oxide; Formaldehyde; Isobutyl Alcohol; Lead; Mercury; Methanol; Methoxychlor; Methyl Chloride; Methyl Chloroformate; Methyl Ethyl Ketone; Methyl Ethyl Ketone Peroxide; Methyl Isobutyl Ketone; Methyl Methacrylate; Methylene Bromide; Methylene Chloride; Naphthalene; Nickel; Nitrobenzene; Nitroglycerine; p-Toluidine; Phenol; Phthalic Anhydride; Polychlorinated Biphenyls; Propargyl Alcohol; Pyridine; Safrole; Selenium; Silver; Styrene; Tetrachloroethylene; Tetrahydrofuran; Thallium; Toluene; Toluene 2,4 Diisocyanate; Trichloroethylene; Trichloroflouromethane; Vanadium; Vinyl Chloride; Warfarin; Xylene; Zinc.

(g) This section applies only to the facility commonly referred to as the OSi Specialties Plant, located on State Route 2, Sistersville, West Virginia ("Sistersville Plant").

(1)(i) No later than 18 months from the date the Sistersville Plant receives written notification of revocation of the temporary deferral for the Sistersville Plant under paragraph (f) of this section, the Sistersville Plant shall, in accordance with the implementation schedule submitted to EPA under paragraph (g)(1)(ii) of this section, either come into compliance with all requirements of this subpart which had been deferred by paragraph (f)(1)(i) of this section, or complete a facility or process modification such that the requirements of §264.1085 are no longer applicable to the two hazardous waste surface impoundments. In any event, the Sistersville Plant must complete the requirements of the previous sentence no later than the MON Compliance Date; if the Sistersville Plant receives written notification of revocation of the temporary deferral after the date 18 months prior to the MON Compliance Date, the date by which the Sistersville Plant must complete the requirements of the previous sentence will be the MON Compliance Date, which would be less than 18 months from the date of notification of revocation.

(ii) Within 30 days from the date the Sistersville Plant receives written no-

tification of revocation under paragraph (f)(3)(iv) of this section, the Sistersville Plant shall enter and maintain in the facility operating record an implementation schedule. The implementation schedule shall demonstrate that within 18 months from the date the Sistersville Plant receives written notification of revocation under paragraph (f)(3)(iv) of this section (but no later than the MON Compliance Date), the Sistersville Plant shall either come into compliance with the regulatory requirements that had been deferred by paragraph (f)(1)(i) of this section, or complete a facility or process modification such that the requirements of §264.1085 are no longer applicable to the two hazardous waste surface impoundments. Within 30 days from the date the Sistersville Plant receives written notification of revocation under paragraph (f)(3)(iv) of this section, the Sistersville Plant shall submit a copy of the implementation schedule to the EPA and WVDEP Project XL contacts identified in paragraph (f)(2)(viii)(H) of this section. The implementation schedule shall reflect the Sistersville Plant's effort to come into compliance as soon as practicable (but no later than 18 months after the date the Sistersville Plant receives written notification of revocation, or the MON Compliance Date, whichever is sooner) with all regulatory requirements that had been deferred under paragraph (f)(1)(i) of this section, or to complete a facility or process modification as soon as practicable (but no later than 18 months after the date the Sistersville Plant receives written notification of revocation, or the MON Compliance Date, whichever is sooner) such that the requirements of §264.1085 are no longer applicable to the two hazardous waste surface impoundments.

(iii) The implementation schedule shall include the information described in either paragraph (g)(1)(iii)(A) or (B) of this section.

(A) Specific calendar dates for: Award of contracts or issuance of purchase orders for the control equipment required by those regulatory requirements that had been deferred by paragraph (f)(1)(i) of this section; initiation of on-site installation of such control

537

equipment; completion of the control equipment installation; performance of any testing to demonstrate that the installed control equipment meets the applicable standards of this subpart; initiation of operation of the control equipment; and compliance with all regulatory requirements that had been deferred by paragraph (f)(1)(i) of this section.

(B) Specific calendar dates for the purchase, installation, performance testing and initiation of operation of equipment to accomplish a facility or process modification such that the requirements of § 264.1085 are no longer applicable to the two hazardous waste surface impoundments.

(2) Nothing in paragraphs (f) or (g) of this section shall affect any regulatory requirements not referenced in paragraph (f)(2)(i) or (ii) of this section, as applicable to the Sistersville Plant.

(3) In the event that a notification of revocation is issued pursuant to paragraph (f)(3)(iv) of this section, the requirements referenced in paragraphs (f)(1)(iii) and (f)(1)(iv) of this section are temporarily deferred, with respect to the two hazardous waste surface impoundments, provided that the Sistersville Plant is in compliance with the requirements of paragraphs (f)(2)(ii), (f)(2)(iii), (f)(2)(iv), (f)(2)(v), (f)(2)(vi) and (g) of this section, except as provided under paragraph (g)(4) of this section. The temporary deferral of the previous sentence shall be effective beginning on the date the Sistersville Plant receives written notification of revocation, and subject to paragraph (g)(5) of this section, shall continue to be effective for a maximum period of 18 months from that date, provided that the Sistersville Plant is in compliance with the requirements of paragraphs (f)(2)(ii), (f)(2)(iii), (f)(2)(iv), (f)(2)(v), (f)(2)(vi) and (g) of this section at all times during that 18-month period.

(4) In the event that a notification of revocation is issued pursuant to paragraph (f)(3)(iv) of this section as a result of the permanent removal of the capper unit from methyl capped polyether production service, the requirements referenced in paragraphs (f)(1)(iii) and (f)(1)(iv) of this section are temporarily deferred, with respect to the two hazardous waste surface impoundments, provided that the Sistersville Plant is in compliance with the requirements of paragraphs (f)(2)(vi), and (g) of this section. The temporary deferral of the previous sentence shall be effective beginning on the date the Sistersville Plant receives written notification of revocation, and subject to paragraph (g)(5) of this section, shall continue to be effective for a maximum period of 18 months from that date, provided that the Sistersville Plant is in compliance with the requirements of paragraphs (f)(2)(vi) and (g) of this section at all times during that 18-month period.

(5) In no event shall the temporary deferral provided under paragraph (g)(3) or (g)(4) of this section be effective after the MON Compliance Date.

[59 FR 62927, Dec. 6, 1994, as amended at 60 FR 26828, May 19, 1995; 60 FR 50428, Sept. 29, 1995; 60 FR 56953, Nov. 13, 1995; 61 FR 28509, June 5, 1996; 61 FR 59952, Nov. 25, 1996; 62 FR 52642, Oct. 8, 1997; 62 FR 64658, Dec. 8, 1997; 63 FR 11131, Mar. 6, 1998; 63 FR 19838, Apr. 22, 1998; 63 FR 49392, Sept. 15, 1998; 63 FR 53847, Oct. 7, 1998; 64 FR 3389, Jan. 21, 1999; 71 FR 40274, July 14, 2006]

§ 264.1081  Definitions.

As used in this subpart, all terms shall have the meaning given to them in 40 CFR 265.1081, the Act, and parts 260 through 266 of this chapter.

§ 264.1082  Standards: General.

(a) This section applies to the management of hazardous waste in tanks, surface impoundments, and containers subject to this subpart.

(b) The owner or operator shall control air pollutant emissions from each hazardous waste management unit in accordance with standards specified in §§ 264.1084 through 264.1087 of this subpart, as applicable to the hazardous waste management unit, except as provided for in paragraph (c) of this section.

(c) A tank, surface impoundment, or container is exempt from standards specified in § 264.1084 through § 264.1087 of this subpart, as applicable, provided that the waste management unit is one of the following:

(1) A tank, surface impoundment, or container for which all hazardous waste entering the unit has an average

VO concentration at the point of waste origination of less than 500 parts per million by weight (ppmw). The average VO concentration shall be determined using the procedures specified in §264.1083(a) of this subpart. The owner or operator shall review and update, as necessary, this determination at least once every 12 months following the date of the initial determination for the hazardous waste streams entering the unit.

(2) A tank, surface impoundment, or container for which the organic content of all the hazardous waste entering the waste management unit has been reduced by an organic destruction or removal process that achieves any one of the following conditions:

(i) A process that removes or destroys the organics contained in the hazardous waste to a level such that the average VO concentration of the hazardous waste at the point of waste treatment is less than the exit concentration limit ($C_t$) established for the process. The average VO concentration of the hazardous waste at the point of waste treatment and the exit concentration limit for the process shall be determined using the procedures specified in §264.1083(b) of this subpart.

(ii) A process that removes or destroys the organics contained in the hazardous waste to a level such that the organic reduction efficiency (R) for the process is equal to or greater than 95 percent, and the average VO concentration of the hazardous waste at the point of waste treatment is less than 100 ppmw. The organic reduction efficiency for the process and the average VO concentration of the hazardous waste at the point of waste treatment shall be determined using the procedures specified in §264.1083(b) of this subpart.

(iii) A process that removes or destroys the organics contained in the hazardous waste to a level such that the actual organic mass removal rate (MR) for the process is equal to or greater than the required organic mass removal rate (RMR) established for the process. The required organic mass removal rate and the actual organic mass removal rate for the process shall be determined using the procedures specified in §264.1083(b) of this subpart.

(iv) A biological process that destroys or degrades the organics contained in the hazardous waste, such that either of the following conditions is met:

(A) The organic reduction efficiency (R) for the process is equal to or greater than 95 percent, and the organic biodegradation efficiency ($R_{bio}$) for the process is equal to or greater than 95 percent. The organic reduction efficiency and the organic biodegradation efficiency for the process shall be determined using the procedures specified in §264.1083(b) of this subpart.

(B) The total actual organic mass biodegradation rate ($MR_{bio}$) for all hazardous waste treated by the process is equal to or greater than the required organic mass removal rate (RMR). The required organic mass removal rate and the actual organic mass biodegradation rate for the process shall be determined using the procedures specified in §264.1083(b) of this subpart.

(v) A process that removes or destroys the organics contained in the hazardous waste and meets all of the following conditions:

(A) From the point of waste origination through the point where the hazardous waste enters the treatment process, the hazardous waste is managed continuously in waste management units which use air emission controls in accordance with the standards specified in §264.1084 through §264.1087 of this subpart, as applicable to the waste management unit.

(B) From the point of waste origination through the point where the hazardous waste enters the treatment process, any transfer of the hazardous waste is accomplished through continuous hard-piping or other closed system transfer that does not allow exposure of the waste to the atmosphere. The EPA considers a drain system that meets the requirements of 40 CFR part 63, subpart RR—National Emission Standards for Individual Drain Systems to be a closed system.

(C) The average VO concentration of the hazardous waste at the point of waste treatment is less than the lowest average VO concentration at the point of waste origination determined for each of the individual waste streams entering the process or 500 ppmw, whichever value is lower. The average VO concentration of each individual waste stream at the point of waste origination shall be determined using the procedures specified in § 264.1083(a) of this subpart. The average VO concentration of the hazardous waste at the point of waste treatment shall be determined using the procedures specified in § 264.1083(b) of this subpart.

(vi) A process that removes or destroys the organics contained in the hazardous waste to a level such that the organic reduction efficiency (R) for the process is equal to or greater than 95 percent and the owner or operator certifies that the average VO concentration at the point of waste origination for each of the individual waste streams entering the process is less than 10,000 ppmw. The organic reduction efficiency for the process and the average VO concentration of the hazardous waste at the point of waste origination shall be determined using the procedures specified in § 264.1083(b) and § 264.1083(a) of this subpart, respectively.

(vii) A hazardous waste incinerator for which the owner or operator has either:

(A) Been issued a final permit under 40 CFR part 270 which implements the requirements of subpart O of this part; or

(B) Has designed and operates the incinerator in accordance with the interim status requirements of 40 CFR part 265, subpart O.

(viii) A boiler or industrial furnace for which the owner or operator has either:

(A) Been issued a final permit under 40 CFR part 270 which implements the requirements of 40 CFR part 266, subpart H, or

(B) Has designed and operates the boiler or industrial furnace in accordance with the interim status requirements of 40 CFR part 266, subpart H.

(ix) For the purpose of determining the performance of an organic destruction or removal process in accordance with the conditions in each of paragraphs (c)(2)(i) through (c)(2)(vi) of this section, the owner or operator shall account for VO concentrations determined to be below the limit of detec-

tion of the analytical method by using the following VO concentration:

(A) If Method 25D in 40 CFR part 60, appendix A is used for the analysis, one-half the blank value determined in the method at section 4.4 of Method 25D in 40 CFR part 60, appendix A, or a value of 25 ppmw, whichever is less.

(B) If any other analytical method is used, one-half the sum of the limits of detection established for each organic constituent in the waste that has a Henry's law constant value at least 0.1 mole-fraction-in-the-gas-phase/mole-fraction-in-the-liquid-phase (0.1 Y/X) [which can also be expressed as $1.8 \times 10^{-6}$ atmospheres/gram-mole/m$^3$] at 25 degrees Celsius.

(3) A tank or surface impoundment used for biological treatment of hazardous waste in accordance with the requirements of paragraph (c)(2)(iv) of this section.

(4) A tank, surface impoundment, or container for which all hazardous waste placed in the unit either:

(i) Meets the numerical concentration limits for organic hazardous constituents, applicable to the hazardous waste, as specified in 40 CFR part 268—Land Disposal Restrictions under Table "Treatment Standards for Hazardous Waste" in 40 CFR 268.40; or

(ii) The organic hazardous constituents in the waste have been treated by the treatment technology established by the EPA for the waste in 40 CFR 268.42(a), or have been removed or destroyed by an equivalent method of treatment approved by EPA pursuant to 40 CFR 268.42(b).

(5) A tank used for bulk feed of hazardous waste to a waste incinerator and all of the following conditions are met:

(i) The tank is located inside an enclosure vented to a control device that is designed and operated in accordance with all applicable requirements specified under 40 CFR part 61, subpart FF—National Emission Standards for Benzene Waste Operations for a facility at which the total annual benzene quantity from the facility waste is equal to or greater than 10 megagrams per year;

(ii) The enclosure and control device serving the tank were installed and began operation prior to November 25, 1996 and

(iii) The enclosure is designed and operated in accordance with the criteria for a permanent total enclosure as specified in "Procedure T—Criteria for and Verification of a Permanent or Temporary Total Enclosure" under 40 CFR 52.741, appendix B. The enclosure may have permanent or temporary openings to allow worker access; passage of material into or out of the enclosure by conveyor, vehicles, or other mechanical or electrical equipment; or to direct air flow into the enclosure. The owner or operator shall perform the verification procedure for the enclosure as specified in Section 5.0 to "Procedure T—Criteria for and Verification of a Permanent or Temporary Total Enclosure" annually.

(d) The Regional Administrator may at any time perform or request that the owner or operator perform a waste determination for a hazardous waste managed in a tank, surface impoundment, or container exempted from using air emission controls under the provisions of this section as follows:

(1) The waste determination for average VO concentration of a hazardous waste at the point of waste origination shall be performed using direct measurement in accordance with the applicable requirements of §264.1083(a) of this subpart. The waste determination for a hazardous waste at the point of waste treatment shall be performed in accordance with the applicable requirements of §264.1083(b) of this subpart.

(2) In performing a waste determination pursuant to paragraph (d)(1) of this section, the sample preparation and analysis shall be conducted as follows:

(i) In accordance with the method used by the owner or operator to perform the waste analysis, except in the case specified in paragraph (d)(2)(ii) of this section.

(ii) If the Regional Administrator determines that the method used by the owner or operator was not appropriate for the hazardous waste managed in the tank, surface impoundment, or container, then the Regional Administrator may choose an appropriate method.

(3) In a case when the owner or operator is requested to perform the waste determination, the Regional Adminis-

trator may elect to have an authorized representative observe the collection of the hazardous waste samples used for the analysis.

(4) In a case when the results of the waste determination performed or requested by the Regional Administrator do not agree with the results of a waste determination performed by the owner or operator using knowledge of the waste, then the results of the waste determination performed in accordance with the requirements of paragraph (d)(1) of this section shall be used to establish compliance with the requirements of this subpart.

(5) In a case when the owner or operator has used an averaging period greater than 1 hour for determining the average VO concentration of a hazardous waste at the point of waste origination, the Regional Administrator may elect to establish compliance with this subpart by performing or requesting that the owner or operator perform a waste determination using direct measurement based on waste samples collected within a 1-hour period as follows:

(i) The average VO concentration of the hazardous waste at the point of waste origination shall be determined by direct measurement in accordance with the requirements of §264.1083(a) of this subpart.

(ii) Results of the waste determination performed or requested by the Regional Administrator showing that the average VO concentration of the hazardous waste at the point of waste origination is equal to or greater than 500 ppmw shall constitute noncompliance with this subpart except in a case as provided for in paragraph (d)(5)(iii) of this section.

(iii) For the case when the average VO concentration of the hazardous waste at the point of waste origination previously has been determined by the owner or operator using an averaging period greater than 1 hour to be less than 500 ppmw but because of normal operating process variations the VO concentration of the hazardous waste determined by direct measurement for any given 1-hour period may be equal to or greater than 500 ppmw, information that was used by the owner or operator to determine the average VO concentration of the hazardous waste (e.g., test results, measurements, calculations, and other documentation) and recorded in the facility records in accordance with the requirements of § 264.1083(a) and § 264.1089 of this subpart shall be considered by the Regional Administrator together with the results of the waste determination performed or requested by the Regional Administrator in establishing compliance with this subpart.

[61 FR 59953, Nov. 25, 1996, as amended at 62 FR 64658, Dec. 8, 1997]

### § 264.1083 Waste determination procedures.

(a) Waste determination procedure to determine average volatile organic (VO) concentration of a hazardous waste at the point of waste origination.

(1) An owner or operator shall determine the average VO concentration at the point of waste origination for each hazardous waste placed in a waste management unit exempted under the provisions of § 264.1082(c)(1) of this subpart from using air emission controls in accordance with standards specified in § 264.1084 through § 264.1087 of this subpart, as applicable to the waste management unit.

(i) An initial determination of the average VO concentration of the waste stream shall be made before the first time any portion of the material in the hazardous waste stream is placed in a waste management unit exempted under the provisions of § 264.1082(c)(1) of this subpart from using air emission controls, and thereafter an initial determination of the average VO concentration of the waste stream shall be made for each averaging period that a hazardous waste is managed in the unit; and

(ii) Perform a new waste determination whenever changes to the source generating the waste stream are reasonably likely to cause the average VO concentration of the hazardous waste to increase to a level that is equal to or greater than the applicable VO concentration limits specified in § 264.1082 of this subpart.

(2) For a waste determination that is required by paragraph (a)(1) of this section, the average VO concentration of a hazardous waste at the point of waste origination shall be determined in accordance with the procedures specified in 40 CFR 265.1084(a)(2) through (a)(4).

(b) Waste determination procedures for treated hazardous waste.

(1) An owner or operator shall perform the applicable waste determinations for each treated hazardous waste placed in waste management units exempted under the provisions of § 264.1082(c)(2)(i) through (c)(2)(vi) of this subpart from using air emission controls in accordance with standards specified in §§ 264.1084 through 264.1087 of this subpart, as applicable to the waste management unit.

(i) An initial determination of the average VO concentration of the waste stream shall be made before the first time any portion of the material in the treated waste stream is placed in the exempt waste management unit, and thereafter update the information used for the waste determination at least once every 12 months following the date of the initial waste determination; and

(ii) Perform a new waste determination whenever changes to the process generating or treating the waste stream are reasonably likely to cause the average VO concentration of the hazardous waste to increase to a level such that the applicable treatment conditions specified in § 264.1082 (c)(2) of this subpart are not achieved.

(2) The waste determination for a treated hazardous waste shall be performed in accordance with the procedures specified in 40 CFR 265.1084 (b)(2) through (b)(9), as applicable to the treated hazardous waste.

(c) Procedure to determine the maximum organic vapor pressure of a hazardous waste in a tank.

(1) An owner or operator shall determine the maximum organic vapor pressure for each hazardous waste placed in a tank using Tank Level 1 controls in accordance with standards specified in § 264.1084(c) of this subpart.

(2) The maximum organic vapor pressure of the hazardous waste may be determined in accordance with the procedures specified in 40 CFR 265.1084 (c)(2) through (c)(4).

(d) The procedure for determining no detectable organic emissions for the purpose of complying with this subpart shall be conducted in accordance with the procedures specified in 40 CFR 265.1084(d).

[61 FR 59954, Nov. 25, 1996, as amended at 62 FR 64658, Dec. 8, 1997; 64 FR 3389, Jan. 21, 1999]

§264.1084   Standards: Tanks.

(a) The provisions of this section apply to the control of air pollutant emissions from tanks for which §264.1082(b) of this subpart references the use of this section for such air emission control.

(b) The owner or operator shall control air pollutant emissions from each tank subject to this section in accordance with the following requirements as applicable:

(1) For a tank that manages hazardous waste that meets all of the conditions specified in paragraphs (b)(1)(i) through (b)(1)(iii) of this section, the owner or operator shall control air pollutant emissions from the tank in accordance with the Tank Level 1 controls specified in paragraph (c) of this section or the Tank Level 2 controls specified in paragraph (d) of this section.

(i) The hazardous waste in the tank has a maximum organic vapor pressure which is less than the maximum organic vapor pressure limit for the tank's design capacity category as follows:

(A) For a tank design capacity equal to or greater than 151 m³, the maximum organic vapor pressure limit for the tank is 5.2 kPa.

(B) For a tank design capacity equal to or greater than 75 m³ but less than 151 m³, the maximum organic vapor pressure limit for the tank is 27.6 kPa.

(C) For a tank design capacity less than 75 m³, the maximum organic vapor pressure limit for the tank is 76.6 kPa.

(ii) The hazardous waste in the tank is not heated by the owner or operator to a temperature that is greater than the temperature at which the maximum organic vapor pressure of the hazardous waste is determined for the purpose of complying with paragraph (b)(1)(i) of this section.

(iii) The hazardous waste in the tank is not treated by the owner or operator

using a waste stabilization process, as defined in 40 CFR 265.1081.

(2) For a tank that manages hazardous waste that does not meet all of the conditions specified in paragraphs (b)(1)(i) through (b)(1)(iii) of this section, the owner or operator shall control air pollutant emissions from the tank by using Tank Level 2 controls in accordance with the requirements of paragraph (d) of this section. Examples of tanks required to use Tank Level 2 controls include: A tank used for a waste stabilization process; and a tank for which the hazardous waste in the tank has a maximum organic vapor pressure that is equal to or greater than the maximum organic vapor pressure limit for the tank's design capacity category as specified in paragraph (b)(1)(i) of this section.

(c) Owners and operators controlling air pollutant emissions from a tank using Tank Level 1 controls shall meet the requirements specified in paragraphs (c)(1) through (c)(4) of this section:

(1) The owner or operator shall determine the maximum organic vapor pressure for a hazardous waste to be managed in the tank using Tank Level 1 controls before the first time the hazardous waste is placed in the tank. The maximum organic vapor pressure shall be determined using the procedures specified in §264.1083(c) of this subpart. Thereafter, the owner or operator shall perform a new determination whenever changes to the hazardous waste managed in the tank could potentially cause the maximum organic vapor pressure to increase to a level that is equal to or greater than the maximum organic vapor pressure limit for the tank design capacity category specified in paragraph (b)(1)(i) of this section, as applicable to the tank.

(2) The tank shall be equipped with a fixed roof designed to meet the following specifications:

(i) The fixed roof and its closure devices shall be designed to form a continuous barrier over the entire surface area of the hazardous waste in the tank. The fixed roof may be a separate cover installed on the tank (e.g., a removable cover mounted on an open-top tank) or may be an integral part of the tank structural design (e.g., a horizontal cylindrical tank equipped with a hatch).

(ii) The fixed roof shall be installed in a manner such that there are no visible cracks, holes, gaps, or other open spaces between roof section joints or between the interface of the roof edge and the tank wall.

(iii) Each opening in the fixed roof, and any manifold system associated with the fixed roof, shall be either:

(A) Equipped with a closure device designed to operate such that when the closure device is secured in the closed position there are no visible cracks, holes, gaps, or other open spaces in the closure device or between the perimeter of the opening and the closure device; or

(B) Connected by a closed-vent system that is vented to a control device. The control device shall remove or destroy organics in the vent stream, and shall be operating whenever hazardous waste is managed in the tank, except as provided for in paragraphs (c)(2)(iii)(B) (1) and (2) of this section.

(1) During periods when it is necessary to provide access to the tank for performing the activities of paragraph (c)(2)(iii)(B)(2) of this section, venting of the vapor headspace underneath the fixed roof to the control device is not required, opening of closure devices is allowed, and removal of the fixed roof is allowed. Following completion of the activity, the owner or operator shall promptly secure the closure device in the closed position or reinstall the cover, as applicable, and resume operation of the control device.

(2) During periods of routine inspection, maintenance, or other activities needed for normal operations, and for removal of accumulated sludge or other residues from the bottom of the tank.

(iv) The fixed roof and its closure devices shall be made of suitable materials that will minimize exposure of the hazardous waste to the atmosphere, to the extent practical, and will maintain the integrity of the fixed roof and closure devices throughout their intended service life. Factors to be considered when selecting the materials for and designing the fixed roof and closure devices shall include: Organic vapor permeability, the effects of any contact with the hazardous waste or its vapors managed in the tank; the effects of outdoor exposure to wind, moisture, and sunlight; and the operating practices used for the tank on which the fixed roof is installed.

(3) Whenever a hazardous waste is in the tank, the fixed roof shall be installed with each closure device secured in the closed position except as follows:

(i) Opening of closure devices or removal of the fixed roof is allowed at the following times:

(A) To provide access to the tank for performing routine inspection, maintenance, or other activities needed for normal operations. Examples of such activities include those times when a worker needs to open a port to sample the liquid in the tank, or when a worker needs to open a hatch to maintain or repair equipment. Following completion of the activity, the owner or operator shall promptly secure the closure device in the closed position or reinstall the cover, as applicable, to the tank.

(B) To remove accumulated sludge or other residues from the bottom of tank.

(ii) Opening of a spring-loaded pressure-vacuum relief valve, conservation vent, or similar type of pressure relief device which vents to the atmosphere is allowed during normal operations for the purpose of maintaining the tank internal pressure in accordance with the tank design specifications. The device shall be designed to operate with no detectable organic emissions when the device is secured in the closed position. The settings at which the device opens shall be established such that the device remains in the closed position whenever the tank internal pressure is within the internal pressure operating range determined by the owner or operator based on the tank manufacturer recommendations, applicable regulations, fire protection and prevention codes, standard engineering codes and practices, or other requirements for the safe handling of flammable, ignitable, explosive, reactive, or hazardous materials. Examples of normal operating conditions that may require these devices to open are during those times when the tank internal pressure exceeds the internal pressure operating range for the tank as a result of loading operations or diurnal ambient temperature fluctuations.

(iii) Opening of a safety device, as defined in 40 CFR 265.1081, is allowed at any time conditions require doing so to avoid an unsafe condition.

(4) The owner or operator shall inspect the air emission control equipment in accordance with the following requirements.

(i) The fixed roof and its closure devices shall be visually inspected by the owner or operator to check for defects that could result in air pollutant emissions. Defects include, but are not limited to, visible cracks, holes, or gaps in the roof sections or between the roof and the tank wall; broken, cracked, or otherwise damaged seals or gaskets on closure devices; and broken or missing hatches, access covers, caps, or other closure devices.

(ii) The owner or operator shall perform an initial inspection of the fixed roof and its closure devices on or before the date that the tank becomes subject to this section. Thereafter, the owner or operator shall perform the inspections at least once every year except under the special conditions provided for in paragraph (l) of this section.

(iii) In the event that a defect is detected, the owner or operator shall repair the defect in accordance with the requirements of paragraph (k) of this section.

(iv) The owner or operator shall maintain a record of the inspection in accordance with the requirements specified in §264.1089(b) of this subpart.

(d) Owners and operators controlling air pollutant emissions from a tank using Tank Level 2 controls shall use one of the following tanks:

(1) A fixed-roof tank equipped with an internal floating roof in accordance with the requirements specified in paragraph (e) of this section;

(2) A tank equipped with an external floating roof in accordance with the requirements specified in paragraph (f) of this section;

(3) A tank vented through a closed-vent system to a control device in accordance with the requirements specified in paragraph (g) of this section;

(4) A pressure tank designed and operated in accordance with the requirements specified in paragraph (h) of this section; or

(5) A tank located inside an enclosure that is vented through a closed-vent system to an enclosed combustion control device in accordance with the requirements specified in paragraph (i) of this section.

(e) The owner or operator who controls air pollutant emissions from a tank using a fixed roof with an internal floating roof shall meet the requirements specified in paragraphs (e)(1) through (e)(3) of this section.

(1) The tank shall be equipped with a fixed roof and an internal floating roof in accordance with the following requirements:

(i) The internal floating roof shall be designed to float on the liquid surface except when the floating roof must be supported by the leg supports.

(ii) The internal floating roof shall be equipped with a continuous seal between the wall of the tank and the floating roof edge that meets either of the following requirements:

(A) A single continuous seal that is either a liquid-mounted seal or a metallic shoe seal, as defined in 40 CFR 265.1081; or

(B) Two continuous seals mounted one above the other. The lower seal may be a vapor-mounted seal.

(iii) The internal floating roof shall meet the following specifications:

(A) Each opening in a noncontact internal floating roof except for automatic bleeder vents (vacuum breaker vents) and the rim space vents is to provide a projection below the liquid surface.

(B) Each opening in the internal floating roof shall be equipped with a gasketed cover or a gasketed lid except for leg sleeves, automatic bleeder vents, rim space vents, column wells, ladder wells, sample wells, and stub drains.

(C) Each penetration of the internal floating roof for the purpose of sampling shall have a slit fabric cover that covers at least 90 percent of the opening.

(D) Each automatic bleeder vent and rim space vent shall be gasketed.

545

(E) Each penetration of the internal floating roof that allows for passage of a ladder shall have a gasketed sliding cover.

(F) Each penetration of the internal floating roof that allows for passage of a column supporting the fixed roof shall have a flexible fabric sleeve seal or a gasketed sliding cover.

(2) The owner or operator shall operate the tank in accordance with the following requirements:

(i) When the floating roof is resting on the leg supports, the process of filling, emptying, or refilling shall be continuous and shall be completed as soon as practical.

(ii) Automatic bleeder vents are to be set closed at all times when the roof is floating, except when the roof is being floated off or is being landed on the leg supports.

(iii) Prior to filling the tank, each cover, access hatch, gauge float well or lid on any opening in the internal floating roof shall be bolted or fastened closed (i.e., no visible gaps). Rim space vents are to be set to open only when the internal floating roof is not floating or when the pressure beneath the rim exceeds the manufacturer's recommended setting.

(3) The owner or operator shall inspect the internal floating roof in accordance with the procedures specified as follows:

(i) The floating roof and its closure devices shall be visually inspected by the owner or operator to check for defects that could result in air pollutant emissions. Defects include, but are not limited to: The internal floating roof is not floating on the surface of the liquid inside the tank; liquid has accumulated on top of the internal floating roof; any portion of the roof seals have detached from the roof rim; holes, tears, or other openings are visible in the seal fabric; the gaskets no longer close off the hazardous waste surface from the atmosphere; or the slotted membrane has more than 10 percent open area.

(ii) The owner or operator shall inspect the internal floating roof components as follows except as provided in paragraph (e)(3)(iii) of this section:

(A) Visually inspect the internal floating roof components through openings on the fixed-roof (e.g., man-holes and roof hatches) at least once every 12 months after initial fill, and

(B) Visually inspect the internal floating roof, primary seal, secondary seal (if one is in service), gaskets, slotted membranes, and sleeve seals (if any) each time the tank is emptied and degassed and at least every 10 years.

(iii) As an alternative to performing the inspections specified in paragraph (e)(3)(ii) of this section for an internal floating roof equipped with two continuous seals mounted one above the other, the owner or operator may visually inspect the internal floating roof, primary and secondary seals, gaskets, slotted membranes, and sleeve seals (if any) each time the tank is emptied and degassed and at least every 5 years.

(iv) Prior to each inspection required by paragraph (e)(3)(ii) or (e)(3)(iii) of this section, the owner or operator shall notify the Regional Administrator in advance of each inspection to provide the Regional Administrator with the opportunity to have an observer present during the inspection. The owner or operator shall notify the Regional Administrator of the date and location of the inspection as follows:

(A) Prior to each visual inspection of an internal floating roof in a tank that has been emptied and degassed, written notification shall be prepared and sent by the owner or operator so that it is received by the Regional Administrator at least 30 calendar days before refilling the tank except when an inspection is not planned as provided for in paragraph (e)(3)(iv)(B) of this section.

(B) When a visual inspection is not planned and the owner or operator could not have known about the inspection 30 calendar days before refilling the tank, the owner or operator shall notify the Regional Administrator as soon as possible, but no later than 7 calendar days before refilling of the tank. This notification may be made by telephone and immediately followed by a written explanation for why the inspection is unplanned. Alternatively, written notification, including the explanation for the unplanned inspection, may be sent so that it is received by the Regional Administrator at least 7 calendar days before refilling the tank.

(v) In the event that a defect is detected, the owner or operator shall repair the defect in accordance with the requirements of paragraph (k) of this section.

(vi) The owner or operator shall maintain a record of the inspection in accordance with the requirements specified in §264.1089(b) of this subpart.

(4) Safety devices, as defined in 40 CFR 265.1081, may be installed and operated as necessary on any tank complying with the requirements of paragraph (e) of this section.

(f) The owner or operator who controls air pollutant emissions from a tank using an external floating roof shall meet the requirements specified in paragraphs (f)(1) through (f)(3) of this section.

(1) The owner or operator shall design the external floating roof in accordance with the following requirements:

(i) The external floating roof shall be designed to float on the liquid surface except when the floating roof must be supported by the leg supports.

(ii) The floating roof shall be equipped with two continuous seals, one above the other, between the wall of the tank and the roof edge. The lower seal is referred to as the primary seal, and the upper seal is referred to as the secondary seal.

(A) The primary seal shall be a liquid-mounted seal or a metallic shoe seal, as defined in 40 CFR 265.1081. The total area of the gaps between the tank wall and the primary seal shall not exceed 212 square centimeters (cm²) per meter of tank diameter, and the width of any portion of these gaps shall not exceed 3.8 centimeters (cm). If a metallic shoe seal is used for the primary seal, the metallic shoe seal shall be designed so that one end extends into the liquid in the tank and the other end extends a vertical distance of at least 61 centimeters above the liquid surface.

(B) The secondary seal shall be mounted above the primary seal and cover the annular space between the floating roof and the wall of the tank. The total area of the gaps between the tank wall and the secondary seal shall not exceed 21.2 square centimeters (cm²) per meter of tank diameter, and

the width of any portion of these gaps shall not exceed 1.3 centimeters (cm).

(iii) The external floating roof shall meet the following specifications:

(A) Except for automatic bleeder vents (vacuum breaker vents) and rim space vents, each opening in a noncontact external floating roof shall provide a projection below the liquid surface.

(B) Except for automatic bleeder vents, rim space vents, roof drains, and leg sleeves, each opening in the roof shall be equipped with a gasketed cover, seal, or lid.

(C) Each access hatch and each gauge float well shall be equipped with a cover designed to be bolted or fastened when the cover is secured in the closed position.

(D) Each automatic bleeder vent and each rim space vent shall be equipped with a gasket.

(E) Each roof drain that empties into the liquid managed in the tank shall be equipped with a slotted membrane fabric cover that covers at least 90 percent of the area of the opening.

(F) Each unslotted and slotted guide pole well shall be equipped with a gasketed sliding cover or a flexible fabric sleeve seal.

(G) Each unslotted guide pole shall be equipped with a gasketed cap on the end of the pole.

(H) Each slotted guide pole shall be equipped with a gasketed float or other device which closes off the liquid surface from the atmosphere.

(I) Each gauge hatch and each sample well shall be equipped with a gasketed cover.

(2) The owner or operator shall operate the tank in accordance with the following requirements:

(i) When the floating roof is resting on the leg supports, the process of filling, emptying, or refilling shall be continuous and shall be completed as soon as practical.

(ii) Except for automatic bleeder vents, rim space vents, roof drains, and leg sleeves, each opening in the roof shall be secured and maintained in a closed position at all times except when the closure device must be open for access.

(iii) Covers on each access hatch and each gauge float well shall be bolted or fastened when secured in the closed position.

(iv) Automatic bleeder vents shall be set closed at all times when the roof is floating, except when the roof is being floated off or is being landed on the leg supports.

(v) Rim space vents shall be set to open only at those times that the roof is being floated off the roof leg supports or when the pressure beneath the rim seal exceeds the manufacturer's recommended setting.

(vi) The cap on the end of each unslotted guide pole shall be secured in the closed position at all times except when measuring the level or collecting samples of the liquid in the tank.

(vii) The cover on each gauge hatch or sample well shall be secured in the closed position at all times except when the hatch or well must be opened for access.

(viii) Both the primary seal and the secondary seal shall completely cover the annular space between the external floating roof and the wall of the tank in a continuous fashion except during inspections.

(3) The owner or operator shall inspect the external floating roof in accordance with the procedures specified as follows:

(i) The owner or operator shall measure the external floating roof seal gaps in accordance with the following requirements:

(A) The owner or operator shall perform measurements of gaps between the tank wall and the primary seal within 60 calendar days after initial operation of the tank following installation of the floating roof and, thereafter, at least once every 5 years.

(B) The owner or operator shall perform measurements of gaps between the tank wall and the secondary seal within 60 calendar days after initial operation of the tank following installation of the floating roof and, thereafter, at least once every year.

(C) If a tank ceases to hold hazardous waste for a period of 1 year or more, subsequent introduction of hazardous waste into the tank shall be considered an initial operation for the purposes of paragraphs (f)(3)(i)(A) and (f)(3)(i)(B) of this section.

(D) The owner or operator shall determine the total surface area of gaps in the primary seal and in the secondary seal individually using the following procedure:

(1) The seal gap measurements shall be performed at one or more floating roof levels when the roof is floating off the roof supports.

(2) Seal gaps, if any, shall be measured around the entire perimeter of the floating roof in each place where a 0.32-centimeter (cm) diameter uniform probe passes freely (without forcing or binding against the seal) between the seal and the wall of the tank and measure the circumferential distance of each such location.

(3) For a seal gap measured under paragraph (f)(3) of this section, the gap surface area shall be determined by using probes of various widths to measure accurately the actual distance from the tank wall to the seal and multiplying each such width by its respective circumferential distance.

(4) The total gap area shall be calculated by adding the gap surface areas determined for each identified gap location for the primary seal and the secondary seal individually, and then dividing the sum for each seal type by the nominal diameter of the tank. These total gap areas for the primary seal and secondary seal are then compared to the respective standards for the seal type as specified in paragraph (f)(1)(ii) of this section.

(E) In the event that the seal gap measurements do not conform to the specifications in paragraph (f)(1)(ii) of this section, the owner or operator shall repair the defect in accordance with the requirements of paragraph (k) of this section.

(F) The owner or operator shall maintain a record of the inspection in accordance with the requirements specified in § 264.1089(b) of this subpart.

(ii) The owner or operator shall visually inspect the external floating roof in accordance with the following requirements:

(A) The floating roof and its closure devices shall be visually inspected by the owner or operator to check for defects that could result in air pollutant emissions. Defects include, but are not limited to: Holes, tears, or other openings in the rim seal or seal fabric of the floating roof; a rim seal detached from the floating roof; all or a portion of the floating roof deck being submerged below the surface of the liquid in the tank; broken, cracked, or otherwise damaged seals or gaskets on closure devices; and broken or missing hatches, access covers, caps, or other closure devices.

(B) The owner or operator shall perform an initial inspection of the external floating roof and its closure devices on or before the date that the tank becomes subject to this section. Thereafter, the owner or operator shall perform the inspections at least once every year except for the special conditions provided for in paragraph (l) of this section.

(C) In the event that a defect is detected, the owner or operator shall repair the defect in accordance with the requirements of paragraph (k) of this section.

(D) The owner or operator shall maintain a record of the inspection in accordance with the requirements specified in §264.1089(b) of this subpart.

(iii) Prior to each inspection required by paragraph (f)(3)(i) or (f)(3)(ii) of this section, the owner or operator shall notify the Regional Administrator in advance of each inspection to provide the Regional Administrator with the opportunity to have an observer present during the inspection. The owner or operator shall notify the Regional Administrator of the date and location of the inspection as follows:

(A) Prior to each inspection to measure external floating roof seal gaps as required under paragraph (f)(3)(i) of this section, written notification shall be prepared and sent by the owner or operator so that it is received by the Regional Administrator at least 30 calendar days before the date the measurements are scheduled to be performed.

(B) Prior to each visual inspection of an external floating roof in a tank that has been emptied and degassed, written notification shall be prepared and sent by the owner or operator so that it is received by the Regional Administrator at least 30 calendar days before refilling the tank except when an inspection is not planned as provided for in paragraph (f)(3)(iii)(C) of this section.

(C) When a visual inspection is not planned and the owner or operator could not have known about the inspection 30 calendar days before refilling the tank, the owner or operator shall notify the Regional Administrator as soon as possible, but no later than 7 calendar days before refilling of the tank. This notification may be made by telephone and immediately followed by a written explanation for why the inspection is unplanned. Alternatively, written notification, including the explanation for the unplanned inspection, may be sent so that it is received by the Regional Administrator at least 7 calendar days before refilling the tank.

(4) Safety devices, as defined in 40 CFR 265.1081, may be installed and operated as necessary on any tank complying with the requirements of paragraph (f) of this section.

(g) The owner or operator who controls air pollutant emissions from a tank by venting the tank to a control device shall meet the requirements specified in paragraphs (g)(1) through (g)(3) of this section.

(1) The tank shall be covered by a fixed roof and vented directly through a closed-vent system to a control device in accordance with the following requirements:

(i) The fixed roof and its closure devices shall be designed to form a continuous barrier over the entire surface area of the liquid in the tank.

(ii) Each opening in the fixed roof not vented to the control device shall be equipped with a closure device. If the pressure in the vapor headspace underneath the fixed roof is less than atmospheric pressure when the control device is operating, the closure devices shall be designed to operate such that when the closure device is secured in the closed position there are no visible cracks, holes, gaps, or other open spaces in the closure device or between the perimeter of the cover opening and the closure device. If the pressure in the vapor headspace underneath the fixed roof is equal to or greater than atmospheric pressure when the control device is operating, the closure device shall be designed to operate with no detectable organic emissions.

(iii) The fixed roof and its closure devices shall be made of suitable materials that will minimize exposure of the hazardous waste to the atmosphere, to the extent practical, and will maintain the integrity of the fixed roof and closure devices throughout their intended service life. Factors to be considered when selecting the materials for and designing the fixed roof and closure devices shall include: Organic vapor permeability, the effects of any contact with the liquid and its vapor managed in the tank; the effects of outdoor exposure to wind, moisture, and sunlight; and the operating practices used for the tank on which the fixed roof is installed.

(iv) The closed-vent system and control device shall be designed and operated in accordance with the requirements of § 264.1087 of this subpart.

(2) Whenever a hazardous waste is in the tank, the fixed roof shall be installed with each closure device secured in the closed position and the vapor headspace underneath the fixed roof vented to the control device except as follows:

(i) Venting to the control device is not required, and opening of closure devices or removal of the fixed roof is allowed at the following times:

(A) To provide access to the tank for performing routine inspection, maintenance, or other activities needed for normal operations. Examples of such activities include those times when a worker needs to open a port to sample liquid in the tank, or when a worker needs to open a hatch to maintain or repair equipment. Following completion of the activity, the owner or operator shall promptly secure the closure device in the closed position or reinstall the cover, as applicable, to the tank.

(B) To remove accumulated sludge or other residues from the bottom of a tank.

(ii) Opening of a safety device, as defined in 40 CFR 265.1081, is allowed at any time conditions require doing so to avoid an unsafe condition.

(3) The owner or operator shall inspect and monitor the air emission control equipment in accordance with the following procedures:

(i) The fixed roof and its closure devices shall be visually inspected by the owner or operator to check for defects that could result in air pollutant emissions. Defects include, but are not limited to, visible cracks, holes, or gaps in the roof sections or between the roof and the tank wall; broken, cracked, or otherwise damaged seals or gaskets on closure devices; and broken or missing hatches, access covers, caps, or other closure devices.

(ii) The closed-vent system and control device shall be inspected and monitored by the owner or operator in accordance with the procedures specified in § 264.1087 of this subpart.

(iii) The owner or operator shall perform an initial inspection of the air emission control equipment on or before the date that the tank becomes subject to this section. Thereafter, the owner or operator shall perform the inspections at least once every year except for the special conditions provided for in paragraph (l) of this section.

(iv) In the event that a defect is detected, the owner or operator shall repair the defect in accordance with the requirements of paragraph (k) of this section.

(v) The owner or operator shall maintain a record of the inspection in accordance with the requirements specified in § 264.1089(b) of this subpart.

(h) The owner or operator who controls air pollutant emissions by using a pressure tank shall meet the following requirements.

(1) The tank shall be designed not to vent to the atmosphere as a result of compression of the vapor headspace in the tank during filling of the tank to its design capacity.

(2) All tank openings shall be equipped with closure devices designed to operate with no detectable organic emissions as determined using the procedure specified in § 264.1083(d) of this subpart.

(3) Whenever a hazardous waste is in the tank, the tank shall be operated as a closed system that does not vent to the atmosphere except under either or the following conditions as specified in paragraph (h)(3)(i) or (h)(3)(ii) of this section.

(i) At those times when opening of a safety device, as defined in §265.1081 of this subpart, is required to avoid an unsafe condition.

(ii) At those times when purging of inerts from the tank is required and the purge stream is routed to a closed-vent system and control device designed and operated in accordance with the requirements of §264.1087 of this subpart.

(i) The owner or operator who controls air pollutant emissions by using an enclosure vented through a closed-vent system to an enclosed combustion control device shall meet the requirements specified in paragraphs (i)(1) through (i)(4) of this section.

(1) The tank shall be located inside an enclosure. The enclosure shall be designed and operated in accordance with the criteria for a permanent total enclosure as specified in "Procedure T—Criteria for and Verification of a Permanent or Temporary Total Enclosure" under 40 CFR 52.741, appendix B. The enclosure may have permanent or temporary openings to allow worker access; passage of material into or out of the enclosure by conveyor, vehicles, or other mechanical means; entry of permanent mechanical or electrical equipment; or direct airflow into the enclosure. The owner or operator shall perform the verification procedure for the enclosure as specified in Section 5.0 to "Procedure T—Criteria for and Verification of a Permanent or Temporary Total Enclosure" initially when the enclosure is first installed and, thereafter, annually.

(2) The enclosure shall be vented through a closed-vent system to an enclosed combustion control device that is designed and operated in accordance with the standards for either a vapor incinerator, boiler, or process heater specified in §264.1087 of this subpart.

(3) Safety devices, as defined in 40 CFR 265.1081, may be installed and operated as necessary on any enclosure, closed-vent system, or control device used to comply with the requirements of paragraphs (i)(1) and (i)(2) of this section.

(4) The owner or operator shall inspect and monitor the closed-vent system and control device as specified in §264.1087 of this subpart.

(j) The owner or operator shall transfer hazardous waste to a tank subject to this section in accordance with the following requirements:

(1) Transfer of hazardous waste, except as provided in paragraph (j)(2) of this section, to the tank from another tank subject to this section or from a surface impoundment subject to §264.1085 of this subpart shall be conducted using continuous hard-piping or another closed system that does not allow exposure of the hazardous waste to the atmosphere. For the purpose of complying with this provision, an individual drain system is considered to be a closed system when it meets the requirements of 40 CFR part 63, subpart RR—National Emission Standards for Individual Drain Systems.

(2) The requirements of paragraph (j)(1) of this section do not apply when transferring a hazardous waste to the tank under any of the following conditions:

(i) The hazardous waste meets the average VO concentration conditions specified in §264.1082(c)(1) of this subpart at the point of waste origination.

(ii) The hazardous waste has been treated by an organic destruction or removal process to meet the requirements in §264.1082(c)(2) of this subpart.

(iii) The hazardous waste meets the requirements of §264.1082(c)(4) of this subpart.

(k) The owner or operator shall repair each defect detected during an inspection performed in accordance with the requirements of paragraph (c)(4), (e)(3), (f)(3), or (g)(3) of this section as follows:

(1) The owner or operator shall make first efforts at repair of the defect no later than 5 calendar days after detection, and repair shall be completed as soon as possible but no later than 45 calendar days after detection except as provided in paragraph (k)(2) of this section.

(2) Repair of a defect may be delayed beyond 45 calendar days if the owner or operator determines that repair of the defect requires emptying or temporary removal from service of the tank and no alternative tank capacity is available at the site to accept the hazardous waste normally managed in the tank. In this case, the owner or operator

shall repair the defect the next time the process or unit that is generating the hazardous waste managed in the tank stops operation. Repair of the defect shall be completed before the process or unit resumes operation.

(l) Following the initial inspection and monitoring of the cover as required by the applicable provisions of this subpart, subsequent inspection and monitoring may be performed at intervals longer than 1 year under the following special conditions:

(1) In the case when inspecting or monitoring the cover would expose a worker to dangerous, hazardous, or other unsafe conditions, then the owner or operator may designate a cover as an "unsafe to inspect and monitor cover" and comply with all of the following requirements:

(i) Prepare a written explanation for the cover stating the reasons why the cover is unsafe to visually inspect or to monitor, if required.

(ii) Develop and implement a written plan and schedule to inspect and monitor the cover, using the procedures specified in the applicable section of this subpart, as frequently as practicable during those times when a worker can safely access the cover.

(2) In the case when a tank is buried partially or entirely underground, an owner or operator is required to inspect and monitor, as required by the applicable provisions of this section, only those portions of the tank cover and those connections to the tank (e.g., fill ports, access hatches, gauge wells, etc.) that are located on or above the ground surface.

[61 FR 59955, Nov. 25, 1996, as amended at 62 FR 64659, Dec. 8, 1997; 64 FR 3389, Jan. 21, 1999]

## § 264.1085 Standards: Surface impoundments.

(a) The provisions of this section apply to the control of air pollutant emissions from surface impoundments for which § 264.1082(b) of this subpart references the use of this section for such air emission control.

(b) The owner or operator shall control air pollutant emissions from the surface impoundment by installing and operating either of the following:

(1) A floating membrane cover in accordance with the provisions specified in paragraph (c) of this section; or

(2) A cover that is vented through a closed-vent system to a control device in accordance with the provisions specified in paragraph (d) of this section.

(c) The owner or operator who controls air pollutant emissions from a surface impoundment using a floating membrane cover shall meet the requirements specified in paragraphs (c)(1) through (c)(3) of this section.

(1) The surface impoundment shall be equipped with a floating membrane cover designed to meet the following specifications:

(i) The floating membrane cover shall be designed to float on the liquid surface during normal operations and form a continuous barrier over the entire surface area of the liquid.

(ii) The cover shall be fabricated from a synthetic membrane material that is either:

(A) High density polyethylene (HDPE) with a thickness no less than 2.5 millimeters (mm); or

(B) A material or a composite of different materials determined to have both organic permeability properties that are equivalent to those of the material listed in paragraph (c)(1)(ii)(A) of this section and chemical and physical properties that maintain the material integrity for the intended service life of the material.

(iii) The cover shall be installed in a manner such that there are no visible cracks, holes, gaps, or other open spaces between cover section seams or between the interface of the cover edge and its foundation mountings.

(iv) Except as provided for in paragraph (c)(1)(v) of this section, each opening in the floating membrane cover shall be equipped with a closure device designed to operate such that when the closure device is secured in the closed position there are no visible cracks, holes, gaps, or other open spaces in the closure device or between the perimeter of the cover opening and the closure device.

(v) The floating membrane cover may be equipped with one or more emergency cover drains for removal of stormwater. Each emergency cover drain shall be equipped with a slotted membrane fabric cover that covers at least 90 percent of the area of the opening or a flexible fabric sleeve seal.

(vi) The closure devices shall be made of suitable materials that will minimize exposure of the hazardous waste to the atmosphere, to the extent practical, and will maintain the integrity of the closure devices throughout their intended service life. Factors to be considered when selecting the materials of construction and designing the cover and closure devices shall include: Organic vapor permeability; the effects of any contact with the liquid and its vapor managed in the surface impoundment; the effects of outdoor exposure to wind, moisture, and sunlight; and the operating practices used for the surface impoundment on which the floating membrane cover is installed.

(2) Whenever a hazardous waste is in the surface impoundment, the floating membrane cover shall float on the liquid and each closure device shall be secured in the closed position except as follows:

(i) Opening of closure devices or removal of the cover is allowed at the following times:

(A) To provide access to the surface impoundment for performing routine inspection, maintenance, or other activities needed for normal operations. Examples of such activities include those times when a worker needs to open a port to sample the liquid in the surface impoundment, or when a worker needs to open a hatch to maintain or repair equipment. Following completion of the activity, the owner or operator shall promptly replace the cover and secure the closure device in the closed position, as applicable.

(B) To remove accumulated sludge or other residues from the bottom of surface impoundment.

(ii) Opening of a safety device, as defined in 40 CFR 265.1081, is allowed at any time conditions require doing so to avoid an unsafe condition.

(3) The owner or operator shall inspect the floating membrane cover in accordance with the following procedures:

(i) The floating membrane cover and its closure devices shall be visually inspected by the owner or operator to check for defects that could result in air pollutant emissions. Defects include, but are not limited to, visible cracks, holes, or gaps in the cover section seams or between the interface of the cover edge and its foundation mountings; broken, cracked, or otherwise damaged seals or gaskets on closure devices; and broken or missing hatches, access covers, caps, or other closure devices.

(ii) The owner or operator shall perform an initial inspection of the floating membrane cover and its closure devices on or before the date that the surface impoundment becomes subject to this section. Thereafter, the owner or operator shall perform the inspections at least once every year except for the special conditions provided for in paragraph (g) of this section.

(iii) In the event that a defect is detected, the owner or operator shall repair the defect in accordance with the requirements of paragraph (f) of this section.

(iv) The owner or operator shall maintain a record of the inspection in accordance with the requirements specified in §264.1089(c) of this subpart.

(d) The owner or operator who controls air pollutant emissions from a surface impoundment using a cover vented to a control device shall meet the requirements specified in paragraphs (d)(1) through (d)(3) of this section.

(1) The surface impoundment shall be covered by a cover and vented directly through a closed-vent system to a control device in accordance with the following requirements:

(i) The cover and its closure devices shall be designed to form a continuous barrier over the entire surface area of the liquid in the surface impoundment.

(ii) Each opening in the cover not vented to the control device shall be equipped with a closure device. If the pressure in the vapor headspace underneath the cover is less than atmospheric pressure when the control device is operating, the closure devices shall be designed to operate such that when the closure device is secured in the closed position there are no visible cracks, holes, gaps, or other open spaces in the closure device or between the perimeter of the cover opening and the closure device. If the pressure in the vapor headspace underneath the cover is equal to or greater than atmospheric pressure when the control device is operating, the closure device shall be designed to operate with no detectable organic emissions using the procedure specified in § 264.1083(d) of this subpart.

(iii) The cover and its closure devices shall be made of suitable materials that will minimize exposure of the hazardous waste to the atmosphere, to the extent practical, and will maintain the integrity of the cover and closure devices throughout their intended service life. Factors to be considered when selecting the materials of construction and designing the cover and closure devices shall include: Organic vapor permeability; the effects of any contact with the liquid or its vapors managed in the surface impoundment; the effects of outdoor exposure to wind, moisture, and sunlight; and the operating practices used for the surface impoundment on which the cover is installed.

(iv) The closed-vent system and control device shall be designed and operated in accordance with the requirements of § 264.1087 of this subpart.

(2) Whenever a hazardous waste is in the surface impoundment, the cover shall be installed with each closure device secured in the closed position and the vapor headspace underneath the cover vented to the control device except as follows:

(i) Venting to the control device is not required, and opening of closure devices or removal of the cover is allowed at the following times:

(A) To provide access to the surface impoundment for performing routine inspection, maintenance, or other activities needed for normal operations. Examples of such activities include those times when a worker needs to open a port to sample liquid in the surface impoundment, or when a worker needs to open a hatch to maintain or repair equipment. Following completion of the activity, the owner or operator shall promptly secure the closure device in the closed position or reinstall the cover, as applicable, to the surface impoundment.

(B) To remove accumulated sludge or other residues from the bottom of the surface impoundment.

(ii) Opening of a safety device, as defined in 40 CFR 265.1081, is allowed at any time conditions require doing so to avoid an unsafe condition.

(3) The owner or operator shall inspect and monitor the air emission control equipment in accordance with the following procedures:

(i) The surface impoundment cover and its closure devices shall be visually inspected by the owner or operator to check for defects that could result in air pollutant emissions. Defects include, but are not limited to, visible cracks, holes, or gaps in the cover section seams or between the interface of the cover edge and its foundation mountings; broken, cracked, or otherwise damaged seals or gaskets on closure devices; and broken or missing hatches, access covers, caps, or other closure devices.

(ii) The closed-vent system and control device shall be inspected and monitored by the owner or operator in accordance with the procedures specified in § 264.1087 of this subpart.

(iii) The owner or operator shall perform an initial inspection of the air emission control equipment on or before the date that the surface impoundment becomes subject to this section. Thereafter, the owner or operator shall perform the inspections at least once every year except for the special conditions provided for in paragraph (g) of this section.

(iv) In the event that a defect is detected, the owner or operator shall repair the defect in accordance with the requirements of paragraph (f) of this section.

(v) The owner or operator shall maintain a record of the inspection in accordance with the requirements specified in § 264.1089(c) of this subpart.

(e) The owner or operator shall transfer hazardous waste to a surface impoundment subject to this section in accordance with the following requirements:

(1) Transfer of hazardous waste, except as provided in paragraph (e)(2) of this section, to the surface impoundment from another surface impoundment subject to this section or from a

tank subject to §264.1084 of this subpart shall be conducted using continuous hard-piping or another closed system that does not allow exposure of the waste to the atmosphere. For the purpose of complying with this provision, an individual drain system is considered to be a closed system when it meets the requirements of 40 CFR part 63, subpart RR—National Emission Standards for Individual Drain Systems.

(2) The requirements of paragraph (e)(1) of this section do not apply when transferring a hazardous waste to the surface impoundment under either of the following conditions:

(i) The hazardous waste meets the average VO concentration conditions specified in §264.1082(c)(1) of this subpart at the point of waste origination.

(ii) The hazardous waste has been treated by an organic destruction or removal process to meet the requirements in §264.1082(c)(2) of this subpart.

(iii) The hazardous waste meets the requirements of §264.1082(c)(4) of this subpart.

(f) The owner or operator shall repair each defect detected during an inspection performed in accordance with the requirements of paragraph (c)(3) or (d)(3) of this section as follows:

(1) The owner or operator shall make first efforts at repair of the defect no later than 5 calendar days after detection and repair shall be completed as soon as possible but no later than 45 calendar days after detection except as provided in paragraph (f)(2) of this section.

(2) Repair of a defect may be delayed beyond 45 calendar days if the owner or operator determines that repair of the defect requires emptying or temporary removal from service of the surface impoundment and no alternative capacity is available at the site to accept the hazardous waste normally managed in the surface impoundment. In this case, the owner or operator shall repair the defect the next time the process or unit that is generating the hazardous waste managed in the surface impoundment stops operation. Repair of the defect shall be completed before the process or unit resumes operation.

(g) Following the initial inspection and monitoring of the cover as required by the applicable provisions of this subpart, subsequent inspection and monitoring may be performed at intervals longer than 1 year in the case when inspecting or monitoring the cover would expose a worker to dangerous, hazardous, or other unsafe conditions. In this case, the owner or operator may designate the cover as an "unsafe to inspect and monitor cover" and comply with all of the following requirements:

(1) Prepare a written explanation for the cover stating the reasons why the cover is unsafe to visually inspect or to monitor, if required.

(2) Develop and implement a written plan and schedule to inspect and monitor the cover using the procedures specified in the applicable section of this subpart as frequently as practicable during those times when a worker can safely access the cover.

[61 FR 59960, Nov. 25, 1996, as amended at 62 FR 64659, Dec. 8, 1997]

### §264.1086 Standards: Containers.

(a) The provisions of this section apply to the control of air pollutant emissions from containers for which §264.1082(b) of this subpart references the use of this section for such air emission control.

(b) *General requirements.* (1) The owner or operator shall control air pollutant emissions from each container subject to this section in accordance with the following requirements, as applicable to the container, except when the special provisions for waste stabilization processes specified in paragraph (b)(2) of this section apply to the container.

(i) For a container having a design capacity greater than 0.1 m³ and less than or equal to 0.46 m³, the owner or operator shall control air pollutant emissions from the container in accordance with the Container Level 1 standards specified in paragraph (c) of this section.

(ii) For a container having a design capacity greater than 0.46 m³ that is not in light material service, the owner or operator shall control air pollutant emissions from the container in accordance with the Container Level 1 standards specified in paragraph (c) of this section.

(iii) For a container having a design capacity greater than 0.46 m³ that is in light material service, the owner or operator shall control air pollutant emissions from the container in accordance with the Container Level 2 standards specified in paragraph (d) of this section.

(2) When a container having a design capacity greater than 0.1 m³ is used for treatment of a hazardous waste by a waste stabilization process, the owner or operator shall control air pollutant emissions from the container in accordance with the Container Level 3 standards specified in paragraph (e) of this section at those times during the waste stabilization process when the hazardous waste in the container is exposed to the atmosphere.

(c) *Container Level 1 standards.* (1) A container using Container Level 1 controls is one of the following:

(i) A container that meets the applicable U.S. Department of Transportation (DOT) regulations on packaging hazardous materials for transportation as specified in paragraph (f) of this section.

(ii) A container equipped with a cover and closure devices that form a continuous barrier over the container openings such that when the cover and closure devices are secured in the closed position there are no visible holes, gaps, or other open spaces into the interior of the container. The cover may be a separate cover installed on the container (e.g., a lid on a drum or a suitably secured tarp on a roll-off box) or may be an integral part of the container structural design (e.g., a "portable tank" or bulk cargo container equipped with a screw-type cap).

(iii) An open-top container in which an organic-vapor suppressing barrier is placed on or over the hazardous waste in the container such that no hazardous waste is exposed to the atmosphere. One example of such a barrier is application of a suitable organic-vapor suppressing foam.

(2) A container used to meet the requirements of paragraph (c)(1)(ii) or (c)(1)(iii) of this section shall be equipped with covers and closure devices, as applicable to the container, that are composed of suitable materials to minimize exposure of the hazardous waste to the atmosphere and to maintain the equipment integrity, for as long as the container is in service. Factors to be considered in selecting the materials of construction and designing the cover and closure devices shall include: Organic vapor permeability; the effects of contact with the hazardous waste or its vapor managed in the container; the effects of outdoor exposure of the closure device or cover material to wind, moisture, and sunlight; and the operating practices for which the container is intended to be used.

(3) Whenever a hazardous waste is in a container using Container Level 1 controls, the owner or operator shall install all covers and closure devices for the container, as applicable to the container, and secure and maintain each closure device in the closed position except as follows:

(i) Opening of a closure device or cover is allowed for the purpose of adding hazardous waste or other material to the container as follows:

(A) In the case when the container is filled to the intended final level in one continuous operation, the owner or operator shall promptly secure the closure devices in the closed position and install the covers, as applicable to the container, upon conclusion of the filling operation.

(B) In the case when discrete quantities or batches of material intermittently are added to the container over a period of time, the owner or operator shall promptly secure the closure devices in the closed position and install covers, as applicable to the container, upon either the container being filled to the intended final level; the completion of a batch loading after which no additional material will be added to the container within 15 minutes; the person performing the loading operation leaving the immediate vicinity of the container; or the shutdown of the process generating the material being added to the container, whichever condition occurs first.

(ii) Opening of a closure device or cover is allowed for the purpose of removing hazardous waste from the container as follows:

(A) For the purpose of meeting the requirements of this section, an empty container as defined in 40 CFR 261.7(b) may be open to the atmosphere at any time (i.e., covers and closure devices are not required to be secured in the closed position on an empty container).

(B) In the case when discrete quantities or batches of material are removed from the container but the container does not meet the conditions to be an empty container as defined in 40 CFR 261.7(b), the owner or operator shall promptly secure the closure devices in the closed position and install covers, as applicable to the container, upon the completion of a batch removal after which no additional material will be removed from the container within 15 minutes or the person performing the unloading operation leaves the immediate vicinity of the container, whichever condition occurs first.

(iii) Opening of a closure device or cover is allowed when access inside the container is needed to perform routine activities other than transfer of hazardous waste. Examples of such activities include those times when a worker needs to open a port to measure the depth of or sample the material in the container, or when a worker needs to open a manhole hatch to access equipment inside the container. Following completion of the activity, the owner or operator shall promptly secure the closure device in the closed position or reinstall the cover, as applicable to the container.

(iv) Opening of a spring-loaded pressure-vacuum relief valve, conservation vent, or similar type of pressure relief device which vents to the atmosphere is allowed during normal operations for the purpose of maintaining the internal pressure of the container in accordance with the container design specifications. The device shall be designed to operate with no detectable organic emissions when the device is secured in the closed position. The settings at which the device opens shall be established such that the device remains in the closed position whenever the internal pressure of the container is within the internal pressure operating range determined by the owner or operator based on container manufacturer recommendations, applicable regulations, fire protection and prevention codes, standard engineering codes and practices, or other requirements for the safe handling of flammable, ignitable, explosive, reactive, or hazardous materials. Examples of normal operating conditions that may require these devices to open are during those times when the internal pressure of the container exceeds the internal pressure operating range for the container as a result of loading operations or diurnal ambient temperature fluctuations.

(v) Opening of a safety device, as defined in 40 CFR 265.1081, is allowed at any time conditions require doing so to avoid an unsafe condition.

(4) The owner or operator of containers using Container Level 1 controls shall inspect the containers and their covers and closure devices as follows:

(i) In the case when a hazardous waste already is in the container at the time the owner or operator first accepts possession of the container at the facility and the container is not emptied within 24 hours after the container is accepted at the facility (i.e., does not meet the conditions for an empty container as specified in 40 CFR 261.7(b)), the owner or operator shall visually inspect the container and its cover and closure devices to check for visible cracks, holes, gaps, or other open spaces into the interior of the container when the cover and closure devices are secured in the closed position. The container visual inspection shall be conducted on or before the date that the container is accepted at the facility (i.e., the date the container becomes subject to the subpart CC container standards). For purposes of this requirement, the date of acceptance is the date of signature that the facility owner or operator enters on Item 20 of the Uniform Hazardous Waste Manifest in the appendix to 40 CFR part 262 (EPA Forms 8700–22 and 8700–22A), as required under subpart E of this part, at 40 CFR 264.71. If a defect is detected, the owner or operator shall repair the defect in accordance with the requirements of paragraph (c)(4)(iii) of this section.

(ii) In the case when a container used for managing hazardous waste remains at the facility for a period of 1 year or more, the owner or operator shall visually inspect the container and its cover and closure devices initially and thereafter, at least once every 12 months, to check for visible cracks, holes, gaps, or other open spaces into the interior of the container when the cover and closure devices are secured in the closed position. If a defect is detected, the owner or operator shall repair the defect in accordance with the requirements of paragraph (c)(4)(iii) of this section.

(iii) When a defect is detected for the container, cover, or closure devices, the owner or operator shall make first efforts at repair of the defect no later than 24 hours after detection and repair shall be completed as soon as possible but no later than 5 calendar days after detection. If repair of a defect cannot be completed within 5 calendar days, then the hazardous waste shall be removed from the container and the container shall not be used to manage hazardous waste until the defect is repaired.

(5) The owner or operator shall maintain at the facility a copy of the procedure used to determine that containers with capacity of 0.46 m³ or greater, which do not meet applicable DOT regulations as specified in paragraph (f) of this section, are not managing hazardous waste in light material service.

(d) *Container Level 2 standards.* (1) A container using Container Level 2 controls is one of the following:

(i) A container that meets the applicable U.S. Department of Transportation (DOT) regulations on packaging hazardous materials for transportation as specified in paragraph (f) of this section.

(ii) A container that operates with no detectable organic emissions as defined in 40 CFR 265.1081 and determined in accordance with the procedure specified in paragraph (g) of this section.

(iii) A container that has been demonstrated within the preceding 12 months to be vapor-tight by using 40 CFR part 60, appendix A, Method 27 in accordance with the procedure specified in paragraph (h) of this section.

(2) Transfer of hazardous waste in or out of a container using Container Level 2 controls shall be conducted in such a manner as to minimize exposure of the hazardous waste to the atmosphere, to the extent practical, considering the physical properties of the hazardous waste and good engineering and safety practices for handling flammable, ignitable, explosive, reactive, or other hazardous materials. Examples of container loading procedures that the EPA considers to meet the requirements of this paragraph include using any one of the following: A submerged-fill pipe or other submerged-fill method to load liquids into the container; a vapor-balancing system or a vapor-recovery system to collect and control the vapors displaced from the container during filling operations; or a fitted opening in the top of a container through which the hazardous waste is filled and subsequently purging the transfer line before removing it from the container opening.

(3) Whenever a hazardous waste is in a container using Container Level 2 controls, the owner or operator shall install all covers and closure devices for the container, and secure and maintain each closure device in the closed position except as follows:

(i) Opening of a closure device or cover is allowed for the purpose of adding hazardous waste or other material to the container as follows:

(A) In the case when the container is filled to the intended final level in one continuous operation, the owner or operator shall promptly secure the closure devices in the closed position and install the covers, as applicable to the container, upon conclusion of the filling operation.

(B) In the case when discrete quantities or batches of material intermittently are added to the container over a period of time, the owner or operator shall promptly secure the closure devices in the closed position and install covers, as applicable to the container, upon either the container being filled to the intended final level; the completion of a batch loading after which no additional material will be added to the container within 15 minutes; the person performing the loading operation leaving the immediate vicinity of the container; or the shutdown of the process generating the material being added to the container, whichever condition occurs first.

(ii) Opening of a closure device or cover is allowed for the purpose of removing hazardous waste from the container as follows:

(A) For the purpose of meeting the requirements of this section, an empty container as defined in 40 CFR 261.7(b) may be open to the atmosphere at any time (i.e., covers and closure devices are not required to be secured in the closed position on an empty container).

(B) In the case when discrete quantities or batches of material are removed from the container but the container does not meet the conditions to be an empty container as defined in 40 CFR 261.7(b), the owner or operator shall promptly secure the closure devices in the closed position and install covers, as applicable to the container, upon the completion of a batch removal after which no additional material will be removed from the container within 15 minutes or the person performing the unloading operation leaves the immediate vicinity of the container, whichever condition occurs first.

(iii) Opening of a closure device or cover is allowed when access inside the container is needed to perform routine activities other than transfer of hazardous waste.

Examples of such activities include those times when a worker needs to open a port to measure the depth of or sample the material in the container, or when a worker needs to open a manhole hatch to access equipment inside the container. Following completion of the activity, the owner or operator shall promptly secure the closure device in the closed position or reinstall the cover, as applicable to the container.

(iv) Opening of a spring-loaded, pressure-vacuum relief valve, conservation vent, or similar type of pressure relief device which vents to the atmosphere is allowed during normal operations for the purpose of maintaining the internal pressure of the container in accordance with the container design specifications. The device shall be designed to operate with no detectable organic emission when the device is secured in the closed position. The settings at which the device opens shall be established such that the device remains in the closed position whenever the internal pressure of the container is within the internal pressure operating range determined by the owner or operator based on container manufacturer recommendations, applicable regulations, fire protection and prevention codes, standard engineering codes and practices, or other requirements for the safe handling of flammable, ignitable, explosive, reactive, or hazardous materials. Examples of normal operating conditions that may require these devices to open are during those times when the internal pressure of the container exceeds the internal pressure operating range for the container as a result of loading operations or diurnal ambient temperature fluctuations.

(v) Opening of a safety device, as defined in 40 CFR 265.1081, is allowed at any time conditions require doing so to avoid an unsafe condition.

(4) The owner or operator of containers using Container Level 2 controls shall inspect the containers and their covers and closure devices as follows:

(i) In the case when a hazardous waste already is in the container at the time the owner or operator first accepts possession of the container at the facility and the container is not emptied within 24 hours after the container is accepted at the facility (i.e., does not meet the conditions for an empty container as specified in 40 CFR 261.7(b)), the owner or operator shall visually inspect the container and its cover and closure devices to check for visible cracks, holes, gaps, or other open spaces into the interior of the container when the cover and closure devices are secured in the closed position. The container visual inspection shall be conducted on or before the date that the container is accepted at the facility (i.e., the date the container becomes subject to the subpart CC container standards). For purposes of this requirement, the date of acceptance is the date of signature that the facility owner or operator enters on Item 20 of the Uniform Hazardous Waste Manifest in the appendix to 40 CFR part 262 (EPA Forms 8700-22 and 8700-22A), as required under subpart E of this part, at 40 CFR 264.71. If a defect is detected, the owner or operator shall repair the defect in accordance with the requirements of paragraph (d)(4)(iii) of this section.

(ii) In the case when a container used for managing hazardous waste remains at the facility for a period of 1 year or more, the owner or operator shall visually inspect the container and its cover and closure devices initially and thereafter, at least once every 12 months, to check for visible cracks, holes, gaps, or other open spaces into the interior of the container when the cover and closure devices are secured in the closed position. If a defect is detected, the owner or operator shall repair the defect in accordance with the requirements of paragraph (d)(4)(iii) of this section.

(iii) When a defect is detected for the container, cover, or closure devices, the owner or operator shall make first efforts at repair of the defect no later than 24 hours after detection, and repair shall be completed as soon as possible but no later than 5 calendar days after detection. If repair of a defect cannot be completed within 5 calendar days, then the hazardous waste shall be removed from the container and the container shall not be used to manage hazardous waste until the defect is repaired.

(e) *Container Level 3 standards.* (1) A container using Container Level 3 controls is one of the following:

(i) A container that is vented directly through a closed-vent system to a control device in accordance with the requirements of paragraph (e)(2)(ii) of this section.

(ii) A container that is vented inside an enclosure which is exhausted through a closed-vent system to a control device in accordance with the requirements of paragraphs (e)(2)(i) and (e)(2)(ii) of this section.

(2) The owner or operator shall meet the following requirements, as applicable to the type of air emission control equipment selected by the owner or operator:

(i) The container enclosure shall be designed and operated in accordance with the criteria for a permanent total enclosure as specified in "Procedure T—Criteria for and Verification of a Permanent or Temporary Total Enclosure" under 40 CFR 52.741, appendix B.

The enclosure may have permanent or temporary openings to allow worker access; passage of containers through the enclosure by conveyor or other mechanical means; entry of permanent mechanical or electrical equipment; or direct airflow into the enclosure. The owner or operator shall perform the verification procedure for the enclosure as specified in Section 5.0 to "Procedure T—Criteria for and Verification of a Permanent or Temporary Total Enclosure" initially when the enclosure is first installed and, thereafter, annually.

(ii) The closed-vent system and control device shall be designed and operated in accordance with the requirements of § 264.1087 of this subpart.

(3) Safety devices, as defined in 40 CFR 265.1081, may be installed and operated as necessary on any container, enclosure, closed-vent system, or control device used to comply with the requirements of paragraph (e)(1) of this section.

(4) Owners and operators using Container Level 3 controls in accordance with the provisions of this subpart shall inspect and monitor the closed-vent systems and control devices as specified in § 264.1087 of this subpart.

(5) Owners and operators that use Container Level 3 controls in accordance with the provisions of this subpart shall prepare and maintain the records specified in § 264.1089(d) of this subpart.

(6) Transfer of hazardous waste in or out of a container using Container Level 3 controls shall be conducted in such a manner as to minimize exposure of the hazardous waste to the atmosphere, to the extent practical, considering the physical properties of the hazardous waste and good engineering and safety practices for handling flammable, ignitable, explosive, reactive, or other hazardous materials. Examples of container loading procedures that the EPA considers to meet the requirements of this paragraph include using any one of the following: A submerged-fill pipe or other submerged-fill method to load liquids into the container; a vapor-balancing system or a vapor-recovery system to collect and control the vapors displaced from the container during filling operations; or a

fitted opening in the top of a container through which the hazardous waste is filled and subsequently purging the transfer line before removing it from the container opening.

(f) For the purpose of compliance with paragraph (c)(1)(i) or (d)(1)(i) of this section, containers shall be used that meet the applicable U.S. Department of Transportation (DOT) regulations on packaging hazardous materials for transportation as follows:

(1) The container meets the applicable requirements specified in 49 CFR part 178—Specifications for Packaging or 49 CFR part 179—Specifications for Tank Cars.

(2) Hazardous waste is managed in the container in accordance with the applicable requirements specified in 49 CFR part 107, subpart B—Exemptions; 49 CFR part 172—Hazardous Materials Table, Special Provisions, Hazardous Materials Communications, Emergency Response Information, and Training Requirements; 49 CFR part 173—Shippers—General Requirements for Shipments and Packages; and 49 CFR part 180—Continuing Qualification and Maintenance of Packagings.

(3) For the purpose of complying with this subpart, no exceptions to the 49 CFR part 178 or part 179 regulations are allowed except as provided for in paragraph (f)(4) of this section.

(4) For a lab pack that is managed in accordance with the requirements of 49 CFR part 178 for the purpose of complying with this subpart, an owner or operator may comply with the exceptions for combination packagings specified in 49 CFR 173.12(b).

(g) To determine compliance with the no detectable organic emissions requirement of paragraph (d)(1)(ii) of this section, the procedure specified in §264.1083(d) of this subpart shall be used.

(1) Each potential leak interface (i.e., a location where organic vapor leakage could occur) on the container, its cover, and associated closure devices, as applicable to the container, shall be checked. Potential leak interfaces that are associated with containers include, but are not limited to: The interface of the cover rim and the container wall; the periphery of any opening on the container or container cover and its as-

sociated closure device; and the sealing seat interface on a spring-loaded pressure-relief valve.

(2) The test shall be performed when the container is filled with a material having a volatile organic concentration representative of the range of volatile organic concentrations for the hazardous wastes expected to be managed in this type of container. During the test, the container cover and closure devices shall be secured in the closed position.

(h) Procedure for determining a container to be vapor-tight using Method 27 of 40 CFR part 60, appendix A for the purpose of complying with paragraph (d)(1)(iii) of this section.

(1) The test shall be performed in accordance with Method 27 of 40 CFR part 60, appendix A of this chapter.

(2) A pressure measurement device shall be used that has a precision of ±2.5 mm water and that is capable of measuring above the pressure at which the container is to be tested for vapor tightness.

(3) If the test results determined by Method 27 indicate that the container sustains a pressure change less than or equal to 750 Pascals within 5 minutes after it is pressurized to a minimum of 4,500 Pascals, then the container is determined to be vapor-tight.

[61 FR 59962, Nov. 25, 1996, as amended at 62 FR 64659, Dec. 8, 1997; 64 FR 3389, Jan. 21, 1999]

## §264.1087 Standards: Closed-vent systems and control devices.

(a) This section applies to each closed-vent system and control device installed and operated by the owner or operator to control air emissions in accordance with standards of this subpart.

(b) The closed-vent system shall meet the following requirements:

(1) The closed-vent system shall route the gases, vapors, and fumes emitted from the hazardous waste in the waste management unit to a control device that meets the requirements specified in paragraph (c) of this section.

(2) The closed-vent system shall be designed and operated in accordance with the requirements specified in §264.1033(k) of this part.

(3) In the case when the closed-vent system includes bypass devices that could be used to divert the gas or vapor stream to the atmosphere before entering the control device, each bypass device shall be equipped with either a flow indicator as specified in paragraph (b)(3)(i) of this section or a seal or locking device as specified in paragraph (b)(3)(ii) of this section. For the purpose of complying with this paragraph, low leg drains, high point bleeds, analyzer vents, open-ended valves or lines, spring loaded pressure relief valves, and other fittings used for safety purposes are not considered to be bypass devices.

(i) If a flow indicator is used to comply with paragraph (b)(3) of this section, the indicator shall be installed at the inlet to the bypass line used to divert gases and vapors from the closed-vent system to the atmosphere at a point upstream of the control device inlet. For this paragraph, a flow indicator means a device which indicates the presence of either gas or vapor flow in the bypass line.

(ii) If a seal or locking device is used to comply with paragraph (b)(3) of this section, the device shall be placed on the mechanism by which the bypass device position is controlled (e.g., valve handle, damper lever) when the bypass device is in the closed position such that the bypass device cannot be opened without breaking the seal or removing the lock. Examples of such devices include, but are not limited to, a car-seal or a lock-and-key configuration valve. The owner or operator shall visually inspect the seal or closure mechanism at least once every month to verify that the bypass mechanism is maintained in the closed position.

(4) The closed-vent system shall be inspected and monitored by the owner or operator in accordance with the procedure specified in § 264.1033(l).

(c) The control device shall meet the following requirements:

(1) The control device shall be one of the following devices:

(i) A control device designed and operated to reduce the total organic content of the inlet vapor stream vented to the control device by at least 95 percent by weight;

(ii) An enclosed combustion device designed and operated in accordance with the requirements of § 264.1033(c) of this part; or

(iii) A flare designed and operated in accordance with the requirements of § 264.1033(d) of this part.

(2) The owner or operator who elects to use a closed-vent system and control device to comply with the requirements of this section shall comply with the requirements specified in paragraphs (c)(2)(i) through (c)(2)(vi) of this section.

(i) Periods of planned routine maintenance of the control device, during which the control device does not meet the specifications of paragraphs (c)(1)(i), (c)(1)(ii), or (c)(1)(iii) of this section, as applicable, shall not exceed 240 hours per year.

(ii) The specifications and requirements in paragraphs (c)(1)(i), (c)(1)(ii), and (c)(1)(iii) of this section for control devices do not apply during periods of planned routine maintenance.

(iii) The specifications and requirements in paragraphs (c)(1)(i), (c)(1)(ii), and (c)(1)(iii) of this section for control devices do not apply during a control device system malfunction.

(iv) The owner or operator shall demonstrate compliance with the requirements of paragraph (c)(2)(i) of this section (i.e., planned routine maintenance of a control device, during which the control device does not meet the specifications of paragraphs (c)(1)(i), (c)(1)(ii), or (c)(1)(iii) of this section, as applicable, shall not exceed 240 hours per year) by recording the information specified in § 264.1089(e)(1)(v) of this subpart.

(v) The owner or operator shall correct control device system malfunctions as soon as practicable after their occurrence in order to minimize excess emissions of air pollutants.

(vi) The owner or operator shall operate the closed-vent system such that gases, vapors, or fumes are not actively vented to the control device during periods of planned maintenance or control device system malfunction (i.e., periods when the control device is not operating or not operating normally) except in cases when it is necessary to vent the gases, vapors, and/or fumes to avoid an unsafe condition or to implement malfunction corrective actions or planned maintenance actions.

(3) The owner or operator using a carbon adsorption system to comply with paragraph (c)(1) of this section shall operate and maintain the control device in accordance with the following requirements:

(i) Following the initial startup of the control device, all activated carbon in the control device shall be replaced with fresh carbon on a regular basis in accordance with the requirements of §264.1033(g) or §264.1033(h) of this part.

(ii) All carbon that is a hazardous waste and that is removed from the control device shall be managed in accordance with the requirements of 40 CFR 264.1033(n), regardless of the average volatile organic concentration of the carbon.

(4) An owner or operator using a control device other than a thermal vapor incinerator, flare, boiler, process heater, condenser, or carbon adsorption system to comply with paragraph (c)(1) of this section shall operate and maintain the control device in accordance with the requirements of §264.1033(j) of this part.

(5) The owner or operator shall demonstrate that a control device achieves the performance requirements of paragraph (c)(1) of this section as follows:

(i) An owner or operator shall demonstrate using either a performance test as specified in paragraph (c)(5)(iii) of this section or a design analysis as specified in paragraph (c)(5)(iv) of this section the performance of each control device except for the following:

(A) A flare;

(B) A boiler or process heater with a design heat input capacity of 44 megawatts or greater;

(C) A boiler or process heater into which the vent stream is introduced with the primary fuel;

(D) A boiler or industrial furnace burning hazardous waste for which the owner or operator has been issued a final permit under 40 CFR part 270 and has designed and operates the unit in accordance with the requirements of 40 CFR part 266, subpart H; or

(E) A boiler or industrial furnace burning hazardous waste for which the owner or operator has designed and op-

erates in accordance with the interim status requirements of 40 CFR part 266, subpart H.

(ii) An owner or operator shall demonstrate the performance of each flare in accordance with the requirements specified in §264.1033(e).

(iii) For a performance test conducted to meet the requirements of paragraph (c)(5)(i) of this section, the owner or operator shall use the test methods and procedures specified in §264.1034(c)(1) through (c)(4).

(iv) For a design analysis conducted to meet the requirements of paragraph (c)(5)(i) of this section, the design analysis shall meet the requirements specified in §264.1035(b)(4)(iii).

(v) The owner or operator shall demonstrate that a carbon adsorption system achieves the performance requirements of paragraph (c)(1) of this section based on the total quantity of organics vented to the atmosphere from all carbon adsorption system equipment that is used for organic adsorption, organic desorption or carbon regeneration, organic recovery, and carbon disposal.

(6) If the owner or operator and the Regional Administrator do not agree on a demonstration of control device performance using a design analysis then the disagreement shall be resolved using the results of a performance test performed by the owner or operator in accordance with the requirements of paragraph (c)(5)(iii) of this section. The Regional Administrator may choose to have an authorized representative observe the performance test.

(7) The closed-vent system and control device shall be inspected and monitored by the owner or operator in accordance with the procedures specified in 40 CFR 264.1033(f)(2) and 40 CFR 264.1033(l). The readings from each monitoring device required by 40 CFR 264.1033(f)(2) shall be inspected at least once each operating day to check control device operation. Any necessary corrective measures shall be immediately implemented to ensure the control device is operated in compliance with the requirements of this section.

[59 FR 62927, Dec. 6, 1994, as amended at 61 FR 4913, Feb. 9, 1996; 61 FR 59965, Nov. 25, 1996; 62 FR 64660, Dec. 8, 1997]

## § 264.1088 Inspection and monitoring requirements.

(a) The owner or operator shall inspect and monitor air emission control equipment used to comply with this subpart in accordance with the applicable requirements specified in § 264.1084 through § 264.1087 of this subpart.

(b) The owner or operator shall develop and implement a written plan and schedule to perform the inspections and monitoring required by paragraph (a) of this section. The owner or operator shall incorporate this plan and schedule into the facility inspection plan required under 40 CFR 264.15.

[61 FR 59966, Nov. 25, 1996]

## § 264.1089 Recordkeeping requirements.

(a) Each owner or operator of a facility subject to requirements of this subpart shall record and maintain the information specified in paragraphs (b) through (j) of this section, as applicable to the facility. Except for air emission control equipment design documentation and information required by paragraphs (i) and (j) of this section, records required by this section shall be maintained in the operating record for a minimum of 3 years. Air emission control equipment design documentation shall be maintained in the operating record until the air emission control equipment is replaced or otherwise no longer in service. Information required by paragraphs (i) and (j) of this section shall be maintained in the operating record for as long as the waste management unit is not using air emission controls specified in §§ 264.1084 through 264.1087 of this subpart in accordance with the conditions specified in § 264.1080(d) or § 264.1080(b)(7) of this subpart, respectively.

(b) The owner or operator of a tank using air emission controls in accordance with the requirements of § 264.1084 of this subpart shall prepare and maintain records for the tank that include the following information:

(1) For each tank using air emission controls in accordance with the requirements of § 264.1084 of this subpart, the owner or operator shall record:

(i) A tank identification number (or other unique identification description as selected by the owner or operator).

(ii) A record for each inspection required by § 264.1084 of this subpart that includes the following information:

(A) Date inspection was conducted.

(B) For each defect detected during the inspection: The location of the defect, a description of the defect, the date of detection, and corrective action taken to repair the defect. In the event that repair of the defect is delayed in accordance with the requirements of § 264.1084 of this subpart, the owner or operator shall also record the reason for the delay and the date that completion of repair of the defect is expected.

(2) In addition to the information required by paragraph (b)(1) of this section, the owner or operator shall record the following information, as applicable to the tank:

(i) The owner or operator using a fixed roof to comply with the Tank Level 1 control requirements specified in § 264.1084(c) of this subpart shall prepare and maintain records for each determination for the maximum organic vapor pressure of the hazardous waste in the tank performed in accordance with the requirements of § 264.1084(c) of this subpart. The records shall include the date and time the samples were collected, the analysis method used, and the analysis results.

(ii) The owner or operator using an internal floating roof to comply with the Tank Level 2 control requirements specified in § 264.1084(e) of this subpart shall prepare and maintain documentation describing the floating roof design.

(iii) Owners and operators using an external floating roof to comply with the Tank Level 2 control requirements specified in § 264.1084(f) of this subpart shall prepare and maintain the following records:

(A) Documentation describing the floating roof design and the dimensions of the tank.

(B) Records for each seal gap inspection required by § 264.1084(f)(3) of this subpart describing the results of the seal gap measurements. The records shall include the date that the measurements were performed, the raw data obtained for the measurements, and the calculations of the total gap surface area. In the event that the seal gap measurements do not conform to the specifications in § 264.1084(f)(1) of this subpart, the records shall include a description of the repairs that were made, the date the repairs were made, and the date the tank was emptied, if necessary.

(iv) Each owner or operator using an enclosure to comply with the Tank Level 2 control requirements specified in § 264.1084(i) of this subpart shall prepare and maintain the following records:

(A) Records for the most recent set of calculations and measurements performed by the owner or operator to verify that the enclosure meets the criteria of a permanent total enclosure as specified in "Procedure T—Criteria for and Verification of a Permanent or Temporary Total Enclosure" under 40 CFR 52.741, appendix B.

(B) Records required for the closed-vent system and control device in accordance with the requirements of paragraph (e) of this section.

(c) The owner or operator of a surface impoundment using air emission controls in accordance with the requirements of § 264.1085 of this subpart shall prepare and maintain records for the surface impoundment that include the following information:

(1) A surface impoundment identification number (or other unique identification description as selected by the owner or operator).

(2) Documentation describing the floating membrane cover or cover design, as applicable to the surface impoundment, that includes information prepared by the owner or operator or provided by the cover manufacturer or vendor describing the cover design, and certification by the owner or operator that the cover meets the specifications listed in § 264.1085(c) of this subpart.

(3) A record for each inspection required by § 264.1085 of this subpart that includes the following information:

(i) Date inspection was conducted.

(ii) For each defect detected during the inspection the following information: The location of the defect, a description of the defect, the date of detection, and corrective action taken to repair the defect. In the event that re-

pair of the defect is delayed in accordance with the provisions of § 264.1085(f) of this subpart, the owner or operator shall also record the reason for the delay and the date that completion of repair of the defect is expected.

(4) For a surface impoundment equipped with a cover and vented through a closed-vent system to a control device, the owner or operator shall prepare and maintain the records specified in paragraph (e) of this section.

(d) The owner or operator of containers using Container Level 3 air emission controls in accordance with the requirements of § 264.1086 of this subpart shall prepare and maintain records that include the following information:

(1) Records for the most recent set of calculations and measurements performed by the owner or operator to verify that the enclosure meets the criteria of a permanent total enclosure as specified in "Procedure T—Criteria for and Verification of a Permanent or Temporary Total Enclosure" under 40 CFR 52.741, appendix B.

(2) Records required for the closed-vent system and control device in accordance with the requirements of paragraph (e) of this section.

(e) The owner or operator using a closed-vent system and control device in accordance with the requirements of § 264.1087 of this subpart shall prepare and maintain records that include the following information:

(1) Documentation for the closed-vent system and control device that includes:

(i) Certification that is signed and dated by the owner or operator stating that the control device is designed to operate at the performance level documented by a design analysis as specified in paragraph (e)(1)(ii) of this section or by performance tests as specified in paragraph (e)(1)(iii) of this section when the tank, surface impoundment, or container is or would be operating at capacity or the highest level reasonably expected to occur.

(ii) If a design analysis is used, then design documentation as specified in 40 CFR 264.1035(b)(4). The documentation shall include information prepared by the owner or operator or provided by the control device manufacturer or vendor that describes the control device design in accordance with 40 CFR 264.1035(b)(4)(iii) and certification by the owner or operator that the control equipment meets the applicable specifications.

(iii) If performance tests are used, then a performance test plan as specified in 40 CFR 264.1035(b)(3) and all test results.

(iv) Information as required by 40 CFR 264.1035(c)(1) and 40 CFR 264.1035(c)(2), as applicable.

(v) An owner or operator shall record, on a semiannual basis, the information specified in paragraphs (e)(1)(v)(A) and (e)(1)(v)(B) of this section for those planned routine maintenance operations that would require the control device not to meet the requirements of § 264.1087(c)(1)(i), (c)(1)(ii), or (c)(1)(iii) of this subpart, as applicable.

(A) A description of the planned routine maintenance that is anticipated to be performed for the control device during the next 6-month period. This description shall include the type of maintenance necessary, planned frequency of maintenance, and lengths of maintenance periods.

(B) A description of the planned routine maintenance that was performed for the control device during the previous 6-month period. This description shall include the type of maintenance performed and the total number of hours during those 6 months that the control device did not meet the requirements of § 264.1087 (c)(1)(i), (c)(1)(ii), or (c)(1)(iii) of this subpart, as applicable, due to planned routine maintenance.

(vi) An owner or operator shall record the information specified in paragraphs (e)(1)(vi)(A) through (e)(1)(vi)(C) of this section for those unexpected control device system malfunctions that would require the control device not to meet the requirements of § 264.1087 (c)(1)(i), (c)(1)(ii), or (c)(1)(iii) of this subpart, as applicable.

(A) The occurrence and duration of each malfunction of the control device system.

(B) The duration of each period during a malfunction when gases, vapors, or fumes are vented from the waste management unit through the closed-vent system to the control device while

the control device is not properly functioning.

(C) Actions taken during periods of malfunction to restore a malfunctioning control device to its normal or usual manner of operation.

(vii) Records of the management of carbon removed from a carbon adsorption system conducted in accordance with § 264.1087(c)(3)(ii) of this subpart.

(f) The owner or operator of a tank, surface impoundment, or container exempted from standards in accordance with the provisions of § 264.1082(c) of this subpart shall prepare and maintain the following records, as applicable:

(1) For tanks, surface impoundments, and containers exempted under the hazardous waste organic concentration conditions specified in § 264.1082(c)(1) or §§ 264.1082(c)(2)(i) through (c)(2)(vi) of this subpart, the owner or operator shall record the information used for each waste determination (e.g., test results, measurements, calculations, and other documentation) in the facility operating log. If analysis results for waste samples are used for the waste determination, then the owner or operator shall record the date, time, and location that each waste sample is collected in accordance with applicable requirements of § 264.1083 of this subpart.

(2) For tanks, surface impoundments, or containers exempted under the provisions of § 264.1082(c)(2)(vii) or § 264.1082(c)(2)(viii) of this subpart, the owner or operator shall record the identification number for the incinerator, boiler, or industrial furnace in which the hazardous waste is treated.

(g) An owner or operator designating a cover as "unsafe to inspect and monitor" pursuant to § 264.1084(l) or § 264.1085(g) of this subpart shall record in a log that is kept in the facility operating record the following information: The identification numbers for waste management units with covers that are designated as "unsafe to inspect and monitor," the explanation for each cover stating why the cover is unsafe to inspect and monitor, and the plan and schedule for inspecting and monitoring each cover.

(h) The owner or operator of a facility that is subject to this subpart and to the control device standards in 40 CFR part 60, subpart VV, or 40 CFR part 61, subpart V, may elect to demonstrate compliance with the applicable sections of this subpart by documentation either pursuant to this subpart, or pursuant to the provisions of 40 CFR part 60, subpart VV or 40 CFR part 61, subpart V, to the extent that the documentation required by 40 CFR parts 60 or 61 duplicates the documentation required by this section.

(i) For each tank or container not using air emission controls specified in §§264.1084 through 264.1087 of this subpart in accordance with the conditions specified in §264.1080(d) of this subpart, the owner or operator shall record and maintain the following information:

(1) A list of the individual organic peroxide compounds manufactured at the facility that meet the conditions specified in §264.1080(d)(1).

(2) A description of how the hazardous waste containing the organic peroxide compounds identified in paragraph (i)(1) of this section are managed at the facility in tanks and containers. This description shall include:

(i) For the tanks used at the facility to manage this hazardous waste, sufficient information shall be provided to describe for each tank: A facility identification number for the tank; the purpose and placement of this tank in the management train of this hazardous waste; and the procedures used to ultimately dispose of the hazardous waste managed in the tanks.

(ii) For containers used at the facility to manage these hazardous wastes, sufficient information shall be provided to describe: A facility identification number for the container or group of containers; the purpose and placement of this container, or group of containers, in the management train of this hazardous waste; and the procedures used to ultimately dispose of the hazardous waste handled in the containers.

(3) An explanation of why managing the hazardous waste containing the organic peroxide compounds identified in paragraph (i)(1) of this section in the tanks and containers as described in paragraph (i)(2) of this section would create an undue safety hazard if the air emission controls, as required under

§§264.1084 through 264.1087 of this subpart, are installed and operated on these waste management units. This explanation shall include the following information:

(i) For tanks used at the facility to manage these hazardous wastes, sufficient information shall be provided to explain: How use of the required air emission controls on the tanks would affect the tank design features and facility operating procedures currently used to prevent an undue safety hazard during the management of this hazardous waste in the tanks; and why installation of safety devices on the required air emission controls, as allowed under this subpart, will not address those situations in which evacuation of tanks equipped with these air emission controls is necessary and consistent with good engineering and safety practices for handling organic peroxides.

(ii) For containers used at the facility to manage these hazardous wastes, sufficient information shall be provided to explain: How use of the required air emission controls on the containers would affect the container design features and handling procedures currently used to prevent an undue safety hazard during the management of this hazardous waste in the containers; and why installation of safety devices on the required air emission controls, as allowed under this subpart, will not address those situations in which evacuation of containers equipped with these air emission controls is necessary and consistent with good engineering and safety practices for handling organic peroxides.

(j) For each hazardous waste management unit not using air emission controls specified in §§264.1084 through 264.1087 of this subpart in accordance with the requirements of §264.1080(b)(7) of this subpart, the owner and operator shall record and maintain the following information:

(1) Certification that the waste management unit is equipped with and operating air emission controls in accordance with the requirements of an applicable Clean Air Act regulation codified under 40 CFR part 60, part 61, or part 63.

(2) Identification of the specific requirements codified under 40 CFR part

60, part 61, or part 63 with which the waste management unit is in compliance.

[61 FR 59966, Nov. 25, 1996, as amended at 62 FR 64660, Dec. 8, 1997]

### § 264.1090 Reporting requirements.

(a) Each owner or operator managing hazardous waste in a tank, surface impoundment, or container exempted from using air emission controls under the provisions of § 264.1082(c) of this subpart shall report to the Regional Administrator each occurrence when hazardous waste is placed in the waste management unit in noncompliance with the conditions specified in § 264.1082 (c)(1) or (c)(2) of this subpart, as applicable. Examples of such occurrences include placing in the waste management unit a hazardous waste having an average VO concentration equal to or greater than 500 ppmw at the point of waste origination; or placing in the waste management unit a treated hazardous waste of which the organic content has been reduced by an organic destruction or removal process that fails to achieve the applicable conditions specified in § 264.1082 (c)(2)(i) through (c)(2)(vi) of this subpart. The owner or operator shall submit a written report within 15 calendar days of the time that the owner or operator becomes aware of the occurrence. The written report shall contain the EPA identification number, facility name and address, a description of the noncompliance event and the cause, the dates of the noncompliance, and the actions taken to correct the noncompliance and prevent recurrence of the noncompliance. The report shall be signed and dated by an authorized representative of the owner or operator.

(b) Each owner or operator using air emission controls on a tank in accordance with the requirements § 264.1084(c) of this subpart shall report to the Regional Administrator each occurrence when hazardous waste is managed in the tank in noncompliance with the conditions specified in § 264.1084(b) of this subpart. The owner or operator shall submit a written report within 15 calendar days of the time that the owner or operator becomes aware of the occurrence. The written report shall contain the EPA identification

number, facility name and address, a description of the noncompliance event and the cause, the dates of the noncompliance, and the actions taken to correct the noncompliance and prevent recurrence of the noncompliance. The report shall be signed and dated by an authorized representative of the owner or operator.

(c) Each owner or operator using a control device in accordance with the requirements of § 264.1087 of this subpart shall submit a semiannual written report to the Regional Administrator excepted as provided for in paragraph (d) of this section. The report shall describe each occurrence during the previous 6-month period when either: (1) A control device is operated continuously for 24 hours or longer in noncompliance with the applicable operating values defined in § 264.1035(c)(4); or (2) A flare is operated with visible emissions for 5 minutes or longer in a two-hour period, as defined in § 264.1033(d). The written report shall include the EPA identification number, facility name and address, and an explanation why the control device could not be returned to compliance within 24 hours, and actions taken to correct the noncompliance. The report shall be signed and dated by an authorized representative of the owner or operator.

(d) A report to the Regional Administrator in accordance with the requirements of paragraph (c) of this section is not required for a 6-month period during which all control devices subject to this subpart are operated by the owner or operator such that:

(1) During no period of 24 hours or longer did a control device operate continuously in noncompliance with the applicable operating values defined in § 264.1035(c)(4); and

(2) No flare was operated with visible emissions for 5 minutes or longer in a two-hour period, as defined in § 264.1033(d).

[59 FR 62927, Dec. 6, 1994, as amended at 61 FR 4913, Feb. 9, 1996; 61 FR 59968, Nov. 25, 1996; 71 FR 40274, July 14, 2006]

§264.1091 [Reserved]

## Subpart DD—Containment Buildings

SOURCE: 57 FR 37265, Aug. 18, 1992, unless otherwise noted.

### §264.1100 Applicability.

The requirements of this subpart apply to owners or operators who store or treat hazardous waste in units designed and operated under §264.1101 of this subpart. The owner or operator is not subject to the definition of land disposal in RCRA section 3004(k) provided that the unit:

(a) Is a completely enclosed, self-supporting structure that is designed and constructed of manmade materials of sufficient strength and thickness to support themselves, the waste contents, and any personnel and heavy equipment that operate within the unit, and to prevent failure due to pressure gradients, settlement, compression, or uplift, physical contact with the hazardous wastes to which they are exposed; climatic conditions; and the stresses of daily operation, including the movement of heavy equipment within the unit and contact of such equipment with containment walls;

(b) Has a primary barrier that is designed to be sufficiently durable to withstand the movement of personnel, wastes, and handling equipment within the unit;

(c) If the unit is used to manage liquids, has:

(1) A primary barrier designed and constructed of materials to prevent migration of hazardous constituents into the barrier;

(2) A liquid collection system designed and constructed of materials to minimize the accumulation of liquid on the primary barrier; and

(3) A secondary containment system designed and constructed of materials to prevent migration of hazardous constituents into the barrier, with a leak detection and liquid collection system capable of detecting, collecting, and removing leaks of hazardous constituents at the earliest practicable time, unless the unit has been granted a variance from the secondary containment system requirements under §264.1101(b)(4);

(d) Has controls sufficient to prevent fugitive dust emissions to meet the no visible emission standard in §264.1101(c)(1)(iv); and

(e) Is designed and operated to ensure containment and prevent the tracking of materials from the unit by personnel or equipment.

[57 FR 37265, Aug. 18, 1992, as amended at 71 FR 16907, Apr. 4, 2006]

### §264.1101 Design and operating standards.

(a) All containment buildings must comply with the following design standards:

(1) The containment building must be completely enclosed with a floor, walls, and a roof to prevent exposure to the elements, (e.g., precipitation, wind, run-on), and to assure containment of managed wastes.

(2) The floor and containment walls of the unit, including the secondary containment system if required under paragraph (b) of this section, must be designed and constructed of materials of sufficient strength and thickness to support themselves, the waste contents, and any personnel and heavy equipment that operate within the unit, and to prevent failure due to pressure gradients, settlement, compression, or uplift, physical contact with the hazardous wastes to which they are exposed; climatic conditions; and the stresses of daily operation, including the movement of heavy equipment within the unit and contact of such equipment with containment walls. The unit must be designed so that it has sufficient structural strength to prevent collapse or other failure. All surfaces to be in contact with hazardous wastes must be chemically compatible with those wastes. EPA will consider standards established by professional organizations generally recognized by the industry such as the American Concrete Institute (ACI) and the American Society of Testing Materials (ASTM) in judging the structural integrity requirements of this paragraph. If appropriate to the nature of the waste management operation to take place in the unit, an exception to the structural strength requirement may be made for light-weight doors and windows that meet these criteria:

569

(i) They provide an effective barrier against fugitive dust emissions under paragraph (c)(1)(iv); and

(ii) The unit is designed and operated in a fashion that assures that wastes will not actually come in contact with these openings.

(3) Incompatible hazardous wastes or treatment reagents must not be placed in the unit or its secondary containment system if they could cause the unit or secondary containment system to leak, corrode, or otherwise fail.

(4) A containment building must have a primary barrier designed to withstand the movement of personnel, waste, and handling equipment in the unit during the operating life of the unit and appropriate for the physical and chemical characteristics of the waste to be managed.

(b) For a containment building used to manage hazardous wastes containing free liquids or treated with free liquids (the presence of which is determined by the paint filter test, a visual examination, or other appropriate means), the owner or operator must include:

(1) A primary barrier designed and constructed of materials to prevent the migration of hazardous constituents into the barrier (e.g., a geomembrane covered by a concrete wear surface).

(2) A liquid collection and removal system to minimize the accumulation of liquid on the primary barrier of the containment building:

(i) The primary barrier must be sloped to drain liquids to the associated collection system; and

(ii) Liquids and waste must be collected and removed to minimize hydraulic head on the containment system at the earliest practicable time.

(3) A secondary containment system including a secondary barrier designed and constructed to prevent migration of hazardous constituents into the barrier, and a leak detection system that is capable of detecting failure of the primary barrier and collecting accumulated hazardous wastes and liquids at the earliest practicable time.

(i) The requirements of the leak detection component of the secondary containment system are satisfied by installation of a system that is, at a minimum:

(A) Constructed with a bottom slope of 1 percent or more; and

(B) Constructed of a granular drainage material with a hydraulic conductivity of $1 \times 10^{-2}$ cm/sec or more and a thickness of 12 inches (30.5 cm) or more, or constructed of synthetic or geonet drainage materials with a transmissivity of $3 \times 10^{-5}$ m²/sec or more.

(ii) If treatment is to be conducted in the building, an area in which such treatment will be conducted must be designed to prevent the release of liquids, wet materials, or liquid aerosols to other portions of the building.

(iii) The secondary containment system must be constructed of materials that are chemically resistant to the waste and liquids managed in the containment building and of sufficient strength and thickness to prevent collapse under the pressure exerted by overlaying materials and by any equipment used in the containment building. (Containment buildings can serve as secondary containment systems for tanks placed within the building under certain conditions. A containment building can serve as an external liner system for a tank, provided it meets the requirements of § 264.193(e)(1). In addition, the containment building must meet the requirements of § 264.193(b) and §§ 264.193(c) (1) and (2) to be considered an acceptable secondary containment system for a tank.)

(4) For existing units other than 90-day generator units, the Regional Administrator may delay the secondary containment requirement for up to two years, based on a demonstration by the owner or operator that the unit substantially meets the standards of this subpart. In making this demonstration, the owner or operator must:

(i) Provide written notice to the Regional Administrator of their request by November 16, 1992. This notification must describe the unit and its operating practices with specific reference to the performance of existing containment systems, and specific plans for retrofitting the unit with secondary containment;

(ii) Respond to any comments from the Regional Administrator on these plans within 30 days; and

(iii) Fulfill the terms of the revised plans, if such plans are approved by the Regional Administrator.

(c) Owners or operators of all containment buildings must:

(1) Use controls and practices to ensure containment of the hazardous waste within the unit; and, at a minimum:

(i) Maintain the primary barrier to be free of significant cracks, gaps, corrosion, or other deterioration that could cause hazardous waste to be released from the primary barrier;

(ii) Maintain the level of the stored/treated hazardous waste within the containment walls of the unit so that the height of any containment wall is not exceeded;

(iii) Take measures to prevent the tracking of hazardous waste out of the unit by personnel or by equipment used in handling the waste. An area must be designated to decontaminate equipment and any rinsate must be collected and properly managed; and

(iv) Take measures to control fugitive dust emissions such that any openings (doors, windows, vents, cracks, etc.) exhibit no visible emissions (see 40 CFR part 60, appendix A, Method 22—Visual Determination of Fugitive Emissions from Material Sources and Smoke Emissions from Flares). In addition, all associated particulate collection devices (e.g., fabric filter, electrostatic precipitator) must be operated and maintained with sound air pollution control practices (see 40 CFR part 60 subpart 292 for guidance). This state of no visible emissions must be maintained effectively at all times during routine operating and maintenance conditions, including when vehicles and personnel are entering and exiting the unit.

(2) Obtain and keep on-site a certification by a qualified Professional Engineer that the containment building design meets the requirements of paragraphs (a), (b), and (c) of this section.

(3) Throughout the active life of the containment building, if the owner or operator detects a condition that could lead to or has caused a release of hazardous waste, the owner or operator must repair the condition promptly, in accordance with the following procedures.

(i) Upon detection of a condition that has led to a release of hazardous waste (e.g., upon detection of leakage from the primary barrier) the owner or operator must:

(A) Enter a record of the discovery in the facility operating record;

(B) Immediately remove the portion of the containment building affected by the condition from service;

(C) Determine what steps must be taken to repair the containment building, remove any leakage from the secondary collection system, and establish a schedule for accomplishing the cleanup and repairs; and

(D) Within 7 days after the discovery of the condition, notify the Regional Administrator of the condition, and within 14 working days, provide a written notice to the Regional Administrator with a description of the steps taken to repair the containment building, and the schedule for accomplishing the work.

(ii) The Regional Administrator will review the information submitted, make a determination regarding whether the containment building must be removed from service completely or partially until repairs and cleanup are complete, and notify the owner or operator of the determination and the underlying rationale in writing.

(iii) Upon completing all repairs and cleanup the owner or operator must notify the Regional Administrator in writing and provide a verification, signed by a qualified, registered professional engineer, that the repairs and cleanup have been completed according to the written plan submitted in accordance with paragraph (c)(3)(i)(D) of this section.

(4) Inspect and record in the facility's operating record, at least once every seven days, except for Performance Track member facilities that must inspect at least once each month, upon approval by the Director, data gathered from monitoring and leak detection equipment as well as the containment building and the area immediately surrounding the containment building to detect signs of releases of hazardous waste. To apply for reduced inspection frequency, the Performance

Track member facility must follow the procedures described in § 264.15(b)(5).

(d) For a containment building that contains both areas with and without secondary containment, the owner or operator must:

(1) Design and operate each area in accordance with the requirements enumerated in paragraphs (a) through (c) of this section;

(2) Take measures to prevent the release of liquids or wet materials into areas without secondary containment; and

(3) Maintain in the facility's operating log a written description of the operating procedures used to maintain the integrity of areas without secondary containment.

(e) Notwithstanding any other provision of this subpart the Regional Administrator may waive requirements for secondary containment for a permitted containment building where the owner operator demonstrates that the only free liquids in the unit are limited amounts of dust suppression liquids required to meet occupational health and safety requirements, and where containment of managed wastes and liquids can be assured without a secondary containment system.

[57 FR 37265, Aug. 18, 1992, as amended at 71 FR 16907, Apr. 4, 2006; 71 FR 40274, July 14, 2006]

§ 264.1102 Closure and post-closure care.

(a) At closure of a containment building, the owner or operator must remove or decontaminate all waste residues, contaminated containment system components (liners, etc.) contaminated subsoils, and structures and equipment contaminated with waste and leachate, and manage them as hazardous waste unless § 261.3(d) of this chapter applies. The closure plan, closure activities, cost estimates for closure, and financial responsibility for containment buildings must meet all of the requirements specified in subparts G and H of this part.

(b) If, after removing or decontaminating all residues and making all reasonable efforts to effect removal or decontamination of contaminated components, subsoils, structures, and equipment as required in paragraph (a) of this section, the owner or operator finds that not all contaminated subsoils can be practicably removed or decontaminated, he must close the facility and perform post-closure care in accordance with the closure and post-closure requirements that apply to landfills (§ 264.310). In addition, for the purposes of closure, post-closure, and financial responsibility, such a containment building is then considered to be a landfill, and the owner or operator must meet all of the requirements for landfills specified in subparts G and H of this part.

[57 FR 37265, Aug. 18, 1992, as amended at 71 FR 40274, July 14, 2006]

§§ 264.1103–264.1110 [Reserved]

## Subpart EE—Hazardous Waste Munitions and Explosives Storage

SOURCE: 62 FR 6652, Feb. 12, 1997, unless otherwise noted.

§ 264.1200 Applicability.

The requirements of this subpart apply to owners or operators who store munitions and explosive hazardous wastes, except as § 264.1 provides otherwise. (NOTE: Depending on explosive hazards, hazardous waste munitions and explosives may also be managed in other types of storage units, including containment buildings (40 CFR part 264, subpart DD), tanks (40 CFR part 264, subpart J), or containers (40 CFR part 264, subpart I); See 40 CFR 266.205 for storage of waste military munitions).

§ 264.1201 Design and operating standards.

(a) Hazardous waste munitions and explosives storage units must be designed and operated with containment systems, controls, and monitoring, that:

(1) Minimize the potential for detonation or other means of release of hazardous waste, hazardous constituents, hazardous decomposition products, or contaminated run-off, to the soil, ground water, surface water, and atmosphere;

(2) Provide a primary barrier, which may be a container (including a shell)

or tank, designed to contain the hazardous waste;

(3) For wastes stored outdoors, provide that the waste and containers will not be in standing precipitation;

(4) For liquid wastes, provide a secondary containment system that assures that any released liquids are contained and promptly detected and removed from the waste area, or vapor detection system that assures that any released liquids or vapors are promptly detected and an appropriate response taken (e.g., additional containment, such as overpacking, or removal from the waste area); and

(5) Provide monitoring and inspection procedures that assure the controls and containment systems are working as designed and that releases that may adversely impact human health or the environment are not escaping from the unit.

(b) Hazardous waste munitions and explosives stored under this subpart may be stored in one of the following:

(1) *Earth-covered magazines.* Earth-covered magazines must be:

(i) Constructed of waterproofed, reinforced concrete or structural steel arches, with steel doors that are kept closed when not being accessed;

(ii) Designed and constructed:

(A) To be of sufficient strength and thickness to support the weight of any explosives or munitions stored and any equipment used in the unit;

(B) To provide working space for personnel and equipment in the unit; and

(C) To withstand movement activities that occur in the unit; and

(iii) Located and designed, with walls and earthen covers that direct an explosion in the unit in a safe direction, so as to minimize the propagation of an explosion to adjacent units and to minimize other effects of any explosion.

(2) *Above-ground magazines.* Above-ground magazines must be located and designed so as to minimize the propagation of an explosion to adjacent units and to minimize other effects of any explosion.

(3) *Outdoor or open storage areas.* Outdoor or open storage areas must be located and designed so as to minimize the propagation of an explosion to adjacent units and to minimize other effects of any explosion.

(c) Hazardous waste munitions and explosives must be stored in accordance with a Standard Operating Procedure specifying procedures to ensure safety, security, and environmental protection. If these procedures serve the same purpose as the security and inspection requirements of 40 CFR 264.14, the preparedness and prevention procedures of 40 CFR part 264, subpart C, and the contingency plan and emergency procedures requirements of 40 CFR part 264, subpart D, then these procedures will be used to fulfill those requirements.

(d) Hazardous waste munitions and explosives must be packaged to ensure safety in handling and storage.

(e) Hazardous waste munitions and explosives must be inventoried at least annually.

(f) Hazardous waste munitions and explosives and their storage units must be inspected and monitored as necessary to ensure explosives safety and to ensure that there is no migration of contaminants out of the unit.

§264.1202 Closure and post-closure care.

(a) At closure of a magazine or unit which stored hazardous waste under this subpart, the owner or operator must remove or decontaminate all waste residues, contaminated containment system components, contaminated subsoils, and structures and equipment contaminated with waste, and manage them as hazardous waste unless §261.3(d) of this chapter applies. The closure plan, closure activities, cost estimates for closure, and financial responsibility for magazines or units must meet all of the requirements specified in subparts G and H of this part, except that the owner or operator may defer closure of the unit as long as it remains in service as a munitions or explosives magazine or storage unit.

(b) If, after removing or decontaminating all residues and making all reasonable efforts to effect removal or decontamination of contaminated components, subsoils, structures, and equipment as required in paragraph (a) of this section, the owner or operator finds that not all contaminated subsoils can be practicably removed or decontaminated, he or she must close the facility and perform post-closure care in accordance with the closure and post-closure requirements that apply to landfills (§ 264.310).

### APPENDIX I TO PART 264—RECORDKEEPING INSTRUCTIONS

The recordkeeping provisions of § 264.73 specify that an owner or operator must keep a written operating record at his facility. This appendix provides additional instructions for keeping *portions* of the operating record. See § 264.73(b) for additional recordkeeping requirements.

The following information must be recorded, as it becomes available, and maintained in the operating record until closure of the facility in the following manner:

Records of each hazardous waste received, treated, stored, or disposed of at the facility which include the following:

(1) A description by its common name and the EPA Hazardous Waste Number(s) from part 261 of this chapter which apply to the waste. The waste description also must include the waste's physical form, i.e., liquid, sludge, solid, or contained gas. If the waste is not listed in part 261, subpart D, of this chapter, the description also must include the process that produced it (for example, solid filter cake from production of ——, EPA Hazardous Waste Number W051).

Each hazardous waste listed in part 261, subpart D, of this chapter, and each hazardous waste characteristic defined in part 261, subpart C, of this chapter, has a four-digit EPA Hazardous Waste Number assigned to it. This number must be used for recordkeeping and reporting purposes. Where a hazardous waste contains more than one listed hazardous waste, or where more than one hazardous waste characteristic applies to the waste, the waste description must include all applicable EPA Hazardous Waste Numbers.

(2) The estimated or manifest-reported weight, or volume and density, where applicable, in one of the units of measure specified in Table 1;

### TABLE 1

| Unit of measure | Code[1] |
|---|---|
| Gallons | G |
| Gallons per Hour | E |
| Gallons per Day | U |
| Liters | L |
| Liters per Hour | H |
| Liters per Day | V |
| Short Tons per Hour | D |
| Metric Tons per Hour | W |
| Short Tons per Day | N |
| Metric Tons per Day | S |
| Pounds per Hour | J |

### TABLE 1—Continued

| Unit of measure | Code[1] |
|---|---|
| Kilograms per Hour | R |
| Cubic Yards | Y |
| Cubic Meters | C |
| Acres | B |
| Acre-feet | A |
| Hectares | Q |
| Hectare-meter | F |
| Btu's per Hour | I |
| Pounds | P |
| Short tons | T |
| Kilograms | K |
| Tons | M |

[1] Single digit symbols are used here for data processing purposes.

(3) The method(s) (by handling code(s) as specified in Table 2) and date(s) of treatment, storage, or disposal.

### Table 2—Handling Codes for Treatment, Storage and Disposal Methods

Enter the handling code(s) listed below that most closely represents the technique(s) used at the facility to treat, store or dispose of each quantity of hazardous waste received.

#### 1. Storage

S01   Container (barrel, drum, etc.)
S02   Tank
S03   Waste Pile
S04   Surface Impoundment
S05   Drip Pad
S06   Containment Building (Storage)
S99   Other Storage (specify)

#### 2. Treatment

(a)  Thermal Treatment—

T06   Liquid injection incinerator
T07   Rotary kiln incinerator
T08   Fluidized bed incinerator
T09   Multiple hearth incinerator
T10   Infrared furnace incinerator
T11   Molten salt destructor
T12   Pyrolysis
T13   Wet air oxidation
T14   Calcination
T15   Microwave discharge
T18   Other (specify)

(b)  Chemical Treatment—

T19   Absorption mound
T20   Absorption field
T21   Chemical fixation
T22   Chemical oxidation
T23   Chemical precipitation
T24   Chemical reduction
T25   Chlorination
T26   Chlorinolysis
T27   Cyanide destruction
T28   Degradation
T29   Detoxification
T30   Ion exchange
T31   Neutralization
T32   Ozonation

T33 Photolysis
T34 Other (specify)

(c) Physical Treatment—

    (1) Separation of components:

T35 Centrifugation
T36 Clarification
T37 Coagulation
T38 Decanting
T39 Encapsulation
T40 Filtration
T41 Flocculation
T42 Flotation
T43 Foaming
T44 Sedimentation
T45 Thickening
T46 Ultrafiltration
T47 Other (specify)

    (2) Removal of Specific Components:

T48 Absorption-molecular sieve
T49 Activated carbon
T50 Blending
T51 Catalysis
T52 Crystallization
T53 Dialysis
T54 Distillation
T55 Electrodialysis
T56 Electrolysis
T57 Evaporation
T58 High gradient magnetic separation
T59 Leaching
T60 Liquid ion exchange
T61 Liquid-liquid extraction
T62 Reverse osmosis
T63 Solvent recovery
T64 Stripping
T65 Sand filter
T66 Other (specify)

(d) Biological Treatment

T67 Activated sludge
T68 Aerobic lagoon
T69 Aerobic tank
T70 Anaerobic tank
T71 Composting
T72 Septic tank
T73 Spray irrigation
T74 Thickening filter
T75 Trickling filter
T76 Waste stabilization pond
T77 Other (specify)
T78–T79 [Reserved]

(e) Boilers and Industrial Furnaces

T80 Boiler
T81 Cement Kiln
T82 Lime Kiln
T83 Aggregate Kiln
T84 Phosphate Kiln
T85 Coke Oven
T86 Blast Furnace
T87 Smelting, Melting, or Refining Furnace
T88 Titanium Dioxide Chloride Process Oxidation Reactor
T89 Methane Reforming Furnace
T90 Pulping Liquor Recovery Furnace

T91 Combustion Device Used in the Recovery of Sulfur Values from Spent Sulfuric Acid
T92 Halogen Acid Furnaces
T93 Other Industrial Furnaces Listed in 40 CFR 260.10 (specify)

(f) Other Treatment

T94 Containment Building (Treatment)

### 3. Disposal

D79 Underground Injection
D80 Landfill
D81 Land Treatment
D82 Ocean Disposal
D83 Surface Impoundment (to be closed as a landfill)
D99 Other Disposal (specify)

### 4. Miscellaneous (Subpart X)

X01 Open Burning/Open Detonation
X02 Mechanical Processing
X03 Thermal Unit
X04 Geologic Repository
X99 Other Subpart X (specify)

[45 FR 33221, May 19, 1980, as amended at 59 FR 13891, Mar. 24, 1994; 71 FR 40274, July 14, 2006]

Appendixes II–III to Part 264
[Reserved]

Appendix IV to Part 264—Cochran's Approximation to the Behrens-Fisher Students' t-Test

Using all the available background data ($n_b$ readings), calculate the background mean ($X_b$) and background variance ($s_b{}^2$). For the single monitoring well under investigation ($n_m$ reading), calculate the monitoring mean ($X_m$) and monitoring variance ($s_m{}^2$).

For any set of data ($X_1, X_2, \ldots, X_n$) the mean is calculated by:

$$\overline{X} = \frac{X_1 + X_2 \cdots + X_n}{n}$$

and the variance is calculated by:

$$s^2 = \frac{\left(X_1 - \overline{X}\right)^2 + \left(X_2 - \overline{X}\right)^2 \cdots + \left(X_n - \overline{X}\right)^2}{n-1}$$

where "n" denotes the number of observations in the set of data.

The t-test uses these data summary measures to calculate a t-statistic ($t^*$) and a comparison t-statistic ($t_c$). The $t^*$ value is compared to the $t_c$ value and a conclusion reached as to whether there has been a statistically significant change in any indicator parameter.

The t-statistic for all parameters except pH and similar monitoring parameters is:

$$t^* = \frac{X_m - \overline{X}_s}{\sqrt{\dfrac{S_m^2}{n_m} + \dfrac{S_b^2}{n_b}}}$$

If the value of this t-statistic is negative then there is no significant difference between the monitoring data and background data. It should be noted that significantly small negative values may be indicative of a failure of the assumption made for test validity or errors have been made in collecting the background data.

The t-statistic ($t_c$), against which $t^*$ will be compared, necessitates finding $t_b$ and $t_m$ from standard (one-tailed) tables where,

$t_b$=t-tables with ($n_b - 1$) degrees of freedom, at the 0.05 level of significance.

$t_m$=t-tables with ($n_m - 1$) degrees of freedom, at the 0.05 level of significance.

Finally, the special weightings $W_b$ and $W_m$ are defined as:

$$W_B = \frac{S_b^2}{n_b} \quad \text{and} \quad W_m = \frac{S_m^2}{n_m}$$

and so the comparison t-statistic is:

$$t_c = \frac{W_b t_b + W_m t_m}{W_b + W_m}$$

The t-statistic ($t^*$) is now compared with the comparison t-statistic ($t_c$) using the following decision-rule:

If $t^*$ *is equal to or larger than* $t_c$, then conclude that there most likely *has been a significant increase* in this specific parameter. If $t^*$ *is less than* $t_c$, then conclude that most likely *there has not been a change* in this specific parameter.

The t-statistic for testing pH and similar monitoring parameters is constructed in the same manner as previously described except the negative sign (if any) is discarded and the caveat concerning the negative value is ignored. The standard (two-tailed) tables are used in the construction $t_c$ for pH and similar monitoring parameters.

If $t^*$ is equal to or larger than $t_c$, then conclude that there most likely *has been a significant increase* (if the initial $t^*$ had been negative, this would imply a significant decrease). If $t^*$ is less than $t_c$, then conclude that there most likely has been no change.

A further discussion of the test may be found in *Statistical Methods* (6th Edition, Section 4.14) by G. W. Snedecor and W. G. Cochran, or *Principles and Procedures of Statistics* (1st Edition, Section 5.8) by R. G. D. Steel and J. H. Torrie.

STANDARD T—TABLES 0.05 LEVEL OF SIGNIFICANCE

| Degrees of freedom | t-values (one-tail) | t-values (two-tail) |
|---|---|---|
| 1 | 6.314 | 12.706 |
| 2 | 2.920 | 4.303 |
| 3 | 2.353 | 3.182 |
| 4 | 2.132 | 2.776 |
| 5 | 2.015 | 2.571 |
| 6 | 1.943 | 2.447 |
| 7 | 1.895 | 2.365 |
| 8 | 1.860 | 2.306 |
| 9 | 1.833 | 2.262 |
| 10 | 1.812 | 2.228 |
| 11 | 1.796 | 2.201 |
| 12 | 1.782 | 2.179 |
| 13 | 1.771 | 2.160 |
| 14 | 1.761 | 2.145 |
| 15 | 1.753 | 2.131 |
| 16 | 1.746 | 2.120 |
| 17 | 1.740 | 2.110 |
| 18 | 1.734 | 2.101 |
| 19 | 1.729 | 2.093 |
| 20 | 1.725 | 2.086 |
| 21 | 1.721 | 2.080 |
| 22 | 1.717 | 2.074 |
| 23 | 1.714 | 2.069 |
| 24 | 1.711 | 2.064 |
| 25 | 1.708 | 2.060 |
| 30 | 1.697 | 2.042 |
| 40 | 1.684 | 2.021 |

Adopted from Table III of "Statistical Tables for Biological, Agricultural, and Medical Research" (1947, R. A. Fisher and F. Yates).

[47 FR 32367, July 26, 1982]

APPENDIX V TO PART 264—EXAMPLES OF POTENTIALLY INCOMPATIBLE WASTE

Many hazardous wastes, when mixed with other waste or materials at a hazardous waste facility, can produce effects which are harmful to human health and the environment, such as (1) heat or pressure, (2) fire or explosion, (3) violent reaction, (4) toxic dusts, mists, fumes, or gases, or (5) flammable fumes or gases.

Below are examples of potentially incompatible wastes, waste components, and materials, along with the harmful consequences which result from mixing materials in one group with materials in another group. The list is intended as a guide to owners or operators of treatment, storage, and disposal facilities, and to enforcement and permit granting officials, to indicate the need for special precautions when managing these potentially incompatible waste materials or components.

This list is not intended to be exhaustive. An owner or operator must, as the regulations require, adequately analyze his wastes so that he can avoid creating uncontrolled substances or reactions of the type listed below, whether they are listed below or not.

It is possible for potentially incompatible wastes to be mixed in a way that precludes a reaction (e.g., adding acid to water rather than water to acid) or that neutralizes them (e.g., a strong acid mixed with a strong base), or that controls substances produced (e.g., by generating flammable gases in a closed tank equipped so that ignition cannot occur, and burning the gases in an incinerator).

In the lists below, the mixing of a Group A material with a Group B material may have the potential consequence as noted.

GROUP 1-A

Acetylene sludge
Alkaline caustic liquids
Alkaline cleaner
Alkaline corrosive liquids
Alkaline corrosive battery fluid
Caustic wastewater
Lime sludge and other corrosive alkalies
Lime wastewater
Lime and water
Spent caustic

GROUP 1-B

Acid sludge
Acid and water
Battery acid
Chemical cleaners
Electrolyte, acid
Etching acid liquid or solvent
Pickling liquor and other corrosive acids
Spent acid
Spent mixed acid
Spent sulfuric acid
   Potential consequences: Heat generation; violent reaction.

GROUP 2-A

Aluminum
Beryllium
Calcium
Lithium
Magnesium
Potassium
Sodium
Zinc powder
Other reactive metals and metal hydrides

GROUP 2-B

Any waste in Group 1-A or 1-B
   Potential consequences: Fire or explosion; generation of flammable hydrogen gas.

GROUP 3-A

Alcohols
Water

GROUP 3-B

Any concentrated·waste in Groups 1-A or 1-B
Calcium
Lithium
Metal hydrides
Potassium

$SO_2 Cl_2$, $SOCl_2$, $PCl_3$, $CH_3 SiCl_3$
Other water-reactive waste
   Potential consequences: Fire, explosion, or heat generation; generation of flammable or toxic gases.

GROUP 4-A

Alcohols
Aldehydes
Halogenated hydrocarbons
Nitrated hydrocarbons
Unsaturated hydrocarbons
Other reactive organic compounds and solvents

GROUP 4-B

Concentrated Group 1-A or 1-B wastes
Group 2-A wastes
   Potential consequences: Fire, explosion, or violent reaction.

GROUP 5-A

Spent cyanide and sulfide solutions

GROUP 5-B

Group 1-B wastes
   Potential consequences: Generation of toxic hydrogen cyanide or hydrogen sulfide gas.

GROUP 6-A

Chlorates
Chlorine
Chlorites
Chromic acid
Hypochlorites
Nitrates
Nitric acid, fuming
Perchlorates
Permanganates
Peroxides
Other strong oxidizers

GROUP 6-B

Acetic acid and other organic acids
Concentrated mineral acids
Group 2-A wastes
Group 4-A wastes
Other flammable and combustible wastes
   Potential consequences: Fire, explosion, or violent reaction.

SOURCE: "Law, Regulations, and Guidelines for Handling of Hazardous Waste." California Department of Health, February 1975.

[46 FR 2872, Jan. 12, 1981]

---

[1] These include counties, city-county consolidations, and independent cities. In the case of Alaska, the political jurisdictions are election districts, and, in the case of Hawaii, the political jurisdiction listed is the island of Hawaii.

APPENDIX VI TO PART 264—POLITICAL JURISDICTIONS[1] IN WHICH COMPLIANCE WITH § 264.18(a) MUST BE DEMONSTRATED

#### ALASKA

| | |
|---|---|
| Aleutian Islands | Kodiak |
| Anchorage | Lynn Canal-Icy |
| Bethel | Straits |
| Bristol Bay | Palmer-Wasilla- |
| Cordova-Valdez | Talkeena |
| Fairbanks-Fort | Seward |
| Yukon | Sitka |
| Juneau | Wade Hampton |
| Kenai-Cook Inlet | Wrangell Petersburg |
| Ketchikan-Prince of | Yukon-Kuskokwim |
| Wales | |

#### ARIZONA

| | |
|---|---|
| Cochise | Greenlee |
| Graham | Yuma |

#### CALIFORNIA

All

#### COLORADO

| | |
|---|---|
| Archuleta | Mineral |
| Conejos | Rio Grande |
| Hinsdale | Saguache |

#### HAWAII

Hawaii

#### IDAHO

| | |
|---|---|
| Bannock | Franklin |
| Bear Lake | Fremont |
| Bingham | Jefferson |
| Bonneville | Madison |
| Caribou | Oneida |
| Cassia | Power |
| Clark | Teton |

#### MONTANA

| | |
|---|---|
| Beaverhead | Lewis and Clark |
| Broadwater | Madison |
| Cascade | Meagher |
| Deer Lodge | Missoula |
| Flathead | Park |
| Gallatin | Powell |
| Granite | Sanders |
| Jefferson | Silver Bow |
| Lake | |

| | |
|---|---|
| Stillwater | Teton |
| Sweet Grass | Wheatland |

#### NEVADA

All

#### NEW MEXICO

| | |
|---|---|
| Bernalillo | Sante Fe |
| Catron | Sierra |
| Grant | Socorro |
| Hidalgo | Taos |
| Los Alamos | Torrance |
| Rio Arriba | Valencia |
| Sandoval | |

#### UTAH

| | |
|---|---|
| Beaver | Piute |
| Box Elder | Rich |
| Cache | Salt Lake |
| Carbon | Sanpete |
| Davis | Sevier |
| Duchesne | Summit |
| Emery | Tooele |
| Garfield | Utah |
| Iron | Wasatch |
| Juab | Washington |
| Millard | Wayne |
| Morgan | Weber |

#### WASHINGTON

| | |
|---|---|
| Chelan | Mason |
| Clallam | Okanogan |
| Clark | Pacific |
| Cowlitz | Pierce |
| Douglas | San Juan Islands |
| Ferry | Skagit |
| Grant | Skamania |
| Grays Harbor | Snohomish |
| Jefferson | Thurston |
| King | Wahkiakum |
| Kitsap | Whatcom |
| Kittitas | Yakima |
| Lewis | |

#### WYOMING

| | |
|---|---|
| Fremont | Teton |
| Lincoln | Uinta |
| Park | Yellowstone National |
| Sublette | Park |

[46 FR 57285, Nov. 23, 1981; 47 FR 953, Jan. 8, 1982]

#### APPENDIXES VII–VIII TO PART 264

[RESERVED]

#### APPENDIX IX TO PART 264—GROUND-WATER MONITORING LIST

##### GROUND-WATER MONITORING LIST

| Common name [1] | CAS RN [2] | Chemical abstracts service index name [3] |
|---|---|---|
| Acenaphthene | 83–32–9 | Acenaphthylene, 1,2-dihydro- |
| Acenaphthylene | 208–96–8 | Acenaphthylene |
| Acetone | 67–64–1 | 2-Propanone |

GROUND-WATER MONITORING LIST—Continued

| Common name [1] | CAS RN [2] | Chemical abstracts service index name [3] |
|---|---|---|
| Acetophenone | 98–86–2 | Ethanone, 1-phenyl- |
| Acetonitrile; Methyl cyanide | 75–05–8 | Acetonitrile |
| 2-Acetylaminofluorene; 2-AAF | 53–96–3 | Acetamide, N–9H-fluoren-2-yl- |
| Acrolein | 107–02–8 | 2-Propenal |
| Acrylonitrile | 107–13–1 | 2-Propenenitrile |
| Aldrin | 309–00–2 | 1,4:5,8-Dimethanonaphthalene, 1,2,3,4,10,10-hexachloro-1,4,4a,5,8,8a-hexahydro- $(1\alpha,4\alpha,4a\beta,5\alpha,8\alpha,8a\beta)$- |
| Allyl chloride | 107–05–1 | 1-Propene, 3-chloro- |
| 4-Aminobiphenyl | 92–67–1 | [1,1′-Biphenyl]-4-amine |
| Aniline | 62–53–3 | Benzenamine |
| Anthracene | 120–12–7 | Anthracene |
| Antimony | (Total) | Antimony |
| Aramite | 140–57–8 | Sulfurous acid, 2-chloroethyl 2-[4-(1,1-dimethylethyl) phenoxy]-1-methylethyl ester |
| Arsenic | (Total) | Arsenic |
| Barium | (Total) | Barium |
| Benzene | 71–43–2 | Benzene |
| Benzo[a]anthracene; Benzanthracene | 56–55–3 | Benz[a]anthracene |
| Benzo[b]fluoranthene | 205–99–2 | Benz[e]acephenanthrylene |
| Benzo[k]fluoranthene | 207–08–9 | Benzo[k]fluoranthene |
| Benzo[ghi]perylene | 191–24–2 | Benzo[ghi]perylene |
| Benzo[a]pyrene | 50–32–8 | Benzo[a]pyrene |
| Benzyl alcohol | 100–51–6 | Benzenemethanol |
| Beryllium | (Total) | Beryllium |
| alpha-BHC | 319–84–6 | Cyclohexane, 1,2,3,4,5,6-hexachloro- ,$(1\alpha,2\alpha,3\beta,4\beta,5\beta,6\beta)$- |
| beta-BHC | 319–85–7 | Cyclohexane, 1,2,3,4,5,6-hexachloro- ,$(1\alpha,2\beta,3\alpha,4\beta,5\alpha,6\beta)$- |
| delta-BHC | 319–86–8 | Cyclohexane, 1,2,3,4,5,6-hexachloro- ,$(1\alpha,2\alpha,3\alpha,4\beta,5\alpha,6\beta)$- |
| gamma-BHC; Lindane | 58–89–9 | Cyclohexane, 1,2,3,4,5,6-hexachloro- ,$(1\alpha,2\alpha,3\beta,4\alpha,5\alpha,6\beta)$- |
| Bis(2-chloroethoxy)methane | 111–91–1 | Ethane, 1,1′-[methylenebis(oxy)]bis [2-chloro- |
| Bis(2-chloroethyl)ether | 111–44–4 | Ethane, 1,1′-oxybis[2-chloro- |
| Bis(2-chloro-1-methylethyl) ether; 2,2′-Dichlorodiisopropyl ether. | 108–60–1 | Propane, 2,2′-oxybis[1-chloro- |
| Bis(2-ethylhexyl) phthalate | 117–81–7 | 1,2-Benzenedicarboxylic acid, bis(2-ethylhexyl)ester |
| Bromodichloromethane | 75–27–4 | Methane, bromodichloro- |
| Bromoform; Tribromomethane | 75–25–2 | Methane, tribromo- |
| 4-Bromophenyl phenyl ether | 101–55–3 | Benzene, 1-bromo-4-phenoxy- |
| Butyl benzyl phthalate; Benzyl butyl phthalate | 85–68–7 | 1,2-Benzenedicarboxylic acid, butyl phenylmethyl ester |
| Cadmium | (Total) | Cadmium |
| Carbon disulfide | 75–15–0 | Carbon disulfide |
| Carbon tetrachloride | 56–23–5 | Methane, tetrachloro- |
| Chlordane | 57–74–9 | 4,7-Methano-1H-indene, 1,2,4,5,6,7,8,8-octachloro-2,3,3a,4,7,7a -hexahydro- |
| p-Chloroaniline | 106–47–8 | Benzenamine, 4-chloro- |
| Chlorobenzene | 108–90–7 | Benzene, chloro- |
| Chlorobenzilate | 510–15–6 | Benzeneacetic acid, 4-chloro-α-(4-chlorophenyl)-α-hydroxy-, ethyl ester |
| p-Chloro-m-cresol | 59–50–7 | Phenol, 4-chloro-3-methyl- |
| Chloroethane; Ethyl chloride | 75–00–3 | Ethane, chloro- |
| Chloroform | 67–66–3 | Methane, trichloro- |
| 2-Chloronaphthalene | 91–58–7 | Naphthalene, 2-chloro- |
| 2-Chlorophenol | 95–57–8 | Phenol, 2-chloro- |
| 4-Chlorophenyl phenyl ether | 7005–72–3 | Benzene, 1-chloro-4-phenoxy- |
| Chloroprene | 126–99–8 | 1,3-Butadiene,2-chloro- |
| Chromium | (Total) | Chromium |
| Chrysene | 218–01–9 | Chrysene |
| Cobalt | (Total) | Cobalt |
| Copper | (Total) | Copper |
| m-Cresol | 108–39–4 | Phenol, 3-methyl- |
| o-Cresol | 95–48–7 | Phenol, 2-methyl- |
| p-Cresol | 106–44–5 | Phenol, 4-methyl- |
| Cyanide | 57–12–5 | Cyanide |
| 2,4-D; 2,4-Dichlorophenoxyacetic acid | 94–75–7 | Acetic acid, (2,4-dichlorophenoxy)- |
| 4,4′-DDD | 72–54–8 | Benzene 1,1′-(2,2-dichloroethylidene) bis[4-chloro- |
| 4,4′-DDE | 72–55–9 | Benzene, 1,1′-(dichloroethenylidene) bis[4-chloro- |
| 4,4′-DDT | 50–29–3 | Benzene, 1,1′-(2,2,2-trichloroethylidene) bis[4-chloro- |

GROUND-WATER MONITORING LIST—Continued

| Common name [1] | CAS RN [2] | Chemical abstracts service index name [3] |
|---|---|---|
| Diallate | 2303–16–4 | Carbamothioic acid, bis(1-methylethyl)- , S- (2,3-dichloro-2-propenyl) ester |
| Dibenz[a,h]anthracene | 53–70–3 | Dibenz[a,h]anthracene |
| Dibenzofuran | 132–64–9 | Dibenzofuran |
| Dibromochloromethane; Chlorodibromomethane | 124–48–1 | Methane, dibromochloro- |
| 1,2-Dibromo-3-chloropropane; DBCP | 96–12–8 | Propane, 1,2-dibromo-3-chloro- |
| 1,2-Dibromoethane; Ethylene dibromide | 106–93–4 | Ethane, 1,2-dibromo- |
| Di-n-butyl phthalate | 84–74–2 | 1,2-Benzenedicarboxylic acid, dibutyl ester |
| o-Dichlorobenzene | 95–50–1 | Benzene, 1,2-dichloro- |
| m-Dichlorobenzene | 541–73–1 | Benzene, 1,3-dichloro- |
| p-Dichlorobenzene | 106–46–7 | Benzene, 1,4-dichloro- |
| 3,3′-Dichlorobenzidine | 91–94–1 | [1,1′-Biphenyl]-4,4′-diamine, 3,3′-dichloro- |
| trans-1,4-Dichloro-2-butene | 110–57–6 | 2-Butene, 1,4-dichloro-, (E)- |
| Dichlorodifluoromethane | 75–71–8 | Methane, dichlorodifluoro- |
| 1,1-Dichloroethane | 75–34–3 | Ethane, 1,1-dichloro- |
| 1,2-Dichloroethane; Ethylene dichloride | 107–06–2 | Ethane, 1,2-dichloro- |
| 1,1-Dichloroethylene; Vinylidene chloride | 75–35–4 | Ethene, 1,1-dichloro- |
| trans-1,2-Dichloroethylene | 156–60–5 | Ethene, 1,2-dichloro-, (E)- |
| 2,4-Dichlorophenol | 120–83–2 | Phenol, 2,4-dichloro- |
| 2,6-Dichlorophenol | 87–65–0 | Phenol, 2,6-dichloro- |
| 1,2-Dichloropropane | 78–87–5 | Propane, 1,2-dichloro- |
| cis-1,3-Dichloropropene | 10061–01–5 | 1-Propene, 1,3-dichloro-, (Z)- |
| trans-1,3-Dichloropropene | 10061–02–6 | 1-Propene, 1,3-dichloro-, (E)- |
| Dieldrin | 60–57–1 | 2,7:3,6-Dimethanonaphth [2,3-b]oxirene, 3,4,5,6,9,9-hexachloro-1a,2,2a,3,6,6a,7,7a-octahydro-, (1aα,2β,2aα,3β,6β;,6aα,7β,7aα)- |
| Diethyl phthalate | 84–66–2 | 1,2-Benzenedicarboxylic acid, diethyl ester |
| O,O-Diethyl O-2-pyrazinyl phosphorothioate; Thionazin. | 297–97–2 | Phosphorothioic acid, O,O-diethyl O-pyrazinyl ester |
| Dimethoate | 60–51–5 | Phosphorodithioic acid, O,O-dimethyl S-[2-(methylamino)-2-oxoethyl] ester |
| p-(Dimethylamino)azobenzene | 60–11–7 | Benzenamine, N,N-dimethyl-4-(phenylazo)- |
| 7,12-Dimethylbenz[a]anthracene | 57–97–6 | Benz[a]anthracene, 7,12-dimethyl- |
| 3,3′-Dimethylbenzidine | 119–93–7 | [1,1′-Biphenyl]-4,4′-diamine, 3,3′-dimethyl- |
| alpha, alpha-Dimethylphenethylamine | 122–09–8 | Benzeneethanamine, α,α-dimethyl- |
| 2,4-Dimethylphenol | 105–67–9 | Phenol, 2,4-dimethyl- |
| Dimethyl phthalate | 131–11–3 | 1,2-Benzenedicarboxylic acid, dimethyl ester |
| m-Dinitrobenzene | 99–65–0 | Benzene, 1,3-dinitro- |
| 4,6-Dinitro-o-cresol | 534–52–1 | Phenol, 2-methyl-4,6-dinitro- |
| 2,4-Dinitrophenol | 51–28–5 | Phenol, 2,4-dinitro- |
| 2,4-Dinitrotoluene | 121–14–2 | Benzene, 1-methyl-2,4-dinitro- |
| 2,6-Dinitrotoluene | 606–20–2 | Benzene, 2-methyl-1,3-dinitro- |
| Dinoseb; DNBP; 2-sec-Butyl-4,6-dinitrophenol | 88–85–7 | Phenol, 2-(1-methylpropyl)-4,6-dinitro- |
| Di-n-octyl phthalate | 117–84–0 | 1,2-Benzenedicarboxylic acid, dioctyl ester |
| 1,4-Dioxane | 123–91–1 | 1,4-Dioxane |
| Diphenylamine | 122–39–4 | Benzenamine, N-phenyl- |
| Disulfoton | 298–04–4 | Phosphorodithioic acid, O,O-diethyl S-[2-(ethylthio)ethyl]ester |
| Endosulfan I | 959–98–8 | 6,9-Methano-2,4,3- benzodioxathiepin, 6,7,8,9,10,10-hexachloro-1,5,5a,6,9,9a-hexahydro-, 3-oxide,(3α,5aβ,6α,9α,9aβ)- |
| Endosulfan II | 33213–65–9 | 6,9-Methano-2,4,3- benzodioxathiepin, 6,7,8,9,10,10-hexachloro-1,5,5a,6,9,9a-hexahydro-, 3-oxide, (3α,5aα,6β,9β,9aα)- |
| Endosulfan sulfate | 1031–07–8 | 6,9-Methano-2,4,3- benzodioxathiepin, 6,7,8,9,10,10-hexachloro-1,5,5a,6,9,9a-hexahydro-, 3,3-dioxide |
| Endrin | 72–20–8 | 2,7:3,6-Dimethanonaphth[2,3-b]oxirene, 3,4,5,6,9,9-hexachloro-,1a,2,2a,3,6,6a,7,7a-octahydro-, (1aα,2β,2aβ,3α,6α,6aβ,7β, 7aα)- |
| Endrin aldehyde | 7421–93–4 | 1,2,4- Methenocyclopenta[cd] pentalene-5-carboxaldehyde, 2,2a,3,3,4,7-hexachlorodecahydro-,(1α,2β,2aβ,4β,4aβ,5β,6aβ,6bβ,7R*)- |
| Ethylbenzene | 100–41–4 | Benzene, ethyl- |
| Ethyl methacrylate | 97–63–2 | 2-Propenoic acid, 2-methyl-, ethyl ester |
| Ethyl methanesulfonate | 62–50–0 | Methanesulfonic acid, ethyl ester |
| Famphur | 52–85–7 | Phosphorothioic acid, O-[4-[(dimethylamino)sulfonyl]phenyl]-O,O-dimethyl ester |
| Fluoranthene | 206–44–0 | Fluoranthene |
| Fluorene | 86–73–7 | 9H-Fluorene |

GROUND-WATER MONITORING LIST—Continued

| Common name [1] | CAS RN [2] | Chemical abstracts service index name [3] |
|---|---|---|
| Heptachlor | 76–44–8 | 4,7-Methano-1H-indene, 1,4,5,6,7,8,8-heptachloro-3a,4,7,7a-tetrahydro- |
| Heptachlor epoxide | 1024–57–3 | 2,5-Methano-2H-indeno[1,2-b] oxirene, 2,3,4,5,6,7,7-heptachloro-1a,1b,5,5a,6,6a,-hexahydro-, (1aα,1bβ,2α,5α,5aβ,6β,6aα) |
| Hexachlorobenzene | 118–74–1 | Benzene, hexachloro- |
| Hexachlorobutadiene | 87–68–3 | 1,3-Butadiene, 1,1,2,3,4,4-hexachloro- |
| Hexachlorocyclopentadiene | 77–47–4 | 1,3-Cyclopentadiene, 1,2,3,4,5,5-hexachloro- |
| Hexachloroethane | 67–72–1 | Ethane, hexachloro- |
| Hexachlorophene | 70–30–4 | Phenol, 2,2'-methylenebis[3,4,6-trichloro- |
| Hexachloropropene | 1888–71–7 | 1-Propene, 1,1,2,3,3,3-hexachloro- |
| 2-Hexanone | 591–78–6 | 2-Hexanone |
| Indeno(1,2,3-cd)pyrene | 193–39–5 | Indeno[1,2,3-cd]pyrene |
| Isobutyl alcohol | 78–83–1 | 1-Propanol, 2-methyl- |
| Isodrin | 465–73–6 | 1,4,5,8-Dimethanonaphthalene,1,2,3,4,1 0,10-hexachloro-1,4,4a,5,8,8a hexahydro-(1α, 4α, 4aβ, 5β, 8β, 8aβ)- |
| Isophorone | 78–59–1 | 2-Cyclohexen-1-one, 3,5,5-trimethyl- |
| Isosafrole | 120–58–1 | 1,3-Benzodioxole, 5-(1-propenyl)- |
| Kepone | 143–50–0 | 1,3,4-Metheno-2H-cyclobuta-[cd]pentalen-2-one, 1,1a,3,3a,4,5,5,5a,5b,6-decachlorooctahydro- |
| Lead | (Total) | Lead |
| Mercury | (Total) | Mercury |
| Methacrylonitrile | 126–98–7 | 2-Propenenitrile, 2-methyl- |
| Methapyrilene | 91–80–5 | 1,2,Ethanediamine,N,N-dimethyl-N'-2-pyridinyl-N'-(2-thienylmethyl)- |
| Methoxychlor | 72–43–5 | Benzene, 1,1'-(2,2,2,trichloroethylidene)bis [4-methoxy- |
| Methyl bromide; Bromomethane | 74–83–9 | Methane, bromo- |
| Methyl chloride; Chloromethane | 74–87–3 | Methane, chloro- |
| 3-Methylcholanthrene | 56–49–5 | Benz[j]aceanthrylene, 1,2-dihydro-3-methyl- |
| Methylene bromide; Dibromomethane | 74–95–3 | Methane, dibromo- |
| Methylene chloride; Dichloromethane | 75–09–2 | Methane, dichloro- |
| Methyl ethyl ketone; MEK; | 78–93–3 | 2-Butanone |
| Methyl iodide; Iodomethane | 74–88–4 | Methane, iodo- |
| Methyl methacrylate | 80–62–6 | 2-Propenoic acid, 2-methyl-, methyl ester |
| Methyl methanesulfonate | 66–27–3 | Methanesulfonic acid, methyl ester |
| 2-Methylnaphthalene | 91–57–6 | Naphthalene, 2-methyl- |
| Methyl parathion; Parathion methyl | 298–00–0 | Phosphorothioic acid, O,O-dimethyl O-(4-nitrophenyl) ester |
| 4-Methyl-2-pentanone; Methyl isobutyl ketone | 108–10–1 | 2-Pentanone, 4-methyl- |
| Naphthalene | 91–20–3 | Naphthalene |
| 1,4-Naphthoquinone | 130–15–4 | 1,4-Naphthalenedione |
| 1-Naphthylamine | 134–32–7 | 1-Naphthalenamine |
| 2-Naphthylamine | 91–59–8 | 2-Naphthalenamine |
| Nickel | (Total) | Nickel |
| o-Nitroaniline | 88–74–4 | Benzenamine, 2-nitro- |
| m-Nitroaniline | 99–09–2 | Benzenamine, 3-nitro- |
| p-Nitroaniline | 100–01–6 | Benzenamine, 4-nitro- |
| Nitrobenzene | 98–95–3 | Benzene, nitro- |
| o-Nitrophenol | 88–75–5 | Phenol, 2-nitro- |
| p-Nitrophenol | 100–02–7 | Phenol, 4-nitro- |
| 4-Nitroquinoline 1-oxide | 56–57–5 | Quinoline, 4-nitro, 1-oxide |
| N-Nitrosodi-n-butylamine | 924–16–3 | 1-Butanamine, N-butyl-N-nitroso- |
| N-Nitrosodiethylamine | 55–18–5 | Ethanamine, N-ethyl-N-nitroso- |
| N-Nitrosodimethylamine | 62–75–9 | Methanamine, N-methyl-N-nitroso- |
| N-Nitrosodiphenylamine | 86–30–6 | Benzenamine, N-nitroso-N-phenyl- |
| N-Nitrosodipropylamine;Di-n-propylnitrosamine | 621–64–7 | 1-Propanamine, N-nitroso-N-propyl- |
| N-Nitrosomethylethalamine | 10595–95–6 | Ethanamine, N-methyl-N-nitroso- |
| N-Nitrosomorpholine | 59–89–2 | Morpholine, 4-nitroso- |
| N-Nitrosopiperidine | 100–75–4 | Piperidine, 1-nitroso- |
| N-Nitrosopyrrolidine | 930–55–2 | Pyrrolidine, 1-nitroso- |
| 5-Nitro-o-toluidine | 99–55–8 | Benzenamine, 2-methyl-5-nitro- |
| Parathion | 56–38–2 | Phosphorothioic acid, O,O-diethyl-O-(4-nitrophenyl) ester |
| Polychlorinated biphenyls; PCBs | See footnote 4 | 1,1'-Biphenyl, chloro derivatives |
| Polychlorinated dibenzo-p-dioxins; PCDDs | See footnote 5 | Dibenzo[b,e][1,4]dioxin, chloro derivatives |
| Polychlorinated dibenzofurans; PCDFs | See footnote 6 | Dibenzofuran, chloro derivatives |
| Pentachlorobenzene | 608–93–5 | Benzene, pentachloro- |
| Pentachloroethane | 76–01–7 | Ethane, pentachloro- |
| Pentachloronitrobenzene | 82–68–8 | Benzene, pentachloronitro- |
| Pentachlorophenol | 87–86–5 | Phenol, pentachloro- |

GROUND-WATER MONITORING LIST—Continued

| Common name [1] | CAS RN [2] | Chemical abstracts service index name [3] |
|---|---|---|
| Phenacetin | 62-44-2 | Acetamide, N-(4-ethoxyphenyl) |
| Phenanthrene | 85-01-8 | Phenanthrene |
| Phenol | 108-95-2 | Phenol |
| p-Phenylenediamine | 106-50-3 | 1,4-Benzenediamine |
| Phorate | 298-02-2 | Phosphorodithioic acid, O,O-diethyl S-[(ethylthio)methyl] ester |
| 2-Picoline | 109-06-8 | Pyridine, 2-methyl- |
| Pronamide | 23950-58-5 | Benzamide, 3,5-dichloro-N-(1,1-dimethyl-2-propynyl)- |
| Propionitrile; Ethyl cyanide | 107-12-0 | Propanenitrile |
| Pyrene | 129-00-0 | Pyrene |
| Pyridine | 110-86-1 | Pyridine |
| Safrole | 94-59-7 | 1,3-Benzodioxole, 5-(2-propenyl)- |
| Selenium | (Total) | Selenium |
| Silver | (Total) | Silver |
| Silvex; 2,4,5-TP | 93-72-1 | Propanoic acid, 2-(2,4,5- trichlorophenoxy)- |
| Styrene | 100-42-5 | Benzene, ethenyl- |
| Sulfide | 18496-25-8 | Sulfide |
| 2,4,5-T; 2,4,5-Trichlorophenoxyacetic acid | 93-76-5 | Acetic acid, (2,4,5-trichlorophenoxy)- |
| 2,3,7,8-TCDD; 2,3,7,8-Tetrachlorodibenzo-p-dioxin | 1746-01-6 | Dibenzo[b,e][1,4]dioxin, 2,3,7,8-tetrachloro- |
| 1,2,4,5-Tetrachlorobenzene | 95-94-3 | Benzene, 1,2,4,5-tetrachloro- |
| 1,1,1,2-Tetrachloroethane | 630-20-6 | Ethane, 1,1,1,2-tetrachloro- |
| 1,1,2,2-Tetrachloroethane | 79-34-5 | Ethane, 1,1,2,2-tetrachloro- |
| Tetrachloroethylene; Perchloroethylene; Tetrachloroethene. | 127-18-4 | Ethene, tetrachloro- |
| 2,3,4,6-Tetrachlorophenol | 58-90-2 | Phenol, 2,3,4,6-tetrachloro- |
| Tetraethyl dithiopyrophosphate; Sulfotepp | 3689-24-5 | Thiodiphosphoric acid ([(HO)$_2$ P(S)]$_2$ O), tetraethyl ester |
| Thallium | (Total) | Thallium |
| Tin | (Total) | Tin |
| Toluene | 108-88-3 | Benzene, methyl- |
| o-Toluidine | 95-53-4 | Benzenamine, 2-methyl- |
| Toxaphene | 8001-35-2 | Toxaphene |
| 1,2,4-Trichlorobenzene | 120-82-1 | Benzene, 1,2,4-trichloro- |
| 1,1,1-Trichloroethane; Methylchloroform | 71-55-6 | Ethane, 1,1,1-trichloro- |
| 1,1,2-Trichloroethane | 79-00-5 | Ethane, 1,1,2-trichloro- |
| Trichloroethylene; Trichloroethene | 79-01-6 | Ethene, trichloro- |
| Trichlorofluoromethane | 75-69-4 | Methane, trichlorofluoro- |
| 2,4,5-Trichlorophenol | 95-95-4 | Phenol, 2,4,5-trichloro- |
| 2,4,6-Trichlorophenol | 88-06-2 | Phenol, 2,4,6-trichloro- |
| 1,2,3-Trichloropropane | 96-18-4 | Propane, 1,2,3-trichloro- |
| O,O,O-Triethyl phosphorothioate | 126-68-1 | Phosphorothioic acid, O,O,O-triethyl ester |
| sym-Trinitrobenzene | 99-35-4 | Benzene, 1,3,5-trinitro- |
| Vanadium | (Total) | Vanadium |
| Vinyl acetate | 108-05-4 | Acetic acid, ethenyl ester |
| Vinyl chloride | 75-01-4 | Ethene, chloro- |
| Xylene (total) | 1330-20-7 | Benzene, dimethyl- |
| Zinc | (Total) | Zinc |

[1] Common names are those widely used in government regulations, scientific publications, and commerce; synonyms exist for many chemicals.

[2] Chemical Abstracts Service registry number. Where "Total" is entered, all species in the ground water that contain this element are included.

[3] CAS index names are those used in the 9th Cumulative Index.

[4] Polychlorinated biphenyls (CAS RN 1336-36-3); this category contains congener chemicals, including constituents of Aroclor-1016 (CAS RN 12674-11-2), Aroclor-1221 (CAS RN 11104-28-2), Aroclor-1232 (CAS RN 11141-16-5), Aroclor-1242 (CAS RN 53469-21-9), Aroclor-1248 (CAS RN 12672-29-6), Aroclor-1254 (CAS RN 11097-69-1), and Aroclor-1260 (CAS RN 11096-82-5).

[5] This category contains congener chemicals, including tetrachlorodibenzo-p-dioxins (see also 2,3,7,8-TCDD), pentachlorodibenzo-p-dioxins, and hexachlorodibenzo-p-dioxins.

[6] This category contains congener chemicals, including tetrachlorodibenzofurans, pentachlorodibenzofurans, and hexachlorodibenzofurans.

[70 FR 34582, June 14, 2005, as amended at 70 FR 44151, Aug. 1, 2005]

# PART 265—INTERIM STATUS STANDARDS FOR OWNERS AND OPERATORS OF HAZARDOUS WASTE TREATMENT, STORAGE, AND DISPOSAL FACILITIES

AUTHORITY: 42 U.S.C. 6905, 6906, 6912, 6922, 6923, 6924, 6925, 6935, 6936, and 6937.

SOURCE: 45 FR 33232, May 19, 1980, unless otherwise noted.

## Subpart A—General

### §265.1 Purpose, scope, and applicability.

(a) The purpose of this part is to establish minimum national standards that define the acceptable management of hazardous waste during the period of interim status and until certification of final closure or, if the facility is subject to post-closure requirements, until

585

post-closure responsibilities are fulfilled.

(b) Except as provided in § 265.1080(b), the standards of this part, and of 40 CFR 264.552, 264.553, and 264.554, apply to owners and operators of facilities that treat, store or dispose of hazardous waste who have fully complied with the requirements for interim status under section 3005(e) of RCRA and § 270.10 of this chapter until either a permit is issued under section 3005 of RCRA or until applicable part 265 closure and post-closure responsibilities are fulfilled, and to those owners and operators of facilities in existence on November 19, 1980 who have failed to provide timely notification as required by section 3010(a) of RCRA and/or failed to file Part A of the permit application as required by 40 CFR 270.10 (e) and (g). These standards apply to all treatment, storage and disposal of hazardous waste at these facilities after the effective date of these regulations, except as specifically provided otherwise in this part or part 261 of this chapter.

[Comment: As stated in section 3005(a) of RCRA, after the effective date of regulations under that section (i.e., parts 270 and 124 of this chapter), the treatment, storage and disposal of hazardous waste is prohibited except in accordance with a permit. Section 3005(e) of RCRA provides for the continued operation of an existing facility that meets certain conditions, until final administrative disposition of the owner's and operator's permit application is made.]

(c) The requirements of this part do not apply to:

(1) A person disposing of hazardous waste by means of ocean disposal subject to a permit issued under the Marine Protection, Research, and Sanctuaries Act;

[Comment: These part 265 regulations do apply to the treatment or storage of hazardous waste before it is loaded onto an ocean vessel for incineration or disposal at sea, as provided in paragraph (b) of this section.]

(2) [Reserved]

(3) The owner or operator of a POTW which treats, stores, or disposes of hazardous waste;

[Comment: The owner or operator of a facility under paragraphs (c)(1) through (3) of this section is subject to the requirements of part 264 of this chapter to the extent they are included in a permit by rule granted to such a person under part 122 of this chapter, or are required by § 144.14 of this chapter.]

(4) A person who treats, stores, or disposes of hazardous waste in a State with a RCRA hazardous waste program authorized under subpart A or B of part 271 of this chapter, except that the requirements of this part will continue to apply:

(i) If the authorized State RCRA program does not cover disposal of hazardous waste by means of underground injection; or

(ii) To a person who treats, stores, or disposes of hazardous waste in a State authorized under subpart A or B of part 271 of this chapter if the State has not been authorized to carry out the requirements and prohibitions applicable to the treatment, storage, or disposal of hazardous waste at his facility which are imposed pursuant to the Hazardous and Solid Waste Act Amendments of 1984. The requirements and prohibitions that are applicable until a State receives authorization to carry them out include all Federal program requirements identified in § 271.1(j);

(5) The owner or operator of a facility permitted, licensed, or registered by a State to manage municipal or industrial solid waste, if the only hazardous waste the facility treats, stores, or disposes of is excluded from regulation under this part by § 261.5 of this chapter;

(6) The owner or operator of a facility managing recyclable materials described in § 261.6 (a)(2), (3), and (4) of this chapter (except to the extent they are referred to in part 279 or subparts C, F, G, or H of part 266 of this chapter).

(7) A generator accumulating waste on-site in compliance with § 262.34 of this chapter, except to the extent the requirements are included in § 262.34 of this chapter;

(8) A farmer disposing of waste pesticides from his own use in compliance with § 262.70 of this chapter; or

(9) The owner or operator of a totally enclosed treatment facility, as defined in § 260.10.

(10) The owner or operator of an elementary neutralization unit or a wastewater treatment unit as defined

in §260.10 of this chapter, provided that if the owner or operator is diluting hazardous ignitable (D001) wastes (other than the D001 High TOC Subcategory defined in §268.40 of this chapter, Table Treatment Standards for Hazardous Wastes), or reactive (D003) waste, to remove the characteristic before land disposal, the owner/operator must comply with the requirements set out in §265.17(b).

(11)(i) Except as provided in paragraph (c)(11)(ii) of this section, a person engaged in treatment or containment activities during immediate response to any of the following situations:

(A) A discharge of a hazardous waste;

(B) An imminent and substantial threat of a discharge of a hazardous waste;

(C) A discharge of a material which, when discharged, becomes a hazardous waste.

(D) An immediate threat to human health, public safety, property, or the environment, from the known or suspected presence of military munitions, other explosive material, or an explosive device, as determined by an explosive or munitions emergency response specialist as defined in 40 CFR 260.10.

(ii) An owner or operator of a facility otherwise regulated by this part must comply with all applicable requirements of subparts C and D.

(iii) Any person who is covered by paragraph (c)(11)(i) of this section and who continues or initiates hazardous waste treatment or containment activities after the immediate response is over is subject to all applicable requirements of this part and parts 122 through 124 of this chapter for those activities.

(iv) In the case of an explosives or munitions emergency response, if a Federal, State, Tribal or local official acting within the scope of his or her official responsibilities, or an explosives or munitions emergency response specialist, determines that immediate removal of the material or waste is necessary to protect human health or the environment, that official or specialist may authorize the removal of the material or waste by transporters who do not have EPA identification numbers and without the preparation of a mani-

fest. In the case of emergencies involving military munitions, the responding military emergency response specialist's organizational unit must retain records for three years identifying the dates of the response, the responsible persons responding, the type and description of material addressed, and its disposition.

(12) A transporter storing manifested shipments of hazardous waste in containers meeting the requirements of 40 CFR 262.30 at a transfer facility for a period of ten days or less.

(13) The addition of absorbent material to waste in a container (as defined in §260.10 of this chapter) or the addition of waste to the absorbent material in a container provided that these actions occur at the time waste is first placed in the containers; and §§265.17(b), 265.171, and 265.172 are complied with.

(14) Universal waste handlers and universal waste transporters (as defined in 40 CFR 260.10) handling the wastes listed below. These handlers are subject to regulation under 40 CFR part 273, when handling the below listed universal wastes.

(i) Batteries as described in 40 CFR 273.2;

(ii) Pesticides as described in §273.3 of this chapter;

(iii) Mercury-containing equipment as described in §273.4 of this chapter; and

(iv) Lamps as described in §273.5 of this chapter.

(15) A New York State Utility central collection facility consolidating hazardous waste in accordance with 40 CFR 262.90.

(d) The following hazardous wastes must not be managed at facilities subject to regulation under this part.

(1) EPA Hazardous Waste Nos. FO20, FO21, FO22, FO23, FO26, or FO27 unless:

(i) The wastewater treatment sludge is generated in a surface impoundment as part of the plant's wastewater treatment system;

(ii) The waste is stored in tanks or containers;

(iii) The waste is stored or treated in waste piles that meet the requirements of §264.250(c) as well as all other applicable requirements of subpart L of this part;

(iv) The waste is burned in incinerators that are certified pursuant to the standards and procedures in § 265.352; or

(v) The waste is burned in facilities that thermally treat the waste in a device other than an incinerator and that are certified pursuant to the standards and procedures in § 265.383.

(e) The requirements of this part apply to owners or operators of all facilities which treat, store or dispose of hazardous waste referred to in 40 CFR part 268, and the 40 CFR part 268 standards are considered material conditions or requirements of the part 265 interim status standards.

(f) Section 266.205 of this chapter identifies when the requirements of this part apply to the storage of military munitions classified as solid waste under § 266.202 of this chapter. The treatment and disposal of hazardous waste military munitions are subject to the applicable permitting, procedural, and technical standards in 40 CFR parts 260 through 270.

[45 FR 33232, May 19, 1980]

EDITORIAL NOTE: For FEDERAL REGISTER citations affecting § 265.1, see the List of CFR Sections Affected, which appears in the Finding Aids section of the printed volume and on GPO Access.

### §§ 265.2–265.3  [Reserved]

### § 265.4  Imminent hazard action.

Notwithstanding any other provisions of these regulations, enforcement actions may be brought pursuant to section 7003 of RCRA.

## Subpart B—General Facility Standards

### § 265.10  Applicability.

The regulations in this subpart apply to owners and operators of all hazardous waste facilities, except as § 265.1 provides otherwise.

### § 265.11  Identification number.

Every facility owner or operator must apply to EPA for an EPA identification number in accordance with the EPA notification procedures (45 FR 12746).

### § 265.12  Required notices.

(a)(1) The owner or operator of a facility that has arranged to receive hazardous waste from a foreign source must notify the Regional Administrator in writing at least four weeks in advance of the date the waste is expected to arrive at the facility. Notice of subsequent shipments of the same waste from the same foreign source is not required.

(2) The owner or operator of a recovery facility that has arranged to receive hazardous waste subject to 40 CFR part 262, subpart H must provide a copy of the tracking document bearing all required signatures to the notifier, to the Office of Enforcement and Compliance Assurance, Office of Compliance, Enforcement Planning, Targeting and Data Division (2222A), Environmental Protection Agency, 1200 Pennsylvania Ave., NW., Washington, DC 20460 and to the competent authorities of all other concerned countries within three working days of receipt of the shipment. The original of the signed tracking document must be maintained at the facility for at least three years.

(b) Before transferring ownership or operation of a facility during its operating life, or of a disposal facility during the post-closure care period, the owner or operator must notify the new owner or operator in writing of the requirements of this part and part 270 of this chapter. (Also see § 270.72 of this chapter.)

[Comment: An owner's or operator's failure to notify the new owner or operator of the requirements of this part in no way relieves the new owner or operator of his obligation to comply with all applicable requirements.]

[45 FR 33232, May 19, 1980, as amended at 48 FR 14295, Apr. 1, 1983; 50 FR 4514, Jan. 31, 1985; 61 FR 16315, Apr. 12, 1996; 71 FR 40274, July 14, 2006]

EFFECTIVE DATE NOTE: At 75 FR 1260, Jan. 8, 2010, § 265.12 was amended by revising paragraph (a)(2), effective July 7, 2010. For the convenience of the user, the revised text is set forth as follows:

### § 265.12  Required notices.

(a) * * *

(2) The owner or operator of a recovery facility that has arranged to receive hazardous waste subject to 40 CFR part 262, subpart H must provide a copy of the movement document bearing all required signatures to the

foreign exporter; to the Office of Enforcement and Compliance Assurance, Office of Federal Activities, International Compliance Assurance Division (2254A), Environmental Protection Agency, 1200 Pennsylvania Avenue, NW., Washington, DC 20460; and to the competent authorities of all other countries concerned within three (3) working days of receipt of the shipment. The original of the signed movement document must be maintained at the facility for at least three (3) years. In addition, such owner or operator shall, as soon as possible, but no later than thirty (30) days after the completion of recovery and no later than one (1) calendar year following the receipt of the hazardous waste, send a certificate of recovery to the foreign exporter and to the competent authority of the country of export and to EPA's Office of Enforcement and Compliance Assurance at the above address by mail, e-mail without a digital signature followed by mail, or fax followed by mail.

\*　　\*　　\*　　\*　　\*

### §265.13　General waste analysis.

(a)(1) Before an owner or operator treats, stores, or disposes of any hazardous wastes, or nonhazardous wastes if applicable under §265.113(d), he must obtain a detailed chemical and physical analysis of a representative sample of the wastes. At a minimum, the analysis must contain all the information which must be known to treat, store, or dispose of the waste in accordance with this part and part 268 of this chapter.

(2) The analysis may include data developed under part 261 of this chapter, and existing published or documented data on the hazardous waste or on waste generated from similar processes.

*Comment:* For example, the facility's records of analyses performed on the waste before the effective date of these regulations, or studies conducted on hazardous waste generated from processes similar to that which generated the waste to be managed at the facility, may be included in the data base required to comply with paragraph (a)(1) of this section. The owner or operator of an off-site facility may arrange for the generator of the hazardous waste to supply part of the information required by paragraph (a)(1) of this section, except as otherwise specified in 40 CFR 268.7 (b) and (c). If the generator does not supply the information, and the owner or operator chooses to accept a hazardous waste, the owner or operator is responsible for obtaining the information required to comply with this section.]

(3) The analysis must be repeated as necessary to ensure that it is accurate and up to date. At a minimum, the analysis must be repeated:

(i) When the owner or operator is notified, or has reason to believe, that the process or operation generating the hazardous wastes or non-hazardous wastes, if applicable, under §265.113(d) has changed; and

(ii) For off-site facilities, when the results of the inspection required in paragraph (a)(4) of this section indicate that the hazardous waste received at the facility does not match the waste designated on the accompanying manifest or shipping paper.

(4) The owner or operator of an off-site facility must inspect and, if necessary, analyze each hazardous waste movement received at the facility to determine whether it matches the identity of the waste specified on the accompanying manifest or shipping paper.

(b) The owner or operator must develop and follow a written waste analysis plan which describes the procedures which he will carry out to comply with paragraph (a) of this section. He must keep this plan at the facility. At a minimum, the plan must specify:

(1) The parameters for which each hazardous waste, or non-hazardous waste if applicable under §265.113(d), will be analyzed and the rationale for the selection of these parameters (i.e., how analysis for these parameters will provide sufficient information on the waste's properties to comply with paragraph (a) of this section);

(2) The test methods which will be used to test for these parameters;

(3) The sampling method which will be used to obtain a representative sample of the waste to be analyzed. A representative sample may be obtained using either:

(i) One of the sampling methods described in appendix I of part 261 of this chapter; or

(ii) An equivalent sampling method.

[*Comment:* See §260.20(c) of this chapter for related discussion.]

589

(4) The frequency with which the initial analysis of the waste will be reviewed or repeated to ensure that the analysis is accurate and up to date;

(5) For off-site facilities, the waste analyses that hazardous waste generators have agreed to supply; and

(6) Where applicable, the methods that will be used to meet the additional waste analysis requirements for specific waste management methods as specified in §§ 265.200, 265.225, 265.252, 265.273, 265.314, 265.341, 265.375, 265.402, 265.1034(d), 265.1063(d), 265.1084, and 268.7 of this chapter.

(7) For surface impoundments exempted from land disposal restrictions under § 268.4(a) of this chapter, the procedures and schedule for:

(i) The sampling of impoundment contents;

(ii) The analysis of test data; and,

(iii) The annual removal of residues which are not delisted under § 260.22 of this chapter or which exhibit a characteristic of hazardous waste and either:

(A) Do not meet applicable treatment standards of part 268, subpart D; or

(B) Where no treatment standards have been established;

(1) Such residues are prohibited from land disposal under § 268.32 or RCRA section 3004(d); or

(2) Such residues are prohibited from land disposal under § 268.33(f).

(8) For owners and operators seeking an exemption to the air emission standards of Subpart CC of this part in accordance with § 265.1083—

(i) If direct measurement is used for the waste determination, the procedures and schedules for waste sampling and analysis, and the results of the analysis of test data to verify the exemption.

(ii) If knowledge of the waste is used for the waste determination, any information prepared by the facility owner or operator or by the generator of the hazardous waste, if the waste is received from off-site, that is used as the basis for knowledge of the waste.

(c) For off-site facilities, the waste analysis plan required in paragraph (b) of this section must also specify the procedures which will be used to inspect and, if necessary, analyze each movement of hazardous waste received

at the facility to ensure that it matches the identity of the waste designated on the accompanying manifest or shipping paper. At a minimum, the plan must describe:

(1) The procedures which will be used to determine the identity of each movement of waste managed at the facility; and

(2) The sampling method which will be used to obtain a representative sample of the waste to be identified, if the identification method includes sampling.

(3) The procedures that the owner or operator of an off-site landfill receiving containerized hazardous waste will use to determine whether a hazardous waste generator or treater has added a biodegradable sorbent to the waste in the container.

[45 FR 33232, May 19, 1980, as amended at 50 FR 4514, Jan. 31, 1985; 50 FR 18374, Apr. 30, 1985; 51 FR 40638, Nov. 7, 1986; 52 FR 25788, July 8, 1987; 54 FR 33396, Aug. 14, 1989; 55 FR 22685, June 1, 1990; 55 FR 25506, June 21, 1990; 56 FR 19290, Apr. 26, 1991; 57 FR 8088, Mar. 6, 1992; 57 FR 54461, Nov. 18, 1992; 59 FR 62935, Dec. 6, 1994; 61 FR 4913, Feb. 9, 1996]

## § 265.14  Security.

(a) The owner or operator must prevent the unknowing entry, and minimize the possibility for the unauthorized entry, of persons or livestock onto the active portion of his facility, *unless:*

(1) Physical contact with the waste, structures, or equipment with the active portion of the facility will not injure unknowing or unauthorized persons or livestock which may enter the active portion of a facility, and

(2) Disturbance of the waste or equipment, by the unknowing or unauthorized entry of persons or livestock onto the active portion of a facility, will not cause a violation of the requirements of this part.

(b) Unless exempt under paragraphs (a)(1) and (2) of this section, a facility must have:

(1) A 24-hour surveillance system (e.g., television monitoring or surveillance by guards or facility personnel) which continuously monitors and controls entry onto the active portion of the facility; or

(2)(i) An artificial or natural barrier (e.g., a fence in good repair or a fence combined with a cliff), which completely surrounds the active portion of the facility; and

(ii) A means to control entry, at all times, through the gates or other entrances to the active portion of the facility (e.g., an attendant, television monitors, locked entrance, or controlled roadway access to the facility).

[*Comment:* The requirements of paragraph (b) of this section are satisfied if the facility or plant within which the active portion is located itself has a surveillance system, or a barrier and a means to control entry, which complies with the requirements of paragraph (b)(1) or (2) of this section.]

(c) Unless exempt under paragraphs (a)(1) and (a)(2) of this section, a sign with the legend, "Danger—Unauthorized Personnel Keep Out," must be posted at each entrance to the active portion of a facility, and at other locations, in sufficient numbers to be seen from any approach to this active portion. The legend must be written in English and in any other language predominant in the area surrounding the facility (e.g., facilities in counties bordering the Canadian province of Quebec must post signs in French; facilities in counties bordering Mexico must post signs in Spanish), and must be legible from a distance of at least 25 feet. Existing signs with a legend other than "Danger—Unauthorized Personnel Keep Out" may be used if the legend on the sign indicates that only authorized personnel are allowed to enter the active portion, and that entry onto the active portion can be dangerous.

[*Comment:* See §265.117(b) for discussion of security requirements at disposal facilities during the post-closure care period.]

[45 FR 33232, May 19, 1980, as amended at 71 FR 40274, July 14, 2006]

### §265.15 General inspection requirements.

(a) The owner or operator must inspect his facility for malfunctions and deterioration, operator errors, and discharges which may be causing—or may lead to: (1) Release of hazardous waste constituents to the environment or (2) a threat to human health. The owner or operator must conduct these inspections often enough to identify problems in time to correct them before they harm human health or the environment.

(b)(1) The owner or operator must develop and follow a written schedule for inspecting all monitoring equipment, safety and emergency equipment, security devices, and operating and structural equipment (such as dikes and sump pumps) that are important to preventing, detecting, or responding to environmental or human health hazards.

(2) He must keep this schedule at the facility.

(3) The schedule must identify the types of problems (e.g., malfunctions or deterioration) which are to be looked for during the inspection (e.g., inoperative sump pump, leaking fitting, eroding dike, etc.).

(4) The frequency of inspection may vary for the items on the schedule. However, the frequency should be based on the rate of deterioration of the equipment and the probability of an environmental or human health incident if the deterioration, malfunction, or operator error goes undetected between inspections. Areas subject to spills, such as loading and unloading areas, must be inspected daily when in use, except for Performance Track member facilities, that must inspect at least once each month, upon approval by the Director, as described in paragraph (b)(5) of this section. At a minimum, the inspection schedule must include the items and frequencies called for in §§265.174, 265.193, 265.195, 265.226, 265.260, 265.278, 265.304, 265.347, 265.377, 265.403, 265.1033, 265.1052, 265.1053, 265.1058, and 265.1084 through 265.1090, where applicable.

(5) Performance Track member facilities that choose to reduce inspection frequencies must:

(i) Submit an application to the Director. The application must identify the facility as a member of the National Environmental Performance Track Program and identify the management units for reduced inspections and the proposed frequency of inspections. Inspections must be conducted at least once each month.

(ii) Within 60 days, the Director will notify the Performance Track member facility, in writing, if the application is approved, denied, or if an extension to the 60-day deadline is needed. This notice must be placed in the facility's operating record. The Performance Track member facility should consider the application approved if the Director does not: (1) Deny the application; or (2) notify the Performance Track member facility of an extension to the 60-day deadline. In these situations, the Performance Track member facility must adhere to the revised inspection schedule outlined in its application and maintain a copy of the application in the facility's operating record.

(iii) Any Performance Track member facility that discontinues its membership or is terminated from the program must immediately notify the Director of its change in status. The facility must place in its operating record a dated copy of this notification and revert back to the non-Performance Track inspection frequencies within seven calendar days.

(c) The owner or operator must remedy any deterioration or malfunction of equipment or structures which the inspection reveals on a schedule which ensures that the problem does not lead to an environmental or human health hazard. Where a hazard is imminent or has already occurred, remedial action must be taken immediately.

(d) The owner or operator must record inspections in an inspection log or summary. He must keep these records for at least three years from the date of inspection. At a minimum, these records must include the date and time of the inspection, the name of the inspector, a notation of the observations made, and the date and nature of any repairs or other remedial actions.

[45 FR 33232, May 19, 1980, as amended at 50 FR 4514, Jan. 31, 1985; 57 FR 3491, Jan. 29, 1992; 59 FR 62935, Dec. 6, 1994; 62 FR 64661, Dec. 8, 1997; 71 FR 16908, Apr. 4, 2006]

## § 265.16  Personnel training.

(a)(1) Facility personnel must successfully complete a program of classroom instruction or on-the-job training that teaches them to perform their duties in a way that ensures the facility's compliance with the requirements of this part. The owner or operator must ensure that this program includes all the elements described in the document required under paragraph (d)(3) of this section.

(2) This program must be directed by a person trained in hazardous waste management procedures, and must include instruction which teaches facility personnel hazardous waste management procedures (including contingency plan implementation) relevant to the positions in which they are employed.

(3) At a minimum, the training program must be designed to ensure that facility personnel are able to respond effectively to emergencies by familiarizing them with emergency procedures, emergency equipment, and emergency systems, including where applicable:

(i) Procedures for using, inspecting, repairing, and replacing facility emergency and monitoring equipment;

(ii) Key parameters for automatic waste feed cut-off systems;

(iii) Communications or alarm systems;

(iv) Response to fires or explosions;

(v) Response to ground-water contamination incidents; and

(vi) Shutdown of operations.

(4) For facility employees that receive emergency response training pursuant to Occupational Safety and Health Administration (OSHA) regulations 29 CFR 1910.120(p)(8) and 1910.120(q), the facility is not required to provide separate emergency response training pursuant to this section, provided that the overall facility training meets all the requirements of this section.

(b) Facility personnel must successfully complete the program required in paragraph (a) of this section within six months after the effective date of these regulations or six months after the date of their employment or assignment to a facility, or to a new position at a facility, whichever is later. Employees hired after the effective date of these regulations must not work in unsupervised positions until they have completed the training requirements of paragraph (a) of this section.

(c) Facility personnel must take part in an annual review of the initial training required in paragraph (a) of this section.

(d) The owner or operator must maintain the following documents and records at the facility:

(1) The job title for each position at the facility related to hazardous waste management, and the name of the employee filling each job;

(2) A written job description for each position listed under paragraph (d)(1) of this Section. This description may be consistent in its degree of specificity with descriptions for other similar positions in the same company location or bargaining unit, but must include the requisite skill, education, or other qualifications, and duties of facility personnel assigned to each position;

(3) A written description of the type and amount of both introductory and continuing training that will be given to each person filling a position listed under paragraph (d)(1) of this section;

(4) Records that document that the training or job experience required under paragraphs (a), (b), and (c) of this section has been given to, and completed by, facility personnel.

(e) Training records on current personnel must be kept until closure of the facility. Training records on former employees must be kept for at least three years from the date the employee last worked at the facility. Personnel training records may accompany personnel transferred within the same company.

[45 FR 33232, May 19, 1980, as amended at 50 FR 4514, Jan. 31, 1985; 71 FR 16908, Apr. 4, 2006; 71 FR 40274, July 14, 2006]

### §265.17 General requirements for ignitable, reactive, or incompatible wastes.

(a) The owner or operator must take precautions to prevent accidental ignition or reaction of ignitable or reactive waste. This waste must be separated and protected from sources of ignition or reaction including but not limited to: Open flames, smoking, cutting and welding, hot surfaces, frictional heat, sparks (static, electrical, or mechanical), spontaneous ignition (e.g., from heat-producing chemical reactions), and radiant heat. While ignitable or reactive waste is being handled, the owner or operator must confine smoking and open flame to specially designated locations. "No Smoking" signs must be conspicuously placed wherever there is a hazard from ignitable or reactive waste.

(b) Where specifically required by other sections of this part, the treatment, storage, or disposal of ignitable or reactive waste, and the mixture or commingling of incompatible wastes, or incompatible wastes and materials, must be conducted so that it does not:

(1) Generate extreme heat or pressure, fire or explosion, or violent reaction;

(2) Produce uncontrolled toxic mists, fumes, dusts, or gases in sufficient quantities to threaten human health;

(3) Produce uncontrolled flammable fumes or gases in sufficient quantities to pose a risk of fire or explosions;

(4) Damage the structural integrity of the device or facility containing the waste; or

(5) Through other like means threaten human health or the environment.

### §265.18 Location standards.

The placement of any hazardous waste in a salt dome, salt bed formation, underground mine or cave is prohibited, except for the Department of Energy Waste Isolation Pilot Project in New Mexico.

[50 FR 28749, July 15, 1985]

### §265.19 Construction quality assurance program.

(a) *CQA program.* (1) A construction quality assurance (CQA) program is required for all surface impoundment, waste pile, and landfill units that are required to comply with §§265.221(a), 265.254, and 265.301(a). The program must ensure that the constructed unit meets or exceeds all design criteria and specifications in the permit. The program must be developed and implemented under the direction of a CQA officer who is a registered professional engineer.

(2) The CQA program must address the following physical components, where applicable:

(i) Foundations;

(ii) Dikes;

(iii) Low-permeability soil liners;

(iv) Geomembranes (flexible membrane liners);

(v) Leachate collection and removal systems and leak detection systems; and

(vi) Final cover systems.

(b) *Written CQA plan.* Before construction begins on a unit subject to the CQA program under paragraph (a) of this section, the owner or operator must develop a written CQA plan. The plan must identify steps that will be used to monitor and document the quality of materials and the condition and manner of their installation. The CQA plan must include:

(1) Identification of applicable units, and a description of how they will be constructed.

(2) Identification of key personnel in the development and implementation of the CQA plan, and CQA officer qualifications.

(3) A description of inspection and sampling activities for all unit components identified in paragraph (a)(2) of this section, including observations and tests that will be used before, during, and after construction to ensure that the construction materials and the installed unit components meet the design specifications. The description must cover: Sampling size and locations; frequency of testing; data evaluation procedures; acceptance and rejection criteria for construction materials; plans for implementing corrective measures; and data or other information to be recorded and retained in the operating record under § 265.73.

(c) *Contents of program.* (1) The CQA program must include observations, inspections, tests, and measurements sufficient to ensure:

(i) Structural stability and integrity of all components of the unit identified in paragraph (a)(2) of this section;

(ii) Proper construction of all components of the liners, leachate collection and removal system, leak detection system, and final cover system, according to permit specifications and good engineering practices, and proper installation of all components (e.g., pipes) according to design specifications;

(iii) Conformity of all materials used with design and other material specifications under §§ 264.221, 264.251, and 264.301 of this chapter.

(2) The CQA program shall include test fills for compacted soil liners, using the same compaction methods as in the full-scale unit, to ensure that the liners are constructed to meet the hydraulic conductivity requirements of §§ 264.221(c)(1), 264.251(c)(1), and 264.301(c)(1) of this chapter in the field. Compliance with the hydraulic conductivity requirements must be verified by using in-situ testing on the constructed test fill. The test fill requirement is waived where data are sufficient to show that a constructed soil liner meets the hydraulic conductivity requirements of §§ 264.221(c)(1), 264.251(c)(1), and 264.301(c)(1) of this chapter in the field.

(d) *Certification.* The owner or operator of units subject to § 265.19 must submit to the Regional Administrator by certified mail or hand delivery, at least 30 days prior to receiving waste, a certification signed by the CQA officer that the CQA plan has been successfully carried out and that the unit meets the requirements of §§ 265.221(a), 265.254, or 265.301(a). The owner or operator may receive waste in the unit after 30 days from the Regional Administrator's receipt of the CQA certification unless the Regional Administrator determines in writing that the construction is not acceptable, or extends the review period for a maximum of 30 more days, or seeks additional information from the owner or operator during this period. Documentation supporting the CQA officer's certification must be furnished to the Regional Administrator upon request.

[57 FR 3491, Jan. 29, 1992, as amended at 71 FR 40274, July 14, 2006]

## Subpart C—Preparedness and Prevention

### § 265.30  Applicability.

The regulations in this subpart apply to owners and operators of all hazardous waste facilities, except as § 265.1 provides otherwise.

### § 265.31  Maintenance and operation of facility.

Facilities must be maintained and operated to minimize the possibility of a fire, explosion, or any unplanned sudden or non-sudden release of hazardous waste or hazardous waste constituents to air, soil, or surface water which could threaten human health or the environment.

## §265.32 Required equipment.

All facilities must be equipped with the following, *unless* none of the hazards posed by waste handled at the facility could require a particular kind of equipment specified below:

(a) An internal communications or alarm system capable of providing immediate emergency instruction (voice or signal) to facility personnel;

(b) A device, such as a telephone (immediately available at the scene of operations) or a hand-held two-way radio, capable of summoning emergency assistance from local police departments, fire departments, or State or local emergency response teams;

(c) Portable fire extinguishers, fire control equipment (including special extinguishing equipment, such as that using foam, inert gas, or dry chemicals), spill control equipment, and decontamination equipment; and

(d) Water at adequate volume and pressure to supply water hose streams, or foam producing equipment, or automatic sprinklers, or water spray systems.

## §265.33 Testing and maintenance of equipment.

All facility communications or alarm systems, fire protection equipment, spill control equipment, and decontamination equipment, where required, must be tested and maintained as necessary to assure its proper operation in time of emergency.

## §265.34 Access to communications or alarm system.

(a) Whenever hazardous waste is being poured, mixed, spread, or otherwise handled, all personnel involved in the operation must have immediate access to an internal alarm or emergency communication device, either directly or through visual or voice contact with another employee, *unless* such a device is not required under §265.32.

(b) If there is ever just one employee on the premises while the facility is operating, he must have immediate access to a device, such as a telephone (immediately available at the scene of operation) or a hand-held two-way radio, capable of summoning external emergency assistance, *unless* such a device is not required under §265.32.

## §265.35 Required aisle space.

The owner or operator must maintain aisle space to allow the unobstructed movement of personnel, fire protection equipment, spill control equipment, and decontamination equipment to any area of facility operation in an emergency, *unless* aisle space is not needed for any of these purposes.

## §265.36 [Reserved]

## §265.37 Arrangements with local authorities.

(a) The owner or operator must attempt to make the following arrangements, as appropriate for the type of waste handled at his facility and the potential need for the services of these organizations:

(1) Arrangements to familiarize police, fire departments, and emergency response teams with the layout of the facility, properties of hazardous waste handled at the facility and associated hazards, places where facility personnel would normally be working, entrances to roads inside the facility, and possible evacuation routes;

(2) Where more than one police and fire department might respond to an emergency, agreements designating primary emergency authority to a specific police and a specific fire department, and agreements with any others to provide support to the primary emergency authority;

(3) Agreements with State emergency response teams, emergency response contractors, and equipment suppliers; and

(4) Arrangements to familiarize local hospitals with the properties of hazardous waste handled at the facility and the types of injuries or illnesses which could result from fires, explosions, or releases at the facility.

(b) Where State or local authorities decline to enter into such arrangements, the owner or operator must document the refusal in the operating record.

## Subpart D—Contingency Plan and Emergency Procedures

### § 265.50 Applicability.

The regulations in this subpart apply to owners and operators of all hazardous waste facilities, except as § 265.1 provides otherwise.

### § 265.51 Purpose and implementation of contingency plan.

(a) Each owner or operator must have a contingency plan for his facility. The contingency plan must be designed to minimize hazards to human health or the environment from fires, explosions, or any unplanned sudden or non-sudden release of hazardous waste or hazardous waste constituents to air, soil, or surface water.

(b) The provisions of the plan must be carried out immediately whenever there is a fire, explosion, or release of hazardous waste or hazardous waste constituents which could threaten human health or the environment.

[45 FR 33232, May 19, 1980, as amended at 50 FR 4514, Jan. 31, 1985]

### § 265.52 Content of contingency plan.

(a) The contingency plan must describe the actions facility personnel must take to comply with §§ 265.51 and 265.56 in response to fires, explosions, or any unplanned sudden or non-sudden release of hazardous waste or hazardous waste constituents to air, soil, or surface water at the facility.

(b) If the owner or operator has already prepared a Spill Prevention, Control, and Countermeasures (SPCC) Plan in accordance with Part 112 of this chapter, or some other emergency or contingency plan, he need only amend that plan to incorporate hazardous waste management provisions that are sufficient to comply with the requirements of this Part. The owner or operator may develop one contingency plan which meets all regulatory requirements. EPA recommends that the plan be based on the National Response

Team's Integrated Contingency Plan Guidance ("One Plan"). When modifications are made to non-RCRA provisions in an integrated contingency plan, the changes do not trigger the need for a RCRA permit modification.

(c) The plan must describe arrangements agreed to by local police departments, fire departments, hospitals, contractors, and State and local emergency response teams to coordinate emergency services, pursuant to § 265.37.

(d) The plan must list names, addresses, and phone numbers (office and home) of all persons qualified to act as emergency coordinator (see § 265.55), and this list must be kept up to date. Where more than one person is listed, one must be named as primary emergency coordinator and others must be listed in the order in which they will assume responsibility as alternates.

(e) The plan must include a list of all emergency equipment at the facility (such as fire extinguishing systems, spill control equipment, communications and alarm systems (internal and external), and decontamination equipment), where this equipment is required. This list must be kept up to date. In addition, the plan must include the location and a physical description of each item on the list, and a brief outline of its capabilities.

(f) The plan must include an evacuation plan for facility personnel where there is a possibility that evacuation could be necessary. This plan must describe signal(s) to be used to begin evacuation, evacuation routes, and alternate evacuation routes (in cases where the primary routes could be blocked by releases of hazardous waste or fires).

[45 FR 33232, May 19, 1980, as amended at 46 FR 27480, May 20, 1981; 50 FR 4514, Jan. 31, 1985; 71 FR 16908, Apr. 4, 2006; 75 FR 13005, Mar. 18, 2010]

### § 265.53 Copies of contingency plan.

A copy of the contingency plan and all revisions to the plan must be:

(a) Maintained at the facility; and

(b) Submitted to all local police departments, fire departments, hospitals, and State and local emergency response teams that may be called upon to provide emergency services.

[45 FR 33232, May 19, 1980, as amended at 50 FR 4514, Jan. 31, 1985]

### §265.54 Amendment of contingency plan.

The contingency plan must be reviewed, and immediately amended, if necessary, whenever:

(a) Applicable regulations are revised;

(b) The plan fails in an emergency;

(c) The facility changes—in its design, construction, operation, maintenance, or other circumstances—in a way that materially increases the potential for fires, explosions, or releases of hazardous waste or hazardous waste constituents, or changes the response necessary in an emergency;

(d) The list of emergency coordinators changes; or

(e) The list of emergency equipment changes.

[45 FR 33232, May 19, 1980, as amended at 50 FR 4514, Jan. 31, 1985]

### §265.55 Emergency coordinator.

At all times, there must be at least one employee either on the facility premises or on call (i.e., available to respond to an emergency by reaching the facility within a short period of time) with the responsibility for coordinating all emergency response measures. This emergency coordinator must be thoroughly familiar with all aspects of the facility's contingency plan, all operations and activities at the facility, the location and characteristics of waste handled, the location of all records within the facility, and the facility layout. In addition, this person must have the authority to commit the resources needed to carry out the contingency plan.

[*Comment:* The emergency coordinator's responsibilities are more fully spelled out in §265.56. Applicable responsibilities for the emergency coordinator vary, depending on factors such as type and variety of waste(s) handled by the facility, and type and complexity of the facility.]

### §265.56 Emergency procedures.

(a) Whenever there is an imminent or actual emergency situation, the emergency coordinator (or his designee when the emergency coordinator is on call) must immediately:

(1) Activate internal facility alarms or communication systems, where applicable, to notify all facility personnel; and

(2) Notify appropriate State or local agencies with designated response roles if their help is needed.

(b) Whenever there is a release, fire, or explosion, the emergency coordinator must immediately identify the character, exact source, amount, and areal extent of any released materials. He may do this by observation or review of facility records or manifests and, if necessary, by chemical analysis.

(c) Concurrently, the emergency coordinator must assess possible hazards to human health or the environment that may result from the release, fire, or explosion. This assessment must consider both direct and indirect effects of the release, fire, or explosion (e.g., the effects of any toxic, irritating, or asphyxiating gases that are generated, or the effects of any hazardous surface water run-offs from water or chemical agents used to control fire and heat-induced explosions).

(d) If the emergency coordinator determines that the facility has had a release, fire, or explosion which could threaten human health, or the environment, outside the facility, he must report his findings as follows:

(1) If his assessment indicates that evacuation of local areas may be advisable, he must immediately notify appropriate local authorities. He must be available to help appropriate officials decide whether local areas should be evacuated; and

(2) He must immediately notify either the government official designated as the on-scene coordinator for that geographical area, or the National Response Center (using their 24-hour toll free number 800/424–8802). The report must include:

(i) Name and telephone number of reporter;

(ii) Name and address of facility;

(iii) Time and type of incident (e.g., release, fire);

(iv) Name and quantity of material(s) involved, to the extent known;

(v) The extent of injuries, if any; and

(vi) The possible hazards to human health, or the environment, outside the facility.

(e) During an emergency, the emergency coordinator must take all reasonable measures necessary to ensure that fires, explosions, and releases do not occur, recur, or spread to other hazardous waste at the facility. These measures must include, where applicable, stopping processes and operations, collecting and containing released waste, and removing or isolating containers.

(f) If the facility stops operations in response to a fire, explosion or release, the emergency coordinator must monitor for leaks, pressure buildup, gas generation, or ruptures in valves, pipes, or other equipment, wherever this is appropriate.

(g) Immediately after an emergency, the emergency coordinator must provide for treating, storing, or disposing of recovered waste, contaminated soil or surface water, or any other material that results from a release, fire, or explosion at the facility.

[Comment: Unless the owner or operator can demonstrate, in accordance with § 261.3(c) or (d) of this chapter, that the recovered material is not a hazardous waste, the owner or operator becomes a generator of hazardous waste and must manage it in accordance with all applicable requirements of parts 262, 263, and 265 of this chapter.]

(h) The emergency coordinator must ensure that, in the affected area(s) of the facility:

(1) No waste that may be incompatible with the released material is treated, stored, or disposed of until cleanup procedures are completed; and

(2) All emergency equipment listed in the contingency plan is cleaned and fit for its intended use before operations are resumed.

(i) The owner or operator must note in the operating record the time, date, and details of any incident that requires implementing the contingency plan. Within 15 days after the incident, he must submit a written report on the incident to the Regional Administrator. The report must include:

(1) Name, address, and telephone number of the owner or operator;

(2) Name, address, and telephone number of the facility;

(3) Date, time, and type of incident (e.g., fire, explosion);

(4) Name and quantity of material(s) involved;

(5) The extent of injuries, if any;

(6) An assessment of actual or potential hazards to human health or the environment, where this is applicable; and

(7) Estimated quantity and disposition of recovered material that resulted from the incident.

[45 FR 33232, May 19, 1980, as amended at 50 FR 4514, Jan. 31, 1985; 71 FR 16908, Apr. 4, 2006; 71 FR 40274, July 14, 2006; 75 FR 13006, Mar. 18, 2010]

## Subpart E—Manifest System, Recordkeeping, and Reporting

### § 265.70  Applicability.

(a) The regulations in this subpart apply to owners and operators of both on-site and off-site facilities, except as § 265.1 provides otherwise. Sections 265.71, 265.72, and 265.76 do not apply to owners and operators of on-site facilities that do not receive any hazardous waste from off-site sources, nor to owners and operators of off-site facilities with respect to waste military munitions exempted from manifest requirements under 40 CFR 266.203(a).

(b) The revised Manifest form and procedures in 40 CFR 260.10, 261.7, 265.70, 265.71. 265.72, and 265.76, shall not apply until September 5, 2006. The Manifest form and procedures in 40 CFR 260.10, 261.7, 265.70, 265.71. 265.72, and 265.76, contained in the 40 CFR, parts 260 to 265, edition revised as of July 1, 2004, shall be applicable until September 5, 2006.

[70 FR 10823, Mar. 4, 2005]

### § 265.71  Use of manifest system.

(a)(1) If a facility receives hazardous waste accompanied by a manifest, the owner, operator or his/her agent must sign and date the manifest as indicated in paragraph (a)(2) of this section to certify that the hazardous waste covered by the manifest was received, that the hazardous waste was received except as noted in the discrepancy space of the manifest, or that the hazardous waste was rejected as noted in the manifest discrepancy space.

(2) If a facility receives a hazardous waste shipment accompanied by a manifest, the owner, operator or his/her agent must:

(i) Sign and date, by hand, each copy of the manifest;

(ii) Note any discrepancies (as defined in §265.72(a)) on each copy of the manifest;

(iii) Immediately give the transporter at least one copy of the manifest;

(iv) Within 30 days of delivery, send a copy of the manifest to the generator; and

(v) Retain at the facility a copy of each manifest for at least three years from the date of delivery.

(3) If a facility receives hazardous waste imported from a foreign source, the receiving facility must mail a copy of the manifest to the following address within 30 days of delivery: International Compliance Assurance Division, OFA/OECA (2254A), U.S. Environmental Protection Agency, Ariel Rios Building, 1200 Pennsylvania Avenue, NW., Washington, DC 20460.

(b) If a facility receives, from a rail or water (bulk shipment) transporter, hazardous waste which is accompanied by a shipping paper containing all the information required on the manifest (excluding the EPA identification numbers, generator's certification, and signatures), the owner or operator, or his agent, must:

(1) Sign and date each copy of the manifest or shipping paper (if the manifest has not been received) to certify that the hazardous waste covered by the manifest or shipping paper was received;

(2) Note any significant discrepancies (as defined in §265.72(a)) in the manifest or shipping paper (if the manifest has not been received) on each copy of the manifest or shipping paper;

[*Comment:* The Agency does not intend that the owner or operator of a facility whose procedures under §265.13(c) include waste analysis must perform that analysis before signing the shipping paper and giving it to the transporter. Section 265.72(b), however, requires reporting an unreconciled discrepancy discovered during later analysis.]

(3) Immediately give the rail or water (bulk shipment) transporter at least one copy of the manifest or shipping

paper (if the manifest has not been received);

(4) Within 30 days after the delivery, send a copy of the signed and dated manifest or a signed and dated copy of the shipping paper .(if the manifest has not been received within 30 days after delivery) to the generator; and

[*Comment:* Section 262.23(c) of this chapter requires the generator to send three copies of the manifest to the facility when hazardous waste is sent by rail or water (bulk shipment).]

(5) Retain at the facility a copy of the manifest and shipping paper (if signed in lieu of the manifest at the time of delivery) for at least three years from the date of delivery.

(c) Whenever a shipment of hazardous waste is initiated from a facility, the owner or operator of that facility must comply with the requirements of part 262 of this chapter.

[*Comment:* The provisions of §262.34 are applicable to the on-site accumulation of hazardous wastes by generators. Therefore, the provisions of §262.34 only apply to owners or operators who are shipping hazardous waste which they generated at that facility.]

(d) Within three working days of the receipt of a shipment subject to 40 CFR part 262, subpart H, the owner or operator of facility must provide a copy of the tracking document bearing all required signatures to the notifier, to the Office of Enforcement and Compliance Assurance, Office of Compliance, Enforcement Planning, Targeting and Data Division (2222A), Environmental Protection Agency, 1200 Pennsylvania Ave., NW., Washington, DC 20460, and to competent authorities of all other concerned countries. The original copy of the tracking document must be maintained at the facility for at least three years from the date of signature.

(e) A facility must determine whether the consignment state for a shipment regulates any additional wastes (beyond those regulated Federally) as hazardous wastes under its state hazardous waste program. Facilities must also determine whether the consignment state or generator state requires the facility to submit any copies of the manifest to these states.

[45 FR 33232, May 19, 1980, as amended at 45 FR 86970, 86974, Dec. 31, 1980; 50 FR 4514, Jan. 31, 1985; 61 FR 16315, Apr. 12, 1996; 70 FR 10823, Mar. 4, 2005]

EFFECTIVE DATE NOTE: At 75 FR 1260, Jan. 8, 2010, § 265.71 was amended by revising paragraphs (a)(3) and (d), effective July 7, 2010. For the convenience of the user, the revised text is set forth as follows:

### § 265.71 Use of manifest system.

(a) * * *

(3) If a facility receives hazardous waste imported from a foreign source, the receiving facility must mail a copy of the manifest and documentation confirming EPA's consent to the import of hazardous waste to the following address within thirty (30) days of delivery: Office of Enforcement and Compliance Assurance, Office of Federal Activities, International Compliance Assurance Division (2254A), Environmental Protection Agency, 1200 Pennsylvania Avenue, NW., Washington, DC 20460.

\*　　\*　　\*　　\*　　\*

(d) Within three (3) working days of the receipt of a shipment subject to 40 CFR part 262, subpart H, the owner or operator of a facility must provide a copy of the movement document bearing all required signatures to the exporter, to the Office of Enforcement and Compliance Assurance, Office of Federal Activities, International Compliance Assurance Division (2254A), Environmental Protection Agency, 1200 Pennsylvania Avenue, NW., Washington, DC 20460, and to competent authorities of all other countries concerned. The original copy of the movement document must be maintained at the facility for at least three (3) years from the date of signature.

\*　　\*　　\*　　\*　　\*

### § 265.72 Manifest discrepancies.

(a) Manifest discrepancies are:

(1) Significant differences (as defined by paragraph (b) of this section) between the quantity or type of hazardous waste designated on the manifest or shipping paper, and the quantity and type of hazardous waste a facility actually receives;

(2) Rejected wastes, which may be a full or partial shipment of hazardous waste that the TSDF cannot accept; or

(3) Container residues, which are residues that exceed the quantity limits for "empty" containers set forth in 40 CFR 261.7(b).

(b) Significant differences in quantity are: For bulk waste, variations greater than 10 percent in weight; for batch waste, any variation in piece count, such as a discrepancy of one drum in a truckload. Significant differences in type are obvious differences which can be discovered by inspection or waste analysis, such as waste solvent substituted for waste acid, or toxic constituents not reported on the manifest or shipping paper.

(c) Upon discovering a significant difference in quantity or type, the owner or operator must attempt to reconcile the discrepancy with the waste generator or transporter (e.g., with telephone conversations). If the discrepancy is not resolved within 15 days after receiving the waste, the owner or operator must immediately submit to the Regional Administrator a letter describing the discrepancy and attempts to reconcile it, and a copy of the manifest or shipping paper at issue.

(d)(1) Upon rejecting waste or identifying a container residue that exceeds the quantity limits for "empty" containers set forth in 40 CFR 261.7(b), the facility must consult with the generator prior to forwarding the waste to another facility that can manage the waste. If it is impossible to locate an alternative facility that can receive the waste, the facility may return the rejected waste or residue to the generator. The facility must send the waste to the alternative facility or to the generator within 60 days of the rejection or the container residue identification.

(2) While the facility is making arrangements for forwarding rejected wastes or residues to another facility under this section, it must ensure that either the delivering transporter retains custody of the waste, or the facility must provide for secure, temporary custody of the waste, pending delivery of the waste to the first transporter designated on the manifest prepared under paragraph (e) or (f) of this section.

(e) Except as provided in paragraph (e)(7) of this section, for full or partial load rejections and residues that are to be sent off-site to an alternate facility, the facility is required to prepare a new manifest in accordance with §262.20(a) of this chapter and the following instructions:

(1) Write the generator's U.S. EPA ID number in Item 1 of the new manifest. Write the generator's name and mailing address in Item 5 of the new manifest. If the mailing address is different from the generator's site address, then write the generator's site address in the designated space in Item 5.

(2) Write the name of the alternate designated facility and the facility's U.S. EPA ID number in the designated facility block (Item 8) of the new manifest.

(3) Copy the manifest tracking number found in Item 4 of the old manifest to the Special Handling and Additional Information Block of the new manifest, and indicate that the shipment is a residue or rejected waste from the previous shipment.

(4) Copy the manifest tracking number found in Item 4 of the new manifest to the manifest reference number line in the Discrepancy Block of the old manifest (Item 18a).

(5) Write the DOT description for the rejected load or the residue in Item 9 (U.S. DOT Description) of the new manifest and write the container types, quantity, and volume(s) of waste.

(6) Sign the Generator's/Offeror's Certification to certify, as the offeror of the shipment, that the waste has been properly packaged, marked and labeled and is in proper condition for transportation, and mail a signed copy of the manifest to the generator identified in Item 5 of the new manifest.

(7) For full load rejections that are made while the transporter remains present at the facility, the facility may forward the rejected shipment to the alternate facility by completing Item 18b of the original manifest and supplying the information on the next destination facility in the Alternate Facility space. The facility must retain a copy of this manifest for its records, and then give the remaining copies of the manifest to the transporter to accompany the shipment. If the original manifest is not used, then the facility must use a new manifest and comply with paragraphs (e)(1), (2), (3), (4), (5), and (6) of this section.

(f) Except as provided in paragraph (f)(7) of this section, for rejected wastes and residues that must be sent back to the generator, the facility is required to prepare a new manifest in accordance with §262.20(a) of this chapter and the following instructions:

(1) Write the facility's U.S. EPA ID number in Item 1 of the new manifest. Write the facility's name and mailing address in Item 5 of the new manifest. If the mailing address is different from the facility's site address, then write the facility's site address in the designated space for Item 5 of the new manifest.

(2) Write the name of the initial generator and the generator's U.S. EPA ID number in the designated facility block (Item 8) of the new manifest.

(3) Copy the manifest tracking number found in Item 4 of the old manifest to the Special Handling and Additional Information Block of the new manifest, and indicate that the shipment is a residue or rejected waste from the previous shipment,

(4) Copy the manifest tracking number found in Item 4 of the new manifest to the manifest reference number line in the Discrepancy Block of the old manifest (Item 18a),

(5) Write the DOT description for the rejected load or the residue in Item 9 (U.S. DOT Description) of the new manifest and write the container types, quantity, and volume(s) of waste.

(6) Sign the Generator's/Offeror's Certification to certify, as offeror of the shipment, that the waste has been properly packaged, marked and labeled and is in proper condition for transportation,

(7) For full load rejections that are made while the transporter remains at the facility, the facility may return the shipment to the generator with the original manifest by completing Item 18a and 18b of the manifest and supplying the generator's information in the Alternate Facility space. The facility must retain a copy for its records and then give the remaining copies of the manifest to the transporter to accompany the shipment. If the original manifest is not used, then the facility must use a new manifest and comply with paragraphs (f)(1), (2), (3), (4), (5), (6), and (8) of this section.

(8) For full or partial load rejections and container residues contained in non-empty containers that are returned to the generator, the facility must also comply with the exception reporting requirements in § 262.42(a).

(g) If a facility rejects a waste or identifies a container residue that exceeds the quantity limits for "empty" containers set forth in 40 CFR 261.7(b) after it has signed, dated, and returned a copy of the manifest to the delivering transporter or to the generator, the facility must amend its copy of the manifest to indicate the rejected wastes or residues in the discrepancy space of the amended manifest. The facility must also copy the manifest tracking number from Item 4 of the new manifest to the discrepancy space of the amended manifest, and must resign and date the manifest to certify to the information as amended. The facility must retain the amended manifest for at least three years from the date of amendment, and must within 30 days, send a copy of the amended manifest to the transporter and generator that received copies prior to their being amended.

[70 FR 10823, Mar. 4, 2005, as amended at 70 FR 35041, June 16, 2005; 75 FR 13006, Mar. 18, 2010]

## § 265.73 Operating record.

(a) The owner or operator must keep a written operating record at his facility.

(b) The following information must be recorded, as it becomes available, and maintained in the operating record for three years unless noted below:

(1) A description and the quantity of each hazardous waste received, and the method(s) and date(s) of its treatment, storage, or disposal at the facility as required by Appendix I to part 265. This information must be maintained in the operating record until closure of the facility;

(2) The location of each hazardous waste within the facility and the quantity at each location. For disposal facilities, the location and quantity of each hazardous waste must be recorded on a map or diagram of each cell or disposal area. For all facilities, this information must include cross-references to manifest document numbers if the waste was accompanied by a manifest. This information must be maintained in the operating record until closure of the facility;

[Comment: See §§ 265.119, 265.279, and 265.309 for related requirements.]

(3) Records and results of waste analysis, waste determinations, and trial tests performed as specified in §§ 265.13, 265.200, 265.225, 265.252, 265.273, 265.314, 265.341, 265.375, 265.402, 265.1034, 265.1063, 265.1084, 268.4(a), and 268.7 of this chapter.

(4) Summary reports and details of all incidents that require implementing the contingency plan as specified in § 265.56(j);

(5) Records and results of inspections as required by § 265.15(d) (except these data need be kept only three years);

(6) Monitoring, testing or analytical data, and corrective action where required by subpart F of this part and by §§ 265.19, 265.94, 265.191, 265.193, 265.195, 265.224, 265.226, 265.255, 265.260, 265.276, 265.278, 265.280(d)(1), 265.302, 265.304, 265.347, 265.377, 265.1034(c) through 265.1034(f), 265.1035, 265.1063(d) through 265. 265.1063(i), 265.1064, and 265.1083 through 265.1090. Maintain in the operating record for three years, except for records and results pertaining to ground-water monitoring and cleanup, and response action plans for surface impoundments, waste piles, and landfills, which must be maintained in the operating record until closure of the facility.

[Comment: As required by § 265.94, monitoring data at disposal facilities must be kept throughout the post-closure period.]

(7) All closure cost estimates under § 265.142 and, for disposal facilities, all post-closure cost estimates under § 265.144 must be maintained in the operating record until closure of the facility.

(8) Records of the quantities (and date of placement) for each shipment of hazardous waste placed in land disposal units under an extension to the effective date of any land disposal restriction granted pursuant to § 268.5 of this chapter, monitoring data required pursuant to a petition under § 268.6 of this chapter, or a certification under § 268.8 of this chapter, and the applicable notice required by a generator

under §268.7(a) of this chapter. All of this information must be maintained in the operating record until closure of the facility.

(9) For an off-site treatment facility, a copy of the notice, and the certification and demonstration if applicable, required by the generator or the owner or operator under §268.7 or §268.8;

(10) For an on-site treatment facility, the information contained in the notice (except the manifest number), and the certification and demonstration if applicable, required by the generator or the owner or operator under §268.7 or §268.8;

(11) For an off-site land disposal facility, a copy of the notice, and the certification and demonstration if applicable, required by the generator or the owner or operator of a treatment facility under §268.7 or §268.8;

(12) For an on-site land disposal facility, the information contained in the notice (except the manifest number), and the certification and demonstration if applicable, required by the generator or the owner or operator of a treatment facility under §268.7 or §268.8.

(13) For an off-site storage facility, a copy of the notice, and the certification and demonstration if applicable, required by the generator or the owner or operator under §268.7 or §268.8; and

(14) For an on-site storage facility, the information contained in the notice (except the manifest number), and the certification and demonstration if applicable, required by the generator or the owner or operator of a treatment facility under §268.7 or §268.8.

(15) Monitoring, testing or analytical data, and corrective action where required by §§265.90, 265.93(d)(2), and 265.93(d)(5), and the certification as required by §265.196(f) must be maintained in the operating record until closure of the facility.

[45 FR 33232, May 19, 1980, as amended at 50 FR 4514, Jan. 31, 1985; 50 FR 18374, Apr. 30, 1985; 51 FR 40638, Nov. 7, 1986; 53 FR 31211, Aug. 17, 1988; 54 FR 26648, June 23, 1989; 55 FR 25507, June 21, 1990; 56 FR 19290, Apr. 26, 1991; 57 FR 3492, Jan. 29, 1992; 59 FR 62935, Dec. 6, 1994; 62 FR 64661, Dec. 8, 1997; 71 FR 16908, Apr. 4, 2006]

### §265.74 Availability, retention, and disposition of records.

(a) All records, including plans, required under this part must be furnished upon request, and made available at all reasonable times for inspection, by any officer, employee, or representative of EPA who is duly designated by the Administrator.

(b) The retention period for all records required under this part is extended automatically during the course of any unresolved enforcement action regarding the facility or as requested by the Administrator.

(c) A copy of records of waste disposal locations and quantities under §265.73(b)(2) must be submitted to the Regional Administrator and local land authority upon closure of the facility (see §265.119).

[45 FR 33232, May 19, 1980, as amended at 50 FR 4514, Jan. 31, 1985]

### §265.75 Biennial report.

The owner or operator must prepare and submit a single copy of a biennial report to the Regional Administrator by March 1 of each even numbered year. The biennial report must be submitted on EPA Form 8700–13B. The report must cover facility activities during the previous calendar year and must include the following information:

(a) The EPA identification number, name, and address of the facility;

(b) The calendar year covered by the report;

(c) For off-site facilities, the EPA identification number of each hazardous waste generator from which the facility received a hazardous waste during the year; for imported shipments, the report must give the name and address of the foreign generator;

(d) A description and the quantity of each hazardous waste the facility received during the year. For off-site facilities, this information must be listed by EPA identification number of each generator;

(e) The method of treatment, storage, or disposal for each hazardous waste;

(f) Monitoring data under §265.94(a)(2)(ii) and (iii), and (b)(2), where required;

(g) The most recent closure cost estimate under §265.142, and, for disposal facilities, the most recent post-closure cost estimate under § 265.144; and

(h) For generators who treat, store, or dispose of hazardous waste on-site, a description of the efforts undertaken during the year to reduce the volume and toxicity of waste generated.

(i) For generators who treat, store, or dispose of hazardous waste on-site, a description of the changes in volume and toxicity of waste actually achieved during the year in comparison to previous years to the extent such information is available for the years prior to 1984.

(j) The certification signed by the owner or operator of the facility or his authorized representative.

[45 FR 33232, May 19, 1980, as amended at 48 FR 3982, Jan. 28, 1983; 50 FR 4514, Jan. 31, 1985; 51 FR 28556, Aug. 8, 1986]

## § 265.76 Unmanifested waste report.

(a) If a facility accepts for treatment, storage, or disposal any hazardous waste from an off-site source without an accompanying manifest, or without an accompanying shipping paper as described by § 263.20(e) of this chapter, and if the waste is not excluded from the manifest requirement by this chapter, then the owner or operator must prepare and submit a letter to the Regional Administrator within fifteen days after receiving the waste. The unmanifested waste report must contain the following information:

(1) The EPA identification number, name and address of the facility;

(2) The date the facility received the waste;

(3) The EPA identification number, name and address of the generator and the transporter, if available;

(4) A description and the quantity of each unmanifested hazardous waste the facility received;

(5) The method of treatment, storage, or disposal for each hazardous waste;

(6) The certification signed by the owner or operator of the facility or his authorized representative; and

(7) A brief explanation of why the waste was unmanifested, if known.

(b) [Reserved]

[70 FR 10824, Mar. 4, 2005]

## § 265.77 Additional reports.

In addition to submitting the biennial report and unmanifested waste reports described in §§ 265.75 and 265.76, the owner or operator must also report to the Regional Administrator:

(a) Releases, fires, and explosions as specified in § 265.56(j);

(b) Ground-water contamination and monitoring data as specified in §§ 265.93 and 265.94; and

(c) Facility closure as specified in § 265.115.

(d) As otherwise required by Subparts AA, BB, and CC of this part.

[45 FR 33232, May 19, 1980, as amended at 48 FR 3982, Jan. 28, 1983; 55 FR 25507, June 21, 1990; 59 FR 62935, Dec. 6, 1994]

## Subpart F—Ground-Water Monitoring

### § 265.90 Applicability.

(a) Within one year after the effective date of these regulations, the owner or operator of a surface impoundment, landfill, or land treatment facility which is used to manage hazardous waste must implement a ground-water monitoring program capable of determining the facility's impact on the quality of ground water in the uppermost aquifer underlying the facility, except as § 265.1 and paragraph (c) of this section provide otherwise.

(b) Except as paragraphs (c) and (d) of this section provide otherwise, the owner or operator must install, operate, and maintain a ground-water monitoring system which meets the requirements of § 265.91, and must comply with §§ 265.92 through 265.94. This ground-water monitoring program must be carried out during the active life of the facility, and for disposal facilities, during the post-closure care period as well.

(c) All or part of the ground-water monitoring requirements of this subpart may be waived if the owner or operator can demonstrate that there is a low potential for migration of hazardous waste or hazardous waste constituents from the facility via the uppermost aquifer to water supply wells (domestic, industrial, or agricultural) or to surface water. This demonstration must be in writing, and must be kept at the facility. This demonstration must be certified by a qualified geologist or geotechnical engineer and must establish the following:

(1) The potential for migration of hazardous waste or hazardous waste constituents from the facility to the uppermost aquifer, by an evaluation of:

(i) A water balance of precipitation, evapotranspiration, runoff, and infiltration; and

(ii) Unsaturated zone characteristics (i.e., geologic materials, physical properties, and depth to ground water); and

(2) The potential for hazardous waste or hazardous waste constituents which enter the uppermost aquifer to migrate to a water supply well or surface water, by an evaluation of:

(i) Saturated zone characteristics (i.e., geologic materials, physical properties, and rate of ground-water flow); and

(ii) The proximity of the facility to water supply wells or surface water.

(d) If an owner or operator assumes (or knows) that ground-water monitoring of indicator parameters in accordance with §§265.91 and 265.92 would show statistically significant increases (or decreases in the case of pH) when evaluated under §265.93(b), he may install, operate, and maintain an alternate ground-water monitoring system (other than the one described in §§265.91 and 265.92). If the owner or operator decides to use an alternate ground-water monitoring system he must:

(1) Within one year after the effective date of these regulations, develop a specific plan, certified by a qualified geologist or geotechnical engineer, which satisfies the requirements of §265.93(d)(3), for an alternate ground-water monitoring system. This plan is to be placed in the facility's operating record and maintained until closure of the facility.

(2) Not later than one year after the effective date of these regulations, initiate the determinations specified in §265.93(d)(4);

(3) Prepare a report in accordance with §265.93(d)(5) and place it in the facility's operating record and maintain until closure of the facility.

(4) Continue to make the determinations specified in §265.93(d)(4) on a quarterly basis until final closure of the facility; and

(5) Comply with the recordkeeping and reporting requirements in §265.94(b).

(e) The ground-water monitoring requirements of this subpart may be waived with respect to any surface impoundment that (1) Is used to neutralize wastes which are hazardous solely because they exhibit the corrosivity characteristic under §261.22 of this chapter or are listed as hazardous wastes in subpart D of part 261 of this chapter only for this reason, and (2) contains no other hazardous wastes, if the owner or operator can demonstrate that there is no potential for migration of hazardous wastes from the impoundment. The demonstration must establish, based upon consideration of the characteristics of the wastes and the impoundment, that the corrosive wastes will be neutralized to the extent that they no longer meet the corrosivity characteristic before they can migrate out of the impoundment. The demonstration must be in writing and must be certified by a qualified professional.

(f) The Regional Administrator may replace all or part of the requirements of this subpart applying to a regulated unit (as defined in 40 CFR 264.90), with alternative requirements developed for groundwater monitoring set out in an approved closure or post-closure plan or in an enforceable document (as defined in 40 CFR 270.1(c)(7)), where Regional Administrator determines that:

(1) A regulated unit is situated among solid waste management units (or areas of concern), a release has occurred, and both the regulated unit and one or more solid waste management unit(s) (or areas of concern) are likely to have contributed to the release; and

(2) It is not necessary to apply the requirements of this subpart because the alternative requirements will protect human health and the environment. The alternative standards for the regulated unit must meet the requirements of 40 CFR 264.101(a).

[45 FR 33232, May 19, 1980, as amended at 47 FR 1255, Jan. 11, 1982; 50 FR 4514, Jan. 31, 1985; 63 FR 56734, Oct. 22, 1998; 71 FR 16909, Apr. 4, 2006; 71 FR 40274, July 14, 2006]

## § 265.91   Ground-water monitoring system.

(a) A ground-water monitoring system must be capable of yielding ground-water samples for analysis and must consist of:

(1) Monitoring wells (at least one) installed hydraulically upgradient (i.e., in the direction of increasing static head) from the limit of the waste management area. Their number, locations, and depths must be sufficient to yield ground-water samples that are:

(i) Representative of background ground-water quality in the uppermost aquifer near the facility; and

(ii) Not affected by the facility; and

(2) Monitoring wells (at least three) installed hydraulically downgradient (i.e., in the direction of decreasing static head) at the limit of the waste management area. Their number, locations, and depths must ensure that they immediately detect any statistically significant amounts of hazardous waste or hazardous waste constituents that migrate from the waste management area to the uppermost aquifer.

(3) The facility owner or operator may demonstrate that an alternate hydraulically downgradient monitoring well location will meet the criteria outlined below. The demonstration must be in writing and kept at the facility. The demonstration must be certified by a qualified ground-water scientist and establish that:

(i) An existing physical obstacle prevents monitoring well installation at the hydraulically downgradient limit of the waste management area; and

(ii) The selected alternate downgradient location is as close to the limit of the waste management area as practical; and

(iii) The location ensures detection that, given the alternate location, is as early as possible of any statistically significant amounts of hazardous waste or hazardous waste constituents that migrate from the waste management area to the uppermost aquifer.

(iv) Lateral expansion, new, or replacement units are not eligible for an alternate downgradient location under this paragraph.

(b) Separate monitoring systems for each waste management component of a facility are not required provided that provisions for sampling upgradient and downgradient water quality will detect any discharge from the waste management area.

(1) In the case of a facility consisting of only one surface impoundment, landfill, or land treatment area, the waste management area is described by the waste boundary (perimeter).

(2) In the case of a facility consisting of more than one surface impoundment, landfill, or land treatment area, the waste management area is described by an imaginary boundary line which circumscribes the several waste management components.

(c) All monitoring wells must be cased in a manner that maintains the integrity of the monitoring well bore hole. This casing must be screened or perforated, and packed with gravel or sand where necessary, to enable sample collection at depths where appropriate aquifer flow zones exist. The annular space (i.e., the space between the bore hole and well casing) above the sampling depth must be sealed with a suitable material (e.g., cement grout or bentonite slurry) to prevent contamination of samples and the ground water.

[45 FR 33232, May 19, 1980, as amended at 50 FR 4514, Jan. 31, 1985; 56 FR 66369, Dec. 23, 1991]

## § 265.92   Sampling and analysis.

(a) The owner or operator must obtain and analyze samples from the installed ground-water monitoring system. The owner or operator must develop and follow a ground-water sampling and analysis plan. He must keep this plan at the facility. The plan must include procedures and techniques for:

(1) Sample collection;

(2) Sample preservation and shipment;

(3) Analytical procedures; and

(4) Chain of custody control.

[*Comment:* See "Procedures Manual For Ground-water Monitoring At Solid Waste Disposal Facilities," EPA–530/SW–611, August 1977 and "Methods for Chemical Analysis of Water and Wastes," EPA–600/4–79–020, March 1979 for discussions of sampling and analysis procedures.]

(b) The owner or operator must determine the concentration or value of the following parameters in ground-water samples in accordance with paragraphs (c) and (d) of this section:

(1) Parameters characterizing the suitability of the ground water as a drinking water supply, as specified in appendix III.

(2) Parameters establishing ground-water quality:

(i) Chloride
(ii) Iron
(iii) Manganese
(iv) Phenols
(v) Sodium
(vi) Sulfate

[*Comment:* These parameters are to be used as a basis for comparison in the event a ground-water quality assessment is required under §265.93(d).]

(3) Parameters used as indicators of ground-water contamination:

(i) pH
(ii) Specific Conductance
(iii) Total Organic Carbon
(iv) Total Organic Halogen

(c)(1) For all monitoring wells, the owner or operator must establish initial background concentrations or values of all parameters specified in paragraph (b) of this section. He must do this quarterly for one year.

(2) For each of the indicator parameters specified in paragraph (b)(3) of this section, at least four replicate measurements must be obtained for each sample and the initial background arithmetic mean and variance must be determined by pooling the replicate measurements for the respective parameter concentrations or values in samples obtained from upgradient wells during the first year.

(d) After the first year, all monitoring wells must be sampled and the samples analyzed with the following frequencies:

(1) Samples collected to establish ground-water quality must be obtained and analyzed for the parameters specified in paragraph (b)(2) of this section at least annually.

(2) Samples collected to indicate ground-water contamination must be obtained and analyzed for the parameters specified in paragraph (b)(3) of this section at least semi-annually.

(e) Elevation of the ground-water surface at each monitoring well must be determined each time a sample is obtained.

[45 FR 33232, May 19, 1980, as amended at 50 FR 4514, Jan. 31, 1985]

## §265.93 Preparation, evaluation, and response.

(a) Within one year after the effective date of these regulations, the owner or operator must prepare an *outline* of a ground-water quality assessment program. The outline must describe a more comprehensive ground-water monitoring program (than that described in §§265.91 and 265.92) capable of determining:

(1) Whether hazardous waste or hazardous waste constituents have entered the ground water;

(2) The rate and extent of migration of hazardous waste or hazardous waste constituents in the ground water; and

(3) The concentrations of hazardous waste or hazardous waste constituents in the ground water.

(b) For each indicator parameter specified in §265.92(b)(3), the owner or operator must calculate the arithmetic mean and variance, based on at least four replicate measurements on each sample, for each well monitored in accordance with §265.92(d)(2), and compare these results with its initial background arithmetic mean. The comparison must consider individually each of the wells in the monitoring system, and must use the Student's t-test at the 0.01 level of significance (see appendix IV) to determine statistically significant increases (and decreases, in the case of pH) over initial background.

(c)(1) If the comparisons for the *upgradient* wells made under paragraph (b) of this section show a significant increase (or pH decrease), the owner or operator must submit this information in accordance with §265.94(a)(2)(ii).

(2) If the comparisons for *downgradient* wells made under paragraph (b) of this section show a significant increase (or pH decrease), the owner or operator must then immediately obtain additional ground-water samples from those downgradient wells where a significant difference was detected, split the samples in two, and

obtain analyses of all additional samples to determine whether the significant difference was a result of laboratory error.

(d)(1) If the analyses performed under paragraph (c)(2) of this section confirm the significant increase (or pH decrease), the owner or operator must provide written notice to the Regional Administrator—within seven days of the date of such confirmation—that the facility may be affecting ground-water quality.

(2) Within 15 days after the notification under paragraph (d)(1) of this section, the owner or operator must develop a specific plan, based on the outline required under paragraph (a) of this section and certified by a qualified geologist or geotechnical engineer, for a ground-water quality assessment at the facility. This plan must be placed in the facility operating record and be maintained until closure of the facility.

(3) The plan to be submitted under § 265.90(d)(1) or paragraph (d)(2) of this section must specify:

(i) The number, location, and depth of wells;

(ii) Sampling and analytical methods for those hazardous wastes or hazardous waste constituents in the facility;

(iii) Evaluation procedures, including any use of previously-gathered ground-water quality information; and

(iv) A schedule of implementation.

(4) The owner or operator must implement the ground-water quality assessment plan which satisfies the requirements of paragraph (d)(3) of this section, and, at a minimum, determine:

(i) The rate and extent of migration of the hazardous waste or hazardous waste constituents in the ground water; and

(ii) The concentrations of the hazardous waste or hazardous waste constituents in the ground water.

(5) The owner or operator must make his first determination under paragraph (d)(4) of this section, as soon as technically feasible, and prepare a report containing an assessment of ground-water quality. This report must be placed in the facility operating record and be maintained until closure of the facility.

(6) If the owner or operator determines, based on the results of the first determination under paragraph (d)(4) of this section, that no hazardous waste or hazardous waste constituents from the facility have entered the ground water, then he may reinstate the indicator evaluation program described in § 265.92 and paragraph (b) of this section. If the owner or operator reinstates the indicator evaluation program, he must so notify the Regional Administrator in the report submitted under paragraph (d)(5) of this section.

(7) If the owner or operator determines, based on the first determination under paragraph (d)(4) of this section, that hazardous waste or hazardous waste constituents from the facility have entered the ground water, then he:

(i) Must continue to make the determinations required under paragraph (d)(4) of this section on a quarterly basis until final closure of the facility, if the ground-water quality assessment plan was implemented prior to final closure of the facility; or

(ii) May cease to make the determinations required under paragraph (d)(4) of this section, if the ground-water quality assessment plan was implemented during the post-closure care period.

(e) Notwithstanding any other provision of this subpart, any ground-water quality assessment to satisfy the requirements of § 265.93(d)(4) which is initiated prior to final closure of the facility must be completed and reported in accordance with § 265.93(d)(5).

(f) Unless the ground water is monitored to satisfy the requirements of § 265.93(d)(4), at least annually the owner or operator must evaluate the data on ground-water surface elevations obtained under § 265.92(e) to determine whether the requirements under § 265.91(a) for locating the monitoring wells continues to be satisfied. If the evaluation shows that § 265.91(a) is no longer satisfied, the owner or operator must immediately modify the number, location, or depth of the monitoring wells to bring the ground-water monitoring system into compliance with this requirement.

[45 FR 33232, May 19, 1980, as amended at 50 FR 4514, Jan. 31, 1985; 71 FR 16909, Apr. 4, 2006]

### §265.94 Recordkeeping and reporting.

(a) Unless the ground water is monitored to satisfy the requirements of §265.93(d)(4), the owner or operator must:

(1) Keep records of the analyses required in §265.92(c) and (d), the associated ground-water surface elevations required in §265.92(e), and the evaluations required in §265.93(b) throughout the active life of the facility, and, for disposal facilities, throughout the post-closure care period as well; and

(2) Report the following ground-water monitoring information to the Regional Administrator:

(i) During the first year when initial background concentrations are being established for the facility: concentrations or values of the parameters listed in §265.92(b)(1) for each ground-water monitoring well within 15 days after completing each quarterly analysis. The owner or operator must separately identify for each monitoring well any parameters whose concentration or value has been found to exceed the maximum contaminant levels listed in appendix III.

(ii) Annually: Concentrations or values of the parameters listed in §265.92(b)(3) for each ground-water monitoring well, along with the required evaluations for these parameters under §265.93(b). The owner or operator must separately identify any significant differences from initial background found in the upgradient wells, in accordance with §265.93(c)(1). During the active life of the facility, this information must be submitted no later than March 1 following each calendar year.

(iii) No later than March 1 following each calendar year: Results of the evaluations of ground-water surface elevations under §265.93(f), and a description of the response to that evaluation, where applicable.

(b) If the ground water is monitored to satisfy the requirements of §265.93(d)(4), the owner or operator must:

(1) Keep records of the analyses and evaluations specified in the plan, which satisfies the requirements of §265.93(d)(3), throughout the active life of the facility, and, for disposal facilities, throughout the post-closure care period as well; and

(2) Annually, until final closure of the facility, submit to the Regional Administrator a report containing the results of his or her ground-water quality assessment program which includes, but is not limited to, the calculated (or measured) rate of migration of hazardous waste or hazardous waste constituents in the ground water during the reporting period. This information must be submitted no later than March 1 following each calendar year.

[45 FR 33232, May 19, 1980, as amended at 48 FR 3982, Jan. 28, 1983; 50 FR 4514, Jan. 31, 1985]

## Subpart G—Closure and Post-Closure

SOURCE: 51 FR 16451, May 2, 1986, unless otherwise noted.

### §265.110 Applicability.

Except as §265.1 provides otherwise:

(a) Sections 265.111 through 265.115 (which concern closure) apply to the owners and operators of all hazardous waste management facilities; and

(b) Sections 265.116 through 265.120 (which concern post-closure care) apply to the owners and operators of:

(1) All hazardous waste disposal facilities;

(2) Waste piles and surface impoundments for which the owner or operator intends to remove the wastes at closure to the extent that these sections are made applicable to such facilities in §265.228 or §265.258;

(3) Tank systems that are required under §265.197 to meet requirements for landfills; and

(4) Containment buildings that are required under §265.1102 to meet the requirement for landfills.

(c) Section 265.121 applies to owners and operators of units that are subject to the requirements of 40 CFR 270.1(c)(7) and are regulated under an enforceable document (as defined in 40 CFR 270.1(c)(7)).

(d) The Regional Administrator may replace all or part of the requirements of this subpart (and the unit-specific standards in § 265.111(c)) applying to a regulated unit (as defined in 40 CFR 264.90), with alternative requirements for closure set out in an approved closure or post-closure plan, or in an enforceable document (as defined in 40 CFR 270.1(c)(7)), where the Regional Administrator determines that:

(1) A regulated unit is situated among solid waste management units (or areas of concern), a release has occurred, and both the regulated unit and one or more solid waste management unit(s) (or areas of concern) are likely to have contributed to the release, and

(2) It is not necessary to apply the closure requirements of this subpart (and/or those referenced herein) because the alternative requirements will protect human health and the environment, and will satisfy the closure performance standard of § 265.111 (a) and (b).

[51 FR 16451, May 2, 1986, as amended at 51 FR 25479, July 14, 1986; 53 FR 34086, Sept. 2, 1988; 57 FR 37267, Aug. 18, 1992; 63 FR 56734, Oct. 22, 1998; 71 FR 40274, July 14, 2006]

§ 265.111  Closure performance standard.

The owner or operator must close the facility in a manner that:

(a) Minimizes the need for further maintenance, and

(b) Controls, minimizes or eliminates, to the extent necessary to protect human health and the environment, post-closure escape of hazardous waste, hazardous constituents, leachate, contaminated run-off, or hazardous waste decomposition products to the ground or surface waters or to the atmosphere, and

(c) Complies with the closure requirements of this subpart, including, but not limited to, the requirements of §§ 265.197, 265.228, 265.258, 265.280, 265.310, 265.351, 265.381, 265.404, and 265.1102.

[51 FR 16451, May 2, 1986, as amended at 57 FR 37267, Aug. 18, 1992; 71 FR 40275, July 14, 2006]

§ 265.112  Closure plan; amendment of plan.

(a) *Written plan.* By May 19, 1981, or by six months after the effective date of the rule that first subjects a facility to provisions of this section, the owner or operator of a hazardous waste management facility must have a written closure plan. Until final closure is completed and certified in accordance with § 265.115, a copy of the most current plan must be furnished to the Regional Administrator upon request, including request by mail. In addition, for facilities without approved plans, it must also be provided during site inspections, on the day of inspection, to any officer, employee, or representative of the Agency who is duly designated by the Administrator.

(b) *Content of plan.* The plan must identify steps necessary to perform partial and/or final closure of the facility at any point during its active life. The closure plan must include, at least:

(1) A description of how each hazardous waste management unit at the facility will be closed in accordance with § 265.111; and

(2) A description of how final closure of the facility will be conducted in accordance with § 265.111. The description must identify the maximum extent of the operation which will be unclosed during the active life of the facility; and

(3) An estimate of the maximum inventory of hazardous wastes ever on-site over the active life of the facility and a detailed description of the methods to be used during partial and final closure, including, but not limited to methods for removing, transporting, treating, storing or disposing of all hazardous waste, identification of and the type(s) of off-site hazardous waste management unit(s) to be used, if applicable; and

(4) A detailed description of the steps needed to remove or decontaminate all hazardous waste residues and contaminated containment system components, equipment, structures, and soils during partial and final closure including, but not limited to, procedures for cleaning equipment and removing contaminated soils, methods for sampling

and testing surrounding soils, and criteria for determining the extent of decontamination necessary to satisfy the closure performance standard; and

(5) A detailed description of other activities necessary during the partial and final closure periods to ensure that all partial closures and final closure satisfy the closure performance standards, including, but not limited to, ground-water monitoring, leachate collection, and run-on and run-off control; and

(6) A schedule for closure of each hazardous waste management unit and for final closure of the facility. The schedule must include, at a minimum, the total time required to close each hazardous waste management unit and the time required for intervening closure activities which will allow tracking of the progress of partial and final closure. (For example, in the case of a landfill unit, estimates of the time required to treat or dispose of all hazardous waste inventory and of the time required to place a final cover must be included.); and

(7) An estimate of the expected year of final closure for facilities that use trust funds to demonstrate financial assurance under § 265.143 or § 265.145 and whose remaining operating life is less than twenty years, and for facilities without approved closure plans.

(8) For facilities where the Regional Administrator has applied alternative requirements at a regulated unit under §§ 265.90(f), 265.110(d), and/or 265.140(d), either the alternative requirements applying to the regulated unit, or a reference to the enforceable document containing those alternative requirements.

(c) *Amendment of plan.* The owner or operator may amend the closure plan at any time prior to the notification of partial or final closure of the facility. An owner or operator with an approved closure plan must submit a written request to the Regional Administrator to authorize a change to the approved closure plan. The written request must include a copy of the amended closure plan for approval by the Regional Administrator.

(1) The owner or operator must amend the closure plan whenever:

(i) Changes in operating plans or facility design affect the closure plan, or

(ii) There is a change in the expected year of closure, if applicable, or

(iii) In conducting partial or final closure activities, unexpected events require a modification of the closure plan.

(iv) The owner or operator requests the Regional Administrator to apply alternative requirements to a regulated unit under §§ 265.90(f), 265.110(d), and/or 265.140(d).

(2) The owner or operator must amend the closure plan at least 60 days prior to the proposed change in facility design or operation, or no later than 60 days after an unexpected event has occurred which has affected the closure plan. If an unexpected event occurs during the partial or final closure period, the owner or operator must amend the closure plan no later than 30 days after the unexpected event. These provisions also apply to owners or operators of surface impoundments and waste piles who intended to remove all hazardous wastes at closure, but are required to close as landfills in accordance with § 265.310.

(3) An owner or operator with an approved closure plan must submit the modified plan to the Regional Administrator at least 60 days prior to the proposed change in facility design or operation, or no more than 60 days after an unexpected event has occurred which has affected the closure plan. If an unexpected event has occurred during the partial or final closure period, the owner or operator must submit the modified plan no more than 30 days after the unexpected event. These provisions also apply to owners or operators of surface impoundments and waste piles who intended to remove all hazardous wastes at closure but are required to close as landfills in accordance with § 265.310. If the amendment to the plan is a Class 2 or 3 modification according to the criteria in § 270.42, the modification to the plan will be approved according to the procedures in § 265.112(d)(4).

(4) The Regional Administrator may request modifications to the plan under the conditions described in paragraph (c)(1) of this section. An owner or operator with an approved closure plan

must submit the modified plan within 60 days of the request from the Regional Administrator, or within 30 days if the unexpected event occurs during partial or final closure. If the amendment is considered a Class 2 or 3 modification according to the criteria in § 270.42, the modification to the plan will be approved in accordance with the procedures in § 265.112(d)(4).

(d) *Notification of partial closure and final closure.* (1) The owner or operator must submit the closure plan to the Regional Administrator at least 180 days prior to the date on which he expects to begin closure of the first surface impoundment, waste pile, land treatment, or landfill unit, or final closure if it involves such a unit, whichever is earlier. The owner or operator must submit the closure plan to the Regional Administrator at least 45 days prior to the date on which he expects to begin partial or final closure of a boiler or industrial furnace. The owner or operator must submit the closure plan to the Regional Administrator at least 45 days prior to the date on which he expects to begin final closure of a facility with only tanks, container storage, or incinerator units. Owners or operators with approved closure plans must notify the Regional Administrator in writing at least 60 days prior to the date on which he expects to begin closure of a surface impoundment, waste pile, landfill, or land treatment unit, or final closure of a facility involving such a unit. Owners or operators with approved closure plans must notify the Regional Administrator in writing at least 45 days prior to the date on which he expects to begin partial or final closure of a boiler or industrial furnace. Owners or operators with approved closure plans must notify the Regional Administrator in writing at least 45 days prior to the date on which he expects to begin final closure of a facility with only tanks, container storage, or incinerator units.

(2) The date when he "expects to begin closure" must be either:

(i) Within 30 days after the date on which any hazardous waste management unit receives the known final volume of hazardous wastes, or, if there is a reasonable possibility that the hazardous waste management unit will receive additional hazardous wastes, no later than one year after the date on which the unit received the most recent volume of hazardous waste. If the owner or operator of a hazardous waste management unit can demonstrate to the Regional Administrator that the hazardous waste management unit or facility has the capacity to receive additional hazardous wastes and he has taken, and will continue to take, all steps to prevent threats to human health and the environment, including compliance with all interim status requirements, the Regional Administrator may approve an extension to this one-year limit; or

(ii) For units meeting the requirements of § 265.113(d), no later than 30 days after the date on which the hazardous waste management unit receives the known final volume of nonhazardous wastes, or if there is a reasonable possibility that the hazardous waste management unit will receive additional nonhazardous wastes, no later than one year after the date on which the unit received the most recent volume of nonhazardous wastes. If the owner or operator can demonstrate to the Regional Administrator that the hazardous waste management unit has the capacity to receive additional nonhazardous wastes and he has taken, and will continue to take, all steps to prevent threats to human health and the environment, including compliance with all applicable interim status requirements, the Regional Administrator may approve an extension to this one-year limit.

(3) The owner or operator must submit his closure plan to the Regional Administrator no later than 15 days after:

(i) Termination of interim status except when a permit is issued simultaneously with termination of interim status; or

(ii) Issuance of a judicial decree or final order under section 3008 of RCRA to cease receiving hazardous wastes or close.

(4) The Regional Administrator will provide the owner or operator and the public, through a newspaper notice, the opportunity to submit written comments on the plan and request modifications to the plan no later than 30

days from the date of the notice. He will also, in response to a request or at his own discretion, hold a public hearing whenever such a hearing might clarify one or more issues concerning a closure plan. The Regional Administrator will give public notice of the hearing at least 30 days before it occurs. (Public notice of the hearing may be given at the same time as notice of the opportunity for the public to submit written comments, and the two notices may be combined.) The Regional Administrator will approve, modify, or disapprove the plan within 90 days of its receipt. If the Regional Administrator does not approve the plan he shall provide the owner or operator with a detailed written statement of reasons for the refusal and the owner or operator must modify the plan or submit a new plan for approval within 30 days after receiving such written statement. The Regional Administrator will approve or modify this plan in writing within 60 days. If the Regional Administrator modifies the plan, this modified plan becomes the approved closure plan. The Regional Administrator must assure that the approved plan is consistent with §§ 265.111 through 265.115 and the applicable requirements of subpart F of this part, and §§ 265.197, 265.228, 265.258, 265.280, 265.310, 265.351, 265.381, 265.404, and 265.1102. A copy of the modified plan with a detailed statement of reasons for the modifications must be mailed to the owner or operator.

(e) *Removal of wastes and decontamination or dismantling of equipment.* Nothing in this section shall preclude the owner or operator from removing hazardous wastes and decontaminating or dismantling equipment in accordance with the approved partial or final closure plan at any time before or after notification of partial or final closure.

[51 FR 16451, May 2, 1986, as amended at 54 FR 37935, Sept. 28, 1988; 56 FR 7207, Feb. 21, 1991; 56 FR 42512, Aug. 27, 1991; 57 FR 37267, Aug. 18, 1992; 63 FR 56734, Oct. 22, 1998; 71 FR 40275, July 14, 2006]

**§ 265.113  Closure; time allowed for closure.**

(a) Within 90 days after receiving the final volume of hazardous wastes, or the final volume of nonhazardous

wastes if the owner or operator complies with all applicable requirements in paragraphs (d) and (e) of this section, at a hazardous waste management unit or facility, or within 90 days after approval of the closure plan, whichever is later, the owner or operator must treat, remove from the unit or facility, or dispose of on-site, all hazardous wastes in accordance with the approved closure plan. The Regional Administrator may approve a longer period if the owner or operator demonstrates that:

(1)(i) The activities required to comply with this paragraph will, of necessity, take longer than 90 days to complete; or

(ii)(A) The hazardous waste management unit or facility has the capacity to receive additional hazardous wastes, or has the capacity to receive non-hazardous wastes if the facility owner or operator complies with paragraphs (d) and (e) of this section; and

(B) There is a reasonable likelihood that he or another person will recommence operation of the hazardous waste management unit or the facility within one year; and

(C) Closure of the hazardous waste management unit or facility would be incompatible with continued operation of the site; and

(2) He has taken and will continue to take all steps to prevent threats to human health and the environment, including compliance with all applicable interim status requirements.

(b) The owner or operator must complete partial and final closure activities in accordance with the approved closure plan and within 180 days after receiving the final volume of hazardous wastes, or the final volume of nonhazardous wastes if the owner or operator complies with all applicable requirements in paragraphs (d) and (e) of this section, at the hazardous waste management unit or facility, or 180 days after approval of the closure plan, if that is later. The Regional Administrator may approve an extension to the closure period if the owner or operator demonstrates that:

(1)(i) The partial or final closure activities will, of necessity, take longer than 180 days to complete; or (ii)(A) The hazardous waste management unit or facility has the capacity to receive additional hazardous wastes, or has the capacity to receive non-hazardous wastes if the facility owner or operator complies with paragraphs (d) and (e) of this section; and

(B) There is reasonable likelihood that he or another person will recommence operation of the hazardous waste management unit or the facility within one year; and

(C) Closure of the hazardous waste management unit or facility would be incompatible with continued operation of the site; and

(2) He has taken and will continue to take all steps to prevent threats to human health and the environment from the unclosed but not operating hazardous waste management unit or facility, including compliance with all applicable interim status requirements.

(c) The demonstrations referred to in paragraphs (a)(1) and (b)(1) of this section must be made as follows:

(1) The demonstrations in paragraph (a)(1) of this section must be made at least 30 days prior to the expiration of the 90-day period in paragraph (a) of this section; and

(2) The demonstration in paragraph (b)(1) of this section must be made at least 30 days prior to the expiration of the 180-day period in paragraph (b) of this section, unless the owner or operator is otherwise subject to the deadlines in paragraph (d) of this section.

(d) The Regional Administrator may allow an owner or operator to receive non-hazardous wastes in a landfill, land treatment, or surface impoundment unit after the final receipt of hazardous wastes at that unit if:

(1) The owner or operator submits an amended part B application, or a part B application, if not previously required, and demonstrates that:

(i) The unit has the existing design capacity as indicated on the part A application to receive non-hazardous wastes; and

(ii) There is a reasonable likelihood that the owner or operator or another person will receive non-hazardous wastes in the unit within one year after the final receipt of hazardous wastes; and

(iii) The non-hazardous wastes will not be incompatible with any remaining wastes in the unit or with the facility design and operating requirements of the unit or facility under this part; and

(iv) Closure of the hazardous waste management unit would be incompatible with continued operation of the unit or facility; and

(v) The owner or operator is operating and will continue to operate in compliance with all applicable interim status requirements; and

(2) The part B application includes an amended waste analysis plan, groundwater monitoring and response program, human exposure assessment required under RCRA section 3019, and closure and post-closure plans, and updated cost estimates and demonstrations of financial assurance for closure and post-closure care as necessary and appropriate to reflect any changes due to the presence of hazardous constituents in the non-hazardous wastes, and changes in closure activities, including the expected year of closure if applicable under § 265.112(b)(7), as a result of the receipt of non-hazardous wastes following the final receipt of hazardous wastes; and

(3) The part B application is amended, as necessary and appropriate, to account for the receipt of non-hazardous wastes following receipt of the final volume of hazardous wastes; and

(4) The part B application and the demonstrations referred to in paragraphs (d)(1) and (d)(2) of this section are submitted to the Regional Administrator no later than 180 days prior to the date on which the owner or operator of the facility receives the known final volume of hazardous wastes, or no later than 90 days after the effective date of this rule in the state in which the unit is located, whichever is later.

(e) In addition to the requirements in paragraph (d) of this section, an owner or operator of a hazardous waste surface impoundment that is not in compliance with the liner and leachate collection system requirements in 42 U.S.C. 3004(o)(1) and 3005(j)(1) or 42 U.S.C. 3004(o)(2) or (3) or 3005(j) (2), (3), (4) or (13) must:

(1) Submit with the part B application:

(i) A contingent corrective measures plan; and

(ii) A plan for removing hazardous wastes in compliance with paragraph (e)(2) of this section; and

(2) Remove all hazardous wastes from the unit by removing all hazardous liquids and removing all hazardous sludges to the extent practicable without impairing the integrity of the liner(s), if any.

(3) Removal of hazardous wastes must be completed no later than 90 days after the final receipt of hazardous wastes. The Regional Administrator may approve an extension to this deadline if the owner or operator demonstrates that the removal of hazardous wastes will, of necessity, take longer than the allotted period to complete and that an extension will not pose a threat to human health and the environment.

(4) If a release that is a statistically significant increase (or decrease in the case of pH) in hazardous constituents over background levels is detected in accordance with the requirements in subpart F of this part, the owner or operator of the unit:

(i) Must implement corrective measures in accordance with the approved contingent corrective measures plan required by paragraph (e)(1) of this section no later than one year after detection of the release, or approval of the contingent corrective measures plan, whichever is later;

(ii) May receive wastes at the unit following detection of the release only if the approved corrective measures plan includes a demonstration that continued receipt of wastes will not impede corrective action; and

(iii) May be required by the Regional Administrator to implement corrective measures in less than one year or to cease receipt of wastes until corrective measures have been implemented if necessary to protect human health and the environment.

(5) During the period of corrective action, the owner or operator shall provide annual reports to the Regional Administrator describing the progress of the corrective action program, compile all ground-water monitoring data, and evaluate the effect of the continued receipt of non-hazardous wastes on the effectiveness of the corrective action.

(6) The Regional Administrator may require the owner or operator to commence closure of the unit if the owner or operator fails to implement corrective action measures in accordance with the approved contingent corrective measures plan within one year as required in paragraph (e)(4) of this section, or fails to make substantial progress in implementing corrective action and achieving the facility's background levels.

(7) If the owner or operator fails to implement corrective measures as required in paragraph (e)(4) of this section, or if the Regional Administrator determines that substantial progress has not been made pursuant to paragraph (e)(6) of this section he shall:

(i) Notify the owner or operator in writing that the owner or operator must begin closure in accordance with the deadline in paragraphs (a) and (b) of this section and provide a detailed statement of reasons for this determination, and

(ii) Provide the owner or operator and the public, through a newspaper notice, the opportunity to submit written comments on the decision no later than 20 days after the date of the notice.

(iii) If the Regional Administrator receives no written comments, the decision will become final five days after the close of the comment period. The Regional Administrator will notify the owner or operator that the decision is final, and that a revised closure plan, if necessary, must be submitted within 15 days of the final notice and that closure must begin in accordance with the deadlines in paragraphs (a) and (b) of this section.

(iv) If the Regional Administrator receives written comments on the decision, he shall make a final decision within 30 days after the end of the comment period, and provide the owner or operator in writing and the public through a newspaper notice, a detailed statement of reasons for the final decision. If the Regional Administrator determines that substantial progress has not been made, closure must be initiated in accordance with the deadlines in paragraphs (a) and (b) of this section.

(v) The final determinations made by the Regional Administrator under paragraphs (e)(7) (iii) and (iv) of this section are not subject to administrative appeal.

[51 FR 16451, May 2, 1986, as amended at 54 FR 33396, Aug. 14, 1989; 56 FR 42512, Aug. 27, 1991; 71 FR 16909, Apr. 4, 2006; 71 FR 40275, July 14, 2006]

## § 265.114  Disposal or decontamination of equipment, structures and soils.

During the partial and final closure periods, all contaminated equipment, structures and soil must be properly disposed of, or decontaminated unless specified otherwise in §§ 265.197, 265.228, 265.258, 265.280, or 265.310. By removing all hazardous wastes or hazardous constituents during partial and final closure, the owner or operator may become a generator of hazardous waste and must handle that hazardous waste in accordance with all applicable requirements of part 262 of this chapter.

[51 FR 16451, May 2, 1986, as amended at 53 FR 34086, Sept. 2, 1988]

## § 265.115  Certification of closure.

Within 60 days of completion of closure of each hazardous waste surface impoundment, waste pile, land treatment, and landfill unit, and within 60 days of completion of final closure, the owner or operator must submit to the Regional Administrator, by registered mail, a certification that the hazardous waste management unit or facility, as applicable, has been closed in accordance with the specifications in the approved closure plan. The certification must be signed by the owner or operator and by a qualified Professional Engineer. Documentation supporting the Professional Engineer's certification must be furnished to the Regional Administrator upon request until he releases the owner or operator from the financial assurance requirements for closure under § 265.143(h).

[71 FR 16909, Apr. 4, 2006]

## § 265.116  Survey plat.

No later than the submission of the certification of closure of each hazardous waste disposal unit, an owner or operator must submit to the local zoning authority, or the authority with jurisdiction over local land use, and to the Regional Administrator, a survey plat indicating the location and dimensions of landfill cells or other hazardous waste disposal units with respect to permanently surveyed benchmarks. This plat must be prepared and certified by a professional land surveyor. The plat filed with the local zoning authority, or the authority with jurisdiction over local land use must contain a note, prominently displayed, which states the owner's or operator's obligation to restrict disturbance of the hazardous waste disposal unit in accordance with the applicable subpart G regulations.

## § 265.117  Post-closure care and use of property.

(a)(1) Post-closure care for each hazardous waste management unit subject to the requirements of §§ 265.117 through 265.120 must begin after completion of closure of the unit and continue for 30 years after that date. It must consist of at least the following:

(i) Monitoring and reporting in accordance with the requirements of subparts F, K, L, M, and N of this part; and

(ii) Maintenance and monitoring of waste containment systems in accordance with the requirements of subparts F, K, L, M, and N of this part.

(2) Any time preceding closure of a hazardous waste management unit subject to post-closure care requirements or final closure, or any time during the post-closure period for a particular hazardous waste disposal unit, the Regional Administrator may:

(i) Shorten the post-closure care period applicable to the hazardous waste management unit, or facility, if all disposal units have been closed, if he finds that the reduced period is sufficient to protect human health and the environment (e.g., leachate or ground-water monitoring results, characteristics of the hazardous waste, application of advanced technology, or alternative disposal, treatment, or re-use techniques indicate that the hazardous waste management unit or facility is secure); or

(ii) Extend the post-closure care period applicable to the hazardous waste management unit or facility, if he finds that the extended period is necessary to protect human health and the environment (e.g., leachate or groundwater monitoring results indicate a potential for migration of hazardous wastes at levels which may be harmful to human health and the environment).

(b) The Regional Administrator may require, at partial and final closure, continuation of any of the security requirements of §265.14 during part or all of the post-closure period when:

(1) Hazardous wastes may remain exposed after completion of partial or final closure; or

(2) Access by the public or domestic livestock may pose a hazard to human health.

(c) Post-closure use of property on or in which hazardous wastes remain after partial or final closure must never be allowed to disturb the integrity of the final cover, liner(s), or any other components of the containment system, or the function of the facility's monitoring systems, unless the Regional Administrator finds that the disturbance:

(1) Is necessary to the proposed use of the property, and will not increase the potential hazard to human health or the environment; or

(2) Is necessary to reduce a threat to human health or the environment.

(d) All post-closure care activities must be in accordance with the provisions of the approved post-closure plan as specified in §265.118.

[51 FR 16451, May 2, 1986, as amended at 71 FR 40275, July 14, 2006]

### §265.118 Post-closure plan; amendment of plan.

(a) *Written plan.* By May 19, 1981, the owner or operator of a hazardous waste disposal unit must have a written post-closure plan. An owner or operator of a surface impoundment or waste pile that intends to remove all hazardous wastes at closure must prepare a post-closure plan and submit it to the Regional Administrator within 90 days of the date that the owner or operator or Regional Administrator determines that the hazardous waste management unit or facility must be closed as a landfill, subject to the requirements of §§265.117 through 265.120.

(b) Until final closure of the facility, a copy of the most current post-closure plan must be furnished to the Regional Administrator upon request, including request by mail. In addition, for facilities without approved post-closure plans, it must also be provided during site inspections, on the day of inspection, to any officer, employee or representative of the Agency who is duly designated by the Administrator. After final closure has been certified, the person or office specified in §265.118(c)(3) must keep the approved post-closure plan during the post-closure period.

(c) For each hazardous waste management unit subject to the requirements of this section, the post-closure plan must identify the activities that will be carried on after closure of each disposal unit and the frequency of these activities, and include at least:

(1) A description of the planned monitoring activities and frequencies at which they will be performed to comply with subparts F, K, L, M, and N of this part during the post-closure care period; and

(2) A description of the planned maintenance activities, and frequencies at which they will be performed, to ensure:

(i) The integrity of the cap and final cover or other containment systems in accordance with the requirements of subparts K, L, M, and N of this part; and

(ii) The function of the monitoring equipment in accordance with the requirements of subparts F, K, L, M, and N of this part; and

(3) The name, address, and phone number of the person or office to contact about the hazardous waste disposal unit or facility during the post-closure care period.

(4) For facilities subject to §265.121, provisions that satisfy the requirements of §265.121(a)(1) and (3).

(5) For facilities where the Regional Administrator has applied alternative requirements at a regulated unit under §§265.90(f), 265.110(d), and/or 265.140(d), either the alternative requirements that apply to the regulated unit, or a reference to the enforceable document containing those requirements.

(d) *Amendment of plan.* The owner or operator may amend the post-closure plan any time during the active life of the facility or during the post-closure care period. An owner or operator with an approved post-closure plan must submit a written request to the Regional Administrator to authorize a change to the approved plan. The written request must include a copy of the amended post-closure plan for approval by the Regional Administrator.

(1) The owner or operator must amend the post-closure plan whenever:

(i) Changes in operating plans or facility design affect the post-closure plan, or

(ii) Events which occur during the active life of the facility, including partial and final closures, affect the post-closure plan.

(iii) The owner or operator requests the Regional Administrator to apply alternative requirements to a regulated unit under §§ 265.90(f), 265.110(d), and/or 265.140(d).

(2) The owner or operator must amend the post-closure plan at least 60 days prior to the proposed change in facility design or operation, or no later than 60 days after an unexpected event has occurred which has affected the post-closure plan.

(3) An owner or operator with an approved post-closure plan must submit the modified plan to the Regional Administrator at least 60 days prior to the proposed change in facility design or operation, or no more than 60 days after an unexpected event has occurred which has affected the post-closure plan. If an owner or operator of a surface impoundment or a waste pile who intended to remove all hazardous wastes at closure in accordance with § 265.228(b) or § 265.258(a) is required to close as a landfill in accordance with § 265.310, the owner or operator must submit a post-closure plan within 90 days of the determination by the owner or Regional Administrator that the unit must be closed as a landfill. If the amendment to the post-closure plan is a Class 2 or 3 modification according to the criteria in § 270.42, the modification to the plan will be approved according to the procedures in § 265.118(f).

(4) The Regional Administrator may request modifications to the plan under the conditions described in paragraph (d)(1) of this section. An owner or operator with an approved post-closure plan must submit the modified plan no later than 60 days of the request from the Regional Administrator. If the amendment to the plan is considered a Class 2 or 3 modification according to the criteria in § 270.42, the modifications to the post-closure plan will be approved in accordance with the procedures in § 265.118(f). If the Regional Administrator determines that an owner or operator of a surface impoundment or waste pile who intended to remove all hazardous wastes at closure must close the facility as a landfill, the owner or operator must submit a post-closure plan for approval to the Regional Administrator within 90 days of the determination.

(e) The owner or operator of a facility with hazardous waste management units subject to these requirements must submit his post-closure plan to the Regional Administrator at least 180 days before the date he expects to begin partial or final closure of the first hazardous waste disposal unit. The date he "expects to begin closure" of the first hazardous waste disposal unit must be either within 30 days after the date on which the hazardous waste management unit receives the known final volume of hazardous waste or, if there is a reasonable possibility that the hazardous waste management unit will receive additional hazardous wastes, no later than one year after the date on which the unit received the most recent volume of hazardous wastes. The owner or operator must submit the post-closure plan to the Regional Administrator no later than 15 days after:

(1) Termination of interim status (except when a permit is issued to the facility simultaneously with termination of interim status); or

(2) Issuance of a judicial decree or final orders under section 3008 of RCRA to cease receiving wastes or close.

(f) The Regional Administrator will provide the owner or operator and the public, through a newspaper notice, the opportunity to submit written comments on the post-closure plan and request modifications to the plan no later than 30 days from the date of the notice. He will also, in response to a request or at his own discretion, hold a public hearing whenever such a hearing might clarify one or more issues concerning a post-closure plan. The Regional Administrator will give public notice of the hearing at least 30 days before it occurs. (Public notice of the hearing may be given at the same time as notice of the opportunity for the public to submit written comments, and the two notices may be combined.) The Regional Administrator will approve, modify, or disapprove the plan within 90 days of its receipt. If the Regional Administrator does not approve the plan he shall provide the owner or operator with a detailed written statement of reasons for the refusal and the owner or operator must modify the plan or submit a new plan for approval within 30 days after receiving such written statement. The Regional Administrator will approve or modify this plan in writing within 60 days. If the Regional Administrator modifies the plan, this modified plan becomes the approved post-closure plan. The Regional Administrator must ensure that the approved post-closure plan is consistent with §§265.117 through 265.120. A copy of the modified plan with a detailed statement of reasons for the modifications must be mailed to the owner or operator.

(g) The post-closure plan and length of the post-closure care period may be modified any time prior to the end of the post-closure care period in either of the following two ways:

(1) The owner or operator or any member of the public may petition the Regional Administrator to extend or reduce the post-closure care period applicable to a hazardous waste management unit or facility based on cause, or alter the requirements of the post-closure care period based on cause.

(i) The petition must include evidence demonstrating that:

(A) The secure nature of the hazardous waste management unit or facility makes the post-closure care requirement(s) unnecessary or supports reduction of the post-closure care period specified in the current post-closure plan (e.g., leachate or ground-water monitoring results, characteristics of the wastes, application of advanced technology, or alternative disposal, treatment, or re-use techniques indicate that the facility is secure), or

(B) The requested extension in the post-closure care period or alteration of post-closure care requirements is necessary to prevent threats to human health and the environment (e.g., leachate or ground-water monitoring results indicate a potential for migration of hazardous wastes at levels which may be harmful to human health and the environment).

(ii) These petitions will be considered by the Regional Administrator only when they present new and relevant information not previously considered by the Regional Administrator. Whenever the Regional Administrator is considering a petition, he will provide the owner or operator and the public, through a newspaper notice, the opportunity to submit written comments within 30 days of the date of the notice. He will also, in response to a request or at his own discretion, hold a public hearing whenever a hearing might clarify one or more issues concerning the post-closure plan. The Regional Administrator will give the public notice of the hearing at least 30 days before it occurs. (Public notice of the hearing may be given at the same time as notice of the opportunity for written public comments, and the two notices may be combined.) After considering the comments, he will issue a final determination, based upon the criteria set forth in paragraph (g)(1) of this section.

(iii) If the Regional Administrator denies the petition, he will send the petitioner a brief written response giving a reason for the denial.

(2) The Regional Administrator may tentatively decide to modify the post-closure plan if he deems it necessary to prevent threats to human health and the environment. He may propose to extend or reduce the post-closure care period applicable to a hazardous waste management unit or facility based on cause or alter the requirements of the post-closure care period based on cause.

(i) The Regional Administrator will provide the owner or operator and the affected public, through a newspaper notice, the opportunity to submit written comments within 30 days of the date of the notice and the opportunity for a public hearing as in paragraph (g)(1)(ii) of this section. After considering the comments, he will issue a final determination.

(ii) The Regional Administrator will base his final determination upon the same criteria as required for petitions under paragraph (g)(1)(i) of this section. A modification of the post-closure plan may include, where appropriate, the temporary suspension rather than permanent deletion of one or more post-closure care requirements. At the end of the specified period of suspension, the Regional Administrator would then determine whether the requirement(s) should be permanently discontinued or reinstated to prevent threats to human health and the environment.

[51 FR 16451, May 2, 1986, as amended at 53 FR 37935, Sept. 28, 1988; 63 FR 56734, Oct. 22, 1998]

### § 265.119  Post-closure notices.

(a) No later than 60 days after certification of closure of each hazardous waste disposal unit, the owner or operator must submit to the local zoning authority, or the authority with jurisdiction over local land use, and to the Regional Administrator, a record of the type, location, and quantity of hazardous wastes disposed of within each cell or other disposal unit of the facility. For hazardous wastes disposed of before January 12, 1981, the owner or operator must identify the type, location and quantity of the hazardous wastes to the best of his knowledge and in accordance with any records he has kept.

(b) Within 60 days of certification of closure of the first hazardous waste disposal unit and within 60 days of certification of closure of the last hazardous waste disposal unit, the owner or operator must:

(1) Record, in accordance with State law, a notation on the deed to the facility property—or on some other instrument which is normally examined during title search—that will in perpetuity notify any potential purchaser of the property that:

(i) The land has been used to manage hazardous wastes; and

(ii) Its use is restricted under 40 CFR part 265, subpart G regulations; and

(iii) The survey plat and record of the type, location, and quantity of hazardous wastes disposed of within each cell or other hazardous waste disposal unit of the facility required by §§ 265.116 and 265.119(a) have been filed with the local zoning authority or the authority with jurisdiction over local land use and with the Regional Administrator; and

(2) Submit a certification signed by the owner or operator that he has recorded the notation specified in paragraph (b)(1) of this section and a copy of the document in which the notation has been placed, to the Regional Administrator.

(c) If the owner or operator or any subsequent owner of the land upon which a hazardous waste disposal unit was located wishes to remove hazardous wastes and hazardous waste residues, the liner, if any, and all contaminated structures, equipment, and soils, he must request a modification to the approved post-closure plan in accordance with the requirements of § 265.118(g). The owner or operator must demonstrate that the removal of hazardous wastes will satisfy the criteria of § 265.117(c). By removing hazardous waste, the owner or operator may become a generator of hazardous waste and must manage it in accordance with all applicable requirements of this chapter. If the owner or operator is granted approval to conduct the removal activities, the owner or operator may request that the Regional Administrator approve either:

(1) The removal of the notation on the deed to the facility property or other instrument normally examined during title search, or

(2) The addition of a notation to the deed or instrument indicating the removal of the hazardous waste.

[51 FR 16451, May 2, 1986, as amended at 71 FR 40275, July 14, 2006]

**§ 265.120 Certification of completion of post-closure care.**

No later than 60 days after the completion of the established post-closure care period for each hazardous waste disposal unit, the owner or operator must submit to the Regional Administrator, by registered mail, a certification that the post-closure care period for the hazardous waste disposal unit was performed in accordance with the specifications in the approved post-closure plan. The certification must be signed by the owner or operator and a qualified Professional Engineer. Documentation supporting the Professional Engineer's certification must be furnished to the Regional Administrator upon request until he releases the owner or operator from the financial assurance requirements for post-closure care under § 265.145(h).

[71 FR 16909, Apr. 4, 2006]

**§ 265.121 Post-closure requirements for facilities that obtain enforceable documents in lieu of post-closure permits.**

(a) Owners and operators who are subject to the requirement to obtain a post-closure permit under 40 CFR 270.1(c), but who obtain enforceable documents in lieu of post-closure permits, as provided under 40 CFR 270.1(c)(7), must comply with the following requirements:

(1) The requirements to submit information about the facility in 40 CFR 270.28;

(2) The requirements for facility-wide corrective action in § 264.101 of this chapter;

(3) The requirements of 40 CFR 264.91 through 264.100.

(b)(1) The Regional Administrator, in issuing enforceable documents under § 265.121 in lieu of permits, will assure a meaningful opportunity for public involvement which, at a minimum, includes public notice and opportunity for public comment:

(i) When the Agency becomes involved in a remediation at the facility as a regulatory or enforcement matter;

(ii) On the proposed preferred remedy and the assumptions upon which the remedy is based, in particular those related to land use and site characterization; and

(iii) At the time of a proposed decision that remedial action is complete at the facility. These requirements must be met before the Regional Administrator may consider that the facility has met the requirements of 40 CFR 270.1(c)(7), unless the facility qualifies for a modification to these public involvement procedures under paragraph (b)(2) or (3) of this section.

(2) If the Regional Administrator determines that even a short delay in the implementation of a remedy would adversely affect human health or the environment, the Regional Administrator may delay compliance with the requirements of paragraph (b)(1) of this section and implement the remedy immediately. However, the Regional Administrator must assure involvement of the public at the earliest opportunity, and, in all cases, upon making the decision that additional remedial action is not needed at the facility.

(3) The Regional Administrator may allow a remediation initiated prior to October 22, 1998 to substitute for corrective action required under a post-closure permit even if the public involvement requirements of paragraph (b)(1) of this section have not been met so long as the Regional Administrator assures that notice and comment on the decision that no further remediation is necessary to protect human health and the environment takes place at the earliest reasonable opportunity after October 22, 1998.

[63 FR 56734, Oct. 22, 1998]

## Subpart H—Financial Requirements

SOURCE: 47 FR 15064, Apr. 7, 1982, unless otherwise noted.

**§ 265.140 Applicability.**

(a) The requirements of §§ 265.142, 265.143 and 265.147 through 265.150 apply to owners or operators of all hazardous waste facilities, except as provided otherwise in this section or in § 265.1.

(b) The requirements of §§ 265.144 and 265.145 apply only to owners and operators of:

(1) Disposal facilities;

(2) Tank systems that are required under § 265.197 of this chapter to meet the requirements for landfills; and

(3) Containment buildings that are required under § 265.1102 to meet the requirements for landfills.

(c) States and the Federal government are exempt from the requirements of this subpart.

(d) The Regional Administrator may replace all or part of the requirements of this subpart applying to a regulated unit with alternative requirements for financial assurance set out in the permit or in an enforceable document (as defined in 40 CFR 270.1(c)(7)), where the Regional Administrator:

(1) Prescribes alternative requirements for the regulated unit under § 265.90(f) and/or 265.110(d), and

(2) Determines that it is not necessary to apply the requirements of this subpart because the alternative financial assurance requirements will protect human health and the environment.

[47 FR 15064, Apr. 7, 1982, as amended at 51 FR 16455, May 2, 1986; 51 FR 25479, July 14, 1986; 57 FR 37267, Aug. 18, 1992; 63 FR 56734, Oct. 22, 1998; 71 FR 40275, July 14, 2006]

### § 265.141   Definitions of terms as used in this subpart.

(a) *Closure plan* means the plan for closure prepared in accordance with the requirements of § 265.112.

(b) *Current closure cost estimate* means the most recent of the estimates prepared in accordance with § 265.142 (a), (b), and (c).

(c) *Current post-closure cost estimate* means the most recent of the estimates prepared in accordance with § 265.144 (a), (b), and (c).

(d) *Parent corporation* means a corporation which directly owns at least 50 percent of the voting stock of the corporation which is the facility owner or operator; the latter corporation is deemed a "subsidiary" of the parent corporation.

(e) *Post-closure plan* means the plan for post-closure care prepared in accordance with the requirements of §§ 265.117 through 265.120.

(f) The following terms are used in the specifications for the financial tests for closure, post-closure care, and liability coverage. The definitions are intended to assist in the understanding of these regulations and are not intended to limit the meanings of terms in a way that conflicts with generally accepted accounting practices.

*Assets* means all existing and all probable future economic benefits obtained or controlled by a particular entity.

*Current assets* means cash or other assets or resources commonly identified as those which are reasonably expected to be realized in cash or sold or consumed during the normal operating cycle of the business.

*Current liabilities* means obligations whose liquidation is reasonably expected to require the use of existing resources properly classifiable as current assets or the creation of other current liabilities.

*Current plugging and abandonment cost estimate* means the most recent of the estimates prepared in accordance with § 144.62(a), (b), and (c) of this title.

*Independently audited* refers to an audit performed by an independent certified public accountant in accordance with generally accepted auditing standards.

*Liabilities* means probable future sacrifices of economic benefits arising from present obligations to transfer assets or provide services to other entities in the future as a result of past transactions or events.

*Net working capital* means current assets minus current liabilities.

*Net worth* means total assets minus total liabilities and is equivalent to owner's equity.

*Tangible net worth* means the tangible assets that remain after deducting liabilities; such assets would not include intangibles such as goodwill and rights to patents or royalties.

(g) In the liability insurance requirements the terms *bodily injury* and *property damage* shall have the meanings given these terms by applicable State law. However, these terms do not include those liabilities which, consistent with standard industry practice, are excluded from coverage in liability policies for bodily injury and property damage. The Agency intends the meanings of other terms used in the liability insurance requirements to be consistent with their common meanings within the insurance industry. The definitions given below of several of the terms are intended to assist in the understanding of these regulations and are not intended to limit their meanings in a way that conflicts with general insurance industry usage.

*Accidental occurrence* means an accident, including continuous or repeated exposure to conditions, which results in bodily injury or property damage neither expected nor intended from the standpoint of the insured.

*Legal defense costs* means any expenses that an insurer incurs in defending against claims of third parties brought under the terms and conditions of an insurance policy.

*Nonsudden accidental occurrence* means an occurrence which takes place over time and involves continuous or repeated exposure.

*Sudden accidental occurrence* means an occurrence which is not continuous or repeated in nature.

(h) *Substantial business relationship* means the extent of a business relationship necessary under applicable State law to make a guarantee contract issued incident to that relationship valid and enforceable. A "substantial business relationship" must arise from a pattern of recent or ongoing business transactions, in addition to the guarantee itself, such that a currently existing business relationship between the guarantor and the owner or operator is demonstrated to the satisfaction of the applicable EPA Regional Administrator.

[47 FR 16558, Apr. 16, 1982, as amended at 51 FR 16456, May 2, 1986; 53 FR 33959, Sept. 1, 1988]

### §265.142  Cost estimate for closure.

(a) The owner or operator must have a detailed written estimate, in current dollars, of the cost of closing the facility in accordance with the requirements in §§265.111 through 265.115 and applicable closure requirements in §§265.197, 265.228, 265.258, 265.280, 265.310, 265.351, 265.381, 265.404, and 265.1102.

(1) The estimate must equal the cost of final closure at the point in the facility's active life when the extent and manner of its operation would make closure the most expensive, as indicated by its closure plan (see §265.112(b)); and

(2) The closure cost estimate must be based on the costs to the owner or operator of hiring a third party to close the facility. A third party is a party who is neither a parent nor a subsidiary of the owner or operator. (See definition of parent corporation in §265.141(d).) The owner or operator may use costs for on-site disposal if he can demonstrate that on-site disposal capacity will exist at all times over the life of the facility.

(3) The closure cost estimate may not incorporate any salvage value that may be realized with the sale of hazardous wastes, or non-hazardous wastes if applicable under §265.113(d), facility structures or equipment, land, or other assets associated with the facility at the time of partial or final closure.

(4) The owner or operator may not incorporate a zero cost for hazardous wastes, or non-hazardous wastes if applicable under §265.113(d), that might have economic value.

(b) During the active life of the facility, the owner or operator must adjust the closure cost estimate for inflation within 60 days prior to the anniversary date of the establishment of the financial instrument(s) used to comply with §265.143. For owners and operators using the financial test or corporate guarantee, the closure cost estimate must be updated for inflation within 30 days after the close of the firm's fiscal year and before submission of updated information to the Regional Administrator as specified in §265.143(e)(3). The adjustment may be made by recalculating the closure cost estimate in current dollars, or by using an inflation factor derived from the most recent Implicit Price Deflator for Gross National Product published by the U.S. Department of Commerce in its *Survey of Current Business*, as specified in paragraphs (b)(1) and (2) of this section. The inflation factor is the result of dividing the latest published annual Deflator by the Deflator for the previous year.

(1) The first adjustment is made by multiplying the closure cost estimate by the inflation factor. The result is the adjusted closure cost estimate.

623

(2) Subsequent adjustments are made by multiplying the latest adjusted closure cost estimate by the latest inflation factor.

(c) During the active life of the facility, the owner or operator must revise the closure cost estimate no later than 30 days after a revision has been made to the closure plan which increases the cost of closure. If the owner or operator has an approved closure plan, the closure cost estimate must be revised no later than 30 days after the Regional Administrator has approved the request to modify the closure plan, if the change in the closure plan increases the cost of closure. The revised closure cost estimate must be adjusted for inflation as specified in § 265.142(b).

(d) The owner or operator must keep the following at the facility during the operating life of the facility: The latest closure cost estimate prepared in accordance with §§ 265.142 (a) and (c) and, when this estimate has been adjusted in accordance with § 265.142(b), the latest adjusted closure cost estimate.

[47 FR 15064, Apr. 7, 1982, as amended at 50 FR 4514, Jan. 31, 1985; 51 FR 16456, May 2, 1986; 54 FR 33397, Aug. 14, 1989; 57 FR 37267, Aug. 18, 1992; 71 FR 40275, July 14, 2006]

### § 265.143 Financial assurance for closure.

By the effective date of these regulations, an owner or operator of each facility must establish financial assurance for closure of the facility. He must choose from the options as specified in paragraphs (a) through (e) of this section.

(a) *Closure trust fund.* (1) An owner or operator may satisfy the requirements of this section by establishing a closure trust fund which conforms to the requirements of this paragraph and submitting an originally signed duplicate of the trust agreement to the Regional Administrator. The trustee must be an entity which has the authority to act as a trustee and whose trust operations are regulated and examined by a Federal or State agency.

(2) The wording of the trust agreement must be identical to the wording specified in § 264.151(a)(1), and the trust agreement must be accompanied by a formal certification of acknowledgment (for example, see § 264.151(a)(2)).

Schedule A of the trust agreement must be updated within 60 days after a change in the amount of the current closure cost estimate covered by the agreement.

(3) Payments into the trust fund must be made annually by the owner or operator over the 20 years beginning with the effective date of these regulations or over the remaining operating life of the facility as estimated in the closure plan, whichever period is shorter; this period is hereafter referred to as the "pay-in period." The payments into the closure trust fund must be made as follows:

(i) The first payment must be made by the effective date of these regulations, except as provided in paragraph (a)(5) of this section. The first payment must be at least equal to the current closure cost estimate, except as provided in § 265.143(f), divided by the number of years in the pay-in period.

(ii) Subsequent payments must be made no later than 30 days after each anniversary date of the first payment. The amount of each subsequent payment must be determined by this formula:

$$\text{Next payment} = \frac{CE - CV}{Y}$$

where CE is the current closure cost estimate, CV is the current value of the trust fund, and Y is the number of years remaining in the pay-in period.

(4) The owner or operator may accelerate payments into the trust fund or he may deposit the full amount of the current closure cost estimate at the time the fund is established. However, he must maintain the value of the fund at no less than the value that the fund would have if annual payments were made as specified in paragraph (a)(3) of this section.

(5) If the owner or operator establishes a closure trust fund after having used one or more alternate mechanisms specified in this section, his first payment must be in at least the amount that the fund would contain if the trust fund were established initially and annual payments made as specified in paragraph (a)(3) of this section.

(6) After the pay-in period is completed, whenever the current closure

cost estimate changes, the owner or operator must compare the new estimate with the trustee's most recent annual valuation of the trust fund. If the value of the fund is less than the amount of the new estimate, the owner or operator, within 60 days after the change in the cost estimate, must either deposit an amount into the fund so that its value after this deposit at least equals the amount of the current closure cost estimate, or obtain other financial assurance as specified in this section to cover the difference.

(7) If the value of the trust fund is greater than the total amount of the current closure cost estimate, the owner or operator may submit a written request to the Regional Administrator for release of the amount in excess of the current closure cost estimate.

(8) If an owner or operator substitutes other financial assurance as specified in this section for all or part of the trust fund, he may submit a written request to the Regional Administrator for release of the amount in excess of the current closure cost estimate covered by the trust fund.

(9) Within 60 days after receiving a request from the owner or operator for release of funds as specified in paragraph (a) (7) or (8) of this section, the Regional Administrator will instruct the trustee to release to the owner or operator such funds as the Regional Administrator specifies in writing.

(10) After beginning partial or final closure, an owner or operator or another person authorized to conduct partial or final closure may request reimbursements for partial or final closure expenditures by submitting itemized bills to the Regional Administrator. The owner or operator may request reimbursements for partial closure only if sufficient funds are remaining in the trust fund to cover the maximum costs of closing the facility over its remaining operating life. No later than 60 days after receiving bills for partial or final closure activities, the Regional Administrator will instruct the trustee to make reimbursements in those amounts as the Regional Administrator specifies in writing, if the Regional Administrator determines that the partial or final clo-

sure expenditures are in accordance with the approved closure plan, or otherwise justified. If the Regional Administrator has reason to believe that the maximum cost of closure over the remaining life of the facility will be significantly greater than the value of the trust fund, he may withhold reimbursements of such amounts as he deems prudent until he determines, in accordance with §265.143(h) that the owner or operator is no longer required to maintain financial assurance for final closure of the facility. If the Regional Administrator does not instruct the trustee to make such reimbursements, he will provide to the owner or operator a detailed written statement of reasons.

(11) The Regional Administrator will agree to termination of the trust when:

(i) An owner or operator substitutes alternate financial assurance as specified in this section; or

(ii) The Regional Administrator releases the owner or operator from the requirements of this section in accordance with §265.143(h).

(b) *Surety bond guaranteeing payment into a closure trust fund.* (1) An owner or operator may satisfy the requirements of this section by obtaining a surety bond which conforms to the requirements of this paragraph and submitting the bond to the Regional Administrator. The surety company issuing the bond must, at a minimum, be among those listed as acceptable sureties on Federal bonds in Circular 570 of the U.S. Department of the Treasury.

(2) The wording of the surety bond must be identical to the wording specified in §264.151(b).

(3) The owner or operator who uses a surety bond to satisfy the requirements of this section must also establish a standby trust fund. Under the terms of the bond, all payments made thereunder will be deposited by the surety directly into the standby trust fund in accordance with instructions from the Regional Administrator. This standby trust fund must meet the requirements specified in §265.143(a), except that:

(i) An originally signed duplicate of the trust agreement must be submitted to the Regional Administrator with the surety bond; and

(ii) Until the standby trust fund is funded pursuant to the requirements of this section, the following are not required by these regulations:

(A) Payments into the trust fund as specified in § 265.143(a);

(B) Updating of Schedule A of the trust agreement (see § 264.151(a)) to show current closure cost estimates;

(C) Annual valuations as required by the trust agreement; and

(D) Notices of nonpayment as required by the trust agreement.

(4) The bond must guarantee that the owner or operator will:

(i) Fund the standby trust fund in an amount equal to the penal sum of the bond before the beginning of final closure of the facility; or

(ii) Fund the standby trust fund in an amount equal to the penal sum within 15 days after an administrative order to begin final closure issued by the Regional Administrator becomes final, or within 15 days after an order to begin final closure is issued by a U.S. district court or other court of competent jurisdiction; or

(iii) Provide alternate financial assurance as specified in this section, and obtain the Regional Administrator's written approval of the assurance provided, within 90 days after receipt by both the owner or operator and the Regional Administrator of a notice of cancellation of the bond from the surety.

(5) Under the terms of the bond, the surety will become liable on the bond obligation when the owner or operator fails to perform as guaranteed by the bond.

(6) The penal sum of the bond must be in an amount at least equal to the current closure cost estimate, except as provided in § 265.143(f).

(7) Whenever the current closure cost estimate increases to an amount greater than the penal sum, the owner or operator, within 60 days after the increase, must either cause the penal sum to be increased to an amount at least equal to the current closure cost estimate and submit evidence of such increase to the Regional Administrator, or obtain other financial assurance as specified in this section to cover the increase. Whenever the current closure cost estimate decreases,

the penal sum may be reduced to the amount of the current closure cost estimate following written approval by the Regional Administrator.

(8) Under the terms of the bond, the surety may cancel the bond by sending notice of cancellation by certified mail to the owner or operator and to the Regional Administrator. Cancellation may not occur, however, during the 120 days beginning on the date of receipt of the notice of cancellation by both the owner or operator and the Regional Administrator, as evidenced by the return receipts.

(9) The owner or operator may cancel the bond if the Regional Administrator has given prior written consent based on his receipt of evidence of alternate financial assurance as specified in this section.

(c) *Closure letter of credit.* (1) An owner or operator may satisfy the requirements of this section by obtaining an irrevocable standby letter of credit which conforms to the requirements of this paragraph and submitting the letter to the Regional Administrator. The issuing institution must be an entity which has the authority to issue letters of credit and whose letter-of-credit operations are regulated and examined by a Federal or State agency.

(2) The wording of the letter of credit must be identical to the wording specified in § 264.151(d).

(3) An owner or operator who uses a letter of credit to satisfy the requirements of this section must also establish a standby trust fund. Under the terms of the letter of credit, all amounts paid pursuant to a draft by the Regional Administrator will be deposited by the issuing institution directly into the standby trust fund in accordance with instructions from the Regional Administrator. This standby trust fund must meet the requirements of the trust fund specified in § 265.143(a), except that:

(i) An originally signed duplicate of the trust agreement must be submitted to the Regional Administrator with the letter of credit; and

(ii) Unless the standby trust fund is funded pursuant to the requirements of this section, the following are not required by these regulations:

(A) Payments into the trust fund as specified in §265.143(a);

(B) Updating of Schedule A of the trust agreement (see §264.151(a)) to show current closure cost estimates;

(C) Annual valuations as required by the trust agreement; and

(D) Notices of nonpayment as required by the trust agreement.

(4) The letter of credit must be accompanied by a letter from the owner or operator referring to the letter of credit by number, issuing institution, and date, and providing the following information: The EPA Identification Number, name, and address of the facility, and the amount of funds assured for closure of the facility by the letter of credit.

(5) The letter of credit must be irrevocable and issued for a period of at least 1 year. The letter of credit must provide that the expiration date will be automatically extended for a period of at least 1 year unless, at least 120 days before the current expiration date, the issuing institution notifies both the owner or operator and the Regional Administrator by certified mail of a decision not to extend the expiration date. Under the terms of the letter of credit, the 120 days will begin on the date when both the owner or operator and the Regional Administrator have received the notice, as evidenced by the return receipts.

(6) The letter of credit must be issued in an amount at least equal to the current closure cost estimate, except as provided in §265.143(f).

(7) Whenever the current closure cost estimate increases to an amount greater than the amount of the credit, the owner or operator, within 60 days after the increase, must either cause the amount of the credit to be increased so that it at least equals the current closure cost estimate and submit evidence of such increase to the Regional Administrator, or obtain other financial assurance as specified in this section to cover the increase. Whenever the current closure cost estimate decreases, the amount of the credit may be reduced to the amount of the current closure cost estimate following written approval by the Regional Administrator.

(8) Following a final administrative determination pursuant to section 3008 of RCRA that the owner or operator has failed to perform final closure in accordance with the approved closure plan when required to do so, the Regional Administrator may draw on the letter of credit.

(9) If the owner or operator does not establish alternate financial assurance as specified in this section and obtain written approval of such alternate assurance from the Regional Administrator within 90 days after receipt by both the owner or operator and the Regional Administrator of a notice from the issuing institution that it has decided not to extend the letter of credit beyond the current expiration date, the Regional Administrator will draw on the letter of credit. The Regional Administrator may delay the drawing if the issuing institution grants an extension of the term of the credit. During the last 30 days of any such extension the Regional Administrator will draw on the letter of credit if the owner or operator has failed to provide alternate financial assurance as specified in this section and obtain written approval of such assurance from the Regional Administrator.

(10) The Regional Administrator will return the letter of credit to the issuing institution for termination when:

(i) An owner or operator substitutes alternate financial assurance as specified in this section; or

(ii) The Regional Administrator releases the owner or operator from the requirements of this section in accordance with §265.143(h).

(d) *Closure insurance.* (1) An owner or operator may satisfy the requirements of this section by obtaining closure insurance which conforms to the requirements of this paragraph and submitting a certificate of such insurance to the Regional Administrator. By the effective date of these regulations the owner or operator must submit to the Regional Administrator a letter from an insurer stating that the insurer is considering issuance of closure insurance conforming to the requirements of this paragraph to the owner or operator. Within 90 days after the effective date of these regulations, the owner or

627

operator must submit the certificate of insurance to the Regional Administrator or establish other financial assurance as specified in this section. At a minimum, the insurer must be licensed to transact the business of insurance, or eligible to provide insurance as an excess or surplus lines insurer, in one or more States.

(2) The wording of the certificate of insurance must be identical to the wording specified in § 264.151(e).

(3) The closure insurance policy must be issued for a face amount at least equal to the current closure cost estimate, except as provided in § 265.143(f). The term "face amount" means the total amount the insurer is obligated to pay under the policy. Actual payments by the insurer will not change the face amount, although the insurer's future liability will be lowered by the amount of the payments.

(4) The closure insurance policy must guarantee that funds will be available to close the facility whenever final closure occurs. The policy must also guarantee that once final closure begins, the insurer will be responsible for paying out funds, up to an amount equal to the face amount of the policy, upon the direction of the Regional Administrator, to such party or parties as the Regional Administrator specifies.

(5) After beginning partial or final closure, an owner or operator or any other person authorized to conduct closure may request reimbursements for closure expenditures by submitting itemized bills to the Regional Administrator. The owner or operator may request reimbursements for partial closure only if the remaining value of the policy is sufficient to cover the maximum costs of closing the facility over its remaining operating life. Within 60 days after receiving bills for closure activities, the Regional Administrator will instruct the insurer to make reimbursements in such amounts as the Regional Administrator specifies in writing if the Regional Administrator determines that the partial or final closure expenditures are in accordance with the approved closure plan or otherwise justified. If the Regional Administrator has reason to believe that the maximum cost of closure over the remaining life of the facility will be sig-

nificantly greater than the face amount of the policy, he may withhold reimbursement of such amounts as he deems prudent until he determines, in accordance with § 265.143(h), that the owner or operator is no longer required to maintain financial assurance for final closure of the particular facility. If the Regional Administrator does not instruct the insurer to make such reimbursements, he will provide to the owner or operator a detailed written statement of reasons.

(6) The owner or operator must maintain the policy in full force and effect until the Regional Administrator consents to termination of the policy by the owner or operator as specified in paragraph (d)(10) of this section. Failure to pay the premium, without substitution of alternate financial assurance as specified in this section, will constitute a significant violation of these regulations, warranting such remedy as the Regional Administrator deems necessary. Such violation will be deemed to begin upon receipt by the Regional Administrator of a notice of future cancellation, termination, or failure to renew due to nonpayment of the premium, rather than upon the date of expiration.

(7) Each policy must contain a provision allowing assignment of the policy to a successor owner or operator. Such assignment may be conditional upon consent of the insurer, provided such consent is not unreasonably refused.

(8) The policy must provide that the insurer may not cancel, terminate, or fail to renew the policy except for failure to pay the premium. The automatic renewal of the policy must, at a minimum, provide the insured with the option of renewal at the face amount of the expiring policy. If there is a failure to pay the premium, the insurer may elect to cancel, terminate, or fail to renew the policy by sending notice by certified mail to the owner or operator and the Regional Administrator. Cancellation, termination, or failure to renew may not occur, however, during the 120 days beginning with the date of receipt of the notice by both the Regional Administrator and the owner or operator, as evidenced by the return receipts. Cancellation, termination, or failure to renew may not occur and the

policy will remain in full force and effect in the event that on or before the date of expiration:

(i) The Regional Administrator deems the facility abandoned; or

(ii) Interim status is terminated or revoked; or

(iii) Closure is ordered by the Regional Administrator or a U.S. district court or other court of competent jurisdiction; or

(iv) The owner or operator is named as debtor in a voluntary or involuntary proceeding under Title 11 (Bankruptcy), U.S. Code; or

(v) The premium due is paid.

(9) Whenever the current closure cost estimate increases to an amount greater than the face amount of the policy, the owner or operator, within 60 days after the increase, must either cause the face amount to be increased to an amount at least equal to the current closure cost estimate and submit evidence of such increase to the Regional Administrator, or obtain other financial assurance as specified in this section to cover the increase. Whenever the current closure cost estimate decreases, the face amount may be reduced to the amount of the current closure cost estimate following written approval by the Regional Administrator.

(10) The Regional Administrator will give written consent to the owner or operator that he may terminate the insurance policy when:

(i) An owner or operator substitutes alternate financial assurance as specified in this section; or

(ii) The Regional Administrator releases the owner or operator from the requirements of this section in accordance with §265.143(h).

(e) *Financial test and corporate guarantee for closure.* (1) An owner or operator may satisfy the requirements of this section by demonstrating that he passes a financial test as specified in this paragraph. To pass this test the owner or operator must meet the criteria of either paragraph (e)(1)(i) or (ii) of this section:

(i) The owner or operator must have:

(A) Two of the following three ratios: A ratio of total liabilities to net worth less than 2.0; a ratio of the sum of net income plus depreciation, depletion,

and amortization to total liabilities greater than 0.1; and a ratio of current assets to current liabilities greater than 1.5; and

(B) Net working capital and tangible net worth each at least six times the sum of the current closure and post-closure cost estimates and the current plugging and abandonment cost estimates; and

(C) Tangible net worth of at least $10 million; and

(D) Assets located in the United States amounting to at least 90 percent of total assets or at least six times the sum of the current closure and post-closure cost estimates and the current plugging and abandonment cost estimates.

(ii) The owner or operator must have:

(A) A current rating for his most recent bond issuance of AAA, AA, A, or BBB as issued by Standard and Poor's or Aaa, Aa, A, or Baa as issued by Moody's; and

(B) Tangible net worth at least six times the sum of the current closure and post-closure cost estimates and the current plugging and abandonment cost estimates; and

(C) Tangible net worth of at least $10 million; and

(D) Assets located in the United States amounting to at least 90 percent of total assets or at least six times the sum of the current closure and post-closure cost estimates and the current plugging and abandonment cost estimates.

(2) The phrase "current closure and post-closure cost estimates" as used in paragraph (e)(1) of this section refers to the cost estimates required to be shown in paragraphs 1–4 of the letter from the owner's or operator's chief financial officer (§264.151(f)). The phrase "current plugging and abandonment cost estimates" as used in paragraph (e)(1) of this section refers to the cost estimates required to be shown in paragraphs 1–4 of the letter from the owner's or operator's chief financial officer (§144.70(f) of this title).

(3) To demonstrate that he meets this test, the owner or operator must submit the following items to the Regional Administrator:

(i) A letter signed by the owner's or operator's chief financial officer and worded as specified in § 264.151(f); and

(ii) A copy of the independent certified public accountant's report on examination of the owner's or operator's financial statements for the latest completed fiscal year; and

(iii) A special report from the owner's or operator's independent certified public accountant to the owner or operator stating that:

(A) He has compared the data which the letter from the chief financial officer specifies as having been derived from the independently audited, year-end financial statements for the latest fiscal year with the amounts in such financial statements; and

(B) In connection with that procedure, no matters came to his attention which caused him to believe that the specified data should be adjusted.

(4) The owner or operator may obtain an extension of the time allowed for submission of the documents specified in paragraph (e)(3) of this section if the fiscal year of the owner or operator ends during the 90 days prior to the effective date of these regulations and if the year-end financial statements for that fiscal year will be audited by an independent certified public accountant. The extension will end no later than 90 days after the end of the owner's or operator's fiscal year. To obtain the extension, the owner's or operator's chief financial officer must send, by the effective date of these regulations, a letter to the Regional Administrator of each Region in which the owner's or operator's facilities to be covered by the financial test are located. This letter from the chief financial officer must:

(i) Request the extension;

(ii) Certify that he has grounds to believe that the owner or operator meets the criteria of the financial test;

(iii) Specify for each facility to be covered by the test the EPA Identification Number, name, address, and current closure and post-closure cost estimates to be covered by the test;

(iv) Specify the date ending the owner's or operator's last complete fiscal year before the effective date of these regulations;

(v) Specify the date, no later than 90 days after the end of such fiscal year, when he will submit the documents specified in paragraph (e)(3) of this section; and

(vi) Certify that the year-end financial statements of the owner or operator for such fiscal year will be audited by an independent certified public accountant.

(5) After the initial submission of items specified in paragraph (e)(3) of this section, the owner or operator must send updated information to the Regional Administrator within 90 days after the close of each succeeding fiscal year. This information must consist of all three items specified in paragraph (e)(3) of this section.

(6) If the owner or operator no longer meets the requirements of paragraph (e)(1) of this section, he must send notice to the Regional Administrator of intent to establish alternate financial assurance as specified in this section. The notice must be sent by certified mail within 90 days after the end of the fiscal year for which the year-end financial data show that the owner or operator no longer meets the requirements. The owner or operator must provide the alternate financial assurance within 120 days after the end of such fiscal year.

(7) The Regional Administrator may, based on a reasonable belief that the owner or operator may no longer meet the requirements of paragraph (e)(1) of this section, require reports of financial condition at any time from the owner or operator in addition to those specified in paragraph (e)(3) of this section. If the Regional Administrator finds, on the basis of such reports or other information, that the owner or operator no longer meets the requirements of paragraph (e)(1) of this section, the owner or operator must provide alternate financial assurance as specified in this section within 30 days after notification of such a finding.

(8) The Regional Administrator may disallow use of this test on the basis of qualifications in the opinion expressed by the independent certified public accountant in his report on examination of the owner's or operator's financial statements (see paragraph (e)(3)(ii) of this section). An adverse opinion or a

disclaimer of opinion will be cause for disallowance. The Regional Administrator will evaluate other qualifications on an individual basis. The owner or operator must provide alternate financial assurance as specified in this section within 30 days after notification of the disallowance.

(9) The owner or operator is no longer required to submit the items specified in paragraph (e)(3) of this section when:

(i) An owner or operator substitutes alternate financial assurance as specified in this section; or

(ii) The Regional Administrator releases the owner or operator from the requirements of this section in accordance with §265.143(h).

(10) An owner or operator may meet the requirements of this section by obtaining a written guarantee. The guarantor must be the direct or higher-tier parent corporation of the owner or operator, a firm whose parent corporation is also the parent corporation of the owner or operator, or a firm with a "substantial business relationship" with the owner or operator. The guarantor must meet the requirements for owners or operators in paragraphs (e)(1) through (8) of this section and must comply with the terms of the guarantee. The wording of the guarantee must be identical to the wording specified in §264.151(h). A certified copy of the guarantee must accompany the items sent to the Regional Administrator as specified in paragraph (e)(3) of this section. One of these items must be the letter from the guarantor's chief financial officer. If the guarantor's parent corporation is also the parent corporation of the owner or operator, the letter must describe the value received in consideration of the guarantee. If the guarantor is a firm with a "substantial business relationship" with the owner or operator, this letter must describe this "substantial business relationship" and the value received in consideration of the guarantee. The terms of the guarantee must provide that:

(i) If the owner or operator fails to perform final closure of a facility covered by the corporate guarantee in accordance with the closure plan and other interim status requirements whenever required to do so, the guar-

antor will do so or establish a trust fund as specified in §265.143(a) in the name of the owner or operator.

(ii) The corporate guarantee will remain in force unless the guarantor sends notice of cancellation by certified mail to the owner or operator and to the Regional Administrator. Cancellation may not occur, however, during the 120 days beginning on the date of receipt of the notice of cancellation by both the owner or operator and the Regional Administrator, as evidenced by the return receipts.

(iii) If the owner or operator fails to provide alternate financial assurance as specified in this section and obtain the written approval of such alternate assurance from the Regional Administrator within 90 days after receipt by both the owner or operator and the Regional Administrator of a notice of cancellation of the corporate guarantee from the guarantor, the guarantor will provide such alternate financial assurance in the name of the owner or operator.

(f) *Use of multiple financial mechanisms.* An owner or operator may satisfy the requirements of this section by establishing more than one financial mechanism per facility. These mechanisms are limited to trust funds, surety bonds, letters of credit, and insurance. The mechanisms must be as specified in paragraphs (a) through (d), respectively, of this section, except that it is the combination of mechanisms, rather than the single mechanism, which must provide financial assurance for an amount at least equal to the current closure cost estimate. If an owner or operator uses a trust fund in combination with a surety bond or a letter of credit, he may use the trust fund as the standby trust fund for the other mechanisms. A single standby trust fund may be established for two or more mechanisms. The Regional Administrator may use any or all of the mechanisms to provide for closure of the facility.

(g) *Use of a financial mechanism for multiple facilities.* An owner or operator may use a financial assurance mechanism specified in this section to meet the requirements of this section for more than one facility. Evidence of financial assurance submitted to the Regional Administrator must include a list showing, for each facility, the EPA Identification Number, name, address, and the amount of funds for closure assured by the mechanism. If the facilities covered by the mechanism are in more than one Region, identical evidence of financial assurance must be submitted to and maintained with the Regional Administrators of all such Regions. The amount of funds available through the mechanism must be no less than the sum of funds that would be available if a separate mechanism had been established and maintained for each facility. In directing funds available through the mechanism for closure of any of the facilities covered by the mechanism, the Regional Administrator may direct only the amount of funds designated for that facility, unless the owner or operator agrees to the use of additional funds available under the mechanism.

(h) *Release of the owner or operator from the requirements of this section.* Within 60 days after receiving certifications from the owner or operator and a qualified Professional Engineer that final closure has been completed in accordance with the approved closure plan, the Regional Administrator will notify the owner or operator in writing that he is no longer required by this section to maintain financial assurance for final closure of the facility, unless the Regional Administrator has reason to believe that final closure has not been in accordance with the approved closure plan. The Regional Administrator shall provide the owner or operator a detailed written statement of any such reason to believe that closure has not been in accordance with the approved closure plan.

[47 FR 15064, Apr. 7, 1982, as amended at 51 FR 16456, May 2, 1986; 57 FR 42843, Sept. 16, 1992; 71 FR 16909, Apr. 4, 2006]

## § 265.144  Cost estimate for post-closure care.

(a) The owner or operator of a hazardous waste disposal unit must have a detailed written estimate, in current dollars, of the annual cost of post-closure monitoring and maintenance of the facility in accordance with the applicable post-closure regulations in §§ 265.117 through 265.120, 265.228, 265.258, 265.280, and 265.310.

(1) The post-closure cost estimate must be based on the costs to the owner or operator of hiring a third party to conduct post-closure care activities. A third party is a party who is neither a parent nor subsidiary of the owner or operator. (See definition of parent corporation in § 265.141(d).)

(2) The post-closure cost estimate is calculated by multiplying the annual post-closure cost estimate by the number of years of post-closure care required under § 265.117.

(b) During the active life of the facility, the owner or operator must adjust the post-closure cost estimate for inflation within 60 days prior to the anniversary date of the establishment of the financial instrument(s) used to comply with § 265.145. For owners or operators using the financial test or corporate guarantee, the post-closure care cost estimate must be updated for inflation no later than 30 days after the close of the firm's fiscal year and before submission of updated information to the Regional Administrator as specified in § 265.145(d)(5). The adjustment may be made by recalculating the post-closure cost estimate in current dollars or by using an inflation factor derived from the most recent Implicit Price Deflator for Gross National Product published by the U.S. Department of Commerce in its *Survey of Current Business* as specified in § 265.145 (b)(1) and (2). The inflation factor is the result of dividing the latest published annual Deflator by the Deflator for the previous year.

(1) The first adjustment is made by multiplying the post-closure cost estimate by the inflation factor. The result is the adjusted post-closure cost estimate.

(2) Subsequent adjustments are made by multiplying the latest adjusted post-closure cost estimate by the latest inflation factor.

(c) During the active life of the facility, the owner or operator must revise the post-closure cost estimate no later than 30 days after a revision to the post-closure plan which increases the cost of post-closure care. If the owner or operator has an approved post-closure plan, the post-closure cost estimate must be revised no later than 30 days after the Regional Administrator has approved the request to modify the plan, if the change in the post-closure plan increases the cost of post-closure care. The revised post-closure cost estimate must be adjusted for inflation as specified in § 265.144(b).

(d) The owner or operator must keep the following at the facility during the operating life of the facility: the latest post-closure cost estimate prepared in accordance with § 265.144 (a) and (c) and, when this estimate has been adjusted in accordance with § 265.144(b), the latest adjusted post-closure cost estimate.

[47 FR 15064, Apr. 7, 1982, as amended at 50 FR 4514, Jan. 31, 1985; 51 FR 16457, May 2, 1986]

**§ 265.145   Financial assurance for post-closure care.**

By the effective date of these regulations, an owner or operator of a facility with a hazardous waste disposal unit must establish financial assurance for post-closure care of the disposal unit(s).

(a) *Post-closure trust fund.* (1) An owner or operator may satisfy the requirements of this section by establishing a post-closure trust fund which conforms to the requirements of this paragraph and submitting an originally signed duplicate of the trust agreement to the Regional Administrator. The trustee must be an entity which has the authority to act as a trustee and whose trust operations are regulated and examined by a Federal or State agency.

(2) The wording of the trust agreement must be identical to the wording specified in § 264.151(a)(1), and the trust agreement must be accompanied by a formal certification of acknowledgment (for example, see § 264.151(a)(2)). Schedule A of the trust agreement must be updated within 60 days after a change in the amount of the current post-closure cost estimate covered by the agreement.

(3) Payments into the trust fund must be made annually by the owner or operator over the 20 years beginning with the effective date of these regulations or over the remaining operating life of the facility as estimated in the closure plan, whichever period is shorter; this period is hereafter referred to as the "pay-in period." The payments into the post-closure trust fund must be made as follows:

(i) The first payment must be made by the effective date of these regulations, except as provided in paragraph (a)(5) of this section. The first payment must be at least equal to the current post-closure cost estimate, except as provided in § 265.145(f), divided by the number of years in the pay-in period.

(ii) Subsequent payments must be made no later than 30 days after each anniversary date of the first payment. The amount of each subsequent payment must be determined by this formula:

$$\text{Next payment} = \frac{CE - CV}{Y}$$

where CE is the current post-closure cost estimate, CV is the current value of the trust fund, and Y is the number of years remaining in the pay-in period.

(4) The owner or operator may accelerate payments into the trust fund or he may deposit the full amount of the current post-closure cost estimate at the time the fund is established. However, he must maintain the value of the fund at no less than the value that the fund would have if annual payments were made as specified in paragraph (a)(3) of this section.

(5) If the owner or operator establishes a post-closure trust fund after having used one or more alternate mechanisms specified in this section, his first payment must be in at least the amount that the fund would contain if the trust fund were established initially and annual payments made as specified in paragraph (a)(3) of this section.

(6) After the pay-in period is completed, whenever the current post-closure cost estimate changes during the operating life of the facility, the owner or operator must compare the new estimate with the trustee's most recent annual valuation of the trust fund. If the value of the fund is less than the amount of the new estimate, the owner or operator, within 60 days after the

change in the cost estimate, must either deposit an amount into the fund so that its value after this deposit at least equals the amount of the current post-closure cost estimate, or obtain other financial assurance as specified in this section to cover the difference.

(7) During the operating life of the facility, if the value of the trust fund is greater than the total amount of the current post-closure cost estimate, the owner or operator may submit a written request to the Regional Administrator for release of the amount in excess of the current post-closure cost estimate.

(8) If an owner or operator substitutes other financial assurance as specified in this section for all or part of the trust fund, he may submit a written request to the Regional Administrator for release of the amount in excess of the current post-closure cost estimate covered by the trust fund.

(9) Within 60 days after receiving a request from the owner or operator for release of funds as specified in paragraph (a) (7) or (8) of this section, the Regional Administrator will instruct the trustee to release to the owner or operator such funds as the Regional Administrator specifies in writing.

(10) During the period of post-closure care, the Regional Administrator may approve a release of funds if the owner or operator demonstrates to the Regional Administrator that the value of the trust fund exceeds the remaining cost of post-closure care.

(11) An owner or operator or any other person authorized to conduct post-closure care may request reimbursements for post-closure expenditures by submitting itemized bills to the Regional Administrator. Within 60 days after receiving bills for post-closure care activities, the Regional Administrator will instruct the trustee to make reimbursements in those amounts as the Regional Administrator specifies in writing, if the Regional Administrator determines that the post-closure expenditures are in accordance with the approved post-closure plan or otherwise justified. If the Regional Administrator does not instruct the trustee to make such reimbursements, he will provide the owner

or operator with a detailed written statement of reasons.

(12) The Regional Administrator will agree to termination of the trust when:

(i) An owner or operator substitutes alternate financial assurance as specified in this section; or

(ii) The Regional Administrator releases the owner or operator from the requirements of this section in accordance with § 265.145(h).

(b) *Surety bond guaranteeing payment into a post-closure trust fund.* (1) An owner or operator may satisfy the requirements of this section by obtaining a surety bond which conforms to the requirements of this paragraph and submitting the bond to the Regional Administrator. The surety company issuing the bond must, at a minimum, be among those listed as acceptable sureties on Federal bonds in Circular 570 of the U.S. Department of the Treasury.

(2) The wording of the surety bond must be identical to the wording specified in § 264.151(b).

(3) The owner or operator who uses a surety bond to satisfy the requirements of this section must also establish a standby trust fund. Under the terms of the bond, all payments made thereunder will be deposited by the surety directly into the standby trust fund in accordance with instructions from the Regional Administrator. This standby trust fund must meet the requirements specified in § 265.145(a), except that:

(i) An originally signed duplicate of the trust agreement must be submitted to the Regional Administrator with the surety bond; and

(ii) Until the standby trust fund is funded pursuant to the requirements of this section, the following are not required by these regulations:

(A) Payments into the trust fund as specified in § 265.145(a);

(B) Updating of Schedule A of the trust agreement (see § 264.151(a)) to show current post-closure cost estimates;

(C) Annual valuations as required by the trust agreement; and

(D) Notices of nonpayment as required by the trust agreement.

(4) The bond must guarantee that the owner or operator will:

(i) Fund the standby trust fund in an amount equal to the penal sum of the bond before the beginning of final closure of the facility; or

(ii) Fund the standby trust fund in an amount equal to the penal sum within 15 days after an administrative order to begin final closure issued by the Regional Administrator becomes final, or within 15 days after an order to begin final closure is issued by a U.S. district court or other court of competent jurisdiction; or

(iii) Provide alternate financial assurance as specified in this section, and obtain the Regional Administrator's written approval of the assurance provided, within 90 days after receipt by both the owner or operator and the Regional Administrator of a notice of cancellation of the bond from the surety.

(5) Under the terms of the bond, the surety will become liable on the bond obligation when the owner or operator fails to perform as guaranteed by the bond.

(6) The penal sum of the bond must be in an amount at least equal to the current post-closure cost estimate, except as provided in §265.145(f).

(7) Whenever the current post-closure cost estimate increases to an amount greater than the penal sum, the owner or operator, within 60 days after the increase, must either cause the penal sum to be increased to an amount at least equal to the current post-closure cost estimate and submit evidence of such increase to the Regional Administrator, or obtain other financial assurance as specified in this section to cover the increase. Whenever the current post-closure cost estimate decreases, the penal sum may be reduced to the amount of the current post-closure cost estimate following written approval by the Regional Administrator.

(8) Under the terms of the bond, the surety may cancel the bond by sending notice of cancellation by certified mail to the owner or operator and to the Regional Administrator. Cancellation may not occur, however, during the 120 days beginning on the date of receipt of the notice of cancellation by both the owner or operator and the Regional Ad-

ministrator, as evidenced by the return receipts.

(9) The owner or operator may cancel the bond if the Regional Administrator has given prior written consent based on his receipt of evidence of alternate financial assurance as specified in this section.

(c) *Post-closure letter of credit.* (1) An owner or operator may satisfy the requirements of this section by obtaining an irrevocable standby letter of credit which conforms to the requirements of this paragraph and submitting the letter to the Regional Administrator. The issuing institution must be an entity which has the authority to issue letters of credit and whose letter-of-credit operations are regulated and examined by a Federal or State agency.

(2) The wording of the letter of credit must be identical to the wording specified in §264.151(d).

(3) An owner or operator who uses a letter of credit to satisfy the requirements of this section must also establish a standby trust fund. Under the terms of the letter of credit, all amounts paid pursuant to a draft by the Regional Administrator will be deposited by the issuing institution directly into the standby trust fund in accordance with instructions from the Regional Administrator. This standby trust fund must meet the requirements of the trust fund specified in §265.145(a), except that:

(i) An originally signed duplicate of the trust agreement must be submitted to the Regional Administrator with the letter of credit; and

(ii) Unless the standby trust fund is funded pursuant to the requirements of this section, the following are not required by these regulations:

(A) Payments into the trust fund as specified in §265.145(a);

(B) Updating of Schedule A of the trust agreement (see §264.151(a)) to show current post-closure cost estimates;

(C) Annual valuations as required by the trust agreement; and

(D) Notices of nonpayment as required by the trust agreement.

(4) The letter of credit must be accompanied by a letter from the owner or operator referring to the letter of credit by number, issuing institution,

and date, and providing the following information: The EPA Identification Number, name, and address of the facility, and the amount of funds assured for post-closure care of the facility by the letter of credit.

(5) The letter of credit must be irrevocable and issued for a period of at least 1 year. The letter of credit must provide that the expiration date will be automatically extended for a period of at least 1 year unless, at least 120 days before the current expiration date, the issuing institution notifies both the owner or operator and the Regional Administrator by certified mail of a decision not to extend the expiration date. Under the terms of the letter of credit, the 120 days will begin on the date when both the owner or operator and the Regional Administrator have received the notice, as evidenced by the return receipts.

(6) The letter of credit must be issued in an amount at least equal to the current post-closure cost estimate, except as provided in § 265.145(f).

(7) Whenever the current post-closure cost estimate increases to an amount greater than the amount of the credit during the operating life of the facility, the owner or operator, within 60 days after the increase, must either cause the amount of the credit to be increased so that it at least equals the current post-closure cost estimate and submit evidence of such increase to the Regional Administrator, or obtain other financial assurance as specified in this section to cover the increase. Whenever the current post-closure cost estimate decreases during the operating life of the facility, the amount of the credit may be reduced to the amount of the current post-closure cost estimate following written approval by the Regional Administrator.

(8) During the period of post-closure care, the Regional Administrator may approve a decrease in the amount of the letter of credit if the owner or operator demonstrates to the Regional Administrator that the amount exceeds the remaining cost of post-closure care.

(9) Following a final administrative determination pursuant to section 3008 of RCRA that the owner or operator has failed to perform post-closure care

in accordance with the approved post-closure plan and other permit requirements, the Regional Administrator may draw on the letter of credit.

(10) If the owner or operator does not establish alternate financial assurance as specified in this section and obtain written approval of such alternate assurance from the Regional Administrator within 90 days after receipt by both the owner or operator and the Regional Administrator of a notice from the issuing institution that it has decided not to extend the letter of credit beyond the current expiration date, the Regional Administrator will draw on the letter of credit. The Regional Administrator may delay the drawing if the issuing institution grants an extension of the term of the credit. During the last 30 days of any such extension the Regional Administrator will draw on the letter of credit if the owner or operator has failed to provide alternate financial assurance as specified in this section and obtain written approval of such assurance from the Regional Administrator.

(11) The Regional Administrator will return the letter of credit to the issuing institution for termination when:

(i) An owner or operator substitutes alternate financial assurance as specified in this section; or

(ii) The Regional Administrator releases the owner or operator from the requirements of this section in accordance with § 265.145(h).

(d) *Post-closure insurance.* (1) An owner or operator may satisfy the requirements of this section by obtaining post-closure insurance which conforms to the requirements of this paragraph and submitting a certificate of such insurance to the Regional Administrator. By the effective date of these regulations the owner or operator must submit to the Regional Administrator a letter from an insurer stating that the insurer is considering issuance of post-closure insurance conforming to the requirements of this paragraph to the owner or operator. Within 90 days after the effective date of these regulations, the owner or operator must submit the certificate of insurance to the Regional

Administrator or establish other financial assurance as specified in this section. At a minimum, the insurer must be licensed to transact the business of insurance, or eligible to provide insurance as an excess or surplus lines insurer, in one or more States.

(2) The wording of the certificate of insurance must be identical to the wording specified in §264.151(e).

(3) The post-closure insurance policy must be issued for a face amount at least equal to the current post-closure cost estimate, except as provided in §265.145(f). The term "face amount" means the total amount the insurer is obligated to pay under the policy. Actual payments by the insurer will not change the face amount, although the insurer's future liability will be lowered by the amount of the payments.

(4) The post-closure insurance policy must guarantee that funds will be available to provide post-closure care of the facility whenever the post-closure period begins. The policy must also guarantee that once post-closure care begins the insurer will be responsible for paying out funds, up to an amount equal to the face amount of the policy, upon the direction of the Regional Administrator, to such party or parties as the Regional Administrator specifies.

(5) An owner or operator or any other person authorized to perform post-closure care may request reimbursement for post-closure care expenditures by submitting itemized bills to the Regional Administrator. Within 60 days after receiving bills for post-closure care activities, the Regional Administrator will instruct the insurer to make reimbursements in those amounts as the Regional Administrator specifies in writing, if the Regional Administrator determines that the post-closure expenditures are in accordance with the approved post-closure plan or otherwise justified. If the Regional Administrator does not instruct the insurer to make such reimbursements, he will provide a detailed written statement of reasons.

(6) The owner or operator must maintain the policy in full force and effect until the Regional Administrator consents to termination of the policy by the owner or operator as specified in paragraph (d)(11) of this section. Failure to pay the premium, without substitution of alternate financial assurance as specified in the section, will constitute a significant violation of these regulations, warranting such remedy as the Regional Administrator deems necessary. Such violation will be deemed to begin upon receipt by the Regional Administrator of a notice of future cancellation, termination, or failure to renew due to nonpayment of the premium, rather than upon the date of expiration.

(7) Each policy most contain a provision allowing assignment of the policy to a successor owner or operator. Such assignment may be conditional upon consent of the insurer, provided such consent is not unreasonably refused.

(8) The policy must provide that the insurer may not cancel, terminate, or fail to renew the policy except for failure to pay the premium. The automatic renewal of the policy must, at a minimum, provide the insured with the option of renewal at the face amount of the expiring policy. If there is a failure to pay the premium, the insurer may elect to cancel, terminate, or fail to renew the policy by sending notice by certified mail to the owner or operator and the Regional Administrator. Cancellation, termination, or failure to renew may not occur, however, during the 120 days beginning with the date of receipt of the notice by both the Regional Administrator and the owner or operator, as evidenced by the return receipts. Cancellation, termination, or failure to renew may not occur and the policy will remain in full force and effect in the event that on or before the date of expiration:

(i) The Regional Administrator deems the facility abandoned; or

(ii) Interim status is terminated or revoked; or

(iii) Closure is ordered by the Regional Administrator or a U.S. district court or other court of competent jurisdiction; or

(iv) The owner or operator is named as debtor in a voluntary or involuntary proceeding under Title 11 (Bankruptcy), U.S. Code; or

(v) The premium due is paid.

(9) Whenever the current post-closure cost estimate increases to an amount

greater than the face amount of the policy during the operating life of the facility, the owner or operator, within 60 days after the increase, must either cause the face amount to be increased to an amount at least equal to the current post-closure cost estimate and submit evidence of such increase to the Regional Administrator, or obtain other financial assurance as specified in this section to cover the increase. Whenever the current post-closure cost estimate decreases during the operating life of the facility, the face amount may be reduced to the amount of the current post-closure cost estimate following written approval by the Regional Administrator.

(10) Commencing on the date that liability to make payments pursuant to the policy accrues, the insurer will thereafter annually increase the face amount of the policy. Such increase must be equivalent to the face amounts of the policy, less any payments made, multiplied by an amount equivalent to 85 percent of the most recent investment rate or of the equivalent coupon-issue yield announced by the U.S. Treasury for 26-week Treasury securities.

(11) The Regional Administrator will give written consent to the owner or operator that he may terminate the insurance policy when:

(i) An owner or operator substitutes alternate financial assurance as specified in this section; or

(ii) The Regional Administrator releases the owner or operator from the requirements of this section in accordance with § 265.145(h).

(e) *Financial test and corporate guarantee for post-closure care.* (1) An owner or operator may satisfy the requirements of this section by demonstrating that he passes a financial test as specified in this paragraph. To pass this test the owner or operator must meet the criteria either of paragraph (e)(1)(i) or (ii) of this section:

(i) The owner or operator must have:

(A) Two of the following three ratios: a ratio of total liabilities to net worth less than 2.0; a ratio of the sum of net income plus depreciation, depletion, and amortization to total liabilities greater than 0.1; and a ratio of current

assets to current liabilities greater than 1.5; and

(B) Net working capital and tangible net worth each at least six times the sum of the current closure and post-closure cost estimates and the current plugging and abandonment cost estimates; and

(C) Tangible net worth of at least $10 million; and

(D) Assets in the United States amounting to at least 90 percent of his total assets or at least six times the sum of the current closure and post-closure cost estimates and the current plugging and abandonment cost estimates.

(ii) The owner or operator must have:

(A) A current rating for his most recent bond issuance of AAA, AA, A, or BBB as issued by Standard and Poor's or Aaa, Aa, A, or Baa as issued by Moody's; and

(B) Tangible net worth at least six times the sum of the current closure and post-closure cost estimates and the current plugging and abandonment cost estimates; and

(C) Tangible net worth of at least $10 million; and

(D) Assets located in the United States amounting to at least 90 percent of his total assets or at least six times the sum of the current closure and post-closure cost estimates and the current plugging and abandonment cost estimates.

(2) The phrase "current closure and post-closure cost estimates" as used in paragraph (e)(1) of this section refers to the cost estimates required to be shown in paragraphs 1-4 of the letter from the owner's or operator's chief financial officer (§ 264.151(f)). The phrase "current plugging and abandonment cost estimates" as used in paragraph (e)(1) of this section refers to the cost estimates required to be shown in paragraphs 1-4 of the letter from the owner's or operator's chief financial officer (§ 144.70(f) of this title).

(3) To demonstrate that he meets this test, the owner or operator must submit the following items to the Regional Administrator:

(i) A letter signed by the owner's or operator's chief financial officer and worded as specified in § 264.151(f); and

(ii) A copy of the independent certified public accountant's report on examination of the owner's or operator's financial statements for the latest completed fiscal year; and

(iii) A special report from the owner's or operator's independent certified public accountant to the owner or operator stating that:

(A) He has compared the data which the letter from the chief financial officer specifies as having been derived from the independently audited, year-end financial statements for the latest fiscal year with the amounts in such financial statements; and

(B) In connection with that procedure, no matters came to his attention which caused him to believe that the specified data should be adjusted.

(4) The owner or operator may obtain an extension of the time allowed for submission of the documents specified in paragraph (e)(3) of this section if the fiscal year of the owner or operator ends during the 90 days prior to the effective date of these regulations and if the year-end financial statements for that fiscal year will be audited by an independent certified public accountant. The extension will end no later than 90 days after the end of the owner's or operator's fiscal year. To obtain the extension, the owner's or operator's chief financial officer must send, by the effective date of these regulations, a letter to the Regional Administrator of each Region in which the owner's or operator's facilities are covered by the financial test are located. This letter from the chief financial officer must:

(i) Request the extension;

(ii) Certify that he has grounds to believe that the owner or operator meets the criteria of the financial test;

(iii) Specify for each facility to be covered by the test the EPA Identification Number, name, address, and the current closure and post-closure cost estimates to be covered by the test;

(iv) Specify the date ending the owner's or operator's latest complete fiscal year before the effective date of these regulations;

(v) Specify the date, no later than 90 days after the end of such fiscal year, when he will submit the documents specified in paragraph (e)(3) of this section; and

(vi) Certify that the year-end financial statements of the owner or operator for such fiscal year will be audited by an independent certified public accountant.

(5) After the initial submission of items specified in paragraph (e)(3) of this section, the owner or operator must send updated information to the Regional Administrator within 90 days after the close of each succeeding fiscal year. This information must consist of all three items specified in paragraph (e)(3) of this section.

(6) If the owner or operator no longer meets the requirements of paragraph (e)(1) of this section, he must send notice to the Regional Administrator of intent to establish alternate financial assurance as specified in this section. The notice must be sent by certified mail within 90 days after the end of the fiscal year for which the year-end financial data show that the owner or operator no longer meets the requirements. The owner or operator must provide the alternate financial assurance within 120 days after the end of such fiscal year.

(7) The Regional Administrator may, based on a reasonable belief that the owner or operator may no longer meet the requirements of paragraph (e)(1) of this section, require reports of financial condition at any time from the owner or operator in addition to those specified in paragraph (e)(3) of this section. If the Regional Administrator finds, on the basis of such reports or other information, that the owner or operator no longer meets the requirements of paragraph (e)(1) of this section, the owner or operator must provide alternate financial assurance as specified in this section within 30 days after notification of such a finding.

(8) The Regional Administrator may disallow use of this test on the basis of qualifications in the opinion expressed by the independent certified public accountant in his report on examination of the owner's or operator's financial statements (see paragraph (e)(3)(ii) of this section). An adverse opinion or a disclaimer of opinion will be cause for

disallowance. The Regional Administrator will evaluate other qualifications on an individual basis. The owner or operator must provide alternate financial assurance as specified in this section within 30 days after notification of the disallowance.

(9) During the period of post-closure care, the Regional Administrator may approve a decrease in the current post-closure cost estimate for which this test demonstrates financial assurance if the owner or operator demonstrates to the Regional Administrator that the amount of the cost estimate exceeds the remaining cost of post-closure care.

(10) The owner or operator is no longer required to submit the items specified in paragraph (e)(3) of this section when:

(i) An owner or operator substitutes alternate financial assurance as specified in this section; or

(ii) The Regional Administrator releases the owner or operator from the requirements of this section in accordance with § 265.145(h).

(11) An owner or operator may meet the requirements of this section by obtaining a written guarantee. The guarantor must be the direct or higher-tier parent corporation of the owner or operator, a firm whose parent corporation is also the parent corporation of the owner or operator, or a firm with a "substantial business relationship" with the owner or operator. The guarantor must meet the requirements for owners or operators in paragraphs (e)(1) through (9) of this section and must comply with the terms of the guarantee. The wording of the guarantee must be identical to the wording specified in § 264.151(h). A certified copy of the guarantee must accompany the items sent to the Regional Administrator as specified in paragraph (e)(3) of this section. One of these items must be the letter from the guarantor's chief financial officer. If the guarantor's parent corporation is also the parent corporation of the owner or operator, the letter must describe the value received in consideration of the guarantee. If the guarantor is a firm with a "substantial business relationship" with the owner or operator, this letter must describe this "substantial business relationship" and the value received in

consideration of the guarantee. The terms of the guarantee must provide that:

(i) If the owner or operator fails to perform post-closure care of a facility covered by the corporate guarantee in accordance with the post-closure plan and other interim status requirements whenever required to do so, the guarantor will do so or establish a trust fund as specified in § 265.145(a) in the name of the owner or operator.

(ii) The corporate guarantee will remain in force unless the guarantor sends notice of cancellation by certified mail to the owner or operator and to the Regional Administrator. Cancellation may not occur, however, during the 120 days beginning on the date of receipt of the notice of cancellation by both the owner or operator and the Regional Administrator, as evidenced by the return receipts.

(iii) If the owner or operator fails to provide alternate financial assurance as specified in this section and obtain the written approval of such alternate assurance from the Regional Administrator within 90 days after receipt by both the owner or operator and the Regional Administrator of a notice of cancellation of the corporate guarantee from the guarantor, the guarantor will provide such alternate financial assurance in the name of the owner or operator.

(f) *Use of multiple financial mechanisms.* An owner or operator may satisfy the requirements of this section by establishing more than one financial mechanism per facility. These mechanisms are limited to trust funds, surety bonds, letters of credit, and insurance. The mechanisms must be as specified in paragraphs (a) through (d), respectively, of this section, except that it is the combination of mechanisms, rather than the single mechanism, which must provide financial assurance for an amount at least equal to the current post-closure cost estimate. If an owner or operator uses a trust fund in combination with a surety bond or a letter of credit, he may use the trust fund as the standby trust fund for the other mechanisms. A single standby trust fund may be established for two or more mechanisms. The Regional Administrator may use any or all of the mechanisms to provide for post-closure care of the facility.

(g) *Use of a financial mechanism for multiple facilities.* An owner or operator may use a financial assurance mechanism specified in this section to meet the requirements of this section for more than one facility. Evidence of financial assurance submitted to the Regional Administrator must include a list showing, for each facility, the EPA Identification Number, name, address, and the amount of funds for post-closure care assured by the mechanism. If the facilities covered by the mechanism are in more than one Region, identical evidence of financial assurance must be submitted to and maintained with the Regional Administrators of all such Regions. The amount of funds available through the mechanism must be no less than the sum of funds that would be available if a separate mechanism had been established and maintained for each facility. In directing funds available through the mechanism for post-closure care of any of the facilities covered by the mechanism, the Regional Administrator may direct only the amount of funds designated for that facility, unless the owner or operator agrees to the use of additional funds available under the mechanism.

(h) *Release of the owner or operator from the requirements of this section.* Within 60 days after receiving certifications from the owner or operator and a qualified Professional Engineer that the post-closure care period has been completed for a hazardous waste disposal unit in accordance with the approved plan, the Regional Administrator will notify the owner or operator in writing that he is no longer required to maintain financial assurance for post-closure care of that unit, unless the Regional Administrator has reason to believe that post-closure care has not been in accordance with the approved post-closure plan. The Regional Administrator shall provide the owner or operator a detailed written statement of any such reason to believe that post-closure care has not been in ac-

cordance with the approved post-closure plan.

[47 FR 15064, Apr. 7, 1982, as amended at 51 FR 16457, May 2, 1986; 57 FR 42843, Sept. 16, 1992; 71 FR 16909, Apr. 4, 2006; 71 FR 40275, July 14, 2006]

## § 265.146 Use of a mechanism for financial assurance of both closure and post-closure care.

An owner or operator may satisfy the requirements for financial assurance for both closure and post-closure care for one or more facilities by using a trust fund, surety bond, letter of credit, insurance, financial test, or corporate guarantee that meets the specifications for the mechanism in both §§ 265.143 and 265.145. The amount of funds available through the mechanism must be no less than the sum of funds that would be available if a separate mechanism had been established and maintained for financial assurance of closure and of post-closure care.

## § 265.147 Liability requirements.

(a) *Coverage for sudden accidental occurrences.* An owner or operator of a hazardous waste treatment, storage, or disposal facility, or a group of such facilities, must demonstrate financial responsibility for bodily injury and property damage to third parties caused by sudden accidental occurrences arising from operations of the facility or group of facilities. The owner or operator must have and maintain liability coverage for sudden accidental occurrences in the amount of at least $1 million per occurrence with an annual aggregate of at least $2 million, exclusive of legal defense costs. This liability coverage may be demonstrated as specified in paragraphs (a) (1), (2), (3), (4), (5), or (6) of this section:

(1) An owner or operator may demonstrate the required liability coverage by having liability insurance as specified in this paragraph.

(i) Each insurance policy must be amended by attachment of the Hazardous Waste Facility Liability Endorsement, or evidenced by a Certificate of Liability Insurance. The wording of the endorsement must be identical to the wording specified in § 264.151(i). The wording of the certificate of insurance must be identical to the wording specified in § 264.151(j). The owner or operator must submit a signed duplicate original of the endorsement or the certificate of insurance to the Regional Administrator, or Regional Administrator if the facilities are located in more than one Region. If requested by a Regional Administrator, the owner or operator must provide a signed duplicate original of the insurance policy.

(ii) Each insurance policy must be issued by an insurer which, at a minimum, is licensed to transact the business of insurance, or eligible to provide insurance as an excess or surplus lines insurer, in one or more States.

(2) An owner or operator may meet the requirements of this section by passing a financial test or using the guarantee for liability coverage as specified in paragraphs (f) and (g) of this section.

(3) An owner or operator may meet the requirements of this section by obtaining a letter of credit for liability coverage as specified in paragraph (h) of this section.

(4) An owner or operator may meet the requirements of this section by obtaining a surety bond for liability coverage as specified in paragraph (i) of this section.

(5) An owner or operator may meet the requirements of this section by obtaining a trust fund for liability coverage as specified in paragraph (j) of this section.

(6) An owner or operator may demonstrate the required liability coverage through the use of combinations of insurance, financial test, guarantee, letter of credit, surety bond, and trust fund, except that the owner or operator may not combine a financial test covering part of the liability coverage requirement with a guarantee unless the financial statement of the owner or operator is not consolidated with the financial statement of the guarantor. The amounts of coverage demonstrated must total at least the minimum amounts required by this section. If the owner or operator demonstrates the required coverage through the use of a combination of financial assurances under this paragraph, the owner or operator shall specify at least one such assurance as "primary" coverage

and shall specify other assurance as "excess" coverage.

(7) An owner or operator shall notify the Regional Administrator in writing within 30 days whenever:

(i) A claim results in a reduction in the amount of financial assurance for liability coverage provided by a financial instrument authorized in paragraphs (a)(1) through (a)(6) of this section; or

(ii) A Certification of Valid Claim for bodily injury or property damages caused by a sudden or non-sudden accidental occurrence arising from the operation of a hazardous waste treatment, storage, or disposal facility is entered between the owner or operator and third-party claimant for liability coverage under paragraphs (a)(1) through (a)(6) of this section; or

(iii) A final court order establishing a judgment for bodily injury or property damage caused by a sudden or non-sudden accidental occurrence arising from the operation of a hazardous waste treatment, storage, or disposal facility is issued against the owner or operator or an instrument that is providing financial assurance for liability coverage under paragraphs (a)(1) through (a)(6) of this section.

(b) *Coverage for nonsudden accidental occurrences.* An owner or operator of a surface impoundment, landfill, or land treatment facility which is used to manage hazardous waste, or a group of such facilities, must demonstrate financial responsibility for bodily injury and property damage to third parties caused by nonsudden accidental occurrences arising from operations of the facility or group of facilities. The owner or operator must have and maintain liability coverage for nonsudden accidental occurrences in the amount of at least $3 million per occurrence with an annual aggregate of at least $6 million, exclusive of legal defense costs. An owner or operator who must meet the requirements of this section may combine the required per-occurrence coverage levels for sudden and nonsudden accidental occurrences into a single per-occurrence level, and combine the required annual aggregate coverage levels for sudden and nonsudden accidental occurrences into a single annual aggregate level. Owners or operators who combine coverage levels for sudden and nonsudden accidental occurrences must maintain liability coverage in the amount of at least $4 million per occurrence and $8 million annual aggregate. This liability coverage may be demonstrated as specified in paragraph (b) (1), (2), (3), (4), (5), or (6) of this section:

(1) An owner or operator may demonstrate the required liability coverage by having liability insurance as specified in this paragraph.

(i) Each insurance policy must be amended by attachment of the Hazardous Waste Facility Liability Endorsement or evidenced by a Certificate of Liability Insurance. The wording of the endorsement must be identical to the wording specified in § 264.151(i). The wording of the certificate of insurance must be identical to the wording specified in § 264.151(j). The owner or operator must submit a signed duplicate original of the endorsement or the certificate of insurance to the Regional Administrator, or Regional Administrators if the facilities are located in more than one Region. If requested by a Regional Administrator, the owner or operator must provide a signed duplicate original of the insurance policy.

(ii) Each insurance policy must be issued by an insurer which, at a minimum, is licensed to transact the business of insurance, or eligible to provide insurance as an excess or surplus lines insurer, in one or more States.

(2) An owner or operator may meet the requirements of this section by passing a financial test or using the guarantee for liability coverage as specified in paragraphs (f) and (g) of this section.

(3) An owner or operator may meet the requirements of this section by obtaining a letter of credit for liability coverage as specified in paragraph (h) of this section.

(4) An owner or operator may meet the requirements of this section by obtaining a surety bond for liability coverage as specified in paragraph (i) of this section.

(5) An owner or operator may meet the requirements of this section by obtaining a trust fund for liability coverage as specified in paragraph (j) of this section.

(6) An owner or operator may demonstrate the required liability coverage through the use of combinations of insurance, financial test, guarantee, letter of credit, surety bond, and trust fund, except that the owner or operator may not combine a financial test covering part of the liability coverage requirement with a guarantee unless the financial statement of the owner or operator is not consolidated with the financial statement of the guarantor. The amounts of coverage demonstrated must total at least the minimum amounts required by this section. If the owner or operator demonstrates the required coverage through the use of a combination of financial assurances under this paragraph, the owner or operator shall specify at least one such assurance as "primary" coverage and shall specify other assurance as "excess" coverage.

(7) An owner or operator shall notify the Regional Administrator in writing within 30 days whenever:

(i) A claim results in a reduction in the amount of financial assurance for liability coverage provided by a financial instrument authorized in paragraphs (b)(1) through (b)(6) of this section; or

(ii) A Certification of Valid Claim for bodily injury or property damages caused by a sudden or non-sudden accidental occurrence arising from the operation of a hazardous waste treatment, storage, or disposal facility is entered between the owner or operator and third-party claimant for liability coverage under paragraphs (b)(1) through (b)(6) of this section; or

(iii) A final court order establishing a judgment for bodily injury or property damage caused by a sudden or non-sudden accidental occurrence arising from the operation of a hazardous waste treatment, storage, or disposal facility is issued against the owner or operator or an instrument that is providing financial assurance for liability coverage under paragraphs (b)(1) through (b)(6) of this section.

(c) *Request for variance.* If an owner or operator can demonstrate to the satisfaction of the Regional Administrator that the levels of financial responsibility required by paragraph (a) or (b) of this section are not consistent with the degree and duration of risk associated with treatment, storage, or disposal at the facility or group of facilities, the owner or operator may obtain a variance from the Regional Administrator. The request for a variance must be submitted in writing to the Regional Administrator. If granted, the variance will take the form of an adjusted level of required liability coverage, such level to be based on the Regional Administrator's assessment of the degree and duration of risk associated with the ownership or operation of the facility or group of facilities. The Regional Administrator may require an owner or operator who requests a variance to provide such technical and engineering information as is deemed necessary by the Regional Administrator to determine a level of financial responsibility other than that required by paragraph (a) or (b) of this section. The Regional Administrator will process a variance request as if it were a permit modification request under § 270.41(a)(5) of this chapter and subject to the procedures of § 124.5 of this chapter. Notwithstanding any other provision, the Regional Administrator may hold a public hearing at his discretion or whenever he finds, on the basis of requests for a public hearing, a significant degree of pubic interest in a tentative decision to grant a variance.

(d) *Adjustments by the Regional Administrator.* If the Regional Administrator determines that the levels of financial responsibility required by paragraph (a) or (b) of this section are not consistent with the degree and duration of risk associated with treatment, storage, or disposal at the facility or group of facilities, the Regional Administrator may adjust the level of financial responsibility required under paragraph (a) or (b) of this section as may be necessary to protect human health and the environment. This adjusted level will be based on the Regional Administrator's assessment of the degree and duration of risk associated with the ownership or operation of the facility or group of facilities. In addition, if the Regional Administrator determines that there is a significant risk to

human health and the environment from nonsudden accidental occurrences resulting from the operations of a facility that is not a surface impoundment, landfill, or land treatment facility, he may require that an owner or operator of the facility comply with paragraph (b) of this section. An owner or operator must furnish to the Regional Administrator, within a reasonable time, any information which the Regional Administrator requests to determine whether cause exists for such adjustments of level or type of coverage. The Regional Administrator will process an adjustment of the level of required coverage as if it were a permit modification under § 270.41(a)(5) of this chapter and subject to the procedures of § 124.5 of this chapter. Notwithstanding any other provision, the Regional Administrator may hold a public hearing at his discretion or whenever he finds, on the basis of requests for a public hearing, a significant degree of public interest in a tentative decision to adjust the level or type of required coverage.

(e) *Period of coverage.* Within 60 days after receiving certifications from the owner or operator and a qualified Professional Engineer that final closure has been completed in accordance with the approved closure plan, the Regional Administrator will notify the owner or operator in writing that he is no longer required by this section to maintain liability coverage for that facility, unless the Regional Administrator has reason to believe that closure has not been in accordance with the approved closure plan.

(f) *Financial test for liability coverage.* (1) An owner or operator may satisfy the requirements of this section by demonstrating that he passes a financial test as specified in this paragraph. To pass this test the owner or operator must meet the criteria of paragraph (f)(1) (i) or (ii) of this section:

(i) The owner or operator must have:

(A) Net working capital and tangible net worth each at least six times the amount of liability coverage to be demonstrated by this test; and

(B) Tangible net worth of at least $10 million; and

(C) Assets in the United States amounting to either: (1) At least 90 percent of his total assets; or (2) at least six times the amount of liability coverage to be demonstrated by this test.

(ii) The owner or operator must have:

(A) A current rating for his most recent bond issuance of AAA, AA, A, or BBB as issued by Standard and Poor's, or Aaa, Aa, A, or Baa as issued by Moody's; and

(B) Tangible net worth of at least $10 million; and

(C) Tangible net worth at least six times the amount of liability coverage to be demonstrated by this test; and

(D) Assets in the United States amounting to either: (1) At least 90 percent of his total assets; or (2) at least six times the amount of liability coverage to be demonstrated by this test.

(2) The phrase "amount of liability coverage" as used in paragraph (f)(1) of this section refers to the annual aggregate amounts for which coverage is required under paragraphs (a) and (b) of this section.

(3) To demonstrate that he meets this test, the owner or operator must submit the following three items to the Regional Administrator:

(i) A letter signed by the owner's or operator's chief financial officer and worded as specified in §264.151(g). If an owner or operator is using the financial test to demonstrate both assurance for closure or post-closure care, as specified by §§264.143(f), 264.145(f), 265.143(e), and 265.145(e), and liability coverage, he must submit the letter specified in §264.151(g) to cover both forms of financial responsibility; a separate letter as specified in §264.151(f) is not required.

(ii) A copy of the independent certified public accountant's report on examination of the owner's or operator's financial statements for the latest completed fiscal year.

(iii) A special report from the owner's or operator's independent certified public accountant to the owner or operator stating that:

(A) He has compared the data which the letter from the chief financial officer specifies as having been derived from the independently audited, year-end financial statements for the latest fiscal year with the amounts in such financial statements; and

(B) In connection with that procedure, no matters came to his attention which caused him to believe that the specified data should be adjusted.

(4) The owner or operator may obtain a one-time extension of the time allowed for submission of the documents specified in paragraph (f)(3) of this section if the fiscal year of the owner or operator ends during the 90 days prior to the effective date of these regulations and if the year-end financial statements for that fiscal year will be audited by an independent certified public accountant. The extension will end no later than 90 days after the end of the owner's or operator's fiscal year. To obtain the extension, the owner's or operator's chief financial officer must send, by the effective date of these regulations, a letter to the Regional Administrator of each Region in which the owner's or operator's facilities to be covered by the financial test are located. This letter from the chief financial officer must:

(i) Request the extension;

(ii) Certify that he has grounds to believe that the owner or operator meets the criteria of the financial test;

(iii) Specify for each facility to be covered by the test the EPA Identification Number, name, address, the amount of liability coverage and, when applicable, current closure and post-closure cost estimates to be covered by the test;

(iv) Specify the date ending the owner's or operator's last complete fiscal year before the effective date of these regulations;

(v) Specify the date, no later than 90 days after the end of such fiscal year, when he will submit the documents specified in paragraph (f)(3) of this section; and

(vi) Certify that the year-end financial statements of the owner or operator for such fiscal year will be audited by an independent certified public accountant.

(5) After the initial submission of items specified in paragraph (f)(3) of this section, the owner or operator must send updated information to the Regional Administrator within 90 days after the close of each succeeding fiscal year. This information must consist of

all three items specified in paragraph (f)(3) of this section.

(6) If the owner or operator no longer meets the requirements of paragraph (f)(1) of this section, he must obtain insurance, a letter of credit, a surety bond, a trust fund, or a guarantee for the entire amount of required liability coverage as specified in this section. Evidence of liability coverage must be submitted to the Regional Administrator within 90 days after the end of the fiscal year for which the year-end financial data show that the owner or operator no longer meets the test requirements.

(7) The Regional Administrator may disallow use of this test on the basis of qualifications in the opinion expressed by the independent certified public accountant in his report on examination of the owner's or operator's financial statements (see paragraph (f)(3)(ii) of this section). An adverse opinion or a disclaimer of opinion will be cause for disallowance. The Regional Administrator will evaluate other qualifications on an individual basis. The owner or operator must provide evidence of insurance for the entire amount of required liability coverage as specified in this section within 30 days after notification of disallowance.

(g) *Guarantee for liability coverage.* (1) Subject to paragraph (g)(2) of this section, an owner or operator may meet the requirements of this section by obtaining a written guarantee, hereinafter referred to as "guarantee." The guarantor must be the direct or higher-tier parent corporation of the owner or operator, a firm whose parent corporation is also the parent corporation of the owner or operator, or a firm with a "substantial business relationship" with the owner or operator. The guarantor must meet the requirements for owners or operators in paragraphs (f)(1) through (f)(6) of this section. The wording of the guarantee must be identical to the wording specified in § 264.151(h)(2) of this chapter. A certified copy of the guarantee must accompany the items sent to the Regional Administrator as specified in paragraph (f)(3) of this section. One of these items must be the letter from the guarantor's chief financial officer. If the guarantor's parent corporation is

also the parent corporation of the owner or operator, this letter must describe the value received in consideration of the guarantee. If the guarantor is a firm with a "substantial business relationship" with the owner or operator, this letter must describe this "substantial business relationship" and the value received in consideration of the guarantee.

(i) If the owner or operator fails to satisfy a judgment based on a determination of liability for bodily injury or property damage to third parties caused by sudden or nonsudden accidental occurrences (or both as the case may be), arising from the operation of facilities covered by this corporate guarantee, or fails to pay an amount agreed to in settlement of claims arising from or alleged to arise from such injury or damage, the guarantor will do so up to the limits of coverage.

(ii) [Reserved]

(2)(i) In the case of corporations incorporated in the United States, a guarantee may be used to satisfy the requirements of this section only if the Attorneys General or Insurance Commissioners of (A) the State in which the guarantor is incorporated, and (B) each State in which a facility covered by the guarantee is located have submitted a written statement to EPA that a guarantee executed as described in this section and § 264.151(h)(2) is a legally valid and enforceable obligation in that State.

(ii) In the case of corporations incorporated outside the United States, a guarantee may be used to satisfy the requirements of this section only if (A) the non-U.S. corporation has identified a registered agent for service of process in each State in which a facility covered by the guarantee is located and in the State in which it has its principal place of business, and if (B) the Attorney General or Insurance Commissioner of each State in which a facility covered by the guarantee is located and the State in which the guarantor corporation has its principal place of business, has submitted a written statement to EPA that a guarantee executed as described in this section and § 264.151(h)(2) is a legally valid and enforceable obligation in that State.

(h) *Letter of credit for liability coverage.* (1) An owner or operator may satisfy the requirements of this section by obtaining an irrevocable standby letter of credit that conforms to the requirements of this paragraph and submitting a copy of the letter of credit to the Regional Administrator.

(2) The financial institution issuing the letter of credit must be an entity that has the authority to issue letters of credit and whose letter of credit operations are regulated and examined by a Federal or State agency.

(3) The wording of the letter of credit must be identical to the wording specified in §264.151(k) of this chapter.

(4) An owner or operator who uses a letter of credit to satisfy the requirements of this section may also establish a standby trust fund. Under the terms of such a letter of credit, all amounts paid pursuant to a draft by the trustee of the standby trust will be deposited by the issuing institution into the standby trust in accordance with instructions from the trustee. The trustee of the standby trust fund must be an entity which has the authority to act as a trustee and whose trust operations are regulated and examined by a Federal or State agency.

(5) The wording of the standby trust fund must be identical to the wording specified in §264.151(n).

(i) *Surety bond for liability coverage.* (1) An owner or operator may satisfy the requirements of this section by obtaining a surety bond that conforms to the requirements of this paragraph and submitting a copy of the bond to the Regional Administrator.

(2) The surety company issuing the bond must be among those listed as acceptable sureties on Federal bonds in the most recent Circular 570 of the U.S. Department of the Treasury.

(3) The wording of the surety bond must be identical to the wording specified in §264.151(l) of this chapter.

(4) A surety bond may be used to satisfy the requirements of this section only if the Attorneys General or Insurance Commissioners of (i) the State in which the surety is incorporated, and (ii) each State in which a facility covered by the surety bond is located have submitted a written statement to EPA that a surety bond executed as described in this section and §264.151(l) of this chapter is a legally valid and enforceable obligation in that State.

(j) *Trust fund for liability coverage.* (1) An owner or operator may satisfy the requirements of this section by establishing a trust fund that conforms to the requirements of this paragraph and submitting an originally signed duplicate of the trust agreement to the Regional Administrator.

(2) The trustee must be an entity which has the authority to act as a trustee and whose trust operations are regulated and examined by a Federal or State agency.

(3) The trust fund for liability coverage must be funded for the full amount of the liability coverage to be provided by the trust fund before it may be relied upon to satisfy the requirements of this section. If at any time after the trust fund is created the amount of funds in the trust fund is reduced below the full amount of the liability coverage to be provided, the owner or operator, by the anniversary date of the establishment of the Fund, must either add sufficient funds to the trust fund to cause its value to equal the full amount of liability coverage to be provided, or obtain other financial assurance as specified in this section to cover the difference. For purposes of this paragraph, "the full amount of the liability coverage to be provided" means the amount of coverage for sudden and/or nonsudden occurrences required to be provided by the owner or operator by this section, less the amount of financial assurance for liability coverage that is being provided by other financial assurance mechanisms being used to demonstrate financial assurance by the owner or operator.

(4) The wording of the trust fund must be identical to the wording specified in §264.151(m) of this part.

(k) Notwithstanding any other provision of this part, an owner or operator using liability insurance to satisfy the requirements of this section may use, until October 16, 1982, a Hazardous Waste Facility Liability Endorsement or Certificate of Liability Insurance that does not certify that the insurer is licensed to transact the business of insurance, or eligible as an excess or surplus lines insurer, in one or more States.

[47 FR 16558, Apr. 16, 1982, as amended at 47 FR 28627, July 1, 1982; 47 FR 30447, July 13, 1982; 48 FR 30115, June 30, 1983; 51 FR 16458, May 2, 1986; 51 FR 25355, July 11, 1986; 52 FR 44321, Nov. 18, 1987; 53 FR 33959, Sept. 1, 1988; 56 FR 30200, July 1, 1991; 56 FR 47912, Sept. 23, 1991; 57 FR 42843, Sept. 16, 1992; 71 FR 16910, Apr. 4, 2006; 71 FR 40275, July 14, 2006]

§ 265.148 Incapacity of owners or operators, guarantors, or financial institutions.

(a) An owner or operator must notify the Regional Administrator by certified mail of the commencement of a voluntary or involuntary proceeding under Title 11 (Bankruptcy), U.S. Code, naming the owner or operator as debtor, within 10 days after commencement of the proceeding. A guarantor of a corporate guarantee as specified in §§ 265.143(e) and 265.145(e) must make such a notification if he is named as debtor, as required under the terms of the corporate guarantee (§ 264.151(h)).

(b) An owner or operator who fulfills the requirements of § 265.143, § 265.145, or § 265.147 by obtaining a trust fund, surety bond, letter of credit, or insurance policy will be deemed to be without the required financial assurance or liability coverage in the event of bankruptcy of the trustee or issuing institution, or a suspension or revocation of the authority of the trustee institution to act as trustee or of the institution issuing the surety bond, letter of credit, or insurance policy to issue such instruments. The owner or operator must establish other financial assurance or liability coverage within 60 days after such an event.

§ 265.149 Use of State-required mechanisms.

(a) For a facility located in a State where EPA is administering the requirements of this subpart but where the State has hazardous waste regulations that include requirements for financial assurance of closure or post-closure care or liability coverage, an owner or operator may use State-required financial mechanisms to meet the requirements of § 265.143, § 265.145, or § 265.147 if the Regional Adminis-

trator determines that the State mechanisms are at least equivalent to the financial mechanisms specified in this subpart. The Regional Administrator will evaluate the equivalency of the mechanisms principally in terms of (1) certainty of the availability of funds for the required closure or post-closure care activities or liability coverage and (2) the amount of funds that will be made available. The Regional Administrator may also consider other factors as he deems appropriate. The owner or operator must submit to the Regional Administrator evidence of the establishment of the mechanism together with a letter requesting that the State-required mechanism be considered acceptable for meeting the requirements of this subpart. The submission must include the following information: The facility's EPA Identification Number, name, and address, and the amount of funds for closure or post-closure care or liability coverage assured by the mechanism. The Regional Administrator will notify the owner or operator of his determination regarding the mechanism's acceptability in lieu of financial mechanisms specified in this subpart. The Regional Administrator may require the owner or operator to submit additional information as is deemed necessary to make this determination. Pending this determination, the owner or operator will be deemed to be in compliance with the requirements of § 265.143, § 265.145, or § 265.147, as applicable.

(b) If a State-required mechanism is found acceptable as specified in paragraph (a) of this section except for the amount of funds available, the owner or operator may satisfy the requirements of this subpart by increasing the funds available through the State-required mechanism or using additional financial mechanisms as specified in this subpart. The amount of funds available through the State and Federal mechanisms must at least equal the amount required by this subpart.

§ 265.150 State assumption of responsibility.

(a) If a State either assumes legal responsibility for an owner's or operator's compliance with the closure, post-closure care, or liability requirements of this part or assures that funds will be available from State sources to cover those requirements, the owner or operator will be in compliance with the requirements of §265.143, §265.145, or §265.147 if the Regional Administrator determines that the State's assumption of responsibility is at least equivalent to the financial mechanisms specified in this subpart. The Regional Administrator will evaluate the equivalency of State guarantees principally in terms of (1) certainty of the availability of funds for the required closure or post-closure care activities or liability coverage and (2) the amount of funds that will be made available. The Regional Administrator may also consider other factors as he deems appropriate. The owner or operator must submit to the Regional Administrator a letter from the State describing the nature of the State's assumption of responsibility together with a letter from the owner or operator requesting that the State's assumption of responsibility be considered acceptable for meeting the requirements of this subpart. The letter from the State must include, or have attached to it, the following information: The facility's EPA Identification Number, name, and address, and the amount of funds for closure or post-closure care or liability coverage that are guaranteed by the State. The Regional Administrator will notify the owner or operator of his determination regarding the acceptability of the State's guarantee in lieu of financial mechanisms specified in this subpart. The Regional Administrator may require the owner or operator to submit additional information as is deemed necessary to make this determination. Pending this determination, the owner or operator will be deemed to be in compliance with the requirements of §§265.143, §265.145, or §265.147, as applicable.

(b) If a State's assumption of responsibility is found acceptable as specified in paragraph (a) of this section except for the amount of funds available, the owner or operator may satisfy the requirements of this subpart by use of both the State's assurance *and* additional financial mechanisms as specified in this subpart. The amount of funds available through the State and Federal mechanisms must at least equal the amount required by this subpart.

## Subpart I—Use and Management of Containers

### §265.170 Applicability.

The regulations in this subpart apply to owners and operators of all hazardous waste facilities that store containers of hazardous waste, except as §265.1 provides otherwise.

### §265.171 Condition of containers.

If a container holding hazardous waste is not in good condition, or if it begins to leak, the owner or operator must transfer the hazardous waste from this container to a container that is in good condition, or manage the waste in some other way that complies with the requirements of this part.

### §265.172 Compatibility of waste with container.

The owner or operator must use a container made of or lined with materials which will not react with, and are otherwise compatible with, the hazardous waste to be stored, so that the ability of the container to contain the waste is not impaired.

### §265.173 Management of containers.

(a) A container holding hazardous waste must always be closed during storage, except when it is necessary to add or remove waste.

(b) A container holding hazardous waste must not be opened, handled, or stored in a manner which may rupture the container or cause it to leak.

[*Comment:* Re-use of containers in transportation is governed by U.S. Department of Transportation regulations, including those set forth in 49 CFR 173.28.]

[45 FR 33232, May 19, 1980, as amended at 45 FR 78529, Nov. 25, 1980]

### §265.174 Inspections.

At least weekly, the owner or operator must inspect areas where containers are stored, except for Performance Track member facilities, that must conduct inspections at least once each month, upon approval by the Director. To apply for reduced inspection frequency, the Performance Track member facility must follow the procedures described in § 265.15(b)(5) of this part. The owner or operator must look for leaking containers and for deterioration of containers caused by corrosion or other factors.

[Comment: See § 265.171 for remedial action required if deterioration or leaks are detected.]

[71 FR 16910, Apr. 4, 2006, as amended at 71 FR 40275, July 14, 2006]

## § 265.175  [Reserved]

## § 265.176  Special requirements for ignitable or reactive waste.

Containers holding ignitable or reactive waste must be located at least 15 meters (50 feet) from the facility's property line.

[Comment: See § 265.17(a) for additional requirements.]

## § 265.177  Special requirements for incompatible wastes.

(a) Incompatible wastes, or incompatible wastes and materials, (see appendix V for examples) must not be placed in the same container, unless § 265.17(b) is complied with.

(b) Hazardous waste must not be placed in an unwashed container that previously held an incompatible waste or material (see appendix V for examples), unless § 265.17(b) is complied with.

(c) A storage container holding a hazardous waste that is incompatible with any waste or other materials stored nearby in other containers, piles, open tanks, or surface impoundments must be separated from the other materials or protected from them by means of a dike, berm, wall, or other device.

[Comment: The purpose of this is to prevent fires, explosions, gaseous emissions, leaching, or other discharge of hazardous waste or hazardous waste constituents which could result from the mixing of incompatible wastes or materials if containers break or leak.]

## § 265.178  Air emission standards.

The owner or operator shall manage all hazardous waste placed in a container in accordance with the applicable requirements of subparts AA, BB, and CC of this part.

[61 FR 59968, Nov. 25, 1996]

## Subpart J—Tank Systems

SOURCE: 51 FR 25479, July 14, 1986, unless otherwise noted.

## § 265.190  Applicability.

The requirements of this subpart apply to owners and operators of facilities that use tank systems for storing or treating hazardous waste except as otherwise provided in paragraphs (a), (b), and (c) of this section or in § 265.1 of this part.

(a) Tank systems that are used to store or treat hazardous waste which contains no free liquids and are situated inside a building with an impermeable floor are exempted from the requirements in § 265.193. To demonstrate the absence or presence of free liquids in the stored/treated waste, the following test must be used: Method 9095B (Paint Filter Liquids Test) as described in "Test Methods for Evaluating Solid Waste, Physical/Chemical Methods," EPA Publication SW–846, as incorporated by reference in § 260.11 of this chapter.

(b) Tank systems, including sumps, as defined in § 260.10, that serve as part of a secondary containment system to collect or contain releases of hazardous wastes are exempted from the requirements in § 265.193(a).

(c) Tanks, sumps, and other collection devices used in conjunction with drip pads, as defined in § 260.10 of this chapter and regulated under 40 CFR part 265 subpart W, must meet the requirements of this subpart.

[51 FR 25479, July 14, 1986, as amended at 53 FR 34087, Sept. 2, 1988; 55 FR 50486, Dec. 6, 1990; 58 FR 46050, Aug. 31, 1993; 70 FR 34585, June 14, 2005]

## § 265.191  Assessment of existing tank system's integrity.

(a) For each existing tank system that does not have secondary containment meeting the requirements of § 265.193, the owner or operator must determine that the tank system is not leaking or is unfit for use. Except as provided in paragraph (c) of this section, the owner or operator must obtain and keep on file at the facility a written assessment reviewed and certified by a qualified Professional Engineer in accordance with §270.11(d) of this chapter, that attests to the tank system's integrity by January 12, 1988.

(b) This assessment must determine that the tank system is adequately designed and has sufficient structural strength and compatibility with the waste(s) to be stored or treated to ensure that it will not collapse, rupture, or fail. At a minimum, this assessment must consider the following:

(1) Design standard(s), if available, according to which the tank and ancillary equipment were constructed;

(2) Hazardous characteristics of the waste(s) that have been or will be handled;

(3) Existing corrosion protection measures;

(4) Documented age of the tank system, if available, (otherwise, an estimate of the age); and

(5) Results of a leak test, internal inspection, or other tank integrity examination such that:

(i) For non-enterable underground tanks, this assessment must consist of a leak test that is capable of taking into account the effects of temperature variations, tank end deflection, vapor pockets, and high water table effects,

(ii) For other than non-enterable underground tanks and for ancillary equipment, this assessment must be either a leak test, as described above, or an internal inspection and/or other tank integrity examination certified by a qualified Professional Engineer in accordance with §270.11(d) of this chapter that addresses cracks, leaks, corrosion, and erosion.

[NOTE: The practices described in the American Petroleum Institute (API) Publication, Guide for Inspection of Refinery Equipment, Chapter XIII, "Atmospheric and Low-Pressure Storage Tanks," 4th edition, 1981, may be used, where applicable, as guidelines in conducting the integrity examination of an other than non-enterable underground tank system.]

(c) Tank systems that store or treat materials that become hazardous wastes subsequent to July 14, 1986 must conduct this assessment within 12 months after the date that the waste becomes a hazardous waste.

(d) If, as a result of the assessment conducted in accordance with paragraph (a) of this section, a tank system is found to be leaking or unfit for use, the owner or operator must comply with the requirements of §265.196.

[51 FR 25479, July 14, 1986, as amended at 71 FR 16910, Apr. 4, 2006]

§ 265.192 **Design and installation of new tank systems or components.**

(a) Owners or operators of new tank systems or components must ensure that the foundation, structural support, seams, connections, and pressure controls (if applicable) are adequately designed and that the tank system has sufficient structural strength, compatibility with the waste(s) to be stored or treated, and corrosion protection so that it will not collapse, rupture, or fail. The owner or operator must obtain a written assessment reviewed and certified by a qualified Professional Engineer in accordance with §270.11(d) of this chapter attesting that the system has sufficient structural integrity and is acceptable for the storing and treating of hazardous waste. This assessment must include the following information:

(1) Design standard(s) according to which the tank(s) and ancillary equipment is or will be constructed.

(2) Hazardous characteristics of the waste(s) to be handled.

(3) For new tank systems or components in which the external shell of a metal tank or any external metal component of the tank system is or will be in contact with the soil or with water, a determination by a corrosion expert of:

(i) Factors affecting the potential for corrosion, including but not limited to:

(A) Soil moisture content;

(B) Soil pH;

(C) Soil sulfides level;

(D) Soil resistivity;

(E) Structure to soil potential;

(F) Influence of nearby underground metal structures (e.g., piping);

(G) Stray electric current; and,

(H) Existing corrosion-protection measures (e.g., coating, cathodic protection), and

651

(ii) The type and degree of external corrosion protection that are needed to ensure the integrity of the tank system during the use of the tank system or component, consisting of one or more of the following:

(A) Corrosion-resistant materials of construction such as special alloys or fiberglass-reinforced plastic;

(B) Corrosion-resistant coating (such as epoxy or fiberglass) with cathodic protection (e.g., impressed current or sacrificial anodes); and

(C) Electrical isolation devices such as insulating joints and flanges.

NOTE: The practices described in the National Association of Corrosion Engineers (NACE) standard, "Recommended Practice (RP–02–85)—Control of External Corrosion on Metallic Buried, Partially Buried, or Submerged Liquid Storage Systems," and the American Petroleum Institute (API) Publication 1632, "Cathodic Protection of Underground Petroleum Storage Tanks and Piping Systems," may be used, where applicable, as guidelines in providing corrosion protection for tank systems.

(4) For underground tank system components that are likely to be affected by vehicular traffic, a determination of design or operational measures that will protect the tank system against potential damage; and

(5) Design considerations to ensure that:

(i) Tank foundations will maintain the load of a full tank;

(ii) Tank systems will be anchored to prevent flotation or dislodgement where the tank system is placed in a saturated zone, or is located within a seismic fault zone; and

(iii) Tank systems will withstand the effects of frost heave.

(b) The owner or operator of a new tank system must ensure that proper handling procedures are adhered to in order to prevent damage to the system during installation. Prior to covering, enclosing, or placing a new tank system or component in use, an independent, qualified installation inspector or a qualified Professional Engineer, either of whom is trained and experienced in the proper installation of tank systems, must inspect the system or component for the presence of any of the following items:

(1) Weld breaks;

(2) Punctures;

(3) Scrapes of protective coatings;

(4) Cracks;

(5) Corrosion;

(6) Other structural damage or inadequate construction or installation.

All discrepancies must be remedied before the tank system is covered, enclosed, or placed in use.

(c) New tank systems or components and piping that are placed underground and that are backfilled must be provided with a backfill material that is a noncorrosive, porous, homogeneous substance and that is carefully installed so that the backfill is placed completely around the tank and compacted to ensure that the tank and piping are fully and uniformly supported.

(d) All new tanks and ancillary equipment must be tested for tightness prior to being covered, enclosed or placed in use. If a tank system is found not to be tight, all repairs necessary to remedy the leak(s) in the system must be performed prior to the tank system being covered, enclosed, or placed in use.

(e) Ancillary equipment must be supported and protected against physical damage and excessive stress due to settlement, vibration, expansion or contraction.

NOTE: The piping system installation procedures described in American Petroleum Institute (API) Publication 1615 (November 1979), "Installation of Underground Petroleum Storage Systems," or ANSI Standard B31.3, "Petroleum Refinery System," may be used, where applicable, as guidelines for proper installation of piping systems.

(f) The owner or operator must provide the type and degree of corrosion protection necessary, based on the information provided under paragraph (a)(3) of this section, to ensure the integrity of the tank system during use of the tank system. The installation of a corrosion protection system that is field fabricated must be supervised by an independent corrosion expert to ensure proper installation.

(g) The owner or operator must obtain and keep on file at the facility written statements by those persons required to certify the design of the tank system and supervise the installation of the tank system in accordance with the requirements of paragraphs (b) through (f) of this section to attest that the tank system was properly designed and installed and that repairs, pursuant to paragraphs (b) and (d) of this section were performed. These written statements must also include the certification statement as required in §270.11(d) of this chapter.

[51 FR 25479, July 14, 1986; 51 FR 29430, Aug. 15, 1986; 71 FR 16910, Apr. 4, 2006]

## §265.193 Containment and detection of releases.

(a) In order to prevent the release of hazardous waste or hazardous constituents to the environment, secondary containment that meets the requirements of this section must be provided (except as provided in paragraphs (f) and (g) of this section):

(1) For all new and existing tank systems or components, prior to their being put into service.

(2) For tank systems that store or treat materials that become hazardous wastes, within 2 years of the hazardous waste listing, or when the tank system has reached 15 years of age, whichever comes later.

(b) Secondary containment systems must be:

(1) Designed, installed, and operated to prevent any migration of wastes or accumulated liquid out of the system to the soil, ground water, or surface water at any time during the use of the tank system; and

(2) Capable of detecting and collecting releases and accumulated liquids until the collected material is removed.

(c) To meet the requirements of paragraph (b) of this section, secondary containment systems must be at a minimum:

(1) Constructed of or lined with materials that are compatible with the waste(s) to be placed in the tank system and must have sufficient strength and thickness to prevent failure due to pressure gradients (including static head and external hydrological forces), physical contact with the waste to which they are exposed, climatic conditions, the stress of installation, and the stress of daily operation (including stresses from nearby vehicular traffic);

(2) Placed on a foundation or base capable of providing support to the secondary containment system and resist-

ance to pressure gradients above and below the system and capable of preventing failure due to settlement, compression, or uplift;

(3) Provided with a leak detection system that is designed and operated so that it will detect the failure of either the primary and secondary containment structure or any release of hazardous waste or accumulated liquid in the secondary containment system within 24 hours, or at the earliest practicable time if the existing detection technology or site conditions will not allow detection of a release within 24 hours;

(4) Sloped or otherwise designed or operated to drain and remove liquids resulting from leaks, spills, or precipitation. Spilled or leaked waste and accumulated precipitation must be removed from the secondary containment system within 24 hours, or in as timely a manner as is possible to prevent harm to human health or the environment, if removal of the released waste or accumulated precipitation cannot be accomplished within 24 hours.

NOTE: If the collected material is a hazardous waste under part 261 of this chapter, it is subject to management as a hazardous waste in accordance with all applicable requirements of parts 262 through 265 of this chapter. If the collected material is discharged through a point source to waters of the United States, it is subject to the requirements of sections 301, 304, and 402 of the Clean Water Act, as amended. If discharged to Publicly Owned Treatment Works (POTWs), it is subject to the requirements of section 307 of the Clear Water Act, as amended. If the collected material is released to the environment, it may be subject to the reporting requirements of 40 CFR part 302.

(d) Secondary containment for tanks must include one or more of the following devices:

(1) A liner (external to the tank);

(2) A vault;

(3) A double-walled tank; or

(4) An equivalent device as approved by the Regional Administrator.

(e) In addition to the requirements of paragraphs (b), (c), and (d) of this section, secondary containment systems must satisfy the following requirements:

(1) External liner systems must be:

(i) Designed or operated to contain 100 percent of the capacity of the largest tank within its boundary;

(ii) Designed or operated to prevent run-on or infiltration of precipitation into the secondary containment system unless the collection system has sufficient excess capacity to contain run-on or infiltration. Such additional capacity must be sufficient to contain precipitation from a 25-year, 24-hour rainfall event;

(iii) Free of cracks or gaps; and

(iv) Designed and installed to completely surround the tank and to cover all surrounding earth likely to come into contact with the waste if released from the tank(s) (i.e., capable of preventing lateral as well as vertical migration of the waste).

(2) Vault systems must be:

(i) Designed or operated to contain 100 percent of the capacity of the largest tank within its boundary;

(ii) Designed or operated to prevent run-on or infiltration of precipitation into the secondary containment system unless the collection system has sufficient excess capacity to contain run-on or infiltration. Such additional capacity must be sufficient to contain precipitation from a 25-year, 24-hour rainfall event;

(iii) Constructed with chemical-resistant water stops in place at all joints (if any);

(iv) Provided with an impermeable interior coating or lining that is compatible with the stored waste and that will prevent migration of waste into the concrete;

(v) Provided with a means to protect against the formation of and ignition of vapors within the vault, if the waste being stored or treated:

(A) Meets the definition of ignitable waste under § 261.21 of this chapter, or

(B) Meets the definition of reactive waste under § 261.23 of this chapter and may form an ignitable or explosive vapor; and

(vi) Provided with an exterior moisture barrier or be otherwise designed or operated to prevent migration of moisture into the vault if the vault is subject to hydraulic pressure.

(3) Double-walled tanks must be:

(i) Designed as an integral structure (i.e., an inner tank within an outer

shell) so that any release from the inner tank is contained by the outer shell;

(ii) Protected, if constructed of metal, from both corrosion of the primary tank interior and the external surface of the outer shell; and

(iii) Provided with a built-in, continuous leak detection system capable of detecting a release within 24 hours or at the earliest practicable time, if the owner or operator can demonstrate to the Regional Administrator, and the Regional Administrator concurs, that the existing leak detection technology or site conditions will not allow detection of a release within 24 hours.

NOTE: The provisions outlined in the Steel Tank Institute's (STI) "Standard for Dual Wall Underground Steel Storage Tank" may be used as guidelines for aspects of the design of underground steel double-walled tanks.

(f) Ancillary equipment must be provided with full secondary containment (e.g., trench, jacketing, double-walled piping) that meets the requirements of paragraphs (b) and (c) of this section except for:

(1) Aboveground piping (exclusive of flanges, joints, valves, and connections) that are visually inspected for leaks on a daily basis;

(2) Welded flanges, welded joints, and welded connections that are visually inspected for leaks on a daily basis;

(3) Sealless or magnetic coupling pumps and sealless valves, that are visually inspected for leaks on a daily basis; and

(4) Pressurized aboveground piping systems with automatic shut-off devices (e.g., excess flow check valves, flow metering shutdown devices, loss of pressure actuated shut-off devices) that are visually inspected for leaks on a daily basis.

(g) The owner or operator may obtain a variance from the requirements of this Section if the Regional Administrator finds, as a result of a demonstration by the owner or operator, either: that alternative design and operating practices, together with location characteristics, will prevent the migration of hazardous waste or hazardous constituents into the ground water or surface water at least as effectively as secondary containment during the active life of the tank system *or* that in the event of a release that does migrate to ground water or surface water, no substantial present or potential hazard will be posed to human health or the environment. New underground tank systems may not, per a demonstration in accordance with paragraph (g)(2) of this section, be exempted from the secondary containment requirements of this section. Application for a variance as allowed in paragraph (g) of this section does not waive compliance with the requirements of this subpart for new tank systems.

(1) In deciding whether to grant a variance based on a demonstration of equivalent protection of ground water and surface water, the Regional Administrator will consider:

(i) The nature and quantity of the waste;

(ii) The proposed alternate design and operation;

(iii) The hydrogeologic setting of the facility, including the thickness of soils between the tank system and ground water; and

(iv) All other factors that would influence the quality and mobility of the hazardous constituents and the potential for them to migrate to ground water or surface water.

(2) In deciding whether to grant a variance, based on a demonstration of no substantial present or potential hazard, the Regional Administrator will consider:

(i) The potential adverse effects on ground water, surface water, and land quality taking into account:

(A) The physical and chemical characteristics of the waste in the tank system, including its potential for migration,

(B) The hydrogeological characteristics of the facility and surrounding land,

(C) The potential for health risks caused by human exposure to waste constituents,

(D) The potential for damage to wildlife, crops, vegetation, and physical structures caused by exposure to waste constituents, and

(E) The persistence and permanence of the potential adverse effects;

(ii) The potential adverse effects of a release on ground-water quality, taking into account:

(A) The quantity and quality of ground water and the direction of ground-water flow,

(B) The proximity and withdrawal rates of water in the area,

(C) The current and future uses of ground water in the area, and

(D) The existing quality of ground water, including other sources of contamination and their cumulative impact on the ground-water quality;

(iii) The potential adverse effects of a release on surface water quality, taking into account:

(A) The quantity and quality of ground water and the direction of ground-water flow,

(B) The patterns of rainfall in the region,

(C) The proximity of the tank system to surface waters,

(D) The current and future uses of surface waters in the area and any water quality standards established for those surface waters, and

(E) The existing quality of surface water, including other sources of contamination and the cumulative impact on surface-water quality; and

(iv) The potential adverse effects of a release on the land surrounding the tank system, taking into account:

(A) The patterns of rainfall in the region, and

(B) The current and future uses of the surrounding land.

(3) The owner or operator of a tank system, for which a variance from secondary containment had been granted in accordance with the requirements of paragraph (g)(1) of this section, at which a release of hazardous waste has occurred from the primary tank system but has not migrated beyond the zone of engineering control (as established in the variance), must:

(i) Comply with the requirements of §265.196, except paragraph (d); and

(ii) Decontaminate or remove contaminated soil to the extent necessary to:

(A) Enable the tank system, for which the variance was granted, to resume operation with the capability for the detection of and response to releases at least equivalent to the capability it had prior to the release, and

(B) Prevent the migration of hazardous waste or hazardous constituents to ground water or surface water; and

(iii) If contaminated soil cannot be removed or decontaminated in accordance with paragraph (g)(3)(ii) of this section, comply with the requirements of § 265.197(b);

(4) The owner or operator of a tank system, for which a variance from secondary containment had been granted in accordance with the requirements of paragraph (g)(1) of this section, at which a release of hazardous waste has occurred from the primary tank system and has migrated beyond the zone of engineering control (as established in the variance), must:

(i) Comply with the requirements of § 265.196(a), (b), (c), and (d); and

(ii) Prevent the migration of hazardous waste or hazardous constituents to ground water or surface water, if possible, and decontaminate or remove contaminated soil. If contaminated soil cannot be decontaminated or removed, or if ground water has been contaminated, the owner or operator must comply with the requirements of § 265.197(b);

(iii) If repairing, replacing, or reinstalling the tank system, provide secondary containment in accordance with the requirements of paragraphs (a) through (f) of this section or reapply for a variance from secondary containment and meet the requirements for new tank systems in § 265.192 if the tank system is replaced. The owner or operator must comply with these requirements even if contaminated soil can be decontaminated or removed, and ground water or surface water has not been contaminated.

(h) The following procedures must be followed in order to request a variance from secondary containment:

(1) The Regional Administrator must be notified in writing by the owner or operator that he intends to conduct and submit a demonstration for a variance from secondary containment as allowed in paragraph (g) of this section according to the following schedule:

(i) For existing tank systems, at least 24 months prior to the date that

secondary containment must be provided in accordance with paragraph (a) of this section; and

(ii) For new tank systems, at least 30 days prior to entering into a contract for installation of the tank system.

(2) As part of the notification, the owner or operator must also submit to the Regional Administrator a description of the steps necessary to conduct the demonstration and a timetable for completing each of the steps. The demonstration must address each of the factors listed in paragraph (g)(1) or paragraph (g)(2) of this section.

(3) The demonstration for a variance must be completed and submitted to the Regional Administrator within 180 days after notifying the Regional Administrator of intent to conduct the demonstration.

(4) The Regional Administrator will inform the public, through a newspaper notice, of the availability of the demonstration for a variance. The notice shall be placed in a daily or weekly major local newspaper of general circulation and shall provide at least 30 days from the date of the notice for the public to review and comment on the demonstration for a variance. The Regional Administrator also will hold a public hearing, in response to a request or at his own discretion, whenever such a hearing might clarify one or more issues concerning the demonstration for a variance. Public notice of the hearing will be given at least 30 days prior to the date of the hearing and may be given at the same time as notice of the opportunity for the public to review and comment on the demonstration. These two notices may be combined.

(5) The Regional Administrator will approve or disapprove the request for a variance within 90 days of receipt of the demonstration from the owner or operator and will notify in writing the owner or operator and each person who submitted written comments or requested notice of the variance decision. If the demonstration for a variance is incomplete or does not include sufficient information, the 90-day time period will begin when the Regional Administrator receives a complete demonstration, including all information

necessary to make a final determination. If the public comment period in paragraph (h)(4) of this section is extended, the 90-day time period will be similarly extended.

(i) All tank systems, until such time as secondary containment meeting the requirements of this section is provided, must comply with the following:

(1) For non-enterable underground tanks, a leak test that meets the requirements of § 265.191(b)(5) must be conducted at least annually;

(2) For other than non-enterable underground tanks, and for all ancillary equipment, the owner or operator must either conduct a leak test as in paragraph (i)(1) of this section or an internal inspection or other tank integrity examination by a qualified Professional Engineer that addresses cracks, leaks, and corrosion or erosion at least annually. The owner or operator must remove the stored waste from the tank, if necessary, to allow the condition of all internal tank surfaces to be assessed.

NOTE: The practices described in the American Petroleum Institute (API) Publication Guide for Inspection of Refining Equipment, Chapter XIII, "Atmospheric and Low Pressure Storage Tanks," 4th edition, 1981, may be used, when applicable, as guidelines for assessing the overall condition of the tank system.

(3) The owner or operator must maintain on file at the facility a record of the results of the assessments conducted in accordance with paragraphs (i)(1) through (i)(3) of this section.

(4) If a tank system or component is found to be leaking or unfit-for-use as a result of the leak test or assessment in paragraphs (i)(1) through (i)(3) of this section, the owner or operator must comply with the requirements of § 265.196.

[51 FR 25479, July 14, 1986; 51 FR 29430, Aug. 15, 1986, as amended at 53 FR 34087, Sept. 2, 1988; 71 FR 16910, Apr. 4, 2006; 71 FR 40275, July 14, 2006]

### § 265.194 General operating requirements.

(a) Hazardous wastes or treatment reagents must not be placed in a tank system if they could cause the tank, its ancillary equipment, or the secondary containment system to rupture, leak, corrode, or otherwise fail.

(b) The owner or operator must use appropriate controls and practices to prevent spills and overflows from tank or secondary containment systems. These include at a minimum:

(1) Spill prevention controls (e.g., check valves, dry disconnect couplings);

(2) Overfill prevention controls (e.g., level sensing devices, high level alarms, automatic feed cutoff, or bypass to a standby tank); and

(3) Maintenance of sufficient freeboard in uncovered tanks to prevent overtopping by wave or wind action or by precipitation.

(c) The owner or operator must comply with the requirements of § 265.196 if a leak or spill occurs in the tank system.

[51 FR 25479, July 14, 1986, as amended at 71 FR 40275, July 14, 2006]

### § 265.195 Inspections.

(a) The owner or operator must inspect, where present, at least once each operating day, data gathered from monitoring and leak detection equipment (e.g., pressure or temperature gauges, monitoring wells) to ensure that the tank system is being operated according to its design.

NOTE: Section 265.15(c) requires the owner or operator to remedy any deterioration or malfunction he finds. Section 265.196 requires the owner or operator to notify the Regional Administrator within 24 hours of confirming a release. Also, 40 CFR part 302 may require the owner or operator to notify the National Response Center of a release.

(b) Except as noted under the paragraph (c) of this section, the owner or operator must inspect at least once each operating day:

(1) Overfill/spill control equipment (e.g., waste-feed cutoff systems, bypass systems, and drainage systems) to ensure that it is in good working order;

(2) Above ground portions of the tank system, if any, to detect corrosion or releases of waste; and

(3) The construction materials and the area immediately surrounding the externally accessible portion of the tank system, including the secondary

containment system (e.g., dikes) to detect erosion or signs of releases of hazardous waste (e.g., wet spots, dead vegetation).

(c) Owners or operators of tank systems that either use leak detection equipment to alert facility personnel to leaks, or implement established workplace practices to ensure leaks are promptly identified, must inspect at least weekly those areas described in paragraphs (b)(1) through (3) of this section. Use of the alternate inspection schedule must be documented in the facility's operating record. This documentation must include a description of the established workplace practices at the facility. .

(d) Performance Track member facilities may inspect on a less frequent basis, upon approval by the Director, but must inspect at least once each month. To apply for a less than weekly inspection frequency, the Performance Track member facility must follow the procedures described in § 265.15(b)(5).

(e) Ancillary equipment that is not provided with secondary containment, as described in § 265.193(f)(1) through (4), must be inspected at least once each operating day.

(f) The owner or operator must inspect cathodic protection systems, if present, according to, at a minimum, the following schedule to ensure that they are functioning properly:

(1) The proper operation of the cathodic protection system must be confirmed within six months after initial installation, and annually thereafter; and

(2) All sources of impressed current must be inspected and/or tested, as appropriate, at least bimonthly (i.e., every other month).

NOTE: The practices described in the National Association of Corrosion Engineers (NACE) standard, "Recommended Practice (RP–02–85)—Control of External Corrosion on Metallic Buried, Partially Buried, or Submerged Liquid Storage Systems," and the American Petroleum Institute (API) Publication 1632, "Cathodic Protection of Underground Petroleum Storage Tanks and Piping Systems," may be used, where applicable, as guidelines in maintaining and inspecting cathodic protection systems.

(g) The owner or operator must document in the operating record of the facility an inspection of those items in paragraphs (a) and (b) of this section.

[51 FR 25479, July 14, 1986; 51 FR 29430, Aug. 15, 1986, as amended at 71 FR 16910, Apr. 4, 2006]

§ 265.196  Response to leaks or spills and disposition of leaking or unfit-for-use tank systems.

A tank system or secondary containment system from which there has been a leak or spill, or which is unfit for use, must be removed from service immediately, and the owner or operator must satisfy the following requirements:

(a) *Cessation of use; prevent flow or addition of wastes.* The owner or operator must immediately stop the flow of hazardous waste into the tank system or secondary containment system and inspect the system to determine the cause of the release.

(b) *Removal of waste from tank system or secondary containment system.* (1) If the release was from the tank system, the owner or operator must, within 24 hours after detection of the leak or, if the owner or operator demonstrates that that is not possible, at the earliest practicable time remove as much of the waste as is necessary to prevent further release of hazardous waste to the environment and to allow inspection and repair of the tank system to be performed.

(2) If the release was to a secondary containment system, all released materials must be removed within 24 hours or in as timely a manner as is possible to prevent harm to human health and the environment.

(c) *Containment of visible releases to the environment.* The owner or operator must immediately conduct a visual inspection of the release and, based upon that inspection:

(1) Prevent further migration of the leak or spill to soils or surface water; and

(2) Remove, and properly dispose of, any visible contamination of the soil or surface water.

(d) *Notifications, reports.* (1) Any release to the environment, except as provided in paragraph (d)(2) of this section, must be reported to the Regional Administrator within 24 hours of detection. If the release has been reported pursuant to 40 CFR part 302, that report will satisfy this requirement.

(2) A leak or spill of hazardous waste that is:

(i) Less than or equal to a quantity of one (1) pound, and

(ii) Immediately contained and cleaned-up is exempted from the requirements of this paragraph.

(3) Within 30 days of detection of a release to the environment, a report containing the following information must be submitted to the Regional Administrator:

(i) Likely route of migration of the release;

(ii) Characteristics of the surrounding soil (soil composition, geology, hydrogeology, climate);

(iii) Results of any monitoring or sampling conducted in connection with the release, (if available). If sampling or monitoring data relating to the release are not available within 30 days, these data must be submitted to the Regional Administrator as soon as they become available;

(iv) Proximity to downgradient drinking water, surface water, and population areas; and

(v) Description of response actions taken or planned.

(e) *Provision of secondary containment, repair, or closure.* (1) Unless the owner or operator satisfies the requirements of paragraphs (e) (2) through (4) of this section, the tank system must be closed in accordance with §265.197.

(2) If the cause of the release was a spill that has not damaged the integrity of the system, the owner/operator may return the system to service as soon as the released waste is removed and repairs, if necessary, are made.

(3) If the cause of the release was a leak from the primary tank system into the secondary containment system, the system must be repaired prior to returning the tank system to service.

(4) If the source of the release was a leak to the environment from a component of a tank system without secondary containment, the owner/operator must provide the component of the system from which the leak occurred with secondary containment that satisfies the requirements of §265.193 before it can be returned to

service, unless the source of the leak is an aboveground portion of a tank system. If the source is an aboveground component that can be inspected visually, the component must be repaired and may be returned to service without secondary containment as long as the requirements of paragraph (f) of this section are satisfied. If a component is replaced to comply with the requirements of this subparagraph, that component must satisfy the requirements for new tank systems or components in §§265.192 and 265.193. Additionally, if a leak has occurred in any portion of a tank system component that is not readily accessible for visual inspection (e.g., the bottom of an inground or onground tank), the entire component must be provided with secondary containment in accordance with §265.193 prior to being returned to use.

(f) *Certification of major repairs.* If the owner/operator has repaired a tank system in accordance with paragraph (e) of this section, and the repair has been extensive (e.g., installation of an internal liner; repair of a ruptured primary containment or secondary containment vessel), the tank system must not be returned to service unless the owner/operator has obtained a certification by a qualified Professional Engineer in accordance with §270.11(d) that the repaired system is capable of handling hazardous wastes without release for the intended life of the system. This certification is to be placed in the operating record and maintained until closure of the facility.

NOTE: The Regional Administrator may, on the basis of any information received that there is or has been a release of hazardous waste or hazardous constituents into the environment, issue an order under RCRA section 3004(v), 3008(h), or 7003(a) requiring corrective action or such other response as deemed necessary to protect human health or the environment.

NOTE: See §265.15(c) for the requirements necessary to remedy a failure. Also, 40 CFR Part 302 requires the owner or operator to notify the National Response Center of a release of any "reportable quantity."

[51 FR 25479, July 14, 1986, as amended at 53 FR 34087, Sept. 2, 1988; 71 FR 16911, Apr. 4, 2006]

## § 265.197 Closure and post-closure care.

(a) At closure of a tank system, the owner or operator must remove or decontaminate all waste residues, contaminated containment system components (liners, etc.), contaminated soils, and structures and equipment contaminated with waste, and manage them as hazardous waste, unless § 261.3(d) of this Chapter applies. The closure plan, closure activities, cost estimates for closure, and financial responsibility for tank systems must meet all of the requirements specified in subparts G and H of this part.

(b) If the owner or operator demonstrates that not all contaminated soils can be practically removed or decontaminated as required in paragraph (a) of this section, then the owner or operator must close the tank system and perform post-closure care in accordance with the closure and post-closure care requirements that apply to landfills (§ 265.310). In addition, for the purposes of closure, post-closure, and financial responsibility, such a tank system is then considered to be a landfill, and the owner or operator must meet all of the requirements for landfills specified in subparts G and H of this part.

(c) If an owner or operator has a tank system which does not have secondary containment that meets the requirements of § 265.193(b) through (f) and which is not exempt from the secondary containment requirements in accordance with § 265.193(g), then,

(1) The closure plan for the tank system must include both a plan for complying with paragraph (a) of this section and a contingent plan for complying with paragraph (b) of this section.

(2) A contingent post-closure plan for complying with paragraph (b) of this section must be prepared and submitted as part of the permit application.

(3) The cost estimates calculated for closure and post-closure care must reflect the costs of complying with the contingent closure plan and the contingent post-closure plan, if these costs are greater than the costs of complying with the closure plan prepared for the

expected closure under paragraph (a) of this section.

(4) Financial assurance must be based on the cost estimates in paragraph (c)(3) of this section.

(5) For the purposes of the contingent closure and post-closure plans, such a tank system is considered to be a landfill, and the contingent plans must meet all of the closure, post-closure, and financial responsibility requirements for landfills under subparts G and H of this part.

[51 FR 25479, July 14, 1986, as amended at 71 FR 40275, July 14, 2006]

## § 265.198 Special requirements for ignitable or reactive wastes.

(a) Ignitable or reactive waste must not be placed in a tank system, unless:

(1) The waste is treated, rendered, or mixed before or immediately after placement in the tank system so that:

(i) The resulting waste, mixture, or dissolved material no longer meets the definition of ignitable or reactive waste under §§ 261.21 or 261.23 of this chapter; and

(ii) Section 265.17(b) is complied with; or

(2) The waste is stored or treated in such a way that it is protected from any material or conditions that may cause the waste to ignite or react; or

(3) The tank system is used solely for emergencies.

(b) The owner or operator of a facility where ignitable or reactive waste is stored or treated in tanks must comply with the requirements for the maintenance of protective distances between the waste management area and any public ways, streets, alleys, or an adjoining property line that can be built upon as required in Tables 2–1 through 2–6 of the National Fire Protection Association's "Flammable and Combustible Liquids Code," (1977 or 1981), (incorporated by reference, see § 260.11).

## § 265.199 Special requirements for incompatible wastes.

(a) Incompatible wastes, or incompatible waste and materials, must not be placed in the same tank system, unless § 265.17(b) is complied with.

(b) Hazardous waste must not be placed in a tank system that has not been decontaminated and that previously held an incompatible waste or material, unless §265.17(b) is complied with.

## §265.200 Waste analysis and trial tests.

In addition to performing the waste analysis required by §265.13, the owner or operator must, whenever a tank system is to be used to treat chemically or to store a hazardous waste that is substantially different from waste previously treated or stored in that tank system; or treat chemically a hazardous waste with a substantially different process than any previously used in that tank system:

(a) Conduct waste analyses and trial treatment or storage tests (e.g., bench-scale or pilot-plant scale tests); or

(b) Obtain written, documented information on similar waste under similar operating conditions to show that the proposed treatment or storage will meet the requirements of §265.194(a).

NOTE: Section 265.13 requires the waste analysis plan to include analyses needed to comply with §§265.198 and 265.199. Section 265.73 requires the owner or operator to place the results from each waste analysis and trial test, or the documented information, in the operating record of the facility.

## §265.201 Special requirements for generators of between 100 and 1,000 kg/mo that accumulate hazardous waste in tanks.

(a) The requirements of this section apply to small quantity generators of more than 100 kg but less than 1,000 kg of hazardous waste in a calendar month, that accumulate hazardous waste in tanks for less than 180 days (or 270 days if the generator must ship the waste greater than 200 miles), and do not accumulate over 6,000 kg on-site at any time.

(b) Generators of between 100 and 1,000 kg/mo hazardous waste must comply with the following general operating requirements:

(1) Treatment or storage of hazardous waste in tanks must comply with §265.17(b).

(2) Hazardous wastes or treatment reagents must not be placed in a tank if they could cause the tank or its inner liner to rupture, leak, corrode, or otherwise fail before the end of its intended life.

(3) Uncovered tanks must be operated to ensure at least 60 centimeters (2 feet) of freeboard, unless the tank is equipped with a containment structure (e.g., dike or trench), a drainage control system, or a diversion structure (e.g., standby tank) with a capacity that equals or exceeds the volume of the top 60 centimeters (2 feet) of the tank.

(4) Where hazardous waste is continuously fed into a tank, the tank must be equipped with a means to stop this inflow (e.g., waste feed cutoff system or by-pass system to a stand-by tank).

NOTE: These systems are intended to be used in the event of a leak or overflow from the tank due to a system failure (e.g., a malfunction in the treatment process, a crack in the tank, etc.).

(c) Except as noted in paragraph (d) of this section, generators who accumulate between 100 and 1,000 kg/mo of hazardous waste in tanks must inspect, where present:

(1) Discharge control equipment (e.g., waste feed cutoff systems, by-pass systems, and drainage systems) at least once each operating day, to ensure that it is in good working order;

(2) Data gathered from monitoring equipment (e.g., pressure and temperature gauges) at least once each operating day to ensure that the tank is being operated according to its design;

(3) The level of waste in the tank at least once each operating day to ensure compliance with §265.201(b)(3);

(4) The construction materials of the tank at least weekly to detect corrosion or leaking of fixtures or seams; and

(5) The construction materials of, and the area immediately surrounding, discharge confinement structures (e.g., dikes) at least weekly to detect erosion or obvious signs of leakage (e.g., wet spots or dead vegetation).

NOTE: As required by §265.15(c), the owner or operator must remedy any deterioration or malfunction he finds.

(d) Generators who accumulate between 100 and 1,000 kg/mo of hazardous waste in tanks or tank systems that have full secondary containment and that either use leak detection equipment to alert facility personnel to leaks, or implement established workplace practices to ensure leaks are promptly identified, must inspect at least weekly, where applicable, the areas identified in paragraphs (c)(1) through (5) of this section. Use of the alternate inspection schedule must be documented in the facility's operating record. This documentation must include a description of the established workplace practices at the facility.

(e) Performance Track member facilities may inspect on a less frequent basis, upon approval by the Director, but must inspect at least once each month. To apply for a less than weekly inspection frequency, the Performance Track member facility must follow the procedures described in § 265.15(b)(5).

(f) Generators of between 100 and 1,000 kg/mo accumulating hazardous waste in tanks must, upon closure of the facility, remove all hazardous waste from tanks, discharge control equipment, and discharge confinement structures.

NOTE: At closure, as throughout the operating period, unless the owner or operator can demonstrate, in accordance with § 261.3(c) or (d) of this chapter, that any solid waste removed from his tank is not a hazardous waste, the owner or operator becomes a generator of hazardous waste and must manage it in accordance with all applicable requirements of parts 262, 263, and 265 of this chapter.

(g) Generators of between 100 and 1,000 kg/mo must comply with the following special requirements for ignitable or reactive waste:

(1) Ignitable or reactive waste must not be placed in a tank, unless:

(i) The waste is treated, rendered, or mixed before or immediately after placement in a tank so that (A) the resulting waste, mixture, or dissolution of material no longer meets the definition of ignitable or reactive waste under § 261.21 or § 261.23 of this chapter, and (B) § 265.17(b) is complied with; or

(ii) The waste is stored or treated in such a way that it is protected from any material or conditions that may cause the waste to ignite or react; or

(iii) The tank is used solely for emergencies.

(2) The owner or operator of a facility which treats or stores ignitable or reactive waste in covered tanks must comply with the buffer zone requirements for tanks contained in Tables 2–1 through 2–6 of the National Fire Protection Association's "Flammable and Combustible Liquids Code," (1977 or 1981) (incorporated by reference, see § 260.11).

(h) Generators of between 100 and 1,000 kg/mo must comply with the following special requirements for incompatible wastes:

(1) Incompatible wastes, or incompatible wastes and materials, (see appendix V for examples) must not be placed in the same tank, unless § 265.17(b) is complied with.

(2) Hazardous waste must not be placed in an unwashed tank which previously held an incompatible waste or material, unless § 265.17(b) is complied with.

[51 FR 25479, July 14, 1986, as amended at 53 FR 34087, Sept. 2, 1988; 71 FR 16911, Apr. 4, 2006; 71 FR 40275, July 14, 2006]

§ 265.202  Air emission standards.

The owner or operator shall manage all hazardous waste placed in a tank in accordance with the applicable requirements of subparts AA, BB, and CC of this part.

[61 FR 59968, Nov. 25, 1996]

## Subpart K—Surface Impoundments

§ 265.220  Applicability.

The regulations in this subpart apply to owners and operators of facilities that use surface impoundments to treat, store, or dispose of hazardous waste, except as § 265.1 provides otherwise.

§ 265.221  Design and operating requirements.

(a) The owner or operator of each new surface impoundment unit, each lateral expansion of a surface impoundment unit, and each replacement of an existing surface impoundment unit must install two or more liners, and a leachate collection and removal system between the liners, and operate the leachate collection and removal system, in accordance with § 264.221(c), unless exempted under § 264.221(d), (e), or (f) of this Chapter.

(b) The owner or operator of each unit referred to in paragraph (a) of this section must notify the Regional Administrator at least sixty days prior to receiving waste. The owner or operator of each facility submitting notice must file a part B application within six months of the receipt of such notice.

(c) The owner or operator of any replacement surface impoundment unit is exempt from paragraph (a) of this section if:

(1) The existing unit was constructed in compliance with the design standards of § 3004(o)(1)(A)(i) and (o)(5) of the Resource Conservation and Recovery Act; and

(2) There is no reason to believe that the liner is not functioning as designed.

(d) The double liner requirement set forth in paragraph (a) of this section may be waived by the Regional Administrator for any monofill, if:

(1) The monofill contains only hazardous wastes from foundry furnace emission controls or metal casting molding sand, and such wastes do not contain constituents which would render the wastes hazardous for reasons other than the Toxicity Characteristic in § 261.24 of this chapter, with EPA Hazardous Waste Numbers D004 through D017; and

(2)(i)(A) The monofill has at least one liner for which there is no evidence that such liner is leaking. For the purposes of this paragraph the term "liner" means a liner designed, constructed, installed, and operated to prevent hazardous waste from passing into the liner at any time during the active life of the facility, or a liner designed, constructed, installed, and operated to prevent hazardous waste from migrating beyond the liner to adjacent subsurface soil, ground water, or surface water at any time during the active life of the facility. In the case of any surface impoundment which has been exempted from the requirements of paragraph (a) of this section on the basis of a liner designed, constructed, installed, and operated to prevent hazardous waste from passing beyond the liner, at the closure of such impoundment the owner or operator must remove or decontaminate all waste residues, all contaminated liner material,

and contaminated soil to the extent practicable. If all contaminated soil is not removed or decontaminated, the owner or operator of such impoundment must comply with appropriate post-closure requirements, including but not limited to ground-water monitoring and corrective action;

(B) The monofill is located more than one-quarter mile from an "underground source of drinking water" (as that term is defined in 40 CFR 270.2); and

(C) The monofill is in compliance with generally applicable ground-water monitoring requirements for facilities with permits under RCRA section 3005(c); or

(ii) The owner or operator demonstrates that the monofill is located, designed and operated so as to assure that there will be no migration of any hazardous constituent into ground water or surface water at any future time.

(e) In the case of any unit in which the liner and leachate collection system has been installed pursuant to the requirements of paragraph (a) of this section and in good faith compliance with paragraph (a) of this section and with guidance documents governing liners and leachate collection systems under paragraph (a) of this section, no liner or leachate collection system which is different from that which was so installed pursuant to paragraph (a) of this section will be required for such unit by the Regional Administrator when issuing the first permit to such facility, except that the Regional Administrator will not be precluded from requiring installation of a new liner when the Regional Administrator has reason to believe that any liner installed pursuant to the requirements of paragraph (a) of this section is leaking.

(f) A surface impoundment must maintain enough freeboard to prevent any overtopping of the dike by overfilling, wave action, or a storm. Except as provided in paragraph (b) of this section, there must be at least 60 centimeters (two feet) of freeboard.

(g) A freeboard level less than 60 centimeters (two feet) may be maintained if the owner or operator obtains certification by a qualified engineer that alternate design features or operating plans will, to the best of his knowledge and opinion, prevent overtopping of the dike. The certification, along with a written identification of alternate design features or operating plans preventing overtopping, must be maintained at the facility.

(h) Surface impoundments that are newly subject to RCRA section 3005(j)(1) due to the promulgation of additional listings or characteristics for the identification of hazardous waste must be in compliance with paragraphs (a), (c) and (d) of this section not later than 48 months after the promulgation of the additional listing or characteristic. This compliance period shall not be cut short as the result of the promulgation of land disposal prohibitions under part 268 of this chapter or the granting of an extension to the effective date of a prohibition pursuant to § 268.5 of this chapter, within this 48-month period.

[50 FR 16048, Apr. 23, 1985. Redesignated at 57 FR 3492, Jan. 29, 1992. 50 FR 28749, July 15, 1985, as amended at 55 FR 11876, Mar. 29, 1990; 57 FR 3492, Jan. 29, 1992; 57 FR 37267, Aug. 18, 1992; 71 FR 16911, Apr. 4, 2006; 71 FR 40275, July 14, 2006]

§ 265.222  Action leakage rate.

(a) The owner or operator of surface impoundment units subject to § 265.221(a) must submit a proposed action leakage rate to the Regional Administrator when submitting the notice required under § 265.221(b). Within 60 days of receipt of the notification, the Regional Administrator will: Establish an action leakage rate, either as proposed by the owner or operator or modified using the criteria in this section; or extend the review period for up to 30 days. If no action is taken by the Regional Administrator before the original 60 or extended 90 day review periods, the action leakage rate will be approved as proposed by the owner or operator.

(b) The Regional Administrator shall approve an action leakage rate for surface impoundment units subject to § 265.221(a). The action leakage rate is the maximum design flow rate that the leak detection system (LDS) can remove without the fluid head on the bottom liner exceeding 1 foot. The action leakage rate must include an adequate safety margin to allow for uncertainties in the design (e.g., slope, hydraulic conductivity, thickness of drainage material), construction, operation, and location of the LDS, waste and leachate characteristics, likelihood and amounts of other sources of liquids in the LDS, and proposed response actions (e.g., the action leakage rate must consider decreases in the flow capacity of the system over time resulting from siltation and clogging, rib layover and creep of synthetic components of the system, overburden pressures, etc.).

(c) To determine if the action leakage rate has been exceeded, the owner or operator must convert the weekly or monthly flow rate from the monitoring data obtained under § 265.226(b), to an average daily flow rate (gallons per acre per day) for each sump. Unless the Regional Administrator approves a different calculation, the average daily flow rate for each sump must be calculated weekly during the active life and closure period, and if the unit closes in accordance with § 265.228(a)(2), monthly during the post-closure care period when monthly monitoring is required under § 265.226(b).

[57 FR 3492, Jan. 29, 1992]

§ 265.223  Containment system.

All earthen dikes must have a protective cover, such as grass, shale, or rock, to minimize wind and water erosion and to preserve their structural integrity.

§ 265.224  Response actions.

(a) The owner or operator of surface impoundment units subject to § 265.221(a) must develop and keep on site until closure of the facility a response action plan. The response action plan must set forth the actions to be taken if the action leakage rate has been exceeded. At a minimum, the response action plan must describe the actions specified in paragraph (b) of this section.

(b) If the flow rate into the leak detection system exceeds the action leakage rate for any sump, the owner or operator must:

(1) Notify the Regional Administrator in writing of the exceedance within 7 days of the determination;

(2) Submit a preliminary written assessment to the Regional Administrator within 14 days of the determination, as to the amount of liquids, likely sources of liquids, possible location, size, and cause of any leaks, and short-term actions taken and planned;

(3) Determine to the extent practicable the location, size, and cause of any leak;

(4) Determine whether waste receipt should cease or be curtailed, whether any waste should be removed from the unit for inspection, repairs, or controls, and whether or not the unit should be closed;

(5) Determine any other short-term and longer-term actions to be taken to mitigate or stop any leaks; and

(6) Within 30 days after the notification that the action leakage rate has been exceeded, submit to the Regional Administrator the results of the analyses specified in paragraphs (b)(3), (4), and (5) of this section, the results of actions taken, and actions planned. Monthly thereafter, as long as the flow rate in the leak detection system exceeds the action leakage rate, the owner or operator must submit to the Regional Administrator a report summarizing the results of any remedial actions taken and actions planned.

(c) To make the leak and/or remediation determinations in paragraphs (b)(3), (4), and (5) of this section, the owner or operator must:

(1)(i) Assess the source of liquids and amounts of liquids by source,

(ii) Conduct a fingerprint, hazardous constituent, or other analyses of the liquids in the leak detection system to identify the source of liquids and possible location of any leaks, and the hazard and mobility of the liquid; and

(iii) Assess the seriousness of any leaks in terms of potential for escaping into the environment; or

(2) Document why such assessments are not needed.

[57 FR 3492, Jan. 29, 1992. Redesignated and amended at 71 FR 16911, Apr. 4, 2006; 71 FR 40275, July 14, 2006]

§ 265.225  **Waste analysis and trial tests.**

(a) In addition to the waste analyses required by § 265.13, whenever a surface impoundment is to be used to:

(1) Chemically treat a hazardous waste which is substantially different from waste previously treated in that impoundment; or

(2) Chemically treat hazardous waste with a substantially different process than any previously used in that impoundment; the owner or operator must, before treating the different waste or using the different process:

(i) Conduct waste analyses and trial treatment tests (e.g., bench scale or pilot plant scale tests); or

(ii) Obtain written, documented information on similar treatment of similar waste under similar operating conditions; to show that this treatment will comply with § 265.17(b).

[*Comment:* As required by § 265.13, the waste analysis plan must include analyses needed to comply with §§ 265.229 and 265.230. As required by § 265.73, the owner or operator must place the results from each waste analysis and trial test, or the documented information, in the operating record of the facility.]

(b) [Reserved]

§ 265.226  **Monitoring and inspection.**

(a) The owner or operator must inspect:

(1) The freeboard level at least once each operating day to ensure compliance with § 265.222, and

(2) The surface impoundment, including dikes and vegetation surrounding the dike, at least once a week to detect any leaks, deterioration, or failures in the impoundment.

(b)(1) An owner or operator required to have a leak detection system under § 265.221(a) must record the amount of liquids removed from each leak detection system sump at least once each week during the active life and closure period.

(2) After the final cover is installed, the amount of liquids removed from each leak detection system sump must be recorded at least monthly. If the liquid level in the sump stays below the pump operating level for two consecutive months, the amount of liquids in the sumps must be recorded at least quarterly. If the liquid level in the sump stays below the pump operating level for two consecutive quarters, the amount of liquids in the sumps must be recorded at least semi-annually. If at any time during the post-closure care

period the pump operating level is exceeded at units on quarterly or semiannual recording schedules, the owner or operator must return to monthly recording of amounts of liquids removed from each sump until the liquid level again stays below the pump operating level for two consecutive months.

(3) "Pump operating level" is a liquid level proposed by the owner or operator and approved by the Regional Administrator based on pump activation level, sump dimensions, and level that avoids backup into the drainage layer and minimizes head in the sump. The timing for submission and approval of the proposed "pump operating level" will be in accordance with § 265.222(a).

[*Comment:* As required by § 265.15(c), the owner or operator must remedy any deterioration or malfunction he finds.]

[45 FR 33232, May 19, 1980, as amended at 57 FR 3493, Jan. 29, 1992]

## § 265.227 [Reserved]

## § 265.228 Closure and post-closure care.

(a) At closure, the owner or operator must:

(1) Remove or decontaminate all waste residues, contaminated containment system components (liners, etc.), contaminated subsoils, and structures and equipment contaminated with waste and leachate, and manage them as hazardous waste unless § 261.3(d) of this chapter applies; or

(2) Close the impoundment and provide post-closure care for a landfill under subpart G and § 265.310, including the following:

(i) Eliminate free liquids by removing liquid wastes or solidifying the remaining wastes and waste residues;

(ii) Stabilize remaining wastes to a bearing capacity sufficient to support the final cover; and

(iii) Cover the surface impoundment with a final cover designed and constructed to:

(A) Provide long-term minimization of the migration of liquids through the closed impoundment;

(B) Function with minimum maintenance;

(C) Promote drainage and minimize erosion or abrasion of the cover;

(D) Accommodate settling and subsidence so that the cover's integrity is maintained; and

(E) Have a permeability less than or equal to the permeability of any bottom liner system or natural subsoils present.

(b) In addition to the requirements of subpart G, and § 265.310, during the post-closure care period, the owner or operator of a surface impoundment in which wastes, waste residues, or contaminated materials remain after closure in accordance with the provisions of paragraph (a)(2) of this section must:

(1) Maintain the integrity and effectiveness of the final cover, including making repairs to the cover as necessary to correct the effects of settling, subsidence, erosion, or other events;

(2) Maintain and monitor the leak detection system in accordance with §§ 264.221(c)(2)(iv) and (3) of this chapter and 265.226(b) and comply with all other applicable leak detection system requirements of this part;

(3) Maintain and monitor the groundwater monitoring system and comply with all other applicable requirements of subpart F of this part; and

(4) Prevent run-on and run-off from eroding or otherwise damaging the final cover.

[52 FR 8708, Mar. 19, 1987, as amended at 57 FR 3493, Jan. 29, 1992; 71 FR 40275, July 14, 2006]

## § 265.229 Special requirements for ignitable or reactive waste.

Ignitable or reactive waste must not be placed in a surface impoundment, unless the waste and impoundment satisfy all applicable requirements of 40 CFR part 268, and:

(a) The waste is treated, rendered, or mixed before or immediately after placement in the impoundment so that:

(1) The resulting waste, mixture, or dissolution of material no longer meets the definition of ignitable or reactive waste under § 261.21 or § 261.23 of this chapter; and

(2) Section 265.17(b) is complied with; or

(b)(1) The waste is managed in such a way that it is protected from any material or conditions which may cause it to ignite or react; and

(2) The owner or operator obtains a certification from a qualified chemist or engineer that, to the best of his knowledge and opinion, the design features or operating plans of the facility will prevent ignition or reaction; and

(3) The certification and the basis for it are maintained at the facility; or

(c) The surface impoundment is used solely for emergencies.

[50 FR 16048, Apr. 23, 1985, as amended at 55 FR 22685, June 1, 1990; 71 FR 40275, July 14, 2006]

§ 265.230 **Special requirements for incompatible wastes.**

Incompatible wastes, or incompatible wastes and materials, (see appendix V for examples) must not be placed in the same surface impoundment, unless § 265.17(b) is complied with.

§ 265.231 **Air emission standards.**

The owner or operator shall manage all hazardous waste placed in a surface impoundment in accordance with the applicable requirements of subparts BB and CC of this part.

[61 FR 59968, Nov. 25, 1996]

## Subpart L—Waste Piles

§ 265.250 **Applicability.**

The regulations in this subpart apply to owners and operators of facilities that treat or store hazardous waste in piles, except as § 265.1 provides otherwise. Alternatively, a pile of hazardous waste may be managed as a landfill under subpart N.

§ 265.251 **Protection from wind.**

The owner or operator of a pile containing hazardous waste which could be subject to dispersal by wind must cover or otherwise manage the pile so that wind dispersal is controlled.

§ 265.252 **Waste analysis.**

In addition to the waste analyses required by § 265.13, the owner or operator must analyze a representative sample of waste from each incoming movement before adding the waste to any existing pile, *unless* (1) The only wastes the facility receives which are amenable to piling are compatible with each other, or (2) the waste received is

compatible with the waste in the pile to which it is to be added. The analysis conducted must be capable of differentiating between the types of hazardous waste the owner or operator places in piles, so that mixing of incompatible waste does not inadvertently occur. The analysis must include a visual comparison of color and texture.

[*Comment:* As required by § 265.13, the waste analysis plan must include analyses needed to comply with §§ 265.256 and 265.257. As required by § 265.73, the owner or operator must place the results of this analysis in the operating record of the facility.]

(b) [Reserved]

§ 265.253 **Containment.**

If leachate or run-off from a pile is a hazardous waste, then either:

(a)(1) The pile must be placed on an impermeable base that is compatible with the waste under the conditions of treatment or storage;

(2) The owner or operator must design, construct, operate, and maintain a run-on control system capable of preventing flow onto the active portion of the pile during peak discharge from at least a 25-year storm;

(3) The owner or operator must design, construct, operate, and maintain a run-off management system to collect and control at least the water volume resulting from a 24-hour, 25-year storm; and

(4) Collection and holding facilities (e.g., tanks or basins) associated with run-on and run-off control systems must be emptied or otherwise managed expeditiously to maintain design capacity of the system; or

(b)(1) The pile must be protected from precipitation and run-on by some other means; and

(2) No liquids or wastes containing free liquids may be placed in the pile.

[*Comment:* If collected leachate or run-off is discharged through a point source to waters of the United States, it is subject to the requirements of section 402 of the Clean Water Act, as amended.]

[45 FR 33232, May 19, 1980, as amended at 47 FR 32367, July 26, 1982]

§ 265.254 Design and operating requirements.

The owner or operator of each new waste pile on which construction commences after January 29, 1992, each lateral expansion of a waste pile unit on which construction commences after July 29, 1992, and each such replacement of an existing waste pile unit that is to commence reuse after July 29, 1992 must install two or more liners and a leachate collection and removal system above and between such liners, and operate the leachate collection and removal systems, in accordance with § 264.251(c), unless exempted under § 264.251(d), (e), or (f), of this chapter; and must comply with the procedures of § 265.221(b). "Construction commences" is as defined in § 260.10 of this chapter under "existing facility".

[57 FR 3493, Jan. 29, 1992]

§ 265.255 Action leakage rates.

(a) The owner or operator of waste pile units subject to § 265.254 must submit a proposed action leakage rate to the Regional Administrator when submitting the notice required under § 265.254. Within 60 days of receipt of the notification, the Regional Administrator will: Establish an action leakage rate, either as proposed by the owner or operator or modified using the criteria in this section; or extend the review period for up to 30 days. If no action is taken by the Regional Administrator before the original 60 or extended 90 day review periods, the action leakage rate will be approved as proposed by the owner or operator.

(b) The Regional Administrator shall approve an action leakage rate for waste pile units subject to § 265.254. The action leakage rate is the maximum design flow rate that the leak detection system (LDS) can remove without the fluid head on the bottom liner exceeding 1 foot. The action leakage rate must include an adequate safety margin to allow for uncertainties in the design (e.g., slope, hydraulic conductivity, thickness of drainage material), construction, operation, and location of the LDS, waste and leachate characteristics, likelihood and amounts of other sources of liquids in the LDS, and proposed response actions (e.g., the action leakage rate must consider decreases in the flow capacity of the system over time resulting from siltation and clogging, rib layover and creep of synthetic components of the system, overburden pressures, etc.).

(c) To determine if the action leakage rate has been exceeded, the owner or operator must convert the weekly flow rate from the monitoring data obtained under § 265.260, to an average daily flow rate (gallons per acre per day) for each sump. Unless the Regional Administrator approves a different calculation, the average daily flow rate for each sump must be calculated weekly during the active life and closure period.

[57 FR 3493, Jan. 29, 1992, as amended at 71 FR 40275, July 14, 2006]

§ 265.256 Special requirements for ignitable or reactive waste.

(a) Ignitable or reactive waste must not be placed in a pile unless the waste and pile satisfy all applicable requirements of 40 CFR part 268, and:

(1) Addition of the waste to an existing pile (i) results in the waste or mixture no longer meeting the definition of ignitable or reactive waste under § 261.21 or § 261.23 of this chapter, and (ii) complies with § 265.17(b); or

(2) The waste is managed in such a way that it is protected from any material or conditions which may cause it to ignite or react.

(b) [Reserved]

[45 FR 33232, May 19, 1980, as amended at 55 FR 22685, June 1, 1990]

§ 265.257 Special requirements for incompatible wastes.

(a) Incompatible wastes, or incompatible wastes and materials, (see appendix V for examples) must not be placed in the same pile, unless § 265.17(b) is complied with.

(b) A pile of hazardous waste that is incompatible with any waste or other material stored nearby in other containers, piles, open tanks, or surface impoundments must be separated from the other materials, or protected from them by means of a dike, berm, wall, or other device.

[*Comment:* The purpose of this is to prevent fires, explosions, gaseous emissions, leaching, or other discharge of hazardous waste or hazardous waste constituents which could result from the contact or mixing of incompatible wastes or materials.]

(c) Hazardous waste must not be piled on the same area where incompatible wastes or materials were previously piled, unless that area has been decontaminated sufficiently to ensure compliance with § 265.17(b).

**§ 265.258 Closure and post-closure care.**

(a) At closure, the owner or operator must remove or decontaminate all waste residues, contaminated containment system components (liners, etc.), contaminated subsoils, and structures and equipment contaminated with waste and leachate, and manage them as hazardous waste unless § 261.3(d) of this chapter applies; or

(b) If, after removing or decontaminating all residues and making all reasonable efforts to effect removal or decontamination of contaminated components, subsoils, structures, and equipment as required in paragraph (a) of this section, the owner or operator finds that not all contaminated subsoils can be practicably removed or decontaminated, he must close the facility and perform post-closure care in accordance with the closure and post-closure requirements that apply to landfills (§ 265.310).

[47 FR 32368, July 26, 1982]

**§ 265.259 Response actions.**

(a) The owner or operator of waste pile units subject to § 265.254 must develop and keep on-site until closure of the facility a response action plan. The response action plan must set forth the actions to be taken if the action leakage rate has been exceeded. At a minimum, the response action plan must describe the actions specified in paragraph (b) of this section.

(b) If the flow rate into the leak determination system exceeds the action leakage rate for any sump, the owner or operator must:

(1) Notify the Regional Administrator in writing of the exceedance within 7 days of the determination;

(2) Submit a preliminary written assessment to the Regional Administrator within 14 days of the determination, as to the amount of liquids, likely sources of liquids, possible location, size, and cause of any leaks, and short-term actions taken and planned;

(3) Determine to the extent practicable the location, size, and cause of any leak;

(4) Determine whether waste receipts should cease or be curtailed, whether any waste should be removed from the unit for inspection, repairs, or controls, and whether or not the unit should be closed;

(5) Determine any other short-term and longer-term actions to be taken to mitigate or stop any leaks; and

(6) Within 30 days after the notification that the action leakage rate has been exceeded, submit to the Regional Administrator the results of the analyses specified in paragraphs (b)(3), (4), and (5) of this section, the results of actions taken, and actions planned. Monthly thereafter, as long as the flow rate in the leak detection system exceeds the action leakage rate, the owner or operator must submit to the Regional Administrator a report summarizing the results of any remedial actions taken and actions planned.

(c) To make the leak and/or remediation determinations in paragraphs (b)(3), (4), and (5) of this section, the owner or operator must:

(1)(i) Assess the source of liquids and amounts of liquids by source,

(ii) Conduct a fingerprint, hazardous constituent, or other analyses of the liquids in the leak detection system to identify the source of liquids and possible location of any leaks, and the hazard and mobility of the liquid; and

(iii) Assess the seriousness of any leaks in terms of potential for escaping into the environment; or

(2) Document why such assessments are not needed.

[57 FR 3494, Jan. 29, 1992, as amended at 71 FR 16911, Apr. 4, 2006; 71 FR 40275, July 14, 2006]

**§ 265.260 Monitoring and inspection.**

An owner or operator required to have a leak detection system under § 265.254 must record the amount of liquids removed from each leak detection

669

system sump at least once each week during the active life and closure period.

[57 FR 3494, Jan. 29, 1992]

## Subpart M—Land Treatment

**§ 265.270 Applicability.**

The regulations in this subpart apply to owners and operators of hazardous waste land treatment facilities, except as § 265.1 provides otherwise.

**§ 265.271 [Reserved]**

**§ 265.272 General operating requirements.**

(a) Hazardous waste must not be placed in or on a land treatment facility unless the waste can be made less hazardous or nonhazardous by degradation, transformation, or immobilization processes occurring in or on the soil.

(b) The owner or operator must design, construct, operate, and maintain a run-on control system capable of preventing flow onto the active portions of the facility during peak discharge from at least a 25-year storm.

(c) The owner or operator must design, construct, operate, and maintain a run-off management system capable of collecting and controlling a water volume at least equivalent to a 24-hour, 25-year storm.

(d) Collection and holding facilities (e.g., tanks or basins) associated with run-on and run-off control systems must be emptied or otherwise managed expeditiously after storms to maintain design capacity of the system.

(e) If the treatment zone contains particulate matter which may be subject to wind dispersal, the owner or operator must manage the unit to control wind dispersal.

[45 FR 33232, May 19, 1980, as amended at 47 FR 32368, July 26, 1982; 50 FR 16048, Apr. 23, 1985]

**§ 265.273 Waste analysis.**

In addition to the waste analyses required by § 265.13, before placing a hazardous waste in or on a land treatment facility, the owner or operator must:

(a) Determine the concentrations in the waste of any substances which

equal or exceed the maximum concentrations contained in Table 1 of § 261.24 of this chapter that cause a waste to exhibit the Toxicity Characteristic;

(b) For any waste listed in part 261, subpart D, of this chapter, determine the concentrations of any substances which caused the waste to be listed as a hazardous waste; and

(c) If food chain crops are grown, determine the concentrations in the waste of each of the following constituents: arsenic, cadmium, lead, and mercury, *unless* the owner or operator has written, documented data that show that the constituent is not present.

[*Comment:* Part 261 of this chapter specifies the substances for which a waste is listed as a hazardous waste. As required by § 265.13, the waste analysis plan must include analyses needed to comply with §§ 265.281 and 265.282. As required by § 265.73, the owner or operator must place the results from each waste analysis, or the documented information, in the operating record of the facility.]

[45 FR 33232, May 19, 1980, as amended at 55 FR 11876, Mar. 29, 1990]

**§§ 265.274–265.275 [Reserved]**

**§ 265.276 Food chain crops.**

(a) An owner or operator of a hazardous waste land treatment facility on which food chain crops are being grown, or have been grown and will be grown in the future, must notify the Regional Administrator within 60 days after the effective date of this part.

[*Comment:* The growth of food chain crops at a facility which has never before been used for this purpose is a significant change in process under § 122.72(c) of this chapter. Owners or operators of such land treatment facilities who propose to grow food chain crops after the effective date of this part must comply with § 122.72(c) of this chapter.]

(b)(1) Food chain crops must not be grown on the treated area of a hazardous waste land treatment facility unless the owner or operator can demonstrate, based on field testing, that any arsenic, lead, mercury, or other constituents identified under § 265.273(b):

(i) Will not be transferred to the food portion of the crop by plant uptake or direct contact, and will not otherwise be ingested by food chain animals (e.g., by grazing); or

(ii) Will not occur in greater concentrations in the crops grown on the land treatment facility than in the same crops grown on untreated soils under similar conditions in the same region.

(2) The information necessary to make the demonstration required by paragraph (b)(1) of this section must be kept at the facility and must, at a minimum:

(i) Be based on tests for the specific waste and application rates being used at the facility; and

(ii) Include descriptions of crop and soil characteristics, sample selection criteria, sample size determination, analytical methods, and statistical procedures.

(c) Food chain crops must not be grown on a land treatment facility receiving waste that contains cadmium unless all requirements of paragraphs (c)(1) (i) through (iii) of this section or all requirements of paragraphs (c)(2) (i) through (iv) of this section are met.

(1)(i) The pH of the waste and soil mixture is 6.5 or greater at the time of each waste application, except for waste containing cadmium at concentrations of 2 mg/kg (dry weight) or less;

(ii) The annual application of cadmium from waste does not exceed 0.5 kilograms per hectare (kg/ha) on land used for production of tobacco, leafy vegetables, or root crops grown for human consumption. For other food chain crops, the annual cadmium application rate does not exceed:

| Time period | Annual Cd application rate (kg/ha) |
|---|---|
| Present to June 30, 1984 | 2.0 |
| July 1, 1984 to December 31, 1986 | 1.25 |
| Beginning January 1, 1987 | 0.5 |

(iii) The cumulative application of cadmium from waste does not exceed the levels in either paragraph (c)(1)(iii)(A) or (B) of this section.

(A)

| Soil caption exchange capacity (meq/100g) | Maximum cumulative application (kg/ha) | |
|---|---|---|
| | Background soil pH less than 6.5 | Background soil pH greater than 6.5 |
| Less than 5 | 5 | 5 |
| 5 to 15 | 5 | 10 |
| Greater than 15 | 5 | 20 |

(B) For soils with a background pH of less than 6.5, the cumulative cadmium application rate does not exceed the levels below: *Provided*, that the pH of the waste and soil mixture is adjusted to and maintained at 6.5 or greater whenever food chain crops are grown.

| Soil caption exchange capacity (meq/100g) | Maximum cumulative application (kg/ha) |
|---|---|
| Less than 5 | 5 |
| 5 to 15 | 10 |
| Greater than 15 | 20 |

(2)(i) The only food chain crop produced is animal feed.

(ii) The pH of the waste and soil mixture is 6.5 or greater at the time of waste application or at the time the crop is planted, whichever occurs later, and this pH level is maintained whenever food chain crops are grown.

(iii) There is a facility operating plan which demonstrates how the animal feed will be distributed to preclude ingestion by humans. The facility operating plan describes the measures to be taken to safeguard against possible health hazards from cadmium entering the food chain, which may result from alternative land uses.

(iv) Future property owners are notified by a stipulation in the land record or property deed which states that the property has received waste at high cadmium application rates and that food chain crops must not be grown except in compliance with paragraph (c)(2) of this section.

[*Comment:* As required by §265.73, if an owner or operator grows food chain crops on his land treatment facility, he must place the information developed in this section in the operating record of the facility.]

[45 FR 33232, May 19, 1980, as amended at 47 FR 32368, July 26, 1982; 48 FR 14295, Apr. 1, 1983]

§ 265.277 [Reserved]

§ 265.278 Unsaturated zone (zone of aeration) monitoring.

(a) The owner or operator must have in writing, and must implement, an unsaturated zone monitoring plan which is designed to:

(1) Detect the vertical migration of hazardous waste and hazardous waste constituents under the active portion of the land treatment facility, and

(2) Provide information on the background concentrations of the hazardous waste and hazardous waste constituents in similar but untreated soils nearby; this background monitoring must be conducted before or in conjunction with the monitoring required under paragraph (a)(1) of this section.

(b) The unsaturated zone monitoring plan must include, at a minimum:

(1) Soil monitoring using soil cores, and

(2) Soil-pore water monitoring using devices such as lysimeters.

(c) To comply with paragraph (a)(1) of this section, the owner or operator must demonstrate in his unsaturated zone monitoring plan that:

(1) The depth at which soil and soil-pore water samples are to be taken is below the depth to which the waste is incorporated into the soil;

(2) The number of soil and soil-pore water samples to be taken is based on the variability of:

(i) The hazardous waste constituents (as identified in § 265.273(a) and (b)) in the waste and in the soil; and

(ii) The soil type(s); and

(3) The frequency and timing of soil and soil-pore water sampling is based on the frequency, time, and rate of waste application, proximity to ground water, and soil permeability.

(d) The owner or operator must keep at the facility his unsaturated zone monitoring plan, and the rationale used in developing this plan.

(e) The owner or operator must analyze the soil and soil-pore water samples for the hazardous waste constituents that were found in the waste during the waste analysis under § 265.273 (a) and (b).

[Comment: As required by § 265.73, all data and information developed by the owner or oper-ator under this section must be placed in the operating record of the facility.]

§ 265.279 Recordkeeping.

The owner or operator must include hazardous waste application dates and rates in the operating record required under § 265.73.

[47 FR 32368, July 26, 1982]

§ 265.280 Closure and post-closure.

(a) In the closure plan under § 265.112 and the post-closure plan under § 265.118, the owner or operator must address the following objectives and indicate how they will be achieved:

(1) Control of the migration of hazardous waste and hazardous waste constituents from the treated area into the ground water;

(2) Control of the release of contaminated run-off from the facility into surface water;

(3) Control of the release of airborne particulate contaminants caused by wind erosion; and

(4) Compliance with § 265.276 concerning the growth of food-chain crops.

(b) The owner or operator must consider at least the following factors in addressing the closure and post-closure care objectives of paragraph (a) of this section:

(1) Type and amount of hazardous waste and hazardous waste constituents applied to the land treatment facility;

(2) The mobility and the expected rate of migration of the hazardous waste and hazardous waste constituents;

(3) Site location, topography, and surrounding land use, with respect to the potential effects of pollutant migration (e.g., proximity to ground water, surface water and drinking water sources);

(4) Climate, including amount, frequency, and pH of precipitation;

(5) Geological and soil profiles and surface and subsurface hydrology of the site, and soil characteristics, including cation exchange capacity, total organic carbon, and pH;

(6) Unsaturated zone monitoring information obtained under § 265.278; and

(7) Type, concentration, and depth of migration of hazardous waste constituents in the soil as compared to their background concentrations.

(c) The owner or operator must consider at least the following methods in addressing the closure and post-closure care objectives of paragraph (a) of this section:

(1) Removal of contaminated soils;

(2) Placement of a final cover, considering:

(i) Functions of the cover (e.g., infiltration control, erosion and run-off control, and wind erosion control); and

(ii) Characteristics of the cover, including material, final surface contours, thickness, porosity and permeability, slope, length of run of slope, and type of vegetation on the cover; and

(3) Monitoring of ground water.

(d) In addition to the requirements of subpart G of this part, during the closure period the owner or operator of a land treatment facility must:

(1) Continue unsaturated zone monitoring in a manner and frequency specified in the closure plan, except that soil pore liquid monitoring may be terminated 90 days after the last application of waste to the treatment zone;

(2) Maintain the run-on control system required under §265.272(b);

(3) Maintain the run-off management system required under §265.272(c); and

(4) Control wind dispersal of particulate matter which may be subject to wind dispersal.

(e) For the purpose of complying with §265.115, when closure is completed the owner or operator may submit to the Regional Administrator certification both by the owner or operator and by an independent, qualified soil scientist, in lieu of a qualified Professional Engineer, that the facility has been closed in accordance with the specifications in the approved closure plan.

(f) In addition to the requirements of §265.117, during the post-closure care period the owner or operator of a land treatment unit must:

(1) Continue soil-core monitoring by collecting and analyzing samples in a manner and frequency specified in the post-closure plan;

(2) Restrict access to the unit as appropriate for its post-closure use;

(3) Assure that growth of food chain crops complies with §265.276; and

(4) Control wind dispersal of hazardous waste.

[45 FR 33232, May 19, 1980, as amended at 47 FR 32368, July 26, 1982; 71 FR 16911, Apr. 4, 2006; 71 FR 40275, July 14, 2006]

### §265.281 Special requirements for ignitable or reactive waste.

The owner or operator must not apply ignitable or reactive waste to the treatment zone unless the waste and treatment zone meet all applicable requirements of 40 CFR part 268, and:

(a) The waste is immediately incorporated into the soil so that:

(1) The resulting waste, mixture, or dissolution of material no longer meets the definition of ignitable or reactive waste under §261.21 or §261.23 of this chapter; and

(2) Section 264.17(b) is complied with; or

(b) The waste is managed in such a way that it is protected from any material or conditions which may cause it to ignite or react.

[47 FR 32368, July 26, 1982, as amended at 55 FR 22686, June 1, 1990; 71 FR 40275, July 14, 2006]

### §265.282 Special requirements for incompatible wastes.

Incompatible wastes, or incompatible wastes and materials (see appendix V for examples), must not be placed in the same land treatment area, unless §265.17(b) is complied with.

## Subpart N—Landfills

### §265.300 Applicability.

The regulations in this subpart apply to owners and operators of facilities that dispose of hazardous waste in landfills, except as §265.1 provides otherwise. A waste pile used as a disposal facility is a landfill and is governed by this subpart.

### §265.301 Design and operating requirements.

(a) The owner or operator of each new landfill unit, each lateral expansion of a landfill unit, and each replacement of an existing landfill unit must install two or more liners and a leachate collection and removal system above and between such liners, and operate the leachate collection and removal system, in accordance with § 264.301(c), unless exempted under § 264.301(d), (e), or (f) of this chapter.

(b) The owner or operator of each unit referred to in paragraph (a) of this section must notify the Regional Administrator at least sixty days prior to receiving waste. The owner or operator of each facility submitting notice must file a part B application within six months of the receipt of such notice.

(c) The owner or operator of any replacement landfill unit is exempt from paragraph (a) of this section if:

(1) The existing unit was constructed in compliance with the design standards of section 3004(o)(1)(A)(i) and (o)(5) of the Resource Conservation and Recovery Act; and

(2) There is no reason to believe that the liner is not functioning as designed.

(d) The double liner requirement set forth in paragraph (a) of this section may be waived by the Regional Administrator for any monofill, if:

(1) The monofill contains only hazardous wastes from foundry furnace emission controls or metal casting molding sand, and such wastes do not contain constituents which would render the wastes hazardous for reasons other than the Toxicity Characteristic in § 261.24 of this chapter, with EPA Hazardous Waste Numbers D004 through D017; and

(2)(i)(A) The monofill has at least one liner for which there is no evidence that such liner is leaking;

(B) The monofill is located more than one-quarter mile from an "underground source of drinking water" (as that term is defined in 40 CFR 270.2); and

(C) The monofill is in compliance with generally applicable ground-water monitoring requirements for facilities with permits under RCRA section 3005(c); or

(ii) The owner or operator demonstrates that the monofill is located, designed and operated so as to assure that there will be no migration of any hazardous constituent into ground water or surface water at any future time.

(e) In the case of any unit in which the liner and leachate collection system has been installed pursuant to the requirements of paragraph (a) of this section and in good faith compliance with paragraph (a) of this section and with guidance documents governing liners and leachate collection systems under paragraph (a) of this section, no liner or leachate collection system which is different from that which was so installed pursuant to paragraph (a) of this section will be required for such unit by the Regional Administrator when issuing the first permit to such facility, except that the Regional Administrator will not be precluded from requiring installation of a new liner when the Regional Administrator has reason to believe that any liner installed pursuant to the requirements of paragraph (a) of this section is leaking.

(f) The owner or operator must design, construct, operate, and maintain a run-on control system capable of preventing flow onto the active portion of the landfill during peak discharge from at least a 25-year storm.

(g) The owner or operator must design, construct, operate and maintain a run-off management system to collect and control at least the water volume resulting from a 24-hour, 25-year storm.

(h) Collection and holding facilities (e.g., tanks or basins) associated with run-on and run-off control systems must be emptied or otherwise managed expeditiously after storms to maintain design capacity of the system.

(i) The owner or operator of a landfill containing hazardous waste which is subject to dispersal by wind must cover or otherwise manage the landfill so that wind dispersal of the hazardous waste is controlled.

[Comment: As required by § 265.13, the waste analysis plan must include analyses needed to comply with §§ 265.312, 265.313, and 265.314. As required by § 265.73, the owner or operator must place the results of these analyses in the operating record of the facility.]

[45 FR 33232, May 19, 1980, as amended at 47 FR 32368, July 26, 1982; 50 FR 18374, Apr. 30, 1985. Redesignated from § 265.302 at 57 FR 3494, Jan. 29, 1992; 50 FR 28750, July 15, 1985, as amended at 57 FR 3494, Jan. 29, 1992; 57 FR 30658, July 10, 1992; 71 FR 16911, Apr. 4, 2006; 71 FR 40275, July 14, 2006]

§ 265.302   Action leakage rate.

(a) The owner or operator of landfill units subject to § 265.301(a) must submit a proposed action leakage rate to the Regional Administrator when submitting the notice required under § 265.301(b). Within 60 days of receipt of the notification, the Regional Administrator will: Establish an action leakage rate, either as proposed by the owner or operator or modified using the criteria in this section; or extend the review period for up to 30 days. If no action is taken by the Regional Administrator before the original 60 or extended 90 day review periods, the action leakage rate will be approved as proposed by the owner or operator.

(b) The Regional Administrator shall approve an action leakage rate for land fill units subject to § 265.301(a). The action leakage rate is the maximum design flow rate that the leak detection system (LDS) can remove without the fluid head on the bottom liner exceeding 1 foot. The action leakage rate must include an adequate safety margin to allow for uncertainties in the design (e.g., slope, hydraulic conductivity, thickness of drainage material), construction, operation, and location of the LDS, waste and leachate characteristics, likelihood and amounts of other sources of liquids in the LDS, and proposed response actions (e.g., the action leakage rate must consider decreases in the flow capacity of the system over time resulting from siltation and clogging, rib layover and creep of synthetic components of the system, overburden pressures, etc.).

(c) To determine if the action leakage rate has been exceeded, the owner or operator must convert the weekly or monthly flow rate from the monitoring data obtained under § 265.304 to an average daily flow rate (gallons per acre per day) for each sump. Unless the Regional Administrator approves a different calculation, the average daily flow rate for each sump must be calculated weekly during the active life and closure period, and monthly during the post-closure care period when monthly monitoring is required under § 265.304(b).

[57 FR 3494, Jan. 29, 1992, as amended at 71 FR 40276, July 14, 2006]

§ 265.303   Response actions.

(a) The owner or operator of landfill units subject to § 265.301(a) must develop and keep on site until closure of the facility a response action plan. The response action plan must set forth the actions to be taken if the action leakage rate has been exceeded. At a minimum, the response action plan must describe the actions specified in paragraph (b) of this section.

(b) If the flow rate into the leak detection system exceeds the action leakage rate for any sump, the owner or operator must:

(1) Notify the Regional Administrator in writing of the exceedance within 7 days of the determination;

(2) Submit a preliminary written assessment to the Regional Administrator within 14 days of the determination, as to the amount of liquids, likely sources of liquids, possible location, size, and cause of any leaks, and short-term actions taken and planned;

(3) Determine to the extent practicable the location, size, and cause of any leak;

(4) Determine whether waste receipt should cease or be curtailed, whether any waste should be removed from the unit for inspection, repairs, or controls, and whether or not the unit should be closed;

(5) Determine any other short-term and longer-term actions to be taken to mitigate or stop any leaks; and

(6) Within 30 days after the notification that the action leakage rate has been exceeded, submit to the Regional Administrator the results of the analyses specified in paragraphs (b)(3), (4), and (5) of this section, the results of actions taken, and actions planned. Monthly thereafter, as long as the flow rate in the leak detection system exceeds the action leakage rate, the owner or operator must submit to the Regional Administrator a report summarizing the results of any remedial actions taken and actions planned.

(c) To make the leak and/or remediation determinations in paragraphs (b)(3), (4), and (5) of this section, the owner or operator must:

(1)(i) Assess the source of liquids and amounts of liquids by source,

(ii) Conduct a fingerprint, hazardous constituent, or other analyses of the liquids in the leak detection system to identify the source of liquids and possible location of any leaks, and the hazard and mobility of the liquid; and

(iii) Assess the seriousness of any leaks in terms of potential for escaping into the environment; or

(2) Document why such assessments are not needed.

[57 FR 3494, Jan. 29, 1992, as amended at 71 FR 16912, Apr. 4, 2006; 71 FR 40276, July 14, 2006]

### § 265.304 Monitoring and inspection.

(a) An owner or operator required to have a leak detection system under § 265.301(a) must record the amount of liquids removed from each leak detection system sump at least once each week during the active life and closure period.

(b) After the final cover is installed, the amount of liquids removed from each leak detection system sump must be recorded at least monthly. If the liquid level in the sump stays below the pump operating level for two consecutive months, the amount of liquids in the sumps must be recorded at least quarterly. If the liquid level in the sump stays below the pump operating level for two consecutive quarters, the amount of liquids in the sumps must be recorded at least semi-annually. If at any time during the post-closure care period the pump operating level is exceeded at units on quarterly or semi-annual recording schedules, the owner or operator must return to monthly recording of amounts of liquids removed from each sump until the liquid level again stays below the pump operating level for two consecutive months.

(c) "Pump operating level" is a liquid level proposed by the owner or operator and approved by the Regional Administrator based on pump activation level, sump dimensions, and level that avoids backup into the drainage layer and minimizes head in the sump. The timing for submission and approval of the proposed "pump operating level" will be in accordance with § 265.302(a).

[57 FR 3495, Jan. 29, 1992]

### §§ 265.305–265.308 [Reserved]

### § 265.309 Surveying and recordkeeping.

The owner or operator of a landfill must maintain the following items in the operating record required in § 265.73:

(a) On a map, the exact location and dimensions, including depth, of each cell with respect to permanently surveyed benchmarks; and

(b) The contents of each cell and the approximate location of each hazardous waste type within each cell.

### § 265.310 Closure and post-closure care.

(a) At final closure of the landfill or upon closure of any cell, the owner or operator must cover the landfill or cell with a final cover designed and constructed to:

(1) Provide long-term minimization of migration of liquids through the closed landfill;

(2) Function with minimum maintenance;

(3) Promote drainage and minimize erosion or abrasion of the cover;

(4) Accommodate settling and subsidence so that the cover's integrity is maintained; and

(5) Have a permeability less than or equal to the permeability of any bottom liner system or natural subsoils present.

(b) After final closure, the owner or operator must comply with all post-closure requirements contained in §§ 265.117 through 265.120 including maintenance and monitoring throughout the post-closure care period. The owner or operator must:

(1) Maintain the integrity and effectiveness of the final cover, including making repairs to the cover as necessary to correct the effects of settling, subsidence, erosion, or other events;

(2) Maintain and monitor the leak detection system in accordance with §§ 264.301(c)(3)(iv) and (4) of this chapter and 265.304(b), and comply with all other applicable leak detection system requirements of this part;

(3) Maintain and monitor the groundwater monitoring system and comply with all other applicable requirements of subpart F of this part;

(4) Prevent run-on and run-off from eroding or otherwise damaging the final cover; and

(5) Protect and maintain surveyed benchmarks used in complying with § 265.309.

[50 FR 16048, Apr. 23, 1985, as amended at 57 FR 3495, Jan. 29, 1992]

**§ 265.311 [Reserved]**

**§ 265.312 Special requirements for ignitable or reactive waste.**

(a) Except as provided in paragraph (b) of this section, and in § 265.316, ignitable or reactive waste must not be placed in a landfill, unless the waste and landfill meets all applicable requirements of 40 CFR part 268, and:

(1) The resulting waste, mixture, or dissolution of material no longer meets the definition of ignitable or reactive waste under § 261.21 or § 261.23 of this chapter; and

(2) Section 265.17(b) is complied with.

(b) Except for prohibited wastes which remain subject to treatment standards in subpart D of part 268, ignitable wastes in containers may be landfilled without meeting the requirements of paragraph (a) of this section, provided that the wastes are disposed of in such a way that they are protected from any material or conditions which may cause them to ignite. At a minimum, ignitable wastes must be disposed of in non-leaking containers which are carefully handled and placed so as to avoid heat, sparks, rupture, or any other condition that might cause ignition of the wastes; must be covered daily with soil or other non-combustible material to minimize the potential for ignition of the wastes; and must not be disposed of in cells that contain or will contain other wastes which may generate heat sufficient to cause ignition of the waste.

[47 FR 32368, July 26, 1982, as amended at 55 FR 22686, June 1, 1990; 71 FR 40276, July 14, 2006]

**§ 265.313 Special requirements for incompatible wastes.**

Incompatible wastes, or incompatible wastes and materials, (see appendix V for examples) must not be placed in the same landfill cell, unless § 265.17(b) is complied with.

**§ 265.314 Special requirements for bulk and containerized liquids.**

(a) The placement of bulk or non-containerized liquid hazardous waste or hazardous waste containing free liquids (whether or not sorbents have been added) in any landfill is prohibited.

(b) Containers holding free liquids must not be placed in a landfill unless:

(1) All free-standing liquid,

(i) has been removed by decanting, or other methods,

(ii) has been mixed with sorbent or solidified so that free-standing liquid is no longer observed; or

(iii) had been otherwise eliminated; or

(2) The container is very small, such as an ampule; or

(3) The container is designed to hold free liquids for use other than storage, such as a battery or capacitor; or

(4) The container is a lab pack as defined in § 265.316 and is disposed of in accordance with § 265.316.

(c) To demonstrate the absence or presence of free liquids in either a containerized or a bulk waste, the following test must be used: Method 9095B (Paint Filter Liquids Test) as described in "Test Methods for Evaluating Solid Waste, Physical/Chemical Methods," EPA Publication SW-846, as incorporated by reference in § 260.11 of this chapter.

(d) The date for compliance with paragraph (a) of this section is November 19, 1981. The date for compliance with paragraph (c) of this section is March 22, 1982.

(e) Sorbents used to treat free liquids to be disposed of in landfills must be nonbiodegradable. Nonbiodegradable sorbents are: materials listed or described in paragraph (e)(1) of this section; materials that pass one of the tests in paragraph (e)(2) of this section; or materials that are determined by EPA to be nonbiodegradable through the Part 260 petition process.

(1) Nonbiodegradable sorbents. (i) Inorganic minerals, other inorganic materials, and elemental carbon (e.g., aluminosilicates, clays, smectites, Fuller's earth, bentonite, calcium bentonite, montmorillonite, calcined montmorillonite, kaolinite, micas (illite), vermiculites, zeolites; calcium

carbonate (organic free limestone); oxides/hydroxides, alumina, lime, silica (sand), diatomaceous earth; perlite (volcanic glass); expanded volcanic rock; volcanic ash; cement kiln dust; fly ash; rice hull ash; activated charcoal/activated carbon); or

(ii) High molecular weight synthetic polymers (e.g., polyethylene, high density polyethylene (HDPE), polypropylene, polystyrene, polyurethane, polyacrylate, polynorborene, polyisobutylene, ground synthetic rubber, cross-linked allylstyrene and tertiary butyl copolymers). This does not include polymers derived from biological material or polymers specifically designed to be degradable; or

(iii) Mixtures of these nonbiodegradable materials.

(2) Tests for nonbiodegradable sorbents. (i) The sorbent material is determined to be nonbiodegradable under ASTM Method G21–70 (1984a)—Standard Practice for Determining Resistance of Synthetic Polymer Materials to Fungi; or

(ii) The sorbent material is determined to be nonbiodegradable under ASTM Method G22–76 (1984b)—Standard Practice for Determining Resistance of Plastics to Bacteria; or

(iii) The sorbent material is determined to be non-biodegradable under OECD test 301B: [$CO_2$ Evolution (Modified Sturm Test)].

(f) The placement of any liquid which is not a hazardous waste in a landfill is prohibited unless the owner or operator of such landfill demonstrates to the Regional Administrator or the Regional Administrator determines that:

(1) The only reasonably available alternative to the placement in such landfill is placement in a landfill or unlined surface impoundment, whether or not permitted or operating under interim status, which contains, or may reasonably be anticipated to contain, hazardous waste; and

(2) Placement in such owner or operator's landfill will not present a risk of contamination of any "underground source of drinking water" (as that term is defined in 40 CFR 270.2).

[45 FR 33232, May 19, 1980, as amended at 47 FR 12318, Mar. 22, 1982; 47 FR 32369, July 26, 1982; 50 FR 18374, Apr. 30, 1985; 50 FR 28750, July 15, 1985; 51 FR 19177, May 28, 1986; 57 FR 54461, Nov. 18, 1992; 58 FR 46050, Aug. 31, 1993; 60 FR 35705, July 11, 1995; 70 FR 34585, June 14, 2005; 71 FR 16912, Apr. 4, 2006; 71 FR 40276, July 14, 2006; 75 FR 13006, Mar. 18, 2010]

§ 265.315  Special requirements for containers.

Unless they are very small, such as an ampule, containers must be either:

(a) At least 90 percent full when placed in the landfill; or

(b) Crushed, shredded, or similarly reduced in volume to the maximum practical extent before burial in the landfill.

[50 FR 16048, Apr. 23, 1985]

§ 265.316  Disposal of small containers of hazardous waste in overpacked drums (lab packs).

Small containers of hazardous waste in overpacked drums (lab packs) may be placed in a landfill if the following requirements are met:

(a) Hazardous waste must be packaged in non-leaking inside containers. The inside containers must be of a design and constructed of a material that will not react dangerously with, be decomposed by, or be ignited by the waste held therein. Inside containers must be tightly and securely sealed. The inside containers must be of the size and type specified in the Department of Transportation (DOT) hazardous materials regulations (49 CFR parts 173, 178 and 179), if those regulations specify a particular inside container for the waste.

(b) The inside containers must be overpacked in an open head DOT-specification metal shipping container (49 CFR parts 178 and 179) of no more than 416-liter (110 gallon) capacity and surrounded by, at a minimum, a sufficient quantity of sorbent material, determined to be nonbiodegradable in accordance with § 265.314(e), to completely sorb all of the liquid contents of the inside containers. The metal outer container must be full after it has been packed with inside containers and sorbent material.

(c) The sorbent material used must not be capable of reacting dangerously with, being decomposed by, or being ignited by the contents of the inside containers in accordance with §265.17(b).

(d) Incompatible wastes, as defined in §260.10 of this chapter, must not be placed in the same outside container.

(e) Reactive waste, other than cyanide- or sulfide-bearing waste as defined in §261.23(a)(5) of this chapter, must be treated or rendered non-reactive prior to packaging in accordance with paragraphs (a) through (d) of this section. Cyanide- and sulfide-bearing reactive waste may be packaged in accordance with paragraphs (a) through (d) of this section without first being treated or rendered non-reactive.

(f) Such disposal is in compliance with the requirements of 40 CFR part 268. Persons who incinerate lab packs according to the requirements in 40 CFR 268.42(c)(1) may use fiber drums in place of metal outer containers. Such fiber drums must meet the DOT specifications in 49 CFR 173.12 and be overpacked according to the requirements in paragraph (b) of this section.

[46 FR 56596, Nov. 17, 1981, as amended at 55 FR 22686, June 1, 1990; 57 FR 54461, Nov. 18, 1992; 71 FR 40276, July 14, 2006; 75 FR 13006, Mar. 18, 2010]

## Subpart O—Incinerators

SOURCE: 46 FR 7680, Jan. 23, 1981, unless otherwise noted.

### §265.340  Applicability.

(a) The regulations of this subpart apply to owners and operators of hazardous waste incinerators (as defined in §260.10 of this chapter), except as §265.1 provides otherwise.

(b) *Integration of the MACT standards.* (1) Except as provided by paragraphs (b)(2) and (b)(3) of this section, the standards of this part no longer apply when an owner or operator demonstrates compliance with the maximum achievable control technology (MACT) requirements of part 63, subpart EEE, of this chapter by conducting a comprehensive performance test and submitting to the Administrator a Notification of Compliance under §§63.1207(j) and 63.1210(d) of this chapter documenting compliance with the requirements of part 63, subpart EEE, of this chapter.

(2) The MACT standards do not replace the closure requirements of §264.351 or the applicable requirements of subparts A through H, BB and CC of this part.

(3) Section 265.345 generally prohibiting burning of hazardous waste during startup and shutdown remains in effect if you elect to comply with §270.235(b)(1)(i) of this chapter to minimize emissions of toxic compounds from startup and shutdown.

(c) Owners and operators of incinerators burning hazardous waste are exempt from all of the requirements of this subpart, except §265.351 (Closure), provided that the owner or operator has documented, in writing, that the waste would not reasonably be expected to contain any of the hazardous constituents listed in part 261, appendix VIII, of this chapter, and such documentation is retained at the facility, if the waste to be burned is:

(1) Listed as a hazardous waste in part 261, subpart D, of this chapter solely because it is ignitable (Hazard Code I), corrosive (Hazard Code C), or both; or

(2) Listed as a hazardous waste in part 261, subpart D, of this chapter solely because it is reactive (Hazard Code R) for characteristics other than those listed in §261.23(a) (4) and (5), and will not be burned when other hazardous wastes are present in the combustion zone; or

(3) A hazardous waste solely because it possesses the characteristic of ignitability, corrosivity, or both, as determined by the tests for characteristics of hazardous wastes under part 261, subpart C, of this chapter; or

(4) A hazardous waste solely because it possesses the reactivity characteristics described by §261.23(a) (1), (2), (3), (6), (7), or (8) of this chapter, and will not be burned when other hazardous wastes are present in the combustion zone.

[47 FR 27533, June 24, 1982 and 50 FR 666, Jan. 4, 1985, as amended at 50 FR 49203, Nov. 29, 1985; 56 FR 7208, Feb. 21, 1991; 64 FR 53075, Sept. 30, 1999; 67 FR 6816, Feb. 13, 2002; 70 FR 59575, Oct. 12, 2005]

## § 265.341 Waste analysis.

In addition to the waste analyses required by § 265.13, the owner or operator must sufficiently analyze any waste which he has not previously burned in his incinerator to enable him to establish steady state (normal) operating conditions (including waste and auxiliary fuel feed and air flow) and to determine the type of pollutants which might be emitted. At a minimum, the analysis must determine:

(a) Heating value of the waste;

(b) Halogen content and sulfur content in the waste; and

(c) Concentrations in the waste of lead and mercury, unless the owner or operator has written, documented data that show that the element is not present.

[*Comment:* As required by § 265.73, the owner or operator must place the results from each waste analysis, or the documented information, in the operating record of the facility.]

## §§ 265.342–265.344 [Reserved]

## § 265.345 General operating requirements.

During start-up and shut-down of an incinerator, the owner or operator must not feed hazardous waste unless the incinerator is at steady state (normal) conditions of operation, including steady state operating temperature and air flow.

## § 265.346 [Reserved]

## § 265.347 Monitoring and inspections.

The owner or operator must conduct, as a minimum, the following monitoring and inspections when incinerating hazardous waste:

(a) Existing instruments which relate to combustion and emission control must be monitored at least every 15 minutes. Appropriate corrections to maintain steady state combustion conditions must be made immediately either automatically or by the operator. Instruments which relate to combustion and emission control would normally include those measuring waste feed, auxiliary fuel feed, air flow, incinerator temperature, scrubber flow, scrubber pH, and relevant level controls.

(b) The complete incinerator and associated equipment (pumps, valves, conveyors, pipes, etc.) must be inspected at least daily for leaks, spills, and fugitive emissions, and all emergency shutdown controls and system alarms must be checked to assure proper operation.

[46 FR 7678, Jan. 23, 1981, as amended at 47 FR 27533, June 24, 1982]

## §§ 265.348–265.350 [Reserved]

## § 265.351 Closure.

At closure, the owner or operator must remove all hazardous waste and hazardous waste residues (including but not limited to ash, scrubber waters, and scrubber sludges) from the incinerator.

[*Comment:* At closure, as throughout the operating period, unless the owner or operator can demonstrate, in accordance with § 261.3(d) of this chapter, that the residue removed from his incinerator is not a hazardous waste, the owner or operator becomes a generator of hazardous waste and must manage it in accordance with all applicable requirements of parts 262 through 266 of this chapter.]

## § 265.352 Interim status incinerators burning particular hazardous wastes.

(a) Owners or operators of incinerators subject to this subpart may burn EPA Hazardous Wastes FO20, FO21, FO22, FO23, FO26, or FO27 if they receive a certification from the Assistant Administrator for Solid Waste and Emergency Response that they can meet the performance standards of subpart O of part 264 when they burn these wastes.

(b) The following standards and procedures will be used in determining whether to certify an incinerator:

(1) The owner or operator will submit an application to the Assistant Administrator for Solid Waste and Emergency Response containing applicable information in §§ 270.19 and 270.62 demonstrating that the incinerator can meet the performance standards in subpart O of part 264 when they burn these wastes.

(2) The Assistant Administrator for Solid Waste and Emergency Response will issue a tentative decision as to whether the incinerator can meet the

performance standards in subpart O of part 264. Notification of this tentative decision will be provided by newspaper advertisement and radio broadcast in the jurisdiction where the incinerator is located. The Assistant Administrator for Solid Waste and Emergency Response will accept comment on the tentative decision for 60 days. The Assistant Administrator for Solid Waste and Emergency Response also may hold a public hearing upon request or at his discretion.

(3) After the close of the public comment period, the Assistant Administrator for Solid Waste and Emergency Response will issue a decision whether or not to certify the incinerator.

[50 FR 2005, Jan. 14, 1985]

§§265.353–265.369 [Reserved]

## Subpart P—Thermal Treatment

§265.370 Other thermal treatment.

The regulations in this subpart apply to owners or operators of facilities that thermally treat hazardous waste in devices other than enclosed devices using controlled flame combustion, except as §265.1 provides otherwise. Thermal treatment in enclosed devices using controlled flame combustion is subject to the requirements of subpart O if the unit is an incinerator, and subpart H of part 266, if the unit is a boiler or an industrial furnace as defined in §260.10.

[50 FR 666, Jan. 4, 1985, as amended at 56 FR 32692, July 17, 1991]

§§265.371–265.372 [Reserved]

§265.373 General operating requirements.

Before adding hazardous waste, the owner or operator must bring his thermal treatment process to steady state (normal) conditions of operation—including steady state operating temperature—using auxiliary fuel or other means, unless the process is a non-continuous (batch) thermal treatment process which requires a complete thermal cycle to treat a discrete quantity of hazardous waste.

§265.374 [Reserved]

§265.375 Waste analysis.

In addition to the waste analyses required by §265.13, the owner or operator must sufficiently analyze any waste which he has not previously treated in his thermal process to enable him to establish steady state (normal) or other appropriate (for a non-continuous process) operating conditions (including waste and auxiliary fuel feed) and to determine the type of pollutants which might be emitted. At a minimum, the analysis must determine:

(a) Heating value of the waste;

(b) Halogen content and sulfur content in the waste; and

(c) Concentrations in the waste of lead and mercury, *unless* the owner or operator has written, documented data that show that the element is not present.

[Comment: As required by §265.73, the owner or operator must place the results from each waste analysis, or the documented information, in the operating record of the facility.]

§265.376 [Reserved]

§265.377 Monitoring and inspections.

(a) The owner or operator must conduct, as a minimum, the following monitoring and inspections when thermally treating hazardous waste:

(1) Existing instruments which relate to temperature and emission control (if an emission control device is present) must be monitored at least every 15 minutes. Appropriate corrections to maintain steady state or other appropriate thermal treatment conditions must be made immediately either automatically or by the operator. Instruments which relate to temperature and emission control would normally include those measuring waste feed, auxiliary fuel feed, treatment process temperature, and relevant process flow and level controls.

(2) The stack plume (emissions), where present, must be observed visually at least hourly for normal appearance (color and opacity). The operator must immediately make any indicated operating corrections necessary to return any visible emissions to their normal appearance.

681

(3) The complete thermal treatment process and associated equipment (pumps, valves, conveyors, pipes, etc.) must be inspected at least daily for leaks, spills, and fugitive emissions, and all emergency shutdown controls and system alarms must be checked to assure proper operation.

(b) [Reserved]

## §§ 265.378–265.380　[Reserved]

## § 265.381　Closure.

At closure, the owner or operator must remove all hazardous waste and hazardous waste residues (including, but not limited to, ash) from the thermal treatment process or equipment.

[Comment: At closure, as throughout the operating period, unless the owner or operator can demonstrate, in accordance with § 261.3 (c) or (d) of this chapter, that any solid waste removed from his thermal treatment process or equipment is not a hazardous waste, the owner or operator becomes a generator of hazardous waste and must manage it in accordance with all applicable requirements of parts 262, 263, and 265 of this chapter.]

## § 265.382　Open burning; waste explosives.

Open burning of hazardous waste is prohibited except for the open burning and detonation of waste explosives. Waste explosives include waste which has the potential to detonate and bulk military propellants which cannot safely be disposed of through other modes of treatment. Detonation is an explosion in which chemical transformation passes through the material faster than the speed of sound (0.33 kilometers/second at sea level). Owners or operators choosing to open burn or detonate waste explosives must do so in accordance with the following table and in a manner that does not threaten human health or the environment.

| Pounds of waste explosives or propellants | Minimum distance from open burning or detonation to the property of others |
|---|---|
| 0 to 100 | 204 meters (670 feet). |
| 101 to 1,000 | 380 meters (1,250 feet). |
| 1,001 to 10,000 | 530 meters (1,730 feet). |
| 10,001 to 30,000 | 690 meters (2,260 feet). |

## § 265.383　Interim status thermal treatment devices burning particular hazardous waste.

(a) Owners or operators of thermal treatment devices subject to this subpart may burn EPA Hazardous Wastes FO20, FO21, FO22, FO23, FO26, or FO27 if they receive a certification from the Assistant Administrator for Solid Waste and Emergency Response that they can meet the performance standards of subpart O of part 264 when they burn these wastes.

(b) The following standards and procedures will be used in determining whether to certify a thermal treatment unit:

(1) The owner or operator will submit an application to the Assistant Administrator for Solid Waste and Emergency Response containing the applicable information in §§ 270.19 and 270.62 demonstrating that the thermal treatment unit can meet the performance standard in subpart O of part 264 when they burn these wastes.

(2) The Assistant Administrator for Solid Waste and Emergency Response will issue a tentative decision as to whether the thermal treatment unit can meet the performance standards in subpart O of part 264. Notification of this tentative decision will be provided by newspaper advertisement and radio broadcast in the jurisdiction where the thermal treatment device is located. The Assistant Administrator for Solid Waste and Emergency Response will accept comment on the tentative decision for 60 days. The Assistant Administrator for Solid Waste and Emergency Response also may hold a public hearing upon request or at his discretion.

(3) After the close of the public comment period, the Assistant Administrator for Solid Waste and Emergency Response will issue a decision whether or not to certify the thermal treatment unit.

[50 FR 2005, Jan. 14, 1985]

## Subpart Q—Chemical, Physical, and Biological Treatment

### § 265.400　Applicability.

The regulations in this subpart apply to owners and operators of facilities which treat hazardous wastes by chemical, physical, or biological methods in other than tanks, surface impoundments, and land treatment facilities, except as §265.1 provides otherwise. Chemical, physical, and biological treatment of hazardous waste in tanks, surface impoundments, and land treatment facilities must be conducted in accordance with subparts J, K, and M, respectively.

### §265.401 General operating requirements.

(a) Chemical, physical, or biological treatment of hazardous waste must comply with §265.17(b).

(b) Hazardous wastes or treatment reagents must not be placed in the treatment process or equipment if they could cause the treatment process or equipment to rupture, leak, corrode, or otherwise fail before the end of its intended life.

(c) Where hazardous waste is continuously fed into a treatment process or equipment, the process or equipment must be equipped with a means to stop this inflow (e.g., a waste feed cut-off system or by-pass system to a standby containment device).

[*Comment:* These systems are intended to be used in the event of a malfunction in the treatment process or equipment.]

### §265.402 Waste analysis and trial tests.

(a) In addition to the waste analysis required by §265.13, whenever:

(1) A hazardous waste which is substantially different from waste previously treated in a treatment process or equipment at the facility is to be treated in that process or equipment, or

(2) A substantially different process than any previously used at the facility is to be used to chemically treat hazardous waste;

the owner or operator must, before treating the different waste or using the different process or equipment:

(i) Conduct waste analyses and trial treatment tests (e.g., bench scale or pilot plant scale tests); or

(ii) Obtain written, documented information on similar treatment of similar waste under similar operating conditions;

to show that this proposed treatment will meet all applicable requirements of §265.401 (a) and (b).

[(b) [Reserved]

[*Comment:* As required by §265.13, the waste analysis plan must include analyses needed to comply with §§265.405 and 265.406. As required by §265.73, the owner or operator must place the results from each waste analysis and trial test, or the documented information, in the operating record of the facility.]

### §265.403 Inspections.

(a) The owner or operator of a treatment facility must inspect, where present:

(1) Discharge control and safety equipment (e.g., waste feed cut-off systems, by-pass systems, drainage systems, and pressure relief systems) at least once each operating day, to ensure that it is in good working order;

(2) Data gathered from monitoring equipment (e.g., pressure and temperature gauges), at least once each operating day, to ensure that the treatment process or equipment is being operated according to its design;

(3) The construction materials of the treatment process or equipment, at least weekly, to detect corrosion or leaking of fixtures or seams; and

(4) The construction materials of, and the area immediately surrounding, discharge confinement structures (e.g., dikes), at least weekly, to detect erosion or obvious signs of leakage (e.g., wet spots or dead vegetation).

(b) [Reserved]

[*Comment:* As required by §265.15(c), the owner or operator must remedy any deterioration or malfunction he finds.]

### §265.404 Closure.

At closure, all hazardous waste and hazardous waste residues must be removed from treatment processes or equipment, discharge control equipment, and discharge confinement structures.

[*Comment:* At closure, as throughout the operating period, unless the owner or operator can demonstrate, in accordance with §261.3 (c) or (d) of this chapter, that any solid waste removed from his treatment process or equipment is not a hazardous waste, the owner or operator becomes a generator of hazardous waste and must manage it in accordance with all applicable requirements of parts 262, 263, and 265 of this chapter.]

**§ 265.405 Special requirements for ignitable or reactive waste.**

(a) Ignitable or reactive waste must not be placed in a treatment process or equipment unless:

(1) The waste is treated, rendered, or mixed before or immediately after placement in the treatment process or equipment so that (i) the resulting waste, mixture, or dissolution of material no longer meets the definition of ignitable or reactive waste under § 261.21 or 261.23 of this chapter, and (ii) § 265.17(b) is complied with; or

(2) The waste is treated in such a way that it is protected from any material or conditions which may cause the waste to ignite or react.

(b) [Reserved]

[45 FR 33232, May 19, 1980, as amended at 71 FR 40276, July 14, 2006]

**§ 265.406 Special requirements for incompatible wastes.**

(a) Incompatible wastes, or incompatible wastes and materials, (see appendix V for examples) must not be placed in the same treatment process or equipment, unless § 265.17(b) is complied with.

(b) Hazardous waste must not be placed in unwashed treatment equipment which previously held an incompatible waste or material, unless § 265.17(b) is complied with.

## Subpart R—Underground Injection

**§ 265.430 Applicability.**

Except as § 265.1 provides otherwise:

(a) The owner or operator of a facility which disposes of hazardous waste by underground injection is excluded from the requirements of subparts G and H of this part.

(b) The requirements of this subpart apply to owners and operators of wells used to dispose of hazardous waste which are classified as Class I under § 144.6(a) of this chapter and which are classified as Class IV under § 144.6(d) of this chapter.

[*Comment:* In addition to the requirements of subparts A through E of this part, the owner or operator of a facility which disposes of hazardous waste by underground injection ultimately must comply with the requirements of §§ 265.431 through 265.437. These sections are reserved at this time. The Agency will propose regulations that would establish those requirements.]

[45 FR 33232, May 19, 1980, as amended at 48 FR 30115, June 30, 1983]

## Subparts S-V [Reserved]

## Subpart W—Drip Pads

Source: 55 FR 50486, Dec. 6, 1990, unless otherwise noted.

**§ 265.440 Applicability.**

(a) The requirements of this subpart apply to owners and operators of facilities that use new or existing drip pads to convey treated wood drippage, precipitation, and/or surface water run-off to an associated collection system. Existing drip pads are those constructed before December 6, 1990 and those for which the owner or operator has a design and has entered into binding financial or other agreements for construction prior to December 6, 1990. All other drip pads are new drip pads. The requirement at § 265.443(b)(3) to install a leak collection system applies only to those drip pads that are constructed after December 24, 1992 except for those constructed after December 24, 1992 for which the owner or operator has a design and has entered into binding financial or other agreements for construction prior to December 24, 1992.

(b) The owner or operator of any drip pad that is inside or under a structure that provides protection from precipitation so that neither run-off nor run-on is generated is not subject to regulation under § 265.443(e) or § 265.443(f), as appropriate.

(c) The requirements of this subpart are not applicable to the management of infrequent and incidental drippage in storage yards provided that:

(1) The owner or operator maintains and complies with a written contingency plan that describes how the owner or operator will respond immediately to the discharge of such infrequent and incidental drippage. At a minimum, the contingency plan must describe how the facility will do the following:

(i) Clean up the drippage;

(ii) Document the cleanup of the drippage;

(iii) Retain documents regarding cleanup for three years; and

(iv) Manage the contaminated media in a manner consistent with Federal regulations.

[55 FR 50486, Dec. 6, 1990, as amended by 56 FR 30198, July 1, 1991; 57 FR 61503, Dec. 24, 1992]

### §265.441 Assessment of existing drip pad integrity.

(a) For each existing drip pad as defined in §265.440, the owner or operator must evaluate the drip pad and determine that it meets all of the requirements of this subpart, except the requirements for liners and leak detection systems of §265.443(b). No later than the effective date of this rule, the owner or operator must obtain and keep on file at the facility a written assessment of the drip pad, reviewed and certified by a qualified Professional Engineer that attests to the results of the evaluation. The assessment must be reviewed, updated, and re-certified annually until all upgrades, repairs, or modifications necessary to achieve compliance with all the standards of §265.443 are complete. The evaluation must document the extent to which the drip pad meets each of the design and operating standards of §265.443, except the standards for liners and leak detection systems, specified in §265.443(b).

(b) The owner or operator must develop a written plan for upgrading, repairing, and modifying the drip pad to meet the requirements of §265.443(b), and submit the plan to the Regional Administrator no later than 2 years before the date that all repairs, upgrades, and modifications are complete. This written plan must describe all changes to be made to the drip pad in sufficient detail to document compliance with all the requirements of §265.443. The plan must be reviewed and certified by a qualified Professional Engineer.

(c) Upon completion of all repairs and modifications, the owner or operator must submit to the Regional Administrator or State Director, the as-built drawings for the drip pad together with a certification by a qualified Professional Engineer attesting that the drip pad conforms to the drawings.

(d) If the drip pad is found to be leaking or unfit for use, the owner or operator must comply with the provisions of §265.443(m) of this subpart or close the drip pad in accordance with §265.445 of this subpart.

[55 FR 50486, Dec. 6, 1990, as amended at 57 FR 61504, Dec. 24, 1992; 71 FR 16912, Apr. 4, 2006; 71 FR 40276, July 14, 2006]

### §265.442 Design and installation of new drip pads.

Owners and operators of new drip pads must ensure that the pads are designed, installed, and operated in accordance with one of the following:

(a) All of the applicable requirements of §§265.443 (except §265.443(a)(4)), 265.444 and 265.445 of this subpart, or

(b) All of the applicable requirements of §§265.443 (except §265.443(b)), 265.444 and 265.445 of this subpart.

[57 FR 61504, Dec. 24, 1992]

### §265.443 Design and operating requirements.

(a) Drip pads must:

(1) Be constructed of non-earthen materials, excluding wood and non-structurally supported asphalt;

(2) Be sloped to free-drain treated wood drippage, rain and other waters, or solutions of drippage and water or other wastes to the associated collection system;

(3) Have a curb or berm around the perimeter;

(4)(i) Have a hydraulic conductivity of less than or equal to $1\times10^{-7}$ centimeters per second, e.g., existing concrete drip pads must be sealed, coated, or covered with a surface material with a hydraulic conductivity of less than or equal to $1\times10^{-7}$ centimeters per second such that the entire surface where drippage occurs or may run across is capable of containing such drippage and mixtures of drippage and precipitation, materials, or other wastes while being routed to an associated collection system. This surface material must be maintained free of cracks and gaps that could adversely affect its hydraulic conductivity, and the material must be chemically compatible with the preservatives that contact the drip pad. The requirements of this provision apply only to existing drip pads and those drip pads for which the owner or

685

operator elects to comply with § 265.442(b) instead of § 265.442(a).

(ii) The owner or operator must obtain and keep on file at the facility a written assessment of the drip pad, reviewed and certified by a qualified Professional Engineer that attests to the results of the evaluation. The assessment must be reviewed, updated and recertified annually. The evaluation must document the extent to which the drip pad meets the design and operating standards of this section, except for paragraph (b) of this section.

(5) Be of sufficient structural strength and thickness to prevent failure due to physical contact, climatic conditions, the stress of installation, and the stress of daily operations, e.g., variable and moving loads such as vehicle traffic, movement of wood, etc.

NOTE: EPA will generally consider applicable standards established by professional organizations generally recognized by industry such as the American Concrete Institute (ACI) and the American Society of Testing Materials (ASTM) in judging the structural integrity requirement of this paragraph.

(b) If an owner/operator elects to comply with § 265.442(a) instead of § 265.442(b), the drip pad must have:

(1) A synthetic liner installed below the drip pad that is designed, constructed, and installed to prevent leakage from the drip pad into the adjacent subsurface soil or groundwater or surface water at any time during the active life (including the closure period) of the drip pad. The liner must be constructed of materials that will prevent waste from being absorbed into the liner and prevent releases into the adjacent subsurface soil or ground water or surface water during the active life of the facility. The liner must be:

(i) Constructed of materials that have appropriate chemical properties and sufficient strength and thickness to prevent failure due to pressure gradients (including static head and external hydrogeologic forces), physical contact with the waste or drip pad leakage to which they are exposed, climatic conditions, the stress of installation, and the stress of daily operation (including stresses from vehicular traffic on the drip pad);

(ii) Placed upon a foundation or base capable of providing support to the liner and resistance to pressure gradients above and below the liner to prevent failure of the liner due to settlement, compression or uplift; and

(iii) Installed to cover all surrounding earth that could come in contact with the waste or leakage; and

(2) A leakage detection system immediately above the liner that is designed, constructed, maintained and operated to detect leakage from the drip pad. The leakage detection system must be:

(i) Constructed of materials that are:

(A) Chemically resistant to the waste managed in the drip pad and the leakage that might be generated; and

(B) Of sufficient strength and thickness to prevent collapse under the pressures exerted by overlaying materials and by any equipment used at the drip pad; and

(ii) Designed and operated to function without clogging through the scheduled closure of the drip pad.

(iii) Designed so that it will detect the failure of the drip pad or the presence of a release of hazardous waste or accumulated liquid at the earliest practicable time.

(3) A leakage collection system immediately above the liner that is designed, constructed, maintained and operated to collect leakage from the drip pad such that it can be removed from below the drip pad. The date, time, and quantity of any leakage collected in this system and removed must be documented in the operating log.

(c) Drip pads must be maintained such that they remain free of cracks, gaps, corrosion, or other deterioration that could cause hazardous waste to be released from the drip pad.

NOTE: See § 265.443(m) for remedial action required if deterioration or leakage is detected.

(d) The drip pad and associated collection system must be designed and operated to convey, drain, and collect liquid resulting from drippage or precipitation in order to prevent run-off.

(e) Unless protected by a structure, as described in § 265.440(b) of this subpart, the owner or operator must design, construct, operate and maintain a run-on control system capable of preventing flow onto the drip pad during peak discharge from at least a 24-hour, 25-year storm unless the system has sufficient excess capacity to contain any run-on that might enter the system, or the drip pad is protected by a structure or cover, as described in §265.440(b) of this subpart.

(f) Unless protected by a structure or cover, as described in §265.440(b) of this subpart, the owner or operator must design, construct, operate and maintain a run-off management system to collect and control at least the water volume resulting from a 24-hour, 25-year storm.

(g) The drip pad must be evaluated to determine that it meets the requirements of paragraphs (a) through (f) of this section and the owner or operator must obtain a statement from a qualified Professional Engineer certifying that the drip pad design meets the requirements of this section.

(h) Drippage and accumulated precipitation must be removed from the associated collection system as necessary to prevent overflow onto the drip pad.

(i) The drip pad surface must be cleaned thoroughly in a manner and frequency such that accumulated residues of hazardous waste or other materials are removed, with residues being properly managed as hazardous waste, so as to allow weekly inspections of the entire drip pad surface without interference or hindrance from accumulated residues of hazardous waste or other materials on the drip pad. The owner or operator must document the date and time of each cleaning and the cleaning procedure used in the facility's operating log.

(j) Drip pads must be operated and maintained in a manner to minimize tracking of hazardous waste or hazardous waste constituents off the drip pad as a result of activities by personnel or equipment.

(k) After being removed from the treatment vessel, treated wood from pressure and non-pressure processes must be held on the drip pad until drippage has ceased. The owner or operator must maintain records sufficient to document that all treated wood is held on the pad following treatment in accordance with this requirement.

(l) Collection and holding units associated with run-on and run-off control systems must be emptied or otherwise managed as soon as possible after storms to maintain design capacity of the system.

(m) Throughout the active life of the drip pad, if the owner or operator detects a condition that may have caused or has caused a release of hazardous waste, the condition must be repaired within a reasonably prompt period of time following discovery, in accordance with the following procedures:

(1) Upon detection of a condition that may have caused or has caused a release of hazardous waste (e.g., upon detection of leakage by the leak detection system), the owner or operator must:

(i) Enter a record of the discovery in the facility operating log;

(ii) Immediately remove the portion of the drip pad affected by the condition from service;

(iii) Determine what steps must be taken to repair the drip pad, remove any leakage from below the drip pad, and establish a schedule for accomplishing the clean up and repairs;

(iv) Within 24 hours after discovery of the condition, notify the Regional Administrator of the condition and, within 10 working days, provide a written notice to the Regional Administrator with a description of the steps that will be taken to repair the drip pad, and clean up any leakage, and the schedule for accomplishing this work.

(2) The Regional Administrator will review the information submitted, make a determination regarding whether the pad must be removed from service completely or partially until repairs and clean up are complete, and notify the owner or operator of the determination and the underlying rationale in writing.

(3) Upon completing all repairs and clean up, the owner or operator must notify the Regional Administrator in writing and provide a certification, signed by an independent qualified, registered professional engineer, that the repairs and clean up have been completed according to the written plan submitted in accordance with paragraph (m)(1)(iv) of this section.

(n) The owner or operator must maintain, as part of the facility operating log, documentation of past operating and waste handling practices. This must include identification of preservative formulations used in the past, a description of drippage management practices, and a description of treated wood storage and handling practices.

[55 FR 50486, Dec. 6, 1990, as amended at 56 FR 30198, July 1, 1991; 57 FR 5861, Feb. 18, 1992; 57 FR 61504, Dec. 24, 1992; 71 FR 16912, Apr. 4, 2006; 71 FR 40276, July 14, 2006]

§ 265.444   Inspections.

(a) During construction or installation, liners and cover systems (e.g., membranes, sheets, or coatings) must be inspected for uniformity, damage and imperfections (e.g., holes, cracks, thin spots, or foreign materials). Immediately after construction or installation, liners must be inspected and certified as meeting the requirements of § 265.443 by a qualified Professional Engineer. This certification must be maintained at the facility as part of the facility operating record. After installation, liners and covers must be inspected to ensure tight seams and joints and the absence of tears, punctures, or blisters.

(b) While a drip pad is in operation, it must be inspected weekly and after storms to detect evidence of any of the following:

(1) Deterioration, malfunctions or improper operation of run-on and run-off control systems;

(2) The presence of leakage in and proper functioning of leakage detection system.

(3) Deterioration or cracking of the drip pad surface.

NOTE: See § 265.443(m) for remedial action required if deterioration or leakage is detected.

[55 FR 50486, Dec. 6, 1990, as amended at 71 FR 16912, Apr. 4, 2006]

§ 265.445   Closure.

(a) At closure, the owner or operator must remove or decontaminate all waste residues, contaminated containment system components (pad, liners, etc.), contaminated subsoils, and structures and equipment contaminated with waste and leakage, and manage them as hazardous waste.

(b) If, after removing or decontaminating all residues and making all reasonable efforts to effect removal or decontamination of contaminated components, subsoils, structures, and equipment as required in paragraph (a) of this section, the owner or operator finds that not all contaminated subsoils can be practically removed or decontaminated, he must close the facility and perform post-closure care in accordance with closure and post-closure care requirements that apply to landfills (§ 265.310). For permitted units, the requirement to have a permit continues throughout the post-closure period.

(c)(1) The owner or operator of an existing drip pad, as defined in § 265.440 of this subpart, that does not comply with the liner requirements of § 265.443(b)(1) must:

(i) Include in the closure plan for the drip pad under § 265.112 both a plan for complying with paragraph (a) of this section and a contingent plan for complying with paragraph (b) of this section in case not all contaminated subsoils can be practicably removed at closure; and

(ii) Prepare a contingent post-closure plan under § 265.118 of this part for complying with paragraph (b) of this section in case not all contaminated subsoils can be practicably removed at closure.

(2) The cost estimates calculated under §§ 265.112 and 265.144 of this part for closure and post-closure care of a drip pad subject to this paragraph must include the cost of complying with the contingent closure plan and the contingent post-closure plan, but are not required to include the cost of expected closure under paragraph (a) of this section.

[55 FR 50486, Dec. 6, 1990, as amended at 71 FR 40276, July 14, 2006]

## Subparts X–Z [Reserved]

## Subpart AA—Air Emission Standards for Process Vents

SOURCE: 55 FR 25507, June 21, 1990, unless otherwise noted.

**§265.1030 Applicability.**

(a) The regulations in this subpart apply to owners and operators of facilities that treat, store, or dispose of hazardous wastes (except as provided in §265.1).

(b) Except for §§265.1034, paragraphs (d) and (e), this subpart applies to process vents associated with distillation, fractionation, thin-film evaporation, solvent extraction, or air or steam stripping operations that manage hazardous wastes with organic concentrations of at least 10 ppmw, if these operations are conducted in one of the following:

(1) A unit that is subject to the permitting requirements of 40 CFR part 270, or

(2) A unit (including a hazardous waste recycling unit) that is not exempt from permitting under the provisions of 40 CFR 262.34(a) (i.e., a hazardous waste recycling unit that is not a 90-day tank or container) and that is located at a hazardous waste management facility otherwise subject to the permitting requirements of 40 CFR part 270, or

(3) A unit that is exempt from permitting under the provisions of 40 CFR 262.34(a) (i.e., a "90-day" tank or container) and is not a recycling unit under the requirements of 40 CFR 261.6.

NOTE: The requirements of §§265.1032 through 265.1036 apply to process vents on hazardous waste recycling units previously exempt under paragraph 261.6(c)(1). Other exemptions under §§261.4, and 265.1(c) are not affected by these requirements.]

(c) The requirements of this subpart do not apply to the pharmaceutical manufacturing facility, commonly referred to as the Stonewall Plant, located at Route 340 South, Elkton, Virginia, provided that facility is operated in compliance with the requirements contained in a Clean Air Act permit issued pursuant to 40 CFR 52.2454. The requirements of this subpart shall apply to the facility upon termination of the Clean Air Act permit issued pursuant to 40 CFR 52.2454.

(d) The requirements of this subpart do not apply to the process vents at a facility where the facility owner or operator certifies that all of the process vents that would otherwise be subject to this subpart are equipped with and operating air emission controls in accordance with the process vent requirements of an applicable Clean Air Act regulation codified under 40 CFR part 60, part 61, or part 63. The documentation of compliance under regulations at 40 CFR part 60, part 61, or part 63 shall be kept with, or made readily available with, the facility operating record.

[55 FR 25507, June 21, 1990, as amended at 56 FR 19290, Apr. 26, 1991; 61 FR 59968, Nov. 25, 1996; 62 FR 52642, Oct. 8, 1997; 62 FR 64661, Dec. 8, 1997]

**§265.1031 Definitions.**

As used in this subpart, all terms shall have the meaning given them in §264.1031, the Act, and parts 260–266.

**§265.1032 Standards: Process vents.**

(a) The owner or operator of a facility with process vents associated with distillation, fractionation, thin-film evaporation, solvent extraction or air or steam stripping operations managing hazardous wastes with organic concentrations at least 10 ppmw shall either:

(1) Reduce total organic emissions from all affected process vents at the facility below 1.4 kg/h (3 lb/h) and 2.8 Mg/yr (3.1 tons/yr), or

(2) Reduce, by use of a control device, total organic emissions from all affected process vents at the facility by 95 weight percent.

(b) If the owner or operator installs a closed-vent system and control device to comply with the provisions of paragraph (a) of this section, the closed-vent system and control device must meet the requirements of §265.1033.

(c) Determinations of vent emissions and emission reductions or total organic compound concentrations achieved by add-on control devices may be based on engineering calculations or performance tests. If performance tests are used to determine vent emissions, emission reductions, or total organic compound concentrations achieved by add-on control devices, the performance tests must conform with the requirements of §265.1034(c).

(d) When an owner or operator and the Regional Administrator do not agree on determinations of vent emissions and/or emission reductions or

689

total organic compound concentrations achieved by add-on control devices based on engineering calculations, the test methods in § 265.1034(c) shall be used to resolve the disagreement.

§ 265.1033 Standards: Closed-vent systems and control devices.

(a)(1) Owners or operators of closed-vent systems and control devices used to comply with provisions of this part shall comply with the provisions of this section.

(2)(i) The owner or operator of an existing facility who cannot install a closed-vent system and control device to comply with the provisions of this subpart on the effective date that the facility becomes subject to the requirements of this subpart must prepare an implementation schedule that includes dates by which the closed-vent system and control device will be installed and in operation. The controls must be installed as soon as possible, but the implementation schedule may allow up to 30 months after the effective date that the facility becomes subject to this subpart for installation and startup.

(ii) Any unit that begins operation after December 21, 1990, and is subject to the requirements of this subpart when operation begins, must comply with the rules immediately (i.e., must have control devices installed and operating on startup of the affected unit); the 30-month implementation schedule does not apply.

(iii) The owner or operator of any facility in existence on the effective date of a statutory or EPA regulatory amendment that renders the facility subject to this subpart shall comply with all requirements of this subpart as soon as practicable but no later than 30 months after the amendment's effective date. When control equipment required by this subpart can not be installed and begin operation by the effective date of the amendment, the facility owner or operator shall prepare an implementation schedule that includes the following information: Specific calendar dates for award of contracts or issuance of purchase orders for the control equipment, initiation of on-site installation of the control equipment, completion of the control equipment installation, and perform-ance of any testing to demonstrate that the installed equipment meets the applicable standards of this subpart. The owner or operator shall enter the implementation schedule in the operating record or in a permanent, readily available file located at the facility.

(iv) Owners and operators of facilities and units that become newly subject to the requirements of this subpart after December 8, 1997, due to an action other than those described in paragraph (a)(2)(iii) of this section must comply with all applicable requirements immediately (i.e., must have control devices installed and operating on the date the facility or unit becomes subject to this subpart; the 30-month implementation schedule does not apply).

(b) A control device involving vapor recovery (e.g., a condenser or adsorber) shall be designed and operated to recover the organic vapors vented to it with an efficiency of 95 weight percent or greater unless the total organic emission limits of § 265.1032(a)(1) for all affected process vents can be attained at an efficiency less than 95 weight percent.

(c) An enclosed combustion device (e.g., a vapor incinerator, boiler, or process heater) shall be designed and operated to reduce the organic emissions vented to it by 95 weight percent or greater; to achieve a total organic compound concentration of 20 ppmv, expressed as the sum of the actual compounds, not carbon equivalents, on a dry basis corrected to 3 percent oxygen; or to provide a minimum residence time of 0.50 seconds at a minimum temperature of 760 °C. If a boiler or process heater is used as the control device, then the vent stream shall be introduced into the flame combustion zone of the boiler or process heater.

(d)(1) A flare shall be designed for and operated with no visible emissions as determined by the methods specified in paragraph (e)(1) of this section, except for periods not to exceed a total of 5 minutes during any 2 consecutive hours.

(2) A flare shall be operated with a flame present at all times, as determined by the methods specified in paragraph (f)(2)(iii) of this section.

(3) A flare shall be used only if the net heating value of the gas being combusted is 11.2 MJ/scm (300 Btu/scf) or greater, if the flare is steam-assisted or air-assisted; or if the net heating value of the gas being combusted is 7.45 MJ/scm (200 Btu/scf) or greater if the flare is nonassisted. The net heating value of the gas being combusted shall be determined by the methods specified in paragraph (e)(2) of this section.

(4)(i) A steam-assisted or nonassisted flare shall be designed for and operated with an exit velocity, as determined by the methods specified in paragraph (e)(3) of this section, of less than 18.3 m/s (60 ft/s), except as provided in paragraphs (d)(4) (ii) and (iii) of this section.

(ii) A steam-assisted or nonassisted flare designed for and operated with an exit velocity, as determined by the methods specified in paragraph (e)(3) of this section, equal to or greater than 18.3 m/s (60 ft/s) but less than 122 m/s (400 ft/s) is allowed if the net heating value of the gas being combusted is greater than 37.3 MJ/scm (1,000 Btu/scf).

(iii) A steam-assisted or nonassisted flare designed for and operated with an exit velocity, as determined by the methods specified in paragraph (e)(3) of this section, less than the velocity, $V_{max}$, as determined by the method specified in paragraph (e)(4) of this section, and less than 122 m/s (400 ft/s) is allowed.

(5) An air-assisted flare shall be designed and operated with an exit velocity less than the velocity, $V_{max}$, as determined by the method specified in paragraph (e)(5) of this section.

(6) A flare used to comply with this section shall be steam-assisted, air-assisted, or nonassisted.

(e)(1) Reference Method 22 in 40 CFR part 60 shall be used to determine the compliance of a flare with the visible emission provisions of this subpart. The observation period is 2 hours and shall be used according to Method 22.

(2) The net heating value of the gas being combusted in a flare shall be calculated using the following equation:

$$H_T = K \left[ \sum_{i=1}^{n} C_i H_i \right]$$

where:

$H_T$=Net heating value of the sample, MJ/scm; where the net enthalpy per mole of offgas is based on combustion at 25 °C and 760 mm Hg, but the standard temperature for determining the volume corresponding to 1 mol is 20 °C;

K=Constant, $1.74 \times 10^{-7}$ (1/ppm) (g mol/scm) (MJ/kcal) where standard temperature for (g mol/scm) is 20 °C;

$C_i$=Concentration of sample component i in ppm on a wet basis, as measured for organics by Reference Method 18 in 40 CFR part 60 and measured for hydrogen and carbon monoxide by ASTM D 1946–82 (incorporated by reference as specified in §260.11); and

$H_i$=Net heat of combustion of sample component i, kcal/g mol at 25 °C and 760 mm Hg. The heats of combustion may be determined using ASTM D 2382–83 (incorporated by reference as specified in §260.11) if published values are not available or cannot be calculated.

(3) The actual exit velocity of a flare shall be determined by dividing the volumetric flow rate (in units of standard temperature and pressure), as determined by Reference Methods 2, 2A, 2C, or 2D in 40 CFR part 60 as appropriate, by the unobstructed (free) cross-sectional area of the flare tip.

(4) The maximum allowed velocity in m/s, $V_{max}$, for a flare complying with paragraph (d)(4)(iii) of this section shall be determined by the following equation:

$$\text{Log}_{10}(V_{max}) = (H_T + 28.8)/31.7$$

where:

$H_T$=The net heating value as determined in paragraph (e)(2) of this section.
28.8=Constant,
31.7=Constant.

(5) The maximum allowed velocity in m/s, $V_{max}$, for an air-assisted flare shall be determined by the following equation:

$$V_{max} = 8.706 + 0.7084 \, (H_T)$$

where:

8.706 = Constant.
0.7084 = Constant.
$H_T$ = The net heating value as determined in paragraph (e)(2) of this section.

(f) The owner or operator shall monitor and inspect each control device required to comply with this section to ensure proper operation and maintenance of the control device by implementing the following requirements:

691

(1) Install, calibrate, maintain, and operate according to the manufacturer's specifications a flow indicator that provides a record of vent stream flow from each affected process vent to the control device at least once every hour. The flow indicator sensor shall be installed in the vent stream at the nearest feasible point to the control device inlet, but before being combined with other vent streams.

(2) Install, calibrate, maintain, and operate according to the manufacturer's specifications a device to continuously monitor control device operation as specified below:

(i) For a thermal vapor incinerator, a temperature monitoring device equipped with a continuous recorder. The device shall have an accuracy of ±1 percent of the temperature being monitored in ° C or ±0.5 ° C, whichever is greater. The temperature sensor shall be installed at a location in the combustion chamber downstream of the combustion zone.

(ii) For a catalytic vapor incinerator, a temperature monitoring device equipped with a continuous recorder. The device shall be capable of monitoring temperature at two locations and have an accuracy of ±1 percent of the temperature being monitored in ° C or ±0.5 ° C, whichever is greater. One temperature sensor shall be installed in the vent stream at the nearest feasible point to the catalyst bed inlet and a second temperature sensor shall be installed in the vent stream at the nearest feasible point to the catalyst bed outlet.

(iii) For a flare, a heat sensing monitoring device equipped with a continuous recorder that indicates the continuous ignition of the pilot flame.

(iv) For a boiler or process heater having a design heat input capacity less than 44 MW, a temperature monitoring device equipped with a continuous recorder. The device shall have an accuracy of ±1 percent of the temperature being monitored in ° C or ±0.5 ° C, whichever is greater. The temperature sensor shall be installed at a location in the furnace downstream of the combustion zone.

(v) For a boiler or process heater having a design heat input capacity greater than or equal to 44 MW, a moni-

toring device equipped with a continuous recorder to measure a parameter(s) that indicates good combustion operating practices are being used.

(vi) For a condenser, either:

(A) A monitoring device equipped with a continuous recorder to measure the concentration level of the organic compounds in the exhaust vent stream from the condenser; or

(B) A temperature monitoring device equipped with a continuous recorder. The device shall be capable of monitoring temperature with an accuracy of ±1 percent of the temperature being monitored in degrees Celsius ( °C) or ±0.5 °C, whichever is greater. The temperature sensor shall be installed at a location in the exhaust vent stream from the condenser exit (i.e., product side).

(vii) For a carbon adsorption system such as a fixed-bed carbon adsorber that regenerates the carbon bed directly in the control device, either:

(A) A monitoring device equipped with a continuous recorder to measure the concentration level of the organic compounds in the exhaust vent stream from the carbon bed, or

(B) A monitoring device equipped with a continuous recorder to measure a parameter that indicates the carbon bed is regenerated on a regular, predetermined time cycle.

(3) Inspect the readings from each monitoring device required by paragraphs (f) (1) and (2) of this section at least once each operating day to check control device operation and, if necessary, immediately implement the corrective measures necessary to ensure the control device operates in compliance with the requirements of this section.

(g) An owner or operator using a carbon adsorption system such as a fixed-bed carbon adsorber that regenerates the carbon bed directly onsite in the control device, shall replace the existing carbon in the control device with fresh carbon at a regular, predetermined time interval that is no longer than the carbon service life established as a requirement of § 265.1035(b)(4)(iii)(F).

(h) An owner or operator using a carbon adsorption system such as a carbon canister that does not regenerate the carbon bed directly onsite in the control device shall replace the existing carbon in the control device with fresh carbon on a regular basis by using one of the following procedures:

(1) Monitor the concentration level of the organic compounds in the exhaust vent stream from the carbon adsorption system on a regular schedule and replace the existing carbon with fresh carbon immediately when carbon breakthrough is indicated. The monitoring frequency shall be daily or at an interval no greater than 20 percent of the time required to consume the total carbon working capacity established as a requirement of §265.1035(b)(4)(iii)(G), whichever is longer.

(2) Replace the existing carbon with fresh carbon at a regular, predetermined time interval that is less than the design carbon replacement interval established as a requirement of §265.1035(b)(4)(iii)(G).

(i) An owner or operator of an affected facility seeking to comply with the provisions of this part by using a control device other than a thermal vapor incinerator, catalytic vapor incinerator, flare, boiler, process heater, condenser, or carbon adsorption system is required to develop documentation including sufficient information to describe the control device operation and identify the process parameter or parameters that indicate proper operation and maintenance of the control device.

(j) A closed-vent system shall meet either of the following design requirements:

(1) A closed-vent system shall be designed to operate with no detectable emissions, as indicated by an instrument reading of less than 500 ppmv above background as determined by the procedure in §265.1034(b) of this subpart, and by visual inspections; or

(2) A closed-vent system shall be designed to operate at a pressure below atmospheric pressure. The system shall be equipped with at least one pressure gauge or other pressure measurement device that can be read from a readily accessible location to verify that negative pressure is being maintained in the closed-vent system when the control device is operating.

(k) The owner or operator shall monitor and inspect each closed-vent system required to comply with this section to ensure proper operation and maintenance of the closed-vent system by implementing the following requirements:

(1) Each closed-vent system that is used to comply with paragraph (j)(1) of this section shall be inspected and monitored in accordance with the following requirements:

(i) An initial leak detection monitoring of the closed-vent system shall be conducted by the owner or operator on or before the date that the system becomes subject to this section. The owner or operator shall monitor the closed-vent system components and connections using the procedures specified in §265.1034(b) of this subpart to demonstrate that the closed-vent system operates with no detectable emissions, as indicated by an instrument reading of less than 500 ppmv above background.

(ii) After initial leak detection monitoring required in paragraph (k)(1)(i) of this section, the owner or operator shall inspect and monitor the closed-vent system as follows:

(A) Closed-vent system joints, seams, or other connections that are permanently or semi-permanently sealed (e.g., a welded joint between two sections of hard piping or a bolted and gasketed ducting flange) shall be visually inspected at least once per year to check for defects that could result in air pollutant emissions. The owner or operator shall monitor a component or connection using the procedures specified in §265.1034(b) of this subpart to demonstrate that it operates with no detectable emissions following any time the component is repaired or replaced (e.g., a section of damaged hard piping is replaced with new hard piping) or the connection is unsealed (e.g., a flange is unbolted).

(B) Closed-vent system components or connections other than those specified in paragraph (k)(1)(ii)(A) of this section shall be monitored annually and at other times as requested by the Regional Administrator, except as provided for in paragraph (n) of this section, using the procedures specified in

§ 265.1034(b) of this subpart to demonstrate that the components or connections operate with no detectable emissions.

(iii) In the event that a defect or leak is detected, the owner or operator shall repair the defect or leak in accordance with the requirements of paragraph (k)(3) of this section.

(iv) The owner or operator shall maintain a record of the inspection and monitoring in accordance with the requirements specified in § 265.1035 of this subpart.

(2) Each closed-vent system that is used to comply with paragraph (j)(2) of this section shall be inspected and monitored in accordance with the following requirements:

(i) The closed-vent system shall be visually inspected by the owner or operator to check for defects that could result in air pollutant emissions. Defects include, but are not limited to, visible cracks, holes, or gaps in ductwork or piping or loose connections.

(ii) The owner or operator shall perform an initial inspection of the closed-vent system on or before the date that the system becomes subject to this section. Thereafter, the owner or operator shall perform the inspections at least once every year.

(iii) In the event that a defect or leak is detected, the owner or operator shall repair the defect in accordance with the requirements of paragraph (k)(3) of this section.

(iv) The owner or operator shall maintain a record of the inspection and monitoring in accordance with the requirements specified in § 265.1035 of this subpart.

(3) The owner or operator shall repair all detected defects as follows:

(i) Detectable emissions, as indicated by visual inspection, or by an instrument reading greater than 500 ppmv above background, shall be controlled as soon as practicable, but not later than 15 calendar days after the emission is detected, except as provided for in paragraph (k)(3)(iii) of this section.

(ii) A first attempt at repair shall be made no later than 5 calendar days after the emission is detected.

(iii) Delay of repair of a closed-vent system for which leaks have been detected is allowed if the repair is technically infeasible without a process unit shutdown, or if the owner or operator determines that emissions resulting from immediate repair would be greater than the fugitive emissions likely to result from delay of repair. Repair of such equipment shall be completed by the end of the next process unit shutdown.

(iv) The owner or operator shall maintain a record of the defect repair in accordance with the requirements specified in § 265.1035 of this subpart.

(l) Closed-vent systems and control devices used to comply with provisions of this subpart shall be operated at all times when emissions may be vented to them.

(m) The owner or operator using a carbon adsorption system to control air pollutant emissions shall document that all carbon that is a hazardous waste and that is removed from the control device is managed in one of the following manners, regardless of the average volatile organic concentration of the carbon:

(1) Regenerated or reactivated in a thermal treatment unit that meets one of the following:

(i) The owner or operator of the unit has been issued a final permit under 40 CFR part 270 which implements the requirements of 40 CFR part 264 subpart X; or

(ii) The unit is equipped with and operating air emission controls in accordance with the applicable requirements of subparts AA and CC of either this part or of 40 CFR part 264; or

(iii) The unit is equipped with and operating air emission controls in accordance with a national emission standard for hazardous air pollutants under 40 CFR part 61 or 40 CFR part 63.

(2) Incinerated in a hazardous waste incinerator for which the owner or operator either:

(i) Has been issued a final permit under 40 CFR part 270 which implements the requirements of 40 CFR part 264, subpart O; or

(ii) Has designed and operates the incinerator in accordance with the interim status requirements of subpart O of this part.

(3) Burned in a boiler or industrial furnace for which the owner or operator either:

(i) Has been issued a final permit under 40 CFR part 270 which implements the requirements of 40 CFR part 266, subpart H; or

(ii) Has designed and operates the boiler or industrial furnace in accordance with the interim status requirements of 40 CFR part 266, subpart H.

(n) Any components of a closed-vent system that are designated, as described in §265.1035(c)(9) of this subpart, as unsafe to monitor are exempt from the requirements of paragraph (k)(1)(ii)(B) of this section if:

(1) The owner or operator of the closed-vent system determines that the components of the closed-vent system are unsafe to monitor because monitoring personnel would be exposed to an immediate danger as a consequence of complying with paragraph (k)(1)(ii)(B) of this section; and

(2) The owner or operator of the closed-vent system adheres to a written plan that requires monitoring the closed-vent system components using the procedure specified in paragraph (k)(1)(ii)(B) of this section as frequently as practicable during safe-to-monitor times.

[59 FR 62935, Dec. 6, 1994, as amended at 61 FR 4913, Feb. 9, 1996; 61 FR 59969, Nov. 25, 1996; 62 FR 64661, Dec. 8, 1997; 71 FR 40276, July 14, 2006]

### §265.1034 Test methods and procedures.

(a) Each owner or operator subject to the provisions of this subpart shall comply with the test methods and procedures requirements provided in this section.

(b) When a closed-vent system is tested for compliance with no detectable emissions, as required in §265.1033(k) of this subpart, the test shall comply with the following requirements:

(1) Monitoring shall comply with Reference Method 21 in 40 CFR part 60.

(2) The detection instrument shall meet the performance criteria of Reference Method 21.

(3) The instrument shall be calibrated before use on each day of its use by the procedures specified in Reference Method 21.

(4) Calibration gases shall be:

(i) Zero air (less than 10 ppm of hydrocarbon in air).

(ii) A mixture of methane or n-hexane and air at a concentration of approximately, but less than, 10,000 ppm methane or n-hexane.

(5) The background level shall be determined as set forth in Reference Method 21.

(6) The instrument probe shall be traversed around all potential leak interfaces as close to the interface as possible as described in Reference Method 21.

(7) The arithmetic difference between the maximum concentration indicated by the instrument and the background level is compared with 500 ppm for determining compliance.

(c) Performance tests to determine compliance with §265.1032(a) and with the total organic compound concentration limit of §265.1033(c) shall comply with the following:

(1) Performance tests to determine total organic compound concentrations and mass flow rates entering and exiting control devices shall be conducted and data reduced in accordance with the following reference methods and calculation procedures:

(i) Method 2 in 40 CFR part 60 for velocity and volumetric flow rate.

(ii) Method 18 or Method 25A in 40 CFR part 60, appendix A, for organic content. If Method 25A is used, the organic HAP used as the calibration gas must be the single organic HAP representing the largest percent by volume of the emissions. The use of Method 25A is acceptable if the response from the high-level calibration gas is at least 20 times the standard deviation of the response from the zero calibration gas when the instrument is zeroed on the most sensitive scale.

(iii) Each performance test shall consist of three separate runs; each run conducted for at least 1 hour under the conditions that exist when the hazardous waste management unit is operating at the highest load or capacity level reasonably expected to occur. For the purpose of determining total organic compound concentrations and mass flow rates, the average of results of all runs shall apply. The average shall be computed on a time-weighted basis.

(iv) Total organic mass flow rates shall be determined by the following equation:

(A) For sources utilizing Method 18.

$$E_h = Q_{2sd} \left\{ \sum_{i=1}^{n} C_i MW_i \right\} [0.0416] [10^{-6}]$$

Where:

$E_h$ = Total organic mass flow rate, kg/h;
$Q_{2sd}$ = Volumetric flow rate of gases entering or exiting control device, as determined by Method 2, dscm/h;
$n$ = Number of organic compounds in the vent gas;
$C_i$ = Organic concentration in ppm, dry basis, of compound i in the vent gas, as determined by Method 18;
$MW_i$ = Molecular weight of organic compound i in the vent gas, kg/kg-mol;
0.0416 = Conversion factor for molar volume, kg-mol/m3 (@ 293 K and 760 mm Hg);
$10^{-6}$ = Conversion from ppm

(B) For sources utilizing Method 25A.

$E_h = (Q)(C)(MW)(0.0416)(10^{-6})$

Where:

$E_h$ = Total organic mass flow rate, kg/h;
$Q$ = Volumetric flow rate of gases entering or exiting control device, as determined by Method 2, dscm/h;
$C$ = Organic concentration in ppm, dry basis, as determined by Method 25A;
$MW$ = Molecular weight of propane, 44;
0.0416 = Conversion factor for molar volume, kg-mol/m3 (@ 293 K and 760 mm Hg);
$10^{-6}$ = Conversion from ppm.

(v) The annual total organic emission rate shall be determined by the following equation:

$E_A = (E_h) (H)$

where:

$E_A$ = Total organic mass emission rate, kg/y;
$E_h$ = Total organic mass flow rate for the process vent, kg/h;
$H$ = Total annual hours of operations for the affected unit, h.

(vi) Total organic emissions from all affected process vents at the facility shall be determined by summing the hourly total organic mass emission rates ($E_h$, as determined in paragraph (c)(1)(iv) of this section) and by summing the annual total organic mass emission rates ($E_A$, as determined in paragraph (c)(1)(v) of this section) for all affected process vents at the facility.

(2) The owner or operator shall record such process information as may be necessary to determine the conditions of the performance tests. Operations during periods of startup, shutdown, and malfunction shall not constitute representative conditions for the purpose of a performance test.

(3) The owner or operator of an affected facility shall provide, or cause to be provided, performance testing facilities as follows:

(i) Sampling ports adequate for the test methods specified in paragraph (c)(1) of this section.

(ii) Safe sampling platform(s).

(iii) Safe access to sampling platform(s).

(iv) Utilities for sampling and testing equipment.

(4) For the purpose of making compliance determinations, the time-weighted average of the results of the three runs shall apply. In the event that a sample is accidentally lost or conditions occur in which one of the three runs must be discontinued because of forced shutdown, failure of an irreplaceable portion of the sample train, extreme meteorological conditions, or other circumstances beyond the owner or operator's control, compliance may, upon the Regional Administrator's approval, be determined using the average of the results of the two other runs.

(d) To show that a process vent associated with a hazardous waste distillation, fractionation, thin-film evaporation, solvent extraction, or air or steam stripping operation is not subject to the requirements of this subpart, the owner or operator must make an initial determination that the time-weighted, annual average total organic concentration of the waste managed by the waste management unit is less than 10 ppmw using one of the following two methods:

(1) Direct measurement of the organic concentration of the waste using the following procedures:

(i) The owner or operator must take a minimum of four grab samples of waste for each waste stream managed in the affected unit under process conditions expected to cause the maximum waste organic concentration.

(ii) For waste generated onsite, the grab samples must be collected at a point before the waste is exposed to the atmosphere such as in an enclosed pipe or other closed system that is used to transfer the waste after generation to the first affected distillation fractionation, thin-film evaporation, solvent extraction, or air or steam stripping operation. For waste generated offsite, the grab samples must be collected at the inlet to the first waste management unit that receives the waste provided the waste has been transferred to the facility in a closed system such as a tank truck and the waste is not diluted or mixed with other waste.

(iii) Each sample shall be analyzed and the total organic concentration of the sample shall be computed using Method 9060A (incorporated by reference under § 260.11 of this chapter) of "Test Methods for Evaluating Solid Waste, Physical/Chemical Methods," EPA Publication SW–846; or analyzed for its individual organic constituents.

(iv) The arithmetic mean of the results of the analyses of the four samples shall apply for each waste stream managed in the unit in determining the time-weighted, annual average total organic concentration of the waste. The time-weighted average is to be calculated using the annual quantity of each waste stream processed and the mean organic concentration of each waste stream managed in the unit.

(2) Using knowledge of the waste to determine that its total organic concentration is less than 10 ppmw. Documentation of the waste determination is required. Examples of documentation that shall be used to support a determination under this provision include production process information documenting that no organic compounds are used, information that the waste is generated by a process that is identical to a process at the same or another facility that has previously been demonstrated by direct measurement to generate a waste stream having a total organic content less than 10 ppmw, or prior speciation analysis results on the same waste stream where it can also be documented that no process changes have occurred since that analysis that could affect the waste total organic concentration.

(e) The determination that distillation fractionation, thin-film evapo-

ration, solvent extraction, or air or steam stripping operations manage hazardous wastes with time-weighted annual average total organic concentrations less than 10 ppmw shall be made as follows:

(1) By the effective date that the facility becomes subject to the provisions of this subpart or by the date when the waste is first managed in a waste management unit, whichever is later; and

(2) For continuously generated waste, annually; or

(3) Whenever there is a change in the waste being managed or a change in the process that generates or treats the waste.

(f) When an owner or operator and the Regional Administrator do not agree on whether a distillation, fractionation, thin-film evaporation, solvent extraction, or air or steam stripping operation manages a hazardous waste with organic concentrations of at least 10 ppmw based on knowledge of the waste, the dispute may be resolved using direct measurement as specified at paragraph (d)(1) of this section.

[55 FR 25507, June 21, 1990, as amended at 56 FR 19290, Apr. 26, 1991; 61 FR 59970, Nov. 25, 1996; 62 FR 32463, June 13, 1997; 70 FR 34586, June 14, 2005]

## § 265.1035 Recordkeeping requirements.

(a)(1) Each owner or operator subject to the provisions of this subpart shall comply with the recordkeeping requirements of this section.

(2) An owner or operator of more than one hazardous waste management unit subject to the provisions of this subpart may comply with the recordkeeping requirements for these hazardous waste management units in one recordkeeping system if the system identifies each record by each hazardous waste management unit.

(b) Owners and operators must record the following information in the facility operating record:

(1) For facilities that comply with the provisions of § 265.1033(a)(2), an implementation schedule that includes dates by which the closed-vent system and control device will be installed and in operation. The schedule must also

include a rationale of why the installation cannot be completed at an earlier date. The implementation schedule must be in the facility operating record by the effective date that the facility becomes subject to the provisions of this subpart.

(2) Up-to-date documentation of compliance with the process vent standards in § 265.1032, including:

(i) Information and data identifying all affected process vents, annual throughput and operating hours of each affected unit, estimated emission rates for each affected vent and for the overall facility (i.e., the total emissions for all affected vents at the facility), and the approximate location within the facility of each affected unit (e.g., identify the hazardous waste management units on a facility plot plan); and

(ii) Information and data supporting determinations of vent emissions and emission reductions achieved by add-on control devices based on engineering calculations or source tests. For the purpose of determining compliance, determinations of vent emissions and emission reductions must be made using operating parameter values (e.g., temperatures, flow rates or vent stream organic compounds and concentrations) that represent the conditions that result in maximum organic emissions, such as when the waste management unit is operating at the highest load or capacity level reasonably expected to occur. If the owner or operator takes any action (e.g., managing a waste of different composition or increasing operating hours of affected waste management units) that would result in an increase in total organic emissions from affected process vents at the facility, then a new determination is required.

(3) Where an owner or operator chooses to use test data to determine the organic removal efficiency or total organic compound concentration achieved by the control device, a performance test plan. The test plan must include:

(i) A description of how it is determined that the planned test is going to be conducted when the hazardous waste management unit is operating at the highest load or capacity level reasonably expected to occur. This shall include the estimated or design flow rate and organic content of each vent stream and define the acceptable operating ranges of key process and control device parameters during the test program.

(ii) A detailed engineering description of the closed-vent system and control device including:

(A) Manufacturer's name and model number of control device.

(B) Type of control device.

(C) Dimensions of the control device.

(D) Capacity.

(E) Construction materials.

(iii) A detailed description of sampling and monitoring procedures, including sampling and monitoring locations in the system, the equipment to be used, sampling and monitoring frequency, and planned analytical procedures for sample analysis.

(4) Documentation of compliance with § 265.1033 shall include the following information:

(i) A list of all information references and sources used in preparing the documentation.

(ii) Records, including the dates, of each compliance test required by § 265.1033(j).

(iii) If engineering calculations are used, a design analysis, specifications, drawings, schematics, and piping and instrumentation diagrams based on the appropriate sections of "APTI Course 415: Control of Gaseous Emissions" (incorporated by reference as specified in § 260.11) or other engineering texts acceptable to the Regional Administrator that present basic control device design information. Documentation provided by the control device manufacturer or vendor that describes the control device design in accordance with paragraphs (b)(4)(iii)(A) through (b)(4)(iii)(G) of this section may be used to comply with this requirement. The design analysis shall address the vent stream characteristics and control device operation parameters as specified below.

(A) For a thermal vapor incinerator, the design analysis shall consider the vent stream composition, constituent concentrations, and flow rate. The design analysis shall also establish the design minimum and average temperature in the combustion zone and the combustion zone residence time.

(B) For a catalytic vapor incinerator, the design analysis shall consider the vent stream composition, constituent concentrations, and flow rate. The design analysis shall also establish the design minimum and average temperatures across the catalyst bed inlet and outlet.

(C) For a boiler or process heater, the design analysis shall consider the vent stream composition, constituent concentrations, and flow rate. The design analysis shall also establish the design minimum and average flame zone temperatures, combustion zone residence time, and description of method and location where the vent stream is introduced into the combustion zone.

(D) For a flare, the design analysis shall consider the vent stream composition, constituent concentrations, and flow rate. The design analysis shall also consider the requirements specified in §265.1033(d).

(E) For a condenser, the design analysis shall consider the vent stream composition, constituent concentrations, flow rate, relative humidity, and temperature. The design analysis shall also establish the design outlet organic compound concentration level, design average temperature of the condenser exhaust vent stream, and design average temperatures of the coolant fluid at the condenser inlet and outlet.

(F) For a carbon adsorption system such as a fixed-bed adsorber that regenerates the carbon bed directly onsite in the control device, the design analysis shall consider the vent stream composition, constituent concentrations, flow rate, relative humidity, and temperature. The design analysis shall also establish the design exhaust vent stream organic compound concentration level, number and capacity of carbon beds, type and working capacity of activated carbon used for carbon beds, design total steam flow over the period of each complete carbon bed regeneration cycle, duration of the carbon bed steaming and cooling/drying cycles, design carbon bed temperature after regeneration, design carbon bed regeneration time, and design service life of carbon.

(G) For a carbon adsorption system such as a carbon canister that does not regenerate the carbon bed directly onsite in the control device, the design analysis shall consider the vent stream composition, constituent concentrations, flow rate, relative humidity, and temperature. The design analysis shall also establish the design outlet organic concentration level, capacity of carbon bed, type and working capacity of activated carbon used for carbon bed, and design carbon replacement interval based on the total carbon working capacity of the control device and source operating schedule.

(iv) A statement signed and dated by the owner or operator certifying that the operating parameters used in the design analysis reasonably represent the conditions that exist when the hazardous waste management unit is or would be operating at the highest load or capacity level reasonably expected to occur.

(v) A statement signed and dated by the owner or operator certifying that the control device is designed to operate at an efficiency of 95 percent or greater unless the total organic concentration limit of §265.1032(a) is achieved at an efficiency less than 95 weight percent or the total organic emission limits of §265.1032(a) for affected process vents at the facility can be attained by a control device involving vapor recovery at an efficiency less than 95 weight percent. A statement provided by the control device manufacturer or vendor certifying that the control equipment meets the design specifications may be used to comply with this requirement.

(vi) If performance tests are used to demonstrate compliance, all test results.

(c) Design documentation and monitoring, operating, and inspection information for each closed-vent system and control device required to comply with the provisions of this part shall be recorded and kept up-to-date in the facility operating record. The information shall include:

(1) Description and date of each modification that is made to the closed-vent system or control device design.

(2) Identification of operating parameter, description of monitoring device, and diagram of monitoring sensor location or locations used to comply with § 265.1033(f)(1) and (f)(2).

(3) Monitoring, operating and inspection information required by paragraphs (f) through (k) of § 265.1033 of this subpart.

(4) Date, time, and duration of each period that occurs while the control device is operating when any monitored parameter exceeds the value established in the control device design analysis as specified below:

(i) For a thermal vapor incinerator designed to operate with a minimum residence time of 0.50 seconds at a minimum temperature of 760 °C, period when the combustion temperature is below 760 °C.

(ii) For a thermal vapor incinerator designed to operate with an organic emission reduction efficiency of 95 percent or greater, period when the combustion zone temperature is more than 28 °C below the design average combustion zone temperature established as a requirement of paragraph (b)(4)(iii)(A) of this section.

(iii) For a catalytic vapor incinerator, period when:

(A) Temperature of the vent stream at the catalyst bed inlet is more than 28 °C below the average temperature of the inlet vent stream established as a requirement of paragraph (b)(4)(iii)(B) of this section; or

(B) Temperature difference across the catalyst bed is less than 80 percent of the design average temperature difference established as a requirement of paragraph (b)(4)(iii)(B) of this section.

(iv) For a boiler or process heater, period when:

(A) Flame zone temperature is more than 28 °C below the design average flame zone temperature established as a requirement of paragraph (b)(4)(iii)(C) of this section; or

(B) Position changes where the vent stream is introduced to the combustion zone from the location established as a requirement of paragraph (b)(4)(iii)(C) of this section.

(v) For a flare, period when the pilot flame is not ignited.

(vi) For a condenser that complies with § 265.1033(f)(2)(vi)(A), period when the organic compound concentration level or readings of organic compounds in the exhaust vent stream from the condenser are more than 20 percent greater than the design outlet organic compound concentration level established as a requirement of paragraph (b)(4)(iii)(E) of this section.

(vii) For a condenser that complies with § 265.1033(f)(2)(vi)(B), period when:

(A) Temperature of the exhaust vent stream from the condenser is more than 6 °C above the design average exhaust vent stream temperature established as a requirement of paragraph (b)(4)(iii)(E) of this section; or

(B) Temperature of the coolant fluid exiting the condenser is more than 6 °C above the design average coolant fluid temperature at the condenser outlet established as a requirement of paragraph (b)(4)(iii)(E) of this section.

(viii) For a carbon adsorption system such as a fixed-bed carbon adsorber that regenerates the carbon bed directly onsite in the control device and complies with § 265.1033(f)(2)(vii)(A), period when the organic compound concentration level or readings of organic compounds in the exhaust vent stream from the carbon bed are more than 20 percent greater than the design exhaust vent stream organic compound concentration level established as a requirement of paragraph (b)(4)(iii)(F) of this section.

(ix) For a carbon adsorption system such as a fixed-bed carbon adsorber that regenerates the carbon bed directly onsite in the control device and complies with § 265.1033(f)(2)(vii)(B), period when the vent stream continues to flow through the control device beyond the predetermined carbon bed regeneration time established as a requirement of paragraph (b)(4)(iii)(F) of this section.

(5) Explanation for each period recorded under paragraph (c)(4) of this section of the cause for control device operating parameter exceeding the design value and the measures implemented to correct the control device operation.

(6) For carbon adsorption systems operated subject to requirements specified in § 265.1033(g) or § 265.1033(h)(2), date when existing carbon in the

control device is replaced with fresh carbon.

(7) For carbon adsorption systems operated subject to requirements specified in §265.1033(h)(1), a log that records:

(i) Date and time when control device is monitored for carbon breakthrough and the monitoring device reading.

(ii) Date when existing carbon in the control device is replaced with fresh carbon.

(8) Date of each control device start-up and shutdown.

(9) An owner or operator designating any components of a closed-vent system as unsafe to monitor pursuant to §265.1033(n) of this subpart shall record in a log that is kept in the facility operating record the identification of closed-vent system components that are designated as unsafe to monitor in accordance with the requirements of §265.1033(n) of this subpart, an explanation for each closed-vent system component stating why the closed-vent system component is unsafe to monitor, and the plan for monitoring each closed-vent system component.

(10) When each leak is detected as specified in §265.1033(k) of this subpart, the following information shall be recorded:

(i) The instrument identification number, the closed-vent system component identification number, and the operator name, initials, or identification number.

(ii) The date the leak was detected and the date of first attempt to repair the leak.

(iii) The date of successful repair of the leak.

(iv) Maximum instrument reading measured by Method 21 of 40 CFR part 60, appendix A after it is successfully repaired or determined to be nonrepairable.

(v) "Repair delayed" and the reason for the delay if a leak is not repaired within 15 calendar days after discovery of the leak.

(A) The owner or operator may develop a written procedure that identifies the conditions that justify a delay of repair. In such cases, reasons for delay of repair may be documented by citing the relevant sections of the written procedure.

(B) If delay of repair was caused by depletion of stocked parts, there must be documentation that the spare parts were sufficiently stocked on-site before depletion and the reason for depletion.

(d) Records of the monitoring, operating, and inspection information required by paragraphs (c)(3) through (c)(10) of this section shall be maintained by the owner or operator for at least 3 years following the date of each occurrence, measurement, maintenance, corrective action, or record.

(e) For a control device other than a thermal vapor incinerator, catalytic vapor incinerator, flare, boiler, process heater, condenser, or carbon adsorption system, monitoring and inspection information indicating proper operation and maintenance of the control device must be recorded in the facility operating record.

(f) Up-to-date information and data used to determine whether or not a process vent is subject to the requirements in §265.1032 including supporting documentation as required by §265.1034(d)(2) when application of the knowledge of the nature of the hazardous waste stream or the process by which it was produced is used, shall be recorded in a log that is kept in the facility operating record.

[55 FR 25507, June 21, 1990, as amended at 56 FR 19290, Apr. 26, 1991; 61 FR 59970, Nov. 25, 1996; 71 FR 40276, July 14, 2006]

## §§ 265.1036–265.1049 [Reserved]

## Subpart BB—Air Emission Standards for Equipment Leaks

SOURCE: 55 FR 25512, June 21, 1990, unless otherwise noted.

### §265.1050 Applicability.

(a) The regulations in this subpart apply to owners and operators of facilities that treat, store, or dispose of hazardous wastes (except as provided in §265.1).

(b) Except as provided in §265.1064(k), this subpart applies to equipment that contains or contacts hazardous wastes with organic concentrations of at least 10 percent by weight that are managed in one of the following:

(1) A unit that is subject to the permitting requirements of 40 CFR part 270, or

(2) A unit (including a hazardous waste recycling unit) that is not exempt from permitting under the provisions of 40 CFR 262.34(a) (i.e., a hazardous waste recycling unit that is not a 90-day tank or container) and that is located at a hazardous waste management facility otherwise subject to the permitting requirements of 40 CFR part 270, or

(3) A unit that is exempt from permitting under the provisions of 40 CFR 262.34(a) (i.e., a "90-day" tank or container) and is not a recycling unit under the provisions of 40 CFR 261.6.

(c) Each piece of equipment to which this subpart applies shall be marked in such a manner that it can be distinguished readily from other pieces of equipment.

(d) Equipment that is in vacuum service is excluded from the requirements of § 265.1052 to § 265.1060 if it is identified as required in § 265.1064(g)(5).

(e) Equipment that contains or contacts hazardous waste with an organic concentration of at least 10 percent by weight for less than 300 hours per calendar year is excluded from the requirements of §§ 265.1052 through 265.1060 of this subpart if it is identified, as required in § 265.1064(g)(6) of this subpart.

(f) The requirements of this subpart do not apply to the pharmaceutical manufacturing facility, commonly referred to as the Stonewall Plant, located at Route 340 South, Elkton, Virginia, provided that facility is operated in compliance with the requirements contained in a Clean Air Act permit issued pursuant to 40 CFR 52.2454. The requirements of this subpart shall apply to the facility upon termination of the Clean Air Act permit issued pursuant to 40 CFR 52.2454.

(g) Purged coatings and solvents from surface coating operations subject to the national emission standards for hazardous air pollutants (NESHAP) for the surface coating of automobiles and light-duty trucks at 40 CFR part 63, subpart IIII, are not subject to the requirements of this subpart.

[NOTE: The requirements of §§ 265.1052 through 265.1064 apply to equipment associated with hazardous waste recycling units previously exempt under paragraph 261.6(c)(1). Other exemptions under §§ 261.4 and 265.1(c) are not affected by these requirements.]

[55 FR 25512, June 21, 1990, as amended at 61 FR 59970, Nov. 25, 1996; 62 FR 52642, Oct. 8, 1997; 62 FR 64661, Dec. 8, 1997; 69 FR 22661, Apr. 26, 2004]

§ 265.1051   Definitions.

As used in this subpart, all terms shall have the meaning given them in § 264.1031, the Act, and parts 260–266.

§ 265.1052   Standards: Pumps in light liquid service.

(a)(1) Each pump in light liquid service shall be monitored monthly to detect leaks by the methods specified in § 265.1063(b), except as provided in paragraphs (d), (e), and (f) of this section.

(2) Each pump in light liquid service shall be checked by visual inspection each calendar week for indications of liquids dripping from the pump seal.

(b)(1) If an instrument reading of 10,000 ppm or greater is measured, a leak is detected.

(2) If there are indications of liquids dripping from the pump seal, a leak is detected.

(c)(1) When a leak is detected, it shall be repaired as soon as practicable, but not later than 15 calendar days after it is detected, except as provided in § 265.1059.

(2) A first attempt at repair (e.g., tightening the packing gland) shall be made no later than 5 calendar days after each leak is detected.

(d) Each pump equipped with a dual mechanical seal system that includes a barrier fluid system is exempt from the requirements of paragraph (a), *provided* the following requirements are met:

(1) Each dual mechanical seal system must be:

(i) Operated with the barrier fluid at a pressure that is at all times greater than the pump stuffing box pressure, or

(ii) Equipped with a barrier fluid degassing reservoir that is connected by a closed-vent system to a control device that complies with the requirements of § 265.1060, or

(iii) Equipped with a system that purges the barrier fluid into a hazardous waste stream with no detectable emissions to the atmosphere.

(2) The barrier fluid system must not be a hazardous waste with organic concentrations 10 percent or greater by weight.

(3) Each barrier fluid system must be equipped with a sensor that will detect failure of the seal system, the barrier fluid system or both.

(4) Each pump must be checked by visual inspection, each calendar week, for indications of liquids dripping from the pump seals.

(5)(i) Each sensor as described in paragraph (d)(3) of this section must be checked daily or be equipped with an audible alarm that must be checked monthly to ensure that it is functioning properly.

(ii) The owner or operator must determine, based on design considerations and operating experience, a criterion that indicates failure of the seal system, the barrier fluid system, or both.

(6)(i) If there are indications of liquids dripping from the pump seal or the sensor indicates failure of the seal system, the barrier fluid system, or both based on the criterion determined in paragraph (d)(5)(ii) of this section, a leak is detected.

(ii) When a leak is detected, it shall be repaired as soon as practicable, but not later than 15 calendar days after it is detected, except as provided in §265.1059.

(iii) A first attempt at repair (e.g., relapping the seal) shall be made no later than 5 calendar days after each leak is detected.

(e) Any pump that is designated, as described in §265.1064(g)(2), for no detectable emissions, as indicated by an instrument reading of less than 500 ppm above background, is exempt from the requirements of paragraphs (a), (c), and (d) of this section if the pump meets the following requirements:

(1) Must have no externally actuated shaft penetrating the pump housing.

(2) Must operate with no detectable emissions as indicated by an instrument reading of less than 500 ppm above background as measured by the methods specified in §265.1063(c).

(3) Must be tested for compliance with paragraph (e)(2) of this section initially upon designation, annually,

and at other times as requested by the Regional Administrator.

(f) If any pump is equipped with a closed-vent system capable of capturing and transporting any leakage from the seal or seals to a control device that complies with the requirements of §265.1060, it is exempt from the requirements of paragraphs (a) through (e) of this section.

[55 FR 25512, June 21, 1990, as amended at 56 FR 19290, Apr. 26, 1991]

## §265.1053 Standards: Compressors.

(a) Each compressor shall be equipped with a seal system that includes a barrier fluid system and that prevents leakage of total organic emissions to the atmosphere, except as provided in paragraphs (h) and (i) of this section.

(b) Each compressor seal system as required in paragraph (a) of this section shall be:

(1) Operated with the barrier fluid at a pressure that is at all times greater than the compressor stuffing box pressure, or

(2) Equipped with a barrier fluid system that is connected by a closed-vent system to a control device that complies with the requirements of §265.1060, or

(3) Equipped with a system that purges the barrier fluid into a hazardous waste stream with no detectable emissions to atmosphere.

(c) The barrier fluid must not be a hazardous waste with organic concentrations 10 percent or greater by weight.

(d) Each barrier fluid system as described in paragraphs (a) through (c) of this section shall be equipped with a sensor that will detect failure of the seal system, barrier fluid system, or both.

(e)(1) Each sensor as required in paragraph (d) of this section shall be checked daily or shall be equipped with an audible alarm that must be checked monthly to ensure that it is functioning properly unless the compressor is located within the boundary of an unmanned plant site, in which case the sensor must be checked daily.

(2) The owner or operator shall determine, based on design considerations and operating experience, a criterion

703

that indicates failure of the seal system, the barrier fluid system or both.

(f) If the sensor indicates failure of the seal system, the barrier fluid system, or both based on the criterion determined under paragraph (e)(2) of this section, a leak is detected.

(g)(1) When a leak is detected, it shall be repaired as soon as practicable, but not later than 15 calendar days after it is detected, except as provided in § 265.1059.

(2) A first attempt at repair (e.g., tightening the packing gland) shall be made no later than 5 calendar days after each leak is detected.

(h) A compressor is exempt from the requirements of paragraphs (a) and (b) of this section if it is equipped with a closed-vent system capable of capturing and transporting any leakage from the seal to a control device that complies with the requirements of § 265.1060, except as provided in paragraph (i) of this section.

(i) Any compressor that is designated, as described in § 265.1064(g)(2), for no detectable emission as indicated by an instrument reading of less than 500 ppm above background is exempt from the requirements of paragraphs (a) through (h) of this section if the compressor:

(1) Is determined to be operating with no detectable emissions, as indicated by an instrument reading of less than 500 ppm above background, as measured by the method specified in § 265.1063(c).

(2) Is tested for compliance with paragraph (i)(1) of this section initially upon designation, annually, and at other times as requested by the Regional Administrator.

§ 265.1054  Standards: Pressure relief devices in gas/vapor service.

(a) Except during pressure releases, each pressure relief device in gas/vapor service shall be operated with no detectable emissions, as indicated by an instrument reading of less than 500 ppm above background, as measured by the method specified in § 265.1063(c).

(b)(1) After each pressure release, the pressure relief device shall be returned to a condition of no detectable emissions, as indicated by an instrument reading of less than 500 ppm above

background, as soon as practicable, but no later than 5 calendar days after each pressure release, except as provided in § 265.1059.

(2) No later than 5 calendar days after the pressure release, the pressure relief device shall be monitored to confirm the condition of no detectable emissions, as indicated by an instrument reading of less than 500 ppm above background, as measured by the method specified in § 265.1063(c).

(c) Any pressure relief device that is equipped with a closed-vent system capable of capturing and transporting leakage from the pressure relief device to a control device as described in § 265.1060 is exempt from the requirements of paragraphs (a) and (b) of this section.

§ 265.1055  Standards: Sampling connection systems.

(a) Each sampling connection system shall be equipped with a closed-purge, closed-loop, or closed-vent system. This system shall collect the sample purge for return to the process or for routing to the appropriate treatment system. Gases displaced during filling of the sample container are not required to be collected or captured.

(b) Each closed-purge, closed-loop, or closed-vent system as required in paragraph (a) of this section shall:

(1) Return the purged process fluid directly to the process line; or

(2) Collect and recycle the purged process fluid; or

(3) Be designed and operated to capture and transport all the purged process fluid to a waste management unit that complies with the applicable requirements of § 265.1085 through § 265.1087 of this subpart or a control device that complies with the requirements of § 265.1060 of this subpart.

(c) *In-situ* sampling systems and sampling systems without purges are exempt from the requirements of paragraphs (a) and (b) of this section.

[61 FR 59971, Nov. 25, 1996]

§ 265.1056  Standards:        Open-ended valves or lines.

(a)(1) Each open-ended valve or line shall be equipped with a cap, blind flange, plug, or a second valve.

(2) The cap, blind flange, plug, or second valve shall seal the open end at all times except during operations requiring hazardous waste stream flow through the open-ended valve or line.

(b) Each open-ended valve or line equipped with a second valve shall be operated in a manner such that the valve on the hazardous waste stream end is closed before the second valve is closed.

(c) When a double block and bleed system is being used, the bleed valve or line may remain open during operations that require venting the line between the block valves but shall comply with paragraph (a) of this section at all other times.

## § 265.1057 Standards: Valves in gas/vapor service or in light liquid service.

(a) Each valve in gas/vapor or light liquid service shall be monitored monthly to detect leaks by the methods specified in § 265.1063(b) and shall comply with paragraphs (b) through (e) of this section, except as provided in paragraphs (f), (g), and (h) of this section, and §§ 265.1061 and 265.1062.

(b) If an instrument reading of 10,000 ppm or greater is measured, a leak is detected.

(c)(1) Any valve for which a leak is not detected for two successive months may be monitored the first month of every succeeding quarter, beginning with the next quarter, until a leak is detected.

(2) If a leak is detected, the valve shall be monitored monthly until a leak is not detected for 2 successive months.

(d)(1) When a leak is detected, it shall be repaired as soon as practicable, but no later than 15 calendar days after the leak is detected, except as provided in § 265.1059.

(2) A first attempt at repair shall be made no later than 5 calendar days after each leak is detected.

(e) First attempts at repair include, but are not limited to, the following best practices where practicable:

(1) Tightening of bonnet bolts.

(2) Replacement of bonnet bolts.

(3) Tightening of packing gland nuts.

(4) Injection of lubricant into lubricated packing.

(f) Any valve that is designated, as described in § 265.1064(g)(2), for no detectable emissions, as indicated by an instrument reading of less than 500 ppm above background, is exempt from the requirements of paragraph (a) of this section if the valve:

(1) Has no external actuating mechanism in contact with the hazardous waste stream.

(2) Is operated with emissions less than 500 ppm above background as determined by the method specified in § 265.1063(c).

(3) Is tested for compliance with paragraph (f)(2) of this section initially upon designation, annually, and at other times as requested by the Regional Administrator.

(g) Any valve that is designated, as described in § 265.1064(h)(1), as an unsafe-to-monitor valve is exempt from the requirements of paragraph (a) of this section if:

(1) The owner or operator of the valve determines that the valve is unsafe to monitor because monitoring personnel would be exposed to an immediate danger as a consequence of complying with paragraph (a) of this section.

(2) The owner or operator of the valve adheres to a written plan that requires monitoring of the valve as frequently as practicable during safe-to-monitor times.

(h) Any valve that is designated, as described in § 265.1064(h)(2), as a difficult-to-monitor valve is exempt from the requirements of paragraph (a) of this section if:

(1) The owner or operator of the valve determines that the valve cannot be monitored without elevating the monitoring personnel more than 2 meters above a support surface.

(2) The hazardous waste management unit within which the valve is located was in operation before June 21, 1990.

(3) The owner or operator of the valve follows a written plan that requires monitoring of the valve at least once per calendar year.

## § 265.1058 Standards: Pumps and valves in heavy liquid service, pressure relief devices in light liquid or heavy liquid service, and flanges and other connectors.

(a) Pumps and valves in heavy liquid service, pressure relief devices in light liquid or heavy liquid service, and flanges and other connectors shall be monitored within 5 days by the method specified in § 265.1063(b) if evidence of a potential leak is found by visual, audible, olfactory, or any other detection method.

(b) If an instrument reading of 10,000 ppm or greater is measured, a leak is detected.

(c)(1) When a leak is detected, it shall be repaired as soon as practicable, but not later than 15 calendar days after it is detected, except as provided in § 265.1059.

(2) The first attempt at repair shall be made no later than 5 calendar days after each leak is detected.

(d) First attempts at repair include, but are not limited to, the best practices described under § 265.1057(e).

(e) Any connector that is inaccessible or is ceramic or ceramic-lined (e.g., porcelain, glass, or glass-lined) is exempt from the monitoring requirements of paragraph (a) of this section and from the recordkeeping requirements of § 265.1064 of this subpart.

[55 FR 25512, June 21, 1990, as amended at 61 FR 59971, Nov. 25, 1996]

### § 265.1059  Standards: Delay of repair.

(a) Delay of repair of equipment for which leaks have been detected will be allowed if the repair is technically infeasible without a hazardous waste management unit shutdown. In such a case, repair of this equipment shall occur before the end of the next hazardous waste management unit shutdown.

(b) Delay of repair of equipment for which leaks have been detected will be allowed for equipment that is isolated from the hazardous waste management unit and that does not continue to contain or contact hazardous waste with organic concentrations at least 10 percent by weight.

(c) Delay of repair for valves will be allowed if:

(1) The owner or operator determines that emissions of purged material resulting from immediate repair are greater than the emissions likely to result from delay of repair.

(2) When repair procedures are effected, the purged material is collected and destroyed or recovered in a control device complying with § 265.1060.

(d) Delay of repair for pumps will be allowed if:

(1) Repair requires the use of a dual mechanical seal system that includes a barrier fluid system.

(2) Repair is completed as soon as practicable, but not later than 6 months after the leak was detected.

(e) Delay of repair beyond a hazardous waste management unit shutdown will be allowed for a valve if valve assembly replacement is necessary during the hazardous waste management unit shutdown, valve assembly supplies have been depleted, and valve assembly supplies had been sufficiently stocked before the supplies were depleted. Delay of repair beyond the next hazardous waste management unit shutdown will not be allowed unless the next hazardous waste management unit shutdown occurs sooner than 6 months after the first hazardous waste management unit shutdown.

### § 265.1060  Standards: Closed-vent systems and control devices.

(a) Owners and operators of closed-vent systems and control devices subject to this subpart shall comply with the provisions of § 265.1033 of this part.

(b)(1) The owner or operator of an existing facility who can not install a closed-vent system and control device to comply with the provisions of this subpart on the effective date that the facility becomes subject to the provisions of this subpart must prepare an implementation schedule that includes dates by which the closed-vent system and control device will be installed and in operation. The controls must be installed as soon as possible, but the implementation schedule may allow up to 30 months after the effective date that the facility becomes subject to this subpart for installation and startup.

(2) Any units that begin operation after December 21, 1990, and are subject to the provisions of this subpart when operation begins, must comply with the rules immediately (i.e., must have control devices installed and operating on startup of the affected unit); the 30-month implementation schedule does not apply.

(3) The owner or operator of any facility in existence on the effective date of a statutory or EPA regulatory amendment that renders the facility subject to this subpart shall comply with all requirements of this subpart as soon as practicable but no later than 30 months after the amendment's effective date. When control equipment required by this subpart can not be installed and begin operation by the effective date of the amendment, the facility owner or operator shall prepare an implementation schedule that includes the following information: Specific calendar dates for award of contracts or issuance of purchase orders for the control equipment, initiation of on-site installation of the control equipment, completion of the control equipment installation, and performance of any testing to demonstrate that the installed equipment meets the applicable standards of this subpart. The owner or operator shall enter the implementation schedule in the operating record or in a permanent, readily available file located at the facility.

(4) Owners and operators of facilities and units that become newly subject to the requirements of this subpart after December 8, 1997 due to an action other than those described in paragraph (b)(3) of this section must comply with all applicable requirements immediately (i.e., must have control devices installed and operating on the date the facility or unit becomes subject to this subpart; the 30-month implementation schedule does not apply).

[62 FR 64662, Dec. 8, 1997]

**§265.1061  Alternative standards for valves in gas/vapor service or in light liquid service: percentage of valves allowed to leak.**

(a) An owner or operator subject to the requirements of §265.1057 may elect to have all valves within a hazardous waste management unit comply with an alternative standard which allows no greater than 2 percent of the valves to leak.

(b) The following requirements shall be met if an owner or operator decides to comply with the alternative standard of allowing 2 percent of valves to leak:

(1) A performance test as specified in paragraph (c) of this section shall be conducted initially upon designation, annually, and at other times requested by the Regional Administrator.

(2) If a valve leak is detected, it shall be repaired in accordance with §265.1057 (d) and (e).

(c) Performance tests shall be conducted in the following manner:

(1) All valves subject to the requirements in §265.1057 within the hazardous waste management unit shall be monitored within 1 week by the methods specified in §265.1063(b).

(2) If an instrument reading of 10,000 ppm or greater is measured, a leak is detected.

(3) The leak percentage shall be determined by dividing the number of valves subject to the requirements in §265.1057 for which leaks are detected by the total number of valves subject to the requirements in §265.1057 within the hazardous waste management unit.

[55 FR 25512, June 21, 1990, as amended at 71 FR 16912, Apr. 4, 2006]

**§265.1062  Alternative standards for valves in gas/vapor service or in light liquid service: skip period leak detection and repair.**

(a) An owner or operator subject to the requirements of §265.1057 may elect for all valves within a hazardous waste management unit to comply with one of the alternative work practices specified in paragraphs (b) (2) and (3) of this section.

(b)(1) An owner or operator shall comply with the requirements for valves, as described in §265.1057, except as described in paragraphs (b)(2) and (b)(3) of this section.

(2) After two consecutive quarterly leak detection periods with the percentage of valves leaking equal to or less than 2 percent, an owner or operator may begin to skip one of the quarterly leak detection periods (i.e., monitor for leaks once every six months) for the valves subject to the requirements in §265.1057 of this subpart.

(3) After five consecutive quarterly leak detection periods with the percentage of valves leaking equal to or less than 2 percent, an owner or operator may begin to skip three of the quarterly leak detection periods (i.e.,

monitor for leaks once every year) for the valves subject to the requirements in § 265.1057 of this subpart.

(4) If the percentage of valves leaking is greater than 2 percent, the owner or operators shall monitor monthly in compliance with the requirements in § 265.1057, but may again elect to use this section after meeting the requirements of § 265.1057(c)(1).

[55 FR 25512, June 21, 1990, as amended at 62 FR 64662, Dec. 8, 1997; 71 FR 16912, Apr. 4, 2006]

§ 265.1063 Test methods and procedures.

(a) Each owner or operator subject to the provisions of this subpart shall comply with the test methods and procedures requirements provided in this section.

(b) Leak detection monitoring, as required in §§ 265.1052 through 265.1062, shall comply with the following requirements:

(1) Monitoring shall comply with Reference Method 21 in 40 CFR part 60.

(2) The detection instrument shall meet the performance criteria of Reference Method 21.

(3) The instrument shall be calibrated before use on each day of its use by the procedures specified in Reference Method 21.

(4) Calibration gases shall be:

(i) Zero air (less than 10 ppm of hydrocarbon in air).

(ii) A mixture of methane or n-hexane and air at a concentration of approximately, but less than, 10,000 ppm methane or n-hexane.

(5) The instrument probe shall be traversed around all potential leak interfaces as close to the interface as possible as described in Reference Method 21.

(c) When equipment is tested for compliance with no detectable emissions, as required in §§ 265.1052(e), 265.1053(i), 265.1054, and 265.1057(f), the test shall comply with the following requirements:

(1) The requirements of paragraphs (b) (1) through (4) of this section shall apply.

(2) The background level shall be determined, as set forth in Reference Method 21.

(3) The instrument probe shall be traversed around all potential leak interfaces as close to the interface as possible as described in Reference Method 21.

(4) The arithmetic difference between the maximum concentration indicated by the instrument and the background level is compared with 500 ppm for determining compliance.

(d) In accordance with the waste analysis plan required by § 265.13(b), an owner or operator of a facility must determine, for each piece of equipment, whether the equipment contains or contacts a hazardous waste with organic concentration that equals or exceeds 10 percent by weight using the following:

(1) Methods described in ASTM Methods D 2267–88, E 169–87, E 168–88, E 260–85 (incorporated by reference under § 260.11);

(2) Method 9060A (incorporated by reference under § 260.11 of this chapter) of "Test Methods for Evaluating Solid Waste," EPA Publication SW–846 or analyzed for its individual organic constituents; or

(3) Application of the knowledge of the nature of the hazardous waste stream or the process by which it was produced. Documentation of a waste determination by knowledge is required. Examples of documentation that shall be used to support a determination under this provision include production process information documenting that no organic compounds are used, information that the waste is generated by a process that is identical to a process at the same or another facility that has previously been demonstrated by direct measurement to have a total organic content less than 10 percent, or prior speciation analysis results on the same waste stream where it can also be documented that no process changes have occurred since that analysis that could affect the waste total organic concentration.

(e) If an owner or operator determines that a piece of equipment contains or contacts a hazardous waste with organic concentrations at least 10 percent by weight, the determination can be revised only after following the procedures in paragraph (d)(1) or (d)(2) of this section.

(f) When an owner or operator and the Regional Administrator do not agree on whether a piece of equipment contains or contacts a hazardous waste with organic concentrations at least 10 percent by weight, the procedures in paragraph (d)(1) or (d)(2) of this section can be used to resolve the dispute.

(g) Samples used in determining the percent organic content shall be representative of the highest total organic content hazardous waste that is expected to be contained in or contact the equipment.

(h) To determine if pumps or valves are in light liquid service, the vapor pressures of constituents may be obtained from standard reference texts or may be determined by ASTM D–2879–86 (incorporated by reference under §260.11).

(i) Performance tests to determine if a control device achieves 95 weight percent organic emission reduction shall comply with the procedures of §265.1034 (c)(1) through (c)(4).

[55 FR 25512, June 21, 1990, as amended at 62 FR 32463, June 13, 1997; 70 FR 34586, June 14, 2005; 71 FR 40276, July 14, 2006]

## §265.1064 Recordkeeping requirements.

(a)(1) Each owner or operator subject to the provisions of this subpart shall comply with the recordkeeping requirements of this section.

(2) An owner or operator of more than one hazardous waste management unit subject to the provisions of this subpart may comply with the recordkeeping requirements for these hazardous waste management units in one recordkeeping system if the system identifies each record by each hazardous waste management unit.

(b) Owners and operators must record the following information in the facility operating record:

(1) For each piece of equipment to which subpart BB of part 265 applies:

(i) Equipment identification number and hazardous waste management unit identification.

(ii) Approximate locations within the facility (e.g., identify the hazardous waste management unit on a facility plot plan).

(iii) Type of equipment (e.g., a pump or pipeline valve).

(iv) Percent-by-weight total organics in the hazardous waste stream at the equipment.

(v) Hazardous waste state at the equipment (e.g., gas/vapor or liquid).

(vi) Method of compliance with the standard (e.g., "monthly leak detection and repair" or "equipped with dual mechanical seals").

(2) For facilities that comply with the provisions of §265.1033(a)(2), an implementation schedule as specified in §265.1033(a)(2).

(3) Where an owner or operator chooses to use test data to demonstrate the organic removal efficiency or total organic compound concentration achieved by the control device, a performance test plan as specified in §265.1035(b)(3).

(4) Documentation of compliance with §265.1060, including the detailed design documentation or performance test results specified in §265.1035(b)(4).

(c) When each leak is detected as specified in §§265.1052, 265.1053, 265.1057, and 265.1058, the following requirements apply:

(1) A weatherproof and readily visible identification, marked with the equipment identification number, the date evidence of a potential leak was found in accordance with §265.1058(a), and the date the leak was detected, shall be attached to the leaking equipment.

(2) The identification on equipment, except on a valve, may be removed after it has been repaired.

(3) The identification on a valve may be removed after it has been monitored for 2 successive months as specified in §265.1057(c) and no leak has been detected during those 2 months.

(d) When each leak is detected as specified in §§265.1052, 265.1053, 265.1057, and 265.1058, the following information shall be recorded in an inspection log and shall be kept in the facility operating record:

(1) The instrument and operator identification numbers and the equipment identification number.

(2) The date evidence of a potential leak was found in accordance with §265.1058(a).

(3) The date the leak was detected and the dates of each attempt to repair the leak.

(4) Repair methods applied in each attempt to repair the leak.

(5) "Above 10,000" if the maximum instrument reading measured by the methods specified in § 265.1063(b) after each repair attempt is equal to or greater than 10,000 ppm.

(6) "Repair delayed" and the reason for the delay if a leak is not repaired within 15 calendar days after discovery of the leak.

(7) Documentation supporting the delay of repair of a valve in compliance with § 265.1059(c).

(8) The signature of the owner or operator (or designate) whose decision it was that repair could not be effected without a hazardous waste management unit shutdown.

(9) The expected date of successful repair of the leak if a leak is not repaired within 15 calendar days.

(10) The date of successful repair of the leak.

(e) Design documentation and monitoring, operating, and inspection information for each closed-vent system and control device required to comply with the provisions of § 265.1060 shall be recorded and kept up-to-date in the facility operating record as specified in § 265.1035(c). Design documentation is specified in § 265.1035 (c)(1) and (c)(2) and monitoring, operating, and inspection information in § 265.1035 (c)(3)–(c)(8).

(f) For a control device other than a thermal vapor incinerator, catalytic vapor incinerator, flare, boiler, process heater, condenser, or carbon adsorption system, monitoring and inspection information indicating proper operation and maintenance of the control device must be recorded in the facility operating record.

(g) The following information pertaining to all equipment subject to the requirements in §§ 265.1052 through 265.1060 shall be recorded in a log that is kept in the facility operating record:

(1) A list of identification numbers for equipment (except welded fittings) subject to the requirements of this subpart.

(2)(i) A list of identification numbers for equipment that the owner or operator elects to designate for no detectable emissions, as indicated by an instrument reading of less than 500 ppm

above background, under the provisions of §§ 265.1052(e), 265.1053(i), and 265.1057(f).

(ii) The designation of this equipment as subject to the requirements of §§ 265.1052(e), 265.1053(i), or 265.1057(f) shall be signed by the owner or operator.

(3) A list of equipment identification numbers for pressure relief devices required to comply with § 265.1054(a).

(4)(i) The dates of each compliance test required in §§ 265.1052(e), 265.1053(i), 265.1054, and 265.1057(f).

(ii) The background level measured during each compliance test.

(iii) The maximum instrument reading measured at the equipment during each compliance test.

(5) A list of identification numbers for equipment in vacuum service.

(6) Identification, either by list or location (area or group) of equipment that contains or contacts hazardous waste with an organic concentration of at least 10 percent by weight for less than 300 hours per calendar year.

(h) The following information pertaining to all valves subject to the requirements of § 265.1057 (g) and (h) shall be recorded in a log that is kept in the facility operating record:

(1) A list of identification numbers for valves that are designated as unsafe to monitor, an explanation for each valve stating why the valve is unsafe to monitor, and the plan for monitoring each valve.

(2) A list of identification numbers for valves that are designated as difficult to monitor, an explanation for each valve stating why the valve is difficult to monitor, and the planned schedule for monitoring each valve.

(i) The following information shall be recorded in the facility operating record for valves complying with § 265.1062:

(1) A schedule of monitoring.

(2) The percent of valves found leaking during each monitoring period.

(j) The following information shall be recorded in a log that is kept in the facility operating record:

(1) Criteria required in §§ 265.1052 (d)(5)(ii) and 265.1053(e)(2) and an explanation of the criteria.

(2) Any changes to these criteria and the reasons for the changes.

(k) The following information shall be recorded in a log that is kept in the facility operating record for use in determining exemptions as provided in the applicability section of this subpart and other specific subparts:

(1) An analysis determining the design capacity of the hazardous waste management unit.

(2) A statement listing the hazardous waste influent to and effluent from each hazardous waste management unit subject to the requirements in §§265.1052 through 265.1060 and an analysis determining whether these hazardous wastes are heavy liquids.

(3) An up-to-date analysis and the supporting information and data used to determine whether or not equipment is subject to the requirements in §§265.1052 through 265.1060. The record shall include supporting documentation as required by §265.1063(d)(3) when application of the knowledge of the nature of the hazardous waste stream or the process by which it was produced is used. If the owner or operator takes any action (e.g., changing the process that produced the waste) that could result in an increase in the total organic content of the waste contained in or contacted by equipment determined not to be subject to the requirements in §§265.1052 through 265.1060, then a new determination is required.

(l) Records of the equipment leak information required by paragraph (d) of this section and the operating information required by paragraph (e) of this section need be kept only 3 years.

(m) The owner or operator of any facility with equipment that is subject to this subpart and to leak detection, monitoring, and repair requirements under regulations at 40 CFR part 60, part 61, or part 63 may elect to determine compliance with this subpart either by documentation pursuant to §265.1064 of this subpart, or by documentation of compliance with the regulations at 40 CFR part 60, part 61, or part 63 pursuant to the relevant provisions of the regulations at 40 part 60, part 61, or part 63. The documentation of compliance under regulation at 40 CFR part 60, part 61, or part 63 shall be kept with or made readily available with the facility operating record.

[55 FR 25512, June 21, 1990, as amended at 56 FR 19290, Apr. 26, 1991; 61 FR 59971, Nov. 25, 1996; 62 FR 64662, Dec. 8, 1997]

## §§ 265.1065–265.1079 [Reserved]

## Subpart CC—Air Emission Standards for Tanks, Surface Impoundments, and Containers

SOURCE: 59 FR 62935, Dec. 6, 1994, unless otherwise noted.

### § 265.1080 Applicability.

(a) The requirements of this subpart apply to owners and operators of all facilities that treat, store, or dispose of hazardous waste in tanks, surface impoundments, or containers subject to either subpart I, J, or K of this part except as §265.1 and paragraph (b) of this section provide otherwise.

(b) The requirements of this subpart do not apply to the following waste management units at the facility:

(1) A waste management unit that holds hazardous waste placed in the unit before December 6, 1996, and in which no hazardous waste is added to the unit on or after December 6, 1996.

(2) A container that has a design capacity less than or equal to 0.1 m$^3$.

(3) A tank in which an owner or operator has stopped adding hazardous waste and the owner or operator has begun implementing or completed closure pursuant to an approved closure plan.

(4) A surface impoundment in which an owner or operator has stopped adding hazardous waste (except to implement an approved closure plan) and the owner or operator has begun implementing or completed closure pursuant to an approved closure plan.

(5) A waste management unit that is used solely for on-site treatment or storage of hazardous waste that is placed in the unit as a result of implementing remedial activities required under the corrective action authorities of RCRA sections 3004(u), 3004(v), or 3008(h); CERCLA authorities; or similar Federal or State authorities.

(6) A waste management unit that is used solely for the management of radioactive mixed waste in accordance

with all applicable regulations under the authority of the Atomic Energy Act and the Nuclear Waste Policy Act.

(7) A hazardous waste management unit that the owner or operator certifies is equipped with and operating air emission controls in accordance with the requirements of an applicable Clean Air Act regulation codified under 40 CFR part 60, part 61, or part 63. For the purpose of complying with this paragraph, a tank for which the air emission control includes an enclosure, as opposed to a cover, must be in compliance with the enclosure and control device requirements of § 265.1085(i), except as provided in § 265.1083(c)(5).

(8) A tank that has a process vent as defined in 40 CFR 264.1031.

(c) For the owner and operator of a facility subject to this subpart who has received a final permit under RCRA section 3005 prior to December 6, 1996, the following requirements apply:

(1) The requirements of 40 CFR part 264, subpart CC shall be incorporated into the permit when the permit is reissued in accordance with the requirements of 40 CFR 124.15 or reviewed in accordance with the requirements of 40 CFR 270.50(d).

(2) Until the date when the permit is reissued in accordance with the requirements of 40 CFR 124.15 or reviewed in accordance with the requirements of 40 CFR 270.50(d), the owner and operator is subject to the requirements of this subpart.

(d) The requirements of this subpart, except for the recordkeeping requirements specified in § 265.1090(i) of this subpart, are administratively stayed for a tank or a container used for the management of hazardous waste generated by organic peroxide manufacturing and its associated laboratory operations when the owner or operator of the unit meets all of the following conditions:

(1) The owner or operator identifies that the tank or container receives hazardous waste generated by an organic peroxide manufacturing process producing more than one functional family of organic peroxides or multiple organic peroxides within one functional family, that one or more of these organic peroxides could potentially undergo self-accelerating thermal decomposition at or below ambient temperatures, and that organic peroxides are the predominant products manufactured by the process. For the purpose of meeting the conditions of this paragraph, "organic peroxide" means an organic compound that contains the bivalent -O-O- structure and which may be considered to be a structural derivative of hydrogen peroxide where one or both of the hydrogen atoms has been replaced by an organic radical.

(2) The owner or operator prepares documentation, in accordance with the requirements of § 265.1090(i) of this subpart, explaining why an undue safety hazard would be created if air emission controls specified in §§ 265.1085 through 265.1088 of this subpart are installed and operated on the tanks and containers used at the facility to manage the hazardous waste generated by the organic peroxide manufacturing process or processes meeting the conditions of paragraph (d)(1) of this section.

(3) The owner or operator notifies the Regional Administrator in writing that hazardous waste generated by an organic peroxide manufacturing process or processes meeting the conditions of paragraph (d)(1) of this section are managed at the facility in tanks or containers meeting the conditions of paragraph (d)(2) of this section. The notification shall state the name and address of the facility, and be signed and dated by an authorized representative of the facility owner or operator.

(e)(1) Except as provided in paragraph (e)(2) of this section, the requirements of this subpart do not apply to the pharmaceutical manufacturing facility, commonly referred to as the Stonewall Plant, located at Route 340 South, Elkton, Virginia, provided that facility is operated in compliance with the requirements contained in a Clean Air Act permit issued pursuant to 40 CFR 52.2454. The requirements of this subpart shall apply to the facility upon termination of the Clean Air Act permit issued pursuant to 40 CFR 52.2454.

(2) Notwithstanding paragraph (e)(1) of this section, any hazardous waste surface impoundment operated at the Stonewall Plant is subject to the

standards in §265.1086 and all requirements related to hazardous waste surface impoundments that are referenced in or by §265.1086, including the closed-vent system and control device requirements of §265.1088 and the recordkeeping requirements of §265.1090(c).

(f) This section applies only to the facility commonly referred to as the OSi Specialties Plant, located on State Route 2, Sistersville, West Virginia ("Sistersville Plant").

(1)(i) Provided that the Sistersville Plant is in compliance with the requirements of paragraph (f)(2) of this section, the requirements referenced in paragraph (f)(1)(iii) of this section are temporarily deferred, as specified in paragraph (f)(3) of this section, with respect to the two hazardous waste surface impoundments at the Sistersville Plant. Beginning on the date that paragraph (f)(1)(ii) of this section is first implemented, the temporary deferral of this paragraph shall no longer be effective.

(ii)(A) In the event that a notice of revocation is issued pursuant to paragraph (f)(3)(iv) of this section, the requirements referenced in paragraph (f)(1)(iii) of this section are temporarily deferred, with respect to the two hazardous waste surface impoundments, provided that the Sistersville Plant is in compliance with the requirements of paragraphs (f)(2)(ii), (f)(2)(iii), (f)(2)(iv), (f)(2)(v), (f)(2)(vi) and (g) of this section, except as provided under paragraph (f)(1)(ii)(B) of this section. The temporary deferral of the previous sentence shall be effective beginning on the date the Sistersville Plant receives written notification of revocation, and continuing for a maximum period of 18 months from that date, provided that the Sistersville Plant is in compliance with the requirements of paragraphs (f)(2)(ii), (f)(2)(iii), (f)(2)(iv), (f)(2)(v), (f)(2)(vi) and (g) of this section at all times during that 18-month period. In no event shall the temporary deferral continue to be effective after the MON Compliance Date.

(B) In the event that a notification of revocation is issued pursuant to paragraph (f)(3)(iv) of this section as a result of the permanent removal of the capper unit from methyl capped polyether production service, the requirements referenced in paragraph (f)(1)(iii) of this section are temporarily deferred, with respect to the two hazardous waste surface impoundments, provided that the Sistersville Plant is in compliance with the requirements of paragraphs (f)(2)(vi), and (g) of this section. The temporary deferral of the previous sentence shall be effective beginning on the date the Sistersville Plant receives written notification of revocation, and continuing for a maximum period of 18 months from that date, provided that the Sistersville Plant is in compliance with the requirements of paragraphs (f)(2)(vi) and (g) of this section at all times during that 18-month period. In no event shall the temporary deferral continue to be effective after the MON Compliance Date.

(iii) The standards in §265.1086 of this part, and all requirements referenced in or by §265.1086 that otherwise would apply to the two hazardous waste surface impoundments, including the closed-vent system and control device requirements of §265.1088 of this part.

(2) Notwithstanding the effective period and revocation provisions in paragraph (f)(3) of this section, the temporary deferral provided in paragraph (f)(1)(i) of this section is effective only if the Sistersville Plant meets the requirements of paragraph (f)(2) of this section.

(i) The Sistersville Plant shall install an air pollution control device on the polyether methyl capper unit ("capper unit"), implement a methanol recovery operation, and implement a waste minimization/pollution prevention ("WMPP") project. The installation and implementation of these requirements shall be conducted according to the schedule described in paragraphs (f)(2)(i) and (f)(2)(vi) of this section.

(A) The Sistersville Plant shall complete the initial start-up of a thermal incinerator on the capper unit's process vents from the first stage vacuum pump, from the flash pot and surge tank, and from the water stripper, no later than April 1, 1998.

(B) The Sistersville Plant shall provide to the EPA and the West Virginia Department of Environmental Protection, written notification of the actual date of initial start-up of the thermal incinerator, and commencement of the methanol recovery operation. The Sistersville Plant shall submit this written notification as soon as practicable, but in no event later than 15 days after such events.

(ii) The Sistersville Plant shall install and operate the capper unit process vent thermal incinerator according to the requirements of paragraphs (f)(2)(ii)(A) through (f)(2)(ii)(D) of this section.

(A) Capper unit process vent thermal incinerator.

(1) Except as provided under paragraph (f)(2)(ii)(D) of this section, the Sistersville Plant shall operate the process vent thermal incinerator such that the incinerator reduces the total organic compounds ("TOC") from the process vent streams identified in paragraph (f)(2)(i)(A) of this section, by 98 weight-percent, or to a concentration of 20 parts per million by volume, on a dry basis, corrected to 3 percent oxygen, whichever is less stringent.

(i) Prior to conducting the initial performance test required under paragraph (f)(2)(ii)(B) of this section, the Sistersville Plant shall operate the thermal incinerator at or above a minimum temperature of 1600 Fahrenheit.

(ii) After the initial performance test required under paragraph (f)(2)(ii)(B) of this section, the Sistersville Plant shall operate the thermal incinerator at or above the minimum temperature established during that initial performance test.

(iii) The Sistersville Plant shall operate the process vent thermal incinerator at all times that the capper unit is being operated to manufacture product.

(2) The Sistersville Plant shall install, calibrate, and maintain all air pollution control and monitoring equipment described in paragraphs (f)(2)(i)(A) and (f)(2)(ii)(B)(3) of this section, according to the manufacturer's specifications, or other written procedures that provide adequate assurance that the equipment can reasonably be expected to control and monitor accurately, and in a manner consistent with good engineering practices during all periods when emissions are routed to the unit.

(B) The Sistersville Plant shall comply with the requirements of paragraphs (f)(2)(ii)(B)(1) through (f)(2)(ii)(B)(3) of this section for performance testing and monitoring of the capper unit process vent thermal incinerator.

(1) Within 120 days after thermal incinerator initial start-up, the Sistersville Plant shall conduct a performance test to determine the minimum temperature at which compliance with the emission reduction requirement specified in paragraph (f)(4) of this section is achieved. This determination shall be made by measuring TOC minus methane and ethane, according to the procedures specified in paragraph (f)(2)(ii)(B) of this section.

(2) The Sistersville Plant shall conduct the initial performance test in accordance with the standards set forth in paragraph (f)(4) of this section.

(3) Upon initial start-up, the Sistersville Plant shall install, calibrate, maintain and operate, according to manufacturer's specifications and in a manner consistent with good engineering practices, the monitoring equipment described in paragraphs (f)(2)(ii)(B)(3)(i) through (f)(2)(ii)(B)(3)(iii) of this section.

(i) A temperature monitoring device equipped with a continuous recorder. The temperature monitoring device shall be installed in the firebox or in the duct work immediately downstream of the firebox in a position before any substantial heat exchange is encountered.

(ii) A flow indicator that provides a record of vent stream flow to the incinerator at least once every fifteen minutes. The flow indicator shall be installed in the vent stream from the process vent at a point closest to the inlet of the incinerator.

(iii) If the closed-vent system includes bypass devices that could be used to divert the gas or vapor stream to the atmosphere before entering the control device, each bypass device shall be equipped with either a bypass flow indicator or a seal or locking device as specified in this paragraph. For the purpose of complying with this paragraph, low leg drains, high point bleeds, analyzer vents, open-ended valves or lines, spring-loaded pressure

714

relief valves, and other fittings used for safety purposes are not considered to be bypass devices. If a bypass flow indicator is used to comply with this paragraph, the bypass flow indicator shall be installed at the inlet to the bypass line used to divert gases and vapors from the closed-vent system to the atmosphere at a point upstream of the control device inlet. If a seal or locking device (e.g. car-seal or lock-and-key configuration) is used to comply with this paragraph, the device shall be placed on the mechanism by which the bypass device position is controlled (e.g., valve handle, damper levels) when the bypass device is in the closed position such that the bypass device cannot be opened without breaking the seal or removing the lock. The Sistersville Plant shall visually inspect the seal or locking device at least once every month to verify that the bypass mechanism is maintained in the closed position.

(C) The Sistersville Plant shall keep on-site an up-to-date, readily accessible record of the information described in paragraphs (f)(2)(ii)(C)(1) through (f)(2)(ii)(C)(4) of this section.

(1) Data measured during the initial performance test regarding the firebox temperature of the incinerator and the percent reduction of TOC achieved by the incinerator, and/or such other information required in addition to or in lieu of that information by the WVDEP in its approval of equivalent test methods and procedures.

(2) Continuous records of the equipment operating procedures specified to be monitored under paragraph (f)(2)(ii)(B)(3) of this section, as well as records of periods of operation during which the firebox temperature falls below the minimum temperature established under paragraph (f)(2)(ii)(A)(1) of this section.

(3) Records of all periods during which the vent stream has no flow rate to the extent that the capper unit is being operated during such period.

(4) Records of all periods during which there is flow through a bypass device.

(D) The Sistersville Plant shall comply with the start-up, shutdown, maintenance and malfunction requirements contained in paragraphs (f)(2)(ii)(D)(1)

through (f)(2)(ii)(D)(6) of this section, with respect to the capper unit process vent incinerator.

(1) The Sistersville Plant shall develop and implement a Start-up, Shutdown and Malfunction Plan as required by the provisions set forth in paragraph (f)(2)(ii)(D) of this section. The plan shall describe, in detail, procedures for operating and maintaining the thermal incinerator during periods of start-up, shutdown and malfunction, and a program of corrective action for malfunctions of the thermal incinerator.

(2) The plan shall include a detailed description of the actions the Sistersville Plant will take to perform the functions described in paragraphs (f)(2)(ii)(D)(2)(i) through (f)(2)(ii)(D)(2)(iii) of this section.

(i) Ensure that the thermal incinerator is operated in a manner consistent with good air pollution control practices.

(ii) Ensure that the Sistersville Plant is prepared to correct malfunctions as soon as practicable after their occurrence in order to minimize excess emissions.

(iii) Reduce the reporting requirements associated with periods of start-up, shutdown and malfunction.

(3) During periods of start-up, shutdown and malfunction, the Sistersville Plant shall maintain the process unit and the associated thermal incinerator in accordance with the procedures set forth in the plan.

(4) The plan shall contain record keeping requirements relating to periods of start-up, shutdown or malfunction, actions taken during such periods in conformance with the plan, and any failures to act in conformance with the plan during such periods.

(5) During periods of maintenance or malfunction of the thermal incinerator, the Sistersville Plant may continue to operate the capper unit, provided that operation of the capper unit without the thermal incinerator shall be limited to no more than 240 hours each calendar year.

(6) For the purposes of paragraph (f)(2)(iii)(D) of this section, the Sistersville Plant may use its operating procedures manual, or a plan developed for other reasons, provided

715

that plan meets the requirements of paragraph (f)(2)(iii)(D) of this section for the start-up, shutdown and malfunction plan.

(iii) The Sistersville Plant shall operate the closed-vent system in accordance with the requirements of paragraphs (f)(2)(iii)(A) through (f)(2)(iii)(D) of this section.

(A) Closed-vent system.

(1) At all times when the process vent thermal incinerator is operating, the Sistersville Plant shall route the vent streams identified in paragraph (f)(2)(i) of this section from the capper unit to the thermal incinerator through a closed-vent system.

(2) The closed-vent system will be designed for and operated with no detectable emissions, as defined in paragraph (f)(6) of this section.

(B) The Sistersville Plant will comply with the performance standards set forth in paragraph (f)(2)(iii)(A)(1) of this section on and after the date on which the initial performance test referenced in paragraph (f)(2)(ii)(B) of this section is completed, but no later than sixty (60) days after the initial start-up date.

(C) The Sistersville Plant shall comply with the monitoring requirements of paragraphs (f)(2)(iii)(C)(1) through (f)(2)(iii)(C)(3) of this section, with respect to the closed-vent system.

(1) At the time of the performance test described in paragraph (f)(2)(ii)(B) of this section, the Sistersville Plant shall inspect the closed-vent system as specified in paragraph (f)(5) of this section.

(2) At the time of the performance test described in paragraph (f)(2)(ii)(B) of this section, and annually thereafter, the Sistersville Plant shall inspect the closed-vent system for visible, audible, or olfactory indications of leaks.

(3) If at any time a defect or leak is detected in the closed-vent system, the Sistersville Plant shall repair the defect or leak in accordance with the requirements of paragraphs (f)(2)(iii)(C)(3)(i) and (f)(2)(iii)(C)(3)(ii) of this section.

(i) The Sistersville Plant shall make first efforts at repair of the defect no later than five (5) calendar days after detection, and repair shall be completed as soon as possible but no later than forty-five (45) calendar days after detection.

(ii) The Sistersville Plant shall maintain a record of the defect repair in accordance with the requirements specified in paragraph (f)(2)(iii)(D) of this section.

(D) The Sistersville Plant shall keep on-site up-to-date, readily accessible records of the inspections and repairs required to be performed by paragraph (f)(2)(iii) of this section.

(iv) The Sistersville Plant shall operate the methanol recovery operation in accordance with paragraphs (f)(2)(iv)(A) through (f)(2)(iv)(C) of this section.

(A) The Sistersville Plant shall operate the condenser associated with the methanol recovery operation at all times during which the capper unit is being operated to manufacture product.

(B) The Sistersville Plant shall comply with the monitoring requirements described in paragraphs (f)(2)(B)(1) through (f)(2)(B)(3) of this section, with respect to the methanol recovery operation.

(1) The Sistersville Plant shall perform measurements necessary to determine the information described in paragraphs (f)(2)(iv)(B)(1)(i) and (f)(2)(iv)(B)(1)(ii) of this section to demonstrate the percentage recovery by weight of the methanol contained in the influent gas stream to the condenser.

(i) Information as is necessary to calculate the annual amount of methanol generated by operating the capper unit.

(ii) The annual amount of methanol recovered by the condenser associated with the methanol recovery operation.

(2) The Sistersville Plant shall install, calibrate, maintain and operate according to manufacturer specifications, a temperature monitoring device with a continuous recorder for the condenser associated with the methanol recovery operation, as an indicator that the condenser is operating.

(3) The Sistersville Plant shall record the dates and times during which the capper unit and the condenser are operating.

(C) The Sistersville Plant shall keep on-site up-to-date, readily-accessible records of the parameters specified to be monitored under paragraph (f)(2)(iv)(B) of this section.

(v) The Sistersville Plant shall comply with the requirements of paragraphs (f)(2)(v)(A) through (f)(2)(v)(C) of this section for the disposition of methanol collected by the methanol recovery operation.

(A) On an annual basis, the Sistersville Plant shall ensure that a minimum of 95% by weight of the methanol collected by the methanol recovery operation (also referred to as the "collected methanol") is utilized for reuse, recovery, or thermal recovery/treatment. The Sistersville Plant may use the methanol on-site, or may transfer or sell the methanol for reuse, recovery, or thermal recovery/treatment at other facilities.

(1) Reuse. To the extent reuse of all of the collected methanol destined for reuse, recovery, or thermal recovery is not economically feasible, the Sistersville Plant shall ensure the residual portion is sent for recovery, as defined in paragraph (f)(6) of this section, except as provided in paragraph (f)(2)(v)(A)(2) of this section.

(2) Recovery. To the extent that reuse or recovery of all the collected methanol destined for reuse, recovery, or thermal recovery is not economically feasible, the Sistersville Plant shall ensure that the residual portion is sent for thermal recovery/treatment, as defined in paragraph (f)(6) of this section.

(3) The Sistersville Plant shall ensure that, on an annual basis, no more than 5% of the methanol collected by the methanol recovery operation is subject to bio-treatment.

(4) In the event the Sistersville Plant receives written notification of revocation pursuant to paragraph (f)(3)(iv) of this section, the percent limitations set forth under paragraph (f)(2)(v)(A) of this section shall no longer be applicable, beginning on the date of receipt of written notification of revocation.

(B) The Sistersville Plant shall perform such measurements as are necessary to determine the pounds of collected methanol directed to reuse, recovery, thermal recovery/treatment and bio-treatment, respectively, on a monthly basis.

(C) The Sistersville Plant shall keep on-site up-to-date, readily accessible records of the amounts of collected methanol directed to reuse, recovery, thermal recovery/treatment and bio-treatment necessary for the measurements required under paragraph (f)(2)(iv)(B) of this section.

(vi) The Sistersville Plant shall perform a WMPP project in accordance with the requirements and schedules set forth in paragraphs (f)(2)(vi)(A) through (f)(2)(vi)(C) of this section.

(A) In performing the WMPP Project, the Sistersville Plant shall use a Study Team and an Advisory Committee as described in paragraphs (f)(2)(vi)(A)(1) through (f)(2)(vi)(A)(6) of this section.

(1) At a minimum, the multi-functional Study Team shall consist of Sistersville Plant personnel from appropriate plant departments (including both management and employees) and an independent contractor. The Sistersville Plant shall select a contractor that has experience and training in WMPP in the chemical manufacturing industry.

(2) The Sistersville Plant shall direct the Study Team such that the team performs the functions described in paragraphs (f)(2)(vi)(A)(2)(i) through (f)(2)(vi)(A)(2)(v) of this section.

(i) Review Sistersville Plant operations and waste streams.

(ii) Review prior WMPP efforts at the Sistersville Plant.

(iii) Develop criteria for the selection of waste streams to be evaluated for the WMPP Project.

(iv) Identify and prioritize the waste streams to be evaluated during the study phase of the WMPP Project, based on the criteria described in paragraph (f)(2)(vi)(A)(2)(iii) of this section.

(v) Perform the WMPP Study as required by paragraphs (f)(2)(vi)(A)(3) through (f)(2)(vi)(A)(5), paragraph (f)(2)(vi)(B), and paragraph (f)(2)(vi)(C) of this section.

(3)(i) The Sistersville Plant shall establish an Advisory Committee consisting of a representative from EPA, a representative from WVDEP, the Sistersville Plant Manager, the Sistersville Plant Director of Safety, Health and Environmental Affairs, and a stakeholder representative(s).

(*ii*) The Sistersville Plant shall select the stakeholder representative(s) by mutual agreement of EPA, WVDEP and the Sistersville Plant no later than 20 days after receiving from EPA and WVDEP the names of their respective committee members.

(*4*) The Sistersville Plant shall convene a meeting of the Advisory Committee no later than thirty days after selection of the stakeholder representatives, and shall convene meetings periodically thereafter as necessary for the Advisory Committee to perform its assigned functions. The Sistersville Plant shall direct the Advisory Committee to perform the functions described in paragraphs (f)(2)(vi)(A)(*4*)(*i*) through (f)(2)(vi)(A)(*4*)(*iii*) of this section.

(*i*) Review and comment upon the Study Team's criteria for selection of waste streams, and the Study Team's identification and prioritization of the waste streams to be evaluated during the WMPP Project.

(*ii*) Review and comment upon the Study Team progress reports and the draft WMPP Study Report.

(*iii*) Periodically review the effectiveness of WMPP opportunities implemented as part of the WMPP Project, and, where appropriate, WMPP opportunities previously determined to be infeasible by the Sistersville Plant but which had potential for feasibility in the future.

(*5*) Beginning on January 15, 1998, and every ninety (90) days thereafter until submission of the final WMPP Study Report required by paragraph (f)(2)(vi)(C) of this section, the Sistersville Plant shall direct the Study Team to submit a progress report to the Advisory Committee detailing its efforts during the prior ninety (90) day period.

(B) The Sistersville Plant shall ensure that the WMPP Study and the WMPP Study Report meet the requirements of paragraphs (f)(2)(vi)(B)(*1*) through (f)(2)(vi)(B)(*3*) of this section.

(*1*) The WMPP Study shall consist of a technical, economic, and regulatory assessment of opportunities for source reduction and for environmentally sound recycling for waste streams identified by the Study Team.

(*2*) The WMPP Study shall evaluate the source, nature, and volume of the waste streams; describe all the WMPP opportunities identified by the Study Team; provide a feasibility screening to evaluate the technical and economical feasibility of each of the WMPP opportunities; identify any cross-media impacts or any anticipated transfers of risk associated with each feasible WMPP opportunity; and identify the projected economic savings and projected quantitative waste reduction estimates for each WMPP opportunity identified.

(*3*) No later than October 19, 1998, the Sistersville Plant shall prepare and submit to the members of the Advisory Committee a draft WMPP Study Report which, at a minimum, includes the results of the WMPP Study, identifies WMPP opportunities the Sistersville Plant determines to be feasible, discusses the basis for excluding other opportunities as not feasible, and makes recommendations as to whether the WMPP Study should be continued. The members of the Advisory Committee shall provide any comments to the Sistersville Plant within thirty (30) days of receiving the WMPP Study Report.

(C) Within thirty (30) days after receipt of comments from the members of the Advisory Committee, the Sistersville Plant shall submit to EPA and WVDEP a final WMPP Study Report which identifies those WMPP opportunities the Sistersville Plant determines to be feasible and includes an implementation schedule for each such WMPP opportunity. The Sistersville Plant shall make reasonable efforts to implement all feasible WMPP opportunities in accordance with the priorities identified in the implementation schedule.

(*1*) For purposes of this section, a WMPP opportunity is feasible if the Sistersville Plant considers it to be technically feasible (taking into account engineering and regulatory factors, product line specifications and customer needs) and economically practical (taking into account the full environmental costs and benefits associated with the WMPP opportunity and the company's internal requirements for approval of capital projects). For purposes of the WMPP Project, the Sistersville Plant shall use "An Introduction to Environmental Accounting as a Business Management Tool," (EPA 742/R–95/001) as one tool to identify the full environmental costs and benefits of each WMPP opportunity.

(2) In implementing each WMPP opportunity, the Sistersville Plant shall, after consulting with the other members of the Advisory Committee, develop appropriate protocols and methods for determining the information required by paragraphs (f)(2)(vi)(2)(i) through (f)(2)(vi)(2)(iii) of this section.

(i) The overall volume of wastes reduced.

(ii) The quantities of each constituent identified in paragraph (f)(8) of this section reduced in the wastes.

(iii) The economic benefits achieved.

(3) No requirements of paragraph (f)(2)(vi) of this section are intended to prevent or restrict the Sistersville Plant from evaluating and implementing any WMPP opportunities at the Sistersville Plant in the normal course of its operations or from implementing, prior to the completion of the WMPP Study, any WMPP opportunities identified by the Study Team.

(vii) The Sistersville Plant shall maintain on-site each record required by paragraph (f)(2) of this section, through the MON Compliance Date.

(viii) The Sistersville Plant shall comply with the reporting requirements of paragraphs (f)(2)(viii)(A) through (f)(2)(viii)(G) of this section.

(A) At least sixty days prior to conducting the initial performance test of the thermal incinerator, the Sistersville Plant shall submit to EPA and WVDEP copies of a notification of performance test, as described in 40 CFR 63.7(b). Following the initial performance test of the thermal incinerator, the Sistersville Plant shall submit to EPA and WVDEP copies of the performance test results that include the information relevant to initial performance tests of thermal incinerators contained in 40 CFR 63.7(g)(1), 40 CFR 63.117(a)(4)(i), and 40 CFR 63.117(a)(4)(ii).

(B) Beginning in 1999, on January 31 of each year, the Sistersville Plant shall submit a semiannual written report to the EPA and WVDEP, with re-

spect to the preceding six month period ending on December 31, which contains the information described in paragraphs (f)(2)(viii)(B)(1) through (f)(2)(viii)(B)(10) of this section.

(1) Instances of operating below the minimum operating temperature established for the thermal incinerator under paragraph (f)(2)(ii)(A)(1) of this section which were not corrected within 24 hours of onset.

(2) Any periods during which the capper unit was being operated to manufacture product while the flow indicator for the vent streams to the thermal incinerator showed no flow.

(3) Any periods during which the capper unit was being operated to manufacture product while the flow indicator for any bypass device on the closed vent system to the thermal incinerator showed flow.

(4) Information required to be reported during that six month period under the preconstruction permit issued under the state permitting program approved under subpart XX of 40 CFR Part 52—Approval and Promulgation of Implementation Plans for West Virginia.

(5) Any periods during which the capper unit was being operated to manufacture product while the condenser associated with the methanol recovery operation was not in operation.

(6) The amount (in pounds and by month) of methanol collected by the methanol recovery operation during the six month period.

(7) The amount (in pounds and by month) of collected methanol utilized for reuse, recovery, thermal recovery/ treatment, or bio-treatment, respectively, during the six month period.

(8) The calculated amount (in pounds and by month) of methanol generated by operating the capper unit.

(9) The status of the WMPP Project, including the status of developing the WMPP Study Report.

(10) Beginning in the year after the Sistersville Plant submits the final WMPP Study Report required by paragraph (f)(2)(vi)(C) of this section, and continuing in each subsequent Semiannual Report required by paragraph (f)(2)(viii)(B) of this section, the Sistersville Plant shall report on the

progress of the implementation of feasible WMPP opportunities identified in the WMPP Study Report. The Semiannual Report required by paragraph (f)(2)(viii)(B) of this section shall identify any cross-media impacts or impacts to worker safety or community health issues that have occurred as a result of implementation of the feasible WMPP opportunities.

(C) Beginning in 1999, on July 31 of each year, the Sistersville Plant shall provide an Annual Project Report to the EPA and WVDEP Project XL contacts containing the information required by paragraphs (f)(2)(viii)(C)(*1*) through (f)(2)(viii)(C)(*8*) of this section.

(*1*) The categories of information required to be submitted under paragraphs (f)(2)(viii)(B)(*1*) through (f)(2)(viii)(B)(*8*) of this section, for the preceding 12 month period ending on June 30.

(*2*) An updated Emissions Analysis for January through December of the preceding calendar year. The Sistersville Plant shall submit the updated Emissions Analysis in a form substantially equivalent to the previous Emissions Analysis prepared by the Sistersville Plant to support Project XL. The Emissions Analysis shall include a comparison of the volatile organic emissions associated with the capper unit process vents and the wastewater treatment system (using the EPA Water 8 model or other model agreed to by the Sistersville Plant, EPA and WVDEP) under Project XL with the expected emissions from those sources absent Project XL during that period.

(*3*) A discussion of the Sistersville Plant's performance in meeting the requirements of this section, specifically identifying any areas in which the Sistersville Plant either exceeded or failed to achieve any such standard.

(*4*) A description of any unanticipated problems in implementing the XL Project and any steps taken to resolve them.

(*5*) A WMPP Implementation Report that contains the information contained in paragraphs paragraphs (f)(2)(viii)(C)(5)(i) through (viii)(C)(5)(vi) of this section.

(*i*) A summary of the WMPP opportunities selected for implementation.

(*ii*) A description of the WMPP opportunities initiated and/or completed.

(*iii*) Reductions in volume of waste generated and amounts of each constituent reduced in wastes including any constituents identified in paragraph (f)(8) of this section.

(*iv*) An economic benefits analysis.

(*v*) A summary of the results of the Advisory Committee's review of implemented WMPP opportunities.

(*vi*) A reevaluation of WMPP opportunities previously determined to be infeasible by the Sistersville Plant but which had potential for future feasibility.

(*6*) An assessment of the nature of, and the successes or problems associated with, the Sistersville Plant's interaction with the federal and state agencies under the Project.

(*7*) An update on stakeholder involvement efforts.

(*8*) An evaluation of the Project as implemented against the Project XL Criteria and the baseline scenario.

(D) The Sistersville Plant shall submit to the EPA and WVDEP Project XL contacts a written Final Project Report covering the period during which the temporary deferral was effective, as described in paragraph (f)(3) of this section.

(*1*) The Final Project Report shall contain the information required to be submitted for the Semiannual Report required under paragraph (f)(2)(viii)(B) of this section, and the Annual Project Report required under paragraph (f)(2)(viii)(C) of this section.

(*2*) The Sistersville Plant shall submit the Final Project Report to EPA and WVDEP no later than 180 days after the temporary deferral of paragraph (f)(1) of this section is revoked, or 180 days after the MON Compliance Date, whichever occurs first.

(E)(*1*) The Sistersville Plant shall retain on-site a complete copy of each of the report documents to be submitted to EPA and WVDEP in accordance with requirements under paragraph (f)(2) of this section. The Sistersville Plant shall retain this record until 180 days after the MON Compliance Date. The Sistersville Plant shall provide to stakeholders and interested parties a written notice of availability (to be mailed to all persons on the Project mailing list and to be provided to at least one local newspaper of general circulation) of each such document, and provide a copy of each document to any such person upon request, subject to the provisions of 40 CFR part 2.

(2) Any reports or other information submitted to EPA or WVDEP may be released to the public pursuant to the Federal Freedom of Information Act (42 U.S.C. 552 *et seq.*), subject to the provisions of 40 CFR part 2.

(F) The Sistersville Plant shall make all supporting monitoring results and records required under paragraph (f)(2) of this section available to EPA and WVDEP within a reasonable amount of time after receipt of a written request from those Agencies, subject to the provisions of 40 CFR Part 2.

(G) Each report submitted by the Sistersville Plant under the requirements of paragraph (f)(2) of this section shall be certified by a Responsible Corporate Officer, as defined in 40 CFR 270.11(a)(1).

(H) For each report submitted in accordance with paragraph (f)(2) of this section, the Sistersville Plant shall send one copy each to the addresses in paragraphs (f)(2)(viii) (H)(1) through (H)(3) of this section.

(1) U.S. EPA Region 3, 1650 Arch Street, Philadelphia, PA 19103–2029, Attention Tad Radzinski, Mail Code 3WC11.

(2) U.S. EPA, 1200 Pennsylvania Ave., NW., Washington, DC 20460, Attention L. Nancy Birnbaum, Mail Code 1812.

(3) West Virginia Division of Environmental Protection, Office of Air Quality, 1558 Washington Street East, Charleston, WV 25311–2599, Attention John H. Johnston.

(3) Effective period and revocation of temporary deferral.

(i) The temporary deferral contained in this section is effective from April 1, 1998, and shall remain effective until the MON Compliance Date. The temporary deferral contained in this section may be revoked prior to the MON Compliance Date, as described in paragraph (f)(3)(iv) of this section.

(ii) On the MON Compliance Date, the temporary deferral contained in this section will no longer be effective.

(iii) The Sistersville Plant shall come into compliance with those require-

ments deferred by this section no later than the MON Compliance Date. No later than 18 months prior to the MON Compliance Date, the Sistersville Plant shall submit to EPA an implementation schedule that meets the requirements of paragraph (g)(1)(iii) of this section.

(iv) The temporary deferral contained in this section may be revoked for cause, as determined by EPA, prior to the MON Compliance Date. The Sistersville Plant may request EPA to revoke the temporary deferral contained in this section at any time. The revocation shall be effective on the date that the Sistersville Plant receives written notification of revocation from EPA.

(v) Nothing in this section shall affect the provisions of the MON, as applicable to the Sistersville Plant.

(vi) Nothing in paragrahs (f) or (g) of this section shall affect any regulatory requirements not referenced in paragraph (f)(1)(iii) of this section, as applicable to the Sistersville Plant.

(4) The Sistersville Plant shall conduct the initial performance test required by paragraph (f)(2)(ii)(B) of this section using the procedures in paragraph (f)(4) of this section. The organic concentration and percent reduction shall be measured as TOC minus methane and ethane, according to the procedures specified in paragraph (f)(4) of this section.

(i) Method 1 or 1A of 40 CFR part 60, appendix A, as appropriate, shall be used for selection of the sampling sites.

(A) To determine compliance with the 98 percent reduction of TOC requirement of paragraph (f)(2)(ii)(A)(1) of this section, sampling sites shall be located at the inlet of the control device after the final product recovery device, and at the outlet of the control device.

(B) To determine compliance with the 20 parts per million by volume TOC limit in paragraph (f)(2)(ii)(A)(1) of this section, the sampling site shall be located at the outlet of the control device.

(ii) The gas volumetric flow rate shall be determined using Method 2, 2A, 2C, or 2D of 40 CFR part 60, appendix A, as appropriate.

(iii) To determine compliance with the 20 parts per million by volume TOC limit in paragraph (f)(2)(ii)(A)(*1*) of this section, the Sistersville Plant shall use Method 18 of 40 CFR part 60, appendix A to measure TOC minus methane and ethane. Alternatively, any other method or data that has been validated according to the applicable procedures in Method 301 of 40 CFR part 63, appendix A, may be used. The following procedures shall be used to calculate parts per million by volume concentration, corrected to 3 percent oxygen:

(A) The minimum sampling time for each run shall be 1 hour in which either an integrated sample or a minimum of four grab samples shall be taken. If grab sampling is used, then the samples shall be taken at approximately equal intervals in time, such as 15 minute intervals during the run.

(B) The concentration of TOC minus methane and ethane ($C_{TOC}$) shall be calculated as the sum of the concentrations of the individual components, and shall be computed for each run using the following equation:

$$C_{TOC} = \sum_{i=1}^{x} \frac{\left( \sum_{j=1}^{n} C_{ji} \right)}{x}$$

Where:

$C_{TOC}$=Concentration of TOC (minus methane and ethane), dry basis, parts per million by volume.

$C_{ji}$=Concentration of sample components j of sample i, dry basis, parts per million by volume.

n=Number of components in the sample.

x=Number of samples in the sample run.

(C) The concentration of TOC shall be corrected to 3 percent oxygen if a combustion device is the control device.

(*1*) The emission rate correction factor or excess air, integrated sampling and analysis procedures of Method 3B of 40 CFR part 60, appendix A shall be used to determine the oxygen concentration ($\%O_{2d}$). The samples shall be taken during the same time that the TOC (minus methane or ethane) samples are taken.

(*2*) The concentration corrected to 3 percent oxygen ($C_c$) shall be computed using the following equation:

$$C_c = C_m \left( \frac{17.9}{20.9 - \%O_{2d}} \right)$$

Where:

$C_c$=Concentration of TOC corrected to 3 percent oxygen, dry basis, parts per million by volume.

$C_m$=Concentration of TOC (minus methane and ethane), dry basis, parts per million by volume.

$\%O_{2d}$=Concentration of oxygen, dry basis, percent by volume.

(iv) To determine compliance with the 98 percent reduction requirement of paragraph (f)(2)(ii)(A)(*1*) of this section, the Sistersville Plant shall use Method 18 of 40 CFR part 60, appendix A; alternatively, any other method or data that has been validated according to the applicable procedures in Method 301 of 40 CFR part 63, appendix A may be used. The following procedures shall be used to calculate percent reduction efficiency:

(A) The minimum sampling time for each run shall be 1 hour in which either an integrated sample or a minimum of four grab samples shall be taken. If grab sampling is used, then the samples shall be taken at approximately equal intervals in time such as 15 minute intervals during the run.

(B) The mass rate of TOC minus methane and ethane ($E_i$, $E_o$) shall be computed. All organic compounds (minus methane and ethane) measured by Method 18 of 40 CFR part 60, Appendix A are summed using the following equations:

$$E_i = K_2 \left( \sum_{j=1}^{n} C_{ij} M_{ij} \right) Q_i$$

$$E_o = K_2 \left( \sum_{j=1}^{n} C_{oj} M_{oj} \right) Q_o$$

Where:

$C_{ij}$, $C_{oj}$=Concentration of sample component j of the gas stream at the inlet and outlet of the control device, respectively, dry basis, parts per million by volume.

$E_i$, $E_o$=Mass rate of TOC (minus methane and ethane) at the inlet and outlet of the control device, respectively, dry basis, kilogram per hour.

$M_{ij}$, $M_{oj}$=Molecular weight of sample component j of the gas stream at the inlet and outlet of the control device, respectively, gram/gram-mole.

$Q_i$, $Q_o$=Flow rate of gas stream at the inlet and outlet of the control device, respectively, dry standard cubic meter per minute.

$K_2$=Constant, $2.494 \times 10^{-6}$ (parts per million)$^{-1}$ (gram-mole per standard cubic meter) (kilogram/gram) (minute/hour), where standard temperature (gram-mole per standard cubic meter) is 20 °C.

(C) The percent reduction in TOC (minus methane and ethane) shall be calculated as follows:

$$R = \frac{E_i E_o}{E_i} (100)$$

where:

R=Control efficiency of control device, percent.

$E_i$=Mass rate of TOC (minus methane and ethane) at the inlet to the control device as calculated under paragraph (f)(4)(iv)(B) of this section, kilograms TOC per hour.

$E_o$=Mass rate of TOC (minus methane and ethane) at the outlet of the control device, as calculated under paragraph (f)(5)(iv)(B) of this section, kilograms TOC per hour.

(5) At the time of the initial performance test of the process vent thermal incinerator required under paragraph (f)(2)(ii)(B) of this section, the Sistersville Plant shall inspect each closed vent system according to the procedures specified in paragraphs (f)(5)(i) through (f)(5)(vi) of this section.

(i) The initial inspections shall be conducted in accordance with Method 21 of 40 CFR part 60, appendix A.

(ii)(A) Except as provided in paragraph (f)(5)(ii)(B) of this section, the detection instrument shall meet the performance criteria of Method 21 of 40 CFR part 60, appendix A, except the instrument response factor criteria in section 3.1.2(a) of Method 21 of 40 CFR part 60, appendix A shall be for the average composition of the process fluid not each individual volatile organic compound in the stream. For process streams that contain nitrogen, air, or other inerts which are not organic hazardous air pollutants or volatile organic compounds, the average stream response factor shall be calculated on an inert-free basis.

(B) If no instrument is available at the plant site that will meet the performance criteria specified in para-

graph (f)(5)(ii)(A) of this section, the instrument readings may be adjusted by multiplying by the average response factor of the process fluid, calculated on an inert-free basis as described in paragraph (f)(5)(ii)(A) of this section.

(iii) The detection instrument shall be calibrated before use on each day of its use by the procedures specified in Method 21 of 40 CFR part 60, appendix A.

(iv) Calibration gases shall be as follows:

(A) Zero air (less than 10 parts per million hydrocarbon in air); and

(B) Mixtures of methane in air at a concentration less than 10,000 parts per million. A calibration gas other than methane in air may be used if the instrument does not respond to methane or if the instrument does not meet the performance criteria specified in paragraph (f)(5)(ii)(A) of this section. In such cases, the calibration gas may be a mixture of one or more of the compounds to be measured in air.

(v) The Sistersville Plant may elect to adjust or not adjust instrument readings for background. If the Sistersville Plant elects to not adjust readings for background, all such instrument readings shall be compared directly to the applicable leak definition to determine whether there is a leak. If the Sistersville Plant elects to adjust instrument readings for background, the Sistersville Plant shall measure background concentration using the procedures in 40 CFR 63.180(b) and (c). The Sistersville Plant shall subtract background reading from the maximum concentration indicated by the instrument.

(vi) The arithmetic difference between the maximum concentration indicated by the instrument and the background level shall be compared with 500 parts per million for determining compliance.

(6) Definitions of terms as used in paragraphs (f) and (g) of this section.

(i) Closed vent system is defined as a system that is not open to the atmosphere and that is composed of piping, connections and, if necessary, flow-inducing devices that transport gas or vapor from the capper unit process vent to the thermal incinerator.

(ii) No detectable emissions means an instrument reading of less than 500 parts per million by volume above background as determined by Method 21 in 40 CFR part 60.

(iii) Reuse includes the substitution of collected methanol (without reclamation subsequent to its collection) for virgin methanol as an ingredient (including uses as an intermediate) or as an effective substitute for a commercial product.

(iv) Recovery includes the substitution of collected methanol for virgin methanol as an ingredient (including uses as an intermediate) or as an effective substitute for a commercial product following reclamation of the methanol subsequent to its collection.

(v) Thermal recovery/treatment includes the use of collected methanol in fuels blending or as a feed to any combustion device to the extent permitted by federal and state law.

(vi) Bio-treatment includes the treatment of the collected methanol through introduction into a biological treatment system, including the treatment of the collected methanol as a waste stream in an on-site or off-site wastewater treatment system. Introduction of the collected methanol to the on-site wastewater treatment system will be limited to points downstream of the surface impoundments, and will be consistent with the requirements of federal and state law.

(vii) Start-up shall have the meaning set forth at 40 CFR 63.2.

(viii) Flow indicator means a device which indicates whether gas flow is present in the vent stream, and, if required by the permit for the thermal incinerator, which measures the gas flow in that stream.

(ix) Continuous Recorder means a data recording device that records an instantaneous data value at least once every fifteen minutes.

(x) MON means the National Emission Standards for Hazardous Air Pollutants for the source category Miscellaneous Organic Chemical Production and Processes ("MON"), promulgated under the authority of Section 112 of the Clean Air Act.

(xi) MON Compliance Date means the date 3 years after the effective date of the National Emission Standards for Hazardous Air Pollutants for the source category Miscellaneous Organic Chemical Production and Processes ("MON").

(7) OSi Specialties, Incorporated, a subsidiary of Witco Corporation ("OSi"), may seek to transfer its rights and obligations under this section to a future owner of the Sistersville Plant in accordance with the requirements of paragraphs (f)(7)(i) through (f)(7)(iii) of this section.

(i) OSi will provide to EPA a written notice of any proposed transfer at least forty-five days prior to the effective date of any such transfer. The written notice will identify the proposed transferee.

(ii) The proposed transferee will provide to EPA a written request to assume the rights and obligations under this section at least forty-five days prior to the effective date of any such transfer. The written request will describe the transferee's financial and technical capability to assume the obligations under this section, and will include a statement of the transferee's intention to fully comply with the terms of this section and to sign the Final Project Agreement for this XL Project as an additional party.

(iii) Within thirty days of receipt of both the written notice and written request described in paragraphs (f)(7)(i) and (f)(7)(ii) of this section, EPA will determine, based on all relevant information, whether to approve a transfer of rights and obligations under this section from OSi to a different owner.

(8) The constituents to be identified by the Sistersville Plant pursuant to paragraphs (f)(2)(vi)(C)(2)(ii) and (f)(2)(viii)(C)(5)(iii) of this section are: 1 Naphthalenamine; 1, 2, 4 Trichlorobenzene; 1,1 Dichloroethylene; 1,1,1 Trichloroethane; 1,1,2 Tetrachloroethane; 1,1,2 Trichloro 1,2,2 Triflouroethane; 1,1,2 Trichloroethane; 1,1,2,2 Tetrachloroethane; 1,2 Dichlorobenzene; 1,2 Dichloroethane; 1,2 Dichloropropane; 1,2 Dichloropropanone; 1,2 Transdichloroethene; 1,2, Trans— Dichloroethene; 1,2,4,5 Tetrachlorobenzine; 1,3 Dichlorobenzene; 1,4 Dichloro 2 butene; 1,4 Dioxane; 2 Chlorophenol; 2 Cyclohexyl 4,6 dinitrophenol; 2 Methyl

Pyridine; 2 Nitropropane; 2, 4-Di-nitrotoluene; Acetone; Acetonitrile; Acrylonitrile; Allyl Alcohol; Aniline; Antimony; Arsenic; Barium; Benzene; Benzotrichloride; Benzyl Chloride; Beryllium; Bis (2 ethyl Hexyl) Phthalate; Butyl Alcohol, n; Butyl Benzyl Phthalate; Cadmium; Carbon Disulfide; Carbon Tetrachloride; Chlorobenzene; Chloroform; Chloromethane; Chromium; Chrysene; Copper; Creosol; Creosol, m-; Creosol, o; Creosol, p; Cyanide; Cyclohexanone; Di-n-octyl phthalate; Dichlorodiflouromethane; Diethyl Phthalate; Dihydrosafrole; Dimethylamine; Ethyl Acetate; Ethyl benzene; Ethyl Ether; Ethylene Glycol Ethyl Ether; Ethylene Oxide; Formaldehyde; Isobutyl Alcohol; Lead; Mercury; Methanol; Methoxychlor; Methyl Chloride; Methyl Chloroformate; Methyl Ethyl Ketone; Methyl Ethyl Ketone Peroxide; Methyl Isobutyl Ketone; Methyl Methacrylate; Methylene Bromide; Methylene Chloride; Naphthalene; Nickel; Nitrobenzene; Nitroglycerine; p-Toluidine; Phenol; Phthalic Anhydride; Polychlorinated Biphenyls; Propargyl Alcohol; Pyridine; Safrole; Selenium; Silver; Styrene; Tetrachloroethylene; Tetrahydrofuran; Thallium; Toluene; Toluene 2,4 Diisocyanate; Trichloroethylene; Trichloroflouromethane; Vanadium; Vinyl Chloride; Warfarin; Xylene; Zinc.

(g) This section applies only to the facility commonly referred to as the OSi Specialties Plant, located on State Route 2, Sistersville, West Virginia ("Sistersville Plant").

(1)(i) No later than 18 months from the date the Sistersville Plant receives written notification of revocation of the temporary deferral for the Sistersville Plant under paragraph (f) of this section, the Sistersville Plant shall, in accordance with the implementation schedule submitted to EPA under paragraph (g)(1)(ii) of this section, either come into compliance with all requirements of this subpart which had been deferred by paragraph (f)(1)(i) of this section, or complete a facility or process modification such that the requirements of §265.1086 are no longer applicable to the two hazardous waste surface impoundments. In any event, the Sistersville Plant must complete the requirements of the previous sen-

tence no later than the MON Compliance Date; if the Sistersville Plant receives written notification of revocation of the temporary deferral after the date 18 months prior to the MON Compliance Date, the date by which the Sistersville Plant must complete the requirements of the previous sentence will be the MON Compliance Date, which would be less than 18 months from the date of notification of revocation.

(ii) Within 30 days from the date the Sistersville Plant receives written notification of revocation under paragraph (f)(3)(iv) of this section, the Sistersville Plant shall enter and maintain in the facility operating record an implementation schedule. The implementation schedule shall demonstrate that within 18 months from the date the Sistersville Plant receives written notification of revocation under paragraph (f)(3)(iv) of this section (but no later than the MON Compliance Date), the Sistersville Plant shall either come into compliance with the regulatory requirements that had been deferred by paragraph (f)(1)(i) of this section, or complete a facility or process modification such that the requirements of §265.1086 are no longer applicable to the two hazardous waste surface impoundments. Within 30 days from the date the Sistersville Plant receives written notification of revocation under paragraph (f)(3)(iv) of this section, the Sistersville Plant shall submit a copy of the implementation schedule to the EPA and WVDEP Project XL contacts identified in paragraph (f)(2)(viii)(H) of this section. The implementation schedule shall reflect the Sistersville Plant's effort to come into compliance as soon as practicable (but no later than 18 months after the date the Sistersville Plant receives written notification of revocation, or the MON Compliance Date, whichever is sooner) with all regulatory requirements that had been deferred under paragraph (f)(1)(i) of this section, or to complete a facility or process modification as soon as practicable (but no later than 18 months after the date the Sistersville Plant receives written notification of revocation, or the MON Compliance Date, whichever is sooner) such that the requirements of § 265.1086 are no longer applicable to the two hazardous waste surface impoundments.

(iii) The implementation schedule shall include the information described in either paragraph (g)(1)(iii)(A) or (B) of this section.

(A) Specific calendar dates for: award of contracts or issuance of purchase orders for the control equipment required by those regulatory requirements that had been deferred by paragraph (f)(1)(i) of this section; initiation of on-site installation of such control equipment; completion of the control equipment installation; performance of any testing to demonstrate that the installed control equipment meets the applicable standards of this subpart; initiation of operation of the control equipment; and compliance with all regulatory requirements that had been deferred by paragraph (f)(1)(i) of this section.

(B) Specific calendar dates for the purchase, installation, performance testing and initiation of operation of equipment to accomplish a facility or process modification such that the requirements of § 265.1086 are no longer applicable to the two hazardous waste surface impoundments.

(2) Nothing in paragraphs (f) or (g) of this section shall affect any regulatory requirements not referenced in paragraph (f)(2)(i) or (ii) of this section, as applicable to the Sistersville Plant.

(3) In the event that a notification of revocation is issued pursuant to paragraph (f)(3)(iv) of this section, the requirements referenced in paragraph (f)(1)(iii) of this section are temporarily deferred, with respect to the two hazardous waste surface impoundments, provided that the Sistersville Plant is in compliance with the requirements of paragraphs (f)(2)(ii), (f)(2)(iii), (f)(2)(iv), (f)(2)(v), (f)(2)(vi) and (g) of this section, except as provided under paragraph (g)(4) of this section. The temporary deferral of the previous sentence shall be effective beginning on the date the Sistersville Plant receives written notification of revocation, and subject to paragraph (g)(5) of this section, shall continue to be effective for a maximum period of 18 months from that date, provided that the Sistersville Plant is in compliance with the requirements of paragraphs

(f)(2)(ii), (f)(2)(iii), (f)(2)(iv), (f)(2)(v), (f)(2)(vi) and (g) of this section at all times during that 18-month period.

(4) In the event that a notification of revocation is issued pursuant to paragraph (f)(3)(iv) of this section as a result of the permanent removal of the capper unit from methyl capped polyether production service, the requirements referenced in paragraph (f)(1)(iii) of this section are temporarily deferred, with respect to the two hazardous waste surface impoundments, provided that the Sistersville Plant is in compliance with the requirements of paragraphs (f)(2)(vi), and (g) of this section. The temporary deferral of the previous sentence shall be effective beginning on the date the Sistersville Plant receives written notification of revocation, and subject to paragraph (g)(5) of this section, shall continue to be effective for a maximum period of 18 months from that date, provided that the Sistersville Plant is in compliance with the requirements of paragraphs (f)(2)(vi) and (g) of this section at all times during that 18-month period.

(5) In no event shall the temporary deferral provided under paragraph (g)(3) or (g)(4) of this section be effective after the MON Compliance Date.

[59 FR 62935, Dec. 6, 1994]

EDITORIAL NOTE: For FEDERAL REGISTER citations affecting § 265.1030, see the List of CFR Sections Affected, which appears in the Finding Aids section of the printed volume and on GPO Access.

§ 265.1081   Definitions.

As used in this subpart, all terms not defined herein shall have the meaning given to them in the Act and parts 260 through 266 of this chapter.

*Average volatile organic concentration* or *average VO concentration* means the mass-weighted average volatile organic concentration of a hazardous waste as determined in accordance with the requirements of § 265.1084 of this subpart.

*Closure device* means a cap, hatch, lid, plug, seal, valve, or other type of fitting that blocks an opening in a cover such that when the device is secured in the closed position it prevents or reduces air pollutant emissions to the atmosphere. Closure devices include devices that are detachable from the cover (e.g., a sampling port cap), manually operated (e.g., a hinged access lid or hatch), or automatically operated (e.g., a spring-loaded pressure relief valve).

*Continuous seal* means a seal that forms a continuous closure that completely covers the space between the edge of the floating roof and the wall of a tank. A continuous seal may be a vapor-mounted seal, liquid-mounted seal, or metallic shoe seal. A continuous seal may be constructed of fastened segments so as to form a continuous seal.

*Cover* means a device that provides a continuous barrier over the hazardous waste managed in a unit to prevent or reduce air pollutant emissions to the atmosphere. A cover may have openings (such as access hatches, sampling ports, gauge wells) that are necessary for operation, inspection, maintenance, and repair of the unit on which the cover is used. A cover may be a separate piece of equipment which can be detached and removed from the unit or a cover may be formed by structural features permanently integrated into the design of the unit.

*Enclosure* means a structure that surrounds a tank or container, captures organic vapors emitted from the tank or container, and vents the captured vapors through a closed-vent system to a control device.

*External floating roof* means a pontoon-type or double-deck type cover that rests on the surface of the material managed in a tank with no fixed roof.

*Fixed roof* means a cover that is mounted on a unit in a stationary position and does not move with fluctuations in the level of the material managed in the unit.

*Floating membrane cover* means a cover consisting of a synthetic flexible membrane material that rests upon and is supported by the hazardous waste being managed in a surface impoundment.

*Floating roof* means a cover consisting of a double deck, pontoon single deck, or internal floating cover which rests upon and is supported by the material being contained, and is equipped with a continuous seal.

*Hard-piping* means pipe or tubing that is manufactured and properly installed in accordance with relevant standards and good engineering practices.

*In light material service* means the container is used to manage a material for which both of the following conditions apply: The vapor pressure of one or more of the organic constituents in the material is greater than 0.3 kilopascals (kPa) at 20 °C; and the total concentration of the pure organic constituents having a vapor pressure greater than 0.3 kPa at 20 °C is equal to or greater than 20 percent by weight.

*Internal floating roof* means a cover that rests or floats on the material surface (but not necessarily in complete contact with it) inside a tank that has a fixed roof.

*Liquid-mounted seal* means a foam or liquid-filled primary seal mounted in contact with the hazardous waste between the tank wall and the floating roof continuously around the circumference of the tank.

*Malfunction* means any sudden, infrequent, and not reasonably preventable failure of air pollution control equipment, process equipment, or a process to operate in a normal or usual manner. Failures that are caused in part by poor maintenance or careless operation are not malfunctions.

*Maximum organic vapor pressure* means the sum of the individual organic constituent partial pressures exerted by the material contained in a tank, at the maximum vapor pressure-causing conditions (i.e., temperature, agitation, pH effects of combining wastes, etc.) reasonably expected to occur in the tank. For the purpose of this subpart, maximum organic vapor pressure is determined using the procedures specified in §265.1084(c) of this subpart.

*Metallic shoe seal* means a continuous seal that is constructed of metal sheets which are held vertically against the wall of the tank by springs, weighted levers, or other mechanisms and is connected to the floating roof by braces or other means. A flexible coated fabric (envelope) spans the annular space between the metal sheet and the floating roof.

*No detectable organic emissions* means no escape of organics to the atmosphere as determined using the procedure specified in § 265.1084(d) of this subpart.

*Point of waste origination* means as follows:

(1) When the facility owner or operator is the generator of the hazardous waste, the point of waste origination means the point where a solid waste produced by a system, process, or waste management unit is determined to be a hazardous waste as defined in 40 CFR part 261.

NOTE: In this case, this term is being used in a manner similar to the use of the term "point of generation" in air standards established for waste management operations under authority of the Clean Air Act in 40 CFR parts 60, 61, and 63.]

(2) When the facility owner and operator are not the generator of the hazardous waste, point of waste origination means the point where the owner or operator accepts delivery or takes possession of the hazardous waste.

*Point of waste treatment* means the point where a hazardous waste to be treated in accordance with § 265.1083(c)(2) of this subpart exits the treatment process. Any waste determination shall be made before the waste is conveyed, handled, or otherwise managed in a manner that allows the waste to volatilize to the atmosphere.

*Safety device* means a closure device such as a pressure relief valve, frangible disc, fusible plug, or any other type of device which functions exclusively to prevent physical damage or permanent deformation to a unit or its air emission control equipment by venting gases or vapors directly to the atmosphere during unsafe conditions resulting from an unplanned, accidental, or emergency event. For the purpose of this subpart, a safety device is not used for routine venting of gases or vapors from the vapor headspace underneath a cover such as during filling of the unit or to adjust the pressure in this vapor headspace in response to normal daily diurnal ambient temperature fluctuations. A safety device is designed to remain in a closed position during normal operations and open only when the internal pressure, or an-

other relevant parameter, exceeds the device threshold setting applicable to the air emission control equipment as determined by the owner or operator based on manufacturer recommendations, applicable regulations, fire protection and prevention codes, standard engineering codes and practices, or other requirements for the safe handling of flammable, ignitable, explosive, reactive, or hazardous materials.

*Single-seal system* means a floating roof having one continuous seal. This seal may be vapor-mounted, liquid-mounted, or a metallic shoe seal.

*Vapor-mounted seal* means a continuous seal that is mounted such that there is a vapor space between the hazardous waste in the unit and the bottom of the seal.

*Volatile organic concentration* or *VO concentration* means the fraction by weight of the volatile organic compounds contained in a hazardous waste expressed in terms of parts per million (ppmw) as determined by direct measurement or by knowledge of the waste in accordance with the requirements of § 265.1084 of this subpart. For the purpose of determining the VO concentration of a hazardous waste, organic compounds with a Henry's law constant value of at least 0.1 mole-fraction-in-the-gas-phase/mole-fraction-in the liquid-phase (0.1 Y/X) (which can also be expressed as $1.8 \times 10^{-6}$ atmospheres/gram-mole/m³) at 25 degrees Celsius must be included. Appendix VI of this subpart presents a list of compounds known to have a Henry's law constant value less than the cutoff level.

*Waste determination* means performing all applicable procedures in accordance with the requirements of § 265.1084 of this subpart to determine whether a hazardous waste meets standards specified in this subpart. Examples of a waste determination include performing the procedures in accordance with the requirements of § 265.1084 of this subpart to determine the average VO concentration of a hazardous waste at the point of waste origination; the average VO concentration of a hazardous waste at the point of waste treatment and comparing the results to the exit concentration limit specified for the process used to treat the hazardous waste; the organic reduction efficiency and the organic biodegradation efficiency for a biological process used to treat a hazardous waste and comparing the results to the applicable standards; or the maximum volatile organic vapor pressure for a hazardous waste in a tank and comparing the results to the applicable standards.

*Waste stabilization process* means any physical or chemical process used to either reduce the mobility of hazardous constituents in a hazardous waste or eliminate free liquids as determined by Test Method 9095B (Paint Filter Liquids Test) in "Test Methods for Evaluating Solid Waste, Physical/Chemical Methods," EPA Publication SW–846, as incorporated by reference in §260.11. A waste stabilization process includes mixing the hazardous waste with binders or other materials, and curing the resulting hazardous waste and binder mixture. Other synonymous terms used to refer to this process are "waste fixation" or "waste solidification." This does not include the adding of absorbent materials to the surface of a waste, without mixing, agitation, or subsequent curing, to absorb free liquid.

[59 FR 62935, Dec. 6, 1994, as amended at 61 FR 4914, Feb. 9, 1996; 61 FR 59971, Nov. 25, 1996; 62 FR 64662, Dec. 8, 1997; 70 FR 34586, June 14, 2005]

### §265.1082 Schedule for implementation of air emission standards.

(a) Owners or operators of facilities existing on December 6, 1996 and subject to subparts I, J, and K of this part shall meet the following requirements:

(1) Install and begin operation of all control equipment or waste management units required to comply with this subpart and complete modifications of production or treatment processes to satisfy exemption criteria in accordance with §265.1083(c) of this subpart by December 6, 1996, except as provided for in paragraph (a)(2) of this section.

(2) When control equipment or waste management units required to comply with this subpart cannot be installed and in operation or modifications of production or treatment processes to satisfy exemption criteria in accordance with §265.1083(c) of this subpart

cannot be completed by December 6, 1996, the owner or operator shall:

(i) Install and begin operation of the control equipment and waste management units, and complete modifications of production or treatment processes as soon as possible but no later than December 8, 1997.

(ii) Prepare an implementation schedule that includes the following information: specific calendar dates for award of contracts or issuance of purchase orders for control equipment, waste management units, and production or treatment process modifications; initiation of on-site installation of control equipment or waste management units, and modifications of production or treatment processes; completion of control equipment or waste management unit installation, and production or treatment process modifications; and performance of testing to demonstrate that the installed equipment or waste management units, and modified production or treatment processes meet the applicable standards of this subpart.

(iii) For facilities subject to the recordkeeping requirements of §265.73 of this part, the owner or operator shall enter the implementation schedule specified in paragraph (a)(2)(ii) of this section in the operating record no later than December 6, 1996.

(iv) For facilities not subject to §265.73 of this part, the owner or operator shall enter the implementation schedule specified in paragraph (a)(2)(ii) of this section in a permanent, readily available file located at the facility no later than December 6, 1996.

(b) Owners or operators of facilities and units in existence on the effective date of a statutory or EPA regulatory amendment that renders the facility subject to subparts I, J, or K of this part shall meet the following requirements:

(1) Install and begin operation of control equipment or waste management units required to comply with this subpart, and complete modifications of production or treatment processes to satisfy exemption criteria of §265.1083(c) of this subpart by the effective date of the amendment, except as provided for in paragraph (b)(2) of this section.

(2) When control equipment or waste management units required to comply with this subpart cannot be installed and begin operation, or when modifications of production or treatment processes to satisfy exemption criteria of § 265.1083(c) of this subpart cannot be completed by the effective date of the amendment, the owner or operator shall:

(i) Install and begin operation of the control equipment or waste management unit, and complete modification of production or treatment processes as soon as possible but no later than 30 months after the effective date of the amendment.

(ii) For facilities subject to the recordkeeping requirements of § 265.73 of this part, enter and maintain the implementation schedule specified in paragraph (a)(2)(ii) of this section in the operating record no later than the effective date of the amendment, or

(iii) For facilities not subject to § 265.73 of this part, the owner or operator shall enter and maintain the implementation schedule specified in paragraph (a)(2)(ii) of this section in a permanent, readily available file located at the facility site no later than the effective date of the amendment.

(c) Owners and operators of facilities and units that become newly subject to the requirements of this subpart after December 8, 1997 due to an action other than those described in paragraph (b) of this section must comply with all applicable requirements immediately (i.e., must have control devices installed and operating on the date the facility or unit becomes subject to this subpart; the 30-month implementation schedule does not apply).

(d) The Regional Administrator may elect to extend the implementation date for control equipment at a facility, on a case by case basis, to a date later than December 8, 1997, when special circumstances that are beyond the facility owner's or operator's control delay installation or operation of control equipment, and the owner or operator has made all reasonable and prudent attempts to comply with the requirements of this subpart.

[62 FR 64662, Dec. 8, 1997]

§ 265.1083   Standards: General.

(a) This section applies to the management of hazardous waste in tanks, surface impoundments, and containers subject to this subpart.

(b) The owner or operator shall control air pollutant emissions from each hazardous waste management unit in accordance with standards specified in §§ 265.1085 through 265.1088 of this subpart, as applicable to the hazardous waste management unit, except as provided for in paragraph (c) of this section.

(c) A tank, surface impoundment, or container is exempt from standards specified in § 265.1085 through § 265.1088 of this subpart, as applicable, provided that the waste management unit is one of the following:

(1) A tank, surface impoundment, or container for which all hazardous waste entering the unit has an average VO concentration at the point of waste origination of less than 500 parts per million by weight (ppmw). The average VO concentration shall be determined using the procedures specified in § 265.1084(a) of this subpart. The owner or operator shall review and update, as necessary, this determination at least once every 12 months following the date of the initial determination for the hazardous waste streams entering the unit.

(2) A tank, surface impoundment, or container for which the organic content of all the hazardous waste entering the waste management unit has been reduced by an organic destruction or removal process that achieves any one of the following conditions:

(i) A process that removes or destroys the organics contained in the hazardous waste to a level such that the average VO concentration of the hazardous waste at the point of waste treatment is less than the exit concentration limit ($C_t$) established for the process. The average VO concentration of the hazardous waste at the point of waste treatment and the exit concentration limit for the process shall be determined using the procedures specified in § 265.1084(b) of this subpart.

(ii) A process that removes or destroys the organics contained in the hazardous waste to a level such that the organic reduction efficiency (R) for the process is equal to or greater than 95 percent, and the average VO concentration of the hazardous waste at the point of waste treatment is less than 100 ppmw. The organic reduction efficiency for the process and the average VO concentration of the hazardous waste at the point of waste treatment shall be determined using the procedures specified in § 265.1084(b) of this subpart.

(iii) A process that removes or destroys the organics contained in the hazardous waste to a level such that the actual organic mass removal rate (MR) for the process is equal to or greater than the required organic mass removal rate (RMR) established for the process. The required organic mass removal rate and the actual organic mass removal rate for the process shall be determined using the procedures specified in § 265.1084(b) of this subpart.

(iv) A biological process that destroys or degrades the organics contained in the hazardous waste, such that either of the following conditions is met:

(A) The organic reduction efficiency (R) for the process is equal to or greater than 95 percent, and the organic biodegradation efficiency ($R_{bio}$) for the process is equal to or greater than 95 percent. The organic reduction efficiency and the organic biodegradation efficiency for the process shall be determined using the procedures specified in § 265.1084(b) of this subpart.

(B) The total actual organic mass biodegradation rate ($MR_{bio}$) for all hazardous waste treated by the process is equal to or greater than the required organic mass removal rate (RMR). The required organic mass removal rate and the actual organic mass degradation rate for the process shall be determined using the procedures specified in § 265.1084(b) of this subpart.

(v) A process that removes or destroys the organics contained in the hazardous waste and meets all of the following conditions:

(A) From the point of waste origination through the point where the hazardous waste enters the treatment process, the hazardous waste is managed continuously in waste management units which use air emission controls in accordance with the standards specified in § 265.1085 through § 265.1088 of this subpart, as applicable to the waste management unit.

(B) From the point of waste origination through the point where the hazardous waste enters the treatment process, any transfer of the hazardous waste is accomplished through continuous hard-piping or other closed system transfer that does not allow exposure of the waste to the atmosphere. The EPA considers a drain system that meets the requirements of 40 CFR part 63, subpart RR—National Emission Standards for Individual Drain Systems to be a closed system.

(C) The average VO concentration of the hazardous waste at the point of waste treatment is less than the lowest average VO concentration at the point of waste origination determined for each of the individual waste streams entering the process or 500 ppmw, whichever value is lower. The average VO concentration of each individual waste stream at the point of waste origination shall be determined using the procedures specified in § 265.1084(a) of this subpart. The average VO concentration of the hazardous waste at the point of waste treatment shall be determined using the procedures specified in § 265.1084(b) of this subpart.

(vi) A process that removes or destroys the organics contained in the hazardous waste to a level such that the organic reduction efficiency (R) for the process is equal to or greater than 95 percent and the owner or operator certifies that the average VO concentration at the point of waste origination for each of the individual waste streams entering the process is less than 10,000 ppmw. The organic reduction efficiency for the process and the average VO concentration of the hazardous waste at the point of waste origination shall be determined using the procedures specified in § 265.1084(b) and § 265.1084(a) of this subpart, respectively.

(vii) A hazardous waste incinerator for which the owner or operator has either:

(A) Been issued a final permit under 40 CFR part 270 which implements the requirements of 40 CFR part 264, subpart O; or

(B) Has designed and operates the incinerator in accordance with the interim status requirements of subpart O of this part.

(viii) A boiler or industrial furnace for which the owner or operator has either:

(A) Been issued a final permit under 40 CFR part 270 which implements the requirements of 40 CFR part 266, subpart H, or

(B) Has designed and operates the boiler or industrial furnace in accordance with the interim status requirements of 40 CFR part 266, subpart H.

(ix) For the purpose of determining the performance of an organic destruction or removal process in accordance with the conditions in each of paragraphs (c)(2)(i) through (c)(2)(vi) of this section, the owner or operator shall account for VO concentrations determined to be below the limit of detection of the analytical method by using the following VO concentration:

(A) If Method 25D in 40 CFR part 60, appendix A is used for the analysis, one-half the blank value determined in the method at section 4.4 of Method 25D in 40 CFR part 60, appendix A, or a value of 25 ppmw, whichever is less.

(B) If any other analytical method is used, one-half the sum of the limits of detection established for each organic constituent in the waste that has a Henry's law constant value at least 0.1 mole-fraction-in-the-gas-phase/mole-fraction-in-the-liquid-phase (0.1 Y/X) [which can also be expressed as $1.8 \times 10^{-6}$ atmospheres/gram-mole/m³] at 25 degrees Celsius.

(3) A tank or surface impoundment used for biological treatment of hazardous waste in accordance with the requirements of paragraph (c)(2)(iv) of this section.

(4) A tank, surface impoundment, or container for which all hazardous waste placed in the unit either:

(i) Meets the numerical concentration limits for organic hazardous constituents, applicable to the hazardous waste, as specified in 40 CFR part 268—Land Disposal Restrictions under Table "Treatment Standards for Hazardous Waste" in 40 CFR 268.40; or

(ii) The organic hazardous constituents in the waste have been treated by the treatment technology established

by the EPA for the waste in 40 CFR 268.42(a), or have been removed or destroyed by an equivalent method of treatment approved by EPA pursuant to 40 CFR 268.42(b).

(5) A tank used for bulk feed of hazardous waste to a waste incinerator and all of the following conditions are met:

(i) The tank is located inside an enclosure vented to a control device that is designed and operated in accordance with all applicable requirements specified under 40 CFR part 61, subpart FF—National Emission Standards for Benzene Waste Operations for a facility at which the total annual benzene quantity from the facility waste is equal to or greater than 10 megagrams per year;

(ii) The enclosure and control device serving the tank were installed and began operation prior to November 25, 1996; and

(iii) The enclosure is designed and operated in accordance with the criteria for a permanent total enclosure as specified in "Procedure T—Criteria for and Verification of a Permanent or Temporary Total Enclosure" under 40 CFR 52.741, Appendix B. The enclosure may have permanent or temporary openings to allow worker access; passage of material into or out of the enclosure by conveyor, vehicles, or other mechanical or electrical equipment; or to direct air flow into the enclosure. The owner or operator shall perform the verification procedure for the enclosure as specified in Section 5.0 to "Procedure T—Criteria for and Verification of a Permanent or Temporary Total Enclosure" annually.

(d) The Regional Administrator may at any time perform or request that the owner or operator perform a waste determination for a hazardous waste managed in a tank, surface impoundment, or container exempted from using air emission controls under the provisions of this section as follows:

(1) The waste determination for average VO concentration of a hazardous waste at the point of waste origination shall be performed using direct measurement in accordance with the applicable requirements of § 265.1084(a) of this subpart. The waste determination for a hazardous waste at the point of waste treatment shall be performed in

accordance with the applicable requirements of §265.1084(b) of this subpart.

(2) In performing a waste determination pursuant to paragraph (d)(1) of this section, the sample preparation and analysis shall be conducted as follows:

(i) In accordance with the method used by the owner or operator to perform the waste analysis, except in the case specified in paragraph (d)(2)(ii) of this section.

(ii) If the Regional Administrator determines that the method used by the owner or operator was not appropriate for the hazardous waste managed in the tank, surface impoundment, or container, then the Regional Administrator may choose an appropriate method.

(3) In a case when the owner or operator is requested to perform the waste determination, the Regional Administrator may elect to have an authorized representative observe the collection of the hazardous waste samples used for the analysis.

(4) In a case when the results of the waste determination performed or requested by the Regional Administrator do not agree with the results of a waste determination performed by the owner or operator using knowledge of the waste, then the results of the waste determination performed in accordance with the requirements of paragraph (d)(1) of this section shall be used to establish compliance with the requirements of this subpart.

(5) In a case when the owner or operator has used an averaging period greater than 1 hour for determining the average VO concentration of a hazardous waste at the point of waste origination, the Regional Administrator may elect to establish compliance with this subpart by performing or requesting that the owner or operator perform a waste determination using direct measurement based on waste samples collected within a 1-hour period as follows:

(i) The average VO concentration of the hazardous waste at the point of waste origination shall be determined by direct measurement in accordance with the requirements of §265.1084(a) of this subpart.

(ii) Results of the waste determination performed or requested by the Regional Administrator showing that the average VO concentration of the hazardous waste at the point of waste origination is equal to or greater than 500 ppmw shall constitute noncompliance with this subpart except in a case as provided for in paragraph (d)(5)(iii) of this section.

(iii) For the case when the average VO concentration of the hazardous waste at the point of waste origination previously has been determined by the owner or operator using an averaging period greater than 1 hour to be less than 500 ppmw but because of normal operating process variations the VO concentration of the hazardous waste determined by direct measurement for any given 1-hour period may be equal to or greater than 500 ppmw, information that was used by the owner or operator to determine the average VO concentration of the hazardous waste (e.g., test results, measurements, calculations, and other documentation) and recorded in the facility records in accordance with the requirements of §265.1084(a) and §265.1090 of this subpart shall be considered by the Regional Administrator together with the results of the waste determination performed or requested by the Regional Administrator in establishing compliance with this subpart.

[61 FR 59972, Nov. 25, 1996, as amended at 62 FR 64663, Dec. 8, 1997]

### §265.1084 Waste determination procedures.

(a) Waste determination procedure to determine average volatile organic (VO) concentration of a hazardous waste at the point of waste origination.

(1) An owner or operator shall determine the average VO concentration at the point of waste origination for each hazardous waste placed in a waste management unit exempted under the provisions of §265.1083(c)(1) of this subpart from using air emission controls in accordance with standards specified in §265.1085 through §265.1088 of this subpart, as applicable to the waste management unit.

(i) An initial determination of the average VO concentration of the waste stream shall be made before the first time any portion of the material in the hazardous waste stream is placed in a waste management unit exempted under the provisions of § 265.1083(c)(1) of this subpart from using air emission controls, and thereafter an initial determination of the average VO concentration of the waste stream shall be made for each averaging period that a hazardous waste is managed in the unit; and

(ii) Perform a new waste determination whenever changes to the source generating the waste stream are reasonably likely to cause the average VO concentration of the hazardous waste to increase to a level that is equal to or greater than the VO concentration limit specified in § 265.1083(c)(1) of this subpart.

(2) For a waste determination that is required by paragraph (a)(1) of this section, the average VO concentration of a hazardous waste at the point of waste origination shall be determined using either direct measurement as specified in paragraph (a)(3) of this section or by knowledge as specified in paragraph (a)(4) of this section.

(3) Direct measurement to determine average VO concentration of a hazardous waste at the point of waste origination.

(i) Identification. The owner or operator shall identify and record the point of waste origination for the hazardous waste.

(ii) Sampling. Samples of the hazardous waste stream shall be collected at the point of waste origination in a manner such that volatilization of organics contained in the waste and in the subsequent sample is minimized and an adequately representative sample is collected and maintained for analysis by the selected method.

(A) The averaging period to be used for determining the average VO concentration for the hazardous waste stream on a mass-weighted average basis shall be designated and recorded. The averaging period can represent any time interval that the owner or operator determines is appropriate for the hazardous waste stream but shall not exceed 1 year.

(B) A sufficient number of samples, but no less than four samples, shall be collected and analyzed for a hazardous

waste determination. All of the samples for a given waste determination shall be collected within a one-hour period. The average of the four or more sample results constitutes a waste determination for the waste stream. One or more waste determinations may be required to represent the complete range of waste compositions and quantities that occur during the entire averaging period due to normal variations in the operating conditions for the source or process generating the hazardous waste stream. Examples of such normal variations are seasonal variations in waste quantity or fluctuations in ambient temperature.

(C) All samples shall be collected and handled in accordance with written procedures prepared by the owner or operator and documented in a site sampling plan. This plan shall describe the procedure by which representative samples of the hazardous waste stream are collected such that a minimum loss of organics occurs throughout the sample collection and handling process, and by which sample integrity is maintained. A copy of the written sampling plan shall be maintained on-site in the facility operating records. An example of acceptable sample collection and handling procedures for a total volatile organic constituent concentration may be found in Method 25D in 40 CFR part 60, appendix A.

(D) Sufficient information, as specified in the "site sampling plan" required under paragraph (a)(3)(ii)(C) of this section, shall be prepared and recorded to document the waste quantity represented by the samples and, as applicable, the operating conditions for the source or process generating the hazardous waste represented by the samples.

(iii) Analysis. Each collected sample shall be prepared and analyzed in accordance with Method 25D in 40 CFR part 60, appendix A for the total concentration of volatile organic constituents, or using one or more methods when the individual organic compound concentrations are identified and summed and the summed waste concentration accounts for and reflects all organic compounds in the waste with Henry's law constant values at least 0.1 mole-fraction-in-the-gas-phase/mole-fraction-in-the-liquid-phase (0.1 Y/X) [which can also be expressed as 1.8 × $10^{-6}$ atmospheres/gram-mole/m³] at 25 degrees Celsius. At the owner or operator's discretion, the owner or operator may adjust test data obtained by any appropriate method to discount any contribution to the total volatile organic concentration that is a result of including a compound with a Henry's law constant value of less than 0.1 Y/X at 25 degrees Celsius. To adjust these data, the measured concentration of each individual chemical constituent contained in the waste is multiplied by the appropriate constituent-specific adjustment factor ($f_{m25D}$). If the owner or operator elects to adjust test data, the adjustment must be made to all individual chemical constituents with a Henry's law constant value greater than or equal to 0.1 Y/X at 25 degrees Celsius contained in the waste. Constituent-specific adjustment factors ($f_{m25D}$) can be obtained by contacting the Waste and Chemical Processes Group, Office of Air Quality Planning and Standards, Research Triangle Park, NC 27711. Other test methods may be used if they meet the requirements in paragraph (a)(3)(iii)(A) or (B) of this section and provided the requirement to reflect all organic compounds in the waste with Henry's law constant values greater than or equal to 0.1 Y/X [which can also be expressed as 1.8 × $10^{-6}$ atmospheres/gram-mole/ m³] at 25 degrees Celsius, is met.

(A) Any EPA standard method that has been validated in accordance with "Alternative Validation Procedure for EPA Waste and Wastewater Methods," 40 CFR part 63, appendix D.

(B) Any other analysis method that has been validated in accordance with the procedures specified in Section 5.1 or Section 5.3, and the corresponding calculations in Section 6.1 or Section 6.3, of Method 301 in 40 CFR part 63, appendix A. The data are acceptable if they meet the criteria specified in Section 6.1.5 or Section 6.3.3 of Method 301. If correction is required under section 6.3.3 of Method 301, the data are acceptable if the correction factor is within the range 0.7 to 1.30. Other sections of Method 301 are not required.

(iv) Calculations.

(A) The average VO concentration ($\bar{C}$) on a mass-weighted basis shall be calculated by using the results for all waste determinations conducted in accordance with paragraphs (a)(3) (ii) and (iii) of this section and the following equation:

$$\bar{C} = \frac{1}{Q_T} \times \sum_{i=1}^{\eta} \left( Q_i \times C_i \right)$$

where:

$\bar{C}$ = Average VO concentration of the hazardous waste at the point of waste origination on a mass-weighted basis, ppmw.

i = Individual waste determination "i" of the hazardous waste.

n = Total number of waste determinations of the hazardous waste conducted for the averaging period (not to exceed 1 year).

$Q_i$ = Mass quantity of hazardous waste stream represented by $C_i$, kg/hr.

$Q_T$ = Total mass quantity of hazardous waste during the averaging period, kg/hr.

$C_i$ = Measured VO concentration of waste determination "i" as determined in accordance with the requirements of paragraph (a)(3)(iii) of this section (i.e. the average of the four or more samples specified in paragraph (a)(3)(ii)(B) of this section), ppmw.

(B) For the purpose of determining $C_i$, for individual waste samples analyzed in accordance with paragraph (a)(3)(iii) of this section, the owner or operator shall account for VO concentrations determined to be below the limit of detection of the analytical method by using the following VO concentration:

(1) If Method 25D in 40 CFR part 60, Appendix A is used for the analysis, one-half the blank value determined in the method at section 4.4 of Method 25D in 40 CFR part 60, appendix A.

(2) If any other analytical method is used, one-half the sum of the limits of detection established for each organic constituent in the waste that has a Henry's law constant values at least 0.1 mole-fraction-in-the-gas-phase/mole-fraction-in-the-liquid-phase (0.1 Y/X) [which can also be expressed as 1.8×$10^{-6}$ atmospheres/gram-mole/m³] at 25 degrees Celsius.

(v) Provided that the test method is appropriate for the waste as required under paragraph (a)(3)(iii) of this section, the EPA will determine compliance based on the test method used by the owner or operator as recorded pursuant to § 265.1090(f)(1) of this subpart.

(4) Use of owner or operator knowledge to determine average VO concentration of a hazardous waste at the point of waste origination.

(i) Documentation shall be prepared that presents the information used as the basis for the owner's or operator's knowledge of the hazardous waste stream's average VO concentration. Examples of information that may be used as the basis for knowledge include: Material balances for the source or process generating the hazardous waste stream; constituent-specific chemical test data for the hazardous waste stream from previous testing that are still applicable to the current waste stream; previous test data for other locations managing the same type of waste stream; or other knowledge based on information included in manifests, shipping papers, or waste certification notices.

(ii) If test data are used as the basis for knowledge, then the owner or operator shall document the test method, sampling protocol, and the means by which sampling variability and analytical variability are accounted for in the determination of the average VO concentration. For example, an owner or operator may use organic concentration test data for the hazardous waste stream that are validated in accordance with Method 301 in 40 CFR part 63, appendix A as the basis for knowledge of the waste.

(iii) An owner or operator using chemical constituent-specific concentration test data as the basis for knowledge of the hazardous waste may adjust the test data to the corresponding average VO concentration value which would have been obtained had the waste samples been analyzed using Method 25D in 40 CFR part 60, appendix A. To adjust these data, the measured concentration for each individual chemical constituent contained in the waste is multiplied by the appropriate constituent-specific adjustment factor ($f_{m25D}$).

(iv) In the event that the Regional Administrator and the owner or operator disagree on a determination of the average VO concentration for a hazardous waste stream using knowledge,

then the results from a determination of average VO concentration using direct measurement as specified in paragraph (a)(3) of this section shall be used to establish compliance with the applicable requirements of this subpart. The Regional Administrator may perform or request that the owner or operator perform this determination using direct measurement. The owner or operator may choose one or more appropriate methods to analyze each collected sample in accordance with the requirements of paragraph (a)(3)(iii) of this section.

(b) Waste determination procedures for treated hazardous waste.

(1) An owner or operator shall perform the applicable waste determination for each treated hazardous waste placed in a waste management unit exempted under the provisions of § 265.1083 (c)(2)(i) through (c)(2)(vi) of this subpart from using air emission controls in accordance with standards specified in §§ 265.1085 through 265.1088 of this subpart, as applicable to the waste management unit.

(i) An initial determination of the average VO concentration of the waste stream shall be made before the first time any portion of the material in the treated waste stream is placed in a waste management unit exempted under the provisions of § 265.1083(c)(2), § 265.1083(c)(3), or § 265.1083(c)(4) of this subpart from using air emission controls, and thereafter update the information used for the waste determination at least once every 12 months following the date of the initial waste determination; and

(ii) Perform a new waste determination whenever changes to the process generating or treating the waste stream are reasonably likely to cause the average VO concentration of the hazardous waste to increase to a level such that the applicable treatment conditions specified in § 265.1083(c)(2), § 265.1083(c)(3), or § 265.1083(c)(4) of this subpart are not achieved.

(2) The owner or operator shall designate and record the specific provision in § 265.1083(c)(2) of this subpart under which the waste determination is being performed. The waste determination for the treated hazardous waste shall

be performed using the applicable procedures specified in paragraphs (b)(3) through (b)(9) of this section.

(3) Procedure to determine the average VO concentration of a hazardous waste at the point of waste treatment.

(i) Identification. The owner or operator shall identify and record the point of waste treatment for the hazardous waste.

(ii) Sampling. Samples of the hazardous waste stream shall be collected at the point of waste treatment in a manner such that volatilization of organics contained in the waste and in the subsequent sample is minimized and an adequately representative sample is collected and maintained for analysis by the selected method.

(A) The averaging period to be used for determining the average VO concentration for the hazardous waste stream on a mass-weighted average basis shall be designated and recorded. The averaging period can represent any time interval that the owner or operator determines is appropriate for the hazardous waste stream but shall not exceed 1 year.

(B) A sufficient number of samples, but no less than four samples, shall be collected and analyzed for a hazardous waste determination. All of the samples for a given waste determination shall be collected within a one-hour period. The average of the four or more sample results constitutes a waste determination for the waste stream. One or more waste determinations may be required to represent the complete range of waste compositions and quantities that occur during the entire averaging period due to normal variations in the operating conditions for the process generating or treating the hazardous waste stream. Examples of such normal variations are seasonal variations in waste quantity or fluctuations in ambient temperature.

(C) All samples shall be collected and handled in accordance with written procedures prepared by the owner or operator and documented in a site sampling plan. This plan shall describe the procedure by which representative samples of the hazardous waste stream are collected such that a minimum loss of organics occurs throughout the sample collection and handling process,

and by which sample integrity is maintained. A copy of the written sampling plan shall be maintained on-site in the facility operating records. An example of acceptable sample collection and handling procedures for a total volatile organic constituent concentration may be found in Method 25D in 40 CFR part 60, appendix A.

(D) Sufficient information, as specified in the "site sampling plan" required under paragraph (C) of (b)(3)(ii)this section, § 265.1084(b)(3)(ii), shall be prepared and recorded to document the waste quantity represented by the samples and, as applicable, the operating conditions for the process treating the hazardous waste represented by the samples.

(iii) *Analysis.* Each collected sample shall be prepared and analyzed in accordance with Method 25D in 40 CFR part 60, appendix A for the total concentration of volatile organic constituents, or using one or more methods when the individual organic compound concentrations are identified and summed and the summed waste concentration accounts for and reflects all organic compounds in the waste with Henry's law constant values at least 0.1 mole-fraction-in-the-gas-phase/mole-fraction-in-the-liquid-phase (0.1 Y/X) [which can also be expressed as 1.8 × 10⁻⁶ atmospheres/gram-mole/m³] at 25 degrees Celsius. When the owner or operator is making a waste determination for a treated hazardous waste that is to be compared to an average VO concentration at the point of waste origination or the point of waste entry to the treatment system to determine if the conditions of § 264.1082(c)(2)(i) through (c)(2)(vi) of this chapter, or § 265.1083(c)(2)(i) through (c)(2)(vi) of this subpart are met, then the waste samples shall be prepared and analyzed using the same method or methods as were used in making the initial waste determinations at the point of waste origination or at the point of entry to the treatment system. At the owner or operator's discretion, the owner or operator may adjust test data obtained by any appropriate method to discount any contribution to the total volatile organic concentration that is a result of including a compound with a Henry's law constant value less than

0.1 Y/X at 25 degrees Celsius. To adjust these data, the measured concentration of each individual chemical constituent in the waste is multiplied by the appropriate constituent-specific adjustment factor ($f_{m25D}$). If the owner or operator elects to adjust test data, the adjustment must be made to all individual chemical constituents with a Henry's law constant value greater than or equal to 0.1 Y/X at 25 degrees Celsius contained in the waste. Constituent-specific adjustment factors ($f_{m25D}$) can be obtained by contacting the Waste and Chemical Processes Group, Office of Air Quality Planning and Standards, Research Triangle Park, NC 27711. Other test methods may be used if they meet the requirements in paragraph (a)(3)(iii)(A) or (B) of this section and provided the requirement to reflect all organic compounds in the waste with Henry's law constant values greater than or equal to 0.1 Y/X [which can also be expressed as $1.8 \times 10^{-6}$ atmospheres/gram-mole/m³] at 25 degrees Celsius, is met.

(A) Any EPA standard method that has been validated in accordance with "Alternative Validation Procedure for EPA Waste and Wastewater Methods," 40 CFR part 63, appendix D.

(B) Any other analysis method that has been validated in accordance with the procedures specified in Section 5.1 or Section 5.3, and the corresponding calculations in Section 6.1 or Section 6.3, of Method 301 in 40 CFR part 63, appendix A. The data are acceptable if they meet the criteria specified in Section 6.1.5 or Section 6.3.3 of Method 301. If correction is required under section 6.3.3 of Method 301, the data are acceptable if the correction factor is within the range 0.7 to 1.30. Other sections of Method 301 are not required.

(iv) *Calculations.* The average VO concentration ($\bar{C}$) on a mass-weighted basis shall be calculated by using the results for all waste determinations conducted in accordance with paragraphs (b)(3)(ii) and (iii) of this section and the following equation:

$$\bar{C} = \frac{1}{Q_T} \times \sum_{i=1}^{\eta} \left( Q_i \times C_i \right)$$

where:

$\bar{C}$=Average VO concentration of the hazardous waste at the point of waste treatment on a mass-weighted basis, ppmw.
i=Individual waste determination "i" of the hazardous waste.
n=Total number of waste determinations of the hazardous waste conducted for the averaging period (not to exceed 1 year).
$Q_i$=Mass quantity of hazardous waste stream represented by $C_i$, kg/hr.
$Q_T$=Total mass quantity of hazardous waste during the averaging period, kg/hr.
$C_i$=Measured VO concentration of waste determination "i" as determined in accordance with the requirements of paragraph (b)(3)(iii) of this section (i.e. the average of the four or more samples specified in paragraph (b)(3)(ii)(B) of this section), ppmw.

(v) Provided that the test method is appropriate for the waste as required under paragraph (b)(3)(iii) of this section, compliance shall be determined based on the test method used by the owner or operator as recorded pursuant to § 265.1090(f)(1) of this subpart.

(4) Procedure to determine the exit concentration limit ($C_t$) for a treated hazardous waste.

(i) The point of waste origination for each hazardous waste treated by the process at the same time shall be identified.

(ii) If a single hazardous waste stream is identified in paragraph (b)(4)(i) of this section, then the exit concentration limit ($C_t$) shall be 500 ppmw.

(iii) If more than one hazardous waste stream is identified in paragraph (b)(4)(i) of this section, then the average VO concentration of each hazardous waste stream at the point of waste origination shall be determined in accordance with the requirements of paragraph (a) of this section. The exit concentration limit ($C_t$) shall be calculated by using the results determined for each individual hazardous waste stream and the following equation:

$$C_t = \frac{\sum_{x=1}^{m}\left( Q_x \times \overline{C_x} \right) + \sum_{y=1}^{n}\left( Q_y \times 500 \text{ ppmw} \right)}{\sum_{x=1}^{m} Q_x + \sum_{y=1}^{n} Q_y}$$

Where:

$C_t$ = Exit concentration limit for treated hazardous waste, ppmw.

x = Individual hazardous waste stream "x" that has an average VO concentration less than 500 ppmw at the point of waste origination as determined in accordance with the requirements of §265.1084(a) of this subpart.

y = Individual hazardous waste stream "y" that has an average VO concentration equal to or greater than 500 ppmw at the point of waste origination as determined in accordance with the requirements of §265.1084(a) of this subpart.

m = Total number of "x" hazardous waste streams treated by process.

n = Total number of "y" hazardous waste streams treated by process.

$Q_x$ = Annual mass quantity of hazardous waste stream "x," kg/yr.

$Q_y$ = Annual mass quantity of hazardous waste stream "y," kg/yr.

$\bar{C}_x$ = Average VO concentration of hazardous waste stream "x" at the point of waste origination as determined in accordance with the requirements of §265.1084(a) of this subpart, ppmw.

(5) Procedure to determine the organic reduction efficiency (R) for a treated hazardous waste.

(i) The organic reduction efficiency (R) for a treatment process shall be determined based on results for a minimum of three consecutive runs.

(ii) All hazardous waste streams entering the treatment process and all hazardous waste streams exiting the treatment process shall be identified. The owner or operator shall prepare a sampling plan for measuring these streams that accurately reflects the retention time of the hazardous waste in the process.

(iii) For each run, information shall be determined for each hazardous waste stream identified in paragraph (b)(5)(ii) of this section using the following procedures:

(A) The mass quantity of each hazardous waste stream entering the process ($Q_b$) and the mass quantity of each hazardous waste stream exiting the process ($Q_a$) shall be determined.

(B) The average VO concentration at the point of waste origination of each hazardous waste stream entering the process ($\bar{C}_b$) during the run shall be determined in accordance with the requirements of paragraph (a)(3) of this section. The average VO concentration at the point of waste treatment of each waste stream exiting the process ($\bar{C}_a$) during the run shall be determined in

accordance with the requirements of paragraph (b)(3) of this section.

(iv) The waste volatile organic mass flow entering the process ($E_b$) and the waste volatile organic mass flow exiting the process ($E_a$) shall be calculated by using the results determined in accordance with paragraph (b)(5)(iii) of. this section and the following equations:

$$E_b = \frac{1}{10^6} \sum_{j=1}^{m} \left( Q_{bj} \times \overline{C}_{bj} \right)$$

$$E_a = \frac{1}{10^6} \sum_{j=1}^{m} \left( Q_{aj} \times \overline{C}_{aj} \right)$$

Where:

$E_a$ = Waste volatile organic mass flow exiting process, kg/hr.

$E_b$ = Waste volatile organic mass flow entering process, kg/hr.

m = Total number of runs (at least 3)

j = Individual run "j"

$Q_b$ = Mass quantity of hazardous waste entering process during run "j," kg/hr.

$Q_a$ = Average mass quantity of hazardous waste exiting process during run "j," kg/hr.

$\bar{C}_a$ = Average VO concentration of hazardous waste exiting process during run "j" as determined in accordance with the requirements of §265.1084(b)(3) of this subpart, ppmw.

$\bar{C}_b$ = Average VO concentration of hazardous waste entering process during run "j" as determined in accordance with the requirements of §265.1084(a)(3) of this subpart, ppmw

(v) The organic reduction efficiency of the process shall be calculated by using the results determined in accordance with paragraph (b)(5)(iv) of this section and the following equation:

$$R = \frac{E_b - E_a}{E_b} \times 100\%$$

Where:

R = Organic reduction efficiency, percent.

$E_b$ = Waste volatile organic mass flow entering process as determined in accordance with the requirements of paragraph (b)(5)(iv) of this section, kg/hr.

$E_a$ = Waste volatile organic mass flow exiting process as determined in accordance with the requirements of paragraph (b)(5)(iv) of this section, kg/hr.

(6) Procedure to determine the organic biodegradation efficiency ($R_{bio}$) for a treated hazardous waste.

(i) The fraction of organics biodegraded ($F_{bio}$) shall be determined using the procedure specified in 40 CFR part 63, appendix C of this chapter.

(ii) The $R_{bio}$ shall be calculated by using the following equation:

$$R_{bio} = F_{bio} \times 100\%$$

Where:

$R_{bio}$ = Organic biodegradation efficiency, percent.

$F_{bio}$ = Fraction of organic biodegraded as determined in accordance with the requirements of paragraph (b)(6)(i) of this section.

(7) Procedure to determine the required organic mass removal rate (RMR) for a treated hazardous waste.

(i) All of the hazardous waste streams entering the treatment process shall be identified.

(ii) The average VO concentration of each hazardous waste stream at the point of waste origination shall be determined in accordance with the requirements of paragraph (a) of this section.

(iii) For each individual hazardous waste stream that has an average VO concentration equal to or greater than 500 ppmw at the point of waste origination, the average volumetric flow rate and the density of the hazardous waste stream at the point of waste origination shall be determined.

(iv) The RMR shall be calculated by using the average VO concentration, average volumetric flow rate, and density determined for each individual hazardous waste stream, and the following equation:

$$RMR = \sum_{y=1}^{n} \left[ V_y \times k_y \times \frac{\left( \overline{C_y} - 500 \text{ ppmw} \right)}{10^6} \right]$$

Where:

RMR = Required organic mass removal rate, kg/hr.

y = Individual hazardous waste stream "y" that has an average VO concentration equal to or greater than 500 ppmw at the point of waste origination as determined in accordance with the requirements of § 265.1084(a) of this subpart.

n = Total number of "y" hazardous waste streams treated by process.

$V_y$ = Average volumetric flow rate of hazardous waste stream "y" at the point of waste origination, m³/hr.

$k_y$ = Density of hazardous waste stream "y," kg/m³

$\bar{C}_y$ = Average VO concentration of hazardous waste stream "y" at the point of waste origination as determined in accordance with the requirements of § 265.1084(a) of this subpart, ppmw.

(8) Procedure to determine the actual organic mass removal rate (MR) for a treated hazardous waste.

(i) The MR shall be determined based on results for a minimum of three consecutive runs. The sampling time for each run shall be 1 hour.

(ii) The waste volatile organic mass flow entering the process ($E_b$) and the waste volatile organic mass flow exiting the process ($E_a$) shall be determined in accordance with the requirements of paragraph (b)(5)(iv) of this section.

(iii) The MR shall be calculated by using the mass flow rate determined in accordance with the requirements of paragraph (b)(8)(ii) of this section and the following equation:

$$MR = E_b - E_a$$

Where:

MR = Actual organic mass removal rate, kg/hr.

$E_b$ = Waste volatile organic mass flow entering process as determined in accordance with the requirements of paragraph (b)(5)(iv) of this section, kg/hr.

$E_a$ = Waste volatile organic mass flow exiting process as determined in accordance with the requirements of paragraph (b)(5)(iv) of this section, kg/hr.

(9) Procedure to determine the actual organic mass biodegradation rate ($MR_{bio}$) for a treated hazardous waste.

(i) The $MR_{bio}$ shall be determined based on results for a minimum of three consecutive runs. The sampling time for each run shall be 1 hour.

(ii) The waste organic mass flow entering the process ($E_b$) shall be determined in accordance with the requirements of paragraph (b)(5)(iv) of this section.

(iii) The fraction of organic biodegraded ($F_{bio}$) shall be determined using the procedure specified in 40 CFR part 63, appendix C of this chapter.

(iv) The $MR_{bio}$ shall be calculated by using the mass flow rates and fraction of organic biodegraded determined in accordance with the requirements of paragraphs (b)(9)(ii) and (b)(9)(iii) of this section, respectively, and the following equation:

$$MR_{bio} = E_b \times F_{bio}$$

Where:

$MR_{bio}$=Actual organic mass biodegradation rate, kg/hr.
$E_b$=Waste organic mass flow entering process as determined in accordance with the requirements of paragraph (b)(5)(iv) of this section, kg/hr.
$F_{bio}$=Fraction of organic biodegraded as determined in accordance with the requirements of paragraph (b)(9)(iii) of this section.

(c) Procedure to determine the maximum organic vapor pressure of a hazardous waste in a tank.

(1) An owner or operator shall determine the maximum organic vapor pressure for each hazardous waste placed in a tank using Tank Level 1 controls in accordance with the standards specified in § 265.1085(c) of this subpart.

(2) An owner or operator shall use either direct measurement as specified in paragraph (c)(3) of this section or knowledge of the waste as specified in paragraph (c)(4) of this section to determine the maximum organic vapor pressure which is representative of the hazardous waste composition stored or treated in the tank.

(3) Direct measurement to determine the maximum organic vapor pressure of a hazardous waste.

(i) Sampling. A sufficient number of samples shall be collected to be representative of the waste contained in the tank. All samples shall be collected and handled in accordance with written procedures prepared by the owner or operator and documented in a site sampling plan. This plan shall describe the procedure by which representative samples of the hazardous waste are collected such that a minimum loss of organics occurs throughout the sample collection and handling process and by which sample integrity is maintained. A copy of the written sampling plan shall be maintained on-site in the facility operating records. An example of acceptable sample collection and handling procedures may be found in Method 25D in 40 CFR part 60, appendix A.

(ii) Analysis. Any appropriate one of the following methods may be used to analyze the samples and compute the maximum organic vapor pressure of the hazardous waste:

(A) Method 25E in 40 CFR part 60 appendix A;

(B) Methods described in American Petroleum Institute Publication 2517, Third Edition, February 1989, "Evaporative Loss from External Floating-Roof Tanks," (incorporated by reference—refer to § 260.11 of this chapter);

(C) Methods obtained from standard reference texts;

(D) ASTM Method 2879–92 (incorporated by reference—refer to § 260.11 of this chapter); and

(E) Any other method approved by the Regional Administrator.

(4) Use of knowledge to determine the maximum organic vapor pressure of the hazardous waste. Documentation shall be prepared and recorded that presents the information used as the basis for the owner's or operator's knowledge that the maximum organic vapor pressure of the hazardous waste is less than the maximum vapor pressure limit listed in § 265.1085(b)(1)(i) of this subpart for the applicable tank design capacity category. An example of information that may be used is documentation that the hazardous waste is generated by a process for which at other locations it previously has been determined by direct measurement that the waste maximum organic vapor pressure is less than the maximum vapor pressure limit for the appropriate tank design capacity category.

(d) Procedure for determining no detectable organic emissions for the purpose of complying with this subpart:

(1) The test shall be conducted in accordance with the procedures specified in Method 21 of 40 CFR part 60, appendix A. Each potential leak interface (i.e., a location where organic vapor leakage could occur) on the cover and associated closure devices shall be checked. Potential leak interfaces that are associated with covers and closure devices include, but are not limited to: The interface of the cover and its foundation mounting; the periphery of any opening on the cover and its associated closure device; and the sealing seat interface on a spring-loaded pressure relief valve.

(2) The test shall be performed when the unit contains a hazardous waste having an organic concentration representative of the range of concentrations for the hazardous waste expected to be managed in the unit. During the test, the cover and closure devices shall be secured in the closed position.

(3) The detection instrument shall meet the performance criteria of Method 21 of 40 CFR part 60, appendix A, except the instrument response factor criteria in section 3.1.2(a) of Method 21 shall be for the average composition of the organic constituents in the hazardous waste placed in the waste management unit, not for each individual organic constituent.

(4) The detection instrument shall be calibrated before use on each day of its use by the procedures specified in Method 21 of 40 CFR part 60, appendix A.

(5) Calibration gases shall be as follows:

(i) Zero air (less than 10 ppmv hydrocarbon in air), and

(ii) A mixture of methane or n-hexane and air at a concentration of approximately, but less than, 10,000 ppmv methane or n-hexane.

(6) The background level shall be determined according to the procedures in Method 21 of 40 CFR part 60, appendix A.

(7) Each potential leak interface shall be checked by traversing the instrument probe around the potential leak interface as close to the interface as possible, as described in Method 21 of 40 CFR part 60, appendix A. In the case when the configuration of the cover or closure device prevents a com-

plete traverse of the interface, all accessible portions of the interface shall be sampled. In the case when the configuration of the closure device prevents any sampling at the interface and the device is equipped with an enclosed extension or horn (e.g., some pressure relief devices), the instrument probe inlet shall be placed at approximately the center of the exhaust area to the atmosphere.

(8) The arithmetic difference between the maximum organic concentration indicated by the instrument and the background level shall be compared with the value of 500 ppmv except when monitoring a seal around a rotating shaft that passes through a cover opening, in which case the comparison shall be as specified in paragraph (d)(9) of this section. If the difference is less than 500 ppmv, then the potential leak interface is determined to operate with no detectable organic emissions.

(9) For the seals around a rotating shaft that passes through a cover opening, the arithmetic difference between the maximum organic concentration indicated by the instrument and the background level shall be compared with the value of 10,000 ppmv. If the difference is less than 10,000 ppmw, then the potential leak interface is determined to operate with no detectable organic emissions.

[61 FR 59974, Nov. 25, 1996, as amended at 62 FR 64664, Dec. 8, 1997; 64 FR 3390, January 21, 1999; 70 FR 34586, June 14, 2005]

## § 265.1085 Standards: Tanks.

(a) The provisions of this section apply to the control of air pollutant emissions from tanks for which § 265.1083(b) of this subpart references the use of this section for such air emission control.

(b) The owner or operator shall control air pollutant emissions from each tank subject to this section in accordance with the following requirements, as applicable:

(1) For a tank that manages hazardous waste that meets all of the conditions specified in paragraphs (b)(1)(i) through (b)(1)(iii) of this section, the owner or operator shall control air pollutant emissions from the tank in accordance with the Tank Level 1 controls specified in paragraph (c) of this

section or the Tank Level 2 controls specified in paragraph (d) of this section.

(i) The hazardous waste in the tank has a maximum organic vapor pressure which is less than the maximum organic vapor pressure limit for the tank's design capacity category as follows:

(A) For a tank design capacity equal to or greater than 151 m³, the maximum organic vapor pressure limit for the tank is 5.2 kPa.

(B) For a tank design capacity equal to or greater than 75 m³ but less than 151 m³, the maximum organic vapor pressure limit for the tank is 27.6 kPa.

(C) For a tank design capacity less than 75 m³, the maximum organic vapor pressure limit for the tank is 76.6 kPa.

(ii) The hazardous waste in the tank is not heated by the owner or operator to a temperature that is greater than the temperature at which the maximum organic vapor pressure of the hazardous waste is determined for the purpose of complying with paragraph (b)(1)(i) of this section.

(iii) The hazardous waste in the tank is not treated by the owner or operator using a waste stabilization process, as defined in §265.1081 of this subpart.

(2) For a tank that manages hazardous waste that does not meet all of the conditions specified in paragraphs (b)(1)(i) through (b)(1)(iii) of this section, the owner or operator shall control air pollutant emissions from the tank by using Tank Level 2 controls in accordance with the requirements of paragraph (d) of this section. Examples of tanks required to use Tank Level 2 controls include: A tank used for a waste stabilization process; and a tank for which the hazardous waste in the tank has a maximum organic vapor pressure that is equal to or greater than the maximum organic vapor pressure limit for the tank's design capacity category as specified in paragraph (b)(1)(i) of this section.

(c) Owners and operators controlling air pollutant emissions from a tank using Tank Level 1 controls shall meet the requirements specified in paragraphs (c)(1) through (c)(4) of this section:

(1) The owner or operator shall determine the maximum organic vapor pressure for a hazardous waste to be managed in the tank using Tank Level 1 controls before the first time the hazardous waste is placed in the tank. The maximum organic vapor pressure shall be determined using the procedures specified in §265.1084(c) of this subpart. Thereafter, the owner or operator shall perform a new determination whenever changes to the hazardous waste managed in the tank could potentially cause the maximum organic vapor pressure to increase to a level that is equal to or greater than the maximum organic vapor pressure limit for the tank design capacity category specified in paragraph (b)(1)(i) of this section, as applicable to the tank.

(2) The tank shall be equipped with a fixed roof designed to meet the following specifications:

(i) The fixed roof and its closure devices shall be designed to form a continuous barrier over the entire surface area of the hazardous waste in the tank. The fixed roof may be a separate cover installed on the tank (e.g., a removable cover mounted on an open-top tank) or may be an integral part of the tank structural design (e.g., a horizontal cylindrical tank equipped with a hatch).

(ii) The fixed roof shall be installed in a manner such that there are no visible cracks, holes, gaps, or other open spaces between roof section joints or between the interface of the roof edge and the tank wall.

(iii) Each opening in the fixed roof, and any manifold system associated with the fixed roof, shall be either:

(A) Equipped with a closure device designed to operate such that when the closure device is secured in the closed position there are no visible cracks, holes, gaps, or other open spaces in the closure device or between the perimeter of the opening and the closure device; or

(B) Connected by a closed-vent system that is vented to a control device. The control device shall remove or destroy organics in the vent stream, and shall be operating whenever hazardous waste is managed in the tank, except as provided for in paragraphs (c)(2)(iii)(B)(*1*) and (*2*) of this section.

(*1*) During periods it is necessary to provide access to the tank for performing the activities of paragraph (c)(2)(iii)(B)(*2*) of this section, venting of the vapor headspace underneath the fixed roof to the control device is not required, opening of closure devices is allowed, and removal of the fixed roof is allowed. Following completion of the activity, the owner or operator shall promptly secure the closure device in the closed position or reinstall the cover, as applicable, and resume operation of the control device.

(*2*) During periods of routine inspection, maintenance, or other activities needed for normal operations, and for the removal of accumulated sludge or other residues from the bottom of the tank.

(iv) The fixed roof and its closure devices shall be made of suitable materials that will minimize exposure of the hazardous waste to the atmosphere, to the extent practical, and will maintain the integrity of the fixed roof and closure devices throughout their intended service life. Factors to be considered when selecting the materials for and designing the fixed roof and closure devices shall include: Organic vapor permeability, the effects of any contact with the hazardous waste or its vapors managed in the tank; the effects of outdoor exposure to wind, moisture, and sunlight; and the operating practices used for the tank on which the fixed roof is installed.

(3) Whenever a hazardous waste is in the tank, the fixed roof shall be installed with each closure device secured in the closed position except as follows:

(i) Opening of closure devices or removal of the fixed roof is allowed at the following times:

(A) To provide access to the tank for performing routine inspection, maintenance, or other activities needed for normal operations. Examples of such activities include those times when a worker needs to open a port to sample the liquid in the tank, or when a worker needs to open a hatch to maintain or repair equipment. Following completion of the activity, the owner or operator shall promptly secure the closure device in the closed position or re-

install the cover, as applicable, to the tank.

(B) To remove accumulated sludge or other residues from the bottom of tank.

(ii) Opening of a spring-loaded pressure-vacuum relief valve, conservation vent, or similar type of pressure relief device which vents to the atmosphere is allowed during normal operations for the purpose of maintaining the tank internal pressure in accordance with the tank design specifications. The device shall be designed to operate with no detectable organic emissions when the device is secured in the closed position. The settings at which the device opens shall be established such that the device remains in the closed position whenever the tank internal pressure is within the internal pressure operating range determined by the owner or operator based on the tank manufacturer recommendations, applicable regulations, fire protection and prevention codes, standard engineering codes and practices, or other requirements for the safe handling of flammable, ignitable, explosive, reactive, or hazardous materials. Examples of normal operating conditions that may require these devices to open are during those times when the tank internal pressure exceeds the internal pressure operating range for the tank as a result of loading operations or diurnal ambient temperature fluctuations.

(iii) Opening of a safety device, as defined in § 265.1081 of this subpart, is allowed at any time conditions require doing so to avoid an unsafe condition.

(4) The owner or operator shall inspect the air emission control equipment in accordance with the following requirements.

(i) The fixed roof and its closure devices shall be visually inspected by the owner or operator to check for defects that could result in air pollutant emissions. Defects include, but are not limited to, visible cracks, holes, or gaps in the roof sections or between the roof and the tank wall; broken, cracked, or otherwise damaged seals or gaskets on closure devices; and broken or missing hatches, access covers, caps, or other closure devices.

(ii) The owner or operator shall perform an initial inspection of the fixed roof and its closure devices on or before the date that the tank becomes subject to this section. Thereafter, the owner or operator shall perform the inspections at least once every year except under the special conditions provided for in paragraph (l) of this section.

(iii) In the event that a defect is detected, the owner or operator shall repair the defect in accordance with the requirements of paragraph (k) of this section.

(iv) The owner or operator shall maintain a record of the inspection in accordance with the requirements specified in §265.1090(b) of this subpart.

(d) Owners and operators controlling air pollutant emissions from a tank using Tank Level 2 controls shall use one of the following tanks:

(1) A fixed-roof tank equipped with an internal floating roof in accordance with the requirements specified in paragraph (e) of this section;

(2) A tank equipped with an external floating roof in accordance with the requirements specified in paragraph (f) of this section;

(3) A tank vented through a closed-vent system to a control device in accordance with the requirements specified in paragraph (g) of this section;

(4) A pressure tank designed and operated in accordance with the requirements specified in paragraph (h) of this section; or

(5) A tank located inside an enclosure that is vented through a closed-vent system to an enclosed combustion control device in accordance with the requirements specified in paragraph (i) of this section.

(e) The owner or operator who controls air pollutant emissions from a tank using a fixed-roof with an internal floating roof shall meet the requirements specified in paragraphs (e)(1) through (e)(3) of this section.

(1) The tank shall be equipped with a fixed roof and an internal floating roof in accordance with the following requirements:

(i) The internal floating roof shall be designed to float on the liquid surface except when the floating roof must be supported by the leg supports.

(ii) The internal floating roof shall be equipped with a continuous seal between the wall of the tank and the floating roof edge that meets either of the following requirements:

(A) A single continuous seal that is either a liquid-mounted seal or a metallic shoe seal, as defined in §265.1081 of this subpart; or

(B) Two continuous seals mounted one above the other. The lower seal may be a vapor-mounted seal.

(iii) The internal floating roof shall meet the following specifications:

(A) Each opening in a noncontact internal floating roof except for automatic bleeder vents (vacuum breaker vents) and the rim space vents is to provide a projection below the liquid surface.

(B) Each opening in the internal floating roof shall be equipped with a gasketed cover or a gasketed lid except for leg sleeves, automatic bleeder vents, rim space vents, column wells, ladder wells, sample wells, and stub drains.

(C) Each penetration of the internal floating roof for the purpose of sampling shall have a slit fabric cover that covers at least 90 percent of the opening.

(D) Each automatic bleeder vent and rim space vent shall be gasketed.

(E) Each penetration of the internal floating roof that allows for passage of a ladder shall have a gasketed sliding cover.

(F) Each penetration of the internal floating roof that allows for passage of a column supporting the fixed roof shall have a flexible fabric sleeve seal or a gasketed sliding cover.

(2) The owner or operator shall operate the tank in accordance with the following requirements:

(i) When the floating roof is resting on the leg supports, the process of filling, emptying, or refilling shall be continuous and shall be completed as soon as practical.

(ii) Automatic bleeder vents are to be set closed at all times when the roof is floating, except when the roof is being floated off or is being landed on the leg supports.

(iii) Prior to filling the tank, each cover, access hatch, gauge float well or lid on any opening in the internal floating roof shall be bolted or fastened closed (i.e., no visible gaps). Rim space vents are to be set to open only when

745

the internal floating roof is not floating or when the pressure beneath the rim exceeds the manufacturer's recommended setting.

(3) The owner or operator shall inspect the internal floating roof in accordance with the procedures specified as follows:

(i) The floating roof and its closure devices shall be visually inspected by the owner or operator to check for defects that could result in air pollutant emissions. Defects include, but are not limited to: The internal floating roof is not floating on the surface of the liquid inside the tank; liquid has accumulated on top of the internal floating roof; any portion of the roof seals have detached from the roof rim; holes, tears, or other openings are visible in the seal fabric; the gaskets no longer close off the hazardous waste surface from the atmosphere; or the slotted membrane has more than 10 percent open area.

(ii) The owner or operator shall inspect the internal floating roof components as follows except as provided in paragraph (e)(3)(iii) of this section:

(A) Visually inspect the internal floating roof components through openings in the fixed-roof (e.g., manholes and roof hatches) at least once every 12 months after initial fill, and

(B) Visually inspect the internal floating roof, primary seal, secondary seal (if one is in service), gaskets, slotted membranes, and sleeve seals (if any) each time the tank is emptied and degassed and at least every 10 years.

(iii) As an alternative to performing the inspections specified in paragraph (e)(3)(ii) of this section for an internal floating roof equipped with two continuous seals mounted one above the other, the owner or operator may visually inspect the internal floating roof, primary and secondary seals, gaskets, slotted membranes, and sleeve seals (if any) each time the tank is emptied and degassed and at least every 5 years.

(iv) Prior to each inspection required by paragraph (e)(3)(ii) or (e)(3)(iii) of this section, the owner or operator shall notify the Regional Administrator in advance of each inspection to provide the Regional Administrator with the opportunity to have an observer present during the inspection. The owner or operator shall notify the Regional Administrator of the date and location of the inspection as follows:

(A) Prior to each visual inspection of an internal floating roof in a tank that has been emptied and degassed, written notification shall be prepared and sent by the owner or operator so that it is received by the Regional Administrator at least 30 calendar days before refilling the tank except when an inspection is not planned as provided for in paragraph (e)(3)(iv)(B) of this section.

(B) When a visual inspection is not planned and the owner or operator could not have known about the inspection 30 calendar days before refilling the tank, the owner or operator shall notify the Regional Administrator as soon as possible, but no later than 7 calendar days before refilling of the tank. This notification may be made by telephone and immediately followed by a written explanation for why the inspection is unplanned. Alternatively, written notification, including the explanation for the unplanned inspection, may be sent so that it is received by the Regional Administrator at least 7 calendar days before refilling the tank.

(v) In the event that a defect is detected, the owner or operator shall repair the defect in accordance with the requirements of paragraph (k) of this section.

(vi) The owner or operator shall maintain a record of the inspection in accordance with the requirements specified in § 265.1090(b) of this subpart.

(4) Safety devices, as defined in § 265.1081 of this subpart, may be installed and operated as necessary on any tank complying with the requirements of paragraph (e) of this section.

(f) The owner or operator who controls air pollutant emissions from a tank using an external floating roof shall meet the requirements specified in paragraphs (f)(1) through (f)(3) of this section.

(1) The owner or operator shall design the external floating roof in accordance with the following requirements:

(i) The external floating roof shall be designed to float on the liquid surface except when the floating roof must be supported by the leg supports.

(ii) The floating roof shall be equipped with two continuous seals, one above the other, between the wall of the tank and the roof edge. The lower seal is referred to as the primary seal, and the upper seal is referred to as the secondary seal.

(A) The primary seal shall be a liquid-mounted seal or a metallic shoe seal, as defined in §265.1081 of this subpart. The total area of the gaps between the tank wall and the primary seal shall not exceed 212 square centimeters (cm²) per meter of tank diameter, and the width of any portion of these gaps shall not exceed 3.8 centimeters (cm). If a metallic shoe seal is used for the primary seal, the metallic shoe seal shall be designed so that one end extends into the liquid in the tank and the other end extends a vertical distance of at least 61 centimeters above the liquid surface.

(B) The secondary seal shall be mounted above the primary seal and cover the annular space between the floating roof and the wall of the tank. The total area of the gaps between the tank wall and the secondary seal shall not exceed 21.2 square centimeters (cm²) per meter of tank diameter, and the width of any portion of these gaps shall not exceed 1.3 centimeters (cm).

(iii) The external floating roof shall meet the following specifications:

(A) Except for automatic bleeder vents (vacuum breaker vents) and rim space vents, each opening in a noncontact external floating roof shall provide a projection below the liquid surface.

(B) Except for automatic bleeder vents, rim space vents, roof drains, and leg sleeves, each opening in the roof shall be equipped with a gasketed cover, seal, or lid.

(C) Each access hatch and each gauge float well shall be equipped with a cover designed to be bolted or fastened when the cover is secured in the closed position.

(D) Each automatic bleeder vent and each rim space vent shall be equipped with a gasket.

(E) Each roof drain that empties into the liquid managed in the tank shall be equipped with a slotted membrane fabric cover that covers at least 90 percent of the area of the opening.

(F) Each unslotted and slotted guide pole well shall be equipped with a gasketed sliding cover or a flexible fabric sleeve seal.

(G) Each unslotted guide pole shall be equipped with a gasketed cap on the end of the pole.

(H) Each slotted guide pole shall be equipped with a gasketed float or other device which closes off the liquid surface from the atmosphere.

(I) Each gauge hatch and each sample well shall be equipped with a gasketed cover.

(2) The owner or operator shall operate the tank in accordance with the following requirements:

(i) When the floating roof is resting on the leg supports, the process of filling, emptying, or refilling shall be continuous and shall be completed as soon as practical.

(ii) Except for automatic bleeder vents, rim space vents, roof drains, and leg sleeves, each opening in the roof shall be secured and maintained in a closed position at all times except when the closure device must be open for access.

(iii) Covers on each access hatch and each gauge float well shall be bolted or fastened when secured in the closed position.

(iv) Automatic bleeder vents shall be set closed at all times when the roof is floating, except when the roof is being floated off or is being landed on the leg supports.

(v) Rim space vents shall be set to open only at those times that the roof is being floated off the roof leg supports or when the pressure beneath the rim seal exceeds the manufacturer's recommended setting.

(vi) The cap on the end of each unslotted guide pole shall be secured in the closed position at all times except when measuring the level or collecting samples of the liquid in the tank.

(vii) The cover on each gauge hatch or sample well shall be secured in the closed position at all times except when the hatch or well must be opened for access.

(viii) Both the primary seal and the secondary seal shall completely cover the annular space between the external floating roof and the wall of the tank in a continuous fashion except during inspections.

(3) The owner or operator shall inspect the external floating roof in accordance with the procedures specified as follows:

(i) The owner or operator shall measure the external floating roof seal gaps in accordance with the following requirements:

(A) The owner or operator shall perform measurements of gaps between the tank wall and the primary seal within 60 calendar days after initial operation of the tank following installation of the floating roof and, thereafter, at least once every 5 years.

(B) The owner or operator shall perform measurements of gaps between the tank wall and the secondary seal within 60 calendar days after initial operation of the tank following installation of the floating roof and, thereafter, at least once every year.

(C) If a tank ceases to hold hazardous waste for a period of 1 year or more, subsequent introduction of hazardous waste into the tank shall be considered an initial operation for the purposes of paragraphs (f)(3)(i)(A) and (f)(3)(i)(B) of this section.

(D) The owner or operator shall determine the total surface area of gaps in the primary seal and in the secondary seal individually using the following procedure:

(1) The seal gap measurements shall be performed at one or more floating roof levels when the roof is floating off the roof supports.

(2) Seal gaps, if any, shall be measured around the entire perimeter of the floating roof in each place where a 0.32-centimeter (cm) diameter uniform probe passes freely (without forcing or binding against the seal) between the seal and the wall of the tank and measure the circumferential distance of each such location.

(3) For a seal gap measured under paragraph (f)(3) of this section, the gap surface area shall be determined by using probes of various widths to measure accurately the actual distance from the tank wall to the seal and multiplying each such width by its respective circumferential distance.

(4) The total gap area shall be calculated by adding the gap surface areas

determined for each identified gap location for the primary seal and the secondary seal individually, and then dividing the sum for each seal type by the nominal diameter of the tank. These total gap areas for the primary seal and secondary seal are then compared to the respective standards for the seal type as specified in paragraph (f)(1)(ii) of this section.

(E) In the event that the seal gap measurements do not conform to the specifications in paragraph (f)(1)(ii) of this section, the owner or operator shall repair the defect in accordance with the requirements of paragraph (k) of this section.

(F) The owner or operator shall maintain a record of the inspection in accordance with the requirements specified in § 265.1090(b) of this subpart.

(ii) The owner or operator shall visually inspect the external floating roof in accordance with the following requirements:

(A) The floating roof and its closure devices shall be visually inspected by the owner or operator to check for defects that could result in air pollutant emissions. Defects include, but are not limited to: Holes, tears, or other openings in the rim seal or seal fabric of the floating roof; a rim seal detached from the floating roof; all or a portion of the floating roof deck being submerged below the surface of the liquid in the tank; broken, cracked, or otherwise damaged seals or gaskets on closure devices; and broken or missing hatches, access covers, caps, or other closure devices.

(B) The owner or operator shall perform an initial inspection of the external floating roof and its closure devices on or before the date that the tank becomes subject to this section. Thereafter, the owner or operator shall perform the inspections at least once every year except for the special conditions provided for in paragraph (l) of this section.

(C) In the event that a defect is detected, the owner or operator shall repair the defect in accordance with the requirements of paragraph (k) of this section.

(D) The owner or operator shall maintain a record of the inspection in accordance with the requirements specified in §265.1090(b) of this subpart.

(iii) Prior to each inspection required by paragraph (f)(3)(i) or (f)(3)(ii) of this section, the owner or operator shall notify the Regional Administrator in advance of each inspection to provide the Regional Administrator with the opportunity to have an observer present during the inspection. The owner or operator shall notify the Regional Administrator of the date and location of the inspection as follows:

(A) Prior to each inspection to measure external floating roof seal gaps as required under paragraph (f)(3)(i) of this section, written notification shall be prepared and sent by the owner or operator so that it is received by the Regional Administrator at least 30 calendar days before the date the measurements are scheduled to be performed.

(B) Prior to each visual inspection of an external floating roof in a tank that has been emptied and degassed, written notification shall be prepared and sent by the owner or operator so that it is received by the Regional Administrator at least 30 calendar days before refilling the tank except when an inspection is not planned as provided for in paragraph (f)(3)(iii)(C) of this section.

(C) When a visual inspection is not planned and the owner or operator could not have known about the inspection 30 calendar days before refilling the tank, the owner or operator shall notify the Regional Administrator as soon as possible, but no later than 7 calendar days before refilling of the tank. This notification may be made by telephone and immediately followed by a written explanation for why the inspection is unplanned. Alternatively, written notification, including the explanation for the unplanned inspection, may be sent so that it is received by the Regional Administrator at least 7 calendar days before refilling the tank.

(4) Safety devices, as defined in 40 CFR 265.1081, may be installed and operated as necessary on any tank complying with the requirements of paragraph (f) of this section.

(g) The owner or operator who controls air pollutant emissions from a tank by venting the tank to a control device shall meet the requirements specified in paragraphs (g)(1) through (g)(3) of this section.

(1) The tank shall be covered by a fixed roof and vented directly through a closed-vent system to a control device in accordance with the following requirements:

(i) The fixed roof and its closure devices shall be designed to form a continuous barrier over the entire surface area of the liquid in the tank.

(ii) Each opening in the fixed roof not vented to the control device shall be equipped with a closure device. If the pressure in the vapor headspace underneath the fixed roof is less than atmospheric pressure when the control device is operating, the closure devices shall be designed to operate such that when the closure device is secured in the closed position there are no visible cracks, holes, gaps, or other open spaces in the closure device or between the perimeter of the cover opening and the closure device. If the pressure in the vapor headspace underneath the fixed roof is equal to or greater than atmospheric pressure when the control device is operating, the closure device shall be designed to operate with no detectable organic emissions.

(iii) The fixed roof and its closure devices shall be made of suitable materials that will minimize exposure of the hazardous waste to the atmosphere, to the extent practical, and will maintain the integrity of the fixed roof and closure devices throughout their intended service life. Factors to be considered when selecting the materials for and designing the fixed roof and closure devices shall include: Organic vapor permeability, the effects of any contact with the liquid and its vapor managed in the tank; the effects of outdoor exposure to wind, moisture, and sunlight; and the operating practices used for the tank on which the fixed roof is installed.

(iv) The closed-vent system and control device shall be designed and operated in accordance with the requirements of §265.1088 of this subpart.

(2) Whenever a hazardous waste is in the tank, the fixed roof shall be installed with each closure device secured in the closed position and the

vapor headspace underneath the fixed roof vented to the control device except as follows:

(i) Venting to the control device is not required, and opening of closure devices or removal of the fixed roof is allowed at the following times:

(A) To provide access to the tank for performing routine inspection, maintenance, or other activities needed for normal operations. Examples of such activities include those times when a worker needs to open a port to sample liquid in the tank, or when a worker needs to open a hatch to maintain or repair equipment. Following completion of the activity, the owner or operator shall promptly secure the closure device in the closed position or reinstall the cover, as applicable, to the tank.

(B) To remove accumulated sludge or other residues from the bottom of a tank.

(ii) Opening of a safety device, as defined in § 265.1081 of this subpart, is allowed at any time conditions require doing so to avoid an unsafe condition.

(3) The owner or operator shall inspect and monitor the air emission control equipment in accordance with the following procedures:

(i) The fixed roof and its closure devices shall be visually inspected by the owner or operator to check for defects that could result in air pollutant emissions. Defects include, but are not limited to, visible cracks, holes, or gaps in the roof sections or between the roof and the tank wall; broken, cracked, or otherwise damaged seals or gaskets on closure devices; and broken or missing hatches, access covers, caps, or other closure devices.

(ii) The closed-vent system and control device shall be inspected and monitored by the owner or operator in accordance with the procedures specified in § 265.1088 of this subpart.

(iii) The owner or operator shall perform an initial inspection of the air emission control equipment on or before the date that the tank becomes subject to this section. Thereafter, the owner or operator shall perform the inspections at least once every year except for the special conditions provided for in paragraph (l) of this section.

(iv) In the event that a defect is detected, the owner or operator shall repair the defect in accordance with the requirements of paragraph (k) of this section.

(v) The owner or operator shall maintain a record of the inspection in accordance with the requirements specified in § 265.1090(b) of this subpart.

(h) The owner or operator who controls air pollutant emissions by using a pressure tank shall meet the following requirements.

(1) The tank shall be designed not to vent to the atmosphere as a result of compression of the vapor headspace in the tank during filling of the tank to its design capacity.

(2) All tank openings shall be equipped with closure devices designed to operate with no detectable organic emissions as determined using the procedure specified in § 265.1084(d) of this subpart.

(3) Whenever a hazardous waste is in the tank, the tank shall be operated as a closed system that does not vent to the atmosphere except under either of the following conditions as specified in paragraph (h)(3)(i) or (h)(3)(ii) of this section.

(i) At those times when opening of a safety device, as defined in § 265.1081 of this subpart, is required to avoid an unsafe condition.

(ii) At those times when purging of inerts from the tank is required and the purge stream is routed to a closed-vent system and control device designed and operated in accordance with the requirements of § 265.1088 of this subpart.

(i) The owner or operator who controls air pollutant emissions by using an enclosure vented through a closed-vent system to an enclosed combustion control device shall meet the requirements specified in paragraphs (i)(1) through (i)(4) of this section.

(1) The tank shall be located inside an enclosure. The enclosure shall be designed and operated in accordance with the criteria for a permanent total enclosure as specified in "Procedure T—Criteria for and Verification of a Permanent or Temporary Total Enclosure" under 40 CFR 52.741, appendix B. The enclosure may have permanent or temporary openings to allow worker

access; passage of material into or out of the enclosure by conveyor, vehicles, or other mechanical means; entry of permanent mechanical or electrical equipment; or direct airflow into the enclosure. The owner or operator shall perform the verification procedure for the enclosure as specified in Section 5.0 to "Procedure T—Criteria for and Verification of a Permanent or Temporary Total Enclosure" initially when the enclosure is first installed and, thereafter, annually.

(2) The enclosure shall be vented through a closed-vent system to an enclosed combustion control device that is designed and operated in accordance with the standards for either a vapor incinerator, boiler, or process heater specified in §265.1088 of this subpart.

(3) Safety devices, as defined in §265.1081 of this subpart, may be installed and operated as necessary on any enclosure, closed-vent system, or control device used to comply with the requirements of paragraphs (i)(1) and (i)(2) of this section.

(4) The owner or operator shall inspect and monitor the closed-vent system and control device as specified in §265.1088 of this subpart.

(j) The owner or operator shall transfer hazardous waste to a tank subject to this section in accordance with the following requirements:

(1) Transfer of hazardous waste, except as provided in paragraph (j)(2) of this section, to the tank from another tank subject to this section or from a surface impoundment subject to §265.1086 of this subpart shall be conducted using continuous hard-piping or another closed system that does not allow exposure of the hazardous waste to the atmosphere. For the purpose of complying with this provision, an individual drain system is considered to be a closed system when it meets the requirements of 40 CFR part 63, subpart RR—National Emission Standards for Individual Drain Systems.

(2) The requirements of paragraph (j)(1) of this section do not apply when transferring a hazardous waste to the tank under any of the following conditions:

(i) The hazardous waste meets the average VO concentration conditions specified in §265.1083(c)(1) of this subpart at the point of waste origination.

(ii) The hazardous waste has been treated by an organic destruction or removal process to meet the requirements in §265.1083(c)(2) of this subpart.

(iii) The hazardous waste meets the requirements of §265.1083(c)(4) of this subpart.

(k) The owner or operator shall repair each defect detected during an inspection performed in accordance with the requirements of paragraphs (c)(4), (e)(3), (f)(3), or (g)(3) of this section as follows:

(1) The owner or operator shall make first efforts at repair of the defect no later than 5 calendar days after detection, and repair shall be completed as soon as possible but no later than 45 calendar days after detection except as provided in paragraph (k)(2) of this section.

(2) Repair of a defect may be delayed beyond 45 calendar days if the owner or operator determines that repair of the defect requires emptying or temporary removal from service of the tank and no alternative tank capacity is available at the site to accept the hazardous waste normally managed in the tank. In this case, the owner or operator shall repair the defect the next time the process or unit that is generating the hazardous waste managed in the tank stops operation. Repair of the defect shall be completed before the process or unit resumes operation.

(l) Following the initial inspection and monitoring of the cover as required by the applicable provisions of this subpart, subsequent inspection and monitoring may be performed at intervals longer than 1 year under the following special conditions:

(1) In the case when inspecting or monitoring the cover would expose a worker to dangerous, hazardous, or other unsafe conditions, then the owner or operator may designate a cover as an "unsafe to inspect and monitor cover" and comply with all of the following requirements:

(i) Prepare a written explanation for the cover stating the reasons why the cover is unsafe to visually inspect or to monitor, if required.

(ii) Develop and implement a written plan and schedule to inspect and monitor the cover, using the procedures specified in the applicable section of this subpart, as frequently as practicable during those times when a worker can safely access the cover.

(2) In the case when a tank is buried partially or entirely underground, an owner or operator is required to inspect and monitor, as required by the applicable provisions of this section, only those portions of the tank cover and those connections to the tank (e.g., fill ports, access hatches, gauge wells, etc.) that are located on or above the ground surface.

[61 FR 59979, Nov. 25, 1996, as amended at 62 FR 64666, Dec. 8, 1997; 64 FR 3391, Jan. 21, 1999; 71 FR 40276, July 14, 2006]

### § 265.1086 Standards: Surface impoundments.

(a) The provisions of this section apply to the control of air pollutant emissions from surface impoundments for which § 265.1083(b) of this subpart references the use of this section for such air emission control.

(b) The owner or operator shall control air pollutant emissions from the surface impoundment by installing and operating either of the following:

(1) A floating membrane cover in accordance with the provisions specified in paragraph (c) of this section; or

(2) A cover that is vented through a closed-vent system to a control device in accordance with the requirements specified in paragraph (d) of this section.

(c) The owner or operator who controls air pollutant emissions from a surface impoundment using a floating membrane cover shall meet the requirements specified in paragraphs (c)(1) through (c)(3) of this section.

(1) The surface impoundment shall be equipped with a floating membrane cover designed to meet the following specifications:

(i) The floating membrane cover shall be designed to float on the liquid surface during normal operations and form a continuous barrier over the entire surface area of the liquid.

(ii) The cover shall be fabricated from a synthetic membrane material that is either:

(A) High density polyethylene (HDPE) with a thickness no less than 2.5 millimeters (mm); or

(B) A material or a composite of different materials determined to have both organic permeability properties that are equivalent to those of the material listed in paragraph (c)(1)(ii)(A) of this section and chemical and physical properties that maintain the material integrity for the intended service life of the material.

(iii) The cover shall be installed in a manner such that there are no visible cracks, holes, gaps, or other open spaces between cover section seams or between the interface of the cover edge and its foundation mountings.

(iv) Except as provided for in paragraph (c)(1)(v) of this section, each opening in the floating membrane cover shall be equipped with a closure device designed to operate such that when the closure device is secured in the closed position there are no visible cracks, holes, gaps, or other open spaces in the closure device or between the perimeter of the cover opening and the closure device.

(v) The floating membrane cover may be equipped with one or more emergency cover drains for removal of stormwater. Each emergency cover drain shall be equipped with a slotted membrane fabric cover that covers at least 90 percent of the area of the opening or a flexible fabric sleeve seal.

(vi) The closure devices shall be made of suitable materials that will minimize exposure of the hazardous waste to the atmosphere, to the extent practical, and will maintain the integrity of the closure devices throughout their intended service life. Factors to be considered when selecting the materials of construction and designing the cover and closure devices shall include: Organic vapor permeability; the effects of any contact with the liquid and its vapor managed in the surface impoundment; the effects of outdoor exposure to wind, moisture, and sunlight; and the operating practices used for the surface impoundment on which the floating membrane cover is installed.

(2) Whenever a hazardous waste is in the surface impoundment, the floating

membrane cover shall float on the liquid and each closure device shall be secured in the closed position except as follows:

(i) Opening of closure devices or removal of the cover is allowed at the following times:

(A) To provide access to the surface impoundment for performing routine inspection, maintenance, or other activities needed for normal operations. Examples of such activities include those times when a worker needs to open a port to sample the liquid in the surface impoundment, or when a worker needs to open a hatch to maintain or repair equipment. Following completion of the activity, the owner or operator shall promptly replace the cover and secure the closure device in the closed position, as applicable.

(B) To remove accumulated sludge or other residues from the bottom of surface impoundment.

(ii) Opening of a safety device, as defined in §265.1081 of this subpart, is allowed at any time conditions require doing so to avoid an unsafe condition.

(3) The owner or operator shall inspect the floating membrane cover in accordance with the following procedures:

(i) The floating membrane cover and its closure devices shall be visually inspected by the owner or operator to check for defects that could result in air pollutant emissions. Defects include, but are not limited to, visible cracks, holes, or gaps in the cover section seams or between the interface of the cover edge and its foundation mountings; broken, cracked, or otherwise damaged seals or gaskets on closure devices; and broken or missing hatches, access covers, caps, or other closure devices.

(ii) The owner or operator shall perform an initial inspection of the floating membrane cover and its closure devices on or before the date that the surface impoundment becomes subject to this section. Thereafter, the owner or operator shall perform the inspections at least once every year except for the special conditions provided for in paragraph (g) of this section.

(iii) In the event that a defect is detected, the owner or operator shall repair the defect in accordance with the requirements of paragraph (f) of this section.

(iv) The owner or operator shall maintain a record of the inspection in accordance with the requirements specified in §265.1090(c) of this subpart.

(d) The owner or operator who controls air pollutant emissions from a surface impoundment using a cover vented to a control device shall meet the requirements specified in paragraphs (d)(1) through (d)(3) of this section.

(1) The surface impoundment shall be covered by a cover and vented directly through a closed-vent system to a control device in accordance with the following requirements:

(i) The cover and its closure devices shall be designed to form a continuous barrier over the entire surface area of the liquid in the surface impoundment.

(ii) Each opening in the cover not vented to the control device shall be equipped with a closure device. If the pressure in the vapor headspace underneath the cover is less than atmospheric pressure when the control device is operating, the closure devices shall be designed to operate such that when the closure device is secured in the closed position there are no visible cracks, holes, gaps, or other open spaces in the closure device or between the perimeter of the cover opening and the closure device. If the pressure in the vapor headspace underneath the cover is equal to or greater than atmospheric pressure when the control device is operating, the closure device shall be designed to operate with no detectable organic emissions using the procedure specified in §265.1084(d) of this subpart.

(iii) The cover and its closure devices shall be made of suitable materials that will minimize exposure of the hazardous waste to the atmosphere, to the extent practical, and will maintain the integrity of the cover and closure devices throughout their intended service life. Factors to be considered when selecting the materials of construction and designing the cover and closure devices shall include: Organic vapor permeability; the effects of any contact with the liquid or its vapors managed in the surface impoundment; the effects of outdoor exposure to wind, moisture, and sunlight; and the operating practices used for the surface impoundment on which the cover is installed.

(iv) The closed-vent system and control device shall be designed and operated in accordance with the requirements of § 265.1088 of this subpart.

(2) Whenever a hazardous waste is in the surface impoundment, the cover shall be installed with each closure device secured in the closed position and the vapor headspace underneath the cover vented to the control device except as follows:

(i) Venting to the control device is not required, and opening of closure devices or removal of the cover is allowed at the following times:

(A) To provide access to the surface impoundment for performing routine inspection, maintenance, or other activities needed for normal operations. Examples of such activities include those times when a worker needs to open a port to sample liquid in the surface impoundment, or when a worker needs to open a hatch to maintain or repair equipment. Following completion of the activity, the owner or operator shall promptly secure the closure device in the closed position or reinstall the cover, as applicable, to the surface impoundment.

(B) To remove accumulated sludge or other residues from the bottom of the surface impoundment.

(ii) Opening of a safety device, as defined in § 265.1081 of this subpart, is allowed at any time conditions require doing so to avoid an unsafe condition.

(3) The owner or operator shall inspect and monitor the air emission control equipment in accordance with the following procedures:

(i) The surface impoundment cover and its closure devices shall be visually inspected by the owner or operator to check for defects that could result in air pollutant emissions. Defects include, but are not limited to, visible cracks, holes, or gaps in the cover section seams or between the interface of the cover edge and its foundation mountings; broken, cracked, or otherwise damaged seals or gaskets on closure devices; and broken or missing hatches, access covers, caps, or other closure devices.

(ii) The closed-vent system and control device shall be inspected and monitored by the owner or operator in accordance with the procedures specified in § 265.1088 of this subpart.

(iii) The owner or operator shall perform an initial inspection of the air emission control equipment on or before the date that the surface impoundment becomes subject to this section. Thereafter, the owner or operator shall perform the inspections at least once every year except for the special conditions provided for in paragraph (g) of this section.

(iv) In the event that a defect is detected, the owner or operator shall repair the defect in accordance with the requirements of paragraph (f) of this section.

(v) The owner or operator shall maintain a record of the inspection in accordance with the requirements specified in § 265.1090(c) of this subpart.

(e) The owner or operator shall transfer hazardous waste to a surface impoundment subject to this section in accordance with the following requirements:

(1) Transfer of hazardous waste, except as provided in paragraph (e)(2) of this section, to the surface impoundment from another surface impoundment subject to this section or from a tank subject to § 265.1085 of this subpart shall be conducted using continuous hard-piping or another closed system that does not allow exposure of the waste to the atmosphere. For the purpose of complying with this provision, an individual drain system is considered to be a closed system when it meets the requirements of 40 CFR part 63, subpart RR—National Emission Standards for Individual Drain Systems.

(2) The requirements of paragraph (e)(1) of this section do not apply when transferring a hazardous waste to the surface impoundment under either of the following conditions:

(i) The hazardous waste meets the average VO concentration conditions specified in § 265.1083(c)(1) of this subpart at the point of waste origination.

(ii) The hazardous waste has been treated by an organic destruction or removal process to meet the requirements in § 265.1083(c)(2) of this subpart.

(iii) The hazardous waste meets the requirements of §265.1083(c)(4) of this subpart.

(f) The owner or operator shall repair each defect detected during an inspection performed in accordance with the requirements of paragraph (c)(3) or (d)(3) of this section as follows:

(1) The owner or operator shall make first efforts at repair of the defect no later than 5 calendar days after detection, and repair shall be completed as soon as possible but no later than 45 calendar days after detection except as provided in paragraph (f)(2) of this section.

(2) Repair of a defect may be delayed beyond 45 calendar days if the owner or operator determines that repair of the defect requires emptying or temporary removal from service of the surface impoundment and no alternative capacity is available at the site to accept the hazardous waste normally managed in the surface impoundment. In this case, the owner or operator shall repair the defect the next time the process or unit that is generating the hazardous waste managed in the tank stops operation. Repair of the defect shall be completed before the process or unit resumes operation.

(g) Following the initial inspection and monitoring of the cover as required by the applicable provisions of this subpart, subsequent inspection and monitoring may be performed at intervals longer than 1 year in the case when inspecting or monitoring the cover would expose a worker to dangerous, hazardous, or other unsafe conditions. In this case, the owner or operator may designate the cover as an "unsafe to inspect and monitor cover" and comply with all of the following requirements:

(1) Prepare a written explanation for the cover stating the reasons why the cover is unsafe to visually inspect or to monitor, if required.

(2) Develop and implement a written plan and schedule to inspect and monitor the cover using the procedures specified in the applicable section of this subpart as frequently as practicable during those times when a worker can safely access the cover.

[61 FR 59984, Nov. 25, 1996, as amended at 62 FR 64666, Dec. 8, 1997]

## §265.1087  Standards: Containers.

(a) The provisions of this section apply to the control of air pollutant emissions from containers for which §265.1083(b) of this subpart references the use of this section for such air emission control.

(b) *General requirements.* (1) The owner or operator shall control air pollutant emissions from each container subject to this section in accordance with the following requirements, as applicable to the container, except when the special provisions for waste stabilization processes specified in paragraph (b)(2) of this section apply to the container.

(i) For a container having a design capacity greater than 0.1 m³ and less than or equal to 0.46 m³, the owner or operator shall control air pollutant emissions from the container in accordance with the Container Level 1 standards specified in paragraph (c) of this section.

(ii) For a container having a design capacity greater than 0.46 m³ that is not in light material service, the owner or operator shall control air pollutant emissions from the container in accordance with the Container Level 1 standards specified in paragraph (c) of this section.

(iii) For a container having a design capacity greater than 0.46 m³ that is in light material service, the owner or operator shall control air pollutant emissions from the container in accordance with the Container Level 2 standards specified in paragraph (d) of this section.

(2) When a container having a design capacity greater than 0.1 m³ is used for treatment of a hazardous waste by a waste stabilization process, the owner or operator shall control air pollutant emissions from the container in accordance with the Container Level 3 standards specified in paragraph (e) of this section at those times during the waste stabilization process when the hazardous waste in the container is exposed to the atmosphere.

(c) *Container Level 1 standards.* (1) A container using Container Level 1 controls is one of the following:

(i) A container that meets the applicable U.S. Department of Transportation (DOT) regulations on packaging hazardous materials for transportation as specified in paragraph (f) of this section.

(ii) A container equipped with a cover and closure devices that form a continuous barrier over the container openings such that when the cover and closure devices are secured in the closed position there are no visible holes, gaps, or other open spaces into the interior of the container. The cover may be a separate cover installed on the container (e.g., a lid on a drum or a suitably secured tarp on a roll-off box) or may be an integral part of the container structural design (e.g., a "portable tank" or bulk cargo container equipped with a screw-type cap).

(iii) An open-top container in which an organic-vapor suppressing barrier is placed on or over the hazardous waste in the container such that no hazardous waste is exposed to the atmosphere. One example of such a barrier is application of a suitable organic-vapor suppressing foam.

(2) A container used to meet the requirements of paragraph (c)(1)(ii) or (c)(1)(iii) of this section shall be equipped with covers and closure devices, as applicable to the container, that are composed of suitable materials to minimize exposure of the hazardous waste to the atmosphere and to maintain the equipment integrity for as long as it is in service. Factors to be considered in selecting the materials of construction and designing the cover and closure devices shall include: Organic vapor permeability, the effects of contact with the hazardous waste or its vapor managed in the container; the effects of outdoor exposure of the closure device or cover material to wind, moisture, and sunlight; and the operating practices for which the container is intended to be used.

(3) Whenever a hazardous waste is in a container using Container Level 1 controls, the owner or operator shall install all covers and closure devices for the container, as applicable to the container, and secure and maintain each closure device in the closed position except as follows:

(i) Opening of a closure device or cover is allowed for the purpose of adding hazardous waste or other material to the container as follows:

(A) In the case when the container is filled to the intended final level in one continuous operation, the owner or operator shall promptly secure the closure devices in the closed position and install the covers, as applicable to the container, upon conclusion of the filling operation.

(B) In the case when discrete quantities or batches of material intermittently are added to the container over a period of time, the owner or operator shall promptly secure the closure devices in the closed position and install covers, as applicable to the container, upon either the container being filled to the intended final level; the completion of a batch loading after which no additional material will be added to the container within 15 minutes; the person performing the loading operation leaving the immediate vicinity of the container; or the shutdown of the process generating the material being added to the container, whichever condition occurs first.

(ii) Opening of a closure device or cover is allowed for the purpose of removing hazardous waste from the container as follows:

(A) For the purpose of meeting the requirements of this section, an empty container as defined in 40 CFR 261.7(b) may be open to the atmosphere at any time (i.e., covers and closure devices are not required to be secured in the closed position on an empty container).

(B) In the case when discrete quantities or batches of material are removed from the container but the container does not meet the conditions to be an empty container as defined in 40 CFR 261.7(b), the owner or operator shall promptly secure the closure devices in the closed position and install covers, as applicable to the container, upon the completion of a batch removal after which no additional material will be removed from the container within 15 minutes or the person performing the unloading operation leaves the immediate vicinity of the container, whichever condition occurs first.

(iii) Opening of a closure device or cover is allowed when access inside the container is needed to perform routine activities other than transfer of hazardous waste. Examples of such activities include those times when a worker needs to open a port to measure the depth of or sample the material in the container, or when a worker needs to open a manhole hatch to access equipment inside the container. Following completion of the activity, the owner or operator shall promptly secure the closure device in the closed position or reinstall the cover, as applicable to the container.

(iv) Opening of a spring-loaded, pressure-vacuum relief valve, conservation vent, or similar type of pressure relief device which vents to the atmosphere is allowed during normal operations for the purpose of maintaining the container internal pressure in accordance with the design specifications of the container. The device shall be designed to operate with no detectable organic emissions when the device is secured in the closed position. The settings at which the device opens shall be established such that the device remains in the closed position whenever the internal pressure of the container is within the internal pressure operating range determined by the owner or operator based on container manufacturer recommendations, applicable regulations, fire protection and prevention codes, standard engineering codes and practices, or other requirements for the safe handling of flammable, ignitable, explosive, reactive, or hazardous materials. Examples of normal operating conditions that may require these devices to open are during those times when the internal pressure of the container exceeds the internal pressure operating range for the container as a result of loading operations or diurnal ambient temperature fluctuations.

(v) Opening of a safety device, as defined in §265.1081 of this subpart, is allowed at any time conditions require doing so to avoid an unsafe condition.

(4) The owner or operator of containers using Container Level 1 controls shall inspect the containers and their covers and closure devices as follows:

(i) In the case when a hazardous waste already is in the container at the time the owner or operator first accepts possession of the container at the facility and the container is not emptied within 24 hours after the container is accepted at the facility (i.e., does not meet the conditions for an empty container as specified in 40 CFR 261.7(b)), the owner or operator shall visually inspect the container and its cover and closure devices to check for visible cracks, holes, gaps, or other open spaces into the interior of the container when the cover and closure devices are secured in the closed position. The container visual inspection shall be conducted on or before the date that the container is accepted at the facility (i.e., the date the container becomes subject to the subpart CC container standards). For purposes of this requirement, the date of acceptance is the date of signature that the facility owner or operator enters on Item 20 of the Uniform Hazardous Waste Manifest in the appendix to 40 CFR part 262 (EPA Forms 8700–22 and 8700–22A), as required under subpart E of this part, at 40 CFR 265.71. If a defect is detected, the owner or operator shall repair the defect in accordance with the requirements of paragraph (c)(4)(iii) of this section.

(ii) In the case when a container used for managing hazardous waste remains at the facility for a period of 1 year or more, the owner or operator shall visually inspect the container and its cover and closure devices initially and thereafter, at least once every 12 months, to check for visible cracks, holes, gaps, or other open spaces into the interior of the container when the cover and closure devices are secured in the closed position. If a defect is detected, the owner or operator shall repair the defect in accordance with the requirements of paragraph (c)(4)(iii) of this section.

(iii) When a defect is detected for the container, cover, or closure devices, the owner or operator shall make first efforts at repair of the defect no later than 24 hours after detection, and repair shall be completed as soon as possible but no later than 5 calendar days after detection. If repair of a defect cannot be completed within 5 calendar days, then the hazardous waste shall be removed from the container and the container shall not be used to manage

757

hazardous waste until the defect is repaired.

(5) The owner or operator shall maintain at the facility a copy of the procedure used to determine that containers with capacity of 0.46 m³ or greater, which do not meet applicable DOT regulations as specified in paragraph (f) of this section, are not managing hazardous waste in light material service.

(d) *Container Level 2 standards.* (1) A container using Container Level 2 controls is one of the following:

(i) A container that meets the applicable U.S. Department of Transportation (DOT) regulations on packaging hazardous materials for transportation as specified in paragraph (f) of this section.

(ii) A container that operates with no detectable organic emissions as defined in § 265.1081 of this subpart and determined in accordance with the procedure specified in paragraph (g) of this section.

(iii) A container that has been demonstrated within the preceding 12 months to be vapor-tight by using 40 CFR part 60, appendix A, Method 27 in accordance with the procedure specified in paragraph (h) of this section.

(2) Transfer of hazardous waste in or out of a container using Container Level 2 controls shall be conducted in such a manner as to minimize exposure of the hazardous waste to the atmosphere, to the extent practical, considering the physical properties of the hazardous waste and good engineering and safety practices for handling flammable, ignitable, explosive, reactive or other hazardous materials. Examples of container loading procedures that the EPA considers to meet the requirements of this paragraph include using any one of the following: A submerged-fill pipe or other submerged-fill method to load liquids into the container; a vapor-balancing system or a vapor-recovery system to collect and control the vapors displaced from the container during filling operations; or a fitted opening in the top of a container through which the hazardous waste is filled and subsequently purging the transfer line before removing it from the container opening.

(3) Whenever a hazardous waste is in a container using Container Level 2 controls, the owner or operator shall install all covers and closure devices for the container, and secure and maintain each closure device in the closed position except as follows:

(i) Opening of a closure device or cover is allowed for the purpose of adding hazardous waste or other material to the container as follows:

(A) In the case when the container is filled to the intended final level in one continuous operation, the owner or operator shall promptly secure the closure devices in the closed position and install the covers, as applicable to the container, upon conclusion of the filling operation.

(B) In the case when discrete quantities or batches of material intermittently are added to the container over a period of time, the owner or operator shall promptly secure the closure devices in the closed position and install covers, as applicable to the container, upon either the container being filled to the intended final level; the completion of a batch loading after which no additional material will be added to the container within 15 minutes; the person performing the loading operation leaving the immediate vicinity of the container; or the shutdown of the process generating the material being added to the container, whichever condition occurs first.

(ii) Opening of a closure device or cover is allowed for the purpose of removing hazardous waste from the container as follows:

(A) For the purpose of meeting the requirements of this section, an empty container as defined in 40 CFR 261.7(b) may be open to the atmosphere at any time (i.e., covers and closure devices are not required to be secured in the closed position on an empty container).

(B) In the case when discrete quantities or batches of material are removed from the container but the container does not meet the conditions to be an empty container as defined in 40 CFR 261.7(b), the owner or operator shall promptly secure the closure devices in the closed position and install covers, as applicable to the container, upon the completion of a batch removal after which no additional material will be removed from the container within 15 minutes or the person performing the unloading operation leaves the immediate vicinity of the container, whichever condition occurs first.

(iii) Opening of a closure device or cover is allowed when access inside the container is needed to perform routine activities other than transfer of hazardous waste. Examples of such activities include those times when a worker needs to open a port to measure the depth of or sample the material in the container, or when a worker needs to open a manhole hatch to access equipment inside the container. Following completion of the activity, the owner or operator shall promptly secure the closure device in the closed position or reinstall the cover, as applicable to the container.

(iv) Opening of a spring-loaded, pressure-vacuum relief valve, conservation vent, or similar type of pressure relief device which vents to the atmosphere is allowed during normal operations for the purpose of maintaining the internal pressure of the container in accordance with the container design specifications. The device shall be designed to operate with no detectable organic emission when the device is secured in the closed position. The settings at which the device opens shall be established such that the device remains in the closed position whenever the internal pressure of the container is within the internal pressure operating range determined by the owner or operator based on container manufacturer recommendations, applicable regulations, fire protection and prevention codes, standard engineering codes and practices, or other requirements for the safe handling of flammable, ignitable, explosive, reactive, or hazardous materials. Examples of normal operating conditions that may require these devices to open are during those times when the internal pressure of the container exceeds the internal pressure operating range for the container as a result of loading operations or diurnal ambient temperature fluctuations.

(v) Opening of a safety device, as defined in § 265.1081 of this subpart, is allowed at any time conditions require doing so to avoid an unsafe condition.

(4) The owner or operator of containers using Container Level 2 controls shall inspect the containers and their covers and closure devices as follows:

(i) In the case when a hazardous waste already is in the container at the time the owner or operator first accepts possession of the container at the facility and the container is not emptied within 24 hours after the container is accepted at the facility (i.e., does not meet the conditions for an empty container as specified in 40 CFR 261.7(b)), the owner or operator shall visually inspect the container and its cover and closure devices to check for visible cracks, holes, gaps, or other open spaces into the interior of the container when the cover and closure devices are secured in the closed position. The container visual inspection shall be conducted on or before the date that the container is accepted at the facility (i.e., the date the container becomes subject to the subpart CC container standards). For purposes of this requirement, the date of acceptance is the date of signature that the facility owner or operator enters on Item 20 of the Uniform Hazardous Waste Manifest in the appendix to 40 CFR part 262 (EPA Forms 8700–22 and 8700–22A), as required under subpart E of this part, at § 265.71. If a defect is detected, the owner or operator shall repair the defect in accordance with the requirements of paragraph (d)(4)(iii) of this section.

(ii) In the case when a container used for managing hazardous waste remains at the facility for a period of 1 year or more, the owner or operator shall visually inspect the container and its cover and closure devices initially and thereafter, at least once every 12 months, to check for visible cracks, holes, gaps, or other open spaces into the interior of the container when the cover and closure devices are secured in the closed position. If a defect is detected, the owner or operator shall repair the defect in accordance with the requirements of paragraph (d)(4)(iii) of this section.

(iii) When a defect is detected for the container, cover, or closure devices, the owner or operator shall make first efforts at repair of the defect no later

759

than 24 hours after detection, and repair shall be completed as soon as possible but no later than 5 calendar days after detection. If repair of a defect cannot be completed within 5 calendar days, then the hazardous waste shall be removed from the container and the container shall not be used to manage hazardous waste until the defect is repaired.

(e) *Container Level 3 standards.* (1) A container using Container Level 3 controls is one of the following:

(i) A container that is vented directly through a closed-vent system to a control device in accordance with the requirements of paragraph (e)(2)(ii) of this section.

(ii) A container that is vented inside an enclosure which is exhausted through a closed-vent system to a control device in accordance with the requirements of paragraphs (e)(2)(i) and (e)(2)(ii) of this section.

(2) The owner or operator shall meet the following requirements, as applicable to the type of air emission control equipment selected by the owner or operator:

(i) The container enclosure shall be designed and operated in accordance with the criteria for a permanent total enclosure as specified in "Procedure T—Criteria for and Verification of a Permanent or Temporary Total Enclosure" under 40 CFR 52.741, appendix B. The enclosure may have permanent or temporary openings to allow worker access; passage of containers through the enclosure by conveyor or other mechanical means; entry of permanent mechanical or electrical equipment; or direct airflow into the enclosure. The owner or operator shall perform the verification procedure for the enclosure as specified in Section 5.0 to "Procedure T—Criteria for and Verification of a Permanent or Temporary Total Enclosure" initially when the enclosure is first installed and, thereafter, annually.

(ii) The closed-vent system and control device shall be designed and operated in accordance with the requirements of § 265.1088 of this subpart.

(3) Safety devices, as defined in § 265.1081 of this subpart, may be installed and operated as necessary on any container, enclosure, closed-vent system, or control device used to comply with the requirements of paragraph (e)(1) of this section.

(4) Owners and operators using Container Level 3 controls in accordance with the provisions of this subpart shall inspect and monitor the closed-vent systems and control devices as specified in § 265.1088 of this subpart.

(5) Owners and operators that use Container Level 3 controls in accordance with the provisions of this subpart shall prepare and maintain the records specified in § 265.1090(d) of this subpart.

(6) Transfer of hazardous waste in or out of a container using Container Level 3 controls shall be conducted in such a manner as to minimize exposure of the hazardous waste to the atmosphere, to the extent practical, considering the physical properties of the hazardous waste and good engineering and safety practices for handling flammable, ignitable, explosive, reactive, or other hazardous materials. Examples of container loading procedures that the EPA considers to meet the requirements of this paragraph include using any one of the following: A submerged-fill pipe or other submerged-fill method to load liquids into the container; a vapor-balancing system or a vapor-recovery system to collect and control the vapors displaced from the container during filling operations; or a fitted opening in the top of a container through which the hazardous waste is filled and subsequently purging the transfer line before removing it from the container opening.

(f) For the purpose of compliance with paragraph (c)(1)(i) or (d)(1)(i) of this section, containers shall be used that meet the applicable U.S. Department of Transportation (DOT) regulations on packaging hazardous materials for transportation as follows:

(1) The container meets the applicable requirements specified in 49 CFR part 178—Specifications for Packaging or 49 CFR part 179—Specifications for Tank Cars.

(2) Hazardous waste is managed in the container in accordance with the applicable requirements specified in 49 CFR part 107, subpart B—Exemptions; 49 CFR part 172—Hazardous Materials Table, Special Provisions, Hazardous

Materials Communications, Emergency Response Information, and Training Requirements; 49 CFR part 173—Shippers—General Requirements for Shipments and Packages; and 49 CFR part 180—Continuing Qualification and Maintenance of Packagings.

(3) For the purpose of complying with this subpart, no exceptions to the 49 CFR part 178 or part 179 regulations are allowed except as provided for in paragraph (f)(4) of this section.

(4) For a lab pack that is managed in accordance with the requirements of 49 CFR part 178 for the purpose of complying with this subpart, an owner or operator may comply with the exceptions for combination packagings specified in 49 CFR 173.12(b).

(g) To determine compliance with the no detectable organic emissions requirements of paragraph (d)(1)(ii) of this section, the procedure specified in §265.1084(d) of this subpart shall be used.

(1) Each potential leak interface (i.e., a location where organic vapor leakage could occur) on the container, its cover, and associated closure devices, as applicable to the container, shall be checked. Potential leak interfaces that are associated with containers include, but are not limited to: The interface of the cover rim and the container wall; the periphery of any opening on the container or container cover and its associated closure device; and the sealing seat interface on a spring-loaded pressure-relief valve.

(2) The test shall be performed when the container is filled with a material having a volatile organic concentration representative of the range of volatile organic concentrations for the hazardous wastes expected to be managed in this type of container. During the test, the container cover and closure devices shall be secured in the closed position.

(h) Procedure for determining a container to be vapor-tight using Method 27 of 40 CFR part 60, appendix A for the purpose of complying with paragraph (d)(1)(iii) of this section.

(1) The test shall be performed in accordance with Method 27 of 40 CFR part 60, appendix A of this chapter.

(2) A pressure measurement device shall be used that has a precision of ±2.5 mm water and that is capable of measuring above the pressure at which the container is to be tested for vapor tightness.

(3) If the test results determined by Method 27 indicate that the container sustains a pressure change less than or equal to 750 Pascals within 5 minutes after it is pressurized to a minimum of 4,500 Pascals, then the container is determined to be vapor-tight.

[61 FR 59986, Nov. 25, 1996, as amended at 62 FR 64666, Dec. 8, 1997; 64 FR 3391, Jan. 21, 1999; 71 FR 40276, July 14, 2006]

## §265.1088 Standards: Closed-vent systems and control devices.

(a) This section applies to each closed-vent system and control device installed and operated by the owner or operator to control air emissions in accordance with standards of this subpart.

(b) The closed-vent system shall meet the following requirements:

(1) The closed-vent system shall route the gases, vapors, and fumes emitted from the hazardous waste in the waste management unit to a control device that meets the requirements specified in paragraph (c) of this section.

(2) The closed-vent system shall be designed and operated in accordance with the requirements specified in §265.1033(j) of this part.

(3) In the case when the closed-vent system includes bypass devices that could be used to divert the gas or vapor stream to the atmosphere before entering the control device, each bypass device shall be equipped with either a flow indicator as specified in paragraph (b)(3)(i) of this section or a seal or locking device as specified in paragraph (b)(3)(ii) of this section. For the purpose of complying with this paragraph, low leg drains, high point bleeds, analyzer vents, open-ended valves or lines, spring-loaded pressure relief valves, and other fittings used for safety purposes are not considered to be bypass devices.

(i) If a flow indicator is used to comply with paragraph (b)(3) of this section, the indicator shall be installed at the inlet to the bypass line used to divert gases and vapors from the closed-vent system to the atmosphere at a point upstream of the control device inlet. For this paragraph, a flow indicator means a device which indicates the presence of either gas or vapor flow in the bypass line.

(ii) If a seal or locking device is used to comply with paragraph (b)(3) of this section, the device shall be placed on the mechanism by which the bypass device position is controlled (e.g., valve handle, damper lever) when the bypass device is in the closed position such that the bypass device cannot be opened without breaking the seal or removing the lock. Examples of such devices include, but are not limited to, a car-seal or a lock-and-key configuration valve. The owner or operator shall visually inspect the seal or closure mechanism at least once every month to verify that the bypass mechanism is maintained in the closed position.

(4) The closed-vent system shall be inspected and monitored by the owner or operator in accordance with the procedure specified in 40 CFR 265.1033(k).

(c) The control device shall meet the following requirements:

(1) The control device shall be one of the following devices:

(i) A control device designed and operated to reduce the total organic content of the inlet vapor stream vented to the control device by at least 95 percent by weight;

(ii) An enclosed combustion device designed and operated in accordance with the requirements of § 265.1033(c); or

(iii) A flare designed and operated in accordance with the requirements of § 265.1033(d).

(2) The owner or operator who elects to use a closed-vent system and control device to comply with the requirements of this section shall comply with the requirements specified in paragraphs (c)(2)(i) through (c)(2)(vi) of this section.

(i) Periods of planned routine maintenance of the control device, during which the control device does not meet the specifications of paragraphs (c)(1)(i), (c)(1)(ii), or (c)(1)(iii) of this section, as applicable, shall not exceed 240 hours per year.

(ii) The specifications and requirements in paragraphs (c)(1)(i), (c)(1)(ii), and (c)(1)(iii) of this section for control devices do not apply during periods of planned routine maintenance.

(iii) The specifications and requirements in paragraphs (c)(1)(i), (c)(1)(ii), and (c)(1)(iii) of this section for control devices do not apply during a control device system malfunction.

(iv) The owner or operator shall demonstrate compliance with the requirements of paragraph (c)(2)(i) of this section (i.e., planned routine maintenance of a control device, during which the control device does not meet the specifications of paragraphs (c)(1)(i), (c)(1)(ii), or (c)(1)(iii) of this section, as applicable, shall not exceed 240 hours per year) by recording the information specified in § 265.1090(e)(1)(v) of this subpart.

(v) The owner or operator shall correct control device system malfunctions as soon as practicable after their occurrence in order to minimize excess emissions of air pollutants.

(vi) The owner or operator shall operate the closed-vent system such that gases, vapors, and/or fumes are not actively vented to the control device during periods of planned maintenance or control device system malfunction (i.e., periods when the control device is not operating or not operating normally) except in cases when it is necessary to vent the gases, vapors, or fumes to avoid an unsafe condition or to implement malfunction corrective actions or planned maintenance actions.

(3) The owner or operator using a carbon adsorption system to comply with paragraph (c)(1) of this section shall operate and maintain the control device in accordance with the following requirements:

(i) Following the initial startup of the control device, all activated carbon in the control device shall be replaced with fresh carbon on a regular basis in accordance with the requirements of § 265.1033(g) or § 265.1033(h).

(ii) All carbon that is a hazardous waste and that is removed from the control device shall be managed in accordance with the requirements of 40 CFR 265.1033(m), regardless of the average volatile organic concentration of the carbon.

(4) An owner or operator using a control device other than a thermal vapor

incinerator, flare, boiler, process heater, condenser, or carbon adsorption system to comply with paragraph (c)(1) of this section shall operate and maintain the control device in accordance with the requirements of §265.1033(i).

(5) The owner or operator shall demonstrate that a control device achieves the performance requirements of paragraph (c)(1) of this section as follows:

(i) An owner or operator shall demonstrate using either a performance test as specified in paragraph (c)(5)(iii) of this section or a design analysis as specified in paragraph (c)(5)(iv) of this section the performance of each control device except for the following:

(A) A flare;

(B) A boiler or process heater with a design heat input capacity of 44 megawatts or greater;

(C) A boiler or process heater into which the vent stream is introduced with the primary fuel;

(D) A boiler or industrial furnace burning hazardous waste for which the owner or operator has been issued a final permit under 40 CFR part 270 and has designed and operates the unit in accordance with the requirements of 40 CFR part 266, subpart H; or

(E) A boiler or industrial furnace burning hazardous waste for which the owner or operator has designed and operates in accordance with the interim status requirements of 40 CFR part 266, subpart H.

(ii) An owner or operator shall demonstrate the performance of each flare in accordance with the requirements specified in §265.1033(e).

(iii) For a performance test conducted to meet the requirements of paragraph (c)(5)(i) of this section, the owner or operator shall use the test methods and procedures specified in §265.1034(c)(1) through (c)(4).

(iv) For a design analysis conducted to meet the requirements of paragraph (c)(5)(i) of this section, the design analysis shall meet the requirements specified in §265.1035(b)(4)(iii).

(v) The owner or operator shall demonstrate that a carbon adsorption system achieves the performance requirements of paragraph (c)(1) of this section based on the total quantity of organics vented to the atmosphere from all carbon adsorption system

equipment that is used for organic adsorption, organic desorption or carbon regeneration, organic recovery, and carbon disposal.

(6) If the owner or operator and the Regional Administrator do not agree on a demonstration of control device performance using a design analysis then the disagreement shall be resolved using the results of a performance test performed by the owner or operator in accordance with the requirements of paragraph (c)(5)(iii) of this section. The Regional Administrator may choose to have an authorized representative observe the performance test.

(7) The closed-vent system and control device shall be inspected and monitored by the owner or operator in accordance with the procedures specified in 40 CFR 265.1033(f)(2) and 40 CFR 265.1033(k). The readings from each monitoring device required by 40 CFR 265.1033(f)(2) shall be inspected at least once each operating day to check control device operation. Any necessary corrective measures shall be immediately implemented to ensure the control device is operated in compliance with the requirements of this section.

[59 FR 62935, Dec. 6, 1994, as amended at 61 FR 4915, Feb. 9, 1996; 61 FR 59989, Nov. 25, 1996; 62 FR 64667, Dec. 8, 1997]

### §265.1089 Inspection and monitoring requirements.

(a) The owner or operator shall inspect and monitor air emission control equipment used to comply with this subpart in accordance with the applicable requirements specified in §265.1085 through §265.1088 of this subpart.

(b) The owner or operator shall develop and implement a written plan and schedule to perform the inspections and monitoring required by paragraph (a) of this section. The owner or operator shall incorporate this plan and schedule into the facility inspection plan required under 40 CFR 265.15.

[61 FR 59990, Nov. 25, 1996]

### §265.1090 Recordkeeping requirements.

(a) Each owner or operator of a facility subject to requirements in this subpart shall record and maintain the information specified in paragraphs (b) through (j) of this section, as applicable to the facility. Except for air emission control equipment design documentation and information required by paragraphs (i) and (j) of this section, records required by this section shall be maintained in the operating record for a minimum of 3 years. Air emission control equipment design documentation shall be maintained in the operating record until the air emission control equipment is replaced or otherwise no longer in service. Information required by paragraphs (i) and (j) of this section shall be maintained in the operating record for as long as the waste management unit is not using air emission controls specified in §§ 265.1085 through 265.1088 of this subpart in accordance with the conditions specified in § 265.1080(d) or § 265.1080(b)(7) of this subpart, respectively.

(b) The owner or operator of a tank using air emission controls in accordance with the requirements of § 265.1085 of this subpart shall prepare and maintain records for the tank that include the following information:

(1) For each tank using air emission controls in accordance with the requirements of § 265.1085 of this subpart, the owner or operator shall record:

(i) A tank identification number (or other unique identification description as selected by the owner or operator).

(ii) A record for each inspection required by § 265.1085 of this subpart that includes the following information:

(A) Date inspection was conducted.

(B) For each defect detected during the inspection: The location of the defect, a description of the defect, the date of detection, and corrective action taken to repair the defect. In the event that repair of the defect is delayed in accordance with the provisions of § 265.1085 of this subpart, the owner or operator shall also record the reason for the delay and the date that completion of repair of the defect is expected.

(2) In addition to the information required by paragraph (b)(1) of this section, the owner or operator shall record the following information, as applicable to the tank:

(i) The owner or operator using a fixed roof to comply with the Tank Level 1 control requirements specified in § 265.1085(c) of this subpart shall pre-

pare and maintain records for each determination for the maximum organic vapor pressure of the hazardous waste in the tank performed in accordance with the requirements of § 265.1085(c) of this subpart. The records shall include the date and time the samples were collected, the analysis method used, and the analysis results.

(ii) The owner or operator using an internal floating roof to comply with the Tank Level 2 control requirements specified in § 265.1085(e) of this subpart shall prepare and maintain documentation describing the floating roof design.

(iii) Owners and operators using an external floating roof to comply with the Tank Level 2 control requirements specified in § 265.1085(f) of this subpart shall prepare and maintain the following records:

(A) Documentation describing the floating roof design and the dimensions of the tank.

(B) Records for each seal gap inspection required by § 265.1085(f)(3) of this subpart describing the results of the seal gap measurements. The records shall include the date that the measurements were performed, the raw data obtained for the measurements, and the calculations of the total gap surface area. In the event that the seal gap measurements do not conform to the specifications in § 265.1085(f)(1) of this subpart, the records shall include a description of the repairs that were made, the date the repairs were made, and the date the tank was emptied, if necessary.

(iv) Each owner or operator using an enclosure to comply with the Tank Level 2 control requirements specified in § 265.1085(i) of this subpart shall prepare and maintain the following records:

(A) Records for the most recent set of calculations and measurements performed by the owner or operator to verify that the enclosure meets the criteria of a permanent total enclosure as specified in "Procedure T—Criteria for and Verification of a Permanent or Temporary Total Enclosure" under 40 CFR 52.741, appendix B.

(B) Records required for the closed-vent system and control device in accordance with the requirements of paragraph (e) of this section.

(c) The owner or operator of a surface impoundment using air emission controls in accordance with the requirements of §265.1086 of this subpart shall prepare and maintain records for the surface impoundment that include the following information:

(1) A surface impoundment identification number (or other unique identification description as selected by the owner or operator).

(2) Documentation describing the floating membrane cover or cover design, as applicable to the surface impoundment, that includes information prepared by the owner or operator or provided by the cover manufacturer or vendor describing the cover design, and certification by the owner or operator that the cover meets the specifications listed in §265.1086(c) of this subpart.

(3) A record for each inspection required by §265.1086 of this subpart that includes the following information:

(i) Date inspection was conducted.

(ii) For each defect detected during the inspection the following information: The location of the defect, a description of the defect, the date of detection, and corrective action taken to repair the defect. In the event that repair of the defect is delayed in accordance with the provisions of §265.1086(f) of this subpart, the owner or operator shall also record the reason for the delay and the date that completion of repair of the defect is expected.

(4) For a surface impoundment equipped with a cover and vented through a closed-vent system to a control device, the owner or operator shall prepare and maintain the records specified in paragraph (e) of this section.

(d) The owner or operator of containers using Container Level 3 air emission controls in accordance with the requirements of §265.1087 of this subpart shall prepare and maintain records that include the following information:

(1) Records for the most recent set of calculations and measurements performed by the owner or operator to verify that the enclosure meets the criteria of a permanent total enclosure as specified in "Procedure T—Criteria for and Verification of a Permanent or Temporary Total Enclosure" under 40 CFR 52.741, appendix B.

(2) Records required for the closed-vent system and control device in accordance with the requirements of paragraph (e) of this section.

(e) The owner or operator using a closed-vent system and control device in accordance with the requirements of §265.1088 of this subpart shall prepare and maintain records that include the following information:

(1) Documentation for the closed-vent system and control device that includes:

(i) Certification that is signed and dated by the owner or operator stating that the control device is designed to operate at the performance level documented by a design analysis as specified in paragraph (e)(1)(ii) of this section or by performance tests as specified in paragraph (e)(1)(iii) of this section when the tank, surface impoundment, or container is or would be operating at capacity or the highest level reasonably expected to occur.

(ii) If a design analysis is used, then design documentation as specified in 40 CFR 265.1035(b)(4). The documentation shall include information prepared by the owner or operator or provided by the control device manufacturer or vendor that describes the control device design in accordance with 40 CFR 265.1035(b)(4)(iii) and certification by the owner or operator that the control equipment meets the applicable specifications.

(iii) If performance tests are used, then a performance test plan as specified in 40 CFR 265.1035(b)(3) and all test results.

(iv) Information as required by 40 CFR 265.1035(c)(1) and 40 CFR 265.1035(c)(2), as applicable.

(v) An owner or operator shall record, on a semiannual basis, the information specified in paragraphs (e)(1)(v)(A) and (e)(1)(v)(B) of this section for those planned routine maintenance operations that would require the control device not to meet the requirements of §265.1088 (c)(1)(i), (c)(1)(ii), or (c)(1)(iii) of this subpart, as applicable.

(A) A description of the planned routine maintenance that is anticipated to be performed for the control device during the next 6-month period. This description shall include the type of maintenance necessary, planned frequency of maintenance, and lengths of maintenance periods.

(B) A description of the planned routine maintenance that was performed for the control device during the previous 6-month period. This description shall include the type of maintenance performed and the total number of hours during those 6 months that the control device did not meet the requirements of § 265.1088 (c)(1)(i), (c)(1)(ii), or (c)(1)(iii) of this subpart, as applicable, due to planned routine maintenance.

(vi) An owner or operator shall record the information specified in paragraphs (e)(1)(vi)(A) through (e)(1)(vi)(C) of this section for those unexpected control device system malfunctions that would require the control device not to meet the requirements of § 265.1088 (c)(1)(i), (c)(1)(ii), or (c)(1)(iii) of this subpart, as applicable.

(A) The occurrence and duration of each malfunction of the control device system.

(B) The duration of each period during a malfunction when gases, vapors, or fumes are vented from the waste management unit through the closed-vent system to the control device while the control device is not properly functioning.

(C) Actions taken during periods of malfunction to restore a malfunctioning control device to its normal or usual manner of operation.

(vii) Records of the management of carbon removed from a carbon adsorption system conducted in accordance with § 265.1088(c)(3)(ii) of this subpart.

(f) The owner or operator of a tank, surface impoundment, or container exempted from standards in accordance with the provisions of § 265.1083(c) of this subpart shall prepare and maintain the following records, as applicable:

(1) For tanks, surface impoundments, or containers exempted under the hazardous waste organic concentration conditions specified in § 265.1083(c)(1) or § 265.1083(c)(2)(i) through (c)(2)(vi) of this subpart, the owner or operator shall record the information used for each waste determination (e.g., test results, measurements, calculations, and other documentation) in the facility

operating log. If analysis results for waste samples are used for the waste determination, then the owner or operator shall record the date, time, and location that each waste sample is collected in accordance with applicable requirements of § 265.1084 of this subpart.

(2) For tanks, surface impoundments, or containers exempted under the provisions of § 265.1083(c)(2)(vii) or § 265.1083(c)(2)(viii) of this subpart, the owner or operator shall record the identification number for the incinerator, boiler, or industrial furnace in which the hazardous waste is treated.

(g) An owner or operator designating a cover as "unsafe to inspect and monitor" pursuant to § 265.1085(l) or § 265.1086(g) of this subpart shall record in a log that is kept in the facility operating record the following information: The identification numbers for waste management units with covers that are designated as "unsafe to inspect and monitor," the explanation for each cover stating why the cover is unsafe to inspect and monitor, and the plan and schedule for inspecting and monitoring each cover.

(h) The owner or operator of a facility that is subject to this subpart and to the control device standards in 40 CFR part 60, subpart VV, or 40 CFR part 61, subpart V, may elect to demonstrate compliance with the applicable sections of this subpart by documentation either pursuant to this subpart, or pursuant to the provisions of 40 CFR part 60, subpart VV or 40 CFR part 61, subpart V, to the extent that the documentation required by 40 CFR parts 60 or 61 duplicates the documentation required by this section.

(i) For each tank or container not using air emission controls specified in §§ 265.1085 through 265.1088 of this subpart in accordance with the conditions specified in § 265.1080(d) of this subpart, the owner or operator shall record and maintain the following information:

(1) A list of the individual organic peroxide compounds manufactured at the facility that meet the conditions specified in § 265.1080(d)(1).

(2) A description of how the hazardous waste containing the organic peroxide compounds identified in paragraph (i)(1) of this section are managed

766

at the facility in tanks and containers. This description shall include the following information:

(i) For the tanks used at the facility to manage this hazardous waste, sufficient information shall be provided to describe for each tank: A facility identification number for the tank; the purpose and placement of this tank in the management train of this hazardous waste; and the procedures used to ultimately dispose of the hazardous waste managed in the tanks.

(ii) For containers used at the facility to manage these hazardous wastes, sufficient information shall be provided to describe: A facility identification number for the container or group of containers; the purpose and placement of this container, or group of containers, in the management train of this hazardous waste; and the procedures used to ultimately dispose of the hazardous waste handled in the containers.

(3) An explanation of why managing the hazardous waste containing the organic peroxide compounds identified in paragraph (i)(1) of this section in the tanks and containers as described in paragraph (i)(2) of this section would create an undue safety hazard if the air emission controls, as required under §§265.1085 through 265.1088 of this subpart, are installed and operated on these waste management units. This explanation shall include the following information:

(i) For tanks used at the facility to manage these hazardous wastes, sufficient information shall be provided to explain: How use of the required air emission controls on the tanks would affect the tank design features and facility operating procedures currently used to prevent an undue safety hazard during the management of this hazardous waste in the tanks; and why installation of safety devices on the required air emission controls, as allowed under this subpart, will not address those situations in which evacuation of tanks equipped with these air emission controls is necessary and consistent with good engineering and safety practices for handling organic peroxides.

(ii) For containers used at the facility to manage these hazardous wastes, sufficient information shall be pro-

vided to explain: How use of the required air emission controls on the containers would affect the container design features and handling procedures currently used to prevent an undue safety hazard during the management of this hazardous waste in the containers; and why installation of safety devices on the required air emission controls, as allowed under this subpart, will not address those situations in which evacuation of containers equipped with these air emission controls is necessary and consistent with good engineering and safety practices for handling organic peroxides.

(j) For each hazardous waste management unit not using air emission controls specified in §§265.1085 through 265.1088 of this subpart in accordance with the provisions of §265.1080(b)(7) of this subpart, the owner and operator shall record and maintain the following information:

(1) Certification that the waste management unit is equipped with and operating air emission controls in accordance with the requirements of an applicable Clean Air Act regulation codified under 40 CFR part 60, part 61, or part 63.

(2) Identification of the specific requirements codified under 40 CFR part 60, part 61, or part 63 with which the waste management unit is in compliance.

[61 FR 59990, Nov. 25, 1996, as amended at 62 FR 64667, Dec. 8, 1997; 71 FR 40276, July 14, 2006]

§265.1091  [Reserved]

## Subpart DD—Containment Buildings

SOURCE: 57 FR 37268, Aug. 18, 1992, unless otherwise noted.

§265.1100  Applicability.

The requirements of this subpart apply to owners or operators who store or treat hazardous waste in units designed and operated under §265.1101 of this subpart. The owner or operator is not subject to the definition of land disposal in RCRA section 3004(k) provided that the unit:

(a) Is a completely enclosed, self-supporting structure that is designed and constructed of manmade materials of sufficient strength and thickness to support themselves, the waste contents, and any personnel and heavy equipment that operate within the units, and to prevent failure due to pressure gradients, settlement, compression, or uplift, physical contact with the hazardous wastes to which they are exposed; climatic conditions; and the stresses of daily operation, including the movement of heavy equipment within the unit and contact of such equipment with containment walls;

(b) Has a primary barrier that is designed to be sufficiently durable to withstand the movement of personnel and handling equipment within the unit;

(c) If the unit is used to manage liquids, has:

(1) A primary barrier designed and constructed of materials to prevent migration of hazardous constituents into the barrier;

(2) A liquid collection system designed and constructed of materials to minimize the accumulation of liquid on the primary barrier; and

(3) A secondary containment system designed and constructed of materials to prevent migration of hazardous constituents into the barrier, with a leak detection and liquid collection system capable of detecting, collecting, and removing leaks of hazardous constituents at the earliest possible time, unless the unit has been granted a variance from the secondary containment system requirements under § 265.1101(b)(4);

(d) Has controls as needed to prevent fugitive dust emissions; and

(e) Is designed and operated to ensure containment and prevent the tracking of materials from the unit by personnel or equipment.

[57 FR 37268, Aug. 18, 1992, as amended at 71 FR 16912, Apr. 4, 2006; 71 FR 40276, July 14, 2006]

**§ 265.1101 Design and operating standards.**

(a) All containment buildings must comply with the following design standards:

(1) The containment building must be completely enclosed with a floor, walls, and a roof to prevent exposure to the elements, (e.g., precipitation, wind, run-on), and to assure containment of managed wastes.

(2) The floor and containment walls of the unit, including the secondary containment system if required under paragraph (b) of this section, must be designed and constructed of materials of sufficient strength and thickness to support themselves, the waste contents, and any personnel and heavy equipment that operate within the unit, and to prevent failure due to pressure gradients, settlement, compression, or uplift, physical contact with the hazardous wastes to which they are exposed; climatic conditions; and the stresses of daily operation, including the movement of heavy equipment within the unit and contact of such equipment with containment walls. The unit must be designed so that it has sufficient structural strength to prevent collapse or other failure. All surfaces to be in contact with hazardous wastes must be chemically compatible with those wastes. EPA will consider standards established by professional organizations generally recognized by the industry such as the American Concrete Institute (ACI) and the American Society of Testing Materials (ASTM) in judging the structural integrity requirements of this paragraph. If appropriate to the nature of the waste management operation to take place in the unit, an exception to the structural strength requirement may be made for light-weight doors and windows that meet these criteria:

(i) They provide an effective barrier against fugitive dust emissions under paragraph (c)(1)(iv); and

(ii) The unit is designed and operated in a fashion that assures that wastes will not actually come in contact with these openings.

(3) Incompatible hazardous wastes or treatment reagents must not be placed in the unit or its secondary containment system if they could cause the unit or secondary containment system to leak, corrode, or otherwise fail.

(4) A containment building must have a primary barrier designed to withstand the movement of personnel, waste, and handling equipment in the unit during the operating life of the unit and appropriate for the physical

and chemical characteristics of the waste to be managed.

(b) For a containment building used to manage hazardous wastes containing free liquids or treated with free liquids (the presence of which is determined by the paint filter test, a visual examination, or other appropriate means), the owner or operator must include:

(1) A primary barrier designed and constructed of materials to prevent the migration of hazardous constituents into the barrier (e.g. a geomembrane covered by a concrete wear surface).

(2) A liquid collection and removal system to prevent the accumulation of liquid on the primary barrier of the containment building:

(i) The primary barrier must be sloped to drain liquids to the associated collection system; and

(ii) Liquids and waste must be collected and removed to minimize hydraulic head on the containment system at the earliest practicable time that protects human health and the environment.

(3) A secondary containment system including a secondary barrier designed and constructed to prevent migration of hazardous constituents into the barrier, and a leak detection system that is capable of detecting failure of the primary barrier and collecting accumulated hazardous wastes and liquids at the earliest practicable time.

(i) The requirements of the leak detection component of the secondary containment system are satisfied by installation of a system that is, at a minimum:

(A) Constructed with a bottom slope of 1 percent or more; and

(B) Constructed of a granular drainage material with a hydraulic conductivity of $1 \times 10^{-2}$ cm/sec or more and a thickness of 12 inches (30.5 cm) or more, or constructed of synthetic or geonet drainage materials with a transmissivity of $3 \times 10^{-5}$ m²/sec or more.

(ii) If treatment is to be conducted in the building, an area in which such treatment will be conducted must be designed to prevent the release of liquids, wet materials, or liquid aerosols to other portions of the building.

(iii) The secondary containment system must be constructed of materials that are chemically resistant to the waste and liquids managed in the containment building and of sufficient strength and thickness to prevent collapse under the pressure exerted by overlaying materials and by any equipment used in the containment building. (Containment buildings can serve as secondary containment systems for tanks placed within the building under certain conditions. A containment building can serve as an external liner system for a tank, provided it meets the requirements of §265.193(e)(1). In addition, the containment building must meet the requirements of §265.193 (b) and (c) to be considered an acceptable secondary containment system for a tank.)

(4) For existing units other than 90-day generator units, the Regional Administrator may delay the secondary containment requirement for up to two years, based on a demonstration by the owner or operator that the unit substantially meets the standards of this Subpart. In making this demonstration, the owner or operator must:

(i) Provide written notice to the Regional Administrator of their request by February 18, 1993. This notification must describe the unit and its operating practices with specific reference to the performance of existing containment systems, and specific plans for retrofitting the unit with secondary containment;

(ii) Respond to any comments from the Regional Administrator on these plans within 30 days; and

(iii) Fulfill the terms of the revised plans, if such plans are approved by the Regional Administrator.

(c) Owners or operators of all containment buildings must:

(1) Use controls and practices to ensure containment of the hazardous waste within the unit; and, at a minimum:

(i) Maintain the primary barrier to be free of significant cracks, gaps, corrosion, or other deterioration that could cause hazardous waste to be released from the primary barrier;

(ii) Maintain the level of the stored/treated hazardous waste within the containment walls of the unit so that the height of any containment wall is not exceeded;

(iii) Take measures to prevent the tracking of hazardous waste out of the unit by personnel or by equipment used in handling the waste. An area must be designated to decontaminate equipment and any rinsate must be collected and properly managed; and

(iv) Take measures to control fugitive dust emissions such that any openings (doors, windows, vents, cracks, etc.) exhibit no visible emissions. In addition, all associated particulate collection devices (e.g., fabric filter, electrostatic precipitator) must be operated and maintained with sound air pollution control practices. This state of no visible emissions must be maintained effectively at all times during normal operating conditions, including when vehicles and personnel are entering and exiting the unit.

(2) Obtain and keep on-site a certification by a qualified Professional Engineer that the containment building design meets the requirements of paragraphs (a), (b), and (c) of this section.

(3) Throughout the active life of the containment building, if the owner or operator detects a condition that could lead to or has caused a release of hazardous waste, the owner or operator must repair the condition promptly, in accordance with the following procedures.

(i) Upon detection of a condition that has led to a release of hazardous waste (e.g., upon detection of leakage from the primary barrier) the owner or operator must:

(A) Enter a record of the discovery in the facility operating record;

(B) Immediately remove the portion of the containment building affected by the condition from service;

(C) Determine what steps must be taken to repair the containment building, remove any leakage from the secondary collection system, and establish a schedule for accomplishing the cleanup and repairs; and

(D) Within 7 days after the discovery of the condition, notify the Regional Administrator of the condition, and within 14 working days, provide a written notice to the Regional Administrator with a description of the steps taken to repair the containment build-

ing, and the schedule for accomplishing the work.

(ii) The Regional Administrator will review the information submitted, make a determination regarding whether the containment building must be removed from service completely or partially until repairs and cleanup are complete, and notify the owner or operator of the determination and the underlying rationale in writing.

(iii) Upon completing all repairs and cleanup the owner or operator must notify the Regional Administrator in writing and provide a verification, signed by a qualified, registered professional engineer, that the repairs and cleanup have been completed according to the written plan submitted in accordance with paragraph (c)(3)(i)(D) of this section.

(4) Inspect and record in the facility's operating record at least once every seven days, except for Performance Track member facilities, that must inspect up to once each month, upon approval of the director, data gathered from monitoring and leak detection equipment as well as the containment building and the area immediately surrounding the containment building to detect signs of releases of hazardous waste. To apply for reduced inspection frequency, the Performance Track member facility must follow the procedures described in § 265.15(b)(5).

(d) For a containment building that contains both areas with and without secondary containment, the owner or operator must:

(1) Design and operate each area in accordance with the requirements enumerated in paragraphs (a) through (c) of this section;

(2) Take measures to prevent the release of liquids or wet materials into areas without secondary containment; and

(3) Maintain in the facility's operating log a written description of the operating procedures used to maintain the integrity of areas without secondary containment.

(e) Notwithstanding any other provision of this subpart, the Regional Administrator may waive requirements for secondary containment for a permitted containment building where the owner or operator demonstrates that the only free liquids in the unit are limited amounts of dust suppression liquids required to meet occupational health and safety requirements, and where containment of managed wastes and liquids can be assured without a secondary containment system.

[57 FR 37268, Aug. 18, 1992, as amended at 71 FR 16912, Apr. 4, 2006; 71 FR 40276, July 14, 2006]

### §265.1102 Closure and post-closure care.

(a) At closure of a containment building, the owner or operator must remove or decontaminate all waste residues, contaminated containment system components (liners, etc.), contaminated subsoils, and structures and equipment contaminated with waste and leachate, and manage them as hazardous waste unless §261.3(d) of this chapter applies. The closure plan, closure activities, cost estimates for closure, and financial responsibility for containment buildings must meet all of the requirements specified in subparts G and H of this part.

(b) If, after removing or decontaminating all residues and making all reasonable efforts to effect removal or decontamination of contaminated components, subsoils, structures, and equipment as required in paragraph (a) of this section, the owner or operator finds that not all contaminated subsoils can be practically removed or decontaminated, he must close the facility and perform post-closure care in accordance with the closure and post-closure requirements that apply to landfills (§265.310). In addition, for the purposes of closure, post-closure, and financial responsibility, such a containment building is then considered to be a landfill, and the owner or operator must meet all of the requirements for landfills specified in subparts G and H of this part.

### §§265.1103–265.1110 [Reserved]

## Subpart EE—Hazardous Waste Munitions and Explosives Storage

Source: 62 FR 6653, Feb. 12, 1997, unless otherwise noted.

### §265.1200 Applicability.

The requirements of this subpart apply to owners or operators who store munitions and explosive hazardous wastes, except as §265.1 provides otherwise. (NOTE: Depending on explosive hazards, hazardous waste munitions and explosives may also be managed in other types of storage units, including containment buildings (40 CFR part 265, subpart DD), tanks (40 CFR part 265, subpart J), or containers (40 CFR part 265, subpart I); See 40 CFR 266.205 for storage of waste military munitions).

### §265.1201 Design and operating standards.

(a) Hazardous waste munitions and explosives storage units must be designed and operated with containment systems, controls, and monitoring, that:

(1) Minimize the potential for detonation or other means of release of hazardous waste, hazardous constituents, hazardous decomposition products, or contaminated run-off, to the soil, ground water, surface water, and atmosphere;

(2) Provide a primary barrier, which may be a container (including a shell) or tank, designed to contain the hazardous waste;

(3) For wastes stored outdoors, provide that the waste and containers will not be in standing precipitation;

(4) For liquid wastes, provide a secondary containment system that assures that any released liquids are contained and promptly detected and removed from the waste area, or vapor detection system that assures that any released liquids or vapors are promptly detected and an appropriate response taken (e.g., additional containment, such as overpacking, or removal from the waste area); and

(5) Provide monitoring and inspection procedures that assure the controls and containment systems are working as designed and that releases that may adversely impact human health or the environment are not escaping from the unit.

(b) Hazardous waste munitions and explosives stored under this subpart may be stored in one of the following:

(1) *Earth-covered magazines.* Earth-covered magazines must be:

(i) Constructed of waterproofed, reinforced concrete or structural steel arches, with steel doors that are kept closed when not being accessed;

(ii) Designed and constructed:

(A) To be of sufficient strength and thickness to support the weight of any explosives or munitions stored and any equipment used in the unit;

(B) To provide working space for personnel and equipment in the unit; and

(C) To withstand movement activities that occur in the unit; and

(iii) Located and designed, with walls and earthen covers that direct an explosion in the unit in a safe direction, so as to minimize the propagation of an explosion to adjacent units and to minimize other effects of any explosion.

(2) *Above-ground magazines.* Above-ground magazines must be located and designed so as to minimize the propagation of an explosion to adjacent units and to minimize other effects of any explosion.

(3) *Outdoor or open storage areas.* Outdoor or open storage areas must be located and designed so as to minimize the propagation of an explosion to adjacent units and to minimize other effects of any explosion.

(c) Hazardous waste munitions and explosives must be stored in accordance with a Standard Operating Procedure specifying procedures to ensure safety, security, and environmental protection. If these procedures serve the same purpose as the security and inspection requirements of 40 CFR 265.14, the preparedness and prevention procedures of 40 CFR part 265, subpart C, and the contingency plan and emergency procedures requirements of 40 CFR part 265, subpart D, then these procedures will be used to fulfill those requirements.

(d) Hazardous waste munitions and explosives must be packaged to ensure safety in handling and storage.

(e) Hazardous waste munitions and explosives must be inventoried at least annually.

(f) Hazardous waste munitions and explosives and their storage units must be inspected and monitored as necessary to ensure explosives safety and to ensure that there is no migration of contaminants out of the unit.

§ 265.1202 Closure and post-closure care.

(a) At closure of a magazine or unit which stored hazardous waste under this subpart, the owner or operator must remove or decontaminate all waste residues, contaminated containment system components, contaminated subsoils, and structures and equipment contaminated with waste, and manage them as hazardous waste unless § 261.3(d) of this chapter applies. The closure plan, closure activities, cost estimates for closure, and financial responsibility for magazines or units must meet all of the requirements specified in subparts G and H of this part, except that the owner or operator may defer closure of the unit as long as it remains in service as a munitions or explosives magazine or storage unit.

(b) If, after removing or decontaminating all residues and making all reasonable efforts to effect removal or decontamination of contaminated components, subsoils, structures, and equipment as required in paragraph (a) of this section, the owner or operator finds that not all contaminated subsoils can be practically removed or decontaminated, he or she must close the facility and perform post-closure care in accordance with the closure and post-closure requirements that apply to landfills (40 CFR 264.310).

APPENDIX I TO PART 265—
RECORDKEEPING INSTRUCTIONS

The recordkeeping provisions of § 265.73 specify that an owner or operator must keep a written operating record at his facility. This appendix provides additional instructions for keeping *portions* of the operating record. See § 265.73(b) for additional recordkeeping requirements.

The following information must be recorded, as it becomes available, and maintained in the operating record until closure of the facility in the following manner:

Records of each hazardous waste received, treated, stored, or disposed of at the facility which include the following:

(1) A description by its common name and the EPA Hazardous Waste Number(s) from part 261 of this chapter which apply to the waste. The waste description also must include the waste's physical form, i.e., liquid,

sludge, solid, or contained gas. If the waste is not listed in part 261, subpart D, of this chapter, the description also must include the process that produced it (for example, solid filter cake from production of _____, EPA Hazardous Waste Number W051).

Each hazardous waste listed in part 261, subpart D, of this chapter, and each hazardous waste characteristic defined in part 261, subpart C, of this chapter, has a four-digit EPA Hazardous Waste Number assigned to it. This number must be used for record-keeping and reporting purposes. Where a hazardous waste contains more than one listed hazardous waste, or where more than one hazardous waste characteristic applies to the waste, the waste description must include all applicable EPA Hazardous Waste Numbers.

(2) The estimated or manifest-reported weight, or volume and density, where applicable, in one of the units of measure specified in Table 1; and

TABLE 1

| Unit of measure | Code [1] |
|---|---|
| Gallons | G |
| Gallons per Hour | E |
| Gallons per Day | U |
| Liters | L |
| Liters Per Hour | H |
| Liters Per Day | V |
| Short Tons Per Hour | D |
| Metric Tons Per Hour | W |
| Short Tons Per Day | N |
| Metric Tons Per Day | S |
| Pounds Per Hour | J |
| Kilograms Per Hour | R |
| Cubic Yards | Y |
| Cubic Meters | C |
| Acres | B |
| Acre-feet | A |
| Hectares | Q |
| Hectare-meter | F |
| Btu's per Hour | I |
| Pounds | P |
| Short tons | I |
| Kilograms | K |
| Tons | M |

[1] Single digit symbols are used here for data processing purposes.

(3) The method(s) (by handling code(s) as specified in Table 2) and date(s) of treatment, storage, or disposal.

TABLE 2—HANDLING CODES FOR TREATMENT, STORAGE AND DISPOSAL METHODS

Enter the handling code(s) listed below that most closely represents the technique(s) used at the facility to treat, store or dispose of each quantity of hazardous waste received.

1. Storage

S01  Container (barrel, drum, etc.)
S02  Tank
S03  Waste Pile
S04  Surface Impoundment
S05  Drip Pad
S06  Containment Building (Storage)
S99  Other Storage (specify)

2. Treatment

(a)  Thermal Treatment—

T06  Liquid injection incinerator
T07  Rotary kiln incinerator
T08  Fluidized bed incinerator
T09  Multiple hearth incinerator
T10  Infrared furnace incinerator
T11  Molten salt destructor
T12  Pyrolysis
T13  Wet Air oxidation
T14  Calcination
T15  Microwave discharge
T18  Other (specify)

(b)  Chemical Treatment—

T19  Absorption mound
T20  Absorption field
T21  Chemical fixation
T22  Chemical oxidation
T23  Chemical precipitation
T24  Chemical reduction
T25  Chlorination
T26  Chlorinolysis
T27  Cyanide destruction
T28  Degradation
T29  Detoxification
T30  Ion exchange
T31  Neutralization
T32  Ozonation
T33  Photolysis
T34  Other (specify)

(c)  Physical Treatment—
     (1) Separation of components

T35  Centrifugation
T36  Clarification
T37  Coagulation
T38  Decanting
T39  Encapsulation
T40  Filtration
T41  Flocculation
T42  Flotation
T43  Foaming
T44  Sedimentation
T45  Thickening
T46  Ultrafiltration
T47  Other (specify)

(2) Removal of Specific Components

T48  Absorption-molecular sieve
T49  Activated carbon
T50  Blending
T51  Catalysis
T52  Crystallization
T53  Dialysis
T54  Distillation
T55  Electrodialysis
T56  Electrolysis
T57  Evaporation
T58  High gradient magnetic separation
T59  Leaching
T60  Liquid ion exchange
T61  Liquid-liquid extraction
T62  Reverse osmosis

T63  Solvent recovery
T64  Stripping
T65  Sand filter
T66  Other (specify)

(d)  Biological Treatment

T67  Activated sludge
T68  Aerobic lagoon
T69  Aerobic tank
T70  Anaerobic tank
T71  Composting
T72  Septic tank
T73  Spray irrigation
T74  Thickening filter
T75  Trickling filter
T76  Waste stabilization pond
T77  Other (specify)
T78–T79 [Reserved]

(e)  Boilers and Industrial Furnaces

T80  Boiler
T81  Cement Kiln
T82  Lime Kiln
T83  Aggregate Kiln
T84  Phosphate Kiln
T85  Coke Oven
T86  Blast Furnace
T87  Smelting, Melting, or Refining Furnace
T88  Titanium Dioxide Chloride Process Oxidation Reactor
T89  Methane Reforming Furnace
T90  Pulping Liquor Recovery Furnace
T91  Combustion Device Used in the Recovery of Sulfur Values From Spent Sulfuric Acid
T92  Halogen Acid Furnaces
T93  Other Industrial Furnaces Listed in 40 CFR 260.10 (specify)

(f)  Other Treatment

T94  Containment Building (Treatment)

3. Disposal

D79  Underground Injection
D80  Landfill
D81  Land Treatment
D82  Ocean Disposal
D83  Surface Impoundment (to be closed as a landfill)
D99  Other Disposal (specify)

4. Miscellaneous

X01  Open Burning/Open Detonation
X02  Mechanical Processing
X03  Thermal Unit
X04  Geologic Repository
X99  Other (specify)

[45 FR 33232, May 19, 1980, as amended at 59 FR 13892, Mar. 24, 1994; 71 FR 40276, July 14, 2006]

APPENDIX II TO PART 265 [RESERVED]

APPENDIX III TO PART 265—EPA INTERIM PRIMARY DRINKING WATER STANDARDS

| Parameter | Maximum level (mg/l) |
| --- | --- |
| Arsenic | 0.05 |
| Barium | 1.0 |
| Cadmium | 0.01 |
| Chromium | 0.05 |
| Fluoride | 1.4–2.4 |
| Lead | 0.05 |
| Mercury | 0.002 |
| Nitrate (as N) | 10 |
| Selenium | 0.01 |
| Silver | 0.05 |
| Endrin | 0.0002 |
| Lindane | 0.004 |
| Methoxychlor | 0.1 |
| Toxaphene | 0.005 |
| 2,4-D | 0.1 |
| 2,4,5-TP Silver | 0.01 |
| Radium | 5 pCi/1 |
| Gross Alpha | 15 pCi/1 |
| Gross Beta | 4 millirem/yr |
| Turbidity | 1/TU |
| Coliform Bacteria | 1/100 ml |

[*Comment:* Turbidity is applicable only to surface water supplies.]

APPENDIX IV TO PART 265—TESTS FOR SIGNIFICANCE

As required in § 265.93(b) the owner or operator must use the Student's t-test to determine statistically significant changes in the concentration or value of an indicator parameter in periodic ground-water samples when compared to the initial background concentration or value of that indicator parameter. The comparison must consider individually each of the wells in the monitoring system. For three of the indicator parameters (specific conductance, total organic carbon, and total organic halogen) a single-tailed Student's t-test must be used to test at the 0.01 level of significance for significant increases over background. The difference test for pH must be a two-tailed Student's t-test at the overall 0.01 level of significance.

The student's t-test involves calculation of the value of a t-statistic for each comparison of the mean (average) concentration or value (based on a minimum of four replicate measurements) of an indicator parameter with its initial background concentration or value. The calculated value of the t-statistic must then be compared to the value of the t-statistic found in a table for t-test of significance at the specified level of significance. A calculated value of t which exceeds the value of t found in the table indicates a statistically significant change in the concentration or value of the indicator parameter.

Formulae for calculation of the t-statistic and tables for t-test of significance can be found in most introductory statistics texts.

APPENDIX V TO PART 265—EXAMPLES OF POTENTIALLY INCOMPATIBLE WASTE

Many hazardous wastes, when mixed with other waste or materials at a hazardous waste facility, can produce effects which are harmful to human health and the environment, such as (1) heat or pressure, (2) fire or explosion, (3) violent reaction, (4) toxic dusts, mists, fumes, or gases, or (5) flammable fumes or gases.

Below are examples of potentially incompatible wastes, waste components, and materials, along with the harmful consequences which result from mixing materials in one group with materials in another group. The list is intended as a guide to owners or operators of treatment, storage, and disposal facilities, and to enforcement and permit granting officials, to indicate the need for special precautions when managing these potentially incompatible waste materials or components.

This list is not intended to be exhaustive. An owner or operator must, as the regulations require, adequately analyze his wastes so that he can avoid creating uncontrolled substances or reactions of the type listed below, whether they are listed below or not.

It is possible for potentially incompatible wastes to be mixed in a way that precludes a reaction (e.g., adding acid to water rather than water to acid) or that neutralizes them (e.g., a strong acid mixed with a strong base), or that controls substances produced (e.g., by generating flammable gases in a closed tank equipped so that ignition cannot occur, and burning the gases in an incinerator).

In the lists below, the mixing of a Group A material with a Group B material may have the potential consequence as noted.

| Group 1–A | Group 1–B |
|---|---|
| Acetylene sludge | Acid sludge |
| Alkaline caustic liquids | Acid and water |
| Alkaline cleaner | Battery acid |
| Alkaline corrosive liquids | Chemical cleaners |
| Alkaline corrosive battery fluid | Electrolyte, acid |
| Caustic wastewater | Etching acid liquid or solvent |
| Lime sludge and other corrosive alkalies | |
| Lime wastewater | Pickling liquor and other corrosive acids |
| Lime and water | Spent acid |
| Spent caustic | Spent mixed acid |
| | Spent sulfuric acid |

Potential consequences: Heat generation; violent reaction.

| Group 2–A | Group 2–B |
|---|---|
| Aluminum | Any waste in Group 1–A or 1–B |
| Beryllium | |
| Calcium | |
| Lithium | |
| Magnesium | |
| Potassium | |
| Sodium | |
| Zinc powder | |
| Other reactive metals and metal hydrides | |

Potential consequences: Fire or explosion; generation of flammable hydrogen gas.

| Group 3–A | Group 3–B |
|---|---|
| Alcohols | Any concentrated waste in Groups 1–A or 1–B |
| Water | Calcium |
| | Lithium |
| | Metal hydrides |
| | Potassium |
| | $SO_2Cl_2$, $SOCl_2$, $PCl_3$, $CH_3SiCl_3$ |
| | Other water-reactive waste |

Potential consequences: Fire, explosion, or heat generation; generation of flammable or toxic gases.

| Group 4–A | Group 4–B |
|---|---|
| Alcohols | Concentrated Group 1–A or 1–B wastes |
| Aldehydes | Group 2–A wastes |
| Halogenated hydrocarbons | |
| Nitrated hydrocarbons | |
| Unsaturated hydrocarbons | |
| Other reactive organic compounds and solvents | |

Potential consequences: Fire, explosion, or violent reaction.

| Group 5–A | Group 5–B |
|---|---|
| Spent cyanide and sulfide solutions | Group 1–B wastes |

Potential consequences: Generation of toxic hydrogen cyanide or hydrogen sulfide gas.

| Group 6–A | Group 6–B |
|---|---|
| Chlorates | Acetic acid and other organic acids |
| Chlorine | Concentrated mineral acids |
| Chlorites | Group 2–A wastes |
| Chromic acid | Group 4–A wastes |

| Group 6-A | Group 6-B |
|---|---|
| Hyphochlorites | Other flammable and combustible wastes |
| Nitrates | |
| Nitric acid, fuming | |
| Perchlorates | |
| Permanganates | |
| Peroxides | |
| Other strong oxidizers | |

Potential consequences: Fire, explosion, or violent reaction.

SOURCE: "Law, Regulations, and Guidelines for Handling of Hazardous Waste." California Department of Health, February 1975.

[45 FR 33232, May 19, 1980, as amended at 71 FR 40276, July 14, 2006]

APPENDIX VI TO PART 265—COMPOUNDS WITH HENRY'S LAW CONSTANT LESS THAN 0.1 Y/X

| Compound name | CAS No. |
|---|---|
| Acetaldol | 107-89-1 |
| Acetamide | 60-35-5 |
| 2-Acetylaminofluorene | 53-96-3 |
| 3-Acetyl-5-hydroxypiperidine. | |
| 3-Acetylpiperidine | 618-42-8 |
| 1-Acetyl-2-thiourea | 591-08-2 |
| Acrylamide | 79-06-1 |
| Acrylic acid | 79-10-7 |
| Adenine | 73-24-5 |
| Adipic acid | 124-04-9 |
| Adiponitrile | 111-69-3 |
| Alachlor | 15972-60-8 |
| Aldicarb | 116-06-3 |
| Ametryn | 834-12-8 |
| 4-Aminobiphenyl | 92-67-1 |
| 4-Aminopyridine | 504-24-5 |
| Aniline | 62-53-3 |
| o-Anisidine | 90-04-0 |
| Anthraquinone | 84-65-1 |
| Atrazine | 1912-24-9 |
| Benzenearsonic acid | 98-05-5 |
| Benzenesulfonic acid | 98-11-3 |
| Benzidine | 92-87-5 |
| Benzo(a)anthracene | 56-55-3 |
| Benzo(k)fluoranthene | 207-08-9 |
| Benzoic acid | 65-85-0 |
| Benzo(g,h,i)perylene | 191-24-2 |
| Benzo(a)pyrene | 50-32-8 |
| Benzyl alcohol | 100-51-6 |
| gamma-BHC | 58-89-9 |
| Bis(2-ethylhexyl)phthalate | 117-81-7 |
| Bromochloromethyl acetate. | |
| Bromoxynil | 1689-84-5 |
| Butyric acid | 107-92-6 |
| Caprolactam (hexahydro-2H-azepin-2-one) | 105-60-2 |
| Catechol (o-dihydroxybenzene) | 120-80-9 |
| Cellulose | 9004-34-6 |
| Cell wall. | |
| Chlorhydrin (3-Chloro-1,2-propanediol) | 96-24-2 |
| Chloroacetic acid | 79-11-8 |
| 2-Chloroacetophenone | 93-76-5 |
| p-Chloroaniline | 106-47-8 |
| p-Chlorobenzophenone | 134-85-0 |
| Chlorobenzilate | 510-15-6 |
| p-Chloro-m-cresol (6-chloro-m-cresol) | 59-50-7 |
| 3-Chloro-2,5-diketopyrrolidine. | |
| Chloro-1,2-ethane diol. | |
| 4-Chlorophenol | 106-48-9 |
| Chlorophenol polymers (2-chlorophenol & 4-chlorophenol) | 95-57-8 & 106-48-9 |
| 1-(o-Chlorophenyl)thiourea | 5344-82-1 |
| Chrysene | 218-01-9 |
| Citric acid | 77-92-9 |
| Creosote | 8001-58-9 |
| m-Cresol | 108-39-4 |
| o-Cresol | 95-48-7 |
| p-Cresol | 106-44-5 |

| Compound name | CAS No. |
|---|---|
| Cresol (mixed isomers) | 1319–77–3 |
| 4-Cumylphenol | 27576–86 |
| Cyanide | 57–12–5 |
| 4-Cyanomethyl benzoate. | |
| Diazinon | 333–41–5 |
| Dibenzo(a,h)anthracene | 53–70–3 |
| Dibutylphthalate | 84–74–2 |
| 2,5-Dichloroaniline (N,N'-dichloroaniline) | 95–82–9 |
| 2,6-Dichlorobenzonitrile11 | 1194–65–6 |
| 2,6-Dichloro-4-nitroaniline | 99–30–9 |
| 2,5-Dichlorophenol | 333–41–5 |
| 3,4-Dichlorotetrahydrofuran | 3511–19 |
| Dichlorvos (DDVP) | 62–73–7 |
| Diethanolamine | 111–42–2 |
| N,N-Diethylaniline | 91–66–7 |
| Diethylene glycol | 111–46–6 |
| Diethylene glycol dimethyl ether (dimethyl Carbitol) | 111–96–6 |
| Diethylene glycol monobutyl ether (butyl Carbitol) | 112–34–5 |
| Diethylene glycol monoethyl ether acetate (Carbitol acetate) | 112–15–2 |
| Diethylene glycol monoethyl ether (Carbitol Cellosolve) | 111–90–0 |
| Diethylene glycol monomethyl ether (methyl Carbitol) | 111–77–3 |
| N,N'-Diethylhydrazine | 1615–80–1 |
| Diethyl (4-methylumbelliferyl) thionophosphate | 299–45–6 |
| Diethyl phosphorothioate | 126–75–0 |
| N,N'-Diethylpropionamide | 15299–99–7 |
| Dimethoate | 60–51–5 |
| 2,3-Dimethoxystrychnidin-10-one | 357–57–3 |
| 4-Dimethylaminoazobenzene | 60–11–7 |
| 7,12-Dimethylbenz(a)anthracene | 57–97–6 |
| 3,3-Dimethylbenzidine | 119–93–7 |
| Dimethylcarbamoyl chloride | 79–44–7 |
| Dimethyldisulfide | 624–92–0 |
| Dimethylformamide | 68–12–2 |
| 1,1-Dimethylhydrazine | 57–14–7 |
| Dimethylphthalate | 131–11–3 |
| Dimethylsulfone | 67–71–0 |
| Dimethylsulfoxide | 67–68–5 |
| 4,6-Dinitro-o-cresol | 534–52–1 |
| 1,2-Diphenylhydrazine | 122–66–7 |
| Dipropylene glycol (1,1'-oxydi-2-propanol) | 110–98–5 |
| Endrin | 72–20–8 |
| Epinephrine | 51–43–4 |
| mono-Ethanolamine | 141–43–5 |
| Ethyl carbamate (urethane) | 5–17–96 |
| Ethylene glycol | 107–21–1 |
| Ethylene glycol monobutyl ether (butyl Cellosolve) | 111–76–2 |
| Ethylene glycol monoethyl ether (Cellosolve) | 110–80–5 |
| Ethylene glycol monoethyl ether acetate (Cellosolve acetate) | 111–15–9 |
| Ethylene glycol monomethyl ether (methyl Cellosolve) | 109–86–4 |
| Ethylene glycol monophenyl ether (phenyl Cellosolve) | 122–99–6 |
| Ethylene glycol monopropyl ether (propyl Cellosolve) | 2807–30–9 |
| Ethylene thiourea (2-imidazolidinethione) | 96–45–7 |
| 4-Ethylmorpholine | 100–74–3 |
| 3-Ethylphenol | 620–17–7 |
| Fluoroacetic acid, sodium salt | 62–74–8 |
| Formaldehyde | 50–00–0 |
| Formamide | 75–12–7 |
| Formic acid | 64–18–6 |
| Fumaric acid | 110–17–8 |
| Glutaric acid | 110–94–1 |
| Glycerin (Glycerol) | 56–81–5 |
| Glycidol | 556–52–5 |
| Glycinamide | 598–41–4 |
| Glyphosate | 1071–83–6 |
| Guthion | 86–50–0 |
| Hexamethylene-1,6-diisocyanate (1,6-diisocyanatohexane) | 822–06–0 |
| Hexamethyl phosphoramide | 680–31–9 |
| Hexanoic acid | 142–62–1 |
| Hydrazine | 302–01–2 |
| Hydrocyanic acid | 74–90–8 |
| Hydroquinone | 123–31–9 |
| Hydroxy-2-propionitrile (hydracrylonitrile) | 109–78–4 |
| Indeno (1,2,3-cd) pyrene | 193–39–5 |
| Lead acetate | 301–04–2 |

| Compound name | CAS No. |
|---|---|
| Lead subacetate (lead acetate, monobasic) | 1335–32–6 |
| Leucine | 61–90–5 |
| Malathion | 121–75–5 |
| Maleic acid | 110–16–7 |
| Maleic anhydride | 108–31–6 |
| Mesityl oxide | 141–79–7 |
| Methane sulfonic acid | 75–75–2 |
| Methomyl | 16752–77–5 |
| p-Methoxyphenol | 150–76–5 |
| Methyl acrylate | 96–33–3 |
| 4,4′-Methylene-bis-(2-chloroaniline) | 101–14–4 |
| 4,4′-Methylenediphenyl diisocyanate (diphenyl methane diisocyanate) | 101–68–8 |
| 4,4′-Methylenedianiline | 101–77–9 |
| Methylene diphenylamine (MDA). | |
| 5-Methylfurfural | 620–02–0 |
| Methylhydrazine | 60–34–4 |
| Methyliminoacetic acid. | |
| Methyl methane sulfonate | 66–27–3 |
| 1-Methyl-2-methoxyaziridine. | |
| Methylparathion | 298–00–0 |
| Methyl sulfuric acid (sulfuric acid, dimethyl ester) | 77–78–1 |
| 4-Methylthiophenol | 106–45–6 |
| Monomethylformamide (N-methylformamide) | 123–39–7 |
| Nabam | 142–59–6 |
| alpha-Naphthol | 90–15–3 |
| beta-Naphthol | 135–19–3 |
| alpha-Naphthylamine | 134–32–7 |
| beta-Naphthylamine | 91–59–8 |
| Neopentyl glycol (dimethylpropane) | 126–30–7 |
| Niacinamide | 98–92–0 |
| o-Nitroaniline | 88–74–4 |
| Nitroglycerin | 55–63–0 |
| 2-Nitrophenol | 88–75–5 |
| 4-Nitrophenol | 100–02–7 |
| N-Nitrosodimethylamine | 62–75–9 |
| Nitrosoguanidine | 674–81–7 |
| N-Nitroso-n-methylurea | 684–93–5 |
| N-Nitrosomorpholine (4-nitrosomorpholine) | 59–89–2 |
| Oxalic acid | 144–62–7 |
| Parathion | 56–38–2 |
| Pentaerythritol | 115–77–5 |
| Phenacetin | 62–44–2 |
| Phenol | 108–95–2 |
| Phenylacetic acid | 103–82–2 |
| m-Phenylene diamine | 108–45–2 |
| o-Phenylene diamine | 95–54–5 |
| p-Phenylene diamine | 106–50–3 |
| Phenyl mercuric acetate | 62–38–4 |
| Phorate | 298–02–2 |
| Phthalic anhydride | 85–44–9 |
| alpha-Picoline (2-methyl pyridine) | 109–06–8 |
| 1,3-Propane sultone | 1120–71–4 |
| beta-Propiolactone | 57–57–8 |
| Proporur (Baygon). | |
| Propylene glycol | 57–55–6 |
| Pyrene | 129–00–0 |
| Pyridinium bromide | 39416–48–3 |
| Quinoline | 91–22–5 |
| Quinone (p-benzoquinone) | 106–51–4 |
| Resorcinol | 108–46–3 |
| Simazine | 122–34–9 |
| Sodium acetate | 127–09–3 |
| Sodium formate | 141–53–7 |
| Strychnine | 57–24–9 |
| Succinic acid | 110–15–6 |
| Succinimide | 123–56–8 |
| Sulfanilic acid | 121–47–1 |
| Terephthalic acid | 100–21–0 |
| Tetraethyldithiopyrophosphate | 3689–24–5 |
| Tetraethylenepentamine | 112–57–2 |
| Thiofanox | 39196–18–4 |
| Thiosemicarbazide | 79–19–6 |
| 2,4-Toluenediamine | 95–80–7 |
| 2,6-Toluenediamine | 823–40–5 |

| Compound name | CAS No. |
|---|---|
| 3,4-Toluenediamine | 496–72–0 |
| 2,4-Toluene diisocyanate | 584–84–9 |
| p-Toluic acid | 99–94–5 |
| m-Toluidine | 108–44–1 |
| 1,1,2-Trichloro-1,2,2-trifluoroethane | 76–13–1 |
| Triethanolamine | 102–71–6 |
| Triethylene glycol dimethyl ether. | |
| Tripropylene glycol | 24800–44–0 |
| Warfarin | 81–81–2 |
| 3,4-Xylenol (3,4-dimethylphenol) | 95–65–8 |

[62 FR 64668, Dec. 8, 1997, as amended at 71 FR 40276, July 14, 2006]

# FINDING AIDS

A list of CFR titles, subtitles, chapters, subchapters and parts and an alphabetical list of agencies publishing in the CFR are included in the CFR Index and Finding Aids volume to the Code of Federal Regulations which is published separately and revised annually.

# Table of CFR Titles and Chapters

(Revised as of July 1, 2010)

## Title 1—General Provisions

## Title 2—Grants and Agreements

# Title 5—Administrative Personnel—Continued

# Title 6—Domestic Security

## Title 7—Agriculture

# Title 7—Agriculture—Continued

# Title 8—Aliens and Nationality

# Title 9—Animals and Animal Products

# Title 10—Energy

# Title 18—Conservation of Power and Water Resources—Continued
Chap.

# Title 19—Customs Duties

# Title 20—Employees' Benefits

# Title 21—Food and Drugs

# Title 22—Foreign Relations

# Title 22—Foreign Relations—Continued

# Title 23—Highways

# Title 24—Housing and Urban Development

791

# Title 28—Judicial Administration

# Title 29—Labor

# Title 30—Mineral Resources

# Title 42—Public Health

# Title 43—Public Lands: Interior

# Title 44—Emergency Management and Assistance

# Title 45—Public Welfare

# Title 46—Shipping

# Title 47—Telecommunication

## Title 49—Transportation

## Title 50—Wildlife and Fisheries

## Title 50—Wildlife and Fisheries—Continued

## CFR Index and Finding Aids

# Alphabetical List of Agencies Appearing in the CFR

(Revised as of July 1, 2010)

| Agency | CFR Title, Subtitle or Chapter |
|---|---|
| Administrative Committee of the Federal Register | 1, I |
| Advanced Research Projects Agency | 32, I |
| Advisory Council on Historic Preservation | 36, VIII |
| African Development Foundation | 22, XV |
| Federal Acquisition Regulation | 48, 57 |
| Agency for International Development | 22, II |
| Federal Acquisition Regulation | 48, 7 |
| Agricultural Marketing Service | 7, I, IX, X, XI |
| Agricultural Research Service | 7, V |
| Agriculture Department | 5, LXXIII |
| Agricultural Marketing Service | 7, I, IX, X, XI |
| Agricultural Research Service | 7, V |
| Animal and Plant Health Inspection Service | 7, III; 9, I |
| Chief Financial Officer, Office of | 7, XXX |
| Commodity Credit Corporation | 7, XIV |
| Economic Research Service | 7, XXXVII |
| Energy Policy and New Uses, Office of | 2, IX; 7, XXIX |
| Environmental Quality, Office of | 7, XXXI |
| Farm Service Agency | 7, VII, XVIII |
| Federal Acquisition Regulation | 48, 4 |
| Federal Crop Insurance Corporation | 7, IV |
| Food and Nutrition Service | 7, II |
| Food Safety and Inspection Service | 9, III |
| Foreign Agricultural Service | 7, XV |
| Forest Service | 36, II |
| Grain Inspection, Packers and Stockyards Administration | 7, VIII; 9, II |
| Information Resources Management, Office of | 7, XXVII |
| Inspector General, Office of | 7, XXVI |
| National Agricultural Library | 7, XLI |
| National Agricultural Statistics Service | 7, XXXVI |
| National Institute of Food and Agriculture. | 7, XXXIV |
| Natural Resources Conservation Service | 7, VI |
| Operations, Office of | 7, XXVIII |
| Procurement and Property Management, Office of | 7, XXXII |
| Rural Business-Cooperative Service | 7, XVIII, XLII, L |
| Rural Development Administration | 7, XLII |
| Rural Housing Service | 7, XVIII, XXXV, L |
| Rural Telephone Bank | 7, XVI |
| Rural Utilities Service | 7, XVII, XVIII, XLII, L |
| Secretary of Agriculture, Office of | 7, Subtitle A |
| Transportation, Office of | 7, XXXIII |
| World Agricultural Outlook Board | 7, XXXVIII |
| Air Force Department | 32, VII |
| Federal Acquisition Regulation Supplement | 48, 53 |
| Air Transportation Stabilization Board | 14, VI |
| Alcohol and Tobacco Tax and Trade Bureau | 27, I |
| Alcohol, Tobacco, Firearms, and Explosives, Bureau of | 27, II |
| AMTRAK | 49, VII |
| American Battle Monuments Commission | 36, IV |
| American Indians, Office of the Special Trustee | 25, VII |
| Animal and Plant Health Inspection Service | 7, III; 9, I |
| Appalachian Regional Commission | 5, IX |
| Architectural and Transportation Barriers Compliance Board | 36, XI |

803

| Agency | CFR Title, Subtitle or Chapter |
|---|---|
| Veterans' Employment and Training Service, Office of the Assistant Secretary for | 41, 61; 20, IX |
| Vice President of the United States, Office of | 32, XXVIII |
| Vocational and Adult Education, Office of | 34, IV |
| Wage and Hour Division | 29, V |
| Water Resources Council | 18, VI |
| Workers' Compensation Programs, Office of | 20, I |
| World Agricultural Outlook Board | 7, XXXVIII |

# List of CFR Sections Affected

All changes in this volume of the Code of Federal Regulations that were made by documents published in the FEDERAL REGISTER since January 1, 2001, are enumerated in the following list. Entries indicate the nature of the changes effected. Page numbers refer to FEDERAL REGISTER pages. The user should consult the entries for chapters and parts as well as sections for revisions.

Title 40 was established at 36 FR 12213, June 29, 1971. For the period before January 1, 2001, see the "List of CFR Sections Affected, 1964–1972, 1973–1985, and 1986–2000," published in ten separate volumes.

○

805